BIOACTIVE FOODS IN PROMOTING HEALTH: FRUITS AND VEGETABLES

BIOACTIVE FOODS IN PROMOTING HEALTH: FRUITS AND VEGETABLES

Edited by

RONALD ROSS WATSON
University of Arizona,
Arizona Health Sciences Center,
Tucson, Arizona, USA

VICTOR R. PREEDY
Department of Nutrition and Dietetics,
King's College London,
London, UK

ELSEVIER

AMSTERDAM • BOSTON • HEIDELBERG • LONDON
NEW YORK • OXFORD • PARIS • SAN DIEGO
SAN FRANCISCO • SINGAPORE • SYDNEY • TOKYO
Academic Press is an imprint of Elsevier

Academic Press is an imprint of Elsevier

32 Jamestown Road, London NW1 7BY, UK
30 Corporate Drive, Suite 400, Burlington, MA 01803, USA
525 B Street, Suite 1900, San Diego, CA 92101-4495, USA

First edition 2010

Notice
No responsibility is assumed by the publisher for any injury and/or damage to persons or property as a matter of products liability, negligence or otherwise, or from any use or operation of any methods, products, instructions or ideas contained in the material herein. Because of rapid advances in the medical sciences, in particular, independent verification of diagnoses and drug dosages should be made

Knowledge and best practice in this field are constantly changing. As new research and experience broaden our knowledge, changes in practice, treatment and drug therapy may become necessary or appropriate. Readers are advised to check the most current information provided (i) on procedure feature or (ii) by the manufacturer of each product to be administered, to verify the recommended dose or formula, the method and duration of administration, and contradictions. It is the responsibility of the practitioner, relying on his or her own experience and knowledge of the patient, to make diagnoses, to determine dosages and the best treatment for each individual patient, and to take all appropriate safety precautions. To the fullest extent of the law neither the publisher nor the Authors assume any liability for any injury and/or damage to persons or property arising out of or related to any use of the material contained in this book.

British Library Cataloguing-in-Publication Data
A catalogue record for this book is available from the British Library

Library of Congress Cataloging-in-Publication Data
A catalog record for this book is available from the Library of Congress

ISBN: 978-0-12-374628-3

For information on all Academic Press publications
visit our website at www.elsevierdirect.com

Typeset by Macmillan Publishing Solutions
www.macmillansolutions.com

Printed and bound in United States of America

10   11   12   13   10 9 8 7 6 5 4 3 2 1

Working together to grow
libraries in developing countries

www.elsevier.com | www.bookaid.org | www.sabre.org

ELSEVIER   BOOK AID
             International   Sabre Foundation

Contents

Section B

EFFECTS OF INDIVIDUAL VEGETABLES ON HEALTH 221

Section C

ACTIONS OF INDIVIDUAL FRUITS IN DISEASE AND CANCER PREVENTION AND TREATMENT 457

Preface

Diet and nutrition are vital keys to controlling or promoting morbidity and mortality from chronic diseases. The multitude of biomolecules in dietary fruits and vegetables play a crucial role in health maintenance. They may, therefore, be more effective and certainly could have different actions beyond nutrients.

The U.S. National Cancer Institute reports that only 18% of adults meet the recommended intake of vegetables. Increasingly, Americans, Japanese, and Europeans are turning to the use of dietary vegetables, medicinal herbs, and their extracts or components to prevent or treat disease and cancer. It has been known for decades that those populations with high vegetable consumption have reduced risks of cancer. However, which vegetables or fruits, how much of them, and which extracts or components are best to prevent disease and promote health?

This book brings together experts working on the different aspects of supplementation, foods, and plant extracts, in health promotion and disease prevention. Their expertise and experience provide the most current knowledge to promote future research. Dietary habits need to be altered, for most people. Therefore, the conclusions and recommendations from the various chapters will provide a basis for change.

The basic outline of the book has three sections: (A) Fruit and Vegetables in Health Promotion, (B) Effects of Individual Vegetables on Health, and (C) Actions of Individual Fruits in Disease and Cancer Prevention and Treatment.

Constituents with anticancer activities in the prevention phytochemicals, are described.

Bioavailability of important constituents of fruit and vegetables plays a key role in their effectiveness. Their roles as well as that of whole vegetables in gastrointestinal disease, heart disease, and old age are defined. Each vegetable contains thousands of different biomolecules, some with the potential to promote health or retard disease and cancer. By use of vegetable extracts as well as increased consumption of whole plants, people can dramatically expand their exposure to protective chemicals and thus readily reduce their risk of multiple diseases. Specific foods, individual fruits or vegetables and their by-products are biomedicines with expanded understanding and use. For decades, it has been appreciated that oxidative pathways can lead to tissue damage and contribute to pathology. Fortunately, nature has provided us the mechanisms found predominately in plants to defend against such injury. Antioxidant nutritional agents have consequently attracted major attention and rightfully deserve to be studied carefully for possible beneficial roles. One of the main reasons for the interest in antioxidant agents in dietary vegetables, and their products, is their virtually complete lack of harmful side effects. This stands in stark contrast to many drugs that are promoted and studied for possible disease-preventive activity.

Plant extracts as dietary supplements are now a multi-billion dollar business, built upon extremely little research data. For example, the U.S. Food and Drug Administration are pushing this industry, with the support of Congress, to base its claims and products on scientific research. Since common dietary vegetables and over-the-counter extracts are readily available, this book will be useful to laypersons who apply it to modify their lifestyles, as well as to the growing nutrition, food science, and natural product community. This book focuses on the growing body of knowledge on the role of various dietary plants in reducing disease.

Expert reviews will define and support the actions of bioflavonoids, antioxidants, and similar materials that are part of dietary vegetables, dietary supplements, and nutraceuticals. As nonvitamin minerals with health-promoting activities, nutraceuticals are an increasing body of materials and extracts that may have biological activity. Therefore, their role is a major emphasis, along with discussions of which agents may be the active components.

The overall goal is to provide the most current, concise, scientific appraisal of the efficacy of key foods and constituents medicines in dietary plants in preventing disease and improving the quality of life. While vegetables have traditionally been seen to be good sources of vitamins, the roles of other constituents have only recently become more widely recognized. This book reviews and often presents new hypotheses and conclusions on the effects of different bioactive components of the diet, derived particularly from vegetables, to prevent disease and improve the health of various populations.

Ronald Ross Watson
Victor R. Preedy

Acknowledgments

Special appreciation is extended to the Natural Health Research Institute (non-profit) http://www.naturalhealthresearch.org. Its goal is to educate scientists, government regulators, and the lay public about the role of nutrition, bioactive foods and dietary supplements in health and wellness. The NHRI stimulated this book which was approved by its board and advisory panel. Their contribution to the book is sincerely acknowledged. In addition, the NHRI supported the book's production by providing partial support for Bethany L. Stevens, the project's editorial assistant who was critical to the book's success. Her excellent work with the authors, editors and publisher greatly supported the work. The editors also greatly appreciate her assistance without which the book would not have been possible.

Contributors

Alexy, Ute Research Institute of Child Nutrition, (FKE), Heinstueck 11, Dortmund, Germany

Arjmandi, Bahram H. Department of Nutrition, Food and Exercise Sciences, Florida State University, Tallahassee, FL, USA

Badrie, Neela Department of Food Production, Faculty of Science and Agriculture, University of West Indies, St. Augustine, Republic of Trinidad and Tobago, West Indies

Ball, Kylie Centre for Physical Activity and Nutrition Research, Deakin University, Burwood, Victoria, Australia

Basu, Tapan K. Department of Agriculture, Food and Nutritional Science, Faculty of Agricultural, Environmental and Life Sciences, University of Alberta Edmonton, Alberta, Canada

Blasa, Manuela Department of Biomolecular Sciences, Università di Urbino 'Carlo Bo,' Urbino (PU) Italy

Borek, Carmia Department of Public Health and Family Medicine, Tufts University School of Medicine, Boston, MA, USA

Broomes, Jacklyn Department of Food Production, Faculty of Science and Agriculture, University of West Indies, St. Augustine, Republic of Trinidad and Tobago, West Indies

Calhau, Conceição Department of Biochemistry (U38-FCT) Faculty of Medicine of the University of Porto, University of Porto, Rua do Campo Alegre, 687, Porto, Portugal

Casagrande, Stark Sarah Department of Epidemiology, Johns Hopkins Bloomberg School of Public Health, Baltimore, MD, USA

Chandra, Amar K. Department of Physiology, University College of Science and Technology University of Calcutta, Kolkata, West Bengal, India

Chen, Yu Ming Centre of Research and Promotion of Women's Health, School of Public Health and Primary Care, The Chinese University of Hong Kong, Guangzhou, China

Christensen, Lars P. Institute of Chemical Engineering, Biotechnology and Environmental Technology, Faculty of Engineering, University of Southern Denmark, Odense M, Denmark

Clarke, Stephen L. Nutritional Sciences Department, Oklahoma State University, Stillwater, OK, USA

Clementi, Elisabetta M. CNR-Istituto di Chimica del Riconoscimento Molecolare (ICRM), L.go F. Vito n.1, Rome, Italy

Clifton Peter M. CSIRO Human Nutrition, CSIRO Preventative Health Flagship, Adelaide, SA, Australia

Cordeiro, Luciana N. Department of Physiology and Pharmacology, Federal University of Ceará (UFC), Nunes de Melo, 1127, Fortaleza, Brazil

Crawford, David Centre for Physical Activity and Nutrition Research, Deakin University, Burwood, Victoria, Australia

Crujeiras, Ana B. Department of Nutrition and Food Sciences, Physiology and Toxicology, University of Navarra, Pamplona, Spain

Donato Angelino Department of Biomolecular Sciences, Università di Urbino 'Carlo Bo,' Urbino (PU) Italy

Dumancas, Gerard G. Chemistry Department, Oklahoma State University, Stillwater, OK, USA

Ellinger, Sabine Department of Food and Nutrition Science – Nutritional Physiology, University of Bonn, Endenicher Allee 11-13, Bonn, Germany

Elmadfa, Ibrahim University of Vienna, Institute for Nutritional Sciences, Althanstrasse 14, Vienna, Austria

Encabo, Rosario R. Department of Science and Technology, Food and Nutrition Research Institute, Bicutan, Taguig, Metro Manila, Philippines

Faoro, Franco Dipartimento di Produzione Vegetale, Università di Milano and Istituto di Virologia Vegetale, Dipartimento Agroalimentare, CNR, Milano, Italy

Faria, Ana Department of Biochemistry (U38-FCT) Faculty of Medicine of the University of Porto, University of Porto, Rua do Campo Alegre, 687, Porto, Portugal

Ferguson, A. Ross The New Zealand Institute for Plant and Food Research Limited, Functional Foods and Health, Mt Albert, Auckland, New Zealand

Gary-Webb, Tiffany L. Department of Epidemiology, Johns Hopkins Bloomberg School of Public Health, Baltimore, MD, USA

Gennari, Lorenzo Department of Biomolecular Sciences, Universitàdi Urbino 'Carlo Bo,' Urbino (PU) Italy

Goyenechea, Estibaliz Department of Nutrition and Food Sciences, Physiology and Toxicology, University of Navarra, Pamplona, Spain

Gupta, Sanjay Department of Urology, Case Western Reserve University, Cleveland, OH, USA

Havermans, Remco C. Department of Clinical Psychological Science, Maastricht University, P.O. Box 616, Maastricht, Netherlands

Ho, Suzanne C. School of Public Health, Sun Yat-sen University, Guangzhou, China

Hunter, Denise C. The New Zealand Institute for Plant and Food Research Limited, Functional Foods and Health, 120 Mt Albert Road, Mt Albert, Auckland, New Zealand

Ibrahim, Salam A. North Carolina Agricultural and Technical State University, Human Env/Family Science, Benbow Hall, Greensboro, NC, USA

Iriti, Marcello Dipartimento di Produzione Vegetale, Università di Milano and Istituto di Virologia Vegetale, Dipartimento Agroalimentare, CNR, Milano, Italy

Jariwalla, Raxit J. Dr Rath Research Institute, Santa Clara, CA, USA

Kersting, Mathilde Research Institute of Child Nutrition, (FKE), Heinstueck 11, Dortmund, Germany

Khatib, Soliman Laboratory of Natural Medicinal Compounds, MIGAL – Galilee Technology Center, Kiryat Shmona, Israel

Kopsell, David E. Department of Agriculture, Illinois State University, Normal, IL, USA

Kopsell, Dean A. Plant Sciences Department, University of Tennessee, Knoxville, TN, USA

Laaksonen, Mikko Department of Public Health, University of Helsinki, Helsinki, Finland

Lallukka, Tea Department of Public Health, University of Helsinki, Helsinki, Finland

López-Sobaler, Ana M. Departamento de Nutrición, Universidad Complutense, Facultad de Farmacia, Madrid, Spain

Loyola, Anacleta C. Department of Science and Technology, Food and Nutrition Research Institute, Bicutan, Taguig, Metro Manila, Philippines

Lucas, Edralin A. Nutritional Sciences Department, Oklahoma State University, Stillwater, OK, USA

Maeda, Naoki Laboratory of Food and Nutritional Sciences, Department of Nutritional Science, Kobe-Gakuin University, Nishi-ku, Kobe, Hyogo, Japan

Mallillin, Aida C. Department of Science and Technology, Food and Nutrition Research Institute, Bicutan, Taguig, Metro Manila, Philippines

Martínez, J. Alfredo Department of Nutrition and Food Sciences, Physiology and Toxicology, University of Navarra, Pamplona, Spain

Matos, F. J. A. Department of Physiology and Pharmacology, Federal University of Ceará (UFC), Nunes de Melo, 1127, Fortaleza, Brazil

Menezes, Silvana Magalhães Siqueira Department of Physiology and Pharmacology, Federal University of Ceará (UFC), Nunes de Melo, 1127, Fortaleza, Brazil

Misiti, Francesco Department of Health and Motor Sciences, University of Cassino, V.le Bonomi, Cassino, FR, Italy

Mizushina, Yoshiyuki Laboratory of Food and Nutritional Sciences, Department of Nutritional Science, Kobe-Gakuin University, Nishi-ku, Kobe, Hyogo, Japan

Niedzwiecki, Aleksandra Dr Rath Research Institute, Santa Clara, CA, USA

Obenchain, Janel Nutrition and Food Science Track, Urban Public Health Program, Hunter College, City University of New York, New York, USA

O'Mahony, Rachel Royal College of Physicians, London, Senior Research Fellow, NCC-CC, St. Andrew's Place, Regents Park, London, UK

Ortega, Rosa M. Departamento de Nutrición, Universidad Complutense, Facultad de Farmacia, Madrid, Spain

Paolino Ninfali Department of Biomolecular Sciences, Università di Urbino 'Carlo Bo,' Urbino (PU) Italy

Pollard, Christina M. Curtin University of Technology, Perth, Western Australia, Australia

Rahkonen, Ossi Department of Public Health, University of Helsinki, Helsinki, Finland

Rahman, Khalid School of Pharmacy and Biomolecular Sciences, Liverpool John Moores University, Byrom Street, Liverpool, UK

Rath, Matthias Dr Rath Research Institute, Santa Clara, CA, USA

Reinik, Mari Tartu Laboratory, Health Protection Inspectorate, PK272 Tartu, Estonia

Ribeiro, Sônia Machado Rocha Department of Health and Nutrition, Federal University of Vicosa, CEP: 36.570-000, Vicosa, Minas Gerais State, Brazil

Roasto, Mati Department of Food Science and Hygiene, Estonian University of Life Sciences, Kreutzwaldi 58a, Tartu, Estonia

Rodríguez-Rodríguez, Elena Departamento de Nutrición, Universidad Complutense, Facultad de Farmacia, Madrid, Spain

Rowley, Chris Horticulture Australia Limited, Sydney, NSW, Australia

Sagum, Rosario S. Department of Science and Technology, Food and Nutrition Research Institute, Bicutan, Taguig, Metro Manila, Philippines

Schauss, Alexander G. Natural and Medicinal Foods Division, AIBMR Life Sciences, Inc., Puyallup, WA, USA

Schieber, Andreas Department of Agricultural, Food and Nutritional Science, University of Alberta, Edmonton, AB, Canada

Shibamoto, Takayuki Department of Environmental Toxicology, University of California, Davis, CA, USA

Shukla, Sanjeev Department of Urology, Case Western Reserve University, Cleveland, OH, USA

Skinner, Margot A. The New Zealand Institute for Plant and Food Research Limited, Functional Foods and Health, Mt Albert, Auckland, New Zealand

Smith, Brenda J. Nutritional Sciences Department, Oklahoma State University, Stillwater, OK, USA

Song, Danfeng North Carolina Agricultural and Technical State University, Human Env/Family Science, Benbow Hall, Greensboro, NC, USA

Srichamroen, Anchalee Department of Agriculture, Food and Nutritional Science, Faculty of Agricultural, Environmental and Life Sciences, University of Alberta Edmonton, Alberta, Canada

Stevenson, Lesley M. The New Zealand Institute for Plant and Food Research Limited, Functional Foods and Health, Mt Albert, Auckland, New Zealand

Tamme, Terje Department of Food Science and Hygiene, Estonian University of Life Sciences, Tartu, Estonia

Tang, Guangwen Jean Mayer US Department of Agriculture Human Nutrition Research Center on Aging, Tufts University, Boston, MA, USA

Thompson Henry J. Crops for Health Research Program and the Cancer Prevention Laboratory, Colorado State University, Fort Collins, CO, USA

Thompson, Matthew D. Crops for Health Research Program and the Cancer Prevention Laboratory, Colorado State University, Fort Collins, CO, USA

Trinidad, Trinidad P. Department of Science and Technology, Food and Nutrition Research Institute, Bicutan, Taguig, Metro Manila, Philippines

Vaya, Jacob Laboratory of Natural Medicinal Compounds, MIGAL – Galilee Technology Center, Kiryat Shmona, Israel

Viana, Glauce S.B. Rua Barbosa de Freitas, 1100, Fortaleza, Brazil

Viladrich, Anahi Community Health Education Track, Urban Public Health Program, Hunter College, City University of New York, New York, USA

Wei, Alfreda Department of Molecular Biosciences, University of California, Davis, CA, USA

Wolf, Alexandra Austrian Agency for Health and Food Safety (AGES), Competence Center for Nutrition and Prevention, Vienna, Austria

Yeh, Ming-Chin Nutrition and Food Science Track, Urban Public Health Program, Hunter College, City University of New York, New York, USA

Yoshida, Hiromi Laboratory of Food and Nutritional Sciences, Department of Nutritional Science, Kobe-Gakuin University, Nishi-ku, Kobe, Hyogo, Japan

FRUIT AND VEGETABLES IN HEALTH PROMOTION

1

Botanical Diversity in Vegetable and Fruit Intake: Potential Health Benefits

Matthew D. Thompson and Henry J. Thompson

Crops for Health Research Program and the Cancer Prevention Laboratory, Colorado State University, Fort Collins, CO

1. OVERVIEW

Dietary guidelines are evolving from a primary focus on providing adequate intake of essential nutrients in order to prevent nutritional deficiency to an emphasis on reducing the prevalence of chronic diseases including cardiovascular disease, cancer, type II diabetes, and obesity [1–3]. During this transition, there has been a movement to broaden nutritional terminology such that nutrients are divided into two categories: essential and nonessential [4]. *Essential nutrients* are those substances that cannot be made in the human body but that are required for normal cellular function. The absence of essential dietary nutrients results in defined disease syndromes. *Nonessential nutrients* are not required for life, but they promote health [5]. Many chemical constituents of plant-based foods, i.e. foods which are plants or are derived from plants, are termed nonessential nutrients since they positively impact health; such phytochemicals are also referred to as phytonutrients.

Current dietary recommendations attempt to meet these nutrient requirements and are based on grouping foods using culinary definitions and knowledge of essential nutrient content. Despite the recognition that literally thousands of chemicals exist in plant-based foods and that they are likely to exert a wide range of bioactivities in living systems, dietary guidelines continue to provide limited direction about the specific types of plant-based foods that should be combined to render maximal health benefits. This situation exists for many reasons including: 1) the lack of a systematic approach by which plant-based foods are nutritionally classified; 2) the limited information regarding the chemical profile of each type of plant-based food; 3) the lack of data on the biological activities of plant chemicals; and 4) the paucity of information about the impact of plant-based food combinations on health outcomes. However, technological advances in chromatographic separation and chemical identification of phytochemicals are occurring at a rapid rate [6] and this progress is providing a large amount of information regarding the chemical composition of plant-based foods. This situation has created an unprecedented opportunity to expand our approach to dietary guidelines and menu planning.

The objective of this chapter is to raise the awareness of health care professionals about opportunities to extend dietary guidance about plant-based food intake beyond meeting the recommended servings/day of cereal grains, vegetables, and fruit by incorporating information about the botanical families from which the plant-based foods are selected for menu planning. The approach also has the potential to identify food combinations that reduce chronic disease risk. The remainder of this chapter addresses three topics: Section 2 details the rationale underlying the proposed use of botanical families, Section 3 provides evidence of the potential usefulness of this approach in an effort to reduce chronic disease risk, and Section 4 considers how botanical families can be applied to meal planning.

2. RATIONALE FOR USING BOTANICAL FAMILIES

2.1 Categorizing Vegetables and Fruit

The focus of this chapter is on vegetables and fruit, yet a careful inspection of how these terms are defined and the manner in which they are used reveals a surprising amount of ambiguity about the foods placed in each category. While the term 'vegetable' generally refers to the edible parts of plants, the categorization of foods as vegetables is traditional rather than scientific, varying by cultural customs of food selection and preparation. Moreover, in the biomedical literature, some foods are not classified as vegetables because of their content of starch, e.g. potatoes [7,8], without consideration that these foods, as well

as other staple food crops, are vehicles for the delivery of a wide array of small molecular weight compounds in addition to carbohydrate. The categorization of foods as fruits is no less ambiguous. Strictly, a fruit is the ripened ovary of a plant and its contents. More loosely, the term is extended to the ripened ovary and seeds together with any structure with which they are combined. The botanical definitions for fruit are not uniformly applied in nutrition and dietetics; rather, cultural customs tend to determine what differentiates a fruit from vegetables and grains. Examples include: 1) the apple, a pome, in which the true fruit (core) is surrounded by flesh derived from the floral receptacle; 2) wheat, a fertilized ovule is comprised of an outer coat (testa) that encloses a food store and embryo; seeds of wheat, rice, and oats, which are botanically the fruits of the plant, are classified in food terms as cereal grains, i.e. they are neither vegetable nor fruit; 3) tomato is classified as a vegetable, though it is the ovary of the plant; and 4) legumes, which could be botanically classified as fruits, are sometimes considered vegetables, but if they are consumed as a staple crop, they are categorized in the meat food group. Together, these examples demonstrate the need to acknowledge how we classify plant foods and in categorizing them, may bias ourselves to thinking certain foods are either more or less related, more or less diverse, or more or less likely to provide health benefit. To overcome this bias, we need to acknowledge the different ways we categorize plant-based foods, e.g. scientific and cultural perspectives. The remainder of this discussion is as inclusive as possible, as almost all the plant-based foods we eat could be classified as a fruit or vegetable depending on the organizational scheme. Inclusion allows us to consider what advantages might be gained from using a scientifically-derived botanical and taxonomic scheme as an additional filter through which plant-based foods are categorized.

2.2 Linnaean Taxonomy

Plant taxonomy classifies plants in a hierarchical manner. Table 1.1 shows an example of the taxonomic classification scheme for the apple (*Malus domestica*, Borkh.) using the Linnaean system, which is the most common method of classification for living organisms. Ascertaining groupings of plant-based foods by this taxonomic classification at the level of the botanical family, as shown in Table 1.2, is useful in promoting an understanding of general relationships among food crops which often go unrecognized [9–11]. This classification scheme has been used: 1) to gain insight regarding specific chemical components of foods that may account for health benefits; and 2) to develop functional foods and nutraceutical supplements that emphasize a particular class of chemicals [12–15]. However, little attention has been given to using this information to identify health-promoting combinations of plant-based foods that, when eaten as a regular component of the diet, result in a reduced risk for chronic diseases.

2.3 Chemotaxonomy

To better understand how botanical family classification informs understanding of the

phytochemical composition of various plant-based foods, an additional approach to taxonomy is needed. Chemotaxonomy, also called chemosystematics, is the attempt to identify and classify plants according to differences and similarities in their biochemical components [16]. The products of plant biosynthesis are generally divided into primary and secondary metabolites as shown in Figure 1.1 [17]. Primary plant metabolites, e.g. carbohydrate, protein, and fat, are considered as the essential building blocks for plant growth and development. The production of these macromolecules is under stringent genetic control and while variation among plants in the content of primary metabolites does exist and is of interest to nutritionists, those differences are of limited value in chemotaxonomy. On the other hand, plants have evolved the capacity for the combinatorial chemical synthesis of a vast array of secondary metabolites. The synthesis of secondary metabolites by plants has two main purposes: 1) signaling (e.g. plant hormones); and 2) defense against abiotic (e.g. ultraviolet light) and biotic stressors (e.g. microbes) [16,18,19]. In terms of defense, secondary metabolites protect plants against microbes, pests, and other plants as indicated in Figure 1.1 [19]. In a broad sense, all of these chemicals function as semiochemicals, i.e. 'message carriers' [20], a term often used in chemical ecology. The chemicals that have evolved over the millennia span at least 14 defined chemical classes of compounds (Table 1.3) and in excess of 200,000 chemical structures. Available evidence indicates that all chemical classes have a biological activity that was selected for during evolution and that these biosynthetic strategies sustained the survival of the plant species [17,19,21]. When classical taxonomic information and chemotaxonomic data are overlaid, relationships become apparent; plants within a botanical family tend to have greater chemical similarity than plants in different families, i.e. plants within a botanical family emphasize

TABLE 1.1 Taxonomic Classification Scheme of an Example Food

Taxonomic Rank	Taxon
Kingdom	Plantae
Division	Magnoliophyta
Class	Magnoliopsida
Order	Rosales
Family	**Rosaceae**
Subfamily	Maloideae
Genus	*Malus*
Species	*M. domestica*
Common name	**Apple**

Bolded ranks and taxa indicate chapter emphasis.

TABLE 1.2   Botanical Families

Botanical Family	Common Name	Vegetables, Fruits, Herbs, and Spices
Actinidiaceae	–	Chinese gooseberry, kiwi
Agaricaceae	–	Mushroom
Anacardiaceae	–	Black currant, mango
Annonaceae	Custard apple	Cherimoya, custard apple, pawpaw, sugar apple
Apiaceae (Umbelliferae*)	Carrot	Carrots, celery, chervil, coriander (Chinese parsley), dill, fennel, parsley, parsnips
Asteraceae (Compositae*)	Daisy or sunflower	Artichoke, chamomile, chicory leaf, dandelion, echinacea, endive, lettuce (iceberg, radicchio, red leaf, romaine), sunflower sprouts, tarragon
Brassicaceae (Cruciferae)*	Mustard or cabbage	Broccoli, Brussels sprouts, cabbages (Chinese, green, red), cauliflower, collard greens, cress leaf, daikon, horseradish, kale, kohlrabi, mustard greens, radish, rocket salad (roquette/arugula), rutabaga, turnip, turnip greens, watercress
Bromeliaceae	–	Pineapple
Caricaceae	–	Papaya
Chenopodiaceae	Goosefoot	Beet, beet greens, orach, spinach, Swiss chard
Convolvulaceae	Morning glory	Sweet potato
Cucurbitaceae	Gourd	Cucumber, melon (canteloupe, Crenshaw, honeydew, watermelon), pumpkin, squash (summer, winter, chayote, zucchini)
Ericaceae	Heath	Blueberry, cranberry, huckleberry, lingonberry, oheloberry, wintergreen
Euphorbiaceae	Spurge	Cassava, tapioca
Fabaceae (Leguminosae*)	Pea, bean, or pulse	Soy bean, dry bean (cranberry, kidney, navy, northern, pink, pinto), alfalfa sprouts, peanuts, various beans: fava, garbanzo, green, lima, mung; fenugreek, jicama, lentils, licorice, peas (blackeye, green, pods, split), tamarind
Lamiaceae (Labiatae*)	Mint	Basil, hoarhound, hyssop, lavender, lemon balm, marjoram, mint, oregano, pennyroyal, rosemary, sage, savory, thyme
Lauraceae	Laurel	Avocado, cinnamon, sassafras
Liliaceae	Lily	Aloe vera, asparagus, chives, garlic, leeks, onion, scallion, shallot
Malvaceae	–	Chocolate
Moraceae	Mulberry	Breadfruit, fig. jackfruit, mulberry
Musaceae	Banana	Banana, plantain
Palmae	Palm	Coconut, date, palm heart
Piperaceae	–	Black pepper
Poaceae (Gramineae*)	Grass	Bamboo shoots, corn, lemon grass, wheat, rice
Polygonaceae	Buckwheat	Rhubarb, sorrel
Rosaceae	Rose	Apple, applesauce, apricot, blackberry, cherry, Juneberry, loganberry, nectarine, peach, pear, plum, prune, quince, raspberry, salmonberry, sloes, strawberry
Rutaceae	Citrus	Curry leaf, grapefruit, kumquat, lemon, lime, mandarin orange, orange, tangerine
Sapindaceae	Soapberry	Lychee, ackee, maple syrup
Solanaceae	Nightshade	Eggplant, peppers (chili, green, red), pimiento, potato, tomatillo, tomato
Vitaceae	Grape	Grape, raisin

*Alternate botanical family names.

A. FRUIT AND VEGETABLES IN HEALTH PROMOTION

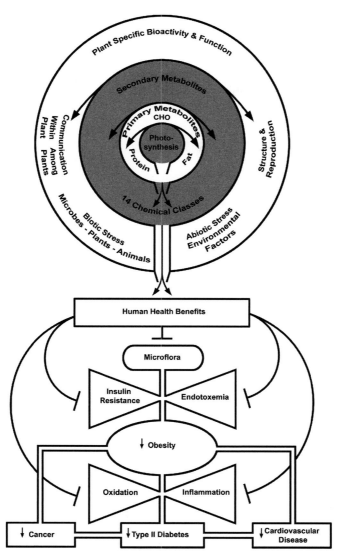

FIGURE 1.1   Integration of plant and animal metabolism for chronic disease prevention in human beings.
Plants fix carbon dioxide during the process of photosynthesis and use the fixed carbon to synthesize primary and secondary metabolites. Whereas primary metabolites, e.g. carbohydrates, proteins, and lipids, provide structural and functional components of the plant, secondary metabolites are used for communication, reproduction, defense, etc. Fourteen classes of chemical compounds are produced as plant secondary metabolites and each class of chemicals exerts biological activities that have the potential to promote human health via their effects on microbial and mammalian metabolism within the individual that ingests these chemicals as plant-based foods. Major chronic diseases that can be affected by plant primary and secondary metabolites are: cancer, cardiovascular disease (CVD), type II diabetes, and obesity, diseases that account for 60% of all deaths worldwide. Common to the pathogenesis of these diseases are altered glucose metabolism, chronic inflammation, increased cellular oxidation, and chronic endotoxemia, metabolic processes that plant metabolites have been shown to affect.

TABLE 1.3   Classes of Plant Secondary Metabolites*

Chemical Classes

Alkaloids

Amines

Cyanogenic glycosides

Diterpenes

Flavonoids

Glucosinolates

Monoterpenes

Non-protein amino acids

Phenylpropanes

Polyacetylenes

Polyketides

Sesquiterpenes

Tetraterpenes

Triterpenes, saponins, steroids

*Source: Wink [17].

biosynthetic pathways for specific classes of chemicals [16,17,22,23], and plants within botanical subfamilies emphasize particular chemical compounds within those subclasses in comparison to plants in other subfamilies [17]. The further apart the botanical families are from one another in the evolutionary tree (Figure 1.2), the more likely they are to differ in the composition of secondary metabolites [17]. In fields such as pharmocognosy, where medicinal plants, crude herbs or extracts, pure natural compounds, and foods are being evaluated for health benefits [18], the goal is to identify chemicals with specific molecular targets. The success of efforts to identify natural products will be based on successfully targeting mammalian proteins involved in cell signaling during the pathogenesis of chronic diseases. The fact that compounds from all chemical classes listed in Table 1.3 have activity in mammalian systems provides considerable support for developing recommendations for using phytochemically diverse food combinations (recipes and menus) as a method to maximize the potential to enhance health and to prevent

chronic diseases in the context of promoting variety and moderation. As more research reveals the level of conservation among plant and animal signaling pathways, the relationships among botanical taxonomy, chemotaxonomy, and the phytochemical composition of plant-based foods will provide a framework for predicting the bioactivity of the foods that are consumed as part of a diverse plant-based diet (Figure 1.1).

2.4 From Botanical Family to Chronic Disease Prevention

A reason for developing dietary guidelines is to reduce the risk of four related chronic diseases. Specifically, cardiovascular disease, cancer, type II diabetes, and obesity are metabolic disorders with shared impairments in both cellular processes and metabolism, although each disease also retains unique characteristics. A better understanding of their interrelationships has come as a result of proteomic investigations providing evidence of a common pathogenic basis for their occurrence [24–28]. At the cellular level, the pathologies associated with each disease display alterations in cell proliferation, blood vessel formation, and cell death. Also common to these diseases are alterations in glucose metabolism, chronic inflammation, and cellular oxidation attributed to a common network of cell signaling events that are perturbed in each of these disease states [24]. In addition, emerging evidence indicates that modulation of gut microflora predisposes an individual to each of the disease processes [27,28]. Microflora appear to be able to exert effects through either biosynthesis of new compounds or chemical transformations of ingested ones, and as a consequence, influence exposure of the host to gut microflora-associated endotoxins [27–29]. By overlaying the chemotaxonomic and bioactivity relationships, i.e. matching of plant-based foods with dysfunctional signaling pathways

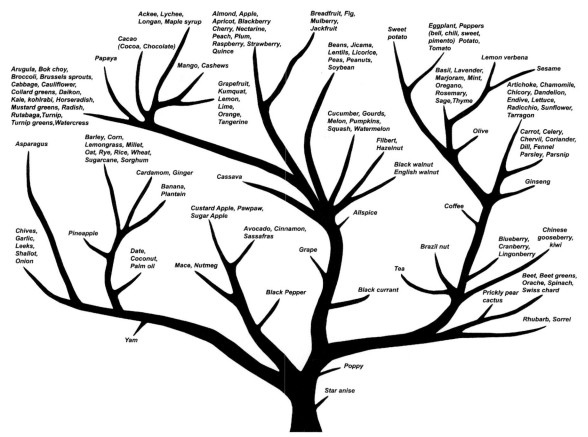

FIGURE 1.2 Evolutionary tree of plant-based foods.

Plants that provide foods consumed by human beings arose at different times during evolution. The tree shows the relationships that exist among commonly eaten plant-based foods. Foods on the same branches of the tree share more similarities, genetically and chemically, than foods that are further separated on the tree. The relationships shown are based on Linnaean taxonomy and were determined based on information available from the Tree of Life web project (http://tolweb.org/tree/). Chemical relationships among botanical families are studied in the field of chemotaxonomy as discussed in Section 2.3 (see cited references). Only angiosperms are shown (does not include gymnosperms, seedless vascular, and seedless nonvascular plants).

associated with chronic disease, a basis for identification and implementation of patterns of plant-based food consumption that inhibit the pathogenesis of chronic diseases will be developed.

2.5 Advantages and Limitations of Using Botanical Families

The use of botanical families to investigate food combinations provides a broader framework upon which to construct diets, as opposed to the reductionist approach that does not take into account the full spectrum of plant chemicals available and essentially violates the variety and moderation axiom. Perpetuation of the quest for a single phytochemical solution, whether it is at the level of a specific chemical (e.g. beta-carotene), a class of compounds (e.g. carotenoids), or a particular food (e.g. carrots), will fail to provide the diversity of the botanical family approach. Use of botanical families requires the consideration, identification, and recommendation of dietary patterns of plant-based food intake for enhanced health benefit that encourages variation and, as a corollary, moderation.

Nonetheless, there are limitations to this approach that have been identified by efforts associated with using chemical composition for taxonomic classification of plants [17,21–23]. One limitation of the botanical family approach is that in some cases, plant-based foods from different botanical families may have more chemical similarities than foods from within a botanical family. This would result from the convergent use of a certain class of chemicals for a specific function (e.g. soil–microbe interactions), environmental factors (e.g. exposed to light versus underground), and plant anatomical origin (e.g. fruit versus stem versus root). Potato versus tomato is a good example of plant-based foods which are chemically quite different from each other [30,31], though they reside within the same botanical family. Tomatoes are exposed to sunlight, do not have to contend with soil fauna and microbes, and are of fundamentally different plant anatomical origin (i.e. tomatoes are the ovaries of the plant while potatoes are underground stems). In considering these within-family disparities, examples of within-family divergent function of chemical classes may also be found. In general, one should always exercise caution in exploring botanical families beyond the commonly consumed foods, i.e. wild versus cultivated plants. As an example, the nightshade family (Solanaceae), which has the well-known members potato and tomato, also has plants known for toxic effects such as belladonna. In general, plant-based foods could be further categorized based on plant anatomical origin (i.e. leaves, roots, stems, fruits) or location of plant-based food components relative to the surroundings (i.e. underground/on ground/above ground or soil/microbes/pests/fauna/light) and should consider known toxicities.

3. EVIDENCE FOR THE VALUE OF USING BOTANICAL FAMILIES

3.1 Dietary Guidelines

Dietary guidelines have been developed by the World Health Organization, most national governments, and other organizations [32]. The guidelines are intended to educate the public regarding healthy food choices, set nutrition policies, and plan menus, and have worked well for controlling nutritional deficiency diseases. The development of dietary guidelines is recognized as an evolving process; as evidence, the Recommended Dietary Allowances have been reformulated as the Recommended Dietary Intakes, and the US Dietary Guidelines for Americans have been updated every 5 years since their inception. The updating process takes into account new information on health and disease and trends in the food supply. The benefit that can result from this process has been illustrated, as chronic disease prevention is complicated by the relatively unknown nature of the dietary factors that control disease occurrence, and updating is required for making progress. The 1995 Dietary Guidelines for Americans focused on the reduction of total fat with little distinction among types of fat or among forms of carbohydrate [33]. Because emerging data provided little evidence that the percentage of total fat in the diet was related to major health outcomes, but that the types of fat, forms of carbohydrate, and sources of protein had important influences on the risk of CVD and type II diabetes, McCullough and Willett developed an alternative dietary pattern named the Alternate Healthy Eating Index that took into account these factors [34]. As reported in 2008, these investigators determined that individuals who followed the 1995 US dietary guidelines did not experience a reduced risk for the occurrence of chronic disease, whereas those individuals whose dietary pattern mirrored the Alternative Healthy Eating Index did experience a reduction in risk [32]. The above situation parallels reports of null effects for increased vegetable and fruit intake on various chronic disease endpoints, where assessment of vegetable and fruit intake can be crude and relatively nonspecific [35]. This prompts the question of whether greater specificity in studying vegetable and fruit consumption would identify a dietary pattern, which is currently masked, that has the potential to reduce chronic disease [32]. Evidence is beginning to emerge in support of the argument that limiting the evaluation of vegetable and fruit intake to total amount consumed rather than using more detailed information, e.g. botanical families, is masking health-related activity of these plant-based foods. In a large prospective study, no relations between total intake of vegetables and fruit combined or the total of each considered separately were observed for lung cancer risk [11]. However, higher consumption of several botanical groups, including Rosaceae, Convolvulaceae, and Umbelliferae, was significantly inversely associated with lung cancer risk in men. Similarly, in several other studies, investigators analyzed the relation of individual plant-based foods and/or plant-based food groups with lung cancer risk; the most consistent findings were for Rosaceae [36–38], Brassicaceae [36,39,40], and Rutaceae [36,40]. A perusal of the literature shows additional examples similar to these exist [10,41], yet there is little discussion of using botanical families as a primary tool to study the effects of plant-based food intake on disease risk nor are we aware of ongoing efforts to develop food consumption instruments validated for the collection of this information or of efforts to identify botanical combinations (patterns) associated with health benefit.

3.2 Botanical Diversity and Oxidative Biomarkers

Our laboratory has investigated the importance of considering not only the amount of vegetable and fruit consumed but also its type, defined by its botanical family, using an intervention approach [42]. In this study, two diets that varied in botanical diversity were evaluated for the ability to reduce oxidative damage of lipids or DNA. The diets provided 8–10 servings of fruits and vegetables per day: the high botanical diversity diet (HBD) included plant-based foods from 18 botanical families and the low botanical diversity diet (LBD) emphasized 5 of the 18 botanical families which had been reported to have high antioxidant activity. In Table 1.4, the major foods and botanical families are listed, with more detail provided elsewhere [9,42]. The 106 women who completed the study on the two different diets, LBD and HBD, did not consume different overall amounts of fruits and vegetables (9.1 ± 2.6 and 8.3 ± 2.1 servings per day, $P > 0.1$). Yet, the HBD diet reduced DNA oxidation ($P < 0.05$). Both diets were associated with a significant reduction in lipid peroxidation ($P < 0.01$), with the effect being greatest in the HBD group. Key findings from the study are presented in Table 1.5. These findings indicate that botanical diversity plays a role in determining the bioactivity of high vegetable and fruit diets and that smaller amounts of many phytochemicals may have greater beneficial effects than larger amounts of a less diverse set of phytochemicals.

4. TRANSLATION OF BOTANICAL FAMILY CONCEPTS TO DIETETIC PRACTICE

The use of botanical families provides a scientific basis by which to plan recipes and menus that maximize the phytochemical diversity

TABLE 1.4 Major Foods and Botanical Families Used in the Clinical Intervention

Botanical Family	Common Foods
Actinidiaceae	Kiwi
Agaricaceae	Mushroom
Bromeliaceae	Pineapple
Chenopodiaceae*	Spinach, Swiss chard, beet
Compositeae	Artichoke, endive, lettuce
Convolvulaceae	Sweet potato
Cruciferae*	Cabbage, broccoli, radish
Cucurbitaceae	Cucumber, zucchini, melon
Ericaceae	Blueberry, cranberry
Gramineae	Corn, bamboo shoots
Leguminosae	Chickpeas, lentils, soybeans
Liliaceae*	Chive, garlic, onion, scallion
Musaceae	Banana, plantain
Rosaceae	Apple, peach, strawberry
Rutaceae*	Grapefruit, orange, lemon, lime
Solanaceae*	Tomato, eggplant, peppers
Umbelliferae	Carrot, celery, parsnip, parsley
Vitaceae	Grape

*Denotes botanical families emphasized in low botanical diversity diet.
Note: Botanical family names are reflective of what is in Thompson et al. [42].

TABLE 1.5 Effect of Botanical Diversity on Oxidative Biomarkers in the Clinical Intervention

Biomarker Class	Low Biodiversity[3–5]	High Biodiversity[3–5]
Lipid oxidation[1]	-0.05 ± 0.02*	-0.13 ± 0.02*
DNA oxidation[2]	-0.03 ± 0.34	-0.81 ± 0.33*

[1]Lipid oxidation (urinary 8-iso-PGF2α μmol/mol creatinine).
[2]DNA oxidation (lymphocyte 8-oxo-dG/10^6dG).
[3]Mean difference between pre- and post-intervention.
[4]*$P < 0.05$.
[5]$N = 53$.

of the diet by promoting the consumption of a wide variety of botanically distinct foods. We have constructed recipes and menus with high levels of botanical diversity which meet established dietary guidelines, e.g. the Recommended Dietary Intakes, US Dietary Guidelines, and US Food Guide Pyramid, thus demonstrating the feasibility of this approach. These diets were successfully prepared either at home [42] or by food retailers [9], and the plans were followed by a large number of individuals ($N > 350$) for a period of up to 2 months. Moreover, given current patterns of food acquisition by the general public, e.g. use of designer foods, convenience foods, and pre-packaged food ingredients, and the high frequency of eating out-of-home, there are many avenues through which the botanical diversity of the diet can be increased. The following goals and supporting rationale illustrate our approach to the development of menus and recipes.

4.1 Rationale and Implementation

Nutrients in foods act in synergy such that the beneficial effects of the phytochemicals in the diet emerge as something other than a simple additive property. The evidence presented in Section 2.3 indicates that each of the 14 classes of phytochemicals listed in Table 1.3 exerts biological effects but their distribution in foods differs across botanical families. Consequently, the goal of consuming many botanical families daily is to maximize potential for synergism among various phytochemicals while minimizing the likelihood of deleterious side effects. This strategy is supported by evidence that dietary phytochemicals exert their biological effects through hormetic relationships, i.e. in a U- or J-shaped dose response [43–45]. Hormesis is an adaptive response to a biological perturbation; in the absence of a health-promoting phytochemical milieu, the system does not function at an arbitrarily defined optimum. By constructing menus comprised of foods from many botanical families which meet established dietary guidelines, variety and moderation can be effectively achieved. The metric for diversity in the diet utilizes the list of botanical families in Table 1.2. The established goal of dietary diversity is attained by following a set of rules to guide menu design, which would consist of specific requirements for eating foods within each botanical family. This approach can be used to design recipes and menus that consistently deliver between 5 and 18 botanical families daily divided among three meals and three

snacks and in as few as 1400 kcal per day [9,42]. A high botanical diversity menu is shown in Table 1.6. Key points illustrated contain the inclusion of multiple botanical families throughout the day and the use of different foods within a botanical family when that family is consumed more than once within a day. The use of multiple staple food crops is also illustrated as recommended in Section 4.2.1. Note that the botanical families used span the evolutionary tree of food origins shown in Figure 1.2.

TABLE 1.6   Example of High Botanical Diversity Menu

Meal	Menu	Food	Botanical Family
		Orange	Rutaceae[1, a]
Breakfast	Waffle and orange juice	Waffle	Poaceae (Gramineae)[2, a]
		Maple	Sapindaceae[3, a]
Snack	Banana	Banana	Musaceae[4, a]
		Spinach	Chenopodiaceae[5, a]
		Pasta	Poaceae (Gramineae)[2, b]
		Red potato	Solanaceae[6, a]
Lunch	Spinach salad and red potato gratin	Tofu	Fabaceae (Leguminosae)[7, a]
		Snow peas	Fabaceae (Leguminosae)[7, b]
		Leek	Alliaceae (Liliaceae)[8, a]
		Bean	Fabaceae (Leguminosae)[7, c]
		Scallions	Alliaceae (Liliaceae)[8, b]
		Garlic	Alliaceae (Liliaceae)[8, c]
Snack	Kiwi	Kiwi	Actinidiaceae[9, a]
		Kidney	Fabaceae (Leguminosae)[7, d]
		Tomato	Solanaceae[6, b]
		Sweet	Convolvulaceae[10, a]
		Corn meal	Poaceae (Gramineae)[2, c]
		Onion	Alliaceae (Liliaceae)[8, d]
Dinner	Vegetarian chili and sweet potato cornbread	Tomato	Solanaceae[6, c]
		Carrots	Apiaceae (Umbelliferae)[11, a]
		Celery	Apiaceae (Umbelliferae)[11, b]
		Garlic	Alliaceae (Liliaceae)[8, c]
		Cumin	Apiaceae (Umbelliferae)[11, c]
		Black	Piperaceae[12, a]
Snack	Hershey's kisses	Chocolate	Malvaceae[13, a]

*Number superscripts (*i*) are indexing the *i*th botanical family used and letter superscripts (*j*) indicate *j*th food of the *i*th family.
Note: Menu ingredients not relevant to discussion of botanical families are not listed in the table, e.g. dairy, etc.

4.2 Other Considerations

4.2.1 Staple Food Crops

Recognizing that food is the primary vehicle by which health-promoting chemicals are delivered to the human body and that staple food crops, i.e. dry beans, corn, rice, wheat, and potatoes, are consumed in large amounts on a daily basis, these plant-based foods should be recognized as prominent delivery vehicles for plant secondary metabolites. Therefore, staple crops should not be neglected in developing approaches that promote botanical diversity. Table 1.7 indicates the lack of botanical family diversity in the typical Western diet; there is a veritable monopoly of the family Poaceae in the staple foods that are eaten. Two approaches are suggested to increase variety. The first is to recognize that legumes, botanical family Fabaceae (Leguminosae), and specifically dry bean (*Phaseolus vulgaris*, L.), serve as a staple food crop in many parts of the world with per capita intake of 1.5 to 3 cups per day, whereas average per capita intake in the United States is less than one-eighth cup per day. Increasing dry bean consumption should be a goal given its reported health benefits [46]. The second approach is to systematically vary the sub-families of Poaceae that are used in menu planning and to include less commonly consumed grains that serve as staple crops in some regions of the world. Additionally, though not feasible for all crops, the varieties (i.e. cultivars) can be rotated, as differential health benefit has been reported for specific varieties of crops [30,46,47].

4.2.2 Factors that can Affect Dietary Phytochemical Exposure

Plants use various mechanisms to defend themselves against microbes, pests, and other plants. As outlined elsewhere [48,49], plants have evolved two major strategies for storing these chemicals: 1) constitutive accumulation of metabolites in the target tissue organelles; and 2) accumulation of precursor compounds and enzymes required for their activation in distinct tissue compartments. This information has implications in food preparation. First, since most of the natural products that are isolated in substantial amounts from plants accumulate in specific organelles, it is essential to determine where the health beneficial phytochemicals are stored in the food and to develop food preparation techniques that maximize their ingestion. Second, since release of bioactive compounds from foods such as onions and garlic depends on mixing of precursor phytochemicals and plant enzymes, food preparation techniques should assure that these reactions can occur and result in maximal benefit to the individual consuming the food.

4.3 Use of Botanical Classification to Identify Food Combinations with Human Health Benefit

A major difficulty in conducting studies of single foods or nutrients in relation to human health is the high degree of correlation among many dietary constituents [50]. Because of this, identifying the effects of a single food or nutrient represents a serious methodological problem. Moreover, the assumption that single foods or nutrients have isolated effects is not likely to be valid [51]. Rather, foods and nutrients, as

TABLE 1.7 Botanical Families of Major Calorie-Contributing Plant-Based Foods to Western Diet

Common Name	Genus	Botanical Family
Wheat	*Triticum*	Poaceae
Corn	*Zea*	Poaceae
Rice	*Oryza*	Poaceae
Potato	*Solanum*	Solanaceae
Sugar*	*Saccharum*	Poaceae

*Plant-derived food component.

outlined in the rationale, are more likely to act in synergy. In response to this situation, efforts have been initiated to identify dietary patterns associated with differences in risk for various chronic diseases [51]. The expectation is that a dietary pattern is more likely to detect the totality of dietary phytochemical exposure, including relevant interactions among the chemicals ingested, in a manner that studies of single nutrients or of individual foods cannot [52]. Given this perspective and the data reported in Table 1.5, we propose that efforts to identify dietary patterns/food combinations that inhibit chronic disease should be extended to include the botanical family.

5. SUMMARY

Variety and moderation in the diet remain the hallmarks of sound nutritional advice with clear application to food-based health promotion and disease prevention. The thesis of this chapter is that most typical dietary patterns fail to capitalize on the wide variety of phytochemicals available from food that, when consumed in appropriate amounts, may provide health benefit. Botanical families from which foods are derived can be used as a metric for quantifying the phytochemical diversity of the diet and to systematically identify plant-based food combinations with health benefit. A framework has been provided for using the botanical family concept as a tool by which to assist the health care professional with creating diets that capitalize on the richness of beneficial chemicals in plant-based foods to maintain a lifestyle that promotes well-being.

ACKNOWLEDGMENTS

This work was supported in part by PHS grants R01-CA125243 and U54-CA116847 from the National Cancer Institute. The authors thank Blair Dorsey and John McGinley for their assistance in the preparation of this chapter.

References

1. USDA. (2008). *Dietary guidance.* Beltsville MD: United States Department of Agriculture.
2. Aggett, P. J., Bresson, J., Haschke, F., Hernell, O., Koletzko, B., Lafeber, H. N., Michaelsen, K. F., Micheli, J., Ormisson, A., Rey, J., Salazar de, S. J., & Weaver, L. (1997). Recommended dietary allowances (RDAs), recommended dietary intakes (RDIs), recommended nutrient intakes (RNIs), and population reference intakes (PRIs) are not 'recommended intakes'. *Journal of Pediatric Gastroenterology and Nutrition, 25,* 236–241.
3. Harris, S. S. (2000). Dietary guidelines for Americans recommendations for the year 2000. *Food Australia, 52,* 212–214.
4. Burlingame, B. (2001). What is a nutrient? *Journal of Food Composition and Analysis, 14,* 1.
5. Charles, J. (2003). Encyclopedia of food and culture, 3 vols. *Library Journal, 128,* 71.
6. Eisenreich, W., & Bacher, A. (2007). Advances of high-resolution NMR techniques in the structural and metabolic analysis of plant biochemistry. *Phytochemistry, 68,* 2799–2815.
7. Johansson, I., Hallmans, G., Wikman, A., Biessy, C., Riboli, E., & Kaaks, R. (2002). Validation and calibration of food-frequency questionnaire measurements in the Northern Sweden Health and Disease cohort. *Public Health Nutrition, 5,* 487–496.
8. Kaaks, R., & Riboli, E. (1997). Validation and calibration of dietary intake measurements in the EPIC project: Methodological considerations. European prospective investigation into cancer and nutrition. *International Journal of Epidemiology, 26*(Suppl. 1), S15–S25.
9. Thompson, H. J., Heimendinger, J., Sedlacek, S., Haegele, A., Diker, A., O'Neill, C., Meinecke, B., Wolfe, P., Zhu, Z., & Jiang, W. (2005). 8-Isoprostane F2alpha excretion is reduced in women by increased vegetable and fruit intake. *American Journal of Clinical Nutrition, 82,* 768–776.
10. Shannon, J., Ray, R., Wu, C., Nelson, Z., Gao, D. L., Li, W., Hu, W., Lampe, J., Horner, N., Satia, J., Patterson, R., Fitzgibbons, D., Porter, P., & Thomas, D. (2005). Food and botanical groupings and risk of breast cancer: A case-control study in Shanghai, China. *Cancer Epidemiology, Biomarkers & Prevention, 14,* 81–90.
11. Wright, M. E., Park, Y., Subar, A. F., Freedman, N. D., Albanes, D., Hollenbeck, A., Leitzmann, M. F., & Schatzkin, A. (2008). Intakes of fruit, vegetables, and

specific botanical groups in relation to lung cancer risk in the NIH-AARP diet and health study. *American Journal of Epidemiology, 168*, 1024–1034.

12. Molyneux, R. J., Lee, S. T., Gardner, D. R., Panter, K. E., & James, L. F. (2007). Phytochemicals: The good, the bad and the ugly? *Phytochemistry, 68*, 2973–2985.

13. Espin, J. C., Garcia-Conesa, M. T., & Tomas-Barberan, F. A. (2007). Nutraceuticals: Facts and fiction. *Phytochemistry, 68*, 2986–3008.

14. Lachance, P. A., & Das, Y. T. (2007). Nutraceuticals. In D. J. Triggle & J. B. Taylor (Eds.), *Comprehensive medicinal chemistry II* (pp. 449–461). Oxford: Elsevier.

15. Marin, F. R., Perez-Alvarez, J. A., & Soler-Rivas, C. (2005). Isoflavones as functional food components. In Atta-ur-Rahman (Ed.), *Studies in natural products chemistry bioactive natural products (Part L)* (pp. 117–207). Oxford: Elsevier.

16. Reynolds, T. (2007). The evolution of chemosystematics. *Phytochemistry, 68*, 2887–2895.

17. Wink, M. (2003). Evolution of secondary metabolites from an ecological and molecular phylogenetic perspective. *Phytochemistry, 64*, 3–19.

18. Phillipson, J. D. (2007). Phytochemistry and pharmacognosy. *Phytochemistry, 68*, 2960–2972.

19. Hartmann, T. (2007). From waste products to ecochemicals: Fifty years research of plant secondary metabolism. *Phytochemistry, 68*, 2831–2846.

20. Macias, F. A., Galindo, J. L. G., & Galindo, J. C. G. (2007). Evolution and current status of ecological phytochemistry. *Phytochemistry, 68*, 2917–2936.

21. Wink, M. (2008). Plant secondary metabolism: Diversity, function and its evolution. *Natural Product Communications, 3*, 1205–1216.

22. Larsson, S. (2007). The 'new' chemosystematics: Phylogeny and phytochemistry. *Phytochemistry, 68*, 2904–2908.

23. Waterman, P. G. (2007). The current status of chemical systematics. *Phytochemistry, 68*, 2896–2903.

24. Marshall, S. (2006). Role of insulin, adipocyte hormones, and nutrient-sensing pathways in regulating fuel metabolism and energy homeostasis: A nutritional perspective of diabetes, obesity, and cancer. *Science's Signal Transduction Knowledge Environment, 346*, re7.

25. Holmes, E., & Nicholson, J. K. (2007). Human metabolic phenotyping and metabolome wide association studies. *Ernst Schering Foundation Symposium Proceedings*, 227–249.

26. Holmes, E., Wilson, I. D., & Nicholson, J. K. (2008). Metabolic phenotyping in health and disease. *Cell, 134*, 714–717.

27. Li, M., Wang, B., Zhang, M., Rantalainen, M., Wang, S., Zhou, H., Zhang, Y., Shen, J., Pang, X., Zhang, M., Wei, H., Chen, Y., Lu, H., Zuo, J., Su, M., Qiu, Y., Jia, W., Xiao, C., Smith, L. M., Yang, S., Holmes, E., Tang, H., Zhao, G., Nicholson, J. K., Li, L., & Zhao, L. (2008). Symbiotic gut microbes modulate human metabolic phenotypes. *Proceedings of the National Academy of Sciences of the USA, 105*, 2117–2122.

28. Martin, F. P., Dumas, M. E., Wang, Y., Legido-Quigley, C., Yap, I. K., Tang, H., Zirah, S., Murphy, G. M., Cloarec, O., Lindon, J. C., Sprenger, N., Fay, L. B., Kochhar, S., van Bladeren, P., Holmes, E., & Nicholson, J. K. (2007). A top-down systems biology view of microbiome-mammalian metabolic interactions in a mouse model. *Molecular Systems Biology, 3*, 112.

29. Nicholson, J. K., Holmes, E., & Elliott, P. (2008). The metabolome-wide association study: A new look at human disease risk factors. *Journal of Proteome Research, 7*, 3637–3638.

30. Stushnoff, C., Holm, D. G., Thompson, M. D., Jiang, W., Thompson, H. J., Joyce, N. I., & Wilson, P. (2008). Antioxidant properties of cultivars and selections from the Colorado potato breeding program. *American Journal of Potato Research, 85*, 267–276.

31. Cox, S. E., Stushnoff, C., & Sampson, D. A. (2003). Relationship of fruit color and light exposure to lycopene content and antioxidant properties of tomato. *Canadian Journal of Plant Science, 83*, 913–919.

32. Willett, W. C., & McCullough, M. L. (2008). Dietary pattern analysis for the evaluation of dietary guidelines. *Asia Pacific Journal of Clinical Nutrition, 17*(Suppl. 1), 75–78.

33. Kennedy, E. (2008). Nutrition policy in the US: 50 years in review. *Asia Pacific Journal of Clinical Nutrition, 17*(Suppl. 1), 340–342.

34. McCullough, M. L., & Willett, W. C. (2006). Evaluating adherence to recommended diets in adults: The alternate healthy eating index. *Public Health Nutrition, 9*, 152–157.

35. WCRF/AICR. (2008). *Food, Nutrition, Physical activity, and the prevention of cancer: A global perspective.* Washington, DC: American Institute for Cancer Research.

36. Feskanich, D., Ziegler, R. G., Michaud, D. S., Giovannucci, E. L., Speizer, F. E., Willett, W. C., & Colditz, G. A. (2000). Prospective study of fruit and vegetable consumption and risk of lung cancer among men and women. *Journal of the National Cancer Institute, 92*, 1812–1823.

37. Fraser, G. E., Beeson, W. L., & Phillips, R. L. (1991). Diet and lung cancer in California Seventh-Day Adventists. *American Journal of Epidemiology, 133*, 683–693.

38. Linseisen, J., Rohrmann, S., Miller, A. B., Bueno-De-Mesquita, H. B., Buchner, F. L., Vineis, P., Agudo,

A., Gram, I. T., Janson, L., Krogh, V., Overvad, K., Rasmuson, T., Schulz, M., Pischon, T., Kaaks, R., Nieters, A., Allen, N. E., Key, T. J., Bingham, S., Khaw, K. T., Amiano, P., Barricarte, A., Martinez, C., Navarro, C., Quiros, R., Clavel-Chapelon, F., Boutron-Ruault, M. C., Touvier, M., Peeters, P. H., Berglund, G., Hallmans, G., Lund, E., Palli, D., Panico, S., Tumino, R., Tjonneland, A., Olsen, A., Trichopoulou, A., Trichopoulos, D., Autier, P., Boffetta, P., Slimani, N., & Riboli, E. (2007). Fruit and vegetable consumption and lung cancer risk: Updated information from the European prospective investigation into cancer and nutrition (EPIC). *International Journal of Cancer, 121*, 1103–1114.

39. Neuhouser, M. L., Patterson, R. E., Thornquist, M. D., Omenn, G. S., King, I. B., & Goodman, G. E. (2003). Fruits and vegetables are associated with lower lung cancer risk only in the placebo arm of the β-Carotene and Retinol Efficacy Trial (CARET). *Cancer Epidemiology, Biomarkers & Prevention, 12*, 350–358.

40. Voorrips, L. E., Goldbohm, R. A., Verhoeven, D. T. H., van Poppel, G. A. F. C., Sturmans, F., Hermus, R. J. J., & van den Brandt, P. A. (2000). Vegetable and fruit consumption and lung cancer risk in the Netherlands Cohort Study on Diet and Cancer. *Cancer Causes Control, 11*, 101–115.

41. Adebamowo, C. A., Cho, E., Sampson, L., Katan, M. B., Spiegelman, D., Willett, W. C., & Holmes, M. D. (2005). Dietary flavonols and flavonol-rich foods intake and the risk of breast cancer. *International Journal of Cancer, 114*, 628–633.

42. Thompson, H. J., Heimendinger, J., Diker, A., O'Neill, C., Haegele, A., Meinecke, B., Wolfe, P., Sedlacek, S., Zhu, Z., & Jiang, W. (2006). Dietary botanical diversity affects the reduction of oxidative biomarkers in women due to high vegetable and fruit intake. *Journal of Nutrition, 136*, 2207–2212.

43. Mattson, M. P. (2008). Hormesis defined. *Ageing Research Reviews, 7*, 1–7.

44. Mattson, M. P. (2008). Dietary factors, hormesis and health. *Ageing Research Reviews, 7*, 43–48.

45. Son, T. G., Camandola, S., & Mattson, M. P. (2008). Hormetic dietary phytochemicals. *Neuromolecular Medicine, 10*, 236–246.

46. Thompson, M. D., Thompson, H. J., Brick, M. A., McGinley, J. N., Jiang, W., Zhu, Z., & Wolfe, P. (2008). Mechanisms associated with dose-dependent inhibition of rat mammary carcinogenesis by dry bean (Phaseolus vulgaris, L.). *Journal of Nutrition, 138*, 2091–2097.

47. Thompson, M. D., Brick, M. A., McGinley, J. N., & Thompson, H. J. (2009). Chemical composition and mammary cancer inhibitory activity of dry bean. *Crop Science, 49*, 179–186.

48. Hartmann, T., Kutchan, T. M., & Strack, D. (2005). Evolution of metabolic diversity. *Phytochemistry, 66*, 1198–1199.

49. Hartmann, T. (2004). Plant-derived secondary metabolites as defensive chemicals in herbivorous insects: A case study in chemical ecology. *Planta, 219*, 1–4.

50. Hu, F. B. (2002). Dietary pattern analysis: A new direction in nutritional epidemiology. *Current Opinion in Lipidology, 13*, 3–9.

51. Jacobs, D. R., Jr., & Steffen, L. M. (2003). Nutrients, foods, and dietary patterns as exposures in research: A framework for food synergy. *American Journal of Clinical Nutrition, 78*, 508S–513S.

52. Flood, A., Rastogi, T., Wirfalt, E., Mitrou, P. N., Reedy, J., Subar, A. F., Kipnis, V., Mouw, T., Hollenbeck, A. R., Leitzmann, M., & Schatzkin, A. (2008). Dietary patterns as identified by factor analysis and colorectal cancer among middle-aged Americans. *American Journal of Clinical Nutrition, 88*, 176–184.

2

Vegetable and Fruit Intake and the Development of Cancer: A Brief Review and Analysis

Henry J. Thompson

Crops for Health Research Program and the Cancer Prevention Laboratory
Colorado State University, Fort Collins, CO

1. OVERVIEW

Although some population studies have identified plant food intake as a potential modulator of chronic disease risk, evidence supporting a beneficial role of vegetable and fruit consumption against cancer is conflicting. While the original report of the World Cancer Research Fund (WCRF)/American Institute for Cancer Research (AICR) [1] indicated that increased consumption of vegetables and fruit was likely to be protective against cancer at many organ sites, a large prospective study, the European Prospective Investigation into Cancer and Nutrition (EPIC), found no association between fruit and vegetable intake and the occurrence of cancer at many of the same organ sites [2]. A recent exhaustive, evidence-based review of the literature on this topic has drawn a similar conclusion [3]. The findings of these epidemiological studies continue to be debated within the scientific community because of recognized problems inherent in the measurement of food intake using food frequency questionnaires and the limited range in the amount of vegetable and fruit consumption and its consistency in many populations [4–7]. While this debate is an important component of the scientific process, it has limited potential to advance the field beyond the current vegetable and fruit–cancer conundrum (VFCC). The goal of this chapter is to provide a different perspective about factors that should be considered in evaluating the role of foods of plant origin in modulating the carcinogenic process.

2. USE OF A PHARMACOLOGY ANALOGY TO REFRAME DISCUSSIONS OF THE VFCC

The goal of constructing this 'pharmacology analogy' is to gain new insight about factors that may account for the VFCC if vegetable and fruit consumption is viewed from the perspective that foods represent drug delivery vehicles. The intent is *not* to portray foods as medicines, but rather to identify the underlying nature of the science that drives current efforts to discover how vegetable and fruit intake is associated with cancer incidence and cancer mortality. Based on extensive reviews of this topic such as [3], it can be argued that the intent of current reductionistic approaches is to identify a specific chemical or set of chemicals in a food that account for the food's ability to inhibit the development of cancer. This intent is also reflected in efforts to develop functional foods [8] and nutraceuticals [9,10] designed to increase consumption of specific phytochemicals in order to decrease cancer risk. Accordingly, for the purposes of this analogy, food is conceptualized as a pharmaceutical agent. In this context it becomes relevant to ask which of the chemical components of foods serve as the delivery vehicle versus the active ingredients that prevent and control the development of cancer.

2.1 Chemical Components of Plants

A distinguishing feature of plants is their ability to capture atmospheric carbon dioxide during the process of photosynthesis. Plants use the carbon that is sequestered to make an array of organic molecules that are classified as either primary or secondary metabolites. Primary metabolites serve structural functions in the plant, and secondary metabolites are involved: 1) in cellular communication within and among plants; 2) in diverse functions related to reproduction; and 3) as the mediators of a plant's response to biotic and abiotic stresses [11].

2.1.1 *Primary Metabolites*

Plant primary metabolites in foods are classified as carbohydrates, fats, nucleic acids and proteins. Of these primary metabolites, three classes are considered nutrients (carbohydrates, fats, and proteins) and they have received an exceptional level of attention relative to their impact on the development of cancer (reviewed in [3]). However, it can be argued that other than in circumstances related to deficiencies of these macronutrients in the diet, there is little compelling evidence and clearly no consensus about the ability of proteins, fats, or carbohydrates, either type or amount consumed, to impact the development of cancer [3]. For this reason, in this pharmacological analogy presented herein, primary plant metabolites are assigned the role of inert fillers, i.e. they are the delivery matrix. Accordingly, the remainder of this section focuses on plant secondary metabolites.

2.1.2 *Secondary Metabolites*

Plant secondary metabolites can be categorized into 14 chemical classes as shown in Table 2.1 [12]. It is estimated that there are in excess of 200,000 chemical structures that are synthesized by plants [13]. Given the essential biochemical functions that these chemicals play, it is not surprising that human beings have discovered that many of these chemicals have medicinal activity. In fact, plant-based preparations account for 70% of remedies used in traditional medicines around the world and are the basis of more than 50% of prescription and/or over the counter drugs used in the Western-type practice of medicine [14]. Specifically as it relates to cancer treatment, over 60% of current therapies are based on plant-derived materials or synthesized as an

TABLE 2.1 Classes of Plant Secondary Metabolites

Chemical Classes*	PubMed Citations**
	Cancer $\times$ Class
Alkaloids	43,465
Amines	36,821
Cyanogenic glycosides	6
Diterpenes	24,029
Flavonoids	7,546
Glucosinolates	164
Monoterpenes	497
Non-protein amino acids	51
Phenylpropanes	5
Polyacetylenes	513
Polyketides	31
Sesquiterpenes	2055
Tetraterpenes (carotenoids)	13,199
Triterpenes, saponins, steroids	1315, 636, 961

*Source: Wink [12].
**Search conducted by the author.

analogue, the core structure of which is a plant secondary metabolite [14]. However, despite these statistics, only a small percentage of the 200,000 chemical structures present in plant tissue are used medicinally. While this information documents that plant secondary metabolites have the capacity for potent biological effects in human beings, it does not directly inform understanding of what chemicals in vegetables and fruits have the capacity to affect the development of cancer.

In our search of the literature, a comprehensive review of the secondary metabolites present in plant-derived foods and their potential to affect the development of cancer was not found; rather most reviews focus on a relatively small number of compounds that are perceived to be the key players in the prevention and control of cancer [15]. While a comprehensive review of this topic is beyond the scope of this chapter, what follows is a brief description of an ongoing process in my laboratory that is building the necessary framework for such an effort. Table 2.1 not only includes the names of the 14 classes of chemicals into which plant secondary metabolites have been categorized, but it also contains the results of a search of the PubMed database using each 'chemical class name' and the word 'cancer' as the search terms. Although the range in the number of citations per class is quite large, it is evident that chemicals from all classes of plant secondary metabolites have been reported to have effects on some aspect of cancer. The point of this exercise is to underscore the need for unbiased assessment, across all classes of plant secondary metabolites, of the potential of chemicals within foods to inhibit the development of cancer, and not to limit the analysis to only a few of the many secondary plant metabolites that have been reported to have bioactivity in mammalian systems. That phytochemicals in each chemical class have biological activity is documented in Table 2.2.

The current approach to classification of foods of plant origin that is used in nutrition and dietetics is based on culinary practice and does not have a strong scientific rationale. As an alternative, plant-derived foods can be grouped according to the botanical families to which they belong [16,17]. Since plants within a botanical family emphasize the same biosynthetic pathways for specific classes of chemicals [12,18–20], foods derived from plants in different botanical families are the more likely to differ in their composition of secondary metabolites [12]. The information presented in Table 2.3 shows prominent chemical classes present in primary botanical families that are the source of plant-derived foods for various populations of the world. Note that chemical classes are not uniformly present across all botanical families, but that all chemical classes are present when the listed botanical families are considered in aggregate. If, in fact, protection against cancer is best afforded by a chemically diverse diet as suggested in [16], then the

TABLE 2.2 Bioactive Compounds in Each Class of Plant Secondary Metabolites

Chemical Classes	Examples of Bioactive Compounds
Alkaloids	Nicotine, nornicotine, anabasine, anatabine, vincristine, vinblastine, symphytine, 7-acetyllycopsamine, 7-acetylintermedine, piperine, coniine, trigonelline, pilocarpine, cytisine, sparteine, pelletierine, hygrine, atropine, ecgonine, scopamine, catuabine, quinine, strychnine, brucine, morphine, codeine, thebaine, papaverine, narcotine, sanguinarine, narceine, hydrastine, berberine, emetine, ephedrine, serotonin, ergine, caffeine, theobromine, capsaicin
Amines	Pyrrolidine (tetrahydropyrrole), piperidine, piperazine
Cyanogenic glycosides	Laetrile (amygdalin), prunasin, sambunigrin, taxiphyllin, dhurrin, linamarin, lotaustralin, vicianin
Diterpenes	Retinoids, retinol, taxol, dihydrogrindelic acid, labd-13E-en-15-oate, dihydrogrindelaldehyde, erythrofordin, norerythrofordin, hedychilactone, hedychinone, phytol
Flavonoids	Apigenin, luteolin, rutin, quercetin, myricetin, baicalein, galangin, naringenin, hesperidin, catechin, wogonin, deguelin, rotenone, cyanidin, pelargonidin, epicatechin, epigallocatechin gallate, delphinidin, fisetin, silymarin, peonidin, petunidin, malvidin, isorhamnetin, kaempferol, tangeritin, epicatechin gallate, pachypodol, rhamnazin, naringin, eriodictyol, genistein, daidzein, glycitein, gallocatechin, coumestrol, biochanin A
Glucosinolates	Sinigrin, gluconapin, glucobrassicin, glucobrassicanapin, progoitrin, glucoiberin, gluconapoleiferin, glucocheirolin, glucoerucin, glucoberteroin, gluconasturtiin
Monoterpenes	Geraniol, limonene, terpineol, carvone, carveol, α-pinene, β-pinene, eucalyptol, linalool, myrcene, carene, citral, citronellal, citronellol, camphor, borneol, eucalyptol
Non-protein amino acids	Alliin, methiin, propiin, isoalliin, etiin, butiin, S-allyl cysteine, djenkolic acid, ethionine, canavanine
Phenylpropanes	Caffeic acid, theaflavin, thearubigin, sesamol, rosavins, resveratrol, pterostilbene, piceatannol
Polyacetylenes	Panaxynol (falcarinol), falcarindiol, falcarinone, panaxydol, panaxytriol, panaxacol, dihydropanaxacol, capillin
Polyketides	Acetogenins (annonacin, uvaricin), erythromycin A, pikromycin, clarithromycin, azithromycin, amphotericin, bullatacin, tetracyclines, discodermolide, aflatoxin, aloenin, aloesin, barbaloin, anthraquinones
Sesquiterpenes	Farnesene, farnesol, artemisinin, caryophyllene, bisabolol, cadinene, vetivazulene, guaiazulene, longifolene, copaene, lactucin, parthenolide
Tetraterpenes	α-carotene, β-carotene, and β-cryptoxanthin, lutein, lycopene, phytoene, crocetin, crocin, annatto, phytofluene, sporopollenin, zeaxanthine, cryptoxanthine
Triterpenes, saponins, steroids	Su1, squalene, beta-sitosterol, lupeol, lanosterol, glycyrrhizin, liquiritic acid, glabrolide, licorice acid, ginsenosides, betulinic acid, ursolic acid, oleanolic acid, lantadene, lantanolic acid, lantic acid

Source: references [58–112].

VFCC may reflect a lack of adequate variety in the human diet as defined via the complementary chemistry associated with all 14 classes of plant secondary metabolites.

Despite the considerable breadth of the information shown in Tables 2.1–2.3, that data is still insufficient to fully understand either a specific food's chemical profile or chemical complementarity among various foods. Table 2.4, which lists specific chemicals in each chemical class and the foods in which they are found, provides a profile of the type of information needed for these purposes. While information of this type for any food is currently incomplete, technical advances in high throughput chemical analysis platforms are rapidly

TABLE 2.3   Prominent Chemical Classes in the Botanical Families from which Plant-Derived Foods are Selected

Botanical Family	Prominent Chemical Classes
Actinidiaceae	Flavonoids, triterpenes, tetraterpenes
Alliaceae (Liliaceae)	Non-protein amino acids, flavonoids, triterpenes, alkaloids
Anacardiaceae	Flavonoids, tetraterpenes
Annonaceae	Alkaloids, flavonoids, polyketides
Apiaceae (Umbelliferae)	Tetraterpenes, monoterpenes, sesquiterpenes, flavonoids, phenylpropanes, polyacetylenes, amines
Araliaceae	Triterpenes, polyacetylenes
Arecaceae (Palmae)	Alkaloids, flavonoids, tetraterpenes
Asparagaceae (Liliaceae)	Flavonoids, triterpenes, alkaloids, non-protein amino acids
Asteraceae (Compositae)	Sesquiterpenes, alkaloids, flavonoids, tetraterpenes, triterpenes, diterpenes, amines, cyanogenic glycosides, non-protein amino acids, monoterpenes, phenylpropanes
Betulaceae	Flavonoids, cyanogenic glycosides, triterpenes
Brassicaceae (Cruciferae)	Glucosinolates, tetraterpenes, alkaloids, flavonoids, non-protein amino acids, phenylpropanes
Bromeliaceae	Flavonoids, alkaloids, tetraterpenes
Cactaceae	Alkaloids, amines, flavonoids
Caricaceae	Glucosinolates
Chenopodiaceae	Alkaloids, amines, non-protein amino acids, flavonoids, tetraterpenes
Convolvulaceae	Alkaloids, flavonoids
Cucurbitaceae	Triterpenes, amines, non-protein amino acids
Dioscoraceae	Triterpenes
Ericaceae	Flavonoids, triterpenes, phenylpropanes
Euphorbiaceae	Glucosinolates, alkaloids, flavonoids, cyanogenic glycosides, non-protein amino acids, phenylpropanes
Fabaceae (Leguminosae)	Alkaloids, flavonoids, non-protein amino acids, diterpenes, triterpenes, amines, cyanogenic glycosides, phenylpropanes
Grossulariaceae	Flavonoids
Illiciaceae	Flavonoids, phenylpropanes
Juglandaceae	Cyanogenic glycosides, flavonoids, phenylpropanes, alkaloids
Lamiaceae (Labiatae)	Alkaloids, monoterpenes, triterpenes, flavonoids, sesquiterpenes, diterpenes, phenylpropanes
Lauraceae	Alkaloids, flavonoids, amines, phenylpropanes, monoterpenes
Lecythidaceae	Triterpenes, flavonoids, non-protein amino acids
Malvaceae	Flavonoids, alkaloids
Moraceae	Flavonoids, triterpenes, alkaloids
Musaceae	Flavonoids, amines, non-protein amino acids, tetraterpenes
Myristicaceae	Phenylpropanes
Myrtaceae	Phenylpropanes, sesquiterpenes, cyanogenic glycosides, monoterpenes
Oleaceae	Monoterpenes, amines, triterpenes, cyanogenic glycosides
Papaveraceae	Alkaloids, cyanogenic glycosides

(Continued)

A. FRUIT AND VEGETABLES IN HEALTH PROMOTION

TABLE 2.3　(Continued)

Botanical Family	Prominent Chemical Classes
Pedaliaceae	Amines, non-protein amino acids, phenylpropanes
Piperaceae	Amines, alkaloids, phenylpropanes, sesquiterpenes
Poaceae (Gramineae)	Alkaloids, amines, cyanogenic glycosides, flavonoids, tetraterpenes, triterpenes, monoterpenes, sesquiterpenes
Polygonaceae	Flavonoids, alkaloids
Rosaceae	Phenylpropanes, flavonoids, tetraterpenes, monoterpenes, sesquiterpenes, diterpenes, alkaloids, amines, cyanogenic glycosides, triterpenes
Rubiaceae	Alkaloids, non-protein amino acids, phenylpropanes
Rutaceae	Alkaloids, flavonoids, triterpenes, monoterpenes, sesquiterpenes, amines, phenylpropanes, tetraterpenes
Sapindaceae	Flavonoids, alkaloids, triterpenes, cyanogenic glycosides, non-protein amino acids
Solanaceae	Alkaloids, sesquiterpenes, phenylpropanes, tetraterpenes, amines, flavonoids
Theaceae	Flavonoids, phenylpropanes, alkaloids
Verbenaceae	Monoterpenes, triterpenes, alkaloids, flavonoids, phenylpropanes
Vitaceae	Flavonoids, phenylpropanes, triterpenes
Zingiberaceae	Flavonoids, phenylpropanes, diterpenes, sesquiterpenes

Source: references [58–112].

providing unbiased chemical composition data that can be added to the relational database that is being developed. This information will provide a framework by which to systematically formulate and test plant-derived food–cancer hypotheses based on profiles of plant secondary metabolites that span the 14 classes of these chemicals present in foods.

In aggregate, the evidence presented in Tables 2.1–2.4 substantiates the perception that bioactive chemicals in plant-derived foods have the potential to modulate the carcinogenic process. Thus, it does not appear that the null results of the epidemiological investigations indicate that plant-derived foods are devoid of chemicals with cancer inhibitory activity. Rather, this information suggests that attention needs to be directed to alternative explanations of the VFCC. They include: 1) with our current understanding of plant-derived foods, as reflected in the ways we assess plant food intake, protection against cancer is not apparent; 2) the plant-derived foods to which consumers have access in the market place are not those with maximum potential for human health benefit; and 3) the way populations currently eat plant-derived foods, protection against cancer is not detected. The pharmacology analogy will be extended in an effort to gain further insights about these candidate explanations.

2.2 Efficacy Is in the Dose

Various national and international dietary guidelines partition foods of plant origin into categories derived from the culinary use of those foods, which can vary within and among geographic regions [21]. These same categories are then generally used to code food frequency questionnaires in order to test specific hypotheses about vegetable and fruit intake and cancer risk or cancer mortality [4,22]. As discussed in Section 2.1, this approach to food classification does not have a strong scientific basis.

TABLE 2.4 Relational Matrix of Secondary Metabolites in Plant-Derived Foods as Classified By Chemical Class, Chemical Compound, and Botanical Family

Chemical Class	Bioactive Compound	Food (Botanical Family)
Alkaloids	Berberine	Berberis (Berberidaceae)
	Caffeine	Coffee (Rubiaceae); tea (Theaceae); guarana (Sapindaceae)
	Capsaicin	Chili pepper (Solanaceae)
	Cytisine	Leguminosae
	Pelletierine	Pomegranate (Punicaceae)
	Pilocarpine	Citrus (Rutaceae)
	Piperine	Black pepper (Piperaceae)
	Serotonin	Banana, plantain (Musaceae); tomato (Solanaceae); walnut (Juglandaceae); pineapple (Bromeliaceae); plum (Rosaceae)
	Trigonelline	Coffee (Rubiaceae)
	Theobromine	Tea (Theaceae); cacao (Malvaceae)
Amines	Piperazine	Black pepper (Piperaceae)
	Piperidine	Black pepper (Piperaceae)
	Pyrrolidine	Carrot (Apiaceae)
Cyanogenic glycosides	Amygdalin	Walnut (Juglandaceae); apricot and black cherry (Rosaceae)
	Dhurrin	Betulaceae; sorghum (Poaceae)
	Linamarin	Cassava (Euphorbiaceae); lima bean (Leguminosae); flax (Linaceae); Compositae
	Lotaustralin	Cassava (Euphorbiaceae); lima bean, white clover (Fabaceae)
	Prunasin	Rosaceae
	Vicianin	Fabaceae
Diterpenes	Carnosic acid	Rosemary and sage (Lamiaceae)
	Carnosol	Rosemary and sage (Lamiaceae)
	Erythrofordin	Leguminosae
	Hedychilactone	Zingiberaceae
	Hedychinone	Zingiberaceae
	Norerythrofordin	Leguminosae
	Rosmanol	Rosemary and sage (Lamiaceae)
	Rosmarinic acid	Rosemary and sage (Lamiaceae)
Flavonoids	Apigenin	Celery and parsley (Apiaceae)
	Catechin	Kiwi (Actinidiaceae); tea (Theaceae); cocoa (Malvaceae); cinnamon (Lauraceae)
	Coumestrol	Alfalfa, soybean, clover (Leguminosae); Brussels sprouts (Brassicaceae); spinach (Chenopodiaceae)
	Cyanidin	Purple asparagus (Asparagaceae); purple-fleshed sweet potato (Convolvulaceae); grape (Vitaceae); blueberry, cranberry (Ericaceae); blackberry, cherry, loganberry, raspberry, apple, plum (Rosaceae); red cabbage (Brassicaceae); red onion (Alliaceae); black currant (Grossulariaceae)
	Daidzein	Soybean (Leguminosae)

(Continued)

A. FRUIT AND VEGETABLES IN HEALTH PROMOTION

TABLE 2.4   (Continued)

Chemical Class	Bioactive Compound	Food (Botanical Family)
	Delphinidin	Grape (Vitaceae); cranberry (Ericaceae); pomegranate (Lythraceae); black currant (Grossulariaceae)
	Epicatechin	Kiwi (Actinidiaceae); cocoa (Malvaceae); tea (Theaceae); cinnamon (Lauraceae)
	Epicatechin gallate	Tea (Theaceae)
	Epigallocatechin gallate	Tea (Theaceae)
	Gallocatechin	Tea (Theaceae); pomegranate (Lythraceae)
	Genistein	Soybean (Leguminosae)
	Glycitein	Soybean (Leguminosae)
	Hesperidin	Citrus (Rutaceae)
	Isorhamnetin	Asparagus (Asparagaceae)
	Kaempferol	Asparagus (Asparagaceae); tea (Theaceae); broccoli (Brassicaceae); grapefruit (Rutaceae)
	Luteolin	Celery (Apiaceae); green pepper (Solanaceae); tea (Theaceae); perilla and sage (Lamiaceae)
	Malvidin	Grapes (Vitaceae)
	Morin	Fig (Moraceae); almond (Rosaceae)
	Myricetin	Grape (Vitaceae); berries (Rosaceae)
	Naringin	Grapefruit (Rutaceae)
	Pelargonidin	Raspberry, strawberry, plum, blackberry (Rosaceae); blueberry and cranberry (Ericaceae); pomegranate (Lythraceae)
	Peonidin	Cranberry (Ericaceae); red onion (Alliaceae)
	Quercetin	Onion (Alliaceae); sugar apple (Annonaceae); asparagus (Asparagaceae); parsley and celery (Apiaceae); citrus (Rutaceae); buckwheat (Polygonaceae); lovage (Apiaceae); apples, chokeberry, cherry, raspberry (Rosaceae); tea (Theaceae); red grape (Vitaceae); tomato (Solanaceae); broccoli (Brassicaceae); cranberry, lingonberry (Ericaceae); prickly pear (Cactaceae)
	Rutin	Citrus (Rutaceae)
	Silymarin	Artichoke (Asteraceae)
	Tangeritin	Tangerine (Rutaceae)
Glucosinolates	Glucobrassicin	Cabbage, broccoli, mustard, cress (Brassicaceae)
	Gluconasturtiin	Horseradish (Brassicaceae)
	Sinagrin	Brussels sprouts, broccoli, black mustard, horseradish (Brassicaceae)
	Sinalbin	White mustard (Brassicaceae)
Monoterpenes	Borneol	Asteraceae (*Artemisia* spp.)
	Camphor	Lauraceae; berries (Rosaceae)
	Citral	Lemongrass (Poaceae); lemon verbena (Verbenaceae)
	Geraniol	Lemongrass (Poaceae)
	Limonene	Rutaceae
	Linalool	Sweet basil (Lamiaceae); sweet orange (Rutaceae); coriander (Apiaceae)

(*Continued*)

TABLE 2.4 (Continued)

Chemical Class	Bioactive Compound	Food (Botanical Family)
Non-protein amino acids	Alliin	Garlic and onion (Alliaceae)
	Butiin	Onion (Alliaceae)
	Canavanine	Leguminosae
	Etiin	Onion (Alliaceae)
	Isoalliin	Onion (Alliaceae)
	Methiin	Onion (Alliaceae)
	Propiin	Onion (Alliaceae)
	S-allyl cysteine	Garlic (Alliaceae)
Phenylpropanes	Caffeic acid	Coffee (Rubiaceae); red clover (Fabaceae); pear, hawthorn, apple (Rosaceae); basil, thyme, oregano, rosemary (Lamiaceae); verbena (Verbenaceae); tarragon (Asteraceae); turmeric (Zingiberaceae)
	Piceatannol	Grape (Vitaceae)
	Pterostilbene	Grape (Vitaceae); blueberry (Ericaceae)
	Resveratrol	Grape (Vitaceae); peanut (Fabaceae)
	Sesamol	Sesame oil (Pedaliaceae)
	Theaflavin	Tea (Theaceae)
Polyacetylenes	Falcarindiol	Carrot, celery, parsley (Apiaceae)
	Falcarinol	Carrot, lovage, celery, parsley (Apiaceae); red ginseng (Araliaceae)
	Panaxydol	Ginseng (Araliaceae); celery (Apiaceae)
Polyketides	Annonacin	Guanabana (Annonaceae)
	Uvaricin	*Uvaria accuminata* (Annonaceae)
Sesquiterpenes	Caryophyllene	Rosemary (Lamiaceae); black pepper (Piperaceae); cloves (Myrtaceae)
	Farnesol	Rosaceae; Poaceae
	Lactucin	Lettuce (Asteraceae)
	Zingiberene	Ginger oil (Zingiberaceae)
Tetraterpenes	Carotene	Banana (Musaceae); carrot and parsley (Apiaceae); sweet potato (Convolvulaceae); melon and pumpkin (Cucurbitaceae); apricot (Rosaceae); spinach (Chenopodiaceae); collard greens and broccoli (Brassicaceae); thyme (Lamiaceae); cassava (Euphorbiaceae)
	Cryptoxanthin	Banana (Musaceae)
	Lutein	Banana (Musaceae); spinach (Chenopodiaceae); kale (Brassicaceae)
	Lycopene	Tomato (Solanaceae); rosehip (Rosaceae)
	Phytofluene	Tomato (Solanaceae)
Triterpenes, saponins, steroids	Betulinic acid	*Caesalpinia paraguariensis* (Leguminosae)
	Ginsenosides	Ginseng (Araliaceae)
	Glabrolide	Licorice (Fabaceae)
	Glycyrrhizin	Licorice (Fabaceae)
	Lantadene	Spanish sage (Verbenaceae)
	Lantanolic acid	Spanish sage (Verbenaceae)
	Lantic acid	Spanish sage (Verbenaceae)

(Continued)

A. FRUIT AND VEGETABLES IN HEALTH PROMOTION

TABLE 2.4 (Continued)

Chemical Class	Bioactive Compound	Food (Botanical Family)
	Licorice acid	Licorice (Fabaceae)
	Liquiritic acid	Licorice (Fabaceae)
	Oleanolic acid	Sage (Lamiaceae); Sapindaceae
	Squalene	Rice bran, wheat germ (Poaceae); olive (Oleaceae)
	Ursolic acid	Sage (Lamiaceae); blueberry (Ericaceae); Spanish sage (Verbenaceae)

Source: references [58–112].

To advance the field beyond the current VFCC, alternative approaches are proposed for identifying how plant-derived food consumption can affect the development of cancer. These approaches were conceived in part based on the pharmacology analogy in which foods are viewed as pharmaceutical agents and the inherent importance is recognized of drug dose in determining treatment efficacy.

In a drug development model, there is an explicit understanding of the active ingredient(s) and of the pharmacokinetics and pharmacodynamics of the ingredient(s) that is (are) associated with treatment efficacy [14]. Given that we currently do not know the identity of the plant secondary metabolites that function as the 'active ingredients' against cancer, let alone what concentrations must be maintained in plasma for efficacy, it must be questioned whether hypothesis testing should be limited to plant-derived foods that meet the culinary definitions of vegetables and fruit. The concern is that foods assigned to these categories are generally not eaten in large amounts or on a consistent, daily basis [23]. There are not many pharmacology models that would predict that a drug administered in varying amounts with different frequencies, including no intake on some days, during the course of a week would exert a consistent or sustained biological effect. Consequently, alternative approaches are proposed for evaluating the effects of plant-derived foods on the development of cancer.

2.2.1 Staple Food Crops

There are food crops that are consumed by the populations of the world in large amounts on a daily basis; they are referred to as staple food crops. Staple food crops include dry beans, corn, rice, wheat, potatoes (white and sweet), and sorghum. In general, none of these staple food crops have been classified as vegetables or fruits in the evaluation of the effects of plant-derived foods on the development of cancer. This situation exists despite the fact that dry beans, corn, rice and wheat actually fit the botanical definition of a fruit, and potato and sorghum fit the botanical definition of a vegetable. In the typical classification scheme, staple crops are not included, at least in part, because they are viewed as sources of 'empty starch calories,' a perspective that merits re-examination [24].

Recent reports indicate that dietary levels of specific types (varieties or cultivars) of dry beans or potatoes, when included as a significant component of dietary calories, i.e. when eaten as a staple, inhibited experimentally induced breast and/or colon cancer [25–27]. In addition, ongoing work in our laboratory indicates that anticancer activity is present in various cultivars of wheat and rice (unpublished).

However, as reported in [28] and reviewed extensively in [24], not all cultivars (varieties) within a staple food crop have equivalent amounts of anticancer activity per gram edible portion. In fact, different cultivars of each staple food crop are likely to contain different bioactive chemicals [24]. This situation reflects a heretofore unappreciated complexity in efforts to evaluate various plant food cancer hypotheses. However, this complexity is also a source of new opportunities to re-examine food consumption databases gathered from broad geographical regions of the world to determine if the consumption of different cultivars of the same staple crop has distinct effects on the development of cancer. While at first glance, there might be concern that globalization of the food supply would prevent examination of differences among varieties within a crop on disease outcomes, this is unlikely to be an insurmountable obstacle. First, it needs to be recognized that different crop cultivars are used to make different products, e.g. different types of wheat cultivars are used to make noodle products versus bread products. Second, certain crop cultivars are characteristically eaten by different populations around the world, e.g. dry beans from the Andean center of domestication are typically eaten in Africa and in various parts of Europe, whereas dry beans from the Middle American center of domestication are eaten in the Americas. Dry beans from these centers of domestication are genetically distinct and have been reported to have different profiles of secondary metabolites and different effects on carcinogenesis [24]. Similar examples can be cited for each staple food crop. Thus what is required is a greater understanding of the uses of each staple food crop and the factors that determine the cultivars of each crop that are characteristically grown in various regions of the world. Failure to account for the existence of cultivar-associated differences in foods is likely to contribute to the VFCC since such differences also exist in crops

that are traditionally classified as vegetables or fruits [24].

That staple crops can and should be viewed as more than a starchy source of calories, specifically as important delivery vehicles for various nutrients, is not a new concept within the nutrition community. The importance of staple crops is probably best reflected in ongoing international efforts to improve human health and well-being through the prevention of nutritional deficiency diseases via the biofortification of staple crops for essential micronutrients [29]. This approach was explicitly developed to capitalize on the regular daily intake of a consistent and large amount of food staples by all family members, and in so doing, address a major public health concern about whether the foods developed are likely to be consumed. Because staple foods predominate in the diets of the poor, this strategy implicitly targets lower-income households.

2.2.2 Vegetable and Fruit Combinations

In Section 2.2.1, a concern was expressed about the relatively small amounts and irregular frequency with which traditionally defined vegetables and fruits are consumed [23,30]. Nonetheless, there may be combinations of vegetable and fruit intake that, when considered in aggregate, would identify food combinations associated with protection against cancer [5–7]. While the identification of food combinations has been proposed for more than a decade, it is only now beginning to gain in its use and acceptance [31]. Initially criticized when proposed because of concerns about biases that can be introduced due to the use of factor analyses to identify food consumption patterns associated with disease risk, the existence of food patterns has been widely studied by those investigating the history of foods [21]. From the contemporary perspective, the null effects not only of food-based studies of cancer risk but also of single and combined agent intervention studies to reduce cancer risk has led to the conclusion that the chemicals that are provided by vegetables and fruit are likely to exert health benefits through their combinatorial effects and as modified by the food matrix [32,33]. The only realistic approach likely to identify meaningful relationships is via the study of food combinations with the level to which the analyses are reduced, that is of foods as they act in combination [6,7]. This approach has considerable appeal in view of the complexity of the set of diseases generally referred to as cancer and of which there are more than 40 types with multiple subtypes at each major organ site [34]. Failure to search for and identify plant-derived food combinations that protect against cancer could underlie the current VFCC. In support of this statement, in a large prospective study, no relations between total intake of vegetables and fruit combined or the total of each considered separately were observed for lung cancer risk [7]. However, higher consumption of several botanical groups, including Rosaceae, Convolvulaceae, and Umbelliferae, was significantly inversely associated with lung cancer risk in men. Similarly, in several other studies, investigators analyzed the relation of individual plant-based foods and/or plant-based food groups with lung cancer risk; the most consistent findings were for Rosaceae, [35–37], Brassicaceae [35,38,39], and Rutaceae [35,39]. A perusal of the literature shows additional examples similar to these exist [5,40], yet there is limited mention of food combinations in ongoing discussions of the VFCC.

3. THE PATHOGENESIS OF THE DISEASE

An important consideration in any effort to understand and resolve the VFCC is to predict when during the carcinogenic process that plant food consumption is likely to disrupt the pathogenesis of the disease. For this discussion, we operationally divide effects on carcinogenesis into primary and secondary events.

3.1 Primary Events

Primary events are defined as those related to the mutation of oncogenes and/or tumor suppressor genes. Thus, the benefit of reducing the occurrence of those events involved in the initial steps of transformation or tumor progression has to be considered in the framework of multi-hit models such as those established for most solid tumors, notably the colon [41]. If in fact, the prevention of these events is thought to be a primary mechanism of benefit, it should be expected that protection would be observed after 10–15 years of establishing a characteristic pattern of consumption, the time-frame that is predicted to be required from an initial transforming mutation until clinically detectable disease is manifest [42]. It is not clear that any of the designs of prospective population-based investigations of the effects of plant-derived food on cancer incidence or cancer mortality have the sensitivity to detect effects on this time scale. If they do not, then the null effects on which the VFCC is based should be anticipated and limitations in experimental designs would in part account for the VFCC.

3.2 Secondary Events

Secondary events that plant food consumption may impact are noted in [42] and involve three major cellular processes misregulated during carcinogenesis: cell proliferation, cell death (and therefore survival), and blood vessel formation. If one attempts to simplify the complex array of signaling events involved in these processes relative to those that plant secondary metabolites have been reported to affect, it becomes clear that signaling reported to be affected by plant metabolites and signaling mis-regulated during carcinogenesis intersect at multiple levels [15,43,44]. The relationships and specific regulation of these pathways by phyto-chemicals has been the subject of a number of excellent reviews and the goal is not to review those details here [32,45]. Rather, the objective is to call attention to the need for predictions about how regulation of altered signaling by a pattern of food consumption would be expected to manifest in terms of effect on vari-ous cancer endpoints. Specifically, what quanti-fiable aspect of the disease would be affected, and how many years of observation would be necessary before a clinically detectable effect on signaling should be anticipated? At issue is when during the lifecycle that plant-derived food consumption is important and whether lifecycle changes in eating behaviors are ade-quately captured using current data collection instruments.

4. DESIGNING A PLANT FOOD RICH DIET TO REDUCE CANCER RISK OR CANCER MORTALITY

For health care professionals in general and the dietitian in particular, the VFCC poses prac-tical problems relative to recommending diets for cancer prevention and control. The perspec-tives gained from this brief review and analy-sis suggest that sound dietary guidance will continue to revolve around the cornerstone con-cept that variety and moderation is the key to health promotion and disease prevention. The opportunity for the future lies in the identifica-tion of food combinations that are tailored to the metabolic needs of the individual. For that purpose, there remains untapped potential in the design of plant food rich diets with a broad base of botanically defined diversity.

The interested reader is referred to another chapter in this book for a detailed discussion of these concepts [28].

5. SUMMARY

It is now becoming clear that the many forms of cancer have specific genotypes, i.e. they are the consequence of specific defects in gene expression and/or protein activity [42]. That level of understanding is now being used in the pharmaceutical industry to develop target-specific drugs. This same information, as it becomes more widely available to the epide-miologist, will provide new tools with which to analyze for food-related effects on pathway-specific development of cancer [46]. Similarly, polymorphisms are also being identified in the systems that regulate the metabolism of both xenobiotics and essential and nonessential nutrients in plant-derived foods [47]. Functional polymorphisms in affected genes will need to be used to identify cohorts that can be pre-dicted to be responsive or nonresponsive to a particular class(es) of plant secondary metabo-lites [44,48–53]. While complex, the opportu-nities for new insights will grow with the availability of an increasing number of high dimensional datasets that are derived from genomic, proteomic and metabolic analyses of human tissues and body fluids as well as the metabolomic profiles of plant foods and of the microbiome that is likely to have a profound yet unappreciated impact on both the com-pounds absorbed from the gut and the activity of key metabolic processes within the mamma-lian system [54–57]. In this context, currently available databases that form the basis of the VFCC are useful in identifying important leads about the design of prospective cohort studies and the formulation of food-based interventions that are population and context defined and mechanism driven. As such, the null effects of today that have propagated the VFCC offer a wealth of opportunity and insight that will inform the studies of tomorrow.

ACKNOWLEDGMENTS

This work was supported in part by PHS grants R01-CA125243 and U54-CA116847 from the National Cancer Institute. The authors thank Blair Dorsey and John McGinley for their assistance in the preparation of this chapter.

References

1. Potter, J. D. (1997). *Food, nutrition and the prevention of cancer: A global perspective.* Washington, DC: American Institute for Cancer Research.
2. van Gils, C. H., Peeters, P. H. T., Bueno-De-Mesquita, H. B., Boshuizen, H. C., Lahmann, P. H., Clavel-Chapelon, F., Thiebaut, A., Kesse, E., Sieri, S., Palli, D., Tumino, R., Panico, S., Vineis, P., Gonzalez, C. A., Ardanaz, E., Sanchez, M. J., Amiano, P., Navarro, C., Quiros, J. R., Key, T. J., Allen, N., Khaw, K. T., Bingham, S. A., Psaltopoulou, T., Koliva, M., Trichopoulou, A., Nagel, G., Linseisen, J., Boeing, H., Berglund, G., Wirfalt, E., Hallmans, G., Lenner, P., Overvad, K., Tjonneland, A., Olsen, A., Lund, E., Engeset, D., Alsaker, E., Norat, T. A., Kaaks, R., Slimani, N., & Riboli, E. (2005). Consumption of vegetables and fruits and risk of breast cancer. *Journal of the American Medical Association, 293,* 183–193.
3. WCRF/AICR. (2008). *Food, nutrition, physical activity, and the prevention of cancer: A global perspective.* Washington, DC: American Institute for Cancer Research.
4. Willett, W. C., & Hu, F. B. (2007). The food frequency questionnaire. *Cancer Epidemiology Biomarkers & Prevention, 16,* 182–183.
5. Shannon, J., Ray, R., Wu, C., Nelson, Z., Gao, D. L., Li, W., Hu, W., Lampe, J., Horner, N., Satia, J., Patterson, R., Fitzgibbons, D., Porter, P., & Thomas, D. (2005). Food and botanical groupings and risk of breast cancer: A case-control study in Shanghai, China. *Cancer Epidemiology Biomarkers & Prevention, 14,* 81–90.
6. Flood, A., Rastogi, T., Wirfalt, E., Mitrou, P. N., Reedy, J., Subar, A. F., Kipnis, V., Mouw, T., Hollenbeck, A. R., Leitzmann, M., & Schatzkin, A. (2008). Dietary patterns as identified by factor analysis and colorectal cancer among middle-aged Americans. *American Journal of Clinical Nutrition, 88,* 176–184.
7. Wright, M. E., Park, Y., Subar, A. F., Freedman, N. D., Albanes, D., Hollenbeck, A., Leitzmann, M. F., & Schatzkin, A. (2008). Intakes of fruit, vegetables, and specific botanical groups in relation to lung cancer risk in the NIH-AARP Diet and Health Study. *American Journal of Epidemiology, 168,* 1024–1034.
8. Hasler, C. M. (2002). Functional foods: benefits, concerns and challenges – a position paper from the American Council on Science and Health. *Journal of Nutrition, 132,* 3772–3781.
9. Lachance, P. A., & Das, Y. T. (2007). Nutraceuticals. In D. J. Triggle, & J. B. Taylor, (Eds.), *Comprehensive medicinal chemistry II* (pp. 449–461). Oxford: Elsevier.
10. Kalra, E. K. (2003). Nutraceutical – definition and introduction. *AAPS Pharmaceutical Science, 5,* E25.
11. Hartmann, T. (2007). From waste products to eco-chemicals: Fifty years research of plant secondary metabolism. *Phytochemistry, 68,* 2831–2846.
12. Wink, M. (2003). Evolution of secondary metabolites from an ecological and molecular phylogenetic perspective. *Phytochemistry, 64,* 3–19.
13. Hartmann, T., Kutchan, T. M., & Strack, D. (2005). Evolution of metabolic diversity. *Phytochemistry, 66,* 1198–1199.
14. Gad, S. C. (2005). Drug discovery handbook, Hoboken: Wiley.
15. Finley, J. W. (2005). Proposed criteria for assessing the efficacy of cancer reduction by plant foods enriched in carotenoids, glucosinolates, polyphenols and selenocompounds. *Annals of Botany (London), 95,* 1075–1096.
16. Thompson, H. J., Heimendinger, J., Diker, A., O'Neill, C., Haegele, A., Meinecke, B., Wolfe, P., Sedlacek, S., Zhu, Z., & Jiang, W. (2006). Dietary botanical diversity affects the reduction of oxidative biomarkers in women due to high vegetable and fruit intake. *Journal of Nutrition, 136,* 2207–2212.
17. Thompson, M. D., & Thompson, H. J. (2009). Botanical diversity in vegetable and fruit intake: Potential health benefits. In R. R. Watson, & V. R. Preedy, (Eds.), *Bioactive foods in promoting health: Fruits and vegetables.* Elsevier.
18. Reynolds, T. (2007). The evolution of chemosystematics. *Phytochemistry, 68,* 2887–2895.
19. Larsson, S. (2007). The 'new' chemosystematics: Phylogeny and phytochemistry. *Phytochemistry, 68,* 2904–2908.
20. Waterman, P. G. (2007). The current status of chemical systematics. *Phytochemistry, 68,* 2896–2903.
21. Charles, J. (2003). Encyclopedia of food and culture, 3 vols. *Library Journal, 128,* 71.
22. Johansson, I., Hallmans, G., Wikman, A., Biessy, C., Riboli, E., & Kaaks, R. (2002). Validation and calibration of food-frequency questionnaire measurements in

the Northern Sweden Health and Disease cohort. *Public Health Nutrition, 5,* 487–496.

23. Thompson, H. J., Heimendinger, J., Haegele, A., Sedlacek, S. M., Gillette, C., O'Neill, C., Wolfe, P., & Conry, C. (1999). Effect of increased vegetable and fruit consumption on markers of oxidative cellular damage. *Carcinogenesis, 20,* 2261–2266.

24. Thompson, M. D., & Thompson, H. J. (2009). Biomedical agriculture: a systematic approach to crop improvement for chronic disease prevention. In D. Sparks, (Ed.), *Advances in agronomy (102).* Philadelphia: Elsevier.

25. Stushnoff, C., Holm, D. G., Thompson, M. D., Jiang, W., Thompson, H. J., Joyce, N. I., & Wilson, P. (2008). Antioxidant properties of cultivars and selections from the Colorado potato breeding program. *American Journal of Potato Research, 85,* 267–276.

26. Thompson, M. D., Thompson, H. J., Brick, M. A., McGinley, J. N., Jiang, W., Zhu, Z., & Wolfe, P. (2008). Mechanisms associated with dose-dependent inhibition of rat mammary carcinogenesis by dry bean (*Phaseolus vulgaris,* L.). *Journal of Nutrition, 138,* 2091–2097.

27. Bobe, G., Barrett, K. G., Mentor-Marcel, R. A., Saffiotti, U., Young, M. R., Colburn, N. H., Albert, P. S., Bennink, M. R., & Lanza, E. (2008). Dietary cooked navy beans and their fractions attenuate colon carcinogenesis in azoxymethane-induced ob/ob mice. *Nutrition and Cancer, 60,* 373–381.

28. Thompson, M. D., Brick, M. A., McGinley, J. N., & Thompson, H. J. (2009). Chemical composition and mammary cancer inhibitory activity of dry bean. *Crop Science, 49,* 179–186.

29. Nestel, P., Bouis, H. E., Meenakshi, J. V., & Pfeiffer, W. (2006). Biofortification of staple food crops. *Journal of Nutrition, 136,* 1064–1067.

30. Stables, G. J., Subar, A. F., Patterson, B. H., Dodd, K., Heimendinger, J., Van Duyn, M. A., & Nebeling, L. (2002). Changes in vegetable and fruit consumption and awareness among US adults: Results of the 1991 and 1997 5 A Day for Better Health Program surveys. *Journal of the American Dietetic Association, 102,* 809–817.

31. Hu, F. B. (2002). Dietary pattern analysis: A new direction in nutritional epidemiology. *Current Opinion in Lipidology, 13,* 3–9.

32. Mayne, S. T. (2003). Antioxidant nutrients and chronic disease: Use of biomarkers of exposure and oxidative stress status in epidemiologic research. *Journal of Nutrition, 133,* 933S–9940S.

33. Seifried, H. E., McDonald, S. S., Anderson, D. E., Greenwald, P., & Milner, J. A. (2003). The antioxidant conundrum in cancer. *Cancer Research, 63,* 4295–4298.

34. American Cancer Society. (2008). *Cancer facts and figures.* Atlanta: The American Cancer Society.

35. Feskanich, D., Ziegler, R. G., Michaud, D. S., Giovannucci, E. L., Speizer, F. E., Willett, W. C., & Colditz, G. A. (2000). Prospective study of fruit and vegetable consumption and risk of lung cancer among men and women. *Journal of the National Cancer Institute, 92,* 1812–1823.

36. Fraser, G. E., Beeson, W. L., & Phillips, R. L. (1991). Diet and lung cancer in California Seventh-Day Adventists. *American Journal of Epidemiology, 133,* 683–693.

37. Linseisen, J., Rohrmann, S., Miller, A. B., Bueno-De-Mesquita, H. B., Buchner, F. L., Vineis, P., Agudo, A., Gram, I. T., Janson, L., Krogh, V., Overvad, K., Rasmuson, T., Schulz, M., Pischon, T., Kaaks, R., Nieters, A., Allen, N. E., Key, T. J., Bingham, S., Khaw, K. T., Amiano, P., Barricarte, A., Martinez, C., Navarro, C., Quiros, R., Clavel-Chapelon, F., Boutron-Ruault, M. C., Touvier, M., Peeters, P. H., Berglund, G., Hallmans, G., Lund, E., Palli, D., Panico, S., Tumino, R., Tjonneland, A., Olsen, A., Trichopoulou, A., Trichopoulos, D., Autier, P., Boffetta, P., Slimani, N., & Riboli, E. (2007). Fruit and vegetable consumption and lung cancer risk: Updated information from the European Prospective Investigation into Cancer and Nutrition (EPIC). *International Journal of Cancer, 121,* 1103–1114.

38. Neuhouser, M. L., Patterson, R. E., Thornquist, M. D., Omenn, G. S., King, I. B., & Goodman, G. E. (2003). Fruits and vegetables are associated with lower lung cancer risk only in the placebo arm of the {beta}-Carotene and Retinol Efficacy Trial (CARET). *Cancer Epidemiology Biomarkers & Prevention, 12,* 350–358.

39. Voorrips, L. E., Goldbohm, R. A., Verhoeven, D. T. H., van Poppel, G. A. F. C., Sturmans, F., Hermus, R. J. J., & van den Brandt, P. A. (2000). Vegetable and fruit consumption and lung cancer risk in the Netherlands Cohort Study on Diet and Cancer. *Cancer Causes Control, 11,* 101–115.

40. Adebamowo, C. A., Cho, E., Sampson, L., Katan, M. B., Spiegelman, D., Willett, W. C., & Holmes, M. D. (2005). Dietary flavonols and flavonol-rich foods intake and the risk of breast cancer. *International Journal of Cancer, 114,* 628–633.

41. Fearon, E. R., & Vogelstein, B. (1990). A genetic model for colorectal tumorigenesis. *Cell, 61,* 759–767.

42. Hanahan, D., & Weinberg, R. A. (2000). The hallmarks of cancer. *Cell, 100,* 57–70.

43. Mattson, M. P. (2008). Dietary factors, hormesis and health. *Ageing Research Reviews, 7,* 43–48.

44. Li, M., Wang, B., Zhang, M., Rantalainen, M., Wang, S., Zhou, H., Zhang, Y., Shen, J., Pang, X., Zhang, M.,

Wei, H., Chen, Y., Lu, H., Zuo, J., Su, M., Qiu, Y., Jia, W., Xiao, C., Smith, L. M., Yang, S., Holmes, E., Tang, H., Zhao, G., Nicholson, J. K., Li, L., & Zhao, L. (2008). Symbiotic gut microbes modulate human metabolic phenotypes. *Proceedings of the National Academy of Sciences*, *105*, 2117–2122.

45. Williams, R. J., Spencer, J. P., & Rice-Evans, C. (2004). Flavonoids: Antioxidants or signalling molecules? *Free Radical Biology & Medicine*, *36*, 838–849.

46. Milner, J. A. (2000). Functional foods: The US perspective. *American Journal of Clinical Nutrition*, *71*, 1654S–1659S.

47. Ambrosone, C. B., Freudenheim, J. L., Thompson, P. A., Bowman, E., Vena, J. E., Marshall, J. R., Graham, S., Laughlin, R., Nemoto, T., & Shields, P. G. (1999). Manganese superoxide dismutase (MnSOD) genetic polymorphisms, dietary antioxidants, and risk of breast cancer. *Cancer Research*, *59*, 602–606.

48. Holmes, E., & Nicholson, J. K. (2007). Human metabolic phenotyping and metabolome wide association studies. *Ernst Schering Foundations Symposium Proceedings*, 227–249.

49. Holmes, E., Wilson, I. D., & Nicholson, J. K. (2008). Metabolic phenotyping in health and disease. *Cell*, *134*, 714–717.

50. Kinross, J. M., von Roon, A. C., Holmes, E., Darzi, A., & Nicholson, J. K. (2008). The human gut microbiome: Implications for future health care. *Current Gastroenterology Reports*, *10*, 396–403.

51. Martin, F. P., Dumas, M. E., Wang, Y., Legido-Quigley, C., Yap, I. K., Tang, H., Zirah, S., Murphy, G. M., Cloarec, O., Lindon, J. C., Sprenger, N., Fay, L. B., Kochhar, S., van, B. P., Holmes, E., & Nicholson, J. K. (2007). A top-down systems biology view of microbiome-mammalian metabolic interactions in a mouse model. *Molecular Systems Biology*, *3*, 112.

52. Nicholson, J. K., Holmes, E., & Elliott, P. (2008). The metabolome-wide association study: A new look at human disease risk factors. *Journal of Proteome Research*, *7*, 3637–3638.

53. Nicholson, J. K., & Lindon, J. C. (2008). Systems biology: Metabonomics. *Nature*, *455*, 1054–1056.

54. Barnes, S., & Kim, H. (2004). Nutriproteomics: Identifying the molecular targets of nutritive and non-nutritive components of the diet. *Journal of Biochemistry and Molecular Biology*, *37*, 59–74.

55. Barton, R. H., Nicholson, J. K., Elliott, P., & Holmes, E. (2008). High-throughput H-1 NMR-based metabolic analysis of human serum and urine for large-scale epidemiological studies: Validation study. *International Journal of Epidemiology*, *37*, 31–40.

56. Cloarec, O., Dumas, M. E., Craig, A., Barton, R. H., Trygg, J., Hudson, J., Blancher, C., Gauguier, D.,

Lindon, J. C., Holmes, E., & Nicholson, J. (2005). Statistical total correlation spectroscopy: An exploratory approach for latent biomarker identification from metabolic 1 H NMR data sets. *Analytical Chemistry*, *77*, 1282–1289.

57. Dumas, M. E., Barton, R. H., Toye, A., Cloarec, O., Blancher, C., Rothwell, A., Fearnside, J., Tatoud, R., Blanc, V., Lindon, J. C., Mitchell, S. C., Holmes, E., McCarthy, M. I., Scott, J., Gauguier, D., & Nicholson, J. K. (2006). Metabolic profiling reveals a contribution of gut microbiota to fatty liver phenotype in insulin-resistant mice. *Proceedings of the National Academy of Sciences of the United States of America*, *103*, 12511–12516.

58. Alasalvar, C., Karamac, M., Amarowicz, R., & Shahidi, F. (2006). Antioxidant and antiradical activities in extracts of hazelnut kernel (*Corylus avellana* L.) and hazelnut green leafy cover. *Journal of Agricultural and Food Chemistry*, *54*, 4826–4832.

59. Angelova, N., Kong, H. W., van der Heijden, R., Yang, S. Y., Choi, Y. H., Kim, H. K., Wang, M., Hankemeier, T., van der Greef, J., Xu, G., & Verpoorte, R. (2008). Recent methodology in the phytochemical analysis of ginseng. *Phytochemical Analysis*, *19*, 2–16.

60. Asl, M. N., & Hosseinzadeh, H. (2008). Review of pharmacological effects of *Glycyrrhiza* sp. and its bioactive compounds. *Phytotherapy Research*, *22*, 709–724.

61. Avente-Garcia, O., & Castillo, J. (2008). Update on uses and properties of citrus flavonoids: New findings in anticancer, cardiovascular, and anti-inflammatory activity. *Journal of Agricultural and Food Chemistry*, *56*, 6185–6205.

62. Barreto, J. C., Trevisan, M. T., Hull, W. E., Erben, G., de Brito, E. S., Pfundstein, B., Wurtele, G., Spiegelhalder, B., & Owen, R. W. (2008). Characterization and quantitation of polyphenolic compounds in bark, kernel, leaves, and peel of mango (*Mangifera indica* L.). *Journal of Agricultural and Food Chemistry*, *56*, 5599–5610.

63. Beier, R. C. (1990). Natural pesticides and bioactive components in foods. *Reviews of Environmental Contamination and Toixicolgy*, *113*, 47–137.

64. Belz, G. G., & Loew, D. (2003). Dose-response related efficacy in orthostatic hypotension of a fixed combination of D-camphor and an extract from fresh crataegus berries and the contribution of the single components. *Phytomedicine*, *10*(Suppl. 4), 61–67.

65. Bohm, V., Frohlich, K., & Bitsch, R. (2003). Rosehip – a 'new' source of lycopene?. *Molecular Aspects of Medicine*, *24*, 385–389.

66. Bordonaba, J. G., & Terry, L. A. (2008). Biochemical profiling and chemometric analysis of seventeen UK-grown black currant cultivars. *Journal of Agricultural and Food Chemistry*, *56*, 7422–7430.

67. Capasso, R., & Mascolo, N. (2003). Inhibitory effect of the plant flavonoid galangin on rat vas deferens *in vitro*. *Life Sciences, 72*, 2993–3001.

68. Christensen, L. P., & Brandt, K. (2006). Bioactive polyacetylenes in food plants of the Apiaceae family: Occurrence, bioactivity and analysis. *Journal of Pharmaceutical and Biomedical Analysis, 41*, 683–693.

69. Christensen, L. P., & Jensen, M. (2009). Biomass and content of ginsenosides and polyacetylenes in American ginseng roots can be increased without affecting the profile of bioactive compounds. *Nature Medicine (Tokyo), 63*, 159–168.

70. Craig, W. J. (1999). Health-promoting properties of common herbs. *American Journal of Clinical Nutrition, 70*, 491S–499S.

71. Crublet, M. L., Long, C., Sevenet, T., Hadi, H. A., & Lavaud, C. (2003). Acylated flavonol glycosides from leaves of Planchonia grandis. *Phytochemistry, 64*, 589–594.

72. De Marino, S., Gala, F., Zollo, F., Vitalini, S., Fico, G., Visioli, F., & Iorizzi, M. (2008). Identification of minor secondary metabolites from the latex of Croton lechleri (Muell-Arg) and evaluation of their antioxidant activity. *Molecules, 13*, 1219–1229.

73. Dembitsky, V. M. (2006). Anticancer activity of natural and synthetic acetylenic lipids. *Lipids, 41*, 883–924.

74. Dini, I., Tenore, G. C., & Dini, A. (2008). S-alkenyl cysteine sulfoxide and its antioxidant properties from Allium cepa var. tropeana (red onion) seeds. *Journal of Natural Products, 71*, 2036–2037.

75. Englberger, L., Schierle, J., Aalbersberg, W., Hofmann, P., Humphries, J., Huang, A., Lorens, A., Levendusky, A., Daniells, J., Marks, G. C., & Fitzgerald, M. H. (2006). Carotenoid and vitamin content of Karat and other Micronesian banana cultivars. *International Journal of Food Sciences and Nutrition, 57*, 399–418.

76. Ferguson, L. R., & Philpott, M. (2008). Nutrition and mutagenesis. *Annual Review of Nutrition, 28*, 313–329.

77. Ferreira, D., Marais, J. P., & Slade, D. (2003). Phytochemistry of the mopane, Colophospermum mopane. *Phytochemistry, 64*, 31–51.

78. Fuentes-Alventosa, J. M., Jaramillo, S., Rodriguez-Gutierrez, G., Cermeno, P., Espejo, J. A., Jimenez-Araujo, A., Guillen-Bejarano, R., Fernandez-Bolanos, J., & Rodriguez-Arcos, R. (2008). Flavonoid profile of green asparagus genotypes. *Journal of Agricultural and Food Chemistry, 56*, 6977–6984.

79. Ghisalberti, E. L. (2000). Lantana camara L. (Verbenaceae). *Fitoterapia, 71*, 467–486.

80. Gleadow, R. M., Haburjak, J., Dunn, J. E., Conn, M. E., & Conn, E. E. (2008). Frequency and distribution of cyanogenic glycosides in Eucalyptus L'Herit. *Phytochemistry, 69*, 1870–1874.

81. Greenberg, D. M. (1980). The case against laetrile: The fraudulent cancer remedy. *Cancer, 45*, 799–807.

82. Hayes, J. D., Kelleher, M. O., & Eggleston, I. M. (2008). The cancer chemopreventive actions of phytochemicals derived from glucosinolates. *European Journal of Nutrition, 47*(Suppl. 2), 73–88.

83. Ito, H., Okuda, T., Fukuda, T., Hatano, T., & Yoshida, T. (2007). Two novel dicarboxylic acid derivatives and a new dimeric hydrolyzable tannin from walnuts. *Journal of Agricultural and Food Chemistry, 55*, 672–679.

84. Itoigawa, M., Ito, C., Tokuda, H., Enjo, F., Nishino, H., & Furukawa, H. (2004). Cancer chemopreventive activity of phenylpropanoids and phytoquinoids from Illicium plants. *Cancer Letters, 214*, 165–169.

85. Jan, K. C., Ho, C. T., & Hwang, L. S. (2009). Elimination and metabolism of sesamol, a bioactive compound in sesame oil, in rats. *Molecular Nutrition & Food Research, 53*, S36–S43.

86. Jiang, S., Liu, Z. H., Sheng, G., Zeng, B. B., Cheng, X. G., Wu, Y. L., & Yao, Z. J. (2002). Mimicry of annonaceous acetogenins: Enantioselective synthesis of a (4R)-hydroxy analogue having potent antitumor activity. *Journal of Organic Chemistry, 67*, 3404–3408.

87. Kamatou, G. P., Makunga, N. P., Ramogola, W. P., & Viljoen, A. M. (2008). South African Salvia species: A review of biological activities and phytochemistry. *Journal of Ethnopharmacology, 119*, 664–672.

88. Khanna, D., Sethi, G., Ahn, K. S., Pandey, M. K., Kunnumakkara, A. B., Sung, B., Aggarwal, A., & Aggarwal, B. B. (2007). Natural products as a gold mine for arthritis treatment. *Current Opinion in Pharmacology, 7*, 344–351.

89. Krishna, S., Bustamante, L., Haynes, R. K., & Staines, H. M. (2008). Artemisinins: Their growing importance in medicine. *Trends in Pharmacological Sciences, 29*, 520–527.

90. Mertens-Talcott, S. U., Rios, J., Jilma-Stohlawetz, P., Pacheco-Palencia, L. A., Meibohm, B., Talcott, S. T., & Derendorf, H. (2008). Pharmacokinetics of anthocyanins and antioxidant effects after the consumption of anthocyanin-rich acai juice and pulp (Euterpe oleracea Mart.) in human healthy volunteers. *Journal of Agricultural and Food Chemistry, 56*, 7796–7802.

91. Montefiori, M., McGhie, T. K., Costa, G., & Ferguson, A. R. (2005). Pigments in the fruit of red-fleshed kiwifruit (Actinidia chinensis and Actinidia deliciosa). *Journal of Agricultural and Food Chemistry, 53*, 9526–9530.

92. Moon, Y. J., Wang, X., & Morris, M. E. (2006). Dietary flavonoids: Effects on xenobiotic and carcinogen metabolism. *Toxicology in Vitro, 20*, 187–210.

93. Neto, C. C. (2007). Cranberry and blueberry: Evidence for protective effects against cancer and vascular diseases. *Molecular Nutrition & Food Research, 51*, 652–664.

A. FRUIT AND VEGETABLES IN HEALTH PROMOTION

94. Noble, C. O., Guo, Z., Hayes, M. E., Marks, J. D., Park, J. W., Benz, C. C., Kirpotin, D. B., & Drummond, D. C. (2009). Characterization of highly stable liposomal and immunoliposomal formulations of vincristine and vinblastine. *Cancer Chemotherapy and Pharmacology*, DOI 10.1007/s00280-008-0923-3.

95. Oboh, I. E., Obasuyi, O., & Akerele, J. O. (2008). Phytochemical and comparative antibacterial studies on the crude ethanol and aqueous extracts of the leaves of Lecaniodiscus cupanoides Planch (Sapindaceae). *Acta Poloniae Pharmaceutica*, 65, 565–569.

96. Ortega, R. M. (2006). Importance of functional foods in the Mediterranean diet. *Public Health Nutrition*, 9, 1136–1140.

97. Panda, S., & Kar, A. (2007). Antidiabetic and antioxidative effects of Annona squamosa leaves are possibly mediated through quercetin-3-O-glucoside. *Biofactors*, 31, 201–210.

98. Peng, X., Cheng, K. W., Ma, J., Chen, B., Ho, C. T., Lo, C., Chen, F., & Wang, M. (2008). Cinnamon bark proanthocyanidins as reactive carbonyl scavengers to prevent the formation of advanced glycation endproducts. *Journal of Agricultural and Food Chemistry*, 56, 1907–1911.

99. Rose, P., Whiteman, M., Moore, P. K., & Zhu, Y. Z. (2005). Bioactive S-alk(en)yl cysteine sulfoxide metabolites in the genus Allium: The chemistry of potential therapeutic agents. *Natural Product Reports*, 22, 351–368.

100. Ross, J. A., & Kasum, C. M. (2002). Dietary flavonoids: Bioavailability, metabolic effects, and safety. *Annual Review of Nutrition*, 22, 19–34.

101. Seigler, D. S., Pauli, G. F., Frohlich, R., Wegelius, E., Nahrstedt, A., Glander, K. E., & Ebinger, J. E. (2005). Cyanogenic glycosides and menisdaurin from Guazuma ulmifolia, Ostrya virginiana, Tiquilia plicata, and Tiquilia canescens.. *Phytochemistry*, 66, 1567–1580.

102. Sotiroudis, T. G., & Kyrtopoulos, S. A. (2008). Anticarcinogenic compounds of olive oil and related biomarkers. *European Journal of Nutrition*, 47(Suppl. 2), 69–72.

103. Subash, S., & Subramanian, P. (2009). Morin a flavonoid exerts antioxidant potential in chronic hyperammonemic rats: A biochemical and histopathological study. *Molecular and Cellular Biochemistry*, 327, 153–161.

104. Syed, D. N., Suh, Y., Afaq, F., & Mukhtar, H. (2008). Dietary agents for chemoprevention of prostate cancer. *Cancer Letters*, 265, 167–176.

105. Topcu, G. (2006). Bioactive triterpenoids from Salvia species. *Journal of Natural Products*, 69, 482–487.

106. Tsao, C. C., Shen, Y. C., Su, C. R., Li, C. Y., Liou, M. J., Dung, N. X., & Wu, T. S. (2008). New diterpenoids and the bioactivity of Erythrophleum fordii. *Bioorganic & Medicinal Chemistry*, 16, 9867–9870.

107. Uzunovic, A., & Vranic, E. (2008). Stability of anthocyanins from commercial black currant juice under simulated gastrointestinal digestion. *The Bosnian Journal of Basic Medical Sciences*, 8, 254–258.

108. van Wyk, B. E., & Albrecht, C. (2008). A review of the taxonomy, ethnobotany, chemistry and pharmacology of Sutherlandia frutescens (Fabaceae). *Journal of Ethnopharmacology*, 119, 620–629.

109. Vetter, J. (2000). Plant cyanogenic glycosides.. *Toxicon*, 38, 11–36.

110. Wagner, K. H., & Elmadfa, I. (2003). Biological relevance of terpenoids. Overview focusing on mono-, di- and tetraterpenes. *Annals of Nutrition and Metabolism*, 47, 95–106.

111. Yogeeswari, P., & Sriram, D. (2005). Betulinic acid and its derivatives: a review on their biological properties. *Current Medicinal Chemistry*, 12, 657–666.

112. Zidorn, C., Johrer, K., Ganzera, M., Schubert, B., Sigmund, E. M., Mader, J., Greil, R., Ellmerer, E. P., & Stuppner, H. (2005). Polyacetylenes from the Apiaceae vegetables carrot, celery, fennel, parsley, and parsnip and their cytotoxic activities. *Journal of Agricultural and Food Chemistry*, 53, 2518–2523.

3

Fruit and Vegetable Antioxidants in Health

Manuela Blasa, Lorenzo Gennari, Donato Angelino, and Paolino Ninfali

Department of Biomolecular Sciences, Università di Urbino 'Carlo Bo,' Urbino (PU) Italy

1. INTRODUCTION

In the history of human nutrition, one of the most widespread alimentary regimens linked to health protection is represented by the Mediterranean diet (MD), evolved naturally in the populations that border the Mediterranean Sea as a consequence of the availability of particular foods that grow well in this area [1]. MD eating patterns consist of a wide use of whole grains, fruits, vegetables, nuts, fish, and olive oil. Red meat is consumed only a few times a month. The MD is therefore very low in saturated fat, alcohol, salt, and food preserved with chemical compounds.

Recently, this traditional diet has received worldwide interest, as several epidemiological studies have indicated that this diet may have healthy benefits for the heart, blood pressure, diabetes, cancer, allergies, and arthritis [2–4]. The MD has also been linked to longer life, and it represents a useful tool in managing the metabolic syndrome, a condition characterized by central obesity, hypertension, and disturbed glucose and insulin metabolism [5].

Most of those diseases generically referred as chronic diseases are linked to the aging process and have been associated with oxidative damage, a constant process that occurs in aerobic organisms. The oxidative damage theory suggests that mitochondria produce reactive oxygen species (ROS), or free radicals, as normal byproducts in the phosphorylating chain of ATP [6,7]. Free radicals are also produced by enzymatic reactions, such as the NADPH oxidase in the phagocytes, to destroy invading microbes, or xanthine oxidase for uric acid formation [8]. In addition, external sources such as pollution, cigarette smoke, and sunlight cause the production of free radicals, which are incorporated by organisms and cells [9]. Free radicals are highly reactive towards other molecules, stealing electrons in order to return to stable states. The molecule from which the original free radical steals the electron becomes another free radical, resulting in a self-perpetuating process [8]. Excessive production of free radicals causes damage to lipids, proteins, carbohydrates, and DNA [9]. Membranes exposed to free radicals lose their ability to properly transport nutrients; lipoproteins are changed into oxidized forms; and damaged DNA has the potential to accumulate consecutive mutations, which can lead to carcinogenesis [10]. Therefore, the theory of oxidative damage strictly recalls the concept that antioxidant molecules, able to slow down the rate of the oxidative process, are very important for the homeostasis of normal body metabolism.

2. FIBER, MINERALS, AND ANTIOXIDANT ENZYMES

Earlier epidemiological studies have suggested a positive role played by whole grains, fruits, and vegetables in the interruption of oxidative stress and therefore in disease prevention. This effect has been initially associated with the presence of fiber, unsaturated fatty acids, oligoelements (minerals), and vitamins. As indigestible polysaccharides, the fiber health benefits were linked to the formation of a gelatinous matrix that increases fecal mass. This contributes to a reduction in the concentration of harmful biliary acids and other potential cancerous compounds in excrement [11]. Actually, the health-protective mechanisms of fiber intake are under close examination in the search for a more complete picture that includes other molecules bound to polysaccharides [12]. Numerous studies have been developed regarding oligoelements and vitamins [13] as factors able to prevent or slow 'oxidative damage.' Oligoelements are transitional metals that have a role in the synthesis and structural stabilization of proteins and nucleic acids [8]. The most important of these metals are magnesium, copper, zinc [14], manganese [15], and selenium [16]. Among these, a major line of research has been devoted to selenium, the consumption

of which is inversely correlated with the risk of developing cancer [17]. Hydrogen selenide, methylselenol, and selenomethionine, the principle selenium metabolites, are able to regulate gene expression, protect DNA from damage, and stimulate the repair and regulation of cell cycles and apoptosis.

Oligoelements also enter into the mechanism of preventive defense against free radicals, performed by antioxidant enzymes (Figure 3.1). These enzymes work in a coordinated and integrated manner, centered on the availability of oligoelements and NADPH, which is the source of reducing equivalents against oxygen radicals [18]. The erythrocytes, which totally depend on the hexose monophosphate shunt pathway for the production of NADPH, are dramatically exposed to oxidative stress when the G6PD is defective [19]. Also, normal myoblasts, in the presence of oxidants, lose NADPH and show damage of membrane lipids [20].

3. PHYTOCHEMICAL ANTIOXIDANTS

The cooperative integration of antioxidant enzymes and various antioxidant molecules and minerals is crucial to suppressing free radical reactions in cellular and extracellular compartments. Other metabolism-derived molecules, such as urate, bilirubin, ceruloplasmin, transferrin, and hormones, also contribute to the antioxidant defense [8], but their concentration is quite stable in human plasma and depends on different metabolic pathways, so that their role as antioxidants is minor. In contrast, the concentration of phytochemical antioxidants in the biological fluids changes from one person to another, depending on diet. The search for phytochemical antioxidants that are bioavailable and therefore able to prevent or retard chronic diseases has become a discipline called chemoprevention [21].

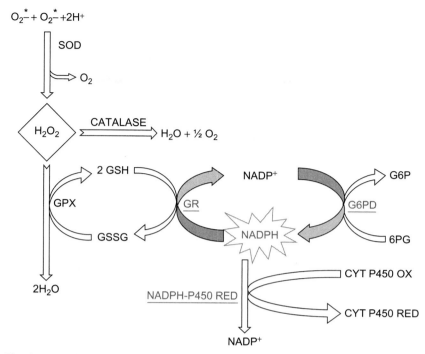

FIGURE 3.1 The integrated system of the antioxidant enzymes supported by NADPH. Abbreviations used: SOD = superoxide dismutase; GPX = glutathione peroxidase; GR = glutathione reductase; G6PD = glucose-6-phosphate dehydrogenase; NADPH-P450 RED = NADPH-P450 reductase; GSH = glutathione (reduced form); GSSG = glutathione (oxidized form).

SOD catalyzes, in the cytosol and mitochondria, the dismutation of two superoxide anions $(O_2^\cdot)$ into hydrogen peroxide (H_2O_2) and molecular oxygen. The cytosolic SOD requires Cu^{2+} and Zn^{2+}, while the mitochondrial SOD requires Mn^{2+}. The H_2O_2, formed by SOD, is converted by CAT to molecular oxygen and water. GPX, which is located both in the cytosol and in the mitochondrial matrix, converts hydrogen or lipid peroxides into water or other less reactive molecules. The cytosolic GPX strictly depends on Se^{2+} and GSH which is used by GPX to provide the electrons for reducing peroxides. GSH is transformed into GSSG, which is in turn regenerated to GSH by a reaction catalyzed by the GR. GR utilizes the NADPH produced by the G6PD. Therefore, it is the NADPH that directly provides the electrons necessary to reduce the hydrogen or lipid peroxides. In addition, NADPH supports the metabolism of toxic insoluble molecules which can damage the DNA and proteins by working as a coenzyme of the NADPH- Cyt P450 reductase.

The primary dietary phytochemical antioxidants are reported in Figure 3.2. Some consideration of their chemical structures and importance in cell physiology and biochemistry is necessary.

3.1 Vitamin C

The chemical structure of vitamin C, also called ascorbic acid (AA), is similar to that of the hexose sugars, with an ene-diol group involving carbons 2 and 3. This group makes AA a strong reducing agent, easily oxidable to dehydroascorbic acid (DHA). In cells, AA and DHA are in chemical equilibrium, and both are endowed with vitaminic activity. The acid character of AA is due to the facility of the –OH group, bound to C2 to release a proton, since the mono-DHA ion that is formed is stabilized by resonance.

Most vegetables synthesize AA, starting from glucose; however, primates are unable to

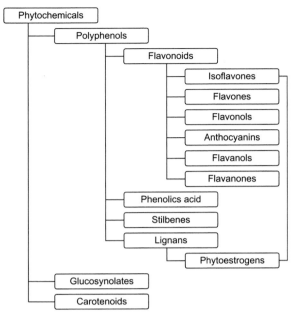

FIGURE 3.2 Different categories of diet phytochemicals with antioxidant properties.

do this, since they are deficient in the enzyme gulonolactone oxidase, which is involved in AA synthesis [22]. Vitamin C is therefore an 'essential' vitamin and must be introduced with fresh vegetables and fruits. At a physiological level, AA is involved in several metabolic processes: corticosteroid hormones, biliary acids, carnitine, prostaglandins, histamine, collagen, iron, thyroxine, and some neurotransmitters. Vitamin C also improves the immune response and favors the elimination of xenobiotics and radicals. In fact, it directly reacts with superoxide anions, hydroxyl radicals, and various lipid hydroperoxides [23].

3.2 Vitamin E

Molecules possessing vitamin E activity are grouped under the term 'tocopherols.' They are all homologous, derived from the 6-hydroxychromane structure, and are classified into two groups: 1) *the tocopherols*, including the α-, β-, γ, δ-tocopherols, which have a long isoprenic saturated side chain at C2 and a variable group of methyl groups bound to the chromanol ring; and 2) *the tocotrienols*, divided into the α-, β-, γ-, δ-tocotrienols, which have the same structure as the related tocopherols, except for the side chain, which is unsaturated with three double bonds [24]. The tocopherols come largely from cereal sprouts, green leafy vegetables, oleic fruits, seeds, and their respective oils. Tocopherols are hydrolyzed in the small intestine, emulsified with the biliary acids, transported in the blood by low density lipoprotein (LDL), stored in the liver, and eliminated through bile and urine [25]. Vitamin E is a powerful lipid-soluble antioxidant, which acts synergistically with selenium to prevent the oxidation of fatty acids, membrane phospholipids, and proteins [26]. In the dentate gyrus of mammalian hippocampus, α- and β-tocopherols have a neuroprotective role by preventing apoptosis and prolonging newborn neuron life [27].

When vitamin E functions as an antioxidant and donates its electron, it cannot function again until it has been 'recharged' by vitamin C. A class of chimeric molecules combining ascorbic acid with the chromanyl head of the tochopherols were synthesized in order to confer higher antioxidant protection in critical conditions like ischemia-reperfusion [28].

3.3 Carotenoids

These are nature's most widespread fat-soluble pigments. More than 600 different carotenoids have been identified in plants, microorganisms, and animals, and approximately 20 of them can be identified in human blood serum after the intake of fruits and vegetables [29]. Carotene occurs in many forms, designated as α-carotene and β-carotene followed by γ-, δ-, and ε-carotene. β-Carotene is composed of two *retinyl* groups and is broken down in the *mucosa* of the *small intestine* by β-carotene dioxygenase to *retinal* and used in the body as both retinoic acids and retinals, which are the active forms of vitamin A [30]. Vitamin A is essential, and its lack may lead to blindness and other health consequences. Carotenoids function as antioxidants protecting lipids against peroxidation by quenching free radicals, particularly the singlet oxygen [31]. Citrus fruits and vegetables, including carrots, sweet potatoes, winter squash, pumpkin, papaya, mango and cantaloupe, are rich sources of carotenoids [32].

Lycopene, a linear carotenoid with 11 double conjugated bonds, is the precursor of all carotenoids because they are formed from the cyclization of its structure and subsequent hydroxylation of specific carbons (Figure 3.3). Tomatoes, watermelon, pink grapefruits, apricots, and pink guavas are the most common sources of lycopene. In humans, differently from β-carotene, lycopene is not transformed into vitamin A, lacking the β-ionone ring required for conversion into retinoids. Lycopene's role has been associated with decreased incidence of prostate cancer [2].

Xanthophylls are yellow carotenoid pigments involved in photosynthesis and found in the leaves of most plants. In the presence of an excess of photochemical energy, they contribute to the non-photochemical extinction of the fluorescence of chlorophyll. Zeaxanthin and lutein are the xanthophylls that are absorbed and bioavailable. They are found in the macula of the eye, where they decrease the risk of developing age-related macular degeneration [33] and cataracts [34].

3.4 Glucosinolates

A unique group of thioglucosides naturally occurring in plants of the Brassicales order are the glucosinolates (GLs) [35]. GLs contain a β-D-glucopyranose moiety, linked to a sulfur

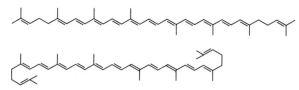

FIGURE 3.3 Molecular structure of lycopene (top) and of the molecule in the cycling phase for the synthesis of other carotenoids (bottom).

atom and either an aliphatic, aromatic, or indole side chain; indolic GLs are derived from tryptophan, while non-indolic GLs derive from other amino acids. The most common GLs present in edible cruciferous vegetables are shown in Table 3.1.

A mixture of indolic and non-indolic GLs is naturally hydrolyzed by the enzyme myrosinase (β-thioglucoside glucohydrolase, EC. 3.2.1.147) into isothiocyanates (ITCs) and indoles (Figure 3.4). These are known to efficiently reduce the risks of cancer and degenerative diseases by inhibiting phase 1 enzymes and by activating phase 2 enzymes [36,37].

3.5 Polyphenols

These molecules are secondary metabolites of plants; they contribute to organoleptic qualities, color, and defense against pathogen attacks.

The chemical structure of phenols possesses one or more aromatic rings, with one or more hydroxyl groups. The radical scavenging activity consists of the ability to directly inactivate the ROS or bind pro-oxidant metal ions by means of their –OH groups [38].

In the former case, the polyphenol transfers one hydrogen to the peroxyl radical as follows:

$$ROO^{\cdot} + ArOH \rightarrow ArO^{\cdot} + ROOH$$

The fenoxyl radical (ArO$^{\cdot}$) formed in this reaction is relatively stable and reacts slowly with other substrates, thus interrupting the

TABLE 3.1   Examples of Aromatic and Aliphatic Glucosinolates Found in Brassica Vegetables

GLUCOIBERIN

$$CH_3SOCH_2CH_2CH_2CH_2 — C \overset{S — glucose}{\underset{NOSO_2O^-}{}}$$

GLUCORAPHANIN

$$CH_3SOCH_2CH_2CH_2CH_2 — C \overset{S — glucose}{\underset{NOSO_2O^-}{}}$$

PROGOITRIN

$$CH_2 = CHCH_2CH_2 — C \overset{S — glucose}{\underset{NOSO_2O^-}{}}$$
OH

SINIGRIN

$$CH_2 = CHCH_2 — C \overset{S — glucose}{\underset{NOSO_2O^-}{}}$$

GLUCONAPIN

$$CH_2 = CHCH_2CH_2 — C \overset{S — glucose}{\underset{NOSO_2O^-}{}}$$

4-HYDROXYGLUCOBRASSICIN

GLUCOBRASSICIN

NEOGLUCOBRASSICIN

After the mirosinase action the aliphatic glucosinolates produces isothyocianates and the aromatic the indole-3-carbinols.

chain of oxidative reactions. At high concentrations, the polyphenol may act as a pro-oxidant, since the amount of formed phenoxyl radicals is able to trigger oxidative reactions [39,40]. Figure 3.2 shows the categories of phenolic compounds.

The phenolic acids comprise two main subgroups: benzoic acids and cinnamic acids. Natural phenolic acids occur in fruits and vegetables either in free or conjugated forms, usually as esters or amides.

Whole cereals are particularly rich in phenolic acids: ferulic, p-coumaric, syringic and vanillic acids are the most common in bran, while those most represented in oat are dihydrocaffeic, synapic and p-hydroxybenzoic acids [12,13].

The stilbenes are phenolic compounds containing two benzenic rings linked by an ethane or ethene bridge. They are widely distributed in higher plants, acting as phytoalexins and growth regulators. Resveratrol (3,4,5-trihydroxystilbene) is the member of this chemical family found

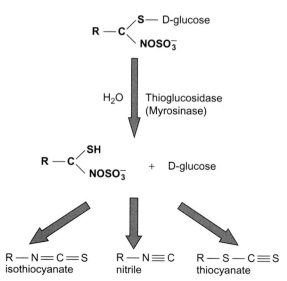

FIGURE 3.4 Hydrolysis of aliphatic glucosinolates. Myrosinase catalyses the hydrolyzation of indolic and non-indolic glucosinolates into isothiocyanate and indoles. The intermediate step shows the momentary dissociation of the glucosinolates into the sugar and the functional group that will be transformed in one of the three azo-groups.

in grapes and wine, and it is reputed to be responsible for preventing heart disease, primarily through the consumption of red wine [41].

The flavonoids' generic structure consists of two aromatic rings, A and B, linked by an oxygenated heterocyclic C ring [42]. Based on the structure of the C ring, as well as on its oxidation state and functional groups, flavonoids are classified as flavonols, flavones, flavanols (catechins), flavanones, anthocyanidins, and isoflavonoids (Figure 3.2). In fruit and vegetables, flavonoids are frequently present as glycosides, since glycosylation makes the molecule less reactive but more soluble. Glucose is the sugar that is most frequently involved in glycoside formation, but we can also find galactose, rhamnose, xylose, and arabinose, and beyond them disaccharides like rutose [42].

The major sources of flavonoids in the Western diet include fruits such as citrus fruits, apricots, cherries, grapes, black currants, bilberries, and apples; among the vegetables we include onions, broccoli, tomatoes, spinach, beet greens, soybeans, and aromatic herbs. In fact, aromatic herbs and spices have a flavonoid concentration that is several times higher than that of common vegetables [43]. Aromatic herbs also contain special antioxidants of phenolic origin. For instance, rosemary contains the phenolic diterpenes, carnosol and rosmanol, which together with carnosic and rosmarinic acids, are the most active phenolic antioxidants of this herb [44]; *Piper nigrum* contains five amides of phenolic acids with marked antioxidant properties [45].

The lignans are basically dimers of the cinnamic alcohol, which cyclizes in different ways, generating a wide range of molecules. The lignans are contained in woody tissues, cereals, and vegetables like carrots, broccoli, and berries. Together with isoflavones, the lignans belong to the class of phytoestrogens, which are protective factors of cardiovascular and immune systems [46].

Tannins are polymers of phenolic acids or flavonoids, present in nature as hydrolyzable and non-hydrolyzable (condensed) tannins. The basic units of hydrolyzable tannins are gallic and ellagic acids, esterified to a core molecule, commonly glucose or a polyphenol such as catechin. Condensed tannins, also called proanthocyanidins, are mainly flavonoid polymers. Tannins are potent antioxidants, but they are scarcely absorbed by the gut and are considered antinutritional factors since they are able to complicate and precipitate proteins and inhibit digestive enzymes [47].

4. PHENOLIC DISTRIBUTION AND CHANGES DURING DEVELOPMENT

The distribution of phenolic compounds varies greatly in the different parts of fruits or vegetables. In mature cherries, the phenolic acids, for instance the derivatives of synaptic acid, are very often localized in the epicarp [48].

In grape skin, the concentration of hydroxyl-cinnamoyltartaric acid is several-fold greater than in the pulp. The skin of apples has a concentration of 5-caffeoylquinic acid and catechins higher than in the parenchyma [48,49].

Interesting changes in phenolic concentration were also described during the developmental stages of vegetables. The levels of hydroxycinnamic acid and catechin in apples and pears generally increase for a short time during the initial development of these fruits, and are extremely high in the raw fruits [49]. However, the concentration decreases to a stable level once the fruits mature. The same pattern has been observed in cherries, apricots, prunes, grapes, and berries [48].

5. ACTION OF PHYTOCHEMICAL ANTIOXIDANTS

The antioxidants present in fruits, vegetables, and grains exhibit various beneficial properties in humans. A number of phenolic compounds, particularly flavonoids, are efficient antiproliferative agents, being able to inhibit proliferation of tumor cells by interfering with cell-cycle proteins or inducing apoptosis [50]. Apple extracts contain phytochemicals that inhibit tumor cell growth *in vitro* by modulating the expression of selected genes [51]. Apples with peel inhibit colon cancer cell proliferation by 43%, although this inhibition was reduced to 29% when peeled apples were tested in one study [52]. The mechanisms by which polyphenols act as antitumoral agents are manifold and have been shown to be connected with the functions of radical scavengers, detoxification agents, cell signaling modulators, inhibitors of cell-cycle phases, and activators of apoptosis [53]. Some flavonoids are able to achieve this effect by inhibiting the enzyme DNA topoisomerase II, which is necessary for the survival and spread of cancerous cells [53].

Broccoli, cabbage, kale, and Brussels sprouts, of the Brassicaceae family, are considered among the most important anticancer vegetables. Studies have demonstrated their anticancer effects due to glucosinolates (Figure 3.4). Brassica vegetables also have a remarkable content of antioxidants like ascorbic acid and carotenoids, and phenolic compounds such as quercetin and kaempferol [54]. Spinach and beet greens, belonging to the Chenopodiaceae family, are also interesting for chemoprevention. A marked antiproliferative activity on HepG2 human liver cancer cells was shown in spinach by Chu et al. [55]. In *Beta Vulgaris cicla*, we demonstrated a strong cytostatic activity towards MCF-7 breast tumor cells, attributable to glycosides of the flavone apigenin [56].

Indeed, phenolic compounds show anti-inflammatory activity, which is mediated by inhibiting the formation of transcription factors closely linked to inflammation, such as NF-κB and enzymes such as xanthine oxidase, cytochrome oxidase, and lipoxygenase, which mediate the inflammatory process [55,57].

The flavonoids of red wine and green tea, particularly quercetin and catechin, exhibit antioxidant and antiatherosclerotic activity through binding to LDL. This binding process reduces their sensitivity to oxidation and their atherogenic potency [58,59].

Flavonoids have also been shown to reverse vascular endothelial dysfunction by increasing endothelium-derived nitric oxide bioactivity [60,61].

Garlic and onions, both from the Liliaceae family, are interesting for the cardiovascular-protective effect of allicin. Metabolism of allicin produces hydrogen sulfide, which relaxes the blood vessels, increases blood flow, and boosts heart health [62]. Allicin is also considered to be responsible for the anticancer activity of garlic extract [63].

Phytochemicals can also have antimicrobial activity [64–66]. For instance, extracts of grape seed or rosemary may be used as food preservatives [67].

Fruits are no less important than vegetables in health protection. A screening of the antiproliferative activities of fruits on HepG2 cells [68] showed the highest effect in cranberries, followed by lemons, apples, red grapes, bananas, grapefruits, and peaches. Apples were shown to be able to prevent mammary cancer in a rat model in a dose-dependent manner [69]. Cranberry powder and juice have been proposed to prevent urinary tract infections, due to the procyanidins that inhibit the adhesion of *Escherichia coli* [70,71].

Another category of vegetable foods that were studied for their antioxidant value are the sprouts. The sprouts are a good source of amino acids, minerals, fiber, and phenolic compounds [72]; interestingly, some sprouts of the Brassicaceae family have a content of glucosinolates and antioxidants higher than that of the mature plant [73].

The list of vegetables and fruits rich in molecules useful for health protection is too long to repeat in its entirety, and only some examples have been mentioned here. The multiplicity of actions expressed by the polyphenols suggests that it is useful to have a number of different fruits and vegetables in the human diet in order to take in a wide range of antioxidants [61,74]. Since at a physiological level some antioxidants work better in hydrophilic conditions and others within a fatty environment, the best health protection is achieved when both antioxidant types are present together in such a way that they work synergistically and possibly regenerate themselves after oxidation [75]. All these antioxidants are best acquired through whole-fruit and vegetable consumption, possibly improved through seasonal selections.

6. METHODS FOR ASSAYING ANTIOXIDANT CAPACITY

The multiplicity of molecules that exploit antioxidant activity, their different chemical and physical characteristics, and the complexity of their interactions, make the use of reliable methods important for determining total antioxidant capacity in foods. For this goal, several methods have been developed through the years and labeled with acronyms such as the following: TRAP (total radical antioxidant power) [76]; FRAP (ferric reducing antioxidant power) [77]; ORAC (oxygen radical absorbance capacity) [78]; and TEAC (Trolox equivalent antioxidant capacity) [79]. An excellent review on the advantages and disadvantages of these methods has been published [80].

Table 3.2 lists some characteristics of the most commonly used methods.

ORAC and TRAP use a fluorescent dye as a target for radicals generated by thermal decomposition of diazocompounds. The antioxidants of the sample function as a protective shield between the fluorescent dye and the radical species. The fluorescence decays slowly during the initial phases of the process when the antioxidants are present at high concentrations, whereas it decreases quickly when antioxidants are almost exhausted. The area under the decay curve in the ORAC assay, or the length of the induction period (lag phase) in the TRAP, is compared to that of an internal standard and then quantitatively related to the antioxidant capacity of the sample.

The FRAP method uses the ferric complex Fe-TPTZ as a probe, which is transformed to a colored ferrous form when it reacts with the antioxidants.

The TEAC test utilizes ABTS, which is oxidized by peroxyl radicals to the stable radical cation (ABTS$^{\cdot +}$), absorbing the light at 734 nm. In the presence of antioxidants, radical formation is inhibited, and the absorbance decay is monitored spectrophotometrically in the presence of the testing sample, or Trolox, at a fixed point in time, and then the antioxidant capacity is expressed as Trolox equivalents.

The TOSC method is based on gas chromatographic measurements of the ethylene

TABLE 3.2 Comparison of different *in vitro* total antioxidant capacity assays

Method	ORAC	TRAP	FRAP	TEAC	TOSC	Crocin
Reagents	AAPH, Trolox, Fluorescein	AAPH, Trolox, R-phycoerythrin	TPTZ, FeCl$_3$, Na-acetate	ABTS, K$_2$SO$_4$, Trolox	KMBA, AAPH,	Crocin, AAPH, Trolox
Instrument	Fluorescent plate reader spectrometer	Fluorescent plate reader spectrometer	Spectro-photometer	Spectro-photometer	Head space gaschromatography	Spectro-photometer
Temperature	37°C	37°C	37°C	37°C	37°C	37°C
Abs (nm)	Ex.485, Em. 520	Ex. 495, Em. 575	593	415 or 734	Depending on detector type	443
pH	7.4	7.4	3.6	7.4	7.4	7.4
Endpoint	End of fluorescence decay	Lag phase	Fixed time (4–10 min)	Fixed time (4–6 min)	End of production of etilene	Fixed time (10 min)
Calibration	Trolox solutions	Trolox solutions	Fe^{3+} standard solutions	Trolox solutions	–	Trolox or α-tocopherol solutions
Calculation of the results	Area of the fluorescence curve decay	Length of the lag phase	Absorbance of final reading minus blank	Absorbance decrease in presence of the sample	Area under the kinetic curve	K$_a$/K$_c$ of antioxidant divided for K$_a$/K$_c$ of Trolox
Lipophylic antioxidants	Yes, in presence of cyclodextrins	Yes, in presence of cyclodextrins	No	Yes	Yes	Yes

ORAC (oxigen radical adsorbance capacity); AAPH or ABAP (Azo-bis(2-amidinopropane); TRAP (total radical-reducing antioxidant potential); FRAP (ferric-reducing antioxidant potential); TPTZ (2.4.6-tripyridyl-s-triazine); TEAC (trolox equivalent antioxidant capacity); ABTS (2.2′-azinobis(3-ethylbenzothiazoline-6-sulfonic acid); TOSC (total oxidant scavenging capacity); KMBA (α-keto-γ-methiolbutyric acid).

produced from the oxidation of α-keto-γ-methiol-butyric acid, caused by the radicals released by the thermal decomposition of AAPH.

Two other methods are commonly used: the first is based on the oxidation of crocin by peroxyl radicals [81], while the second is based on luminol [82]. The old rancidity test against fats and oils is still applied to compare antioxidant capacities of fruit and vegetable extracts [83].

An advance on these methods for chemical antioxidant activity assay is a new test, the cellular *antioxidant* activity (CAA) assay, which is a more biologically relevant method. In fact, it is performed in the presence of cells and accounts for uptake, metabolism, and localization of antioxidant compounds [84]. The CAA

centers on dichlorofluorescin, a probe molecule trapped within cells, which can be easily oxidized to produce fluorescence. The test uses ABAP-generated peroxyl radicals to oxidize dichlorofluorescin. The ability of antioxidants to inhibit this process is measured, and the decrease in cellular fluorescence indicates the antioxidant capacity of the sample.

Different studies have compared data obtained with the popular antioxidant capacity assays; sometimes a significant correlation has been found and sometimes not [80]. The importance of the antioxidant capacity and the huge number of molecules and products to be evaluated have generated a number of data, utilization of which is not easy because of the lack of normalization methods. However, a similar trend among values has frequently been observed [80].

The ORAC assay has been considered by the US Department of Agriculture (USDA) as the official method for antioxidant capacity analysis, and tables of ORAC values for phytochemicals, foods, and single antioxidants are available on the USDA website (http://www.usda.gov/wps/portal/usdahome).

These tables are of great importance, since data on antioxidant capacity have several implications for the *in vivo* health-protective effect of vegetable foods. In fact, the antioxidant capacity of ingested foods is the best way of changing the antioxidant capacity *in vivo* and therefore protecting against ROS-induced diseases.

In our laboratory we applied the ORAC method with a high throughput system, carried out using a plate reader, equipped with a temperature-controlled incubation chamber and an injection pump. The ORAC method results are particularly adaptable for determining the antiradical activity of complex mixtures, and they have also been validated for lipophilic antioxidants in the presence of cyclodextrins [85]. These polymeric molecules are able to incorporate lipophilic substances, leaving them available for radical attack and at the same time making them hydrosoluble.

Using the ORAC method, we provided the antioxidant capacity of both the lipo- (L-ORAC) and hydrosoluble (H-ORAC) fractions of fruits, vegetables, herbs and spices, olive oils and salad dressings, sauces, vegetable soups, botanical extracts, beverages, biscuits, and cookies. Different processing and storage conditions of fruits and vegetables were also tested in order to search for the mildest food processing [86].

7. PLASMA ORAC VALUES

When applied to human plasma, the ORAC method can provide the index of bioavailability of antioxidants after the intake of foods. A correlation between the ORAC units of ingested foods and the antioxidant capacity of plasma has often been detected for finding a correct intake of antioxidants in humans. In the past decade, the daily dose of fruits and vegetables necessary to maintain health-protective effects in young and old people was tentatively ranked in the range 3500–6000 μmol Trolox equivalents (TE) [87]. In 2007, Prior et al. [88] studied a number of subjects exposed to a variable daily food intake and measured the ORAC antioxidant capacity of the plasma in the range of 1–4 hours after a meal. These authors showed that the consumption of dried plums and dried plum juice did not alter either the H-ORAC or the L-ORAC plasmatic value; however, the consumption of cherries, for a total of 4500 μmol TE, increased the L-ORAC but not the H-ORAC plasmatic value; and consumption of blueberries to 12.500 or 39.900 μmol TE and mixed grape juices to 8.600 μmol TE increased both H-ORAC and L-ORAC of the plasma [88].

These researchers concluded that: a) consumption of certain berries and fruits modifies the plasmatic antioxidant capacity; b) after a complete meal, oxidative stress increases with respect to the fasting condition; and c) fruits and vegetables are able to prevent the postprandial increase in oxidative stress [88].

Broadly speaking, the bioavailability of the most common polyphenols varies from 0.3 to 43% depending on the type and ingested amount of food [83,89]. Such a wide range may be explained by the fact that, to be bioavailable, the phenolic compounds should pass several barriers, such as solubility and permeability across the digestive tract, renal filtration, uptake from the tissues, metabolism, elimination, and final excretion in the urine [90]. Glycosylated flavonoids, which are soluble but too big and polar to pass through the enterocyte membrane, have been thought to need the action of intestinal microflora to hydrolyze their sugars and make their phenolic components potentially better absorbed, but this is not the case. In fact, aglycon bioavailability does not increase because its solubility decreases and dramatically reduces its uptake. In some cases the presence of the glucose group on the molecule creates an advantage, since the specific glucose transporters (GLUT 1–5) bind to glucose and transport the whole molecule inside the enterocyte [90,91].

Due to this complex series of stochastic events, the bioefficacy of a vegetable food to increase the plasmatic antioxidant capacity must always be evaluated in terms of both the type of ingested food and the plasma.

8. ANTIOXIDANT CAPACITY OF FRESH AND STORED PRODUCTS

The famous campaign 'five a day,' launched in the USA at the end of the 1990s, suggested eating five servings of fresh fruits and vegetables every day. This is because fruits and vegetables, if eaten fresh, maintain their complete phytochemical patrimony. In fact, collection and transport from field to the storage facility or marketplace do not significantly deplete the nutritional and organoleptic properties or the antioxidant capacity of the produce. This is not the case, however, for produce that undergoes lengthy storage in refrigerated rooms, since vitamins and polyphenol chemical structures are quite sensitive to modifications of enzymatic oxidases and chemical oxidation occurring during storage. The control of environmental conditions and the extent of the storage time of fresh fruits and vegetables represent important steps for maintaining their nutritional quality [86]. Storage requires temperatures in the range 5–8°C, humidity levels of 80%, and atmospheres that are carefully controlled. Under these conditions, some fruits, like apples, can be conserved for more than a year, and vegetables like chicory and cabbage for 3–5 months [86]. During these time periods,

we have observed significant fluctuations of the phenolics and consequently of the antioxidant capacities [86]. Since cells utilize nutrients for their survival, after some months of stored fruits and vegetables have lost most of their phenolic content, yet they are still sold as fresh. Consumers should be informed that the nutritional value of long-stored produce is very often far less than that of fresh produce. However, it is better to eat long-stored produce than no produce at all. The nutritional guidelines of every country suggest that people use vegetables of the current season and cultivated locally, but with globalization this has become an ideal situation, only applicable in certain districts with deep agricultural development, such as California or the Sicilian areas of Italy.

9. ANTIOXIDANT CAPACITY OF PROCESSED FRUITS AND VEGETABLES

Fruits and vegetables, after their arrival from the farm to the industry, are cleaned and immediately processed. The rapidity of processing is very important for optimal quality of the final products, but inevitably the processing challenges the phytochemical molecules. The content of total phenols, or single marker compounds like vitamin C, decreases significantly with processing [86,92]. The antioxidant capacity, when measured on each stock of processed fruits and vegetables, describes the extent of quality loss. Many variables occur in processing, and producers wishing to give consumers high quality products should be aware of total antioxidant capacity and add this value to their packaging labels as proof of their attention to nutritional quality.

Minimally processed products are increasingly consumed nowadays. These are the 'ready-to-eat' fruits and vegetables, which are washed, cleaned, cut, then finally submitted to modified atmosphere packaging (MAP). The cutting is one of the most delicate steps, since it triggers degradative oxidation as a consequence of the decompartmentalization and contact of the phenolic substrates, present in the vacuoles, with the cytoplasmatic oxidases [93]. Therefore, phenolic concentration and antioxidant capacity may change significantly in comparison with the fresh products. Organoleptic characteristics and phenol concentrations strictly depend on type of storage and MAP [94]. These parameters also seem to be relevant in determining the bioavailability of the phenolics in humans [95,96].

Fermentation is another process that may change, either positively or negatively, the chemical structure, the antioxidant activity, and possibly the bioavailability of many phytochemicals. Fermentation is mainly necessary for soy-derived products such as miso, tofu, soy sauce, and yogurt. During this process, hydrolysis of the glycosylated soy isoflavones occurs, as well as the transformation of the aglycones. In preparation of black tea from green tea leaves, fermentation oxidizes the catechins to aromatic substances that make the tea more flavorful but less rich in phenolics. In contrast, however, fermentation may act positively in the release of healthy molecules from waste material, as happens during the fermentation of grapes, during which the stilbenes present in the peel are released in the must [94].

In the processing of freezing or canning vegetables, the heating phase, also called blanching, is the major factor responsible for modifications in phytochemicals [92]. During the blanching process, the industry searches for a subtle equilibrium between the need to inactivate the peroxidases, which is necessary to avoid undesirable smell and taste, and to maintain the flavonoids in their natural state. The heat treatment partially denatures the enzymes and prevents enzymatic degradation of the phytochemicals but exposes them to thermal modifications [92]. Blanching performed with vapor is preferred to that from boiling water, since the latter can cause a phenolic loss of up to 80%, in comparison to an average of only 20% in the former [43]. Blanching of Brassicaceae vegetables, for instance, reduces the functional properties of the glucosinolates by inactivating the myrosinase. Antioxidant capacity, total phenols, ascorbic acid, and minerals are the principal markers for evaluating the dispersion of nutrients during thermal treatments [43,92].

Other flavonoids that are temperature sensitive can also be monitored. For instance, quercetin is very thermal-hardy, as is observed in onions, which lose about 33% of the initial content after frying, 14–20% after boiling, 14% after steaming, and 4% after microwaving [97]. Soy isoflavones are also not destroyed following cooking; however, they are subject to interconversion in different forms, so that modifications of the total content of isoflavones may be negligible [98]. On the contrary, temperature increases the release of lycopene from tomato peel, and the total antioxidant capacity of the final product increases as well [99].

The use of high-pressure technologies for vegetable sterilization is sometimes an excellent alternative to conventional thermal treatments. In carrots, tomatoes, and broccoli, the concentration of lycopene, β-carotene, and vitamin C, in addition to their antioxidant capacity, did not change after 60 min of pressurization [100]. However, structural modifications of phytochemicals may occur, and consequently the bioavailability of the molecules can change [100].

Dehydrated fruits, used in desserts or fruit salads, and dehydrated vegetables, used as powders to aromatize pasta or other commodities, show significant changes in antioxidant capacity before and after processing. ORAC values of lipophilic and hydrophilic components in dehydrated nuts, figs, prunes, and raisins were lower than those of fresh fruits [101]. Phenolic content and antioxidant capacity of vegetables used for aromatized pasta showed a significant decay [86]. The extent of

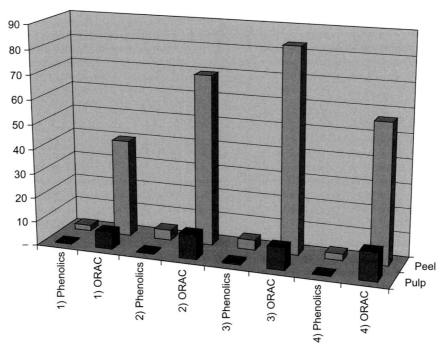

FIGURE 3.5 Phenolics and antioxidant capacity in peel and pulp of two cultivars of apples and pears. Sample number: 1) Apple cv. Golden Delicious; 2) Apple cv. Stark; 3) Pear cv. Abate; 4) Pear cv. Conference. Phenolics are expressed in mg/g of fresh weight, ORAC in μmolTE/g of fresh weight.

the decay depends on the type of dehydration: lyophilization reduces the loss of antioxidants minimally, while a ventilated oven promotes a greater decay. Residual water activity, which remains in fruits and vegetables, may be in the range of 5 to 40% after the dehydration treatment [102], and this value is important both to the food quality and the extent of the shelf-life.

A study performed in our laboratory found significant differences in antioxidant capacities of apples and pears in the presence or absence of the peel. Figure 3.5 shows the results of the phenolic content and ORAC values; it emerges that on the basis of weight, the peel's contribution to the ORAC value is several times higher than that of the pulp. However, from a nutritional standpoint, the contribution of the peel is minimal in fresh fruit, due to the low weight of the peel as compared to the total mass of the fruit, but when the fruit is dehydrated, the loss of weight makes the peel's antioxidant contribution significant [102]. The highest content of antioxidants in the peel suggests that it is better to consume fruits that are both organically cultivated and dehydrated with peels intact. However, organic cultivation, in respect to conventional agriculture, does not show consistent differences in nutritional value, just the difference in content of chemicals on the peels. For example, a 3-year study of phenolic and antioxidant capacity measurements of organic and conventional olives did not show significant differences in the nutritional content in these fruits [103].

Changes induced by technological treatments in the concentration and structure of

TABLE 3.3   Phenolic, Flavonoids and ORAC Values of Fresh and Processed Vegetables

Vegetable	Processing	Total phenols (mg/g)	Flavonoids (mg/g)	ORAC (μmolTE/g)
Beet green	Fresh	0.97±0.07	0.47±0.12	26.24±2.10
	[d]Frozen	0.44±0.05*	0.22±0.08*	11.68±1.10*
Spinach	Fresh	1.01±0.11	0.32±0.10	27.32±2.30
	[d]Frozen	0.81±0.07*	0.28±0.10	16.87±1.30*
Yellow pepper	Fresh	1.54±0.07	0.18±0.01	9.50±0.89
	[c]Grilled	1.45±0.09	0.11±0.01*	6.94±0.81*
Cauliflower	Fresh	0.62±0.09	0.32±0.02	9.25±0.97
	[b]Steamed	0.54±0.06	0.10±0.01*	6.20±0.58*
Aubergine	Fresh	0.26±0.02	0.09±0.01	3.81±0.15
	[b]Steamed	0.19±0.02*	0.04±0.01*	2.45±0.26*
Carrot	Fresh	0.35±0.02	0.13±0.02	6.96±0.70
	[a]Ready-to-eat	0.30±0.02	0.12±0.02	4.99±0.50*
Rocket	Fresh	1.36±0.12	0.46±0.02	23.27±2.10
	[a]Ready-to-eat	0.61±0.03*	0.11±0.01*	10.12±0.90*
Valerian	Fresh	1.40±0.12	1.05±0.12	22.57±2.0
	[a]Ready-to-eat	1.08±0.74*	0.44±0.02*	18.20±1.30*

Samples were obtained from the market. Vegetables were:
[a]cleaned, cut and packaged under modified atmosphere (MAP);
[b]Steamed than packaged under MAP;
[c]Grilled then packaged under MAP;
[d]Vegetables cleaned, blanched than refrigerated and packaged. Values are referred to edible weight and are the mean ± S.D. of four independent determinations.
*Significantly different from the corresponding fresh vegetable by ANOVA ($p < 0.05$).

phytochemicals in fruits and vegetables occur in relation to the type of processing, the phytochemical concentrations, the chemical structure, the redox state of the molecules, and the interactions with other components of the food or the environment. The incidences of these parameters cannot be foreseen, and the phenolic content and antioxidant capacity must be evaluated to create an index of the effective decay in phytochemical antioxidants.

Table 3.3 shows phenolic, flavonoid, and ORAC values of vegetables that undergo different processes and compares these values to those of fresh products.

10. ADDITIVE AND SYNERGIC ANTIOXIDANT EFFECTS

Different species and varieties of fruits, vegetables, and grains have different phytochemical profiles. When an organism consumes a wide variety of grains, fruits, and vegetables, the number of phytochemicals taken up by the organism increases. The combination of many phytochemicals at low concentrations provides that the molecules display additive and very often synergistic effects in their antioxidant properties. This synergism is frequently demonstrated by the antioxidant capacity of the mixture, which is greater than the sum of the antioxidant capacity of the single molecules. This synergism has frequently been demonstrated *in vitro* and very likely also happens under *in vivo* conditions [104,105].

The synergy among phytochemicals is one of the reasons that nutritional guidelines insist on varying the foods in one's diet, particularly fruits and vegetables. As the complete picture of the determinants for health-protective effects of grains, fruits, and vegetables is still unknown, as wide a range of phytochemicals possible must be ingested to provide the most efficacious health-protective effects.

11. NEW FOODS AND THE 'OMICS' DISCIPLINES

Several components of fruits and vegetables were isolated, and the structure was elucidated and finally tested singularly or in mixtures for obtaining the expected effect on cells and animals.

Potentially, all vegetable cell metabolites can be identified through so-called 'metabolomic analysis' [106], which is developing into an essential tool in natural sciences. Metabolomic analysis is linked with both nuclear magnetic resonance (NMR) spectroscopic analysis for structure elucidation and pluricomponent analysis (PCA), a statistical method that evaluates the importance of one parameter in conditioning the others. The manipulation of plants and the search for new plants endowed with great antioxidant potential are conducted in tandem with metabolomic analysis, which is integrated with nutrigenetics and nutrigenomics [107,108]. These disciplines, which have evolved rapidly in recent years, have increased our knowledge of the interactions between life processes and specific components of our diet. Nutrigenetics asks how individual genetic disposition affects susceptibility to diet; nutrigenomics addresses

how diet influences gene transcription, protein expression, and metabolism. The future goal of these disciplines, combined in an integrated approach to nutrition, is to provide a wealth of useful information on the effects of nutrients and food components on metabolic pathways [107,108]. Personalized diet regimens, to prevent the onset of nutrition-related disorders in genetically predisposed individuals, will be set up with the highest precision.

12. FUNCTIONAL FOOD PRODUCTS

Progress in plant antioxidant and bioactive compounds has boosted the dietary industries to search for functional foods, which are modified foods or food ingredients that can provide health benefits beyond those furnished by traditional nutrients. The main interest, at commercial level, concerns fruit juices and isotonic beverages, which are widely consumed by younger people and athletes. Other interesting products are snacks and chocolates enriched in phenolic compounds with herbal extracts in order to increase their antioxidant capacity [109].

In our laboratory we analyzed the antioxidant capacity of isotonic beverages, juices, dressings, biscuits, and cookies fortified with fruit and vegetable extracts. Table 3.4 shows the remarkable increase of antioxidant capacity with respect to the standard food.

13. SUMMARY

Consumers can obtain a wide range of antioxidants from the intake of a large variety of fresh fruits and vegetables. Investigations have shown that the risk of cancer and other chronic diseases is inversely related to the consumption of vegetables and fruits. Results are maximally oriented to attribute the highest

TABLE 3.4 Total Phenols and ORAC Increase of Functional Compared to Standard Products

Product	Type	Total phenols (mg/g)	ORAC (μMolTE/g)
[a]Vegetable soup portion size 200 g	Standard	0.14 ± 0.012	2.30 ± 0.31
	Functional	0.53 ± 0.049	8.70 ± 0.25
[a]Tomato sauce portion size 20 g	Standard	0.28 ± 0.03	5.26 ± 0.27
	Functional	0.54 ± 0.06	10.75 ± 0.77
[a]Sweet each one 4 g	Standard	0.02 ± 0.003	0.08 ± 0.02
	Functional	1.43 ± 0.11	25.56 ± 1.9
[a]Biscuit each one 6 g	Standard	0.41 ± 0.05	12.19 ± 0.6
	Functional	4.61 ± 0.42	76.97 ± 3.7
[b]Yogurt drink portion size 200 g	Standard	0.16 ± 0.02	1.76 ± 0.25
	Functional	0.41 ± 0.05	8.99 ± 1.06
[a]Isotonic beverage portion size 500 g	Standard	0.18 ± 0.05	1.13 ± 0.04
	Functional	0.63 ± 0.08	5.07 ± 0.20
[c]Dark chocolate portion size 100 g	Standard	0.013 ± 0.001	0.102 ± 0.009
	Functional	0.025 ± 0.001	0.156 ± 0.012

[a]Product integrated with spray-dried extract of Green Tea;
[b]Product integrated with catechins;
[c]Product integrated with catechins obtained from dry powder of Cocoa.

protective role to the antioxidant compounds contained in fruits and vegetables.

Isolated pure compounds given as supplements do not appear to be effective since they either lose bioactivity or lack synergistic compounds present in the whole food [110].

Processed fruits and vegetables show a wide range of phytochemical loss. The technology in the food industry should reduce the loss of antioxidants and micronutrients to the minimum level by means of mild processes and the monitoring of each step of the transformation with due control assays.

Functional foods, containing fruit and vegetable juices or extracts, are an important part of the healthful lifestyle, which includes a balanced diet and physical activity. However, to deliver their potential public health benefits, functional foods need to be quality controlled through the collaborative efforts of food-control organizations and the food industry, in order to market only those functional foods that are clearly supported with scientific evidence of nutritional value.

The emerging field of nutrigenomics, or 'personalized nutrition,' provides individual dietary recommendations and may one day have a greater ability to reduce the risk of disease. Biochemical sciences should support the development of foods that address these requirements and provide consumers with the most appropriate and beneficial information for their specific nutritional needs.

However, these expectations should not divert attention from the true problem, which is actually the supply of fresh fruits and vegetables at affordable prices. The huge number of scientific results on the health-protective effects of fruits and vegetables must be able to prime a cultural and political process that brings agricultural fresh products to the center of alimentary education in both families and primary schools. Agricultural products supported by science and technology can provide the most relevant contributions to the long-term health and well-being of consumers.

References

1. Keys, A. (1995). Mediterranean diet and public health: Personal reflections. *The American Journal of Clinical Nutrition, 61*, 1321S–1323S.

2. Itsiopoulos, C., Hodge, A., & Kaimakamis, M. (2008). Can the Mediterranean diet prevent prostate cancer? *Molecular Nutrition Food Research, 53*, 227–239.

3. Simopoulos, A. P. (2004). The traditional diet of Greece and cancer. *European Journal of Cancer Prevention, 13*, 219–230.

4. Kapiszewska, M., Soltys, E., Visioli, F., Cierniak, A., & Zajac, G. (2005). The protective ability of the Mediterranean plant extracts against the oxidative DNA damage. The role of the radical oxygen species and the polyphenol content. *Journal of Physiology and Pharmacology, 56*, 183–197.

5. Salas-Salvadó, J., Fernández-Ballart, J., Ros, E., Martinez-González, M. A., Fitó, M., Estruch, R., Corella, D., Fiol, M., Gómez-Gracia, E., Arós, F., Flores, G., Lapetra, J., Lamuela-Raventós, R., Ruiz-Gutiérrez, V., Bulló, M., Basora, J., & Covas, M. I. (2008). Effect of a Mediterranean diet supplemented with nuts on metabolic syndrome status. *Archives of Internal Medicine, 168*, 2449–2458.

6. Harman, D. (1956). Aging: A theory based on free radical and radiation chemistry. *The Journals of Gerontology, 11*, 298–300.

7. Harman, D. (1972). The biological clock: the mitochondria? *Journal of the American Geriatric Society, 20*, 145–147.

8. Halliwell, B., & Gutteridge, J. M. C. (1989). Role of free radical and catalytic metals ions in human diseases: An overview. *Methods Enzymologia, 186*, 1–85.

9. Dean, R. T., Fu, S., Stocker, R., & Davies, M. J. (1997). Biochemistry and pathology of radical-mediated protein oxidation. *The Biochemical Journal, 15*, 1–18.

10. Ames, B. M., Shigena, M. K., & Hagen, T. M. (1993). Oxidants, antioxidants and the degenerative disease of aging. *Proceedings of the National Academy of Sciences of the United States of America, 90*, 7915–7922.

11. Gallus, S., Bosetti, C., & La Vecchia, C. (2004). Mediterranean diet and cancer risk. *European Journal of Cancer Prevention, 13*, 447–452.

12. Vitaglione, P., Napolitano, A., & Fogliano, V. (2008). Cereal dietary fibre: A natural functional ingredient to deliver phenolic compounds into the gut. *Trends in Food Science and Technology, 19*, 451–463.

13. Galli, C., & Visioli, F. (2001). Antioxidant properties of Mediterranean diet. *International Journal for Vitamin and Nutrition Research, 71*, 185–188.

14. Harris, E. D. (1992). Regulation of antioxidant enzymes. *The FASEB Journal, 6*, 2675–2683.

15. Coassin, M., Ursini, F., & Bindoli, A. (1992). Antioxidant effect of manganese. *Archives of Biochemistry and Biophysics, 299*, 330–333.

16. Brown, K. M., Pickard, K., Nicol, F., Beckett, G. J., Duthie, G. G., & Arthur, J. R. (2000). Effects of organic and inorganic selenium supplementation on selenoenzyme activity in blood lymphocytes, granulocytes, platelets and erythrocytes. *Clinical Science, 98*, 593–599.

17. Jackson, M. I., & Combs, G. F. Jr., (2008). Selenium and anticarcinogenesis: Underlyng mechanisms. *Current Opinion in Clinical Nutrition and Metabolic Care, 11*, 718–726.

18. Ferri, P. E., Biagiotti, P., Ambrogini, S., Santi, P., Del Grande, P., & Ninfali, P. (2005). NADPH-consuming reactions correlate with glucose-6-phosphate dehydrogenase in Purkinje cells: An immunohistochemical and enzyme histochemical study of the rat cerebellar cortex. *Neurosciences Research, 51*, 185–197.

19. Luzzatto, L. (1995). Glucose-6-phosphate dehydrogenase deficency and the pentose phosphate pathways. In R. I. Handin, S. E. Lux, & T. P. Stossel, (Eds.), *Blood: Principles and practice of hematology*. Philadelphia: J.B. Lippincott Company.

20. Ninfali, P., Perini, M. P., Bresolin, N., Aluigi, G., Cambiaggi, C., Ferrali, M., & Pompella, A. (2000). Iron release and oxidant damage in human myoblasts by divicine. *Life Sciences, 66*, 85–91.

21. Sporn, M. B. (1980). Combination chemoprevention of cancer. *Nature, 287*, 107–108.

22. Linster, C. L., & Van Schaftingen, E. (2007). Vitamin C. Biosynthesis, recycling and degradation in mammals. *FEBS Journal, 274*, 1–22.

23. Rose, R. C., & Bode, A. M. (1993). Biology of free radical scavengers: An evaluation of ascorbate. *FASEB Journal, 7*, 1135–1142.

24. Pennock, J. F., Hemming, F. W., & Kerr, J. D. (1964). A reassessment of tocopherol in chemistry. *Biochemical and Biophysical Research Communications, 17*, 542–548.

25. Brigelius-Flohé, R., & Traber, M. G. (1999). Vitamin E: Function and metabolism. *FASEB Journal, 13*, 1145–1156.

26. Islam, K. N., O'Byrne, D., Devaraj, S., Palmer, B., Grundy, S. M., & Jialal, I. (2000). Alpha-tocopherol supplementation decreases the oxidative susceptibility of LDL in renal failure patients on dialysis therapy. *Atherosclerosis, 150*, 217–224.

27. Cuppini, R., Ciaroni, S., Cecchini, T., Ambrogini, P., Ferri, P., Cuppini, C., Ninfali, P., & Del Grande, P. (2002). Tocopherols enhance neurogenesis in dentate gyrus of adult rats. *International Journal for Vitamin and Nutrition Research, 3*, 170–176.

28. Pessina, F., Marazova, K., Ninfali, P., Avanzi, L., Manfredini, S., & Sgaragli, G. (2004). *In vitro* neuroprotection by novel antioxidants in guinea pig urinary bladder subjected to anoxia-glucopenia/reperfusion damage. *Naunyn-Schmiedeberg's Archives of Pharmacology, 370,* 521–528.

29. Mayne, S. T. (1996). Beta-carotene, carotenoids, and disease prevention in humans. *FASEB Journal, 10,* 690–701.

30. Parker, R. S. (1996). Absorption, metabolism, and transport of carotenoids. *FASEB Journal, 10,* 542–551.

31. Foote, C. S., & Denny, R. W. (1968). Chemistry of singlet oxygen. VII. Quenching by b-carotene. *Journal of the American Chemical Society, 90,* 6233–6235.

32. Khachik, F., Beecher, G. R., Goli, M. B., & Lusby, W. R. (1991). Separation, identification, and quantification of carotenoids in fruits, vegetables and human plasma by high performance liquid chromatography. *Pure and Applied Chemistry, 63,* 71–80.

33. Humphries, J. M., & Khachik, F. (2003). Distribution of lutein, zeaxanthin, and related geometrical isomers in fruit, vegetables, wheat, and pasta products. *Journal of Agricultural and Food Chemistry, 51,* 1322–1327.

34. Lyle, B. J., Mares-Perlman, J. A., Klein, B. E., Klein, R., & Greger, J. L. (1999). Antioxidant intake and risk of incident age-related nuclear cataracts in the Beaver Dam Eye Study. *American Journal of Epidemiology, 149,* 801–809.

35. Fenwick, G. R., & Heaney, R. K. (1983). Glucosinolates and their breakdown products in cruciferous crops, foods and feeding stuffs. *Food Chemistry, 11,* 249–271.

36. Canistro, D., Croce, C. D., Iori, R., Bacillari, J., Bronzetti, G., Poi, G., Cini, M., Caltavuturo, L., Perocco, P., & Paolini, M. (2004). Genetic and metabolic effects of gluconasturtiin, a glucosinolate derived from cruciferae. *Mutation Research, 545,* 23–35.

37. Verhoeven, D. T. (1997). A review of mechanisms underlying anticarcinogenicity by brassica vegetables. *Chemico-Biological Interactions, 103,* 79–129.

38. Rice-Evans, C. A., Miller, N. J., & Paganga, G. (1997). Antioxidant properties of phenolic compounds. *Trends Plant Science, 2,* 152–159.

39. Cao, G., Sofic, E., & Prior, R. L. (1997). Antioxidant and prooxidant behavior of flavonoids: Structure-activity relationships. *Free Radical Biology and Medicine, 22,* 749–760.

40. Skibola, C. F., & Smith, M. T. (2000). Potential health impacts of excessive flavonoid intake. *Free Radical Biology and Medicine, 29,* 375–383.

41. Vidavalur, R., Otani, H., Singal, P. K., & Maulik, N. (2006). Significance of wine and resveratrol in cardiovascular disease: French paradox revisited. *Experimental and Clinical Cardiology, 11,* 217–225.

42. Heim, K. E., Tagliaferro, A. R., & Bobilya, D. G. (2002). Flavonoid antioxidants: Chemistry, metabolism and structure-activity relationships. *The Journal of Nutritional Biochemistry, 13,* 572–584.

43. Ninfali, P., Mea, G., Giorgini, S., Rocchi, M., & Bacchiocca, M. (2005). Antioxidant capacity of vegetables, spices and dressings relevant to nutrition. *The British Journal of Nutrition, 93,* 257–266.

44. Zeng, H. H., Tu, P. F., Zhou, K., Wang, H., Wang, B. H., & Lu, J. F. (2001). Antioxidant properties of phenolic diterpenes from *Rosmarinus officinalis*. *Acta Pharmacologica Sinica, 22,* 1094–1098.

45. Nakatani, N., Inatani, R., Ohta, H., & Nishioka, A. (1986). Chemical constituents of peppers (*Piper spp.*) and applications of food preservation: Naturally occurring antioxidative compounds. *Environmental Health Perspectives, 67,* 135–142.

46. Adlercreutz, H. (1995). Phytoestrogenes: Epidemiology and a possible role in cancer protection. *Environmental Health Perspectives, 103,* 103–112.

47. Reed, J. D. (1995). Nutritional toxicology of tannins and related polyphenols in forage legumes. *Journal of Animal Science, 73,* 1516–1528.

48. Stöhr, H., Mosel, H. D., & Herrmann, K. (1975). The phenolics of fruits. VII. The phenolics of cherries and plums and the changes in catechins and hydroxycinnamic acid derivatives during the development of fruits. *Zeitschrift Fur Lebensmittel-Untersuchung Und-Forschung, 159,* 85–91.

49. Mosel, H. D., & Herrmann, K. (1974). Changes in catechins and hydroxycinnamic acid derivatives during development of apples and pears. *Journal of the Science of Food and Agriculture, 25,* 251–256.

50. Yang, C. S., Landau, J. M., Huang, M. T., & Newmark, H. L. (2001). Inhibition of carcinogenesis by dietary polyphenolic compounds. *Annual Review of Nutrition, 21,* 381–406.

51. Veeriah, S., Miene, C., Habermann, N., Hofmann, T., Klenow, S., Sauer, J., Böhmer, F., Wölfl, S., & Pool-Zobel, B. L. (2008). Apple polyphenols modulate expression of selected genes related to toxicological defence and stress response in human colon adenoma cells. *International Journal of Cancer, 22,* 2647–2655.

52. Eberhardt, M. V., Lee, C. Y., & Liu, R. H. (2000). Antioxidant activity of fresh apples. *Nature, 405,* 903–904.

53. Fresco, P., Borges, F., Diniz, C., & Marques, M. P. M. (2006). New insights on the anticancer properties of dietary poliphenols. *Medicinal Research Reviews, 26,* 747–766.

54. Kurilich, A. C., Jeffery, E. H., Juvik, J. A., Wallig, M. A., & Klein, B. P. (2002). Antioxidant capacity

of different broccoli (Brassica oleracea) genotypes using the oxygen radical absorbance capacity (ORAC) assay. *Journal of Agricultural and Food Chemistry, 28,* 5053–5057.

55. Chu, Y. F., Sun, J., Wu, X., & Liu, R. H. (2002). Antioxidant and antiproliferative activities of common vegetables. *Journal of Agricultural and Food Chemistry, 50,* 6910–6916.

56. Ninfali, P., Bacchiocca, M., Antonelli, A., Biagiotti, E., Di Gioacchino, A., Piccoli, G., Stocchi, V., & Brandi, G. (2007). Characterization and biological activity of the main flavonoids from B*eta vulgaris* (subsp. *Cycla*). *Phytomedicine, 14,* 216–221.

57. Read, M. A. (1995). Flavonoids: Naturally occurring anti-inflammatory agents. *The American Journal of Pathology, 147,* 235–237.

58. Hayek, T., Fuhrman, B., Vaya, J., Rosenblat, M., Belinky, P., Coleman, R., Elis, A., & Aviram, M. (1997). Reduced progression of atherosclerosis in apolipoprotein E-deficient mice following consumption of red wine, or its polyphenols quercetin or catechin, is associated with reduced susceptibility of LDL to oxidation and aggregation. *Arteriosclerosis, Thrombosis, and Vascular Biology, 17,* 2744–2752.

59. Vinson, J. A., & Dabbagh, Y. A. (1998). Effect of green and black tea supplementation on lipids, lipid oxidation and fibrinogen in the hamster: mechanisms for the epidemiological benefits of tea drinking. *FEBS Letters, 433,* 44–46.

60. Duffy, S. J., & Vita, J. A. (2003). Effects of phenolics on vascular endothelial function. *Current Opinion in Lipidology, 14,* 21–27.

61. Van Ackers, S. A., Tromp, M. N., Haenen, G. R., van der Vijgh, W. J., & Bast, A. (1995). Flavonoids as scavengers of nitric oxide radical. *Biochemical and Biophysical Research Communications, 214,* 755–759.

62. Benavides, G. A., Squadrito, G. L., Mills, R. W., Patel, H. D., Isbell, T. S., Patel, R. P., Darley-Usmar, V. M., Doeller, J. E., & Kraus, D. W. (2007). Hydrogen sulfide mediates the vasoactivity of garlic. *PNAS, 104,* 17977–17982.

63. Hirsch, K., Danilenko, M., Giat, J., Miron, T., Rabinkov, A., Wilchek, M., Mirelman, D., Levy, J., & Sharoni, Y. (2000). Effect of purified allicin, the major ingredient of freshly crushed garlic, on cancer cell proliferation. *Nutrition and Cancer, 38,* 245–254.

64. Cowan, M. M. (1999). Plant products as antimicrobial agents. *Clinical Microbiology Reviews, 12,* 564–582.

65. Brandi, G., Amagliani, G., Sisti, M., Fraternale, D., Ninfali, P., & Scoccianti, V. (2007). Antibacterial activity of a stable standardized *in vitro* culture of Rubus ulmifolius Schott against food-borne pathogenic bacteria. *Italian Journal of Food Science, 19,* 471–476.

66. Fraternale, D., Bacchiocca, M., Giamperi, L., Ricci, D., Scoccianti, V., & Ninfali, P. (2007). Control of isolates of Fusarium species by using extracts of Rubus ulmifolius micropropagated plantlets. *Journal of the Food Agricultural Environmental, 5,* 442–445.

67. Ahn, J., Grün, I. U., & Mustapha, A. (2007). Effects of plant extracts on microbial growth, color change, and lipid oxidation in cooked beef. *Food Microbiology, 24,* 7–14.

68. Sun, J., Chu, Y. F., Wu, X., & Liu, R. H. (2002). Antioxidant and antiproliferative activities of common fruits. *Journal of Agricultural and Food Chemistry, 50,* 7449–7454.

69. Liu, R. H., Liu, J., & Chen, B. (2005). Apples prevent mammary tumors in rats. *Journal of Agricultural and Food Chemistry, 53,* 2341–2343.

70. Vinson, J. A., Bose, P., Proch, J., Al Kharrat, H., & Samman, N. (2008). Cranberries and cranberry products: Powerful *in vitro, ex vivo,* and *in vivo* sources of antioxidants. *Journal of Agricultural and Food Chemistry, 56,* 5884–5891.

71. Agence Francaise de Securité des Aliments (AFSSA). (April 6, 2004). Notification 2003-SA-0352.

72. Lorenz, K. (1980). Cereal sprouts: Composition, nutritive value, food applications. *Critical Reviews in Food Science, 13,* 353–385.

73. Barillari, J., Cervellati, R., Costa, S., Guerra, M. C., Speroni, E., Utan, A., & Iori, R. (2006). Antioxidant and choleretic properties of *Raphanus sativus L.* sprout (Kaiware Daikon) extract. *Journal of Agricultural and Food Chemistry, 54,* 9773–9778.

74. Tournaire, C., Croux, S., Maurette, M. T., Beck, I., Hocquax, M., Braun, A. M., & Oliveros, E. (1993). Antioxidant activity of flavonoids: Efficiency of singlet oxygen (1 delta g) quenching. *Journal of Photochemistry and Photobiology B, 19,* 205–215.

75. Glascott, P. A., Jr., & Farber, J. L. (1999). Assessment of physiological interaction between vitamin E and vitamin C. *Methods in Enzymologica, 300,* 78–88.

76. Wayner, D. D. M., Burton, G. W., Ingold, K. U., & Locke, S. (1985). Quantitative measurement of the total, peroxyl radical-trapping antioxidant capacity of human blood plasma by controlled peroxidation. *FEBS Letters, 187,* 33–37.

77. Benzie, I. F., & Strain, J. J. (1999). Ferric reducing/antioxidant power assay: Direct measure of total antioxidant activity of biological fluids and modified version for simultaneous measurement of total antioxidant power and ascorbic acid concentration. *Methods in Enzymology, 299,* 15–27.

78. Cao, G., Alessio, H. M., & Cutler, R. G. (1993). Oxygen-radical absorbance capacity assay for antioxidants. *Free Radical Biology and Medicine, 14,* 303–311.

79. Miller, N. J., Rice-Evans, C., Davies, M. J., Gopinathan, V., & Milner, A. (1993). A novel method for measuring antioxidant capacity and its application to monitoring the antioxidant status in premature neonates. *Clinical Science, 84*, 407–412.

80. Prior, R. L., Wu, X., & Schaich, K. (2005). Standardized methods for the determination of antioxidant capacity and phenolics in foods and dietary supplements. *Journal of Agricultural and Food Chemistry, 53*, 4290–4302.

81. Tubaro, F., Ghiselli, A., Rapuzzi, P., Maiorino, M., & Urini, F. (1998). Analysis of plasma antioxidant capacity by competition kinetics. *Free Radical Biology and Medicine, 24*, 1228–1234.

82. Nakamura, M., & Nakamura, S. (1998). One- and two-electron oxidations of luminol by peroxidise systems. *Free Radical Biology and Medicine, 24*, 537–544.

83. Prior, R. L., Hoang, H., Gu, L., Wu, X., Bacchiocca, M., Howard, L., Hampsch-Woodill, M., Huang, D., Ou, B., & Jacob, R. (2003). Assays for hydrophilic and lipophilic antioxidant capacity (oxygen radical absorbance capacity (ORAC(FL)) of plasma and other biological and food samples. *Journal of Agricultural and Food Chemistry, 51*, 3273–3279.

84. Wolfe, K. L., & Liu, R. H. (2007). Cellular antioxidant activity (CAA) assay for assessing antioxidants, foods, and dietary supplements. *Journal of Agricultural and Food Chemistry, 55*, 8896–8907.

85. Huang, D., Ou, B., Hampsch-Woodill, M., Flanagan, J. A., & Deemer, E. K. (2002). Development and validation of oxygen radical absorbance capacity assay for lipophilic antioxidants using randomly methylated beta-cyclodextrin as the solubility enhancer. *Journal of Agricultural and Food Chemistry, 50*, 1815–1821.

86. Ninfali, P., & Bacchiocca, M. (2004). Parameters for the detection of post-harvest quality in fresh or transformed horticultural crops. *Journal of the Food Agricultural Environmental, 2*, 122–127.

87. Cao, G., Booth, S. L., Sadowski, J. A., & Prior, R. L. (1998). Increases in human plasma antioxidant capacity after consumption of controlled diets high in fruit and vegetables. *The American Journal of Clinical Nutrition, 68*, 1081–1087.

88. Prior, R. L., Gu, L., Wu, X., Jacob, R. A., Sotoudeh, G., Kader, A. A., & Cook, R. A. (2007). Plasma antioxidant capacity changes following a meal as a measure of the ability of a food to alter *in vivo* antioxidant status. *Journal of the American College of Nutrition, 26*, 170–181.

89. Manach, C., Williamson, G., Morand, C., Scalbert, A., & Rémésy, C. (2005). Bioavailability and bioefficacy of polyphenols in humans. I. Review of 97 bioavailability studies. *The American Journal of Clinical Nutrition, 81*, 230S–242S.

90. Walle, T. (2004). Absorption and metabolism of flavonoids. *Free Radical Biology and Medicine, 36*, 829–837.

91. Chen, C. H., Hsu, H. J., Huang, Y. J., & Lin, C. J. (2007). Interaction of flavonoids and intestinal facilitated glucose transporters. *Planta Medica, 73*, 348–354.

92. Ninfali, P., & Bacchiocca, M. (2003). Polyphenols and antioxidant capacity of vegetables under fresh and frozen conditions. *Journal of Agricultural and Food Chemistry, 51*, 2222–2226.

93. Varoquaux, P., & Mazollier, J. (2002). Overview of the European fresh-cut produce industry. In O. Lamikanra, (Ed.), *Fresh-cut fruits and vegetables: Science, technology, and market* (p. 21). Boca Raton: CRC Press.

94. Costantini, V., Bellincontro, A., De Santis, D., Botondi, R., & Mencarelli, F. (2006). Metabolic changes of Malvasia grapes for wine production during postharvest drying. *Journal of Agricultural and Food Chemistry, 54*, 3334–3340.

95. Rolle, R. S., & Chrism, C. W. III, (1987). Physiological consequences of minimally processed fruits and vegetables. *Journal of Food Quality, 10*, 157–162.

96. Serafini, M., Bugianesi, R., Salucci, M., Azzini, E., Raguzzini, A., & Maiani, G. (2002). Effect of acute ingestion of fresh and stored lettuce (*Lactuca sativa*) on plasma total antioxidant capacity and antioxidant levels in human subjects. *The British Journal of Nutrition, 88*, 615–623.

97. Lee, S. U., Lee, J. H., Choi, S. H., Lee, J. S., Ohnisi-Kameyama, M., Kozukue, N., Levin, C. E., & Friedman, M. (2008). Flavonoid content in fresh, home-processed, and light-exposed onions and in dehydrated commercial onion products. *Journal of Agricultural and Food Chemistry, 56*, 8541–8548.

98. Coward, L., Smith, M., Kirk, M., & Barnes, S. (1998). Chemical modification of isoflavones in soyfoods during cooking and processing. *The American Journal of Clinical Nutrition, 68*, 1486S–1491S.

99. Gärtner, C., Stahl, W., & Sies, H. (1997). Lycopene is more bioavailable from tomato paste than from fresh tomatoes. *The American Journal of Clinical Nutrition, 66*, 116–122.

100. Yaldagard, M., Mortazavi, S. A., & Tabatabaie, F. (2008). The principles of ultra high pressure technology and its application in food processing/preservation: A review of microbiological and quality aspects. *African Journal of Biotechnology, 7*, 2739–2767.

101. Wu, X., Beecher, G. R., Holden, J. M., Haytowitz, D. B., Gebhardt, S. E., & Prior, R. L. (2004). Lipophilic and hydrophilic antioxidant capacities of

common foods in the United States. *Journal of Agricultural and Food Chemistry, 52*, 4026–4037.

102. Bacchiocca, M., Biagiotti, E., & Ninfali, P. (2006). Nutritional and technological reasons for evaluating the antioxidant capacity of vegetable products. *Italian Journal of Food Science, 2*, 209–217.

103. Ninfali, P., Bacchiocca, M., Biagiotti, E., Esposto, S., Servili, M., Rosati, A., & Montedoro, G. (2008). A 3-year study on quality, nutritional and organoleptic evaluation of organic and conventional extra-virgin olive oils. *Journal of the American Oil Chemists Society, 85*, 151–158.

104. Sun, J., Chu, Y. F., Wu, X., & Liu, R. H. (2002). Antioxidant and antiproliferative activities of common fruits. *Journal of Agricultural and Food Chemistry, 50*, 7449–7454.

105. Chu, Y. F., Sun, J., Wu, X., & Liu, R. H. (2002). Antioxidant and antiproliferative activities of common vegetables. *Journal of Agricultural and Food Chemistry, 50*, 6910–6916.

106. Thongboonkerd, V. (2007). Proteomics. *Forum of Nutrition, 60*, 80–90.

107. Lindsay, D. G. (2000). The nutritional enhancement of plant foods in Europe 'NEODIET'. *Trends in Food Science and Technology, 11*, 145–151.

108. Kussmann, M., & Affolter, M. (2006). Proteomic methods in nutrition. *Current Opinion in Clinical Nutrition and Metabolic Care, 9*, 575–583.

109. Ninfali, P., Gennari, L., Biagiotti, E., Cangi, F., Mattoli, L., & Maidecchi, A. (2009). Improvement in botanical standardization of commercial freeze-dried herbal extracts by using the combination of antioxidant capacity and constituent marker concentrations. *JAOAC International, 92*, 797–805.

110. Omenn, G. S., Goodman, G. E., Thornquist, M. D., Balmes, J., Cullen, M. R., Glass, A., Keogh, J. P., Meyskens, F. L., Valanis, B., Williams, J. H., Barnhart, S., & Hammar, S. (1996). Effects of a combination of beta carotene and vitamin A on lung cancer and cardiovascular disease. *The New England Journal of Medicine, 334*, 1150–1155.

4

Medicinal Activities of Essential Oils: Role in Disease Prevention

Alfreda Wei[1] and Takayuki Shibamoto[2]

[1]Department of Molecular Biosciences, University of California, Davis, CA
[2]Department of Environmental Toxicology, University of California, Davis, CA

1. INTRODUCTION

Essential oils are derived from various parts of the plant, including leaves, flowers, fruits, seeds, rhizomes, roots, and bark. In the plant, these constituents serve several physiological purposes for the plant – protection from pests and microorganisms, attraction of pollinating insects or birds, providing photoprotection to the plant, and allelopathy. The complex mixtures of volatile, aromatic components are secondary plant metabolites, and are usually obtained by distillation (water distillation, steam distillation, water-and-steam distillation, or dry distillation). Mild techniques such as enfleurage, expression, extraction, or fermentation are also used for the isolation of more sensitive compounds [1]. In general, plant essential oils contain constituents predominantly of aromatic (derived via the shikimate pathway) or terpenoid (derived via the deoxyxylulose phosphate pathway) character. The proportion of these individual constituents, the location on the plant from which these chemicals were isolated, variations in climate or growing regions, harvest times, and extraction techniques are all factors that affect oil quality and observed pharmacological properties of the essential oil. The biosynthetic pathways of these chemicals have been described [2].

Essential oils have traditionally been used to impart flavoring or preservative effects to foods, or to instill fragrances in cosmetics and aromatherapy. Since ancient times, numerous civilizations have also valued essential oils for their therapeutic qualities in disease prevention and treatment. The ancient Egyptians buried essential oils with their pharaohs for medicinal uses in the afterlife. Later, the Greeks and Romans absorbed Egyptian practices of using essential oils in aromatherapy and expanded it to their baths for promotion of well-being. For instance, baths infused with the oils of jasmine, lavender, or ylang-ylang stimulated mental relaxation. Similarly, current interest in essential oils arises from the various bioactive effects they display, including antioxidant [3,4], anti-inflammatory [5,6], antimicrobial [7,8], antiviral [9,10], and anticarcinogenic [11]. In developed countries, the benefits derived from using essential oils appear optimistic. Demand for plant essential oils has risen as a consequence of consumers searching for cheaper, more 'natural' alternatives to disease-fighting medications. In food and cosmetic applications, essential oils are considered to be biodegradable, readily available, and 'less toxic' than synthetic preservative agents. As such, this optimism has raised concerns and stimulated studies to evaluate the safety and efficacy of essential oils in various systems in order to better understand their pharmacological properties and roles in health. In this regard, the purpose of this book chapter is to provide an overview of some of the bioactivities demonstrated by essential oils, and the effects that these actions can have on health. The scope of this work will focus on the medicinal activities of essential oils (antioxidant, antiviral, antimicrobial, anticarcinogenic, and anti-inflammatory) and the implications that these bioactivities have on diseases such as aging, skin damage, and cancer.

2. ANTIOXIDANT ACTIVITIES

As organisms in an aerobic environment, humans are constantly exposed to environmental factors (including stress, pollution, and ultraviolet (UV) radiation) that promote the generation of reactive oxygen species (ROS) within our bodies. The deleterious effects of ROS are such because they initiate the cycle of lipid oxidation. Oftentimes, disease progression is associated with lipid oxidation of cellular entities – lipid membranes, DNA bases, and proteins [12–14]. The damage associated with pathological diseased states continues with formation of peroxides and secondary lipid oxidation products, such as malonaldehyde (MA), which can crosslink and bind to proteins and DNA bases to form mutagenic Schiff bases [15]. Schiff bases can then proceed to attack at sites further away from the original site of oxidation.

In humans, sunlight-induced skin disorders may occur as a consequence of oxidation of skin surface lipids. As the major component of human sebum, squalene has been studied for its potential to undergo oxidation due to its high degree of unsaturation in its triterpene structure [16,17] and its possible role in promoting oxidative skin damage. The deleterious effects of squalene hydroperoxides have been associated with acne [18,19], skin roughness and wrinkle formation in hairless mice [20], induction of comedone formation in rabbit ear skin [21], and generation of MA [22,23]. While the body possesses natural antioxidants to prevent oxidation, they are not present at the surface of the skin. Research to find ways to inhibit oxidation of skin lipids is lacking, hence, it is important to find suitable candidates with antioxidant potentials that can prevent sunlight-induced skin disorders such as skin cancer and aging.

The therapeutic effects demonstrated by essential oils may be a consequence of their antioxidant activities. In essential oils, the presence of phenolic compounds with redox properties may allow them to behave as hydrogen donators, singlet oxygen quenchers, free radical scavengers, and reducing agents [24]. While many studies using numerous antioxidant assays have assessed the antioxidant activities of plant extracts, more research to evaluate the antioxidative effects of plant essential oils (or 'volatile plant extracts') is needed. This is particularly important since essential oils are commonly used in cosmetics and skin care products. Since oxidative damage from UV light is implicated in many skin diseases, the abilities of essential oils to inhibit skin lipid oxidation have also been studied.

Wei and Shibamoto [25] reported the antioxidant activities of 13 essential oils. Antioxidant activity measurements were done based on the abilities of the essential oils to scavenge the 1,1-diphenyl-2-picrylhydrazyl (DPPH) radical, to inhibit the cycle of lipid oxidation by preventing the formation of hexanoic acid from hexanal (aldehyde/carboxylic acid assay), or malonaldehyde from UV-irradiated squalene (malonaldehyde/gas chromatography assay). Jasmine, parsley seed, rose, and ylang-ylang essential oils exhibited nearly 100% inhibitory actions in preventing hexanal oxidation at 500 μg/mL after 40 days, comparable to that of the standard α-tocopherol. Moderate inhibitory activities were observed in juniper berry, patchouli, and angelica seed oils (56–72%) at the same concentration. Dose-dependent activities were exhibited by patchouli and angelica seed oils at the concentrations tested for the assay (0, 10, 20, 50, 100, 200, and 500 μg/mL). Modest DPPH radical scavenging abilities were observed for lavender, sandalwood, chamomile, ginger, or peppermint essential oils at 200 μg/mL (48–53%). In contrast, α-tocopherol was 86% effective at scavenging DPPH radicals. Compared to other essential oils at the same concentration (500 μg/mL), parsley seed oil showed greatest inhibitory actions against malonaldehyde formation from UV-irradiated squalene (67%). This was attributed to the high content of myristicin and apiole present in the oil. Pro-oxidant actions (based on the formation of more malonaldehyde than initially present) were observed in some concentrations of patchouli and jasmine essential oils. This was attributed to the complexity of constituents present in these oil mixtures, which can affect the overall biological activities of the essential oil. As such, this finding supports the importance to thoroughly evaluate the pharmacological actions of essential oils and their constituents in various assay systems in order to better understand their potential roles in improving health.

Two essential oils that have demonstrated excellent antioxidant activities in numerous assays are thyme and clove leaf oils. White thyme oil (versus red thyme oil) is a pale yellow liquid that is redistilled from the red thyme oil. Water-and-steam distillation of the partially dried herb (*Thymus vulgaris*, *Thymus zygis*, or related species) yields a brownish-red, orange red, or gray-brown oil with a rich, powerful, sweet, warm-herbal, spicy, distinctly aromatic odor. In contrast, white thyme oil has a similar odor, but is sweeter, less sharp, and less herbaceous [1]. Major constituents that have been isolated from a volatile thyme extract are thymol, carvacrol, linalool, α-terpineol, and 1,8-cineole [26]. Thymol was also shown to inhibit the formation of malonaldehyde from blood plasma oxidation by 43% at 400 μg/mL. In comparison, at the same oncentration, α-tocopherol and butylated hydroxytoluene (BHT) inhibited malonaldehyde formation by 52 and 70%, respectively [27]. Thymol and other major components identified in thyme oil (carvacrol, γ-terpinene, myrcene, linalool, p-cymene, limonene, 1,8-cineole, and α-pinene) also showed inhibitory activities against ferric ion stimulated lipid peroxidation of rat brain homogenates, but the degree of effectiveness for the individual constituents were not as great as the overall thyme oil [28]. Lee and Shibamoto [4] found that thyme volatile extracts inhibited the oxidation of hexanal for 40 days at 10 μg/mL in the aldehyde/carboxylic acid assay. In the conjugated diene assay, they found that thyme extract retarded methyl linoleate deterioration and at 50 μg/mL, the activity it demonstrated was similar to that of BHT and α-tocopherol. Lee et al. [26] found that thymol inhibited hexanal oxidation by almost 100% over 30 days at a concentration of 5 μg/mL, comparable to that of α-tocopherol and BHT.

Clove leaf oil is distilled from the leaves and twigs of the clove tree (*Eugenia caryophyllata*). The crude oil is a dark brown, often violet- or purple-brown oil with some cloudiness or precipitate. In addition to eugenol (80–90%), the other major constituent is caryophyllene, a sesquiterpene. The odor of crude clove leaf oil is somewhat harsh, slightly sweet with a 'burnt' breadlike note. It is usually characterized as 'woody' and 'dry' [1]. The antioxidant activity of clove leaf essential oil has also been tested in various assay systems. Jirovetz et al. [29] found that at a much lower concentration of 0.5 μg/mL, clove leaf essential oil exhibited significantly higher antiradical activity (91%) toward the 2,2-diphenyl-1-picrylhydrazyl (DPPH) radical than 20 μg/mL of eugenol (89%), BHT (82%), or butylated hydroxyanisole (90%). Significant hydroxyl-radical scavenging activity (94%) was also demonstrated by this oil against $\cdot$OH in the deoxyribose assay at 0.2 μg/mL when compared to eugenol (91% at 0.6 μg/mL) and quercetin (78% at 20 μg/mL), suggesting that clove oil had considerable chelate-generating potential against Fe^{3+} as well. Evaluation of antioxidant activity in the linoleic acid model system was done by monitoring for conjugated diene formation and malonaldehyde formation. The ability of clove leaf oil to inhibit conjugated diene formation exceeded that of BHT at a concentration of 0.005% oil. The ability of the oil to inhibit the formation of secondary products of lipid oxidation, such as malonaldehyde, was monitored by the thiobarbituric acid reactive substances (TBARS) method. Comparable activity was demonstrated by clove leaf oil at half the concentration of BHT. This suggested that application of clove leaf oil as an antioxidant could still be effective at later stages of lipid oxidation. The main compounds found in this oil were eugenol (~77%), β-caryophyllene (~17%), α-humulene (~2%), and eugenyl acetate (~1%).

The extent to which thyme and clove leaf oils could inhibit oxidation as a mixture was evaluated by Wei and Shibamoto [30]. Mixtures of thyme or clove leaf oils with rose, celery seed, or parsley seed oils were also evaluated for their abilities to inhibit malonaldehyde formation from UV-irradiated squalene. All mixtures of thyme (50, 100, or 500 μg/mL) with clove leaf oil (500 μg/mL) were over 90% effective at inhibiting

malonaldehyde formation, greater than that of the BHT (85%) or α-tocopherol (76%). Thymol, as well as clove leaf oil, eugenol, were also tested for their antioxidant actions on squalene oxidation. All concentrations of thymol or eugenol demonstrated dose-dependent activities, comparable to BHT. This was also consistent with the findings of Lee and Shibamoto [4]. Mixtures of essential oils at 500 μg/mL demonstrated increasing potencies in the order of thyme/cinnamon leaf (7%) < thyme/parsley seed (71%) < clove leaf/cinnamon leaf (77%) < thyme/parsley seed (83%) < clove leaf/rose clove leaf/parsley seed (87%) < thyme/clove leaf (93%). Pro-oxidant effects were observed when lower concentrations of thyme oil (50 or 100 μg/mL) were mixed with cinnamon or rose oils (500 μg/mL). These studies suggest that essential oils, their individual constituents, and mixtures of essential oils may possess potential for behaving as antioxidants in the prevention of oxidative skin damage and disease. However, further investigations are required to study the complexity of each essential oil. The concentrations at which essential oils and their individual constituents demonstrate antioxidant or pro-oxidant actions, require more research in order for an accurate assessment to be made regarding their use for prevention of UV-induced oxidative skin damage.

3. ANTIVIRAL AND ANTIMICROBIAL ACTIVITIES

While many viruses exist, research on the virucidal effects of essential oils is deficient. However, one of the most commonly studied human pathogens, herpes simplex virus (HSV, type 1 or type 2), is often used because of its prevalence. Typically, herpes simplex type 1 virus (HSV-1) is often used for testing the antiviral actions of potential agents. HSV can cause infections of the mouth, genitals, or central nervous system, and may even result in death for the immunocompromised. Acyclovir has been used to manage herpes infections; however, some strains of the virus have become resistant to it.

Occurrences of drug-resistant viral strains have promoted investigations into the potential application of essential oils as alternatives to synthetic antiviral medications. Schnitzler et al. [9] examined the antiviral activity of eucalyptus oil against type-1 and type-2 herpes simplex viruses *in vitro* on RC-37 cells using a plaque reduction assay. The IC_{50} value for the oil was 0.009% for the herpes simplex virus type-1 and 0.008% for the type-2 virus. The cytotoxic activity of the oil was determined to be 0.03% as based on a neutral red dye uptake assay. Pretreatment of the cells with eucalyptus oil resulted in a reduction in plaque formation, suggesting that the oil could exert an antiviral effect before or during adsorption of the virus to the host cell. These authors also tested the virucidal actions of ginger, thyme, hyssop and sandalwood oils on HSV-1 and the acyclovir resistant HSV-1 [10]. Pretreatment of all cells with each essential oil prior to infection significantly reduced the infectivity of both strains of HSV-1. Since both strains of HSV were inhibited, the mechanism by which essential oils act is suggested to be different than that by acyclovir. Similar to their findings using eucalyptus, the researchers in this study determined that the essential oils inhibited viral replication during the stage before virus adsorption to the host cells, possibly by disruption of the viral envelope structures which are important for adsorption to host cells. Hence, essential oils may inactivate HSV prior to host cell entry whereas acyclovir inhibits viral replication by interfering with host cellular DNA polymerase. Other essential oils that have been tested for their virucidal abilities against HSV include *Melaleuca* species [31], lemongrass [32], sandalwood [33] and *Juniperus* species [34]. Isoborneol, a constituent of some plant essential oils, was also examined for its antiviral

activities against HSV-1 [35]. Exposure of the virus to isoborneol resulted in inactivation of HSV-1. Isoborneol also inhibited virus replication (specifically, viral glycosylation), at a concentration of 0.06% without affecting adsorption of HSV-1. The concentrations at which isoborneol inhibited viral replication were tested to be nontoxic to three human cell lines, suggesting that isoborneol possesses potential as an antiviral therapeutic agent.

In relation to its preservative effects in preventing foodborne pathogens [36], the antimicrobial effects of essential oils are also sought after in treating cutaneous infections. For instance, individuals carrying the methicillin-resistant strain of *S. aureus* have chronic skin lesions that can be potentially spread to other compromised individuals. Tea tree oil (*Melaleuca alternifolia*) has been widely tested and reviewed for its antimicrobial action, particularly as a potential treatment agent against this antibiotic-resistant bacterial strain of *Staphylococcus aureus* [37,38]. The traditional use of tea tree oil on fungal infections such as vaginal candidiasis and dermatophytoses has also prompted investigations into exploring its anti-candidal and antifungal potential [39,40]. Hammer et al. [41] investigated the *in vitro* activities of 24 essential oils, including tea tree oil, against *Candida* species. Tea tree oil was found to inhibit 90% of the isolates for *Candida albicans* and non-*albicans Candida* species at the lowest concentration of 0.25% (v/v) using the broth microdilution method. Application of 0.25% tea tree oil was also determined to be the minimum concentration of oil needed to kill 90% of isolates of *C. albicans*. For non-*albicans Candida* species, the concentration was 0.5% tea tree oil. The sensitivity of *Candida* species to tea tree oil was accomplished using the agar dilution method. For 57 *Candida* isolates tested, the minimum concentration of tea tree oil required to inhibit 90% of isolates was found to be 0.5%.

The application of essential oils as antimicrobial agents for treatment of oral pathogens is also under investigation. Periodontal disease and dental caries (tooth decay or cavities) are common ailments that are also associated with the presence of *Candida albicans* and *Streptococcus mutans*. The potential of using essential oils in the prevention of oral diseases was studied by Botelho et al. [42]. In their study, the essential oil *Lippia sidoides* and its major constituents thymol (57%) and carvacrol (17%) were added by the disk diffusion method to four strains of cariogenic bacteria (*Streptococcus*) and one strain of yeast (*Candida albicans*) in order to determine their antibacterial and antifungal properties. The inhibition zones for all microorganisms tested were dependent on the concentration of *L. sidoides* essential oil, thymol, or carvacrol added. Even though the concentrations of thymol and carvacrol tested were four times more dilute than the essential oil (50 mg/mL versus 217.5 mg/mL), greater inhibitory actions against microbial growth were observed by these constituents. The minimum inhibitory concentrations for thymol and carvacrol ranged from 2.5–5.0 mg/mL whereas it was 5.0–10.0 mg/mL for the *L. sidoides* essential oil. The minimum fungicidal concentration of carvacrol was determined to be 2.5 mg/mL whereas that of the essential oil was 5.0 mg/mL, suggesting the strong fungicidal potential of carvacrol. The minimum bactericidal concentration for the *L. sidoides* oil ranged from 20 to 40 mg/mL, whereas thymol and carvacrol demonstrated values of 10.0 mg/mL and 5.0 mg/mL, respectively. The mechanisms by which essential oil components demonstrate antimicrobial action are based on the lipophilic nature of their hydrocarbon skeleton and hydrophilic character of the functional groups present [43].

4. ANTICARCINOGENIC ACTIVITIES

The therapeutic potential of essential oils has been shown to include chemopreventive and chemotherapeutic activities. Chemoprevention describes the use of essential oils during the initiation phase of carcinogenesis in order to protect against interactions of the carcinogen with DNA. During chemoprevention, phase I and phase II enzymes are induced to detoxify the carcinogen. In contrast, the chemotherapeutic action of essential oils describes their ability to inhibit the expression of tumor cells (by inhibiting tumor proliferation, enhancing tumor death, or inducing tumor cell differentiation) during the promotion phase of carcinogenesis [44]. A review of early works suggested that the inhibitory actions of dietary isoprenoids, particularly monoterpenes in essential oils, were particularly effective toward tumor tissues derived by the mevalonate pathway [45].

Sandalwood oil (*Santalum album*) is one of the oldest perfume materials used. The anticarcinogenic activity of this oil was demonstrated on mouse hepatic tissues [46] and skin papillomas [47]. Major components that have been identified in sandalwood oil include α-santalol (46%), β-santalol (20%), E-β-santalol (7%), and trans-α-bergamotol (5%) [48]. The chemopreventive effect of α-santalol has been demonstrated on skin tumor development in mice [49], and in human epidermoid carcinoma cells [50]. Aruna and Sivaramakrishnan [11] studied the chemopreventive actions of basil oil. They assessed the anticarcinogenic actions of basil oil by its effects on the carcinogen-detoxifying enzyme, glutathione-S-transferase (GST), and on 3,4-benzo(a)pyrene-induced neoplasia in Swiss mice. In the presence of basil oil, GST levels increased and significant inhibition of benzo(a)pyrene-induced squamous cell carcinoma in the stomachs of the mice occurred. Similarly, Zheng et al. [51] tested myristicin, a major constituent of parsley leaf oil, for its ability to induce increased activity of GST in mice tissue. Because increased induction of GST has been correlated to inhibition of tumorigenesis, the observed induction of GST by myristicin in the liver and small intestinal mucosa in this study suggested that myristicin could be a potential chemopreventive agent. Thyme oil has also been studied for its antitumor actions on human ovarian adenocarcinoma cells (IGR-OV1) and its chemoresistant counterparts [52]. Thyme oil showed varying degrees of cytotoxic effects on the different cell lines of ovarian cancer. *In vivo* tumoricidal effects were studied by injecting thyme oil into tumor sites of tumor-bearing DBA-2 (H_2^d) mice. After 30 days, tumor volume increased in the control mice while mice receiving thyme oil injections (10, 30, or 50 μL of thyme oil per mouse) showed inhibited tumor proliferation in a dose-dependent manner.

Chemopreventive actions of plant metabolites other than essential oils have also been studied. For instance, interest in the contribution of ginger and its constituents toward chemoprevention of multistage carcinogenesis has spawned much research into the anticarcinogenic and antitumor effect of ginger. Katiyar et al. [53] used an ethanolic ginger extract to study antitumor effects on a model of mouse skin tumorigenesis. Ginger extract significantly inhibited markers of skin tumor promotion, including inhibition of the induction of epidermal ornithine decarboxylase (ODC) by 12-O-tetradecanoylphorbol-13-acetate (TPA), inhibition of cyclooxygenase activity and lipoxygenase activity, inhibition of ODC mRNA expression, and inhibition of epidermal edemas caused by TPA. Pretreatment of ginger extract to 7,12-dimethylbenz(a)anthracene-initiated SENCAR mice before TPA application resulted in significant protection against skin tumor incidence and its subsequent multiplicity. The antitumor activity of 6-gingerol, a constituent in ginger, was also demonstrated in a two-stage mouse skin carcinogenesis model using ICR mice. Application of 6-gingerol to the backs of mice before TPA treatment significantly inhibited 7,12-dimethylbenz(a)anthracene-induced skin papillomagenesis [54]. The mechanisms by which essential oils or their constituents behave as chemopreventive or chemotherapeutic agents were previously reviewed [55,56].

Essential oils may also play a role as biological agents that potentiate the effects of therapeutic drugs. Since the mechanisms by which they display their pharmacological actions may vary from those of therapeutic drugs, combined application may present a new means for the effective treatment of diseases and cancers. Geraniol, a major constituent of geranium essential oil, was tested by Carnesecchi et al. [57] for its potential synergistic effect with 5-fluorouracil (5-FU) on inhibiting growth of human colon cancer cell lines. Treatment of colon cancer tumors uses 5-FU as a chemotherapy agent, but its success is limited due to variations in sensitivity of the tumor cells to 5-FU. In cancer cells, a two-fold reduction in the expression of thymidylate synthase and thymidine kinase – two enzymes related to the cytotoxicity of 5-FU – was observed with geraniol application but not 5-FU application. In conjunction with 5-FU (20 mg/kg), geraniol (150 mg/kg) reduced the volume of TC-118 human tumors transplanted in Swiss nu/nu mice by 53%, whereas alone, geraniol reduced tumor volume by 26%. No effects were observed in tumor volume when 5-FU was used alone. The mechanisms by which geraniol affected cancer cells were by increasing cell membrane permeability in order to enhance 5-FU uptake by cancer cells and by altering intracellular signaling pathways involved in the metabolism of 5-FU.

5. ANTI-INFLAMMATORY ACTIVITIES

Inflammation, or the inflammatory process, refers to the cascade of events that occurs at the tissue and cellular levels to remove or regenerate new tissue in response to an injury [58]. Previously, the role of ROS in initiating lipid oxidation was discussed. The presence of

ROS can also activate transcription factors that induce the production of signaling molecules and various cytokines to exert inflammatory actions in the body. The assays and markers used for studying inflammation are diverse and their descriptions are beyond the scope of this discussion. Recently however, inflammation has been linked to a growing number of diseases (including hypertension, cancer, and stroke) and it has become of interest to find suitable agents that will minimize inflammation. The traditional use of essential oils as anti-inflammatory agents suggests that they may possess a role in alleviating inflammation. As a consequence, the anti-inflammatory actions of essential oils, their constituents, and mechanisms of action have been the source of much study.

One of the plants most noted for its anti-inflammatory activities is aloe vera (*Barbadensis miller*). While much research exists on the anti-inflammatory activities of aloe vera [59,60], studies on the anti-inflammatory actions of its essential oil are limited. The pale, translucent oil is produced by cold press and is usually used as a carrier oil in aromatherapy. Analysis of the major aroma chemicals isolated and identified from one species of aloe by steam distillation showed (Z)-3-hexenol, (Z)-3-hexenal, (E)-hexenal, 4-methyl-3-pentenol, and butanol [61]. Extracts of aloe vera have demonstrated anti-inflammatory activities in tests on carrageenan-induced edema in the rat paw, and inhibition of cyclooxygenase activity [62]. Enhancement of wound healing by use of aloe vera occurred in diabetic rats [63] and in various cases of dermal ischemia [64]. Chao et al. [65] examined the influence of cinnamon leaf essential oil on the ability of lipopolysaccharide (LPS) activated J774A.1 murine macrophage to stimulate the production of inflammatory cytokines such as tumor necrosis factor-α (TNF-α), interleukin (IL)-1β, and IL-6 proteins. Dose-dependent inhibitory effects on LPS-induced IL-1β protein secretion were observed for the different concentrations of essential oil (0, 5, 10, 30

and 60 µg/mL). Also, the presence of 60 µg/mL of cinnamon leaf oil inhibited the production of prointerleukin-1-beta (proIL-1) protein expression from LPS-stimulated macrophages, as determined by Western blot. At this concentration, cinnamon leaf oil also inhibited approximately 65% of IL-6 protein expression compared to LPS-stimulated macrophages that were not treated with the oil. A reduction in LPS-induced TNF-α secretion from 52 ng/mL in macrophages treated with only LPS to 35 ng/mL in macrophages stimulated with LPS and treated with 60 µg/mL cinnamon leaf oil was also observed by ELISA measurements. In 2008, these authors used similar *in vitro* methodologies to test the anti-inflammatory abilities of cinnamaldehyde, one of the major constituents in cinnamon leaf essential oil [66]. Dose-dependent inhibition of IL-6 production (11, 6, 6, and 6 ng/mL) from LPS-stimulated J774A.1 macrophages was observed when cinnamaldehyde was added at concentrations of 8, 24, 40, or 80 µM, respectively. Interleukin-1, TNF-α, and proIL-1 expression was also reduced when LPS-induced J774A.1 macrophages were treated with cinnamaldehyde. Secretion of TNF and IL-1 was inhibited in a dose-dependent fashion when cinnamaldehyde was added to lipoteichoic acid (LTA) stimulated J774A.1 macrophages (instead of LPS). No significant inhibition of TNF by cinnamaldehyde was observed with polyinosinic-polycytidylic acid (poly-IC) stimulated J774A.1 macrophages. The inhibitory effect of cinnamaldehyde on cytokine production was not specific to cell types tested since a reduction in secretion of LPS-induced TNF and IL-1 cytokine production was also observed in human blood monocytes derived primary macrophages and human THP-1 monocytes. It was suggested that cinnamaldehyde exhibits its inhibitory effects on cytokine production partly by reducing ROS release from LPS-stimulated macrophages and decreasing LPS-mediated phosphorylation of mitogen-activated protein kinases (extracellular signal-regulated kinase 1/2 (ERK1/2) and c-Jun N-terminal kinase 1/2 (JNK1/2)), which help in the regulation of cytokines gene expression.

Another constituent in cinnamon leaf oil, 1,8-cineole, may also contribute to the anti-inflammatory actions demonstrated in Chao et al. [65]. Santos and Rao [67] showed that 1,8-cineole possessed anti-inflammatory actions based on inhibition of carrageenan-induced paw edema and cotton pellet–induced granulomas in rats. In addition, a dose of 200 mg/kg of 1,8-cineole significantly inhibited the promotion of blue dye leakage by acetic acid into the peritoneal cavity of mice by 35% in the acetic acid–induced increase in peritoneal vascular permeability test. Because of the widespread presence of 1,8-cineole in eucalyptus, melaleuca, and rosemary oils, the anti-inflammatory effects of this constituent may explain why some early cultures used these plant essential oils to soothe skin ailments and rheumatism.

Anti-inflammatory studies such as these suggest that the pharmacological properties exhibited by major constituents of an essential oil may play a defining role in the expression of overall biological activity of the essential oil. Research on the use of essential oils for inflammatory diseases requires further investigations into the interactions between individual constituents and their responses to inflammation. Since the inflammatory response is a complex cascade of reactions, the mechanisms by which essential oils and their constituents participate and intervene in reducing inflammation require further elucidation. By doing so, a rational scientific-based approach to recommending essential oils as anti-inflammatory substances can be obtained.

6. SUMMARY

The extensive analysis of essential oils, their constituents, and biological activities is significant due to concerns of human health and safety. Considerable progress is being made to define the role of essential oils in reducing the risk for diseases and to characterize the mechanisms by which they promote these therapeutic effects. The demand for natural protective agents is growing as more antibiotic-resistant infections continue to evolve. However, it is important to realize that further investigations into the deleterious or adverse biological effects of essential oils in *in vivo* models need to be performed. By doing so, we can better understand their mechanisms of action in combating disease, and better evaluate the quantities at which they best exert their beneficial actions to improving human health.

References

1. Arctander, S. (1960). *Perfume and flavor materials of natural origin.* Denmark: Det Hoffensbergske Etablissement.
2. Dewick, P. M. (2001). *Medicinal natural products: A biosynthetic approach.* (2nd ed). Chichester, England: John Wiley & Sons.
3. Dorman, H. J. D., Surai, P., & Deans, S. G. (2000). In vitro antioxidant activity of a number of plant essential oils and phytoconstituents. *Journal of Essential Oil Research, 12,* 241–248.
4. Lee, K. G., & Shibamoto, T. (2002). Determination of antioxidant potential of volatile extracts isolated from various herbs and spices. *Journal of Agricultural and Food Chemistry, 50,* 4947–4952.
5. Vazquez, B., Avila, G., Segura, D., & Escalante, B. (1996). Antiinflammatory activity of extracts from aloe vera gel. *Journal of Ethnopharmacology, 55,* 69–75.
6. Park, K. K., Chun, K. S., Lee, J. M., Lee, S. S., & Surh, Y. J. (1998). Inhibitory effects of [6]-gingerol, a major pungent principle of ginger, on phorbol ester-induced inflammation, epidermal ornithine decarboxylase activity and skin tumor promotion in ICR mice. *Cancer Letters, 129,* 139–144.
7. Elgayyar, M., Draughon, F. A., Golden, D. A., & Mount, J. R. (2001). Antimicrobial activity of essential oils from plants against selected pathogenic and saprophytic microorganisms. *Journal of Food Protection, 64,* 1019–1024.
8. Lima, E. D., Gompertz, O. F., Paulo, M. D., & Giesbrecht, A. M. (1992). *In vitro* antifungal activity of essential oils against clinical isolates of dermatophytes. *Revista de Microbiologia, 23,* 235–238.

9. Schnitzler, P., Schon, K., & Reichling, J. (2001). Antiviral activity of Australian tea tree oil and eucalyptus oil against herpes simplex virus in cell culture. *Pharmazie, 56,* 343–347.

10. Schnitzler, P., Koch, C., & Reichling, J. (2007). Susceptibility of drug-resistant clinical herpes simplex virus type 1 strains to essential oils of ginger, thyme, hyssop, and sandalwood. *Antimicrobial Agents and Chemotherapy, 51,* 1859–1862.

11. Aruna, K., & Sivaramakrishnan, V. M. (1996). Anticarcinogenic effects of the essential oils from cumin, poppy and basil. *Phytotherapy Research, 10,* 577–580.

12. Frankel, E. N., & Neff, W. E. (1983). Formation of malonaldehyde from lipid oxidation products. *Biochimica et Biophysica Acta, 754,* 264–270.

13. Basu, A. K., Marnett, L. J., & Romano, L. J. (1984). Dissociation of malonaldehyde mutagenicity in *Salmonella typhimurium* from its ability to induce interstrand DNA cross-links. *Mutation Research, 129,* 39–46.

14. Nair, V., Cooper, S. C., Vietti, D. E., & Turner, G. A. (1986). The chemistry of lipid peroxidation metabolites: Crosslinking reactions of malonaldehyde. *Lipids, 21,* 6–10.

15. Esterbauer, H., Schaur, R. J., & Zollner, H. (1991). Chemistry and biochemistry of 4-hydroxynonenal, malonaldehyde and related aldehydes. *Free Radical Biology and Medicine, 11,* 81–128.

16. Ekanayake-Mudiyanselage, S., Hamburger, M., Elsner, P., & Thiele, J. J. (2003). Ultraviolet a induces generation of squalene monohydroperoxide isomers in human sebum and skin surface lipids *in vitro* and *in vivo*. *Journal of Investigative Dermatology, 120,* 915–922.

17. Picardo, M., Zompetta, C., Deluca, C., Cirone, M., Faggioni, A., Nazzaroporro, M., Passi, S., & Prota, G. (1991). Role of skin surface-lipids in UV-induced epidermal cell changes. *Archives of Dermatological Research, 283,* 191–197.

18. Saint-Leger, D., Bague, A., Cohen, E., & Chivot, M. (1986). A possible role for squalene in the pathogenesis of acne. I. *In vitro* study of squalene oxidation. *British Journal of Dermatology, 114,* 535–542.

19. Saint-Leger, D., Bague, A., Lefebvre, E., Cohen, E., & Chivot, M. (1986). A possible role for squalene in the pathogenesis of acne. II. *In vivo* study of squalene oxides in skin surface and intra-comedonal lipids of acne patients. *British Journal of Dermatology, 114,* 543–552.

20. Chiba, K., Sone, T., Kawakami, K., & Onoue, M. (1999). Skin roughness and wrinkle formation induced by repeated application of squalene-monohydroperoxide to the hairless mouse. *Experimental Dermatology, 8,* 471–479.

21. Chiba, K., Yoshizawa, K., Makino, I., Kawakami, K., & Onoue, M. (2000). Comedogenicity of squalene monohydroperoxide in the skin after topical application. *Journal de Toxicologie Science, 25,* 77–83.

22. Dennis, K. J., & Shibamoto, T. (1989). Production of malonaldehyde from squalene, a major skin surface lipid, during UV-irradiation. *Photochemistry and Photobiology, 49,* 711–716.

23. Yeo, H. C. H., & Shibamoto, T. (1992). Formation of formaldehyde and malonaldehyde by photooxidation of squalene. *Lipids, 27,* 50–53.

24. Rice-Evans, C. A., Miller, N. T., & Paganga, G. (1997). Antioxidant properties of phenolic compounds. *Trends in Plant Science, 4,* 304–309.

25. Wei, A., & Shibamoto, T. (2007). Antioxidant activities and volatile constituents of various essential oils. *Journal of Agricultural and Food Chemistry, 55,* 1737–1742.

26. Lee, S. J., Umano, K., Shibamoto, T., & Lee, K. G. (2005). Identification of volatile components in basil (*Ocimum basilicum* L.) and thyme leaves (*Thymus vulgaris* L.) and their antioxidant properties. *Food Chemistry, 91,* 131–137.

27. Lee, K. G., & Shibamoto, T. (2001). Inhibition of malonaldehyde formation from blood plasma oxidation by aroma extracts and aroma components isolated from clove and eucalyptus. *Food and Chemical Toxicology, 39,* 1199–1204.

28. Youdim, K. A., Deans, S. G., & Finlayson, H. J. (2002). The antioxidant properties of thyme (*Thymus zygis* L.) essential oil: An inhibitor of lipid peroxidation and a free radical scavenger. *Journal of Essential Oil Research, 14,* 210–215.

29. Jirovetz, L., Buchbauer, G., Stoilova, I., Stoyanova, A., Krastanov, A., & Schmidt, E. (2006). Chemical composition and antioxidant properties of clove leaf essential oil. *Journal of Agricultural and Food Chemistry, 54,* 6303–6307.

30. Wei, A., & Shibamoto, T. (2007). Antioxidant activities of essential oil mixtures toward skin lipid squalene oxidized by UV irradiation. *Cutaneous and Ocular Toxicology, 26,* 227–233.

31. Farag, R. S., Shalaby, A. S., El-Baroty, G. A., Ibrahim, N. A., Ali, M. A., & Hassan, E. M. (2004). Chemical and biological evaluation of the essential oils of different *Melaleuca* species. *Phytotherapy Research, 18,* 30–35.

32. Minami, M., Kita, M., Nakaya, T., Yamamoto, T., Kuriyama, H., & Imanishi, J. (2003). The inhibitory effect of essential oils on herpes simplex virus type-1 replication *in vitro*. *Microbiology and Immunology, 47,* 681–684.

33. Benencia, F., & Courreges, M. C. (1999). Antiviral activity of sandalwood oil against herpes simplex viruses-1 and -2. *Phytomedicine, 6,* 119–123.

34. Marongiu, B., Porcedda, S., Caredda, A., Gioannis, B. D., Vargiu, L., & Colla, P. L. (2003). Extraction of *Juniperus oxycedrus ssp. oxycedrus* essential oil by super-critical carbon dioxide: Influence of some process parameters and biological activity. *Flavour and Fragrance Journal, 18*, 390–397.

35. Armaka, M., Papanikolaou, E., Sivropoulou, A., & Arsenakis, M. (1999). Antiviral properties of isoborneol, a potent inhibitor of herpes simplex virus type 1. *Antiviral Research, 43*, 79–92.

36. Burt, S. (2004). Essential oils: Their antibacterial properties and potential applications in foods – a review. *International Journal of Food Microbiology, 94*, 223–253.

37. Carson, C. F., Hammer, K. A., & Riley, T. V. (2006). *Melaleuca alternifolia* (tea tree) oil: A review of antimicrobial and other medicinal properties. *Clinical Microbiology Reviews, 19*, 50–62.

38. Halcón, L., & Milkus, K. (2004). *Staphylococcus aureus* and wounds: A review of tea tree oil as a promising antimicrobial. *American Journal of Infection Control, 32*, 402–408.

39. Hammer, K. A., Carson, C. F., & Riley, T. V. (2002). *In vitro* activity of *Melaleuca alternifolia* (tea tree) oil against dermatophytes and other filamentous fungi. *Journal of Antimicrobial Chemotherapy, 50*, 195–199.

40. Mondello, F., Bernardis, F. D., Girolamo, A., Cassone, A., & Salvatore, G. (2006). *In vivo* activity of terpinen-4-ol, the main bioactive component of *Melaleuca alternifolia* cheel (tea tree) oil against azole-susceptible and -resistant human pathogenic *Candida* species. *BMC Infectious Diseases, 6*, 158–165.

41. Hammer, K. A., Carson, C. F., & Riley, T. V. (1998). *In-vitro* activity of essential oils, in particular *Melaleuca alternifolia* (tea tree) oil and tea tree oil products, against *Candida* spp. *Journal of Antimicrobial Chemotherapy, 42*, 591–595.

42. Botelho, M. A., Nogueira, N. A. P., Bastos, G. M., Fonseca, S. G. C., Lemos, T. L. G., Matos, F. J. A., Montenegro, D., Heukelbach, J., Rao, V. S., & Brito, G. A. C. (2007). Antimicrobial activity of the essential oil from *Lippia sidoides*, carvacrol and thymol against oral pathogens. *Brazilian Journal of Medical and Biological Research, 40*, 349–356.

43. Kalemba, D., & Kunicka, A. (2003). Antibacterial and antifungal properties of essential oils. *Current Medicinal Chemistry, 10*, 813–829.

44. Morse, M., & Stoner, G. (1993). Cancer chemoprevention: Principle and prospects. *Carcinogenesis, 14*, 1737–1746.

45. Elson, C. E., & Yu, S. G. (1994). The chemoprevention of cancer by mevalonate-derived constituents of fruits and vegetables. *Journal of Nutrition, 124*, 607–614.

46. Banerjee, S., Ecavade, A., & Rao, A. R. (1993). Modulatory influence of sandalwood oil on mouse hepatic glutathione-s-transferase activity and acid-soluble sulfhydryl level. *Cancer Letters, 68*, 105–109.

47. Dwivedi, C., & AbuGhazaleh, A. (1997). Chem-opreventive effects of sandalwood oil on skin papillomas in mice. *European Journal of Cancer Prevention, 6*, 399–401.

48. Marongiu, B., Piras, A., Porcedda, S., & Tuveri, E. (2006). Extraction of *Santalum album* and *Boswellia carterii* volatile oils by supercritical carbon dioxide: Influence of some process parameters. *Flavour and Fragrance Journal, 21*, 718–724.

49. Dwivedi, C., Guan, X. M., Harmsen, W. L., Voss, A. L., Goetz-Parten, D. E., Koopman, E. M., Johnson, K. M., Valluri, H. B., & Matthees, D. P. (2003). Chemopreventive effects of alpha-santalol on skin tumor development in CD-1 and SENCAR mice. *Cancer Epidemiology Biomarkers & Prevention, 12*, 151–156.

50. Kaur, C., & Kapoor, H. C. (2002). Anti-oxidant activity and total phenolic content of some Asian vegetables. *International Journal of Food Sciences and Technology, 37*, 153–161.

51. Zheng, G. Q., Kenney, P. M., & Lam, L. K. T. (1992). Myristicin – a potential cancer chemopreventive agent from parsley leaf oil. *Journal of Agricultural and Food Chemistry, 40*, 107–110.

52. M'Barek, L. A., Mouse, H. A., Jaâfari, A., Aboufatima, R., Benharref, A., Kamal, M., Bénard, J., Abbadi, N. E., Bensalah, M., Gamouh, A., Chait, A., Dalal, A., & Zyad, A. (2007). Cytotoxic effect of essential oil of thyme (*Thymus broussonettii*) on the IGR-OV1 tumor cells resistant to chemotherapy. *Brazilian Journal of Medical and Biological Research, 40*, 1537–1544.

53. Katiyar, S. K., Agarwal, R., & Mukhtar, H. (1996). Inhibition of tumor promotion in SENCAR mouse skin by ethanol extract of *Zingiber officinale* rhizome. *Cancer Research, 56*, 1023–1030.

54. Park, K. K., Chun, K. S., Lee, J. M., Lee, S. S., & Surh, Y. J. (1998). Inhibitory effects of [6]-gingerol, a major pungent principle of ginger, on phorbol ester-induced inflammation, epidermal ornithine decarboxylase activity and skin tumor promotion in ICR mice. *Cancer Letters, 129*, 139–144.

55. Crowell, P. L. (1999). Prevention and therapy of cancer by dietary monoterpenes. *Journal of Nutrition, 129*, 775S–778S.

56. Edris, A. E. (2007). Pharmaceutical and therapeutic potentials of essential oils and their individual volatile constituents: A review. *Phytotherapy Research, 21*, 308–323.

A. FRUIT AND VEGETABLES IN HEALTH PROMOTION

57. Carnesecchi, S., Bras-Gonçalves, R., Bradaia, A., Zeisel, M., Gossé, F., Poupon, M., & Raul, F. (2004). Geraniol, a component of plant essential oils, modulates DNA synthesis and potentiates 5-fluorouracil efficacy on human colon tumor xenografts. *Cancer Letters*, *215*, 53–59.

58. Schmid-Schönbein, G. W. (2006). Analysis of inflammation. *Annual Review of Biomedical Engineering*, *8*, 93–151.

59. Reynolds, T., & Dweck, A. C. (1999). Aloe vera leaf gel: A review update. *Journal of Ethnopharmacology*, *68*, 3–37.

60. Shelton, R. M. (1991). Aloe vera – its chemical and therapeutic properties. *International Journal of Dermatology*, *30*, 679–683.

61. Umano, K., Nakahara, K., Shoji, A., & Shibamoto, T. (1999). Aroma chemicals isolated and identified from leaves of *Aloe arborescens* Mill. Var. *natalensis* Berger. *Journal of Agricultural and Food Chemistry*, *47*, 3702–3705.

62. Vazquez, B., Avila, G., Segura, D., & Escalante, B. (1996). Antiinflammatory activity of extracts from aloe vera gel. *Journal of Ethnopharmacology*, *55*, 69–75.

63. Chithra, P., Sajithlal, G. B., & Chandrakasan, G. (1998). Influence of aloe vera on the healing of dermal wounds in diabetic rats. *Journal of Ethnopharmacology*, *59*, 195–201.

64. Heggers, J. P., Pelley, R. P., & Robson, M. C. (1993). Beneficial effects of aloe in wound healing. *Phytotherapy Research*, *7*, S47–S48.

65. Chao, L. K., Hua, K. F., Hsu, H. Y., Cheng, S. S., Liu, J. Y., & Chang, S. T. (2005). Study on the anti-inflammatory activity of essential oil from leaves of *Cinnamomum osmophloeum*. *Journal of Agricultural and Food Chemistry*, *53*, 7274–7278.

66. Chao, L. K., Hua, K. F., Hsu, H. Y., Cheng, S. S., Lin, I. F., Chen, C. J., Chen, S. T., & Chang, S. T. (2008). Cinnamaldehyde inhibits pro-inflammatory cytokines secretion from monocytes/macrophages through suppression of intracellular signaling. *Food and Chemical Toxicology*, *46*, 220–231.

67. Santos, F. A., & Rao, V. S. (2000). Antiinflammatory and antinociceptive effects of 1,8-cineole, a terpenoid oxide present in many plant essential oils. *Phytotherapy Research*, *14*, 240–244.

5

Emerging Knowledge of the Bioactivity of Foods in the Diets of Indigenous North Americans

Alexander G. Schauss

Natural and Medicinal Products Division, AIBMR Life Sciences, Inc., Puyallup, WA, USA

1. INTRODUCTION

In the foreword to Harriet V. Kuhnlein's and Nancy J. Turner's wonderful book, *Traditional Plant Foods of Canadian Indigenous Peoples*, Laurie Montour with the Assembly of First Nations in Canada, writes: 'I never knew what a weed was, since I was taught that every plant has a purpose on this planet' [1]. As North American natives volunteered or were forced to abandon their dependence on sustenance from the land, the depth of knowledge of the traditional use and benefit of fruits, vegetables, nuts, and grains, from among well over a thousand species of plants, slowly disappeared. This loss of knowledge continues today.

Indigenous people did not separate nutrition and medicine the way our modern world does today. Edible plants often functioned as food for sustenance, and at the same time contained compounds with medicinal properties.

The O'odham Indians live in the northern Sonora Desert that spans southern Arizona, California and northern Mexico. A native O'odham expressed his concern for the effect the transition from traditional foods to modern foods of commerce has had on the health and lifespan of his people:

> The desert food is meant for the Indians to eat. The reason so many Indians die young is because they don't eat the desert food. I worry about what will happen to this new generation of Indians who have become accustomed to present food they buy at the markets. [2, p. 56]

This opinion is expressed by tribal elders throughout North America as they reflect upon the impact modern foods of convenience have had on the health of their tribes, for historically food was not just about nutrition, it was part of the social fiber around which a community organized itself year to year.

Today food scientists are finding that native foods, many of which we are only now beginning to study, are not only nutrient-dense food sources but also rich sources of bioactive phytochemicals. These nutrient- and phytochemical-dense foods have the potential to protect against chronic and degenerative diseases as we have begun to discover. This realization forces us to reflect upon whether our modern diets are protecting us or contributing to so many of the degenerative and aging-related diseases that are prevalent today.

What emerges from studying the chemistry of native foods is the discovery that *if* natives could revert back to their ancestral food sources, they contain not only health-promoting macronutrients such as complex carbohydrates, proteins and beneficial fats, but also numerous bioactive compounds that could protect against and mitigate common diseases such as obesity, diabetes and cardiovascular disease.

Whether we look at the traditional food habits of indigenous people in the arid Sonoran Desert or those living in the lush forested Pacific Northwest of the United States or in Canada's British Columbia, we find an urgency to learn as much as we can about their diets, before this knowledge is forever lost. This effort becomes obvious as discovery after discovery reveals that indigenous people knew far more than they have been given credit for about the foods they selected for themselves.

By looking at the foods consumed from coast to coast, or those foods restricted by climate or geography to certain areas of North America, one marvels at the knowledge indigenous people needed to have in order to select the foods that most contributed to their sustenance and health.

On one trip this author took to Death Valley, California, to learn about the indigenous diet of the Timbisha Shoshone Indians, it was revealed that despite what to most visitors appeared to be a barren land, these indigenous semi-nomadic people had access to over 800 different plant foods, among which several hundred were consumed annually. When in season, berries, roots, and green foods, found at various elevations from below sea level to several thousand feet above the desert floor, were supplemented with a diet of fish, fowl (doves or quail), and game (lizards, jack rabbits and bighorn sheep). Pinyon pine nuts (*Pinus* subsection Cembroides) and mesquite trees (*Prosopis veluntina*) were particularly important to the Shoshone in that they provided a source of food that could be made into flour and when formed into cakes, could provide calories and nourishment that would last well into fall, winter, and early spring.

Pinyon nuts were also a major source of calories. Natives claimed that a family of four could harvest more than 90 kilograms (200 lb) of nuts in a single day. Nutritional studies have found that a pound of shelled pinyon nuts yields nearly 2800 calories [3]. This translates into a harvest of 336,000 calories per day. Hence, if each family member requires 2000 calories a day, a family could in a single day's harvest provide enough calories to last 6 weeks.

The Timbisha Shoshone's traditional indigenous plant foods included root vegetables, green vegetables, fruit, seeds, nuts, grains, and rarely mushrooms, occasionally supplemented with flowers, lichens, algae, and the outer or inner bark of trees. For the indigenous people of the Pacific Northwest, tubers, bulbs, rhizomes, and corms provided foods such as camas (*Camassia quamash*) bulbs, yellow avalanche lily (*Erythronium grandiflorum*), silverweed (*Argentina anserina*), springbank clover rhizomes (*Trifolium wormskioldii*), and knotweed, a member of the buckwheat family (Polygonaceae) [4]. Rich in carbohydrates, the food sources provided much-needed nourishment that could be consumed at harvest or stored for later consumption. Stem and shoot vegetables included Indian celery (*Lomatium nudicaule*), fern fiddleheads, and parsnip (*Pastinaca sativa*), a root vegetable related to the carrot. Leaf vegetables came from watercress (*Nasturtium officinale*), mustard greens (*Brassica juncea*), stinging nettles (*Urtica dioica*) and common lambsquarters (*Chenopodium album*).

Grains from wild rice (*Zizania palustris*), an annual plant native to the Great Lakes region of North America, or maize (*Zea mays*) found growing in the southwestern United States and throughout what is now Mexico, were important sources of food that could be stored. In the case of maize, it could be cultivated even in semi-arid regions to provide crops from year to year.

For Mesoamericans, maize was *the* major staple food. These indigenous ancient peoples learned to grow maize via a system known as the 'Three Sisters' that incorporated cultivation of maize with beans to provide nitrogen from nitrogen-fixing bacteria, and squashes to provide ground cover to inhibit weed growth and evaporation. The Mesoamericans figured out that by adding alkali from ashes or lime (calcium carbonate) to cornmeal, the alkali liberated the B vitamin niacin, which prevented pellagra. Non-indigenous people that settled in the Americas did not have this knowledge and once they adopted maize as a major staple food, parts of the southern United States in particular began to suffer from an epidemic of pellagra, a deficiency of niacin [5].

Indigenous people also learned to combine maize consumption with other sources of protein from plants, such as amaranth (*Amaranthus*) and chia (*Salvia hispanica*), in order to provide a complete range of amino acids needed for protein synthesis.

Although maize is important to meet macronutrient needs, less is known about staple foods found in other parts of North America that were consumed by indigenous people.

To get a snapshot of what was known by hundreds of indigenous communities throughout the continent about at least a thousand different plant foods, three topics were selected to illustrate what science is learning about traditional foods, starting with natives that lived in the Sonoran Desert, then on to natives that lived along and near the borders of what is now Canada and the United States, and finally to the indigenous tribes that lived in the Pacific Northwest. As will become evident in examining the diet of these indigenous groups, there is still much to be learned about the bioactivity and chemical composition of their native foods, which suggests this area of research deserves far more attention by food scientists and others, given the promising knowledge learned to date.

2. NATIVE TRIBES OF THE SONORAN DESERT

One-third of the Earth's land surface is desert. With meager rainfall, this arid land supports sparse vegetation and a limited population of people. The Sonoran Desert of the American Southwest has the most complex desert vegetation on Earth [6].

The Tohono O'odham tribes have lived in the Sonoran Desert for millennia [7]. Through collective experience they learned to adapt to the arid and harsh environment of the desert. Through a semi-nomadic lifestyle that involved the collection and harvesting of wild plants and heat-tolerant crop foods they were able to feed themselves year round [8]. Today the O'odham live on the second largest land reserve in the United States, covering an area of nearly 3 million acres. Subpopulations of the O'odham on this reserve were distinguished by their subsistence patterns.

The colonial Spanish referred to the O'odham as the *Papago* based on the O'odham expression, *papahvi o-otom*, referring to their dietary use of the tepary bean (*Phaseolus acutifolius* var. *latifolius*; *Phaseolus metcalfei*; *Phaseolus ritensis*), an important source of protein, as well as the *Pima* by European settlers based on language similarities [9]. In 1986, the Tohono O'odham Nation voted to discard these colonial/European references and adopted their own name, Tohono O'odham, which means 'desert people.'

That the Spanish referred to the O'odham as the Papagos due to their consumption of *Phaseolus* species helps to illustrate the risk associated with forced or voluntary abandonment of traditional diets.

Today the Tohono O'odham have the highest documented incidence of diabetes of almost any Indian tribe in North America. With the transition from native foods and the adoption of modern foods of commerce, the O'odham no longer consume the exceptionally high fiber diet of legumes such as the teparies and mesquite that past generations did [10]. These legumes combined with chia (*Salvia hispanica*), tansy mustard (*Descurainia pinnata*), various species of *Plantago* and native seeds contributed to maintaining normal post-prandial blood sugar levels [11]. The use of chia for food, medicine and polyunsaturated fatty acids has a long history in the Sonoran Desert stretching back to pre-Columbian Mesoamerica [12]. Chia is rich in fiber, protein and the omega-3 polyunsaturated fatty acid (alpha-linolenic acid), and a complement of vitamins and minerals [13]. We now know that its antioxidant activity comes from flavonol glycosides, chlorogenic acid, caffeic acid, kaempferol, quercetin, and myricetin [14].

One subgroup living in the Sonoran Desert was known as the Sand Papagos before they changed their name to O'odham. Based on the recollection of elders, the Sand's diet included 21 wild and 9 cultivated plant species, and at least 23 animal species [2]. Their diverse diet served them well. They survived in what many would consider a very demanding, arid, and inhospitable desert environment, yet its members found they could meet their caloric and nutritional needs, as long as knowledge of what foods to select, when to harvest, and how to prepare them, was passed on from generation to generation. Knowing what foods to select, and when, and their location was essential to survival. Learning how to prepare some foods and how to maximize their use was equally as important.

One can marvel at the Sand's selection and/or cultivation of different food sources, in light of our knowledge today of their nutritional composition. The Sand's diet included mesquite tree (*Prosopis veluntina*) beans, desert asparagus (broomrape) (*Orobanche sp.*), wild amaranth (*Amaranthus hybridus*), saguaro (*Carnegiea gigantea*) and organ (*Stenocereus thurberi*) cactus fruit. These plant foods were supplemented with ocean (Sea of Cortez) and freshwater river and lake fish, and meat obtained from hunting numerous wild animals, including lizards, rabbits and deer.

Wild amaranth was a favorite vegetable of the Sand's. The leaves of this plant appeared soon after the monsoon season began in the Sonoran Desert. During the hottest days of the year, the Sands would get on their hands and knees to eat the emerging leaves once they reached a certain height without removing them or even touching them, other than to bite the leaves off and ingest them. Literally, they grazed on fresh wild amaranth.

Nutritional studies of wild greens found in the southwestern United States such as *A. hydrides*, have revealed a remarkable difference in nutritional density of greens compared to cultivated greens. For example, 100 g of cultivated and mass produced iceberg lettuce, so commonly used in salads today, lacks the nutritional density of an equal amount of wild amaranth, which has 'three times as many calories (36), eighteen times the amount of vitamin A (6100 I.U.), thirteen times the amount of vitamin C (80 mg.), twenty times the amount of calcium (411 mg.), and almost seven times the amount of iron (3.4 mg.)' compared to iceberg lettuce [2, p. 97]. Wild amaranth was only one of many green Soronan vegetables consumed by natives. Others included lambsquarters (*Chenopodium album*), winter spinach (*Spinacia oleracea*), saltbush greens (*Atriplex hymenelytra, Atriplex lentiformis*), and purslane (*Portulaca oleracea*). Purslane contains many biologically active compounds, including omega-3 fatty acids, coumarins, cardiac glycosides, anthraquinone glycosides, free oxalic acids, and numerous alkaloids and flavonoids [15].

The O'odham calendar was organized around four seasons beginning with the Light-Green Time (*Ce:dagi Masad*) during the months of February and March. During these months light rain brought dormant plants to life. New leaves appeared on bushes and trees, allowing for the gathering of wild plants, particularly the beans from drought-resistant velvet mesquite (*Prosopis velutina, Proscopis glandulosa*) and the Palo Verde tree (*Cercidium microphyllum, Cercidium floridum*).

The bean pods of mesquite were harvested once the pods turn from green to yellow-brown, and just before falling to the ground. Once collected, beans were dried and ground into flour, adding a nutty taste to breads, or used to make jelly, which could later be made into flour, or an alcoholic beverage, such as wine [16]. When ground into a sweet, nutritionally-dense, high protein, high fiber, and gluten-free flour, mesquite beans prove to be a suitable choice for diabetics, compared to eating gluten-rich wheat flour.

The mesquite tree also had medicinal uses. The leaves were infused and used as eye drops for the treatment of conjunctivitis [3]. The sap was traditionally used to soothe sore throats, while the root was made into a salve to treat cuts.

April was the Yellow Time (*Uam Masad*), which allowed for the harvesting of wild greens and wild onions. This was followed by

May, called the Painful Time (*U'us Wihogdag Masad*), when the desert starts to dry out and turn very hot. During May food became scarce as stored food began to run out. During these difficult times the O'odham survived on various cacti and agave plants, and the buds of the cholla (*Opuntia imbricata*), a good source of calcium. Those O'odham living on the Gila River (the Akimel O'odham) harvested wheat and peas, which they shared if necessary when other O'odhams visited them.

June through July was known as the Saguaro-Harvest Time (*Ha:san Bak Masad*). This is the hottest time of the year in the desert. Harvesting of mesquite beans continues. But most important, it is the time when the fruiting of the saguaro (*Carnegiea gigantea*) and organ (*Stenocereus thurberi*) cactus occurs.

With its tall trunk and arms, the Saguaro cactus is the most recognizable plant found in the Sonoran Desert. When fully mature, a saguaro can weigh as much as 9000 kilograms (10 tons). The tallest saguaro is found in Maricopa County, Arizona, and is nearly 14 meters tall with a girth of over 3 meters. To the O'odham, gathering the ruby-colored saguaro fruit (*ha:san*) that matures in late June held importance due to the variety of foodstuffs that it produces for consumption. The fruit contains up to 2000 seeds, which can be ground into flour and stored for later use.

When mature, the sweet strawberry-like fruit was picked off the tops of the trunk or arms of the saguaro using a long gathering pole called a *kuipad*. Once the fruit dropped to the ground and was gathered, the pulp was scooped out and placed into a basket, whereafter it was taken to the O'odham's camp for processing. The fruit was mashed and then boiled and made into a wine (*nawait*) or syrup. When used to make wine, the syrup was taken and put into a ceramic pot called an *ollas*. The pots were then sealed and allowed to ferment into wine. Another option was to use the fruit pulp to make either a jam or just sun dry it for future use. The fruit contains up to 2000 seeds. The small seeds were strained out of the fruit using a plaited basket made of sotol leaves (*Dasylirion wheeleri*) after which they were ground into flour.

Virtually every tribe residing in what are now the states of New Mexico, Arizona and California, and those of northwestern Mexico, also consumed the inner bark of the large tuber of *Proboscidea altheaefolia*, known as the unicorn plant. *P. parviflora* and its dried fruit, named by the Spanish 'devil's claw,' due to its appearance, contained raw seeds that were usually chewed or infrequently cooked.

In the Sonoran Desert, *Panicum sonorum*, better known as panic grass, millet, or sauwi, was frequently cultivated and its range expanded by various groups living in the Sonoran Desert [17]. Each plant yields up to 2000 grains. Families could work together to gather kilos of *P. sonorum* grain and store them in baskets or pots for later consumption.

A demonstration plot organized by the US Soil Conservation Service in 1980 found that one hectare of *P. sonorum* produced 440 kg of the grain. Few southwestern natives grow this sweet grain today, unfortunately, due to water diversion projects completed in the twentieth century that virtually destroyed most of the fields used to plant panic grass.

Wild desert gourds of the genus *Cucurbita*, a genus that includes squashes and pumpkins, were also well known by Sonoran natives [3]. The cucurbitacin compounds found in native Sonoran gourds (*C. sororia, C. moschata, C. foetidissima,* and *C. digitata)* contain tetracyclic triterpenes called cucurbitacins that impart an intense bitter taste. This intense bitterness was put to good use. Rubbed on cultivated plants it kept grazing animals from devouring their crop. Recent studies have found that an extract of *C. foetidissima* can induce contractions of the uterine muscles to hasten childbirth, a use known to natives.

The fan palm (*Washingtonia filifera*) was the only native palm in the western United States that provided food. Sometimes natives would consume the tip of the palm, but without question the most important food was the palm fruit. In wet years, the harvest might be 180 kg (400 lb) of the pea-sized fruit per tree [3]. The fruit was eaten fresh, although the seeds were removed. Sun dried palm fruit was stored in ceramic jars.

3. SUBSISTENCE BERRIES OF NATIVE AMERICAN TRIBAL COMMUNITIES

A consortium of institutions in North America have studied four wild berries for their phytochemical composition and metabolic performance-enhancing activity. These berries were known to be an integral part of the traditional subsistence diet in regions of Canada and the United States among different tribal communities: chokeberry (*Prunus virginiana*), highbush cranberry (*Viburnum trilobum*), Juneberry (*Amelanchier alnifolia*), and silver buffaloberry (*Shepherdia argentea*) [18,19]. They were either eaten raw, sun dried for later use, turned into wines or jellies, or mixed into animal meat to preserve the meat, suggesting the berries had antioxidant properties. Each of these berries had a traditional history of use. For example, *Viburnum edule*, known as 'cramp bark' by Tanaina Indians in Alaska, and a close species of *V. trilobum*, has been listed in numerous reference works as an antispasmodic [18]. A muscle relaxant, the shavings of the bark of *V. edule* are commonly used to treat menstrual cramps. The Upper Cook Inlet people report its value in the treatment of colds, sore throat, and laryngitis. *Viburnum opulus*, a related species, also called highbush cranberry, has been found to contain viopudial and viburtinal, both sesquiterpenes, which affect cholinesterase activity [20,21]. Cholinesterase catalyzes the hydrolysis of acetylcholine into choline and acetic acid which allows a cholinergic neuron to return to its resting state after activation. Both scopoletin and viopudial have been determined *in vivo* to be responsible for the uterine relaxant activity of *V. opulus* and *V. prunifolium* [22,23].

Studies have demonstrated antioxidant activity of *V. trilobum* fruit owing to its polyphenolic composition, as well as the fruit of *A. alnifolia*, *Prunus virginiana*, and *Shepherdia argentea* [24]. The antioxidant polyphenols contribute to preserving meat and its fat. For example, wild blueberries mixed into bison meat produced pemmican which delayed the fat in meat from becoming rancid, a method of meat preservation and prevention of fat rancidity well known to Plains Indians [25].

The Saskatoon berry (*A. alnifolia*), also known as Juneberry or serviceberry, is native to the northern plain provinces of Canada and the western United States. It has an extensive history of use by native cultures both as a food source and for its medicinal properties (e.g. to prevent miscarriages).

A crude extract of *A. alnifolia* berries has been shown to inhibit nitric oxide production in activated macrophages, which would suggest it might have a role in protecting against cardiovascular disease or chronic inflammation. Analytical studies of fresh berries of *A. alnifolia* have found it contains appreciable concentrations of polyphenolic anthocyanins, with 3-galactoside and 3-glucoside conjugates as the predominant cyanidins, and other anthocyanins (cyanin 3-xyloside, perlargonin 3-glucoside, and malvidin derivates), in addition to the phenolic acids, 3-feruloylquinic, 5-feruloylquinic, as well as cholorogenic acids, and the flavonoid compounds avicularin, hyperoside, quercitin, and rutin [26–29].

Natives used chokeberry as a food, either as juice and wine, mixed with meat to preserve it, or as a medicine. Although considerably more work into its chemical composition is needed, a methanol extract of the berry has been found to have antioxidant activity *in vitro* owing to its phenolic acids [30]. Bioassays have also found that chokeberry inhibits aldose reductase activity. There is good evidence that aldose reductase contributes to diabetic complications during disease progression (e.g. reduced nerve conduction, retinopathy, nephropathy), as it is a rate-limiting enzyme in the polyol pathway associated with the conversion of glucose to sorbitol. As sorbitol accumulation increases during tissue depletion of myoinositol content, this results in derangement of sodium-potassium adenosine triphosphate (ATP) activity. Only in a hyperglycemic state, when the enzyme hexokinase is saturated, is aldose reductase activated, resulting in accumulation of sorbitol. Drugs designed to prevent or slow the action of aldose reductase may be a means to prevent or delay the complications of diabetes. Hence, finding evidence that chokeberry inhibits aldose reductase activity may be meaningful especially for populations with a higher incidence of diabetes, such as experienced by native North Americans who have reduced or abandoned their consumption of such indigenous foodstuffs.

The silver buffaloberry (*S. argenta*) was traditionally eaten by natives either fresh, dried, or in jellies, for its medicinal properties (i.e. to treat gastrointestinal problems), and to symbolically offer it to females entering puberty [25,31]. Nutritional and chemical analyses of the berries have found them to contain appreciable amounts of β-carotene, ascorbic acid, leucoanthocyanins, catechols, and flavonols [32,33].

Human immunodeficiency virus (HIV)-1 reverse transcriptase is a DNA polymerase that will use RNA or DNA to contribute to the production of a double-stranded DNA copy of the single-stranded RNA genome that is contained in the HIV-1 virus particle. The reverse transcriptase from HIV-1 is of great interest to drug companies and the medical community because it is a target enzyme for anti-AIDS drugs, which act to cause termination of the polymerase reaction.

Two hydrolyzable tannins, shephagenins A and B, along with hippophaenin A and strictinin, from a leaf extract of *S. argentea*, strongly inhibit HIV-1 reverse transcriptase *in vitro*. This activity was stronger than that of $(-)$-epigallocatechin gallate, used as a positive control [34].

Using a cytotoxicity assay model, a methanolic extract of the closely related species *S. canadensis* (Canadian buffaloberry) has been demonstrated to inhibit the growth of mouse mastocytoma cells [35].

In North America, highbush cranberry (*V. trilobum*) grows in Canada bordering eastern and mid-western states, as well as Massachusetts and Oregon, with an isolated population in New Mexico, and the Canadian provinces from Newfoundland to British Columbia. It can be eaten fresh, made into jellies, or mixed with meat to create pemmican. Native preparations of the berry were used as an astringent to treat swollen glands, while the bark was used as an antispasmodic for relief of asthma, and menstrual and stomach cramps, owing to the bitter compound viburnine.

Taxonomically, highbush cranberry is not a cranberry (*Vaccinium macrocarpon*) at all, despite the strong resemblance in appearance and taste. *V. trilobum* is a member of the Caprifoliaceae, or honeysuckle family, while 'lowbush' cranberry is a member of the Ericaceae or heather family.

Besides being a rich source of vitamin C, the primary anthocyanidins that contribute to the antioxidant capacity of *V. trilobum* fruit include cyanidin 3-arabinosylsambubioside and cyanidin 3-arabinoglucoside [36–38]. However, more research is needed on the chemistry of the fruit, as little is known about its composition compared to *V. macrocarpon*, whose fruit is commonly consumed around the Thanksgiving holiday.

The major phytochemical constituents of the Saskatoon berry, chokeberry, silver buffaloberry, and highbush cranberry have been investigated to evaluate their ability to protect against or mitigate the progression of diabetes, by improving glucose utilization, inhibiting production of aldose reductase, suppressing pro-inflammatory gene expression (i.e. cyclooxygenase-2, interleukin-1 and interleukin-6), and by measuring changing energy expenditure and lipid metabolism [19]. The endpoint measurements selected for these studies relate to the protective health value these berries might provide if consumed and traditionally practiced.

Chemical analysis of the Saskatoon berry revealed the presence of phenolic acids, anthocyanins, and proanthocyanidin polymers, which demonstrated hypoglycemic activity and strong inhibition of IL-1β and COX-2 gene expression *in vitro*. Non-polar constituents, including carotenoids, were found to be potent inhibitors of aldose reductase.

Inhibition of aldose reductase activity was also demonstrated with silverbuffalo berry, despite no detectable anthocyanins, low concentrations of a few phenolic acids and cholorogenic acid, and the presence of several carotenoids. Nevertheless, various water, crude and methanolic extracts of the berry caused improved glycogen accumulation, reduced expression of IL-1β and COX-2, and altered energy expenditure in cells via different mechanisms. Collectively these findings suggested to the authors that consumption of the berry might protect against diabetic microvascular complications and a reduction in chronic inflammation. Since it demonstrated the ability to improve energy expenditure and glucose uptake, the authors believe it may have the potential to counter the symptoms of metabolic syndrome [19].

Bioassays that were performed of highbush cranberry demonstrated the berry had the capacity to inhibit aldose reductase, improve glucose uptake, reduce IL-1β expression, modulate energy expenditure, and significantly improve glycogen accumulation in cells. Despite the author's opinion that these results showed it was not a powerful mediator of inflammation, the compounds in the berry may 'not only provide protection against diabetic microvascular complications but also improve glucose uptake via an insulin-like effect as well as change energy expenditure,' which when taken together, suggest the potential 'to partially modulate mechanisms that may play a role in the development of insulin resistance and metabolic syndrome' [19, p. 658].

Further studies *in vivo* are clearly needed to confirm the benefits of consuming these four berries given their phytochemical composition. Activity-guided fractionation and characterization of the berries will yield further insight into the biological activity and health-giving properties of each of these native berries. The findings will contribute to our understanding of their nutritional and health-giving properties.

Black cherries from the *Prunus serotina* tree can be harvested wherever the trees are found in North America, from Nova Scotia to Minnesota, south to Florida and Texas, and Arizona and Mexico. The fruit is astringent and was used for the treatment of dysentery.

The fruit of *P. serotina* was eaten raw or cooked to make into a jelly [25]. However, the seed contains compounds that can be converted into cyanide upon release of three enzymes: prunasin β-glucoside, amygdalin β-glucoside, and mandelonitrile lyase [39,40]. Although the flesh of the fruit contains cyanogenic glycosides, the enzymes needed to release cyanide are lacking, so the flesh is safe to eat [41]. Young thin bark from *P. serotina* contains the glycoside, prunasin, which is a very toxic hydrocyanic acid, hence only small amounts were used for medicinal purposes to stimulate respiration in asthmatics or to treat indigestion.

Autumn olive (*Elaeagnus umbellata*) berries are found in southern Canada and the eastern United States, from Maine to Alabama and west to Wisconsin. A shrub native to Asia, *E. umbellata* provides a small red fruit that is extremely rich in lycopene (30–55 mg/100 g) with potential health benefits [42]. The red fruit is high in phenolic content, flavanols, and hydrobenzoic acid.

Lycopene is a carotenoid that exhibits antioxidant properties, particularly against singlet oxygen, *in vitro*. Evidence suggests that increased dietary intake of lycopene, which accounts for about 50% of carotenoids in human serum, can reduce the incidence of some cancers (i.e. cervix, gastrointestinal tract, and prostate) and cardiovascular disease [43,44]. *E. umbellata* fruit is 17 times greater in lycopene content than tomatoes [45].

Since the introduction of *E. umbellata* from Asia in the early nineteenth century, the shrub has proven to be invasive and even declared a noxious weed in states like Massachusetts, Connecticut, Rhode Island, New Hampshire, Tennessee, and West Virginia, and listed as invasive in states such as Texas, Virginia, and South Carolina. The plant has nitrogen-fixing ability, and its seed spreads readily. The nitrogen cycle of native plant communities that rely on infertile soil for their existence suffers in the presence of this shrub, as well as native species in open fields it intrudes upon, that rely on maximal sunlight exposure.

Since it is well established in the eastern United States, and given the fruit's high content of lycopene and strong antioxidant capacity, harvesting of this fruit may warrant greater attention by agronomists in increasing its availability among native North Americans given their dietary need for lycopene, and the increased prevalence among natives of cardiovascular disease and various cancers.

4. PACIFIC NORTHWEST NATIVE DIETS

Tanoak trees (*Lithocarpus densilorus*), such as the Oregon white oak, are members of the beech family Fagaceae native to the western United States and Canada. Tanoaks can live over 500 years and reach a height of 20–24 m under the right circumstances, although tanoak trees as tall as 36 m have been measured. At one time, the white oak's range spanned from British Columbia, Canada, to northern California. Today these trees occupy less than one percent of the area they once covered.

Tanoak trees love sun and avoid shade. Unfortunately, given this preference, it is also the same land that commercial timberlands, roads, homes, and businesses favor as well. With increasing encroachment on these lands by lumber companies and settlers, white oaks found themselves competing with large vigorous Douglas firs (*Pseudotsuga menziesii* var. *mensiesii*), Western hemlock (*Tsuga heterophylla*) and Western red cedar (*Thuja plicata*), the former of which can reach heights of 60–75 m, eventually dramatically reducing needed light for production of acorns. The trees also depend on wildlife for their survival, such as the Western gray squirrel (*Sciurus griseus*), which bury the oak's acorns, and thereby help to spread its range.

The seed of tanoak trees is a nut about 2–3 cm long with a diameter of 2 cm. The nut sits in a cup that requires 18 months to mature. The nuts are produced in clusters, although a few grow on a single stem. Despite the nut's bitter taste, owing to its high tannins content, natives utilized the nut as a source of food and medicine. The bitterness was reduced by leaching the nut in a process mastered by natives living in the region.

Natives made flour and soup from the acorns. It has been estimated that upwards of 50% of native northern Californians' diet consisted of acorns [46]. Food scientists have determined that acorns are a good source of nutrients and polyphenolics in addition to having high antioxidant activity because of the presence of hydrolyzable tannins [47].

The quantity and type of tannins in a food, whether condensed tannins, proanthocyanidins or condensed hydrolyzable tannins, affect its biological activity [48]. Tanoak tree acorns are particularly rich in hydrolyzable tannins. These tannins possess anticarcinogenic and antimutagenic activity *in vitro* [49,50].

Procyanins tannins are found in many nut varieties, their activity dependent on chain length in terms of potential health benefits. The phytochemical composition of tanoak acorn cotyledon tissue and the pericarp, including hydrolyzable tannins, has been determined [47]. Condensed tannins up through hexamers and a total of 22 hydrolyzable tannins were identified. The condensed tannins were procyanidins of the B type, while the hydrolyzable tannins were gallic acid and ellagic acid derivatives.

Prior to European and American settlements arriving in the Pacific Northwest, natives periodically burned the prairies and valleys where oak trees thrived to keep the conifers at bay. These burns were carefully controlled and kept at ground level, but were restricted or discouraged by settlers as they encroached upon the harvest areas natives had relied on for millennia. Tribes subsisted in part not only on acorns as a dietary staple, but also on low ground vegetation, particularly the common camas (*Camassia quamash*) plant, which also requires considerable sun. *C. quamash* is a perennial member of the lily family (Liliaceae) with an edible bulb, and grows to 30–75 cm tall with a dense inflorescence.

With the exception of dried salmon, no other food item was more widely desired or traded than *C. quamash* or camas bulbs. The bulbs were dug after flowering in the summer, although they could be harvested in the spring as well. To prevent threatening the plant's production, harvesting of any one area was only done several years apart. Only the largest bulbs were harvested, leaving the smaller ones for the future. Natives could estimate how many camas bulbs might be found on a plot of land by observing flowering in the spring. If a sea of deep blue flowers appeared that stretched across a field, natives knew it would be a rich source of camas bulbs when the quiescent period time came to harvest the bulbs, following seed maturation, foliar senescence, and development of bulb offshoots [51]. Camas bulbs were used medicinally as a cough medicine by boiling the bulbs and mixing the strained juice with honey [1].

Rich in inulin, a naturally occurring polysaccharide belonging to a class of fibers known as fructans typically found in roots or rhizomes, camas bulbs could also be cooked to convert inulin to fructose and thereby reduce the resulting mass to a sweetener which when dried could be added to other foods.

The inulin content of camas bulbs has now been shown to be important to natives not only for its nutritional value but also due to significant health benefits that have only recently been discovered. Inulin increases magnesium and calcium absorption, promotes the growth of desirable intestinal bacteria, contains soluble and insoluble fibers, has minimal impact on blood sugar, and unlike fructose does not cause a rise in insulin or raise the level of triglycerides [52–54]. This makes it particularly suitable for diabetics or those managing sugar-related diseases.

It is estimated that regular consumption of camas bulbs provides up to 10 g of inulin a day [55]. Since inulin is indigestible by the amylase and ptyalin enzymes that break down starch, it passes through the stomach and small intestine intact, and then enters the colon where bacteria metabolize it. This is why inulin is considered a prebiotic. Since inulin also contains soluble and insoluble fibers, it increases stool bulk. The soluble fiber fraction also helps lower glucose levels and cholesterol.

5. SUMMARY

A database search of wild plants growing in North America and their ancestral use by indigenous people suggests the need for a significant increase in funding for research to fill gaps in our understanding of not just their nutritional value but also their phytochemical characterization and bioactivity. The work of Tozer on the uses and cultivation of more than 1200 species in over 500 genera of wild plants provides a useful guide as to many of the plants that warrant further study of the chemistry and bioactivity [56].

The need for the study of native foods is made all the more urgent given the high prevalence of obesity and metabolic syndrome among indigenous populations. Establishing foods that could aid in protecting against or mitigating the progression of diseases such as diabetes concomitant to re-establishing these foods in the diet of indigenous people, could temper the ethnic differences in the prevalence of diagnosed diabetes.

The prevalence of diabetes for American Indian elders in the United States is three times higher than the national average [57], yet this prevalence is not restricted to older Indians. The prevalence of American Indian and Alaska Native children, adolescents and young adults, from 1990 to 1998, with diagnosed diabetes increased by 71%, and prevalence increased by 46% [58]. Among the Ute Indians, the prevalence of diabetes is four times the average of others living in Utah, while the rate of diabetic neuropathy is at least 43 times that of the diabetic population in the general Utah population [59]. Similar differences are found among most native Americans and indigenous populations in Canada.

The reason for focusing attention on the native foods of the O'odham population that live in the Sonoran Desert is due to the high incidence of diabetes *in utero* that contributes to most of the increase in diabetes prevalence in O'odham children that has been reported in recent decades [60]. Various ancestral foods contain bioactive compounds that can play a role in protecting the O'odham from this insidious metabolic disease. The same is evident from an examination of the diets of indigenous people along the US–Canadian border and the Pacific Northwest.

It is hoped that efforts will be made to direct further scientific study of their traditional foods, and that of other indigenous populations. By such an effort these native foods might find their place back in the diet of indigenous people, and maybe benefit that of others living in North America.

Anthropologist, Daniel E. Moerman, has provided an exceptional work that describes the medicinal uses of nearly 3,000 plants used as food or for medicinal purposes, that would be of particular interest to those studying the ethnobotanical and therapeutic applications of such plants among over 200 North American tribes [61].

References

1. Kuhnlein, H. V., & Turner, N. J. (1991). *Traditional plant foods of Canadian indigenous peoples. Nutrition, botany and use.* London: Gordon & Breach Science Publishers.
2. Nabhan, G. P. (1985). *Gathering the desert,* Tucson: University of Arizona Press.
3. Cornett, J. W. (2002). *How Indians used desert plants.* Palm Springs, CA: Nature Trails Press.
4. Martin, A. C., Zim, H. S., & Nelson, A. L. (1951). *American wildlife and plants: A guide to wildlife food habits.* New York: Dove Publications.
5. Sydenstricker, V. P. (1958). The history of pellagra, its recognition as a disorder of nutrition and its conquest. *The American Journal of Clinical Nutrition, 6,* 409–414.
6. Walker, A. S. (2008). *Deserts: geology and resources.* p. 25. Denver: US Department of the Interior, US Geological Survey.
7. Erickson, W. P. (1994). *Sharing the desert: The Tohono O'odham in history.* Tucson: University of Arizona Press.
8. Nabhan, G. P., Hodgson, W., & Fellows, F. (1989). A meager living on lava and sand? Hia Ced O'odham food resources and habitat diversity in oral and documentary histories. *Journal Southwest, 31,* 509–533.
9. McIntyre, A. J. Arizona Historical Society. (2008). *The Tohono O'Odham and Pimeria Alta.* Charleston, SC: Arcadia Publishing.
10. Scheerens, J. C., Tinsley, A. M., Abbas, I. R., Weber, C. W., & Berry, J. W. (1983). The nutritional significance of tepary bean consumption. *Desert Plants, 5,* 11–14 50-56.
11. Leeds, A. R. (1981). Legume diets for diabetics? *Journal Plant Foods, 3,* 219–223.
12. Cahill, J. P. (2003). Ethnobotany of chia, *Salvia hispanica* L. (Lamiaceae). *Economic Botany, 57,* 604–618.
13. Bushway, A. A., Belyea, P. R., & Bushway, R. J. (2006). Chia seed as a source of oil, polysaccharide, and protein. *Journal of Food Science, 46,* 1346–1350.
14. Taga, M. S., Miller, E. E., & Pratt, D. E. (1984). Chia seeds as a source of natural lipid antioxidants. *Journal American Oil Chemists Society, 61,* 928–931.
15. Ezekwe, M. O., Omara Alwala, T. R., & Membrahtu, T. (1999). Nutritive characterization of purslane accessions as influenced by planting date. *Plant Foods for Human Nutrition, 54,* 183–191.
16. Hesse, Z. (1973). *Southwestern Indian recipe book: Apache, Pima, Papago, Pueblo, and Navajo.* Palmer Lake, CO: Filter Press.
17. Nabhan, G., & de Wet, J. M. J. (1984). *Panicum sonorum* in Sonoran desert agriculture. *Economic Botany, 38,* 65–82.
18. Viereck, E. G. (1987). *Alaska's wilderness medicines: Healthful plants of the Far North.* p. 35. Portland, OR: Alaska Northwest Books.
19. Burns-Kraft, T. F., Dey, M., Rogers, R. B., Ribnicky, D. M., Gipp, D. M., Cefalu, W. T., Raskin, I., & Lila, M. A. (2008). Phytochemical composition and metabolic performance-enhancing activity of dietary berries traditionally used by native North Americans. *Journal of Agricultural and Food Chemistry, 56,* 654–660.
20. Nicholson, J. A., Darby, T. D., & Jarboe, C. H. (1972). Viopudial, a hypotensive and smooth muscle antispasmodic from *Viburnum opulus. Proceedings of the Society for Experimental Biology and Medicine, 140,* 457–461.
21. Brayer, J. L., Alazard, J. P., & Thai, C. (1983). A simple total synthesis of viburtinal. *Journal of the Chemical Society Chemical Communications, 6,* 257–258.
22. Jarboe, C. H., Schmidt, C. M., Nicholson, J. A., & Zirvi, K. A. (1966). Uterine relaxant properties of *Viburnum. Nature, 212,* 837.

23. Jarboe, C. H, Zirvi, K. A., Nicholson, J. A., & Schmidt, C. M. (1967). Scopoletin, an antispasmodic component of *Viburnum opulus* and *V. prunifolium*. *Journal of Medicinal Chemistry*, 10, 488–489.

24. Kahkonen, M. P., Hopia, A. I., Vuorela, H. J., Jussi-Pekka, R., Kujala, T. S., & Heinonen, M. (1999). Antioxidant activity of plant extracts containing phenolic compounds. *Journal of Agricultural and Food Chemistry*, 47, 3954–3962.

25. Moerman, D. E. (1998). *Native American Ethnobotany*. pp. 445–448, 528–530. Portland, OR: Timber Press.

26. Sergeeva, N. V., Bandyukova, V. A., Shapiro, D. K., Narizhnaya, T. I., & Anikhimovskaya, L. V. (1980). Phenolic acids from fruits of some species of the *Amelanchier* Medic genus. *Khimiia Prirodnykh Soedinenii*, 5, 726–728.

27. Vereskovskii, V. V., Shapiro, D. K., & Narizhnaya, T. I. (1982). Flavonoids in the fruits of different species of the genus *Amelanchier*. *Khimiia Prirodnykh Soedinenii*, 2, 257.

28. Vereskovskii, V. V., Shapiro, D. K., & Narizhnaya, T. I. (1982). Anthocyanins in the fruits of different species of the genus *Amelanchier*. *Khimiia Prirodnykh Soedinenii*, 4, 522–523.

29. Mazza, G. (1986). Anthocyanins and other phenolic compounds of Saskatoon berries (*Amelanchier alnifolia* Nutt.). *Journal of Food Science*, 51, 1260–1264.

30. Acuna, U. M., Athan, D. E., Ma, J., Nee, M. H., & Kennelly, E. J. (2002). Antioxidant capacities of ten edible North American plants. *Phytotherapy Research*, 16, 63–65.

31. Gilmore, M. R. (1977). *Uses of plants by Indians of the missouri river region*. pp. 36–54. Lincoln: University of Nebraska Press.

32. Boboreko, E. Z., Shapiro, D. K., Anikhimovskaya, L. V., & Narizhnaya, T. I. (1978). *Shepherdia argentea* (Pursh.) Nutt. as a promising vitamin source and decorative plant. *Vestsi Akademica Navuk BSSR, Ser Biyal Navuk*, 4, 89–91.

33. Bekker, N. P., & Glushenkova, A. I. (2001). Components of certain species of the Elaeagnaceae family. *Chemistry Natural Compounds*, 37, 97–116.

34. Yoshida, T., Ito, H., Hatano, T., Kurata, M., Nakanishi, T., Inada, A., Murata, H., Inatomi, T., Matsuura, N., Ono, K., & Nakane, H., et al. (1996). New hydrolyzable tannins, shephagenins A and B, from Shepherdia argentea as HIV-1 reverse transcriptase inhibitors. *Chemical and Pharmaceutical Bulletin*, 44, 1436–1439.

35. Ritch-Krc, E. M., Turner, N. J., & Towers, G. H. (1996). Carrier herbal medicine: An evaluation of the antimicrobial and anticancer activity in some frequently used remedies. *Journal of Ethnopharmacology*, 52, 151–156.

36. Stevens, O. A. (1963). *Handbook of ND plants*. pp. 260–262. Fargo, ND: North Dakota Institute for Regional Studies.

37. Wang, P. L., & Francis, F. J. (1972). New anthocyanin from *Viburnum trilobum*. *Horticultural Science*, 7, 87.

38. Du, C. T., Wang, P. L., & Francis, F. J. (1974). Cyanidin-3-arabinosylsambubioside in *Viburnum trilobum*. *Phytochemical Report*, 13, 1998–1999.

39. Poulton, J. E. (1988). Localization and catabolism of cyanogenic glycosides. *Ciba Foundation Symposium*, 140, 67–91.

40. Yemm, R. S., & Poulton, J. E. (1992). Isolation and characterization of multiple forms of mandelonitrile lyase from mature black cherry (*Prunus serotina* Ehrh.) seeds. *Archives of Biochemistry and Biophysics*, 247, 440–445.

41. Swain, E., Li, C. P., & Poulton, J. E. (1992). Development of the potential for cyanogenesis in maturing black cherry (*Prunus serotina* Ehrh.) fruits. *Plant Physiology*, 98, 1423–1428.

42. Veazie, P. M., Black, B. L., Fordham, I. M., & Howard, L. R. (2005). Lycopene and total phenol content of autumn olive (*Elaeagnus umbellate*) selections. *Hort Science*, 40, 883.

43. Gerster, H. (1997). The potential role of lycopene for human health. *Journal of the American College of Nutrition*, 16, 109–126.

44. Collins, J., Perkins Veazie, P., & Roberts, B. (2006). Lycopene: From plants to humans. *Hort Science*, 41, 1135–1144.

45. Collins, J. K., Arjmandi, B. H., Claypool, P. L., Perkins Veazie, P. M., Baker, R. A., & Clevidence, B. A. (2004). Lycopene from two food sources does not affect antioxidant or cholesterol status of middle-aged adults. *Nutrition Journal*, 3, 15.

46. Heizer, R. F., & Elasser, A. B. (1980). *The natural world of the California Indians*, Berkeley, CA: University of California Press.

47. Meyers, K. J., Swiecki, T. J., & Mitchell, A. E. (2006). Understanding the native California diet: Identification of condensed and hydrolysable tannins in tanoak acorns (*Lithocarpus densiflorus*). *Journal of Agricultural and Food Chemistry*, 54, 7686–7691.

48. Chung, K. T., Wei, C. I., & Johnson, M. G. (1998). Are tannins a double-edged sword in biology and health? *Trends Food Science and Technology*, 9, 168–175.

49. Fujiki, H., Yoshizawa, S., Horiuchi, T., Suganuma, M., Yatsunami, J., Nishiwaki, S., & Okabe, S. (1992). Anticarcinogenic effects of (−)-epigallocatechin gallate. *Preventive Medicine*, 21, 503–509.

50. Howikawa, K., Mohri, T., Tanaka, Y., & Tokiwa, H. (1994). Moderate inhibition of mutagenicity and carcinogenicity of benzo[a]pyrene 1,6-dinitropyrene and 3,9-dinitrofluoranthene by Chinese medical herbs. *Mutagenesis, 9*, 523–526.

51. Guard, J. B. (1995). *Wetland plants of Oregon and Washington.* Redmond, WA: Lone Pine Publishing.

52. Niness, K. R. (1999). Inulin and oligofructose: What are they?. *The Journal of Nutrition, 129*, 1402S–1406S.

53. Coudray, C., Demigne, C., & Rayssiquier, Y. (2003). Effects of dietary fibers on magnesium absorption in animals and humans. *The Journal of Nutrition, 133*, 1–4.

54. Abrams, S. A., Griffin, I. J., Hawthorne, K. M., Liang, L., Gunn, S. K., Darlington, G., & Ellis, K. J. (2005). A combination of prebiotic short- and long-chain inulin-type fructans enhances calcium absorption and bone mineralization in young adolescents. *The American Journal of Clinical Nutrition, 82*, 471–476.

55. Coussement, P. A. (1999). Inulin and oligofructose: Safe intakes and legal status. *The Journal of Nutrition, 129*, 1412S–1417S.

56. Tozer, F. (2007). *The uses of wild plants. Using and growing the wild plants of the United States and Canada.* Santa Cruz, CA: Green Man Publishing.

57. *National Indian Council on Aging.* (2009). < http://www.nicoa.org/Diabetes_Education/home.html > downloaded 23.02.09.

58. Acton, K. J., Burrows, N. R., Moore, K., Querec, L., Geiss, L. S., & Engegau, M. M. (2002). Trends in diabetes prevalence among American Indian and Alaska Native children, adolescents, and young adults. *American Journal of Public Health, 92*, 1485–1490.

59. Giger, J. N., & Davidhizar, R. E. (2008). *Transcultural nursing: Assessment and intervention.* p. 290. St. Louis: Mosby, Elsevier.

60. Dabelea, D., Hanson, R. L., Bennett, P. H., Roumain, J., Knowler, W. C., & Pettitt, D. J. (1998). Increasing prevalence of type II diabetes in American Indian children. *Diabetologia, 41*, 904–910.

61. Moerman, D. E. (2009). *Native American Medicinal Plants: An Ethnobotanical Dictionary.* Portland (OR): Timber Press.

6

Barriers and Facilitating Factors Affecting Fruit and Vegetable Consumption

Ming-Chin Yeh[1,3]*, Janel Obenchain*[1]*, and Anahi Viladrich*[2,3]

[1]Nutrition and Food Science Track, Urban Public Health Program, Hunter College,
City University of New York, New York, NY, USA

[2]Community Health Education Track, Urban Public Health Program, Hunter College,
City University of New York, New York, NY, USA

[3]Immigration and Health Initiative, School of Health Sciences, Hunter College,
City University of New York, New York, NY, USA

1. INTRODUCTION

It is generally agreed that a healthy diet includes eating fruit and vegetables. Yet to date, the majority of Americans do not eat the recommended amounts. Although many people experience barriers inhibiting fruit and vegetable consumption, facilitators also exist that promote fruit and vegetable consumption. This chapter first describes the current fruit and vegetable consumption level in the United States and how it compares to the dietary guidelines promulgated by the US government. We then proceed to provide possible barriers and facilitators to adult fruit and vegetable intake. We also discuss the important role the government plays in establishing guidelines and how governmental food assistance programs can impact fruit and vegetable consumption. Finally, we conclude with recommendations on how to promote fruit and vegetable consumption.

2. CURRENT FRUIT AND VEGETABLE CONSUMPTION IN THE US, AND US DIETARY GUIDANCE

Fruit and vegetable consumption can be measured in many ways, such as number of times eaten daily, number of cups, and number of servings. The US government publishes the *Dietary Guidelines for Americans* [1], sets

85

public health dietary goals in documents such as *Healthy People 2010* [2], conducts nutritional surveillance and surveys to assess if the population is meeting these guidelines, and plans food and nutrition education programs following these recommendations. *Dietary Guidelines for Americans* is jointly published every 5 years by the Department of Health and Human Services (DHHS) and the Department of Agriculture (USDA). Recommendations for an individual's fruit and vegetable consumption are based on the *Dietary Guidelines* and are published by the USDA for public use in the form of the on-line educational tool 'MyPyramid' (previously the 'Food Guide Pyramid') [3].

Current MyPyramid dietary recommendations for fruit and vegetable vary depending on total calories needed for age, gender, and activity level and range from 1 to 2½ cups of fruit and 1–4 cups of vegetables per day for a total of 2 cups of fruit and vegetables to 6½ cups of fruit and vegetables combined. For example, a 2-year-old (of any gender/activity level) should eat 2 combined cups of fruit and vegetables, whereas a physically active 27-year-old male requires 3000 calories per day and should eat 6½ combined cups of fruit and vegetables, but a less active 60-year-old female only requires 1600 calories and should eat 3½ combined cups of fruit and vegetables.

Since previous US dietary guidelines defined a serving as a half-cup for most items, the 2005 recommendations would require adults to generally consume 7–13 servings [4]. Although current US dietary recommendations are given in cups, nutrition surveillance tools and interventions vary in assessment method and are still sometimes conducted in servings. Regardless of methodology, however, the results are unfortunately the same – most Americans simply do not meet current fruit and vegetable consumption recommendations, let alone the previous less ambitious '5-a-Day'

program that recommended eating at least 5 servings of fruit and vegetables a day [5,6].

In 2007, the nutritional surveillance data from the 2005 Behavioral Risk Factor Surveillance System reports that only 32.6% of adults consumed fruit two or more times per day and only 27.2% ate vegetables three or more times per day [7]. It is even more critical to ask how many people eat the recommended amounts of both fruit and vegetables. A study of the National Health and Nutrition Examination Survey (NHANES) data from 1988 to 2002 showed that only 11% of adults meet the combined fruit and vegetable recommendation [8].

In fact, of all age groups, children aged 2–3 are most likely to meet current fruit and vegetable consumption guidelines, with an estimated 48% of children meeting the combined recommendation of 1 cup. The group with the second highest compliance – 17% – is women aged 51–70. Even assuming the minimum requirement necessary to support a sedentary level of activity, all other age–gender groups have an estimated compliance rate under 11%. Adolescent boys aged 14–18 have a particularly low consumption rate, with less than 1% (only 0.7%) meeting their combined recommendation of 5 cups of fruits and vegetables. Furthermore, only 3% of the population consumes 3 or more servings of vegetables with the specific requirement that at least one third of the vegetables consumed are dark green or orange [9].

Fruit and vegetable consumption also vary across other demographic characteristics. Non-Hispanic blacks are significantly less likely to meet USDA guidelines than whites [8]. Latinos have reported higher intakes of fruit and vegetables than whites and blacks [10], but this has also been reported to be a decreasing effect when compared to social acculturation [11].

In general, individuals with higher income and education are more likely to achieve dietary guidance [8,10]. Additional reported demographic factors describing greater likelihood of consumption include being married, having access to a garden, and possessing food security [12].

Worsening of food security [13] and lower household income both show a consistent negative association with fruit and vegetable consumption [12]. An environment-level study has shown a positive association between neighborhood socioeconomic status (SES; a measure evaluating income, occupation, and education) and fruit and vegetable intake: a 1-standard deviation increase in the neighborhood socioeconomic status index was associated with consumption of nearly 2 additional servings of fruit and vegetables per week [14]. Additionally, smokers are also generally less likely to consume fruit and vegetables [15], and low-income smokers in particular are one of the groups at highest risk for inadequate fruit and vegetable consumption [16].

Because dietary patterns emphasizing fruit and vegetable consumption are associated with diabetes prevention and weight management [17–19], treating and preventing metabolic syndrome [20], and decreased risk of certain cancers [21] and cardiovascular disease (CVD) [22–24], understanding barriers, facilitators and predictors of fruit and vegetable consumption is an important public health goal.

3. BARRIERS INHIBITING FRUIT AND VEGETABLE CONSUMPTION

Recognizing the discrepancy between recommended consumption and actual consumption leads to the question: What barriers are keeping the American population and individuals from consuming the recommended amounts of fruit and vegetables?

Barriers can prevent or inhibit fruit and vegetable consumption in a variety of manners from a macro-level to a personal level. For example, at the macro-level there is a very basic, but significant, barrier preventing the population as a whole from consuming the recommended amounts of fruits and vegetables. Put very simply, we do not have enough on hand. The USDA reports per capita food availability (adjusted for loss) as defined by cups. During 2006, the most recent year reported at the time of writing, there was only enough fruit available in any form (including juice) to provide less than one cup (0.91) per person per day, and only enough vegetables in any form (including beans, lentils, legumes, and dehydrated vegetables) to provide less than two cups (1.74) per person per day [25]. Although this data is a rough estimate, which does not account for personal gardens, etc., it does provide important macro-level barrier information. Our food and agricultural policy as implemented approximately every 5 years via the legislation colloquially known as 'the Farm Bill' plays an important role in acting both as a barrier (not enough supply) and as a facilitator (providing food assistance).

However, the inherent limitations of our food supply do not address why any particular individual or group is not consuming recommended amounts of fruit and vegetables. Numerous studies have been conducted at neighborhood and individual levels with the goal of elucidating actual and perceptual barriers. Barriers reported as common themes in the literature include lack of availability, shopping practicalities, cost, time, and taste/preference.

3.1 Lack of Availability

Environmental studies report the presence of fewer grocery stores in minority/mixed census tracts [26–29] and less availability of healthy foods [30]. These quantitative studies provide additional support to the qualitative reporting of lack of availability by African Americans in a focus group study [31]. Other groups have also reported lack of availability as a specific barrier, such as women living on Prince Edward Island, Canada [32], and rural Lakota Indians [33]. Additionally, consumers may perceive that healthy foods are not available or are cost-prohibitive in their local area [34]. Lack of availability extends beyond grocery purchasing. An increasing number of eating occasions are outside of the home, and the lack of availability of fruit and vegetables in food consumption environments such as work cafeterias and take-out restaurants has also been reported as a barrier to consumption [35].

3.2 Shopping Practicalities/Types of Stores Utilized

The logistic requirements of shopping may act as a disincentive to purchase fresh produce [35]. Logistics, such as distance, may be more problematic for low-income people or those with transportation limitations. For example, Food Stamp participants reported distance from home to food source inversely related with fruit use [36]. Shopping practicalities may dictate the type of store utilized. In an urban environment, smaller stores may be more prevalent and more convenient than grocery stores and may have a significant impact on fruit and vegetable consumption. For example, a cross-sectional survey in New Orleans of 102 randomly selected households found that greater fresh vegetable availability within 100 meters was a positive predictor of household vegetable intake and an increase in servings per day was associated with increased shelf space [37]. Unfortunately, small stores (such as bodegas) in lower-income, lower-resource neighborhoods may be less likely to carry healthy food and fresh produce [38]. In an area of central Brooklyn, New York City, 80% of the food stores are convenience stores but only 28% of those stores carried even the basic fruits – apples, oranges and bananas [39].

3.3 Cost

Cost is an often-reported barrier [31,33,35]. A survey of low-income women in Minnesota found that more than one-third cited money concerns as a barrier [40]. Although cost may in some cases be a perceptual barrier, physically finding less costly produce may require a careful selection of grocery stores. A study on the cost of purchasing a fruit and vegetable market-basket satisfying the 2005 Dietary Guidelines found that prices vary across very-low-income areas by 65% in Los Angeles and 76% in Sacramento, with variations even in the same supermarket chain. This study also found that this basket would require the allocation of 43–70% of lower-income families' food-at-home budget, which would require a significant increase on spending for fruit and vegetables and a corresponding reduction in other areas [41]. This supports the finding of another study that for those with inflexible food budgets, the perception that fruit and vegetables were expensive was a relatively intractable barrier [42].

Cost can manifest in a variety of barriers. Cost barriers are not necessarily limited to low-income consumers. For senior citizens (and others who grew up with a garden supplying produce), the objection to cost may also be due to the experience of produce as 'free' [31]. Another component to cost is perception of value. The cost barrier has been reported as a value judgment of quality [33], especially when fruit is out of season and does not taste as appealing [43].

3.4 Time Issues

Time issues are another frequently identified barrier [31,35,43,44]. A survey of low-income women in Minnesota found that half identified time as a barrier [40]. Effort of preparation [32] and lack of preparation time, may be seen as more of an issue for vegetables than fruit [45,46]. The time issue barrier may also vary by age. A survey of 1000 European Union adults found that time issues were more important to younger responders [47].

3.5 Taste/Preference

Some people identify taste as a barrier to healthy eating [47]. Taste is also a predictor of consumption [48]. Many people report a preference for other foods [45] and without a desire to substitute fruit and vegetables for another food item. The ability to simply add fruit and vegetables to the diet may be hindered by a relative homeostasis in overall food and caloric amount intake. Taste and preference may also be associated with meal patterns that may vary across cultures and individuals. For example, for some, vegetables are consumed only at an evening meal [45] and this perceived restriction of eating occasion accordingly limits perceived available opportunities to increase consumption.

Other reported barriers include social norms as portrayed by the media [31], the marketing of unhealthy food [44], the convenience of purchasing pre-packaged foods [31], and fear concerning pesticides [43].

Finally, certain populations such as seniors, children and adolescents may experience additional barriers to fruit and vegetable intake. Seniors may have increased difficulty in shopping and chewing foods, and less sensory motivation due to decreased taste-bud sensitivity. Seniors have reported 'difficulties with digestion,' and 'too much trouble' [49]. Indeed, an analysis of NHANES III data showed that risk factors for lowest consumption of fruit and vegetables by seniors included social isolation and missing pairs of posterior teeth [50]. For children and adolescents, intake of fruit and vegetables may be highly dependent on parental behavior and home and school environment [51].

4. FACILITATORS PROMOTING FRUIT AND VEGETABLE CONSUMPTION

Encouraging the American public to eat more fruit and vegetables is an important public health goal. In addition to publishing dietary recommendations for individuals, the US government has also set goals for increasing fruit and vegetable consumption in the population. Fruit and vegetable consumption goals in Healthy People 2010 include Objective 19-5: to increase the proportion of persons aged 2 years and older who consume at least 2 daily servings of fruit, and Objective 19-6: to increase the proportion of persons aged 2 years and older who consume at least 3 daily servings of vegetables, with at least one-third being orange or dark green [2]. In order to increase consumption at a population and individual level, we need to understand what factors can encourage people to overcome the reported barriers inhibiting consumption. In other words, what factors facilitate consumption?

Commonly reported facilitators include availability, methods of serving, and ability/exposure to growing one's own produce. While some facilitators may specifically address barriers (i.e. cut up and prepared fruit and vegetables addresses time and convenience barriers), other reported facilitators have a more complex relationship with barriers. Commonly reported influencing factors, such as cultural support, and social support and norms, may act as either a facilitator or a barrier. Social support and norms, in particular, are important psychosocial predictors of fruit and vegetable consumption and will be discussed in more detail in the psychosocial section following.

4.1 Availability

Perceived availability has been reported as a measure of influence [52]. Some participants in an intervention providing home delivery of fruit and vegetables reported that the sheer presence of the fruit prompted consumption [45].

Federal food assistance programs, such as school-based programs, the Women, Infants, and Children (WIC) program, and the Farmers' Markets Nutrition Programs play a key role in facilitating consumption through availability. Low-income women in Minnesota who received food from assistance programs perceived that the programs were promoting healthy eating [40]. An evaluation of a federally-funded free fruit distribution program found that Wisconsin students who received the snacks reported an increased willingness to try new fruits and vegetables [53], and an evaluation in Missouri concluded that fruit distribution could be an effective component of a comprehensive approach to change dietary behavior [54].

In 2009, the WIC program updated the food package offered. This update is a significant facilitator of fruit and vegetable consumption, as the original list of eligible foods did not include fruits and vegetables, with the exception of fruit juice. Participants will now receive vouchers ranging from $6 to $10, which can be used to purchase fresh, frozen or canned fruit and vegetables [55].

Additional vouchers are available for specific use at local farmers' markets through a program called the Farmers' Market Nutrition Program [56]. Pilot studies of the program reported that when WIC participants were given vouchers for fresh fruit and vegetables, there was an increased consumption sustained for 6 months after termination, and that the increase was greater in those given farmers' market vouchers than supermarket ones [57]. Similarly, an evaluation of the program in Michigan found a direct relationship between providing coupons and increased fruit and vegetable consumption, but found that the benefits were maximized when nutrition education was provided with the coupons [58]. A similar program providing vouchers to low-income seniors for use at local farmers' markets has also shown an ability to increase fruit and vegetable consumption [59].

4.2 Methods of Serving

Providing a salad bar or serving food buffet or family style can increase consumption of fruit and vegetables. A cross-sectional survey of Los Angeles elementary schoolchildren exposed to a salad bar found a significant increase in fruit and vegetables consumed, almost all during lunch [60], and a Danish study observing worksite canteens using either a buffet style or à la carte found the buffet serving was associated with an average 76 g increased intake of fruit and vegetables [61]. A focus group reported that participants were more likely to eat vegetables if served with a dinner entrée rather than being offered à la carte [31]. Similarly, a convenience sample of a 3-day dietary analysis from 503 college students found that those who participated in the school's pre-paid plan had fruit and vegetable intake closer to recommendations [62].

4.3 Ability to Grow Own Produce/ Exposure to Growing Produce

Exposure to school-based gardening programs has been shown to increase consumption over an intervention without a gardening component [63]. Although growing one's own produce is often discussed in light of cultural immigrant food patterns (such as reinforcing the traditional patterns among first-generation Hmong immigrants) [44], this facilitator is also meaningful to non-immigrant groups, such as rural parents and pre-schoolers. A cross-sectional study found that frequency of eating homegrown produce was associated with increased home availability of produce, increased pre-schooler preference for fruit and vegetables and increased parental role modeling [64].

4.4 Cultural Support

For immigrant populations adjusting to differing food resources, education by members of the cultural group delivered through media and trusted community sources has been found helpful [44]. Cultural support may assist in identifying resources for familiar fruit and vegetables or by providing education about how to incorporate new and unfamiliar ones [31]; for example, street vendors with push carts who are familiar to the community could help facilitate the adoption of new fruits and vegetables [65].

A final note to keep in mind is that a population or individual may have different motivators or facilitators for fruit as compared with vegetables. A survey of Canadian women found that fruit was perceived as a healthy between-meal snack, whereas vegetables were seen as an aesthetic meal component [43]. Similarly, a survey conducted as part of a British intervention found that perceived practical opportunities to increase fruit consumption focused on eating fruit as a dessert or as a between-meal snack, whereas increased vegetable consumption was perceived as possible by having two servings at dinner [35]. A nationally representative survey in the United States found that for African American men, fruit consumption appears to be motivated by perceived benefits and standards set by important people in their lives, whereas vegetable consumption is a function of extrinsic rewards and preferences for high-calorie, fatty foods [66].

5. PSYCHOSOCIAL PREDICTORS OF FRUIT AND VEGETABLE CONSUMPTION

In addition to the above mentioned barriers and facilitating factors, it is also important to understand psychosocial determinants of fruit and vegetable consumption. A literature review of psychosocial predictors of fruit and vegetable consumption co-authored by the first author found strong support for three main predictors – self-efficacy, social support and knowledge [67].

5.1 Self-Efficacy

Self-efficacy refers to situation-specific confidence in one's ability to make and maintain a health behavior change [68]. Positive self-efficacy has been found to be a predictor of fruit and vegetable intake [69,70]. Self-efficacy can manifest in various areas, such as lack of familiarity with available fruit and vegetables [31], or cooking confidence, and the self-efficacy to cook healthy foods has been associated with higher vegetable consumption [71]. Cooking classes have been shown to improve fruit and vegetable intake in youth and adults [72,73].

5.2 Social Support

Social support can exist at various levels in which an individual interacts in society. For example, family was the most common promoter identified by low-income women [40]. Social support provided by churches [31] has been shown to increase fruit and vegetable consumption [74]. Facilitated small groups in an intervention can also function as social support [75], and personal contact can make a difference. Individual attention, including motivational counseling sessions [76], has been reported as increasing consumption across numerous studies.

However, it is important to realize that social support can occur in a positive or negative direction. For example, family and friends may act as both a barrier and facilitator [35] and social pressure has been reported in both directions for vegetable consumption [43]. The direction of influence may also vary for different individuals or populations. In one study, women reported that children and male partners were obstructive to their attempts to eat more fruit and vegetables, while men reported that their partners were supportive of the change [42].

5.3 Knowledge of Dietary Recommendations/Health Literacy

Knowledge is a mediating factor of fruit and vegetable intake [70]. Although adults may be aware that consuming fruits and vegetables is recommended or are familiar with the highly promoted message of '5-a-Day,' this knowledge is not universal and the practicalities of how much to consume may not be well understood [77]. Additionally, the mistaken belief that one already is eating healthily enough is itself a barrier [45]. A UK survey of low-income persons found that less than 5% believed they had problems with healthy eating, but only 18% claimed to be eating the recommended daily amounts of fruits and vegetables [78]. In one report of an intervention concerning senior citizens in Georgia, only 7% were aware prior to the intervention of the recommended daily consumption amounts for fruits and vegetables, and increased intake was found post-intervention to be independently associated with improved knowledge of dietary recommendations [49]. In a baseline survey of adults conducted for the 5-a-Day program, it was found that one of the most important factors in determining intake was the number of servings they thought they should have in a day [48]. Finally, general health knowledge, or health literacy, is also an influencer. A cross-sectional survey of 759 adults found that every point higher on the health literacy scale increased likelihood of eating at least 5 portions of fruit and vegetables a day [79].

6. RECOMMENDATIONS FOR PROMOTING FRUIT AND VEGETABLE CONSUMPTION

Dieticians and physicians who feel stressed about having too little time with their patients still have some avenues to embed methods of encouraging fruit and vegetable consumption into their practice. For example, one study found that it took only 1–2 minutes for medical professionals to write prescriptions for the purchase of discounted produce and hand it over with an explicit '5-a-day' message [80]. In a rural population of African Americans, a low-intensity physician-endorsed self-help dietary intervention in which low-literacy nutrition education materials and personalized dietary feedback were administered by mail and telephone was successful in initiating fruit and vegetable dietary changes at 1 and 6 months post-intervention [81]. The following are additional specific suggestions for designing interventions for promoting fruit and vegetables consumption.

6.1 Be Culturally Competent

Nutrition information, whether presented to an individual as counseling or as a planned intervention for a population, should take into account any specific relevant cultural views about food. Eating patterns, food selection, preparation and even food storage may have underpinnings in a cultural context underlying individual preferences, beliefs and actions. Accordingly, different interventions/strategies may be necessary to address fruit and vegetable consumption barriers in diverse populations. As an example, magazines with culturally tailored health information were found to be more effective at increasing fruit and vegetable consumption than magazines without the cultural component [82]. Level of acculturation is another important factor to consider [11,83,84]. A study of 662 WIC-eligible women found that foreign-born women who had lived in the United States for 4 or fewer years consumed 2.5 more fruit and vegetable servings daily than native-born women but this difference diminished with longer residence [85].

6.2 Be Practical

When assessing needs or designing an intervention, consider what is available for the population of interest. Consider walking distance in urban neighborhoods, challenges and perceptions of low socioeconomic status individuals, and creative solutions such as mobile produce vendors [86]. Specifically encourage fresh produce consumption (particularly in season), but emphasize the benefits of eating fruits and vegetables of any kind.

6.3 Be Holistic

Consider potential multiple health impacts. For example, an intervention that delivered fruit baskets with a newsletter and recipes to home-bound seniors not only increased fruit and vegetable consumption among the seniors, but also provided the ancillary benefit of personal contact for seniors suffering social isolation [87,88]. While the proportion of participants in a collective kitchen program in Canada consuming at least 5 fruit and vegetable servings a day rose from 29% to 47%, the most common reason reported (90%) for joining this program concerned social interactions and support [89].

6.4 Be Inclusive

Multi-component interventions have been reported to provide increased effect on produce consumption [90]. A family-based intervention may provide stronger reinforcement [40] and impact a greater number of people. Additionally, although many studies and interventions are focused on low-income persons, most Americans do not meet dietary guidance for fruit and vegetables. Cost-effective worksite interventions are also important. E-mail and web-based worksite programs have shown the potential to reach a broad adult population [91,92].

6.5 Be Specific

Be aware that fruits and vegetables may have different barriers and facilitators. When designing educational material, consider that information can be tailored in a variety of ways that can increase consumption, such as lifestyle [93], an individual's regulatory approach (motivated by promotion or prevention goals) [94], and an individual's preference for style of expert communication [95]. Individual goal setting, such as making a 'contract for change,' has been shown to help increase consumption [96]. A systematic review of dietary interventions reported that two components seemed particularly effective – goal setting and small groups [97].

6.6 Be Flexible

Unanticipated barriers may arise, even for those willing to make changes in their fruit and vegetable consumption. An intervention studying barriers reported at the end as well as the beginning of a 6-month dietary trial that a flexible action plan allowed participants to adapt [42].

6.7 Be Visual

Demonstrating the amount that should be consumed by providing a visual cue, such as a plate containing the recommended daily amount of fruit and vegetables, is a successful means of education [45].

6.8 Be Personal

A small amount of personal contact can make a difference. Although a systematic review of fruit and vegetable consumption interventions has shown that face-to-face methods are most effective, telephone counseling is seen as an effective alternative [90] and behavior change tool for fruit and vegetable consumption in both healthy and ill people [98,99], with reported long-term success (up to 4 years) in a randomized trial of breast cancer survivors [100].

7. SUMMARY

Most Americans do not consume enough fruit and vegetables. To effectively encourage consumption, medical professionals should be aware of the barriers, facilitators and psychosocial predictors that may impact individuals. References cited in this chapter provide further information for those seeking innovative ways to design interventions and programs, but all medical professionals can utilize social support and personal contact to encourage fruit and vegetable consumption. Providing easy to read nutritional brochures in the waiting room or asking patients a simple question 'Did you eat any fruit and vegetables yesterday?' are examples of effective, but simple, strategies that may be easily adoptable by medical professionals.

References

1. U.S. Department of Health and Human Services and U.S. Department of Agriculture. (2005). *Dietary Guidelines for Americans, 2005.* (6th ed.). Washington, DC: U.S. Government Printing Office.
2. Healthy People 2010. Available at: < http://www.healthypeople.gov/> Accessed 20.12.08.
3. USDA Food Guide Pyramid. Available at: < http://www.mypyramid.gov/> Accessed 20.12.08.
4. More Matters campaign. Available at: < http://www.fruitsandveggiesmorematters.org/> Accessed 20.12.08.
5. Krebs-Smith, S. M., Cook, A., Subar, A. F., Cleveland, L., & Friday, J. (1995). US adults' fruit and vegetable intakes, 1989 to 1991: A revised baseline for the Healthy People 2000 objective. *American Journal Public Health, 85,* 1623–1629.
6. United States Department of Agriculture, ARS. (1997). *Data Tables: Results from USDA's 1994–96 Continuing Survey of Food Intakes by Individuals and 1994–96 Diet and Health Survey.* Maryland: Beltsville Human Nutrition Research Center.
7. Centers for Disease Control and Prevention. (2007). *MMWR. Morbidity and Mortality Weekly Report, 56,* 213–217.
8. Casagrande, S. S., Wang, Y., Anderson, C., & Gary, T. L. (2007). Have Americans increased their fruit and vegetable intake? The trends between 1988 and 2002. *American Journal of Preventive Medicine, 32,* 354–355.
9. Guenther, P. M., Dodd, K. W., Reedy, J., & Krebs-Smith, S. M. (2006). Most Americans eat much less than recommended amounts of fruits and vegetables. *Journal of the American Dietetic Association, 106,* 1371–1379.
10. Thompson, F. E., Midthune, D., Subar, A. F., McNeel, T., Berrigan, D., & Kipnis, V. (2005). Dietary intake estimates in the National Health Interview Survey, 2000: Methodology, results and interpretation. *Journal of the American Dietetic Association, 105,* 352–363.
11. Gregory-Mercado, K. Y., Staten, L. K., Ranger-Moore, J., Thomson, C. A., Will, J. C., Ford, E. S., Guillen, J., Larkey, L. K, Giuliano, A. R., & Marshall, J. (2006). Fruit and vegetable consumption of older Mexican-American women is associated with their acculturation level. *Ethnopharmacol Disease, 16,* 89–95.
12. Kamphuis, C. B., Giskes, K., de Bruijn, G. J., Wendel-Vos, W., Brug, J., & van Lenthe, F. J. (2006). Environmental determinants of fruit and vegetable consumption among adults: A systematic review. *The British Journal of Nutrition, 96,* 620–635.
13. Kendall, A., Olson, C. M., & Frongillo, E. A. Jr., (1996). Relationship of hunger and food insecurity to food availability and consumption.. *Journal of the American Dietetic Association, 96,* 1019–1024.
14. Dubowitz, T., Heron, M., Bird, C. E., Lurie, N., Finch, B. K., Basurto-Dávila, R., Hale, L., & Escarce, J. J.

(2008). Neighborhood socioeconomic status and fruit and vegetable intake among whites, blacks, and Mexican Americans in the United States. *The American Journal of Clinical Nutrition, 87,* 1883–1891.

15. Genkinger, J. M., Platz, E. A., Hoffman, S. C., Comstock, G. W., & Helzlsouer, K. J. (2004). Fruit, vegetable, and antioxidant intake and all-cause, cancer, and cardiovascular disease mortality in a community-dwelling population in Washington County, MD. *American Journal of Epidemiology, 160,* 1223–1233.

16. Nollen, N., Befort, C., Pulvers, K., James, A. S., Kaur, H., Mayo, M. S., Hou, Q., & Ahluwalia, J. S. (2008). Demographic and psychosocial factors associated with increased fruit and vegetable consumption among smokers in public housing enrolled in a randomized trial. *Health Psychology, 27,* S252–S259.

17. Ford, E., & Mokdad, A. (2001). Fruit and vegetable consumption and diabetes mellitus incidence among U.S. adults. *Preventive Medicine, 32,* 33–39.

18. Hodge, A. M., English, D. R., O'Dea, K., & Giles, G. G. (2007). Dietary patterns and diabetes incidence in the Melbourne Collaborative Cohort Study. *American Journal of Epidemiology, 165,* 603–610.

19. Bazzano, L. (2006). The high cost of not consuming fruits and vegetables. *Journal of the American Dietetic Association, 106,* 1364–1368.

20. Feldeisen, S. E., & Tucker, K. L. (2007). Nutritional strategies in the prevention and treatment of metabolic syndrome. *Applied Physiology, Nutrition, and Metabolism, 32,* 46–60.

21. Lee, J. E., Giovannucci, E., Smith-Warner, S. A., Spiegelman, D., Willet, W. C., & Curhan, G. C. (2006). Intakes of fruits, vegetables, vitamin A, C, and E, and carotenoids and risk of renal cell cancer. *Cancer Epidemiology Biomarkers Prevention, 15,* 2445–2452.

22. Ness, A., & Powles, J. (1997). Fruit and vegetables, and cardiovascular disease: A review. *International Journal of Epidemiology, 26,* 1–13.

23. Nikolic, M., Nikic, D., & Petrovic, B. (2008). Fruit and vegetable intake and the risk for developing coronary heart disease. *Central European Journal of Public Health, 16,* 17–20.

24. He, F. J., Nowson, C. A., Lucas, M., & MacGregor, G. A. (2007). Increased consumption of fruit and vegetables is related to a reduced risk of coronary heart disease: meta-analysis of cohort studies. *Journal of Human Hypertension, 21,* 717–728.

25. United States Department of Agriculture, ERS. Available at: < http://www.ers.usda.gov/Data/FoodConsumption/FoodGuideIndex.htm/> Accessed 10.01.09.

26. Zenk, S. N., Schulz, A. J., Israel, B. A., James, S. A., Bao, S., & Wilson, M. L. (2006). Fruit and vegetable access differs by community racial composition and socioeconomic position in Detroit, Michigan. *Ethnicity & Disease, 16,* 275–280.

27. Moore, L. V., & Diez Roux, A. V. (2006). Associations of neighborhood characteristics with the location and type of food stores. *American Journal of Public Health, 96,* 325–331.

28. Morland, K., & Filomena, S. (2007). Disparities in the availability of fruits and vegetables between racially segregated urban neighbourhoods. *Public Health Nutrition, 10,* 1481–1489.

29. Hosler, A. S., Rajulu, D. T., Fredrick, B. L., & Ronsani, A. E. (2008). Assessing retail fruit and vegetable availability in urban and rural underserved communities. *Preventing Chronic Disease, 5,* A123.

30. Baker, E. A., Schootman, M., Barnidge, E., & Kelly, C. (2006). The role of race and poverty in access to foods that enable individuals to adhere to dietary guidelines.. *Preventing Chronic Disease, 3,* A76.

31. Yeh, M. C., Ickes, S., Lowenstein, L., Shuval, K., Ammerman, A., Farris, R., & Katz, D. (2008). Understanding barriers and facilitators of fruit and vegetable consumption among a diverse multi-ethnic population in the USA. *Health Promotion International, 23,* 42–51.

32. Maclellan, D. L., Gottschall-Pass, K., & Larsen, R. (2004). Fruit and vegetable consumption: Benefits and barriers. *Canadian Journal of Dietetic Practice and Research, 65,* 101–105.

33. Harnack, L., Story, M., & Rock, B. H. (1999). Diet and physical activity patterns of Lakota Indian adults. *Journal of the American Dietetic Association, 99,* 829–835.

34. Inglis, V., Ball, K., & Crawford, D. (2008). Socioeconomic variations in women's diets: What is the role of perceptions of the local food environment?. *Journal of Epidemiology and Community Health, 62,* 191–197.

35. Anderson, A. S., Cox, D. N., McKellar, S., Reynolds, J., Lean, M. E., & Mela, D. J. (1998). Take Five, a nutrition education intervention to increase fruit and vegetable intakes: Impact on attitudes towards dietary change. *The British Journal of Nutrition, 80,* 119–120.

36. Rose, D., & Richards, R. (2004). Food store access and household fruit and vegetable use among participants in the US Food Stamp Program. *Public Health Nutrition, 7,* 1081–1088.

37. Bodor, J. N., Rose, D., Farley, T. A., Swalm, C., & Scott, S. K. (2008). Neighbourhood fruit and vegetable availability and consumption: The role of small food stores in an urban environment. *Public Health Nutrition, 11,* 413–420.

38. Horowitz, C. R., Colson, K. A., Herbert, P. L., & Lancaster, K. (2004). Barriers to buying healthy foods

for people with diabetes: Evidence of environmental disparities. *American Journal of Public Health, 94,* 1549–1554.

39. Graham, R., Kaufman, L., Novoa, Z., & Karpata, A. (2006). *Eating in, eating out, eating well: Access to healthy food in North and Central Brooklyn.* New York, NY: New York City Department of Health and Mental Hygiene.

40. Eikenberry, N., & Smith, C. (2004). Healthful eating: Perceptions, motivations, barriers, and promoters in low-income Minnesota communities. *Journal of the American Dietetic Association, 104,* 1158–1161.

41. Cassady, D., Jetter, K. M., & Culp, J. (2007). Is price a barrier to eating more fruits and vegetables for low-income families?. *Journal of the American Dietetic Association, 107,* 1909–1915.

42. John, J. H., & Ziebland, S. (2004). Reported barriers to eating more fruit and vegetables before and after participation in a randomized controlled trial: A qualitative study. *Health Education Research, 19,* 165–174.

43. Paisley, J., & Skrzypczyk, S. (2005). Qualitative investigation of differences in benefits and challenges of eating fruits versus vegetables as perceived by Canadian women. *Journal of Nutrition Education and Behavior, 37,* 77–82.

44. Pham, K. L., Harrison, G. G., & Kagawa-Singer, M. (2007). Perceptions of diet and physical activity among California Hmong adults and youths.. *Preventing Chronic Disease, 4,* A93.

45. Dixon, H., Mullins, R., Wakefield, M., & Hill, D. (2004). Encouraging the consumption of fruit and vegetables by older Australians: An experiential study. *Journal of Nutrition Education and Behavior, 36,* 245–249.

46. Pollard, C., Miller, M., Woodman, R. J., Meng, R., & Binns, C. (2009). Changes in knowledge, beliefs, and behaviors related to fruit and vegetable consumption among Western Australian adults from 1995 to 2004. *American Journal of Public Health, 99,* 355–361.

47. Kearney, J. M., & McElhone, S. (1999). Perceived barriers in trying to eat healthier – results of a pan-EU consumer attitudinal survey. *The British Journal of Nutrition, 81,* S133–S137.

48. Krebs-Smith, S. M., Heimendinger, J., Patterson, B. H., Subar, A. F., Kessler, R., & Pivonka, E. (1995). Psychosocial factors associated with fruit and vegetable consumption. *American Journal of Health Promotion, 10,* 98–104.

49. Hendrix, S. J., Fischer, J. G., Reddy, R. D., Lommel, T. S., Speer, E. M., Stephens, H., Park, S., & Johnson, M. A. (2008). Fruit and vegetable intake and knowledge increased following a community-based intervention in older adults in Georgia senior centers. *Journal of Nutrition for the Elderly, 27,* 155–178.

50. Sahyoun, N. R., Zhang, X. L., & Serdula, M. K. (2005). Barriers to the consumption of fruits and vegetables among older adults. *Journal of Nutrition for the Elderly, 24,* 5–21.

51. Kubik, M. Y., Lytle, L., & Fulkerson, J. A. (2005). Fruits, vegetables, and football: Findings from focus groups with alternative high school students regarding eating and physical activity. *The Journal of Adolescent Health, 36,* 494–500.

52. Maddock, J. E., Silbanuz, A., & Reger-Nash, B. (2008). Formative research to develop a mass media campaign to increase physical activity and nutrition in a multi-ethnic state. *Journal of Health Communication, 13,* 208–215.

53. Jamelske, E., Bica, L. A., McCarty, D. J., & Meinen, A. (2008). Preliminary findings from an evaluation of the USDA Fresh Fruit and Vegetable Program in Wisconsin schools. *WMJ, 107,* 225–230.

54. Centers for Disease Control and Prevention. (2006). Evaluation of a fruit and vegetable distribution program – Mississippi, 2004–05 school year. *Morbidity and Mortality Weekly Report, 55,* 957–961.

55. United States Department of Agriculture, WIC. Available at: < http://www.fns.usda.gov/wic/benefitsandservices/foodpkg.HTM/> Accessed 10.01.09.

56. United States Department of Agriculture, WIC. Available at: < http://www.fns.usda.gov/wic/fmnp/FMNPfaqs.htm#1/> . Accessed 10.01.09.

57. Herman, D. R., Harrison, G. G., Afifi, A. A., & Jenks, E. (2008). Effect of a targeted subsidy on intake of fruits and vegetables among low-income women in the Special Supplemental Nutrition Program for Women, Infants, and Children. *American Journal of Public Health, 98,* 98–105.

58. Anderson, J. V., Bybee, D. I., Brown, R. M., McLean, D. F., Garcia, E. M., Breer, M. L., & Schillo, B. A. (2001). Five a day fruit and vegetable intervention improves consumption in a low income population. *Journal of the American Dietetic Association, 101,* 195–202.

59. Kunkel, M. E., Luccia, B., & Moore, A. C. (2003). Evaluation of the South Carolina seniors farmers' market nutrition education program. *Journal of Dietetic Association, 103,* 880–883.

60. Slusser, W. M., Cumberland, W. G., Browdy, B. L., Lange, L., & Neumann, C. (2007). A school salad bar increases frequency of fruit and vegetable consumption among children living in low-income households. *Public Health Nutrition, 10,* 1490–1496.

61. Lassen, A., Hansen, K., & Trolle, E. (2007). Comparison of buffet and à la carte serving at worksite canteens on nutrient intake and fruit and vegetable consumption. *Public Health Nutrition, 10,* 292–297.

62. Brown, L. B., Dresen, R. K., & Eggett, D. L. (2005). College students can benefit by participating in a prepaid meal plan. *Journal of the American Dietetic Association, 105,* 445–448.

63. McAleese, J. D., & Rankin, L. L. (2007). Garden-based nutrition education affects fruit and vegetable consumption in sixth-grade adolescents. *Journal of the American Dietetic Association, 107,* 662–665.

64. Nanney, M. S., Johnson, S., Elliott, M., & Haire-Joshu, D. (2007). Frequency of eating homegrown produce is associated with higher intake among parents and their preschool-aged children in rural Missouri. *American Dietetic Association, 107,* 577–584.

65. Yeh, M. C., & Katz, D. (2006). Food, nutrition and the health of urban populations. In N. Freudenberg, S. Galea, & D. Vlahov, (Eds.), *Cities and the Health of the Public* (pp. 106–125). Nashville: Vanderbilt University Press.

66. Moser, R. P., Green, V., Weber, D., & Doyle, C. (2005). Psychosocial correlates of fruit and vegetable consumption among African American men. *Journal of Nutrition Education and Behavior, 37,* 306–314.

67. Shaikh, A. R., Yaroch, A. L., Nebeling, L., Yeh, M. C., & Resnicow, K. (2008). Psychosocial predictors of fruit and vegetable consumption in adults a review of the literature. *American Journal of Preventive Medicine, 34,* 535–543.

68. Bandura, A. (1977). Self-efficacy: Toward a unifying theory of behavioral change. *Psychological Review, 81,* 191–215.

69. De Bourdeaudhuij, I., te Velde, S., Brug, J., Due, P., Wind, M., Sandvik, C., Maes, L., Wolf, A., Perez Rodrigo, C., Yngve, A., Thorsdottir, I., Rasmussen, M., Elmadfa, I., Franchini, B., & Klepp, K. I. (2008). Personal, social and environmental predictors of daily fruit and vegetable intake in 11-year-old children in nine European countries. *European Journal of Clinical Nutrition, 62,* 834–841.

70. Campbell, M. K., McLerran, D., Turner-McGrievy, G., Feng, Z., Havas, S., Sorensen, G., Buller, D., Beresford, S. A., & Nebeling, L. (2008). Mediation of adult fruit and vegetable consumption in the National 5 A Day for Better Health community studies. *Annals of Behavioral Medicine, 35,* 49–60.

71. Zehle, K., Smith, B. J., Chey, T., McLean, M., Bauman, A. E., & Cheung, N. W. (2008). Psychosocial factors related to diet among women with recent gestational diabetes: opportunities for intervention.. *The Diabetes Educator, 34,* 807–814.

72. Brown, B. J., & Hermann, J. R. (2005). Cooking classes increase fruit and vegetable intake and food safety behaviors in youth and adults. *Journal of Nutrition Education and Behavior, 37,* 104–105.

73. Wansink, B., & Lee, K. (2004). Cooking habits provide a key to 5 a day success. *Journal of the American Dietetic Association, 104,* 1648–1650.

74. Winett, R. A., Anderson, E. S., Wojcik, J. R., Winett, S. G., & Bowden, T. (2007). Guide to health: Nutrition and physical activity outcomes of a group-randomized trial of an Internet-based intervention in churches. *Annals of Behavioral Medicine, 33,* 251–261.

75. Devine, C. M., Farrell, T. J., & Hartman, R. (2005). Sisters in health: Experiential program emphasizing social interaction increases fruit and vegetable intake among low-income adults. *Journal of Nutrition Education and Behavior, 37,* 265–270.

76. Ahluwalia, J. S., Nollen, N., Kaur, H., James, A. S., Mayo, M. S., & Resnicow, K. (2007). Pathway to health: Cluster-randomized trial to increase fruit and vegetable consumption among smokers in public housing. *Health Psychologica, 26,* 214–221.

77. Welch, N., McNaughton, S. A., Hunter, W., Hume, C., & Crawford, D. (2008). Is the perception of time pressure a barrier to healthy eating and physical activity among women?. *Public Health Nutrition, 23,* 1–8.

78. Dibsdall, L. A., Lambert, N., Bobbin, R. F., & Frewer, L. J. (2003). Low-income consumers' attitudes and behaviour towards access, availability and motivation to eat fruit and vegetables. *Public Health Nutrition, 6,* 159–168.

79. von Wagner, C., Knight, K., Steptoe, A., & Wardle, J. (2007). Functional health literacy and health-promoting behaviour in a national sample of British adults. *Journal of Epidemiology and Community Health, 61,* 1086–1090.

80. Kearney, M., Bradbury, C., Ellahi, B., Hodgson, M., & Thurston, M. (2005). Mainstreaming prevention: Prescribing fruit and vegetables as a brief intervention in primary care. *Public Health, 119,* 981–986.

81. Carcaise-Edinboro, P., McClish, D., Kracen, A. C., Bowen, D., & Fries, E. (2008). Fruit and vegetable dietary behavior in response to a low-intensity dietary intervention: The rural physician cancer prevention project. *The Journal of Rural Health, 24,* 299–305.

82. Kreuter, M. W., Sugg-Skinner, C., Holt, C. L., Clark, E. M., Haire-Joshu, D., Fu, Q., Booker, A. C., Steger-May, K., & Bucholtz, D. (2005). Cultural tailoring for mammography and fruit and vegetable intake among low-income African-American women in urban public health centers. *Preventive Medicine, 41,* 53–62.

83. Neuhouser, M. L., Thompson, B., Coronado, G. D., & Solomon, C. C. (2004). Higher fat intake and lower fruit and vegetables intakes are associated with greater acculturation among Mexicans living in Washington State. *Journal of the American Dietetic Association, 104,* 51–57.

84. Montez, J. K., & Eschbach, K. (2008). Country of birth and language are uniquely associated with intakes of fat, fiber, and fruits and vegetables among Mexican-American women in the United States. *Journal of the American Dietetic Association, 108*, 473–480.

85. Dubowitz, T., Smith-Warner, S. A., Acevedo-Garcia, D., Subramanian, S. V., & Peterson, K. E. (2007). Nativity and duration of time in the United States: differences in fruit and vegetable intake among low-income postpartum women. *American Journal of Public Health, 97*, 1787–1790.

86. Algert, S. J., Agrawal, A., & Lewis, D. S. (2006). Disparities in access to fresh produce in low-income neighborhoods in Los Angeles. *American Journal of Preventive Medicine, 30*, 365–370.

87. Johnson, D. B., Beaudoin, S., Smith, L. T., Beresford, S. A., & LoGerfo, J. P. (2004). Increasing fruit and vegetable intake in homebound elders: The Seattle Senior Farmers' Market Nutrition Pilot Program. *Preventing Chronic Disease, 1*, A03.

88. Smith, L. T., Johnson, D. B., Beaudoin, S., Monsen, E. R., & LoGerfo, J. P. (2004). Qualitative assessment of participant utilization and satisfaction with the Seattle Senior Farmers' Market Nutrition Pilot Program. *Preventing Chronic Disease, 1*, A06.

89. Fano, T. J., Tyminski, S. M., & Flynn, M. A. (2004). Evaluation of a collective kitchens program: Using the Population Health Promotion Model. *Canadian Journal of Dietetic Practice and Research, 65*, 72–80.

90. Pomerleau, J., Lock, K., Knai, C., & McKee, M. (2005). Interventions designed to increase adult fruit and vegetable intake can be effective: A systematic review of the literature. *Journal of Nutrition, 135*, 2486–2495.

91. Franklin, P. D., Rosenbaum, P. F., Carey, M. P., & Roizen, M. F. (2006). Using sequential e-mail messages to promote health behaviors: Evidence of feasibility and reach in a worksite sample. *Journal of Medical Internet Research, 8*, e3.

92. Block, G., Block, T., Wakimoto, P., & Block, C. H. (2004). Demonstration of an E-mailed worksite nutrition intervention program. *Preventing Chronic Disease, 1*, A06.

93. de Vries, H., Kremers, S. P., Smeets, T., Brug, J., & Eijmael, K. (2008). The effectiveness of tailored feedback and action plans in an intervention addressing multiple health behaviors. *American Journal of Health Promotion, 22*, 417–425.

94. Latimer, A. E., Williams-Piehota, P., Katulak, N. A., Cox, A., Mowad, L., Higgins, E. T., & Salovey, P. (2008). Promoting fruit and vegetable intake through messages tailored to individual differences in regulatory focus. *Annals of Behavioral Medicine, 35*, 363–369.

95. Resnicow, K., Davis, R. E., Zhang, G., Konkel, J., Strecher, V. J., Shaikh, A. R., Tolsma, D., Calvi, J., Alexander, G., Anderson, J. P., & Wiese, C. (2008). Tailoring a fruit and vegetable intervention on novel motivational constructs: Results of a randomized study. *Annals of Behavioral Medicine, 35*, 159–169.

96. Heneman, K., Block-Joy, A., Zidenberg-Cherr, S., Donohue, S., Garcia, L., Martin, A., Metz, D., Smith, D., West, E., & Steinberg, F. M. (2005). A 'contract for change' increases produce consumption in low-income women: A pilot study. *Journal of the American Dietetic Association, 105*, 1793–1796.

97. Ammerman, A. S., Lindquist, C. H., Lohr, K. N., & Hersey, J. (2002). The efficacy of behavioral interventions to modify dietary fat and fruit and vegetable intake: A review of the evidence. *Preventive Medicine, 35*, 25–41.

98. Vanwormer, J. J., Boucher, J. L., & Pronk, N. P. (2006). Telephone-based counseling improves dietary fat, fruit, and vegetable consumption: A best-evidence synthesis. *Journal of the American Dietetic Association, 106*, 1434–1444.

99. Newman, V. A., Flatt, S. W., & Pierce, J. P. (2008). Telephone counseling promotes dietary change in healthy adults: Results of a pilot trial. *Journal of the American Dietetic Association, 108*, 1350–1354.

100. Pierce, J. P., Newman, V. A., Natarajan, L., Flatt, S. W., Al-Delaimy, W. K, Caan, B. J., Emond, J. A., Faerber, S., Gold, E. B., Hajek, R. A., Hollenbach, K., Jones, L. A., Karanja, N., Kealey, S., Madlensky, L., Marshall, J., Ritenbaugh, C., Rock, C. L., Stefanick, M. L., Thomson, C., Wasserman, L., & Parker, B. A. (2007). Telephone counseling helps maintain long-term adherence to a high-vegetable dietary pattern. *Journal of Nutrition, 137*, 2291–2296.

7

Healthy Eating: What Is the Role of the Economic Situation?

Tea Lallukka, Mikko Laaksonen, and Ossi Rahkonen

Department of Public Health, University of Helsinki, Finland

1. INTRODUCTION

This chapter focuses on the relationships between income, other indicators of economic situations, and healthy food habits. Numerous epidemiological studies have emphasized the importance of healthy food habits in the prevention of major chronic diseases such as cardiovascular diseases and several types of cancer. With the decline in the prevalence of smoking, food habits may have become the most important determinants of health and well-being in the developed countries [1].

Many studies have examined associations between education or occupation and consumption of fruit and vegetables [2,3]. These socioeconomic factors are also often taken into account in studies focused on the associations between eating and health. However, fewer studies have taken into account the economic situation, which may have an effect over and above the contribution of education and occupational class. This is of importance, because the economic situation may be a key factor explaining the adoption of healthy food habits for the prevention of diseases [4]. Failing to take into account the economic situation, intervention and health education programs aiming to enhance healthy food habits in terms of the promotion of fruit and vegetables consumption are likely to be ineffective. Moreover, if the economics partly underlie the epidemic of obesity as well, it is vital to consider policies that make the healthy choices the easiest choices by taking into account affordability and accessibility to healthy foods across all socioeconomic groups.

In this chapter, as well as in more general terms, healthy eating refers to the adoption of dietary guidelines to meet the recommended intake of nutrients for maintenance of health and well-being. Accordingly, food habits in epidemiological studies usually reflect main segments of dietary guidelines [5]. Typically, healthy eating habits are examined as consumption of fruit and vegetables, and consumption of fruit and vegetables can be seen as an indicator of overall healthiness of diet. These are also foods that are typically associated with a higher

99

socioeconomic position, but there are other more traditional food items of a recommended diet such as bread and potato which show decreasing consumption with an increasing socioeconomic position [6]. Thus, although the price of foods may be one factor influencing their consumption, there are also other aspects in these foods like their cultural appreciation that may affect their consumption in different population groups.

Epidemiological studies focused on the consumption of fruit and vegetables are reviewed in this chapter. In particular, this chapter is focused on several indicators of the economic situation in explaining healthy food habits and consumption of fruit and vegetables. The emphasis is on reviewing the current empirical evidence on the associations between the economic situation and healthy eating among the adult populations in different European regions, the USA, and Australia. The focus is on current reviews and comparative surveys. When applicable, results from good quality epidemiological studies from single countries will also be pointed out. First, household income, economic difficulties, and other material resources are conceptualized with the emphasis on health promotion, and second, their associations with healthy eating are reviewed. Additionally, food cost is discussed as a further external environmental barrier to healthy eating. Living area as well as access and availability are linked to various obstacles to improving food habits.

The questions to be asked when assessing the associations between healthy food habits and the role of the economic situation are as follows:

1. What is the role of income and other indicators of the economic situation with respect to healthy eating? More specifically, what are the key economic constraints to adoption of healthy food habits?
2. Do the effects of the economic situation on healthy eating remain when education and occupational social class are taken into account, i.e. does the economic situation show independent effects on healthy eating irrespective of socioeconomic position?
3. Is the cost of healthy foods higher than that of less healthy food items? Are there differences in prices and availability of healthy foods by country and region? Does the price of food affect the consumption of healthy foods more among low income people? What are the key socioeconomic reasons for different access to healthy eating?
4. With respect to implications and applicability of the current existing evidence, which are the key socioeconomic factors that need to be taken into account when aiming to promote healthy food habits?
5. What actions do health policies need to take to successfully increase consumption of fruit and vegetables among the most disadvantaged people?

2. ECONOMIC SITUATION AND SOCIOECONOMIC POSITION

Socioeconomic position is a broad, multidimensional concept that covers social, financial and material circumstances [7]. Conventional indicators of socioeconomic position comprise education, occupational class, and income. All of these indicators reflect both common ranking in society and particular circumstances according to the specific nature of the socioeconomic indicator in question [8–11]. Moreover, socioeconomic indicators cannot be used interchangeably [12], since a single indicator captures only a specific part of people's socioeconomic circumstances over the life course.

The association between the economic situation and food habits is compounded by other socioeconomic factors like educational qualifications and occupational position which are achieved earlier in one's life course. These could be more fundamental determinants of food habits than the economic situation as such, and offer different interpretations and policy recommendations for the observed associations. Examining the impact of the economic situation independently of education and occupation can help explain inequalities in healthy eating and therefore highlight specific needs for health promotion among the most disadvantaged groups.

Studying income only at one point of time may not, however, be sufficient to describe the full extent of differences in economic resources across the life course. 'Economic situation' captures the childhood economic situation and current income as well as wealth, i.e. long-term accumulation of economic resources. The whole spectrum of economic circumstances over the life course is expected to affect food habits. Accordingly, indicators of the socioeconomic position included in this review are assumed and interpreted as causal and successive.

Individual and household incomes are mainly based on paid employment. The level of income is likely to be dependent on education and occupational class. Income provides individuals and families with the necessary material resources and determines their purchasing power. Furthermore, income is the indicator of the socioeconomic position that can most easily change on a short-term basis [13].

For those with families, the household disposable income rather than individual income indicates their true financial resources. Thus, income reflects the availability of economic and material resources, and directly determines dietary quality by making healthy foods more affordable and readily accessible [14]. Individual income only partly reflects status and prestige. Concerning healthy eating, household income is a more proper measure. Economic difficulties in adulthood refer to everyday financial troubles such as those related to buying food or paying bills [15]. These difficulties reflect available material resources and their influences on food habits.

In addition to income and other indicators of individual material resources, the cost and subsequent affordability of healthy foods as well as area differences in access and availability are also of importance when assessing the complex relationships between an economic situation and adoption and maintenance of healthy eating habits.

3. INCOME AND HEALTHY EATING

Studies across the Western world have consistently shown that people from low-income households tend to consume less healthy foods, smaller quantities of fruit and vegetables, and larger quantities of meat, fat, and sugar [3,12,16–20]. People having low household incomes are less likely to buy foods that are high in fiber and low in fat, salt, and sugar [12]. In addition, people in low-income households purchase less regularly and a more limited range of fruit and vegetables than people from high-income groups. Similar associations have been found when examining the consumption of fruit and vegetables high in vitamin C and folate [3,12]. A nationally representative Finnish study showed that the intake of vitamin C and carotenoids also increases with increasing income level [6]. The intake of these vitamins is likely to reflect consumption of fruit and vegetables.

The UK 2002/03 Expenditure and Food Survey shows that the richest quintile of households consumes 20% more fruit as compared to that consumed in the poorest quintile of households [21]. Moreover, people in poverty or who live on state benefits or the minimum wage often lack money to buy enough or appropriate healthy food, especially if they have to meet other essential expenditures at the same time, such as high housing costs or paying debt. However, an important question on promoting healthy eating is whether there is a threshold under which unhealthy diet is more prevalent or whether unhealthy diet increases gradually with decreasing income. Figure 7.1 shows that consumption of fruit and vegetables gradually declines as net family income decreases. At the same time, the share of income that is spent on food in these families increases. Thus, low consumption of fruit and vegetables is not just due to poverty and absolute material hardship (see Fig. 7.1).

If one wishes to separate the effect of the economic situation from other aspects of the socioeconomic position, it is essential that several measures of the socioeconomic position are examined simultaneously. An Australian study found that the association between education and occupation with fruit and vegetables consumption substantially attenuated to nonsignificant or marginally significant when these socioeconomic indicators were adjusted for each other and income [12]. However, the association between household income and food purchasing remained strong and robust [12]. The same held true in a large Finnish study [19,23], showing that the consumption of vegetables remained strongly related to household income after adjusting for other socioeconomic indicators. The association between individual income and use of vegetables was rather similar, but among men the association disappeared after adjustments [19]. A study examining the association between household income and healthy eating among middle-aged public sector employees found that the association was clear, and among women remained even when other measures of the socioeconomic position (SEP) were controlled for [24]. All these results indicate that income, especially household income, has an independent effect on food habits that is not reducible to educational level and occupational social class.

4. ECONOMIC DIFFICULTIES AND HEALTHY EATING

Economic difficulties indicate that people – although sometimes at work – cannot afford to buy enough food for the needs of their families [15]. However, economic difficulties can exist

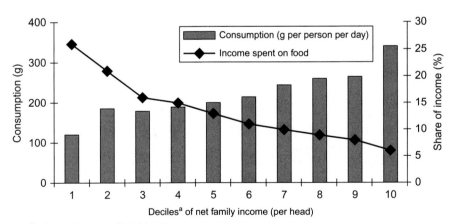

FIGURE 7.1 The relationship of income to consumption of fruit and vegetables and the share of income spent on food in the UK.
Reproduced with permission from DEFRA [22].

at all income levels [25] and thus affect healthy eating irrespective of current income. These difficulties may arise from excessive consumption behaviors, debt, or problems that are unrelated to food purchasing but may limit the opportunities to healthy eating. Dealing with economic difficulties also involves coping skills, since devaluation of monetary success may act as a buffer against feelings of deprivation [15]. Thus, with effective coping, the effects of limited resources may be less detrimental. When facing economic difficulties, different values are also likely to affect the choices where disposable income is used. Additionally, people with limited resources may not choose to use any extra money on healthy eating but rather on other areas of life.

A study among employed middle-aged women and men found that the more often people had economic difficulties, the less often they consumed vegetables, fruit, and other healthy foods (see Figure 7.2). This held true although all the participants were employed and had a constant salary. The association between current economic difficulties and healthy food remained although both SEP and

parental SEP were controlled for [24]. An Australian study used several related measures on food cost concerns (such as 'Sometimes my family cannot afford to buy enough food for our needs'). 'Food cost concern' was related to purchasing healthy food although household income was controlled for [26]. Thus, these results indicate that in all income groups – also in the highest groups – consumption of healthy food decreases with increasing economic difficulties. It is therefore not sufficient to take into account only income when examining the economic situation and healthy eating or assume that high income linearly contributes to healthy eating. Instead, economic difficulties can act as barriers to healthy eating irrespective of current income level.

5. AFFORDABILITY: QUESTIONS OF FOOD COST AND PRICING

In addition to differences in consumption of fruit and vegetables and other healthy foods by income, prices of such foods play a large role in the opportunities for healthy eating.

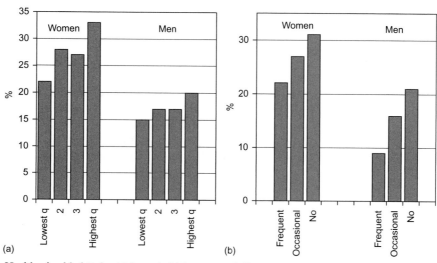

(a)   (b)

FIGURE 7.2   Healthy food habits by (a) household income and (b) economic difficulties among women and men [24].

Cost of healthy food is of importance, as it has risen much faster than that of less healthy and energy dense foods, at least in the USA and Australia, where information on trends is available [27,28]. Between 2004 and 2006 the cost of healthy foods increased by 19.5% in the Seattle metropolitan area [27], whereas the price change of the most energy dense foods was −1.8% during the same period. Similar disproportionate price changes were observed in Australia, where cost of healthier food items increased by 14% in the urban areas and 18% in the rural areas, while cost of unhealthy foods was less affected by the remoteness of area [28]. These changes are likely to widen inequalities in affordability and access to healthy food items between income groups and areas, in particular for fruit and vegetables.

International comparisons provide one source of information when examining the association between food costs and consumption. Additionally, trends in time within countries provide information on the changes in consumption behaviors and the economic situation. A study compared nine European countries taking into account both purchasing power (Gross Domestic Product per capita) and price levels concerning affordability of vegetables in 2001 [29]. The results showed that vegetables are cheaper in southern European countries (Italy and Spain), but also in France and Germany. Instead, prices of vegetables are higher in the Baltic countries as compared to other EU countries, making it more difficult for low-income people to afford frequent consumption. If the general availability of healthy foods at low price is good, this is likely to diminish the effect of the economic situation on food consumption.

Accordingly, cost and affordability of healthy foods may be more important for actual consumption habits than education, which is a marker of knowledge of healthy eating. Healthy eating costs more with fruit and vegetables as the key food items contributing to overall higher cost of a healthy diet [30]. Low-cost foods also provide more energy and thus serve as an easy, affordable option for meeting energy needs among people with limited economic resources [27,31–33]. In particular, among low-income people, the high cost of healthy foods is one of the barriers to increased fruit and vegetable consumption [34,35]. When consuming fruit and vegetables at the habitual level, price is not a barrier, but additional purchases may be perceived as unaffordable among those with low incomes [34]. In the USA, a healthy market basket has been estimated to cost 17–19% more than the standard market basket [35]. The difference was mostly due to higher prices of wholegrain products and lean meat. Additionally, availability of healthier foods was better in high-income neighborhoods, further limiting the opportunities of those with limited economic resources to follow healthier food habits.

Overall, economic constraints are suggested to explain why people with low food budgets tend to select less healthy foods that meet the energy needs with less money [36]. Nonetheless, food cost is likely to more strongly determine energy density of diet than the other way round, i.e. food choices of people with low income are more likely to be based on cost rather than on preference for energy dense foods [36]. This suggests that food choices are based on price more than appreciation of unhealthy foods in place of the recommended choices.

In addition to lower cost, less nutritious foods with high energy density are also more palatable and are also consumed for taste and pleasure, whereas the cheapest vegetables such as cabbage and carrots may be rejected [37]. However, those with low incomes are also found to be less motivated to eat fruit and vegetables and unaware of the discrepancy between their reported frequency of vegetable consumption and the recommended level [34]. Thus, high cost cannot fully explain the low prevalence of fruit and vegetable consumption in the low-income groups. Therefore, health policies that aim to increase affordability of fruit and vegetables among the low-income groups also need to focus on processes that affect actual food purchasing behaviors, i.e. find effective means to increase motivation towards the benefits of healthy eating and make the healthiest choices the easiest choices at all levels; including equal access, affordable prices, and improved availability. An intervention study from the USA showed that among a low-income population, combining education to affect attitudes and perceptions concerning access and cost of healthy foods, and providing food coupons and cash incentives to purchase fruit and vegetables was more effective in increasing consumption than either of the interventions applied separately [38]. Thus, it is necessary to simultaneously increase affordability of fruit and vegetables and affect actual food purchasing to achieve an improvement in fruit and vegetable consumption, at least among the most disadvantaged. In addition to direct costs, perception of prices, in particular concerning those of healthy foods, may also affect dietary behaviors independent of socioeconomic position or the perception of prices may mediate the effects of socioeconomic status on healthy eating [26,39,40].

6. AVAILABILITY AND ACCESS TO HEALTHY FOODS

Even if socioeconomic differences in dietary habits are well documented, the underlying causes for these differences are not well understood. One set of possible mechanisms for the association between the economic situation and food habits relates to differences in neighborhood and local environment, including availability and access to healthy foods. While wealthy people usually live in different areas than the poor, living areas as such can also affect food choices. Studies have examined whether healthy foods are more expensive in low-income areas

and whether availability of healthy foods is lower in low-income areas [16].

In countries with easy access to vegetables such as southern European countries, consumption of vegetables tends to be higher in the lower socioeconomic groups as compared to northern Europe with opposite trends [29,41]. Overall, differences in the level of fruit and vegetable consumption between northern and southern European countries have been narrowing, although disparities remain between countries in the consumption of pulses and olive oil, which represent typical characteristics of southern European healthy food habits [42]. Other studies have found that availability of healthy food is poorer and that it is more expensive in rural than in urban areas [43]. A study examining vegetable use in nine European countries found urban respondents in northern Europe to be more frequent daily vegetable users, but in central and southern Europe vegetable consumption did not vary by the place of residence [29]. A possible explanation for this finding is that the better availability of fruit and vegetables narrows differences by urbanization.

Studies examining small-area variation across urban neighborhoods have also found differences in fruit and vegetable consumption. Those living in low-income neighborhoods have lower consumption of fruit and vegetables than those in wealthier neighborhoods [43,44]. A multilevel study in four US communities showed that people living in the least wealthy areas had the lowest fruit and vegetable intake, irrespective of their race, personal income, or education [45].

Access to foods may have an important influence on differences in dietary habits by neighborhood socioeconomic status and racial composition [46]. The type and number of food stores have been shown to vary according to the racial and income level of neighborhoods, with supermarkets generally more common in white and wealthier areas than in minority and poorer neighborhoods. Supermarkets typically have a wider variety of high-quality food products and lower prices than smaller neighborhood food stores. Greater shelf-space of healthier foods in local food stores has been associated with higher consumption of those foods. Shopping at supermarkets has been associated with increased fruit and vegetable consumption [47].

Differences in supermarket access by socioeconomic status of living area have been documented especially in the USA. A nationwide study reported in 2007 found significant differences by neighborhood income, racial and ethnic characteristics in the availability of food stores between urban areas [48]. Similarly, other studies have also found fewer available supermarkets in low-income neighborhoods. The Atherosclerosis Risk in Communities (ARIC) Study found that fruit and vegetable intake increased with each additional supermarket in a census tract [49]. Also, among participants in the USA Food Stamp Program supermarket access was associated with fruit consumption [50].

However, studies outside the USA have shown less consistent findings on food availability and access to supermarkets between deprived and affluent areas. While some studies have found that there are fewer healthy choices available in stores in deprived areas compared with wealthier areas, other studies have failed to find such differences [51]. Some studies have even found differences favoring more-deprived rather than less-deprived neighborhoods [52,53]. As an explanation for these differences it has been suggested that area, race, and income segregation in the USA may be larger than elsewhere. In addition, if the availability of healthy foods in general is good, the area of living is less likely to make much difference in the use of fruit and vegetables.

Most studies examining how neighborhood food environment affects the use of fruit and vegetables have used access to supermarkets as the measure of their availability. However, access to supermarkets is a crude measure for access to fruit and vegetables with potential sources of bias. Some people may choose to do their shopping elsewhere than where they live. If access to food stores in the living area is poor, it is more likely that people will do their shopping elsewhere. It is therefore plausible that the local food environment has the greatest effect among those who do not own a car. A poor economic situation may force people to use their local food environments, as low-income families are unlikely to have a car. Only a few studies have used more refined measures than food store access to examine food availability directly across different types of neighborhoods [54].

While the availability of healthy foods in the neighborhood may affect their use, it is also possible that the consumption of these products affects their supply. Over time, low demand for fruit and vegetables is likely to decrease their availability in local shops. If the demand is low in low-income neighborhoods, this may create a genuine area effect where decreased availability also affects the use of those with higher income. Studies that have examined how increasing availability affects fruit and vegetable consumption have produced mixed findings. In one study, a small increase in fruit and vegetable consumption was found [55]. Another study found little evidence for increased fruit and vegetable consumption in an area where a large supermarket was established as compared to the control area [56].

7. GENERAL DISCUSSION

There is much evidence that income level, economic difficulties, and affordability and availability of fruit and vegetables contribute to healthy eating. It needs to be noted, however, that one's economic situation only partly captures the complex reasons for healthy eating or following a certain diet. For example, people on a low income are faced with greater structural, material, and economic difficulties when purchasing healthy food [12]. A variety of differences between countries, food cultures, and traditions, tastes, and pleasure, alongside personal variations also modify eating behaviors. Furthermore, differences in fruit and vegetable consumption by socioeconomic indicators may depend on area level factors such as pricing and availability of healthy foods [29]. Moreover, one study reported that an increase in income is unlikely to encourage poor people to purchase more fruit and vegetables [57]. Rather, the additional income is spent on other goods. The study further suggested that taste and preference, time constraints, and other factors are the primary factors influencing low consumption of fruits and vegetables among low-income families [57].

Inequalities in healthy eating are, nevertheless, clear, and following an unhealthy diet directly affects the health of the poor and their chances for a healthy life. Thus, concrete messages on the opportunities to follow healthy food habits at low or moderate cost are needed to enhance healthier food habits among low-income people. Moreover, in contrast to common perceptions, it is also possible to adopt recommendations for healthy eating among low-income groups [26].

The effects of income and economic situation on food choices are of crucial interest for policy makers. Low income and other economic constraints are independently associated with low use of fruit and vegetables, which is a strong incentive to reduce taxation of healthy foods in order to encourage increased consumption by increasing affordability for all. Alongside cost, it is also important to pay attention to the availability of healthy, recommended food items in countries where availability of fruit and vegetables varies by neighborhood, season, or other reasons. Similar conclusions were pointed out in a report by the WHO on food and health in Europe [58] stating that 'People's chances for a healthy diet depend less on individual choices than on what food is available and whether it is affordable.' Thus, in order to succeed in improving access to healthy foods in all population subgroups, action at all policy levels is needed as well as a wider understanding of the economic and other constraints to healthy eating.

7.1 Health Applications and Policy Implications

To enhance healthy eating, promote health and well-being, and prevent diet-related chronic diseases, action at several policy levels is needed. Lower taxation of healthy foods in order to increase affordability of fruit and vegetables may increase consumption in countries where prices are high and the availability of fruit and vegetables is limited. However, motivation, health education, and positive perceptions on healthy eating are also needed to modify actual food purchasing behaviors. All in all, taking into account the economic situation, the healthiest choices need to be the easiest choices in terms of price, availability, and pleasure to be adopted and followed in a regular, everyday diet as much as possible among all population subgroups.

8. SUMMARY

The economic situation has a key role in healthy eating. The motive to focus on an economic situation and healthy eating is to find effective means for healthy promotion and disease prevention. If an economic situation acts as a barrier for healthy eating, health education will be ineffective.

Moreover, income and economic difficulties affect healthy eating over and above the effects of education and occupation. Food costs also play a large role. Thus, improving affordability promotes healthy eating in all population subgroups. Alongside cost, availability is also of importance. In countries with easy access to fruit and vegetables, the level of the household income has a small effect on consumption. When prices are low, and availability is high, people with limited economic means have a better opportunity to include more fruit and vegetables in their diets. To increase the use of fruit and vegetables the healthiest choices need to become the easiest choices across all population subgroups.

References

1. Shahar, D., Shai, I., Vardi, H., Shahar, A., & Fraser, D. (2005). Diet and eating habits in high and low socioeconomic groups. *Nutrition, 21,* 559–566.

2. Roos, E., Talala, K., Laaksonen, M., Helakorpi, S., Rahkonen, O., Uutela, A., & Prättälä, R. (2008). Trends of socioeconomic differences in daily vegetable consumption, 1979–2002. *European Journal of Clinical Nutrition, 62,* 823–833.

3. Giskes, K., Turrell, G., Patterson, C., & Newman, B. (2002). Socioeconomic differences among Australian adults in consumption of fruit and vegetables and intakes of vitamins A, C and folate. *Journal of Human Nutrition and Dietetics, 15,* 375–385 discussion 387–390.

4. Darmon, N., & Drewnowski, A. (2008). Does social class predict diet quality? *American Journal of Clinical Nutrition, 87,* 1107–1117.

5. Roos, E., Lahelma, E., Virtanen, M., Prättälä, R., & Pietinen, P. (1998). Gender, socioeconomic status and family status as determinants of food behaviour. *Social Science & Medicine, 46,* 1519–1529.

6. Roos, E., Prättälä, R., Lahelma, E., Kleemola, P., & Pietinen, P. (1996). Modern and healthy?: Socioeconomic differences in the equality of diet. *European Journal of Clinical Nutrition, 50,* 753–760.

7. Bartley, M., Sacker, A., Firth, D., & Fitzpatrick, R. (1999). Understanding social variation in cardiovascular risk factors in women and men: The advantage of theoretically based measures. *Social Science & Medicine, 49,* 831–845.

8. Laaksonen, M., Rahkonen, O., Martikainen, P., & Lahelma, E. (2005). Socioeconomic position and self-rated health: The contribution of childhood socioeconomic circumstances, adult socioeconomic status, and material resources. *American Journal of Public Health, 95,* 1403–1409.

9. Lahelma, E., Martikainen, P., Laaksonen, M., & Aittomäki, A. (2004). Pathways between socioeconomic determinants of health. *Journal of Epidemiology and Community Health, 58,* 327–332.

10. Liberatos, P., Link, B., & Kelsey, J. (1988). The measurement of social class in epidemiology. *Epidemiologic Reviews, 10,* 87–121.

11. Lynch, J., & Kaplan, G. (2000). Socioeconomic position. In L. F. Berkman, & I. Kawachi, (Eds.), *Social epidemiology* (pp. 13–35). New York: Oxford University Press.

12. Turrell, G., Hewitt, B., Patterson, C., & Oldenburg, B. (2003). Measuring socio-economic position in dietary research: Is choice of socio-economic indicator important? *Public Health Nutrition, 6,* 191–200.

13. Shaw, M., Galobardes, B., Lawlor, D. A., Lynch, J., Wheeler, B., & Davey Smith, G. (2007). *The handbook of inequality and socioeconomic position, concepts and measures.* Bristol: Policy Press.

14. Sooman, A., Macintyre, S., & Anderson, A. (1993). Scotland's health – A more difficult challenge for some? The price and availability of healthy foods in socially contrasting localities in the west of Scotland. *Health Bulletin, 51,* 276–284.

15. Pearlin, L. I., & Schooler, C. (1978). The structure of coping. *Journal of Health and Social Behavior, 19,* 2–21.

16. Kamphuis, C. B., Giskes, K., de Bruijn, G. J., Wendel-Vos, W., Brug, J., & van Lenthe, F. J. (2006). Environmental determinants of fruit and vegetable consumption among adults: A systematic review. *The British Journal of Nutrition, 96,* 620–635.

17. Worsley, A., Blasche, R., Ball, K., & Crawford, D. (2003). Income differences in food consumption in the 1995 Australian National Nutrition Survey. *European Journal of Clinical Nutrition, 57,* 1198–1211.

18. Wandel, M. (1995). Dietary intake of fruits and vegetables in Norway: Influence of life phase and socio-economic factors. *International Journal of Food Sciences and Nutrition, 46,* 291–301.

19. Laaksonen, M., Prättälä, R., Helasoja, V., Uutela, A., & Lahelma, E. (2003). Income and health behaviours. Evidence from monitoring surveys among Finnish adults. *Journal of Epidemiology and Community Health, 57,* 711–717.

20. Kirkpatrick, S., & Tarasuk, V. (2003). The relationship between low income and household food expenditure patterns in Canada. *Public Health Nutrition, 6,* 589–597.

21. Office for National Statistics, Department for Environment, Food and Rural Affairs. (2004). *Expenditure and food survey, 2002–2003.* London: ONS.

22. DEFRA. (2007). National Food Survey 2000. Department for Food, Environment and Rural Affairs. London: The Stationery Office.

23. Ebrahim, S. (2005). Socioeconomic position (again), causes and confounding. *International Journal of Epidemiology, 34,* 237–238.

24. Lallukka, T., Laaksonen, M., Rahkonen, O., Roos, E., & Lahelma, E. (2007). Multiple socio-economic circumstances and healthy food habits. *European Journal of Clinical Nutrition, 61*, 701–710.

25. Laaksonen, E., Martikainen, P., Lahelma, E., Lallukka, T., Rahkonen, O., Head, J., & Marmot, M. (2007). Socioeconomic circumstances and common mental disorders among Finnish and British public sector employees: Evidence from the Helsinki Health Study and the Whitehall II Study. *International Journal of Epidemiology, 36*, 776–786.

26. Turrell, G., & Kavanagh, A. M. (2006). Socio-economic pathways to diet: Modelling the association between socio-economic position and food purchasing behaviour. *Public Health Nutrition, 9*, 375–383.

27. Monsivais, P., & Drewnowski, A. (2007). The rising cost of low-energy-density foods. *Journal of the American Dietetic Association, 107*, 2071–2076.

28. Harrison, M. S., Coyne, T., Lee, A. J., Leonard, D., Lowson, S., Groos, A., & Ashton, B. A. (2007). The increasing cost of the basic foods required to promote health in Queensland. *The Medical Journal of Australia, 186*, 9–14.

29. Prättälä, R., Hakala, S., Roskam, A. R., Roos, E., Helmert, U., Klumbiene, J., Van Oyen, H., Regidor, E., Kunst, A. E. (2009). Association between educational level and vegetable use in nine European countries. *Public Health Nutrition, 30*, 1–9.

30. Cade, J., Upmeier, H., Calvert, C., & Greenwood, D. (1999). Costs of a healthy diet: Analysis from the UK Women's Cohort Study. *Public Health Nutrition, 2*, 505–512.

31. Andrieu, E., Darmon, N., & Drewnowski, A. (2006). Low-cost diets: More energy, fewer nutrients. *European Journal of Clinical Nutrition, 60*, 434–436.

32. Drewnowski, A., & Darmon, N. (2005). Food choices and diet costs: An economic analysis. *The Journal of Nutrition, 135*, 900–904.

33. Drewnowski, A., Darmon, N., & Briend, A. (2004). Replacing fats and sweets with vegetables and fruits – A question of cost. *American Journal of Public Health, 94*, 1555–1559.

34. Dibsdall, L. A., Lambert, N., Bobbin, R. F., & Frewer, L. J. (2003). Low-income consumers' attitudes and behaviour towards access, availability and motivation to eat fruit and vegetables. *Public Health Nutrition, 6*, 159–168.

35. Jetter, K. M., & Cassady, D. L. (2006). The availability and cost of healthier food alternatives. *American Journal of Preventive Medicine, 30*, 38–44.

36. Darmon, N., Ferguson, E., & Briend, A. (2003). Do economic constraints encourage the selection of energy dense diets? *Appetite, 41*, 315–322.

37. Drewnowski, A., & Darmon, N. (2005). The economics of obesity: Dietary energy density and energy cost. *The American Journal of Clinical Nutrition, 82*(Suppl. 1), 265S–273S.

38. Anderson, J. V., Bybee, D. I., Brown, R. M., McLean, D. F., Garcia, E. M., Breer, M. L., & Schillo, B. A. (2001). Five a day fruit and vegetable intervention improves consumption in a low income population. *Journal of the American Dietetic Association., 101*, 195–202.

39. Beydoun, M. A., & Wang, Y. (2008). How do socio-economic status, perceived economic barriers and nutritional benefits affect quality of dietary intake among US adults? *European Journal of Clinical Nutrition, 62*, 303–313.

40. Giskes, K., Van Lenthe, F. J., Brug, J., Mackenbach, J. P., & Turrell, G. (2007). Socioeconomic inequalities in food purchasing: The contribution of respondent-perceived and actual (objectively measured) price and availability of foods. *Preventive Medicine, 45*, 41–48.

41. Roos, G., Johansson, L., Kasmel, A., Klumbiene, J., & Prättälä, R. (2001). Disparities in vegetable and fruit consumption: European cases from the north to the south. *Public Health Nutrition, 4*, 35–43.

42. Naska, A., Fouskakis, D., Oikonomou, E., Almeida, M. D., Berg, M. A., Gedrich, K., Moreiras, O., Nelson, M., Trygg, K., Turrini, A., Remaut, A. M., Volatier, J. L., & Trichopoulou, A. DAFNE participants. (2006). Dietary patterns and their socio-demographic determinants in 10 European countries: Data from the DAFNE databank. *European Journal of Clinical Nutrition, 60*, 181–190.

43. Larson, N. I., Story, M. T., & Nelson, M. C. (2009). Neighborhood environments: Disparities in access to healthy foods in the U.S. *American Journal of Preventive Medicine, 36*, 74–81.

44. Turrell, G., Blakely, T., Patterson, C., & Oldenburg, B. (2004). A multilevel analysis of socioeconomic (small area) differences in household food purchasing behaviour. *Journal of Epidemiology and Community Health, 58*, 208–215.

45. Diez-Roux, A. V., Nieto, F. J., Caulfield, L., Tyroler, H. A., Watson, R. L., & Szklo, M. (1999). Neighbourhood differences in diet: The Atherosclerosis Risk in Communities (ARIC) Study. *Journal of Epidemiology and Community Health, 53*, 55–63.

46. Pearce, J., Hiscock, R., Blakely, T., & Witten, K. (2008). The contextual effects of neighbourhood access to supermarkets and convenience stores on individual fruit and vegetable consumption. *Journal of Epidemiology and Community Health, 62*, 198–201.

47. Zenk, S. N., Schulz, A. J., Hollis-Neely, T., Campbell, R. T., Holmes, N., Watkins, G., Nwankwo, R., &

Odoms-Young, A. (2005). Fruit and vegetable intake in African Americans income and store characteristics. *American Journal of Preventive Medicine, 29,* 1–9.

48. Powell, L. M., Slater, S., Mirtcheva, D., Bao, Y., & Chaloupka, F. J. (2007). Food store availability and neighborhood characteristics in the United States. *Preventive Medicine, 44,* 189–195.

49. Morland, K., Wing, S., & Diez Roux, A. (2002). The contextual effect of the local food environment on residents' diets: The atherosclerosis risk in communities (ARIC) study. *American Journal of Public Health, 92,* 1761–1767.

50. Rose, D., & Richards, R. (2004). Food store access and household fruit and vegetable use among participants in the US Food Stamp Program. *Public Health Nutrition, 7,* 1081–1088.

51. Winkler, E., Turrell, G., & Patterson, C. (2006). Does living in a disadvantaged area entail limited opportunities to purchase fresh fruit and vegetables in terms of price, availability, and variety? Findings from the Brisbane Food Study. *Health Place, 12,* 741–748.

52. Cummins, S., & Macintyre, S. (1999). The location of food stores in urban areas: A case study in Glasgow. *The British Food Journal, 101,* 545–553.

53. van Lenthe, F. J., Brug, J., & Mackenbach, J. P. (2005). Neighbourhood inequalities in physical inactivity: The role of neighbourhood attractiveness, proximity to local facilities and safety in the Netherlands. *Social Science & Medicine, 60,* 763–775.

54. Bodor, J. N., Rose, D., Farley, T. A., Swalm, C., & Scott, S. K. (2008). Neighbourhood fruit and vegetable availability and consumption: The role of small food stores in an urban environment. *Public Health Nutrition, 11,* 413–420.

55. Wrigley, N., Warm, D., & Margetts, B. (2003). Deprivation, diet, and food-retail access: Findings from the Leeds 'food deserts' study. *Environment and Planning A, 35,* 151–188.

56. Cummins, S., Petticrew, M., Higgins, C., Findlay, A., & Sparks, L. (2005). Large scale food retailing as an intervention for diet and health: Quasi-experimental evaluation of a natural experiment. *Journal of Epidemiology and Community Health, 59,* 1035–1040.

57. Stewart, H., Blisard, N., & Jolliffe, D. (2003). Do income constraints inhibit expenditures on fruits and vegetables among low-income households? *Journal of Agricultural and Resource Economics, 28,* 465–480.

58. WHO Regional Publications, European series. (2004). *Food and health in Europe: A new basis for action.* Europe: WHO 96: 1–388.

8

Trends in US Adult Fruit and Vegetable Consumption

Sarah Stark Casagrande[1] and Tiffany L. Gary-Webb[1,2]

[1]Department of Epidemiology, Johns Hopkins Bloomberg School
of Public Health, Baltimore, Maryland, USA
[2]Welch Center for Prevention, Epidemiology, and Clinical Medicine,
Medical Institutions, Baltimore, Maryland, USA

1. EVIDENCE SUMMARY

1.1 Fruit and Vegetable Consumption

Consuming a diet high in fruits and vegetables is associated with a decreased risk of certain chronic diseases including cardiovascular disease [1–4], cancer [1,5], and diabetes [6–9]. Previous literature indicates that US adult fruit and vegetable consumption is below recommendations [10–16]. Beginning in 1985, the United States Department of Agriculture (USDA) *Dietary Guidelines for Healthy Americans* (*Guidelines* hereafter) recommended consuming at least two servings of fruit and three servings of vegetables daily [17,18]. In 1991, the 5-A-Day Program for Better Health was initiated by the National Cancer Institute and the Produce for Better Health Foundation to increase public awareness of the importance of eating at least five fruits and vegetables daily [19]. This was done through advertising campaigns, education, and school and workplace interventions. In 2005, the USDA *Guidelines* were revised to emphasize fruit and vegetable intakes based on individual energy requirements and recommended 5–9 servings per day. Lastly, in 2007, the Produce for Better Health Foundation launched the Fruits and Veggies – More Matters campaign to reflect the 2005 *Guideline* revisions.

Previous research indicates that only a small proportion of Americans meet the USDA *Guidelines* for daily servings of fruits and vegetables [13,20,21]. An evidence table summarizing fruit and vegetable consumption studies is referenced in Table 8.1. Twenty-four hour dietary recall data from the Second National Health and Nutrition Examination Survey (NHANES) (1976–1980) estimated that only 27% of adults consumed the three or more servings of vegetables and 29% of adults

TABLE 8.1   Evidence Summary for Fruit and Vegetable (F&V) Consumption

Reference	Study	Population	Measures	Results
Patterson *et al.* [13]	NHANES II, 1976–1980	11,648 adults	24-hour dietary recall administered by a trained interviewer	45% no fruit servings 22% no vegetable servings 27% consumed 3 + vegetables 29% consumed 2 + fruits 9% consumed both Consumption low
Swanson *et al.* [15]	Population survey in three areas of US	881 African Americans and 1095 white adults	Food frequency data	African Americans were more frequent consumers of F&V than whites
Patterson *et al.* [14]	1987 National Health Interview Survey (NHIS)	20,143 adults	Food Frequency Questionnaire	Females consumed more F&V, fewer meats and high-fat foods Whites consumed a more varied diet than African Americans or Hispanics Hispanics consumed the lowest amounts of high-fat foods; African Americans the most
Basiotis [16]	NHANES 1999–2002	8070 age > 2 yrs	Healthy Eating Index	28% consumed ≥3 servings vegetables 17% consumed ≥2 servings fruit Few met F&V recommendations
Gary *et al.* [12]	Project DIRECT	2172 African American adults	Modified Block Questionnaire	8% met fruit recommendations (≥2) 16% met vegetable recommendations (≥3) Intake increased with higher income and more education
Casagrande *et al.* [11]	NHANES III, 1988–1994 and NHANES 1999–2002	NHANES III: 14,997 adults NHANES 1999–2002: 8910 adults	24-hour dietary recall	11% met USDA Guidelines for F&V in both study periods Non-Hispanic blacks less likely to meet Guidelines compared to non-Hispanic whites

consumed the two or more servings of fruit recommended by USDA and DHHS [13]. Only 9% of adults met both fruit and vegetable recommendations.

Demographic patterns for fruit and vegetable consumption have been varied. Racial differences in consumption patterns are inconsistent depending on the region, sample size, and the method of measurement. One study demonstrated that whites tend to eat a more varied diet than African Americans and Mexican Americans, but that Mexican Americans consume the lowest amounts of high-fat foods while African Americans eat the most [14]. A second study reported that African Americans have low consumption of fruits and vegetables with only 8% meeting the recommendations for fruit and 16% meeting the recommendations for vegetable intake [12]. A third study reported that African Americans consumed more fruits and vegetables (citrus, cruciferous, and vegetables rich in vitamin A and C) and slightly less total and saturated fat than whites [15]. Finally, a fourth study among African Americans reported that fruit intake was higher among participants with higher income and more education [12]. Females have been shown to consume a more diverse diet than males and consume more fruits and vegetables, less meat, and fewer high-fat foods [14].

1.2 Fruit and Vegetable Trends

Table 8.2 summarizes evidence of trends in fruit and vegetable consumption. Using the National Health Interview Survey (NHIS) between 1987 and 1992, the proportion of fruit and vegetable consumption and the mean number of servings per week remained stable over time [10]. Similarly, results from the Behavioral Risk Factor Surveillance System (BRFSS) examining F&V consumption at intervals between 1994 and 2000 found that the proportion of individuals consuming fruits and vegetables five or more times per day remained constant at 25% [21]. The BRFSS survey conducted from 1990 to 1996 indicated 19%, 22%, and 23% of adults in 1990, 1994, and 1996, respectively, consumed at least five fruits and vegetables per day which indicated slightly higher consumption over time [20]. These results indicate a slight increase in consumption over relatively the same time period as the 1994–2000 BRFFS survey.

1.3 Trends by Race/Ethnicity

A cross-sectional study identified quality of diet trends between 1965 and 1991 using data from the Nationwide Food Consumption Surveys and the Continuing Survey of Food Intake by Individuals using the Diet Quality Index [22]. In 1965 whites of high socioeconomic status were eating the least healthful diet and African Americans of low socioeconomic status were eating the most healthful. By 1991, the diets among all groups had improved and were relatively similar.

Cross-sectional data from NHANES I (1971–1975), II (1976–1980), III (1988–1994), and NHANES 2001–2002 indicated that dietary trends were similar between blacks and whites across the study periods [23]. There was a significant trend toward lower fruit consumption among whites and a significant trend toward higher vegetable consumption among black females. Black males and females reported lower intakes of vegetables compared to their white counterparts.

2. DIETARY GUIDELINES

2.1 Background

Since 1980, the *Guidelines* have been jointly published by the Department of Health and Human Services (DHHS) and the United States Department of Agriculture (USDA). The *Guidelines* are reviewed and revised as needed every five years. The purpose of this publication is to provide scientific evidence for beneficial nutrients to promote health and reduce disease risk through nutrition and healthy lifestyles.

The *Guidelines* are developed through a three-stage process involving advisory committees from both DHHS and USDA. An external scientific Advisory Committee is appointed by the two Departments. The Advisory Committee is responsible for compiling new scientific information relating to nutrition, disease, and physical activity and preparing a summarized report of the findings. The report is available for public and Government agencies for comment and then the two Departments join to develop Key

TABLE 8.2 Evidence Summary for Trends in Fruit and Vegetable Consumption

Reference	Study	Population	Measures	Results
Popkin et al. [22]	1965, 1977–78 Nationwide Food Consumption Survey (NFCS) and 1989–1991 Continuing Survey of Food Intake (CSFI)	1965: 6061 1977–78: 16,425 1989–1991: 9920 adults	Diet Quality Index (DQI)	Overall dietary quality improved Racial differences decreased over time
Breslow et al. [50]	1987 and 1992 National Health Interview Survey (NHIS) Cancer Control Supplements	10,000 adult respondents	57-item Food Frequency Questionnaire	Proportion of F&V consumers remained stable over time
Li et al. [20]	1990–1996 Behavioral Risk Factor Surveillance System (BRFFS)	1990: 25,499 1994: 32,076 1996: 37,581 adults	6-item Food Frequency Questionnaire	F&V consumption ≥5 was 19%, 22%, 23% in 1990, 1994, 1996, respectively Slight increase in consumption
Serdula et al. [21]	1994–2000 Behavioral Risk Factor Surveillance System (BRFSS)	1994: 87,582 1996: 96,511 1998: 114,129 2000: 135,899 adults	6-item Food Frequency Questionnaire	F&V consumption ≥5 was 25% for each year No trend in consumption
Kant et al. [23]	NHANES I (1971–1975), II (1976–1980), III (1988–1994), NHANES 2001–2002	Non-Hispanic black ($N = 7099$) and white ($N = 23,314$) adults	24-hour dietary recall administered by a trained interviewer	Black men and women reported lower intakes of vegetables compared to white counterparts Dietary trends were similar between blacks and whites Significant decreased trend in fruit reporting among whites Significant increased trend in vegetable reporting among black women

Recommendations based on the Advisory Committee's initial report and the comments made by other agencies. The final *Guidelines* contain the details of the science-based Key Recommendations and, finally, the two Departments develop messages to communicate the *Guidelines* to the public.

The *Guidelines* are the basis for national nutrition education programs, intervention programs, and federal food programs. Federal dietary guidance publications are, by law, required to follow the *Guidelines*. Examples include the USDA Food Guide, the food label, and the Nutrition Facts Panel. Policies for nutrition-related programs such as the National Child Nutrition Programs or the Women, Infants and Children Program are based on the *Guidelines*.

2.2 History of Dietary Guidelines

The *Guidelines* were first implemented in 1980, and, over the past 25 years, the content, wording, and terminology have significantly evolved [18]. The first two editions, in 1980 and 1985, stated 'Eat food with adequate starch and fiber' to include sources from wholegrain breads and cereals, fruit, vegetables, beans, peas, and nuts. To help consumers implement the *Guidelines* in daily food choices, the 1985 edition outlined suggested servings from each of the major food groups. Specifically, the USDA suggested a minimum of at least two daily servings of fruits and at least three daily servings of vegetables. In 1990 the Advisory Committee revised the statement to 'Choose a diet with plenty of fruits, vegetables, and grain products' which explicitly targeted the food groups to choose from in an effort to make the *Guidelines* clearer to Americans.

2.3 The Food Guide Pyramid

The development of the USDA Food Guide Pyramid spans over six decades. The first National Nutrition Conference, prompted by President Franklin D. Roosevelt, was held in 1941. As a result of this conference, the USDA developed Recommended Dietary Allowances (RDAs) and specified caloric intakes and essential nutrients. In 1943, the USDA announced the 'Basic Seven' which was a special modification of the nutritional guidelines to alleviate the shortage of food supplies during the Second World War. The seven categories included milk, vegetables, fruit, eggs, all meat, cheese, fish, and poultry, cereal and bread, and butter. To simplify, the 'Basic Four' was introduced in 1956 and continued until 1979; categories included milk, vegetable and fruit, meat, and grain. With the rise of chronic diseases, the USDA addressed the roles of unhealthy foods and added a fifth group in the late 1970s: fats, sweets, and alcoholic beverages to be consumed in moderation. Although the USDA's food guide, *A Pattern for Daily Food Choices,* was published annually beginning in the 1980s, it was not well known. Beginning in 1988, the USDA began to represent the *Guidelines* graphically to convey the messages of variety, proportionality, and moderation. The Food Guide Pyramid was released in 1992 and in 1994 the Nutrition Labeling and Education Act required all grocery items to have a nutritional label.

2.4 2005 Guidelines: The Current Guidelines

Unlike past revisions, the 2005 *Guidelines* emphasized nutrient intakes based on individual energy requirements for weight maintenance (Table 8.3). Fiber-rich fruits and vegetables were one of the food groups strongly encouraged in the *Guidelines*. Two cups of fruit and 2½ cups of vegetables per day (9 servings) were recommended as a reference for a 2000-calorie intake; an increase from the recommended five

TABLE 8.3 The 2005 *Dietary Guidelines* for Number of Servings of Fruits and Vegetables, Percentage of Calories from Total and Saturated Fat, and Fiber

Energy (kcal)	Daily Servings of Fruits and Vegetables	% of Energy from Total Fat	% of Energy from Saturated Fat	Fiber (g)
1200	5	≤30	<10	17
2000	9	≤30	<10	28
3200	13	≤30	<10	45

daily servings that began in 1985. These intakes should be adjusted depending on specific energy needs. Intakes would equate to 2½ to 6½ cups of fruits and vegetables per day (5–13 servings) for 1200–3200 calorie levels. The Key Recommendations emphasized choosing 'a wide variety of fruits and vegetables each day' and selecting from 'all five vegetable subgroups (dark green, orange, legumes, starchy vegetables, and other vegetables) several times a week' [17]. The *Guidelines* recommended three cups per week of dark green vegetables, up to 11 cups per week of starchy and other vegetables, and three cups per week of legumes.

2.5 MyPyramid: The Current Food Guide

Along with the release of the 2005 *Dietary Guidelines*, the USDA and DHHS released the revised Food Guide Pyramid. In past years, the Food Guide Pyramid displayed the major food groups proportional to consumption recommendations. Grains and cereals formed the base of the pyramid, fruits and vegetables formed the next layer followed by dairy and meat, beans, and nuts, and the tip of the pyramid included fats, oils and sweets to be used sparingly. The 2005 edition had a similar pyramid shape but contained a vertical rainbow of colors, with each color identifying a specific food group and the proportion of total intake. In addition, a color segment for physical activity was included. The 2005 Food Guide Pyramid was promoted as being interactive on the internet by allowing individuals to tailor their daily intakes depending on their age, gender, and activity levels. Unfortunately, those populations without internet access were disadvantaged in becoming educated about the importance of a healthy diet and ways to incorporate nutritious foods into daily intakes.

3. FRUIT AND VEGETABLE INITIATIVES

The 5-A-Day for Better Health Program was the first large-scale program aimed at increasing fruit and vegetable consumption by promoting the consumption of five fruit and vegetable servings per day; in 2005 the 5-A-Day began to encourage 5–9 servings per day to reflect changes in the *Dietary Guidelines* [24]. The 5-A-Day began in 1991 as a public/private partnership between the National Cancer Institute (NCI) and Produce for Better Health (PBH), a non-profit organization. The mission of the program was to increase public awareness of the importance of eating at least five fruits and vegetables daily through education and advertising campaigns. The 5-A-Day has initiated several successful local intervention initiatives in schools and the workplace [25–28] (Table 8.4). These multicomponent interventions were based on the stages of behavioral change model and included interactive activities involving F&V, education, cooking demonstrations, and involving the family and community. Although many 5-A-Day interventions were successful, whether individuals maintained fruit and vegetable consumption after the intervention was removed has not been well established. Research does indicate that demographic and psychosocial determinants have a strong bearing on sustained consumption [29]; whether 5-A-Day interventions had a positive effect on predictors of consumption has yet to be determined.

The 5-A-Day logo, placed on grocery items, was developed to educate consumers on which grocery items counted towards the 5-A-Day goal. Advertising funding has been low with total funding for public communication under $3 million in 1999, which is a miniscule amount compared to the advertising spent on fast food and junk food [30].

TABLE 8.4   Evidence Summary of Interventions Implemented for the 5-A-Day for Better Health Program

Reference	Population	Measures	Results
Beech et al. [25]	2213 students in parochial school in New Orleans	22 nutrition-related items to assess knowledge of F&V	Knowledge and consumption low Attitudes toward learning about healthier eating practices favorable
Beresford et al. [26]	2828 adults at 28 work-sites in Seattle	Food Frequency Questionnaire, 24-hour recalls, fiber and fat-related diet behavior questions, usual day checklist Intervention followed behavioral stages of change model	At 2 years, significant increase in F&V consumption (0.2 baseline vs. 0.5 follow-up)
Stables et al. [27]	2755 and 2544 adults in 1991 and 1997	Random digit dialing baseline and follow-up survey Questions regarding F&V consumption and attitudes and knowledge about F&V	Mean F&V consumption significantly higher at follow-up; null findings when data adjusted for demographic shifts Weighted only: ≥5 servings F&V, 27% in 1997 and 23% in 1991
Stables et al. [28]	Seven 5-A-Day school-based intervention studies funded by 1-year grants	F&V consumption baseline vs. follow-up; measures varied	Average effect size of 0.4 servings of F&V Results similar to larger scale trials

In March 2007, the Produce for Better Health Foundation launched the Fruit and Veggies – More Matters campaign to replace the 5-A-Day Program [31]. The new campaign relies heavily on internet access but provides a multitude of online resources including recipes, individualized serving calculators, lessons for kids, shopping tips, and the ability to post questions for experts and the community. The Fruit and Veggies – More Matters campaign is part of the Produce for Better Health Foundation's National Action Plan to increase fruit and vegetable consumption among Americans [32]. Marketing to children, targeting worksites, supermarket retailers, fruit and vegetable growers and processors, health-care industries and organizations, communities, research, and federal policies are some of the strategies outlined to meet fruit and vegetable intake goals. We now present a recent fruit and vegetable trend study among US adults.

4. TRENDS IN FRUIT AND VEGETABLE CONSUMPTION: DATA FROM NHANES III (1988–1994) AND NHANES 1999–2002

4.1 Introduction

To determine if national efforts to increase F&V consumption were successful, F&V consumption trends were examined using data from NHANES III (1988–1994) and NHANES 1999–2002. Data from NHANES III (1988–1994) coincided with the initiation of the 5-A-Day Program in 1991, and data from NHANES 1999–2002 were used as the most recently available data for comparison with NHANES III. This analysis compared consumption to the USDA *Dietary Guidelines* that were implemented from 1985 to 2005, which explicitly recommended eating at least two servings of fruits and at least three servings of vegetables daily.

4.2 Methods

The National Health and Nutrition Examination Survey (NHANES) is a stratified multistage probability survey conducted in the non-institutionalized population, aged ≥6 months administered by the National Center for Health Statistics (NCHS). The NHANES survey has two parts: 1) an in-home interview for demographic and basic health information, and 2) a health examination in a mobile examination center (MEC). Household interviews are conducted by trained staff and the MEC is staffed by physicians, medical and health technicians, and dietary and health interviewers. All survey information is confidential and approved by the NCHS Institutional Review Board [33].

4.3 Study Population

This study included 14,997 adults (≥18 years) from 1988 to 1994 and 8910 adults from 1999 to 2002 with complete demographic and dietary data. For 1988–1994, the number of missing values for variables ranged from 108 (0.6% missing) for education to 1582 (9%) for poverty income ratio (PIR). Individuals with missing PIR tended to be slightly older, more often Mexican American, and have less education than individuals not missing PIR (data not shown). For 1999–2002, missing values for variables ranged from 19 (0.2%) for education to 1031 (10%) for PIR. Individuals with missing PIR tended to be slightly older, more often non-Hispanic Black or Mexican American, and have less education than individuals having PIR. There were no differences in F&V consumption overall or by socio-demographic indicators when individuals with missing PIR were included in analysis in either NHANES dataset. Therefore, individuals with missing socio-demographic data were excluded.

4.4 Dietary Measures

As part of the standard NHANES data collection protocol, 24-hour dietary recalls (24HR) were conducted and were used to estimate the number of daily F&V servings. The main F&V variable used in these analyses was the 7- and 8-digit food coding scheme developed by the USDA for NHANES III and NHANES 1999–2002, respectively, which categorized items by food group and subgroup [34]. If the main ingredient of a mixed dish was a fruit or vegetable, the item was categorized according to the main fruit or vegetable. Sweets containing fruit where fruit was not the main ingredient were excluded from analysis. Serving sizes for each recorded F&V were determined using serving size estimations from USDA/DHHS *Dietary Guidelines* [35] and a previous NHANES study [13]. Fruit servings included whole fruit, dried and mixed fruit dishes, and 100% fruit juice. Vegetable servings included white potatoes, fried potatoes, garden vegetables (dark leafy greens, yellow vegetables, tomatoes, green beans, starchy vegetables), salad, and legumes. Although fried potatoes have been criticized as a vegetable, this subcategory was included in analyses to be consistent with previously published NHANES II data [13]. Additional analyses were conducted to estimate the proportion of individuals consuming fried potatoes. Reporting a salad on a single occasion was coded as being equal to one serving rather than multiple servings for each ingredient.

For each vegetable, one serving consisted of 30–149 g and two servings consisted of ≥150 g. For each fruit, one serving consisted of 30–239 g and two servings consisted of ≥240 g. For fruit juice, one serving was 62–371 g (2–12 ounces), two servings 372–587 g (12–18 ounces), and three servings ≥588 g (≥18 ounces). To minimize the impact of reported overestimation, an upper limit of two servings for F&V and an upper limit of three servings for fruit juice were established. The lower limit for one serving was equal to one ounce to capture consumption for individuals with lower energy intakes. Although portions less than one ounce could sum to a serving over the day, and consequently result in underestimation, the bias would be minimal since the lower limit reflects very small intake amounts.

Individuals with total caloric intakes <350 kcal or >7000 kcal were considered to have unreliable consumption patterns and were excluded from our final analysis [N = 195 (1.0%) and N = 67 (0.7%) for 1988–1994 and 1999–2002, respectively] [36,37].

4.5 Statistical Analysis

All analyses were conducted using survey weighting to account for the complex survey design, which consisted of multistage, stratified, clustered samples. Probability sampling weights were used in conjunction with strata and primary sampling units (psu) to weight the analysis. Mean servings of F&V, the proportion of individuals consuming specific types of F&V, and the proportion of individuals meeting the *Dietary Guidelines* (≥2 servings of fruit and ≥3 servings of vegetables based on an average 2000 kcal daily intake) were reported [35]. To explore demographic shifts over time between NHANES datasets, secondary analyses were conducted adjusted for age, ethnicity, and gender. Previously published data from NHANES II, 1976–1980 [13] were used to assess trends across three NHANES surveys.

Logistic regression analysis was used to compare F&V consumption patterns across socio-demographic groups. Mean intakes of total energy, percentage of energy from total and saturated fat, and fiber were stratified by F&V intake and adjusted for age, gender, and ethnicity using multiple regression models. To assess differences in consumption between 1988–1994 and 1999–2002, χ^2 tests for significance were performed. All analyses were conducted using STATA 9.0 statistical software (StatCorp. 2005 *Stata Statistical Software: Release 9*. College Station, TX; StataCorp LP).

4.6 Results

4.6.1 Population Characteristics

Sixty-eight percent and 65% of the study population was <50 years of age in 1988–1994 and 1999–2002, respectively, with a greater number of individuals ≥70 years represented in 1999–2002 (11.2% vs. 9.4%) (P<0.001) (Table 8.5). There were slight differences, though not significant (P = 0.055), in ethnicity between 1988–1994 and 1999–2002 with 76.8% and 72.4%, respectively, of those sampled reporting ethnicity as non-Hispanic White, 10.8% and 10.3% reporting non-Hispanic Black, and 5.0% and 6.9% reporting Mexican American. Fewer participants were poor or near poor in 1988–1994 (12.9%, 4.9%, respectively) compared to participants in 1999–2002 (14.9%, 6.7%, respectively) (P = 0.008) [38]. Education level was significantly different between the two NHANES populations with 41.5% in 1988–1994 reporting an education level > high school compared to 52.5% in 1999–2002 (P<0.001).

4.6.2 Fruit and Vegetable Patterns

Daily recommended fruit consumption (≥2 servings) was similar between 1988–1994 and 1999–2002 (27%, 28%, respectively; P = 0.196) (Table 8.6). There was a slight decrease in daily recommended vegetable consumption (≥3 servings) from 35% in 1988–1994 to 32% in 1999–2002, which was statistically significant (P = 0.026). Eleven percent met the daily recommendations (≥2 servings fruit and ≥3

TABLE 8.5   Adult (≥18 years) Population Characteristics for NHANES III (1988–1994) and NHANES 1999–2002

	NHANES III, 1988–1994 (N = 14,997) % (SE)[a]	NHANES 1999–2002 (N = 8910) % (SE)	P-value[b]
Age			
18–29	25.3 (0.9)	21.6 (0.8)	<0.001
30–39	24.1 (0.7)	20.3 (0.9)	
40–49	18.7 (0.7)	21.6 (0.8)	
50–59	12.0 (0.7)	15.6 (0.6)	
60–69	10.6 (0.5)	9.7 (0.5)	
≥70	9.4 (0.6)	11.2 (0.5)	
Gender			
Female	52.2 (0.5)	51.9 (0.5)	0.693
Ethnicity			
Non-Hispanic White	76.8 (1.3)	72.4 (1.7)	0.055
Non-Hispanic Black	10.8 (0.6)	10.3 (1.2)	
Mexican American	5.0 (0.4)	6.9 (0.9)	
Other	7.4 (0.8)	10.4 (1.7)	
Poverty Income Ratio (PIR)			
≤1.0 (poor)	12.9 (0.8)	14.9 (0.9)	0.008
1.0–1.25 (near poor)	4.9 (0.3)	6.7 (0.7)	
1.25–2.5 (moderately poor)	26.3 (1.0)	22.2 (0.9)	
>2.5 (not poor)	55.9 (1.4)	56.2 (1.8)	
Education			
< High school	23.9 (1.1)	21.3 (0.9)	<0.001
High school completion	34.5 (0.7)	26.2 (1.1)	
> High school	41.5 (1.3)	52.5 (1.5)	

[a]Survey weighted percentage and standard error.
[b]P-value for Pearson χ^2 test for difference between NHANES III and NHANES 1999–2002.

servings vegetables) in 1988–1994 and in 1999–2002 (P = 0.963), indicating no change in consumption. Mean daily serving intakes for fruits and vegetables were 3.06 and 3.04 for 1988–1994 and 1999–2002, respectively (P = 0.754). For both study periods, roughly half of the participants reported no fruit servings, about 25% reported no vegetable servings, and 14% reported no vegetable and no fruit servings (Table 8.7). Fruit and vegetable intakes adjusted for age, sex, and ethnicity did not significantly change consumption patterns (data not shown).

4.6.3 Fruit and Vegetable Patterns by Demographics

In both study periods, recommended consumption frequencies for fruits (≥2) and vegetables (≥3) were significantly different by age, ethnicity, poverty income ratio, and education level (data not shown). In 1988–1994

TABLE 8.6 Mean Servings (SE) of Fruits and Vegetables and Percentage (SE) of Adults (≥ 18 Years) Meeting Daily Serving Recommendations for Fruits and Vegetables

Servings per day	NHANES III, 1988–1994 ($N = 14,997$) % (SE)	NHANES 1999–2002 ($N = 8910$) % (SE)	P-value[a]
Fruit			
Total Fruit			
Mean servings (SE)	0.99 (0.03)	1.07 (0.04)	0.081
Whole Fruit			
0	63.1 (0.9)	61.3 (1.2)	0.290
≥ 1	36.9 (0.9)	38.7 (1.2)	
Fruit Juice			
0	75.4 (0.8)	75.4 (0.9)	0.983
≥ 1	24.6 (0.8)	24.6 (0.9)	
Recommended (≥ 2)	26.7 (0.8)	28.4 (1.2)	0.196
Vegetables			
Total Vegetables			
Mean servings (SE)	2.08 (0.03)	1.97 (0.03)	0.029
White Potatoes			
0	79.3 (0.8)	82.1 (0.6)	0.004
≥ 1	20.7 (0.8)	17.9 (0.6)	
Fried Potatoes			
0	80.2 (0.6)	79.4 (1.0)	0.402
≥ 1	19.8 (0.6)	20.6 (1.0)	
Salad			
0	70.8 (0.9)	72.0 (0.7)	0.335
≥ 1	29.2 (0.8)	28.0 (0.7)	
Garden Vegetables			
0	46.2 (0.9)	48.9 (1.0)	0.033
≥ 1	53.7 (0.9)	51.1 (1.0)	
Legumes			
0	88.7 (0.6)	87.1 (0.5)	0.059
≥ 1	11.2 (0.6)	12.9 (0.5)	
Recommended (≥ 3)	35.0 (0.7)	32.5 (0.7)	0.026
Total Fruit and Vegetable			
Mean Servings of Fruit and Vegetables			
Mean (SE)	3.06 (0.04)	3.04 (0.06)	0.754
Recommended Fruit (≥ 2) and Vegetable (≥ 3)			
Yes	10.9 (0.4)	10.8 (0.6)	0.963
≥ 5 *Servings Fruits and Vegetables*			
Yes	24.3 (0.6)	23.6 (0.8)	0.544

[a]Pearson χ^2 test for difference between NHANES III and NHANES 1999–2002.

TABLE 8.7   Percentage (SE) of Adults (≥ 18 years) by Number of Consumed Fruit and Vegetable Servings

| Fruit Servings | NHANES III, 1988–1994 (N = 14,997) | | | | | NHANES 1999–2002 (N = 8910) | | | | |
| | Vegetable Servings | | | | | Vegetable Servings | | | | |
	0	1	2	≥ 3	Total	0	1	2	≥ 3	Total
0	14 (0.5)	11 (0.5)	10 (0.4)	16 (0.6)	50 (0.9)	14 (0.8)	11 (0.5)	10 (0.6)	14 (0.7)	49 (1.3)
1	5 (0.3)	4 (0.3)	5 (0.3)	8 (0.4)	23 (0.5)	5 (0.3)	5 (0.3)	5 (0.3)	8 (0.4)	22 (0.6)
2	3 (0.2)	3 (0.3)	3 (0.2)	6 (0.3)	15 (0.5)	3 (0.2)	3 (0.2)	3 (0.3)	5 (0.3)	14 (0.6)
≥ 3	2 (0.2)	2 (0.2)	2 (0.2)	5 (0.3)	12 (0.6)	3 (0.2)	3 (0.2)	3 (0.2)	6 (0.4)	15 (0.8)
Total	24 (0.7)	21 (0.5)	20 (0.5)	35 (0.7)	100	25 (0.9)	22 (0.6)	21 (0.6)	32 (0.7)	100

and 1999–2002, older individuals (≥ 50 years) were more likely to consume ≥ 2 fruit servings and were more likely to meet both fruit and vegetable recommendations in both study periods (all $P<0.05$) (Table 8.8).

Females were more likely to consume ≥ 2 fruit servings in both study periods ($P<0.05$) but less likely to consume ≥ 3 servings of vegetables in NHANES III [odds ratio (OR) = 0.83, confidence interval (CI) = 0.77–0.89]. There were no gender differences for meeting both recommendations.

Non-Hispanic Blacks were less likely to meet fruit recommendations compared to non-Hispanic Whites in 1988–1994 (OR = 0.63, CI = 0.57–0.69) and 1999–2002 (OR = 0.68, CI = 0.60–0.77). Mexican Americans were less likely to meet fruit recommendations in 1988–1994 compared to non-Hispanic Whites (OR = 0.75, CI = 0.69–0.83); there was no difference in NHANES 1999–2002. Non-Hispanic Blacks were less likely to meet vegetable recommendations in 1988–1994 and 1999–2002 compared to non-Hispanic Whites (OR = 0.52, 0.0.61, respectively, $P<0.05$); there was no difference among Mexican Americans in both study periods. For meeting both fruit and vegetable recommendations, non-Hispanic Blacks were less likely in 1988–1994 and 1999–2002 (OR = 0.45, 0.57, respectively; $P<0.05$) to meet the recommendations compared to non-Hispanic

Whites. Mexican Americans were less likely to meet fruit and vegetable recommendations compared to non-Hispanic Whites in 1988–1994 (OR = 0.76, CI = 0.68–0.88) but there was no difference in the later study period.

In 1988–1994, a poverty income ratio >1.0 was associated with an increased likelihood of meeting fruit and vegetable recommendations compared to a ratio ≤ 1.0; those with the highest PIR (≥ 2.5) had 2.06 times the odds of meeting the recommendations than those in the lowest PIR. In 1999–2002, only individuals with a PIR ≥ 1.25 had an increased likelihood of meeting the fruit and vegetable recommendations ($P<0.05$).

In 1988–1994 and 1999–2002, those with $\geq$ high school completion were more likely to meet fruit, vegetable, and both fruit and vegetable recommendations (all $P<0.05$). A consistent dose–response relationship for education levels and the likelihood of meeting fruit and vegetable recommendations was observed.

4.6.4 Types of Vegetables Consumed

There was no difference in the number of daily vegetable servings consumed by individuals in NHANES 1988–1994 and NHANES 1999–2002 (data not shown). Among those consuming one vegetable serving, 26% in 1988–1994 and 27% in 1999–2002 consumed fried

TABLE 8.8 Odds Ratios (95% CI) for Meeting Fruit and Vegetable Serving Recommendations Stratified by Age, Gender, Ethnicity, Poverty Income Ratio, and Education for Adults (≥18 Years)

	Recommended Fruit (≥2 servings)		Recommended Vegetables (≥3 servings)		Recommended Fruits (≥2) and Vegetables (≥3)	
	NHANES III 1988–1994[a]	NHANES 1999–2002	NHANES III 1988–1994	NHANES 1999–2002	NHANES III 1988–1994	NHANES 1999–2002
Age						
18–29	Ref	Ref	Ref	Ref	Ref	Ref
30–39	1.18* (1.05–1.33)	1.04 (0.90–1.21)	1.07 (0.97–1.19)	1.15* (1.00–1.33)	1.18 (0.99–1.41)	1.23 (0.99–1.54)
40–49	1.22* (1.07–1.38)	1.12 (0.97–1.30)	1.13* (1.01–1.27)	1.10 (0.95–1.27)	1.22* (1.01–1.48)	1.25 (1.00–1.56)
50–59	1.80* (1.58–2.06)	1.33* (1.14–1.57)	1.10 (0.97–1.24)	1.24* (1.06–1.45)	1.84* (1.51–2.23)	1.45* (1.14–1.83)
60–69	2.11* (1.86–2.39)	1.67* (1.44–1.94)	1.03 (0.92–1.16)	1.27* (1.09–1.46)	1.71* (1.42–2.06)	1.80* (1.45–2.23)
≥70	2.96* (2.64–3.32)	2.10* (1.82–2.41)	0.94 (0.84–1.05)	0.96 (0.83–1.11)	2.10* (1.78–2.49)	1.74* (1.42–2.14)
Gender						
Female (vs. male)	1.12* (1.04–1.20)	1.20* (1.09–1.31)	0.83* (0.77–0.89)	0.95 (0.87–1.04)	0.97 (0.88–1.08)	1.00 (0.88–1.14)
Ethnicity						
Non-Hispanic White	Ref	Ref	Ref	Ref	Ref	Ref
Non-Hispanic Black	0.63* (0.57–0.69)	0.68* (0.60–0.77)	0.52* (0.47–0.57)	0.61* (0.53–0.69)	0.45* (0.39–0.53)	0.57* (0.46–0.70)
Mexican American	0.75* (0.69–0.83)	0.93 (0.83–1.03)	1.04 (0.96–1.13)	1.01 (0.91–1.13)	0.76* (0.68–0.88)	0.95 (0.82–1.11)
Other	1.02 (0.85–1.23)	0.97 (0.82–1.15)	0.90 (0.75–1.08)	0.83* (0.70–1.00)	0.92 (0.71–1.20)	0.94 (0.74–1.21)
Poverty Income Ratio (PIR)						
≤1.0 (poor)	Ref	Ref	Ref	Ref	Ref	Ref
1.0–1.25 (near poor)	1.30* (1.11–1.51)	0.88 (0.73–1.07)	1.01 (0.87–1.17)	1.03 (0.85–1.25)	1.40* (1.10–1.77)	0.89 (0.65–1.22)
1.25–2.5 (moderately poor)	1.40* (1.26–1.56)	1.17* (1.02–1.34)	1.14* (1.03–1.25)	1.22* (1.07–1.40)	1.44* (1.22–1.69)	1.31* (1.06–1.61)
>2.5 (not poor)	1.77* (1.61–1.96)	1.38* (1.22–1.56)	1.50* (1.37–1.64)	1.57* (1.39–1.78)	2.06* (1.77–2.39)	1.65* (1.37–2.00)
Education						
< High school	Ref	Ref	Ref	Ref	Ref	Ref
High school diploma	0.98 (0.90–1.07)	1.12 (0.99–1.27)	1.12* (1.03–1.22)	1.20* (1.06–1.35)	1.14 (1.00–1.31)	1.21* (1.00–1.47)
> High school	1.60* (1.47–1.74)	1.52* (1.37–1.69)	1.31* (1.21–1.43)	1.62* (1.46–1.80)	1.78* (1.57–2.02)	1.90* (1.62–2.22)

[a]NHANES III (N = 14,997); NHANES 1999–2002 (N = 8910).
*P<0.05.

potatoes ($P = 0.51$) and 44% in 1988–1994 and 43% in 1999–2002 consumed garden vegetables ($P = 0.53$) (Fig. 8.1). In both study periods, as vegetable servings increased and vegetable combinations became more varied, garden vegetables and salads were more frequently consumed. For individuals consuming 2 servings of vegetables, roughly 22% consumed a salad and a garden vegetable and roughly 19% consumed 2 servings of garden vegetables. Among individuals consuming ≥3 servings in 1988–1994 and 1999–2002, roughly 12% consumed 3 servings of garden vegetables and 14% consumed 2 servings of garden and a salad

with far fewer consuming any combination containing fried potatoes. The combination of vegetables individuals consumed over the two study periods did not change significantly (data not shown).

4.6.5 Trends in Consumption Between NHANES II, NHANES III, and NHANES 1999–2002

Previously published data from NHANES II (1976–1980) was used to compare fruit and vegetable consumption over the past three NHANES surveys [39] (Fig. 8.2). The

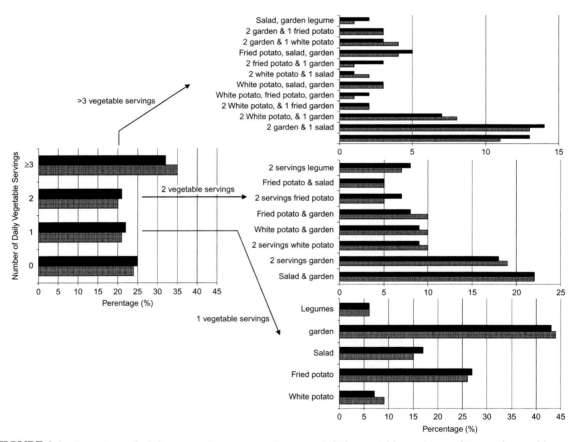

FIGURE 8.1 Percentage of adults consuming zero, one, two, or ≥3 daily vegetable servings and types of vegetables consumed in NHANES 1988–1994 ■ and NHANES 1999–2002 ■.

percentage of individuals meeting fruit and vegetable recommendations increased only slightly from 9% in 1976–1980 to 11% in 1988–1994 and remained constant at 11% in 1999–2002. Fruit consumption decreased slightly from 29% in 1976–1980 to 27% in 1988–1994 and 28% in 1999–2002. Vegetable consumption increased from 27% in 1976–1980 to 35% in 1988–1994 and then decreased slightly to 33% in 1999–2002.

4.6.6 Energy and Macronutrient Consumption by Fruit and Vegetable Intake

The USDA *Dietary Guidelines* recommend 14 g of fiber per 1000 kcal consumed [40]. After adjusting for age, gender, and ethnicity, mean energy and fiber intakes were higher for those consuming more F&V (Table 8.9), but remained below recommendations in both

study periods. The *Dietary Guidelines* recommend 20–35% of energy from total fat and <10% of energy from saturated fat. The mean percentage of energy from total fat was slightly above 30% and the mean percentage of energy from saturated fat was slightly above 10% for all F&V intake levels.

4.6.7 Discussion

Based on 24-hour dietary recall data from NHANES III 1988–1994 and NHANES 1999–2002, approximately 89% of Americans failed to meet USDA *Dietary Guidelines* for F&V. Furthermore, there was no improvement in Americans' fruit consumption during this period and there was a small decrease in vegetable intake. About half of participants did not consume any fruit and a quarter consumed no vegetables. Although 35% in 1988–1994 and

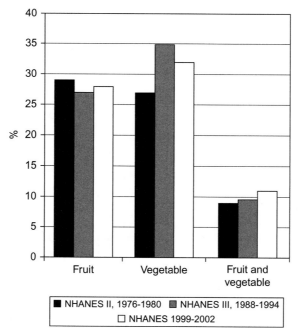

FIGURE 8.2 Percentage of adults (≥18 years) meeting daily recommended servings of fruits (≥2) and vegetables (≥3) by NHANES study periods.
(NHANES II from previously published data; Patterson *et al.* [13]).

TABLE 8.9 Mean Intakes (SE) of Energy, Total and Saturated Fat, and Fiber by Fruit and Vegetable Consumption for American Adults, Adjusted for Age, Gender and Ethnicity

	NHANES III, 1988–994 (N = 14,997)				NHANES 1999–2002 (N = 8910)			
	Energy Intake (kcal)	% of Energy from Total Fat	% of Energy from Saturated Fat	Fiber (g/ 1000 kcal)	Energy Intake (kcal)	% of Energy from Total Fat	% of Energy from Saturated Fat	Fiber (g/ 1000 kcal)
F&V Servings								
0	1818 (37)	32.4 (0.4)	11.4 (0.2)	6.1 (0.1)	1813 (32)	31.9 (0.4)	10.7 (0.1)	5.5 (0.1)
1	1922 (27)	33.1 (0.3)	11.3 (0.1)	6.9 (0.1)	1985 (29)	33.1 (0.3)	10.8 (0.1)	6.4 (0.1)
2	2075 (37)	33.8 (0.3)	11.3 (0.1)	7.7 (0.1)	2090 (27)	33.7 (0.3)	10.9 (0.1)	7.0 (0.1)
3	2091 (28)	32.7 (0.3)	10.8 (0.1)	8.5 (0.1)	2164 (22)	33.1 (0.3)	10.7 (0.1)	7.8 (0.1)
4	2253 (39)	33.3 (0.3)	10.8 (0.1)	9.4 (0.1)	2269 (36)	33.0 (0.5)	10.5 (0.2)	8.1 (0.1)
≥5	2372 (29)	32.4 (0.3)	10.3 (0.1)	10.2 (0.1)	2453 (24)	32.1 (0.4)	10.0 (0.1)	10.0 (0.2)

32% in 1999–2002 consumed ≥3 servings of vegetables per day on average, many individuals consumed multiple servings of the same vegetable, which reflects a lack of balance and variety. These results are similar to other studies that have estimated fruit and vegetable intake among American adults.

In both surveys, non-Hispanic Blacks were less likely to meet F&V *Guidelines* than non-Hispanic Whites (7% vs. 11%). These results are consistent with previous NHANES findings [41], although published literature is varied [12,14,15]. The results that F&V consumption was positively associated with income and education are also consistent with findings in other studies [12,20,21,42,43]. This result suggests that poverty continues to be a barrier for purchasing and consuming F&V, and could be one reason for low F&V consumption. In addition to the general population, behavioral interventions should target demographic groups that are less likely to meet the national *Dietary Guidelines* such as African Americans and low socioeconomic status groups.

In contrast to study expectations, the analysis indicated little change in F&V consumption between 1988–1994 and 1999–2002. A slight increase in consumption was expected as a result of national initiatives such as the 5-A-Day for Better Health Program launch in 1991 and increased media attention on the importance of a healthy, balanced diet, which includes a variety of F&V. Although the 5-A-Day Campaign has produced some immediate successful results in local interventions [44–46], the analyses indicated that consumption at the national level has not changed (Table 8.2). Similar results have been documented in the BRFSS, which examined trends over a shorter period. The BRFSS data collected between 1994 and 2000 showed that the proportion of individuals consuming F&V five or more times per day remained constant at 25% [47]; another BRFSS analysis indicated 19%, 22%, and 23% of adults in 1990, 1994, and 1996, respectively, consumed at least five daily F&V servings [48]. Most recently, a report from 2005 BRFSS data indicated that 32.6% of the US adult population consumed fruit two or more times per day and 27.2% ate vegetables three or more times per day; results for ≥5 servings of fruits and vegetables was not reported [49].

These estimates are slightly higher for fruits and slightly lower for vegetables than reported estimates from NHANES III and NHANES 1999–2002. The differences are likely to be a result of the BRFSS using a brief self-reported fruit and vegetable screener and NHANES using a 24-hour dietary recall as reporting methods. The 1987–1992 NHIS found that F&V consumption and the mean number of servings per week remained stable over time [50]. Although the trends in BRFSS and NHIS are similar to the current study, the difference in the actual proportion meeting the *Guidelines* is likely due to inaccurate portion size estimation and limiting F&V consumption to the specific combination of 2 fruits and 3 vegetables. In these analyses, when any combination of F&V was considered, 24% of individuals consumed ≥5 F&V.

There are limitations of this study worth noting. First, NHANES provides only one 24-hour dietary recall. Four 24-hour recalls are optimal to measure individuals' usual intake, however, some research has indicated that food intake data based on 1-day dietary recall can be reliable measures of usual intakes of population groups [51]. Second, it is likely that there may be inaccuracies in converting grams to servings. Nevertheless, the implemented method follows protocols used in the USDA *Dietary Guidelines* and is interpretable by the public. Even though small portions (<1 ounce) were excluded, which could sum to a serving over the span of a day, it is unlikely that the results were underestimated since a generous lower limit of one ounce was used in calculation of servings. Approximately <1% of individuals had small fruit intakes that could cumulate to a 1 ounce serving over a day; <10% of individuals were estimated to have small vegetable intakes that could cumulate to a 1 ounce serving. On the other hand, the lower limit is equivalent to less than half of a USDA defined serving, which could result in an overestimation of consumption. The latter bias would be

expected to be larger, thus, adult F&V consumption may be, regrettably, lower than reported here. Indeed, additional analyses determined that less than 10% of individuals reported consuming between 2 and 6 ounces of juice and that the sum of these small quantities often did not reach a total of 6 ounces.

5. CONCLUSION

A lack of an improvement in F&V consumption could be attributed to a variety of factors. First, food preferences are often personal and rooted in cultural backgrounds [52–54]. Second, environmental barriers continue to deter individuals from eating the recommended number of F&V. Snack and unhealthy foods are relatively cheap compared to fresh produce due to subsidies, costs in fresh food distribution, and the large US food supply [30]; eating out is common and convenient but facilitates consumption of larger portion sizes with extra energy and fat content [55]; advertising for nutritionally poor foods is much more widespread than for the promotion of F&V [30]; access to fruits and vegetables may be limited in disadvantaged neighborhoods [56]. Third, confusion over implementing the *Dietary Guidelines* into daily practice may deter individuals from trying to meet F&V recommendations [57]. Without formal education or access to pertinent information, many Americans are missing important messages about the health benefits of including F&V in their daily diet.

The 2005 *Dietary Guidelines* incorporate energy intake into F&V *Guidelines* to emphasize that individuals who consume more energy should consume more F&V (≥5 servings for <1800 kcal, ≥9 for 1800–2200 kcal and >9 for >2200 kcal). Previous analyses in NHANES 1999–2002 found that 3.7% met the 2005 *Dietary Guidelines* recommendation of ≥9 F&V servings for a dietary intake of approximately 2000 kcal. Although individuals surveyed during 1999–2002 cannot be expected to meet the 2005 *Guidelines*, these analyses stress the point that the proportion of individuals meeting current *Guidelines* is likely even lower and the need for further nutritional education is warranted.

Consistent reports of low F&V consumption with no indication of improvement as well as consumption disparities across ethnic, income, and educational groups over the past several decades should alarm public health officials and professionals. With two-thirds of the US adult population overweight or obese, the implications of a diet low in F&V are extensive [58]. Actions to improve the availability, accessibility, and affordability of fresh produce and to increase consumption through federally funded food programs, such as the Food Stamp Program, should be components for implementation. Schools and worksites provide an avenue in which to reach the general public about the importance of fruits and vegetables in a healthy diet. Successful components of the 5-A-Day Program, such as information sessions, cooking demonstrations, and community and family involvement, are starting points for future interventions. The new Fruit and Veggies – More Matters campaign aims to increase consumption by targeting the primary household food purchasers and by educating on the health benefits of fruit and vegetables and ways to incorporate them in the daily diet. Despite these campaigns, new strategies may be necessary to help Americans make desirable behavioral changes to consume a healthy diet that includes a variety of fruits and vegetables.

6. SUMMARY

Consumption of fruits and vegetables among US adults is below USDA recommended servings. Low consumption has a negative effect on a healthy diet. A healthy diet that includes a variety of fruits and vegetables may protect against preventable chronic diseases including cardiovascular disease and diabetes. To increase consumption, concerted efforts must address individual behaviors and the social and physical environments in which individuals interact and obtain fruits and vegetables.

References

1. Genkinger, J. M., Platz, E. A., Hoffman, S. C., Comstock, G. W., & Helzlsouer, K. J. (2004). Fruit, vegetable, and antioxidant intake and all-cause, cancer, and cardiovascular disease mortality in a community-dwelling population in Washington County, Maryland. *American Journal of Epidemiology, 160*, 1223–1233.
2. He, F. J., Nowson, C. A., & MacGregor, G. A. (2006). Fruit and vegetable consumption and stroke: Meta-analysis of cohort studies. *Lancet, 367*, 320–326.
3. He, K., Hu, F. B., Colditz, G. A., Manson, J. E., Willett, W. C., & Liu, S. (2004). Changes in intake of fruits and vegetables in relation to risk of obesity and weight gain among middle-aged women. *International Journal of Obesity Related Metabolism Disorder, 28*, 1569–1574.
4. Hung, H. C., Joshipura, K. J., Jiang, R., Hu, F. B., Hunter, D., Smith-Warner, S. A., Colditz, G. A., Rosner, B., Spiegelman, D., & Willett, W. C. (2004). Fruit and vegetable intake and risk of major chronic disease. *Journal of National Cancer Institute, 96*, 1577–1584.
5. Ziegler, R. G., Mason, T. J., Stemhagen, A., Hoover, R., Schoenberg, J. B., Gridley, G., Virgo, P. W., & Fraumeni, J. F., Jr. (1986). Carotenoid intake, vegetables, and the risk of lung cancer among white men in New Jersey. *American Journal of Epidemiology, 123*, 1080–1093.
6. Feskens, E. J., Virtanen, S. M., Rasanen, L., Tuomilehto, J., Stengard, J., Pekkanen, J., Nissinen, A., & Kromhout, D. (1995). Dietary factors determining diabetes and impaired glucose tolerance. A 20-year follow-up of the Finnish and Dutch cohorts of the Seven Countries Study. *Diabetes Care, 18*, 1104–1112.
7. Ford, E. S., & Mokdad, A. H. (2001). Fruit and vegetable consumption and diabetes mellitus incidence among U.S. adults. *Preventive Medicine, 32*, 33–39.
8. Sargeant, L. A., Khaw, K. T., Bingham, S., Day, N. E., Luben, R. N., Oakes, S., Welch, A., & Wareham, N. J. (2001). Fruit and vegetable intake and population glycosylated haemoglobin levels: The EPIC-Norfolk Study. *European Journal of Clinical Nutrition, 55*, 342–348.
9. Williams, D. E., Wareham, N. J., Cox, B. D., Byrne, C. D., Hales, C. N., & Day, N. E. (1999). Frequent salad vegetable consumption is associated with a reduction in the risk of diabetes mellitus. *Journal of Clinical Epidemiology, 52*, 329–335.

10. Breslow, R. A., Subar, A. F., Patterson, B. H., & Block, G. (1997). Trends in food intake: the 1987 and 1992 National Health Interview Surveys. *Nutrition and Cancer, 28,* 86–92.

11. Casagrande, S. S., Wang, Y., Anderson, C., & Gary, T. L. (2007). Have Americans increased their fruit and vegetable intake? The trends between 1988 and 2002. *American Journal of Preventive Medicine, 32,* 257–263.

12. Gary, T. L., Baptiste-Roberts, K., Gregg, E. W., Williams, D. E., Beckles, G. L., Miller, E. J., III, & Engelgau, M. M. (2004). Fruit, vegetable and fat intake in a population-based sample of African Americans. *Journal of National Medicine Association, 96,* 1599–1605.

13. Patterson, B. H., Block, G., Rosenberger, W. F., Pee, D., & Kahle, L. L. (1990). Fruit and vegetables in the American diet: Data from the NHANES II survey. *American Journal of Public Health, 80,* 1443–1449.

14. Patterson, B. H., Harlan, L. C., Block, G., & Kahle, L. (1995). Food choices of whites, blacks, and Hispanics: Data from the 1987 National Health Interview Survey. *Nutrition and Cancer, 23,* 105–119.

15. Swanson, C. A., Gridley, G., Greenberg, R. S., Schoenberg, J. B., Swanson, G. M., Brown, L. M., Hayes, R., Silverman, D., & Pottern, L. (1993). A comparison of diets of blacks and whites in three areas of the United States. *Nutrition and Cancer, 20,* 153–165.

16. Basiotis, P. P. (2004). *Healthy eating index: NHANES 1999–2002.* Washington, DC: USDA Center for Nutrition and Policy Promotion.

17. HHS/USDA (2005). *Dietary guidelines for Americans* (pp. 1–84). Washington, DC: United States Department of Human Services and US Department of Agriculture.

18. Davis, C., & Saltos, E. (1998). *Dietary recommendations and how they have changed over time.* (pp. 33–50). Washington, DC: United States Department of Agriculture, Center for Nutrition Policy Promotion.

19. Heimendinger, J., Van Duyn, M. A., Chapelsky, D., Foerster, S., & Stables, G. (1996). The national 5 A Day for Better Health Program: A large-scale nutrition intervention. *Journal of Public Health Management Practice, 2,* 27–35.

20. Li, R., Serdula, M., Bland, S., Mokdad, A., Bowman, B., & Nelson, D. (2000). Trends in fruit and vegetable consumption among adults in 16 US states: Behavioral Risk Factor Surveillance System, 1990–1996. *American Journal of Public Health, 90,* 777–781.

21. Serdula, M. K., Gillespie, C., Kettel-Khan, L., Farris, R., Seymour, J., & Denny, C. (2004). Trends in fruit and vegetable consumption among adults in the United States: Behavioral risk factor surveillance system, 1994–2000. *American Journal of Public Health, 94,* 1014–1018.

22. Popkin, B. M., Siega-Riz, A. M., & Haines, P. S. (1996). A comparison of dietary trends among racial and socioeconomic groups in the United States. *New England Journal of Medicine, 335,* 716–720.

23. Kant, A. K., Graubard, B. I., & Kumanyika, S. K. (2007). Trends in black-white differentials in dietary intakes of U.S. adults, 1971–2002. *American Journal of Preventive Medicine, 32,* 264–272.

24. The National Cancer Institute. (2006). The 5-A-day for better health program. <http://www.5aday.gov>.

25. Beech, B. M., Rice, R., Myers, L., Johnson, C., & Nicklas, T. A. (1999). Knowledge, attitudes, and practices related to fruit and vegetable consumption of high school students. *Journal of Adolescent Health, 24,* 244–250.

26. Beresford, S. A., Thompson, B., Feng, Z., Christianson, A., McLerran, D., & Patrick, D. L. (2001). Seattle 5 a Day worksite program to increase fruit and vegetable consumption. *Preventive Medicine, 32,* 230–238.

27. Stables, G. J., Subar, A. F., Patterson, B. H., Dodd, K., Heimendinger, J., Van Duyn, M. A., & Nebeling, L. (2002). Changes in vegetable and fruit consumption and awareness among US adults: Results of the 1991 and 1997 5 A Day for Better Health Program surveys. *Journal of the American Dietetic Association, 102,* 809–817.

28. Stables, G. J., Young, E. M., Howerton, M. W., Yaroch, A. L., Kuester, S., Solera, M. K., Cobb, K., & Nebeling, L. (2005). Small school-based effectiveness trials increase vegetable and fruit consumption among youth. *Journal of the American Dietetic Association, 105,* 252–256.

29. Trudeau, E., Kristal, A. R., Li, S., & Patterson, R. E. (1998). Demographic and psychosocial predictors of fruit and vegetable intakes differ: Implications for dietary interventions. *Journal of the American Dietetic Association, 98,* 1412–1417.

30. Nestle, M. (2002). *Food politics: How the food industry influences nutrition and health.* (pp. 130–132). Berkeley and Los Angeles, CA: University of California Press.

31. Produce for Better Health Foundation. (2008). Fruit and veggies – more matters.

32. Produce for a Better Health Foundation. (2005). National action plan: To promote health through increased fruit and vegetable consumption. Produce for a Better Health Foundation..

33. National Center for Health Statistics. (1999). *National health and nutrition examination survey: Overview 1999–2002.* Hyattsville, MD: NCHS.

34. US Department of Agriculture. (2004). *USDA food and nutrient database for dietary studies, 1.0-documentation and user guide.* Beltsville, MD: Agricultural Research Service, Food Surveys Research Group.

35. Cronin, F., Shaw, A., Krebs-Smith, S. M., Marsland, P., & Light, L. (1987). Developing a food guidance system to implement the dietary guidelines. *Journal of Nutrition Education, 19,* 281–302.

36. Krebs-Smith, S. M., Cleveland, L. E., Ballard-Barbash, R., Cook, D. A., & Kahle, L. L. (1997). Characterizing food intake patterns of American adults. *American Journal of Clinical Nutrition, 65*, 1264S–1268S.

37. Thompson, F. E., Midthune, D., Subar, A. F., McNeel, T., Berrigan, D., & Kipnis, V. (2005). Dietary intake estimates in the National Health Interview Survey, 2000: Methodology, results, and interpretation. *Journal of the American Dietetic Association, 105*, 352–363.

38. US Census Bureau. (2005). *Poverty income ratio.* Washington, DC: U.S. Department of Commerce.

39. Patterson, B. H., Block, G., Rosenberger, W. F., Pee, D., & Kahle, L. L. (1990). Fruit and vegetables in the American diet: Data from the NHANES II survey. *American Journal of Public Health, 80*, 1443–1449.

40. US Department of Health and Human Services and US Department of Agriculture. (2006). *Dietary guidelines for Americans, 2005.* Washington, DC: US Government Printing Office.

41. Patterson, B. H., Block, G., Rosenberger, W. F., Pee, D., & Kahle, L. L. (1990). Fruit and vegetables in the American diet: Data from the NHANES II survey. *American Journal of Public Health, 80*, 1443–1449.

42. Dibsdall, L. A., Lambert, N., Bobbin, R. F., & Frewer, L. J. (2003). Low-income consumers' attitudes and behaviour towards access, availability and motivation to eat fruit and vegetables. *Public Health Nutrition, 6*, 159–168.

43. Miller, R. R., Sales, A. E., Kopjar, B., Fihn, S. D., & Bryson, C. L. (2005). Adherence to heart-healthy behaviors in a sample of the U.S. population. *Prevention Chronical Disease, 2*, A18.

44. Beresford, S. A., Thompson, B., Feng, Z., Christianson, A., McLerran, D., & Patrick, D. L. (2001). Seattle 5 a Day worksite program to increase fruit and vegetable consumption. *Preventive Medicine, 32*, 230–238.

45. Stables, G. J., Subar, A. F., Patterson, B. H., Dodd, K., Heimendinger, J., Van Duyn, M. A., & Nebeling, L. (2002). Changes in vegetable and fruit consumption and awareness among US adults: Results of the 1991 and 1997 5 A Day for Better Health Program surveys. *Journal of the American Dietetic Association, 102*, 809–817.

46. Stables, G. J., Young, E. M., Howerton, M. W., Yaroch, A. L., Kuester, S., Solera, M. K., Cobb, K., & Nebeling, L. (2005). Small school-based effectiveness trials increase vegetable and fruit consumption among youth. *Journal of the American Dietetic Association, 105*, 252–256.

47. Serdula, M. K., Gillespie, C., Kettel-Khan, L., Farris, R., Seymour, J., & Denny, C. (2004). Trends in fruit and vegetable consumption among adults in the United States: Behavioral risk factor surveillance system, 1994–2000. *American Journal of Public Health, 94*, 1014–1018.

48. Li, R., Serdula, M., Bland, S., Mokdad, A., Bowman, B., & Nelson, D. (2000). Trends in fruit and vegetable consumption among adults in 16 US states: Behavioral Risk Factor Surveillance System, 1990–1996. *American Journal of Public Health, 90*, 777–781.

49. Blanck, H. M., Galuska, D. A., Gillespie, C., Kettel Kahn, L., Serdula, M. K., Solera, M. K., & Mokdad, A. K. (2007). Fruit and vegetable consumption among adults – United States 2005. *MMWR, 56*, 213–217.

50. Breslow, R. A., Subar, A. F., Patterson, B. H., & Block, G. (1997). Trends in food intake: The 1987 and 1992 National Health Interview Surveys. *Nutrition Cancer, 28*, 86–92.

51. Basiotis, P. P., Welsh, S. O., Cronin, F. J., Kelsay, J. L., & Mertz, W. (1987). Number of days of food intake records required to estimate individual and group nutrient intakes with defined confidence. *Journal of Nutrition, 117*, 1638–1641.

52. Shepherd, R. (2005). Influences on food choice and dietary behavior. *Forum Nutrition, 57*, 36–43.

53. Tucker, K. L., Maras, J., Champagne, C., Connell, C., Goolsby, S., Weber, J., Zaghloul, S., Carithers, T., & Bogle, M. L. (2005). A regional food-frequency questionnaire for the US Mississippi Delta. *Public Health Nutrition, 8*, 87–96.

54. Van Duyn, M. A., Kristal, A. R., Dodd, K., Campbell, M. K., Subar, A. F., Stables, G., Nebeling, L., & Glanz, K. (2001). Association of awareness, intrapersonal and interpersonal factors, and stage of dietary change with fruit and vegetable consumption: A national survey. *American Journal of Health Promotion, 16*, 69–78.

55. French, S. A., Harnack, L., & Jeffery, R. W. (2000). Fast food restaurant use among women in the Pound of Prevention study: Dietary, behavioral and demographic correlates. *International Journal of Obesity Related Metabolism Disorder, 24*, 1353–1359.

56. Morland, K., Wing, S., Diez, R. A., & Poole, C. (2002). Neighborhood characteristics associated with the location of food stores and food service places. *American Journal of Preventive Medicine, 22*, 23–29.

57. Brown, D. (2005). New dietary guidelines need dietetic interpretation. *Journal of the American Dietetic Association, 105*, 1356–1357.

58. Hedley, A. A., Ogden, C. L., Johnson, C. L., Carroll, M. D., Curtin, L. R., & Flegal, K. M. (2004). Prevalence of overweight and obesity among US children, adolescents, and adults, 1999–2002. *JAMA, 291*, 2847–2850.

9

Fruit and Vegetables in the Optimized Mixed Diet

Ute Alexy and Mathilde Kersting
Research Institute of Child Nutrition (FKE), Dortmund, Germany

Unhealthy diet and physical inactivity increase the risk for several chronic diseases. Since dietary habits during childhood and adolescence continue to a moderate degree until adulthood [1] and unfavorable dietary practices during youth have a negative effect on later health [2], dietary interventions at an early age are desirable.

To improve dietary practice, there is a clear need for so-called 'food-based dietary guidelines' (FBDG). Such FBDGs translate scientific nutritional population goals, e.g. fat intake below 30% of energy, into comprehensible recommendations for the total diet. In 1998, the Food and Agriculture Organization of the United Nations and World Health Organization (FAO/WHO) published a report on the preparation and use of FBDG. FBDG should cover the total daily diet and should not be restricted to single meals (e.g. breakfast) or food groups (e.g. fruit and vegetables). FBDG should be population based and take into account nutrition-related health problems within a country. They

should also be developed on a national level considering traditional food consumption patterns and should be practical, simple and transformed into slogans [3].

In the FAO/WHO report, children < 2 years of age were not mentioned initially. However, it cannot be tacitly assumed that FBDG for adults are, per se, also appropriate for children or that dietary problems within a population are the same for all age groups. Therefore there is a need for specific FBDG that consider their special nutritional needs for children and adolescents, especially the increasing energy intake per day and the simultaneously decreasing energy requirement per kg body weight, as well as specific food patterns and preferences during this period. Such FBDG should be country specific even for the young, since food and meal patterns, for instance weaning habits, differ even between neighboring countries.

For Germany, the Research Institute of Child Nutrition Dortmund (FKE) in 1993 developed FBDG termed the Optimized Mixed Diet

(OMD) for children and adolescents aged 1–18 years. Development and preparation were in accordance with the FAO/WHO guidelines [4]. The scientific panel of the European Food Safety Authority (EFSA) accentuated that in the OMD, 'a core of quantified food groups can be adequate for age groups between 1 and 18 years' (www.efsa.europa.eu; 2008).

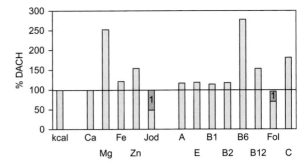

¹Use of fortified table salt (100 mg folic acid, 13.8 mg iodine per kg)

FIGURE 9.1 Intake of energy, minerals and vitamins in the OMD (% of dietary reference intakes).

1. THE DEVELOPMENT OF THE OMD

The basis of the OMD is 7-day daily menus, which were conceived taking into account:

1. the typical meal patterns of families in Germany;
2. the use of only common, easy available and non-fortified foods;
3. the typical food preferences of children and adolescents, including, for example, fast-food dishes and sweets.

This approach ensured the traditional food pattern as demanded by the FAO/WHO [3]. Fortified foods were avoided since fortification practice in Germany is very heterogeneous: the range of fortification varies even within a product group, and fortification with several vitamins and/or minerals is customary [5,6].

In a second step, these 7-day menus were optimized, i.e. food amounts and/or food selections were modified by trained dieticians, until the mean energy and nutrient intake over the 7 days was in accordance with German reference values [7]. In total, the OMD provided 100% of energy requirements, and reached or exceeded the reference values for most vitamins and minerals (Fig 9.1). Protein constitutes 13.8% of energy intake, fat 32.8%, and carbohydrates 53.4%. Percentages of energy from saturated, monounsaturated and polyunsaturated fatty acids are 10.3%, 15.2% and 7.3%. The mean

energy density (incl. beverages) is 71 kcal/100 g. Per 100 g, the OMD includes 17 g dietary fiber.

In spite of the high intake of vegetables, the most important source for folic acid, the reference values for folic acid could not be achieved without fortification. The same applied for iodine. Therefore, in the OMD the use of iodized table salt fortified with folic acid (100 mg/per 100 g) is recommended.

In the German reference values, six age groups, and subgroups by gender are indicated. For the 7-day menus, the 4–6-year-old age group was chosen as an example. Since reference nutrient densities of 1–18-year-old children and adolescents remained fairly constant [4], by adapting food amounts in the menus the altered energy and nutrient demands of other age groups could be achieved using the same food items, but with varying food group amounts (Table 9.1). Food group proportions in the OMD are independent of age and sex.

In a third step, the individual foods of the optimized menus were categorized into 11 food groups. The criteria for aggregating these food groups were their specific nutrient properties or their use in the traditional German meal pattern. Weight proportions of the 11 food groups in the total diet, and proportions of energy and nutrients were calculated.

TABLE 9.1   Recommended Daily Food Group Amounts by Age and Sex in the OMD

		\multicolumn{7}{c}{Age (Years)}		% Total Diet
		1	2–3	4–6	7–9	10–12	13–14 Boys/Girls	15–18 Boys/Girls	
Beverages	mL/d	600	700	800	900	1000	1200/1300	1400/1500	38.5
Bread, grain, potatoes	g/d	200	260	350	420	520	630/520	700/580	19.3
Vegetables	g/d	120	150	200	220	250	260/300	300/350	10
Fruits	g/d	120	150	200	220	250	260/300	300/350	10
Dairy, cheese	g/d	300	330	350	400	420	425/450	450/500	13.7
Meat, fish, eggs	g/d	45	52	63	77	93	110/100	120/110	3.1
Oil, fat	g/d	15	20	25	30	35	35/40	40/45	1.2
Confectionery[a]	kcal/d	100	110	150	180	220	270/220	310/250	3.5

[a]Maximum 10% of daily energy intake.

2. IMPLEMENTATION OF THE OMD IN GERMANY

For public use, three simple messages were derived from the aggregated food groups:

- Consume abundantly: beverages and plant foods (e.g. fruit and vegetables, bread, potatoes);
- Consume moderately: foods of animal origin (e.g. dairy, meat);
- Consume sparingly: fatty and sugary foods (e.g. oil, fat, confectionery, soft drinks).

Using the three traffic light colors (green for abundantly, yellow for moderately, and red for sparingly), these rules are a simple and catchy tool for nutrition education.

The OMD is now widely accepted as the standard of a healthy diet during growth by pediatric and nutritional professionals. For example, the OMD is used for dietary adiposity and obesity intervention programs [8], as reference for food group intake in representative dietary studies [9], and as a basis for recommendations for communal feeding.

3. FRUIT AND VEGETABLES IN THE OMD

Plant food and beverages account for 78% of the OMD, followed by foods of animal origin (17%) and fatty and sugary foods (5%). Fruit and vegetables each account for 10%. As Table 9.2 shows, the impact of these food groups on energy was disproportionately low. The low energy density of fruit and especially vegetables compared to other food groups limits their proportion in the diet during growth. Other plant food groups with a higher energy density are necessary to fulfill the high energy needs in this age with an acceptable food volume. In contrast to energy, fruit and especially vegetables are important sources for many essential nutrients, e.g. fiber, potassium, folic acid, and vitamin C. Vegetables additionally yield magnesium, vitamin K, and vitamin A via β-carotene (Table 9.2).

TABLE 9.2   Percent (%) Food Consumption, Energy and Nutrients Provided by Food Groups in the Optimized Mixed Diet

% of Daily Intake	Beverages and Plant Food				Food of Animal Origin		Fatty and Sugary Food
	Fruit	Vegetables	Beverages	Potatoes, Bread, Grain	Dairy, Cheese	Meat, Fish, Eggs	Fat, Oils, Confectionery
Consumption	9.9	10.1	38.5	19.3	13.7	3.1	5.5
Energy	6.8	2.7	0.2	42.4	12.7	7.9	27.3
Protein	3.4	6.1	0.0	40.0	24.5	21.6	4.5
Fat	1.1	1.0	0.1	12.7	16.1	14.8	54.3
Carbohydrates	11.2	2.8	0.3	61.5	7.5	0.1	16.6
Fiber	13.6	15.1	0.0	65.7	0.3	0.1	5.3
Water	10.0	11.3	46.2	13.2	14.1	2.5	2.7
Potassium	16.5	20.7	1.2	34.6	15.5	6.1	5.6
Magnesium	10.8	12.4	4.2	50.3	11.6	5.2	5.6
Carotene	3.0	79.7	0.0	1.8	1.3	0.2	14.0
Vitamin K	9.0	65.6	0.0	7.5	0.7	3.6	13.7
Vitamin B1	6.6	9.7	2.4	55.2	7.8	16.0	2.4
Folic acid	13.2	30.5	0.0	43.1	7.1	4.2	1.9
Vitamin C	35.2	48.7	0.0	8.9	2.7	2.6	1.9

4. MEAL-BASED RECOMMENDATIONS OF THE OMD

A specialty of the OMD is the addition of meal-based recommendations to the total diet concept. Food groups of the optimized 7-day menus were also aggregated by the five meals that occur in the menus: first and second breakfast, lunch, afternoon snack and dinner. Since the food patterns of the first breakfast and dinner, as well as second breakfast and afternoon snack respectively were similar, the guidelines for these meals were combined, resulting in three meal-based guidelines: cold main meals (i.e. breakfast and dinner), snacks (in the mid morning and afternoon), and lunch.

All three meal guidelines are common with a great proportion of fruit or vegetables. Thus, the OMD is in accordance with 5-a-day campaign to promote fruit and vegetable intake worldwide.

The basis of the cold main meal is dairy products (41% of food intake), fruit/vegetables (28%), and grain (21%), completed by small amounts of meat and eggs (4%) and fats/oils (2%). The OMD lunch consists predominantly of potatoes, rice, pasta and legumes (41% of food intake), followed by fruit/vegetables (28%). Three times a week a lunch with a small portion of meat should be served; once a week with a portion of fish. In OMD snacks the percentage of fruit and vegetables is highest (40%); dairy (21%) and grain (18%) are the other main components.

Figures 9.2–9.4 show the three OMD meal pyramids, which display the composition of meals.

5. THE OMD IN COMMUNAL FEEDING

The meal-based guidelines of the OMD are an important tool for dietary recommendations in communal feeding. The calculated food group amounts provide an informative basis for portion sizes in school lunches. The daily food

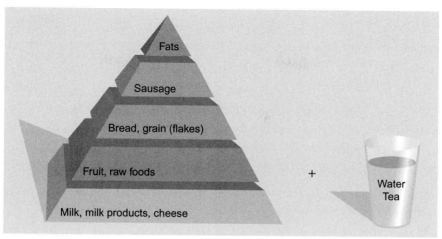

FIGURE 9.2 Composition of breakfast and dinner in the OMD.

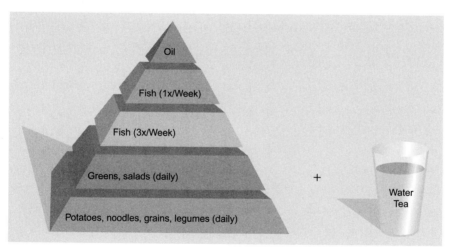

FIGURE 9.3 Composition of lunch in the OMD.

group amounts recommend a portion size of vegetables in school lunches (Table 9.1), and according to the OMD this varies with age. For 7–9-year-old children 110 g vegetables per meal are designated. For 15–18-year-old boys the vegetable portion size is doubled. Since consumed amounts of raw vegetables are often smaller than those of more soft, cooked vegetables, the amounts of uncooked vegetables may be reduced by one-third. Two studies on the feasibility of the OMD recommendations in communal feeding in schools showed that kitchen staff prefer concrete guidelines for their work. Therefore the FKE transcribed the OMD food amounts for lunch into a couple of recipes. Compared with common recipes, these optimized recipes contained 50% more vegetables, less meat, more wholegrain products, and had a better fat quality. Sensory tests with pupils showed good acceptance of these modified recipes compared with the customary dishes, especially in younger children [10].

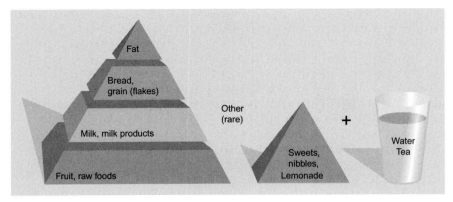

FIGURE 9.4   Composition of snacks in the OMD.

TABLE 9.3   Mean Intakes (Grams per Day) of the Three Food Items with the Highest Intakes per Food Group and Percentage of Total Food Group Intake Attained by these Food Items [12]

	Children (4–6 years)			Adolescents (13–14 years)	
	Boys (g/d)	Girls (g/d)		Boys (g/d)	Girls (g/d)
Fruits			**Fruits**		
Apple	38.3	33.3	Apple	56.5	61.2
Banana	20.9	21.2	Banana	22.0	20.0
Pears	7.8	7.6	Strawberry	13.0	10.5[a]
			Pears	3.8[a]	11.0
%[b]	59.0	53.8	%[b]	59.2	55.8
Vegetables			**Vegetables**		
Carrots	9.6	9.9	Creamed spinach[c]	10.6	6.2[a]
Cucumber	8.0	7.6	Carrot	8.8	9.6
Tomato	6.0	8.1	Tomato	8.5	13.6
			Cucumber	7.5[a]	9.2
%[b]	33.8	35.6	%[b]	24.2	23.4

[a]These food items do not originally belong to the three food items with the highest intake per food group, but were listed for comparison with the corresponding amount of the other sex group.
[b]Percentage of total food group intake attained by these food items.
[c]Convenience product, frozen.

6. FRUIT AND VEGETABLE INTAKE IN GERMANY COMPARED WITH THE OMD

Dietary surveys in Germany showed some clear discrepancies of food group intake compared with the OMD (see Table 9.3). Predominantly, children and adolescents should be encouraged to consume more plant foods and beverages. Figure 9.5 shows fruit and vegetables intake in 4–18-year-old participants of the *DOrtmund Nutritional and Anthropometric Longitudinally Designed*

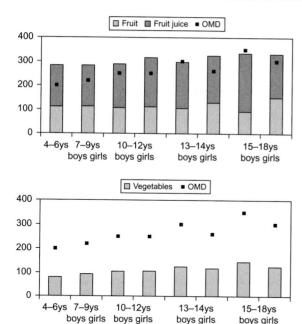

FIGURE 9.5 Intake of fruit and vegetables in 4–18-year-old participants of the Dortmund Nutritional and Anthropometric Longitudinally Designed (DONALD) Study compared with the recommended food amounts of the OMD.

Table 9.3 shows a customary selection of fruit and vegetable varieties in the DONALD Study. Those three varieties with the highest consumption amounts during the 3 days of recording were displayed. Apples and bananas are the most popular fruits; carrots, tomatoes, and cucumbers the most popular vegetables [12]. The three most consumed fruit varieties accounted for more than half of fruit intake. For vegetables, the most consumed varieties only accounted for 23–36% of total intake, indicating a higher diversity of consumption. Whereas food group intake differs between sex and age groups, selection of different foods in the DONALD Study is very similar, i.e. the most important food items of each food group are almost the same in all four subgroups [12].

7. FRUIT AND VEGETABLES IN CONVENIENCE (PRE-PREPARED) FOOD

Consumer researchers have reported an increasing trend toward the use of convenience food (CF) products in Germany during the last decades. This development is attributed to several social changes: the increasing number of single or small family households; the increasing number of women in employment; a social trend toward more leisure time; the break-up of the traditional family mealtimes; and an overall increase in the demand for easily prepared and just-in-time meals. Simultaneously, the technical equipment in household kitchens (freezer, microwave) has increased, and food preparation skills have decreased. However, little is known about consumption patterns and composition of convenience products. In an evaluation of those convenience foods consumed in families of the DONALD Study 2003–2006, in 86% of all records at least one convenience product was recorded [13]. Vegetable dishes, e.g. canned or frozen vegetables, with spices, cream or fat additives or vegetarian dishes, accounted

(DONALD) Study compared with the recommended food amounts of the OMD. At first sight, intake of fruits was in good accordance with the recommended intake of the OMD. However, more than half of the fruit intake was consumed as fruit juice. High fruit juice intake is suspected to promote the development of obesity in adolescence [11]. Additionally, fruit juice consumption increased the risk of caries and the fiber content of fruit juice is lower than that of the original fruits. Therefore, at most one portion of fruit should be consumed as juice, and the consumption of unprocessed foods has to be increased.

Vegetable consumption in the DONALD study marginally increased with age in spite of clearly increasing energy and food requirements. Hence, the recommended vegetable amounts of the OMD were not reached in any age group.

for 10% of all recorded convenience products (69 different products). Median energy density of these products was 80 kcal/100 g; 55% of energy content came from fat, 31% from carbohydrates, and 14% from protein [13], indicating a selection of convenience vegetable products with additives of fat and/or cream. The important advantages of vegetable intake, the low energy density and fat content, are lost in this way. Therefore, consumers should be encouraged to read CF ingredients labels to get an idea of the composition. Current political efforts to enforce the defining of nutrient profiles of foods could be a significant step to simplify the choice of healthy CF products by the consumer. Vice versa, improved consumer demand could encourage the food industry to offer improved products.

8. STRATEGIES FOR AN INCREASED FRUIT AND VEGETABLE INTAKE IN CHILDREN AND ADOLESCENTS

The family is the central unit for the development of food preferences and dietary habits [14], not only in childhood but also in adolescence. The development of preferences has already started during pregnancy and continues during lactation. Children are more likely to accept foods that are eaten by the mother and hence introduced via the amniotic fluid and breast milk. This early exposure lays the foundation for the acceptance of a culturally determined diet.

During infancy and early childhood, familiarity is a major determinant of food preferences [14]. Parents should be encouraged to offer a variety of flavors. Food preferences are also acquired by social learning. Food intake is modeled on those people around us, particularly those whom we respect. For this reason, the example of parents is of great importance. Several studies have shown a high correlation

of food patterns within families. Such similarity is not only found in children but also during adolescence, although the influence of peers is suspected to be more powerful in this age group [15].

The learning of preferences is influenced by the emotional atmosphere at a meal. In a positive atmosphere, food preferences and food consumption amounts increase [16].

A common strategy to get children to eat a 'healthy' but disliked food is to lure them with a reward, e.g. a dessert. However, this leads to a further decline in the preference for this rejected food. On the other hand, the preference for foods given as a reward enhances. Therefore, parents and caregivers should reflect how they present food to children and establish food preferences by the way they do it [14].

It is not only families that are responsible for food selection and intake; politicians should also be aware of the need to create positive frameworks. Food selection is determined to a certain degree by food costs. A calculation on the costs of the OMD showed that the German concept of calculating social welfare for children and adolescents did not result in a sufficient amount for this healthy diet, because energy requirements were not sufficiently attended to [17]. Per 1000 kcal, a diet corresponding to the OMD costs on average €1.67, using foods only from a discounter. Buying in supermarkets raises the cost by €1 per 1000 kcal. Fruit and vegetables account for approximately 36% of food costs in the OMD, followed by cereals (12–22%), meat and fish (12–18%) and dairy (9–10%), depending on the type of store (discounter, supermarket, healthy food shop) [17].

Schools are important locations to reach almost all children and adolescents and therefore give the opportunity for interventions to improve fruit and vegetable intake. It should be self-evident that school lunches should include sufficient (see above) daily portions of vegetables. Unfortunately, in Germany there are no binding standards on school lunch quality [10].

To improve school breakfasts, the distribution of school milk, at least until the fourth school year, is traditional in Germany and government subsidized. Especially for those children who fail to eat breakfast, this is an important tool to improve dietary patterns. However, an additional school fruit program should be established.

9. SUMMARY

The OMD as an example of FBDG for children and adolescents recommends daily amounts of fruit and vegetables, which are on average not reached in the actual diet. To improve dietary patterns, different strategies are necessary, not only on an individual basis in families but also at governmental level. Parents (and caregivers) should be encouraged to:

- offer fruit or vegetables at every daily meal in a positive emotional atmosphere;
- give a good example by showing a healthy food pattern and be aware that preferences are impressed very early;
- offer a variety of healthy and palatable foods but let the child decide what and how much to eat of this.

References

1. Lake, A. A., Mathers, J. C., Rugg-Gunn, A. J., & Adamson, A. J. (2006). Longitudinal change in food habits between adolescence (11–12 years) and adulthood (32–33 years): the ASH30 Study. *Journal of Public Health (Oxford), 28,* 10–16.
2. American Academy of Pediatrics. Committee on Nutrition. (1998). Cholesterol in childhood. *Pediatrics, 101,* 141–147.
3. FAO/WHO. (1998). *Preparation and use of food-based dietary guidelines.* Geneva: WHO.
4. Kersting, M., Alexy, U., & Clausen, K. (2005). Using the concept of food based dietary guidelines to develop an optimized mixed diet (OMD) for German children and adolescents. *Journal of Pediatrics, Gastroenterology, Nutrition, 40,* 301–308.
5. Sichert-Hellert, W., Kersting, M., & Schöch, G. (1999). Consumption of fortified food between 1985 and 1996 in 2- to 14-year-old German children and adolescents. *International Journal of Food Science Nutrition, 50,* 65–72.
6. Sichert-Hellert, W., Kersting, M., & Manz, F. (2001). Changes in time-trends of nutrient intake from fortified and non-fortified food in German children and adolescents – 15 year results of the DONALD study. Dortmund nutritional and anthropometric longitudinally designed study. *European Journal of Nutrition, 40,* 49–55.
7. German Society of Nutrition (Deutsche Gesellschaft für Ernährung, DGE). (2000). *Referenzwerte für die nährstoffzufuhr [nutrient reference values].* Frankfurt: Umschau-Verlag.
8. Alexy, U., Reinehr, T., Sichert-Hellert, W., Wollenhaupt, A., Kersting, M., & Andler, W. (2006). Positive changes of dietary habits after an outpatient training for overweight children. *Nutrition Research, 26,* 202–208.
9. Mensink, G. B., Bauch, A., Vohmann, C., Stahl, A., Six, J., Kohler, S., Fischer, J., & Heseker, H. (2007). [EsKiMo – the nutrition module in the German health interview and examination survey for children and adolescents (KiGGS)]. *Bundesgesundheitsblatt Gesundheitsforschung Gesundheitsschutz, 50,* 902–908.
10. Kersting, M., & Clausen, K. (2007). School feeding at whole-day schools. *Ernährungs-Umschau., 54,* 114–119.
11. Libuda, L., Alexy, U., Remer, T., Stehle, P., Schoenau, E., & Kersting, M. (2008). Association between long-term consumption of soft drinks and variables of bone modeling and remodeling in a sample of healthy German children and adolescents. *American Journal of Clinical Nutrition, 88,* 1670–1677.
12. Alexy, U., Sichert-Hellert, W., Kersting, M., & Manz, F. (2001). The foods most consumed by German children and adolescents. *Annals of Nutrition and Metabolism, 45,* 128–134.
13. Alexy, U., Sichert-Hellert, W., Rode, T., & Kersting, M. (2008). Convenience food in the diet of children and adolescents: consumption and composition. *British Journal of Nutrition, 99,* 345–351.
14. Benton, D. (2004). Role of parents in the determination of the food preferences of children and the development of obesity. *International Journal of Obesity Related Metabolism Disorder, 28,* 858–869.
15. Story, M., Neumark-Sztainer, D., & French, S. (2002). Individual and environmental influences on adolescent eating behaviors. *Journal of the American Dietetic Association, 102,* S40–S51.
16. Birch, L. L. (1999). Development of food preferences. *Annual Review of Nutrition, 19,* 41–62.
17. Kersting, M., & Clausen, K. (2007). What is the price of a healthy diet for children and adolescents? *Ernährungs-Umschau., 54,* 508–513.

10

The Antibacterial Properties of Dietary Fruit

Rachel O'Mahony

NCC-CC, Royal College of Physicians, Regents Park, London, UK

1. INTRODUCTION

Bacteria pose a significant problem for the health of humans, animals, and plants, and they also infect the produce that we eat. They have therefore been the target of eradication since their discovery over 100 years ago as causative agents of infection and disease. Since then, with the discovery of antibiotics, antibacterial agents, and vaccines, bacterial infection became less of a widespread problem and a number of common diseases such as pneumonia, tuberculosis (TB) and cholera were practically eradicated from Western society. However, in the last 30 years there has been a drastic global increase in the use of broad-spectrum antibiotics to treat bacterial infection, and the addition of antibacterial agents to everyday hygiene and cleaning products. As a result, problems have arisen and we are beginning to see, in the early twenty-first century, the gradual reappearance of diseases that were previously thought to be under control (such as TB and pneumonia) [1–3].

This problem has arisen because antibiotics and antibacterial agents kill bacteria. Because bacteria have a very short generation time, they are able to produce rapid responses to environmental changes. When the selective pressure of an antibacterial agent is introduced into their environment, bacteria respond by becoming resistant, that is they remain capable of reproducing and surviving even in the presence of the agent. This is known as acquired resistance. Widespread use of antibiotics and antibacterials has therefore led to a huge selective pressure, and strains of bacteria around the world have rapidly developed resistance to these agents faster than we are able to develop new ones.

2. BACTERIAL RESISTANCE

Acquired resistance is a consequence of changes in the cellular structure and physiology

141

of a bacterium and is brought about by modifications in its normal genetic makeup. Most modifications are caused by genetic mutations, by the transfer of genes from one organism to another enabling the acquisition of resistance genes, or by a combination of both.

There are three main mechanisms by which bacteria may develop resistance to the effects of an agent [4].These are:

1. Prevention of intracellular drug accumulation by:
    *   changes in the outer bacterial membrane so it can no longer bind;
    *   terminating active transport of the drug across the membrane; or
    *   up-regulating active efflux mechanisms to pump the drug out before it causes damage.
2. Modification of the part of the microbe that the drug targets, resulting in ineffective levels of drug binding to the target site.
3. Production of a drug-inactivating enzyme that reduces or even eliminates the ability of the drug to kill the microorganism.

However many new antibiotics are discovered that have static effects or can kill bacteria, it is almost certain that the bacteria will develop resistance to all the agents developed. Additionally, the speed at which new agents are made available for use (discovered and tested for safety and efficacy) is unlikely to be able to keep up with the rate at which these organisms develop resistance to those in current use, in order to provide replacements. It is therefore necessary to look for new sources of antibacterial agents as well as to develop novel, alternative strategies to successfully treat infection. Such strategies must circumvent the mechanisms that induce bacteria to develop resistance (namely interference with growth, metabolism or killing).

3. DISCOVERY OF NEW METHODS TO COMBAT BACTERIAL INFECTION

Most current antibacterial agents work in the following ways:

1. bacteriostatic activity – inhibits the growth of the organism.
2. bactericidal activity – kills the organism, often through lysis.

These therefore lead to the development of resistant bacteria. However, to combat this a number of new strategies and agents have been devised to overcome bacterial resistance mechanisms. These include:

1. Inhibitors of bacterial efflux pumps [5–8]
2. Specific targeting of antimicrobial drugs via:
    *   probiotics [9,10];
    *   photodynamic therapy [11];
    *   bioadhesives and mucolytic agents (aids drug targeting at the mucosal surface) [12–14];
    *   immune enhancement and vaccine development; and
    *   inhibition of bacterial adhesion [15,16].

Inhibition of adhesion has been an important strategy as it is an alternative method to the use of antibiotics that does not involve 'killing' the target organism as most of the other alternative strategies do. The principle behind the method is to target the first step of bacterial infection: to inhibit the adhesion of the bacterium to the host tissue [17]. Because inhibition of adhesion sterically blocks bacteria from attaching to host tissue rather than having a bactericidal effect, the selection of inhibitor-resistant strains is unlikely to occur [18]. If the bacterium can be prevented from binding, then it is less likely to invade and establish infection, resulting in tissue damage and host debilitation. Inhibitors may also be able to 'remove' organisms that are already bound, thus providing a therapeutic as well as prophylactic application. This has been shown for *Helicobacter pylori* [19,20].

Bacteria adhere to tissue by specific ligand–receptor interactions. Therefore, by using molecules (analogues) that mimic the microbial adhesin or its complementary host cell receptor, or by using antibodies that target the adhesin or receptor, it is possible to block microbial adhesion to the host tissue (Figure 10.1).

4. DISCOVERY OF NEW SOURCES OF ANTIMICROBIALS

In addition to new strategies to fight infection, it is important that new sources of antimicrobials are found, particularly those which may have limited adverse effects on the 'patients' who are using them. A number of new sources have recently been investigated and these include:

1. Animal products such as:
   - colostrum and milk from humans and animals [21–23]
   - products such as propolis or honey (both produced by bees) [24–26].
2. Food products such as:
   - melanoidins which form in heat-treated foods [27]
   - glycopeptides in buttermilk [28].
3. Plants and plant products [29].

Plants are one of the most up and coming as well as promising natural sources of antibacterial substances and anti-adhesives. Investigations based on ethnobotanical studies have found that many plant products have cidal, static and anti-adhesive effects on a huge variety of different microorganisms [29,30]. In particular, a number of studies have shown that certain dietary fruits are very good sources of antibacterial agents.

The major focus of this chapter will be to look at dietary fruits as novel sources of antimicrobials, in particular, to investigate which bacteria they target, which components are responsible for their activity, and most importantly what evidence there is for their effectiveness in humans.

5. THE ANTIBACTERIAL EFFECTS OF DIETARY FRUIT

For thousands of years, fruits have been used by cultures all over the world as an integral part of medicinal treatments. Although some uses were based on superstition, many fruits did in fact have healing properties. With the modern trend toward using complementary, alternative, and natural medicines, many of these lost 'secrets' are being rediscovered and numerous studies and articles have been conducted to address the topic. This chapter is a review of the literature on the antibacterial effects of fruits (see box for definition) that are eaten in the Western world, and will give an overview of the experimental and clinical studies that have been conducted for each fruit in turn. Unless specifically

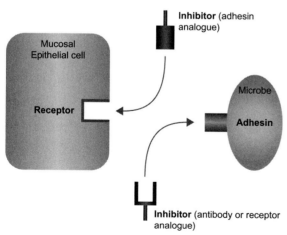

FIGURE 10.1 Principles of inhibition of microbial adhesion Block the first step of infection, the adhesion of the microbe to the host cell, by designing inhibitors (analogues) that mimic the adhesin or receptor.

DEFINITION OF A FRUIT

In botanical terms, a 'true' fruit is the part of the plant that develops from the ovary after fertilization, its flesh constituting the pericarp walls. Several 'true' fruit (such as the avocado, tomato and olive) are commonly and mistakenly called vegetables and used to form savory dishes in Western society.

mentioned, it is the 'fruit' of the plant that is referred to in this chapter, rather than other aerial parts of the plant itself.

5.1 Apricot (*Prunus armeniaca*)

Apricot extracts have the potential to be used as an alternative source of antibacterial agents, as they have been shown to be effective against a wide range of Gram-positive (G+) and Gram-negative (G−) bacteria. The *in vitro* experiment found that two flavonoid glycosides (from butanolic fractions of dried apricot extracts) were effective inhibitors of bacterial growth [31]. They were particularly effective against G+ bacteria, including species of *Staphylococcus* and *Streptococcus*, clinical isolates of methicillin-resistant *Staphylococcus aureus* (MRSA) and non-tuberculous mycobacteria. The G− bacteria included *Escherichia coli (E. coli)*, species of *Salmonella, Klebsiella pneumoniae, Pseudomonas, Shigella dysenteriae* and *Proteus* species. A second study has shown that other active antibacterial components from the apricot include triterpene acid and its glycoside [32].

5.2 Berries

Berries are known to be naturally high in antioxidants and therefore would probably be an obvious choice when looking for potential antibacterial agents. True to this, a number of studies have tested a variety of berries for their bioactive properties against bacteria.

5.2.1 Raspberry (*Rubus idaeus*)

Two studies have looked at the effects of raspberry juice and cordial on the growth inhibition of bacteria *in vitro*. These studies found that they reduced the growth of five human pathogenic bacteria (including *Salmonella, Shigella* and *E. coli*) [33] and a number of other G+ and G− bacteria [34]. In a third study [35], the seed extracts of the black raspberry were shown to be bactericidal against *E. coli* and to inhibit the growth of *Listeria monocytogenes* in fish oils. Raspberry extracts could therefore have a use as food preservatives as well as antibacterial agents against human pathogenic bacteria. Raspberries contain complex phenolic compounds such as ellagitannins which are known to have antibacterial properties on common human gastrointestinal (GI) pathogens such as *Staphylococcus* and *Salmonella* [36]. Pre-treatment of the fruit with enzymes may also increase their activity (as for bilberries and strawberries – see later sections) by releasing the active components from the cell walls of the fruit [36].

5.2.2 Strawberry (*Fragaria*)

Although the strawberry is a very commonly eaten fruit, few studies have investigated its potential as a source of antibacterial compounds. Complex phenolic compounds and polymers such as ellagitannins are present in strawberries and these are known to inhibit the growth and possibly have anti-adhesive effects against the common human GI pathogens, *Staphylococcus* and *Salmonella* [36]. Pre-treatment of the fruit with enzymes may increase their activity by releasing the active components (as for bilberries and raspberries) [36]. Strawberries may therefore be useful as anti-microbials in the food industry as using them in fruit products may increase their shelf-life.

5.2.3 Bilberry (Vaccinium myrtillus)

The bilberry (in particular phenolics isolated from the cell wall) has been shown to have antibacterial activity (growth inhibition) against *Salmonella enterica* and *Staphylococcus aureus* [37]. The antibacterial activity can be increased by pre-treating juices, press cakes and berry mashes with enzymes that liberate the cell wall phenolics, especially under acidic conditions, which is the natural environment of these products (particularly the juices). The bacteria that were tested in these experimental studies are involved in food spoilage and thus pre-treatment of the fruit with enzymes could increase the shelf-life of these products, making them useful as antimicrobials in the food industry.

5.2.4 Blackberry (Rubus fruticosus) and blackcurrant (Ribes nigrum)

Fresh blackberries and blackcurrants, as well as commercial culinary cordials, have been shown to inhibit the growth of a number of G + and G − bacteria, but *Mycobacterium phlei* (a bacterium involved in food spoilage) was particularly susceptible [34]. The most active product was blackcurrant cordial, which inhibited the growth of all 12 strains of bacteria. Like the bilberry, blackberries and blackcurrants may be useful if incorporated into food products, to enhance their shelf-life. They may also be useful for treating *H. pylori* infection as one study has demonstrated that acidic high molecular weight galactans from blackcurrant seeds can inhibit the adhesion of *H. pylori* to human gastric mucosa tissue sections [38].

5.2.5 Cranberry (Vaccinium macrocarpon)

Since the publication of early *in vitro* studies showing that the cranberry had antibacterial properties, this fruit has been the subject of numerous experimental and clinical studies. Cranberry and its constituents have been shown to have bactericidal, bacteriostatic and anti-adhesive effects on the pathogenic bacteria that infect the stomach, the mouth, and the urinary tract.

Helicobacter pylori is a bacterium which infects the stomach of over half the population worldwide, leading to the development of severe gastroduodenal diseases such as ulcers and cancer in 10–20% of those infected. Several studies have demonstrated the antibacterial effects of cranberry extracts on *H. pylori*. A high molecular weight non-dialyzable constituent has been shown to inhibit *H. pylori* adhesion [39–41], phenolic extracts to inhibit its growth and its urease activity [42], and the juice (especially in combination with triple therapy − antibiotics + proton pump inhibitors) to increase the eradication rate of *H. pylori* in infected women [43]. Interestingly, many clinicians tell their patients to drink cranberry juice as part of their treatment program for *H. pylori* infection.

The non-dialyzable material (NDM) of cranberry, which was shown to inhibit *H. pylori* adhesion, has also been tested for its effectiveness against oral pathogens. It was found to reduce the ability of *S. mutans* to form biofilms [44] and to adhere to saliva-coated hydroxyapatite [45]. Additionally, when given as a mouthwash in a clinical trial, there was a significant reduction in the number of salivary *S. mutans* and total bacteria compared to the placebo mouthwash [45].

Most studies on cranberry have assessed the ability of the juice, raw fruit, or various extracts as a treatment for UTIs (urinary tract infections) and the bacteria that are known to cause them. Studies have shown that cranberry has anti-adhesive effects, can reduce bacterial growth, and has some therapeutic effect on UTIs (Table 10.1) [46–54]. However, the clinical study data provides limited evidence since most of the studies and trials had various sources of bias (most had small sample sizes and were thus likely to be underpowered, some were open-label and the results were mixed). The studies also varied hugely in treatment regimen (different cranberry extracts, doses, and outcome measures). Larger, longer-term studies using extracts at a variety of doses are

TABLE 10.1   Urinary Tract Infection (UTI) Studies Involving Cranberry

Cranberry Component	Methods	Result	Reference
H. pylori			
High molecular weight non-dialyzable constituent	Anti-adhesion to gastric monolayers, gastric mucus and human erythrocytes	Anti-adhesion	[39–41]
Phenolic phytochemical extracts	*In vitro*	Inhibited growth and urease; synergistic effect when mixed with oregano	[42]
Juice	Clinical trial (RCT): Patients with *H. pylori* infection given cranberry juice vs. cranberry + triple therapy vs. placebo twice/day for 2 weeks	NS difference between the groups for eradication rate. However eradication was significantly higher for cranberry + triple therapy vs. placebo in women	[43]
Oral bacteria			
Non-dialyzable material (NDM)	Clinical trial: mouthwash containing NDM vs. placebo – daily use for 6 weeks	Significant reduction in salivary mutans streptococci and total bacteria compared to placebo. No change in plaque and gingival index	[45]
Non-dialyzable material (NDM)	*In vitro*: Anti-adhesion of *S. sobrinus* to saliva-coated hydroxyapatite	Inhibited adhesion	[45]
Non-dialyzable material (NDM)	*In vitro*: Anti-adhesion of *S. sobrinus* to artificial dental biofilm	Pre-coating bacteria with NDM inhibited adhesion. NDM also prompted desorption of *S. sobrinus* from the biofilm	[44]
UTI			
Juice	Clinical study: UTI patients drink cranberry juice after antibiotic treatment finished	0% patients treated with cranberry had relapse of symptoms vs. 21% of controls (no juice)	[46]

(Continued)

TABLE 10.1 (Continued)

Cranberry Component	Methods	Result	Reference
Juice	Clinical trial (RCT): Older people in hospital given cranberry juice vs. placebo to see if it reduced incidence of UTIs	NS difference between the groups for incidence of UTIs. Significant reduction in number of *E. coli* infections (cranberry vs. placebo)	[47]
Juice	Clinical trial: Cranberry juice vs. mixed berry juice in geriatric patients	No reduction in incidence of UTIs	[48]
Juice	Clinical study: P-fimbriated *E. coli* from women with UTIs incubated with urine from healthy women after consuming cranberry juice ($N = 39$ isolates tested)	Inhibited adhesion of P-fimbriated *E. coli* to uroepithelial cells (79% of antibiotic resistant strains and 80% of non-resistant strains)	[49]
Cranberry supplement	Clinical trial (RCT): cranberry supplement vs. placebo for UTI prevention in patients with neurogenic bladders. Three tablets/day for 4 weeks	NS difference between the groups for bacterial count in urine	[50]
Cranberry capsules	Clinical trial (RCT): *E. coli* cultured from urine of healthy people 12 hours after consumption of cranberry capsules. Measured adhesion of *E. coli* to epithelial cells and ability of *E. coli* to kill *C. elegans*	*E. coli* – reduced ability to kill *C. elegans* after growth in urine from people taking cranberry capsules. Significant reduction of adhesion compared to controls	[51]
Dried cranberries	Clinical study: UTI patients ate dried cranberries vs. raisins vs. controls – urine collected after consumption of each and incubated with uropathogenic *E. coli*. These bacteria then tested for ability to inhibit adhesion of P-fimbriated *E. coli*	Some anti-adhesion activity by urine after consumption of cranberries. None after consumption of raisins or control	[52]
Concentrated cranberry preparation (30% phenolics)	Clinical study: 1 capsule given twice/day for 12 weeks to women with history of recurrent UTIs. Two-year follow-up	During the 4 months no women had UTIs. Over the 2-year follow-up the $N = 8$ women who continued to take cranberry had no UTI infections	[53]
Fractions rich in anthocyanins and proanthocyanidins	*In vitro*: 9 bacterial strains (G+ and G−) tested for inhibition of growth.	Some strains susceptible to some extracts (*S. aureus*, *Enterofaecalis* and *Micrococcus luteus* susceptible to proanthocyanin-rich extracts). *S. mutans*, *E. coli* and *P. aeruginosa* not susceptible to any fractions	[54]
Cranberry proanthocyanidin extract	Clinical study: P-fimbriated *E. coli* from women with UTIs incubated with urine from healthy women after consuming cranberry extract ($N = 39$ isolates tested)	Inhibited adhesion of 100% of P-fimbriated *E. coli* (antibiotic resistant and non-resistant strains) to uroepithelial cells	[49]

RCT, randomized controlled trial; NS, non-significant.

needed to have a better idea of the effectiveness of cranberry at treating urinary tract infections (UTIs) or other infections.

However, despite the inconsistent evidence from published trials, in clinical practice cranberry juice (in conjunction with antibiotics) is often recommended to UTI patients by their doctor, and seems to provide help, especially as many UTI-causing bacteria are becoming resistant to the antibiotics currently used.

5.3 Citrus fruit

5.3.1 Grapefruit (Citrus paradisi)

Most studies that have investigated the antibacterial properties of grapefruit have focused on extracts of the seeds. Seed extracts have been shown to inhibit the growth of [55] and have bactericidal effects [56] on numerous G+ and G− bacteria and on multiple strains of *Legionella pneumophila* [57]. However the extracts may be toxic at high concentrations to human cells [56]. Seed and pulp extracts have also been found to inhibit the growth of 20 bacteria, being most effective against *Salmonella enteridis* [58]. The extracts contained flavanones, naringin and hesperidin, indicating that these may be the active components.

Interestingly, grapefruit seeds have been tested in a basic clinical study [59], given orally for 2 weeks to patients with UTIs caused by different antibiotic-resistant strains of bacteria (*P. aeruginosa, Klebsiella, S. aureus,* and *E. coli*). The results showed that dried or fresh seeds had comparable efficacy to that of proven antibacterial drugs, and there was a satisfactory response in all patients except for the one whose UTI was caused by *P. aeruginosa*. Bacterial growth was reduced over time and a reversal of antibiotic resistance pattern was seen. However, grapefruit seed extracts may have some toxic effects when taken orally [60]; this needs to be tested and confirmed in further studies.

In addition to the seed extracts, the volatile oils of grapefruit have also been shown to have antibacterial activity, inhibiting the growth of 10 bacteria (particularly *S. aureus, S. faecalis, Shigella flexneri,* and *Mycobacterium smegmatis*) [61].

The grapefruit therefore has the potential to be a good source of several antibacterial components and a possible alternative treatment for UTIs.

5.3.2 Lemon (Citrus limon)

Oils from the lemon have been shown by several studies to have antibacterial activity. The essential oil and volatile oil were able to inhibit the growth of a number of G+ and G− bacteria [61,62] with the lemon volatile oil being more effective than those found in orange and grapefruit. When tested against foodborne pathogens, lemon oils citral and linalool were found to be effective against bacteria found in apple juice (*E. coli* and *S. enterica*) [63]. Additionally, the vapor of lemon oils (particularly citral and linalool) has been shown to inhibit the growth of G+ and G− bacteria (*Campylobacter jejuni, Escherichia coli, Listeria monocytogenes,* and *Staphylococcus aureus*) both *in vitro* and when tested in food systems [64].

Lemon juice has also been shown to be an effective antibacterial agent, inhibiting the growth of *Vibrio* species and antibiotic-resistant clinical isolates of *Staphylococcus aureus, E. coli, Klebsiella aerogenes,* and *Klebsiella pneumoniae* in two *in vitro* studies [65,66]. Interestingly, the active component in both studies was identified as citric acid.

These studies therefore show that the oils of lemon may be useful in food preservation, and the juice (particularly citric acid) and oils as antibacterial agents.

5.3.3 Orange (Citrus sinensis)

Several different components of regular sweet oranges (*C. sinensis*) have been shown to have antibacterial activity. Two studies found that the volatile oil inhibited the growth of various G+ and G− bacteria (*Staphylococcus aureus, Streptococcus faecalis, Shigella flexneri, Mycobacterium smegmatis, E. coli, Klebsiella pneumoniae, Pseudomonas aeruginosa,* and *Proteus vulgaris*) [61,62], and a third study [67] found that orange oils were antimicrobial against G+ *S. epidermis* and G− *E. coli*. A fourth study [68] found that essential oils (which contained high concentrations of limonene) from the peel of both the sweet orange and the bitter orange (*C. aurantium*), inhibited the growth of *Propionibacterium acnes* but had little effect at low concentrations on *S. epidermis*. The oil also had low cytotoxicity to human skin cells, despite its antioxidant activity.

The rind of the mandarin orange (*C. reticulata* and *C. deliciosa*) has also been shown to have antibacterial activity. Particularly active components against G+ and G− bacteria were identified as four polymethoxyflavones and two flavone glycosides [69,70]. However, in another study [71], the crude extract from the rind of the sweet orange was shown to have no or little effect against multidrug-resistant, β-lactamase-producing MRSA and methicillin-sensitive *Staphylococcus aureus* (MSSA) isolates. Other extracts, however (the exact part of the fruit is not mentioned), have been shown to be active against drug-resistant *Mycobacterium tuberculosis* isolates, and not cytotoxic to human cells [72].

Bitter orange (*C. aurantium*) extracts have also been found to strongly inhibit the growth of *E. coli* and *S. aureus* [73].

There is therefore the potential for several different types and parts of oranges to provide a new source of antibacterial agents.

5.3.4 Lime (Citrus aurantifolia)

Extracts from the lime have been shown to inhibit the growth of a number of different bacteria. This includes methanol extracts against *E. coli* and *S. aureus* [13], the citric acid component against various strains of *Vibrio* [65], and several extracts against *M. tuberculosis* (exact part of fruit not mentioned) which were also shown not to be toxic to human cells [72]. One study [62] also found that the antibacterial activity of the lime was superior to that of the orange or lemon when tested against various G+ and G− bacteria (*E. coli, Klebsiella pneumoniae, Pseudomonas aeruginosa, Proteus vulgaris,* and *Staphylococcus aureus*). The lime may therefore be a useful source of antibacterial agents to treat food poisoning (*E. coli* infection), respiratory tract infections including pneumonia, UTIs, wound infections, and nosocomial infections.

5.4 Exotic and tropical fruits (that are commonly eaten in Western society)

5.4.1 Avocado (Persea americana)

Although the avocado is commonly eaten as a savory food, it is actually classed as a fruit. Botanically speaking it is a type of berry or drupe because it develops from the ovary of the plant, with its flesh comprising the pericarp.

Active compounds (long-chain aliphatics and their derivatives) have been isolated from the avocado and shown to have antibacterial effects on a number of different bacteria [74]. In particular, this study found that the most active compound was 1,2,4-trihydroxy-n-hepadeca-16-en, which inhibited the growth of, and in some cases was bactericidal against, several G+ (*Bacillus* species, *S. aureus*) and G− bacteria (*Salmonella typhi, Shigella dysenteriae*). So, as well as being a good source of over 20 vitamins and minerals, and healthy fats, the avocado may also be an effective source of antibacterial compounds.

5.4.2 Banana (Musa)

The dwarf banana (*Musa acuminata*) found in East Asia has been shown *in vitro* to have antibacterial activity against drug-resistant variants of *Mycobacterium tuberculosis* [72]. Isothiocyanates from a relative of the banana, the plantain (*M. paradisiaca*), have also been shown to possess antibacterial activity, having cidal effects on strains of *E. coli* when tested *in vitro* [75]. However, one study [76] found that the unripe pulp of plantain may have antagonistic antimicrobial effects against *E. coli* when given in combination with the antibiotic ciprofloxacin. Species of the banana may therefore have some use as agents for treating TB and the bacteria that cause food poisoning (*E. coli*).

5.4.3 Fig (Ficus)

Extracts from several species of Egyptian fig. (*F. sycomorus* L., *F. benjamina* L., *F. bengalensis* L., and *F. religiosa* L) have been shown *in vitro* to have significant antibacterial activity. However, two species (*F. bengalensis* and *F. religiosa*) may also be toxic to cells [77], and therefore may have limited use as antibacterial agents in humans, but may still be useful as disinfectants.

5.4.4 Guava (Psidium guajava)

Several studies have investigated the antibacterial effects of the guava plant. The leaf extract has been shown to inhibit the growth of cultured bacteria from acne lesions (*P. acnes*) and to have a significantly greater effect than tea tree oil [78]. The same study showed that when compared to the effects of the antibiotics doxycycline or clindamycin, guava extracts were equally effective for *Staphylococcus* species and less effective for *P. acnes*. The extract was also found to have anti-inflammatory activity and is therefore a good alternative treatment for acne to both antibiotics and tea tree oil, both of which are currently used widely and are probably inducing the development of antibiotic-resistant strains [79,80].

Guava fruit extracts have also been assessed in a number of *in vitro* studies. Growth

inhibition has been demonstrated by boiled water extracts (on penicillin-resistant *S. aureus* and *E. coli*) [81], methanol extracts (on 11 wound isolates) [82], and ethanol extracts (against 21 strains of foodborne pathogens). The ethanol extracts were found to be particularly effective against G+ bacteria, but none were effective against *E. coli* or *Salmonella enteritidis* [83]. The methanol extracts were also shown to increase wound healing *in vivo* and were significantly better than antibiotics or control (water) groups [82].

A close relative of the guava is the pineapple guava (*Feijoa sellowinana*), which has also been shown to have antibacterial properties *in vitro*. One study [84] found that the essential oils were active against bacteria and the major-constituents of the oil were: limonene (30%), β-caryophylliene (28%), α-pinenem β-pinene (8%), isocaryophyllene and estragole (all <5%). A second study [85] found that aqueous extracts of the fruit inhibited the growth of a number of G+ and G− bacteria and also had antioxidant properties. The most sensitive bacteria were *Pseudomonas aeruginosa*, *Enterobacter aerogenes* and *Enterobacter cloacae* – opportunistic bacteria often involved in nosocomial infections (*Enterobacter* species) or affecting immunocompromised patients (*P. aeruginosa*).

The leaves and fruit of the guava therefore have the potential to be used as an alternative treatment for acne, for wound healing, and against foodborne diseases and *Staphylococcus* infections. Additionally, its relative the pineapple guava could be used against opportunistic and nosocomial infections, where it may be put to good use as an alternative disinfectant for catheters.

5.4.5 Pineapple (Ananas sativus)

A hydroalcoholic extract of the pineapple fruit (formulated as a medicinal drink) has been tested against *M. tuberculosis*, but was found not to have bactericidal activity even after 24 hours of exposure to the drink [86]. However, the enzyme bromelain from the pineapple has been shown in a number of other studies [87–90] to increase the absorption of antibiotics. Treatment with bromelain may therefore indirectly enhance the antibacterial activity of antibiotics, as has been shown in two clinical studies [91,92]. Further studies are needed to see whether various components of the fruit have antibacterial effects on other types of bacteria.

5.4.6 Kiwi (Actinidia chinensis)

Extracts of the pulp, skin, and seeds of the kiwi fruit have been shown to have bacteriostatic effects on eight G+ and G− bacteria, the seeds being most active [93]. Although extracts of kiwi may inhibit the growth of these bacteria, other extracts have also been shown to be inactive against H. pylori and cytotoxic to human cells [94,95].

5.4.7 Papaya (Carica papaya)

Three studies have shown that extracts of papaya are active against E. coli and other G− or G+ bacteria (S. aureus, Streptococcus faecalis, Bacillus cereus, P. vulgaris, P. aeruginosa, and Shigella flexneri) [96–98]. The active parts were identified as chitinase (which inhibited bacterial growth) [98], and seed extracts from the ripe and unripe epicarp and endocarp [96,97] (which were bactericidal, especially against G+ bacteria) [96]. The epicarp and endocarp themselves did not inhibit bacterial growth [97]. Extracts from the seeds of papaya and the enzyme chitinase may therefore be useful agents against: opportunistic, nosocomial, UTI and wound infections; food poisoning; dysentery; toxic shock syndrome (TSS) and septicemia.

5.4.8 Pomegranate (Punica granatum)

The pomegranate has been the subject of many studies investigating its antibacterial effects. This is probably because of the publicity in Western society hailing it as a 'superfood', as well as its widespread use in traditional Middle Eastern and Indian medicine, where it is used for its antibacterial, antiinflammatory and alleviating properties, as well as for curing ailments such as mouth ulcers. Most studies performed in vitro have shown that extracts of the pomegranate are able to inhibit the growth of a number of different bacteria (Table 10.2). Most of the extracts tested were obtained by alcohol extraction, and the rest were either water extracts or the crude material itself. The major active components identified in these studies were phenols [71,99–101] and flavonoids [71,100–102], which are highly concentrated in the peel, while several other studies indicated sterols, triterpenes, tannins and ellagitannins, alkaloids and saponins [71,100,101,103].

Some extracts of pomegranate (particularly alcohol extracts) have also been shown to have bactericidal effects on MRSA (some acting synergistically with antibiotics) [104], on enterohemorrhagic E. coli [100], and on H. pylori [105]. In addition, one study also showed that the methanol extract of pomegranate peel was able to increase wound healing in vivo in rats, when applied topically as a water-soluble gel, and it gave better results than a commercial topical antibacterial agent. The gel was found to be high in phenolic compounds and the major active components responsible for its antibacterial activity were gallic acid and catechin. As well as wound healing, a hydroalcoholic extract from the pomegranate has been found in a clinical trial to be as equally effective as chlorhexidine at reducing dental plaque bacteria, when used by healthy people as a mouth rinse [106].

The pomegranate could therefore have multiple uses as an antibacterial agent or disinfectant against many types of bacteria that cause a wide range of diseases. The future of using this fruit as a new source of such agents now lies in further testing being done, tests for toxicity and tests for its effectiveness in human trials.

TABLE 10.2 Studies Assessing the Antibacterial Effects of Pomegranate – Bacteria Tested and the Diseases They Cause

Bacteria	Examples of Diseases Caused	Reference
S. aureus, MRSA, MSSA	Nosocomial infections, toxic shock syndrome (TSS) and septicemia	[73,81,101,102,117–120]
Bacillus species (B. cereus, B. megaterium, and B. subtilis)	B. cereus – food poisoning; others non-pathogenic	[102,117,121]
P. vulgaris	UTIs and wound infections	[121]
K. pneumoniae	Respiratory tract infections including pneumonia	[121]
E. coli (including EPEC and EHEC)	Food poisoning	[73,81,100–102,117,120,122,123]
Salmonella typhi and other species	Food poisoning	[101,102,122,124–126]
H. pylori	Gastroduodenal ulcers and cancer	[105]
Shigella dysenteriae and other species	Shigellosis – dysentery–diarrhea, fever	[101,102,122,126]
Vibrio cholerae	Cholera	[102,122]
Corynebacteria species	Diphtheria, diseases from infection of intravascular and prosthetic devices (e.g. endocarditis)	[102]
Streptococcus species	Toxic shock syndrome, skin infections, oral and throat infections, tooth decay, pneumonia	[102]
Proteus vulgaris	UTIs and wound infections	[117,126]
Enterobacter aerogenes	Opportunistic and nosocomial infections (e.g. UTIs, wound infections, endocarditis)	[117]

EPEC, enteropathogenic *E. coli*; EHEC, enterohemorrhagic *E. coli*; UTI, urinary tract infection.

5.5 Olive (*Olea europaea*)

Like the avocado, the olive is eaten as a savory food, but it is actually a fruit. A number of studies have looked at various products of olive fruit for their antibacterial properties. These include virgin olive oil (which inhibited the growth of foodborne pathogens, due to its phenolic compounds) [107], green olive brines (which were bactericidal against 10 strains of a non-pathogenic probiotic bacteria, *Lactobacillus plantarum*) [108], and olive mill waste extracts (which had broad-spectrum antibacterial and antioxidant activity against *S. aureus*, *E. coli*, and *Pseudomonas aeruginosa*) [109].

Specific compounds within the olive have been shown to be responsible for its antibacterial activity. For example, aliphatic long-chain aldehydes were active against the G+ and G− bacteria that cause human intestinal and respiratory tract infection [110]. Most studies, however, have found that phenols and phenolic compounds are the main active components, inhibiting bacterial growth [107,109,111] as well as having bactericidal (lytic) effects [107,108]. Olive extracts have also been shown to have antibacterial effects against drug-resistant variants of *Mycobacterium tuberculosis* (which causes TB) and to be non-toxic to human cells [72].

Extracts of olive, particularly phenolic compounds, may therefore be useful in the future to treat TB, dermatitis, and nosocomial infections as well as the infections and diseases caused by foodborne and other human pathogenic bacteria.

5.6 Tomato (*Solanum lycopersicum*)

The tomato is often mistakenly called a vegetable and is nearly always eaten as such, being a common component of savory dishes and having been nutritionally categorized as a vegetable. However, strictly speaking it is a fruit, and more specifically a berry.

Lycopene, the red pigment found in tomatoes, has become the subject of much investigation. This is probably due to its properties as an antioxidant and having been thought to help reduce the risk of developing certain types of cancer (although the evidence for this is weak, as reported by the US Food and Drug Administration, FDA) [112]. Lycopene has been shown *in vivo* (chronic bacterial prostatitis, CBP, rat model) to have significant antibacterial effects. It was found to reduce the number of bacteria cultured from the prostate and urine, and to improve prostatic inflammation [113]. When given in combination with the antibiotic ciprofloxacin, its effects were synergistically enhanced. Further studies are clearly needed to assess its effectiveness and adverse effects in humans; however, lycopene from the tomato has the potential to be used as an alternative treatment for CBP and to enhance the effects of current antibiotic treatment for this condition.

5.7 Grapes (*Vitis*)

The 70% aqueous propane extract (P70) of red grape (particularly from the skin) has been shown to have antibacterial activity against G + oral bacteria. When tested using *S. mutans*, it was found to be bacteriostatic, to prevent bacterial acid production, and to inhibit its adhesion to glass [114]. The active component responsible is likely to be 'resveratol' which is found in the skins of red fruit such as red grapes, and has been shown in other studies to have antibacterial, antioxidant, and anticarcinogenic effects [115,116]. Flour made from the seeds of white grapes (Chardonnay) has also been shown to have antibacterial effects, thought to be due to its high phenolic content [35]. The flour was found to be bactericidal

against *E. coli*, to inhibit the growth of *Listeria monocytogenes*, and to have preservative effects on fish oils [35]. Interestingly, when grapes have been processed into red and white wine, their antibacterial effects still remain. One study [107] found that red and white wine killed a variety of foodborne bacteria including *Salmonella enteritidis* and *Listeria monocytogenes*. However, whether this is due to the alcohol rather than the grape component of the wine remains to be seen.

A pilot clinical trial has assessed the effect of dried grapes (raisins) on UTIs. Consumption of raisins by UTI patients was found not to elicit bacterial anti-adhesion activity in their urine when tested against uropathogenic *E. coli* [52].

Although raisins may not be effective at inhibiting the adhesion of UTI bacteria, the other studies have shown that grapes and grape products may be a new source of antibacterial agents to treat food poisoning and dental plaque, and be used as a preservative in food products.

6. CONCLUSION

Investigations into plant materials as alternative sources of antimicrobials have become more common over the past few years, due to the increased rate of development of antibiotic-resistant organisms. New strategies to combat infection are also being sought such as the use of 'anti-adhesive' molecules, targeting the primary step of infection – adhesion of the organism to host tissue [18], and plant extracts have been shown to do this as well [29,38,40].

The health benefits of plants and plant products such as fresh fruits may extend beyond their nutritional value (sources of fiber, vitamins, and minerals) and be far more wide-reaching as they contain components that can kill, inhibit the growth, and inhibit adhesion of a wide variety of bacterial species including human pathogens. Many fruit extracts have been shown to be effective at killing or inhibiting the growth of bacteria. Despite this success, bacterial resistance can develop to these extracts too. However, some fruit extracts have been shown to inhibit the adhesion of bacteria. These include blackcurrant [38], grapes [114], and cranberry [39–41,44,45,49,51,52]. Because inhibition of adhesion works on the principle of sterically blocking bacteria from attaching to host tissue, the likelihood of resistance developing in bacteria, which occurs when the organism is killed, is less likely. These fruit extracts are therefore most likely to be the best potential new sources of alternative antibacterial agents.

Most investigations into the antibacterial properties of fruit have only been conducted at the basic experimental stage *in vitro* or in food systems and still need validation in human trials. However, even for those that have been tested in patients, the clinical study data provides limited evidence since most of the studies and trials had methodological limitations including very small sample sizes or lack of blinding. Additionally studies varied widely in their treatment regimen and so consistency of results is a problem. Despite the lack of trial data, cranberry and pomegranate are two examples of fruit that have been tested in clinical studies and cranberry is certainly used by clinicians today to help treat UTI and *H. pylori* infections.

In order to have a better idea of the effectiveness of fruit extracts at treating bacterial diseases, larger, well-conducted RCTs are needed which assess different doses, as it may be that at higher doses the extract is more effective than the dose currently tested. In addition, studies need to be carried out for longer periods of time and compare different extracts or products of each fruit in several trials, since published data so far have been short and not consistent in their use of extract under scrutiny. Toxicity studies are also required because, although the fruit themselves are edible, high concentrations of particular extracted components may have toxic effects in human or animal systems. Additionally, for many fruits or fruit extracts, the active component(s) still remains to be identified.

One of the problems with currently used broad-spectrum antibiotics is that they also kill the normal members of the microflora within the human gut – the 'good' bacteria. Fruit extracts also need to be tested to see whether they are specific in killing pathogenic bacteria rather than members of the normal microflora. For example, some experiments have shown that extracts from some fruit (including lime, orange, and pomegranate) killed both pathogenic and non-pathogenic bacteria such as lactobacilli and *Bacillus subtilis* [62,102,117].

Despite all these limitations, including the time taken for further experiments and trials to be conducted, it does not negate the fact that fruit and fruit extracts do contain antibacterial compounds which can kill, inhibit the growth of, and in some cases inhibit the adhesion of, bacteria that are the causes of major health problems in the world today. If further experiments are conducted and the active components identified, dietary fruit may comprise one of the prominent sources of antibacterial agents that will be regularly available to us in the future.

7. SUMMARY

Bacteria are rapidly becoming resistant to common antibiotics and antibacterial agents, and previously controlled diseases are beginning to re-emerge. New sources are therefore required and one such source is plants and plant products including dietary fruit.

Components and extracts from dietary fruit have been shown to kill, inhibit the growth, and inhibit the adhesion of a wide range of pathogenic bacteria.

Further studies are still required for many fruits to identify the active components and to test their effectiveness and safety in humans.

Despite this, dietary fruits are a potential source of new antibacterial agents which is desperately needed in the world today, where a number of bacteria seem to have the upper hand in the fight against infection.

References

1. Cassell, G. H., & Mekalanos, J. (2001). Development of antimicrobial agents in the era of new and reemerging infectious diseases and increasing antibiotic resistance. *JAMA, 285*, 601–605.

2. MMWR (1999). Geographic variation in penicillin resistance in *Streptococcus pneumoniae* – selected sites, United States, 1997. *MMWR Morbidity Mortality Weekly Report, 48*, 656–661.

3. Morens, D. M., Folkers, G. K., & Fauci, A. S. (2004). The challenge of emerging and re-emerging infectious diseases. *Nature, 430*, 242–249.

4. Sanders, C. C., & Sanders, W. E. (1995). Resistance to antibacterial agents. In D. I. Jungkind, J. E. Mortensen, H. S. Fraimow, & G. B. Calandra, (Eds.), Antimicrobial resistance, a crisis in health care. *Advances in experimental medicine and biology* New York: Plenum Press.

5. Lechner, D., Gibbons, S., & Bucar, F. (2008). Plant phenolic compounds as ethidium bromide efflux inhibitors in *Mycobacterium smegmatis*. *Journal of Antimicrobial Chemotherapy, 62*, 345–348.

6. Kvist, M., Hancock, V., & Klemm, P. (2008). Inactivation of efflux pumps abolishes bacterial biofilm formation. *Applied and Environmental Microbiology, 74*, 7376–7382.

7. Amaral, L., Martins, M., Viveiros, M., Molnar, J., & Kristiansen, J. E. (2008). Promising therapy of XDR-TB/MDR-TB with thioridazine an inhibitor of bacterial efflux pumps. *Current Drug Targets, 9*, 816–819.

8. Kamicker, B. J., Sweeney, M. T., Kaczmarek, F., Dib-Hajj, F., Shang, W., Crimin, K., Duignan, J., & Gootz, T. D. (2008). Bacterial efflux pump inhibitors. *Methods Molecular Medicine, 142*, 187–204.

9. Coconnier, M. H., Lievin, V., Hemery, E., & Servin, A. L. (1998). Antagonistic activity against Helicobacter infection *in vitro* and *in vivo* by the human Lactobacillus acidophilus strain LB. *Applied and Environmental Microbiology, 64*, 4573–4580.

10. Midolo, P. D., Lambert, J. R., Hull, R., Luo, F., & Grayson, M. L. (1995). *In vitro* inhibition of *Helicobacter pylori* NCTC 11637 by organic acids and lactic acid bacteria. *Journal of Applied Bacteriology, 79*, 475–479.

11. Demidova, T. N., & Hamblin, M. R. (2004). Photodynamic therapy targeted to pathogens. *International Journal of Immunopathology and Pharmacology, 17*, 245–254.

12. Conway, B. R. (2005). Drug delivery strategies for the treatment of *Helicobacter pylori* infections. *Current Pharmaceutical Design, 11*, 775–790.

13. Gotoh, A., Akamatsu, T., Shimizu, T., Shimodaira, K., Kaneko, T., Kiyosawa, K., Ishida, K., Ikeno, T., Sugiyama, A., Kawakami, Y., Ota, H., & Katsuyama, T. (2002). Additive effect of pronase on the efficacy of eradication therapy against *Helicobacter pylori*. *Helicobacter, 7*, 183–191.

14. Higo, S., Ori, K., Takeuchi, H., Yamamoto, H., Hino, T., & Kawashima, Y. (2004). A novel evaluation method of gastric mucoadhesive property *in vitro* and the mucoadhesive mechanism of tetracycline-sucralfate acidic complex for eradication of *Helicobacter pylori*. *Pharmaceutical Research, 21*, 413–419.

15. Vajdy, M., Singh, M., Ugozzoli, M., Briones, M., Soenawan, E., Cuadra, L., Kazzaz, J., Ruggiero, P., Peppoloni, S., Norelli, F., Del Giudice, G., & O'Hagan, D. (2003). Enhanced mucosal and systemic immune responses to *Helicobacter pylori* antigens through mucosal priming followed by systemic boosting immunizations. *Immunology, 110*, 86–94.

16. Del Giudice, G., & Michetti, P. (2004). Inflammation, immunity and vaccines for *Helicobacter pylori*. *Helicobacter, 9*(1 Suppl), 23–28.

17. Kelly, C. G., & Younson, J. S. (2000). Anti-adhesive strategies in the prevention of infectious disease at mucosal surfaces. *Expert Opinion on Investigational Drugs, 9*, 1711–1721.

18. Sharon, N., & Ofek, I. (2002). Fighting infectious diseases with inhibitors of microbial adhesion to host tissues. *Critical Reviews in Food Science and Nutrition, 42*, 267–272.

19. O'Mahony, R., Basset, C., Holton, J., Vaira, D., & Roitt, I. (2005). Comparison of image analysis software packages in the assessment of adhesion of microorganisms to mucosal epithelium using confocal laser scanning microscopy. *Journal of Microbiological Methods, 61*, 105–126.

20. Younsen, J., O'Mahony, R. Novel inhibitors of *H. pylori* binding to fucosylated blood group antigens. (Submitted for publication).

21. Clare, D. A., Catignani, G. L., & Swaisgood, H. E. (2003). Biodefense properties of milk: The role of antimicrobial proteins and peptides. *Current Pharmaceutical Design, 9*, 1239–1255.

22. Gopal, P. K., & Gill, H. S. (2000). Oligosaccharides and glycoconjugates in bovine milk and colostrum. *British Journal of Nutrition, 84*(1 Suppl), S69–S74.

23. van Hooijdonk, A. C., Kussendrager, K. D., & Steijns, J. M. (2000). *In vivo* antimicrobial and antiviral activity of components in bovine milk and colostrum involved in non-specific defence. *British Journal of Nutrition, 84*(1 Suppl), S127–S134.

24. al Somal, N., Coley, K. E., Molan, P. C., & Hancock, B. M. (1994). Susceptibility of *Helicobacter pylori* to the antibacterial activity of manuka honey. *Journal of the Royal Society of Medicine, 87*, 9–12.

25. Koo, H., Gomes, B. P., Rosalen, P. L., Ambrosano, G. M., Park, Y. K., & Cury, J. A. (2000). *In vitro* antimicrobial activity of propolis and *Arnica montana* against oral pathogens. *Archives of Oral Biology, 45*, 141–148.

26. Nostro, A., Cellini, L., Di Bartolomeo, S., Di Campli, E., Grande, R., Cannatelli, M. A., Marzio, L., & Alonzo, V. (2005). Antibacterial effect of plant extracts against *Helicobacter pylori*. *Phytotherapy Research, 19*, 198–202.

27. Hiramoto, S., Itoh, K., Shizuuchi, S., Kawachi, Y., Morishita, Y., Nagase, M., Suzuki, Y., Nobuta, Y., Sudou, Y., Nakamura, O., Kagaya, I., Goshima, H., Kodama, Y., Icatro, F. C., Koizumi, W., Saigenji, K., Miura, S., Sugiyama, T., & Kimura, N. (2004). Melanoidin, a food protein-derived advanced maillard reaction product, suppresses *Helicobacter pylori in vitro* and *in vivo*. *Helicobacter, 9*, 429–435.

28. Matsumoto, M., Hara, K., Kimata, H., Benno, Y., & Shimamoto, C. (2005). Exfoliation of *Helicobacter pylori* from gastric mucin by glycopolypeptides from buttermilk. *Journal of Dairy Science, 88*, 49–54.

29. O'Mahony, R., Al Khtheeri, H., Weerasekera, D., Fernando, N., Vaira, D., Holton, J., & Basset, C. (2005). Bactericidal and anti-adhesive properties of culinary and medicinal plants against *Helicobacter pylori*. *World Journal of Gastroenterology, 11*, 7499–7507.

30. Cowan, M. M. (1999). Plant products as antimicrobial agents. *Clinical Microbiology Reviews, 12*, 564–582.

31. Rashid, F., Ahmed, R., Mahmood, A., Ahmad, Z., Bibi, N., & Kazmi, S. U. (2007). Flavonoid glycosides from *Prunus armeniaca* and the antibacterial activity of a crude extract. *Archives of Pharmacal Research, 30*(8), 932–937.

32. Rashid, F., Bibi, N., Ahmed, R., & Kazmi, S. U. (2006). Triterpene acid and its glycoside from *Prunus armeniaca*, and antibacterial and antioxidant activities of fruit extracts. *Journal of Tropical Medicinal Plants, 6*, 31–35.

33. Ryan, T., Wilkinson, J. M., & Cavanagh, H. M. (2001). Antibacterial activity of raspberry cordial *in vitro*. *Research in Veterinary Science, 71*, 155–159.

34. Cavanagh, H. M., Hipwell, M., & Wilkinson, J. M. (2003). Antibacterial activity of berry fruits used for culinary purposes. *Journal of Medicine Food, 6*, 57–61.

35. Luther, M., Parry, J., Moore, J., Meng, J., Zhang, Y., Cheng, Z., & Yu, L. (2007). Inhibitory effect of Chardonnay and black raspberry seed extracts on lipid oxidation in fish oil and their radical scavenging and antimicrobial properties. *Food Chemistry, 104*, 1065–1073.

36. Puupponen-Pimia, R., Nohynek, L., Alakomi, H. L., & Oksman-Caldentey, K. M. (2005). The action of berry phenolics against human intestinal pathogens. *Biofactors, 23*, 243–251.

37. Puupponen-Pimia, R., Nohynek, L., Ammann, S., Oksman-Caldentey, K. M., & Buchert, J. (2008). Enzyme-assisted processing increases antimicrobial and antioxidant activity of bilberry. *Journal of Agricultural Food Chemistry, 56*, 681–688.

38. Lengsfeld, C., Deters, A., Faller, G., & Hensel, A. (2004). High molecular weight polysaccharides from black currant seeds inhibit adhesion of *Helicobacter pylori* to human gastric mucosa. *Planta Medica, 70*, 620–626.

39. Shmuely, H., Burger, O., Neeman, I., Yahav, J., Samra, Z., Niv, Y., Sharon, N., Weiss, E., Athamna, A., Tabak, M., & Ofek, I. (2004). Susceptibility of *Helicobacter pylori* isolates to the antiadhesion activity of a high-molecular-weight constituent of cranberry. *Diagnostic Microbiology and Infectious Disease, 50*, 231–235.

40. Burger, O., Ofek, I., Tabak, M., Weiss, E. I., Sharon, N., & Neeman, I. (2000). A high molecular mass constituent of cranberry juice inhibits *Helicobacter pylori* adhesion to human gastric mucus. *FEMS Immunology and Medical Microbiology, 29*, 295–301.

41. Burger, O., Weiss, E., Sharon, N., Tabak, M., Neeman, I., & Ofek, I. (2002). Inhibition of *Helicobacter pylori* adhesion to human gastric mucus by a high-molecular-weight constituent of cranberry juice. *Critical Reviews in Food Science and Nutrition, 42*, 279–284.

42. Lin, Y. T., Kwon, Y. I., Labbe, R. G., & Shetty, K. (2005). Inhibition of *Helicobacter pylori* and associated urease by oregano and cranberry phytochemical synergies. *Applied and Environmental Microbiology, 71*, 8558–8564.

43. Shmuely, H., Yahav, J., Samra, Z., Chodick, G., Koren, R., Niv, Y., & Ofek, I. (2007). Effect of cranberry juice on eradication of *Helicobacter pylori* in patients treated with antibiotics and a proton pump inhibitor. *Molecular Nutritional Food Research, 51*, 746–751.

44. Steinberg, D., Feldman, M., Ofek, I., & Weiss, E. I. (2005). Cranberry high molecular weight constituents promote *Streptococcus sobrinus* desorption from artificial biofilm. *International Journal of Antimicrobial Agents, 25*, 247–251.

45. Weiss, E. I., Kozlovsky, A., Steinberg, D., Lev-Dor, R., Greenstein, R. B. N., Feldman, M., Sharon, N., & Ofek, I. (2004). A high molecular mass cranberry constituent reduces mutans streptococci level in saliva and inhibits *in vitro* adhesion to hydroxyapatite. *FEMS Microbiology Letters, 232,* 89–92.

46. Park, S. J., Yoon, H. N., & Shim, B. S. (2005). Prevention of relapse with the cranberry juice in chronic pelvic pain syndrome. [Korean]. *Korean Journal of Urology, 46,* 63–67.

47. McMurdo, M. E., Bissett, L. Y., Price, R. J., Phillips, G., & Crombie, I. K. (2005). Does ingestion of cranberry juice reduce symptomatic urinary tract infections in older people in hospital? A double-blind, placebo-controlled trial. *Age Ageing, 34,* 256–261.

48. Kirchhoff, M., Renneberg, J., Damkjaer, K., Pietersen, I., & Schroll, M. (2001). [Can ingestion of cranberry juice reduce the incidence of urinary tract infections in a department of geriatric medicine?]. [Danish]. *Ugeskr. Laeger, 163,* 2782–2786.

49. Howell, A. B., & Foxman, B. (2002). Cranberry juice and adhesion of antibiotic-resistant uropathogens [6]. *JAMA, 287,* 3082–3083.

50. Linsenmeyer, T. A., Harrison, B., Oakley, A., Kirshblum, S., Stock, J. A., & Millis, S. R. (2004). Evaluation of cranberry supplement for reduction of urinary tract infections in individuals with neurogenic bladders secondary to spinal cord injury. A prospective, double-blinded, placebo-controlled, crossover study. *Journal of Spinal Cord Medicine, 27,* 29–34.

51. Lavigne, J.. P., Bourg, G., Combescure, C., Botto, H., & Sotto, A. (2008). *In-vitro* and *in-vivo* evidence of dose-dependent decrease of uropathogenic *Escherichia coli* virulence after consumption of commercial Vaccinium macrocarpon (cranberry) capsules. *Clinical Microbiology and Infection, 14,* 350–355.

52. Greenberg, J. A., Newmann, S. J., & Howell, A. B. (2005). Consumption of sweetened dried cranberries versus unsweetened raisins for inhibition of uropathogenic *Escherichia coli* adhesion in human urine: a pilot study. *Journal of Alternative and Complementary Medicine, 11,* 875–878.

53. Bailey, D. T., Dalton, C., Joseph, D. F., & Tempesta, M. S. (2007). Can a concentrated cranberry extract prevent recurrent urinary tract infections in women? A pilot study. *Phytomedicine, 14,* 237–241.

54. Leitao, D. P., Polizello, A. C., Ito, I. Y., & Spadaro, A. C. (2005). Antibacterial screening of anthocyanic and proanthocyanic fractions from cranberry juice. *Journal of Medicine Food, 8,* 36–40.

55. Reagor, L., Gusman, J., McCoy, L., Carino, E., & Heggers, J. P. (2002). The effectiveness of processed grapefruit-seed extract as an antibacterial agent: I. an *in vitro* agar assay. *Journal of Alternative and Complementary Medicine – New York, 8,* 325–332.

56. Heggers, J. P., Cottingham, J., Gusman, J., Reagor, L., McCoy, L., Carino, E., Cox, R., & Zhao, J. G. (2002). The effectiveness of processed grapefruit-seed extract as an antibacterial agent: II. mechanism of action and *in vitro* toxicity. *Journal of Alternative and Complementary Medicine – New York, 8,* 333–340.

57. Furuhata, K., Dogasaki, C., Hara, M., & Fukuyama, M. (2003). Inactivation of *Legionella pneumophila* from whirlpool bath waters by grapefruit (*Citrus paradisi*) seed extract. *Biocontrol Science, 8,* 129–132.

58. Cvetnic, Z., & Vladimir-Knezevic, S. (2004). Antimicrobial activity of grapefruit seed and pulp ethanolic extract. *Acta Pharmaceutical, 54,* 243–250.

59. Oyelami, O. A., Agbakwuru, E. A., Adeyemi, L. A., & Adedeji, G. B. (2005). The effectiveness of grapefruit (*Citrus paradisi*) seeds in treating urinary tract infections. *Journal of Alternative and Complementary Medicine – New York, 11,* 369–371.

60. Brandin, H., Myrberg, O., Rundlof, T., Arvidsson, A. K., & Brenning, G. (2007). Adverse effects by artificial grapefruit seed extract products in patients on warfarin therapy. *European Journal of Clinical Pharmacology, 63,* 565–570.

61. Mihele, D., Gird, C. E., Pop, A., & Al Borsh, M. A. (2008). Study regarding the antimicrobian and antifungic activity of the volatile oils from the citrus species. *Archives of Balkan Medicine Union, 43,* 11–14.

62. Prabuseenivasan, S., Jayakumar, M., & Ignacimuthu, S. (2006). *In vitro* antibacterial activity of some plant essential oils. *BMC Complementary Alternative Medicine, 6,* Article Number 39.

63. Friedman, M., Henika, P. R., Levin, C. E., & Mandrell, R. E. (2004). Antibacterial activities of plant essential oils and their components against *Escherichia coli* O157:H7 and *Salmonella enterica* in apple juice. *Journal of Agricultural Food Chemistry, 52,* 6042–6048.

64. Fisher, K., & Phillips, C. A. (2006). The effect of lemon, orange and bergamot essential oils and their components on the survival of *Campylobacter jejuni*, *Escherichia coli* O157, *Listeria* monocytogenes, *Bacillus cereus* and *Staphylococcus aureus in vitro* and in food systems. *Journal of Applied Microbiology, 101,* 1232–1240.

65. Tomotake, H., Koga, T., Yamato, M., Kassu, A., & Ota, F. (2006). Antibacterial activity of citrus fruit juices against Vibrio species. *Journal of Nutritional Science and Vitaminology, 52,* 157–160.

66. Tumane, P. M., Wadher, B. J., Gomashe, A. V., & Deshmukh, S. R. (2007). Antibacterial activity of citrus

limon fruit juice against clinical isolates of human pathogens. *Asian Journal of Microbiological Biotechnology Environment Science, 9*, 129–132.

67. Schelz, Z., Molnar, J., & Hohmann, J. (2006). Antimicrobial and antiplasmid activities of essential oils. *Fitoterapia, 77*, 279–285.

68. Baik, J. S., Kim, S. S., Lee, J. A., Oh, T. H., Kim, J. Y., Lee, N. H., & Hyun, C. G. (2008). Chemical composition and biological activities of essential oils extracted from Korean endemic citrus species. *Journal of Microbiol Biotechnology, 18*, 74–79.

69. El Shafa, A. M. (2002). Bioactive polymethoxyflavones and flavanone glycosides from the peels of Citrus deliciosa. *Chinese Pharmaceutical Journal, 54*, 199–206.

70. Jayaprakasha, G. K., Negi, P. S., Sikder, S., Rao, L. J., & Sakariah, K. K. (2000). Antibacterial activity of Citrus reticulata peel extracts. *Z. Naturforsch. C. Journal of Biosciences, 55*, 1030–1034.

71. Aqil, F., Khan, M. S., Owais, M., & Ahmad, I. (2005). Effect of certain bioactive plant extracts on clinical isolates of beta-lactamase producing methicillin resistant *Staphylococcus aureus*. *Journal of Basic Microbiology, 45*, 106–114.

72. Camacho-Corona, M. D. R., Ramirez-Cabrera, M. A., Gonzalez-Santiago, O., Garza-Gonzalez, E., Palacios, I. D. P., & Luna-Herrera, J. (2008). Activity against drug resistant-tuberculosis strains of plants used in Mexican traditional medicine to treat tuberculosis and other respiratory diseases. *Phytotherapy Research, 22*, 82–85.

73. Melendez, P. A., & Capriles, V. A. (2006). Antibacterial properties of tropical plants from Puerto Rico. *Phytomedicine, 13*, 272–276.

74. Neeman, I., Lifshitz, A., & Kashman, Y. (1970). New antibacterial agent isolated from the avocado pear. *Applied Microbiology, 19*, 470–473.

75. Ono, H., Tesaki, S., Tanabe, S., & Watanabe, M. (1998). 6-Methylsulfinylhexyl isothiocyanate and its homologues as food-originated compounds with antibacterial activity against *Escherichia coli* and *Staphylococcus aureus*. *Bioscience Biotechnology and Biochemistry, 62*, 363–365.

76. Nwafor, S. V., Esimone, C. O., Amadi, C. A., & Nworu, C. S. (2003). *In vivo* interaction between ciprofloxacin hydrochloride and the pulp of unripe plantain (*Musa paradisiaca*). *European Journal of Drug Metabolism Pharmacokinetics, 28*, 253–258.

77. Mousa, O., Vuorela, P., Kiviranta, J., Wahab, S. A., Hiltunen, R., & Vuorela, H. (1994). Bioactivity of certain Egyptian Ficus species. *Journal of Ethnopharmacology, 41*, 71–76.

78. Qa'dan, F., Thewaini, A.. J., Ali, D. A., Afifi, R., Elkhawad, A., & Matalka, K. Z. (2005). The antimicrobial activities of *Psidium guajava* and *Juglans regia* leaf extracts to acne-developing organisms. *American Journal of Chinese Medicine, 33*, 197–204.

79. McMahon, M. A., Blair, I. S., Moore, J. E., & McDowell, D. A. (2007). Habituation to sub-lethal concentrations of tea tree oil (*Melaleuca alternifolia*) is associated with reduced susceptibility to antibiotics in human pathogens. *Journal of Antimicrobial Chemotherapy, 59*, 125–127.

80. McMahon, M. A., Tunney, M. M., Moore, J. E., Blair, I. S., Gilpin, D. F., & McDowell, D. A. (2008). Changes in antibiotic susceptibility in staphylococci habituated to sub-lethal concentrations of tea tree oil (*Melaleuca alternifolia*). *Letters in Applied Microbiology, 47*, 263–268.

81. Anesini, C., & Perez, C. (1993). Screening of plants used in Argentine folk medicine for antimicrobial activity. *Journal of Ethnopharmacology, 39*, 119–128.

82. Chah, K. F., Eze, C. A., Emuelosi, C. E., & Esimone, C. O. (2006). Antibacterial and wound healing properties of methanolic extracts of some Nigerian medicinal plants. *Journal of Ethnopharmacology, 104*, 164–167.

83. Mahfuzul, H., Bari, M. L., Inatsu, Y., Juneja, V. K., & Kawamoto, S. (2007). Antibacterial activity of guava (*Psidium guajava* L.) and neem (*Azadirachta indica* A. Juss.) extracts against foodborne pathogens and spoilage bacteria. *Foodborne Pathogenicity Disorder, 4*, 481–488.

84. Saj, O. P., Roy, R. K., & Savitha, S. V. (2008). Chemical composition and antimicrobial properties of essential oil of *Feijoa sellowiana* O. Berg. (Pineapple guava). *Journal of Pure Applied Microbiology, 2*, 227–230.

85. Vuotto, M. L., Basile, A., Moscatiello, V., De Sole, P., Castaldo-Cobianchi, R., Laghi, E., & Ielpo, M. T. (2000). Antimicrobial and antioxidant activities of Feijoa sellowiana fruit. *International Journal of Antimicrobial Agents, 13*, 197–201.

86. Oliveira, D. G., Prince, K. A., Higuchi, C. T., Santos, A. C. B., Lopes, L. M. X., Simoes, M. J. S., & Leite, C. Q. F. (2007). Antimycobacterial activity of some Brazilian indigenous medicinal drinks. *Revista de Ciencias Farmaceuticas Basica e Aplicada, 28*, 165–169.

87. Friesen, A., Schilling, A., Gobstetter, A., & Adam, D. (1987). Tetracyclin-Konzentration im Prostata-Sekret. *Z. antimikrob. antineoplast. Chirurgie, 2*, 61–65.

88. Renzini, G., & Varengo, M. (1972). [Absorption of tetracycline in presence of bromelain after oral administration]. *Arzneimittelforschung, 22*, 410–412.

89. Tinozzi, S., & Venegoni, A. (1978). Effect of bromelain on serum and tissue levels of amoxycillin. *Drugs Under Experimental and Clinical Research, 1*, 39–44.

90. Maurer, H. R. (2001). Bromelain: biochemistry, pharmacology and medical use. *Cellular and Molecular Life Sciences, 58*, 234–245.

91. Nuebauer, R. A. (1961). A plant protease for potentiation of and possible replacement of antibiotics. *Experimental Medicine and Surgery, 19,* 143–160.

92. Ryan, R. E. (1967). A double-blind clinical evaluation of bromelains in the treatment of acute sinusitis. *Headache, 7,* 13–17.

93. Basile, A., Vuotto, M. L., Violante, U., Sorbo, S., Martone, G., & Castaldo-Cobianchi, R. (1997). Antibacterial activity in *Actinidia chinensis, Feijoa sellowiana* and *Aberia caffra. International Journal of Antimicrobial Agents, 8,* 199–203.

94. Motohashi, N., Shirataki, Y., Kawase, M., Tani, S., Sakagami, H., Satoh, K., Kurihara, T., Nakashima, H., Wolfard, K., Miskolci, C., & Molnar, J. (2001). Biological activity of kiwifruit peel extracts. *Phytotherapy Research, 15,* 337–343.

95. Motohashi, N., Shirataki, Y., Kawase, M., Tani, S., Sakagami, H., Satoh, K., Kurihara, T., Nakashima, H., Mucsi, I., Varga, A., & Molnar, J. (2002). Cancer prevention and therapy with kiwifruit in Chinese folklore medicine: a study of kiwifruit extracts. *Journal of Ethnopharmacology, 81,* 357–364.

96. Emeruwa, A. C. (1982). Antibacterial substance from *Carica papaya* fruit extract. *Journal of Natural Products, 45,* 123–127.

97. Dawkins, G., Hewitt, H., Wint, Y., Obiefuna, P. C., & Wint, B. (2003). Antibacterial effects of *Carica papaya* fruit on common wound organisms. *West Indian Medical Journal, 52,* 290–292.

98. Chen, Y. T., Hsu, L. H., Huang, I. P., Tsai, T. C., Lee, G. C., & Shaw, J. F. (2007). Gene cloning and characterization of a novel recombinant antifungal chitinase from papaya (*Carica papaya*). *Journal of Agricultural Food Chemistry, 55,* 714–722.

99. Murthy, K. N., Reddy, V. K., Veigas, J. M., & Murthy, U. D. (2004). Study on wound healing activity of *Punica granatum* peel. *Journal of Medicine Food, 7,* 256–259.

100. Voravuthikunchai, S. P., Sririrak, T., Limsuwan, S., Supawita, T., Iida, T., & Honda, T. (2005). Inhibitory effects of active compounds from *Punica granatum* pericarp on verocytotoxin production by enterohemorrhagic *Escherichia coli* O157:H7. *Journal of Health Science, 51,* 590–596.

101. Ahmad, I., & Beg, A. Z. (2001). Antimicrobial and phytochemical studies on 45 Indian medicinal plants against multi-drug resistant human pathogens. *Journal of Ethnopharmacology, 74,* 113–123.

102. Naz, S., Siddiqi, R., Ahmad, S., Rasool, S. A., & Sayeed, S. A. (2007). Antibacterial activity directed isolation of compounds from *Punica granatum. Journal of Food Science, 72,* 341–345.

103. Machado, T. B., Pinto, A. V., Pinto, M. C. F. R., Leal, I. C. R., Silva, M. G., Amaral, A. C. F., Kuster, R. M., & Netto-dosSantos, K. R. (2003). *In vitro* activity of Brazilian medicinal plants, naturally occurring naphthoquinones and their analogues, against methicillin-resistant *Staphylococcus aureus. International Journal of Antimicrobial Agents, 21,* 279–284.

104. Braga, L. C., Leite, A. A. M., Xavier, K. G. S., Takahashi, J. A., Bemquerer, M. P., Chartone-Souza, E., & Nascimento, A. M. A. (2005). Synergic interaction between pomegranate extract and antibiotics against *Staphylococcus aureus. Canadian Journal of Microbiology, 51,* 541–547.

105. Voravuthikunchai, S. P., Limsuwan, S., & Mitchell, H. (2006). Effects of *Punica granatum* pericarps and *Quercus infectoria* nutgalls on cell surface hydrobicity and cell survival of *Helicobacter pylori. Journal of Health Science, 52,* 154–159.

106. Menezes, S. M., Cordeiro, L. N., & Viana, G. S. (2006). *Punica granatum* (pomegranate) extract is active against dental plaque. *Journal of Herb Pharmacotherapy, 6,* 79–92.

107. Medina, E., Romero, C., Brenes, M., & De Castro, A. (2007). Antimicrobial activity of olive oil, vinegar, and various beverages against foodborne pathogens. *Journal of Food Protection, 70,* 1194–1199.

108. Ruiz-Barba, J. L., Rios-Sanchez, R. M., Fedriani-Iriso, C., Olias, J. M., Rios, J. L., & Jimenez-Diaz, R. (1990). Bactericidal effect of phenolic compounds from green olives on *Lactobacillus plantarum. Systematic and Applied Microbiology, 13,* 199–205.

109. Obied, H. K., Bedgood, D. R., Jr., Prenzler, P. D., & Robards, K. (2007). Bioscreening of Australian olive mill waste extracts: Biophenol content, antioxidant, antimicrobial and molluscicidal activities. *Food and Chemical Toxicology, 45,* 1238–1248.

110. Bisignano, G., Lagana, M. G., Trombetta, D., Arena, S., Nostro, A., Uccella, N., Mazzanti, G., & Saija, A. (2001). *In vitro* antibacterial activity of some aliphatic aldehydes from *Olea europaea* L. *FEMS Microbiology Letters, 198,* 9–13.

111. Aziz, N. H., Farag, S. E., Mousa, L. A., & Abo-Zaid, M. A. (1998). Comparative antibacterial and antifungal effects of some phenolic compounds. *Microbios, 93,* 43–54.

112. Kavanaugh, C. J., Trumbo, P. R., & Ellwood, K. C. (2007). The U.S. Food and Drug Administration's evidence-based review for qualified health claims: tomatoes, lycopene, and cancer. *Journal of the National Cancer Institute, 99,* 1074–1085.

113. Han, C. H., Yang, C. H., Sohn, D. W., Kim, S. W., Kang, S. H., & Cho, Y.. H. (2008). Synergistic effect between lycopene and ciprofloxacin on a chronic bacterial prostatitis rat model. *International Journal of Antimicrobial Agents, 31,* 102–107.

114. Smullen, J., Koutsou, G. A., Foster, H. A., Zumbe, A., & Storey, D. M. (2007). The antibacterial activity

of plant extracts containing polyphenols against *Streptococcus mutans*. *Caries Research, 41*, 342–349.

115. Zdrojewicz, Z., & Belowska-Bien, K. (2005). Resveratrol – Its activity and clinical role. [Polish]. *Advances in Clinical Experimental Medicine, 14*, 1051–1056.

116. Pervaiz, S. (2004). Chemotherapeutic potential of the chemopreventive phytoalexin resveratrol. *Drug Resistance Updates, 7*, 333–344.

117. Ghosh, A., Das, B. K., Roy, A., Mandal, B., & Chandra, G. (2008). Antibacterial activity of some medicinal plant extracts. *Nature Medicine, 62*, 259–262.

118. Braga, L. C., Shupp, J. W., Cummings, C., Jett, M., Takahashi, J. A., Carmo, L. S., Chartone-Souza, E., & Nascimento, A. M. (2005). Pomegranate extract inhibits *Staphylococcus aureus* growth and subsequent enterotoxin production. *Journal of Ethnopharmacology, 96*, 335–339.

119. Korban, S. S., Krasnyanski, S. F., & Buetow, D. E. (2002). Foods as production and delivery vehicles for human vaccines. *Journal of the American College of Nutrition, 21*(3 Suppl), 212S–217S.

120. Aqil, F., & Ahmad, I. (2007). Antibacterial properties of traditionally used Indian medicinal plants. *Methods and Findings in Experimental Clinical Pharmacology, 29*, 79–92.

121. Nair, R. R., & Chanda, S. V. (2005). *Punica granatum*: A potential source as antibacterial drug. *Asian Journal of Microbiology and Biotechnology Environmental Science, 7*, 625–628.

122. Shajahan, A., & Ramesh, S. (2004). Antimicrobial activity of crude ectocarp extract of pomegranate (*Punica granatum* L.) against some selected enteropathogenic bacteria. *Asian Journal of Microbiology and Biotechnology Environmental Science, 6*, 647–648.

123. Alanis, A. D., Calzada, F., Cervantes, J. A., Torres, J., & Ceballos, G. M. (2005). Antibacterial properties of some plants used in Mexican traditional medicine for the treatment of gastrointestinal disorders. *Journal of Ethnopharmacology, 100*, 153–157.

124. Perez, C., & Anesini, C. (1994). *In vitro* antibacterial activity of Argentine folk medicinal plants against *Salmonella typhi*. *Journal of Ethnopharmacology, 44*, 41–46.

125. Rani, P., & Khullar, N. (2004). Antimicrobial evaluation of some medicinal plants for their anti-enteric potential against multi-drug resistant *Salmonella typhi*. *Phytotherapy Research, 18*, 670–673.

126. Pradeep, B. V., Manojbabu, M. K., & Palaniswamy, M. (2008). Antibacterial activity of *Punica granatum* L against gastrointestinal tract infection causing organisms. *Ethnobotannical Leaflets, 12*, 1085–1089.

11

Fruit and Vegetable Intake of Mothers in Europe: Risks/Benefits

Alexandra Wolf[1] and Ibrahim Elmadfa[2]

[1]Austrian Agency for Health and Food Safety (AGES), Competence Center for Nutrition and Prevention, Austria
[2]Institute for Nutritional Sciences, University of Vienna, Austria

1. INTRODUCTION

Epidemiological and clinical studies in the nutrition field have shown the health benefits of adequate fruit and vegetable consumption [1,2] especially with regards to the prevention of chronic diseases [3–6].

Recommendations for adequate fruit and vegetable intake levels have been set by international health agencies and most European countries. National food-based dietary guidelines (FBDG) vary between 400 and 750 g fruit and vegetables per day [7], yet most of the national guidelines suggest at least 400 g as a minimum intake amount, which is also in line with the World Health Organization (WHO) recommendation [8,9].

While diets high in fruit and vegetables have been associated with beneficial health effects, a large proportion of the European population does not meet these recommendations [3,7,10–12]. Furthermore, adequate consumption of fruit and vegetables in mothers is of utmost importance, not only for the health of the mothers but also for the development and health of their children. Beside the health benefits it is worth emphasizing that mothers' fruit and vegetable consumption patterns play an essential role in shaping the food preferences and the fruit and vegetable consumption pattern of their children. Mothers are therefore an important target group for intervention programs promoting fruit and vegetable intake in Europe.

The aim of this paper is to review available and comparable intake data in Europe and to provide insight into risks and benefits of maternal fruit and vegetable consumption.

2. DATA ON FRUIT AND VEGETABLE INTAKE IN EUROPE

As fruit and vegetable intake is one of the main nutrition quality indicators [13], it is of particular importance to monitor the intake across Europe. Differences in consumption patterns are useful in order to identify important target groups for future interventions. Some attempts have been made to review data from national studies to make international comparisons [14,15]. The comparability of such national surveys is however limited, due to methodological differences including variations in food categorization systems and food composition databases [16]. At the time of writing, comparable data from cross-European studies using the same standardized instruments is limited.

Relevant studies collecting data using standardized instruments for measuring fruit and vegetable intake are the European Prospective Investigation in Cancer and Nutrition (EPIC) study and the Pro Children study. The EPIC study provided cross-European data of fruit and vegetable intake in women [17,18], whereas the Pro Children study was the first that provided detailed information on fruit and vegetable intake of European mothers [7,19].

2.1 Fruit and Vegetable Intake in European Women – Results from the EPIC Study

The European Prospective Investigation in Cancer and Nutrition (EPIC) study was not specifically designed to measure fruit and vegetable intake [17,18]. While focusing on risk factors for cancer, food consumption patterns were measured generally by using standardized computer-assisted 24-hour diet recalls. As a result, information on fruit and vegetable intake was obtained from 35,955 participants, of whom 22,924 were women, in 27 administrative centers in 10 European countries. Unfortunately, the publication did not indicate how many mothers were included in the analysis [3].

The results reported that fruit and vegetable intake levels in European women were not sufficient in many countries. A south–north gradient for fruit and vegetable intake could be observed in both sexes, but was less clearly defined among women.

The highest fruit consumption in women was found in Spain (400 g/d in Ragusa), the lowest in Sweden (151 g/d in Malmö). The highest vegetable consumption was shown in the South of France (261 g/d), while the lowest intake levels were observed in Norway (118 g/d) and surprisingly also in the North of Spain (103 g/d in Asturias). The highest consumption of fruit and vegetable juice was observed in Germany, followed by Dutch women [3].

Additionally, differences in consumption patterns could be observed with regards to type and preparation of vegetables. Relatively large amounts of leafy vegetables were consumed in the Mediterranean countries, whereas root vegetables were more common in Scandinavian countries, while the intake of cabbage was highest in the UK, the Netherlands, and in Germany. The proportion of cooked vegetables was highest in Greece, France, some Italian regions, the UK, and the Netherlands, while people in Norway, Sweden, and Germany tended to consume more raw vegetables [3].

2.2 Fruit and Vegetable Intake in European Mothers – Results from the Pro Children Study

Data on the fruit and vegetable intake of European mothers was provided by the Pro Children study, carried out in nine European countries in 2003. The main objective of the study was to develop strategies to promote adequate consumption levels of fruit and vegetables among European schoolchildren and their parents. In order to achieve that aim the project was designed to collect information on actual fruit and vegetable consumption in all participating countries, to identify factors associated with the consumption patterns, and to develop effective intervention strategies [20]. A simple, self-administered survey instrument for fruit and vegetable intake [21] was used to collect the data of 9070 mothers of 11-year-old children who participated in the survey.

The Pro Children study results showed that in all participating countries, fruit and vegetable intake was below the recommended level. The consumption of vegetables was consistently lower than that of fruit, while a high percentage of mothers reported eating fruit and vegetables less than once a day [7].

As illustrated in Figure 11.1, comparatively high average fruit intake levels could be identified in Portugal, Denmark, and Sweden (211, 203 and 194 g/d), while the lowest intake was found in Iceland (97 g/d). High vegetable intake levels were shown in Portugal and Belgium (169 and 150 g/d). The lowest intake was observed in Spain (88 g/d), as consistent

with the findings from the EPIC study. In this study, the highest intake levels of fruit juice were observed in Austria and again in the Netherlands. Since Spain was the country with the lowest vegetable intake level, a south–north gradient could not be observed in the Pro Children data [7].

Differences between countries were again shown for the preparation of vegetables (Table 11.1). More than two-thirds of the total amounts were eaten as cooked vegetables in Portugal, Belgium, and the Netherlands. Vegetable preparation in Portugal and Belgium was additionally characterized by a high intake of vegetable soup. A high consumption level of raw vegetables was again observed in Sweden, but also in Denmark and Iceland [7].

Despite the fact that Portugal and Denmark showed the highest total consumptions of fruit and vegetables with an average intake of 380 g/d

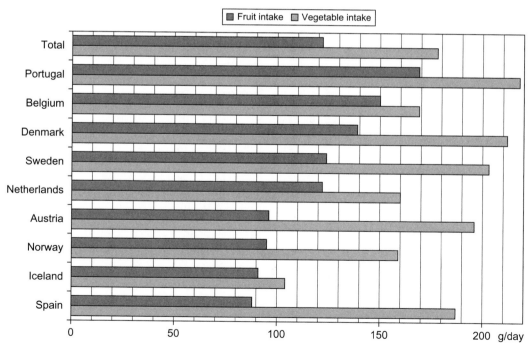

FIGURE 11.1 Average fruit and vegetable intake (g/d) of European mothers by country. Results from the Pro Children study. Adapted from Wolf et al. [7].

and 342 g/d, respectively, European mothers are far from meeting the international recommendations. Notably, vegetable consumption was low in most of the surveyed countries.

TABLE 11.1   Percentage of Total Vegetable Consumption per Preparation Method and Country (Results from the Pro Children Study)

	% of Total Vegetable Consumption per Country			
	Salad	**Raw Vegetables**	**Cooked Vegetables**	**Vegetable Soup**
Portugal	16.7	7.2	31.2	44.9
Belgium	13.7	16.3	33.2	36.8
Denmark	18.7	50.0	24.7	6.6
Sweden	28.6	44.9	21.7	4.8
Netherlands	16.5	17.9	50.2	15.4
Austria	25.5	29.1	26.6	18.8
Norway	15.7	40.0	39.9	6.4
Iceland	26.3	42.4	24.1	7.2
Spain	27.1	17.7	31.9	23.3
Total	20.0	27.8	30.6	21.6

Many recommendations do not specify whether fruit juice should be included as a part of total fruit consumption or not. Therefore juice intake was analyzed separately. With the inclusion of fruit juice, the WHO population goal for total fruit and vegetable consumption of ≥ 400 g/d was met by most countries surveyed, with the exception of Norway and Iceland. When fruit juice was excluded average consumption did not reach the recommendations in any country.

As shown in Figure 11.2, the consumption levels of both fruit and vegetables (without fruit juice) were in line with WHO recommendations for only 27% of all participating mothers.

3. INTERPRETATION OF AVAILABLE DATA ON FRUIT AND VEGETABLE INTAKE IN EUROPE

International nutritional surveys clearly indicated that fruit and vegetable intake is below

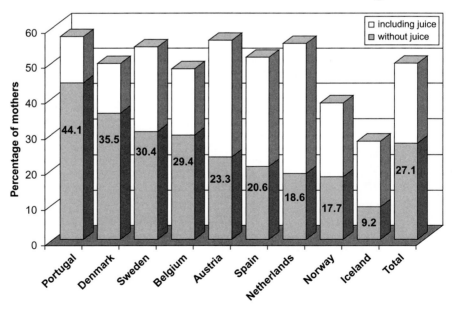

FIGURE 11.2   Percentage of mothers meeting WHO recommendations for fruit and vegetable intake of ≥ 400 g/d Results from the Pro Children study. Adapted from Wolf et al. [7].

the recommended amount in many European countries. In particular, low intake levels were found in the northern European countries [3,7,12]. Therefore, a south–north gradient in fruit and vegetable intake could be observed in some studies [3] and is also reflected in some agricultural statistics [12].

This south–north gradient for fruit and vegetable intake was to some extent also shown in European women [3], but could not be observed in mothers of 11-year-old children. In fact, the lowest intake level of vegetables was identified in Spanish mothers [7].

However, a tendency towards dietary uniformity has been widely observed in Europe over the last decade [22,23]. That might be one reason why a decrease in fruit and vegetable variety could also be identified in the southern European countries [19]. Therefore, interventions should not only promote the increase in fruit and vegetable intake but also encourage variety in the diets of European mothers.

4. BENEFITS AND RISKS OF FRUIT AND VEGETABLE INTAKE IN MOTHERS

Fruit and vegetables are low in energy but relatively rich in vitamins (especially vitamin C and folate), minerals, and other bioactive components which impart beneficial health effects. Convincing evidence exists for the role of fruit and vegetables in the prevention of chronic diseases such as cardiovascular diseases [24–26], stroke [27–29], and hypertension [30] and probable evidence for the protection against some cancers [31]. Furthermore, there is evidence that high fruit and vegetable intake is associated with a reduced risk of osteoporosis [32,33] and some eye diseases such as cataracts [34] and age-related nuclear lens opacities [35]. Besides that, fruit and vegetable intake may play an important role in weight management

and thereby in the prevention of overweight and obesity [1,36–38].

It is important to emphasize that aside from these general health effects, adequate fruit and vegetable consumption levels in mothers would positively affect both the mothers' and children's health.

4.1 Health-relevant Aspects of Fruit and Vegetable Intake in Women of Childbearing Age

High fruit and vegetable intake in pregnancy has been shown to be positively associated with birth weight [39]. In contrast, inadequate fruit and vegetable consumption may increase the risk of adverse pregnancy complications and outcomes [40]. Low fruit and vegetable consumption and the resulting low intake of nutrients such as folate and lutein/zeaxanthin may increase the risk of sporadic retinoblastoma in the infant [41]. Low concentrations of dietary and circulating folate have further been associated with preterm delivery, low birth weight and fetal growth retardation [40]. Folate also plays an essential role in the prevention of neural tube defects. Thus, adequate folate intake is already crucial prior to conception and throughout pregnancy [42].

Low intake of fruit and vegetables, especially green leafy vegetables, is a primary cause of folate deficiency [43], which is associated with an elevation of blood homocysteine. High maternal homocystein concentrations were in turn identified as a risk factor for habitual spontaneous abortion and pregnancy complications such as placental abruption and preeclampsia [40]. The long-term high consumption of vegetables has the potential to improve folate status and is therefore associated with a reduced risk for folate deficiency during pregnancy [42].

Fruit and vegetables are important sources of micronutrients that are crucial for the health of the mother and also for the development of the unborn child. High consumption of fruit and vegetables to optimize the folate status is therefore recommended by many nutritional guidelines [42]. Despite this, a significant proportion of women of reproductive age consume diets that are low in fruits and vegetables [7] and subsequently also low in micronutrients such as folate [40]. So far these recommendations alone have not shown any positive effects in reducing the occurrence of neural tube defects, therefore strategies for supplementation and food fortification are still a contentious issue in the nutritional and medical field [44].

5. MATERNAL INFLUENCE ON CHILDREN'S FRUIT AND VEGETABLE CONSUMPTION

Healthy diets in childhood are important as they shape food preferences and eating habits in later life and protect against early manifestations of nutrition-related diseases. Nevertheless, in many European countries, children consume far less than the minimum recommended amount [10].

Children's food choice is strongly guided by their taste preferences [45–48]. Therefore it is of the utmost importance to promote healthy food preferences and thereby healthy eating habits in children. Generally, children's nearest social environment plays an important role in shaping food preferences [49]. In particular, maternal eating behavior has an important function in the development of their children's eating preferences and behavior. People in close proximity to children's social environments can influence children's eating habits either directly by creating the child's food environment (by influencing exposure and accessibility to food) and by acting as positive role models or indirectly through the transmission of attitudes and values [50,51].

5.1 Early Exposure and Food Preferences

Several studies noted that fruit and vegetable preferences are either the only significant or the most important predictor of fruit and vegetable consumption in children [47,52–54]. Children's food preferences, on the other hand, are strongly associated with their consumption pattern. They are determined by the children's food experiences and the extent to which food is familiar to the children [45,55]. The innate preference for a sweet taste is modified from the very beginning, initiated by early experiences with food [47,56]. The early exposure to different tastes, either to fruits and vegetables or to high-energy dense foods, may have an important role in establishing a hierarchy of food selection later on [47]. Therefore research suggests that the earlier and broader the experience, the healthier the child's diet [45].

Consequently, it was hypothesized that the acceptability of flavors to infants increases if previously exposed via amniotic fluid or breast milk. Breastfeeding in particular might provide an advantage in the initial acceptability of food, especially when mothers consume the same food regularly [57].

Indeed, evidence suggests that prenatal and early postnatal exposure to a flavor, for example that of carrots, may promote infants' enjoyment and acceptability of that flavor in solid foods during weaning [47,57,58]. It was further indicated that even the fruit consumption of 2- to 6-year-old children was positively affected by breastfeeding and the early introduction to fruit [59]. Frequent exposure to a wide variety of fruits in the first two years was shown to predict fruit variety in school-aged children [60].

Based on these findings, mothers should be encouraged to eat plentiful amounts, as well as a variety, of fruit and vegetables during pregnancy and lactation, not only to guarantee satisfactory nutrient intake levels (for example that of folate), but also with regards to the early introduction of these flavors to their infants.

Mothers should be made aware of the importance of early and repeated food-related experiences for the acceptability of a variety of fruits and vegetables later on [60]. They should continue to offer new flavors such as those of unfamiliar fruits and vegetables, as often as possible. This is in order to familiarize their children with new tastes and to allow their children to integrate the new food into their diets [45,60].

5.2 Family Food Environment and Parental Facilitation

Parents occupy a key position in their children's food environment. They are the 'gate-keepers' of home food supply (selection of menus, variety of foods, cooking or preparation practices) and play an important role in promoting the availability and accessibility of fruits and vegetables [61]. Mothers can influence children's fruit and vegetable consumption patterns by controlling fruit and vegetable availability, by providing a supportive eating environment, and by establishing clear meal structures [62]. Mothers can encourage their children's fruit and vegetable consumption; for example, by eating fruit and vegetables together with their children, by cutting up fruit and vegetables between meals, and by allowing their children to eat as much fruit or as many vegetables as they like. They can further facilitate their children's intake by acting as a positive role model [62,63].

5.3 Model Learning

Children's eating behaviors are particularly influenced by the behavior of people in their social environment; for example, by their parents and especially by their mothers, who are traditionally responsible for their children's care.

As they act as potential role models for their children's health and eating behavior, their own fruit and vegetable consumption becomes a significant predictor of their children's consumption pattern [63–68]. Literature suggests that children are likely to develop similar taste preferences and eating patterns to their parents [69]. Children's food intake often mirrors their parents' intake [70]. Positive associations between parents' and children's intake could be found for nutrients [71,72] as well as for fruits and vegetables [63,72]. The observed effects were stronger between mothers and children than between fathers and children [71].

Children who perceived their role models – in particular their mother, their father, and their best friends – to be daily fruit and vegetable consumers were more likely to have positive attitudes towards fruit and vegetables [73] and to be daily fruit and vegetable consumers themselves [52,74].

Thus, a number of studies show positive associations between healthy consumption patterns of mothers and consumption patterns of children or adolescents, respectively [51,75].

It was further shown that mothers reporting higher levels of concern for healthful eating had higher intakes of fruits, vegetables, and dairy products. Additionally, these mothers were more likely to create a supportive home food environment for their children. Consequently, adolescent perception of maternal concern for healthy eating was shown to be positively associated with adolescent fruit and vegetable intake [51]. These findings again indicated that maternal modeling effects have a consistent impact on children's eating behavior. However, it was also shown that children and adolescents do not only imitate the eating behaviors of their role models, but also pick up preferences, attitudes, and beliefs, as well as norms and expectations towards food from their social environment [76].

Mothers should recognize that their own fruit and vegetable intake, as well as attitudes regarding fruit and vegetable intake, have the potential to influence their children's intake levels. Furthermore, mothers should be made aware of the importance of adequate fruit and vegetable consumption and should be encouraged to share these health beliefs with their children.

6. SUMMARY

Despite the importance of fruit and vegetable consumption in the health of mothers in particular, the results of European cross-sectional studies show that intake levels of fruit and vegetables in women are still below the recommended amount. As a high proportion of women of a reproductive age consume diets that are low in fruit and vegetables, adequate intake of essential nutrients such as folate, vitamin C and β-carotene cannot be guaranteed, and might have an adverse effect on the health and development of their unborn children. Unfortunately, a high percentage of mothers report that they eat fruit and vegetables less than once a day. Above all, mothers should be made aware of the great importance of their own fruit and vegetable intake and the association of their food preferences with the fruit and vegetable intake patterns of their children.

Published data show that national and international interventions are necessary to promote fruit and especially vegetable consumption in the European population of mothers.

ACKNOWLEDGMENT

The Pro Children survey was carried out with financial support from the Commission of the European Communities, specific RTD program 'Quality of Life and Management of Living Resources,' QLK1-2001-00547 'Promoting and sustaining health through increased vegetable and fruit consumption among European school-children.' It does not necessarily reflect the Commission's views and in no way anticipates its future policy in this area.

Special thanks to the Pro Children consortium and to all mothers who took the time to participate in the Pro Children survey.

References

1. Bes-Rastrollo, M., Martinez-Gonzalez, M. A., Sanchez-Villegas, A., de la Fuente Arrillaga, C., & Martinez, J. A. (2006). Association of fiber intake and fruit/vegetable consumption with weight gain in a mediterranean population. *Nutrition, 22,* 504–511.

2. Van Duyn, M. S., & Pivonka, E. (2000). Overview of the health benefits of fruit and vegetable consumption for the dietetics professional: Selected literature. *Journal of the American Dietetic Association, 100,* 1511–1521.

3. Agudo, A., Slimani, N., Ocke, M. C., Naska, A., Miller, A. B., Kroke, A., Bamia, C., Karalis, D., Vineis, P., Palli, D., Bueno-de-Mesquita, H. B., Peeters, P. H., Engeset, D., Hjartaker, A., Navarro, C., Martinez Garcia, C., Wallstrom, P., Zhang, J. X., Welch, A. A., Spencer, E., Stripp, C., Overvad, K., Clavel-Chapelon, F., Casagrande, C., & Riboli, E. (2002). Consumption of vegetables, fruit and other plant foods in the European Prospective Investigation into Cancer and Nutrition (EPIC) cohorts from 10 European countries. *Public Health Nutrition, 5,* 1179–1196.

4. Trichopoulou, A., Naska, A., Antoniou, A., Friel, S., Trygg, K., & Turrini, A. (2003). Vegetable and fruit: The evidence in their favour and the public health perspective. *International Journal for Vitamin and Nutrition Research, 73,* 63–69.

5. van't Veer, P., Jansen, M. C., Klerk, M., & Kok, F. J. (2000). Fruits and vegetables in the prevention of cancer and cardiovascular disease. *Public Health Nutrition, 3,* 103–107.

6. Thorsdottir, I., & Ramel, A. (2003). Dietary intake of 10- to 16-year-old children and adolescents in central and northern Europe and association with the incidence of type 1 diabetes. *Annals of Nutrition and Metabolism, 47,* 267–275.

7. Wolf, A., Yngve, A., Elmadfa, I., Poortvliet, E., Ehrenblad, B., Perez-Rodrigo, C., Thorsdottir, I., Haraldsdottir, J., Brug, J., Maes, L., Vaz de Almeida, M. D., Krolner, R., & Klepp, K. I. (2005). Fruit and vegetable intake of mothers of 11-year-old children in nine European countries: The pro children cross-sectional survey. *Annals of Nutrition and Metabolism, 49,* 246–254.

8. WHO (2002). *Food-based dietary guidelines in WHO European member states*. Copenhagen: WHO Regional Office for Europe.
9. WHO (2004). Global strategy on diet, physical activity and health – what are the challenges for follow-up in Europe? Workshop 1. *European Journal of Public Health*, 14(Suppl 1), 8.
10. Yngve, A., Wolf, A., Poortvliet, E., Elmadfa, I., Brug, J., Ehrenblad, B., Franchini, B., Haraldsdottir, J., Krolner, R., Maes, L., Perez-Rodrigo, C., Sjostrom, M., Thorsdottir, I., & Klepp, K. I. (2005). Fruit and vegetable intake in a sample of 11-year-old children in nine European countries: The pro children cross-sectional survey. *Annals of Nutrition and Metabolism*, 49, 236–245.
11. Trichopoulou, A., Orfanos, P., Norat, T., Bueno-de-Mesquita, B., Ocke, M. C., Peeters, P. H., van der Schouw, Y. T., Boeing, H., Hoffmann, K., Boffetta, P., Nagel, G., Masala, G., Krogh, V., Panico, S., Tumino, R., Vineis, P., Bamia, C., Naska, A., Benetou, V., Ferrari, P., Slimani, N., Pera, G., Martinez-Garcia, C., Navarro, C., Rodriguez-Barranco, M. Dorronsoro, M., Spencer, E. A., Key, T. J., Bingham, S., Khaw, K. T., Kesse, E., Clavel-Chapelon, F., Boutron-Ruault, M. C., Berglund, G., Wirfalt, E., Hallmans, G., Johansson, I., Tjonneland, A., Olsen, A., Overvad, K., Hundborg, H. H., Riboli, E., & Trichopoulos, D. (2005). Modified Mediterranean diet and survival: EPIC-elderly prospective cohort study. *BMJ*, 330, 991.
12. Elmadfa, I., Weichselbaum, E., Konig, J., de Winter, A. M. R., Trolle, E., Haapala, I., Uusitalo, U., Mennen, L., Hercberg, S., Wolfram, G., Trichopoulou, A., Naska, A., Benetou, V., Kritsellis, E., Rodler, I., Zajkas, G., Branca, F., D'Acapito, P., Klepp, K. I., Ali-Madar, A., De Almeida, M. D., Alves, E., Rodrigues, S., Sarra-Majem, L., Roman, B., Sjostrom, M., Poortvliet, E., & Margetts, B. (2005). European nutrition and health report 2004. *Forum Nutrition*, 1–220.
13. Steingrimsdottir, L., Ovesen, L., Moreiras, O., & Jacob, S. (2002). Selection of relevant dietary indicators for health. *European Journal of Clinical Nutrition*, 56 (Suppl 2), 8–11.
14. Irala-Estevez, J. D., Groth, M., Johansson, L., Oltersdorf, U., Prattala, R., & Martinez-Gonzalez, M. A. (2000). A systematic review of socio-economic differences in food habits in Europe: Consumption of fruit and vegetables. *European Journal of Clinical Nutrition*, 54, 706–714.
15. Beer-Borst, S., Hercberg, S., Morabia, A., Bernstein, M. S., Galan, P., Galasso, R, Giampaoli, S., McCrum, E., Panico, S., Preziosi, P., Ribas, L., Serra-Majem, L., Vescio, M. F., Vitek, O., Yarnell, J., & Northridge, M. E. (2000). Dietary patterns in six european populations: results from EURALIM, a collaborative European data harmonization and information campaign. *European Journal of Clinical Nutrition*, 54, 253–262.
16. Verger, P., Ireland, J., Moller, A., Abravicius, J. A., De Henauw, S., & Naska, A. (2002). Improvement of comparability of dietary intake assessment using currently available individual food consumption surveys. *European Journal of Clinical Nutrition*, 56(Suppl 2), 18–24.
17. Slimani, N., Ferrari, P., Ocke, M., Welch, A., Boeing, H., Liere, M., Pala, V., Amiano, P., Lagiou, A., Mattisson, I., Stripp, C., Engeset, D., Charrondiere, R., Buzzard, M., Staveren, W., & Riboli, E. (2000). Standardization of the 24-hour diet recall calibration method used in the european prospective investigation into cancer and nutrition (EPIC): general concepts and preliminary results. *European Journal of Clinical Nutrition*, 54, 900–917.
18. Slimani, N., Deharveng, G., Unwin, I., Southgate, D. A., Vignat, J., Skeie, G., Salvini, S., Parpinel, M., Moller, A., Ireland, J., Becker, W., Farran, A., Westenbrink, S., Vasilopoulou, E., Unwin, J., Borgejordet, A., Rohrmann, S., Church, S., Gnagnarella, P., Casagrande, C., van Bakel, M., Niravong, M., Boutron-Ruault, M. C., Stripp, C., Tjonneland, A., Trichopoulou, A., Georga, K., Nilsson, S., Mattisson, I., Ray, J., Boeing, H., Ocke, M., Peeters, P. H., Jakszyn, P., Amiano, P., Engeset, D., Lund, E., de Magistris, M. S., Sacerdote, C., Welch, A., Bingham, S., Subar, A. F., & Riboli, E. (2007). The EPIC nutrient database project (ENDB): A first attempt to standardize nutrient databases across the 10 European countries participating in the EPIC study. *European Journal of Clinical Nutrition*, 61, 1037–1056.
19. Klepp, K.. I., Perez-Rodrigo, C., Thorsdottir, I., Due, P., Vaz De Almeida, M. D., Elmadfa, I., Wolf, A., Haraldsdottir, J., Brug, J., Sjostrom, M., Yngve, A., & De Bourdeaudhuij, I. (2005). Promoting and sustaining health through increased vegetable and fruit consumption among European schoolchildren: The pro children project. *Journal of Public Health*, 13, 97–101.
20. Klepp, K. I., Perez-Rodrigo, C., De Bourdeaudhuij, I., Due, P. P., Elmadfa, I., Haraldsdottir, J., Konig, J., Sjostrom, M., Thorsdottir, I., Vaz de Almeida, M. D., Yngve, A., & Brug, J. (2005). Promoting fruit and vegetable consumption among European schoolchildren: rationale, conceptualization and design of the pro children project. *Annals of Nutrition and Metabolism*, 49, 212–220.
21. Kristjansdottir, A. G., Andersen, L. F., Haraldsdottir, J., de Almeida, M. D., & Thorsdottir, I. (2006). Validity of a questionnaire to assess fruit and vegetable intake in adults. *European Journal of Clinical Nutrition*, 60, 408–415.

22. Hill, M. J. (1997). Changing pattern of diet in Europe. *European Journal of Cancer Prevention, 6*(Suppl 1), 11–13.

23. Rumm-Kreuter, D. (2001). Comparison of the eating and cooking habits of northern Europe and the mediterranean countries in the past, present and future. *International Journal for Vitamin and Nutrition Research, 71*, 141–148.

24. Bazzano, L.A. (2005). Dietary intake of fruits and vegetables and risk of diabetes mellitus and cardiovascular disease. Background paper for the joint FAO/WHO workshop on fruit and vegetables for health, 1–3 September 2004, Kobe, Japan..

25. Dauchet, L., Amouyel, P., Hercberg, S., & Dallongeville, J. (2006). Fruit and vegetable consumption and risk of coronary heart disease: A meta-analysis of cohort studies. *Journal of Nutrition, 136*, 2588–2593.

26. He, F. J., Nowson, C. A., Lucas, M., & MacGregor, G. A. (2007). Increased consumption of fruit and vegetables is related to a reduced risk of coronary heart disease: Meta-analysis of cohort studies. *Journal of Human Hypertension, 21*, 717–728.

27. Dauchet, L., Amouyel, P., & Dallongeville, J. (2005). Fruit and vegetable consumption and risk of stroke: A meta-analysis of cohort studies. *Neurology, 65*, 1193–1197.

28. He, F. J., Nowson, C. A., & MacGregor, G. A. (2006). Fruit and vegetable consumption and stroke: Meta-analysis of cohort studies. *Lancet, 367*, 320–326.

29. Howard, B. V., Van Horn, L., Hsia, J., Manson, J. E., Stefanick, M. L., Wassertheil-Smoller, S., Kuller, L. H., LaCroix, A. Z., Langer, R. D., Lasser, N. L., Lewis, C. E., Limacher, M. C., Margolis, K. L., Mysiw, W. J., Ockene, J. K., Parker, L. M., Perri, M. G., Phillips, L., Prentice, R. L., Robbins, J., Rossouw, J. E., Sarto, G. E., Schatz, I. J., Snetselaar, L. G., Stevens, V. J., Tinker, L. F., Trevisan, M., Vitolins, M. Z., Anderson, G. L., Assaf, A. R., Bassford, T., Beresford, S. A., Black, H. R., Brunner, R. L., Brzyski, R. G., Caan, B., Chlebowski, R. T., Gass, M., Granek, I., Greenland, P., Hays, J., Heber, D., Heiss, G., Hendrix, S. L., Hubbell, F. A., Johnson, K. C., & Kotchen, J. M. (2006). Low-fat dietary pattern and risk of cardiovascular disease: The women's health initiative randomized controlled dietary modification trial. *JAMA, 295*, 655–666.

30. Dauchet, L., Kesse-Guyot, E., Czernichow, S., Bertrais, S., Estaquio, C., Peneau, S., Vergnaud, A. C., Chat-Yung, S., Castetbon, K., Deschamps, V., Brindel, P., & Hercberg, S. (2007). Dietary patterns and blood pressure change over 5-y follow-up in the SU.VI.MAX cohort. *American Journal of Clinical Nutrition, 85*, 1650–1656.

31. World Cancer Research Fund (WCRF) and American Institute for Cancer Research (AICR). (2007). *Food, Nutrition and Prevention of Cancer: a Global Perspective.* Washington, DC: AICR.

32. Bell, J. A., & Whiting, S. J. (2004). Effect of fruit on net acid and urinary calcium excretion in an acute feeding trial of women. *Nutrition, 20*, 492–493.

33. Lin, P. H., Ginty, F., Appel, L. J., Aickin, M., Bohannon, A., Garnero, P., Barclay, D., & Svetkey, L. P. (2003). The DASH diet and sodium reduction improve markers of bone turnover and calcium metabolism in adults. *Journal of Nutrition, 133*, 3130–3136.

34. Christen, W. G., Liu, S., Schaumberg, D. A., & Buring, J. E. (2005). Fruit and vegetable intake and the risk of cataract in women. *American Journal of Clinical Nutrition, 81*, 1417–1422.

35. Moeller, S. M., Taylor, A., Tucker, K. L., McCullough, M. L., Chylack, L. T., Jr., Hankinson, S. E., Willett, W. C., & Jacques, P. F. (2004). Overall adherence to the dietary guidelines for Americans is associated with reduced prevalence of early age-related nuclear lens opacities in women. *Journal of Nutrition, 134*, 1812–1819.

36. Newby, P. K., Muller, D., Hallfrisch, J., Qiao, N., Andres, R., & Tucker, K. L. (2003). Dietary patterns and changes in body mass index and waist circumference in adults. *American Journal of Clinical Nutrition, 77*, 1417–1425.

37. Rolls, B. J., Ello-Martin, J. A., & Tohill, B. C. (2004). What can intervention studies tell us about the relationship between fruit and vegetable consumption and weight management? *Nutrition Reviews, 62*, 1–17.

38. WHO/FAO. (2003). Diet, nutrition and the prevention of chronic diseases. Report of a joint WHO/FAO expert consultation, January 28 – February 1, 2002, Geneva. *World Health Organization Technical Report Services, 916*, 1–149.

39. Mikkelsen, T. B., Osler, M., Orozova-Bekkevold, I., Knudsen, V. K., & Olsen, S. F. (2006). Association between fruit and vegetable consumption and birth weight: A prospective study among 43,585 Danish women. *Scandinavian Journal of Public Health, 34*, 616–622.

40. Scholl, T. O., & Johnson, W. G. (2000). Folic acid: Influence on the outcome of pregnancy. *American Journal of Clinical Nutrition, 71*, 1295S–1303S.

41. Orjuela, M. A., Titievsky, L., Liu, X., Ramirez-Ortiz, M., Ponce-Castaneda, V., Lecona, E., Molina, E., Beaverson, K., Abramson, D. H., & Mueller, N. E. (2005). Fruit and vegetable intake during pregnancy and risk for development of sporadic retinoblastoma. *Cancer Epidemiology Biomarkers and Prevention, 14*, 1433–1440.

42. Koebnick, C., Heins, U. A., Hoffmann, I., Dagnelie, P. C., & Leitzmann, C. (2001). Folate status during pregnancy in women is improved by long-term high vegetable intake compared with the average western diet. *Journal of Nutrition, 131*, 733–739.

43. Allen, L. H. (2008). Causes of vitamin B12 and folate deficiency. *Food and Nutrition Bulletin, 29*, S20–S34; discussion S35–S37.

44. Botto, L. D., Lisi, A., Robert-Gnansia, E., Erickson, J. D., Vollset, S. E., Mastroiacovo, P., Botting, B., Cocchi, G., de Vigan, C., de Walle, H., Feijoo, M., Irgens, L. M., McDonnell, B., Merlob, P., Ritvanen, A., Scarano, G., Siffel, C., Metneki, J., Stoll, C., Smithells, R., & Goujard, J. (2005). International retrospective cohort study of neural tube defects in relation to folic acid recommendations: Are the recommendations working? *BMJ, 330*, 571.

45. Cooke, L. (2007). The importance of exposure for healthy eating in childhood: A review. *Journal of Human Nutrition and Dietetics, 20*, 294–301.

46. Ramos, M., & Stein, L. M. (2000). Development children's eating behavior. *Journal of Pediatric (Rio J), 76* (Suppl 3), S229–S237.

47. Birch, L. L., & Fisher, J. O. (1998). Development of eating behaviors among children and adolescents. *Pediatrics, 101*, 539–549.

48. Baranowski, T., Cullen, K. W., Nicklas, T., Thompson, D., & Baranowski, J. (2003). Are current health behavioral change models helpful in guiding prevention of weight gain efforts? *Obesity Research, 11*(Suppl), 23S–43S.

49. Hart, K. H., Herriot, A., Bishop, J. A., & Truby, H. (2003). Promoting healthy diet and exercise patterns amongst primary school children: A qualitative investigation of parental perspectives. *Journal of Human Nutrition and Dietetics, 16*, 89–96.

50. Benton, D. (2004). Role of parents in the determination of the food preferences of children and the development of obesity. *International Journal of Obesity and Related Metabolic Disorders, 28*, 858–869.

51. Boutelle, K. N., Birkeland, R. W., Hannan, P. J., Story, M., & Neumark-Sztainer, D. (2007). Associations between maternal concern for healthful eating and maternal eating behaviors, home food availability, and adolescent eating behaviors. *Journal of Nutrition Education and Behavior, 39*, 248–256.

52. Brug, J., Tak, N. I., te Velde, S. J., Bere, E., & de Bourdeaudhuij, I. (2008). Taste preferences, liking and other factors related to fruit and vegetable intakes among schoolchildren: results from observational studies. *British Journal of Nutrition, 99*(Suppl 1), S7–S14.

53. Perez-Rodrigo, C., Ribas, L., Serra-Majem, L., & Aranceta, J. (2003). Food preferences of Spanish children and young people: The enKid study. *European Journal of Clinical Nutrition, 57*(Suppl 1), S45–S48.

54. Neumark-Sztainer, D., Wall, M., Perry, C., & Story, M. (2003). Correlates of fruit and vegetable intake among adolescents. Findings from project EAT. *Preventive Medicine, 37*, 198–208.

55. Cooke, L. J., & Wardle, J. (2005). Age and gender differences in children's food preferences. *British Journal of Nutrition, 93*, 741–746.

56. Westenhoefer, J. (2002). Establishing dietary habits during childhood for long-term weight control. *Annals of Nutrition and Metabolism, 46*(Suppl 1), 18–23.

57. Forestell, C. A., & Mennella, J. A. (2007). Early determinants of fruit and vegetable acceptance. *Pediatrics, 120*, 1247–1254.

58. Mennella, J. A., Jagnow, C. P., & Beauchamp, G. K. (2001). Prenatal and postnatal flavor learning by human infants. *Pediatrics, 107*, E88.

59. Cooke, L. J., Wardle, J., Gibson, E. L., Sapochnik, M., Sheiham, A., & Lawson, M. (2004). Demographic, familial and trait predictors of fruit and vegetable consumption by pre-school children. *Public Health Nutrition, 7*, 295–302.

60. Skinner, J. D., Carruth, B. R., Bounds, W., Ziegler, P., & Reidy, K. (2002). Do food-related experiences in the first 2 years of life predict dietary variety in school-aged children? *Journal of Nutrition Education and Behavior, 34*, 310–315.

61. Cullen, K. W., Baranowski, T., Owens, E., Marsh, T., Rittenberry, L., & de Moor, C. (2003). Availability, accessibility, and preferences for fruit, 100% fruit juice, and vegetables influence children's dietary behavior. *Health Education and Behavior, 30*, 615–626.

62. Nicklas, T. A., Baranowski, T., Baranowski, J. C., Cullen, K., Rittenberry, L., & Olvera, N. (2001). Family and child-care provider influences on preschool children's fruit, juice, and vegetable consumption. *Nutrition Reviews, 59*, 224–235.

63. Blanchette, L., & Brug, J. (2005). Determinants of fruit and vegetable consumption among 6-12-year-old children and effective interventions to increase consumption. *Journal of Human Nutrition and Dietetics, 18*, 431–443.

64. Gibson, E. L., Wardle, J., & Watts, C. J. (1998). Fruit and vegetable consumption, nutritional knowledge and beliefs in mothers and children. *Appetite, 31*, 205–228.

65. Young, E. M., Fors, S. W., & Hayes, D. M. (2004). Associations between perceived parent behaviors and middle school student fruit and vegetable consumption. *Journal of Nutrition Education and Behavior, 36*, 2–8.

66. Bere, E., & Klepp, K. I. (2004). Correlates of fruit and vegetable intake among Norwegian schoolchildren: parental and self-reports. *Public Health Nutrition, 7*, 991–998.

67. Fisher, J. O., Mitchell, D. C., Smiciklas-Wright, H., & Birch, L. L. (2002). Parental influences on young girls' fruit and vegetable, micronutrient, and fat intakes. *Journal of the American Dietetic Association, 102*, 58–64.

68. Rasmussen, M., Krolner, R., Klepp, K. I., Lytle, L., Brug, J., Bere, E., & Due, P. (2006). Determinants of fruit and vegetable consumption among children and adolescents: A review of the literature. Part I: Quantitative studies. *International Journal of Behavioral Nutrition and Physical Activity, 3*, 22.

69. Belton, T. (2003). Exploring attitudes to eating fruits and vegetables, Berlin: Springer.

70. Elfhag, K., Tholin, S., & Rasmussen, F. (2008). Consumption of fruit, vegetables, sweets and soft drinks are associated with psychological dimensions of eating behaviour in parents and their 12-year-old children. *Public Health Nutrition, 11*, 914–923.

71. Oliveria, S. A., Ellison, R. C., Moore, L. L., Gillman, M. W., Garrahie, E. J., & Singer, M. R. (1992). Parent-child relationships in nutrient intake: The Framingham Children's Study. *American Journal of Clinical Nutrition, 56*, 593–598.

72. Thorsdottir, I., Gunnarsdottir, I., Ingolfsdottir, S. E., & Palsson, G. (2006). Fruit and vegetable intake: Vitamin C and β-carotene intake and serum concentrations in six-year-old children and their parents. *Scandinavian Journal of Food and Nutrition, 50*, 71–76.

73. Wind, M., Bobelijn, K., De Bourdeaudhuij, I., Klepp, K. I., & Brug, J. (2005). A qualitative exploration of determinants of fruit and vegetable intake among 10- and 11-year-old schoolchildren in the low countries. *Annals of Nutrition and Metabolism, 49*, 228–235.

74. De Bourdeaudhuij, I., te Velde, S., Brug, J., Due, P., Wind, M., Sandvik, C., Maes, L., Wolf, A., Perez Rodrigo, C., Yngve, A., Thorsdottir, I., Rasmussen, M., Elmadfa, I., Franchini, B., & Klepp, K. I. (2008). Personal, social and environmental predictors of daily fruit and vegetable intake in 11-year-old children in nine European countries. *European Journal of Clinical Nutrition, 62*, 834–841.

75. Hill, A. J., & Franklin, J. A. (1998). Mothers, daughters and dieting: investigating the transmission of weight control. *British Journal of Clinical Psychology, 37*, 3–13.

76. Cullen, K. W., Baranowski, T., Rittenberry, L., & Olvera, N. (2000). Social-environmental influences on children's diets: Results from focus groups with African-, Euro- and Mexican-American children and their parents. *Health Education Research, 15*, 581–590.

12

Fruit, Vegetables, and Bone Health

Yu Ming Chen[1,2] and Suzanne C. Ho[1]

[1]Centre of Research and Promotion of Women's Health, School of Public Health and Primary Care, the Chinese University of Hong Kong

[2]School of Public Health, Sun Yat-sen University, Guangzhou, China

1. INTRODUCTION

Osteoporosis is defined as a metabolic bone disease characterized by low bone mass and microarchitectural deterioration of bone tissue, leading to enhanced bone fragility and a consequent increase in fracture risk. Osteoporosis has become a significant public health problem and has received great attention in industrialized countries. It affects an estimated 44 million Americans or 55% of people 50 years of age or older [1]. Another 34 million Americans are estimated to have low bone mass, meaning that they are at an increased risk of osteoporosis [1]. Bone mineral density (BMD) is affected by genetic, endocrine, mechanical, and nutritional factors, with extensive interactions between these factors [2]. Nutritional factors are considered to be of particular importance to bone health because they are potentially modifiable.

Previous studies have pointed toward the important role of nutritional factors in the maintenance of optimal bone health. Until recently, focuses have primarily been placed on calcium and a few isolated nutrients. Nutrients such as alkaline ions (potassium and magnesium), vitamin K, and vitamin C, found abundantly in fruit and vegetables, have been found to be associated with greater bone mass and lower urinary calcium excretion in human studies [3]. These findings have led to the examination of linkages between fruit and vegetable consumption and bone health. A few studies have reported the beneficial associations between fruit and vegetable consumption and BMD and/or bone size among children and adults [3–11]. An increase in the consumption of fruits and vegetables to 4.5 servings a day has been advocated by the US Department of Agriculture and the American Heart Association Nutrition Committee on the assumption that such a change would reduce the incidences of both cancer and cardiovascular disease [12,13]. This chapter will summarize the association between the consumption of fresh fruits and vegetables and bone health, and the potential mechanisms.

2. ASSOCIATION OF FRUIT AND VEGETABLE INTAKE AND BONE HEALTH

2.1 Animal or *In Vitro* Studies

In 1999, Mühlbauer and Li first reported that short-term treatments (10 days) with 14 vegetables commonly consumed by humans (onion, leek, garlic, wild garlic, common parsley, Italian parsley, dill, lettuce, tomato, cucumber, arugula, broccoli, Brussels sprouts, Chinese cabbage, and red cabbage) can significantly inhibit bone resorption in rats [14,15]. The effects of one gram of the dried test foodstuff on a urinary bone resorption marker (tetracycline) were similar to or even higher than those of calcitonin at doses (1.25 or 2.5 IU/kg body weight) used to treat postmenopausal osteoporosis [14,15] (Figure 12.1). Their study group further examined many other fruits, herbs, nuts, beans, and seeds, mushrooms, carbohydrate sources, and beverages [16]. Figure 12.2 summarizes the ranked potency of inhibition of bone resorption in rats by 24 fresh food items and red wine residue when administered at a dosage of 1 g/d of fresh weight [16]. For vegetables and bone health, a summary of animal studies strongly suggested the positive effects of onion on bone. A study using short-term treatment (10 days) with onion (0.03–1.5 g/rat/day) has shown a decrease in the ovariectomy-induced bone resorption rate in a dose-dependent manner [16]. Mühlbauer and his colleagues noted that relative to controls, rats fed 1 g dry onion per day for 4 weeks had an increase in bone mineral content (BMC) by 17.7% [standard deviation (SD) 6.4%; P<0.05], in mean cortical thickness by 14.8% (SD, 7.6%), and in BMD of trabecular bone by 13.5% (SD 3.1%; P<0.05) [14]. Another study showed that onion-enriched diets could counteract ovariectomy-induced osteopenia in young adult rats in a dose-dependent manner. The efficacies of

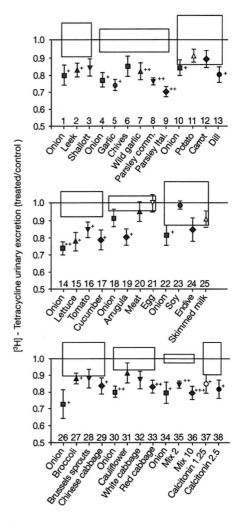

FIGURE 12.1 Effect of foodstuffs and the hormone calcitonin on bone resorption as assessed by the urinary excretion of previously administered radiolabeled tetracycline.

Data are plotted as the ratio of treated/untreated control [± standard error of the mean (SEM); N=5 per group] over 10 days. The 95% confidence interval of the untreated control groups (N=5–6; 10 experiments) is shaded. Onion was used as a positive control for all foodstuffs. All rats received the same total daily amount of food, including 1 g of the dried test foodstuff. Fresh foods were air- or freeze-dried and ground. Calcitonin was injected daily at doses of 1.25 or 2.5 IU per kg body mass at the optimal time. *Cooked before drying; +P = 0.05, ++P<0.01, +++P<0.001 [14,15].

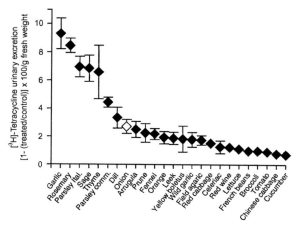

FIGURE 12.2 Inhibition of bone resorption in rats by 24 fresh food items and the red wine residue.
To allow a comparison of the activity of various food items as they are consumed in real life, the results for items that inhibited bone resorption when administered at a dosage of 1 g/d of dry matter were converted to fresh weight using their recorded loss of weight (water and ethanol) during drying. The values are means ± SEM and are ranked according to potency [16].

the highest dose of onion (14%) in preventing bone loss were similar to those of alendronate treatment in many bone indices [17]. In *in vitro* studies, Tang et al. [18] found that water-soluble extracts of onion inhibited osteoclastogenesis from co-cultures of bone marrow stromal cells and macrophage cells via attenuation of the receptor activator of nuclear factor kappa B ligand (RANKL)-induced extracellular signal-regulated kinase (ERK), p38 and NF-κB activation. A substance, γ-L-glutamyl-*trans*-S-1-propenyl-L-cysteine sulfoxide (GPCS), isolated from onion (*Allium cepa* L.) by bioassay-guided fractionation, has a molecular mass of 306 Da and was found to inhibit dose-dependently the resorption activity of osteoclasts. The minimal effective dose was approximately 2 mM [19].

Several studies have examined the effects of fruits on bone indices. Among fruits, dried plum was extensively studied by Arjmandi and colleagues. In 2001, they first reported that dried plum increases indices of bone

formation, i.e. serum total alkaline phosphatase and bone-specific alkaline phosphatase activity, and insulin-like growth factor I (IGF-I) [20]. Further studies found that dried plum, as low as 5% (g/100 g diet), was effective in restoring femoral and tibial bone density in an osteopenic rat: 5%, 15%, and 25% dried plum could dose-dependently reverse the loss of bone density and the deterioration of microarchitectural properties in an ovariectomized rat model of postmenopausal osteoporosis [21]. A 25% prune-enriched diet was found to have an equally protective effect as 17β-estradiol E$_2$ on bone volume ratio (bone volume/total volume, BV/TV) [21]. Other studies have noted the effect of dried plum on preventing osteopenia in androgen-deficient male rats [22], and improving bone mass, trabecular and cortical bone microarchitecture, and strength in orchidectomized (ORX) rats [23]. Dried plum had approximately 50% of the effect of parathyroid hormone (PTH) on vertebral bone density, but a greater effect (i.e. 60%) on bone biomechanical properties. These findings suggest that dried plum may positively affect components of overall bone strength, including bone turnover rate, protein matrix synthesis, and regulation of osteoblast and osteoclast activities [23]. These beneficial effects of dried plum may partly be attributable to a decrease in osteoclastogenesis via the down-regulation of nuclear factor kappa B ligand and nuclear factor for activated T cell expression as well as nitric oxide (NO) and tumor necrosis factor (TNF)-α production; and an enhancement of osteoblast activity and function by up-regulating Runx2, osterix, and IGF-I and increasing lysyl oxidase expression [22,24,25]. The restoration in some of the bone structural and biomechanical parameters might also share some similarities with PTH [23].

Only a few studies have assessed the effects of other fruits or fruit juices on bone indices. Mühlbauer et al. reported that a 10-day treatment with prune and orange, but not banana or apple, could significantly inhibit bone resorption in rats [16]. Deyhim et al. [26] observed that orange juice and grapefruit juice reversed ORX-induced antioxidant suppression, decreased alkaline phosphatase and acid phosphatase activities, moderately restored femoral density, increased femoral strength, significantly delayed time-induced femoral fracture, and decreased urinary excretion of hydroxyproline. Similar positive effects of grapefruit juice and grapefruit pulp on bone quality were also observed in ORX rats by enhancing bone mineral deposition and slowing down bone turnover [27,28] in non-ORX rats via an undefined mechanism. Moreover, hesperidin, one of the main flavonoids present in oranges, was reported to improve bone quality in rodents and prevented ovariectomy-induced bone loss in mice and rats [29,30]. In addition, Devareddy et al. [31] found that blueberry was effective in preventing bone loss caused by ovarian hormone deficiency as seen from whole body, tibial, and femoral BMD values. However, such a benefit was not observed in orchidectomized male rats fed cranberry juice [32].

2.2 Human Studies

Many observational studies have directly examined the association of fruit and/or vegetable consumption with indices of bone health [33–45]. Most of these studies are cross-sectional in design; three are prospective cohort studies [34,38,40]. A summary of these results is presented in Table 12.1.

2.2.1 Adult Studies

In 1997, New et al. [33] first reported the association of fruit and vegetables with bone health in a cross-sectional study. They found that BMD at the lumbar spine, femoral neck, greater trochanter, and Ward's area was significantly lower by 3.4–4.8% in middle-aged (44–50 y) women who reported a low intake of fruit in early adulthood than in those who reported a medium or high intake. Tucker et al. [34] observed a similar favorable effect of fruit and vegetable consumption on BMD in elderly men and women (mean age 75 y) of the Framingham Heart Study cohort population in a cross-sectional examination. One serving increase in baseline consumption of fruits and vegetables was associated with 0.0049 ($P<0.1$) and 0.0053 ($P<0.05$) (g/cm^2) increases in subsequent 4-year change in BMD in men after adjusting for baseline BMD, and other dietary and lifestyle factors (age, body mass index, physical activity score, smoking status, alcohol use, calcium supplement use, vitamin D supplement use, season of BMD measurement, energy intake, dietary calcium intake, and dietary vitamin D intake) [34]. Similar positive associations between fruit and vegetable intake and BMD, BMC, or low fracture rate were also observed in middle-aged Hong Kong postmenopausal Chinese women [41], in population-based Mainland Chinese adult men and women [44], and in old women in the UK [42]. However, inconsistent results were also found in other studies. In cross-sectional studies, Prynne et al. [42] did not observe this positive link between fruit and vegetable consumption and BMD among 130 young women and 70 elderly men. A similar result was found in 57 men aged 39–42 years after controlling for lean body mass and energy intake [37]. In prospective cohort studies, there was no association between baseline fruit and vegetable intake and longitudinal change in BMD in 562 women in the Framingham study (34), and in 2–5 year hip BMD loss rates in 470 white men and 474 white women aged 67–79 years in a prospective population-based diet and cancer study (EPIC-Norfolk) in Eastern England (38). In another cohort study of 1865 peri- and postmenopausal women (Adventist Health Study) who completed two lifestyle surveys 25 years apart, increasing fruit and vegetables was not

TABLE 12.1 A Summary of Results of Studies on Fruit and Vegetables on Bone Indices

Source	Country	Study Design	Recruitment Methods	Study Size	Age, Mean (SD) or Range	Dietary Measurements	Results: Fruit and Vegetable and Bone Indices
New [33]	UK	CS	Random	F: 994	47.1(1.4)	FFQ, self-adm.	Fru at age 20–30 y and current BMD at LS, FN, Troch.: positive
Tucker [34]	US	CS cohort	Framingham subjects	M: 345 F: 562	75(4.9)	FFQ, Self-adm.	M: B (BMD in g/cm2/serving F&V): 0.0086* (FN), 0.011* (Ward's) 0.0043* (Radius); high V&F, lower 4-y Ward's bone loss F: B: 0.0056*(Troch) and 0.0049** (Radius), NS for bone loss rate
Frassetto [140]	World	ES		F: 33 countries	Elderly women		Veg. Protein & hip fracture, r = − 0.37* Veg. protein/animal protein & hip fracture, r = − 0.84***
Jones [35]	AU	CS	Clinical subjects	M: 215 F: 116	8.2(0.3)	FFQ by mother or guardian	Fru and BMD (TB, FN, LS): NS Veg and TB BMD: r = − 0.15*; Veg & FN, LS BMD: NS
Whiting [37]	Ca	CS	Random	M: 57	39–42	FFQ, self-adm.	F&V and BMD (TB, Hip, LS): NS
Kaptoge [38]	US	Cohort	Pop.-based	M: 470 F: 474	67–79	7-d FD	V&F and BMD loss: NS
McGartland [39]	UK	CS	Random	M: 594 F: 747	12,15	Diet history method	12y Girls: Fru and heel BMD: positive; Fru and radius BMD: NS 15y girls and 12 or 15y boys: F&V and BMD: NS
Tylavsky [45]	US	CS	Volunteers	F: 56	8–12	FD	F&V and TB bone area: positive; F&V and urine calcium: negative F&V and TB BMD, BMC: NS
Vatanparast [40]	Ca	Cohort	Pop.-based	M: 85 F: 67	8–20	Multi-24-h recalls	Boys: F&V and TB BMC increases: positive; Girls: NS
Prynne [42]	UK	CS	Volunteers	a. F: 125 b. M: 132 c. F: 130 d. M: 70 e. F: 73	a. 16–18 b. 16–18 c. 23–37 d. 60–83 e. 60–83	7-day FD	a. Fru, F&V and BMD/BMC at TB, LS (not Hip, FN): positive: b. Fru, F&V and BMD/BMC at TB, LS, hip, FN: positive: c. Fru or Veg and BMD/BMC at TB, LS, hip, FN: NS d. Fru or Veg and BMD/BMC at TB, LS, hip, FN: NS e. Fru & LS BMC: Positive; F and BMD at TB, LS, hip, FN: NS
Chen [41]	HK	CS	Volunteers	F: 670	48–63	FFQ, interview	F&V and BMD at TB, LS, hip (but not FN): positive, Fru > veg

(Continued)

TABLE 12.1   (Continued)

Source	Country	Study Design	Recruitment Methods	Study Size	Age, Mean (SD) or Range	Dietary Measurements	Results: Fruit and Vegetable and Bone Indices
Okubo [141]	Japan	CS	Com.-based	F: 291	46.4(3.7)	147-item FFQ	Dietary pattern with high Veg, Fru, mushroom, fish and BMD at radius and ulna: positive
Zalloua [44]	China	CS	Pop.-based	M: 5848 F: 6207	24–67	Short FFQ	Fru and BMD at TB and Hip: positive, Veg and BMD: NS
Thorpe [43]	US	Cohort	Volunteers	F: 1406	52–53	FFQ	F&V and all minor trauma fractures: NS

UK, United Kingdom; US, United States; AU, Australia; HK, Hong Kong; CS, cross-sectional study; cohort, prospective cohort study; ES, ecological study; Pop.-based, population based; Com.-based, community-based; FFQ, food frequency questionnaire; Self-adm., self-administrated; FD, food diary; Fru, fruits; Veg, vegetables; F&V, fruit and vegetables; TB, total body; LS, lumbar spine; FN, femur neck; Troch, trochanter; BMD, bone mineral density; BMC, bone mineral content; NS, no significant association.
*,** and ***: $P<0.05$, $P<0.01$ and $P<0.001$.

associated with a reduced risk of all fractures due to minor trauma.

2.2.2 Childhood Studies

Tylavsky et al. [45] showed fruit and vegetable intake to be a significant independent predictor of bone area but not BMD or BMC in 56 girls aged 8–13 years. Children who reported consuming ≥ 3 servings of fruits and vegetables per day had more bone area of the whole body (6.0%; $P = 0.03$) and radius (8.3%; $P = 0.03$), lower urinary calcium excretion and lower PTH than children who reported consuming <3 serving of fruit and vegetables per day. Prynne et al. [42] found that greater fruit and vegetable intake was associated with higher BMD and BMC in the whole body and spine, and BMC at the hip in 111 boys and 101 girls (mean age 17 y). Similar positive association of fruits and vegetables with heel BMD was found by McGartland et al. [39] in 378 girls (but not in 324 boys) aged 12 years. In a 7-year follow-up study, Vatanparast et al. [40] found that every additional serving of vegetables and fruits was associated with an increase of 5.4 g (SE 1.3) of total body BMC accrual in 85 boys

aged 8–20 years, but no significant effect was observed in 67 girls of similar age.

Up to now, only one randomized controlled trial has directly examined the effect of fruits and vegetables on bone indices. In the trial, a 2-year supplementation of 300 g self-selected fruit and vegetables per day in 52 healthy postmenopausal women did not significantly persistently reduce bone turnover or prevent BMD loss over 2 years as compared with 47 controls (Figure 12.3) [46]. A 3-month Dietary Approaches to Stopping Hypertension (DASH) intervention study involving 23 76-year-old men and women showed that a diet high in fruit and vegetables significantly reduced serum osteocalcin (OC) by 8–11% and C-terminal telopeptide of type I collagen (CTX) by 16–18% (both $P<0.001$) [47]. As compared with a typical American diet, the DASH diet emphasizes fruits, vegetables, and low fat dairy foods, and includes whole grains, poultry, fish, and nuts, with reduced fats, red meat, sweets, and sugar-containing beverages. The DASH diet thus contains reduced amounts of total fat, saturated fat, and cholesterol, and increased amounts of potassium, calcium, magnesium, dietary fiber, and protein. Since the calcium

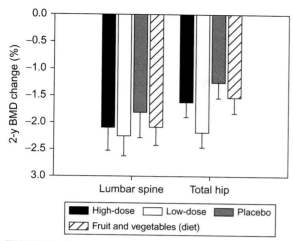

FIGURE 12.3 Mean (±SEM) percentage change over the duration of the study (2 y) in bone mineral density (BMD) at the lumbar spine (LS) ($N = 60$, 61, 66, and 63 in the high-dose potassium citrate, low-dose potassium citrate, placebo, and fruit and vegetables groups, respectively) and mean total hip ($N = 58$, 59, 65, and 60 in the high-dose potassium citrate, low-dose potassium citrate, placebo, and fruit and vegetables groups, respectively). $P = 0.88$ for LS, $P = 0.14$ for mean total hip [one-way analysis of variance (ANOVA)]. The results were similar after adjustment for confounders (i.e. age, weight, height, and social deprivation category). $P = 0.90$ for LS, $P = 0.21$ for mean total hip [analysis of covariance (ANCOVA)] [46].

content is substantially higher in the DASH diet, it is not certain if the benefits of the DASH diet on bone turnover can be attributed to the increases in fruit and vegetable consumption. More well-controlled and larger trials are needed to clarify if fruit and vegetables are beneficial to bone health and the extent of that benefit, if any.

In summary, human studies on the effects of fruit and vegetables on bone health have yielded inconsistent results. Although some observational studies have found a favorable association of fruit and vegetable consumption with bone health, results with null effects may be unreported. We are thus not able to conclude that greater fruit and vegetable intake improves bone health. Moreover, most of the reported studies are cross-sectional in design,

have relatively small sample size, and outcomes are mostly based on BMD and/or bone biomarkers rather than on osteoporotic fractures. Further evidence is required to substantiate the bone health effects of fruits and vegetables in human populations.

3. MECHANISMS OF ACTION

3.1 Acid–base Homeostasis and Bone Health

The maintenance of a stable physiologic systemic pH is of critical importance to the survival of mammals. Almost every biological process in the human body is dependent on the acid–base balance, including bone metabolism. Acid–base homeostasis is tightly regulated in the extracellular fluid at pH 7.4 (±0.05). Bone contributes to the acid–base homeostasis as it delivers cations such as potassium, calcium, magnesium, and sodium, which can be associated with alkali salts such as citrate or carbonate. It was suggested that osteoporosis may, in part, be caused by the continual release of alkaline salts from bone for acid–base balance [48].

The importance of acid–base homeostasis to bone health has been extensively addressed in previous reviews [48–51]. Theoretical considerations of the role that alkaline bone minerals may play in the defense against acidosis date back as far as the early nineteenth century [48,52]. The fundamental concepts were established in the 1960s to 1970s. A large number of human studies in the past decade provided evidence that net endogenous acid production is associated with low BMD [48,53–56]. Dietary intake and nutritional state can strongly influence the acid–base balance in humans. The organic acids that are produced during metabolism and the hepatic oxidation of sulfur-containing amino acids (cysteine and methionine) lower blood pH through increased production of hydrogen ions. Alkaline dietary salts contain the cations (potassium, calcium, and magnesium) and act as buffers for organic acids that have the potential to raise the pH. Insufficient buffering capacity from alkaline salts, in a state of long-term negative acid–base balance, may result in chronic low-grade metabolic acidosis even in healthy persons [48,53–56]. Even mild forms of long-term low-grade metabolic acidosis can impair skeletal architecture and stability because the skeleton is a large but not an endless alkaline reservoir.

3.1.1 In Vitro Studies

In vitro studies have shown that metabolic acidosis induces a calcium efflux from bone. Neuman and Neuman [57] found that a reduction in medium pH produced a marked increase in hydroxyapatite solubility from bone. Dominguez and Raisz [58] observed that acidic medium induced the release of ^{45}Ca from prelabeled fetal rat limb bones in organ culture. Bushinsky et al. noted a net calcium efflux from the calvaria and a fall in mineral sodium, potassium, carbonate, and phosphate, when medium pH was lowered (pH <7.40) by decreasing the bicarbonate ion concentration during acute incubations (3 h) in an *in vitro* bone organ culture system [59,60] (Figure 12.4). Initially, metabolic acidosis stimulates physicochemical mineral dissolution and subsequently cell-mediated bone resorption [51,61]. Any reduction in extracellular pH enhances osteoclastic activity. Arnett and Dempster [62] found that reduction in medium pH from 7.4 to 6.8 resulted in a 14-fold increase in the mean area resorbed per bone slice after a 24-hour incubation. Five- and 9-fold increases were seen at pH 7.2 and 7.0 [62]. Similar results were observed in other *in vitro* studies [63–65]. In addition, many studies found that acidosis also suppresses the activity of the bone-resorbing cells, osteoblasts, and decreases gene expression of specific matrix proteins and alkaline phosphatase activity [51,66–68].

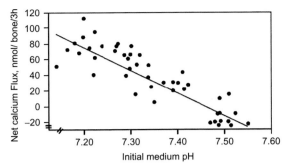

FIGURE 12.4 Effect of initial medium pH on net calcium flux in calvariae cultured for three hours.
A positive flux indicates net calcium movement from the calvariae into the medium. pH was adjusted for the 3-hour incubation with concentrated HCl or NaOH at a Pco_2 of 40 mmHg. Calvariae were preincubated in control medium for 24 hours prior to this 3-hour incubation. $r = 0.890$, $N = 46$, $P < 0.001$. (Reproduced with permission from Bushinsky et al. [60].)

3.1.2 Animal and Human Studies

A few animal and human studies have provided evidence that natural, pathologic, and experimental states of acid loading and/or acidosis have been associated with negative calcium balance and lower bone loss [56,69]. Many epidemiological studies have examined the association of dietary acid load and bone metabolism or bone mass. In observational studies, dietary acid load was estimated by using acid-forming (protein, particular animal protein) or base-forming (potassium and magnesium) nutrients, net endogenous acid production (NEAP) or dietary potential renal acid load (PRAL).

3.2 Dietary Protein and Bone Health

Mixed effects of dietary protein on bone health have been observed [70,71]. Urinary calcium is influenced by the acid–base status of the total diet, increases with acid-forming foods, such as meat, fish, eggs, and cereal, and is lower with plant foods [72,73]. The high protein content of the Western diet is often cited as a risk factor for osteoporosis or bone fractures [74,75]. Diets based on high animal protein might have a greater negative effect on skeletal health than do vegetable-based diets because animal protein induces a greater increase in urinary calcium excretion than vegetable protein [24]. However, a large prospective study of a cohort of Iowa women aged 55–69 years with 104,338 person-years showed that increasing quartiles of animal protein intake but not vegetable protein were associated with a lower risk of hip fracture [RR for Q4 vs. Q1: 0.31 (95% CI 0.10, 0.93)] after the adjustment for potential confounders [76]. A statewide case-control study in Utah including 1167 cases (831 women, 336 men) and 1334 controls (885 women, 449 men) also noted higher total protein intake was associated with a reduced risk of hip fracture [OR for Q4 vs. Q1: 0.35 (0.21–0.59)] in men and women aged 50–69, years but not in men and women 70–89 years of age [77]. A review by Heaney and Layman [71] on how the amount and type of protein influences bone health concluded that optimal protein intake for bone health is likely to be higher than the current recommended

intakes, particularly in the elderly. Concerns about dietary protein increasing urinary calcium and acid production appear to be offset by increases in intakes of calcium, fruit, and vegetables [48].

3.3 Base-forming Nutrients and Bone Health

3.3.1 Potassium and Bone Health

Potassium is a major base-forming nutrient in the diet. Fruits and vegetables are the major source of potassium. Many studies have examined the association of potassium and bone health. In 1997, New et al. [33] first reported a significant association of higher lumbar spine (LS) BMD with higher intake of potassium in 994 healthy premenopausal women aged 45–49 years (Figure 12.5). Similar positive associations between potassium and BMD or BMD changes were observed in some observational epidemiological studies in adults [34,37,53,78–82] and in

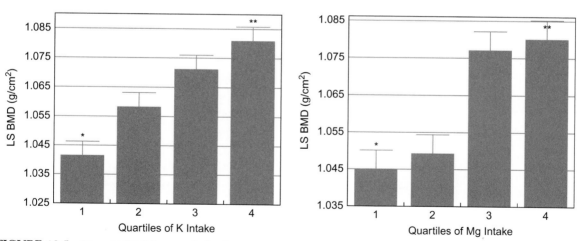

FIGURE 12.5 Mean (±SEM) increase in lumbar spine BMD with quartiles of energy-adjusted magnesium or potassium intake in 994 healthy premenopausal women aged 45–49 years.
All P for linear trend <0.005. *Significantly different from quartile 4, P<0.01 (ANOVA).
**Significantly different from quartile 1, P<0.02 (Mg) and P<0.06 (K) (ANCOVA: adjusted for age, weight, height, physical activity, smoking, and social status) [33].

prepubertal children [35], but not in a study in older men and women [38].

Most short-term intervention studies show a favorable effect of potassium on calcium or bone metabolism. A 18-day randomized controlled trial (RCT) in 18 postmenopausal women showed that potassium bicarbonate given orally in doses of 60–120 mmol/d resulted in no significant change in net intestinal absorption of calcium, but improved the calcium balance ($+56 \pm 76$ mg) with a reduction in urinary calcium excretion (Figure 12.6) [83]. The net renal acid excretion also decreased from 70.9 (SD: 10.1) in the control period to 12.8 (21.8) mmol/d during the supplementation period, indicating that endogenous acid was almost completely neutralized during treatment [83]. Another RCT in 60 postmenopausal women showed that the addition of oral potassium citrate (90 mmol/d) to a high-salt (225 mmol/d sodium) diet prevented the increased excretion of urinary calcium and bone resorption marker (N-telopeptide, NTx) caused by a high salt intake [84]. Many other short-term studies also showed that the administration of potassium bicarbonate ($KHCO_3$) resulted in lower urinary calcium excretion, improved calcium balance, and decreased bone resorption [85–88].

Previously, only two RCTs had examined the long-term effect of potassium on bone health. A 1-year RCT showed that interventions

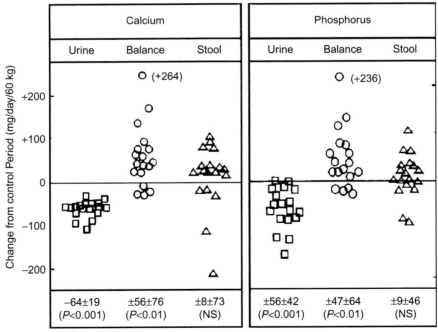

FIGURE 12.6 Effect of potassium bicarbonate supplementation on calcium and phosphorus excretion in urine, external calcium and phosphorus balance, and calcium and phosphorus excretion in stool in 18 postmenopausal women.
The values shown at the bottom of the figure are the average ($\pm$SD) potassium bicarbonate-induced changes from the control period (before supplementation). To convert calcium values to millimoles per day per 60 kg, divide by 40; to convert phosphorus values to millimoles per day per 60 kg, divide by 31. NS denotes not significant. The P values are for the comparisons between the control period and the supplementation period [83].

of 30 mEq/d of oral potassium citrate significantly improved BMD at the lumbar spine, left hip, and total hip (all $P<0.001$) compared with 30 mEq/d of potassium chloride (KCl) in 161 postmenopausal women (age 58.6 ± 4.8 y) with low bone mass (T score -1 to -4) [89]. These results suggested that increasing daily alkali intake as citrate significantly improved bone health, and the effect was independent of reported skeletal effects of potassium. In another randomized placebo-controlled trial, long-term supplementation with alkali-providing potassium citrate did not persistently influence bone turnover and BMD changes in 276 postmenopausal women (55–65 y) with a mean daily calcium intake of ~900 mg/d [46]. Potassium citrate 18.5 or 55.6 mEq/d for 2 years did not result in any significant changes in serum C-telopeptide (CTX), serum N-propeptide (NTx) of type I collagen (P1NP), and urinary free deoxypyridoline cross-links relative to creatinine (fDPD/Cr) at 3, 6, 12, 18, and 24 months, and did not significantly influence 2-year BMD changes (see Figure 12.3) [46]. However urinary pH was significantly and persistently increased in both of the potassium intervention groups but only a transient reduction in fDPD/Cr was noted in the high-dose potassium citrate group at 4–6 weeks, and that value returned to baseline at 3 months [46]. Similar negative results were observed in an animal study in which potassium intervention for 19 months caused a decrease in calcium excretion but had no effect on long-term bone markers, BMD, or bone strength in male rats with either a normal or a high-protein diet [90]. Due to differences in designs and populations, and limited sample size, further larger studies are needed to address the long-term effects of potassium on bone health.

3.3.2 Magnesium and Bone Health

Magnesium (Mg) is the second most abundant intracellular cation in the body. Mg exists in macronutrient quantities in bone (0.5–1% bone ash) and dietary Mg deficiency has been implicated as a risk factor for osteoporosis. Impaired bone growth, decreased bone formation, increased bone resorption, and increased skeletal fragility have been observed in rats with selective dietary Mg depletion [91–95].

Epidemiological studies in humans have generally shown a positive correlation between dietary Mg intake and bone density and/or an increased rate of bone loss with low dietary Mg intake [33,34,81,96,97]. New et al. found a significant correlation of BMD of the lumbar spine with Mg intake in a study of 994 premenopausal women [33]. Yano et al. [80] reported a positive correlation between Mg intake and appendicular BMD in 912 women aged 43–80 years, but not in 1208 men aged 61–81 years. Tucker et al. [34] found that Mg intake was associated with greater BMD at one hip site for both men and women and in the forearm for men in the Framingham Heart Study subjects aged 69–97 years. Greater Mg intake was also associated with less 4-year decline in BMD at two hip sites in men but not in women [34]. Wang et al. [98] observed that dietary Mg intake was positively related to quantitative ultrasound (QUS) properties of calcaneus in pre-adolescent girls. These results suggested that Mg was important in skeletal growth, development, and maintenance.

Human interventional studies examining Mg effects on bone are limited and the results of short-term Mg intervention on bone turnover are conflicting. Dimai et al. [99] found that daily oral supplementation of a moderate dose of 360 mg/d Mg for 30 days significantly reduced serum levels of biochemical markers of both bone formation and resorption in 12 normal young males (non-Mg-deficient). However, the effects were significant only during the first 5–10 days of the study. In a similar trial with a double-blind, placebo-controlled randomized crossover design, increasing Mg intake from the usual level (11 mmol/d) to

22 mmol/d for 28 days did not significantly affect serum osteocalcin, bone-specific alkaline phosphatase, or urinary pyridinoline and Dpd excretion in 26 young women [100]. The effect of dietary Mg intervention on bone mass has not been extensively studied. In an uncontrolled study of postmenopausal osteoporotic women, Mg supplementation was associated with BMD increases in 60% of those treated [101]. A placebo-controlled study of patients with gluten-sensitive enteropathy demonstrated increased BMD after 6 months of Mg supplementation (504–576 mg $MgCl_2$ or Mg lactate daily), compared with placebo-supplemented subjects [102]. Another 1-year double-blind, placebo-controlled, randomized trial showed that a daily dose of 300 mg Mg given orally for 12 months significantly increased accrual ($P = 0.05$) in integrated hip BMC but had no significant effect on bone turnover biomarkers in 60 healthy 8–14-year-old Caucasian girls with dietary Mg intake of less than 220 mg/d [103]. Thus, Mg intervention studies to date have shown positive effects on bone mass but further larger long-term studies are needed to confirm the beneficial effect of Mg.

3.3.3 Dietary Acid Load and Bone Health

Western diets consumed by adults are estimated to generate 50–100 mEq acid/d [104,105]. Renal net acid excretion can be measured directly with 24-hour urine collection, but such measurement is impractical in large population studies. Alternative approaches have been developed to examine the net acid content of the diet [106]. Frassetto et al. [107] have found that protein to potassium ratio predicts net endogenous acid production (NEAP) [NEAP = 54.5 × (protein intake/potassium intake) − 10.2]. Another validated measurement was developed by Remer et al. [108] and estimates the dietary potential renal acid load (PRAL) using the formula: PRAL (mEq/d) = [phosphorus (mg/d) × 0.037 + protein (g/d) ×

0.49] − [potassium (mg/d) × 0.021 + magnesium (mg/d) × 0.0263 + calcium (mg/d) × 0.013] [108].

Many studies have examined and found an inverse correlation between the dietary acid load and bone health in humans. In 1993, a cross-sectional survey of 764 middle-aged and elderly women with markedly different dietary patterns and lifestyles found that urinary calcium was positively correlated with urinary acids, including titratable acid ($r = 0.46$, $P < 0.0001$), ammonia ($r = 0.42$, $P < 0.0001$), and sulfate ($r = 0.52$, $P < 0.0001$) [73]. New et al. [109] observed a significantly inverse association between energy-adjusted NEAP and BMD at the spine, hip, and forearm ($P < 0.02$ to $P < 0.05$) in 1056 premenopausal or perimenopausal women aged 45–54 years. Hip and forearm bone mass decreased significantly across increasing quartiles of energy-adjusted NEAP ($P < 0.02$ to $P < 0.03$), and trends at the spine were similar ($P < 0.09$) (Figure 12.7). Lower estimates of energy-adjusted NEAP were also correlated with lower excretion of deoxypyridinoline and were significant predictors of spine and forearm bone mass [109]. Welch et al. [110] found that PRAL was inversely associated with calcaneal broadband ultrasound attenuation (BUA) in 8188 women ($P = 0.002$) but not in 6375 men ($P = 0.78$) aged 42–82 years in Norfolk (UK) after adjusting for potential confounders. Similar inverse associations between dietary acid load and bone indices were also observed in healthy children and adolescents [111,112], in perimenopausal and early postmenopausal women [53] and in ambulatory Swiss elderly women [55]. These studies suggested high dietary acid load might have an adverse effect on bone health.

Most of the short-term intervention studies using metabolic diets have shown that an alkali diet can improve calcium and bone metabolism. Bleich and coworkers [113] fed normal subjects protein, ammonium chloride (NH_4Cl), and sodium bicarbonate ($NaHCO_3$) and measured urinary calcium excretion.

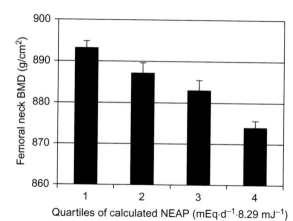

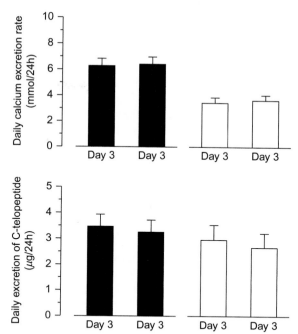

FIGURE 12.7 Mean ($\pm$SEM) femoral neck bone mineral density (BMD) by quartile (Q; $N=264$) of calculated net rate of endogenous non-carbonic acid production (NEAP; mEq/d/8.29 MJ) in 1056 women aged 45–54 years. *P for linear trend* <0.02. Mean BMD values for Q1–Q4 were 852, 848, 841, and 824 g/cm², respectively. *Significantly different from Q4, *P*<0.04 (ANOVA with Tukey's test, adjusted for age, weight, height, and menopausal status). [109]

FIGURE 12.8 Average ($\pm$SEM) daily excretion rates of calcium and C-telopeptide according to diet and study day. A) acid-forming diet; B) base-forming diet. Day 3: no calcium supplement; day 4: 1 g calcium supplement. Diet effect on calcium: $P=0.0002$; on C-telopeptide: $P=0.01$; no significant effect of calcium supplement. [114]

Protein and NH_4Cl, which are metabolized into metabolic acids, led to a marked increase, while the base $NaHCO_3$ led to a decrease, in renal calcium excretion. There was no measurable change in intestinal calcium absorption with any treatment. Another four-phase short-term double-crossover trial in eight healthy male volunteers showed that compared with an alkali diet, an acid diet significantly increased urinary calcium excretion by 74% and urinary CTx excretion by 19% (Figure 12.8) [114]. Similar results were observed in other studies. Short-term administration of bicarbonate or potassium has resulted in decreases in urinary acid, NTX [84,85], and urinary calcium excretion [85–87]. However, long-term supplementation of an alkali diet (18.5 or 55.6 mEq potassium citrate, or 300 g fruits and vegetables) did not significantly improve bone mass and bone biomarkers, except for a transient reduction in fDPD/Cr at 4–6 weeks in an RCT in 276 postmenopausal women (55–65 years) [46]. More stringent longer-term RCTs with larger sample sizes are needed to demonstrate the acid–base hypothesis.

3.4 Other Potential Mechanisms

Except for a major source of base cations, fruit and vegetables are high in calcium, vitamin K, numerous antioxidant compounds, and phytochemicals such as polyphenols, carotenoids, tocopherols, tocotrienols, glutathione, and ascorbic acid, as well as enzymes with antioxidant activity [115]. Thousands of biologically active phytochemicals have been identified in fruits and vegetables. These compounds may also contribute to the favorable effects of fruit and vegetables on bone health. Mühlbauer et al. [116] showed that the effect of vegetables, salads, and herbs, which inhibit bone resorption in the rat, is not mediated by their base excess but possibly by a pharmacologically active compound(s). Since many articles have reviewed the effects of calcium and vitamin K on bone health, we will only summarize the effects of antioxidants and other bioactive phytochemicals on bone health.

3.4.1 Oxidative Stress, Antioxidants, and Bone Health

Reactive oxygen species (ROS) are oxygen-containing molecules produced during normal metabolism [117]. The organism has enzymatic and non-enzymatic antioxidant systems to neutralize the harmful effects of the endogenous ROS products. Oxidative damage can result when the critical balance between free radical generation and antioxidant defenses is unfavorable. Oxidative stress has been hypothesized to play a key role in cardiovascular diseases, cancer initiation, cataract formation, the aging process, inflammatory diseases, and a variety of neurological disorders [118]. A few studies have shown that oxidative damage may also contribute to the progress of osteoporosis [119–122].

In vitro studies suggested that osteoclast-generated superoxide directly contributes to bone degradation. The presence of superoxide production at the osteoclast–bone interface suggests a direct effect of superoxide in osteoclastic bone resorption [119,120]. Oxidative stress inhibits osteoblastic differentiation of bone cells [121]. In addition, it has been demonstrated that osteoblasts produce antioxidants such as glutathione peroxidase (GPx) to protect against ROS [123], and inhibition of osteoclastic superoxide availability results in a reduction in bone resorption [124]. *In vivo* studies by Altindag et al. [122] found that plasma total oxidative status (TOS) and oxidative stress index (OSI) were significantly higher, and plasma total antioxidant status (TAS) level was lower, in osteoporosis patients than in healthy controls

($P<0.001$ for all). A significant negative correlation between oxidative stress index and BMD was found in the lumbar and femoral neck region ($r = -0.63$, $P<0.001$; $r = 0.40$, $P = 0.018$) [122]. These data suggest oxidative stress may increase osteoporosis risk in humans.

Fruits and vegetables are rich sources of dietary antioxidants, such as vitamin C, β-carotene, and other carotenoids, including lutein, zeaxanthin, and lycopene. These antioxidants have the highest singlet oxygen-quenching properties [125] and are thought to protect against oxidative stress and prevent osteoporosis. Many studies have examined dietary antioxidant intake or plasma antioxidant levels and bone health. Maggio et al. [126] observed that plasma antioxidant vitamins (C, E, and A) and the enzymatic activities of superoxide dismutase (SOD) in plasma and erythrocytes and of glutathione peroxidase in plasma were consistently lower in osteoporotic than in control women [126]. An inverse dose–response association between intakes of vitamin E, β-carotene, and selenium and the risk of hip fracture was observed among ever smokers in the elderly Utah population [127]. Maggio et al. noted that plasma levels of carotenoids (zeaxanthin, β-cryptoxanthin, lycopene, α-carotene, β-carotene) and retinol were significantly lower in 45 osteoporotic than in 45 control elderly women [128]. The dietary antioxidant lycopene reduced oxidative stress and the levels of bone turnover markers in 33 postmenopausal women [129]. Several [33,126,127,130,131], but not all [132,133], studies found a positive association between vitamin C and BMD or lower risk of osteoporosis in human. These data suggest that antioxidants rich in fruit and vegetables may explain, at least in part, the beneficial effects of fruits and vegetables on bone health. However, a large cross-sectional study in the Women's Health Initiative Study observed dietary antioxidants (vitamin A, retinol, β-carotene, vitamin C, vitamin E, and selenium) had no independent associations with

BMD [133]. Further high-quality large RCTs are needed to confirm the favorable effect of antioxidants on bone health.

3.4.2 Effects of Other Biologically Active Compounds on Bone Health

Fruits and vegetables contain an abundance of other biologically active compounds with potential bone health effects. Vegetables are rich sources of calcium, which is an essential material for bone structure and also itself an alkali salt [48]. Dark green leafy vegetables are the major sources of dietary vitamin K, which has been demonstrated to improve bone health [134]. Many phytochemicals, such as flavonoids and polyphenols, have been found in both *in vitro* and animal studies to have benefits to bone health. Studies have noted that quercetin, a flavonol and also a phytoestrogen commonly found in onions and apples, may increase bone formation [135], suppress bone resorption by decreasing the differentiation of osteoclast progenitor cells and inhibiting the activity of mature osteoclasts [136], and increase bone minerals in rats [137]. Rutin and hesperidin, flavonoids rich in fruits and vegetables, have also been found to inhibit trabecular bone loss caused by estrogen deficiency in ovariectomized rats [14,30,138]. Bu et al. found that dried plum polyphenols enhance osteoblast activity and function by up-regulating Runx2, osterix and IGF-I and increasing lysyl oxidase expression, and at the same time attenuate osteoclastogenesis signaling [24,25]. Citrus bioactive compounds (limonin and naringin) improve bone quality in orchidectomized rats [139]. A γ-L-glutamyl-*trans*-S-1-propenyl-L-cysteine sulfoxide (GPCS) was isolated from onion (*Allium cepa* L.). It inhibits dose-dependently the resorption activity of osteoclasts [19]. These results show that multiple biologically active compounds contained in fruits and vegetables may contribute to the improvement of bone health.

4. SUMMARY

Animal and observational epidemiological human studies have generally found that fruits and vegetables have a beneficial effect on bone health. Most observational human studies are cross-sectional in design and have a relatively small sample size. Only one relatively small RCT directly examined the association between fruit and vegetable intervention and bone health, and revealed a null effect on bone indices. At present, no sufficient solid evidence is available to conclude that greater fruit and vegetable intake improves bone health. Further well-designed longitudinal studies examining the effect of fruit and vegetable consumption on BMD changes and risk of osteoporotic fractures are needed to examine this hypothesis.

Several mechanisms are proposed for the potential effects of fruits and vegetables on bone health. Among them, acid–base homeostasis theory plays a key role in the beneficial effect of fruits and vegetables. Alkaline bone mineral may play a part in the defense against metabolic acidosis. Base cations (potassium, calcium, and magnesium), rich in fruits and vegetables, act as buffers for organic acids and thus prevent acidosis-induced bone loss. Many human studies have found that greater intakes of potassium and magnesium and lower dietary endogenous acid load are associated with better bone indices. Due to the many limitations in the studies on fruits and vegetables and bone health, better-quality and longer-term RCTs are needed to demonstrate the acid–base hypothesis. Apart from being a major source of base cations, fruits and vegetables are also rich sources of calcium, vitamin K, numerous antioxidant compounds, and other biologically active phytochemicals. The favorable effect of calcium and vitamin K on bone has long been established. An increasing number of studies have suggested that oxidative stress might play a role in the progression of osteoporosis, and dietary antioxidants from fruits and vegetables might reduce oxidative stress and prevent osteoporosis. A few biologically active phytochemicals, such as quercetin, rutin, and hesperidin, have been identified from fruits and vegetables and found to have a beneficial effect on bone health in animal and *in vitro* studies. Further human studies are needed to examine the effects of these phytochemicals on bone indices.

In conclusion, fruits and vegetables might have a favorable effect on bone health, but more solid evidence is required to demonstrate this hypothesis and test the potential mechanisms of action.

References

1. Qaseem, A., Snow, V., Shekelle, P., Hopkins, R., Jr., Forciea, M. A., & Owens, D. K. (2008). Pharmacologic treatment of low bone density or osteoporosis to prevent fractures: a clinical practice guideline from the American College of Physicians. *Annals of Internal Medicine, 149,* 404–415.
2. McGuigan, F. E, Murray, L., & Gallagher, A., et al. (2002). Genetic and environmental determinants of peak bone mass in young men and women. *Journal of Bone and Mineral Research, 17,* 1273–1279.
3. Tucker, K. L., Hannan, M. T., Chen, H., Cupples, L. A., Wilson, P. W., & Kiel, D. P. (1999). Potassium, magnesium, and fruit and vegetable intakes are associated with greater bone mineral density in elderly men and women. *American Journal of Clinical Nutrition, 69,* 727–736.
4. Tucker, K. L., Chen, H., & Hannan, M. T., et al. (2002). Bone mineral density and dietary patterns in older adults: The Framingham Osteoporosis Study. *American Journal of Clinical Nutrition, 76,* 245–252.
5. Tucker, K. L., Hannan, M. T., & Kiel, D. P. (2001). The acid-base hypothesis: diet and bone in the Framingham Osteoporosis Study. *European Journal of Nutrition, 40,* 231–237.
6. Tylavsky, F.A., Holliday, K., Danish, R., Womack, C., Norwood, J., & Carbone, L. (2004). Fruit and vegetable intakes are an independent predictor of bone size in early pubertal children. *American Journal of Clinical Nutrition, 79,* 311–317.
7. Macdonald, H. M., New, S. A., Golden, M. H. N., Campbell, M. K., & Reid, D. M. (2004). Nutritional associations with bone loss during the menopausal transition: evidence of a beneficial effect of calcium, alcohol, and fruit and vegetable nutrients and of a detrimental effect of fatty acids. *American Journal of Clinical Nutrition, 79,* 155–165.
8. McGartland, C. P., Robson, P. J., & Murray, L. J., et al. (2004). Fruit and vegetable consumption and bone mineral density: The Northern Ireland Young Hearts Project. *American Journal of Clinical Nutrition, 80,* 1019–1023.
9. New, S. A., Bolton-Smith, C., Grubb, D. A., & Reid, D. M. (1997). Nutritional influences on bone mineral density: A cross-sectional study in premenopausal women. *American Journal of Clinical Nutrition, 65,* 1831–1839.
10. New, S. A., & Millward, D. J. (2003). Calcium, protein, and fruit and vegetables as dietary determinants of bone health. *American Journal of Clinical Nutrition, 77,* 1340–1341.
11. New, S. A., Robins, S. P., & Campbell, M. K., et al. (2000). Dietary influences on bone mass and bone metabolism: further evidence of a positive link between fruit and vegetable consumption and bone health? *American Journal of Clinical Nutrition, 71,* 142–151.
12. Lichtenstein, A. H., Appel, L. J., & Brands, M., et al. (2006). Diet and lifestyle recommendations revision 2006: a scientific statement from the American Heart Association Nutrition Committee. *Circulation, 114,* 82–96.
13. US Department of Agriculture. (2005). Dietary guidelines for Americans 2005. (6th ed.). Washington, DC: US Government Printing Office.
14. Mühlbauer, R. C., & Li, F. (1999). Effect of vegetables on bone metabolism. *Nature, 401,* 343–344.
15. Mühlbauer, R. C., Li, F., Lozano, A., Reinli, A., & Tschudi, I. (2000). Some vegetables (commonly consumed by humans) efficiently modulate bone metabolism. *Journal of Musculoskeletal and Neuronal Interactions, 1,* 137–140.
16. Mühlbauer, R. C., Lozano, A., Reinli, A., & Wetli, H. (2003). Various selected vegetables, fruits, mushrooms and red wine residue inhibit bone resorption in rats. *Journal of Nutrition, 133,* 3592–3597.
17. Huang, T. H., Mühlbauer, R. C., & Tang, C. H., et al. (2008). Onion decreases the ovariectomy-induced osteopenia in young adult rats. *Bone, 42,* 1154–1163.
18. Tang, C. H., Huang, T. H., Chang, C. S., Fu, W. M., & Yang, R. S. (2009). Water solution of onion crude powder inhibits RANKL-induced osteoclastogenesis through ERK, p38 and NF-kappaB pathways. *Osteoporosis International, 20,* 93–103.
19. Wetli, H. A., Brenneisen, R., & Tschudi, I., et al. (2005). A gamma-glutamyl peptide isolated from

onion (Allium cepa L.) by bioassay-guided fractionation inhibits resorption activity of osteoclasts. *Journal of Agricultural and Food Chemistry, 53*, 3408–3414.

20. Arjmandi, B. H., Lucas, E. A., & Juma, S., et al. (2001). Prune prevents ovariectomy-induced bone loss in rats. *JANA, 4*, 50–56.

21. Deyhim, F., Stoecker, B. J., Brusewitz, G. H., Devareddy, L., & Arjmandi, B. H. (2005). Dried plum reverses bone loss in an osteopenic rat model of osteoporosis. *Menopause, 12*, 755–762.

22. Franklin, M., Bu, S. Y., & Lerner, M. R., et al. (2006). Dried plum prevents bone loss in a male osteoporosis model via IGF-I and the RANK pathway. *Bone, 39*, 1331–1342.

23. Bu, S. Y., Lucas, E. A., & Franklin, M., et al. (2007). Comparison of dried plum supplementation and intermittent PTH in restoring bone in osteopenic orchidectomized rats. *Osteoporosis International, 18*, 931–942.

24. Bu, S. Y., Hunt, T. S., & Smith, B. J. (2009). Dried plum polyphenols attenuate the detrimental effects of TNF-alpha on osteoblast function coincident with upregulation of Runx2, Osterix and IGF-I. *Journal of Nutrition Biochemistry, 20*, 35–44.

25. Bu, S. Y., Lerner, M., & Stoecker, B. J., et al. (2008). Dried plum polyphenols inhibit osteoclastogenesis by downregulating NFATc1 and inflammatory mediators. *Calcified Tissue International, 82*, 475–488.

26. Deyhim, F., Garica, K., & Lopez, E., et al. (2006). Citrus juice modulates bone strength in male senescent rat model of osteoporosis. *Nutrition, 22*, 559–563.

27. Deyhim, F., Mandadi, K., Faraji, B., & Patil, B. S. (2008). Grapefruit juice modulates bone quality in rats. *Journal of Medicinal Food, 11*, 99–104.

28. Deyhim, F., Mandadi, K., Patil, B. S., & Faraji, B. (2008). Grapefruit pulp increases antioxidant status and improves bone quality in orchidectomized rats. *Nutrition, 24*, 1039–1044.

29. Horcajada, M. N., Habauzit, V., & Trzeciakiewicz, A., et al. (2008). Hesperidin inhibits ovariectomized-induced osteopenia and shows differential effects on bone mass and strength in young and adult intact rats. *Journal of Applied Physiology, 104*, 648–654.

30. Chiba, H., Uehara, M., & Wu, J., et al. (2003). Hesperidin, a citrus flavonoid, inhibits bone loss and decreases serum and hepatic lipids in ovariectomized mice. *Journal of Nutrition, 133*, 1892–1897.

31. Devareddy, L., Hooshmand, S., Collins, J. K., Lucas, E. A., Chai, S. C., & Arjmandi, B. H. (2008). Blueberry prevents bone loss in ovariectomized rat model of postmenopausal osteoporosis. *Journal of Nutrition Biochemistry, 19*, 694–699.

32. Villarreal, A., Stoecker, B. J., & Garcia, C., et al. (2007). Cranberry juice improved antioxidant status

without affecting bone quality in orchidectomized male rats. *Phytomedicine, 14*, 815–820.

33. New, S. A., Bolton-Smith, C., Grubb, D. A., & Reid, D. M. (1997). Nutritional influences on bone mineral density: a cross-sectional study in premenopausal women. *American Journal of Clinical Nutrition, 65*, 1831–1839.

34. Tucker, K. L., Hannan, M. T., Chen, H., Cupples, L. A., Wilson, P. W., & Kiel, D. P. (1999). Potassium, magnesium, and fruit and vegetable intakes are associated with greater bone mineral density in elderly men and women. *American Journal of Clinical Nutrition, 69*, 727–736.

35. Jones, G., Riley, M. D., & Whiting, S. (2001). Association between urinary potassium, urinary sodium, current diet, and bone density in prepubertal children. *American Journal of Clinical Nutrition, 73*, 839–844.

36. Tucker, K. L., Chen, H., & Hannan, M. T., et al. (2002). Bone mineral density and dietary patterns in older adults: The Framingham Osteoporosis Study. *American Journal of Clinical Nutrition, 76*, 245–252.

37. Whiting, S. J., Boyle, J. L., Thompson, A., Mirwald, R. L., & Faulkner, R. A. (2002). Dietary protein, phosphorus and potassium are beneficial to bone mineral density in adult men consuming adequate dietary calcium. *Journal of American College Nutrition, 21*, 402–409.

38. Kaptoge, S., Welch, A., & McTaggart, A., et al. (2003). Effects of dietary nutrients and food groups on bone loss from the proximal femur in men and women in the 7th and 8th decades of age. *Osteoporosis International, 14*, 418–428.

39. McGartland, C. P., Robson, P. J., & Murray, L. J., et al. (2004). Fruit and vegetable consumption and bone mineral density: The Northern Ireland Young Hearts Project. *American Journal of Clinical Nutrition, 80*, 1019–1023.

40. Vatanparast, H., Baxter-Jones, A., Faulkner, R. A., Bailey, D. A., & Whiting, S. J. (2005). Positive effects of vegetable and fruit consumption and calcium intake on bone mineral accrual in boys during growth from childhood to adolescence: The University of Saskatchewan Pediatric Bone Mineral Accrual Study. *American Journal of Clinical Nutrition, 82*, 700–706.

41. Chen, Y. M., Ho, S. C., & Woo, J. L. (2006). Greater fruit and vegetable intake is associated with increased bone mass among postmenopausal Chinese women. *British Journal of Nutrition, 96*, 745–751.

42. Prynne, C. J., Mishra, G. D., & O'Connell, M. A., et al. (2006). Fruit and vegetable intakes and bone mineral status: A cross sectional study in 5 age and sex cohorts. *American Journal of Clinical Nutrition, 83*, 1420–1428.

43. Thorpe, D. L., Knutsen, S. F., Beeson, W. L., Rajaram, S., & Fraser, G. E. (2008). Effects of meat consumption and vegetarian diet on risk of wrist fracture over 25 years in a cohort of peri- and postmenopausal women. *Public Health Nutrition, 11*, 564–572.

44. Zalloua, P. A., Hsu, Y. H., & Terwedow, H., et al. (2007). Impact of seafood and fruit consumption on bone mineral density. *Maturitas, 56*, 1–11.

45. Tylavsky, F. A., Holliday, K., Danish, R., Womack, C., Norwood, J., & Carbone, L. (2004). Fruit and vegetable intakes are an independent predictor of bone size in early pubertal children. *American Journal of Clinical Nutrition, 79*, 311–317.

46. Macdonald, H. M., Black, A. J., & Aucott, L., et al. (2008). Effect of potassium citrate supplementation or increased fruit and vegetable intake on bone metabolism in healthy postmenopausal women: A randomized controlled trial. *American Journal of Clinical Nutrition, 88*, 465–474.

47. Lin, P. H., Ginty, F., & Appel, L. J., et al. (2003). The DASH diet and sodium reduction improve markers of bone turnover and calcium metabolism in adults. *Journal of Nutrition, 133*, 3130–3136.

48. New, S. A. (2003). Intake of fruit and vegetables: implications for bone health. *Proceedings of the Nutrition Society, 62*, 889–899.

49. Lemann, J., Jr., Bushinsky, D. A., & Hamm, L. L. (2003). Bone buffering of acid and base in humans. *American Journal of Physiology Renal Physiology, 285*, F811–F832.

50. Bushinsky, D. A., & Frick, K. K. (2000). The effects of acid on bone. *Current Opinion in Nephrology and Hypertension, 9*, 369–379.

51. Krieger, N. S., Frick, K. K., & Bushinsky, D. A. (2004). Mechanism of acid-induced bone resorption. *Current Opinion in Nephrology and Hypertension, 13*, 423–436.

52. Barzel, U. S. (1995). The skeleton as an ion exchange system: Implications for the role of acid-base imbalance in the genesis of osteoporosis. *Journal of Bone and Mineral Research, 10*, 1431–1436.

53. Macdonald, H. M., New, S. A., Fraser, W. D., Campbell, M. K., & Reid, D. M. (2005). Low dietary potassium intakes and high dietary estimates of net endogenous acid production are associated with low bone mineral density in premenopausal women and increased markers of bone resorption in postmenopausal women. *American Journal of Clinical Nutrition, 81*, 923–933.

54. Chan, R. S., Woo, J., Chan, D. C., Cheung, C. S., & Lo, D. H. (2009). Estimated net endogenous acid production and intake of bone health-related nutrients in Hong Kong Chinese adolescents. *European Journal of Clinical Nutrition, 63*, 505–512.

55. Wynn, E., Lanham-New, S. A., Krieg, M. A., Whittamore, D. R., & Burckhardt, P. (2008). Low estimates of dietary acid load are positively associated with bone ultrasound in women older than 75 years of age with a lifetime fracture. *Journal of Nutrition, 138*, 1349–1354.

56. Lemann, J. J., Litzow, J. R., & Lennon, E. J. (1966). The effects of chronic acid load in normal man: further evidence for the participation of bone mineral in the defense against chronic metabolic acidosis. *The Journal of Clinical Investigation, 45*, 1608–1614.

57. Neuman, W. F., & Neuman, M. W. (1958). The chemical dynamics of bone mineral, Chicago: University Chicago Press.

58. Dominguez, J. H., & Raisz, L. G. (1979). Effects of changing hydrogen ion, carbonic acid, and bicarbonate concentrations on bone resorption *in vitro. Calcified Tissue International, 29*, 7–13.

59. Bushinsky, D. A., Krieger, N. S., Geisser, D. I., Grossman, E. B., & Coe, F. L. (1983). Effects of pH on bone calcium and proton fluxes *in vitro. American Journal of Physiology, 245*, F204–F209.

60. Bushinsky, D. A., Goldring, J. M., & Coe, F. L. (1985). Cellular contribution to pH-mediated calcium flux in neonatal mouse calvariae. *American Journal of Physiology, 248*, F785–F789.

61. Bushinsky, D. A., & Lechleider, R. J. (1987). Mechanism of proton-induced bone calcium release: Calcium carbonate-dissolution. *American Journal of Physiology, 253*, F998–F1005.

62. Arnett, T. R., & Dempster, D. W. (1986). Effect of pH on bone resorption by rat osteoclasts *in vitro. Endocrinology, 119*, 119–124.

63. Baron, R., Neff, L., Louvard, D., & Courtoy, P. J. (1985). Cell-mediated extracellular acidification and bone resorption: evidence for a low pH in resorbing lacunae and localization of a 100-kD lysosomal membrane protein at the osteoclast ruffled border. *The Journal of Cell Biology, 101*, 2210–2222.

64. Meghji, S., Morrison, M. S., Henderson, B., & Arnett, T. R. (2001). pH dependence of bone resorption: mouse calvarial osteoclasts are activated by acidosis. *American Journal of Physiology Endocrinology Metabolism, 280*, E112–E119.

65. Walsh, C. A., Dawson, W. E., Birch, M. A., & Gallagher, J. A. (1990). The effects of extracellular pH on bone resorption by avian osteoclasts in vitro. *Journal of Bone and Mineral Research, 5*, 1243–1247.

66. Frick, K. K., & Bushinsky, D. A. (1999). *In vitro* metabolic and respiratory acidosis selectively inhibit osteoblastic matrix gene expression. *American Journal of Physiology, 277*, F750–F755.

67. Frick, K. K., Jiang, L., & Bushinsky, D. A. (1997). Acute metabolic acidosis inhibits the induction of osteoblastic egr-1 and type 1 collagen. *American Journal of Physiology, 272,* C1450–C1456.

68. Krieger, N. S., Sessler, N. E., & Bushinsky, D. A. (1992). Acidosis inhibits osteoblastic and stimulates osteoclastic activity in vitro. *American Journal of Physiology, 262,* F442–F448.

69. Lemann, J. J., Adams, N. D., & Gray, R. W. (1979). Urinary calcium excretion in human beings. *The New England Journal of Medicine, 301,* 535–541.

70. Heaney, R. P. (2001). Protein intake and bone health: the influence of belief systems on the conduct of nutritional science. *American Journal of Clinical Nutrition, 73,* 5–6.

71. Heaney, R. P., & Layman, D. K. (2008). Amount and type of protein influences bone health. *American Journal of Clinical Nutrition, 87,* 1567S–1570S.

72. New, S. A., & Millward, D. J. (2003). Calcium, protein, and fruit and vegetables as dietary determinants of bone health. *American Journal of Clinical Nutrition, 77,* 1340–1 (author reply 1341)..

73. Hu, J. F., Zhao, X. H., Parpia, B., & Campbell, T. C. (1993). Dietary intakes and urinary excretion of calcium and acids: a cross-sectional study of women in China. *American Journal of Clinical Nutrition, 58,* 398–406.

74. Barzel, U. S., & Massey, L. K. (1998). Excess dietary protein can adversely affect bone. *Journal of Nutrition, 128,* 1051–1053.

75. Feskanich, D., Willett, W. C., Stampfer, M. J., & Colditz, G. A. (1996). Protein consumption and bone fractures in women. *American Journal of Epidemiology, 143,* 472–479.

76. Munger, R. G., Cerhan, J. R., & Chiu, B. C. (1999). Prospective study of dietary protein intake and risk of hip fracture in postmenopausal women. *American Journal of Clinical Nutrition, 69,* 147–152.

77. Wengreen, H. J., Munger, R. G., & West, N. A., et al. (2004). Dietary protein intake and risk of osteoporotic hip fracture in elderly residents of Utah. *Journal of Bone and Mineral Research, 19,* 537–545.

78. Macdonald, H. M., New, S. A., Golden, M. H., Campbell, M. K., & Reid, D. M. (2004). Nutritional associations with bone loss during the menopausal transition: evidence of a beneficial effect of calcium, alcohol, and fruit and vegetable nutrients and of a detrimental effect of fatty acids. *American Journal of Clinical Nutrition, 79,* 155–165.

79. Yaegashi, Y., Onoda, T., Tanno, K., Kuribayashi, T., Sakata, K., & Orimo, H. (2008). Association of hip fracture incidence and intake of calcium, magnesium, vitamin D, and vitamin K. *European Journal of Epidemiology, 23,* 219–225.

80. Yano, K., Heilbrun, L. K., Wasnich, R. D., Hankin, J. H., & Vogel, J. M. (1985). The relationship between diet and bone mineral content of multiple skeletal sites in elderly Japanese-American men and women living in Hawaii. *American Journal of Clinical Nutrition, 42,* 877–888.

81. Ryder, K. M., Shorr, R. I., & Bush, A. J., et al. (2005). Magnesium intake from food and supplements is associated with bone mineral density in healthy older white subjects. *Journal of the American Geriatrics Society, 53,* 1875–1880.

82. Tucker, K. L., Hannan, M. T., & Kiel, D. P. (2001). The acid-base hypothesis: diet and bone in the Framingham Osteoporosis Study. *European Journal of Nutrition, 40,* 231–237.

83. Sebastian, A., Harris, S. T., Ottaway, J. H., Todd, K. M., & Morris, R. C. Jr., (1994). Improved mineral balance and skeletal metabolism in postmenopausal women treated with potassium bicarbonate. *The New England Journal of Medicine, 330,* 1776–1781.

84. Sellmeyer, D. E., Schloetter, M., & Sebastian, A. (2002). Potassium citrate prevents increased urine calcium excretion and bone resorption induced by a high sodium chloride diet. *Journal of Clinical Endocrinology and Metabolism, 87,* 2008–2012.

85. Maurer, M., Riesen, W., Muser, J., Hulter, H. N., & Krapf, R. (2003). Neutralization of Western diet inhibits bone resorption independently of K intake and reduces cortisol secretion in humans. *American Journal of Physiology Renal Physiology, 284,* F32–F40.

86. Lemann, J., Jr., Gray, R. W., & Pleuss, J. A. (1989). Potassium bicarbonate, but not sodium bicarbonate, reduces urinary calcium excretion and improves calcium balance in healthy men. *Kidney International, 35,* 688–695.

87. Lemann, J., Jr., Pleuss, J. A., & Gray, R. W. (1993). Potassium causes calcium retention in healthy adults. *Journal of Nutrition, 123,* 1623–1626.

88. Lemann, J., Jr., Pleuss, J. A., Gray, R. W., & Hoffmann, R. G. (1991). Potassium administration reduces and potassium deprivation increases urinary calcium excretion in healthy adults [corrected]. *Kidney International, 39,* 973–983.

89. Jehle, S., Zanetti, A., Muser, J., Hulter, H. N., & Krapf, R. (2006). Partial neutralization of the acidogenic Western diet with potassium citrate increases bone mass in postmenopausal women with osteopenia. *Journal of American Society of Nephrology, 17,* 3213–3222.

90. Mardon, J., Habauzit, V., & Trzeciakiewicz, A., et al. (2008). Long-term intake of a high-protein diet with or

without potassium citrate modulates acid-base metabolism, but not bone status, in male rats. *Journal of Nutrition, 138*, 718–724.

91. Vormann, J., Forster, C., & Zippel, U., et al. (1997). Effects of magnesium deficiency on magnesium and calcium content in bone and cartilage in developing rats in correlation to chondrotoxicity. *Calcified Tissue International, 61*, 230–238.

92. Rude, R. K., Kirchen, M. E., Gruber, H. E., Stasky, A. A., & Meyer, M. H. (1998). Magnesium deficiency induces bone loss in the rat. *Mineral and Electrolyte Metabolism, 24*, 314–320.

93. Rude, R. K., Gruber, H. E., Norton, H. J., Wei, L. Y., Frausto, A., & Mills, B. G. (2004). Bone loss induced by dietary magnesium reduction to 10% of the nutrient requirement in rats is associated with increased release of substance P and tumor necrosis factor-alpha. *Journal of Nutrition, 134*, 79–85.

94. Stendig-Lindberg, G., Koeller, W., Bauer, A., & Rob, P. M. (2004). Prolonged magnesium deficiency causes osteoporosis in the rat. *Journal of American College Nutrition, 23*, 704S–711S.

95. Rude, R. K., Gruber, H. E., Norton, H. J., Wei, L. Y., Frausto, A., & Kilburn, J. (2006). Reduction of dietary magnesium by only 50% in the rat disrupts bone and mineral metabolism. *Osteoporosis International, 17*, 1022–1032.

96. New, S. A., Robins, S. P., & Campbell, M. K., et al. (2000). Dietary influences on bone mass and bone metabolism: further evidence of a positive link between fruit and vegetable consumption and bone health?. *American Journal of Clinical Nutrition, 71*, 142–151.

97. Tranquilli, A. L., Lucino, E., Garzetti, G. G., & Romanini, C. (1994). Calcium, phosphorus and magnesium intakes correlate with bone mineral content in postmenopausal women. *Gynecological Endocrinology, 8*, 55–58.

98. Wang, M. C., Moore, E. C., & Crawford, P. B., et al. (1999). Influence of pre-adolescent diet on quantitative ultrasound measurements of the calcaneus in young adult women. *Osteoporosis International, 9*, 532–535.

99. Dimai, H. P., Porta, S., & Wirnsberger, G., et al. (1998). Daily oral magnesium supplementation suppresses bone turnover in young adult males. *Journal of Clinical Endocrinology and Metabolism, 83*, 2742–2748.

100. Doyle, L., Flynn, A., & Cashman, K. (1999). The effect of magnesium supplementation on biochemical markers of bone metabolism or blood pressure in healthy young adult females. *European Journal of Clinical Nutrition, 53*, 255–261.

101. Stendig-Lindberg, G., Tepper, R., & Leichter, I. (1993). Trabecular bone density in a two year controlled trial of peroral magnesium in osteoporosis. *Magnesium Research, 6*, 155–163.

102. Rude, R. K., & Olerich, M. (1996). Magnesium deficiency: possible role in osteoporosis associated with gluten-sensitive enteropathy. *Osteoporosis International, 6*, 453–461.

103. Carpenter, T. O., DeLucia, M. C., & Zhang, J. H., et al. (2006). A randomized controlled study of effects of dietary magnesium oxide supplementation on bone mineral content in healthy girls. *Journal of Clinical Endocrinology and Metabolism, 91*, 4866–4872.

104. Remer, T., & Manz, F. (1994). Estimation of the renal net acid excretion by adults consuming diets containing variable amounts of protein. *American Journal of Clinical Nutrition, 59*, 1356–1361.

105. Gannon, R. H., Millward, D. J., & Brown, J. E., et al. (2008). Estimates of daily net endogenous acid production in the elderly UK population: analysis of the National Diet and Nutrition Survey (NDNS) of British adults aged 65 years and over. *British Journal of Nutrition, 100*, 615–623.

106. Frassetto, L. A., Lanham-New, S. A., & Macdonald, H. M., et al. (2007). Standardizing terminology for estimating the diet-dependent net acid load to the metabolic system. *Journal of Nutrition, 137*, 1491–1492.

107. Frassetto, L. A., Todd, K. M., Morris, R. C., Jr., & Sebastian, A. (1998). Estimation of net endogenous noncarbonic acid production in humans from diet potassium and protein contents. *American Journal of Clinical Nutrition, 68*, 576–583.

108. Remer, T., & Manz, F. (1995). Potential renal acid load of foods and its influence on urine pH. *Journal of American Dietetic Association, 95*, 791–797.

109. New, S. A., MacDonald, H. M., & Campbell, M. K., et al. (2004). Lower estimates of net endogenous noncarbonic acid production are positively associated with indexes of bone health in premenopausal and perimenopausal women. *American Journal of Clinical Nutrition, 79*, 131–138.

110. Welch, A. A., Bingham, S. A., Reeve, J., & Khaw, K. T. (2007). More acidic dietary acid-base load is associated with reduced calcaneal broadband ultrasound attenuation in women but not in men: results from the EPIC-Norfolk cohort study. *American Journal of Clinical Nutrition, 85*, 1134–1141.

111. Alexy, U., Remer, T., Manz, F., Neu, C. M., & Schoenau, E. (2005). Long-term protein intake and dietary potential renal acid load are associated with bone modeling and remodeling at the proximal radius in healthy children. *American Journal of Clinical Nutrition, 82*, 1107–1114.

112. Chan, R. S., Woo, J., Chan, D. C., Lo, D. H., & Cheung, C. S. (2008). Bone mineral status and its relation with dietary estimates of net endogenous acid production in Hong Kong Chinese adolescents. *British Journal of Nutrition, 100*, 1283–1290.

113. Bleich, H. L., Moore, M. J., Lemann, J., Jr., Adams, N. D., & Gray, R. W. (1979). Urinary calcium excretion in human beings. *The New England Journal of Medicine, 301*, 535–541.

114. Buclin, T., Cosma, M., & Appenzeller, M., et al. (2001). Diet acids and alkalis influence calcium retention in bone. *Osteoporosis International, 12*, 493–499.

115. Benzie, I. F. (2003). Evolution of dietary antioxidants. *Comparative Biochemistry and Physiology A-Molecular & Integrative Physiology, 136*, 113–126.

116. Mühlbauer, R. C., Lozano, A., & Reinli, A. (2002). Onion and a mixture of vegetables, salads, and herbs affect bone resorption in the rat by a mechanism independent of their base excess. *Journal of Bone and Mineral Research, 17*, 1230–1236.

117. Galli, F., Piroddi, M., Annetti, C., Aisa, C., Floridi, E., & Floridi, A. (2005). Oxidative stress and reactive oxygen species. *Contributions to Nephrology, 149*, 240–260.

118. Blomhoff, R. (2005). Dietary antioxidants and cardiovascular disease. *Current Opinion in Lipidology, 16*, 47–54.

119. Key, L. L., Jr., Ries, W. L., Taylor, R. G., Hays, B. D., & Pitzer, B. L. (1990). Oxygen derived free radicals in osteoclasts: the specificity and location of the nitroblue tetrazolium reaction. *Bone, 11*, 115–119.

120. Yang, S., Madyastha, P., Bingel, S., Ries, W., & Key, L. (2001). A new superoxide-generating oxidase in murine osteoclasts. *Journal of Biological Chemistry, 276*, 5452–5458.

121. Bai, X. C., Lu, D., & Bai, J., et al. (2004). Oxidative stress inhibits osteoblastic differentiation of bone cells by ERK and NF-kappaB. *Biochemical and Biophysical Research Communications, 314*, 197–207.

122. Altindag, O., Erel, O., Soran, N., Celik, H., & Selek, S. (2008). Total oxidative/anti-oxidative status and relation to bone mineral density in osteoporosis. *Rheumatology International, 28*, 317–321.

123. Fuller, K., Lean, J. M., Bayley, K. E., Wani, M. R., & Chambers, T. J. (2000). A role for TGFbeta(1) in osteoclast differentiation and survival. *Journal of Cell Science, 113*(Pt 13), 2445–2453.

124. Darden, A. G., Ries, W. L., Wolf, W. C., Rodriguiz, R. M., & Key, L. L. Jr., (1996). Osteoclastic superoxide production and bone resorption: stimulation and inhibition by modulators of NADPH oxidase. *Journal of Bone and Mineral Research, 11*, 671–675.

125. Halliwell, B., & Gutteridge, J. M. C. (1989). Free radicals in biology and medicine, (2nd ed.). Oxford: Clarendon Press.

126. Maggio, D., Barabani, M., & Pierandrei, M., et al. (2003). Marked decrease in plasma antioxidants in aged osteoporotic women: results of a cross-sectional study. *Journal of Clinical Endocrinology and Metabolism, 88*, 1523–1527.

127. Zhang, J., Munger, R. G., West, N. A., Cutler, D. R., Wengreen, H. J., & Corcoran, C. D. (2006). Antioxidant intake and risk of osteoporotic hip fracture in Utah: an effect modified by smoking status. *American Journal of Epidemiology, 163*, 9–17.

128. Maggio, D., Polidori, M. C., & Barabani, M., et al. (2006). Low levels of carotenoids and retinol in involutional osteoporosis.. *Bone, 38*, 244–248.

129. Rao, L. G., Mackinnon, E. S., Josse, R. G., Murray, T. M., Strauss, A., & Rao, A. V. (2007). Lycopene consumption decreases oxidative stress and bone resorption markers in postmenopausal women. *Osteoporosis International, 18*, 109–115.

130. Hall, S. L., & Greendale, G. A. (1998). The relation of dietary vitamin C intake to bone mineral density: results from the PEPI study. *Calcified Tissue International, 63*, 183–189.

131. Morton, D. J., Barrett-Connor, E. L., & Schneider, D. L. (2001). Vitamin C supplement use and bone mineral density in postmenopausal women. *Journal of Bone and Mineral Research, 16*, 135–140.

132. Leveille, S. G., LaCroix, A. Z., Koepsell, T. D., Beresford, S. A., Van Belle, G., & Buchner, D. M. (1997). Dietary vitamin C and bone mineral density in postmenopausal women in Washington State, USA. *Journal of Epidemiology and Community Health, 51*, 479–485.

133. Wolf, R. L., Cauley, J. A., & Pettinger, M., et al. (2005). Lack of a relation between vitamin and mineral antioxidants and bone mineral density: results from the Women's Health Initiative. *American Journal of Clinical Nutrition, 82*, 581–588.

134. Cockayne, S., Adamson, J., Lanham-New, S., Shearer, M. J., Gilbody, S., & Torgerson, D. J. (2006). Vitamin K and the prevention of fractures: systematic review and meta-analysis of randomized controlled trials. *Archives of Internal Medicine, 166*, 1256–1261.

135. Wong, R. W., & Rabie, A. B. (2008). Effect of quercetin on bone formation. *Journal of Orthopaedic Research, 26*, 1061–1066.

136. Woo, J. T., Nakagawa, H., & Notoya, M., et al. (2004). Quercetin suppresses bone resorption by inhibiting the differentiation and activation of osteoclasts. *Biological & Pharmaceutical Bulletin, 27*, 504–509.

137. Kanter, M., Altan, M. F., Donmez, S., Ocakci, A., & Kartal, M. E. (2007). The effects of quercetin on bone minerals, biomechanical behavior, and structure in streptozotocin-induced diabetic rats. *Cell Biochemistry Function, 25*, 747–752.

138. Mühlbauer, R. C. (2001). Rutin cannot explain the effect of vegetables on bone metabolism. *Journal of Bone and Mineral Research, 16*, 970–971.

139. Mandadi, K., Ramirez, M., & Jayaprakasha, G. K., et al. (2009). Citrus bioactive compounds improve bone quality and plasma antioxidant activity in orchidectomized rats. *Phytomedicine, 16*, 513–520.

140. Frassetto, L. A., Todd, K. M., Morris, R. C., Jr., & Sebastian, A. (2000). Worldwide incidence of hip fracture in elderly women: relation to consumption of animal and vegetable foods. *Journal of Gerontology Series A: Biological Sciences and Medical Sciences, 55*, M585–M592.

141. Okubo, H., Sasaki, S., & Horiguchi, H., et al. (2006). Dietary patterns associated with bone mineral density in premenopausal Japanese farmwomen. *American Journal of Clinical Nutrition, 83*, 1185–1192.

13

Socioeconomic Inequalities in Fruit and Vegetable Intakes

Kylie Ball and David Crawford

Centre for Physical Activity and Nutrition Research, Deakin University, Burwood, Victoria, Australia

1. INTRODUCTION

Socioeconomic position (SEP) refers to an individual's social and economic ranking within society based on access to resources (such as material and social assets, including income, wealth, and educational credentials) and prestige (i.e. an individual's status in a social hierarchy, linked for instance to their occupation, income, or education level) [1]. Commonly used proxy indicators for SEP include education level; own or household income; and occupational status. Socioeconomic position can also be assessed at the area (as opposed to individual or household) level, for example, using indicators based on the proportion of residents with particular socioeconomic characteristics residing within neighborhoods.

Compared with those of high SEP, individuals of low SEP have been demonstrated with reasonable consistency to be at increased risk of low or inadequate fruit and vegetable consumption (for example, not meeting recommendations of two daily servings of fruit, and five of vegetables, currently promoted in many developed countries). That this has been reported across a range of studies focusing on different aged target groups, and using different indicators of SEP, attests to the robustness of this finding. The following section provides an overview of evidence demonstrating socioeconomic inequalities in fruit and vegetable consumption in children, adolescents, and adults. These are reported with consideration to the particular measure of SEP used. This is important, since evidence suggests that, while associations of different SEP indicators with diet are similar, they are also independent, potentially reflecting distinct underlying social processes impacting on diet [2,3]. The examination of separate socioeconomic indices therefore provides additional insights into the likely mechanisms underlying socioeconomic variations in diet. These mechanisms are considered further in Section 3.

195

2. OVERVIEW OF EVIDENCE ON SOCIOECONOMIC INEQUALITIES IN FRUIT AND VEGETABLE CONSUMPTION

The following overview of existing evidence on socioeconomic inequalities in fruit and vegetable consumption is restricted to evidence from developed countries, primarily because these countries typically have more established systems of nutrition monitoring and reporting of inequalities in diet. Further, factors affecting food supply, diets, and socioeconomic variations in diet are likely to be quite different between developed and developing countries.

2.1 Children and Adolescents

A child's SEP is largely determined by that of his or her parents, and thus SEP for children is typically characterized using measures such as parental education or income level. Similarly, the SEP of adolescents is most often based on parental SEP. There is evidence that children's diets do vary according to SEP. For example, using the UK Avon Longitudinal Study of Parents and Children (ALSPAC), Northstone and Emmett [4] found that among 4–7-year-olds, a dietary pattern described as being based on 'junk' type foods (which was low in fruits and vegetables) was more common among children of mothers with lower levels of education. Conversely, among the ALSPAC children, a 'health conscious' dietary pattern (which included fruits and vegetables) was more common among children of mothers with higher levels of education. In their review of fruit and vegetable consumption among children and adolescents, Rasmussen et al. [5] identified 46 papers that examined associations of these dietary factors with SEP. They concluded that, despite SEP being operationalized differently across these studies, low SEP was consistently associated with low or less frequent

intake of fruit and vegetables, and this was especially the case when SEP was indicated by family income.

2.2 Adults

Many studies of adult populations show differences by SEP in the quantity and/or variety of fruits and vegetables consumed. For example, a systematic review of socioeconomic variations in diet across Europe showed that higher SEP, assessed using both education and occupation, was consistently associated with greater consumption (grams/day) of both fruits and vegetables [6]. Similarly, socioeconomic gradients in fruit and vegetable consumption among adults have been reported in studies in the US [7], Canada [8], the UK [9] and Australia [10–12], using measures of occupation [9,10], education [8], income [8,11,12], welfare benefit status [9], and neighborhood SEP [7]. Consistent with these findings, studies have also shown parallel socioeconomic differentials in intakes of macro- and micronutrients found in fruits and vegetables, such as fiber, vitamin C and folate among adults [11,13]. Of particular concern are findings from prospective studies showing that socioeconomic inequalities in adults' fruit and vegetable consumption are not decreasing [14] and may actually be widening [15] over time.

2.3 Associations of Multiple SEP Indicators with Fruit and Vegetable Intakes

As noted above, different etiological pathways may underlie associations of different indicators of SEP with fruit and vegetable intakes. For example, education may reflect an individual's knowledge and attitudes that might in turn impact on his or her decision or ability to adopt health-promoting dietary behaviors such as eating fruits and vegetables (such mechanisms are considered further in Section 3). However, relatively few studies have attempted to disentangle the relationships between divergent indicators of SEP and fruit and vegetable consumption. In 2007, Lallukka et al. [16] examined the inter-relationships between seven different indicators of SEP and an index of healthy food habits, which included consumption of fresh fruit and vegetables. That study found that two indicators of childhood SEP – parental education and childhood economic difficulties – were not associated with current healthy eating. However, the remaining indices – education, occupational class, household income, home ownership, and economic difficulties – were all associated with fruit and vegetable consumption in the hypothesized direction. These associations were attenuated but mostly remained significant when all SEP indicators were considered simultaneously, suggesting that these indicators have primarily independent effects on diet.

2.4 Summary

The evidence summarized above demonstrates the existence of socioeconomic inequalities in fruit and vegetable consumption, such that persons of low SEP are at increased risk of consuming relatively lesser quantities and varieties of fruits and vegetables than their higher SEP peers. These socioeconomic inequalities have been observed in samples of children, adolescents, and adults. The associations of SEP with fruit and vegetable consumption appear robust across indicators of SEP, and in fact different indicators appear to be independently associated with fruit and vegetable intakes, suggesting divergent underlying etiological pathways. These findings are a cause for concern, since the wide-ranging benefits to health of frequent consumption of fruits and vegetables are well recognized. Consuming lower than ideal

amounts of fruits and vegetables places persons of low SEP at increased risk of nutritional deficiencies and a range of associated adverse health outcomes, including obesity, cardiovascular problems, and certain cancers. Socioeconomic inequalities in fruit and vegetable consumption parallel inequalities in health outcomes, and represent one potential pathway by which low SEP might lead to poorer health. In order to redress this situation, an understanding of the mechanisms underlying socioeconomic inequalities in fruit and vegetable consumption is required.

3. MECHANISMS UNDERLYING SOCIOECONOMIC INEQUALITIES IN FRUIT AND VEGETABLE CONSUMPTION

While there has now accumulated a reasonable body of research evidence on the determinants of fruit and vegetable intakes, data on the determinants of socioeconomic inequalities in these intakes have been slower to emerge. That is, we do not yet have a good understanding of why people of lower SEP tend to have lower intakes of fruits and vegetables. Qualitative studies and quantitative descriptive studies have shed some light on possible explanations, but until recently very few quantitative studies had empirically tested the contribution of different factors to explaining or mediating socioeconomic inequalities in fruit and vegetable consumption, using appropriate research designs. In recent years, several quantitative studies have capitalized on advances in understanding and application of statistical methods of testing mediating pathways to begin to provide insights into the pathways by which low SEP might lead to less favorable consumption of fruits and vegetables.

An overview of the evidence on potential mediators of socioeconomic gradients in fruit and vegetable intakes, drawing from both qualitative and quantitative research designs, is provided below. This review considers mediating factors in an approach consistent with the conceptual framework of social–ecological models of health behavior [17], which posit the importance of influences within intrapersonal, social, and physical environmental domains.

3.1 Children

To our knowledge, no studies have aimed to explicitly test the mediators of socioeconomic variations in children's fruit and vegetable consumption. However, it is plausible that the mediators that apply to adults' intakes are also likely to impact the intakes of children, given that children's SEP and also their dietary intakes are strongly determined by parental factors. There is some evidence that mothers of low SEP consider costs more often, and health less often, when making food-purchasing choices; however, these factors did not explain socioeconomic (educational) variations in diet [18].

3.2 Adolescents

To our knowledge, only two studies to date have empirically tested potential mediating pathways linking SEP to fruit and/or vegetable consumption among adolescents. We previously tested the role of constructs derived from social cognitive theory (i.e. self-efficacy, the perceived importance of health-promoting behaviors, social observation of mother and of best friend, social support for healthy eating from family and from friends, and availability of fruits, vegetables, and energy-dense snack foods in the home) in explaining socioeconomic variations in fruit intakes among 2529 Australian adolescents [19]. With the exception of social support for healthy eating from friends, all of the constructs tested contributed to explaining socioeconomic variations in

adolescents' fruit intakes. The cognitive factors, particularly the perceived importance of healthy behaviors, appeared to be the strongest mediators. In a longitudinal study of 896 Norwegian adolescents [20], both income and education were positively associated with fruit and vegetable intakes. Perceived accessibility of fruit and vegetables at home was the strongest mediator of these associations.

3.3 Adults

Slightly more evidence is available concerning the potential mediators of socioeconomic discrepancies in fruit and vegetable intakes among adults. An overview of this evidence, categorized according to social–ecological domain (intrapersonal, social, or physical environmental), is provided below.

3.3.1 Intrapersonal Mediators

Lower levels of nutrition knowledge, and less consideration of health as a priority when making food choices, have been implicated as potential explanatory factors in the association of SEP with diet quality among adults [21,22]. For example, Ball et al. [21] showed that women with lower levels of education had less knowledge about the nutrient sources and health effects of different foods, and reported giving less consideration to health when making food-purchasing choices, and these factors partly explained their lower intakes of fruits and vegetables.

Some have suggested that the poorer diets of adults of low SEP are attributable to a lack of cooking skills or interest in cooking among these groups, but evidence of socioeconomic differentials in these constructs is equivocal [23,24]. Other potential explanations for the lower intakes of fruits and vegetables among adults of low SEP include lower perceived palatability of fruits and vegetables [25], apathy toward nutrition messages [26], a lack of motivation, or misperceptions about the adequacy of one's diet [27], but again little empirical evidence directly attests to the role of these factors in mediating socioeconomic variations in fruit and vegetable intakes.

It is also important to consider the possibility of over-reporting of fruit and vegetable intakes by those of high SEP as a potential explanatory factor for socioeconomic gradients in intakes. That is, people of high SEP may report higher than actual intakes of fruits and vegetables, because, for example, they may be more aware of the social desirability of frequent consumption of these foods, and/or more eager to present their diets in a favorable light. However, it is unlikely that such a reporting bias would completely account for the relatively consistent findings across multiple studies, and using a variety of measures of SEP and fruit and vegetable consumption.

3.3.2 Social and Physical Environmental Mediators

More recently, attention has shifted from a primary focus on intrapersonal factors, to encompass broader social and physical environmental factors as possible mediators of socioeconomic inequalities in diet. However, evidence for the importance of social and physical environmental factors in explaining socioeconomic discrepancies in consumption of fruits and vegetables remains patchy and inconsistent. Some data suggest that people of low SEP receive less social support from their families to eat healthily [28], which may impede their efforts to eat more fruits and vegetables. Time pressures associated with long or inflexible working hours have also been implicated as barriers to shopping for and preparing healthy foods in low SEP groups [28].

In terms of broader environmental factors, some data suggest that over the past few decades the prices of fruits and vegetables have increased disproportionately more than those of foods high in sugar or fat [29], and several authors have argued that the poorer diet quality among individuals of low SEP is attributable to the relatively higher costs of a healthy diet [25]. However, this hypothesis is not supported by all available evidence. For example, in a quasi-experimental study, Inglis et al. [30] showed that socioeconomic inequalities in the healthfulness of food choices (including fruits and vegetables) were not reduced through (theoretically) manipulating the food budgets available to low- and high-income women. In another study [31], objectively-assessed food prices did not impact on food-purchasing decisions or explain inequalities in purchasing of healthy and less healthy foods (although this study did not access fruits and vegetables specifically).

Similarly, there is mixed evidence concerning the impact of availability and accessibility to food stores in explaining socioeconomic inequalities in fruit and vegetable consumption. Some research, primarily from the USA, indicates that there are fewer large supermarkets [32–39], or fewer healthier choices available in stores [33,40], in more socioeconomically deprived neighborhoods. These findings might imply that the lower intakes of fruits and vegetables among lower SEP individuals are at least partly attributable to poorer access to these foods in local neighborhoods. However, other studies from the USA [35,41], UK [42,43], Canada [44–47], New Zealand [48], and Australia [49–51] found mixed evidence, or few differences in food availability or access to supermarkets between deprived and affluent areas, or otherwise differences favoring more, rather than less, deprived neighborhoods. Qualitative studies have also demonstrated that availability of and access to good quality healthy foods are not perceived as significant barriers to healthy eating among low SEP groups [28,52]. Less than a handful of studies have explicitly tested the role of food availability and access in mediating associations of SEP with diet, and those that have indicated that the less healthful diets of individuals of low SEP were not explained by socioeconomic differences in food availability or access [21,53].

4. IMPLICATIONS FOR FUTURE RESEARCH

This overview of the scientific literature shows that there are consistent socioeconomic differentials in intakes of fruits and vegetables, by which individuals of low SEP are at increased risk of inadequate consumption of these foods. These findings are relatively robust across age groups, and across different SEP indicators. However, much less is known about the mechanisms underlying associations of SEP with fruit and vegetable consumption. There is a particular dearth of information on mediating factors among children and adolescents. Existing studies of mediators among adults are few in number and findings are inconsistent. More research is required to examine the inter-relationships among different indicators of SEP, and how these might independently or jointly impact on fruit and vegetable consumption. There are also very few longitudinal studies in this area, and hence a lack of evidence on likely temporal associations among SEP, mediating factors, and fruit and vegetable consumption. Clearly, further epidemiological research, in particular longitudinal studies and studies focused on children and adolescents, is required to elucidate more clearly the reasons why persons of low SEP are at increased risk of low consumption of fruits and vegetables. Evidence from intervention studies would also be of great value, since these studies can shed light on the causes of socioeconomic inequalities in fruit and vegetable consumption, and on the most effective strategies for reducing these inequalities.

5. SUMMARY: IMPLICATIONS FOR PRACTICE

The available literature on the existence of socioeconomic differentials in fruit and vegetable consumption among children, adolescents, and adults points to several practice implications. In particular, there is good evidence that persons of low SEP are likely to require additional assistance to enable them to better meet health recommendations regarding consumption of fruits and vegetables. However, in contrast, the factors that mediate socioeconomic inequalities in fruit and vegetable intakes and that might therefore be targeted in nutrition promotion interventions, remain poorly understood. In the absence of strong evidence of mediating factors, it is difficult to recommend public health strategies or policies that might be implemented in order to reduce socioeconomic discrepancies in fruit and vegetable consumption. However, the limited existing evidence suggests several strategies that could be considered, at least for adolescents and adults. Among adolescents, strategies aimed at increasing self-efficacy, at promoting the importance of healthy eating, and at promoting increased availability of fruits and vegetables in the home may help in supporting low SEP adolescents to consume more fruits and vegetables. Among adults, such strategies might include nutrition education and messages aimed at promoting the importance of health when making food-purchasing choices; advice on engaging family and garnering support for making healthy food choices; and tips on time-efficient preparation of fruits and vegetables. While broader policies ensuring the equitable provision of healthy foods in all neighborhoods across the socioeconomic spectrum are clearly important, further research is necessary before advocating for this as a specific strategy for reducing socioeconomic inequalities in fruit and vegetable consumption. In any case, available evidence suggests that such environmental strategies should be supplemented with education and support to enable individuals to make healthy dietary choices, regardless of their socioeconomic position.

References

1. Krieger, N., Williams, D. R., & Moss, N. E. (1997). Measuring social class in US public health research: Concepts, methodologies, and guidelines. *Annual Review of Public Health, 18,* 341–378.
2. Galobardes, B., Morabia, A., & Bernstein, M. (2001). Diet and socioeconomic position: Does the use of different indicators matter? *International Journal of Epidemiology, 30,* 334–340.
3. Turrell, G., Hewitt, B., Patterson, C., & Oldenburg, B. (2003). Measuring socio-economic position in dietary research: Is choice of socio-economic indicator important? *Public Health Nutrition, 6,* 191–201.
4. Northstone, K., & Emmett, P. (2005). Multivariate analysis of diet in children at four and seven years of age and associations with socio-demographic characteristics. *European Journal of Clinical Nutrition, 59,* 751–760.
5. Rasmussen, M., Krolner, R., Klepp, K. I., Lytle, L., Brug, J., Bere, E., & Due, P. (2006). Determinants of fruit and vegetable consumption among children and adolescents: A review of the literature. Part I: Quantitative studies. *International Journal of Behavioral Nutrition Physical Activity, 3,* 22.
6. De Irala-Estevez, J., & Groth, M. (2000). A systematic review of socio-economic differences in food habits in Europe: consumption of fruit and vegetables. *European Journal of Clinical Nutrition, 54,* 706.
7. Dubowitz, T., Heron, M., Bird, C. E., Lurie, N., Finch, B. K., Basurto-Davila, R., Hale, L., & Escarce, J. J. (2008). Neighborhood socioeconomic status and fruit and vegetable intake among whites, blacks, and Mexican Americans in the United States. *American Journal of Clinical Nutrition, 87,* 1883–1891.
8. Riediger, N. D., & Moghadasian, M. H. (2008). Patterns of fruit and vegetable consumption and the influence of sex, age and socio-demographic factors among Canadian elderly. *Journal of the American College of Nutr, 27,* 306–313.
9. Billson, H., Pryer, J. A., & Nichols, R. (1999). Variation in fruit and vegetable consumption among adults in Britain. An analysis from the dietary and nutritional survey of British adults. *European Journal of Clinical Nutrition, 53,* 946–952.
10. Ball, K., Mishra, G. D., Thane, C. W., & Hodge, A. (2004). How well do Australian women comply with dietary guidelines? *Public Health Nutrition, 7,* 443–452.
11. Giskes, K., Turrell, G., Patterson, C., & Newman, B. (2002). Socioeconomic differences among Australian adults in consumption of fruit and vegetables and intakes of vitamins A, C and folate. *Journal of Human Nutrition and Dietetics, 15,* 375–385.
12. Giskes, K., Turrell, G., Patterson, C., & Newman, B. (2002). Socio-economic differences in fruit and vegetable consumption among Australian adolescents and adults. *Public Health Nutrition, 5,* 663–669.
13. Stallone, D., Brunner, E., Bingham, S., & Marmot, M. (1997). Dietary assessment in Whitehall 11: The influence of reporting bias on apparent socioeconomic variation in nutrient intakes. *European Journal of Clinical Nutrition, 51,* 815–825.
14. Perrin, A. E., Simon, C., Hedelin, G., Arveiler, D., Schaffer, P., & Schlienger, J. L. (2002). Ten-year trends of dietary intake in a middle-aged French population: Relationship with educational level. *European Journal of Clinical Nutrition, 56,* 393–401.
15. Wrieden, W. L., Connaghan, J., Morrison, C., & Tunstall-Pedoe, H. (2004). Secular and socio-economic trends in compliance with dietary targets in the north Glasgow MONICA population surveys 1986–1995: Did social gradients widen? *Public Health Nutrition, 7,* 835–842.
16. Lallukka, T., Laaksonen, M., Rahkonen, O., Roos, E., & Lahelma, E. (2007). Multiple socio-economic circumstances and healthy food habits. *European Journal of Clinical Nutrition, 61,* 701–710.
17. Stokols, D. (1996). Translating social ecological theory into guidelines for community health promotion. *American Journal of Health Promotion, 10,* 282–298.
18. Hupkens, C., Knibbe, R., & Drop, M. (2000). Social class differences in food consumption: The explanatory value of permissiveness and health and cost considerations. *European Journal of Public Nutrition, 10,* 108–113.
19. Ball, K., Macfarlane, A., Crawford, D., Savige, G., Andrianopoulos, N., & Worsley, A. (2009). Can social cognitive theory constructs explain socio-economic variations in adolescent eating behaviours? A mediation analysis. *Health Education Research, 24,* 496–506.
20. Bere, E., vanLenthe, F., Klepp, K. I., & Brug, J. (2008). Why do parents' education level and income affect the amount of fruits and vegetables adolescents eat? *European Journal of Public Health, 18,* 611–615.
21. Ball, K., Crawford, D., & Mishra, G. (2006). Socio-economic inequalities in women's fruit and vegetable intakes: A multilevel study of individual, social and

environmental mediators. *Public Health Nutrition, 9,* 623–630.

22. Turrell, G. (1997). Educational differences in dietary guideline food practices: Are they associated with educational differences in food and nutrition knowledge? *Australian Journal of Nutrition and Dietetics, 54,* 25–33.

23. Caraher, M., & Lang, T. (1999). Can't cook, won't cook: A review of cooking skills and their relevance to health promotion. *International Journal of Health Promotional Education, 37,* 89–100.

24. McLaughlin, C., Tarasuk, V., & Kreiger, N. (2003). An examination of at-home food preparation activity among low-income, food-insecure women. *Journal of the American Dietetic Association, 103,* 1506–1512.

25. Darmon, N., & Drewnowski, A. (2008). Does social class predict diet quality? *American Journal of Clinical Nutrition, 87,* 1107–1117.

26. Patterson, R. E., Satia, J. A., Kristal, A. R., Neuhouser, M. L., & Drewnowski, A. (2001). Is there a consumer backlash against the diet and health message? *Journal of the American Dietetic Association, 101,* 37–41.

27. Dibsdall, L. A., Lambert, N., Bobbin, R. F., & Frewer, L. J. (2003). Low-income consumers' attitudes and behaviour towards access, availability and motivation to eat fruit and vegetables. *Public Health Nutrition, 6,* 159–168.

28. Inglis, V., Ball, K., & Crawford, D. (2005). Why do women of low socioeconomic status have poorer dietary behaviours than women of higher socioeconomic status? A qualitative exploration. *Appetite, 45,* 334–343.

29. Drewnowski, A., & Darmon, N. (2005). The economics of obesity: Dietary energy density and energy cost. *American Journal of Clinical Nutrition, 82,* 265S–273S.

30. Inglis, V., Ball, K., & Crawford, D. (2009). Does modifying the household food budget predict changes in the healthfulness of purchasing choices among low- and high-income women? *Appetite, 52,* 273–279.

31. Giskes, K., Van Lenthe, F. J., Brug, J., Mackenbach, J. P., & Turrell, G. (2007). Socioeconomic inequalities in food purchasing: The contribution of respondent-perceived and actual (objectively measured) price and availability of foods. *Preventive Medicine, 45,* 41–48.

32. Alwitt, L. F., & Donley, T. D. (1997). Retail stores in poor urban neighbourhoods. *The Journal of Consumer Affairs, 31,* 139–164.

33. Baker, E. A., Schootman, M., & Barnidge, E. (2006). The role of race and poverty in access to foods that enable individuals to adhere to dietary guidelines. *Preventive Chronic Disease, 5,* [serial online]. Available from: URL: http://www.cdc.gov/pcd/issues/2006/jul/05_0217.htm July.

34. Block, D., & Kouba, J. (2005). A comparison of the availability and affordability of a market basket in two communities in the Chicago area. *Public Health Nutrition, 9,* 837–845.

35. Chung, C., & Myers, S. L. Jr. (1999). Do the poor pay more for food? An analysis of grocery store availability and food price disparities. *The Journal of Consumer Affairs, 33,* 276–296.

36. Moore, L. V., & Diez Roux, A. V. (2006). Associations of neighborhood characteristics with the location and type of food stores. *American Journal of Public Health, 96,* 325–331.

37. Morland, K., Wing, S., Diez Roux, A., & Poole, C. (2002). Neighborhood characteristics associated with the location of food stores and food service places. *American Journal of Preventive Medicine, 22,* 23–29.

38. Powell, L. M., Slater, S., Mirtcheva, D., Bao, Y., & Chaloupka, F. J. (2007). Food store availability and neighborhood characteristics in the United States. *Preventive Medicine, 44,* 189–195.

39. Zenk, S. N., Schulz, A. J., Israel, B. A., James, S. A., Bao, S., & Wilson, M. L. (2005). Neighborhood racial composition, neighborhood poverty, and the spatial accessibility of supermarkets in metropolitan Detroit. *American Journal of Public Health, 95,* 660–667.

40. Horowitz, C. R., Colson, K. A., Hebert, P. L., & Lancaster, K. (2004). Barriers to buying healthy foods for people with diabetes: Evidence of environmental disparities. *American Journal of Public Health, 94,* 1549–1554.

41. Cassady, D., Jetter, K. M., & Culp, J. (2007). Is price a barrier to eating more fruits and vegetables for low-income families? *Journal of the American Dietetic Association, 107,* 1909–1915.

42. Cummins, S., & MacIntyre, S. (1999). The location of food stores in urban areas: A case study in Glasgow. *British Food Journal, 101,* 545.

43. Cummins, S., & Macintyre, S. (2002). A systematic study of an urban foodscape: The price and availability of food in greater Glasgow. *Urban Studies, 39,* 2115–2130.

44. Apparicio, P., Cloutier, M., & Shearmur, R. (2007). The case of Montreal's missing food deserts: Evaluation of accessibility to food supermarkets. *International Journal of Health Geography, 6,* 4.

45. Latham, J., & Moffat, T. (2007). Determinants of variation in food cost and availability in two socioeconomically contrasting neighbourhoods of Hamilton, Ontario, Canada. *Health Place, 13,* 273–287.

46. Smoyer-Tomic, K. E., Spence, J. C., & Amrhein, C. (2006). Food deserts in the Prairies? Supermarket accessibility and neighborhood need in Edmonton, Canada. *The Professional Geographer, 58,* 307–326.

47. Smoyer-Tomic, K. E., Spence, J. C., & Raine, K. D., et al. (2008). The association between neighbourhood

socioeconomic status and exposure to supermarkets and fast food outlets. *Health Place, 14,* 740–754.

48. Pearce, J., Blakely, T., Witten, K., & Bartie, P. (2007). Neighborhood deprivation and access to fast-food retailing: A national study. *American Journal of Preventive Medicine, 32,* 375–382.

49. Ball, K., Timperio, A., & Crawford, D. (2009). Neighbourhood socioeconomic inequalities in food access and affordability. *Health Place, 15,* 578–585.

50. Winkler, E., Turrell, G., & Patterson, C. (2006). Does living in a disadvantaged area mean fewer opportunities to purchase fresh fruit and vegetables in the area? Findings from the Brisbane food study. *Health Place, 12,* 306–319.

51. Winkler, E., Turrell, G., & Patterson, C. (2006). Does living in a disadvantaged area entail limited opportunities to purchase fresh fruit and vegetables in terms of price, availability, and variety? Findings from the Brisbane Food Study. *Health Place, 12,* 741–748.

52. Dibsdall, L. A., Lambert, N., & Frewer, L. J. (2002). Using interpretative phenomenology to understand the food-related experiences and beliefs of a select group of low-income UK women. *Journal of Nutrition Education and Behavior, 34,* 298–309.

53. Giskes, K., van Lenthe, F., Kamphuis, C., Huisman, M., Brug, J., & Mackenbach, J. P. (2009). Household and food shopping environments: Do they play a role in socioeconomic inequalities in fruit and vegetable consumption? A multilevel study among Dutch adults. *Journal of Epidemiology and Community Health, 63,* 113–120.

14

Working with Industry for the Promotion of Fruit and Vegetable Consumption

Christina M. Pollard[1] and Chris Rowley[2]
[1]Curtin University of Technology, Perth, Western Australia
[2]Horticulture Australia Limited, Sydney, New South Wales, Australia

This chapter is about working with industry for the promotion of fruit and vegetable consumption. The term 'industry' is broad and open to interpretation. It usually refers to the commercial food supply chain; however, for the purpose of building partnerships to promote fruit and vegetable consumption, there are many participants who have the potential to influence. Potential partners cross sectors and include the public sector (for example, Health, Education, Agriculture, Transport, Public works, Commerce and Trade, Local government, Environment departments), private sector (agriculture suppliers, producers, fruit and vegetable processing, packaging, transport, marketing, retailers, wholesalers and importer industries, media, financial institutions, worksites, schools, and hospitals), non-government organizations (community groups, consumer groups, non-government health organizations, professional associations – dieticians, parents,

teachers, community or religious leaders) and international bodies [World Health Organization (WHO), Food and Agricultural Organisation (FAO), United Nations Children's Fund (UNICEF), etc., consultative and advocacy groups] [1–3].

1. INCREASING FRUIT AND VEGETABLE CONSUMPTION

Many factors influence what we eat. Interventions to change any behavior need to be aimed at those things that can be changed or modified. Identification of the modifiable determinants of fruit and vegetable consumption assists the development of effective interventions. Modifiable factors should be considered when designing strategies and developing the appropriate partnerships to deliver interventions. Age, gender, and ethnicity are not modifiable determinants of consumption; however, they need to be considered in developing interventions as they may affect other characteristics relating to food intake. Examples of this could include age-related dietary needs, food knowledge, or the likelihood of wanting to eat more.

1.1 Factors Influencing Fruit and Vegetable Consumption

Both personal (demographic and individual) and environment factors influence human behavior and need to be considered when developing interventions to achieve health outcomes [4]. Although fruit and vegetables are often treated as one food category, there are broad differences in their sensory qualities, cultural uses, and attributes which suggest that different factors may influence consumption [5,6]. The factors influencing fruit consumption may be similar to those for vegetables, however the extent of influence differs [7]. Determinants of consumption are also specific for certain forms and type of fruit and vegetables, for example, the barriers to eating cooked vegetables are different to those to eating salad vegetables or fresh fruit [6]. There is a lack of consumer knowledge about what fruit and vegetables are available and what to do with them [8]. People eat fewer fruit and vegetables when they perceive more barriers [9], consider them expensive [6,9], perceive there is an additional cost to adding more fruit and vegetables to the usual food budget [10], or incorrectly perceive they have an adequate or high current intake [6,9]. Children's taste preferences [9], and mother's beliefs concerning disease prevention also influence consumption [9].

1.1.1 Personal Factors

Demographic factors associated with fruit and vegetable consumption include age, gender, socioeconomic status, ethnicity, education, geographic location, and employment. Socioeconomic status includes household income, educational attainment, occupation, marital status, social class, and area of residence [11]. Individual factors including knowledge, attitudes, beliefs, food budgeting, preparation and cooking skills, life course events and experiences (food upbringing, dietary changes for health, social roles, food skills, practice of food traditions) and intentions are associated with fruit and vegetable consumption [10,12–14].

It has been estimated that demographic factors account for about 10% of the variation in fruit and vegetable consumption compared to individual psychosocial factors which account for about 25% [15].

Modifying the individual level factors has been the main focus of many of the interventions to increase consumption, for example providing information and campaigning to improve knowledge, attitudes and beliefs about fruits and vegetables, and taking steps to increase food purchase and preparation skills. The focus of these interventions is usually associated with positive attributes (their health benefits, or taste, etc.).

1.1.2 Environmental Factors

Environmental influences on health behavior have been defined as all factors external to the individual [7]. Environmental factors influence food availability, marketing, and promotion. These influences include, but are not limited to, the proximity of food outlets [16], media marketing and promotion of foods [4], workforce food service, variety of foods available, and portion sizes [17]. Cost and availability seem to play an important role in fruit and vegetable intake [9,18], however the influence on fruit and vegetable consumption varies [19,20]. A system-wide approach is required to modify any environmental influences on fruit and vegetable consumption.

1.2 Addressing the Influences on Consumption of Fruit and Vegetables

Consumption is influenced by structural factors impacting on supply [7]. Anecdotal evidence suggests that retailers are not marketing fruit and vegetables as aggressively as they were a decade ago. Public health and policy makers need to collaborate with industry to assist them to communicate appropriate messages to consumers [21].

A comprehensive approach to increasing fruit and vegetable consumption is required [22]. Interventions need to address socio-cultural, economic, educational and technical challenges to expand and address the fruit and vegetable supply chain. Individual eating behaviors lie alongside these supply factors [23,24]. Interventions need to be based on scientific evidence, evaluated to determine effectiveness and unintended consequences, and messages need to be integrated into dietary guidelines and programs, and be culturally specific [1]. Strategies required include increasing:

- consumer awareness of the benefits of healthy eating, motivation and skills to increase consumption
- fruit and vegetable production and availability
- understanding of the components of consumer food choice attributes (taste, texture, form, price, convenience, quality, and safety)
- innovation and the development of fruit and vegetable based food products
- opportunities for consumption in various settings, for example worksites
- the implementation and evaluation of educational campaigns integrated with efforts to increase availability of and access to fruit and vegetables [1].

Cost components of accessibility occur throughout the value chain: cost of production, processing, transport, wholesale and retail marketing practices [21]. Changes in any of these cost components can have an impact on either final price or cost effectiveness of selling in some locations, particularly rural and remote areas [25].

Policy interventions include guidelines, voluntary and enforceable codes of conduct, legislation, and supporting regulations [26]. Policies, strategies, and guidelines can provide the strategic direction and political impetus to assist organizations and governments to take action to address health issues. There is some agreement that public policy could assist local communities to alter the mix of local food retailing to enable access to fresh, good quality, appropriately priced fruit and vegetables [27]. Inter-sectoral food policies implemented at a local government level have resulted in increased access to fruit and vegetables [28–30]. Policies include food security policies, agricultural production policies, education (and school) policies, credit policies, environmental policies, farming policies, labor, land and water policies, commerce and trade policies, food and agriculture taxation policies, horticultural and poverty reduction policies, and bilateral and multilateral donor programs [1]. Regulation and fiscal policies to encourage consumption of fruit and vegetables are dependent on each country's political, social, and environmental situation. Food manufacturers and retailers lobby governments to influence dietary advice to the public in an effort to establish a 'healthy' or 'nutritious' image for their products [31]. The evidence to suggest that these changes will result in healthier food choices is limited, or not available. Regulation to limit advertising and promotion of nutritionally less-preferable foods or facilitation of advertising and promotion of fruit and vegetables through health claims is a possibility [1]. For example, in Australia, there is a consideration of claims related to dietary advice concerning fruit and vegetables that would be allowed, for example 'eat at least two servings of fruit and five of vegetables every day' and 'oranges are good sources of vitamin C.' It is envisaged that these new regulations would allow use of health claims outlining the health benefits of eating more fruit and vegetables on fresh and partially processed fruit and vegetables at point of sale [32,33].

1.3 Partnerships to Increase Fruit and Vegetable Consumption

1.3.1 Health Sector Interest

Communication models to improve eating habits in line with dietary recommendations need to influence opinion leaders in all strata of society [34]. Structural support for such initiatives requires a broad base and should include government, horticulture, and food industry sectors [34]. There is limited financial support available for such initiatives and partnerships can assist in influencing stakeholders to improve public health by well-funded promotions of fruit and vegetables [34].

There are several reasons for health sector leadership in increasing fruit and vegetable consumption:

1. There is growing evidence of the health protection effects of adequate intake of fruit and vegetables.
2. Average population intakes in many countries are significantly lower than recommended levels across all ages and gender groups.
3. Consumers hold an overoptimistic assessment of their current intake.
4. Consumers have a low awareness of recommended levels of consumption.
5. Consumers generally are amenable to change as they already view fruit and vegetables as 'good for them' or 'healthy.'
6. There is little media promotion of fruits and vegetables compared to the heavy

promotion of 'less nutritious' or 'junk' foods (for example, confectionery, sugary breakfast cereals, soft drinks, savory snacks, take-away meals) [35].

To help reduce chronic disease, the WHO and FAO ask nations to conduct targeted campaigns to increase consumption of fruit and vegetables [36], asserting that effective health communication 'has the capacity to create awareness, improve knowledge and induce long-term changes in individual and social behaviors.' Leadership for national campaigns should be either health or agriculture driven, or both [1].

Increasing consumption of fruit and vegetables requires good quality, appropriately priced fruit and vegetables to be available to the majority of consumers. Barriers to increasing consumption related to taste, price, quality, and supply are best resolved through industry food supply channels. Facilitating dietary change to increase fruit and vegetable consumption requires strong partnership between industry, government, wholesalers, and others in the production, distribution, and marketing chain [2,3,15,35,37–39].

The assumption is that horticultural and retail industries would enhance health sector efforts due to their expertise in, and resources for, the promotion of their products.

1.4 Other Interests in Increasing Fruit and Vegetable Consumption

There are numerous economic and societal benefits to increasing fruit and vegetable consumption across a population. In addition to the well-documented health benefits, from an economic perspective, there are the 'bottom line' financial rewards for industry, employment opportunities across many sectors, and long-term health care cost savings. The horticulture and related industries aim to supply produce that will deliver consumer satisfaction and, thus, improve industry profitability. The horticulture sector and related industries influence the supply, demand, and quality of food and are leading domestic market growth in many countries [34]. The food service sector is an important influence as market share is increasing and there is opportunity to increase the provision of fruit and vegetables in some countries [40].

Food distribution impacts the environment – the supply chain results in 'food miles' [41]. Fruit, vegetables, and horticulture overall are probably the least impacting of all agriculture in terms of emissions. There is some suggestion they are actually carbon positive. The impact of trade practices also needs to be considered [42]. Fresh produce and seafood have the most volatile prices due to their distribution systems, transportation paths, product perishability, and seasonality [21].

Public policy has a role in determining which fruits and vegetables are produced and consumed, for example price corrections through taxes or levy systems to assist production. Strategies can be employed to 'improve the long term productivity, diversity and quality of fruit and vegetable production' [1]. National horticultural research and government agriculture agencies should assist the grower community to develop and implement interventions [1].

Interventions can address all aspects of the supply chain, for example transport cost and handling procedures, and how to add value to the current market. The FAO and the WHO encourage countries to engage many stakeholders to ensure links with promotional programs and adopting practices that ensure product quality and safety [1].

Elements which could be included in a framework for action include:

- advocacy, information and decision support
- supporting urban/peri-urban horticulture
- promoting sustainable production and enhancing efficiency of production factors
- preventing post-harvest losses and enhancing value of fresh produce
- ensuring food safety and quality
- strengthening trade intelligence, marketing, processing, and financial services
- promoting research and technology
- investing in nutrition education, e.g. in rural, home gardens, school feeding and gardens. The FAO/WHO workshop, p. 11 [1]

2. THE COMMON OBJECTIVE AND THE POINTS OF DIFFERENCE

The common objective 'to increase consumption of fruit and vegetables' is a good reason for health and industry to work together.

From a health perspective, dietary recommendations encourage eating more fruit and vegetables and eating a wide range of fruit and vegetables, in a variety of forms, particularly fresh. The fruit and vegetable industry is disparate, with many players who are often in direct competition for the same consumer market (for example, growers of carrots versus celery). Industry is often skeptical of a generic fruit and vegetable promotion, with a preference for specific product or retailer promotion [35], reinforced by the need to show commodity producers or business owners a return on the marketing investment made. However, successful promotion of the consumption of fruit and vegetables will necessitate commercial operators and others – some of whom are in direct competition – to collaborate on issues of mutual benefit [43]. In fact, 'collaborative competition' is required.

The right mix of organizational commitment, leadership, relationships, opportunities, and capacity to achieve results is required [35].

A number of factors are considered essential preconditions to successful collaborative health action: necessity, opportunity, capacity to work together, relationships, planned action, evaluation, and sustainability of outcomes [29,35].

2.1 The Food Supply and Health

Many of the dietary factors that influence health are modifiable, therefore leading to the acknowledgement of preventable diet-related disease. Governments and health authorities are actively working on aligning the food supply chain with healthy diets. All players aim to identify leverage points along the supply chain that can lead to increases in consumption. For example, in the effort to address the obesity epidemic, the Chinese government identified a

need to 'enhance scientific guidance in the fields of agriculture, food manufacturing, distribution and marketing to make them play more important roles in improving people's nutrition and health status' [44]. See Figure 14.1 for examples of modifying factors at an individual and environmental level along the agro-food supply chain and health.

2.2 The Fruit and Vegetable Supply

The fruit and vegetable supply chain is complex and interrelated. There are many potential points to influence consumer consumption. In response to calls for increasing the provision of fruits and vegetables, the food industry often expresses the comment, 'we provide what consumers demand.' In relation to the food supply

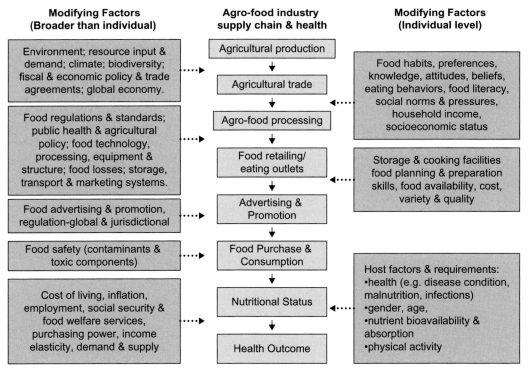

FIGURE 14.1 The food system – identifying agro-food supply and health relationships. Adapted from references 44 and 45.

chain, the 'consumer' context is relevant. It has been suggested that the whole concept of consumer power and consumer choice should be scrutinized in relation to the food supply chain [46]. The processed food manufacturers are the 'consumers' of much of the primary agricultural produce and imports. Supermarkets buy from farms, food processors, and wholesaler. Food services – catering outlets (including fast food chains, public or institutional caterers, e.g. schools and hospitals) – are the purchasers of large amounts of the food passing along the system and influence the types of foods that are on menus. See Figure 14.2 to consider the complex relationships influencing fruit and vegetable supply and demand.

2.3 Industry Trends in Fruit and Vegetable Supply

Supply-side and demand-driven factors contribute to fruit and vegetable consumption. The specific factors vary between countries, and between developed and developing countries. This discussion is based on trends in developed countries. Horticulture industries collect market intelligence to assist in their strategic development. These industries collect information on consumer attitudes to and perceptions of specific fruit and vegetable categories, food purchasing habits, perceptions of quality, taste, and value.

On the supply side, improvements in technology for shipping, handling, plant breeding, and packing mean that produce is less perishable and maintains appearance and quality. Consumers now have more variety available for a much longer period of the year, for example grapes, pears, and strawberries. New varieties with consumer-preferred attributes are available, for example seedless grapes and watermelon [47].

On the demand side, food industry data supports the notion that convenience is an important and increasing factor determining fruit and vegetable intake [47]. The food service

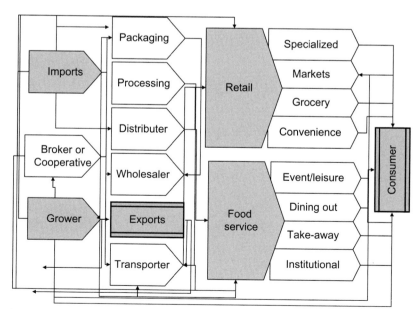

FIGURE 14.2 The fruit and vegetable supply chain

sector is emerging as an important influence as market share is increasing and there is opportunity to increase the provision of fruit and vegetables [40]. Figure 14.3 highlights consumer trends and the main interests of health and industry.

Work in the USA over the last two decades has identified a number of factors that were related to the increase in domestic consumption of fruit and vegetables, specifically relating to the improvements in: availability and quality of produce all year round; addressing the convenience aspect; consumer desire for health benefits [49]. Trends in the increased consumption of specific types of fruit and vegetables point to these factors. Since the mid 1970s food preferences have changed with Americans eating more fresh or frozen and less canned product, and food stores have a greater selection and quantity of fresh produce on display. Consumers are eating more asparagus, broccoli, and types of lettuces. The supply of some products relies on imports, for example, bananas in America.

Consumer preferences, technological innovations, and globalization have affected the volume of sales, price, and quality of fresh fruits and vegetables. Electronic commerce allows for improved grower–retailer communications and marketing of produce. Shelf life and quality of perishable products have improved due to improvements in the transport system. As a result, supermarkets often have a year-round supply of produce varieties, pre-cut produce, and more packaged and branded items. These changes have had profound effects on the way the produce industry is organized and the way it conducts business [42]. Today's supermarkets carry twice as many types of produce than they did a decade ago. Fresh-cut produce

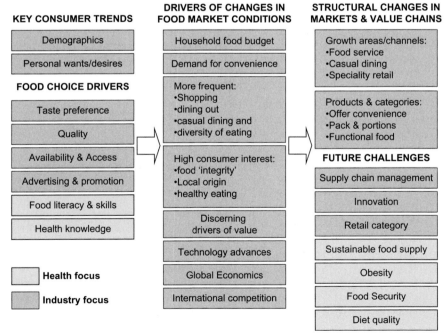

FIGURE 14.3 Health and industry priorities on food supply market trends. Adapted from references 40 and 48.

is growing rapidly, with the proportion of sales of prepackaged salads having doubled. Fruit and vegetables are available all year around in many parts of the world due to improved production, storage, and transportation allowing for the availability of produce during counter seasons in both hemispheres.

3. POINTS OF INFLUENCE IN THE FOOD SUPPLY CHAIN

The objective: 'to increase and sustain access to high quality, safe, affordable fruit and vegetables'

There are many points along the food supply chain where consumption can be influenced (Table 14.1). As well as changes to the supply side factors to influence fruit and vegetable intake, industry partners are key to influencing changes in knowledge, attitudes, and skills to prepare fruit and vegetables. It is important to identify the specific objectives of strategies to increase consumption, for example awareness (increase the proportion of the population aware of the need to increase their consumption of fruit and vegetables); attitude/perceptions (increase the proportion of the population who perceive the benefits of fruit and vegetables in terms of taste, convenience, low relative cost, safety, and health); knowledge (increase the proportion of the population with the knowledge of the recommended minimum intakes of fruits and vegetables); and skills to purchase and prepare (increase the proportion of the population with knowledge, skills, and confidence to select and prepare convenient, low-cost, tasty vegetable and fruit dishes).

One of the main benefits of establishing partnerships to promote fruit and vegetable consumption is that each sector is aware of the barriers and opportunities that exist within

TABLE 14.1 Examples of Leverage Points to Increase Fruit and Vegetable Consumption along the Food Supply Chain

Partnership activities

Agricultural production

Incentives (financial and non-financial)

Planning incentives supporting fruit and vegetables food production

Agricultural trade

Policy

Conduct health impact assessments focused on impact of fruit and vegetable consumption on trade decisions

Agro-food processing

Plant and equipment

Appropriate fruit and vegetable storage facilities and handling practices

Food retailing/eating outlets

Incentives (financial and non-financial)

Incentive or food provision programs to increase access of fruit and vegetables through food services – worksites, childcare, hospitality, retailer, or transport owner accreditation schemes

Community and organizational policy

Local and organizational food and nutrition policies to increase access to fruit and vegetables

Advertising and promotion

Legislation and regulation

Legislate to curb advertising and promotion of foods that displace fruit and vegetables – restrict food advertising directed at children

Communication

Conduct targeted fruit and vegetable social marketing campaigns

Food purchase and consumption

Point of sale

Tastings and demonstrations

Infrastructure support

Information systems

Improve communication of fruit and vegetable marketing and information through the supply chain, e.g. availability, price, quality, volume statistics

Identification and surveillance of determinants

Collect and analyze information to assess fruit and vegetable supply, cost, quality, access, sales/marketing their sector. Sharing this information with other sectors can assist in identifying solutions beyond the reach of one sector alone.

Successful initiatives to increase fruit and vegetable consumption require both effective promotional interventions and initiatives to address the underlying structural factors. Structural supports for public health initiatives include: research to assist the development and assess the potential impact of health and non-health interventions; evaluation and monitoring of change; development and implementation of policy to support initiatives addressing structural barriers; adequate and sustained resourcing (both human resources and funding arrangements); and a strategic management structure to manage coalitions at national, state, and local levels [50]. These are similar to those identified at the first international meeting of 5-A-Day managers to promote fruit and vegetables, Figure 14.4 [51].

Inter-sectoral action is essential for the effective promotion of fruit and vegetable consumption [3,15,35,37–39]. The right mix of organizational commitment, leadership, relationships, opportunities, and capacity to achieve results is required [35]. A number of factors are considered essential preconditions to successful collaborative health action:

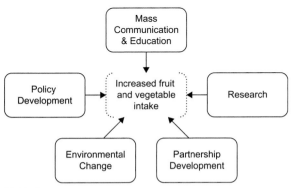

FIGURE 14.4 Effective programs to increase fruit and vegetable intake.
Adapted from Morten Strunge Meyer [51].

necessity, opportunity, capacity to work together, relationships, planned action, evaluation, and sustainability of outcomes [29,35].

3.1 Approaches to Developing Partnerships

Many countries have formed cooperative partnerships to implement strategies to increase fruit and vegetable consumption. The partnerships are formed in recognition of the common agenda to increase fruit and vegetable consumption. The intent of health–industry fruit and vegetable partnerships is clearly to increase the consumption of fruit and vegetables to promote health. These partnerships may underpin population-based campaigns and promotions of a health-based message. As such, the arrangements require an agreement on the objectives of the collaboration, the nature of activities and the recognition of points of difference (if possible).

The International Fruit and Vegetable Alliance (IFAVA) was formed in recognition of the role of health in the promotion of fruits and vegetables globally and the need for industry and health to work together. IFAVA aims to 'foster efforts to increase global fruit and vegetable consumption for better health through promoting efficiencies, facilitating collaboration on shared aims and providing global leadership – all of which is based on sound science.' For more information on the role of IFAVA, visit http://www.ifava.org

The New Zealand '5+ A Day' Fruit and Vegetable Alliance aims to improve public health. The Alliance aims to provide strong leadership, a united voice for promoting the allocation of resources to support a national sustainable campaign, consistent and targeted messages, information sharing, synergy of efforts, reduction in duplication, evidence and evaluation to guide activities, a consumer focus, and an agreement to work at all levels of society (individual, family, interpersonal, organizational, industrial, community, media, and policy).

The Australian *Go for 2&5®* campaign invites partners to undertake a formal agreement to improve public health through a collaborative, coordinated, cooperative approach. The language chosen in the describing of partnerships has implications for how the work is undertaken. For example, in contemporary language, 'collaboration' refers abstractly to all processes wherein people work together – applying both to the work of individuals as well as larger collectives and societies; 'cooperation' refers to the practice of people or entities working with commonly agreed upon goals and possibly methods, instead of working separately in competition; coordination is the regulation of diverse elements into an integrated and harmonious operation – essentially, integrating or linking together different parts of an organization to accomplish a collective set of tasks [52].

3.1.1 Components of Effective Partnerships

There are many models designed to assist in developing effective inter-sectoral partnerships for action. One model identifies important dimensions to inter-sectoral action including: the *necessity* for the sectors or organizations to work together; the factors that are providing the *opportunity* for them to work together; the *capacity* to work together; established *relationships* that will allow them to achieve their goal; the degree of *planning* of the action and the potential for *evaluation*; and the *sustainability* of the action [53]. Common or shared visions, mutual need, shared decisions, shared benefits, and risk have also been identified as important. These types of models for analysis of inter-sectoral action are a constructive tool for identifying potential strengths and weaknesses of the health and industry partnerships.

3.2 Formalizing the Relationship

Campaigns to promote fruit and vegetables are usually supported by formalized licensing agreements to protect the brand or message, and ensure compliance with the health intent. There are many countries in which industry and health work together through formalized relationships to achieve increased fruit and vegetable consumption. Licensing frameworks and campaign materials, including resources to assist the management of the brand and the nutrition aspects of the campaign, need to be established.

> Promotion campaigns, such as USDA's Food Guide Pyramid and the 5-A-Day for Better Health program, improved produce quality, increased variety and year-round availability. They have also boosted consumption of fresh fruit and vegetables.
> Kaufman et al., p. 15 [42]

The purchase of fruit and vegetables in the USA doubled between 1987 and 1997 and consumption increased by 12.3% to 319 pounds per capita [42]. The choice increased, the number of individual produce items increased by 94%, and food service produce sales increased 62% over that time period. The Department of Agriculture found that consumers were more aware of the health benefits of fruit and vegetables, and were responding by increasing consumption.

3.3 Protecting the Brand

Establishment of a brand amplifies the effect of individual promotions and enables expansion of activities across many communities. Formal agreements aim to ensure the brand is protected, that the standard of the campaign is managed, and assist with communication between partners. The national *5 A-Day* logo and licensing scheme, which formalizes the multilevel public and private partnerships, has

---

5 TO 10 A-DAY CANADA

The Canadian fruit and vegetable program is privately funded by the produce industry and non-government organizations and is supported via a trilateral partnership agreement with the two national health organizations: the Canadian Cancer Society and the Heart and Stroke Foundation of Canada. http://www.5to10aday.com

---

been acknowledged as the key to the success of the program in the USA[3].

It has been estimated that there was a four-fold return in promotional value for every dollar invested by the National Cancer Institute in the *5 A-Day* scheme in the USA, and in Australia the return was estimated at $26 AUD of promotional investment for every dollar invested [3,54].

3.4 Protecting the Health Message

Health and horticulture have a common agenda when it comes to encouraging consumers to eat more fruit and vegetables; however, the health sector requires that fruit and vegetables are promoted in the 'total diet' context. Dietary guidelines encourage eating patterns to reduce the risk of diet-related disease and improve population well-being, that is, encouraging a wide variety of fruit and vegetables and other recommended food groups while limiting dietary sodium, fat, alcohol, and added sugar. This remains at odds with many industry-based marketing campaigns that focus on individual products or at best groups of products such as 'Summerfruit' or 'vegetables' or engage in 'cross-promotions' with other less nutritious products, for example those high in fat, added sugar, or salt. It is essential, therefore, that as partnerships are formed they align on the health message and agree to protect the nutritional integrity with any activity.

Most campaigns have some nutrition guidelines or criteria that must be agreed to when being allowed to carry or promote the brand. Usually these criteria promote fruit and vegetables in a way consistent with the country's dietary recommendations. Food companies that wish to develop, label, advertise, and promote food or recipes as 'healthy' or as part of a campaign must demonstrate that product nutrition profiles meet agreed nutrient targets [55]. Specific nutrition criteria in line with health policy were required to assist industry partners to implement campaign messages. The criteria aimed to identify fruit, vegetables, and recipes that could be promoted using the campaign logos. The UK and US *5 A-Day* and the Australian *Go for 2&5*® campaigns supply nutrition criteria to partner organizations to assist their promotion of fruit and vegetables in the context of a healthy diet [56–58].

Dietary guidance encourages consumption of whole foods containing essential nutrients, while limiting dietary sodium, fat, and added sugars. Guidelines recommend an appropriate energy intake to reduce the risk of overweight and obesity but do not assess the 'healthiness' of individual food items. Food selection guides provide a conceptual framework for selecting the types and amounts of foods to meet dietary recommendations. They usually provide food group-based recommendations, specify what constitutes a food group, and give standard serving sizes with visual representations. Campaign nutrition criteria usually require the

5 A-DAY USA

Awareness of the *5 A-Day* message among adults over 18 years in the USA increased from 7.7% in 1991 to 19.2% in 1997 and knowledge of the program increased from 2% to 17.8% [59]. When reflecting on the *5 A-Day* Program in the USA, it was considered that the development of the *5 A-Day* message and logo was one of the best decisions made [3]. Permission to use the 'brand' was integral to licensing arrangements that supported partners to implement the campaign.

GO FOR 2&5® AUSTRALIA

Evaluation of the 2003 Australian *Go for 2&5®* fruit and vegetable campaign found 70% awareness, increases in knowledge of recommended amount of vegetables, and attempts to increase consumption [60]. The industry–government partnership formed to develop this campaign was integral [35]. Campaigns need to be maintained to achieve sustained change [61]. An Australian government review of intervention to promote fruit and vegetables recommended a national social marketing campaign as an effective intervention to promote fruit and vegetables [61]. http://www.gofor2and5.com.au/

The US *5 A-Day* program and the Department of Health in Western Australia's fruit and vegetable campaigns commenced at about the same time in the late 1980s [3,62]. Adequate resources and long-term investment are required to sustain interventions. Experience from these campaigns shows that intervention activity and outcomes are related to the resources invested [3,63]. Partnerships are critical: each partner takes the lead in a range of areas, securing funding, ensuring ongoing activities, identifying new opportunities, and ongoing developments.

inclusion of specific amounts of fruit and/or vegetables in products or recipes to be promoted or endorsed, as well as cut-off levels for nutrients or ingredients to be limited (fat, sugar, alcohol, or salt) [58].

4. FUTURE CHALLENGES

Environmental effects of climate change resulting in changes to natural systems that result in nutrition and diet-related health impacts (e.g. food-producing systems, affecting yields and nutritional quality; food yields, biodiversity of the food supply; reduced food yields and affordability, leading to dietary imbalances and poor nutrition) are just some of the challenges faced by partnerships to encourage increased consumption of fruit and vegetables. Countries need to consider global influences on food and nutrition policy, particularly in relation to food production, distribution, advertising, and promotion [64]. Nutrition policy also has significant implications for agriculture and trade [64]. There is an increasing emphasis on the need for food policy to support the production, access, promotion, and equitable consumption of health-promoting foods [65]. Food security is a priority, that is, food availability, access to food, stability of supply, and safe and healthy food utilization [66]. Climate change, the growing use of food crops as a source of fuel, and soaring food prices threaten to temper efforts to overcome food insecurity and malnutrition.

Inter-sectoral partnerships, working collaboratively to achieve an agreed vision, will provide the foundation, commitment, leadership, and resources required to face these challenges.

References

1. World Health Organization and Food and Agriculture Organization. (2004). *Fruit and vegetables for health. Report of a joint FAO/WHO workshop, 1–3 September 2004, Kobe, Japan.* WHO and FAO.
2. Marshall, D., Anderson, A., Lean, M., & Foster, A. (1995). Eat your greens: The Scottish consumer's perspective on fruit and vegetables. *Health Education Journal, 54,* 186–197.
3. National Institutes of Health and National Cancer Institute. (2001). *5 A Day for Better Health Program Monograph.* Damascus: National Cancer Institute.
4. Swinburn, B., Egger, G., & Raza, F. (1999). Dissecting obesogenic environments: The development and application of a framework for identifying and prioritizing environmental interventions for obesity. *Preventive Medicine, 29,* 563–570.
5. Gibson, E. L., Wardle, J., & Watts, C. J. (1998). Fruit and vegetable consumption, nutritional knowledge and beliefs in mothers and children. *Appetite, 31,* 205–228.
6. Cox, D. N., Anderson, A. S., Lean, M. E., & Mela, D. J. (1998). UK consumer attitudes, beliefs and barriers to increasing fruit and vegetable consumption. *Public Health Nutrition, 1,* 61–68.
7. Kamphuis, C. B. M., Giskes, K., de Bruijn, G.-J., Wendel-Vos, W., Brug, J., & van Lenthe, F. J. (2006). Environmental determinants of fruit and vegetable consumption among adults: A systematic review. *British Journal of Nutrition, 96,* 620–635.
8. Maclellan, D. L., Gottschall-Pass, K., & Larsen, R. (2004). Fruit and vegetable consumption: Benefits and barriers. *Canadian Journal of Dietetic Practice and Research, 65,* 101–105.
9. Bogers, R. P., Assema, P. V., Brug, J., Kester, A. D. M., & Dagnelie, P. C. (2007). Psychosocial predictors of

10. increases in fruit and vegetable consumption. *American Journal of Health Behavior, 31,* 135–145.
10. Dibsdall, L. A., Lambert, N., Bobbin, R. F., & Frewer, L. J. (2003). Low-income consumers' attitudes and behaviour towards access, availability and motivation to eat fruit and vegetables. *Public Health Nutrition, 6,* 159–168.
11. Shohaimi, S., Welch, A., Bingham, S., Luben, R., Day, N., Wareham, N., & Khaw, K.-T. (2004). Residential area deprivation predicts fruit and vegetable consumption independently of individual educational level and occupational social class: A cross sectional population study in the Norfolk cohort of the European Prospective Investigation into Cancer (EPIC-Norfolk). *Journal of Epidemiology and Community Health, 58,* 686–691.
12. Ball, K., Crawford, D., & Mishra, G. (2006). Socioeconomic inequalities in women's fruit and vegetable intakes: A multilevel study of individual, social and environmental mediators. *Public Health Nutrition, 9,* 623–630.
13. Devine, C. M., Wolfe, W. S., Frongillo, E. A., Jr., & Bisogni, C. A. (1999). Life-course events and experiences: Association with fruit and vegetable consumption in 3 ethnic groups. *Journal of the American Dietetic Association, 99,* 309–314.
14. Beydoun, M. A., & Wang, Y. (2008). Do nutrition knowledge and beliefs modify the association of socioeconomic factors and diet quality among US adults? *Preventive Medicine, 46,* 145–153.
15. Miller, M., Shiell, A., & Stafford, H. (2000). *An intervention portfolio to promote fruit and vegetable consumption. Part 1 – the process and portfolio.* Melbourne: National Public Health Partnership.
16. Booth, K. M., Pinkston, M. M., & Poston, W. S. (2005). Obesity and the built environment. *Journal of the American Dietetic Association, 105,* S110–S117.
17. Giskes, K., Kamphuis, C. B. M., van Lenthe, F. J., Kremers, S., Droomers, M., & Brug, J. (2007). A systematic review of associations between environmental factors, energy and fat intakes among adults: Is there evidence for environments that encourage obesogenic dietary intakes? *Public Health Nutrition, 10,* 1005–1017.
18. French, S. A., Story, M., Jeffery, R. W., Snyder, P., Eisenberg, M., Sidebottom, A., & Murray, D. (1997). Pricing strategy to promote fruit and vegetable purchase in high school cafeterias. *Journal of the American Dietetic Association, 97,* 1008–1010.
19. Pearson, T., Russell, J., Campbell, M. J., & Barker, M. E. (2005). Do 'food deserts' influence fruit and vegetable consumption? – A cross-sectional study. *Appetite, 45,* 195–197.
20. Latham, J., & Moffat, T. (2007). Determinants of variation in food cost and availability in two socioeconomically contrasting neighbourhoods of Hamilton, Ontario, Canada. *Health Place, 13,* 273–287.

21. McLaughlin, E. W. (2004). The dynamics of fresh fruit and vegetable pricing in the supermarket channel. *Preventive Medicine, 39*(Suppl 2), S81–S87.

22. Knai, C., Pomerleau, J., Lock, K., & McKee, M. (2006). Getting children to eat more fruit and vegetables: A systematic review. *Preventive Medicine, 42,* 85–95.

23. Giskes, K., Turrell, G., Patterson, C., & Newman, B. (2002). Socioeconomic differences among Australian adults in consumption of fruit and vegetables and intakes of vitamins A, C and folate. *Journal of Human Nutrition and Dietetics, 15,* 375–385; discussion 387–390.

24. Brug, J., Debie, S., van Assema, P., & Weijts, W. (1995). Psychosocial determinants of fruit and vegetable consumption among adults: Results of focus group interviews. *Food Quality and Preference, 6,* 99–107.

25. Lee, A. J., Darcy, A. M., Leonard, D., Groos, A. D., Stubbs, C. O., Lowson, S. K., Dunn, S. M., Coyne, T., & Riley, M. D. (2002). Food availability, cost disparity and improvement in relation to accessibility and remoteness in Queensland. *Australian and New Zealand Journal of Public Health, 26,* 266–272.

26. National Public Health Partnership. (2000). *A planning framework for public health practice. Public health planning and practice improvement.* Melbourne: National Public Health Partnership.

27. Robinson, N., Caraher, M., & Lang, L. (2000). Access to shops: The views of low-income shoppers. *Health Education Journal, 59,* 121–136.

28. NPHP (2001). Hawkesbury district health service, Hawsbury food program. *Food Chain, 6,* 9–11.

29. Webb, K., Hawe, P., & Noort, M. (2001). Collaborative Intersectoral Approaches to Nutrition in a Community on the Urban Fringe. *Health Education & Behavior, 28,* 306–319.

30. *NSW Centre for Public Health Nutrition.* (2003). Food Security Options Paper: A planning framework and menu of options for policy and practice interventions. NSW Department of Health, Australia. < http://www.health.nsw.gov.au/pubs/2003/food_jsecurity.html > Accessed 17.08.09.

31. Nestle, M. (2003). *Food politics: How the food industry influences nutrition and health.* California University of California Press.

32. Food Standards Australia New Zealand (2006). Health claims proposal and sugar levels of fruit. http://www.foodstandards.gov.au/newsroom/factsheets/factsheets2006/healthlaimsproposal3128.cfm Accessed 17.08.09.

33. Food Standards Australia New Zealand. (2007). *Nutritional Information Requirements Australia New Zealand food standards code. Issue 67 Standard 1.2.8 nutrition information requirements.* ANSTAT Pty Ltd. http:// www.foodstandards.gov.au/thecode/foodstandardscode/standardl28nution4235.cfm Accessed 17.08.09.

34. Lea, E., Worsley, A., & Crawford, D. (2005). Australian adult consumers' beliefs about plant foods: A qualitative study. *Health Education & Behavior, 32,* 795–808.

35. Miller, M., & Pollard, C. (2005). Health working with industry to promote fruit and vegetables: A case study of the Western Australian fruit and vegetable campaign with reflection on effectiveness and inter-sectoral action. *Australian and New Zealand Journal of Public Health, 29,* 176–182.

36. World Health Organization. (2003). *World Health Assembly resolution WHA57.17 - Global strategy on diet, physical activity and health.* Geneva: WHO.

37. Glanz, K., & Hoelscher, D. (2004). Increasing fruit and vegetable intake by changing environments, policy and pricing: Restaurant-based research, strategies, and recommendations. *Preventive Medicine, 39*(Suppl 2), S88–S93.

38. (2003). European partnership for fruits, vegetables and better health. http://epbh.org/about/index.shtml Accessed 17.08.09.

39. (2005). The international fruit and vegetables alliance. http://www.ifava.org/ Accessed 17. 08.09.

40. Australian Government Department of Agriculture, Fisheries and Forestry. (2006). FOODmap. A comparative analysis of Australian food distribution channels. *Commonwealth of Australia* http://www.daff.gov.au/_data/assets/pdf_file/0003/298002/foodmap-full.pdf Accessed 17.08.09.

41. Lea, E. J. (2005). Food, health, the environment and consumers' dietary choices. *Australian Journal of Nutrition and Dietetics, 62,* 21–25.

42. Kaufman, P., Handy, C., McLaughlin, E., Park, K., & Green, G. M. (2000). Understanding the dynamics of produce markets: Consumption and consolidation grow. *Agriculture Information Bulletin,* No (AIB758).

43. Pollard, C., & Rowley, C. (2006). *Coordinated, collaborative, cooperative – approach to increasing fruit and vegetable consumption workshop report, February to March 2006.* Strategic Inter-Governmental Nutrition Alliance and Horticulture Australia Ltd.

44. Hawkes, C. (2008). Agro-food industry growth and obesity in China: what role for regulating food advertising and promotion and nutrition labelling? *Obesity Reviews, 9*(Suppl 1), 151–161.

45. Australian Government. (1992). South Australia food and nutrition policy. Department of Human Services. http://www.health.gov.au/internet/main/publishing.nsf/Content/phd-nutrition-fnp-1992 Accessed 17.08.09.

46. Lobstein, T. (2002). Suppose we all ate a healthy diet…could our food supplies cope? *Eurohealth, 10*, 8–12. http://www.sustainweb.org/pdf/afn_m6_)p2.pdf. Accessed 17.08.09.

47. Pollack S. (2005). Consumer Demand for Fruit and Vegetables: The U.S. Example. In Economic Research Service/USDA (Changing Structure of Global Food Consumption and Trade. http://www.ers.usda.gov/publications/wrs011/wrs0llh.pdf Accessed 17.08.09.

48. Australian Government. (2007). *Australian food statistics 2006*. Department of Agriculture, Fisheries and Forestry, Commonwealth of Australia.

49. Nebeling, L., Yaroch, A. L., Seymour, J. D., & Kimmons, J. (2007). Still not enough: Can we achieve our goals for Americans to eat more fruits and vegetables in the future? *The American Journal of Preventive Medicine, 32*, 354–355.

50. Strategic International Nutrition Alliance. (2001). *Eat well Australia: A strategic framework for public health nutrition/national aboriginal & Torres Strait Islander nutrition strategy & action plan*. Canberra: National Public Health Partnership.

51. DiSogra, L., Dudley, P., Strunge Meyer, M. (2004). Effective strategies for enhancing partner collaboration and resource sharing. Workshop for experienced 5 A-Day managers. International 5 A-Day Symposium, Christchurch, New Zealand.

52. Wikipedia®. (2009). *Wikipedia® the free encyclopedia.* Wikimedia Foundation, Inc.

53. Harris, E., Wise, M., & Hawe, P. (1995). *Working together: Intersectoral action for health.* National Centre for Health Promotion & Commonwealth Department of Human Services, Australian Government Publishing Service.

54. Department of Health in Western Australia. (2007). Go for 2&5®: Increasing fruit and vegetable consumption. *Food Chain, 18,* 8.

55. Trichterborn, J., & Harzer, G. (2007). An industry perspective on nutrition profiling in the European environment of public health and nutrition. *Nutrition Bulletin, 32,* 295–302.

56. Produce for Better Health Foundation. (2004). 5 A Day – the color way. www.5aday.com.

57. National Health Service. (2004). 5 A Day. http://www.5aday.nhs/toptips/default.htm.

58. Pollard, C., Nicolson, C., Pulker, C. E., & Binns, C. W. (2009). Translating government policy into recipes for success! Nutrition criteria promoting fruit and vegetables. *Journal of Nutrition Education and Behavior, 41,* 218–226.

59. Stables, G. J., Subar, A. F., Patterson, B. H., Dodd, K., Heimendinger, J., Van Duyn, M. A., & Nebeling, L. (2002). Changes in vegetable and fruit consumption and awareness among US adults: results of the 1991 and 1997 5 A Day for Better Health Program surveys. *Journal of the American Dietetic Association, 102,* 809–817.

60. Woolcott Research Pty Ltd. (2007). *Evaluation of the national go for 2&5® campaign*. Australian Government. Department of Health and Aging, Australia. http://www.health.gov.au/intemethealthyactive/publishing.nsf/content/EAED0B3283A8AlE2CA257259007CFC2D/$File/2&5-eval-jan06.pdf.

61. Miller, M., & Stafford, H. (2000). *An intervention portfolio to promote fruit and vegetable consumption. Part 2 – review of interventions*. Melbourne: National Public Health Partnership.

62. Miller, M., Pollard, C., & Paterson, D. (1996). A public health nutrition campaign to promote fruit and vegetables in Australia. In A. Worsley, (Ed.), *Multidisciplinary approaches to food choice. Proceedings of food choice conference* (pp. 152–158). Adelaide: University of Adelaide.

63. Dixon, H., Borland, R., Segan, C., Stafford, H., & Sindall, C. (1998). Public reaction to Victoria's '2 Fruit 'n' 5 Veg Every Day' campaign and reported consumption of fruit and vegetables. *Preventive Medicine, 27,* 572–582.

64. Norum, K., Johansson, L., Botten, G., Bjorneboe, G., & Oshaug, A. (1997). Nutrition and food policy in Norway: Effects on reduction of coronary heart disease. *Nutrition Reviews, 55,* S32–S39.

65. McMichael, A. J. (2005). Integrating nutrition with ecology: Balancing the health of humans and biosphere. *Public Health Nutrition, 8,* 706–715.

66. Cohen, M., Tirado, C., Aberman, N., & Thompson, B. (2008). *Impact of climate change and bioenergy on nutrition*. International Food Policy Research Institute, FAO.

EFFECTS OF INDIVIDUAL VEGETABLES ON HEALTH

15

Garlic and Aging: Current Knowledge and Future Considerations

Carmia Borek

Department of Public Health and Family Medicine, Tufts University School of Medicine, Boston, MA, USA

1. INTRODUCTION

Of all the herbal remedies consumed for their health benefits, garlic ranks the highest, both in popularity and range of efficacy. Garlic (*Allium sativum*), one of the oldest plants used in medicine, has been an important part of life for centuries, across cultures and millennia. Garlic has been used to spice food, fortify soldiers for war, cure colds, heal infections, and treat ailments ranging from heart disease to cancer and even the plague.

Today, after close to 6000 years of folklore, modern science has confirmed many of garlic's benefits. With its rich source of phytochemicals, largely organosulfur compounds, and high antioxidant activity, garlic has proven to have a broad range of health benefits and anti-aging effects, helping prevent disease and age-related pathological conditions. Scientific and clinical studies have shown that garlic can enhance immunity, protect against infection and inflammation, and help lower the risk of

cancer, heart disease, and dementia, the most common form of which is Alzheimer's disease. Scientific studies also show that garlic does not have to be eaten raw or fresh to be effective, that its potent odor is not needed for its health benefits. Research shows that aged, deodorized garlic extract (kyolic), that is highly rich in anti-oxidants [1–6], often works even better than fresh garlic, without causing digestive disorders and 'garlic breath.'

2. A HISTORY OF GARLIC

Garlic is a hardy perennial plant that belongs to the lily family. Although the exact geographic origin of garlic is not known, modern botanists think it came from Central Asia, some say Siberia, where it was discovered and cultivated by people foraging in the fields for food and healing herbs. Garlic was transported east and west by migrating tribes, becoming native to the Near East, Europe, the Far East, and Africa. Its characteristic white bulb is devoid of smell and flavor when intact, but when cut it is pungent both in taste and smell.

Remnants of garlic have been found in cave dwellings that are over 10 000 years old. Egyptian tombs, dating back to close to 5700 years ago, were found to contain sketches of garlic and clay sculptures of the bulb. The ancient Egyptian text Codex Ebers details formulas with garlic as remedies for heart problems, headaches, tumors and other ailments. Chinese writings dating from 2700 BC describe garlic for treating many ailments and for enhancing vigor. In India, Ayurvedic medicine recommends garlic to boost energy and treat colds and fatigue.

In modern times garlic has become a popular health-promoting herb in the Far and Near Eastern countries, Europe, and the USA. In certain parts of China people eat about 20 grams of garlic a day, approximately eight medium-sized cloves. In Germany, most adults take a daily garlic supplement to promote health. In the USA the use of garlic supplements has been rapidly escalating over the years, with the most popular supplement being the odorless aged garlic extract [1].

3. THE COMPOSITION AND CHEMISTRY OF GARLIC

The composition of garlic is complex, with over 200 different compounds that contribute to its effects. The most important and unique feature of garlic is its high content of organosulfur substances. Garlic contains at least four times more sulfur than other high sulfur vegetables – onion, broccoli, and cauliflower. Garlic also contains fructose-containing carbohydrates, protein, fiber, saponins, phosphorus, potassium, zinc, moderate amounts of selenium and vitamin C, steroidal glycosides, lectins, prostaglandins, essential oil, adenosine, vitamins B1, B2, B6, and E, biotin, nicotinic

acid, fatty acids, glycolipids, phospholipids, anthocyanins, flavonoids, phenolics, and essential amino acids [1–6].

3.1 Organosulfur Compounds

Depending on the conditions of its cultivation, garlic may contain at least 33 different organosulfur compounds. The allyl sulfur constituents in garlic are largely responsible for its health benefits. The major allyl sulfur content in freshly crushed/chopped/cut garlic is allicin, which is unstable and breaks down rapidly to produce odorous oil-soluble di-allyl sulfide, di-allyl disulfide, di-allyl trisulfide, and ajoene. The major allyl sulfur constituents in aged garlic extract (AGE) include S-allyl-cysteine and S-allyl-mercaptocysteine, which are water soluble and formed by the process of natural aging bioconversion [1–6].

From a medicinal point of view, and effectiveness in preventing age-related conditions, the stable water-soluble organosulfur compounds in garlic are highly effective. Present to some extent in fresh garlic, their level is increased by a process of aging [1–4].

4. COMMERCIALLY AVAILABLE GARLIC PREPARATIONS

Modern scientific findings on the medicinal benefits of garlic led to the development of garlic preparations that are available commercially. These include kyolic aged garlic extract (AGE), an odorless supplement produced by extracting and aging organically grown garlic, for 20 months, in an alcohol solution, at room temperature [1–4]. The extraction and aging process produces a high content of water-soluble organosulfur compounds, and converts unstable compounds, such as allicin, to stable compounds, thus increasing the antioxidant content of AGE, compared to fresh garlic [1–5].

The major organosulfur compound in AGE is S-allyl cysteine. It is stable, has high antioxidant activity, and is highly bioavailable, with an absorption of close to 98% into the circulation. S-allyl cystine is used for commercial standardization of the AGE products including tablets, capsules, and liquid forms. S-allyl mercaptocysteine, unique to AGE, is also a water-soluble organosulfur antioxidant compound, formed during the process of extraction and aging [1–3], and similar to S-allyl cysteine has a wide range of health and anti-aging benefits. Aged garlic extract also contains some lipid-soluble organosulfur compounds.

Non-sulfur compounds in AGE include proteins, carbohydrates (sugars, fructans, pectins), saponins, that are steroid substances are shown to have antibacterial and antifungal actions, allixin, and fructosyl arginin [4], that are important antioxidants [1–4].

Other commercially available garlic supplements include garlic powder tablets, that are said to contain a specified amount of allicin, oil of steam-distilled garlic, and oil-macerated garlic, both containing volatile oil-soluble dialyl sulfides [1,3,6].

4.1 Garlic in Modern Medicine

Research on the health benefits of garlic has increased in the last three decades, in part due to the enhanced practice of complementary and alternative medicine. Garlic and some garlic supplements, notably aged garlic extract, the most researched among the supplements with over 600 scientific publications, have been found to play a role in reducing the risk of a wide range of chronic diseases and pathological conditions associated with aging, such as atherosclerosis, heart disease, stroke, cancer, and Alzheimer's disease; helping to boost immunity, and protect against combat fatigue and toxic effects of radiation and certain medications (reviewed in references 1, 5–7).

5. AGING, OXIDANT STRESS AND ANTIOXIDANTS

Aging is a slow wear and tear of the body that is influenced by multiple genetic and environmental factors. Our genes determine the hereditary aspects of life, including susceptibility to disease. However, environmental factors such as diet and lifestyle modify genetic patterns. Aging comes with altered metabolic processes, an accumulation of mutations, and cell death that ultimately result in a variety of age-related conditions, including atherosclerosis and cataracts and diseases such as heart disease, cancer, and dementia [5–13].

The causative link between oxidative stress and aging is one of the most accepted theories of aging, largely observed by studies showing that the intake of fruits and vegetables that are rich in antioxidant is associated with a reduced risk of many chronic diseases and aging. Reactive oxygen species (ROS), that include free radicals, are the major culprits [1,5–13]. Produced as by-products in the generation of energy in normal metabolism, ROS increase during infection and inflammation, exercise, and stress, and in exposure to exogenous sources, including NOx pollutants, smoking, certain drugs (e.g. acetaminophen), and radiation, including sunlight [1].

High levels of ROS with inadequate antioxidants to neutralize them lead to oxidative stress that is a triggering or promoting factor in many age-related pathological conditions; these include cancer, cardiovascular disease, Alzheimer's disease, and several other neurological diseases. Oxidative damage by ROS accumulates with age and its consequences depend on the molecules that are attacked and on the prevailing levels of protective antioxidants, which get depleted by ROS and fall to lower levels with age [6–8,12].

When the target molecule is DNA, the resulting ROS-induced mutations can lead to cancer

[8]. Oxidation of lipids and proteins injures cell membranes, increases blood vessel fragility, damages immune cells, and modifies enzymes, including protective enzymes and many other molecules [7,8,13]. Oxidative modification of low density lipoprotein (LDL) cholesterol increases the risk of atherosclerosis, cardiovascular, and cerebrovascular disease; free radical-producing beta amyloid peptides (Abeta) in the brain trigger neuronal death by apoptosis, increasing the risk of brain atrophy and dementia, including Alzheimer's disease. Stroke that can be triggered by an atherosclerotic clot can be followed by dementia, as brain cells, deprived of oxygen and blood (ischemia) during blockage, die by apoptosis, also called programmed cell death [10–14].

Apoptosis is regulated by the Bcl-2 family of proteins that includes the pro-apoptosis proteins Bax and others and the apoptosis inhibitors Bcl-2, Bag, and others. Apoptosis is associated with characteristic morphological changes in cells and their DNA [14] that result largely from the action of activated cysteine proteinases (caspases). Apoptosis occurs through a mitochondrial-dependent pathway, with the release of cytochrome C, followed by activation of the caspase cascade, with caspase 3 leading cells to their death. Alternately, programmed cell death proceeds via a mitochondrial-independent pathway, with the ligation of death receptors CD95 (Fas/Apo1) and the subsequent recruitment of caspases [14].

Antioxidants neutralize and destroy ROS, and prevent reactive radicals from interacting with tissue molecules, including DNA, proteins, and lipids, and causing genetic changes and apoptosis [7,13]. While internal antioxidants such as glutathione, superoxide dismutase, catalase, and peroxidases provide some protection against excessive oxidative damage, a diet rich in plant products that contain essential nutrients and phytochemicals with antioxidant activity is the optimal assurance for neutralizing free radical-mediated stress that leads to aging and age-related diseases. Garlic represents an important source of antioxidant phytochemicals and has been shown to help prevent diseases associated with aging [1,5,6,13].

6. GARLIC AND CARDIOVASCULAR DISEASE (CVD)

Preclinical and clinical studies have reported the ability of garlic to help prevent cardiovascular disease by multiple effects, reducing cardiovascular risk factors [1,5,6,15–35]. Garlic has been shown to lower blood pressure [23,24], inhibit platelet aggregation and adhesion (thus helping prevent blood clots) [18,19], reduce LDL cholesterol (the bad cholesterol), and elevate high density lipoprotein (HDL; the protective cholesterol) [15,18,20]. Garlic prevents LDL oxidation that is an exacerbating factor in atherogenesis [22]. S-allyl cysteine has been shown to deactivate the cholesterol-synthesizing enzyme, HMA-CoA reductase, by 30–40% [28], having an additive effect with lipid-lowering statins.

A study on elderly subjects found that ingestion of garlic leads to decreased oxidation reactions, which plays a part in the beneficial effects of garlic. Supplementation of garlic extract improves blood lipid profile, strengthens blood antioxidant potential, and results in significant reductions in systolic and diastolic blood pressures. It also leads to a decrease in the level of peroxidation product (MDA) in the blood, which demonstrates reduced oxidation reactions in the body [35].

Another study evaluated the effects of raw garlic consumption on human blood biochemical factors in hyperlipidemic individuals. In this trial 30 volunteer individuals with blood cholesterol higher than 245 mg/dL consumed 5 g raw garlic twice a day for 42 days, followed by the same number of days with no garlic consumption. After 42 days of garlic consumption cholesterol and triglycerides were significantly reduced and HDL cholesterol showed a significant increase. After the 42 days with no garlic consumption total cholesterol and triglycerides significantly increased and HDL was decreased [29].

A double-blind crossover placebo-controlled study on the effect of aged garlic extract on blood lipids was performed in a group of 41 moderately hypercholesterolemic men (cholesterol 220–290 mg/dL). Following 6 months' ingestion of 7.2 g aged garlic extract per day there was a reduction in total serum cholesterol of 6.1% or 7.0% over baseline, and LDL and blood pressure were decreased by 4% when compared with average baseline values and 4.6% in comparison with placebo period concentrations. In addition, there was a 5.5% decrease in systolic blood pressure and a modest reduction of diastolic blood pressure in response to aged garlic extract [36].

Garlic has been found to reduce smoking-related oxidative damage and lower the production of prostaglandins involved in inflammation [6,30]. Aged garlic extract has also been found to reduce levels of homocysteine, a risk factor for CVD [17,20,21,32,33], and reduce the levels of nuclear factor kappa B that is involved in oxidation and inflammation linked to atherosclerosis [34].

A double-blind placebo-controlled randomized clinical study has found that garlic inhibits the progression of coronary artery calcification and potentially reduces the risk of heart attacks [20]. Twenty three patients, who were at high risk for heart disease, received either 1200 mg/d aged garlic extract, for one year, or a placebo. As all patients were on statin and aspirin therapy, any improvement seen under the influence of the extract would be an additional benefit to statin treatment. Using non-invasive electron beam computed tomography to measure calcification, that is a measure of atherosclerosis, investigators found that in patients on placebo coronary artery plaques progressed at a rate of 22% a year, while the addition of aged garlic extract to the diet reduced progression to 7.5%. The garlic extract also elevated HDL and decreased LDL, triglycerides, and homocysteine, as compared to placebo.

6.1 Garlic, Microcirculation, and Endothelial Function

Garlic has been shown to increase blood flow, including microcirculation [1,15,17], and protect endothelial cells from oxidative damage. Improved microcirculation is most important in aging as arteries stiffen with age and microvasculature is damaged, increasing the risk of CVD as well as dementia. Experimental and clinical studies with aged garlic extract show that garlic increases the production of an endothelial relaxing factor, cellular nitric oxide (NO), that improves endothelial function and helps reduce hypertension [17,25].

Elevated plasma homocysteine above 12 μmol/L is associated with increased oxidant stress, a depletion of endogenous antioxidants such as glutathione, and endothelial dysfunction, thus enhancing the risk of CVD [17]. Aged garlic extract is reported to increase glutathione levels and stimulate NO generation in endothelial cells. A placebo-controlled blinded crossover trial in healthy subjects found that aged garlic extract protected endothelial cells from the damage of homocysteine, in part, by preventing a decrease in bioavailable NO and other factors during acute hyperhomocysteinemia [17]. In another study, a randomized double-blind placebo-controlled trial, 15 men with coronary artery disease were treated with 2.4 g/d of aged garlic extract or placebo, for two weeks. Garlic treatment increased endothelial function (dilation and increased blood flow) by 44%, compared to placebo, indicating that short-term treatment with aged garlic extract improves impaired endothelial function in men with coronary artery disease [31].

7. CANCER AND AGING

Cancer, a major cause of human death, increases in frequency with age [8]. Women are most frequently affected by cancers of the breast, lung, colon, uterus, ovary, and skin, while men are most frequently victims of cancers of the prostate, lung, colon, and skin [36]. All cancers involve the malfunction of genes that control cell growth and division. About 5–10% of cancers are hereditary, whereby an inherited defective gene predisposes a person to a particular cancer. The remainder of cancers result from multiple mutations in DNA and promotion events by damage to other molecules, that accumulate with age, following exposure to oxidative injury or cell death from external factors (tobacco, chemicals, ionizing radiation, sunlight) and internal factors (hormones, immune conditions, and oxidative events and inflammation). Reactive oxygen species are believed to play a major role in the initiation and promotion of cancer and aging. Antioxidants that protect against oxidant stress prevent cancer in a wide range of preclinical studies [8,9,37–44]. Some human studies support this concept; for example, the finding that clinical progression of breast cancer is linked to an increase in oxidant stress, as measured by increased levels of peroxidation products [38], and that a high intake of antioxidant-containing plant foods, including garlic, can reduce the risk of certain cancers [7,8,37–44].

7.1 Garlic and Cancer

Epidemiological observations, largely in China, the USA, and Italy show that a high consumption of garlic dramatically lowers the risk of stomach and colorectal cancer, sometimes by 50% [1,40,41,43]. A trend towards an association between garlic consumption and reduced risk of prostate and breast cancer has also been reported [41]. Preclinical studies consistently show a cancer-preventive effect of garlic, aged garlic extract, and their components, inhibiting growth of mammary, colon, stomach, and other cancers and inducing apoptosis in the cancer cells [1,44,45]. Studies also show the ability of aged garlic extract to protect normal cells against apoptosis induced by radiation and chemicals, such as methotrexate, in cancer treatment [1,39,46].

Molecular studies, carried out mostly with aged garlic extract, support existing data that garlic-derived organosulfur compounds reduce cancer risk. S-allyl mercaptocysteine has been shown to inhibit the progression of cancer cells at either the G1/S or G2/M phase and modify redox-sensitive signal pathways, causing mitotic arrest and apoptosis [47], and aged garlic extract was found to inhibit angiogenesis, the step needed for tumor growth [48], and to boost immunity [49].

The antioxidant effects of the water-soluble compounds in garlic and some of the lipid-soluble compounds play an important role in their cancer preventive action [1,8,44]. S-allyl cysteine and S-allyl mercaptocysteine, that is unique to the aged extract, have powerful radical scavenging ability [1–4], a capacity to increase cellular glutathione and inhibit lipid peroxidation, that is associated with cancer development [38]; some lipid-soluble organosulfur compounds in garlic and in aged garlic extract have also been found to exert a protective effect against cancer [50], in part, due to their ability to scavenge ROS and decrease lipid peroxidation [51].

While the protective effects of fresh garlic in significantly lowering the risk of cancer of the gastrointestinal tract are well established [40,41], the role of aged garlic extract as a cancer-preventive agent in humans has now been reported, thus complementing the preclinical studies that have long established the efficacy of the odorless supplement in cancer prevention [44]. In a double-blind randomized clinical trial, patients with precancerous colorectal adenomas received a high dose of kyolic aged garlic extract (2.4 mL/d), while other patients, who served as controls, received a low dose (0.16 mL/d). Using colonoscopy, the number and size of adenomas were measured before aged garlic extract intake and at 6 and 12 months after intake. The study showed that in the control subjects there was a linear increase in the number of adenomas, from the onset of the study (baseline). By contrast, subjects in the group that received the high dose of aged garlic extract had a significant reduction in adenomas. Aged garlic extract decreased both the size and number of colon adenomas after 12 months of treatment ($P = 0.04$), potentially suppressing progression to colorectal cancer [52].

8. GARLIC AND COGNITIVE FUNCTIONS

Cognitive impairment, loss of memory, and the onset of dementia and its most common form Alzheimer's disease (AD), are devastating forms of deterioration that may occur in aging. Alzheimer's disease accounts for more than 70% of all cases of dementia [53]. During the past decade there has been a growing volume of data showing that vascular factors play a role in Alzheimer's disease and that people with cardiovascular risk factors and a history of stroke have an increased risk of both vascular dementia and AD [54].

The antioxidant actions of garlic and aged garlic extract and their broad range of cardiovascular protection may be extended to protecting the brain and helping reduce the risk of dementia, including vascular dementia and AD. Garlic has potential to protect the brain against neurodegenerative conditions by its antioxidant actions, its efficacy in lowering cholesterol, inhibiting inflammation, reducing homocysteine, preventing oxidative brain injury following ischemia, protecting neurons against apoptosis triggered by oxidative stress, and preventing

neurotoxicity by beta amyloid peptide, the hallmark of AD [4]). In addition, aged garlic extract was found to prevent atrophy in the frontal brain of early senescence mice models, improved learning and memory retention, and increased longevity [55]; other studies showed that allixin, a component of garlic and AGE, enhanced the survival of neurons and increased the branching points in axons of hippocampus neurons [56].

8.1 Homocysteine

Elevated homocysteine, a risk factor for cardiovascular and cerebrovascular diseases [57], is also an independent risk factor for AD and other forms of dementia [53,58]. The link between high levels of homocysteine and dementia, found in epidemiological studies, has been confirmed in case-control studies, where people with vascular dementia and AD had higher levels of homocysteine than did healthy people. Investigators provided compelling evidence of a direct link between elevated plasma homocysteine and loss of cognition; they showed that in adults with intact cognition, an increase in plasma homocysteine, over time, is associated with an increased incidence of dementia including AD [53].

Research on the effects of garlic on homocysteine, using AGE, found that AGE reduced homocysteine-related vascular damage [16,17]. In a clinical trial, AGE was found to reduce macro- and microvascular endothelial dysfunction during acute hyperhomocysteinemia [17]; a preclinical trial found that while homocysteine stimulated the uptake of oxidized LDL into macrophages, a step in early atherosclerosis, AGE inhibited the uptake, thus helping prevent formation of early atherosclerotic lesions [16].

In a preclinical study, homocysteine levels were compared following a 4-week feeding of a folate-deficient diet that contained AGE with a feeding of a folate-fortified diet and AGE.

Plasma homocysteine was 30% lower in the folate-deficient animals that received AGE, but not in those with adequate folate. The results suggest that AGE may serve as an added treatment in hyperhomocysteinemia [21]. A clinical study showing that AGE inhibits the progression of coronary artery calcification also showed a trend in a lowering of homocysteine levels [20].

8.2 Protection Against Ischemic/Reperfusion Adverse Effects

Ischemia is a feature of heart diseases, transient ischemic attacks, cerebrovascular accidents, and peripheral artery occlusive disease. The heart, the kidneys, and the brain are among the organs that are the most sensitive to inadequate blood supply. Ischemia in brain tissue, for example due to stroke or head injury, causes a cascade to be unleashed, in which proteolytic enzymes and free radicals damage and kill neurons by apoptosis, sometimes leading to dementia.

Restoring blood flow after ischemia can increase the damage, as the reintroduction of oxygen enhances free radical production, resulting in reperfusion injury and necrosis contributing to the development of dementia following a stroke.

Water-soluble garlic compounds have been found to protect neurons from ischemia/reperfusion injury and increase their survival. In focal ischemia, aged garlic extract, S-allyl cysteine, allyl sulfide (AS) or allyl disulfide (ADS) was administered 30 min prior to ischemic insult. Three days after ischemic insult, water contents of both ischemic and contralateral hemispheres were measured to assess the degree of ischemic damage. Treatment with aged garlic extract and S-allyl cysteine attenuated ROS and prevented brain injury; treatment with S-allyl cysteine reduced infarct volume. None of the lipid-soluble compounds tested had a protective effect [59].

8.3 Garlic Prevents Cognitive Decline

Garlic has been shown to act as a neuroprotective agent, potentially defending against AD [5]. One of the characteristic features of AD is the accumulation of extracellular plaques of aggregated β-amyloid peptide (Abeta) that trigger neuronal death by apoptosis. Reactive oxygen species, produced by Abeta, are thought to play a role in the apoptotic mechanism of Abeta-mediated neurotoxicity [60]. Garlic and S-allyl cysteine have been shown in a number of experimental studies to protect neuronal cells against Abeta toxicity and apoptosis [60–63] and prevent cognitive decline [64–68]. When model neuronal cells (PC12) were exposed to Abeta they showed a significant increase in ROS [60–62]. A number of studies have shown that AGE and S-allyl cysteine prevent Abeta-induced neurotoxicity, suppressing ROS generation, attenuating the activation of the apoptotic enzyme caspase 3, in a dose-related manner, and attenuating DNA fragmentation and apoptosis [61–63].

Preclinical studies in models that are genetically prone to early aging show that garlic has additional anti-aging effects [64,65]. Treatment with AGE or S-allyl cysteine prevented the degeneration of the brain's frontal lobe, improved learning and memory retention, and extended lifespan. Isolated neurons from the hippocampus area, grown in the presence of AGE or S-allyl cysteine, showed an unusual ability to grow and branch, which may be linked to the findings that the extract increases learning and cognition [66].

Studies with AGE and S-allyl cysteine have further confirmed the neuroprotective effects of garlic and its components. The efficacy of S-allyl-l-cysteine in preventing cognitive decline, by protecting neurons from Abeta-induced neuronal apoptosis, was found to be mediated in part by binding to Abeta, inhibiting Abeta fibrillation, and destabilizing preformed Abeta fibrils [67]. Feeding of AGE to mice with accelerated Abeta formation prevented deterioration of hippocampal-based memory tasks in the mice, suggesting that AGE has a potential for preventing AD progression [68].

A study in Alzheimer's Swedish double mutant mouse model (Tg2576), investigated the anti-amyloidogenic, anti-inflammatory, and anti-tangle effects of dietary AGE (2%) and compared its efficacy with that of its prominent constituent, the water-soluble S-allyl-cysteine (20 mg/kg) and the lipid-soluble di-allyl-disulfide (20 mg/kg). The ameliorative effects of dietary interventions were found to be in the order of AGE > SAC > DADS [69]. The authors suggest that dietary intervention with a herbal alternative such as AGE, which has a wide range of benefits, including cholesterol reduction, antioxidant effects, and anti-inflammatory actions and has no adverse side effects, 'may provide greater therapeutic benefit over a single-ingredient synthetic pharmaceutical drug having serious side effects in treating Alzheimer's disease' [69].

9. GARLIC REDUCES FATIGUE

Fatigue and reduced levels of energy are conditions that often accompany aging. Traditional medicine and practices in antiquity used garlic treatment to help increase energy levels [1]. Other studies have shown that garlic, in the form of AGE, can increase endurance in exercise and reduce fatigue [70,71]. Although the underlying mechanism remains unclear, earlier studies showed that garlic's ability to increase vasodilation, blood flow, and consequently oxygen supply may play an important part in its anti-fatigue effects.

10. SUMMARY

Numerous studies point to the health benefits and anti-aging effects of garlic. Rich in antioxidants and in organosulfur compounds, that provide most of its benefits, garlic has been shown to help prevent chronic diseases and conditions associated with aging, notably cardiovascular and cerebrovascular diseases, cancer, and Alzheimer's disease. Garlic has been found to reduce the risk factors for disease and aging, including reducing oxidative stress and inflammation, lowering cholesterol, triglycerides, and homocysteine, inhibiting coronary artery plaque formation, and preventing platelet aggregation. Garlic treatment was found to increase vasodilation and blood flow, lower blood pressure, prevent neuronal cell death and increase memory and cognition, boost immunity, increase glutathione, an important internal antioxidant, increase physical endurance, and ameliorate fatigue. Among the many studies on the anti-aging effects of garlic, the most consistent results, reported in over 600 publications, are from preclinical and clinical research with aged garlic extract (kyolic), an odorless, highly standardized supplement, prepared by a long-term extraction and aging of fresh garlic, which enriches its antioxidant content and renders its stable water-soluble compounds, such as S-allyl cysteine, highly bioavailable [72]. Taken together, a diet rich in garlic and its standardized stable supplement promotes general health, and helps provide protection from chronic diseases and aging.

References

1. Borek, C. (2001). Antioxidant health effects of aged garlic extract. *Journal of Nutrition, 131,* 1010S–1015S.
2. Imai, J., Ide, N., Nagae, S., Moriguchi, T., Matsuura, H., & Itakura, Y. (1994). Antioxidant and radical scavenging effects of aged garlic extract and its constituents. *Planta Medica, 60,* 417–420.
3. Amagase, H., Petesch, B. L., Matsuura, H., Kasuga, S., & Itakura, I. (2001). Intake of garlic and its bioactive components. *Journal of Nutrition, 31,* 955S–962S.
4. Ide, N., Lau, B. H., Ryu, K., Matsuura, H., & Itakura, Y. (1999). Antioxidant effects of fructosyl arginine, a Maillard reaction product in aged garlic extract. *Journal of Nutritional Biochemistry, 10,* 372–376.

5. Borek, C. (2006). Garlic reduces dementia and heart-disease risk. *Journal of Nutrition, 136*, 810S–812S.

6. Rahman, K. (2003). Garlic and aging: New insights into an old remedy. *Ageing Research Reviews, 2*, 39–56.

7. Borek, C. (2006). Aging and antioxidants. Fruits and vegetables are powerful armor. *Advances in Nursing Practice, 14*, 35–38.

8. Borek, C. (2004). Dietary antioxidants and human cancer. *Integrated Cancer Therapy, 3*, 333–341.

9. Borek, C. (1997). Antioxidants and cancer. *Science and Medicine, 4*, 51–61.

10. Richardson, S. J. (1993). Free radicals in the genesis of Alzheimer's disease. *Annals of the New York Academy of Sciences, 695*, 73–76.

11. Margaill, I., Plotkine, M., & Lerouet, D. (2005). Antioxidant strategies in the treatment of stroke. *Free Radical Biology and Medicine, 15*, 29–43.

12. Jama, J. W., Launer, L. J., Witteman, J. C., den Breeijen, J. H., Breteler, M. M., Grobbee, D. E, & Hofman, A. (1996). Dietary antioxidants and cognitive functions in a population based sample of older persons. *American Journal of Epidemiology, 144*, 275–280.

13. Knight, J. A. (2000). The biochemistry of aging. *Advances in Clinical Chemistry, 35*, 1–62.

14. Zimmermann, K. C., Bonbon, C., & Green, D. R. (2001). The machinery of programmed cell death. *Pharmacology and Therapeutics, 92*, 57–70.

15. Ramman, K., & Lowe, G. M. (2006). Garlic and cardiovascular disease: A critical review. *Journal of Nutrition, 136*, 736S–740S.

16. Ide, N., Keller, C., & Weiss, N. (2006). Aged garlic extract inhibits homocysteine-induced CD36 expression and foam cell formation in human macrophages. *Journal of Nutrition, 136*, 755S–758S.

17. Weiss, N., Ide, N., Abahji, T., Nill, L., Keller, C., & Hoffmann, U. (2006). Aged garlic extract improves homocysteine-induced endothelial dysfunction in macro- and microcirculation. *Journal of Nutrition, 136*, 750S–754S.

18. Steiner, M., & Li, W. (2001). Aged garlic extract, a modulator of cardiovascular disease risk factors: A dose finding study of the effects of AGE on platelet function. *Journal of Nutrition, 131*, 980S–984S.

19. Ramman, K. (2007). Effects of garlic on platelet biochemistry and physiology. *Molecular Nutritional Food Research, 51*, 1335–1344.

20. Budoff, M. (2006). Aged garlic extract retards progression of coronary artery calcification. *Journal of Nutrition, 136*, 741S–744S.

21. Yeh, Y. Y., & Yeh, S. M. (2006). Homocysteine-lowering action is another potential cardiovascular protective factor of aged garlic extract. *Journal of Nutrition, 136*, 745S–749S.

22. Lau, B. H. S. (2006). Suppression of LDL oxidation by garlic compounds is a possible mechanism of cardiovascular health benefit. *Journal of Nutrition, 136*, 765S–768S.

23. Ried, K., Frank, O. R., Stocks, N. P., Fakler, P., & Sullivan, T. (2008). Effect of garlic on blood pressure: A systematic review and meta-analysis. *BMC Cardiovascular Disorder, 8*, 13.

24. Harauma, A., & Moriguchi, T. (2006). Aged garlic extract improves blood pressure in spontaneously hypertensive rats more safely than raw garlic. *Journal of Nutrition, 136*, 769S–773S.

25. Al-Qattan, K. K., Thomson, M., Al-Mutawa'a, S., Al-Hajeri, D., Drobiova, H., & Ali, M. (2006). Nitric oxide mediates the blood-pressure lowering effect of garlic in the rat two-kidney, one-clip model of hypertension. *Journal of Nutrition, 136*, 774S–776S.

26. Morihara, N., Sumioka, I., Ide, N., Moriguchi, T., Uda, N., & Kyo, E. (2006). Aged garlic extract maintains cardiovascular homeostasis in mice and rats. *Journal of Nutrition, 136*, 777S–781S.

27. Allison, G. L., Lowe, G. M., & Ramman, K. (2006). Aged garlic extract and its constituents inhibit platelet aggregation through multiple mechanisms. *Journal of Nutrition, 136*, 782S–788S.

28. Liu, L., & Yeh, Y. Y. (2002). Alk(en)yl cysteine of garlic inhibit cholesterol synthesis by deactivating HMA-C.a. reductase in cultured hepatocytes. *Journal of Nutrition, 132*, 1129–1134.

29. Avci, A., Atli, T., Ergüder, I. B., Varli, M., Devrim, E., Aras, S., & Durak, I. (2006). Effects of garlic consumption on plasma and erythrocyte antioxidant parameters in elderly subjects. *Gerontology, 54*, 173–176.

30. Dillon, S. A., Lowe, G. M., Billington, D., & Ramman, A. (2002). Dietary supplementation with aged garlic extract reduces plasma and urine concentration of 8-iso prostaglandin F(2alpha) in smoking and non smoking men and women. *Journal of Nutrition, 132*, 168–171.

31. Williams, M. J., Sutherland, W. H., McCormick, M. P., Yeoman, D. J., & de Jong, S. A. (2005). Aged garlic extract improves endothelial function in men with coronary artery disease. *Phytotherapy Research, 19*, 314–319.

32. Haim, M., Tanne, D., Goldbourt, U., Doolman, R., Boyko, V., Brunner, D., Sela, B. A., & Behar, S. (2007). Serum homocysteine and long-term risk of myocardial infarction and sudden death in patients with coronary heart disease. *Cardiology, 107*, 52–56.

33. Yeh, Y. Y., & Yeh, S. M. (2006). Homocysteine-lowering action is another potential cardiovascular protective factor of aged garlic extract. *Journal of Nutrition, 136*, 745S–749S.

34. Ide, N., & Lau, B. H. (2001). Garlic compounds minimize intracellular oxidative stress and inhibit nuclear factor-kappa b activation. *Journal of Nutrition, 131*, 1020S–1026S.

35. Durak, I., Kavutcu, M., Aytaç, B., Avci, A., Devrim, E., Ozbek, H., & Oztürk, H. S. (2004). Effects of garlic extract consumption on blood lipid and oxidant/antioxidant parameters in humans with high blood cholesterol. *Journal of Nutritional Biochemistry, 15*, 373–377.

36. Steiner, M., Kahn, A. H., Holbert, D., & Lin, R. I. (1996). A double-blind crossover study in moderately hypercholesterolemic men that compared the effect of aged garlic extract and placebo administration on blood lipids. *American Journal of Clinical Nutrition, 64*, 866–870.

37. Greenwald, P. (2005). Lifestyle and medical approaches to cancer prevention. Recent results. *Cancer Research, 166*, 1–15.

38. Khanzode, S. S., Muddeshwar, M. G., Khanzode, S. D., & Dakhale, G. N. (2004). Antioxidant enzymes and lipid peroxidation in different stages of breast cancer. *Free Radical Research, 38*, 81–85.

39. Borek, C. (2004). Antioxidants and radiotherapy. *Journal of Nutrition, 134*, 3207S–3209S.

40. Steinmetz, K. A., & Potter, J. D. (1996). Vegetables, fruit and cancer prevention: A review. *Journal of the American Dietetic Association, 96*, 1027–1039.

41. Fleischauer, A. T., & Arab, L. (2001). Garlic and cancer: A critical review of the epidemiological literature. *Journal of Nutrition, 131*, 1032S–1040S.

42. Galeone, C., Pelucchi, C., Dal Maso, L., Negri, E., Montella, M., Zucchetto, A., Talamini, R., & La Vecchia, C. (2008). Allium vegetables intake and endometrial cancer risk. *Public Health Nutrition, 6*, 1–4.

43. Greenwald, P., Clifford, C. K., & Milner, J. A. (2001). Diet and cancer prevention. *European Journal of Cancer, 37*, 948–965.

44. Milner, J. A. (2006). Preclinical perspectives on garlic and cancer. *Journal of Nutrition, 136*, 827S–831S.

45. Lee, Y. (2008). Induction of apoptosis by S-allylmercapto-L-cysteine, a biotransformed garlic derivative, on a human gastric cancer cell line. *International Journal of Molecular Medicine, 21*, 765–770.

46. Li, T., Ito, K., Sumi, S. I., Fuwa, T., & Horie, T. (2009). Protective effect of aged garlic extract (AGE) on the apoptosis of intestinal epithelial cells caused by methotrexate. *Cancer Chemotherapy and Pharmacology, 63*, 873–880.

47. Pinto, J. T., Krasnikov, B. F., & Cooper, A. J. (2006). Redox-sensitive proteins are potential targets of garlic-derived mercaptocysteine derivatives. *Journal of Nutrition, 136*, 835S–841S.

48. Matsuura, N., Miyamae, Y., Yamane, K., Nagao, Y., Hamada, Y., Kawaguchi, N., Katsuki, T., Hirata, K., Sumi, S., & Ishikawa, H. (2006). Aged garlic extract inhibits angiogenesis and proliferation of colorectal carcinoma cells. *Journal of Nutrition, 136*, 842S–846S.

49. Kyo, E., Uda, N., Kasuga, S., & Itakura, Y. (2001). Immunomodulatory effects of aged garlic extract. *Journal of Nutrition, 131*, 1075S–1079S.

50. Horie, T., Awazu, S., Itakura, Y., & Fuwa, T. (1992). Identified diallyl polysulfides from an aged garlic extract which protects the membranes from lipid peroxidation. *Planta Medica, 58*, 468–469.

51. Seki, T., Hassan, T., Hassan-Fukao, T., Inada, K., Tanaka, R., Ogihara, J., & Ariga, T. (2008). Anticancer effects of diallyl trisulfide derived from garlic. *Asia Pacific Journal of Clinical Nutrition, 17*(1 Suppl), 249–252.

52. Tanaka, S., Haruma, K., Yoshihara, M., Kajiyama, G., Kira, K., Amagase, H., & Chayama, K. (2006). Aged garlic extract has potential suppressive effect on colorectal adenomas in humans. *Journal of Nutrition, 136*, 821S–826S.

53. Seshadri, S., Beiser, A., Selhub, J., Jacques, P. F., Rosenberg, I. H., D'Agostino, R. B., Wilson, P. W., & Wolf, P. A. (2002). Plasma homocysteine as a risk factor for dementia and Alzheimer's disease. *New England Journal of Medicine, 346*, 476–483.

54. Kivipelto, M., Helkala, E. L., Hanninen, T, Laakso, M. P., Hallikainen, M., Alhainen, K., Soininen, H., Tuomilehto, J., & Nissinen, A. (2001). Midlife vascular risk factors and Alzheimer's disease in later life: Longitudinal, population based study. *BMJ, 322*, 1447–1451.

55. Moriguchi, T., Saito, H., & Nishyama, N. (1997). Antiaging effect of aged garlic extract in the inbred brain atrophy mouse model. *Clinical and Experimental Pharmacology and Physiology, 24*, 235–242.

56. Moriguchi, T., Matsuura, H., Itakura, Y., Katsuki, H., Saito, H., & Nishiyama, N. (1997). Allixin, a phytoalexin produced by garlic, and its analogues as novel exogenous substances with neurotrophic activity. *Life Sciences, 61*, 1413–1420.

57. The Homocysteine Studies Collaboration. (2002). Homocysteine and risk of ischemic heart disease and stroke: A meta-analysis. *JAMA, 288*, 2015–2022.

58. Sala, I., Belén Sánchez-Saudinós, M., Molina-Porcel, L., Lázaro, E., Gich, I., Clarimón, J., Blanco-Vaca, F., Blesa, R., Gómez-Isla, T., & Lleó, A. (2008). Homocysteine and cognitive impairment. Relation with diagnosis and neuropsychological performance. *Dementia and Geriatric Cognitive Disorders, 26*, 506–512.

59. Numagami, Y., Sato, S., & Onishi, T. (1996). Attenuation of rat ischemic brain damage by aged garlic extracts: A possible protecting mechanism as an antioxidant. *Neurochemistry International, 29*, 135–143.

60. Brenya, G., Selassie, M., & Gwebie, E. T. (1999). Effect of aged garlic extract on the cytotoxicity of Alzheimer

B. EFFECTS OF INDIVIDUAL VEGETABLES ON HEALTH

beta amyloid peptide in neuronal PC12 cells. *Nutritional Neuroscience, 3,* 139–142.

61. Kosuge, Y., Koen, Y., Ishige, K., Minami, K., Urasawa, H., Saito, H., & Ito, Y. (2003). S-allyl-L-cysteine selectively protects cultured rat hippocampal neurons from amyloid beta-protein- and tunicamycin-induced neuronal death. *Neuroscience, 122,* 885–895.

62. Peng, Q., Buz'Zard, A. R., & Lau, B. H. (2002). Neuroprotective effect of garlic compounds in amyloid-beta peptide-induced apoptosis *in vitro. Medical Science Monitor, 8,* 28–37.

63. Jackson, R., McNeil, B., Taylor, C., Holl, G., Ruff, D., & Gwebu, E. T. (2002). Effect of aged garlic extract on caspase-3 activity, *in vitro. Nutritional Neuroscience, 5,* 287–290.

64. Nishiyama, N., Moriguchi, T., Katsuki, H., & Saito, H. (1996). Effects of aged garlic extract on senescence accelerated mouse and cultured brain cells. Preclinical and clinical strategies for the treatment of neurodegenerative, cerebrovascular and mental disorders. *International Biomedical Drug Research, 11,* 253–258.

65. Moriguchi, T., Saito, H., & Nishiyama, N. (1997). Anti-aging effect of aged garlic extract in the inbred brain atrophy mouse model. *Clinical and Experimental Pharmacology and Physiology, 24,* 235–242.

66. Moriguchi, T., Matsuura, H., Kodera, Y., Itakura, Y., Katsuki, H., Saito, H., & Nishiyama, N. (1997). Neurotrophic activity of organosulfur compounds having a thioallyl group on cultured rat hippocampal neurons. *Neurochemical Research, 22,* 1449–1452.

67. Gupta, V. B., & Rao, K. S. (2007). Anti-amyloidogenic activity of S-allyl-L-cysteine and its activity to destabilize Alzheimer's beta-amyloid fibrils *in vitro. Neuroscience Letters, 429,* 75–80.

68. Chauhan, N. B., & Sandoval, J. (2007). Amelioration of early cognitive deficits by aged garlic extract in Alzheimer's transgenic mice. *Phytotherapy Research, 21,* 629–640.

69. Chauhan, N. B. (2006). Effect of aged garlic extract on APP processing and tau phosphorylation in Alzheimer's transgenic model Tg2576. *Journal of Ethnopharmacology, 108,* 385–394.

70. Morihara, N., Ushijima, M., Kashimoto, N., Sumioka, I., Nishihama, T., Hayama, M., & Takeda, H. (2006). Aged garlic extract ameliorates physical fatigue. *Biological & Pharmaceutical Bulletin, 29,* 962–966.

71. Morihara, N., Nishihama, T., Ushijima, M., Ide, N., Takeda, H., & Hayama, M. (2007). Garlic as an anti-fatigue agent. *Molecular Nutritional Food Research, 11,* 1329–1334.

72. Kasuga, S., Uda, N., Kyo, E., Ushijima, M., Morihara, N., & Itakura, Y. (2001). Pharmacologic activities of aged garlic extract in comparison with other garlic preparations. *Journal of Nutrition, 131,* 1080S–1084S.

16

Garlic and Heart Health

Khalid Rahman

School of Pharmacy and Biomolecular Sciences, Liverpool John Moores University, Liverpool, UK

1. INTRODUCTION

There has been an increase in individual life expectancy throughout the developed countries with parallel increases in chronic diseases such as cardiovascular disease. In addition, there has been an international resurgence in 'greening issues' due to wider coverage of media. Consumers are also more knowledgeable regarding health issues and thus have been more interested in food and its components that can either delay or prevent chronic diseases such as cardiovascular disease [1]. There is also epidemiological evidence for an association between food and chronic diseases. The available evidence indicates that a higher consumption of fruits and vegetables leads to a lower prevalence of risk factors for cardiovascular disease such as hypertension, obesity, type II diabetes, etc. [2,3].

The concept of complementary and alternative medicine (CAM) is now widespread and is now part of the traditional health care systems in many parts of the world. Within CAM, herbs, spices, and components of food with medical properties (nutraceuticals) are being marketed for the treatment and prevention of chronic diseases including cardiovascular disease. In these days of high-tech and high-cost health care, CAM is increasingly being used and is considered to be cost effective thus providing savings in health care [4]. One of the most popular CAMs for the treatment and prevention of cardiovascular disease is garlic.

2. GARLIC

Garlic (*Allium sativum*) originates from Central Asia and has been used universally as a flavoring agent and phytomedicine. The beneficial effects of garlic consumption in treating a wide variety of human diseases and disorders have been known for centuries. In ancient Egypt, it was used to promote general health and was mentioned in the Codex Eber as a treatment for various disorders. In ancient China garlic was used for promoting a healthy digestive system and enhancing male potency, while in Greece soldiers and athletes consumed garlic to enhance their stamina.

The composition of garlic is shown in Table 16.1. The major part of garlic (65%) is made up of water and the rest of the dry

235

weight is composed of mainly fructose-containing carbohydrates, followed by sulfur compounds, protein, fiber, and free amino acids [5]. It also contains high levels of saponins, phosphorus, potassium, sulfur, zinc; moderate levels of selenium, vitamins A and C; and low levels of calcium, magnesium, sodium, iron, manganese, and B-complex vitamins. About 97% of compounds present in garlic are water soluble and between 0.15 and 0.7% of the compounds are oil soluble. Due to the intense interest in garlic, commercial garlic preparations have become widely available over the years and one can buy these from pharmacy outlets and supermarkets. Garlic is reported to have many medicinal properties, however the most widely researched and reported effect of garlic is its ability to either prevent or reduce cardiovascular disease [1,6–8].

TABLE 16.1   General Composition of Garlic

Component	Amount (fresh weight; %)
Water	62–68
Carbohydrates (mainly fructans)	26–30
Protein	1.5–2.1
Amino acids: common	1–1.5
Amino acids: cysteine sulfoxides	0.6–1.9
γ-Glutamylcysteines	0.5–1.6
Lipids	0.1–0.2
Fiber	1.5
Total sulfur compounds[a]	1.1–3.5
Sulfur	0.23–0.37
Nitrogen	0.6–1.3
Minerals	0.7
Vitamins	0.015
Saponins	0.04–0.11
Total oil-soluble compounds	0.15 (whole)–0.7 (cut)
Total water-soluble compounds	97

Source: reference 5.
[a]Excluding protein and inorganic sulfate (0.5%).

3. CARDIOVASCULAR DISEASE

Cardiovascular disease (CVD) remains the leading cause of morbidity and mortality worldwide and is characterized by multiple risk factors. Epidemiological studies have identified some of these as elevated serum lipids (cholesterol and triglycerides), increased plasma fibrinogen and coagulation factors, increased platelet activation and aggregation, alterations in glucose metabolism and smoking [9,10]. Hypertension and the oxidative modification of low density lipoprotein (LDL), free radicals, inflammation of the arteries, and altered calcium handling by the heart are now also considered important mechanisms in the development of CVD [11]. The normalization of abnormal lipids, lipoproteins, and hypertension, inhibition of platelet aggregation, and an increase in antioxidant status and nitric oxide are believed to improve this disease.

Dietary modification is an important part of the strategy of reducing and treating CVD, thus 'designer' diets and phytochemicals, and dietary supplements are being actively promoted as part of the daily diet to maintain a healthy heart. Cardiovascular disease patients are also increasingly demanding natural products to treat their heart disease and consumers are using CAM to maintain a healthy heart. This disease is characterized by multiple factors and is thus open to CAM intervention, and scientific literature shows that many phytochemicals including garlic address the normalization of these factors. There is widespread usage of CAM among National Health Service patients in the UK [12,13] and similar results have been reported for herb and supplement use in the USA adult population [14,15]. There is widespread usage of CAM for maintaining heart health [16–19] and one of the most widely used phytochemicals is garlic. A summary of the mechanisms by which garlic reduces CVD is given in Table 16.2, and the

TABLE 16.2 Summary of the Mechanisms by which Garlic Reduces Cardiovascular Disease

Reduction in serum lipids (cholesterol and triglycerides)
Increase in bioavailability of nitric oxide
Increase in antioxidant status
Decrease in low density lipoprotein (LDL) oxidation
Decrease in platelet aggregation
Increase in elastic properties of blood vessels
Reduction in hypertension

detailed mechanisms by which it maintains a healthy heart are discussed below.

4. EFFECTS OF GARLIC ON SERUM LIPIDS

Hyperlipidemia and lipoprotein abnormalities disrupt serum lipid levels and are well-established risk factors in heart disease. Cholesterol levels are related to nutritional status [20] and heart disease is treated by drugs although the use of natural products such as garlic is also widespread. Numerous studies have reported a reduction in serum lipid levels by garlic [7,8]. *In vitro* studies have indicated that garlic and its constituents can inhibit key enzymes involved in lipid synthesis in cultured rat hepatocytes and human HpG2 cells [21,22]. Cycloalliin, a cyclic sulfur imino acid present in garlic, is reported to reduce serum triacylglycerol levels in rats without affecting hepatic triacylglycerol synthesis and content, suggesting an alteration of lipoprotein assembly and secretion mechanisms in the liver [23]. This was confirmed by another study which indicated that allicin, a component of garlic, may affect atherosclerosis by lipoprotein modification and inhibition of LDL uptake [24]. In another series of experiments di-allyl tetrasulfide, another component of garlic, was administered to rats exposed to cadmium which is associated with distinct pathological changes including dyslipidemia. Di-allyl tetrasulfide administration resulted in a

reduction in cadmium-mediated CVD [25]. Cyclosporin A is a commonly used immunosuppressive drug and one of its side effects is the elevation of plasma LDL. It has been reported that in the Wistar rat model, animals given cyclosporin A and subsequently treated with garlic had reduced LDL levels compared with controls [26]. Garlic has also been subjected to human clinical trials; for a comprehensive review see references 7, 8, and 27. The effect of raw garlic and enteric-coated garlic powder consumption on human blood biochemical parameters in hyperlipidemic individuals has been investigated. Raw garlic (5 g twice per day for 42 days) resulted in a decrease in serum lipids as did enteric-coated garlic tablets (equal to 400 mg/d for 42 days). The study using enteric-coated garlic tablets was a single-blind placebo-controlled study, and the number of subjects for both of these studies was between 30 and 150 [28,29]. The effect of short-term garlic supplementation on lipid metabolism in hypertensive subjects has also been investigated and the results indicated that garlic supplementation could be useful as an adjunct agent in the treatment of hyperlipidemic patients [30]; however, contradictory data has also been reported [31,32]. An analysis of the clinical trials conducted so far does indicate that garlic is effective in lowering serum cholesterol in hypercholesterolemic subjects, and it is likely that this effect is seen at a certain threshold of serum cholesterol [8,27]. The mechanism by which garlic reduces serum cholesterol has not been fully elucidated, however it is likely that garlic reduces cholesterol, at least in part, through competition with dietary and biliary cholesterol for intestinal absorption in mixed micelles and by inhibiting key enzymes involved in cholesterol synthesis. Evidence reported by Calpe-Berdiel et al. suggests that plant sterols and stanols may reduce cholesterol by regulating proteins implicated in cholesterol metabolism both in enterocytes and hepatocytes; this may be one mechanism by which garlic also displays its hypolipidemic property [33].

5. EFFECT OF GARLIC ON LDL OXIDATION

Free radicals or reactive oxygen species (ROS) have also been implicated in the development of CVD [34]. The two most biologically important ROS are superoxide ions and hydrogen peroxide, and both are produced in the vascular cells by a number of oxidases such as NADPH oxidases, xanthine oxidase, lipoxygenase, etc. The production of ROS is counterbalanced by antioxidant enzymes and if this balance is disturbed ROS can accumulate and result in oxidative stress which has been linked to chronic diseases including CVD. ROS can cause the oxidation of low density lipoprotein (OxLDL) that is then taken up by macrophages which have a foamy appearance and become activated leading to a state of chronic inflammation [35] which is now regarded as central to CVD. ROS can also induce platelet aggregation, monocyte adhesion, and vascular smooth muscle cell apoptosis [36], and these can be inactivated by endogenous antioxidant enzymes and by dietary ingestion of phytochemicals with antioxidant properties, such as garlic.

Many studies have investigated the role of garlic as an antioxidant in both *in vitro* and *in vivo* systems. An extract of garlic has been shown to prevent the depletion of intracellular glutathione (GSH) when endothelial cells were incubated with Ox-LDL, and in addition garlic extract prevented the *in vitro* oxidation of isolated human LDL by scavenging ROS and inhibiting the formation of lipid peroxides [37,38]. In another study the antioxidant activity of allicin, a thiosulfinate compound found in fresh garlic, has been reported to act against the oxidation of cumene and methyl linoleate in chlorobenzene [39]. Garlic has also been shown to significantly inhibit the adhesion of monocytes to IL-1α-stimulated endothelial cells indicating

that garlic modulates the expression of intercellular (ICAM) and vascular cell adhesion molecules (VCAM) thus having a beneficial effect in CVD [40]. Garlic components di-allyl disulfide and di-allyl trisulfide have also been reported to suppress oxidized LDL-induced vascular cell adhesion molecule and E-selectin expression through protein kinase A- and B-dependent signaling pathways [41]. Cyclo-oxygenase (COX) enzymes are elevated in inflammatory conditions and garlic has also been reported to inhibit lipid peroxidation and the activity of COX thus further displaying its antioxidant potential [42].

Garlic has also been investigated for its antioxidant properties in *in vivo* systems [8,34], although some conflicting data has been reported. However, the evidence points towards garlic being a strong antioxidant [8,43]. The prostaglandin F_2 isomers, F_2 isoprostanes, have been investigated as reliable markers for enhanced oxidative stress *in vivo*. Urinary excretion of F_2-isoprostane, 8-iso-prostaglandin $F_{2\alpha}$ (8-epi-PGF$_{2\alpha}$) has been proposed as a risk marker in patients with CVD: in fact increased levels have been reported in coronary plaques of patients with heart disease and high levels of oxidative stress. It has also been proposed that 8-epi-PGF$_{2\alpha}$ is a reliable marker of oxidative damage in chronic heart failure [44,45]. Smokers are subjected to increased oxidative stress and thus have elevated levels of F_2-isoprostanes both in their serum and urine. In a study conducted with smokers and non-smokers, dietary supplementation with an aged garlic extract for 14 days reduced plasma and urine concentrations of F2-isoprostanes by 29% and 37% respectively in non-smokers and by 35% and 48% respectively in smokers. Interestingly, 14 days after cessation of the aged garlic extract, plasma and urine concentrations of F_2-isoprostanes in both groups returned to values similar to those at the start of the study. In this study the plasma antioxidant capacity of non-smokers was also twice that of smokers and the antioxidant capacity of smokers significantly increased by 53% following the ingestion of aged garlic extract for 14 days, however this decreased significantly once the aged garlic extract ingestion was discontinued [46]. The effects of garlic consumption on plasma and erythrocyte antioxidant parameters in elderly subjects have been reported. The conclusion of this study was that ingestion of garlic significantly lowered plasma and erythrocyte malondialdehyde (MDA) levels, and increased the activities of the antioxidant enzymes glutathione peroxidase and catalase indicating that garlic consumption may decrease oxidative stress in the elderly [47]. Endothelial dysfunction can also be caused by increases in oxidative stress which can lead to a decrease in the bioavailability of nitric oxide (NO) which plays a critical role in the vascular pathobiology of hyperhomocysteinemia. In a placebo-controlled, blinded crossover trial aged garlic extract improved homocysteinemia-induced endothelial dysfunction in macro- and microcirculation, in part by preventing a decrease in the bioavailable NO [48]. The mechanism by which garlic exerts its antioxidant properties is not clear, and is probably complex due to the complex composition of garlic. It may be acting as an antioxidant by directly scavenging free radicals, chelating metal ions, augmenting endogenous antioxidant systems, and also by stimulating the synthesis of nitric oxide.

6. EFFECT OF GARLIC ON PLATELET AGGREGATION

It is now recognized that platelets play an important role in the etiology of cardiovascular disease. On activation, platelets undergo shape change, aggregate, secrete their contents, and finally liberate arachidonic acid which is converted to prostaglandins and lipoxygenases related products [49]. The platelet plasma membrane contains intrinsic glycoproteins (GPIIb/IIIa) and these act as receptors to which activating and inhibiting agents can bind. On activation the platelets release various mediators into the circulation including transforming growth factor (TGF-β), interleukin 1 beta (IL-1β), prostaglandin E_2 (PHE$_2$), thromboxane A_2 (TXA$_2$), and CD40 ligand (CD40L). There is also a transient increase in free cytoplasmic calcium ions (Ca^{2+}). Thromboxane A_2 stimulates platelet aggregation while the prostaglandin I_2 (PGI$_2$) inhibits platelet function as does an increase in intracellular platelet cyclic adenosine monophosphate (cAMP), cyclic guanine monophosphate (cGMP), and nitric oxide (NO) [10].

The role of garlic in preventing platelet aggregation has been extensively investigated and evidence points to garlic preventing platelet aggregation by multiple effects [50]. Cyclooxygenase (COX) enzymes are elevated in inflamed and cancerous cells and it has been reported that garlic inhibits COX-1 and COX-2 enzymes and lipid peroxidation *in vitro* [42]. In support of this, di-allyl trisulfide (DAT) rich garlic oil displayed anticoagulant properties by inhibition and/or inactivation of thrombin hence preventing thrombus formation. However, at a high dose DAT significantly increased plasma fibrinogen concentration and also affected the levels of other hematological parameters such as erythrocyte count, hemoglobin, and platelets [51]. Platelets express the constitutive isoform of nitric oxide synthetase (cNOS) and activation of this enzyme by calcium influx during platelet aggregation provides an important feedback mechanism which prevents platelet aggregation. Garlic has also been reported to exert its beneficial effect on platelet aggregation by increasing NO availability. Nitric oxide inhibits platelet adhesion, platelet granule secretion, platelet aggregation, and arachidonic acid liberation [49,52]. It has also been reported that low density lipoprotein (LDL) on contact with platelets induces their aggregation [53]; hence garlic may be able to inhibit the pathways responsible for this. Although the majority of the *in vivo* studies indicate that consumption of garlic inhibits platelet aggregation [8], contradictory results have also been reported and could be due to the type and concentration of garlic used and the length of the studies [54,55]. Despite some contradictory data, there is strong evidence that garlic inhibits platelet aggregation by multiple mechanisms such as inhibition of cyclooxygenase activity, suppression of Ca^{2+} mobilization, increases in cAMP, increase in availability of NO, increase in antioxidant levels, and the interaction and inhibition of exposure of GPIIb/IIIa to its fibrinogen receptor [10,56].

7. EFFECT OF GARLIC ON BLOOD PRESSURE

Raised blood pressure (hypertension) is a well-established risk factor for heart disease and there appears to be a strong association between blood pressure and some oxidative stress-related parameters which indicates that oxidative stress may be involved in the pathophysiology of hypertension [57]. Diets rich in phytochemicals such as garlic have been shown to reduce hypertension. Aged garlic extract has been shown to improve endothelial function in men with coronary artery disease, and garlic is reported to improve coronary endothelial function by reducing pulmonary vasoconstriction [58,59] and this may be due to the ability of garlic to activate NO [52]. It has also been reported that garlic exerts its antihypertensive action by displaying prostaglandin-type effects which are known to decrease peripheral vascular resistance [60]. Another mechanism by which garlic may reduce hypertension is by its ability to inhibit angiotensin-converting enzymes and vasodilating effects

[61–63], however no effect of garlic in hypertension has also been reported [8,64].

In 2008, a systematic review and meta-analysis of the effect of garlic on blood pressure has been reported [65]. In this study randomized controlled trials with true placebo groups using garlic-only preparations, and reporting mean systolic and/or diastolic blood pressure and standard deviations were included in the meta-analysis. The conclusion of this review was that garlic preparations are superior to placebo in reducing blood pressure in individuals with hypertension.

8. OTHER PROPERTIES OF GARLIC ON HEART HEALTH

Garlic has been shown to reduce plasma viscosity and increase capillary skin perfusion [66]. In humans, consumption of garlic has shown a decrease in unstable angina, an increase in the elastic properties of the blood vessels, and a decrease in peripheral arterial occlusive disease was also observed [8]. Garlic has been shown to increase peripheral blood flow in healthy subjects and can also inhibit the progression of coronary calcification in patients undergoing statin therapy [67,68]; however, the number of studies within these areas is rather limited and very few clinical trials have been reported.

The majority of studies with garlic indicate that it has a positive effect on heart health; however, some conflicting data has also been reported mainly due to the use of different garlic preparations, the number and type of clinical trials, and the questions addressed by them [8].

9. SUMMARY

Evidence supports the fact that regular consumption of garlic can reduce factors associated with cardiovascular disease; this is summarized

Summary of the Mechanisms by which
Garlic Maintains a Healthy Heart

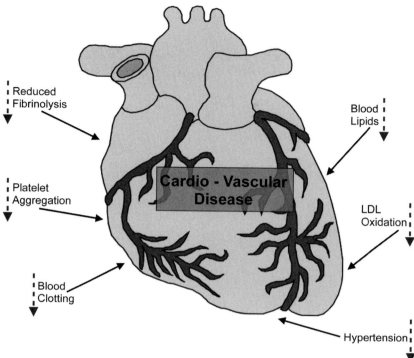

FIGURE 16.1 Summary of the mechanism by which garlic maintains a healthy heart.
Factors which contribute to cardiovascular disease are reduced fibrinolysis, increase in platelet aggregation, blood clotting, elevated serum blood lipids, increase in LDL oxidation, and hypertension. These are indicated by the solid arrows. Garlic reduces the contribution of these as indicated by the broken arrows.

in Figure 16.1 It suggests that garlic can reduce serum lipids in hypercholesterolemic subjects, and in addition, has strong antioxidant properties and can reduce oxidative stress by increasing the availability of NO, which in turn reduces the extent of LDL oxidation. Garlic consumption is also associated with a reduction in hypertension and the most widely reported property of garlic is its ability to inhibit platelet aggregation in healthy subjects and patients with CVD. The mechanisms of action by which garlic prevents CVD are being addressed and its consumption for maintaining a healthy heart should be encouraged.

References

1. Rahman, K. (2003). Garlic and aging: new insights into an old remedy. *Ageing Research Reviews, 2,* 39–56.
2. Bazzano, L. A., Serdula, M. K., & Liu, S. (2003). Dietary intake of fruit and vegetables and risk of cardiovascular disease. *Current Atherosclerosis Reports, 5,* 492–499.
3. Lundberg, J. O., Feelisch, M., Björne, H., Jansson, E. A., & Weitzberg, E. (2006). Cardioprotective effects of vegetables: Is nitrate the answer? *Nitric Oxide, 15,* 359–362.
4. Herman, P. M., Craig, B. M., & Caspi, O. (2005). Is complementary and alternative medicine (CAM) cost-effective? A systematic review. *BMC Complementary and Alternative Medicine, 5,* 11.
5. Lawson, L. D. (1996). The composition and chemistry of garlic cloves and processed garlic. In H. P. Koch, &

L. D. Lawson, (Eds.), *Garlic The Science and Therapeutic Application of Allium sativum L and related species* (pp. 37–107). USA: Williams and Wilkins Press.

6. Tattelman, E. (2005). Health effects of garlic. *American Family Physician, 72*, 103–106.

7. Banerjee, S. K., & Maulik, S. K. (2002). Effect of garlic on cardiovascular disorders: a review. *Nutrition Journal, 1*, 4–14.

8. Rahman, K., & Lowe, G. M. (2006). Garlic and cardiovascular disease: a critical review. *Journal of Nutrition, 136*, 736S–740S.

9. Wood, D. (2001). Established and emerging cardiovascular risk factors. *American Heart Journal, 141*, S49–S57.

10. Rahman, K. (2007). Effects of garlic on platelet biochemistry and physiology. *Molecular Nutrition and Food Research, 51*, 1335–1344.

11. Scott, J. (2004). Pathophysiology and biochemistry of cardiovascular disease. *Current Opinion in Genetics and Development, 14*, 271–279.

12. Sharples, F. M., van Haselen, R., & Fisher, P. (2003). NHS patients' perspective on complementary medicine: a survey. *Complementary Therapies in Medicine, 11*, 243–248.

13. Crawford, N. W., Cincotta, D. R., Lim, A., & Powell, C. V. (2006). A cross-sectional survey of complementary and alternative medicine use by children and adolescents attending the University Hospital of Wales. *BMC Complementary and Alternative Medicine, 6*, 16.

14. Kennedy, J. (2005). Herb and supplement use in the US adult population. *Clinical Therapeutics, 27*, 1847–1858.

15. Buettner, C., Phillips, R. S., Davis, R. B., Gardiner, P., & Mittleman, M. A. (2007). Use of dietary supplements among United States adults with coronary artery disease and atherosclerotic risks. *American Journal of Cardiology, 99*, 661–666.

16. Pharand, C., Ackman, M. L., Jackevicius, C. A., Paradiso-Hardy, F. L., & Pearson, G. J. Canadian Cardiovascular Pharmacists Network. (2003). Use of OTC and herbal products in patients with cardiovascular disease. *Annals of Pharmacotherapy, 37*, 899–904.

17. Yeh, G. Y., Davis, R. B., & Phillips, R. S. (2006). Use of complementary therapies in patients with cardiovascular disease. *American Journal of Cardiology, 98*, 673–680.

18. Edwards, Q. T., Colquist, S., & Maradiegue, A. (2005). What's cooking with garlic: is this complementary and alternative medicine for hypertension? *Journal of the American Academy of Nurse Practitioners, 17*, 381–385.

19. Artz, M. B., Harnack, L. J., Duval, S. J., Armstrong, C., Arnett, D. K., & Luepker, R. V. (2006). Use of non-prescription medications for perceived cardiovascular health. *American Journal of Preventive Medicine, 30*, 78–81.

20. Araujo, J. P., Friões, F., Azevedo, A., Lourenço, P., Rocha-Gonçalves, F., Ferreira, A., & Bettencourt, P. (2008). Cholesterol – A marker of nutritional status in mild to moderate heart failure. *International Journal of Cardiology, 129*, 65–68.

21. Gebhardt, R. (1993). Multiple inhibitory effects of garlic extracts on cholesterol biosynthesis in hepatocytes. *Lipids, 28*, 613–619.

22. Liu, L., & Yeh, Y. Y. (2001). Water-soluble organosulfur compounds of garlic inhibit fatty acid and triglyceride synthesis in cultured rat hepatocytes. *Lipids, 36*, 395–400.

23. Yanagita, T., Han, S., Wang, U., Tsuruta, Y., & Anno, T. (2003). Cycloalliin, a cyclic sulphur imino acid, reduces serum triacylglycerol in rats. *Nutrition, 19*, 140–143.

24. Gonen, A., Harats, D., Rabinkov, A., Miron, T., Mirelman, D., Wilchek, M., Weiner, L., Ulman, E., Levkovitz, H., Ben-Shushan, D., & Shaish, A. (2005). The antiatherogenic effect of allicin: possible modes of action. *Pathobiology, 72*, 325–334.

25. Murugavel, P., & Pari, L. (2007). Diallyl tetrasulfide protects cadmium-induced alterations in lipids and plasma lipoproteins in rats. *Nutrition Research, 27*, 356–361.

26. Taghizadeh, A. A., Shirpoor, A., & Dodangeh, B. E. (2005). The effect of garlic on cyclosporine-a-induced hyperlipidemia in male rats. *Urologic Journals, 2*, 153–156.

27. Gorinstein, S., Jastrzebski, Z., Namiesnik, J., Leontowicz, H., Leontowicz, M., & Trakhtenberg, S. (2007). *Molecular Nutrition and Food Research, 51*, 1365–1381.

28. Mahmoodi, M., Ismail, M. R., Asadi, K. G. R., Khaksari, M., Sahrbghadam, L. A., Hajizadeh, M. R., & Mirzaee, M. R. (2006). Study of the effects of raw garlic consumption on the level of lipids and other blood biochemical factors in hyperlipidemic individuals. *Pakistan Journal of Pharmaceutical Sciences, 19*, 295–298.

29. Kojuri, J., Vosoughi, A. R., & Akrami, M. (2007). Effects of anthum graveolens and garlic on lipid profile in hyperlipidemic patients. *Lipids Health Disease, 6*, 5.

30. Duda, G., Suliburska, J., & Pupek-Musialik, D. (2008). Effects of short-term garlic supplementation on lipid metabolism and antioxidant status in hypertensive adults. *Pharmacology Reports, 60*, 163–170.

31. Zhang, L., Gail, M. H., Wang, Y. Q., Brown, L. M., Pan, K. F., Ma, J. L., Amagase, H., You, W. C., & Moslehi, R. (2006). A randomised factorial study of the effects of long-term garlic and micronutrient supplementation and of 2-wk antibiotic treatment for Helicobacter pylori infection on serum cholesterol and

lipoproteins. *American Journal of Clinical Nutrition, 84,* 912–919.

32. Gardner, C. D., Lawson, L. D., Block, E., Chatterjee, L. M., Kiazand, A., Balise, R. R., & Kraemer, H. C. (2007). Effect of raw garlic vs commercial garlic supplements on plasma lipid concentrations in adults with moderate hypercholesterolemia: a randomised clinical trial. *Archives of Internal Medicine, 167,* 346–353.

33. Calpe-Berdiel, L., Escolá-Gil, J. C., & Blanco-Vaca, F. (2009). New insights into the molecular actions of plant sterols and stanols in cholesterol metabolism. *Atherosclerosis, 203,* 18–31.

34. Rahman, K. (2007). Studies on free radicals, antioxidants, and co-factors. *Clinical Interventions in Aging, 2,* 219–236.

35. Libby, P. (2002). Inflammation in atherosclerosis. *Nature, 420,* 868–874.

36. Papaharalambus, C. A., & Griendling, K. K. (2007). Basic mechanisms of oxidative stress and reactive oxygen species in cardiovascular injury. *Trends in Cardiovascular Medicine, 17,* 48–54.

37. Ide, N., & Lau, B. H. S. (2001). Garlic compounds minimise intracellular oxidative stress and inhibit nuclear factor-κB activation. *Journal of Nutrition, 131,* 1020S–11026S.

38. Dillon, S. A., Burmi, R. S., Lowe, G. M., Billington, D., & Rahman, K. (2003). Antioxidant properties of aged garlic extract: an *in vitro* study incorporating human low density protein. *Life Science, 72,* 1583–1594.

39. Okada, Y., Tanaka, K., Sato, E., & Okajima, H. (2006). Kinetic and mechanistic studies of allicin as an antioxidant. *Organic and Biomolecular Chemistry, 4,* 4113–4117.

40. Rassoul, F., Salvetter, J., Reissig, D., Schneider, W., Thiery, J., & Richter, V. (2006). The influence of garlic (*Allium sativum*) extract on interleukin 1 alpha-induced expression of endothelial intercellular adhesion molecule-1 and vascular cell adhesion molecule-1. *Phytomedicine, 13,* 230–235.

41. Lei, Y. P., Chen, H. W., Sheen, L. Y., & Lii, C. K. (2008). Diallyl disulfide and diallyl trisulfide suppress oxidized LDL-induced vascular cell adhesion molecule and E-selectin expression through protein kinase A- and B-dependent signaling pathways. *Journal of Nutrition, 138,* 996–1003.

42. Raman, P., Dewitt, D. L., & Nair, M. G. (2008). Lipid peroxidation and cyclooxygenase enzyme inhibitory activities of acidic aqueous extracts of some dietary supplements. *Phytotherapy Research, 22,* 204–212.

43. Banerjee, S. K., Mukherjee, P. K., & Maulik, S. K. (2003). Garlic as an antioxidant: The good, the bad and the ugly. *Phytotherapy Research, 17,* 97–106.

44. Nishibe, A., Kijima, Y., Fukunaga, M., Nishiwak, N, Sakai, T, Nakagawa, Y., & Hata, T. (2008). Increased isoprostane content in coronary plaques obtained from vulnerable patients. *Prostaglandins Leukotrienes and Essential Fatty Acids, 78,* 257–263.

45. Radovanovic, S., Krotin, M., Simic, D. V., Mimic-Oka, J., Savic-Radojevic, A., Pijesa-Ercegovac, M., Matic, M, Ninkovic, N., Ivanovic, B., & Simic, T. (2008). Markers of oxidative damage in chronic heart failure: role in disease progression. *Redox Report, 13,* 109–116.

46. Dillon, S. A., Lowe, G. M., Billington, D., & Rahman, K. (2002). Dietary supplementation with aged garlic extract reduces plasma and urine concentrations of 8-iso-prostaglandin $F_{2\alpha}$ in smoking and nonsmoking men and women. *Journal of Nutrition, 132,* 168–171.

47. Avic, A., Atli, T., Ergüder, I. B., Varli, M., Devrim, E., Aras, E., & Durak, I. (2008). Effects of garlic consumption on plasma and erythrocyte antioxidant parameters in elderly subjects. *Gerontology, 54,* 173–176.

48. Wiess, N., Ide, N., Abahji, T., Nill, L., Keller, C., & Hoffmann, U. (2006). Aged garlic extract improves homocysteine-induced endothelial dysfunction in macro- and microcirculation. *Journal of Nutrition, 136,* 750S–754S.

49. Willoughby, S., Holmes, A., & Loscalzo, J. (2002). Platelets and cardiovascular disease. *Eur. Journal of Cardiovascular Nursing, 1,* 273–288.

50. Allison, G. L., Lowe, G. M., & Rahman, K. (2006). Aged garlic extract and its constituents inhibit platelet aggregation through multiple mechanisms. *Journal of Nutrition, 136,* 782S–788S.

51. Chan, K. C., Yin, M. C., & Chao, W. J. (2007). Effect of diallyl trisulfide-rich garlic oil on blood coagulation and plasma activity of anticoagulation factors in rats. *Food and Chemical Toxicology, 45,* 502–507.

52. Ku, D. D., Abdel-Razek, T. T., Dai, J., Kim-Park, S., Fallon, M. B., & Abrams, G. A. (2002). Garlic and its active metabolite allicin produce endothelium- and nitric oxide-dependent relaxation in rat pulmonary arteries. *Clinical and Experimental Pharmacology and Physiology, 29,* 84–91.

53. Akkerman, J. W. (2008). From low-density lipoprotein to platelet activation. *International Journal of Biochemistry and Cell Biology, 40,* 2374–2378.

54. Wojcikowski, K., Myers, S., & Brooks, L. (2007). Effects of garlic oil on platelet aggregation: a double-blind placebo-controlled crossover study. *Platelets, 18,* 29–34.

55. Scharbert, G., Kalb, M. L., Duris, M., Marschalek, C., & Kozek-Langenecker, S. A. (2007). Garlic at dietary doses does not impair platelet function. *Anesthesia and Analgesia, 105,* 1214–1218.

56. Allison, G. L., Lowe, G. M., & Rahman, K. (2006). Aged garlic extract may inhibit aggregation in human

platelets by suppressing calcium mobilization. *Journal of Nutrition, 136*, 789S–792S.

57. Rodrigo, R., Prat, H., Passalacqua, W., Araya, J., Guichard, C., & Bächler, J. P. (2007). Relationship between oxidative stress and essential hypertension. *Hypertension Research, 30*, 1159–1167.

58. Williams, M. J., Sutherland, W. H., McCormick, M. P., Yeoman, D. J., & De Jong, S. A. (2005). Aged garlic extract improves endothelial function in men with coronary artery disease. *Phytotherapy Research, 19*, 314–319.

59. Sun, X., & Ku, D. D. (2006). Allicin in garlic protects against coronary endothelial dysfunction and right heart hypertrophy in pulmonary hypertensive rats. *American Journal Physiology Heart and Circulatory Physiology, 291*, H2431–H2438.

60. Rashid, A., & Khan, H. H. (1985). The mechanism of hypotensive effect of garlic extract. *Journal of Pakistan Medical Association, 35*, 357–362.

61. Al-Qattan, K. K., Khan, I., Alnaqeeb, M. A., & Ali, M. (2003). Mechanism of garlic (Allium sativum) induced reduction of hypertension in 2K-1 C rats: a possible mediation of Na/H exchanger isoform-1. *Prostaglandins Leukotrienes and Essential Fatty Acids, 69*, 217–222.

62. Al-Qattan, K. K., Thomson, M., Al-Mutawa'a, S., Al-Hajeri, D., Drobiova, H., & Ali, M. (2006). Nitric oxide mediates the blood-pressure lowering effect of garlic in the rat two-kidney, one-clip model of hypertension. *Journal of Nutrition, 136*, 774S–776S.

63. Sharif, A. M., Darabi, R., & Akbarloo, N. (2003). Investigation of antihypertensive mechanism of garlic in 2K 1C hypertensive rat. *Journal of Ethnopharmacology, 86*, 219–224.

64. Capraz, M., Dilek, M., & Akpolat, T. (2007). Garlic, hypertension and patient education. *International Journal of Cardiology, 121*, 130–131.

65. Ried, K., Frank, O. R., Stocks, N. P., Fakler, P., & Sullivan, T. (2008). Effect of garlic on blood pressure: A systematic review and meta-analysis. *BMC Cardiovascular Disorders, 8*, 13.

66. Kisewetter, H., Jung, F., Pindur, G., Jung, E. M., Mrowietz, C., & Wenzel, E. (1991). Effect of garlic on the thrombocyte aggregation, microcirculation, and other risk factors. *International Journal of Clinical Pharmacology, Therapy and Toxicology, 29*, 151–155.

67. Anim-Nyame, N., Sooranna, S. R., Johnson, M. R., Gamble, J., & Steer, P. J. (2004). Garlic supplementation increases peripheral blood flow: a role for interleukin-6? *Journal of Nutritional Biochemistry, 15*, 30–36.

68. Budoff, M. J., Takasu, J., Flores, F. R., Nihara, Y., Lu, B., Lau, B. H., Rosen, R. T., & Amagase, H. (2004). Inhibiting progression of coronary calcification using aged garlic extract in patients receiving statin therapy: a preliminary study. *Preventive Medicine, 39*, 985–991.

17

Fig, Carob, Pistachio, and Health

Soliman Khatib[1,2] *and Jacob Vaya*[1,2]

[1]Laboratory of Natural Medicinal Compounds, MIGAL-Galilee Technology Center, Kiryat Shmona, Israel

[2]Department of Biotechnology, Tel-Hai Academic College, Israel

1. INTRODUCTION

1.1 Fig

The word 'fig' usually refers to Ficus, the fig tree and its fruit known as the Common fig (*Ficus carica*). The Common fig is a large, deciduous shrub or small tree native to southwestern Asia and the eastern Mediterranean region (Greece east to Afghanistan). It grows to a height of 3–10 m, with smooth gray bark. The Common fig is widely grown for its edible fruit throughout its natural range in Iran, as well as in the rest of the Mediterranean region, and in other areas of the world with a similar climate. There is evidence that figs, specifically the Common fig (*F. carica*) and Sycamore fig (*F. sycomorus*), were among the first – if not the very first – plant species that were deliberately bred for agriculture in the Middle East, starting more than 11,000 years ago. Nine subfossil *F. carica* figs dated to about 9400–9200 BC were found in the early Neolithic village, Gilgal I (in the Jordan Valley, 13 km north of Jericho). These were a parthenocarpic type, and thus apparently an early cultivar. This finding predates the cultivation of grain in the Middle East by many hundreds of years [1].

The fruit is 3–5 cm long, with a green skin, sometimes ripening towards purple. The sap of the tree's green parts is an irritant to human skin. The fig fruit is actually the flower of the tree, known as an inflorescence (an arrangement of multiple flowers), a false fruit or multiple fruit, in which the flowers and seeds grow together to form a single mass (Figure 17.1A).

Fig fruits can be eaten fresh or dried and are used in jam-making. Most commercial production is in dried or otherwise processed forms since the ripe fruit does not transport well, and once picked does not keep well. It is cooked as a vegetable and is believed to be good for heart ailments.

On the basis of the *Dietary Reference Intakes* (DRI) data published by the Food and Nutrition Board of the US Institute of Medicine, and the *Nutrient Composition Data* published by the US Department of Agriculture (USDA) 2002, dried figs have the best nutrient score among dried fruit. Figs are one of the highest sources of calcium and fiber in plants. Furthermore, dried figs are the richest in fiber, and contain

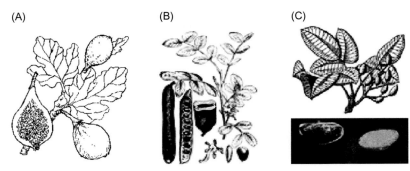

FIGURE 17.1 (A) Fig fruit (bbs.fruit8.com/viewthread.php?tid = 36364). (B) Carob pod (www.thefreedictionary.com/carob+beans). (C) Pistachio nuts (www.thefreedictionary.com/pistachio).

important minerals and vitamins (Table 17.1). Figs have a laxative effect and contain many antioxidants. Moreover, they are a good source of flavonoids and other polyphenols [2]. In one study, a 40 g portion of dried figs (two medium size figs) produced a significant increase in the plasma antioxidant capacity [3].

1.2 Carob

The carob (*Ceratonia siliqua* L.) is a leguminous evergreen shrub or tree of the family Leguminosae (pulse family) native to the eastern Mediterranean, probably the Middle East, where it has been cultivated for at least 4000 years. Its value was recognized by the ancient Greeks, who brought it from its native Middle East to Greece and Italy, and by the Arabs, who disseminated it along the North African coast and north into Spain and Portugal. The fruit is an indehiscent pod, elongated, compressed, straight or curved, thickened at the sutures, 10–30 cm long, 1.5–3.5 cm wide, and about 1 cm thick (Figure 17.1B). In many Mediterranean countries the fruit is used in popular beverages and confectioneries. In Western countries, carob powder is produced by deseeding carob pods. Carob powder is a natural sweetener [4] with a flavor and appearance similar to chocolate; therefore, it is often used as a cocoa substitute, free of caffeine and theobromine.

Carob pods ethanolic extract contains three major carbohydrates: sucrose (437.3 mg/g dry weight), glucose (395.3 mg/g dry weight) and fructose (42.3 mg/g dry weight). Together these three carbohydrates account for 87.54% of the total dry weight extract. Carob pods are also rich in protein (5–8 g protein per 100 g dry weight). Moreover, the pod has vitamins A and B and several important minerals, such as K, P, Ca, and Mg, as major minerals and Fe, Mn, Zn, and Cu as trace minerals. Carob pods are virtually fat-free (0.2–0.6%). Results published by the USDA National Nutrient Database for Standard Reference are presented in Table 17.1 [4–6].

1.3 Pistachio

Pistachio (*Pistacia vera* L., *Anacardiaceae*; sometimes placed in *Pistaciaceae*) is a small tree, reaching up to 10 m tall, and is native to mountainous regions. Pistachio nuts are commonly consumed in western Asia. Pistachios (as part of the *pistacia* genus) have existed for about 80 million years [7]. Archaeological evidence in Turkey indicates that nuts were used as food as early as 7000 BC.

The fruit is a drupe, containing an elongated seed with a hard shell and a striking kernel, which has mauvish skin, light green flesh, and a distinctive flavor. When the fruit ripens, the husk changes from green to an autumnal

TABLE 17.1   Nutritional Value (per 100 g) of Dry Roasted Pistachio Nuts, without Salt Carob Flour, and Dried Uncooked Figs

	Unit	Fig (dried)	Carob	Pistachio
Water	g	30.1	3.6	2.0
Energy	kcal	249.0	222.0	571.0
Protein	g	3.3	4.6	21.4
Total lipid (fat)	g	0.9	0.65	46.0
Fatty acids, total saturated	g	0.1	0.1	5.5
Fatty acids, total monounsaturated	g	0.2	0.2	24.2
Fatty acids, total polyunsaturated	g	0.35	0.2	13.9
Carbohydrate, by difference	g	63.9	88.9	27.7
Fiber, total dietary	g	9.8	39.8	10.3
Sugars, total	g	47.9	49.1	7.8
Minerals and Vitamins				
Calcium (Ca)	mg	162.0	348.0	110.0
Iron (Fe)	mg	2.0	2.9	4.2
Magnesium (Mg)	mg	68.0	54.0	120.0
Phosphorus (P)	mg	67.0	79.0	485.0
Potassium (K)	mg	680.0	827.0	1042.0
Sodium (Na)	mg	10.0	35.0	10.0
Zinc (Zn)	mg	0.5	0.9	2.3
Copper (Cu)	mg	0.3	0.6	1.3
Manganese (Mn)	mg	0.5	0.5	1.3
Selenium (Se)	μg	0.6	5.3	9.3
Vitamin C (total ascorbic acid)	mg	1.2	0.2	2.3
Thiamin (B1)	mg	0.1	0.05	0.8
Riboflavin (B2)	mg	0.1	0.5	0.15
Niacin (B3)	mg	0.6	1.9	1.4
Pantothenic acid (B5)	mg	0.4	0.05	0.5
Vitamin B6	mg	0.1	0.4	1.3
Folate, total (B9)	μg	9.0	29.0	50.0
Vitamin A	μg	0.0	1.0	13.0
Vitamin E (α-tocopherol)	mg	0.3	0.6	1.9
Vitamin K (phylloquinone)	μg	15.6	0.0	13.2

Data from the USDA National Nutrient Database for Standard References.

yellow/red and the shells split partially open (Figure 17.1C). This is known as dehiscence and happens with an audible pop. The kernels are often eaten whole, either fresh, or roasted and salted, and are also used in ice cream and confections. In 2003, the Food and Drug Administration (FDA) approved the first qualified health claim specific to nuts, declaring them to lower the risk of heart disease: 'Scientific evidence suggests, but does not prove, that eating 1.5 ounces (42.5 g) per day of most nuts, such as pistachios, as part of a diet low in saturated fat and cholesterol, may reduce the risk of heart disease.' In a study at Pennsylvania State University, pistachios, in particular, significantly reduced the level of low density lipoprotein in the blood of volunteers [8]. Dry roasted pistachio nuts are rich in carbohydrates (27.65 g total carbohydrates, 7.88 g sugars, and 10.33 g dietary fibers per 100 g), proteins (21.35 g per 100 g) and fat (45.97 g per 100 g; of which 5.5 g are saturated fatty acids and 40.5 g unsaturated fatty acids). In addition, pistachio nuts contain some important vitamins and minerals (see Table 17.1).

2. PHENOLIC AND POLYPHENOL CONTENT AND COMPOSITION IN THE FIG, CAROB AND PISTACHIO

Polyphenol compounds are distributed among many types of fruits and vegetables, where they carry out specific functions and impart sensory properties (flavor and color). The interest in polyphenol compounds has increased over the last few decades, as these agents have become popular among the public, who attribute them several medicinal properties, mostly related to their antioxidant activity [9,10].

Polyphenols are divided into two categories: flavonoidal and non-flavonoidal (Figure 17.2). The non-flavonoidal phenols include phenolic acids, lignans, and stilbenes. The stilbene resveratrol, found in red wine, has recently received great attention regarding its anticarcinogenic properties and neuroprotective effect [11,12].

Flavonoids may be divided into several subclasses, namely anthocyanins, flavonols,

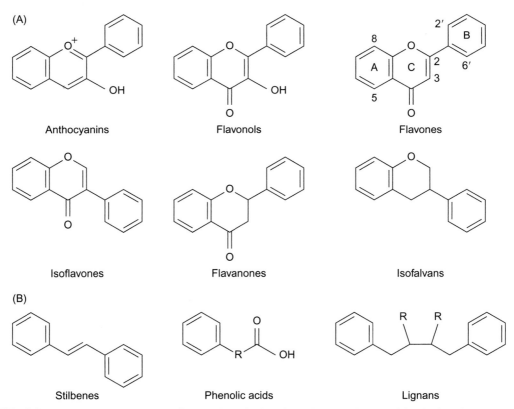

FIGURE 17.2   Polyphenol structures: (A) flavonoidal polyphenols, and (B) non-flavonoidal polyphenols.

flavans, flavanol, flavones, and isoflavones, based on their heterocyclic ring (ring C), degree of oxidation and ring B position (Figure 17.2A). In general, these compounds are constructed from an aromatic ring that is hydroxylated or methoxylated and/or glycosylated. The linked carbohydrate moieties are often glucose or rhamnose. The number of carbohydrate units is commonly 1, but they may also comprise 2 or 3 units bound to the hydroxyl functional group on the aromatic ring at several possible positions [13,14].

Flavonoids are the most widely studied class of polyphenols with respect to their antioxidant and biological properties. They display powerful antioxidant activities *in vitro*, being able to scavenge a wide range of reactive oxygen species (ROS), reactive nitrogen species (RNS), and chlorine species, among them superoxide, hydroxyl, and peroxyl radicals, as well as hypochlorous and peroxynitrous acids. Some flavonoids can also chelate metal ions, such as copper and iron.

Polyphenols constituents, existing in vegetables and fruits, have been shown to have the capacity to delay or prevent diseases associated with oxidative stress (OS), e.g. the neurological diseases Alzheimer's and Parkinson's [14–16], atherosclerosis, cancer, diabetes, and aging [17–24].

It is extremely difficult to estimate the daily average intake of polyphenols, because of the diversity of their chemical structure, making the estimation of their content in foods complex. Moreover, the polyphenol intake level is largely depend on the sort of food consumed, geographical distribution of the specific type, seasonal variations in their content, people's habits (the preferential consumption of tea, coffee, and/or wine), and the analytical methods employed to detect polyphenol levels. It is estimated that in the USA the total intake of polyphenols is 1 g/d, that of anthocyanins is 12.5 mg/d, and the consumption of total flavonoids is 13 mg/d [14,25].

The bioavailability of polyphenols varies widely from one compound to another. It depends on their chemical structure, which determines their absorption rate through the gastrointestinal tract and their metabolism, and thus, their biological bioavailability. Most polyphenolics are poorly absorbed from the intestine and are highly metabolized, or rapidly eliminated. The maximal plasma concentrations of flavonoids are low, usually not more than 1 µM, with a maximum level attained 1–2 h after ingestion. Therefore, the maintenance of a high concentration in the plasma requires repeated ingestion of the polyphenols over time.

Table 17.2 summarizes the phenolic and polyphenol content and composition in the fig, carob, and pistachio.

2.1 Polyphenol Content and Composition in the Fig

The polyphenol antioxidant composition and level in dried fig fruits was determined and analyzed by several investigators. Its total content in fig fruits was analyzed using an oxidation–reduction colorimetric assay (Folin–Ciocalteu reagent) [26], which showed that dark fig fruits were richer in polyphenols than the green varieties, and the skin of the fruit had a higher content of polyphenols than the pulp. As reported by Solomon et al. [27], who used gallic acid as the standard, the total polyphenol content ranged from 49 to 281 mg (gallic acid equivalent)/100 g fruit weight (FW) in the light and the dark fig fruits, respectively; their polyphenol content in the skins varied from 42 to 463 (gallic acid equivalent) mg/100 g FW in the light and the dark ones, respectively; and the polyphenol content in pulps ranged from 37 to 100 mg (gallic acid equivalent)/100 g FW in the light and the dark varieties, respectively. These results have been confirmed by Del Caro and Piga [10], who analyzed the total polyphenol content and

TABLE 17.2   Polyphenol Composition of Fig Fruit, Carob Pods, and Pistachio Nuts

	Fig (mg/100 g FW)	Carob (mg/100 g FW)	Pistachio (mg/100 g FW)
Total polyphenol content	49–281 [3,10,27]	1950 [32]	35 [38]
Gallic acid	0.38 [29]	602 [36]	ND
Ellagic acid	ND	25 [36]	ND
Chlorogenic acid	1.71 [29]	ND	ND
Syringic acid	0.1 [29]	ND	ND
trans-resveratrol	ND	ND	0.709[b] [39–41]
Total flavonoid content	69.7–145.2 [10,27]	ND	ND
Rutin	52.7–107.1 [10,29]	ND	ND
(+)-Catechin	4.03 [29]	50 [36]	ND
(−)-Epicatechin	0.97 [29]	–	ND
Quercetin glycosides	ND	107 [36]	ND
(−)-Epicatechin gallate	ND	30 [36]	ND
(−)-Epigallocatechin gallate	ND	110 [36]	ND
Daidzein	ND	ND	3.68[b] [41]
Genistein	ND	ND	3.4[b] [41]
Total anthocyanin content	3.2–11 [27,31]	ND	ND
Cyanidin (cy)3-rutinoside	48–97%[a] [10,27,31]	ND	ND
Cyanidin (cy)3-glucoside	5–15%[a] [10,27,31]	ND	ND
Cyanidin (cy)3-galactoside	ND	ND	0–42.6 [42]
Proanthocyanidins	ND	290 [32]	268b [41]
Ellagitannins	ND	46 [32]	ND

FW, fruit weight; ND, not detected.
Numbers in brackets are references.
[a]The values are percent of the total anthocyanin content.
[b]The value per 100 g extract and not fruit dry weight.

composition in fig fruit by using the high performance liquid chromatography (HPLC)-diode array detector (DAD) method. These authors showed that polyphenols were localized mostly in the peel, while the pulp had only a low content of polyphenols. In addition, the peel of the dark fig fruit was richer in polyphenols (253 mg/100 g FW) than the peel of the lighter varieties (78 mg/100 g FW).

Vinson et al. [3], who analyzed the total polyphenols in fresh and dried fruits, using the same method as Solomon et al. [27], but with catechin as a standard, reported that the total polyphenol content in fresh fig fruit was 486±218 (mg catechin equivalents per 100 g FW), whereas in dried fruit it was 320±37 (mg catechin equivalents per 100 g FW). Moreover, they testified that dried figs and plums have the best nutrient score among the other dried fruits investigated.

Flavonoids belong to the widespread class of polyphenol compounds. The content and composition of the flavonoids have been analyzed by several methods, such as HPLC–DAD and HPLC–mass spectrometry (MS). The total flavonoid content in fig fruits, as reported by Solomon et al. [27], was measured colorimetrically [28] using a 510 nm wavelength and (+)-catechin as the standard. Similarly, flavonoids are also localized in the skin of the dark fig fruit. Total fruit flavonoids ranged from 21.5 (mg catechin equivalent/100 g FW) for the darker variety to 2.1 (mg catechin equivalent/100 g FW) for the lighter one. The skin was the major tissue that contributed to the total flavonoid content, ranging from 45.6 (mg catechin equivalent/100 g FW) for the darker variety to 2.2 (mg catechin equivalent/100 g FW) for the lighter one. The flavonoid composition of the peel and pulp of two Italian fresh fig cultivars (*Ficus carica* L.) were analyzed by Del Caro and Piga [10], using the HPLC–DAD method. Results showed that all of the flavonoids content is concentrated in the peel in the case of the black cultivar. The total flavonoids content was 145.2 mg/100 g FW for the black peel and 69.7 mg/100 g FW for the green one, where rutin was the major flavonoid constituent, with levels ranging from 52.7 to 107.1 mg/100 g (fresh peel basis) for green and black figs, respectively. Similar results were obtained by Veberic et al. [29], who also analyzed the flavonoids content in the fig fruit, using the HPLC–DAD system, which identified the flavonoids (+)-catechin, (−)-, epicatechin and rutin as well as the polyphenols gallic acid, chlorogenic acid, and syringic acid.

Analysis of polyphenols extracted from light and dark type figs revealed levels of rutin ranging from 28.7 mg per 100 g FW for the dark cultivar to 4.89 mg per 100 g FW for the lighter one, followed by (+)-catechin (up to 4.03 mg per 100 g FW), chlorogenic acid (up to 1.71 mg per 100 g FW), (−)-epicatechin (up to 0.97 mg per 100 g FW), gallic acid (up to 0.38 mg per 100 g FW), and syringic acid (up to 0.10 mg per 100 g FW) [29].

In another study, Vaya and Mahmood [30] analyzed the total flavonoid content of leaf extracts (70% ethanol) from fig (*Ficus carica* L.), carob (*Ceratonia siliqua* L.), and pistachio (*Pistacia lentiscus* L.) plants by reverse phase HPLC and DAD array and electrospray ionization (ESI)-mass spectrometry (MS) detectors. They found that the major flavonoids in the fig leaf are quercetin and luteolin, with a total of 63.1 and 68.1 mg/100 g extract, respectively.

Anthocyanins belong to the widespread class of flavonoid compounds. The content and composition of the anthocyanins were determined by investigators, who employed several methods, such as HPLC–DAD and HPLC–MS. The anthocyanin composition in the fig fruit (*Ficus carica* L.) was analyzed by Duenas et al. using HPLC, reverse phase chromatography coupled to DAD and MS detection [31]. Fifteen anthocyanin pigments were detected, most of them containing cyanidin (Cy), e.g. aglycone and some pelargonidin (Pg) derivatives. Rutinose and glucose were present as substituting sugars, as well as acylation with malonic acid. Minor levels of peonidin 3-rutinose (Pn 3-rutinoside) in the pulp were also detected. According to Duenas et al. [31] the total anthocyanin content in the skin ranged between 3.2 and 9.7 mg per 100 g FW and between 0.15 and 1.5 mg per 100 g FW in the pulp for the light and dark fig fruit, respectively. The main anthocyanin in both parts of the fruit was Cy 3-rutinoside (Figure 17.3A) (48–81% of the total anthocyanins in skin and 68–79% in pulp) followed by Cy 3-glucoside (Figure 17.3B) (5–18% in skin and 10–15% in pulp). Malonyl derivatives were more abundant in the skin (1.2–6.5%) than in the pulp (1.0–2.6%).

The total anthocyanin content described by Del Caro and Piga [10] was 93 (mg per 100 g FW) and 0.49 (mg per 100 g FW) for the peel and the pulp of the dark fig fruit, respectively. In addition, minor amounts of anthocyanins were detected in the pulp of the green fruit (1.7 mg per 100 g FW). As reported by Duenas et al. [31], Del Caro and Piga [10] also stated that cyanidin 3-*O*-rutinoside is the main anthocyanin in the pulp and peel of the fruit (97%), while cyanidin 3-*O*-glucoside was less prevalent, detected only in the peel of the black figs (3%).

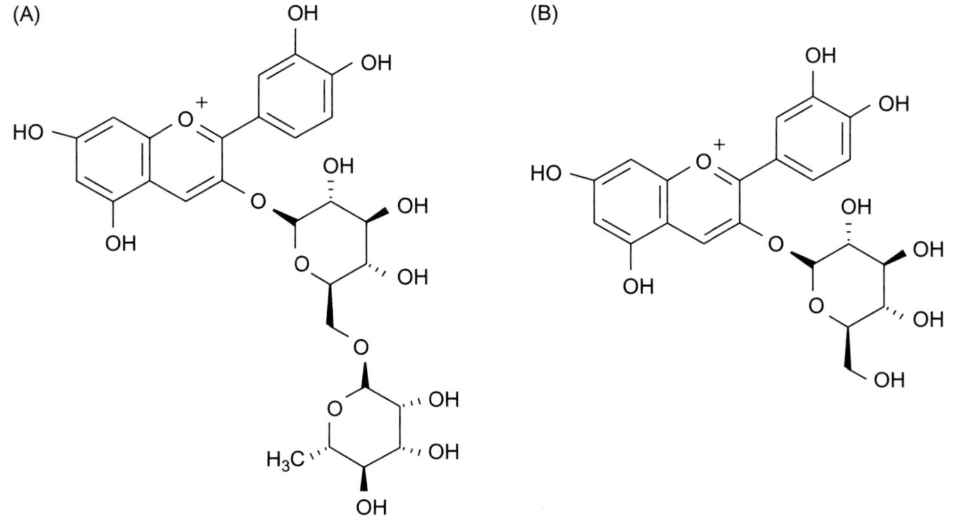

FIGURE 17.3   (A) Cyanidin (Cy) 3-rutinoside, and (B) cyanidin (Cy) 3-glucoside.

Total anthocyanin content was also determined by Solomon et al. [27] using a pH differential method [28]. Absorbance was measured at 520 and 700 nm and expressed as cyanidin-3-glycoside (molar extinction coefficient of $26\,900\,L.m^{-1}.mol^{-1}$ and molecular weight of 449.2) equivalents per 100 g FW or 100 g tissue. The total anthocyanin level was higher in all ripe fruits as compared to that in unripe ones. For example, the anthocyanin level was 3.0 mg/ 100 g FW in the unripe variety and 11.0 mg/ 100 g FW in the ripe one. Furthermore, extracts of darker varieties showed higher contents of anthocyanins, ranging from 11.0 (mg/100 g FW) for the dark varieties to 0.1 (mg/100 g FW) for the light ones. In both ripe and unripe fig fruits, skins had higher anthocyanin levels as compared to pulps. For example, in the dark fig fruit, the skin had 100 times more anthocyanin content, as compared to pulp. Accordingly, dark fruit skins had more anthocyanins, relative to purple or lighter varieties.

Corresponding to the results reported by Duenas et al. [31] and Del Caro and Piga [10], Solomon et al. [27] concluded that cyanidin-3-O-rhamnoglucoside (cyanidin-3-O-rutinoside; C3R) was the main anthocyanin in all fruits (95%), followed by cyanidin 3-O-glucoside.

2.2 Phenol and Polyphenol Content and Composition in the Carob

The polyphenol and tannin contents in an acetone extract of carob pods were analyzed by Avallone et al. [32], using the Folin–Ciocalteu method for the polyphenols [26], and by the vanillin assay [33] for the analysis of the proanthocyanidin quantity. The ellagitannin level was estimated by oxidation with nitrous acid, according to the method described by Bate-Smith [34], and gallotannins were exposed to acid hydrolysis, affording free gallic acid, which was further detected by a rhodanine reagent [35]. The results showed that pure acetone was inefficient in the extraction of polyphenols, and in particular, of tannins from carob pods.

The efficiency of tannin extraction was improved significantly by using a mixture of acetone and water. Thus, 70% acetone extracted 1950 mg of total polyphenols, 290 mg of proanthocyanidins, and 46 mg of ellagitannins in 100 g carob pods (all based on dry weight).

The polyphenol constituents in carob pods were analyzed using an HPLC with a DAD in conjunction with a liquid chromatography LC–MS method for the determination of major polyphenols in vegetables, fruits and teas [36]. It was revealed that carob pods contain the following flavonoids: quercetin glycosides [231 μmole (107 mg)/100 g FW], (+)-catechin [175 μmole (50 mg)/100 g FW], (−)-epicatechin gallate [68 μmole (30 mg)/100 g FW], (−) epigallocatechin gallate [239 μmole (110 mg)/100 g FW]. An abundant quantity of polyphenols was also found, including gallic acid [3540 μmole (602 mg)/100 g FW] and ellagic acid [84 μmole (25 mg)/100 g FW].

Menelaos Papagiannopoulos et al. also analyzed the polyphenol composition (simple phenols, tannins, and flavonoids) of the carob fibers in a similar fashion using an HPLC with a UV detector in conjunction with MS [37]. In carob fibers, which had been prepared from carob pods, 41 individual phenolic compounds could be detected. They included gallic acid, hydrolyzable tannins, and condensed tannins, as well as derivatives of myrictin, quercitin, and kaempferol. The polyphenol content was influenced by carob pod processing. Carob pods (kibbles), carob syrup, carob fiber, and roasted carobs were found to contain 44.8, 108.5, 414.2, and 120.8 mg polyphenols/100 g, respectively. In addition, the polyphenol composition was affected by the carob pod processing. For example, the content of condensed tannins in carob pods (kibbles), carob syrup, carob fiber and roasted carobs was 1.4.8, 0.90, 19.18 and 1.87 mg tannins/100 g, respectively.

The flavonoid content in leaf extracts of carobs (Ceratonia siliqua L.) was also examined [30] by Vaya and Mahmood using HPLC connected to UV/VIS array and ESI-MS detectors; nine different flavonoids were detected, with the major one being myricetin (148.6 mg/100 g extract).

2.3 Phenol and Polyphenol Content and Composition in the Pistachio

The total polyphenol content in pistachio hulls was analyzed by Goli et al. [38] using the Folin–Ciocalteu method [26]. The polyphenols were extracted by two different extraction methods (solvent and ultrasound-assisted) and with a mixture of three different solvents (water, methanol, and ethyl acetate); the total polyphenols were determined and expressed as tannic acid equivalents per 100 g dry weight of sample (TAE/gdw). Higher extraction yields of phenolic compounds were obtained with an increase in the polarity of the solvent. Extraction with water afforded the highest yield (35 mg TAE/gdw), as compared to with ethyl acetate (5.7 mg TAE/gdw) or with methanol (32.8 mg TAE/gdw). There was no significant difference ($P<0.05$) in the extraction yield obtained from the solvent and ultrasound-assisted methods.

The polyphenol composition was determined by either HPLC–DAD, or by an HPLC instrument connected to an MS, as a detector (LC–MS). Sicilian pistachios, harvested on ten different farms in Sicily, were extracted with ethanol:water (80:20), and subsequently their stilbene concentration was determined [39]. Resveratrol derivatives have attracted much attention due to different therapeutic effects. In all tested samples, trans-resveratrol and trans-resveratrol-3-O-b-glucoside (trans-piceid) were detected with no cis isomers. The mean value of trans-resveratrol was 0.012±0.003 mg/100 g, trans-piceid was 0.697±0.055 mg/100 g and the total resveratrol concentration was 0.709±0.054 mg/100 g. Epidemiological and clinical investigations have shown that the polyphenolic phytoalexin resveratrol (3,5,4'-trihydroxystilbene) has been associated with reduced cardiovascular disease and reduced cancer risk.

The polyphenolic phytoalexin resveratrol content in pistachio nuts was determined by investigators by the same methods as mentioned above. Tokusoglu et al. [40] reported that the trans-resveratrol concentration in pistachio varieties is 0.009–0.167 mg/100 g (average = 0.115 mg/100 g), and others [41] claimed that the total amount of trans-resveratrol was 0.012 ± 0.0012 mg/100 g. Exposure of the pistachio extracts to UV light at 366 nm for 1 min converted approximately 20% of the trans-resveratrol to cis-resveratrol [40] according to analysis by the GC–MS method.

Other polyphenols found in pistachio extracts include proanthocyanidins, and the phytoestrogens daidzein and genistein at concentrations of 268.12 ± 8.31, 3.68 ± 0.41, and 3.40 ± 0.37 mg/100 g, respectively [41].

Anthocyanins were extracted from the external skins of pistachio kernels by Bellomo and Fallico [42] using aqueous formic acid (1%) as the solvent, followed by methanol containing 0.1% HCl. Afterward, their structures were elucidated by the HPLC–MS/MS method. Cyanidin-3-galactoside was found to be the major anthocyanin present in pistachio kernels, and in some samples a small amount of cyanidin-3-glucoside was also detected. The level of cyanidine-3-galactoside in pistachios ranged from non-detectable quantities to 42.64 mg/100 g, depending on the ripening progress. Pistachios in the unripe stage contained the lowest anthocyanin content (2.2 mg/100 g), whereas in the intermediate stages the level of anthocyanin was 10.6–14.7 mg/100 g. In ripe samples, it reached 30.0 mg/100 g. The ripening status correlated well with the anthocyanin level with an R^2 of 0.9425.

3. CAROTENOID COMPOSITION IN FIG, CAROB, AND PISTACHIO

Carotenoids are pigments that occur naturally in chromoplasts of plants and in some other photosynthetic organisms, such as algae, some types of fungus, and some bacteria. There are over 600 known carotenoid compounds, which are divided into two classes, xanthophylls and carotenes (Figure 17.4). Carotenoids can act as antioxidants, and in humans β-carotene (Figure 17.4A) is a precursor of vitamin A, a pigment essential to vision.

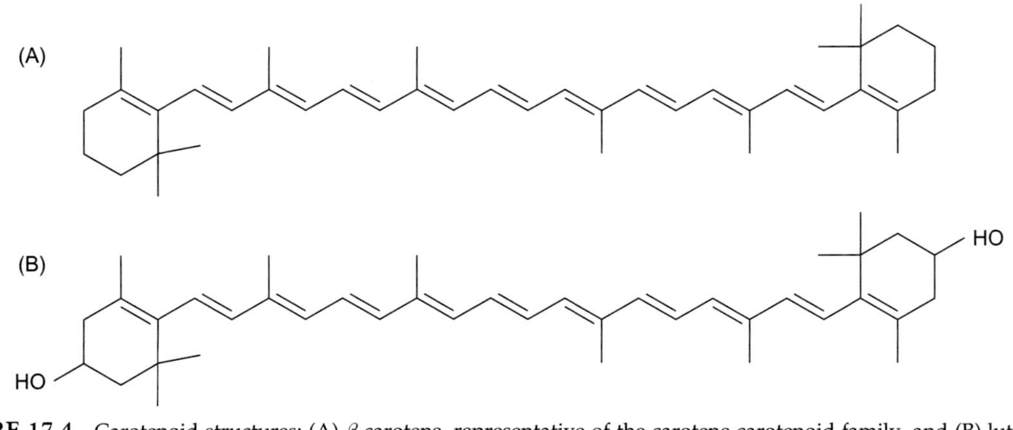

FIGURE 17.4 Carotenoid structures: (A) β-carotene, representative of the carotene carotenoid family, and (B) lutein, representative of the xanthophylls carotenoid family.

Many *in vivo* studies have been carried out to evaluate the contribution of carotenoids to human health, when consumed in diets rich in carotenoids, either owing to high fruit and vegetable content or to diet supplementation with pure compounds, mostly β-carotene and lycopene. The carotenoid content in selected components of the Mediterranean diet has been analyzed [43]. Table 17.3 summarizes the carotenoid content and composition in fig, carob, and pistachio. Major carotenoids present in fig include lutein, cryptoxanthin, lycopene, β-carotene, and α-carotene. Lycopene was the most abundant carotenoid (0.32 mg/100 g FW), followed by lutein and β-carotene (0.08 and 0.04 mg/100 g, respectively). Another study analyzed the carotenoids content in vegetables and fruit commonly consumed in Israel [44]. It was found that carob pods contain 0.2 mg total carotenoids/100 g dry weight, with α- and β-carotene (0.08 mg/100 g dry weight), lycopene (0.03 mg/100 g dry weight), and lutein (0.02 mg/100 g dry weight) as the major constituents. Carotenoids were also extracted from the pistachio kernel by Bellomo and Fallico [42], with acetone:methanol (75:25) as solvents and their concentration determined by an HPLC with a DAD detector.

Lutein was the main carotenoid identified, although a small amount of β-carotene was also present (<0.18 mg/100 g). The lutein level depends on both the degree of ripeness and the source of the kernels, with values ranging from 5.21 ± 0.294 mg/100 g and 4.13 ± 0.164 for the unripe harvested nuts from Turkey and Iran, respectively, to 18.01 ± 0.068 and 3.77 ± 0.20 mg/100 g for the ripe ones harvested in Turkey and Italy, respectively. Giovannini and Condorelli [45,46] claimed that the average level of lutein and β-carotene was 0.79 mg/100 g each in ripe fruit, with a value of 0.94 for β-carotene:lutein, compared to 0.05 for this ratio, as reported by Bellomo and Fallica [42].

4. PHYTOSTEROL COMPOSITION IN THE FIG AND PISTACHIO

Phytosterols are formed in ornamental plants, medicinal herbs, edible plants, shrubs, and trees. It has been widely reported that plant sterols (phytosterols) lower the serum cholesterol level in animals and humans [47–51]. Furthermore, investigators claim that phytosterols have positive effects in the treatment of benign prostatic hyperplasia, rheumatoid arthritis, allergies, and stress-related illness, as well as inhibiting the development of colon cancer [52].

Jeong and Lachance [53] studied the sterol profiles of various 'parts' of fig tree branches, i.e. structural components (bark, stem, pith) and fig fruit, using the GC–MS method. Following saponification (treatment with KOH solution), the weight of the solid concentrate, resulting from the KOH hydrolysis, was 433 mg of solid/100 g dry weight of fig fruit. Four phytosterols were determined in the solid. Sitosterol was the most predominant

TABLE 17.3 Carotenoid Composition of Fig Fruit, Carob Pods, and Pistachio Nuts

	Fig (mg/100 g FW)	Carob (mg/100 g FW)	Pistachio (mg/100 g FW)
Lutein	0.08 [43]	0. 02 [44]	3.77–18 [42]
Lycopene	0.32 [43]	0.03 [44]	ND
α- and β-carotene	0.04 [43]	0.08 [44]	0.18 [42]

FW, fruit weight; ND, not detected.
Numbers in brackets are references.

component (28% of the solid, 121 mg/100 g dry weight of fig fruit), fucosterol (5% of the solid, 21.7 mg/100 g dry weight of fig fruit), stigmastirol (3.2% of the solid, 13.9 mg/100 g dry weight of fig fruit), and campesterol (1.3% of the solid, 5.629 mg/100 g dry weight of fig fruit).

The same procedure was carried out by Arena et al. [54] in the determination of the phytosterol composition in pistachio oil extracted by a Soxhlet extractor in the solvent, petroleum ether, for 6 h. After alkaline hydrolysis of the pistachio oil, nine sterols were identified and quantified. β-Sitosterol was the predominant component in all samples, varying from about 85 to 88%. Also detected were Δ^5-avenasterol (5–9%), campesterol (3–4.5%) and other sterols, such as clerosterol, stigmasterol, $\Delta^{5,24}$-stigmastadienol, Δ^7-stigmastenol, and Δ^7-avenasterol [trace amounts (<1%)].

5. ANTIOXIDANT ACTIVITY AND HEALTH-PROMOTING EFFECTS

The phytochemicals in fruits and vegetables attract the interest of many investigators due to their potential health benefits and largely because of their antioxidant activity. Usually a mixture of compounds contribute to the antioxidant activity of a specific plant, some due to their capacity to donate an electron, others owing to their capacity to chelate transition metals, and still others as a result of their cell signaling effects, which promote the activity of the defense system, including the synthesis of specific enzymes, such as superoxide dismutase, catalase, and glutathione peroxidase [55].

5.1 Antioxidant Activity and Health-promoting Effects of the Fig Fruit

Several studies have been published in the last few years in which the antioxidant level in

fig fruits has been determined. Some of them have shown a high correlation between the polyphenol content and the level of the antioxidant capacity. Solomon et al. [27] measured the antioxidant capacity using the Trolox Equivalent Antioxidant Capacity (TEAC) method. This study demonstrated a correlation between total polyphenols, total flavonoids, or total anthocyanins and antioxidant activity. The antioxidant activities were higher in extracts of the dark fig varieties, with the skin the major contributing tissue, as it had a threefold higher antioxidant capacity as compared to the pulp. In another study, the ability of the water extract of dark fig fruit to protect the low density lipoproteins (LDL) and very low density lipoproteins (VLDL) oxidation was evaluated [3]. The results demonstrated that spiking the plasma with the fig extract (50 μM and 100 μM fig polyphenols, as measured by the Folin assay) increased the lag time of LDL and VLDL oxidation by 25 and 40%, respectively. In the same study, Vinson et al. [3] determined the antioxidant capacity of human plasma (using the TEAC method) after consumption of fig fruit. Five males and five females, aged 25–58, participated in the research, and their plasma antioxidant capacity was analyzed after consumption of either 240 mL of Sprite™ alone (control) or 40 g of dried fig fruit. Results revealed that consumption of Sprite™ decreased the antioxidant capacity, which reached the lowest level at 1 h after the consumption of the soft drink. In contrast, fig fruit consumption increased the antioxidant capacity of the plasma significantly over a period of 4 h.

In another *in vivo* study, the effect of fig leaf extract on diabetic rats was tested [56]. There were four groups of rats: streptozotocin-induced diabetic rats ($N = 10$); diabetic rats, which received a single dose of an aqueous fraction of *Ficus carica* leaf extract ($N = 14$); diabetic rats, which received a single dose of a chloroform fraction of the leaf extract ($N = 10$); and normal rats ($N = 10$). The results demonstrated that both the aqueous and chloroform extract of *Ficus carica* leaves reduced the hyperglycemia in diabetic rats and normalized the fatty acids and vitamin E values, whereas the chloroform extract was more effective in reducing fatty acid levels. These studies confirm that the antioxidant status in rat plasma was affected in the diabetes syndrome, and that *Ficus carica* extracts from leaves tend to normalize it.

The cytotoxicity of fig fruit latex (FFL) against human cancer cells was also examined [57]. It was found that FFL has a strong dose- and time-dependent anticancer activity against human glioma and hepatocellular carcinoma (HCC) cells, with a lower cytotoxicity toward normal liver cells; in short, the HCC cells were more sensitive to FFL than normal liver cells. FFL also inhibits the colony growth potential of cancer cells in a dose-dependent manner *in vitro*. The anticancer activity of original fresh FFL (collected from ripe fig fruits, drop by drop without squeezing) is more potent ($IC_{50} = 0.25$–0.8 $1 \, g/mL$) than that of the hydrophobic resin [6-O-acyl-β-D-glucosyl-β-sitosterols (6-AGS)] extracted from FFL and identified by Rubnov et al. ($IC_{50} = 25 \, \mu g/mL$) [58].

The mechanism of the anticancer activity of FFL against human glioma and HCC cells was investigated by Jing Wang et al. [57]. It was suggested that FFL has a strong antiproliferation effect, since it induced apoptosis, inhibited DNA synthesis, and caused G0/G1 phase arrest of cancer cells.

5.2 Antioxidant Activity and Health-promoting Effects of the Carob Pod

The antioxidant activity of the aqueous carob pod crude polyphenols extract was evaluated by Kumazawa et al. [59] by means of several methods: the β-carotene–linoleic acid system ($10 \, \mu g/mL$ was obtained), 1,1-diphenyl-2-picryl hydrazyl (DPPH) radical scavenging activity ($25 \, \mu g/mL$), rabbit erythrocyte membrane ghost system ($100 \, \mu g/mL$), and rat liver microsomal system ($1 \, mg/mL$). In all of the above methods, the carob pod crude polyphenol extract, with the above concentrations, showed a strong antioxidant activity, in comparison to the well-known polyphenol antioxidants (e.g. catechin, quercetin, and gallic acid), which have been attributed to the carob pod polyphenol content. Similarly, carob fiber, produced by cold water extraction by Papagiannopoulos et al. [37], showed a strong antioxidant activity, when measured by the DPPH method and by the TEAC method. The carob pod's high polyphenolic content and its strong antioxidant activity make this nutrient an interesting potential candidate for health food.

Carob pulp, rich in insoluble fibers, was examined for its ability to lower the total and LDL cholesterol level in hypercholesterolemic patients [60]. Volunteers ($N = 58$) with hypercholesterolemia consumed daily, both bread (two servings) and a fruit bar (one serving), either with ($N = 29$) or without ($N = 29$) a total amount of $15 \, g/d$ of a carob pulp preparation (carob fiber). Total LDL, high density lipoprotein (HDL) cholesterol, and triglyceride concentrations in serum were assessed at baseline and after 4 and 6 weeks. Results showed that the consumption of carob fiber reduced LDL cholesterol by $10.5 \pm 2.2\%$ ($P = 0.010$). The LDL:HDL cholesterol ratio was marginally decreased by $7.9 \pm 2.2\%$ in the carob fiber group compared to the placebo group ($P = 0.058$). Carob fiber consumption also lowered triglycerides in females by $11.3 \pm 4.5\%$ ($P = 0.030$). Lipid-lowering effects were more pronounced in females than in males. Similar research supported the above findings that the daily consumption of food products enriched with carob fiber has beneficial effects on the human blood lipid profile and may be effective in prevention and treatment of hypercholesterolemic patients [61]. These results strengthen the concept that carob fiber affords a beneficial effect on human health. This effect may be attributed to the high content of lignin and polyphenols (especially tannins), and their ability to adsorb bile acids in the chymus.

The postprandial effects of carob fiber, rich in polyphenols, on circulating ghrelin levels, plasma triglycerides, non-esterified fatty acids (NEFA), and substrate utilization in humans were investigated [62]. Ghrelin is a peptide hormone produced and excreted mainly in the stomach [63]. It circulates in two major forms, acylated ghrelin and desacyl ghrelin. Acylated ghrelin acts as an orexigenic signal of the central nervous system, and its levels increase during fasting and are suppressed postprandially. The administration of ghrelin induces body weight gain by promoting food intake and decreasing fat utilization [64]. Gruendel et al. [62] found that consumption of carob fiber extract rich in polyphenols by 20 healthy subjects, aged 22–62 years, lowered acylated ghrelin to 49.1%, triglycerides to 97.2%, and NEFA to 67.2% compared with the effect of the control meal ($P < 0.001$). Total ghrelin and insulin concentrations were not influenced by the consumption of a carob fiber-enriched liquid meal. Postprandial energy expenditure was increased by 42.3%, and respiratory quotient (RQ) was reduced by 99.9% in those who received a liquid meal with carob fiber in comparison to those who were given a control meal ($P < 0.001$). These results indicate that carob fiber might exert beneficial effects on energy intake and body weight.

The ability of carob bean juice in treatment of children with diarrhea was examined [65]. Diarrhea in children may cause dehydration and is treated by the World Health Organization (WHO) standard oral rehydration solution (ORS). The ORS protects children from dehydration and provides effective rehydration, but it does not treat and reduce the diarrhea. Aksit et al. [65] found that the addition of carob bean juice (20 mL per kg body weight of patient) to ORS, shortened the diarrhea duration by 45%, reduced stool output by 44%, and decreased the ORS requirement by 38% relative to children who received ORS alone.

The antiproliferative effects of carob (*Ceratonia siliqua* L.) on the mouse hepatocellular carcinoma cell line (T1) were also investigated [66]. Extracts from pods and leaves of carobs (*Ceratonia siliqua* L.) were tested for their ability to inhibit cell proliferation of the mouse T1 cells. Incubation of these cells for 24 h with carob pod and leaf extracts significantly inhibited their proliferation. The two extracts showed a marked reduction in T1 cell proliferation in a dose-related fashion, reaching the maximal effect at 1 mg/mL. Pod and leaf extracts were also able to induce apoptosis in T1 cell lines, as indicated by the darker staining of nuclei in treated cells. The effect on apoptosis obtained by leaf extracts (0.2 mg/mL) was greater than that obtained with pod extracts (0.4 mg/mL).

In another study, the ability of carob pod fiber to suppress tumor cell growth was investigated by determining selected biological effects in human colon cells [67]. An aqueous carob fiber extract was used to determine its influence on the survival and growth of the HT29 colon adenocarcinoma cell line and the pre-neoplastic adenoma LT97 cell line, as models of late and early cancer stages. The results showed that carob fiber modulates parameters of cell growth differently in human HT29 colon adenocarcinoma cells than in LT97 colon adenoma cells. In both human colon cell lines used, cell growth was inhibited. The inhibition of proliferation was more pronounced in LT97 adenoma than in the less differentiated HT29 adenocarcinoma cells, and the reduction in cell number was distinctly higher in the latter. After 72 h, carob fiber extract reduced survival of rapidly proliferating HT29 cells (by $76.4 \pm 12.9\%$), whereas metabolic activity and DNA synthesis were only transiently impaired. Survival of the slower growing LT97 cells was less decreased (by 21.5±12.9%), but there were marked effects on DNA synthesis (reduction by 95.6±7%, 72 h). The antitumor activity of the carob pod fiber is not due to the main compound found in the fiber, gallic acid.

5.3 Antioxidant Activity and Health-promoting Effects of the Pistachio

The antioxidant activity in polar pistachio hull extract (PHE) was analyzed by Goli et al. [38]. The oxidation of soybean oil in the presence of PHE and synthetic antioxidants (BHA and BHT for comparison) was assessed by measuring the peroxide value (PV) and thiobarbituric acid reactive substances value (TBARS). The PV value is a gauge of the primary products of lipid oxidation, and the TBARS value is a measure of the formation of secondary oxidation products, mainly malondialdehyde, which may contribute to the off flavor or odor of the oxidized oil [68]. According to PV and TBARS, all samples examined in the presence of PHE (0.02–0.06%) were more stable on heating at 60°C than the control. The antioxidant effect of PHE increased with concentration, and at a concentration of 0.06%, the antioxidant activity was not significantly different ($P<0.05$) from that of the synthetic antioxidants (BHA and BHT) at levels of 0.02%.

The activity of the hydrophilic pistachio nut extracts in biological models of lipid oxidation, including bovine liver microsomal membranes and human LDL was investigated by Gentile et al. [41]. When lipid oxidation was induced in microsomes by the hydrophilic radical generator AAPH, in the absence of nut extract, the production of TBARS reached a plateau after a 90 min incubation, while hydrophilic extracts from 0.25 to 1.0 mg pistachio nuts increased the lag time to over 120 min, implying that the antioxidant activity of the nuts slowed down the oxidation rate in a dose-dependent manner. Employing pro-oxidation conditions, such

as the Fe+3/ascorbate system, resulted in a decrease in the lag time, with a maximum of 60 min. Co-incubation with the extracts resulting from 0.5 and 1.0 mg of nuts increased the protective activity observed in the metal-dependent oxidation model. This phenomenon could be derived from the expression of combined peroxyl radical-scavenging abilities and metal-chelating activity of one or more extract components. In addition, the hydrophilic extract of pistachio nuts brought about an increase in the susceptibility of LDL to oxidation. Thus, when LDL was incubated with the hydrophilic nut extract (30–100 µg), a dose-dependent elongation of the lag phase was observed. Five hundred micrograms of the hydrophilic nut extract protected LDL from oxidation for 6 h.

Clinical and epidemiological studies have reported the beneficial effects of pistachio nuts on serum lipid levels. Consumption of pistachio nuts at a level of 15% of the daily caloric intake affected lipid profiles of human subjects with primary and moderate hypercholesterolemia (serum cholesterol greater than 210 mg/dL) [69]. Subjects that consumed pistachio nuts showed statistically significant increases in HDL-C (6%, $P = 0.02$), as well as decreases in total cholesterol (TC):HDL-C ($-9%$, $P = 0.001$), LDL-C: HDL-C ($-14%$, $P = 0.004$) and in B-100:A-1 (apolipoproteins) ($-13%$, $P = 0.009$). Decreases were also seen in LDL-C ($-9%$, $P = 0.06$), with no changes in body mass index or blood pressure observed. It is assumed that changes in ratios of TC:HDL-C and LDL-C:HDL-C are better predictors of coronary heart disease (CHD) risk reduction than changes in each of these specific component levels [70,71]. Others [72] have reported that the apolipoprotein ratio B-100:A-1 is the best lipoprotein-related measure to estimate CHD risk. Similar results have been obtained in another two investigations carried out by Gebauer et al. [73] and Kocyigit et al. [74]. They showed that the consumption of pistachio nuts by healthy volunteers may reduce the risk of cardiovascular disease.

It may be concluded that consumption of pistachio nuts at levels of 15–20% of the daily calories (about 2–3 ounces per day), causes a decrease in oxidative stress and improved total cholesterol and HDL levels in hypercholesterolemic, as well as in healthy volunteers [74], and may reduce the risk of cardiovascular disease.

6. SUMMARY

Fig fruit, carob pods, and pistachio nuts have been grown throughout the Mediterranean region for thousands of years. They have been used as food and their effects on health have been long recognized. As part of the worldwide recognition of the potential health value originating from plants, the antioxidant activity and health-promoting effects of these foods have been extensively investigated. Their consumption brings about an increase in the plasma antioxidant capacity, protects the low (LDL) and very low density lipoproteins from oxidation, reduces lipid levels in plasma, and thus may reduce risk of cardiovascular disease. They have also demonstrated anticancer activities. Studies are continuing to isolate the constituents responsible for such biological effects and elucidate their structure. Although more investigations are necessary to confirm this activity, it is mostly clinical studies that are needed.

References

1. Ronsted, N., Weiblen, G. D., Cook, J. M., Salamin, N., Machado, C. A., & Savolainen, V. (2005). 60 million years of co-divergence in the fig-wasp symbiosis. *Proceedings Biological Sciences*, 272, 2593–2599.
2. Vinson, J. A. (1999). Functional food properties of figs. *Cereal Foods World*, 44, 82–87.
3. Vinson, J. A., Zubik, L., Bose, P., Samman, N., & Proch, J. (2005). Dried fruits: Excellent *in vitro* and *in vivo* antioxidants. *Journal of the American College of Nutrition*, 24, 44–50.
4. Ayaz, F. A., Torun, H., Ayaz, S., Correia, P. J., Alaiz, M., Sanz, C., Gruz, J., & Strand, M. (2007). Determination of chemical composition of Anatolian carob pod (*Ceratonia siliqua* L.): Sugars, amino and organic acids, minerals and phenolic compounds. *Journal of Food Quality*, 30, 1040–1055.
5. Batlle, I., & Tous, J (1997). *Carob Tree*. Ceratonia siliqua L. *Promoting the conservation and use of underutilized and neglected crops, 17.* Rome, Italy: Institute of Plant Genetics and Carob Plant Research, Gatersleben/International Plant Genetic Institute.
6. Morton, J. F. (1987). Carob. In C. F. Dowling, (Ed.), *Fruits of warm climates*. Miami, FL: Morton.
7. Parfitt, D. E., & Maria, L. B. (1997). *Phylogeny of the genus pistacia as determined from analysis of the chloroplast genome*. CA, USA: Department Pomology, University California Davis.
8. Aksoy, N., Aksoy, M., Bagci, C., Gergerlioglu, H. S., Celik, H., Herken, E., Yaman, A., Tarakcioglu, M., Soydinc, S., Sari, I., & Davutoglu, V. (2007). Pistachio intake increases high density lipoprotein levels and inhibits low-density lipoprotein oxidation in rats. *Tohoku Journal of Experimental Medicine*, 212, 43–48.
9. Aviram, M., Vaya, J., & Fuhrman, B. (2004). Licorice root flavonoid antioxidants reduce LDL oxidation and attenuate cardiovascular diseases. *Oxidative Stress and Disease*, 14, 595–614.
10. Del Caro, A., & Piga, A. (2008). Polyphenol composition of peel and pulp of two Italian fresh fig fruits cultivars (*Ficus carica* L.). *European Food Research and Technology*, 226, 715–719.
11. Shan, H., Cuirong, S., & Yuajiang, P. (2008). Red wine polyphenols for cancer prevention. *International Journal of Molecular Sciences*, 9, 842–853.
12. Zykova, T. A., Zhu, F., Zhai, X., Ma, W. Y., Ermakova, S. P., Lee, K. W., Bode, A. M., & Dong, Z. (2008). Resveratrol directly targets COX-2 to inhibit carcinogenesis. *Molecular Carcinogenesis*, 47, 797–805.
13. Rice-Evans, C. A., Miller, N. J., & Paganga, G. (1996). Structure-antioxidant activity relationships of flavonoids and phenolic acids. *Free Radical Biology and Medicine*, 20, 933–956.
14. Singh, M., Arseneault, M., Sanderson, T., Murthy, V., & Ramassamy, C. (2008). Challenges for research on polyphenols from foods in Alzheimer's disease: Bioavailability, metabolism, and cellular and molecular mechanisms. *Journal of Agricultural and Food Chemistry*, 56, 4855–4873.
15. Vaya, J., & Schipper, H. M. (2007). Oxysterols, cholesterol homeostasis, and Alzheimer disease. *Journal of Neurochemistry*, 102, 1727–1737.
16. Vaya, J., Song, W., Khatib, S., Geng, G., & Schipper, H. M. (2007). Effects of heme oxygenase-1 expression on sterol homeostasis in rat astroglia. *Free Radical Biology and Medicine*, 42, 864–871.

17. Aviram, M., Volkova, N., Coleman, R., Dreher, M., Reddy, M. K., Ferreira, D., & Rosenblat, M. (2008). Pomegranate phenolics from the peels, arils, and flowers are antiatherogenic: Studies *in vivo* in atherosclerotic apolipoprotein e-deficient (E 0) mice and *in vitro* in cultured macrophages and lipoproteins. *Journal of Agricultural and Food Chemistry, 56,* 1148–1157.

18. Borochov-Neori, H., Judeinstein, S., Greenberg, A., Fuhrman, B., Attias, J., Volkova, N., Hayek, T., & Aviram, M. (2008). Phenolic antioxidants and anti-atherogenic effects of Marula (*Sclerocarrya birrea* Subsp. *caffra*) fruit juice in healthy humans. *Journal of Agricultural and Food Chemistry, 56,* 9884–9891.

19. Chen, D., & Ping Dou, Q. (2008). Tea polyphenols and their roles in cancer prevention and chemotherapy. *International Journal of Molecular Sciences, 9,* 1196–1206.

20. Dillard, C. J., & Bruce German, J. B. (2000). Phytochemicals: Nutraceuticals and human health. *Journal of the Science of Food and Agriculture, 80,* 1744–1756.

21. Garcia-Closas, R., Gonzalez, C. A., Agudo, A., & Riboli, E. (1999). Intake of specific carotenoids and flavonoids and the risk of gastric cancer in Spain. *Cancer Causes Control, 10,* 71–75.

22. Havsteen, B. H. (2002). The biochemistry and medical significance of the flavonoids. *Pharmacology & Therapeutics, 96,* 67–202.

23. Szuchman, A., Aviram, M., Musa, R., Khatib, S., & Vaya, J. (2008). Characterization of oxidative stress in blood from diabetic vs. hypercholesterolaemic patients, using a novel synthesized marker. *Biomarkers, 13,* 119–131.

24. Vaya, J., Tamir, S., & Somjen, D. (2004). Estrogen-like activity of licorice root extract and its constituents. *Oxidative Stress and Disease, 14,* 615–634.

25. Scalbert, A., & Williamson, G. (2000). Dietary intake and bioavailability of polyphenols. *Journal of Nutrition, 130,* 2073S–2085S.

26. Singleton, V. L., Orthofer, R., & Lamuela-Raventos, R. M. (1999). Analysis of total phenols and other oxidation substrates and antioxidant by means of Folin–Ciocalteu reagent. *Methods in Enzymology, 299,* 152–178.

27. Solomon, A., Golubowicz, S., Yablowicz, Z., Grossman, S., Bergman, M., Gottlieb, H. E., Altman, A., Kerem, Z., & Flaishman, M. A. (2006). Antioxidant activities and anthocyanin content of fresh fruits of common fig (*Ficus carica* L.). *Journal of Agricultural and Food Chemistry, 54,* 7717–7723.

28. Kim, D., Jeong, S. W., & Lee, C. Y. (2003). Antioxidant capacity of phenolic phytochemicals from various cultivars of plums. *Food Chemistry, 81,* 321–326.

29. Veberic, R., Colaric, M., & Stampar, F. (2008). Phenolic acids and flavonoids of fig fruit (*Ficus carica* L.) in the northern Mediterranean region. *Food Chemistry, 106,* 153–157.

30. Vaya, J., & Mahmood, S. (2006). Flavonoid content in leaf extracts of the fig (*Ficus carica* L.), carob (*Ceratonia siliqua* L.) and pistachio (*Pistacia lentiscus* L.). *Biofactors, 28,* 169–175.

31. Duenas, M., Pérez-Alonso, J. J., Santos-Buelga, C., & Escribano-Bailón, T. (2008). Anthocyanin composition in fig (*Ficus carica* L.). *Journal of Food Composition and Analysis, 21,* 107–115.

32. Avallone, R., Plessi, M., Baraldi, M., & Monzani, A. (1997). Determination of chemical composition of carob (*Ceratonia siliqua*): Protein, fat, carbohydrates, and tannins. *Journal of Food Composition and Analysis, 10,* 166–172.

33. Swain, T., & Hillis, W. E. (1959). The phenolic constituents of prunus domestica – I. The quantitative analysis of phynolic constituents. *Journal of the Science of Food and Agriculture, 10,* 63–68.

34. Bate-Smith, E. C. (1972). Detection and determination of ellagitannins. *Phytochemistry, 11,* 1153–1156.

35. Inoue, K. H., & Hagerman, A. E. (1988). Determination of gallotannin with rhodanine. *Analytical Biochemistry, 169,* 363–369.

36. Sakakibara, H., Honda, Y., Nakagawa, S., Ashida, H., & Kanazawa, K. (2003). Simultaneous determination of all polyphenols in vegetables, fruits, and teas. *Journal of Agricultural and Food Chemistry, 51,* 571–581.

37. Papagiannopoulos, M., Wollseifen, H. R., Mellenthin, A., Haber, B., & Galensa, R. (2004). Identification and quantification of polyphenols in carob fruits (*Ceratonia siliqua* L.) and derived products by HPLC-UV-ESI/MSn. *Journal of Agricultural and Food Chemistry, 52,* 3784–3791.

38. Goli, A. H., Barzegar, M., & Sahari, M. A. (2005). Antioxidant activity and total phenolic compounds of pistachio (Pistachia vera) hull extracts. *Food Chemistry, 92,* 521–525.

39. Grippi, F., Crosta, L., Aiello, G., Tolomeo, M., Oliveri, F., Gebbia, N., & Curione, A. (2008). Determination of stilbenes in Sicilian pistachio by high-performance liquid chromatographic diode array (HPLC-DAD/FLD) and evaluation of eventually mycotoxin contamination. *Food Chemistry, 107,* 483–488.

40. Tokusoglu, O., Unal, M. K., & Yemis, F. (2005). Determination of the phytoalexin resveratrol (3,5,4′-trihydroxystilbene) in peanuts and pistachios by high-performance liquid chromatographic diode array (HPLC-DAD) and gas chromatography-mass spectrometry (GC-MS). *Journal of Agricultural and Food Chemistry, 53,* 5003–5009.

B. EFFECTS OF INDIVIDUAL VEGETABLES ON HEALTH

41. Gentile, C., Tesoriere, L., Butera, D., Fazzari, M., Monastero, M., Allegra, M., & Livrea, M. A. (2007). Antioxidant activity of Sicilian pistachio (*Pistacia vera* L. *var. Bronte*) nut extract and its bioactive components. *Journal of Agricultural and Food Chemistry, 55*, 643–648.

42. Bellomo, M. G., & Fallico, B. (2007). Anthocyanins, chlorophylls and xanthophylls in pistachio nuts (*Pistacia vera*) of different geographic origin. *Journal of Food Composition and Analysis, 20*, 352–359.

43. Su, Q., Rowley, K. G., Itsiopoulos, C., & O'Dea, K. (2002). Identification and quantitation of major carotenoids in selected components of the Mediterranean diet: Green leafy vegetables, figs and olive oil. *European Journal of Clinical Nutrition, 56*, 1149–1154.

44. Ben-Amotz, A., & Fishler, R. (1998). Analysis of carotenoids with emphasis on 9-k p-carotene in vegetables and fruits commonly consumed in Israel. *Food Chemistry, 62*, 515–520.

45. Giovannini, E., & Condorelli, G. (1956). Sul contenuto vitaminico dei semi di pistacchio. *Giornale Di Biochimica, 5*, 542–560.

46. Giovannini, E., & Condorelli, G. (1958). Contributo alla conoscenza del metabolismo dei pigmenti cloroplastici e delle loro correlazioni con i tocoferoli. *La Ricerca Scientifica, 28*, 1–10.

47. Andriamiarina, R., Laraki, L., Pelletier, X., & Debry, G. (1989). Effects of stigmasterol-supplemented diets on fecal neutral sterols and bile acid excretion in rats. *Annals of Nutrition and Metabolism, 33*, 297–303.

48. Bhattacharyya, A. K., & Eggen, D. A. (1984). Effects of feeding cholesterol and mixed plant sterols on the fecal excretion of acidic steroids in rhesus monkeys. *Atherosclerosis, 53*, 225–232.

49. Howard, B. V., & Kritchevsky, D. (1997). Phytochemicals and cardiovascular disease. A statement for healthcare professionals from the American Heart Association. *Circulation, 95*, 2591–2593.

50. Laraki, L., Pelletier, X., & Debry, G. (1991). Effects of dietary cholesterol and phytosterol overload on Wistar rat plasma lipids. *Annals of Nutrition and Metabolism, 35*, 221–225.

51. Sklan, D., Budowski, P., & Hurwitz, S. (1974). Effect of soy sterols on intestinal absorption and secretion of cholesterol and bile acids in the chick. *Journal of Nutrition, 104*, 1086–1090.

52. Oomah, B. D., & Mazza, G. (1999). Health benefits of phytochemicals from selected Canadian crops. *Trends Food Science and Technology, 10*, 193–198.

53. Jeong, W.. S., & Lachance, P. A. (2001). Phytosterols and fatty acids in fig (*Ficus carica*, var. Mission) fruit and tree components. *Journal of Food Science, 66*, 278–281.

54. Arena, E., Campisi, S., Fallico, B., & Maccarone, E. (2007). Distribution of fatty acids and phytosterols as a criterion to discriminate geographic origin of pistachio seeds. *Food Chemistry, 104*, 403–408.

55. Vaya, J., & Aviram, M. (2001). Nutritional antioxidants: Mechanisms of action, analyses of activities and medical applications. *Current Medicinal Chemistr Immunology Endocrinology and Metabolism Agents, 1*, 99–117.

56. Perez, C., Canal, J. R., & Torres, M. D. (2003). Experimental diabetes treated with *Ficus carica* extract: Effect on oxidative stress parameters. *Acta Diabetologica, 40*, 3–8.

57. Wang, J., Wang, X., Jiang, S., Lin, P., Zhang, J., Lu, Y., Wang, Q., Xiong, Z., Wu, Y., Ren, J., & Yang, H. (2008). Cytotoxicity of fig fruit latex against human cancer cells. *Food Chemistry Toxicologica, 46*, 1025–1033.

58. Rubnov, S., Kashman, Y., Rabinowitz, R., Schlesinger, M., & Mechoulam, R. (2001). Suppressors of cancer cell proliferation from fig (*Ficus carica*) resin: Isolation and structure elucidation. *Journal of Natural Products, 64*, 993–996.

59. Kumazawa, S., Taniguchi, M., Suzuki, Y., Shimura, M., Kwon, M. S., & Nakayama, T. (2002). Antioxidant activity of polyphenols in carob pods. *Journal of Agricultural and Food Chemistry, 50*, 373–377.

60. Zunft, H. J., Luder, W., Harde, A., Haber, B., Graubaum, H. J., Koebnick, C., & Grunwald, J. (2003). Carob pulp preparation rich in insoluble fibre lowers total and LDL cholesterol in hypercholesterolemic patients. *Europaea Journal of Nutrition, 42*, 235–242.

61. Zunft, H. J., Luder, W., Harde, A., Haber, B., Graubaum, H. J., & Gruenwald, J. (2001). Carob pulp preparation for treatment of hypercholesterolemia. *Advances in Therapy, 18*, 230–236.

62. Gruendel, S., Garcia, A. L., Otto, B., Mueller, C., Steiniger, J., Weickert, M. O., Speth, M., Katz, N., & Koebnick, C. (2006). Carob pulp preparation rich in insoluble dietary fiber and polyphenols enhances lipid oxidation and lowers postprandial acylated ghrelin in humans. *Journal of Nutrition, 136*, 1533–1538.

63. Kojima, M., & Kangawa, K. (2002). Ghrelin, an orexigenic signaling molecule from the gastrointestinal tract. *Current Opinion in Pharmacology, 2*, 665–668.

64. Tschop, M., Smiley, D. L., & Heiman, M. L. (2000). Ghrelin induces adiposity in rodents. *Nature, 407*, 908–913.

65. Aksit, S., Caglayan, S., Cukan, R., & Yaprak, I. (1998). Carob bean juice: A powerful adjunct to oral rehydration solution treatment in diarrhoea. *Paediatric and Perinatal Epidemiology, 12*, 176–181.

66. Corsi, L., Avallone, R., Cosenza, F., Farina, F., Baraldi, C., & Baraldi, M. (2002). Antiproliferative

effects of *Ceratonia siliqua* L. on mouse hepatocellular carcinoma cell line. *Fitoterapia, 73*, 674–684.

67. Klenow, S., Glei, M., Haber, B., Owen, R., & Pool-Zobel, B. L. (2008). Carob fibre compounds modulate parameters of cell growth differently in human HT29 colon adenocarcinoma cells than in LT97 colon adenoma cells. *Food Chemistry Toxicologica, 46*, 1389–1397.

68. Rossel, J. B. (1994). Measurements of rancidity. In J. C. Allen, & R. J. Hamilton, (Eds.), *Rancidity in foods* 3rd ed (pp. 22–53). UK: Blackie.

69. Sheridan, M. J., Cooper, J. N., Erario, M., & Cheifetz, C. E. (2007). Pistachio nut consumption and serum lipid levels. *Journal of the American College of Nutrition, 26*, 141–148.

70. Kinosian, B., Glick, H., & Garland, G. (1994). Cholesterol and coronary heart disease: Predicting risks by levels and ratios. *Annals of Internal Medicine, 121*, 641–647.

71. Natarajan, S., Glick, H., Criqui, M., Horowitz, D., Lipsitz, S. R., & Kinosian, B. (2003). Cholesterol measures to identify and treat individuals at risk for coronary heart disease. *American Journal of Preventive Medicine, 25*, 50–57.

72. Walldius, G., Jungner, I., Aastveit, A. H., Holme, I., Furberg, C. D., & Sniderman, A. D. (2004). The apoB/apoA-I ratio is better than the cholesterol ratios to estimate the balance between plasma proatherogenic and antiatherogenic lipoproteins and to predict coronary risk. *Clinical Chemistry and Laboratory Medicine, 42*, 1355–1363.

73. Gebauer, S. K., West, S. G., Kay, C. D., Alaupovic, P., Bagshaw, D., & Kris-Etherton, P. M. (2008). Effects of pistachios on cardiovascular disease risk factors and potential mechanisms of action: A dose-response study. *American Journal of Clinical Nutrition, 88*, 651–659.

74. Kocyigit, A., Koylu, A. A., & Keles, H. (2006). Effects of pistachio nuts consumption on plasma lipid profile and oxidative status in healthy volunteers. *Nutrition and Metabolism Cardiovascular Disease, 16*, 202–209.

B. EFFECTS OF INDIVIDUAL VEGETABLES ON HEALTH

18

Poi History, Uses, and Role in Health

Salam A. Ibrahim[1] and Danfeng Song[1,2]
[1]North Carolina Agricultural and Technical State University, Greensboro, NC, USA
[2]Gannon University, Erie, PA, USA

1. INTRODUCTION

1.1 *Poi*

Poi is a Hawaiian word for the primary Polynesian staple food made from the corm of the kalo plant (known widely as taro). In the traditional native Hawaiian diet, *poi* is made by first steaming taro which is then mixed with a little water to form a smooth paste (Figure 18.1). *Poi* is usually a bit thick and starchy. When fresh, *poi* tastes rather bland. Most people prefer to eat *poi* warm with other foods or allow it to cool for 2–3 days, after which it develops a strong sour taste. This souring occurs due to the yeast and lactic acid bacteria naturally found on the plant's corm surface [1]. During the 'souring' process acid production changes the pH from 6.3 to 4.5 within 24 hours, and reaches its lowest pH on the fourth or fifth day of fermentation. At this point *poi* is usually discarded. Over the years, innovations in *poi* production have resulted in methods that allow *poi* to stay fresh longer and have a sweeter taste. This has produced a less traditional version group of products generally requiring refrigeration [2].

As early as 1933, Allen and Allen were able to identify the presence of three *Lactobacillus* species and two *Streptococcus* (renamed *Lactococcus*) bacteria, which included the *L. lactis* species, in *poi* [3]. Bilger and Young (1935) later identified the actions of lactic acid, acetic acid, formic acid, alcohol, and acetaldehyde as being the primary agents for why *poi* 'sours' [4]. In one study, Huang et al. (1994) identified the predominant species in sour *poi* as *Lactococcus lactis* [about log10 5.8 colony-forming organisms (CFU)/g] [1]; the major acids are lactic and acetic acid [1].

1.2 Cultivation of *Poi* (Taro)

Taro (*Colocasia esculenta* L.) has been cultivated for many centuries. Originating in Asia, taro is now primarily found in tropical and subtropical regions, where it was historically a major dietary staple on the islands of the Pacific, especially Hawaii, New Zealand, and west to Indonesia (Figure 18.2). Taro became especially important to the Hawaiians who associated it with their gods and the original ancestor of the Hawaiian people, and even used it

FIGURE 18.1   *Poi* is made by first steaming taro and then pounding it until it reaches a nice smooth paste.

FIGURE 18.2   Taro plant and taro root.

for medicinal purposes [5]. The Kanaka Maoli, the indigenous Hawaiian people, are linked closely with the taro plant. They believed that taro had the greatest life force of all foods. According to the Kumulipo, the creation chant of the Kanaka Maoli, taro grew from the first-born son of Wakea (sky father) through Wakea's relationship with his and Papa's (earth mother's) daughter, Ho`ohokulani. The son, Haloa-naka, was stillborn and after he was buried, out of his body grew the taro plant, also called Haloa, which means everlasting breath. From this legend, taro and *poi* became symbols and means of survival for the Hawaiian people, which is celebrated by ritual eating of *poi* as part of a ceremony of life that brings people together and supports a relationship of `ohana (family) and of appreciation of the `aumakua (ancestors).

1.3 Historical Review and Medical Use

The nutrient composition of *poi* primarily includes carbohydrates, along with a few other nutrients [6]. In the first part of the twentieth century, researchers believed that, due to its easy digestibility, *poi* might have beneficial health effects for certain gastrointestinal conditions such as diarrhea, gastroenteritis, irritable bowel syndrome, and inflammatory bowel disease. Other uses began to be explored, including the possibility of *poi* being utilized in medical nutrition therapy for certain health conditions. Early studies in the mid-1960s [7] suggested that *poi* might be useful for the treatment of allergies and failure-to-thrive in infants due to its high content of easily digestible starch. In addition, its probiotic functionality also has led some to conclude that *poi* might also be useful as a cancer treatment and to address depressed immune function.

1.3.1 Digestive Disorders

Poi is easily digested, and this may benefit certain health conditions involving the gastrointestinal tract (Table 18.1) [5]. MacCaughey (1917) recognized how easily *poi* was digested

and attributed this to the small size of the taro starch granule [8]. Langworthy and Deuel (1922) confirmed this finding and further established that the raw starches of rice and taro root were notably more digestible as the result of the smaller size of the starch granules. Further evidence of the high digestibility of *poi* has been demonstrated in human studies which have reported no measure of undigested starch in feces, even when large quantities of *poi* were consumed [9].

The high digestibility of *poi* also appears to be related to the relative ease with which it breaks down. Derstine and Rada (1952), for example, reported that the easy digestibility of *poi* and the high absorbability of its minerals such as calcium and phosphorus appears to be related to its rapid fermentation process [7]. Taro starches have irregular, polygonal shapes and very small granular sizes. The average diameter of taro starches ranges from 2.6 to

TABLE 18.1 Reported Uses of *Poi* [4]

Medical Conditions	Reference	Study Objective	Subject Type	Number of Subjects	Study Length	Treatment Conditions	Results
Digestive disorders	[8,10]	Assess starch granules in *poi*	Plant	N/A	?	N/A	Found taro starch granule size: 1–8 μm
	[9]	Assess digestion of starch found in *poi*	Human	?	?	N/A	98.88% of taro starch was digested with up to 250 g of starch consumed/day
Food allergies	[7]	Observe use of *poi* as a food source	N/A	N/A	N/A	N/A	Found WWII use in hospitals
	[13]	Compare *poi* vs. rice as food source for ill infants	Human	100	6 months	Fed *poi*/rice instead of cereal	Found rice-fed group and *poi*-fed group thrived equally well
	[14]	Compare *poi* vs. cereal for allergic reactions	Human	132	?	Fed *poi*/rice instead of cereal	7% of both groups showed signs of allergy
Failure-to-thrive	[13]	Case study of *poi* as cereal substitute	Human	12	11–45 days	Fed *poi* and formula	All gained enough weight to be discharged
Colorectal cancer	[24]	Evaluation of mutagenic properties of taro	Plant	N/A	?	N/A	Taro fiber composition acts by adsorption of mutagens

N/A, not applicable; WWII, Second World War.

3.76 μm [10]. This small size makes *poi* an excellent food for the patient with digestive disorders.

1.3.2 Infant Allergies

Because of its very low protein content, *poi* is hypoallergenic and as such has been used as a food substitute for people with food allergies [5]. Alverez in 1939 was the first to suggest that *poi* be used as a substitute food for allergic people [11]. During the Second World War, *poi* was used as a substitute starch for people allergic to cereal or grain [7]. Feingold [12] was one of the first to suggest that *poi* be considered a substitute for soy milk in infants allergic to both soy and cow's milk. Physicians in Hawaii were some of the first to investigate *poi* as a substitute for food allergies. In a 1961 paper, Dr Jerome Glaser noted the high use of *poi* for allergic infants and those with gastrointestinal disorders and theorized that infants allergic to grain cereal could eat *poi* as a substitute. Glaser conducted a 6-month study of 100 infants, in which 50 infants were to be fed *poi* and compared with 50 rice-fed babies, and found that both groups of infants thrived equally well [13].

Roth et al. confirmed Glaser's findings after they tested 132 potentially allergic infants and found that breastfed babies remained completely symptom-free. Of the infants fed cow's milk substitutes ($N = 132$), about 7% of the rice-fed infants (4/55) and *poi*-fed infants (5/73) showed signs of allergy. Roth concluded that *poi* was definitely well tolerated by the babies, and *poi* may be a useful alternative when there is a family history of cereal allergy [14].

Glaser et al. also reported two case studies in which *poi* proved to be helpful to allergic infants. One infant boy had a severe multiple food intolerance. He was considered a failure-to-thrive infant. At the age of 9 months, he was started on *poi* as the main dietary carbohydrate, which he had no trouble tolerating. By the age of 19 months, he was at the lower limit of weight for his height. His intake of *poi* was approximately one pound per day. He was last seen at the age of 4 years and 3 months, at which time he had a normal weight and height for his age, appeared healthy, and was still consuming large amounts of *poi* [13]. The other case involved an allergic infant girl. She was experiencing severe gastrointestinal problems that were attributed to cow's milk allergy. She was placed on an elimination diet using *poi* as the cereal alternative. She thrived well on this diet [13]. The lack of gluten in *poi* makes it an ideal substitute for cereals in patients with celiac disease [13].

1.3.3 Failure-to-thrive

Weight gain is often the desired outcome for patients with failure-to-thrive. From the mid 1960s, a few studies have been done on the use of *poi* and failure-to-thrive. Glaser et al. reported in 1967 that preterm infants who consumed *poi* thrived as well as other preterm infants not fed *poi* of comparable weight and size [13]. In a case study of a premature female infant who weighed 1500 g at birth, after being on various formulas and only gaining 100 g in 54 days, her risk of failing to thrive became acute [13]. She was then given *poi* and quickly responded positively. She was discharged from the hospital after being able to maintain a healthy weight (2250–2500 g). Glaser concluded that *poi* can safely be recommended as a food for any very young infant. However, because these studies were conducted a long time ago, further research is warranted for the use of *poi* in infants and children.

1.3.4 Probiotic Effects

Poi can be used as a probiotic in medical nutrition therapy [15]. The predominant bacteria in *poi* are *Lactococcus lactis* (95%) and

Lactobacilli (5%) [1], both of which are lactic acid-producing bacteria. *Poi* contains significantly more of these bacteria per gram than yogurt. The first reported study conducted to investigate the effect of *poi* consumption on gastrointestinal tract bacterial concentration was a crossover clinical study including 18 subjects (a *poi* group of 10 and a control group of 8) by Brown and colleagues in 2005 [16]. They found no significant differences in total bacterial counts following fresh *poi* diet versus following a control diet, nor were significant differences found in counts of specific bacterial species. However, they expected that 'sour *poi*' might have a greater effect than this fresh *poi* as a potential probiotic. More research is needed to confirm that *poi* has a probiotic function.

1.3.5 Colon Cancer

Colorectal cancer continues to be a leading cause of morbidity and mortality in the Western world [17]. Hawaiians tend to have lower incidence rates of colorectal cancer. In addition to the epidemiological data linking *poi* and low colon cancer incidence, *poi* has several properties significant to the decreased risk of carcinogenesis including fiber content, novel phytochemical contents, pH influences, and possible probiotic chemoprotection. Colon cancer prevention has long been associated with plant-rich diets, especially ones supplemented with probiotics. Brown et al. (2005) were first to find out that *poi* extract can have two distinct inhibitory effects towards colon cancer [18]. One is that it can directly inhibit the proliferation of mammalian colon cancer cells by inducing the apoptosis of the cancer cells. The other is that components of *poi* stimulate the immune system by activating lymphocytes which have previously been shown to kill numerous types of colon cancer cells, both in humans and in rodents.

Like many plants, *poi* contains a unique collection of compounds relevant to chemoprotection and anticancer activity. Taro corms have been reported to contain anthocyanins, cyanidin 3-glucoside, pelargonidin 3-glucoside, and cyanidin 3-rhamnoside. These substances have antioxidant and anti-inflammatory properties which could protect the intestine from carcinogens [19]. Kim et al. reported that taro has high 'cancer preventative activity' compared with other vegetables [20,21]. Another study discovered the antioxidant nicotinamide adenine dinucleotide oxidase in taro which has been shown to produce potent inhibition of induced oxygen free radical generation in an animal model, suggesting that taro carries a significant antioxidative cancer preventative potential [22]. Fiber content (*poi* contains 3.7 g of fiber per 100 g) might also contribute to the positive anticancer effects against colon cancers [20,22,23].

1.3.6 Other Potential Health Benefits

Some research suggests that *poi* may be useful to control *Salmonella* typhimurium. Okabe and others (1996) investigated the antimutagenicity effects on the Trp-P2-induced mutagenicity to *Salmonella typhimurium* using different preparation bases. The study investigated the effects on TA 98 of water extracts, ethanol extracts, and gummy materials prepared from four root crops: nagaimo (Chinese yam, *Dioscorea opposita* Thumb.), jinenjo (*D. japonica* Thunb.), satoimo (taro, *Colocasia esculenta* (L.) Schott), and processed taro (freeze-dried *poi*). The gummy processed taro showed the highest inhibition of 60% among the specimens used [24].

It also has been observed that groups who have consumed taro have tended to have clear and conditioned skin. It is believed that poi has some kind of chemical that is good for the skin. However, the components that may be responsible for this property are currently unknown.

2. SUMMARY

Studies from the mid 1960s suggest that *poi* may have great potential for treating certain medical conditions, especially infant food allergies and failure-to-thrive in infants. In addition, the easy digestibility and other characteristics of *poi* might make it a nutritional supplement for promoting weight gain in patients with conditions such as failure-to-thrive, cancer, AIDS, and inflammatory bowel disease. Its combination of antioxidant and anti-inflammatory compounds suggests *poi* may have cancer fighting properties, in particular against colorectal cancer. Finally, *poi* may offer a nutritional supplement unique in its possible probiotic activities and ability to reduce skin damage and conditions that affect complexion and appearance.

2.1 Summary Points

- *Poi* can be a milk substitute for babies born with an allergy to dairy products because of its nutritional value.
- *Poi* can also be used as a baby food for babies with severe food allergies.
- Sour poi has more probiotics than yogurt.
- *Poi* does good to your gastrointestinal system.
- *Poi* has anticancer and antimutagenic properties.

References

1. Huang, A. S., Lam, S. Y., Nakayama, T. M., & Lin, H. (1994). Microbiological and chemical changes in *poi* stored at 20°C. *Journal of Agricultural Food Chemistry*, 42, 45–48.
2. Huang, A. S, Karthik, K., & Liu, X. X. (2002). Textural and sensory properties of α-amylase treated *poi* stored at 4°C. *Journal of Food Processing and Preservation*, 26, 1–10.
3. Allen, O. N., & Allen, E. K. (1933). The manufacture of *poi* from taro in Hawaii: with special emphasis upon its fermentation. *Hawaii Agriculture Experiment Station Bulletin*, (70).
4. Bilger, L. N., & Young, H. Y. (1935). The chemical investigation of the fermentations occurring in the process of *poi* manufacture. *Journal of Agricultural Research*, 51, 45–50.
5. Brown, A. C., & Valiere, A. (2004). The medicinal uses of *poi*. *Nutrition in Clinical Care*, 7, 69–74.
6. Huang, A. S., Titchena, C. A., & Meilleur, B. A. (2001). Nutrient composition of taro corms and breadfruit. *Journal of Food Composition and Analysis*, 13, 859–864.
7. Derstine, V., & Rada, E. (1952). Some dietetic factors influencing the market for *poi* in Hawaii. *Agricultural Economics. (bulletin 3)*, 3, 1–143. University of Hawaii Agricultural Experiment Station, Hawaii.
8. MacCaughey, V. (1917). The Hawaiian taro as food. *Hawaiian Forester Agriculture*, 14, 265–268.
9. Langworthy, C. F., & Deuel, H. J. (1922). Digestibility of raw rice, arrowroot, canna, cassava, taro, tree-fern, and potato starches. *Journal of Biological Chemistry*, 52, 251–261.
10. Jane, J., Shen, L., & Kasemsuwan, T. (1992). Physical and chemical studies of taro starches and flours. *Cereal Chemistry*, 69, 528–535.
11. Alverez, W. C. (1939). Problems of maintaining nutrition in the highly food-sensitive person. *American Journal of Digestive Diseases*, 5, 801–803.
12. Feingold, B. F. (1942). A vegetable milk substitute: taro. *Journal of Allergy*, 13, 488.
13. Glaser, J., Lawrence, R. A., Harrison, A., & Ball, M. R. (1967). *Poi* – its use as a food for normal, allergic and potentially allergic children. *Annals of Allergy*, 25, 496–500.
14. Roth, A., Worth, R. M., & Lichton, I. J. (1967). Use of *poi* in the prevention of allergic disease in potentially allergic infants. *Annals of Allergy*, 25, 501–506.
15. Brown, A. C., Shovic, A., Ibrahim, S. A., Holck, P., & Huang, A. (2005). A non-dairy probiotics's (*poi*) influence on changing the gastrointestinal tract's microflora environment. *Alternative Therapies in Health and Medicine*, 11, 58–64.
16. Brown, A. C., & Valiere, A. (2004). Probiotics and medical nutrition therapy. *Nutrition in Clinical Care*, 7, 56–68.
17. Ries, L., Eisner, M., Kosary, C., Hankey, B., Miller, B., Clegg, L., Mariotto, A., Feuer, E., & Edwards, B. K. (Eds.), (2004). *Seer Cancer Statistics Review, 1975–2001*. Bethesda, MD: National Cancer Institute.
18. Brown, A. C., Reitzenstein, J. E, Liu, J., & Jadus, M. R. (2005). The antiproliferative effect of diluted *poi* (*Colocasia esculenta*) on colonic adenocarcinoma cells *in vitro*. *Phytotherapy Research*, 19, 767–771.

19. Cambie, R. C., & Ferguson, L. R. (2003). Potential functional foods in the traditional Maori diet. *Mutation Research, 523–524*, 109–117.

20. Kim, H. W., Murakami, A., Nakamura, Y., & Ohigashi, H. (2002). Screening of edible Japanese plants for suppressive effects on phorbol ester-induced superoxide generation in differentiated HL-60 cells and AS52 cells. *Cancer Letters, 176*, 7–16.

21. Kim, Y. I. (2000). AGA technical review: impact of dietary fiber on colon epithelial cells in diseases leading to colonic cancer. *Gastroenterol, 118*, 1235.

22. Ferguson, L. R., Roberton, A. M., McKenzie, R. J., Watson, M. E., & Harris, P. J. (1992). Adsorption of a hydrophobic mutagen to dietary fiber from taro (*Colocasia esculenta*), an important food plant of the South Pacific. *Nutrition and Cancer – An International Journal, 17*, 85–95.

23. Bingham, S. A., Day, N. E., Luben, R., Ferrari, P., Slimani, N., Norat, T., Clavel-Chapelon, F., Kesse, E., Nieters, A., & Boeing, H. (2003). Dietary fiber in food and protection against colorectal cancer in the European. Prospective Investigation into Cancer and Nutrition (EPIC): an observational study. *Lancet, 361*, 1496–1501.

24. Okabe, Y., Shinmoto, H., Tsushida, T., & Tokuda, S. (1996). Antimutagenicity of the extracts from four root crops on the Trp-p 2-induced mutagenicity to *Salmonella typhimurium* TA 98. *Journal of the Japanese Society for Food Science and Technology, 43*, 36–39.

B. EFFECTS OF INDIVIDUAL VEGETABLES ON HEALTH

19

Increasing Children's Liking and Intake of Vegetables through Experiential Learning

Remco C. Havermans

Maastricht University, Faculty of Psychology and Neuroscience, Department of Clinical Psychological Science, Maastricht, The Netherlands

1. CHILDREN'S FEAR AND LOATHING OF VEGETABLES

It is broadly accepted that the consumption of ample fruits and vegetables is beneficial to one's health. More specifically, it has been shown to decrease the probability of developing diverse forms of cancer and it protects against the development of obesity (i.e. severe overweight) and hence adverse obesity-related chronic health consequences such as type II diabetes and cardiovascular or coronary heart disease [1,2]. Therefore, various health organizations across the world encourage the consumption of fruits and vegetables. The World Health Organization recommends a minimum intake of 400 g of fruit and vegetables (excluding potatoes and cassava) per day [3]. Although most people are likely unaware of the specific health dangers of not eating vegetables, the general fact that eating fruits and vegetables promotes good health can be considered conventional wisdom. Undoubtedly, the efforts exerted by the health organizations and policy makers in motivating the consumption of fruit and vegetables have contributed to this wisdom, yet growing knowledge concerning healthful dietary habits does not always lead to the desired change in eating behavior [4]. Indeed, most people still consume much less vegetables than is generally recommended and this particularly holds true for children. Many children are reluctant to eat vegetables [5], and this is a matter for concern as dietary habits and food preferences are developed during childhood and appear less susceptible to change in later adulthood [6].

An important question that arises is why so many children are reluctant to eat sufficient amounts of vegetables. Gibson et al. [7] found that vegetable consumption among 9–11-year old children was predicted by their mother's belief that choosing the right food for her child is important for preventing disease. Paradoxically, the children ate less vegetables the more they believed that vegetables are good for one's health and the stronger their mothers' concern for disease prevention. Gibson and colleagues speculate that when parents attempt to push their child to eat vegetables by posing the threat of future health consequences, this only serves to devalue the food, effectively ingraining the cognition that vegetables are good for you but taste bad. This is definitely an effect that one should be careful to avoid as Gibson and colleagues found that an even better predictor of children's vegetable consumption was 'liking' for vegetables. The more a child liked certain vegetables, the more frequently s/he would consume these vegetables. Interestingly, the mothers rated their liking of vegetables higher than their children did and the mothers' liking ratings did not explain their children's vegetable consumption. In other words, the common notion that young children like whatever their parents like does not apply to children's appreciation of vegetables. Gibson and colleagues [7], reviewing a large body of literature, also found that among children and adolescents (6–18 years old) taste preferences (i.e. degree of vegetable liking) are positively correlated with vegetable intake. Overall, one may conclude that the answer to the question of why children tend to eat very few vegetables is relatively straightforward: they do not like the taste of them. So what is the likely origin of this dislike for vegetables in children?

Children's reluctance to eat vegetables has been linked to food neophobia [8]. Food neophobia literally means 'fear of novel foods' and is expressed as a reluctance to eat, or the avoidance of, unfamiliar foods/flavors. It has been suggested that such neophobia should be considered as a phobia in a true clinical sense. In a study, young adults had to rate their willingness to eat certain novel foods and their expected dislike and perceived danger of eating the foods. Perceived danger is known to contribute to the expression of specific phobias such as a spider phobia. Although most participants did not evaluate the foods as particularly dangerous, the ones who did were far less willing to eat these foods [9]. Russell and Worsley [9] demonstrated in an Australian sample of 371 preschool children (2–5 years old) that the degree of food neophobia is negatively correlated with preference of various foods, but especially with the preference of vegetables [10]. In other words, the more neophobic the child, the less it likes vegetables. The challenge then is to overcome this apparent aversion for the flavor of vegetables and this leaves the question how to achieve this.

2. PRACTICE MAKES VEGETABLES TASTE PERFECT

Robert Zajonc [11] showed that repeat mere exposure to certain words in everyday language and communication is correlated with a more positive attitude toward these words. He further showed that repeat exposure to a set of given words actually causes increased liking of these words. This mere exposure effect has been replicated many times across species and across a wide variety of stimulus domains, including taste. Indeed, repeated exposure to a certain taste of food increases liking of that food in adults, and in children it has been found to increase the willingness to eat more of that food [12,13]. Mere exposure to novel foods has thus been suggested as a possible means to reduce food neophobia.

There is ample evidence that repeated exposure to novel foods lessens food neophobia. For example, Pliner and colleagues [10] had one group of participants taste seven different novel foods and another group taste seven similar but familiar foods. At test, all participants were required to make several choices. Each choice comprised the selection to taste one of two foods; one of the two foods being novel (e.g. a bite-size serving of a Peruvian salad) and the other being familiar. Participants who prior to this test had had to taste several other novel foods now proved less neophobic.

The mere exposure effect with regard to neophobia is not limited to adults. Sullivan and Birch [14], for instance, exposed 39 approximately 4–6-year-old preschool children to tofu (a new food to these children) twice a week for a period of 9 weeks and found that the preference for tofu increased with exposure frequency. This effect required about 8–15 exposures, which corroborated earlier findings [12,13,15]. However, Williams et al. [16] noted that in these previous studies the exposure usually comprised just a single food or a very limited number of novel foods. These researchers wondered whether the number of exposures required for acceptance of a novel food decreases with an increase in the number of foods previously accepted. They investigated this in six children being treated for extreme food selectivity or outright food refusal. One particular case concerned a 5-year-old girl with autism whose diet had comprised merely hot dogs, bacon, peanut butter, eggs, toast, and chocolate, but just prior to treatment had stopped eating altogether. Treatment of her food refusal comprised, among other things, regular exposure to novel foods presented in meals and taste sessions. Meals would contain three tablespoons of about three different novel foods (fruits, vegetables, meat, starch, or dairy product), and taste sessions were used to introduce the child to a specific novel food. With the introduction of the first novel food, the 5-year-old girl mentioned above required 27 presentations before she accepted it. Of the 49 different novel foods introduced and accepted across treatment, only the first four foods required 10 or more presentations before acceptance. The same pattern of results was found for the other five patients with the final ten foods introduced at the end of treatment requiring just one to seven exposures. Three months after treatment, the parents of the

abovementioned girl reported that she was still eating a variety of 47 foods. It is not quite clear why the effect of mere exposure is facilitated with an increasing number of foods previously accepted, but one might speculate that this is the result of stimulus generalization. With each introduction of a flavor of a novel food, this flavor is more likely to resemble some other already known and liked flavor. It should be noted that this 'I like it, it tastes like chicken' effect only works if one knows what 'chicken' tastes like and if one likes the taste of 'chicken' [17].

Mere exposure thus appears to be a potentially powerful means to increase children's liking and acceptance of certain foods, but does it also specifically work for vegetables? Williams and colleagues [16] exposed their participants to different novel foods from various food groups, including vegetables. They did not find that the number of exposures required for acceptance of vegetables was any different from other food groups. This suggests that for vegetables too, repeated mere exposure may help to promote acceptance and consumption. A more systematic investigation into this matter was undertaken by Jane Wardle and her co-workers. In one particular study Wardle and colleagues [18] instructed parents to expose their own children (3–7 years of age) to a certain target vegetable (e.g. carrots) every day for a period of 2 weeks. This 'exposure' group was compared with two other groups: one group of parents who received information leaflets containing suggestions on how to promote fruit and vegetable consumption, and one group who were told that they would get advice on promoting healthful dietary habits in their children. At the start and directly after the 2-week period, all children received a taste test in which they were instructed to taste and evaluate six different vegetables. One of the vegetables served as the target vegetable. Subjective liking and intake were measured during these taste tests. At posttest, children from the parents in the 'exposure' group liked the target vegetable better and consumed more of it in comparison to the pretest and relative to the children from the parents in the other two groups [19]. This result was, however, restricted to the parents who managed to expose their child to the target vegetable for at least 10 days. Wardle and colleagues note that the exposure itself was not at all a trouble-free undertaking. Fourteen of the 48 parents did not manage to complete the exposure for at least 10 days and many more parents reported finding the exposure very difficult.

Wardle and colleagues [18] instructed parents to conduct the exposure procedure, but children may also be exposed to vegetables within a school setting. Several such school programs were examined in the Netherlands: Schoolgruiten (a compound name combining the Dutch words for school, fruit, and vegetables) and Pro-Children. Both programs were aimed at increasing children's consumption of fruits and vegetables. The Pro-Children program tested in the Netherlands was part of a larger international European program. Each of these interventions were designed so as to target several determinants of fruit and vegetable consumption, with one of the primary determinants being recognized as liking of fruit and vegetables. Through repeated exposure to vegetables at school, the investigators hoped to improve the schoolchildren's preference of fruit and vegetables and in doing so promote behavior change; that is, increased acceptance and consumption of vegetables. For the duration of each intervention, children received a ready to eat piece of fruit or vegetable twice a week at school. As expected, stable intake of a high frequency of fruits and vegetables or an increase in this frequency was significantly predicted by liking of fruit and vegetables [20].

2.1 Age-related Mere Exposure Effect

It is commonly agreed that food neophobia is strongest between 2 and 8 years of age, so perhaps exposure to vegetables is more effective at increasing vegetable liking in older children. Indeed, Loewen and Pliner [21] demonstrated that 10–12-year-old children could use a single brief exposure to novel but good-tasting foods as a learning experience from which they deduce that novel foods do not necessarily have to taste bad, hence limiting food neophobia. Younger children (i.e. 7–9-year-old children), however, proved unable to use such a cognitive reasoning strategy despite the fact that these children too found the novel foods to be good tasting. This cognitive facility to 'refute' one's food neophobia may well be the reason that food neophobia is less common or less strong among adults [8]. Conducting exposure to vegetables in later childhood may thus be more effective at attenuating food neophobia. However, one might argue that by then potentially established unhealthy dietary habits are likely less amenable to change. With regard to promoting vegetable consumption in children one should perhaps strive to induce a liking for vegetables at the earliest age possible. But what is that 'earliest age'?

Sullivan and Birch [22] exposed 4–6-month-old infants consistently to either a pea or green bean puree during a period of 10 consecutive days and found that the infants demonstrated a clear increase in the amount consumed of specifically the vegetable they had been exposed to. Birch and colleagues [17] exposed 4–7-month-old infants repeatedly to a specific target food and found that acceptance of this food was increased significantly after just a single exposure. Maier et al. [23] also found evidence of a rapid increase in the intake of a previously disliked vegetable puree in approximately 7-month-old infants. Forestell and Mennella [24] evaluated the effects of breastfeeding and repeat exposure on acceptance and consumption of green beans. A sample of 4–8-month-old infants was fed green beans each day for eight consecutive days. Some of the infants were also fed peaches within an hour after having been fed the green beans. As expected, the repeated exposure to (i.e. consumption of) the green beans increased acceptance of the green beans in the infants. This effect was evident to a similar degree in both the breastfed and the formula-fed infants. However, only the infants who had experienced the green beans with the peaches came to display less facial expressions of distaste during feeding of the green beans. At the initial exposure, the infants ate more from the peaches than from the green beans and they clearly liked the peaches. Peaches taste sweet and humans have an innate preference for sweet tastes. Forestell and Mennella speculate that the liking of the green beans was enhanced due to an acquired association with the sweet taste of the peaches. It is certainly true that strong flavor preferences can be established through associative learning (or conditioning) and the results of Forestell and Mennella suggest that such learning may be stronger than (or add to) a mere exposure effect. But how exactly do associative learning mechanisms contribute to the development of a flavor preference?

3. SIGNALS FOR SATISFACTION

As argued above, the appreciation of the flavor of a food is largely based on experience. Next to (or perhaps on top of) a mere exposure effect, associative learning may contribute to the acquisition of food likes [25]. Associative learning in this respect refers specifically to Pavlovian conditioning, the learning of an association between a so-called 'conditioned stimulus' (CS; e.g. a flavor) and a biologically relevant unconditioned stimulus (UCS; e.g. food). Pavlov [26] demonstrated that dogs would come to salivate to a tone after several pairings of the tone (the CS) and the administration of food (UCS). Pavlov further noted that his dogs could differently display a

conditioned salivary reflex dependent upon the content of the UCS. An aversive UCS such as the administration of an electric shock would elicit a defensive salivary reflex, poor in mucin, whereas appetitive UCSs elicited appetitive mucin-rich salivation. Pavlov, however, did not explain this in terms of a hedonic response system. Such a hedonic view on the Pavlovian acquisition of likes and dislikes was not formulated until the 1960s with the work on conditioned flavor aversions by John Garcia (see reference 28).

The so-called 'Garcia effect' refers to the finding that when certain negative events are paired with food consumption, this will typically lead to an aversion for the flavor of that particular food. In a characteristic flavor aversion learning procedure, thirsty rats are given a drink that has been distinctly flavored. After consuming this flavor (the CS), the rats are exposed to gamma radiation or are injected with some poisonous substance (e.g. lithium chloride; the UCS) with the purpose of making the rats feel ill. After recovery, the animals are then given a choice to consume the flavored drink or plain water. Typically, the animals will avoid the flavored drink [28,29]. Similar effects have been demonstrated in human subjects [30–33]. Importantly, such conditioning, or associative learning, may also underlie a positive hedonic shift and as such may be instrumental in promoting liking and consumption of vegetables in children. Two main associative learning processes have been identified that may contribute to the acquisition of food likes: 1) flavor–nutrient learning; and 2) flavor–flavor learning. For both techniques, a neutral or even disliked flavor is paired with a positive UCS; that is, a satisfying post-ingestive stimulus in the case of flavor–nutrient learning, or an already highly liked flavor such as the sweet taste of saccharin in the case of flavor–flavor learning. In both cases, the initially neutral or disliked flavor CS becomes better liked due to the acquired association between this flavor and the satisfying UCS. In other words, this flavor preference learning endows the flavor CS with a signaling value, a signal for satisfaction, and as such becomes better liked itself.

3.1 Flavor–nutrient Learning

Bolles et al. [34] found that rats learn an apparent flavor preference for food paired with high caloric density, suggesting that a positive post-ingestive consequence (e.g. satiation) leads to a positive shift in the hedonic value of the flavor of the food. Such flavor preference learning is a form of conditioning where the preference for an initially neutral, or even disliked flavor increases due to pairings of the flavor of a food (the flavor CS) with the post-ingestive consequences of the macronutrients in that food (the nutrient UCS) [25,35].

Positive shifts in flavor preference due to flavor–nutrient learning have repeatedly been demonstrated in animals, using nutrient UCSs such as ethanol (7 kcal/g), fat (9 kcal/g), protein (4 kcal/g), and carbohydrate (4 kcal/g) [36]. Conditioning of flavor preferences has also been demonstrated in humans. Booth et al. [37] found that adults readily learn a preference for food associated with a high starch content. Mobini et al. [38] showed that in one group of participants, pairing a peach-flavored ice tea with sucrose enhanced liking of this particular flavor, as compared to another group of participants who received a similar but minimally sweetened flavor. Furthermore, this effect was the most prominent among the participants who had received flavor–nutrient training in a hungry state, which is in line with the general notion that flavor–nutrient learning entails the learning of an association between the flavor CS and the satiating consequence of the nutrient UCS [39]. Yeomans et al. [40] further demonstrated in human adults that flavor–nutrient learning may contribute to both enhanced liking and intake of a flavor paired with sucrose. Interestingly, Brunstrom and Mitchell [41] found evidence of flavor–nutrient preference learning in unrestrained but not in restrained eaters. This difference seems at odds with the aforementioned notion that flavor–nutrient learning is most prominent when hungry. Brunstrom and Mitchell, however, speculate that severe long-term or frequent calorie restriction associated with high dietary restraint makes one less sensitive to feelings of appetite, hunger, and satiation, hence debilitating flavor–nutrient learning.

The few human studies on flavor–nutrient learning all seem to suggest that the reinforcing property of the nutrient UCS lies in its property to evoke a subjective post-ingestive signal. It is important to note that this signaling effect of different macronutrients is not directly tied to caloric density. The metabolism of different nutrients occurs at different rates in different tissues and organs, and nutrients may have different satiating effects [42]. Each of these factors can affect the timing and the perceived intensity of the post-ingestive signal provided by macronutrients, and thus the learning of a flavor preference. Sclafani [43], in reviewing studies on flavor–nutrient learning, reported that fat administrations may be less effective in conditioning flavor preferences in rats than the isocaloric administration of carbohydrates. Lucas and Sclafani [44,45] directly compared the reinforcing and satiating effects of carbohydrate and fat in conditioning flavor preferences in rats. Carbohydrate proved to be more effective in conditioning a flavor preference than fat. Lucas and Sclafani [44] also demonstrated that carbohydrates have stronger satiating effects than fat. This may explain the differential reinforcing potency of carbohydrate and fat in conditioning flavor preferences in rats.

In an unpublished study, we tested whether a positive shift in liking of a flavor due to flavor–nutrient learning would be especially pronounced for flavors paired with high carbohydrate content. On a pre-test, participants had to taste, evaluate, and rank order seven different novel (i.e. not commercially available) flavors of yogurt. Subsequently, during 12 conditioning sessions three of these flavors were consistently paired with either a low-energy yogurt, or a high-fat yogurt, or a high-carbohydrate yogurt. At post-test, it was found that only liking of the carbohydrate-paired flavor, as rated on an 11-point Likert scale, had increased relative to pre-test. It should be noted that the flavors were only paired with fat or carbohydrates during training, not at the pre and post-test.

3.1.1 Flavor–nutrient Learning in Children

Flavor preference learning, that is, the acquisition of an association between a flavor and its post-ingestive consequence, extends to children as well. In fact, early failures to demonstrate clear flavor–nutrient learning in human adults led some researchers to speculate that this form of flavor preference learning in humans is restricted to children [41]. Birch et al. [46] demonstrated conditioned flavor preferences in children for drinks paired with a high carbohydrate content [47]. Similar results were found in children when pairing flavors with a high dietary fat content [48,49].

The food consumed in the studies referred to above was usually some type of dessert, and so the question arises whether flavor–nutrient learning is still effective when the test food or flavor comprises the taste and odor of a vegetable. To investigate the potential of flavor–nutrient learning as a means to increase children's liking of vegetables, Zeinstra et al. [50] had 7–8-year-old children consume two different vegetable juices, one of which was paired with a high energy density by adding maltodextrin to it. The children were instructed to consume 150 g of the juice (either high or low caloric) each day for a period of 14 consecutive schooldays. Each

juice was thus drunk seven times. Despite adding non-caloric sweetener to the vegetable juices, the children proved to be extremely hesitant to consume the daily required minimum amount of the juice, which was at least 80% of the serving. The children found the flavor of the juices to be very intense and very unpleasant. Not surprisingly then, no evidence of any flavor–nutrient learning was found. The children consumed far too little of the target vegetable flavors to experience any post-ingestive nutrient effect. This result implies that flavor–nutrient learning may not be an altogether practical and effective method to increase children's liking and acceptance of vegetables.

3.1.2 Flavor–flavor Learning

Eric Holman [51] tested whether a rat's flavor preference could be influenced by the sweetness of saccharin associated with the flavor. In one of his experiments, rats received on alternate days a highly sweetened flavored solution or another less sweetened flavored solution. Flavors were almond and banana and the assignment of one of these flavors to the highly sweetened solution was counterbalanced between the animals. At test, the rats were offered the choice between the almond- and the banana-flavored solution. Holman found that his rats clearly preferred the flavor that previously had been paired with the highest saccharin concentration.

This flavor–flavor learning has also been demonstrated in humans. Baeyens et al. [52] found that pairing a flavor with the aversive taste of polysorbate-20 (the emulsifier Tween) led to decreased liking of that particular flavor. They, however, did not find evidence of positive flavor–flavor learning with pairings of a specific neutral flavor with a sweet taste. This latter failure may be attributed to the fact that overall the sweet taste was only moderately liked. Zellner et al. [53] did observe significant flavor–flavor learning. They had students drink different flavors of tea. Some of these teas had to be tasted several times with sucrose. At test, all participants again had to taste and evaluate the different teas, now left unsweetened, and the participants clearly and specifically liked the previously sweetened teas better than before conditioning. Yeomans and colleagues [40] similarly found evidence of flavor–flavor learning when a specific flavor of dessert was paired with the sweet taste of either sucrose or aspartame. To ensure that the sweet taste (the flavor UCS) was well liked, Yeomans took care to only select self-identified sweet-likers for participation in their study.

Anita Jansen and I discussed the possibility that flavor–flavor learning might be a particularly powerful and practical method to increase children's liking of vegetables. On the basis of previous findings one could argue that successful positive flavor–flavor learning requires far less exposure to the target flavor than is usually the case with a mere exposure procedure and that, unlike flavor–nutrient learning, it does not require abundant ingestion of the target flavor. To test whether such flavor–flavor learning actually increases children's liking of vegetables we conducted an experiment. In this experiment 4–6-year-old children evaluated and rank ordered six different vegetables. Next, they were instructed to repeatedly consume small amounts of two of the six specific vegetables. These two flavors served as the flavor CSs and one of the two vegetables was now sweetened with glucose. After this repeated exposure procedure of approximately one hour's duration, all children again were instructed to evaluate and rank order the six vegetables. At this post-test, the children now specifically ranked the previously sweetened vegetable as better liked than before [54]. This effect was successfully replicated in a recent study by Capaldi and Privitera [55]. In a first experiment they had 2–5-year-old children repeatedly taste grapefruit juice mixed with the sweet taste of sucrose. This led to increased liking of unsweetened grapefruit juice. Moreover, this positive shift in liking proved stable for weeks. In a second experiment, undergraduate students were presented with several occasions in which they were instructed to consume one small stalk of cauliflower and another stalk of broccoli. One of the two vegetables would be sweetened by having it dipped in sugar water. The assignment of vegetable to sugar water was counterbalanced between participants. Capaldi and Privitera found that the pairings of either cauliflower or broccoli increased liking of the taste of these vegetables.

In sum, flavor–flavor learning may be a viable method to increase children's liking of vegetables in both the short and longer term. However, research findings concerning this method are still scarce and hence several questions remain concerning the efficacy and benefits of flavor–flavor learning. One such question is whether the established shift in liking for a particular flavor, as established through flavor–flavor pairings, likely generalizes to other vegetables. In an unpublished study, we investigated this in a sample of 50 8–10-year-old children. In this study, the children received several bite-size servings of bell pepper and tomato. Both the pepper and tomato were offered in two varieties, namely red and yellow. At a pre-test, the children had to taste and evaluate each variety. Next, the children received several exposures to the pepper (red or yellow) and tomato (yellow when the pepper was red and vice versa). In half of the participants, the pepper was sweetened by briefly dipping the vegetable in sugar water. For the other half of the sample, the tomato was sweetened, not the pepper. At a post-test, all children had to evaluate the vegetables once more, now unsweetened. Although the effects were small, the results indicated that at post-test, the initially sweetened vegetable was now better liked than the other vegetable. This effect though was only apparent when considering flavor (pepper versus tomato), not color (red versus yellow), implying that a positive hedonic shift for a vegetable established through flavor–flavor learning appears fairly flavor specific and does not generalize easily to vegetables that have perceptually similar characteristics (e.g. color) but taste differently.

4. SUMMARY

It is important to realize that, generally speaking, children's food preferences and aversions are still fairly limited and certainly not fixed. Therefore, children in particular can still learn to like (or dislike) specific foods and flavors, including the taste of vegetables. This latter finding is noteworthy as children generally dislike the often bitter taste of vegetables – which may be especially prominent in food neophobic children – leading to reluctance to consume healthful amounts of vegetables. The learning referred to in the present chapter concerns learning through direct experience with food, but learning to appreciate certain foods is certainly not limited to such experiential learning. When attending a large banquet where they serve all kinds of foods one does not immediately recognize as such (it can happen), one can learn that a certain food tastes good by carefully observing other food consumers. What foods do they put on their plate? How do they react when they taste a certain food? One could also ask someone attending the banquet how a given food tastes. Such social cues have also been found to affect food choice and food intake in both adults and children [47,56–58], but it is beyond the scope of the present chapter to describe this social learning effect. It suffices to note that at present, it is still unclear whether such social learning also effectively promotes liking, acceptance, and intake of vegetables in children.

As for experientially learning to like the taste of vegetables, mere exposure has proven to effectively increase children's acceptance and liking of vegetables. One prominent practical drawback of this procedure is that it generally requires a large number of exposures to achieve this effect. In this respect, flavor–flavor learning holds great promise. It has been shown that very few exposures to a sweetened vegetable suffice to increase children's liking of that vegetable. More research, however, is needed to determine whether the positive hedonic shift is also associated with more frequent and greater intake of vegetables. Although food liking and food choice/intake are usually strongly correlated, it has been shown that a shift in food liking does not necessarily prompt a relevant change in intake [40].

If one wishes to apply the research reviewed in the present chapter to increase children's liking and intake of vegetables, best practice – when considering the current body of research – would be to repeatedly expose the child to various vegetables. Such exposure should comprise the consumption of at least one spoonful of vegetable, and to facilitate an exposure effect one could mix sugar (e.g. sucrose, glucose, or fructose) with the vegetables for the first few exposures. This is to promote flavor–flavor and potentially flavor–nutrient learning and as such to advance a short- and longer-term positive shift in vegetable liking and intake.

References

1. Steinmetz, K. A., & Potter, J. D. (1996). Vegetables, fruit, and cancer prevention: A review. *Journal of the American Dietetic Association, 96*, 1027–1039.
2. Reddy, K. S., & Katan, M. B. (2004). Diet, nutrition and the prevention of hypertension and cardiovascular diseases. *Public Health Nutrition, 7*, 167–186.
3. WHO/FAO (2003). Diet, nutrition and the prevention of chronic diseases. Report of a joint WHO/FAO Expert Consultation. (WHO Technical report series 916.) Geneva: World Health Organization.

4. Contento, I., Balch, G. I., Bronner, Y. L., Lytle, L. A., & Malony, S. K. (1995). The effectiveness of nutrition education and implications for nutrition education policy, programs, and research: A review of research. *Journal of Nutrition Education, 27*, 278–418.

5. Munoz, K. A., Krebs-Smith, S. M., Ballard-Barbash, R., & Cleveland, L. E. (1997). Food intakes of US children and adolescents compared with recommendations. *Pediatrics, 100*, 323–329.

6. Lyte, L., Seifert, S., Greenstein, J., & McGovern, P. (2000). How do children's eating patterns and food choices change over time? Results from a cohort study. *American Journal of Health Promotion, 14*, 222–228.

7. Gibson, E. L., Wardle, J., & Watts, C. J. (1998). Fruit and vegetable consumption, nutritional knowledge and beliefs in mothers and children. *Appetite, 31*, 205–228.

8. Dovey, T. M., Staples, P. A., Gibson, E. L., & Halford, J. C. G. (2008). Food neophobia and picky/fussy eating in children: A review. *Appetite, 50*, 181–193.

9. Russell, C. G., & Worsley, A. (2008). A population-based study of preschooler's food neophobia and its associations with food preferences. *Journal of Nutrition Education and Behavior, 40*, 11–19.

10. Pliner, P., Pelchat, M., & Grabski, M. (1993). Reduction of neophobia in humans by exposure to novel foods. *Appetite, 20*, 111–123.

11. Zajonc, R. (1968). Attitudinal effects of mere exposure. *Journal of Personality and Social Psychology Monograph Supplement, 9*, 1–27.

12. Pliner, P. (1982). The effect of mere exposure on liking for edible substances. *Appetite, 3*, 283–290.

13. Birch, L. L., & Marlin, D. W. (1982). I don't like it; I never tried it: Effects of exposure on two-year-old children's food preferences. *Appetite, 3*, 353–360.

14. Sullivan, S., & Birch, L. L. (1990). Pass the sugar, pass the salt: Experience dictates preference. *Developmental Psychology, 26*, 546–551.

15. Birch, L. L., McPhee, L., Shoba, B. C., Pirok, E., & Steinberg, L. (1987). What kind of exposure reduces children's food neophobia? *Appetite, 9*, 171–178.

16. Williams, K. E., Paul, C., Pizzo, B., & Riegel, K. (2008). Practice does make perfect: A longitudinal look at repeated taste exposure. *Appetite, 51*, 739–742.

17. Birch, L. L., Gunder, L., & Grimm-Thomas, K. (1998). Infant's consumption of a new food enhances acceptance of similar foods. *Appetite, 30*, 283–295.

18. Wardle, J., Cooke, L. J., Gibson, E. L., Sapochnik, M., Sheiham, A., & Lawson, M. (2003). Increasing children's acceptance of vegetables: A randomized trial of parent-led exposure. *Appetite, 40*, 155–162.

19. Wardle, J., Herrera, M. L., Cooke, L. J., & Gibson, E. L. (2003). Modifying children's food preferences: The effects of exposure and reward on acceptance of an unfamiliar vegetable. *European Journal of Clinical Nutrition, 57*, 341–348.

20. Tak, N. I., Te Velde, S. J., & Brug, J. (2008). Are positive changes in potential determinants associated with increased fruit and vegetable intakes among primary schoolchildren? Results of two intervention studies in the Netherlands: The schoolgruiten project and the pro-children study. *The International Journal of Behavioral Nutrition and Physical Activity, 5*, 21.

21. Loewen, R., & Pliner, P. (1999). Effects of prior exposure to palatable and unpalatable novel foods on children's willingness to taste other novel foods. *Appetite, 32*, 351–366.

22. Sullivan, S. A., & Birch, L. L. (1994). Infant dietary experience and acceptance of solid foods. *Pediatrics, 93*, 271–277.

23. Maier, A., Chabanet, C., Schaal, B., Issanchou, S., & Leathwood, P. (2007). Effects of repeated exposure on acceptance of initially disliked vegetables in 7-month old infants. *Food Quality and Preference, 18*, 1023–1032.

24. Forestell, C. A., & Mennella, J. A. (2007). Early determinants of fruit and vegetable acceptance. *Pediatrics, 120*, 1247–1254.

25. Capaldi, E. D. (1996). Conditioned food preferences. In E. D. Capaldi, (Ed.), *Why we eat what we eat: The psychology of eating* (pp. 53–80). Washington DC: American Psychological Association.

26. Pavlov, I. P. (1927). *Conditioned reflexes.* Oxford: Oxford University Press.

27. Mehiel, R. (1997). The consummatory rat: The psychological hedonism of Robert C. Bolles. In M. E. Bouton, & M. Fanselow, (Eds.), *Learning, motivation and cognition: The functional behaviorism of Robert C. Bolles.* Washington, DC: American Psychological Association.

28. Garcia, J., Kimeldorf, D. J., & Koelling, R. A. (1955). Conditioned aversion to saccharin resulting from exposure to gamma radiation. *Science, 122*, 157–158.

29. Garcia, J., & Koelling, R. A. (1966). Relation of cues to consequence in avoidance learning. *Psychonom Science, 5*, 121–122.

30. Arwas, S., Rolnick, A., & Lubow, R. E. (1989). Conditioned taste aversion in humans using motion-induced sickness as the US: The effects of CS familiarity. *Behaviour Research Therapy, 27*, 295–302.

31. Bernstein, I. L., & Borson, S. (1986). Learned food aversion: A component of anorexia syndromes. *Psychological Review, 93*, 462–472.

32. Cannon, D. S., Best, M. R., Batson, J. D., & Feldman, M. (1983). Taste familiarity and apomorphine-induced taste aversions in humans. *Behaviour Research and Therapy, 21*, 669–673.

33. Havermans, R.C., Salvy, S-J., and Jansen, A. (2009). Single-trial exercise induced taste and odor aversion learning in humans. *Appetite*, Epub ahead of print.

34. Bolles, R. C., Hayward, L., & Crandall, C. (1981). Conditioned taste preferences based on caloric density. *Journal of Experimental Psychology Animal Behaviour Processes*, 7, 59–69.

35. Myers, K. P., & Sclafani, A. (2003). Conditioned acceptance and preference but not alerted taste reactivity responses to bitter and sour flavors paired with intragastric glucose infusion. *Physiology & Behaviour, 78*, 173–183.

36. Fedorchak, P. M. (1997). The nature and strength of caloric conditioning. In M. E. Bouton & M. Fanselow, (Eds.), *Learning, motivation and cognition: The functional behaviorism of Robert C. Bolles* (pp. 255–269). Washington, DC: American Psychological Association.

37. Booth, D., Mather, P., & Fuller, J. (1982). Starch content of ordinary foods associatively conditions human appetite and satiation. *Appetite, 3*, 163–184.

38. Mobini, S., Chambers, L. C., & Yeomans, M. R. (2007). Effects of hunger state on flavour pleasantness conditioning at home: Flavour–nutrient learning vs. flavour–flavour learning. *Appetite, 48*, 20–28.

39. Appleton, K. M., Gentry, R. C., & Shepherd, G. R. (2006). Evidence of a role for conditioning in the development of liking for flavours in humans in everyday life. *Physiology & Behaviour, 87*, 478–486.

40. Yeomans, M. R., Leitch, M., Gould, N. J., & Mobini, S. (2007). Differential hedonic, sensory and behavioral changes associated with flavor-nutrient and flavor-flavor learning. *Physiology & Behaviour, 93*, 798–806.

41. Brunstrom, J. M., & Mitchell, G. L. (2007). Flavour-nutrient learning in restrained and unrestrained eaters. *Physiology Behaviour, 90*, 133–141.

42. Blundell, J. E., Lawton, C. L., Cotton, J. R., & Macdiarmid, J. I. (1996). Control of human appetite: Implications for the intake of dietary fat. *Annual Review of Nutrition, 16*, 285–319.

43. Sclafani, A. (1990). Nutritionally based learned flavor preferences in rats. In E. D. Capaldi & T. L. Powley, (Eds.), *Taste, experience and feeding* (pp. 139–156). Washington DC: American Psychological Association.

44. Lucas, F., & Sclafani, A. (1999). Differential reinforcing and satiating effects of intragastric fat and carbohydrate infusions in rats. *Physiology & Behaviour, 66*, 381–388.

45. Lucas, F., & Sclafani, A. (1999). Flavor preferences conditioned by high-fat versus high-carbohydrate diets vary as a function of session length. *Physiology & Behaviour, 66*, 389–395.

46. Birch, L. L., McPhee, L., Steinberg, L., & Sullivan, S. (1990). Conditioned flavor preferences in young children. *Physiology & Behaviour, 47*, 501–505.

47. Jansen, A., & Tenney, N. (2001). Seeing mum drinking a 'light' product: Is social learning a stronger determinant of taste preference acquisition than caloric conditioning? *European Journal of Clinical Nutrition, 55*, 418–422.

48. Johnson, S., McPhee, L., & Birch, L. L. (1991). Conditioned preferences: Young children prefer flavors associated with high dietary fat. *Physiology & Behaviour, 50*, 1245–1251.

49. Kern, D. L., McPhee, L., Fisher, J., Johnson, S., & Birch, L. L. (1993). The postingestive consequences of fat condition preferences for flavors associated with high dietary fat. *Physiology & Behaviour, 54*, 71–76.

50. Zeinstra, G. G., Koelen, M. A., Kok, F. J., & de Graaf, C. (2009). Children's hard-wired aversion to vegetable tastes. A 'failed' flavour-nutrient learning study. *Appetite, 52*, 528–530.

51. Holman, E. W. (1975). Immediate and delayed reinforcers for flavour preferences in rats. *Learning and Motivation, 6*, 91–100.

52. Baeyens, F., Crombez, G., Hendrickx, H., & Eelen, P. (1995). Parameters of human evaluative flavor-flavor associations in humans. *Learning and Motivation, 26*, 141–160.

53. Zellner, D. A., Rozin, P., Aron, M., & Kulish, C. (1983). Conditioned enhancement of human's liking for flavors by pairing with sweetness. *Learning and Motivation, 14*, 338–350.

54. Havermans, R. C., & Jansen, A. (2007). Increasing children's liking of vegetables through flavour–flavour learning. *Appetite, 48*, 259–262.

55. Capaldi, E. D., & Privitera, G. J. (2008). Decreasing dislike for sour and bitter in children and adults. *Appetite, 50*, 139–145.

56. Baeyens, F., Vansteenwegen, D., de Houwer, J., & Crombez, G. (1996). Observational conditioning of food valence in humans. *Appetite, 27*, 235–250.

57. Pelchat, M. L., & Pliner, P. (1995). 'Try it. You'll like it': Effects of information on willingness to try novel foods. *Appetite, 24*, 153–165.

58. Herman, C. P., & Polivy, J. (2008). External cues in the control of food intake in humans: The sensory-normative distinction. *Physiology & Behaviour, 94*, 722–728.

B. EFFECTS OF INDIVIDUAL VEGETABLES ON HEALTH

20

Bioactivity of Polyacetylenes in Food Plants

Lars P. Christensen

Institute of Chemical Engineering, Biotechnology and Environmental Technology, Faculty of Engineering, University of Southern Denmark, Odense M, Denmark

1. INTRODUCTION

Epidemiological investigations have provided evidence that a diet high in vegetables and fruits is associated with a reduced risk for the development of, for example, cancer and cardiovascular diseases [1–6]. Compounds associated with the health-promoting effects of vegetables are glucosinolates and other organosulfur compounds and their degradation products, phytosterols, vitamins, polyphenols, carotenoids, and dietary fibers [1,6]. These different classes of natural products may only partly explain the health effects of vegetables, and consequently focus has been directed toward other types of potential health-promoting compounds such as glycolipids [7,8] and polyacetylenes [6,9–15].

Naturally-occurring polyacetylenes include all compounds containing two or more conjugated carbon–carbon triple bonds. However, the term 'polyacetylenes' is also often used interchangeably to describe natural products

containing a single acetylenic bond if they are formed from polyacetylene precursors [16,17]. In this review, the latter and broader definition of polyacetylenes will be used. Polyacetylenes are widely distributed, occurring in plants, fungi, lichens, moss, marine algae, and invertebrates [16–18]. Polyacetylenes are often unstable, being sensitive to UV light as well as oxidative and pH-dependent decomposition, which often provides substantial challenges for their isolation and characterization [12,16,18]. The triple bond functionality of polyacetylenes makes these natural products a very interesting group of compounds whose reactivity toward proteins and other natural products may explain their wide variety of bioactivities.

Polyacetylenes are examples of bioactive secondary metabolites that have been considered undesirable in plant foods due to their toxic properties. Some polyacetylenes are known to be potent skin sensitizers, and to be neurotoxic in high concentrations, but are also highly bioactive compounds with potential health-promoting properties. The beneficial effects of

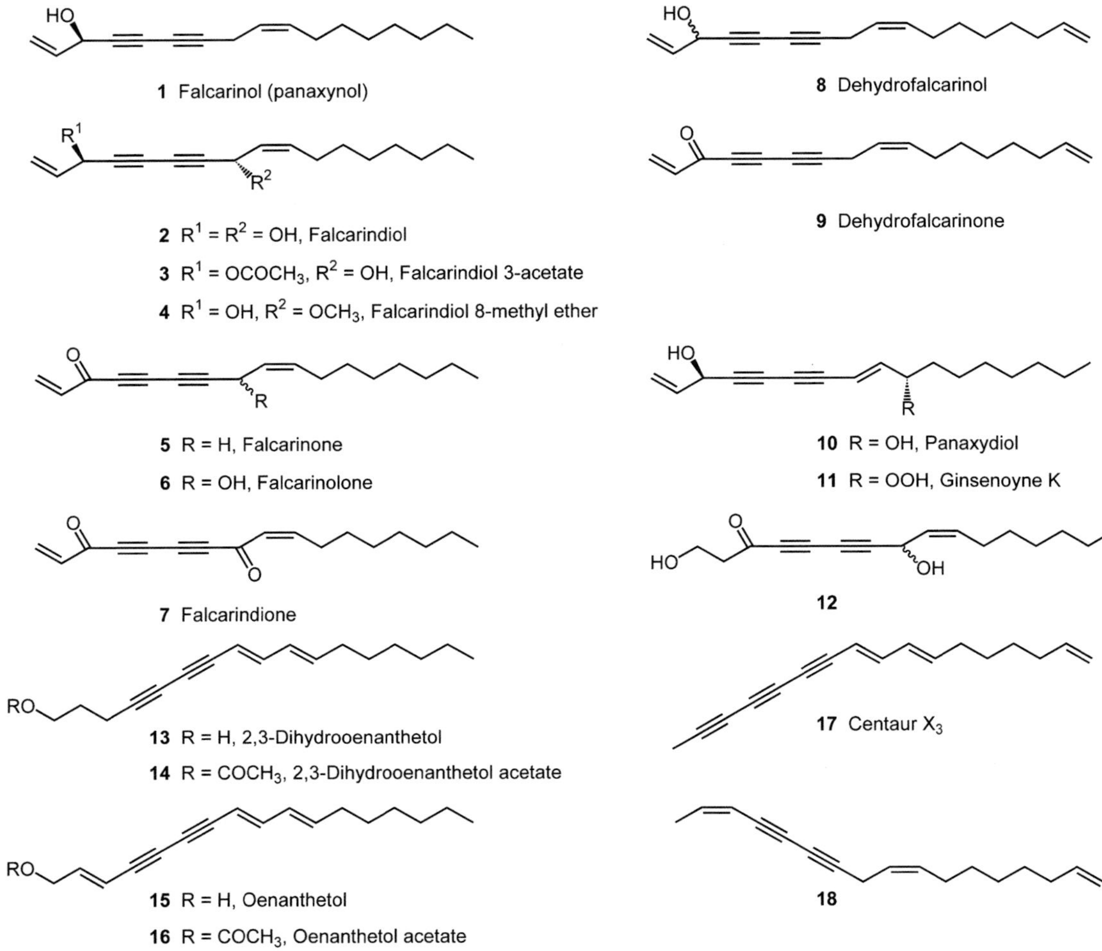

FIGURE 20.1 Aliphatic C_{17}-polyacetylenes isolated from the edible parts of food plants.

most bioactive polyacetylenes from higher plants occur at non-toxic concentrations. Hence, polyacetylenes appear to be an important group of nutraceuticals in vegetable foods that are obvious targets for the development of healthier foods and food products. In particular, aliphatic C_{17}-polyacetylenes such as falcarinol (1)[1] and related metabolites (falcarinol-type polyacetylenes) (Figure 20.1) present in carrots and related vegetables possess interesting bioactivities, including anticancer and anti-inflammatory effects and are now considered as important bioactive compounds that contribute significantly to the health effects of certain vegetable foods [6,9–15].

This review highlights the present state of knowledge on the occurrence of naturally occurring polyacetylenes in the edible parts of food plants with special focus on their bioactivity and possible relevance for human health.

[1]Bold numbers refer to the polyacetylenes illustrated in the figures.

2. POLYACETYLENES IN FOOD PLANTS

More than 2000 polyacetylenes are known, of which the majority have been isolated from higher plants. Polyacetylenes are widespread among species within the botanically related plant families Apiaceae, Araliaceae, and Asteraceae, and have in addition been found to occur sporadically in at least 21 other plant families [16,18–23]. The polyacetylenes isolated among Apiaceae and Araliaceae species are mainly aliphatic polyacetylenes. Plants belonging to the Asteraceae are frequent producers of acetylenes with more than 1100 polyacetylenes in this plant family. The structural diversity of polyacetylenes isolated from Asteraceae species

is considerable and includes thiophenes, thioethers, sulfoxides, alkamides, spiroacetal enol ethers, aromatic and aliphatic polyacetylenes. The structural diversity among polyacetylenes seems to indicate the involvement of many different precursors in their biosynthesis. A comparison of the polyacetylene structures, however, clearly indicates that most polyacetylenes are biosynthesized from unsaturated fatty acids. Feeding experiments with [14]C- and ^{3}H-labeled precursors have confirmed this assumption and further that polyacetylenes are built up from acetate and malonate units [17–26]. The first step in the biosynthesis of polyacetylenes is the dehydrogenation of oleic acid and linoleic acid to the C_{18}-acetylene crepenynic acid (Figure 20.2). Two distinct pathways for the formation of acetylenic bonds

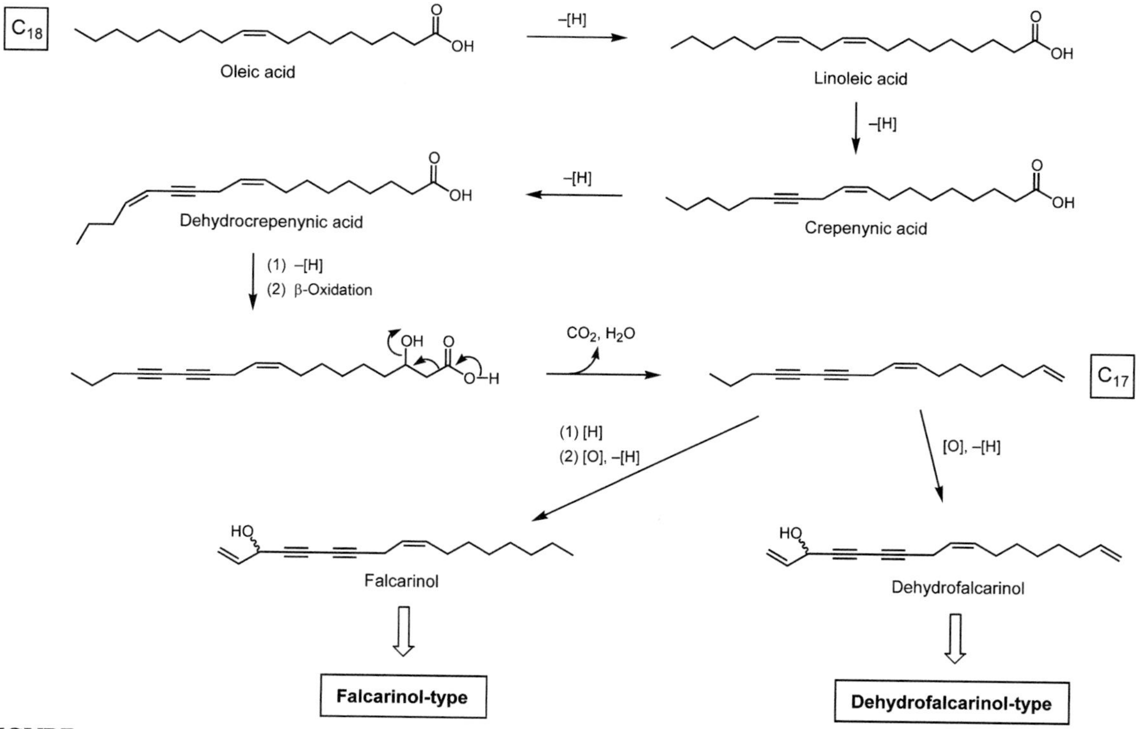

FIGURE 20.2 A possible biosynthetic pathway for aliphatic C_{17}-polyacetylenes of the falcarinol-type and dehydrofalcarinol-type in higher plants.
[O] = oxidation; [H] = reduction; − [H] = iron-catalyzed dehydrogenation with molecular oxygen by the loss of water [17].

have been proposed: desaturation of existing alkene functionality through an iron-catalyzed dehydrogenation with molecular oxygen [25] or the elimination of an activated enol carboxylate intermediate that is driven by CO_2 formation leading to triple bond formation [27]. The former pathway would be operative with full-length acyl lipids whereas the latter would install acetylenic groups during *de novo* fatty acid biosynthesis [17,25,27]. Both hypotheses appear to be valid for the biosynthesis of polyacetylenes discussed in this review. Further dehydrogenation of crepenynic acid leads to diyne and triyne C_{18}-acids that by α- and β-oxidation and other oxidative degradation reactions leads to polyacetylene precursors of various chain lengths, which are then transformed to a large variety of polyacetylenes.

Food plants that are known to produce polyacetylenes in their utilized plant parts include important vegetables such as artichoke, aubergine, carrot, celeriac, lettuce, parsley, parsnip, and tomato (Table 20.1). However, many food plants have not yet been investigated for polyacetylenes, although they belong to plant families that normally produce polyacetylenes. Therefore, it is expected that polyacetylenes are much more common in vegetable foods than described in the present review.

Aliphatic C_{17}-polyacetylenes of the falcarinol-type (e.g. **1**–**7**, **10**, **11**) (Figure 20.1) are widespread amongst Apiaceae and Araliaceae species, and consequently nearly all polyacetylenes found in the edible parts of food plants of the Apiaceae are of the falcarinol-type (Table 20.1). Falcarinol-type polyacetylenes are, however, less common in food plants of other plant families, although falcarinol (**1**) and falcarindiol (**2**) and related C_{14}- and C_{15}-polyacetylenes (**32**, **40**) (Figure 20.3) have been found in tomatoes and aubergines of the Solanaceae (Table 20.1), where they appear to be phytoalexins [28–30]. The biosynthesis of polyacetylenes of the falcarinol-type seems to follow the normal biosynthetic pathway for aliphatic C_{17}-polyacetylenes,

with dehydrogenation of oleic acid leading to the C_{18}-acetylenes crepenynic acid and dehydrocrepenynic acid, which is then transformed to C_{17}-acetylenes by β-oxidation. Further oxidation and dehydrogenation leads to polyacetylenes of the falcarinol-type and/or dehydrofalcarinol-type (**8**, **9**) as outlined in Figure 20.2. Dehydrofalcarinol-type polyacetylenes are characteristic for certain tribes of Asteraceae [16,21–23]. So far, they have only been found in the edible parts of tarragon (**8**) [31] and Jerusalem artichoke (**9**) [32].

From Asteraceae food plants structurally very diverse polyacetylenes have been isolated, including aliphatic polyacetylenes, acetylenic thiophenes, aromatic and spiroacetal enol ethers, in accordance with the structural diversity among polyacetylenes found in this plant family (Table 20.1; Figures 20.1, 20.3, 20.4, 20.5).

Spiroacetal enol ethers are characteristic for species of the tribe Anthemideae (Asteraceae) [16,21], and it is therefore not surprising that the polyacetylenes isolated from the edible parts of *Chrysanthemum coronarium* (garland chrysanthemum) [21,33–35] and *Matricaria chamomilla* (chamomile) [36–40], both belonging to the tribe Anthemideae, are spiroacetal enol ethers (Table 20.1; Figure 20.4). The C_{12}- and C_{13}-spiroacetal enol ethers present in the above-mentioned food plants are most likely biosynthesized from a C_{18}-triyn-ene acid precursor by α-oxidation followed by two β-oxidations leading to a C_{13}-triyn-ene alcohol, which is then transformed into the spiroacetal enol ethers **45** and **46** by further oxidation and ring closure (Figure 20.6). Further oxidation of compounds **45** and **46** leads to the spiroacetal enol ethers **42**–**44**, **47** and **48**, whereas oxidation and decarboxylation followed by addition of CH_3SH or its biochemical equivalent leads to thioether and/or sulfoxide spiroacetal enol ethers such as compounds **49**–**57** (Figures 20.4, 20.6). *Artemisia dracunculus* (tarragon), also belonging to the tribe Anthemideae, is characterized by the

TABLE 20.1   Polyacetylenes in the Edible Parts of Food Plants

Family/species	Common name	Plant part used for foods[a]	Primary use[b]	Polyacetylenes in used plant parts	References
Apiaceae (= Umbelliferae)					
Aegopodium podagraria L.	Bishop's weed, ground elder	L, St	V	1, 2, 12	16,57,83
Anethum graveolens L.	Dill	L, S	C, V	1, 2, 16	16,83
Anthriscus cerefolium (L.) Hoffm.	Chervil, salad chervil, French parsley	L, S	C, V	1, 2	83[c]
A. sylvestris Hoffm.	Cow parsley	L	V	2	c,d
Apium graveolens L. var. Dulce	Celery	L, S	C, V	1, 2	c,e
A. graveolens L. var. Rapaceum	Celeriac, knob celery, celery root	R	V	1, 2, 4 − 6, 10	11,16[f]
Bunium bulbocastanum L.	Great earthnut	T, L, F	C, V	1, 5, 6	16
Carum carvi L.	Caraway	R, L, S	C	1, 2, 6, 7	16,83[g]
Centella asiatica L.	Asiatic or Indian pennywort	L	V	1, 2	c,d
Chaerophyllum bulbosum L.	Turnip-rooted chervil	R, L	V	1, 5	c
Coriandrum sativum L.	Coriander, cilantro	L, S	C, V	1, 2	c,d
Crithmum maritimum L.	Samphire, marine fennel	L	V	1, 2	109
Cryptotaenia canadensis (L.) DC.	Hornwort, white or wild chervil	R, L, St, F	V	1, 2	h
Daucus carota L.	Carrot	R, L	V	1 − 3, 6, 10, 11	9 − 16,19,26, 62 − 65,82[i]
Ferula assa-foetida L.	Asafoetida, giant fennel	R, S, Sh	C	6	16
F. communis L.	Common giant fennel	L, S	C, V	2	93
Foeniculum vulgare Mill.	Fennel	L, S	C, V	1, 2	11,83[d]
Heracleum sphondylium L.	Common cow parsnip, hogweed	L, Sh	V	1, 2	83[c]
Levisticum officinale Koch.	Lovage, garden lovage	L, S	C, V	2, 6	c
Oenanthe javanica (Blume) DC.	Water dropwort, water celery	L, St, Sh	V	1, 2	63[j]
Pastinaca sativa L.	Parsnip	R, L	V	1, 2, 5, 6	11,16,83
Petroselinum crispum (Mill.) Nyman ex A. W Hill. (= P. sativum Hoffm.)	Parsley	L	C, V	1, 2, 5, 6	16,83[c]
P. crispum (Mill.) Nyman ex A. W. Hill. var. tuberosum	Hamburg parsley, turnip-rooted parsley	R, L	C, V	1, 2, 4, 10	11,60
Pimpinella major (L.) Hud.	Greater burnet saxifrage	R, L, S	C	1, 2	83
P. saxifraga L.	Burnet saxifrage	R, L, S	C	31	16[k]
Sium sisarum L.	Skirret, chervil	R	V	6	16[g]
Trachyspermum ammi (L.) Spr.	Ajowan, ajwain	L, S	C	13 − 16	16
Asteraceae (= Compositae)					
Arctium lappa L.	Edible burdock, Gobo	R	V	18, 24 − 29, 36 − 39, 58 − 69	16,20,42 − 45,63,74
Artemisia dracunculus L.	Tarragon, esdragon	L	C, V	8, 19 − 21, 41	31
Bellis perennis L.	Common daisy	L, F	C, V	22, 23	73[l]
Chrysanthemum coronarium L.	Garland chrysanthemum, Shungiku (Japanese), Kor tongho (Chinese)	L	V	42 − 57	21, 33 − 35[c]
Cynara scolymus L.	Globe artichoke	L, F	V	27, 28	16[c]
Cichorium endivia L.	Endive, escarole	L	V	30	c
C. intybus L. var. foliosum	Chicory	R, L	V	30	46
Helianthus tuberosus L.	Jerusalem artichoke	T	V	9	16,32
Lactuca sativa L.	Lettuce	L	V	17	16[m]

(Continued)

B. EFFECTS OF INDIVIDUAL VEGETABLES ON HEALTH

TABLE 20.1 (Continued)

Family/species	Common name	Plant part used for foods[a]	Primary use[b]	Polyacetylenes in used plant parts	References
Matricaria chamomilla L. (= *M. recutita* (L.) Rausch.)	Chamomile, German chamomile	F	C	**45, 46**	36–40
Campanulaceae					
Platycodon grandiflorum (Jacq.) A. DC.	Balloon flower, Chinese bell flower	R, L	V	**33–35**	47,48
Solanaceae					
Lycopersicon esculentum Mill.	Tomato	Fr	V	1^n, 2^n, 32^n	28,29
Solanum melongena L.	Eggplant, aubergine	Fr	V	2^n, 40^n	30

[a]R, roots; T, tubers; L, leaves; St, stems; Sh, shoots; F, flowers; Fr, fruits; P, pods; S, seeds.

[b]V, vegetable; C, condiment or flavoring [12,63].

[c]Christensen, L.P., unpublished results.

[d]Nakano, Y., Matsunaga, H., Saita, T., Mori, M., Katano, M., and Okabe, H. (1998). Antiproliferative constituents in Umbelliferae plants II. Screening for polyacetylenes in some Umbelliferae plants, and isolation of panaxynol and falcarindiol from the root of *Heracleum moellendorffii*. *Biol. Pharm. Bull.* **21**, 257–261.

[e]Avalos, J., Fontan, G.P., and Rodriguez, E. (1995). Simultaneous HPLC quantification of two dermatoxins, 5-methoxypsoralen and falcarinol, in healthy celery. *J. Liq. Chromatogr.* **18**, 2069–2076.

[f]Bohlmann, F. (1967). Notiz über die Inhaltsstoffe von Petersilie- und Sellerie-Wurzeln. *Chem. Ber.* **100**, 3454–3456.

[g]Bohlmann, F., Arndt, C., Bornowski, H., and Kleine, K.-M. (1961). Über Polyine aus der Familie der Umbelliferen. *Chem. Ber.* **94**, 958–967.

[h]Eckenbach, U., Lampman, R. L., Seigler, D. S., Ebinger, J., and Novak, R. J. (1999). Mosquitocidal activity of acetylenic compounds from *Cryptotaenia canadensis*. *J. Chem. Ecol.* **25**, 1885–1893.

[i]Bentley, R.K., Bhattacharjee, D., Jones, E.R.H., and Thaller, V. (1969). Natural acetylenes. Part XXVIII. C_{17}-Polyacetylenic alcohols from the Umbellifer *Daucus carota* L. (carrot): Alkylation of benzene by acetylenyl(vinyl)carbinols in the presence of toluene-*p*-sulphonic acid. *J. Chem. Soc. (C)*, 685–688; Yates, S.G., and England, R.E. (1982). Isolation and analysis of carrot constituents: myristicin, falcarinol, and falcarindiol. *J. Agric. Food Chem.* **30**, 317–320; Lund, E.D. (1992). Polyacetylenic carbonyl compounds in carrots. *Phytochemistry* **31**, 3621–3623; Lund, E.D., and White, J.M. (1990). Polyacetylenes in normal and water-stressed 'Orlando Gold' carrots (*Daucus carota*). *J. Sci. Food Agric.* **51**, 507–516; Czepa, A., and Hofmann, T. (2003). Structural and sensory characterization of compounds contributing to the bitter off-taste of carrots (*Daucus carota* L.) and carrot puree. *J. Agric. Food Chem.* **51**, 3865–3873; Czepa, A., and Hofmann, T. (2004). Quantitative studies and sensory analyses on the influence of cultivar, spatial tissue distribution, and industrial processing on the bitter off-taste of carrots (*Daucus carota* L.) and carrot products. *J. Agric. Food Chem.* **52**, 4508–4514; Kidmose, U., Hansen, S. L., Christensen, L. P., Edelenbos, M., Larsen, E., and Nørbæk, R. (2004). Effects of genotype, processing, storage, and root size on bioactive compounds (carotenoids, polyacetylenes, 6-methoxymellein) in organically grown carrots (*Daucus carota* L.). *J. Food Sci.* **69**, S388–S394; Baranska, M., Schulz, H., Baranski, R., Nothnagel, T., and Christensen, L.P. (2005). *In situ* simultaneous analysis of polyacetylenes, carotenoids and polysaccharides in carrot roots. *J. Agric. Food Chem.* **53**, 6565–6571; Christensen, L.P., and Kreutzmann, S. (2007). Determination of polyacetylenes in carrot roots (*Daucus carota* L.) by high-performance liquid chromatography coupled with diode array detection. *J. Sep. Sci.* **30**, 483–490; Yang, R.-L., Yan, Z.-H., and Lu, Y. (2008). Cytotoxic phenylpropanoids from carrot. *J. Agric. Food Chem.* **56**, 3024–3027.

[j]Fujita, T., Kadoya, Y., Aota, H., and Nakayama, M. (1995). A new phenylpropanoid glucoside and other constituents of *Oenanthe javanica*. *Biosci. Biotech. Biochem.* **59**, 526–528.

[k]Schulte, K.E., Rücker, G., and Backe, W. (1970). Polyacetylenes from *Pimpinella*-species. *Arch. Pharm.* **303**, 912–919.

[l]Avato, P., and Tava, A. (1995). Acetylenes and terpenoids of *Bellis perennis*. *Phytochemistry* **40**, 141–147.

[m]Bentley, R.K., Jones, E.R.H., and Thaller, V. (1969). Natural acetylenes. Part XXX. Polyacetylenes from *Lactuca* (lettuce) species of the Liguliflorae sub family of the Compositae. *J. Chem. Soc. (C)*, 1096–1099.

[n]Mainly isolated from infected plant tissue.

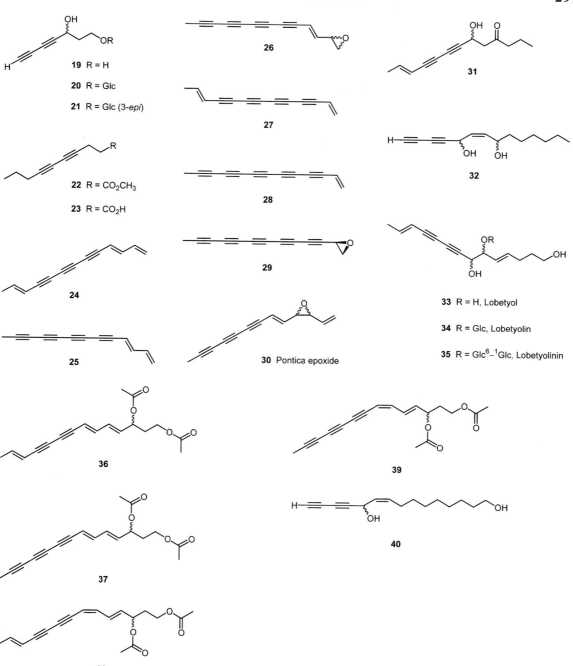

FIGURE 20.3 Aliphatic C_7 to C_{15}-polyacetylenes isolated from the edible parts of food plants. Glc = β-D-glucose.

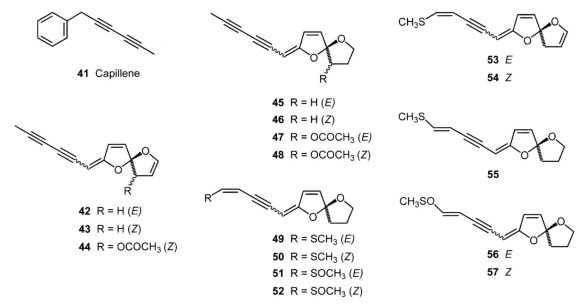

41 Capillene

42 R = H (*E*)
43 R = H (*Z*)
44 R = OCOCH₃ (*Z*)

45 R = H (*E*)
46 R = H (*Z*)
47 R = OCOCH₃ (*E*)
48 R = OCOCH₃ (*Z*)

49 R = SCH₃ (*E*)
50 R = SCH₃ (*Z*)
51 R = SOCH₃ (*E*)
52 R = SOCH₃ (*Z*)

53 *E*
54 *Z*

55

56 *E*
57 *Z*

FIGURE 20.4 Aromatic and spiroacetal enol ether acetylenes derived from polyacetylene precursors and isolated from the edible parts of food plants.

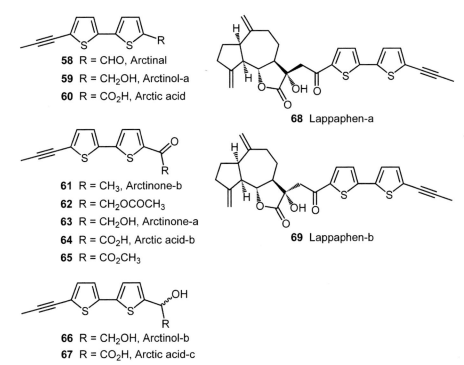

58 R = CHO, Arctinal
59 R = CH₂OH, Arctinol-a
60 R = CO₂H, Arctic acid

61 R = CH₃, Arctinone-b
62 R = CH₂OCOCH₃
63 R = CH₂OH, Arctinone-a
64 R = CO₂H, Arctic acid-b
65 R = CO₂CH₃

66 R = CH₂OH, Arctinol-b
67 R = CO₂H, Arctic acid-c

68 Lappaphen-a

69 Lappaphen-b

FIGURE 20.5 Acetylenic dithiophenes derived from polyacetylene precursors, and isolated from the roots of *Arctium lappa* (edible burdock).

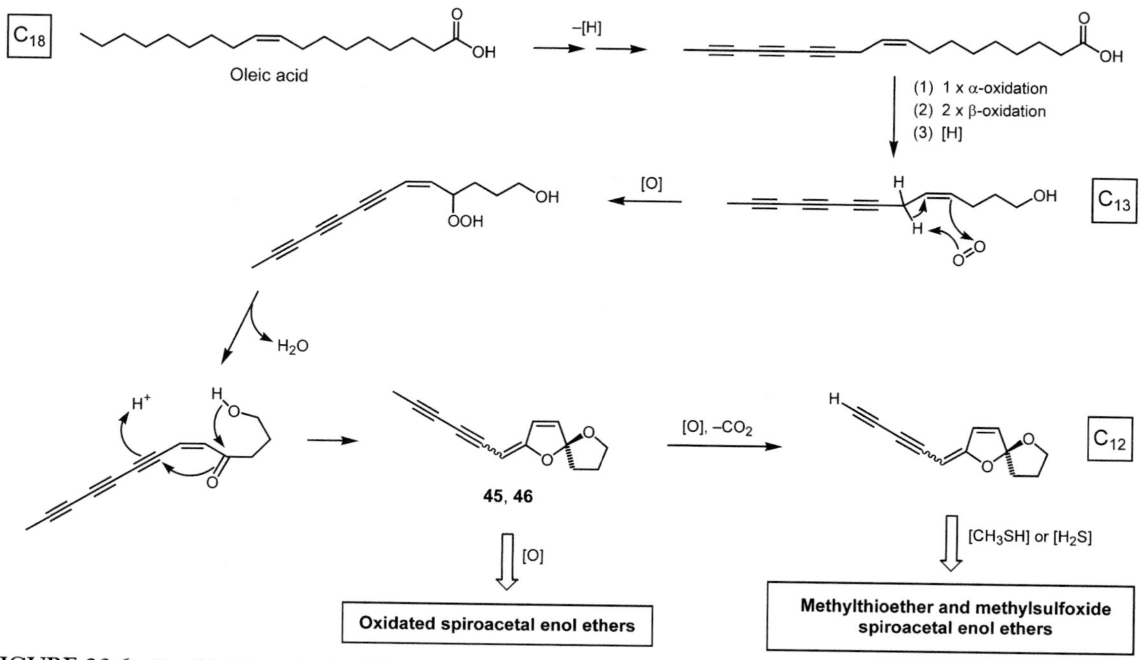

FIGURE 20.6 Possible biosynthesis of C_{12}- and C_{13}-spiroacetal enol ethers isolated from the edible parts of food plants.

presence of aliphatic polyacetylenes as well as aromatic polyacetylenes, especially in the underground parts, which are not used for food [41]. From the aerial parts of tarragon that are used both as a vegetable and condiment the aromatic polyacetylene capillene (**41**) has been isolated together with several aliphatic polyacetylenes (**8**, **19–21**) (Table 20.1; Figures 20.1, 20.3, 20.4). The majority of aromatic acetylenes isolated from higher plants appear to follow almost the same biosynthetic route as the spiroacetal enol ethers [16,18,21].

The roots of *Arctium lappa* (edible burdock), which are used for food in Japan, are especially rich in both aliphatic polyacetylenes and acetylenic dithiophenes (Table 20.1; Figures 20.1, 20.3, 20.5) [42–45]. Most of the aliphatic polyacetylenes isolated from *A. lappa* (**18**, **24–29**, **36–39**) are widely distributed in Asteraceae, whereas most of the dithiophenes (**58–69**) have only been isolated from this plant species [16,18,20–22]. Dithiophenes

are characteristic for certain tribes of the Asteraceae and are most likely biosynthesized from the highly unsaturated and widespread trideca-3,5,7,9,11-pentayn-1-ene (**28**) followed by addition of $2 \times H_2S$ or its biochemical equivalent [16,18,20–22]. A special type of acetylenic dithiophenes (**68**, **69**) found in edible burdock are those linked to guaianolide sesquiterpene lactones. The biosynthesis of these dithiophenes probably involves reaction with arctinal (**58**) and a sesquiterpene lactone to afford oxetane-containing adducts that after cleavage afford diols by hydrolysis, and subsequent oxidation then leads to the ketols **68** and **69** [42].

The aliphatic polyacetylenes isolated from edible burdock and the edible parts of other food plants, including endive (**30**), chicory (**30**), and balloon flower (**33–35**) [46–48] (Table 20.1) are biosynthesized from C_{18}-acetylenic acids by $2 \times \beta$-oxidation or $2 \times \beta$-oxidation and $1 \times \alpha$-oxidation leading to C_{14}-acetylenic and C_{13}-acetylenic precursors, respectively. Further information about the biosynthesis of aliphatic polyacetylenes, including those isolated from food plants, can be found in several reviews on the subject [16–23].

3. BIOACTIVITY OF POLYACETYLENES IN FOOD PLANTS

3.1 Allergenicity

Many plants of the Araliaceae and Apiaceae have been reported to cause allergic contact dermatitis and irritant skin reactions primarily due to occupational exposure (e.g. nursery workers) [49]. The relation between clinical effect and content of polyacetylenes has shown that falcarinol (1) and related C_{18}-polyacetylenes are potent contact allergens, being responsible for many allergic skin reactions caused by plants of the Apiaceae and Araliaceae [49–53]. On the other hand, polyacetylenes of the falcarinol-type such as falcarindiol (2) and falcarinone (5) do not seem to be allergenic [50]. The allergenic properties of falcarinol indicate that it is very reactive toward mercapto and amino groups in proteins, forming a hapten–protein complex (antigen). The reactivity of falcarinol toward proteins is probably due to its hydrophobicity and its ability to form an extremely stable carbocation (resonance stabilized) with the loss of water (Figure 20.7), thereby acting as a very strong alkylating agent toward various biomolecules. This mechanism may explain the anti-inflammatory, anti-platelet-aggregatory and antibacterial effect of falcarinol and related aliphatic C_{17}-polyacetylenes as well as their cytotoxicity and anticancer activity.

Allergic contact dermatitis from common vegetables of the Apiaceae such as carrots is known but rare [54,55], probably due to the relatively low concentrations of falcarinol in food plants compared to ornamental and wild plant

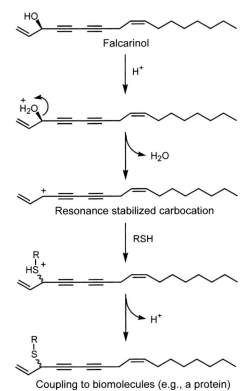

FIGURE 20.7 Possible reaction of falcarinol (1) with biomolecules that may explain its allergenicity. Anti-inflammatory, anti-platelet-aggregatory and/or anti-cancer activity of polyacetylenes of the falcarinol-type and dehydrofalcarinol-type may be explained by a similar mechanism. RSH = thiol residue of a biomolecule.

species [49], or possibly a desensitizing effect of oral intake.

3.2 Antibacterial, Antimycobacterial, and Antifungal Activity

Falcarinol (1) and falcarindiol (2) have been identified as important antifungal compounds in carrots and other Apiaceae plant species, inhibiting spore germination of various fungi in concentrations ranging from 20 to 200 μg/mL [19,56–65]. In many Apiaceae plant species polyacetylenes of the falcarinol-type tend to act as pre-infectional compounds, although some increases can be observed in response to infections [19,58,64,65]. However, in tomatoes and aubergines of the Solanaceae, polyacetylenes of the falcarinol type only seem to be present during infection and hence act as phytoalexins in these food plants [28–30]. Studies have also shown that falcarinol, falcarindiol, and related C_{17}-polyacetylenes have antibacterial effects as well as antimycobacterial effects [66–69]. The antimycobacterial effects toward *Mycobacterium* spp. including *M. aurum*, *M. fortuitum*, and *M. tuberculosis* [68,69], and antibacterial effects toward resistant strains of the gram-positive bacteria *Staphylococcus aureus* [66,68] are particularly interesting. The antistaphylococcal activity and effects against mycoplasma and other antibacterial effects occurred at approximately $10\,\mu g/mL$, i.e. at non-toxic concentrations, and thus represent pharmacological useful properties. This indicates that falcarinol-type polyacetylenes could have positively effects on human health, and may be used to develop new antibiotics.

Several polyacetylenes from the Asteraceae have demonstrated antibacterial effect against gram-positive bacteria (e.g. *Bacillus* spp., *Staphylococcus* spp., *Streptococcus* spp.) and gram-negative bacteria (e.g. *Escherichia* ssp., *Pseudomonas* ssp.) as well as antifungal activity against *Candida albicans* and *Microsporum* spp. [18,56,70–75]. Bacteriostatic and/or fungistatic activity has been demonstrated for the aliphatic polyacetylenes 8, 22, 23, 27, and 28, the spiroacetal enol ethers 44–46, and capillene (41) [18,56,70–75], and this activity in some cases can be enhanced by UV light [18,56,70,76].

3.3 Neurotoxic, Neuritogenic, and Serotonergic Effects

The polyacetylenes cicutoxin and oenanthotoxin (Figure 20.8) occurring in the Apiaceae

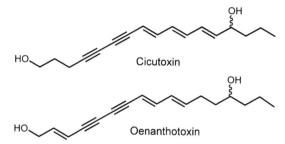

FIGURE 20.8 Chemical structures of highly neurotoxic aliphatic C_{17}-polyacetylenes isolated from non-food Apiaceae plants.

species water-hemlock (*Cicuta virosa*), spotted water-hemlock (*Cicuta maculata*), and hemlock water dropwort (*Oenanthe crocata*), are known to act directly on the central nervous system, causing convulsions and respiratory paralysis, and hence are extremely poisonous [77–79]. Consequently, water-hemlock and related toxic plants have been responsible for the death of numerous human beings and livestock [80]. The mode of action of cicutoxin and oenanthotoxin is probably related to their ability to interact with the γ-aminobutyric acid type A (GABA$_A$) receptors by inhibiting the specific binding of GABA antagonists to GABA-gated chlorine channels of GABA$_A$ receptors as demonstrated in the rat brain [81]. Binding to these ion channels therefore seems to play an important role in the acute toxicity of these compounds and hence their pharmacological mode of action. Structure–activity relationship studies in rats with cicutoxin and related derivatives have shown that the length of the π–bond conjugated system in these polyacetylenes and the geometry of the double bonds are critical for the toxicological effects. Moreover, a terminal hydroxyl group and an allylic alcohol separated a certain distance of approximately 14 carbon atoms appears to be essential for neurotoxicity [79,81]. These findings are supported by the fact that falcarindiol (2) with two hydroxyl groups at a distance of six carbon atoms and falcarinol (1) with only one hydroxyl group do not exert convulsive action *in vitro* [79]. However, the neurotoxicity of falcarinol has been demonstrated upon injection into mice ($LD_{50} = 100\,mg/kg$) [82] whereas falcarindiol does not seem to have any acute effect as demonstrated in rats with an LD_{50} >200 mg/kg [81]. The type of neurotoxic symptoms produced by falcarinol is similar to those of cicutoxin. Cicutoxin with an LD_{50} <3 mg/kg in rats is, however, much more toxic than falcarinol [81]. Poisoning of mammals from voluntary ingestion of natural sources of falcarinol has not been reported and therefore intake of vegetables containing falcarinol is considered to be safe. Dill and/or ajowan contain polyacetylenes [12,16,83], including compounds 13–16 that are closely related to cicutoxin and oenanthotoxin (Table 20.1; Figures 20.1, 20.8). However, as the polyacetylenes 13–16 do not fulfill the requirements for neurotoxicity as described above, these polyacetylenes are not expected to exert any neurotoxic effects and hence their presence in food plants is considered to be safe.

Aliphatic C_{17}-polyacetylenes of the falcarinol-type also seem to have an effect on neuritogenesis of cultured paraneurons. It has for example been demonstrated that falcarinol has a significant neuritogenic effect on paraneurons like PC12 and Neuro2a cells at concentrations above 2 μM [84,85] and a neuroprotective effect on induced neuronal apoptosis [86]. Further, it has been demonstrated that falcarinol improves scopolamine-induced memory deficit in mice, which is probably due to its ability to promote neuritogenesis of paraneurons [84]. Based on the potential neuritogenic and neuroprotective effect of falcarinol together with its good lipophilic character, falcarinol may be considered as a candidate for the prevention and treatment of certain nervous system diseases such as Alzheimer's disease.

Several serotonergic agents, including falcarindiol, have been isolated from the roots of *Angelica sinensis* by serotonin receptor (5-HT$_7$)

binding assay-directed fractionation [87]. This indicates that polyacetylenes of the falcarinol-type may also act on serotonin receptors, and thus may exhibit pharmacological effects that are for example related to improvement of mood and behaviors.

3.4 Anti-inflammatory and Anti-platelet-aggregatory Effects

The inflammatory process attracts leukocytes from the intravascular compartment to the site of damage, a process mediated by the inducible expression of cytokines, chemokines, and cell surface proteins. In addition, inflammation leads to an induced expression and activity of enzymes such as cyclooxygenases (COXs) and lipoxygenases (LOXs), which produce inflammatory mediators like prostaglandin E2. Transcription factor NF-κB plays a key role for the inducible expression of genes mediating pro-inflammatory effects and is thus an important target for the development of anti-inflammatory agents. A large variety of inflammatory signals lead to NF-κB activation, including lipopolysaccharide (LPS), nitric oxide (NO), and pro-inflammatory cytokines such as interleukin-6 (IL-6), IL-1β, and TNF-α. Only a few food plants have been investigated for potential anti-inflammatory effects. However, in a 2008 study it was demonstrated that extracts of purple carrots possess anti-inflammatory activity by decreasing LPS-induced production of IL-6, TNF-α, and NO in macrophage cells in concentrations around 10 μg/mL [14]. Purple carrots are unique due to the presence of very high concentrations of anthocyanins and other polyphenols, which are known to possess anti-inflammatory effects. Bioassay-guided fractionation of carrot extracts, however, revealed that the main anti-inflammatory agents in the extracts were falcarinol (1), falcarindiol (2), and falcarindiol 3-acetate (3), which clearly suggests that polyacetylenes, not polyphenols or carotenoids, are responsible for the anti-inflammatory activity of carrots. Falcarinol and falcarindiol are also strong inhibitors of LOXs (5-, 12- and 15-LOX) involved in tumor progression and atherosclerotic processes [88–92]. Furthermore, falcarindiol is an effective inhibitor of COXs, in particular COX-1, whereas the COX activity of falcarinol does not seem to be pronounced [88,92].

The anti-platelet-aggregatory effects of falcarinol and falcarindiol, and hence their protective effects against development of cardiovascular diseases, are probably related to their anti-inflammatory activity, and in particular their ability to inhibit certain LOXs that are responsible for the production of thromboxanes, such as thromboxane B_2 [88,93–95]. Furthermore, it has been suggested that the anti-platelet-aggregatory activity of falcarinol is related to its ability to modulate prostaglandin catabolism by inhibiting the prostaglandin-catabolizing enzyme 15-hydroxy-prostaglandin dehydrogenase (PGDH) [96]. Falcarinol also has a significant antiproliferative effect on vascular smooth muscle cells (VSMCs) from rats in a concentration of $9 \mu M$ [97]. Abnormal proliferation of VSMCs plays a central role in the pathogenesis of atherosclerosis, and possibly in the development of hypertension [98]. Consequently, the antiproliferative effect of falcarinol on VSMCs further indicates a central role of this polyacetylene in the prevention of atherosclerosis. The anti-inflammatory and anti-platelet-aggegatory

activity of polyacetylenes of the falcarinol-type are most likely due to their ability to react with nucleophiles (Figure 20.7), and hence LOX and COX enzymes as well as NF-κB on a critical nucleophilic site of this transcription factor.

Extracts from the plant *Plagius flosculosus* (Asteraceae) inhibit the induction of NF-κB [99]. The spiroacetal enol ether 45 was identified as the anti-inflammatory agent responsible for this effect, in a concentration of $25 \mu g/mL$, whereas the spiroacetal enol ethers 42 and 47 were found to have no or only a minor anti-inflammatory effect. It has been suggested that the exocyclic double bond of polyacetylene spiroacetal enol ethers has electrophilic properties and hence may react with nucleophiles to form aromatic furan adducts as illustrated in Figure 20.9 [99]. It is therefore possible that NF-κB inhibition occurs by alkylation at a critical point of the transcription factor in the same manner as suggested for falcarinol-type polyacetylenes. The differences in the anti-inflammatory activity of compounds 42, 45, and 47 suggest that the substitution pattern of the saturated furan ring can modulate the electrophilic reactivity of the exo-methylene double bond. The results indicate that spiroacetal enol ether polyacetylenes are powerful inhibitors of NF-κB activity and that the activity of these compounds depends on the substitution pattern of the saturated furan ring. Consequently, chamomile and garland chrysanthemum containing

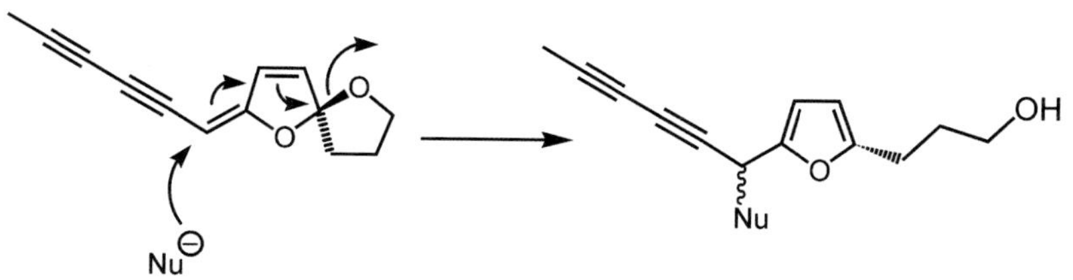

FIGURE 20.9 The possible reaction of spiroacetal enol ethers with nucleophiles explaining their inhibition of, for example, the transcription factor NF-κB, and hence their anti-inflammatory activity. Nu = nucleophile.

compound **45** and other spiroacetal enol ethers (Table 20.1) may be used to treat inflammatory conditions.

Finally, it has been demonstrated that the aromatic polyacetylene capillene (**41**) present in tarragon strongly inhibits induced TGF-β1 apoptosis of primary cultured hepatocytes, which could indicate that this polyacetylene can be used in the treatment of various inflammatory liver diseases [100].

3.5 Cytotoxicity and Anticancer Effect

Focus on the potential anticancer effect of polyacetylenes was initiated early in the 1980s by the discovery of the anticancer activity of petrol extracts of the roots of *Panax ginseng* (Araliaceae) [101,102]. Since then the lipophilic constituents of this plant have been intensively investigated, leading to the isolation and identification of several cytotoxic polyacetylenes of the falcarinol-type, of which falcarinol (**1**), panaxydol, and panaxytriol (Figures 20.1, 20.10) are the best studied polyacetylenes from this plant [12,18,56,102–108]. Falcarinol, panaxydol, and panaxytriol are highly cytotoxic to various cancer cell lines, such as leukemia (L-1210), human gastric adenocarcinoma (MK-1), mouse melanoma (B-16), and mouse fibroblast-derived tumor cells (L-929) [103–105]. The most toxic effect has been observed for MK-1 cells with

ED_{50} values of 0.027 μg/mL (falcarinol), 0.016 μg/mL (panaxydol), and 0.171 μg/mL (panaxytriol) [105]. In addition, these polyacetylenes also inhibit the growth of normal cell cultures such as human fibroblasts (MRC-5), although the ED_{50} against normal cells appears to be around 20 times higher than for cancer cells. In particular, panaxytriol does not inhibit the growth of MRC-5 cells by 50% even at concentrations $>70\,\mu$g/mL [105]. The selective *in vitro* cytotoxicity of falcarinol, panaxydol, and panaxytriol against cancer cells compared to normal cells indicates that they may be useful in the treatment of cancer. Falcarindiol (**2**) also possesses cytotoxic [11,109–111] and antimutagenic [112] activity *in vitro*, although it appears to be less bioactive than falcarinol. Falcarindiol 8-methyl ether (**4**) and panaxydiol (**10**) isolated from celeriac, parsley, and carrots (Table 20.1; Figure 20.1) are examples of further polyacetylenes from food plants exhibiting cytotoxic effects in human cancer and leukemia cell lines [11,113].

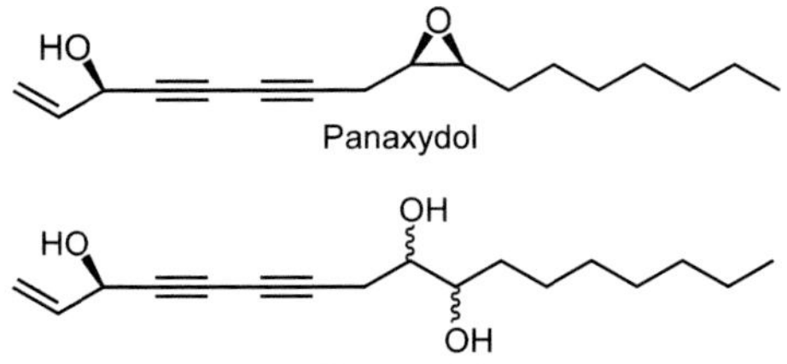

The mechanism for the cytotoxic activity and potential anticancer effect of falcarinol-type polyacetylenes is still not known but may be related to their reactivity towards nucleophiles and hence their ability to interact with various biomolecules (Figure 20.7). This is in accordance with *in vitro* studies showing a suppressive effect of falcarinol on cell proliferation of tumor cells (e.g. K562, Raji, Wish, HeLa, and Calu-1) [114], an effect that is probably related to the ability of falcarinol to arrest the cell cycle progression of the cells at various phases of their cell cycle. This indicates that falcarinol is able to induce apoptosis as demonstrated for Caco-2 cells [13].

Although polyacetylenes of the falcarinol-type have demonstrated interesting *in vitro* effects, only a few studies have been conducted to investigate their effect *in vivo*. From the aerial parts of *Dendropanax arboreus* (Araliaceae), several aliphatic C_{17}-polyacetylenes were isolated, of which falcarinol, falcarindiol, and

FIGURE 20.10 Examples of highly cytotoxic polyacetylenes of the falcarinol-type isolated from ginseng species (Araliaceae) that may contribute to the anticancer effect of ginseng roots.

dehydrofalcarinol (8) (Figure 20.1) were found to exhibit *in vitro* cytotoxicity against human tumor cell lines, with falcarinol showing the strongest activity [110]. Preliminary *in vivo* evaluation of the cytotoxic activity of falcarinol and dehydrofalcarinol using the LOX melanoma mouse xenocraft model demonstrated some potential for *in vivo* antitumor activity of falcarinol and dehydrofalcarinol with the latter showing the strongest therapeutic effect [110]. The most interesting findings on the potential *in vivo* anticancer effect of aliphatic C_{17}-polyacetylenes are from a preclinical study on rats demonstrating inhibitory effects of carrots and falcarinol on the development of induced colon preneoplastic lesions [10].

In the human diet, carrots are the major dietary source of falcarinol, although falcarinol may also be supplied by many other plant food sources (Table 20.1). An *in vitro* study conducted by Reinik et al. showed that falcarinol could stimulate differentiation of primary mammalian cells in concentrations between 0.001 and 0.1 μg/mL falcarinol. Toxic effects were found above >0.5 μg/mL falcarinol (Figure 20.11), while the major carotenoid in carrots, β-carotene, had no effect even at 100 μg/mL [9]. This biphasic effect (hormesis) of falcarinol on cell proliferation has also been demonstrated for cancer cells (Caco-2) [13] and is fully in accordance with the hypothesis that toxic compounds have beneficial effects at certain lower concentrations [115,116]. Therefore, falcarinol appears to be one of the bioactive components in carrots and related vegetables that could explain their health-promoting properties, rather than carotenoids or polyphenols. This hypothesis is further supported by studies on the anti-inflammatory compounds of carrots [14] and bioavailability studies of falcarinol in humans [12]. When falcarinol was administered orally via carrot juice (13.3 mg falcarinol/

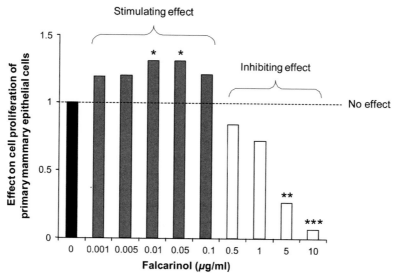

FIGURE 20.11 Effects of increasing concentrations of falcarinol (1) on proliferation, measured by incorporation of [methyl-³H]thymidine into mammary epithelial cells prepared from prepubertal Frisian heifers and grown in three-dimensional collagen gels [9].
No effect on proliferation was observed for β-carotene when tested in the same bioassay [9].
*P<0.09, **P<0.01, ***P<0.001.

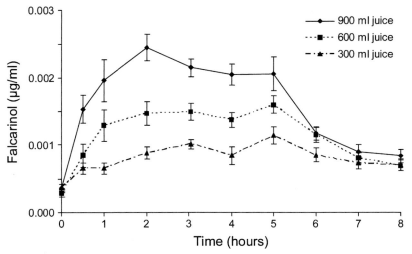

FIGURE 20.12 Concentration of falcarinol (1) in plasma of 14 volunteers as a function of time after ingestion of a break-fast meal with 300, 600 and 900 mL carrot juice, respectively, that contained 13.3 mg falcarinol per liter carrot juice. Mean ± standard error of mean.

L carrot juice) in amounts of 300, 600, and 900 mL, respectively, it was rapidly absorbed, reaching a maximum concentration in serum of $0.0023 \mu g/mL$ and $0.0020 \mu g/mL$ at 2 and 5 hours after administration of 900 mL carrot juice (Figure 20.12). This is within the range where the *in vitro* data indicate a potentially beneficial physiological effect (Figure 20.11), and a possible inhibitive effect on the proliferation of cancer cells.

The possible anticancer effect of falcarinol and carrots has been studied in an established rat model for colon cancer by injections of the carcinogen azoxymethane (AOM) in rats by feeding with carrot or purified falcarinol [10]. Eighteen weeks after the first AOM injection, the rats were killed and the colon examined for tumors and their microscopic precursors, aberrant crypt foci (ACF) [10]. The carrot and falcarinol treatments showed a significant tendency to reduced numbers of (pre)cancerous lesions with increasing size of lesion as shown in Figure 20.13. Although falcarindiol and falcarindiol 3-acetate (**3**), which are also present in carrots (Table 20.1), may have a similar mode of action as falcarinol, they are expected

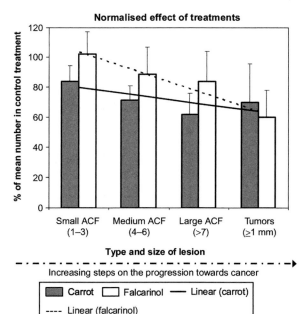

FIGURE 20.13 Effect of treatments with carrot or falcarinol (1) on the average numbers per animal of four types of (pre)cancerous lesions in rat colons.
Smallest tumors correspond to aberrant crypt foci (ACF) size of approximately 20 [10]. The trend for reducing the relative number of lesions with increasing size of lesion was significant ($P<0.03$).

to have less effect than falcarinol. The possibility to generate two active centers for nucleophilic attack in falcarindiol and falcarindiol 3-acetate reduces the lipophilic character of these compounds and hence their reactivity, in accordance with the observed non-allergenic properties of falcarindiol [50] and the less cytotoxic activity observed for falcarindiol compared to falcarinol. So the physiological effects of falcarindiol and falcarindiol 3-acetate are expected to be qualitatively similar but quantitatively less than those of falcarinol, and furthermore, they may even interact with falcarinol in an antagonistic manner thereby affecting its effectiveness. This may also explain the possible, although not significant, differences in the effect and trend observed between the carrot diet and the falcarinol diet shown in Figure 20.13 [10].

The results clearly suggest that the protective effect of carrot against cancer can be explained to a high degree by its content of falcarinol, and not carotenoids or polyphenols as has previously been suggested [117–120]. The cytotoxic activity of polyacetylenes of the falcarinol-type toward cancer cells and possible *in vivo* anticancer effect indicates that they may be valuable in the treatment and/or prevention of different types of cancer, and could contribute to the health-promoting properties of certain food plants.

Cytotoxic activities of aromatic and spiroacetal enol ethers, which have so far not been detected in food plants, have been demonstrated. It is therefore likely that further polyacetylenes may prove to be potential anticancer agents of food plants. It is interesting to note that lipophilic extracts of the roots of *Platycodon grandiflorum* (balloon flower), which is used both as herbal medicine and food in Asia, have anticancer activities [121]. The anticancer principles of balloon flower have not been identified but it is likely that the anticancer agents are identical with the polyacetylenes lobetyol (**33**), lobetyolin (**34**), and lobetyolinin (**35**) isolated from the roots of this plant (Table 20.1).

4. SUMMARY

This review has demonstrated that polyacetylenes in food plants, and in particular aliphatic C_{17}-polyacetylenes of the falcarinol-type, comprise a group of natural products with important bioactivities such as antimycobacterial, antifungal, anti-inflammatory, anti-platelet-aggregatory, neuritogenic, serotonergic, and anticancer activity. The traditional view of polyacetylenes in food as undesirable toxicants therefore needs to be revised and instead these compounds may be regarded as important nutraceuticals that contribute significantly to the health effects of vegetables and in particular those of the Apiaceae family such as carrot, parsley, fennel, and celery. In addition, some polyacetylenes of the falcarinol-type may be used to develop new antibiotics and to prevent and treat cardiovascular diseases, cancer, and certain nervous system diseases such as Alzheimer's disease. The major challenge with regard to bioactive polyacetylenes is to confirm their health-promoting effects *in vivo* in clinical as well as in further preclinical studies.

References

1. Steinmetz, K. A., & Potter, J. D. (1996). Vegetables, fruit, and cancer prevention: A review. *Journal of the American Dietetic Association, 96*, 1027–1039.
2. Block, G., Patterson, B., & Subar, A. (1992). Fruit, vegetables, and cancer prevention: A review of the epidemiological evidence. *Nutrition and Cancer – An International Journal, 18*, 1–29.
3. Greenvald, P., Clifford, C. K., & Milner, J. A. (2001). Diet and cancer prevention. *European Journal of Cancer, 37*, 948–965.
4. Kris-Etherton, P. M. Etherton, T. D., Carlson, J., & Gardner, C. (2002). Recent discoveries in inclusive food-based approaches and dietary patterns for reduction in

risk for cardiovascular disease. *Current Opinion in Lipidology*, 13, 397–407.

5. Maynard, M., Gunnell, D., Emmett, P., Frankel, S., & Smith, G. D. (2003). Fruit, vegetables, and antioxidants in childhood and risk of adult cancer: The Boyd Orr cohort. *Journal of Epidemiology and Community Health*, 57, 218–225.

6. Brandt, K., Christensen, L. P., Hansen-Møller, J., Hansen, S. L., Haraldsdóttir, J., Jespersen, L., Purup, S., Kharazmi, A., Barkholt, V., Frøkiær, H., & Kobæk-Larsen, M. (2004). Health promoting compounds in vegetables and fruits: A systematic approach for identifying plant components with impact on human health. *Trends in Food Science & Technology*, 15, 384–393.

7. Christensen, L. P. (2009). Galactolipids as potential health promoting compounds in vegetable foods. *Recent Patents on Food, Nutrition and Agriculture*, 1, 50–58.

8. Larsen, E., & Christensen, L. P. (2007). Common vegetables and fruits a source of 1,2-di-*O*-α-linolenoyl-3-*O*-β-D-galactopyranosyl-*sn*-glycerol, a potential anti-inflammatory and antitumor agent. *Journal of Food Lipids*, 14, 272–279.

9. Hansen, S. L., Purup, S., & Christensen, L. P. (2003). Bioactivity of falcarinol and the influence of processing and storage on its content in carrots (*Daucus carota* L.). *Journal of the Science of Food and Agriculture*, 83, 1010–1017.

10. Kobæk-Larsen, M., Christensen, L. P., Vach, W., Ritskes-Hoitinga, J., & Brandt, K. (2005). Inhibitory effects of feeding with carrots or (−)-falcarinol on development of azoxymethane-induced preneoplastic lesions in the rat colon. *Journal of Agricultural and Food Chemistry*, 53, 1823–1827.

11. Zidorn, C., Johrer, K., Ganzera, M., Schubert, B., Sigmund, E. M., Mader, J., Greil, R., Ellmerer, E. P., & Stuppner, H. (2005). Polyacetylenes from the Apiaceae vegetables carrot, celery, fennel, parsley, and parsnip and their cytotoxic activities. *Journal of Agricultural and Food Chemistry*, 53, 2518–2523.

12. Christensen, L. P., & Brandt, K. (2006). Bioactive polyacetylenes in food plants of the Apiaceae family: Occurrence, bioactivity and analysis. *Journal of Pharmaceutical and Biomedical Analysis*, 41, 683–693.

13. Young, J. F., Duthie, S. J., Milne, L., Christensen, L. P., Duthie, G. G., & Bestwick, C. S. (2007). Biphasic effect of falcarinol on CaCo-2 cell proliferation, DNA damage, and apoptosis. *Journal of Agricultural and Food Chemistry*, 55, 618–623.

14. Metzger, B. T., Barnes, D. M., & Reed, J. D. (2008). Purple carrot (*Daucus carota* L.) polyacetylenes decrease lipopolysaccharide-induced expression of inflammatory proteins in macrophage and endothelial cells. *Journal of Agricultural and Food Chemistry*, 56, 3554–3560.

15. Young, J. F., Christensen, L. P., Theil, P. K., & Oksbjerg, N. (2008). The polyacetylenes falcarinol and falcarindiol affect stress responses in myotube cultures in a biphasic manner. *Dose-Response*, 6, 239–251.

16. Bohlmann, F., Burkhardt, T., & Zdero, C. (1973). *Naturally occurring acetylenes*. London: Academic Press.

17. Minto, R. E., & Blacklock, B. J. (2008). Biosynthesis and function of polyacetylenes and allied natural products. *Progress in Lipid Research*, 47, 233–306.

18. Christensen, L. P., and Jakobsen, H. B. (2008). Polyacetylenes: Distribution in higher plants, pharmacological effects and analysis. In '*Thin layer chromatography in phytochemistry.*' M. Waksmundzka-Hajnos, J. Sherma and T. Kowalska (Eds.). Chromatographic Science Series. Boca Raton, FL: CRC Press/Taylor & Francis Group Vol. 99, (pp. 757–816).

19. Hansen, L., & Boll, P. M. (1986). Polyacetylenes in Araliaceae: Their chemistry, biosynthesis and biological significance. *Phytochemistry*, 25, 285–293.

20. Christensen, L. P., & Lam, J. (1990). Acetylenes and related compounds in Cynareae. *Phytochemistry*, 29, 2753–2785.

21. Christensen, L. P. (1992). Acetylenes and related compounds in Anthemideae. *Phytochemistry*, 31, 7–49.

22. Christensen, L. P., & Lam, J. (1991). Acetylenes and related compounds in Heliantheae. *Phytochemistry*, 30, 11–49.

23. Christensen, L. P., & Lam, J. (1991). Acetylenes and related compounds in Astereae. *Phytochemistry*, 30, 2453–2476.

24. Bu'Lock, J. D., & Smalley, H. M. (1962). The biosynthesis of polyacetylenes Part V. The role of malonate derivatives and the common origin of fatty acids, polyacetylenes and 'acetate-derived' phenols. *Journal of the Chemical Society*, 4662–4664.

25. Bu'Lock, J. D., & Smith, G. N. (1967). The origin of naturally-occurring acetylenes. *Journal of the Chemical Society(C)*, 332–336.

26. Barley, G. C., Jones, E. R. H., & Thaller, V. (1988). Crepenynate as a precursor of falcarinol in carrot tissue culture. In J. Lam, H. Breteler, T. Arnason & L. Hansen, (Eds.), *Chemistry and biology of naturally-occurring acetylenes and related compounds (NOARC)* (pp. 85–91). Amsterdam: Elsevier.

27. Fleming, I., & Harley-Mason, J. (1963). Enol elimination reactions Part I. A new synthesis of acetylenic acids. *Journal of the Chemical Society (C)*, 4771–4777.

28. De Wit, P. J. G. M., & Kodde, E. (1981). Induction of polyacetylenic phytoalexins in *Lycopersicon esculentum* after inoculation with *Cladosporium fulvum* (syn. *Fulvia fulva*). *Physiologica Plant Pathology*, 18, 143–148.

29. Elgersma, D. M., Weijman, A. C. M., Roeymans, H. J., & van Eijk, G. W. (1984). Occurrence of falcarinol and falcarindiol in tomato plants after infection with *Verticillium alboatrum* and charcterization of four phytoalexins by capillary gas chromatography-mass spectrometry. *Phytopathologische Zeitschrift – Journal of Phytopathology, 109,* 237–240.

30. Imoto, S., & Ohta, Y. (1988). Elicitation of diacetylenic compounds in suspension cultured cells of eggplant. *Plant Physiology, 86,* 176–181.

31. Jakupovic, J., Tan, R. X., Bohlmann, F., Jia, Z. J., & Huneck, S. (1991). Acetylenes and other constituents from *Artemisia dracunculus. Planta Medica, 57,* 450–453.

32. Bohlmann, F., Arndt, C., Bornowski, H., Jastrow, H., & Kleine, K.-M. (1962). Neue Polyine aus dem Tribus Anthemideae. *Chemische Berichte, 95,* 1320–1327.

33. Bohlmann, F., & Fritz, U. (1979). Neue Lyratolester *aus Chrysanthemum coronarium. Phytochemistry, 18,* 1888–1889.

34. Tada, M., & Chiba, K. (1984). Novel plant growth inhibitors and an insect antifeedant from *Chrysanthemum coronarium* (Japanese name: Shungiku). *Agricultural and Biological Chemistry, 48,* 1367–1369.

35. Sanz, J. F., Falcó, E., & Marco, J. A. (1990). New acetylenes from *Chrysanthemum coronarium* L. *Liebigs Annalen der Chemie,* 303–305.

36. Reichling, J., & Becker, H. (1977). Ein Beitrag zur Analytik des ätherischen Öls aus Flores Chamomillae. *Deutsche Apotheker Zeitung, 117,* 275–277.

37. Repcák, M., Halásová, J., Honcariv, R., & Podhradský, D. (1980). The content and composition of the essential oil in the course of anthodium development in wild chamomile (*Matricaria chamomilla* L). *Biologia Plantarum, 22,* 183–189.

38. Redaelli, C., Formentini, L., & Santaniello, E. (1981). High performance liquid chromatography of *cis* and *trans* ene-yne-dicycloethers (spiroethers) in *Matricaria chamomilla* L. flowers and in chamomile extracts. *Journal of Chromatography, 209,* 110–112.

39. Holz, W., & Miething, H. (1994). Wirkstoffe in wässrigen Kamillenaufgüssen. 2, Mitteilung: Freisetzung von Ätherisch-Öl-Komponenten. *Pharmazie, 49,* 53–55.

40. Repcák, M., Imrich, J., & Garcár, J. (1999). Quantitative evaluation of the main sesquiterpenes and polyacetylenes of *Chamomilla recutita* essential oil by high-performance liquid chromatography. *Phytochemical Analysis, 10,* 335–338.

41. Greger, H. (1979). Aromatic acetylenes and dehydrofalcarinone derivatives within the *Artemisia dracunculus* group. *Phytochemistry, 18,* 1319–1322.

42. Washino, T., Kobayashi, H., & Ikawa, Y. (1987). Structures of lappaphen-a and lappahen-b, new guianolides linked with a sulfur-containing acetylenic-compound, from *Arctium lappa* L. *Agricultural and Biological Chemistry, 51,* 1475–1480.

43. Schulte, K. E., Rücker, G., & Boehme, R. (1967). Polyacetylene als Inhaltsstoffe der Klettenwurzeln. *Arzneimittel-Forschung-Drug Research, 17,* 829–833.

44. Washino, T., Yoshikura, M., & Obata, S. (1986). New sulfur-containing acetylenic compounds from *Arctium lappa. Agricultural and Biological Chemistry, 50,* 263–269.

45. Washino, T., Yoshikura, M., & Obata, S. (1986). Polyacetylenic compounds of *Arctium lappa* L. *Nippon Nogeik Kaishi-Journal of the Japan Society for Bioscience Biotechnology and Agrochemistry, 60,* 377–383.

46. Rücker, G., & Noldenn, U. (1991). Polyacetylenes from the underground parts of *Cichorium intybus. Planta Medica, 57,* 97–98.

47. Tada, H., Shimomura, K., & Ishimaru, K. (1995). Polyacetylenes in *Platycodon grandiflorum,* hairy root and Campanulaceous plants. *Journal of Plant Physiology, 145,* 7–10.

48. Ahn, J. C., Hwang, B., Tada, H., Ishimaru, K., Sasaki, K., & Shimomura, K. (1996). Polyacetylenes in hairy roots of *Platycodon grandiflorum. Phytochemistry, 42,* 69–72.

49. Hausen, B. M., & Vieluf, I. K. (2001). *Allergiepflanzen – Pflanzenallergene: Handbuch und Atlas der allergie-induzierenden Wild- und Kulturpflanzen 2. erweit Auflage.* Ecomed Verlagsgesellschaft mbH. Landsberg: München.

50. Hansen, L., Hammershøy, O., & Boll, P. M. (1986). Allergic contact dermatitis from falcarinol isolated from *Schefflera arboricola. Contact Dermatitis, 14,* 91–93.

51. Hausen, B. M., Bröhan, J., König, W. A., Faasch, H., Hahn, H., & Bruhn, G. (1987). Allergic and irritant contact dermatitis from falcarinol and didehydrofalcarinol in common ivy (*Hedera helix* L). *Contact Dermatitis, 17,* 1–9.

52. Oka, K., Saito, F., Yasuhara, T., & Sugimoto, A. (1997). The major allergen of *Dendropanax trifidus* Makino. *Contact Dermatitis, 36,* 252–255.

53. Oka, K., Saito, F., Yasuhara, T., & Sugimoto, A. (1999). The allergens of *Dendropanax trifidus* Makino and *Fatsia japonica* Decne. et Planch. and evaluation of cross-reactions with other plants of the Araliaceae family. *Contact Dermatitis, 40,* 209–213.

54. Murdoch, S. R., & Dempster, J. (2000). Allergic contact dermatitis from carrot. *Contact Dermatitis, 42,* 236.

55. Machado, S., Silva, E., & Massa, A. (2002). Occupational allergic contact dermatitis from falcarinol. *Contact Dermatitis, 47,* 113–114.

56. Christensen, L. P. (1998). Biological activities of naturally occurring acetylenes and related compounds from higher plants. *Recent Research Developments in Phytochemistry, 2,* 227–257.

B. EFFECTS OF INDIVIDUAL VEGETABLES ON HEALTH

57. Kemp, M. S. (1978). Falcarindiol: An antifungal polyacetylene from *Aegopodium podagraria*. *Phytochemistry*, *17*, 1002.

58. Garrod, B., Lewis, B. G., & Coxon, D. T. (1978). *Cis*-heptadeca-1,9-diene-4,6-diyne-3,8-diol, an antifungal polyacetylene from carrot root tissue. *Physiologica Plant Pathology*, *13*, 241–246.

59. Villegas, M., Vargas, D., Msonthi, J. D., Marston, A., & Hostettmann, K. (1988). Isolation of the antifungal compounds falcarindiol and sarisan from *Heteromorpha trifoliate*. *Planta Medica*, *54*, 36–37.

60. Nitz, S., Spraul, M. H., & Drawert, F. (1990). C_{17} Polyacetylenic alcohols as the major constituents in roots of *Petroselinum crispum* Mill. ssp. *tuberosum*. *Journal of Agricultural and Food Chemistry*, *38*, 1445–1447.

61. Lutomski, J., Luan, T. C., & Hoa, T. T. (1992). Polyacetylenes in the Araliaceae family. Part IV. The antibacterial and antifungal activities of two main polyacetylenes from *Panax vietnamensis* Ha et Grushv. and *Polyscias fruticosa* (L.) Harms. *Herba Polonica*, *38*, 137–140.

62. Olsson, K., & Svensson, R. (1996). The influence of polyacetylenes on the susceptibility of carrots to storage diseases. *Journal of Phytopathology*, *144*, 441–447.

63. Christensen, L. P., & Brandt, K. (2006). Acetylenes and psoralens. In A. Crozier, M. N. Clifford, & H. Ashihara, (Eds.), *Plant secondary metabolites: Occurrence structure and role in the human diet* (pp. 137–173). Oxford, UK: Blackwell Publishing Chapter 5.

64. Harding, V. K., & Heale, J. B. (1980). Isolation and identification of the antifungal compounds accumulation in the induced resistance response of carrot slices to *Botrytis cinerea*. *Physiologica Plant Pathology*, *17*, 277–289.

65. Harding, V. K., & Heale, J. B. (1981). The accumulation of inhibitory compounds in the induced resistance response of carrot root slices to *Botrytis cinerea*. *Physiologica Plant Pathology*, *18*, 7–15.

66. Chou, S.-C., Everngam, M. C., Sturtz, G., & Beck, J. J. (2006). Antibacterial activity of components from *Lomatium californicum*. *Phytotherapy Research*, *20*, 153–156.

67. Lechner, D., Stavri, M., Oluwatuyi, M., Pereda-Miranda, R., & Gibbons, S. (2004). The anti-staphylococcal activity of *Angelica dahurica* (Bai Zhi). *Phytochemistry*, *65*, 331–335.

68. Kobaisy, M., Abramowski, Z., Lermer, L, Saxena, G., Hancock, R. E. W., Towers, G. H. N., Doxsee, D., & Stokes, R. W. (1997). Antimycobacterial polyynes of Devil's Club (*Oplopanax horridus*), a North American native medicinal plant. *Journal of Natural Products*, *60*, 1210–1213.

69. Schinkovitz, A., Stavri, M., Gibbons, S., & Bucar, F. (2008). Antimycobacterial polyacetylenes from *Levisticum officinale*. *Phytotherapy Research*, *22*, 681–684.

70. Towers, G. H. N., Wat, C.-K., Graham, E. A., Bandoni, R. J., Mitchell, J. C., & Lam, J. (1977). Ultraviolet-mediated antibiotic activity of species of Compositae caused by polyacetylenic compounds. *Journal of Natural Products*, *40*, 487–498.

71. Towers, G. H. N., & Wat, C.-K. (1978). Biological activity of polyacetylenes. *Revista Latinoamericana de Quimica*, *9*, 162–170.

72. Gonzalez, A. G., Estevez-Reyes, R., Estevez-Braun, A., Ravelo, A. G., Jimenez, I. A., Bazzocchi, I. L., Aguilar, M. A., & Moujir, L. (1997). Biological activities of some *Argyranthemum* species. *Phytochemistry*, *45*, 963–967.

73. Avato, P., Vitali, C., Mongelli, P., & Tava, A. (1997). Antimicrobial activity of polyacetylenes from *Bellis perennis* and their synthetic derivatives. *Planta Medica*, *63*, 503–507.

74. Takasugi, M., Kawashima, S., Katsui, N., & Shirata, A. (1987). Two polyacetylenic phytoalexins from *Arctium lappa*. *Phytochemistry*, *26*, 2957–2958.

75. Wang, Y., Toyota, M., Krause, F., Hamburger, M., & Hostettmann, K. (1990). Polyacetylenes from *Artemisia borealis* and their biological activities. *Phytochemistry*, *29*, 3101–3105.

76. Hudson, J. B., Graham, E. A., Lam, J., & Towers, G. H. N. (1991). Ultraviolet-dependent biological activities of selected polyines of the Asteraceae. *Planta Medica*, *57*, 69–73.

77. Starreveld, E., & Hope, E. (1975). Cicutoxin poisoning. *Neurology*, *25*, 730–734.

78. Wittstock, U., Hadacek, F., Wurz, G., Teuscher, E., & Greger, H. (1995). Polyacetylenes from water hemlock. *Cicuta virosa*. *Planta Medica*, *61*, 439–445.

79. Wittstock, U., Lichtnow, K. H., & Teuscher, E. (1997). Effects of cicutoxin and related polyacetylenes from *Cicuta virosa* on neuronal action potentials: A comparative study on the mechanism of the convulsive action. *Planta Medica*, *63*, 120–124.

80. Anet, E. F. L. J., Lythgoe, B., Silk, M. H., & Trippett, S. (1953). Oenanthotoxin and cicutoxin Isolation and structure. *Journal of the Chemical Society*, 309–322.

81. Uwai, K., Ohashi, K., Takaya, Y., Ohta, T., Tadano, T., Kisara, K., Shibusawa, K., Sakakibara, R., & Oshima, Y. (2000). Exploring the structural basis of neurotoxicity in C_{17}-polyacetylenes isolated from water hemlock. *Journal of Medicinal Chemistry*, *43*, 4508–4515.

82. Crosby, D. G., & Aharonson, N. (1967). The structure of carotatoxin, a natural toxicant from carrot. *Tetrahedron*, *23*, 465–472.

83. Degen, T., Buser, H.-R., & Städler, E. (1999). Patterns of oviposition stimulants for carrot fly in leaves of various host plants. *Journal of Chemical Ecology, 25*, 67–87.

84. Yamazaki, M., Hirakura, K., Miyaichi, Y., Imakura, K., Kita, M., Chiba, K., & Mohri, T. (2001). Effect of polyacetylenes on the neurite outgrowth of neuronal culture cells and scopolamine-induced memory impairment in mice. *Biological & Pharmaceutical Bulletin, 24*, 1434–1436.

85. Wang, Z.-J., Nie, B.-M., Chen, H.-Z., & Lu, Y. (2006). Panaxynol induces neurite outgrowth in PC12D cells via *c*AMP- and MAP kinase-dependent mechanisms. *Chemico-Biological Interactions, 159*, 58–64.

86. Nie, B.-M., Jiang, X.-Y., Cai, J.-X., Fu, S.-L., Yang, L.-M., Lin, L., Hang, Q., Lu, P.-L., & Lu, Y. (2008). Panaxydol and panaxynol protect cultured cortical neurons against A β25-35-induced toxicity. *Neuropharmacology, 54*, 845–853.

87. Deng, S., Chen, S.-N., Yao, P., Nikolic, D., van Breemen, R. B., Bolton, J. L., Fong, H. H. S., Farnsworth, N. R., & Pauli, G. F. (2006). Serotonergic activity-guided phytochemical investigation of the roots of *Angelica sinensis. Journal of Natural Products, 69*, 536–541.

88. Alanko, J., Kurahashi, Y., Yoshimoto, T., Yamamoto, S., & Baba, K. (1994). Panaxynol, a polyacetylene compound isolated from oriental medicines, inhibits mammalian lipoxygenases. *Biochemical Pharmacology, 48*, 1979–1981.

89. Liu, J.-H., Zschocke, S., & Bauer, R. (1998). A polyacetylenic acetate and a coumarin from *Angelica pubescens* f. *biserrata. Phytochemistry, 49*, 211–213.

90. Schneider, I., & Bucar, F. (2005). Lipoxygenase inhibitors from natural plant sources. Medicinal plants with inhibitory activity on arachidonate 5-lipoxygenase and 5-lipoxygenase/cyclooxygenase. *Phytotherapy Research, 19*, 81–102.

91. Schneider, I., & Bucar, F. (2005). Lipoxygenase inhibitors from natural plant sources. Medicinal plants with inhibitory activity on arachidonate 12-lipoxygenase, 15-lipoxygenase and leukotriene receptor antagonists. *Phytotherapy Research, 19*, 263–272.

92. Prior, R. M., Lundgaard, N. H., Light, M. E., Stafford, G. I., van Staden, J., & Jäger, A. K. (2007). The polyacetylene falcarindiol with COX-1 activity isolated from *Aegopodium podagraria* L. *Journal of Ethnopharmacology, 113*, 176–178.

93. Appendino, G., Tagliapietra, S., & Nano, G. M. (1993). An anti-platelet acetylene from the leaves of *Ferula communis. Fitoterapia, 64*, 179.

94. Kuo, S.-C., Teng, C.-M., Lee, J.-C., Ko, F.-N., Chen, S.-C., & Wu, T.-S. (1990). Antiplatelet components in *Panax ginseng. Planta Medica, 56*, 164–167.

95. Teng, C.-M., Kuo, S.-C., Ko, F.-N., Lee, J. C., Lee, L.-G., Chen, S.-C., & Huang, T.-F. (1989). Antiplatelet actions of panaxynol and ginsenosides isolated from ginseng. *Biochimica et Biophysica Acta, 990*, 315–320.

96. Fujimoto, Y., Sakuma, S., Komatsu, S., Sato, D., Nishida, H., Xiao, Y.-Q., Baba, K., & Fujita, T. (1998). Inhibition of 15-hydroxyprostaglandin dehydrogenase activity in rabbit gastric antral mucosa by panaxynol isolated from oriental medicines. *Journal of Pharmacy and Pharmacology, 50*, 1075–1078.

97. Jiang, L.-P., Lu, Y., Nie, B.-M., & Chen, H.-Z. (2008). Antiproliferative effect of panaxynol on RASMCs via inhibition of ERK1/2 and CREB. *Chemico-Biological Interactions, 171*, 348–354.

98. Ross, R. (1990). Mechanisms of atherosclerosis: A review. *Advances in Nephrology from the Necker Hospital, 19*, 79–86.

99. Calzado, M. A., Lüdi, K. S., Fiebich, B., Ben-Neriah, Y., Bacher, S., Munoz, E., Ballero, M., Prosperini, S., Appendino, G., & Schmitz, M. L. (2005). Inhibition of NF-κB activation and expression of inflammatory mediators by polyacetylenes spiroketals from *Plagius flosculosus. Biochimica et Biophysica Acta, 1729*, 88–93.

100. Yamamoto, M., Ogawa, K., Morita, M., Fukuda, K., & Komatsu, K. (1996). The herbal medicine inchinko-to inhibits liver cell apoptosis induced by transforming growth factor beta 1. *Hepatology, 23*, 552–559.

101. Shim, S. C., Koh, H. Y., & Han, B. H. (1983). Polyacetylene compounds from *Panax ginseng* C. A. Meyer. *Bulletin of the Korean Chemical Society, 4*, 183–188.

102. Fujimoto, Y., & Satoh, M. (1988). A new cytotoxic chlorine-containing polyacetylene from the callus of *Panax ginseng. Chemical & Pharmaceutical Bulletin, 36*, 4206–4208.

103. Ahn, B.-Z., & Kim, S.-I. (1988). Beziehung zwischen Struktur und cytotoxischer Aktivität von Panaxydol-Analogen gegen L1210 Zellen. *Archiv Der Pharmazie, 321*, 61–63.

104. Matsunaga, H., Katano, M., Yamamoto, H., Mori, M., & Takata, K. (1989). Studies on the panaxytriol of *Panax ginseng* C. A. Meyer. Isolation, determination and antitumor activity. *Chemical & Pharmaceutical Bulletin, 37*, 1279–1281.

105. Matsunaga, H., Katano, M., Yamamoto, H., Fujito, H., Mori, M., & Takata, K. (1990). Cytotoxic activity of polyacetylene compounds in *Panax ginseng* C. A. Meyer. *Chemical & Pharmaceutical Bulletin, 38*, 3480–3482.

106. Hirakura, K., Takagi, H., Morita, M., Nakajima, K., Niitsu, K., Sasaki, H., Maruno, M., & Okada, M. (2000). Cytotoxic activity of acetylenic compounds from *Panax ginseng. Natural Medicine, 54*, 342–345.

B. EFFECTS OF INDIVIDUAL VEGETABLES ON HEALTH

107. Dembitsky, V. (2006). Anticancer activity of natural and synthetic acetylenic lipids. *Lipids, 41,* 883–924.

108. Siddiq, A., & Dembitsky, V. (2008). Acetylenic anticancer agents. *Anti-Cancer Agents in Medicinal Chemistry, 8,* 132–170.

109. Cunsolo, F., Ruberto, G., Amico, V., & Piattelli, M. (1993). Bioactive metabolites from sicilian marine fennel, *Crithium maritimum. Journal of Natural Products, 56,* 1598–1600.

110. Bernart, M. W., Cardellina, J. H., II, Balaschak, M. S., Alexander, M., Shoemaker, R. H., & Boyd, M. R. (1996). Cytotoxic falcarinol oxylipins from *Dendropanax arboreus. Journal of Natural Products, 59,* 748–753.

111. Fujioka, T., Furumi, K., Fujii, H., Okabe, H., Mihashi, K., Nakano, Y., Matsunaga, H., Katano, M., & Mori, M. (1999). Antiproliferative constituents from Umbelliferae plants. V. A new furanocoumarin and falcarindiol furanocoumarin ethers from the root of *Angelica japonica. Chemical & Pharmaceutical Bulletin, 47,* 96–100.

112. Miyazawa, M., Shimamura, H., Bhuva, R. C., Nakamura, S., & Kameoka, H. (1996). Antimutagenic activity of falcarindiol from *Peucedanum praeruptorum. Journal of Agricultural and Food Chemistry, 44,* 3444–3448.

113. Ahn, B.-Z., & Kim, S.-I. (1988). Heptadeca-1,8*t*-dien-4,6-diin-3,10-diol, ein weiteres, gegen L1210-Zellen cytotoxisches Wirkprinzip aus der Koreanischen Ginseng-wurzel. *Planta Medica, 54,* 183.

114. Kuo, Y.-C., Lin, Y. L., Huang, C.-P., Shu, J.-W., & Tsai, W.-J. (2002). A tumor cell growth inhibitor from *Saposhnikovae divaricata. Cancer Investigation, 20,* 955–964.

115. Calabrese, E. J., & Baldwin, L. A. (2003). The hormetic dose-response model is more common than the threshold model in toxicology. *Toxicological Sciences, 71,* 246–250.

116. Calabrese, E. J. (2004). Hormesis: from marginalization to mainstream. A case for hormesis as the default dose-response model in risk assessment. *Toxicology and Applied Pharmacology, 197,* 125–136.

117. Greenberg, E. R., Baron, J. A., Karagas, M. R., Stukel, T. A., Nierenberg, D. W., Stevens, M. M., Mandel, J. S., & Haile, R. W. (1996). Mortality associated with low plasma concentration of β-carotene and the effect of oral supplementation. *Journal of the American Medical Association, 275,* 699–703.

118. Omenn, G. S., Goodmann, G. E., Thornquist, M. D., Balmes, J., Cullen, M. R., Glass, A., Keogh, J. P., Meyskens, F. L., Jr., Valanis, B., & Williams, J. H. Jr., (1996). Effects of a combination of β-carotene and vitamin A on lung cancer and cardiovascular disease. *New England Journal of Medicine, 334,* 1150–1155.

119. Michaud, D. S., Feskanich, D., Rimm, E. B., Colditz, G. A., Speizer, F. E., Willett, W. C., & Giovannucci, E. (2000). Intake of specific carotenoids and risk of lung cancer in 2 prospective US cohorts. *American Journal of Clinical Nutrition, 72,* 990–997.

120. Wright, M. E., Mayne, S. T., Swanson, C. A., Sinha, R., & Alavanja, M. C. R. (2003). Dietary carotenoids, vegetables, and lung cancer risk in women: The Missouri Women's Health Study (United States). *Cancer Causes Control, 14,* 85–96.

121. Lee, J.-Y., Hwang, W.-I., & Lim, S.-T. (2004). Antioxidant and anticancer activities of organic extracts from *Platycodon grandiflorum* A. De Candole roots. *Journal of Ethnopharmacology, 93,* 409–415.

21

Nitrates and Nitrites in Vegetables: Occurrence and Health Risks

Terje Tamme[1], Mari Reinik[2], and Mati Roasto[1]
[1]Department of Food Science and Hygiene, Estonian University of Life Sciences, Tartu, Estonia
[2]Tartu Laboratory, Health Protection Inspectorate, Tartu, Estonia

1. NITRATE AND NITRITE IN THE ENVIRONMENT

The gaseous form of nitrogen makes up 78% of the troposphere. Its incorporation into terrestrial nitrogenous compounds takes place via different pathways, including microorganisms, plants, and human industrial and agricultural activities.

Nitrogen taken from the air is converted to ammonia by nitrogen-fixing bacteria. In most soils, ammonium is rapidly oxidized to nitrite, and consequently into nitrate in a nitrification process by the action of the aerobic bacteria, such as *Nitrosomonas* and *Nitrobacter*.

The nitrate ion (NO_3^-) is the stable form of nitrogen for oxygenated systems. Although it is chemically unreactive, it can be microbially reduced to the reactive nitrite ion [1]. Nitrate is very soluble and, unless intercepted or taken up by plant roots, leaches down into the soil along with irrigation or rainwater or is carried away by runoff. Under certain conditions nitrate may undergo bacterial conversion to molecular nitrogen (N_2), nitrous oxide (N_2O), or nitric oxide (NO) by a process called denitrification. Presence of nitrates, excess water, and available carbon source are important factors affecting the denitrification process. While nitric oxide easily gets converted to nitrate and is brought down again by precipitation, nitrous oxide escapes to the stratosphere and destroys ozone producing a greenhouse effect [2].

Nitrate and nitrite occur in drinking water mainly as a result of intensive agricultural activities. Nitrate-containing compounds present in soil are generally soluble and readily migrate into groundwater. Contamination of soil with nitrogen-containing fertilizers, including anhydrous ammonia, as well as animal or human natural organic wastes, can raise the concentration of nitrate in water. As nitrite is easily oxidized to nitrate, nitrite levels in water are usually low, and nitrate is the compound predominantly found in groundwater and surface waters. Water in highly polluted wells may also contain nitrites at elevated levels. Drinking water is regarded to be the second-largest source of nitrate in the diet after vegetables [3–5]. To guarantee drinking water safety, maximum permitted concentrations have been established for nitrate and nitrite, being 50 mg/L and 0.5 mg/L, respectively [6].

Being an essential element for plant growth, nitrogen is absorbed by plants in the form of ammonium or nitrate from soil water. Nitrates accumulated in plants form a nitrogen reserve, which is needed for amino acid and protein synthesis [7,8]. Normally, nearly all nitrates absorbed by plant roots are reduced once inside the plant to form ammonia which serves as a precursor for protein synthesis. Every reduction step is catalyzed by a certain enzyme – reductase.

However, under some stress conditions, nitrates can accumulate in plants to high levels. Accumulation of nitrates into the plant is influenced by many factors such as soil composition, growth density of cultivars, lighting, growth temperature, growing under cover or in the open air, harvesting time, occurrence of plant diseases, etc.

As the maximum limit concentrations for nitrates and nitrites in food and drinking water have been established by legislation, a wide scale of analytical methods, including spectrophotometric, high-performance liquid chromatographic (HPLC), ion chromatographic (IC), gas chromatographic (GC), polarographic and capillary electrophoretic (CE), for the determination of nitrate and nitrite content in food or in water have been developed. Use of HPLC methods has gained more popularity in the last decades since they are more rapid than classic methods based on reduction process followed by colorimetry [9].

2. OCCURRENCE OF NITRATES AND NITRITES IN VEGETABLES

Nitrate is an essential nutrient for plants, making vegetables the most important source of nitrates for humans through its accumulation. It has been estimated that vegetables contribute up to 92% of human exposure to nitrates [10–13].

Nitrate levels present in vegetables naturally via the nitrogen cycle are affected by factors such as plant species, climatic and light conditions, soil characteristics, and fertilization regime. Concentrations of nitrates in edible parts of vegetables can vary enormously, ranging from below 10 to up to 10,000 mg/kg [14].

According to nitrate content vegetables can be divided into three groups [15]:

1. Plants with nitrate content higher than 1000 mg/kg – rocket, lettuce, spinach, herbs, beetroot, etc.
2. Plants with average content of nitrate (50–1000 mg/kg) – carrot, green beans, cauliflower, onion, pumpkin, eggplant, potato, etc.
3. Plants with nitrate content lower than 50 mg/kg – berries, fruits, cereals, pod vegetables.

De Martin and Restani showed that leafy green vegetables accumulate high amounts of nitrates, concentrations reaching up to 6000 mg/kg [16]. The highest amounts are absorbed by rocket, a leafy vegetable popular especially in the Mediterranean countries. Increased use of synthetic nitrogen fertilizers and livestock manure in agricultural activities during the last years has led to higher nitrate concentrations in vegetables and drinking water [17]. The nitrite content of most fresh, frozen, or canned vegetables is relatively low and usually in the order of 0–2 mg/kg [18,19]. Levels of nitrates and nitrites in vegetables determined in different countries are presented in Tables 21.1 and 21.2.

In order to protect human health and taking into account the possible health risks, the level of nitrates and nitrites in food should be reduced to as low as reasonably achievable (the ALARA principle). The regulatory limits for nitrates in lettuce, spinach, and vegetable-based infant food

TABLE 21.1 Mean Content of Nitrates in Vegetables in Different Countries, mg/kg

Vegetable Commodity	Belgium[a]	Cyprus[b]	Denmark[c]	Estonia[d]	Finland[e]	France[f]	Germany[g]	Great Britain[h]	Iran[i]	Italy[j]	Korea[k]	New Zealand[l]	Slovenia[m]
Beetroot		656	1493	1446	1800	644	1630	1211		1727			
Cabbage		2408	342	437	607	498	451	338	504	400	725	331	881
Carrot	278			148	264	121	232	97	458	195	316	58	264
Cauliflower				287		214		86		202			
Celery	3135	759	993	565	1057	613			887	1678		1610	
Chinese cabbage				1243							1740		
Cucumber	344	<30		160	240	192			813		212		93
Dill/fennel		1492		2936		1043	1541		313				
Garlic				174		121			229	34	124		
Green beans	585				455	449							298
Leek	841		308			410					56		
Lettuce	2782	1083	2603	2167	1835	1974	1489	1051	1300		2430	1590	1074
Onion	59			55	140	78		48	249	32	23		
Parsley	2690	1320		966		1980				1150			
Potato	154	227	144	94	82	192	93	155	155	81	452	129	158
Pumpkin		137		174							639	67	
Radish	2136			1309		1861	2030		1902	2067	1878		
Rhubarb				201			986						
Spinach	2297	1455	1783	2508		1682	965	1631	748	1845	4259	990	965
Spring onion				477							436		
Sweet (bell) pepper	93				140				623		76		

(Continued)

B. EFFECTS OF INDIVIDUAL VEGETABLES ON HEALTH

TABLE 21.1 (Continued)

Vegetable Commodity	Belgium[a]	Cyprus[b]	Denmark[c]	Estonia[d]	Finland[e]	France[f]	Germany[g]	Great Britain[h]	Iran[i]	Italy[j]	Korea[k]	New Zealand[l]	Slovenia[m]
Tomato	36	<30		41	170		27	17	1681				<6
Turnip				307	908	657		118					

[a]Dejonckheere, W., Streubaut, W., Drieghe, S., Verstraeten, R., and Braeckman, H. (1994). Nitrate in food commodities of vegetable origin and the total diet in Belgium, 1992–1993. *Microbiologie-Aliments-Nutrition* **12**, 359–370.

[b]Ioannou-Kakouri, E., Aletrari, M., Christou E., Procopiou, E., Ralli, A., and Koliou, A. (2004). Levels of nitrates, nitrites and nitrosamines in foodstuffs in Cypros and evaluation of the population exposure to nitrates. COST Action 922 'Health Implications of Dietary Amines.' Larnaca, Cyprus.

[c]Petersen, A., and Stoltze, S. (1999). Nitrate and nitrite in vegetables on the Danish market: content and intake. *Food Additives and Contaminants* **16**, 291–299.

[d]Tamme T., Reinik M., Roasto M., Juhkam K., Tenno T., and Kiis A. (2006). Nitrates and nitrites in vegetables and vegetable-based products and their intakes by the Estonian population. *Food Additives and Contaminants* **23**, 355–361.

[e]Penttilä, P.L. (1995). 'Estimation of food additive and pesticide intakes by means of stepwise method.' Doctoral thesis, University of Turku, Finland.

[f]Menard, C., Heraud, F., Volatier, J.-L., and Leblanc, J.-C. (2008). Assessment of dietary exposure of nitrate and nitrite in France. *Food Additives and Contaminants* **25**, 971–988.

[g]Belitz, H.-D., and Grosch, W. (1999). 'Food Chemistry,' 2nd ed. Springer, Berlin.

[h]Ysart, G., Miller, P., Barret, G., Farrington, D., Lawrance, P., and Harrison, N. (1999). Dietary exposures to nitrate in the UK. *Food Additives and Contaminants* **16**, 521–532.

[i]Shahlaei, A., Ansari, A. N., and Dehkordie, F. S. (2007). Evaluation of nitrate and nitrite content of Iran Southern (Ahwaz) vegetables during winter and spring of 2006. *Asian Journal of Plant Sciences* **6**, 1197–1203.

[j]Santamaria P., Elia A., Serio F., and Todaro E. (1999). A survey of nitrate and oxalate content in fresh vegetables. *Journal of the Science of Food and Agriculture* **79**, 1882–1888.

[k]Chung, S.Y., Kim, J.S., Kim, M., Hong, M.K., Lee, J.O., Kim, C.M., and Song, I.S. (2003). Survey of nitrate and nitrite contents of vegetables grown in Korea. *Food Additives and Contaminants* **20**, 621–628.

[l]Thomson, B. M., Nokes, C. J. and Cressey, P. J. (2007). Intake and risk assessment of nitrate and nitrite from New Zealand foods and drinking water. *Food Additives and Contaminants* **24**(2), 113–121.

[m]Sušin J., Kmecl V., and Gregorcic A. (2006). A survey of nitrate and nitrite content of fruit and vegetables grown in Slovenia during 1996–2002. *Food Additives and Contaminants* **23**, 385–390.

B. EFFECTS OF INDIVIDUAL VEGETABLES ON HEALTH

TABLE 21.2 Mean Content of Nitrites in Vegetables in Different Countries, mg/kg

Vegetable Commodity	China[a]	Denmark[b]	France[c]	Iran[d]	Korea[e]	Slovenia[f]
Beetroot	–	0.91	–	6.93	–	–
Cabbage	0.48	0.16	–	5.23	0.4	0.2
Carrot	–	–	4.15	2.91	0.5	0.2
Cauliflower	–	–	2.00	4.8	–	–
Celery	2.17	–	2.00	3.98	–	–
Chinese cabbage	0.20	0.34	–	–	1.0	–
Cucumber	0.29	–	2.00	5.32	0.3	0.2
Dill/fennel	–	–	7.40	3.1	–	–
Garlic	–	–	2.00	2.77	0.4	–
Green onion	–	–	–	–	0.4	–
Leek	–	0.15	2.53	–	–	–
Lettuce	–	0.20	6.25	4.21	0.7	0.3
Onion	–	–	11.0	5.63	0.3	–
Potato	1.59	0.60	–	1.84	0.6	1.2
Pumpkin	–	–	–	–	0.6	–
Radish	–	–	3.39	4.56	0.7	–
Spinach	–	11.00	18.88	3.42	0.5	–
Sweet (bell) pepper	–	–	–	7.38	0.4	–
Tomato	0.23	–	–	10.13	–	0.3
Turnip	–	–	2.00	3.28	–	–

[a]Zhong, W., Hu, C., and Wang M. (2002). Nitrate and nitrite in vegetables from north China: content and intake. *Food Additives and Contaminants* 19, 1125–1129.
[b]Petersen, A. and Stoltze, S. (1999). Nitrate and nitrite in vegetables on the Danish market: content and intake. *Food Additives and Contaminants* 16, 291–299.
[c]Menard, C., Heraud, F., Volatier, J.-L., and Leblanc, J.-C. (2008). Assessment of dietary exposure of nitrate and nitrite in France. *Food Additives and Contaminants* 25, 971–988.
[d]Shahlaei, A., Ansari, A. N., and Dehkordie, F. S. (2007). Evaluation of nitrate and nitrite content of Iran Southern (Ahwaz) vegetables during winter and spring of 2006. *Asian Journal of Plant Sciences* 6, 1197–1203.
[e]Chung, S.Y., Kim, J.S., Kim, M., Hong, M.K., Lee, J.O., Kim, C.M., and Song, I.S. (2003). Survey of nitrate and nitrite contents of vegetables grown in Korea. *Food Additives and Contaminants* 20, 621–628.
[f]Sušin J., Kmecl V., and Gregorcic A. (2006). A survey of nitrate and nitrite content of fruit and vegetables grown in Slovenia during 1996-2002. *Food Additives and Contaminants* 23, 385–390.

are established by the European Commission Regulation (EC) No. 1881/2006 (Table 21.3) [20].

3. FACTORS AFFECTING NITRATE AND NITRITE CONCENTRATION IN FOOD

The concentration of nitrates in plant products depends on the biological properties of cultivars, plant portion, size of the vegetable, light intensity, soil composition, air temperature, moisture, growth density, maturity of the plant, duration of growth period, harvesting time, nitrogen source, and storage time and conditions [4,21–24].

Plant Portion The accumulation of nitrates in plants differs to a large extent depending on the portion of the plant [4]. Nitrate content in plants reduces as follows: petiole>leaf>root>stem>inflorescence>tuber>bulb, fruit, seed [25]. Accumulation of nitrates can differ between potato breeds. The roots of cabbage contain more nitrates compared to the leaves, which

TABLE 21.3   Maximum Levels for Nitrate [Commission Regulation (EC) No 1881/2006]

Foodstuffs	Maximum Levels (mg NO3 − /kg)	
Fresh spinach (Spinacia oleracea)	Harvested 1 October to 31 March	3000
	Harvested 1 April to 30 September	2500
Preserved, deep-frozen or frozen spinach		2000
	Harvested 1 October to 31 March:	
Fresh lettuce (*Lactuca sativa* L.)	lettuce grown under cover	4500
(protected and open-grown lettuce)	lettuce grown in the open air	4000
excluding lettuce listed next	Harvested 1 April to 30 September:	
	lettuce grown under cover	3500
	lettuce grown in the open air	2500
Iceberg-type lettuce	Lettuce grown under cover	2500
	Lettuce grown in the open air	2000
Processed cereal-based foods and baby foods for infants and young children		200

makes it advisable to avoid cabbage roots in the human diet. Experiments have shown that outer layers of cucumbers and radish contain two to three times more nitrates than inner layers do. To reduce the risk of high nitrate intake, cucumbers, especially early greenhouse cucumbers, should be peeled prior to consumption [26,27].

Temperature This is an important factor influencing the residual nitrate content in vegetables: low temperatures in spring or autumn slow down photosynthesis and favor nitrate accumulation; too high temperatures reduce nitrate reductase activity causing higher nitrate concentrations. In optimal growth temperature conditions stress in plants is avoided and no temperature-related excess nitrate accumulation is observed [23].

Lighting Conditions These influence the activity of the enzyme nitrate reductase. Shaded plants lack sufficient photosynthetic energy to convert nitrate to amino acids. Via the disturbance of nitrate reductase reduced lighting causes disorders in the formation of organic compounds and the concentrations of nitrates will stay high.

Moisture Low humidity conditions in soil are related to an advanced nitrification process which causes excess of nitrate that can accumulate in the plants. Moderate soil moisture conditions (80–90%) favor reasonable plant nitrogen nutrition while wet conditions, like low humidity, increase the nitrate concentrations in plants.

Soil Composition and Fertilization While circumstances such as fertilization and other growth conditions are the same, the lowest nitrate concentrations are detected in vegetables grown in light sandy soils. Higher accumulation of nitrates is reported in clay- and humus-rich soils. Plants grown without excessive nitrogen fertilizer have far less nitrates. Nitrate fertilizer applied shortly before harvest causes the greatest increase in nitrate levels and should be avoided. Slower nitrogen-releasing sources such as animal and green manures enable production of vegetables with significantly lower nitrates. However, plant species, stress factors, and plant growing conditions have been reported to have more influence on nitrate levels in plants than amount of nitrogen fertilizer applied [28].

Growth Density of Cultivars This influences the final nitrate content of vegetables from two aspects. First, higher density will shadow the plants from light and cause decreased growth via enzyme inhibition. The same type of reductase inhibition has also been reported for weedy fields. Second, there is a mutual relationship between growth density, soil fertility, and nitrate content of the plants. In the conditions of a soil rich in plant nutrients and low growth density, over-consumption of nitrogen by plants may occur [29,30].

Plant Maturity and Growth Period Generally, the nitrate content is highest in early plant growth stages and decreases with maturity [30]. Stems of vegetables contain higher amounts of nitrates than leaves. Longer and lighter growth periods favor the reduction of nitrates in plants for time of harvesting.

Storage Time and Conditions Studies have shown that during storage of vegetables in optimal conditions, nitrate content decreases slowly. Studies have also shown that during 6 months of storage, nitrate content in vegetables decreases 1.5–2.0 times [30]. At the same time, nitrite concentrations in vegetables may increase to elevated levels due to bacterial nitrification of nitrate to nitrite when vegetables are stored in rooms with high humidity and poor sanitation. Storage of vegetable salads, raw purees, and juices at room temperatures can increase the nitrite concentrations to potentially hazardous levels [31,32]. For the latter reason raw vegetable-containing food, especially infant food, should preferably be prepared only shortly prior to consumption. Significant changes in nitrate and nitrite content of vegetable products have not been detected following storage in refrigerated conditions.

4. NITRATES AND NITRITES: HEALTH RISKS VERSUS USEFULNESS

Humans are subjected to significant nitrate and nitrite levels, mainly from food and water but also via nitrogen-containing drugs/chemicals used for therapeutic purposes. The fate of nitrate in the human organism has been the subject of numerous studies and the results have been compiled in several reviews [33–36].

In 1990, taking into account the possible health risks, the European Commission's Scientific Committee for Food (SCF) set maximum daily intake (ADI) values for nitrate and nitrite. After reviewing these values in 1995, the present acceptable daily intakes for nitrates 0–3.7 mg/kg body weight (bw) and for nitrites 0–0.06 mg/kg bw were determined [37]. The USA Environmental Protection Agency (EPA) reference dose (RfD) for nitrate is 1.6 mg nitrate nitrogen/kg bw per day and for nitrite 0.1 mg nitrite nitrogen/kg bw per day [1].

Ingested nitrate is effectively absorbed in the upper part of the small intestine [38]. From the proximal small intestine the dietary nitrate is absorbed into the plasma. Plasma nitrite levels are normally lower than nitrate levels because of the lower exposure and due to the rapid oxidation from nitrite to nitrate by oxygenated hemoglobin [39]. Absorbed nitrate is rapidly transported by the blood and selectively secreted by the salivary glands, resulting in high salivary nitrate levels [33,40]. About 25% of ingested nitrate is secreted in the saliva. Approximately 20% of the secreted salivary nitrate, constituting 5–7% of the overall absorbed nitrate dose, is rapidly converted to nitrite by a specialized flora of anaerobic nitrate reducing bacteria on the tongue [33,41,42]. Oral reduction of nitrate is the most important source of nitrite for humans and accounts for 70–80% of the total nitrite exposure. In the stomach nitrite will be transformed to nitric oxide and other metabolites. Most of the absorbed nitrate is ultimately excreted in the urine [34].

In addition to exogenous ingestion, nitrite and nitrate are formed in human saliva and gastrointestinal tract by endogenous synthesis mediated mostly by bacteria [36,41,43–46]. The main source of endogenous nitrate in mammals is the L-arginine-NO pathway; nitric oxide is produced from the amino acid L-arginine and molecular oxygen by nitric oxide synthetase [41,45]. Estimation of the daily amounts of nitrate, nitrite, and N-nitroso compounds formed endogenously is complicated; a contribution of as large as or greater than from the normal diet has been suggested [13,47].

4.1 Health Risks

Nitrate itself is relatively non-toxic. As approximately 5% of all ingested nitrate is converted in saliva and in the gastrointestinal tract into more toxic nitrite, the assessment of the health risk of nitrates should take into account the toxicity of nitrites and also N-nitroso compounds which can be formed in reaction of nitrite with amines or amides in food or endogenously.

4.1.1 Acute Toxicity

The most widely known adverse effect of nitrite in humans is the formation of methemoglobin (metHb) and further symptoms of methemoglobinemia, so-called 'blue baby syndrome,' which occurs probably due to the consumption of high levels of nitrates and nitrites from drinking water or vegetable-based infant foods [5,48]. Clinical symptomatic disease is rare and limited mainly to babies less than 90 days old, but may also be encountered among older children when vitamin C intake is low and in adults with low gastric acidity [35]. Methemoglobin, a derivative of hemoglobin in which the iron has been oxidized from the Fe_2^+ to the Fe_3^+ state, imparts a characteristic brownish color to the blood.

$$NO_2^- + oxyHb(Fe^{2+}) \rightarrow metHb(Fe^{3+}) + NO_3^-$$

As a consequence of the formation of metHb the oxygen delivery to tissues is impaired. Normal levels of methemoglobin in human blood are 1–3%. Once the proportion of metHb reaches 10% of normal Hb levels, clinical symptoms (cyanosis, asphyxia, suffocation) occur [5,24,33,49,50]. Speijers reported in 1996 that the toxic doses for methemoglobin formation ranged from 33 to 350 mg nitrate ion/kg bw, and it could easily lead to death of infants [51]. Babies less than 3 months old are particularly susceptible to methemoglobinemia because of higher levels of fetal oxyHb present in the blood, which oxidizes to metHb more readily than non-fetal oxyHb. Additionally, infants have a higher gastric pH (pH >4), but there are low concentrations of reducing agents to reconvert the metHb back to oxyHb, and their methemoglobin reductase system is immature [1,52]. Intake of nitrate from water used for drinking or food preparation can imply a risk for infants also due to their proportionately higher intake of nitrate through drinking water by body weight.

The controversies over the topics of nitrate and nitrite toxicity have been critically reviewed and analyzed in the book 'Nitrate and Man' [53], where the authors conclude that the regulations enacted in Western Europe and in the USA to control the nitrate content of water are based on faulty epidemiology, and that there is lack of proof of the toxicity of nitrate. The authors suggest that WHO, US, and EU regulations regarding nitrate levels in potable water should be re-examined and indicate that well-water methemoglobinemia cases could be related to unhygienic wells.

4.1.2 Chronic Effects

The second widely known concern is that nitrate, when reduced to nitrite, can react with amines or amides to form N-nitroso compounds [46,50,54]. Nitrosation occurs mainly in two ways: during preparation and storage of food, and in the stomach as a result of the action of salivary nitrite produced through enzymatic reduction of endogenous or exogenous nitrate [55–57]. Antibiotic inhibition studies have shown that some of the broad-spectrum antibiotics significantly reduce the nitrite concentration in the saliva, suggesting an important role of bacteria in the conversion of nitrate to nitrite [58].

N-nitroso compounds have been shown to be carcinogenic in more than 40 animal species and therefore it is quite evident that humans may be included [36,59]. It has been reported that vegetables with high nitrate content in the diet could cause a human to risk gastrointestinal cancer [60]. Several earlier studies have evaluated dietary sources of nitrite in association with pancreatic cancer risk. N-nitroso compounds have been shown to cause tumors in animal species tested and induced cancer in many different organs including the pancreas [61,62]. A study by Coss et al. was especially aimed to determine the association of increased consumption of nitrate and nitrite from drinking water and dietary sources with pancreatic cancer risk [63]. This population-based case-control study in Iowa concluded that long-term exposure to drinking water nitrate at levels below the maximum permitted level of nitrate nitrogen was not associated with pancreatic cancer. However, the authors suggested that the consumption of dietary nitrite from animal products can increase cancer risk. Many other cancer sites in humans are suspected to be related to N-nitroso compounds but unfortunately only a few epidemiological studies exist. Eichholzer and Gutzwiller reported that taking into account the results of different epidemiological studies, a causal relationship between stomach cancer and dietary N-nitroso compounds as well nitrite and nitrate cannot be concluded or excluded [13]. In the book 'Food Safety, Contaminants and Toxins' the authors concluded the need for cohort studies to evaluate properly the effect of dietary N-nitroso compounds, nitrite, and nitrate on human cancer risk [13].

Besides carcinogenesis, the nitrosating substances have also shown teratogenic effects related to DNA alkylation. Malformation of the fetus is induced either by germ-cell mutation or by transplacental transmission of nitroso compounds [58]. However, there is not yet enough data available on dietary exposure and increased risk for congenital malformation. More research is needed to support statements on the teratogenic effects of N-nitroso compounds in humans.

Researchers have presumed some of the N-nitroso compounds, their precursors, and modulators of their metabolism to be risk factors for different types of brain tumors in both adults and children [64]. The causes of brain cancer are not yet well known, and there is still a need for scientific evidence on increased risk of brain tumor regarding intake of vegetables, cured meats, and drinking water. Intake of cured meats and an increased risk for adult brain tumors has been reported, but there is no good proof that consumption of cured meats in childhood or maternal diet during pregnancy can increase the risk of brain tumors in children [13,34,65,66].

4.2 Minimizing the Health Risks

Nitrosamines are found in diverse foods and beverages, primarily those to which nitrates or nitrites have been added. In addition to that, they may also be formed during frying or smoking of food. The most toxic nitrosamine often detected in food is N,N-dimethylnitrosamine (NDMA) [67]. As N-nitroso compounds have shown to be genotoxic carcinogens in animal experiments, there is no threshold dose below which no tumor formation would occur. The levels of these compounds in food should be kept as low as reasonably achievable.

Ascorbic acid and tocopherols have been used to inhibit nitrosation processes in food. The same effect can be attributed to plant polyphenols and tannins, thus plant sources of nitrite may form less N-nitroso compounds than might otherwise be predicted [37]. Chow and Hong reported that there is a protective effect of dietary vitamin E against nitrite and nitrate toxicity which is attributable to its ability to limit the production and availability of superoxide and NO, while dietary selenium enhances the level of selenoenzymes/compounds, which reduce the formation of reactive free radical peroxynitrite [44].

It can be concluded that when nitrate is consumed as a part of a normal vegetable-rich diet, simultaneously consumed bioactive substances such as vitamin C and other plant antioxidants can reduce the amount of nitrosamines formed by up to half [67,68].

4.3 Beneficial Effects

The use of nitrites and nitrates as antimicrobial agents, as well as food preservatives and flavoring/coloring agents in meat and fish products, was already well known at the end of nineteenth century and is still important in everyday practice [53,69,70]. Through the years, the antibacterial effect of nitrite has been demonstrated by food research scientists against the lethal *Clostridium botulinum* and many other bacteria, including enteric pathogens such as *Salmonella* and *Escherichia coli*. By inhibiting bacterial growth, nitrate and nitrite in certain doses inhibit the spoilage of meat and meat products. Nitrate itself, because of its chemical characteristics, has no bactericidal effects. Nitrate metabolites, including nitrite, nitric acid, and NO have antimicrobial activity for prevention of intestinal pathogens [42,71]. A bactericidal effect of nitrite under the acidic conditions present in the mouth and stomach has been verified on some of the most important foodborne pathogens like *Yersinia enterocolitica*, *Salmonella enteritidis*, *Salmonella typhimurium*, *Shigella sonnei*, and *Escherichia coli* [72]. Nitrate displays indirect anti-infective

effects through acidification of nitrite at sites other than the mouth and digestive tract like the skin surface [73], the respiratory tract [74], and the lower urinary tract [75].

It has been reported by many scientists that one of the reduced nitrate products, NO, is beneficial to human health. NO is known as an important signal molecule in regulating many physiological functions in human body and, as reported by Archer, serves as an effective host defense against pathogens [76]. NO is an effective vasodilator and thrombocyte aggregation inhibitor which means that its deficiency can cause arteriosclerosis, diabetes, low blood pressure, and many other diseases [52,63].

Nitrites have been used for medical purposes as a vasodilator, a circulatory depressant to relieve smooth muscle spasm, and an antidote for treating cyanide poisoning [77]. Several toothpastes contain 5% potassium nitrate and are designed to lower dental hypersensitivity to rapid temperature changes. Some curatives against fungal infections (e.g. econazole nitrate) as well as treatment for burns (e.g. 0.5% nitrate aqueous solution) contain nitrate. Here, the nitrate ions are needed to keep the active component of the medicament in solution [78,79].

There is a lot of controversy about the toxic or beneficial effects of nitrite and nitrate in the human body and it seems that it will remain a disputed topic. Vegetables are considered beneficial in human nutrition as a source of fiber, vitamins, and nutrients. They may also contain versatile bioactive molecules, such as antioxidants, which may act as protective agents against chronic diseases, like cardiovascular diseases and diabetes, and even cancer [34]. A large range of these compounds is listed in International Agency for Research on Cancer (IARC) handbooks of cancer prevention [80]. In 1996 Hill reported a 20–40% reduction in deaths from cancer for vegetarians compared to non-vegetarians even when the vegetarians had a nitrate intake three times higher than non-vegetarians; 61 mg/person/d and 185–194 mg/person/d, respectively [35].

However, consumption of large amounts of nitrates in vegetables and drinking water for a prolonged period should be avoided. The carcinogenicity of nitrate, nitrite, and N-nitroso compounds has been and definitely will be reviewed many times regarding ongoing and future research on this very important topic. It can be generally stated that the beneficial and harmful effects of nitrate to human health are dependent on its dose, and reducing nitrate intake could be an acceptable way of preventing certain nitrate-related diseases [41,58].

5. NITRATE AND NITRITE INTAKE

Interest in the dietary intakes of nitrates and nitrites has arisen from concern about their possible adverse effect on health [81,82]. Exposure to nitrate and nitrite is associated mainly with three major dietary sources: vegetables, drinking water, and meat products.

For non-genotoxic food chemicals the estimated exposure is compared to acceptable daily intake (ADI) value. ADI is defined as an estimated maximum amount of an agent, expressed on a body mass basis, to which a subject may be exposed daily over his or her lifetime without appreciable health risk. For the calculation of ADI value, a 100-fold uncertainty factor is applied to the maximum non-effect dose determined in long-term animal studies in order to derive an acceptable intake for humans.

The first international evaluation of the risks associated with nitrate and nitrite ingestion was conducted by the Joint FAO/WHO Expert Committee on Food Additives (JECFA) in 1962 [83]. An acceptable daily intake (ADI) of 0–3.7 mg/kg bw for nitrate was established in 1990 and for nitrite 0–0.06 mg/kg bw in 1995

by the EU Scientific Committee for Food (SCF). In its most recent reviews JECFA reconfirmed an ADI for nitrate and set an ADI of 0–0.07 mg/kg bw for nitrite [33,50].

Differences in dietary habits as well as in water quality between countries result in different nitrate intakes, but vegetables are still the major source of nitrate intake by humans, contributing approximately 40–92% of the average daily intake [10–13]. Up to 20% of the total nitrate intake comes from the consumption of drinking water [84].

According to an oral bioavailability study of nitrate from nitrate-rich vegetables in humans [85], the nitrate from vegetables is absorbed very effectively, resulting in absolute nitrate bioavailability of around 100%, and the authors concluded that reducing the amount of nitrate in vegetables could be an effective measure to lower the systematic nitrate exposure of the general population.

To assess any potential health impact from different vegetable intake scenarios the European Food Safety Authority (EFSA) CONTAM Panel compared the nitrate exposure estimates with the ADI for nitrate of 3.7 mg/kg bw per day, equivalent to 222 mg for a 60 kg adult. Thus, a person eating 400 g of a variety of vegetables, as recommended by the World Health Organization (WHO), at typical median nitrate concentration levels would have a mean dietary exposure to nitrate of 157 mg/d. However, those who consume a high level of vegetables grown under unfavorable production conditions may exceed the ADI approximately 2-fold. Consumption of more than 47 g of rucola at the median nitrate concentration would lead to exceeding the ADI.

Nitrite intake from vegetables is low compared to the amounts resulting from endogenous conversion of nitrate. In consuming the recommended amount of vegetables with a mean nitrite concentration of 0.5 mg/kg, the exposure is 0.003 mg/kg bw per day [34].

6. SUMMARY

The nitrate ion has a low level of acute toxicity, but if transformed into nitrite, it may constitute a health problem. The concern over nitrites in the diet has two aspects: they may create an excess of methemoglobin, possibly leading to toxic effects such as cyanosis; and they can cause the endogenous formation of carcinogenic *N*-nitroso compounds. At the same time, it is generally accepted that a balanced vegetable-rich diet helps to reduce the risk of a number of diseases, including cancer and cardiovascular diseases. Many reports have shown the beneficial effects of nitrates and nitrites to human health, including antimicrobial activity.

Some risk assessments of nitrate and nitrite concluded that exposure to nitrate from vegetables is unlikely to result in appreciable health risks, therefore the beneficial effects of vegetable consumption prevail. However, there may be some specific risk groups, e.g. individuals with a high proportion of nitrate-rich vegetables in their diet or unfavorable local/home production conditions for vegetables that constitute a large part of the diet [34].

References

1. Mensinga, T. T., Speijers, G. J. A., & Meulenbelt, J. (2003). Health implications of exposure to environmental nitrogenous compounds. *Toxicological Reviews, 22,* 41–51.
2. Prakasa Rao, E. V. S., & Puttanna, K. (2000). Nitrates, agriculture and environment. *Current Science, 79,* 1163–1168.
3. Belitz, H.-D., & Grosch, W. (1999). Food chemistry, (2nd ed.). Berlin: Springer.
4. Fytianos, K., & Zarogiannis, P. (1999). Nitrate and nitrite accumulation in fresh vegetables from Greece. *Bulletin of Environment Contamination and Toxicology, 62,* 187–192.
5. Knobeloch, L., Salna, B., Hogan, A., Postle, J., & Anderson, H. (2000). Blue babies and nitrate-contaminated well water. *Environmental Health Perspectives, 108,* 675–678.
6. EC (1998). Council Directive 98/83/EEC of 3rd November 1998 on the quality of water intended for human consumption. *Official Journal of the European Union, L330,* 32–54.
7. Elliott, W., & Elliott, D. (2002). *Biochemistry and molecular biology.* UK: Oxford University Press.
8. Walters, C. L. (1996). Nitrate and nitrite in foods. In M. Hill, (Ed.), *Nitrates and nitrites in food and water* (2nd ed.) (pp. 93–112). Cambridge: Woodhead Publishing.
9. Reinik, M., Tamme, T., & Roasto, M. (2008). Naturally occurring nitrates and nitrites in foods. In J. Gilbert, & Z. Senyuva, (Eds.), *Bioactive compounds in foods* (pp. 227–253). UK: Blackwell Publishing.
10. Penttilä, P. L. (1995). Estimation of food additive and pesticide intakes by means of stepwise method. Doctoral thesis, University of Turku, Finland..
11. Dich, J., Järvinen, R., Knekt, P., & Penttilä, P.-L. (1996). Dietary intakes of nitrate, nitrite and NDMA in the Finnish mobile clinic health examination survey. *Food Additives and Contaminants, 13,* 541–552.
12. Ximenes, M. I. N., Rath, S., & Reyes, F. G. R. (2000). Polargraphic determination of nitrate in vegetables. *Talanta, 51,* 49–56.
13. Eichholzer, M., & Gutzwiller, F. (2003). Dietary nitrates, nitrites and N-nitroso compounds and cancer risk with special emphasis on the epidemiological evidence. In J. P. F. D'Mello, (Ed.), *Food safety, contaminants and toxins* (pp. 217–234). Wallingford, UK: CABI Publishing.
14. WHO (1995). Evaluation of certain food additives and contaminants. Joint FAO/WHO Expert Committee on Food Additives. *WHO Technical Report, 859,* 29–35.
15. Tamme, T., Reinik, M., Roasto, M., Juhkam, K., Tenno, T., & Kiis, A. (2006). Nitrates and nitrites in vegetables and vegetable-based products and their intakes by the Estonian population. *Food Additives and Contaminants, 23,* 355–361.
16. De Martin, S., & Restani, P. (2003). Determination of nitrates by a novel ion chromatographic method: Occurrence in leafy vegetables (organic and conventional) and exposure assessment for Italian consumers. *Food Additives and Contaminants, 20,* 787–792.
17. Santamaria, P. (2006). Nitrate in vegetables: Toxicity, content, intake and EC regulation. *Journal of the Science of Food and Agriculture, 86,* 10–17.
18. Siciliano, J., Krulick, S., Heisler, E. G., Schwartz, J. H., & White, J. W. Jr., (1975). Nitrate and nitrite of some fresh and processed market vegetables. *Journal of Agricultural and Food Chemistry, 23,* 461–464.
19. Corré, W. J., & Breimer, T. (1979). *Nitrate and nitrite in vegetables.* Literature Survey No. 39. Wageningen: Centre for Agricultural Publishing Documentation.

20. EC (2006). Commission Regulation No. 1881/2006 of 19 December 2006 setting maximum levels for certain contaminants in foodstuffs. *Official Journal of the European Communities, L364*, 5–24.

21. Laslo, C., Preda, N., & Bara, V. (2000). Relations between the administration of some nitrate fertilisers and the incidence of nitrates and nitrites in the fond products. *The University of Agricultural Sciences and Veterinary Medicine Cluj-Napoca, 28*, 1–6.

22. Sokolov, O. A. (1989). Hranenie i kulinarnaja obrabotka ovošcei. *Nauka i žizn, 2*, 152–153.

23. Tivo, P. F., & Saskevic, L. A. (1990). *Nitratõ sluhi i realnost.* Byelorussia: Minsk.

24. Walker, R. (1990). Nitrates and N-nitroso compounds: A review of the occurrence in food and diet and the toxicological implications. *Food Additives and Contaminants, 7*, 717–768.

25. Santamaria, P., Elia, A., Serio, F., & Todaro, E. (1999). A survey of nitrate and oxalate content in fresh vegetables. *Journal of the Science of Food and Agriculture, 79*, 1882–1888.

26. Golaszewska, B., & Zalewski, S. (2001). Optimalisation of potato quality in culinary process. *Polish Journal of Food and Nutrition Sciences, 10*, 59–63.

27. Mozolewski, W., & Smoczynski, S. (2004). Effect of culinary processes on the content of nitrates and nitrites in potato. *Pakistan Journal of Nutrition, 3*, 361–375.

28. Hlusek, J., Zrust, J., & Juzl, J. (2000). Nitrate concentration in tubers of early potatoes. *Rostlinna Vyroba, 46*, 17–21.

29. Vulsteke, G., & Biston, R. (1978). Factors affecting nitrate content in field-grown vegetables. *Plant Foods for Human Nutrition, 28*, 71–78.

30. Järvan, M. (1993). Köögiviljade nitraatidesisaldust mõjutavad tegurid. Doctoral thesis. Estonia: Saku.

31. Sokolov, O. A. (1987). Osobennosti raspredelenija nitratov I nitritov v ovošcah. *Kartofel`i i Ovošci, 6.*

32. Chung, S. Y., Kim, J. S., Kim, M., Hong, M. K., Lee, J. O., Kim, C. M., & Song, I. S. (2003). Survey of nitrate and nitrite contents of vegetables grown in Korea. *Food Additives and Contaminants, 20*, 621–628.

33. FAO/WHO (Food and Agriculture Organization of the United Nations/World Health Organization) (2003). *Nitrate (and potential endogenous formation of N-nitroso compounds).* WHO Food Additive series 50. Geneva: World Health Organization. Available at http://www.inchem.org/documents/jecfa/jecmono/v50je06.htm.

34. EFSA (European Food Safety Authority) (2008). Nitrate in vegetables. Scientific Opinion of the Panel on Contaminants in the Food chain. *The EFSA Journal, 689*, 1–79.

35. Hill, M. (1996). *Nitrates and Nitrites in Food and Water.* Cambridge, England: Woodhead Publishing.

36. Gangolli, S. D., van den Brandt, P., Feron, V., Janzowsky, C., Koeman, J., Speijers, G., Speigelhalder, B., Walker, R., & Winshnok, J. (1994). Nitrate, nitrite, and N-nitroso compounds. *European Journal of Pharmacology, 1*, 1–38.

37. EU Scientific Committee for Food (SCF) (1995). *Opinion on Nitrate and Nitrite Expressed on 22 September 1995.* Annex 4 to document III/56/95, CS/CNTM/NO3/20-FINAL. Brussels: European Commission DG III.

38. Bartholomew, B., & Hill, M. J. (1984). The pharmacology of dietary nitrate and the origin of urinary nitrate. *Food and Chemical Toxicology, 22*, 789–795.

39. Lundeberg, J. O., & Weitzberg, E. (2005). NO generation from nitrite and its role in vascular control. *Arteriosclerosis. Thrombosis and Vascular Biology, 25*, 915–922.

40. Lundberg, J. O., Weitzberg, E., Lundberg, J. M., & Alving, K. (1994). Intragastric nitric oxide production in humans: Measurements in expelled air. *Gut, 35*, 1543–1546.

41. Lundberg, J. O., Weitzberg, E., Cole, J. A., & Benjamin, N. (2004). Nitrate, bacteria and human health. *Nature Reviews Microbiology, 2*, 593–602.

42. Duncan, C., Dougall, H., Johnston, P., Green, S., Brogan, R., Leifert, C., Smith, L., Golden, M., & Benjamin, N. (1995). Chemical generation of nitric oxide in the mouth from the enterosalivary circulation of dietary nitrate. *Nature Medicine, 1*, 546–551.

43. Tannenbaum, S. R., Frett, D., Young, V. R., Land, P. D., & Bruce, W. R. (1978). Nitrite and nitrate are formed by endogenous synthesis in the human intestine. *Science, 200*, 1487–1489.

44. Chow, C. K., & Hong, C. B. (2002). Dietary vitamin E and selenium and toxicity of nitrite and nitrate. *Toxicology, 180*, 195–207.

45. Lundberg, J. O., Weitzberg, E., & Gladwin, M. T. (2008). The nitrate-nitrite-nitric oxide pathway in physiology and therapeutics. *Nature Reviews Drug Discovery, 7*, 156–167.

46. Pannala, A. S., Mani, A. R., Spencer, J. P. E., Skinner, V., Bruckdorfer, K. R., Moore, K. P., & Rice-Evans, C. A. (2003). The effect of dietary nitrate on salivary, plasma, and urinary nitrate metabolism in humans. *Free Radical Biology and Medicine, 34*, 576–584.

47. Packer, P. J., & Leach, S. A. (1996). Human exposure, pharmacology and metabolism of nitrate and nitrite. In M. Hill, (Ed.), *Nitrates and nitrites in food and water* (2nd ed) (pp. 131–162). Cambridge: Woodhead Publishing.

B. EFFECTS OF INDIVIDUAL VEGETABLES ON HEALTH

48. Ger, J., Kao, H., Shih, T. S., & Deng, J. F. (1996). Fatal toxic methemoglobinemia due to occupational exposure to methyl nitrite. *Chinese Medical Journal, 57,* S78.

49. Fan, A. M., Willhite, C. C., & Book, S. A. (1987). Evaluation of the nitrate drinking water standard with reference to infant methemoglobinemia and potential reproductive toxicology. *Regulatory Toxicology and Pharmacology, 7,* 135–148.

50. FAO/WHO (Food and Agriculture Organization of the United Nations/World Health Organization). (2003). *Nitrite (and potential endogenous formation of N-nitroso compounds).* WHO Food Additive series 50. Geneva: World Health Organization. Available at <http://www.inchem.org/documents/jecfa/jecmono/v50je05.htm>.

51. Speijers, G. J. A. (1996). Nitrate, toxicological evaluation of certain food additives and contaminants in food. In *WHO Food Additives Series 35* (pp. 325–360). Geneva: World Health Organization.

52. McKnight, G. M., Duncan, C. W., Leifert, L., & Golden, M. H. (1999). Dietary nitrate in man: Friend or foe? *British Journal of Nutrition, 81,* 349–358.

53. L'hirondel, J., & L'hirondel, J.-L. (2002). *Nitrate and Man.* Wallingford, Oxon, UK: CABI Publishing, CAB International.

54. Lijnski, W. (1999). N-Nitroso compounds in the diet. *Mutation Research, 443,* 129–138.

55. Vitozzi, L. (1992). Toxicology of nitrates and nitrites. *Food Additives and Contaminants, 9,* 579–585.

56. Benjamin, N. (2000). Nitrates in the human diet – good or bad? *Annales de Zootechnie, 49,* 207–216.

57. Du, S. T., Zhang, Y. S., & Lin, X. Y. (2007). Accumulation of nitrate in vegetables and its possible implications to human health. *Agricultural Sciences in China, 6,* 1246–1255.

58. Dougall, H. T., Smith, L., Duncan, C., & Benjamin, N. (1995). The effect of amoxicillin on salivary nitrite concentrations: An important mechanism of adverse reactions? *British Journal of Clinical Pharmacology, 39,* 460–462.

59. Turrini, A., Saba, A., & Lintas, C. (1991). Study of the Italian reference diet for monitoring food constituents and contaminants. *Nutrition Research, 11,* 861–873.

60. Slob, W., Van der Berg, R., & van Veen, M. P. (1995). A statistical exposure model applied to nitrate intake in the Dutch population. In *Health Aspects of Nitrates and Its Metabolites,* (pp. 75–82). Strasbourg: Council of Europe Press.

61. Bogovski, P., & Bogovski, S. (1981). Animal species in which *N*-nitroso compounds induce cancer. *International Journal of Cancer, 27,* 471–474.

62. Pour, P. M., Runge, R. G., & Birt, D. (1981). Current knowledge of pancreatic carcinogenesis in the hamster and its relevance to the human disease. *Cancer, 47,* 1573–1587.

63. Coss, A., Cantor, K. P., Reif, J. S., Lynch, C. F., & Ward, M. H. (2004). Pancreatic cancer and drinking water and dietary sources of nitrate and nitrite. *American Journal of Epidemiology, 159,* 693–701.

64. Preston-Martin, S., Pogoda, J. M., Mueller, B. A., Holly, E. A., Lijinsky, W., & Davis, R. L. (1996). Maternal consumption of cured meats and vitamins in relation to pediatric brain tumors. *Cancer Epidemiology Biomarkers and Prevention, 5,* 599–605.

65. Lee, M., Wrensch, M., & Miike, R. (1997). Dietary and tobacco risk factors for adult onset glioma in the San Francisco Bay Area (California, USA). *Cancer Causes Control, 8,* 13–24.

66. Bunin, G. R. (1998). Maternal diet during pregnancy and risk of brain tumors in children. *International Journal of Cancer, 11,* 23–25.

67. Püssa, T. (2008). *Principles of Food Toxicology.* CRC Press, Taylor & Francis Group.

68. Brambilla, G., & Martelli, A. (2007). Genotoxic and carcinogenic risk to humans of drug-nitrite interaction products. *Mutation Research, 635,* 17–52.

69. Bousset, J., & Fournaud, J. (1976). L'emploi des nitrates et des nitrites pour le traitement des produits carnés: Aspects technologiques et microbiologiques. *Annales de la Nutrition et de l Alimentation, 30,* 707–714.

70. Reinik, M., Tamme, T., Roasto, M., Juhkam, K., Jurtšenko, S., Tenno, T., & Kiis, A. (2005). Nitrites, nitrates and N-nitrosoamines in Estonian cured meat products: Intake by Estonian children and adolescents. *Food Additives and Contaminants, 22,* 1098–1105.

71. Tompkin, R. B. (1993). Nitrite. In P. M. Davidson, & A. L. Branen, (Eds.), *Antimicrobials in Foods* (pp. 191–262). New York: Marcel Dekker.

72. Dykhuizen, R. S., Frazer, R., Duncan, C., Smith, C. C., Golden, M., Benjamin, N., & Leifert, C. (1996). Antimicrobial effect of acidified nitrite on gut pathogens: Importance of dietary nitrate in host defense. *Antimicrobial Agents and Chemotherapy, 40,* 1422–1425.

73. Weller, R., Prize, R., Ormerod, A., Benjamin, N., & Leifert, C. (1997). Antimicrobial effect of acidified nitrite on skin commensals and pathogens. *British Journal of Dermatology, 136,* 464.

74. Robbins, R. A., & Rennard, S. I. (1997). Biology of airway epithelial cells. In R. G. Crystal, J. B. West, E. R. Weibel, & P. J. Barnes, (Eds.), *The Lung. Scientific Foundations* (Vol. 1). (pp. 445–457). Philadelphia: Lippincott-Raven.

75. Lundberg, J. O. N., Carlsson, S., Engstrand, L., Morcos, E., Wiklund, N. P., & Weitzberg, E. (1997). Urinary nitrite: More than a marker of infection. *Urology, 50,* 189–191.

76. Archer, J. (2002). How to use nitrates. *Cardiovascular Drugs and Therapy, 16,* 511–514.

77. Robertson, R. M., & Robertson, D. (1996). Drugs used for the treatment of myocardial ichemia. In *Goodman & Gilman's the pharmacological basis of therapeutics,* 9th ed (pp. 759–780). London: MacMillan.

78. Monafo, W. W., Tandson, S. N., Ayvazian, V. H., Tuchschmidt, J., Skimmer, A. M., & Deitz, F. (1976). Cerium nitrate: A new topical antiseptic for extensive burns. *Surgery, 80,* 465–473.

79. Orchardson, R., & Gillam, D. G. (2000). The efficacy of potassium salts as agents for treating dentin hypersensitivity. *Journal of Orofacial Pain, 14,* 9–19.

80. IARC (International Agency for Research on Cancer) (2003). *IARC Handbooks of Cancer Prevention* (Vol. 8). Lyon, France: International Agency for Research on Cancer pp. 1–375.

81. Knekt, P., Jarvinen, R., Dich, J., & Hakulinen, T. (1999). Risk of colorectal and other gastro-intestinal cancers after exposure to nitrate, nitrite and N-nitroso compounds: A follow-up study. *International Journal of Cancer, 80*(6), 852–856.

82. Pegg, R. B., & Shahidi, F. (2000). *Nitrite curing of meat. The N-Nitrosoamine Problem and Nitrite Alternatives.* Trumbull, Connecticut, USA: Food & Nutrition Press.

83. FAO/WHO (Food and Agriculture Organization of the United Nations/World Health Organization) (1962). Evaluation of the toxicity of a number of antimicrobials and antioxidants. Sixth report of the joint FAO/WHO Expert Committee on Food Additives. *World Health Organization Technical Report Series, 228,* 69-75.

84. White, R. J. (1983). Nitrate in British waters. *Aqua, 2,* 51–57.

85. van Velzen, A. G., Sips, A. J. A. M., Schothorst, R. C., Lambers, A. C., & Meulenbelt, J. (2008). The oral bioavailability of nitrate from nitrate-rich vegetables in humans. *Toxicology Letters, 181,* 177–181.

B. EFFECTS OF INDIVIDUAL VEGETABLES ON HEALTH

22

The Essentiality of Nutritional Supplementation in HIV Infection and AIDS: Review of Clinical Studies and Results from a Community Health Micronutrient Program

Raxit J. Jariwalla, Aleksandra Niedwiecki, and Matthias Rath

Dr Rath Research Institute, Santa Clara, CA, USA

1. INTRODUCTION

Human immunodeficiency virus (HIV) infection and acquired immune deficiency syndrome (AIDS) has become a global health crisis. According to the UNAIDS 2008 Report on the Global AIDS Epidemic [1], the number of people living with HIV and AIDS was 33.2 million. Of those, roughly 22.5 million people were living in Africa alone, with an HIV prevalence rate of 5% in the general population. In 2007, approximately 2.5 million people were reported to be infected by HIV. In the same year, more than 2.1 million people died from AIDS [2–4].

Despite the existence of antiretroviral (ARV) drugs, there is no cure, which indicates the need for alternative/complementary therapy.

Scientific evidence gathered to date points to the essentiality of nutrition in the control of HIV infection and AIDS [2–4].

There is an intricate relationship between nutrition, HIV, and AIDS. Very early on in the AIDS epidemic, it was recognized that persons with HIV infection and AIDS have nutritional deficiencies that predispose them to disease progression [2,3,5,6]. Independently, experimental evidence generated prior to the advent of AIDS had established a clear link between nutrition, immunity, and infection [7,8]. Thus, it is well known that nutritional deficiency lowers immunity and predisposes to infection. Conversely, nutritional supplementation modulates the immune system and provides resistance to infection and, hence, may provide the needed solution to this epidemic.

In the case of HIV, infection has been linked to virus-induced enteropathy [9] involving damage to the intestinal mucosa and contributing to nutrient deficiency. In turn, nutrient loss can exacerbate diarrhea, weight loss, and wasting, which are the physical symptoms associated with AIDS. Malnutrition, viral infection, and nutrient loss constitute a triumvirate, creating a vicious circle leading to immune deficiency and increased susceptibility to opportunistic infection. This circle cannot be broken by ARV drugs, which are in themselves immunosuppressive and toxic, as specified in the product information leaflets of ARV drugs provided by the manufacturers [10]. The circle, however, can be remedied by nutritional intervention, as nutrients have been shown to modulate the immune system and suppress viral infection. The role of micronutrients for optimum function of the immune system has been recognized by 9 Nobel prizes and it is a condition, *sine qua non*, in restoring the immune system – a target that is beyond the reach of drug therapy [8,11–13].

This review is aimed at highlighting the essentiality of nutrition in AIDS, focusing specifically on the role of micronutrients in restoring the immune system leading to delay, deceleration, or reversal of disease progression. Here, we summarize the known benefits of nutrients, commonly found in fruits and vegetables, in controlling AIDS. In addition, we describe the results of a micronutrient program conducted in a community-wide setting in South Africa that has unraveled the power of micronutrient supplementation in reducing the severity of physical symptoms of AIDS and improving the health of AIDS patients.

2. CHARACTERISTICS OF AIDS

AIDS is associated with progressive depletion of a specific class of immune cells (CD4+ T lymphocytes) whose loss leads to opportunistic infections and cancer. The defective immunologic responses linked to CD4+ T cell loss include impaired proliferative ability, decreased interleukin-2 (IL-2) secretion, and unresponsiveness to recall antigens or stimulation through the CD3/TCR complex [14].

Since 1984, AIDS has been linked to infection by HIV, which has been implicated as a causative agent of AIDS [15]. Although HIV infects CD4+ T cells, it is found in only a small proportion of such cells in the body. Furthermore, unlike other viruses such as measles, influenza, common cold, and herpes, which have an incubation period of only a few days, the latent period for manifestation of AIDS symptoms following HIV infection is 8–10 years, a remarkably long interval [16]. Considering these facts HIV infection alone can't be the sole cause of AIDS and other factors associated with viral infection, such as nutritional deficiencies, must play an important role in AIDS development [2,3,13].

3. LIMITATIONS OF CURRENT ANTIRETROVIRAL TREATMENT

Conventional treatment is based on the use of ARV drugs that can lower virus load but do not fully restore the immune system or cure AIDS (as the labels on the medication clearly indicate this fact). The claims advanced by medical authorities for a sustained halt of AIDS progression are highly questionable.

There is some evidence that ARV drugs lead to improvements in surrogate markers such as the level of CD4 cells. ARV drugs are capable of lowering viral load in the bloodstream but latent virus persists in a dormant state in lymphoid tissue and macrophage cells from which it can reappear upon reactivation. Furthermore, prolonged ARV treatment is often accompanied by severe side effects as noted below and patients frequently develop resistance to ARV drugs over time, suggesting the need for other treatment modalities.

There is only limited evidence (from uncontrolled surveys or controlled trials) that ARV drugs offer clinical benefits [11,12,17]. The effectiveness of ARV drugs in having any clinical benefits at all depends upon a number of factors, such as the nutritional status of the patient and the level of CD4 cells at the time at which ARV treatment begins. Thus, in an uncontrolled survey of 1219 patients on triple-drug therapy conducted by Hogg et al. [17], it was reported that participants with CD4 cell counts of $<50/\mu L$ were 6.67 times and those with counts of $50-199/\mu L$ were 3.41 times more likely to die than those with counts of at least $200/\mu L$.

Further, in a retrospective cohort study of 394 patients at an HIV referral center in Singapore, it was reported that at the time of starting ARV treatment malnutrition was significantly associated with reduced survival [18]. With respect to improvement of the immune system, the ability of ARV drugs to restore or improve a patient's immune system is limited by their very nature of being toxic to the bone marrow and blood cell system.

In terms of controlled studies, only a few, mostly short-term trials have looked at the effects of ARV drugs on clinical parameters such as survival or disease progression. As stated above, there is no long-term clinical proof for prolonged survival with ARV treatment from controlled studies available in scientific literature.[11, 12].

Among the few controlled studies is the licensing study of AZT (azidothymidine), the first ARV drug to be used in AIDS patients, which provides the only evidence for clinical benefit of ARV drugs [19,20]. However, the benefits were short-term in duration (4-month survival) and were attended by serious adverse reactions.

Thus, approximately 24% of AZT recipients (4-month survivors) developed anemia (reduced red blood cell count or hemoglobin concentration), compared to 4% of those who received a placebo. Furthermore, 21% of the AZT recipients had to be given red cell transfusions to keep them alive compared to only 4% receiving the placebo [19].

Notably, the short-term survival benefit of AZT seen in the initial trial was not maintained in a follow-up 21-month study [11,20], which found a rapid decline in the survival reported for the first 4 months. A few years later, a large, placebo-controlled trial, the British–French Concorde study, conducted in HIV-asymptomatic individuals showed unambiguously that AZT was unable to prevent AIDS in this patient population [12]. Subsequent trials have been mostly uncontrolled surveys or comparative studies that compare one ARV drug regimen with another without inclusion of an ARV-free control group [21–23]. Because there are no other controlled long-term studies, the ARV proponents cite uncontrolled surveys such as the ones by Hogg et al. [17] and Palella et al. [21] to make a case for lowering of disease progression. However, these surveys only show that the conferral of benefits depends on CD4 count at the time of administration of ARV treatment. ARV proponents also rely on surrogate markers to make claims about survival, claiming the drugs to be life saving or life prolonging. Two commonly used surrogate markers are HIV viral load or CD4 count, which have been routinely employed to approve ARV drugs over the last two decades. However, an improvement in surrogate markers without demonstration of corresponding improvement in clinical symptoms is not a significant improvement since the validity of these surrogate markers as predictors of clinical improvement has been questioned [12,24]. Thus, lowering of viral load or improvement in CD4 count has not been unequivocally shown to lead to improvement in survival or life extension.

4. DAMAGING SIDE EFFECTS OF ARV DRUGS

Because of the exclusive emphasis of the medical community on HIV as the etiological agent of AIDS, several different classes of ARV drugs have been developed that are targeted at different steps in the virus lifecycle. These include inhibitors of reverse transcription, polypeptide processing by viral protease, virus fusion, and viral DNA integration. Although all classes of ARV drugs have been employed in clinical practice, latent virus still persists in a dormant state and patients undergo only partial immune recovery. Furthermore, prolonged antiretroviral treatment is often accompanied by severe side effects as noted below and patients frequently develop resistance to ARV drugs over time, suggesting the need for other treatment modalities.

The toxicities of reverse transcriptase (RT) inhibitors, namely nucleoside analogues, non-nucleoside analogues, and protease inhibitors, which form part of the highly active antiretroviral therapy (HAART), are well known [25–33] and documented in patients information leaflet of every commercially available ARV drug [10].

RT inhibitors produce serious side effects and are toxic to all cells in the body. Those organs that have a high rate of cell proliferation (multiplication) are particularly affected by this toxicity. By virtue of their ability to block DNA synthesis, the RT inhibitors cannot discriminate between the replicating virus and the rapidly dividing cells – such as those of the bone marrow – that are essential to keep the human body alive and in optimum health.

Since the bone marrow is the production site of red and white blood cells, an impaired production/function of red blood cells leads to anemia whereas an impaired production/function of white blood cells leads to an impairment of the immune system (drug-induced immune deficiency). The toxicity of ARV drugs (RT inhibitors) on the bone marrow and blood cell system and the known side effects of anemia (loss of red blood cells) or leukopenia (loss of white blood cells) are known to the manufacturers of ARV drugs as this information is part of the mandatory ARV product information and has been reported in published studies [25–28]. Since ARV drugs are known to affect the bone marrow they can further weaken the immune system and pose risk for drug-induced immune deficiency.

Although the doses of NRTIs (nucleoside reverse transcriptase inhibitors) have been lowered in the triple drug combination cocktails such as HAART, the toxicity is still evident as documented by the reports of thrombocytopenia (low platelet count) and anemia in the studies by Carbonara et al. [29] and Mocroft et al. [30].

Additionally, both nucleoside and non-nucleoside reverse transcriptase inhibitors (NRTIs and NNRTIs) contained in triple drug cocktails (e.g. HAART) are hepatotoxic, i.e. cause damage to the liver [31]. Further, NRTIs in HAART have been linked to mitochondrial toxicity [32] and the protease inhibitors in HAART have been associated with a lipodystrophy syndrome, characterized by dyslipidemia, i.e. a severe impairment of the metabolism and storage of fats in the body [33].

Many protease inhibitors also have other serious adverse side effects. There are a number of metabolic complications associated with the use of some ARV drugs, including derangement in glucose and lipid metabolism, bone metabolism, and lactic acidemia (sustained accumulation of lactic acid in the blood), which have been documented in industrialized countries. [32,33]

Two drugs commonly employed in prevention of mother-to-child transmission (PMCT) of HIV are zidovudine (AZT) and nevirapine. Both drugs are highly cytotoxic in the PMCT setting. Nevirapine is reported to damage the liver (hepatotoxicity) and zidovudine has been reported to produce damaging prenatal effects *in utero* and perinatal effects in newborn infants [31,34–37].

4.1 Drug-induced Immune Deficiency

Since ARV drugs are known to affect the bone marrow (precursor of both white and red blood cells), they can further weaken the immune system and pose risk for drug-induced immune deficiency. Based on WHO reports [38], the average annual global mortality rate for all HIV-positive persons (including the minority on ARV drugs) for the year 2000 can be estimated to be ~1.4% (assuming that all newly diagnosed AIDS cases died in the same year). During the years 1998 and 2001, which fall on either side of the WHO year, the rate of death from AIDS in (initially asymptomatic) HIV-positive persons treated with ARV drugs was reported in two uncontrolled surveys conducted in the USA and Canada [17,21] to be 6.7–8.8% – 4–6 times higher than the global mortality rate of HIV-positive persons from the WHO estimation. Thus, it is likely that the use of ARV drugs, while affecting the HIV (viral) load, may actually impair the immune system and cause drug-induced immune deficiencies. Further research in this area must be conducted with high priority in order to avoid a possible failure of current global strategies to control AIDS.

5. NEED FOR NON-TOXIC, NUTRITIONAL THERAPY

The optimum treatment for AIDS should depend on relatively non-toxic, non-immunosuppressive agents or activities that slow disease progression. This includes activities that reduce stress, promote the adoption of healthy lifestyles including proper nutrition (in particular micronutrients that enhance the immune system), cessation of unhealthy habits (e.g. smoking, alcohol, drug use), and prophylaxis of preventable opportunistic infections.

The pathological basis of AIDS is a dysfunctional immune system clinically indicated by abnormally low levels of white blood cells. Micronutrients are essential for both red and white blood cell formation. Of particular importance are vitamins B3, B5, B6, B12, and vitamin C, folic acid, and iron.

ARV drugs have limited ability to restore a patient's immune system. By contrast, micronutrients have been shown to have immune-restoring and disease-retarding potential [39–48].

As discussed above, the median time of progression from HIV infection to AIDS has been estimated by reputable research at 9–10 years, unlike that of only a few days for manifestation of symptoms by measles and influenza. In this period, multivitamins and micronutrients should be used as an effective and economical preventive measure to improve immune system function in HIV infected individuals before exerting aggressive ARV drug therapies [45–47]. The use of ARV drugs in this period, before the progression to AIDS, runs the risk that patients who would otherwise live for an average of 9–10 years without developing AIDS will develop much sooner a drug-induced form of immune deficiency that is indistinguishable from (and as deadly as) AIDS.

Without micronutrients and vitamins, life is not possible and without an optimum daily supply of micronutrients and vitamins in the form of either adequate nutrition or supplementation, the metabolism of cells and the immune system generally is impaired and optimum health is impossible.

Furthermore, analysis of HIV patients in Singapore showed that malnutrition at the time of starting ARV treatment was significantly associated with reduced survival [18].

Protein-calorie malnutrition and specific micronutrient abnormalities are common in HIV and AIDS infection and have been linked to increased disease progression (see section below). Hence, nutritional therapy would be of immense benefit to HIV/AIDS infected patients.

6. NUTRITIONAL DEFICIENCIES AND THEIR IMPACT IN HIV INFECTION AND AIDS

The relationship between HIV and AIDS and poor nutrition has been well established since the onset of the AIDS epidemic [48]. Very early on with the emergence of HIV infection and AIDS, it was established that the disease was associated with protein-calorie malnutrition as well as specific micronutrient abnormalities. Malnutrition is prevalent in the developing world, particularly African and Asian countries.

Micronutrients reported to be altered in HIV/AIDS patients include vitamins A, B12, B6, C, and E; trace minerals selenium and zinc; amino acids cysteine and glutamine; and the tripeptide glutathione.

Infection with HIV exacerbates the impact of poor nourishment, while poor nutrition hastens the progression of HIV infection to AIDS, wasting, and death. Opportunistic infections and their associated symptoms limit food intake and intensify resting energy demands, increasing nutritional needs. HIV-related symptoms such as anorexia, nausea, vomiting, malabsorption, and diarrhea further worsen poor nutrition. For undernourished HIV-infected people, the resulting downward spiral of inadequate nutritional intake, inability to maintain weight and lean tissue mass, micronutrient deficiency, and increased susceptibility to opportunistic infections accelerates the development of AIDS. This decline ultimately leads to malnourished, HIV-infected people who become economically unproductive and unable to control their illness.

It is therefore essential in any treatment of HIV that a comprehensive nutritional program is adopted. Such a program should include, where necessary, supplementary meals and micronutrients. If implemented properly, such regimens will promote and enhance the immune systems of HIV-positive patients.

Optimal nutritional status and maintenance of adequate vitamin and mineral levels delays the progression to AIDS [4,46,47,49]. HIV-positive infants and children face a confluence of three powerful nutritional challenges, namely high nutritional needs to sustain their growth rate, rapid progression to AIDS associated with significant wasting, and an immature, compromised immune system, with increased risk for opportunistic infections and diminished nutritional intake Consequently, all HIV-positive children under the age of 14 should receive nutritional packages consisting of vitamin syrup and a supplement meal.

Providing HIV-infected pregnant women with a multivitamin supplement that contains vitamins B, C, and E along with iron and folate reduces the potential for vertical transmission of HIV.

7. THE ESSENTIALITY OF MICRONUTRIENTS FOR OPTIMAL HEALTH

Micronutrients are essential for sustaining all cellular functions. Vitamins and minerals are needed in smaller amounts than proteins, fats, and sugars but without them cells cannot convert food into biological energy and build different body structures.

The essential micronutrients include: vitamins C, the B-complex vitamins (B1, B2, B3, B5, B6, B12), vitamins E, D, K, A; the minerals calcium magnesium, sodium, potassium; the trace elements selenium and zinc; and the amino acids cysteine, glutamine lysine, proline, and arginine.

Today our food is poor in micronutrients. A comparison of the content of micronutrients in fruits and vegetables between 1985 and 1995 showed that over this 10-year period there was between 14% and 93% decline per 100 g of food in the content of calcium, magnesium, vitamin C, folic acid, and vitamin B6.

Deficiency of micronutrients is difficult to recognize. Whereas the deficit or absence of natural elements such as air (oxygen), water, and food can be felt (by our senses), the deficiency of vitamins or minerals cannot be sensed but has to be measured in blood by biochemical means.

Long-term deficiency of micronutrients leads to many diseases, including failing of the immune system. Disease arises in healthy cells that can affect tissues, leading to sick organs and the spread of disease in the body. Poor nutrition and long-term vitamin deficiency can impair the function of immune cells making a person susceptible to various viral and bacterial infections. In AIDS, there is selective loss of helper (CD4) T cells that leads to a gradual shift from the Th1 (intracellular immunity) to Th2 (extracellular or autoimmune) cytokine shift, increasing vulnerability to opportunistic infection [50].

The critical role of micronutrients for health has been recognized but not implemented. A few years ago UNICEF [51] reported that vitamin and mineral deficiencies affect one-third of the world's people (2 billion persons), debilitating minds, bodies, energies, and the economic prospect of nations. Despite that report, very little has been done by the international agency to alleviate global micronutrient deficiency.

Since AIDS is the major health problem linked to malnutrition and micronutrient deficiency, it presents an excellent opportunity to intervene in this disease with nutrient supplementation. Today we have a real possibility of controlling AIDS by taking advantage of research progress in science-based natural health or cellular medicine by applying micronutrient-based approaches.

8. CLINICAL BENEFITS OF MICRONUTRIENT SUPPLEMENTATION IN HIV/AIDS

Although ARV drugs cannot cure AIDS, there are several studies published in peer-reviewed scientific journals (see sections 8.1 and 8.2 below) which suggest that nutrition, including proper vitamin and micronutrient supplementation, can help to arrest the deterioration of patients with HIV, delay the onset of AIDS and, by increasing CD4 count and reducing viral load, reverse the progression of the disease. This is supported by several clinical and experimental studies conducted with single and multiple micronutrients that are summarized below:

8.1 Effects of Single Nutrients or their Combinations

Specific effects of micronutrients on blocking HIV multiplication have also been demonstrated. Thus, certain vitamins such as vitamin C and amino acid derivatives such as N-acetylcysteine (NAC) were shown to have anti-HIV activity in HIV-infected cells [52–54]. These laboratory studies were validated by clinical improvements seen in HIV-infected patients following supplementation with these nutrients [41,55,56].

In an early observational study, supplementation of AIDS patients with large doses of vitamin C was associated with stabilization of helper T lymphocyte count and amelioration of opportunistic infection [57].

In HIV-infected adults, supplementation with vitamins C and E reversed oxidative damage to

DNA which was promoted by treatment with zidovudine or AZT [58].

In a randomized, placebo-controlled, double-blind trial in HIV-infected subjects, a combination treatment with vitamin C (1000 mg) and vitamin E (800 IU daily) was reported to reduce significantly several markers of oxidative stress and produce a trend towards reduction in HIV viral load after 3 months of supplementation [55].

A pilot study of short-term, combination treatment with high-dose NAC and vitamin C reported improvements of immunological and virological parameters in patients with advanced AIDS [41].

In a cohort of HIV-infected subjects with low CD4 count, subjects with CD4 <200 who took NAC showed significantly longer survival than a comparable group (CD4 <200) who did not supplement with NAC [56].

In a randomized, placebo-controlled, double-blind trial, daily supplementation with selenium in HIV-seropositive men and women (with attained serum selenium levels greater than 26.1 μg/L) was reported to prevent the progressive elevation of HIV-1 viral load and improve CD4 cell count after 9 months of treatment. In contrast, in non-responding selenium-treated subjects (with serum selenium levels of less than or equal to 26.1 μg/L), an elevation in viral load was seen [59].

In 2008, in HIV-positive subjects with a history of unresponsiveness to ARV (viral load >10,000 copies/cm^3), alpha-lipoic acid supplementation was shown to improve lymphocyte function [60].

8.2 Effects of Multivitamins or Multiple Nutrients

Two prospective, longitudinal studies carried out at two well-respected US universities and published in peer-reviewed journals were among the first studies to show that the ingestion of a multivitamin supplement by asymptomatic HIV-positive men was associated with significantly reduced onset of AIDS development [45,46]. These studies demonstrated the important – and acknowledged – role played by vitamins in delaying the onset of AIDS.

In a large Tanzanian trial in pregnant and lactating women [47], multivitamins, mainly vitamins B, C, and E at doses of up to 22 times recommended daily intakes, resulted in significant reductions over the entire period of follow-up in the risk of progression of the disease. The study was a randomized, double-blind, placebo-controlled trial and demonstrated that inexpensive multivitamin treatment was effective in staving off the development of the AIDS disease among HIV-positive women. The key findings were:

a) Multivitamins reduced maternal viral load and all signs of HIV-related complications, and raised maternal CD4 and CD8 counts, compared to no multivitamins.

b) The progression of AIDS to stage 4 (according to the WHO's grading and corresponding to the full-blown manifestation of the disease) was reduced by 50% in patients receiving multivitamins.

c) The relative risk of patients dying from AIDS was lowered by 27% in patients receiving multivitamins.

d) 'Multivitamins also significantly reduced all signs of complications,' including oral lesions (ulcers) reduced by 48%, lip infections reduced by 56%, difficult, painful swallowing reduced by 53%, diarrhea reduced by 25%, and fatigue reduced by 24%.

In a randomized, placebo-controlled, double-blind trial in HIV-infected subjects living in Bangkok, multiple micronutrient supplementation was shown to enhance significantly the survival of individuals with a low CD4 count (<100) [49].

In a randomized, double-blind trial on HIV-infected patients in Kampala, Uganda, daily supplementation with two different combinations of micronutrients was shown to improve CD4 cell count, quality of life, and body weight after 52 weeks [4].

8.3 Advantages of Micronutrients over ARV Drugs

Micronutrients are also non-toxic compared to ARV drugs. There is a fundamental difference between vitamins and ARV drugs with respect to the assessment of 'side effects.' Vitamins are natural substances known to the human body and recognized by millions of its cells; hence, a surplus is generally eliminated by natural means. Thus, perhaps with the exception of a few lipid-soluble vitamins, even relatively high doses of vitamins are known to be safe.

In contrast, ARV drugs are synthetic molecules not recognized by the cells of the body as natural substances. Moreover, the toxicity of these drugs, e.g. their damage to DNA to block its replication, is the desired mechanism for these drugs. The only way for the body to eliminate these synthetic substances is by using a detoxification process in the liver. Since these drugs exert their toxicity even during their elimination, liver failure is an established and frequent side effect of ARV drugs. Moreover, HIV infection is linked to a deficiency of glutathione, a major intracellular antioxidant and detoxifier in the body.

Thus, with respect to side effects, there is no comparison between natural micronutrients and the severe side effects from ARV-induced toxicity (such as organ failure and drug-induced immune deficiencies) which are frequently seen even in the range of the recommended therapeutic doses [19,20,25,27–29]. However, the manufacturers of ARV drugs ignore these facts, with drug patentability being an important economic reason to do so, and continue to draw massive profits from some of the poorest countries in the world [10].

Micronutrients confer substantial clinical benefits at all stages of disease relative to ARV drugs. In short:

a) Multivitamins and micronutrients are essential for optimum immune system function and should be considered the basis of correcting immune deficiencies, including AIDS.

b) There is no significant evidence that once AIDS has developed ARV drugs can deliver proven clinical benefits or are clinically effective, and at best they may deliver improvements in surrogate markers, limited effects on the immune system, or short-term improvement in survival.

c) There is evidence that vitamins and micronutrients can provide clinical benefits once AIDS has developed [13].

d) The side effects from taking ARV drugs are known to be very significant.

e) There are few if any side effects from taking multivitamins and micronutrients and any side effects are trivial when compared to the known side effects of ARV drugs.

Ideally, the case for one treatment being better than another would be made from a comparative study of both treatments in the same trial. However, no such trials have been published or (as far as we are aware) carried out. Nevertheless, the significantly reduced risk of AIDS development at an early non-symptomatic stage of HIV infection from consumption of a daily multivitamin supplement as seen in prospective studies in HIV-positive, asymptomatic men [45,46] was an important finding. Similarly, the delayed progression of HIV disease seen in HIV-positive pregnant women given multivitamin supplements was also an important, significant finding [47]. In contrast, in the Concorde study, AZT was unable to delay the progression of the HIV disease in asymptomatic HIV-positive individuals [12].

Since multivitamin supplementation delayed the progression of HIV to AIDS and vitamins are essentially non-toxic compared to ARV drugs, multivitamin supplementation provides a better alternative to ARV therapy in the early stages of HIV infection.

At later stages of HIV infection, in cohorts of patients with low CD4 counts, multiple micronutrient supplementation was also shown to significantly enhance survival of individuals with low CD4 counts. Those with a count <100 and taking a placebo were 4 times as likely to die compared with those being given micronutrients. Those with higher counts up to 200 on placebo were twice as likely to die [49].

In contrast to the benefits of micronutrient supplementation in HIV patients in the later stages of HIV infection, as discussed above, only short-term improvement in survival with attendant serious side effects was seen from AZT treatment in AIDS patients, a benefit that was not maintained in a longer-term follow-up study [11,19]. Furthermore, the effect of triple-combination ARV treatment on survival in AIDS patients was shown to worsen with the decline in CD4 count ($<200/\mu L$) [17].

In addition, other benefits seen from ingestion of micronutrients include:

a) Immune restoration and retardation of HIV disease progression.

b) Decreases in AIDS-defining symptoms following micronutrient supplementation as seen in our community health nutrient program affecting people with AIDS in South Africa (see section below).

c) Improvement of CD4 count, quality of life and body weight after 52 weeks following daily supplementation with micronutrients [4].

Pending a cure for HIV/AIDS, vitamins and micronutrients are certainly more of an 'answer' than ARV drugs. Whether vitamins and micronutrients will halt the disease's progression to AIDS or delay its onset will depend on individual patient circumstances, but at the pre-AIDS stage they are more effective in the treatment of HIV than ARV drugs.

CD4 counts or HIV viral load have been employed as surrogate markers for AIDS over the last two decades and are routinely used as a biomarker to approve ARV drugs. Insofar as micronutrients are concerned, there is evidence that they not only improve surrogate markers (CD4 count and viral load) but also retard disease progression and decrease AIDS-defining symptoms. In contrast, except for the short-term survival benefit seen with AZT in the Fischl et al. studies [11,19] and limited, relative survival benefit (dependent on CD4 count) reported in the Hogg et al. study [17], there is no significant evidence that ARV drugs cause reversal of the progression of AIDS beyond affecting surrogate markers. In contrast micronutrient supplementation in a community-wide setting in South Africa has been shown to reduce the physical symptoms of AIDS as presented below.

9. OUTCOME FROM A COMMUNITY HEALTH MICRONUTRIENT PROGRAM IN SOUTH AFRICA

9.1 Background and Setting

In 2005, the Dr Rath Health Foundation South Africa, a non-profit organization, donated a micronutrient supplement program to the South African National Civic Organization (SANCO) to support its community-based nutritional health initiative among people affected by AIDS. SANCO organized the distribution of this nutritional supplement in various townships in South Africa, including Cape Town (Townships: Khayelitsha and Western Cape), KwaZulu Natal province (including Durban) and Free State.

9.2 Participants and Methods

9.2.1 Evaluation

Participants in this program were selected by community health professionals and included adult HIV-positive men and non-pregnant women with advanced AIDS symptoms (CDC stage 2 or 3), none of whom were or had been taking ARV drugs.

At the beginning of the program and after periodic intervals of taking the nutritional supplement, the health status of the participants was assessed with the aid of a bilingual questionnaire that graded AIDS-defining symptoms for Africa (according to the Bengui definition of AIDS) and other physical symptoms on a scale of 0 to 4 (0 = no symptoms, 1 = mild, 2 = medium, 3 = advanced, 4 = severe).

Approximately 813 patients were evaluated who had completed all three examinations and questionnaires. The scores from the questionnaires were entered into Excel data sheets and the severity of symptoms after the first three visits (visit 3) was compared to that observed at the first visit (visit 1) which marks the start of nutritional supplement intake by the participants.

Statistical significance of differences in the change of symptom severity between visit 1, visit 2 and visit 3 was determined by the one-tailed paired t test. Statistical significance of the result is indicated by a P-value, where P equal or below 0.05 indicates that the results reached statistical significance. The lower the P-value, the more likely it is that the result will be achieved in a larger population.

9.2.2 Informed Consent

All participants gave their informed consent prior to participation in this program.

9.2.3 Nutritional Supplement

The nutritional supplement consisted of a defined combination of vitamins, minerals, trace elements, amino acids, and polyphenols (from green tea) as well as other nutrients and was supplied in the form of tablets to be taken three times a day with meals. This natural program was declared as a Food/Nutritional supplement for distribution and importation into South Africa in terms of the Foodstuffs, Cosmetics and Disinfectant Act, 1972 (act No 54 of 1972) and allowed into South Africa by the Director, Inspectorate and Law Enforcement of Medicine Regulatory Affairs.

9.3 Results

9.3.1 Khayelitsha Township

The results from evaluation of participants in the Khayelitsha Township near Cape Town have been previously reported [56]. Briefly, changes in severity of symptoms were evaluated after three visits (10–12 weeks) from the beginning of micronutrient supplementation in 56 participants who completed all three examinations and questionnaires. The results showed statistically significant reduction in severity of:

a) All AIDS-defining symptoms that include fever, weight loss, diarrhea, cough, and severity of tuberculosis (TB).
b) Fungal and opportunistic infections.
c) Other physical symptoms associated with AIDS such as skin sores, skin rashes, swollen glands, colds and flu, fatigue, headache, numbness in extremities, and joint pain.

No adverse side effects occurred from intake of the micronutrient supplements.

9.3.2 KwaZulu-Natal

The micronutrient supplement evaluated in Khayelitsha was subsequently rolled out in KwaZulu-Natal by SANCO. Similarly to other sites, the health status of participants was assessed with the aid of a bilingual questionnaire

TABLE 22.1   Severity of AIDS Related Symptoms at Baseline (0 Weeks) and after 4 and 8 Weeks of Taking Nutritional Supplement in Subjects from KwaZulu-Natal (KZN)

Symptom	Number of Patients	Symptom Severity Before Taking Nutritional Supplement	Symptom Severity After Taking Nutritional Supplement					
		Visit 1 (0 weeks)		Visit 2 (4 weeks)		Visit 3 (8 weeks)		
		Score	%	Score	Decrease in %	Score	Decrease in %	P-value
AIDS-defining symptoms								
Cough, breath	421	1.54	100	1.11	28.00	0.96	37.69	<0.0001
Diarrhea	296	1.28	100	0.80	37.47	0.67	48.02	<0.0001
Fever, chills, sweating	473	1.72	100	0.94	45.51	0.89	48.34	<0.0001
Other physical symptoms								
Blurred vision	406	1.07	100	0.91	15.17	0.86	19.1	0.002
Colds, flu	455	1.49	100	0.96	35.74	0.82	45.44	<0.0001
Dizzy, light headed	362	1.24	100	0.69	43.97	0.72	41.96	<0.0001
Dry, itchy skin	440	1.66	100	1.13	31.74	1.00	39.53	<0.0001
Eyes sensitivity	457	1.78	100	1.42	20.30	1.31	26.45	<0.0001
Gum bleeding	276	1.06	100	0.73	30.82	0.67	36.30	<0.0001
Irregular heartbeat	335	1.07	100	0.79	26.94	0.77	28.61	0.0001
Loose teeth	134	0.85	100	0.62	27.19	0.46	46.49	0.001
Skin bruises	297	1.26	100	0.87	30.93	0.75	40.80	<0.0001
Sweating without exertion	429	1.72	100	1.21	29.44	1.06	38.40	<0.0001
Swelling	235	1.16	100	0.68	41.54	0.56	51.47	<0.0001
Unusual thirst	522	2.07	100	1.76	15.00	1.71	17.13	<0.0001
Vomiting, nausea	267	1.10	100	0.64	42.18	0.50	54.42	<0.0001
Pain symptoms								
Headache	511	1.73	100	1.28	25.90	1.16	32.81	<0.0001
Joint pain	476	1.88	100	1.22	34.83	1.06	43.67	<0.0001
Muscle cramps	414	1.56	100	1.15	26.58	0.94	39.88	<0.0001
Pain in mouth, lip, gums	301	1.13	100	0.64	43.40	0.70	38.42	<0.0001
Pains, numbness, tingling	483	1.90	100	1.16	39.28	0.95	50.16	<0.0001

Data express severity of AIDS-related symptoms as mean values of symptom scores in indicated number of patients. These scores were determined with the aid of a questionnaire as described in the text.

A decrease in severity of symptoms after 4 weeks and 8 weeks of taking nutritional supplement was calculated and expressed as percent change. Statistical analysis evaluated significance of change in severity of symptoms between visit 1 (0 weeks) and visit 3 (8 weeks) of micronutrient supplementation. Values with $P<0.05$ were considered statistically significant.

that graded AIDS-defining symptoms for Africa and other physical symptoms on a scale of 0 to 4 (0 = no symptoms, 1 = mild, 2 = medium, 3 = advanced, 4 = severe). The evaluation included assessing the changes in severity of symptoms after three visits (8 weeks) from the beginning of micronutrient supplementation. Five hundred and twenty two participants completed all three examinations and questionnaires.

Results showed that daily micronutrient supplementation was associated with statistically significant reduction in severity of all symptoms (Table 22.1), which included:

a) AIDS-defining symptoms including fever, chills and sweats, diarrhea, and cough (Figure 22.1).
b) Specific clinical symptoms such as blurred vision, colds and flu, unusual thirst, vomiting, and nausea.
c) Other physical symptoms including swelling, skin bruises, irregular heartbeat, and gum bleeding.
d) Pain symptoms such as numbness of extremities, pain in mouth, lip and gums, muscle cramps, joint pain, and headaches.

No adverse side effects occurred from intake of the micronutrient supplements.

9.3.3 Western Cape

The micronutrient supplement evaluated in Khayelitsha and KwaZulu-Natal was also rolled out in Western Cape by SANCO. Similarly to other sites, the health status of participants was assessed with the aid of a bilingual questionnaire that graded AIDS-defining symptoms for Africa and other physical symptoms on a scale of 0 to 4 (0 = no symptoms, 1 = mild, 2 = medium, 3 = advanced, 4 = severe). The evaluation included the changes in severity of symptoms after three visits from the beginning of micronutrient supplementation. One hundred and fifty three participants completed all three examinations and questionnaires.

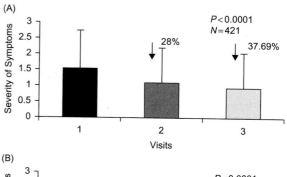
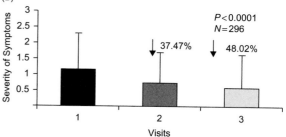
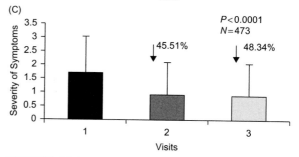

FIGURE 22.1 Changes in severity of cough, shortness of breath (A), diarrhea (B) and fever, chills, and sweating (C) in persons with AIDS before (visit 1) and after visits 2 and 3 of micronutrient supplementation in subjects from KwaZulu-Natal (KZN).
N = number of participants experiencing these symptoms before and during the program; P = statistical significance of the change between visits 3 and 1.

Results showed that daily micronutrient supplementation was associated with statistically significant reduction in severity of multiple physical symptoms (Table 22.2), which included:

a) AIDS-defining symptoms including fever, chills and sweats, diarrhea, and cough (Figure 22.2).

TABLE 22.2 Severity of AIDS-related Symptoms at Baseline (Visit 1) and After Second and Third Visits of Taking Nutritional Supplement in Subjects from Western Cape (WC)

Symptom	Number of Patients	Symptom Severity Before Taking Nutritional Supplements				Symptom Severity After Taking Nutritional Supplements		
		Visit 1 Score	%	Visit 2* Score	Decrease in %	Visit 3* Score	Decrease in %	P-value
AIDS-defining symptoms								
Cough, breath	147	1.84	100	1.11	39.63	0.88	52.22	<0.0001
Diarrhea	70	1.34	100	0.59	56.38	0.29	78.72	<0.0001
Fevers, chills, sweating	153	2.01	100	1.21	39.74	0.97	51.79	<0.0001
Other physical symptoms								
Blurred vision	92	1.82	100	0.67	62.87	0.48	73.65	<0.0001
Colds, flu	153	1.94	100	1.13	41.75	0.84	56.90	<0.0001
Dizzy, light headed	114	1.53	100	0.70	54.02	0.32	78.74	<0.0001
Dry, itchy skin	100	1.94	100	0.72	62.89	0.47	75.77	<0.0001
Eyes sensitivity	102	1.91	100	0.80	57.95	0.49	74.36	<0.0001
Gum bleeding	59	1.32	100	0.71	46.15	0.49	62.82	0.0002
Irregular heartbeat	59	1.42	100	0.80	44.05	0.44	69.05	<0.0001
Skin bruises	78	1.14	100	0.92	19.10	0.63	44.94	0.0037
Sweating without exertion	112	1.51	100	0.83	44.97	0.69	54.44	<0.0001
Swelling	60	1.42	100	0.85	40.00	0.50	64.71	<0.0001
Unusual thirst	128	1.72	100	0.95	45.00	0.55	67.73	<0.0001
Vomiting, nausea	77	1.53	100	0.65	57.63	0.25	83.90	<0.0001
Pain symptoms								
Headache	149	1.50	100	0.95	37.05	0.85	43.30	<0.0001
Joint pain	130	1.79	100	1.11	38.20	0.69	61.37	<0.0001
Muscle cramps	134	1.53	100	1.02	33.17	0.72	53.17	<0.0001
Pain in mouth, lip, gums	63	1.16	100	0.75	35.62	0.41	64.38	<0.0001
Pains, numbness, tingling	146	1.71	100	0.98	42.57	0.82	51.81	<0.0001

Data express severity of AIDS-related symptoms as mean values of symptom scores in indicated number of patients. These scores were determined with the aid of a questionnaire as described in the text.

A decrease in severity of symptoms after second and third visit of taking nutritional supplement was calculated and expressed as percent change. Statistical analysis evaluated significance of change in severity of symptoms between visit 1 and visit 3 of micronutrient supplementation. Values with $P < 0.05$ were considered statistically significant.

*Based on random means taken for 21 subjects, the average time to visit 2 was 17 weeks and that for visit 3 was 40 weeks.

b) Specific clinical symptoms such as colds and flu, blurred vision, dry, itchy skin, and unusual thirst.

c) Other physical symptoms including vomiting and nausea, irregular heartbeat, gum bleeding and swelling.

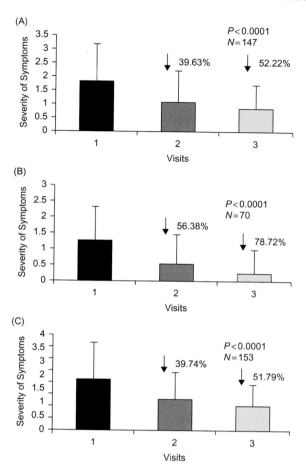

FIGURE 22.2 Changes in severity of cough, shortness of breath (A), diarrhea (B) and fever, chills, and sweating (C) in persons with AIDS before (visit 1) and after visits 2 and 3 of micronutrient supplementation in subjects from Western Cape (WC).
N = number of participants experiencing these symptoms before and during the program; P = statistical significance of the change between visits 3 and 1.

d) Pain symptoms such as numbness of extremities, pain in mouth, lip and gums, joint pain, muscle cramps, and headaches.

No adverse side effects occurred from intake of the micronutrient supplements.

9.3.4 Free State

The micronutrient supplement evaluated in Khayelitsha, Western Cape, and KwaZulu-Natal was also rolled out in Free State by SANCO. Similarly to the other sites, health status of participants was assessed with the aid of a bilingual questionnaire that graded AIDS-defining symptoms for Africa and other physical symptoms on a scale of 0 to 4 (0 = no symptoms, 1 = mild, 2 = medium, 3 = advanced, 4 = severe). The evaluation included assessing the changes in severity of symptoms after 3 visits (8–10 weeks) from the beginning of micronutrient supplementation.

TABLE 22.3 Severity of AIDS-related Symptoms at Baseline (0 weeks) and After Second Visit (4–6 weeks) and Third Visit (8–10 weeks) of Taking Nutritional Supplement in Subjects from Free State (FS)

Symptom	Number of Patients	Symptom Severity Before Taking Nutritional Supplement		Symptom Severity After Taking Nutritional Supplement				
		Visit 1 (0 weeks)			Visit 2 (4–6 weeks)		Visit 3 (8–10 weeks)	
		Score	%	Score	Decrease in %	Score	Decrease in %	P-value
AIDS-defining symptoms								
Cough, breath	73	1.89	100	1.68	10.87	1.45	23.19	0.0237
Fever, chills, sweating	82	2.61	100	2.02	22.43	1.93	26.17	0.0012
Other physical symptoms								
Blurred vision	76	2.84	100	2.05	27.78	1.72	39.35	0.0295
Colds, flu	77	2.40	100	1.78	25.95	1.65	31.35	0.0017
Dizzy, light headed	73	2.25	100	1.73	23.17	1.55	31.10	0.0016
Dry, itchy skin	56	2.07	100	1.54	25.86	1.09	47.41	0.0008
Eyes sensitivity	78	2.40	100	2.12	11.76	1.97	17.65	0.0262
Gum bleeding	50	1.74	100	1.72	1.15	1.22	29.89	0.0313
Irregular heartbeat	71	1.93	100	2.00	−3.65	1.39	27.74	0.0222
Skin bruises	35	1.66	100	1.54	6.90	0.86	48.28	0.0153
Sweating without exertion	73	2.38	100	1.67	29.89	1.75	26.44	0.0074
Unusual thirst	76	2.54	100	2.03	20.21	1.75	31.09	0.0007
Vomiting, nausea	61	1.77	100	1.33	25.00	1.03	41.67	0.0068
Pain symptoms								
Headache	75	2.40	100	2.47	−2.78	1.65	31.11	0.0003
Joint pain	68	2.37	100	1.99	16.15	1.79	24.22	0.0129
Pains, numbness, tingling	79	2.49	100	1.90	23.86	1.61	35.53	0.0001

Data express severity of AIDS-related symptoms as mean values of symptom scores in indicated number of patients. These scores were determined with the aid of a questionnaire as described in the text.

A decrease in severity of symptoms after visit 2 (4–6 weeks) and visit 3 (8–10 weeks) of taking nutritional supplement was calculated and expressed as percent change. Statistical analysis evaluated significance of change in severity of symptoms between visit 1 and visit 3 of micronutrient supplementation. Values with $P<0.05$ were considered statistically significant.

Eighty-two participants completed all three examinations and questionnaires.

Results showed that daily micronutrient supplementation was associated with statistically significant reduction in severity of physical symptoms (Table 22.3), which included:

a) AIDS-defining symptoms including cough, fever, chills, and sweats (Figure 22.3).

b) Specific clinical symptoms such as gum bleeding, irregular heartbeat, and skin bruising.

c) Other physical symptoms including blurred vision, colds and flu, vomiting and nausea, and unusual thirst.

d) Pain symptoms such as numbness of extremities, joint pain, and headaches.

No adverse side effects occurred from intake of the micronutrient supplements.

10. SUMMARY

The evaluation of the effects of a micronutrient program on AIDS-defining symptoms and other symptoms associated with AIDS in different populations of patients living in different townships and different regions of South Africa, which included Cape Town (Khayelitsha and Western Cape), KwaZulu-Natal province (including Durban) and Free State, concludes that:

a) Daily supplementation with a nutritional program containing a defined complex of vitamins, minerals, trace elements, selected amino acids, polyphenols, and other essential nutrients by AIDS patients not taking antiretroviral drugs (ARVs) caused significant reduction in all AIDS-defining and other physical symptoms associated with the progression of AIDS.

b) These observed health improvements in AIDS patients were consistent independent of the township location and region of the country.

This evaluation, along with clinical studies reviewed in this report, suggests that science-based effective, safe and affordable micronutrient supplementation in combination with other measures of improving nutritional status and living conditions can become a basis of public health strategy to fight immune deficiency

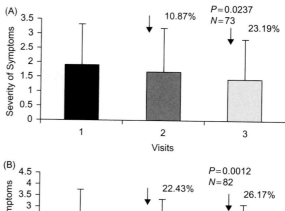

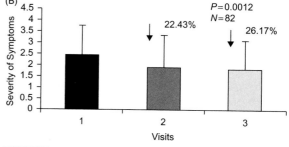

FIGURE 22.3 Changes in severity of cough, shortness of breath (A) and fever, chills, and sweating (B) in persons with AIDS before (visit 1) and after visits 2 and 3 of micronutrient supplementation in subjects from Free State (FS). N = number of participants experiencing these symptoms before and during the program; P = statistical significance of the change between visits 3 and 1.

diseases including AIDS. Implementation of these findings by national governments and international organizations would save millions of lives and represent successful, cost-effective, and sustainable global public health strategies to control AIDS until a cure is found.

ACKNOWLEDGMENTS

We would like to recognize SANCO organization and members of the Rath Health Foundation who assisted in the community health nutrient program. We would like to thank Bhakti Gangapurkar and Anupriya Pandit for their help with data analysis and Cathy Flowers for her comments on the manuscript.

References

1. UNAIDS. (2008). *2008 Report on the global AIDS epidemic.* New York: UNAIDS.

2. Beach, R. S., Mantero-Atienza, E., Shor-Posner, G., Javier, J. J., Szapocznik, J., Morgan, R., Sauberlich, H. E., Cornwell, P. E., Eisdorfer, C., & Baum, M. K. (1992). Specific nutrient abnormalities in asymptomatic HIV-1 infection. *AIDS, 6,* 701–708.

3. Baum, M. K., Shor-Posner, G., Lu, Y., Rosner, B., Sauberlich, H. E., Fletcher, M. A., Szapocznik, J., Eisdorfer, C., Buring, J. E., & Hennekens, C. H. (1995). Micronutrients and HIV-1 disease progression. *AIDS, 9,* 1051–1056.

4. Namulemia, E., Sparling, J., & Foster, H. (2007). Nutritional supplementation can delay the progression of AIDS in HIV-infected patients: Results from a double-blinded, clinical trial at Mengo Hospital, Kampala, Uganda. *Journal of Orthomolecular Medicine, 22*(3), 129–136.

5. Gray, R. H. (1984). Similarities between AIDS and PCM (Letter). *American Journal of Public Health, 73,* 1332.

6. Kotler, D. P., Rosenbaum, K., Wang, J., & Pierson, R. N. (1999). Studies of body composition and fat distribution in HIV-infected and control subjects. *Journal of Acquired Immune Deficiency Syndromes, 20,* 228–237.

7. Scrimshaw, N. S. (2003). Historical concepts of interactions, synergism and antagonism between nutrition and infection. *Journal of Nutrition, 133*(1), 316S–321S.

8. Webb, A. L., & Villamor, E. (2007). Update: Effects of antioxidant and non-antioxidant vitamin supplementation on immune function. *Nutrition Reviews, 65,* 181–217.

9. Ullrich, R., Zeitz, M., Heise, W., L'age, M., Höffken, G., & Riecken, E. O. (1989). Small intestinal structure and function in patients infected with human immunodeficiency virus (HIV): Evidence for HIV-induced enteropathy. *Annals of Internal Medicine, 111,* 15–21.

10. SANCO and Dr Rath Health Foundation Africa. (2007). The pharmaceutical investment business with the AIDS epidemic. In *End AIDS!* (pp. 57–86). Dr Rath Health Foundation <www.dr-rath-foundation.org.za/news/end_aids.html>.

11. Fischl, M. A., Richman, D. D., Causey, D. M., Grieco, M. H., Bryson, Y., Mildvan, D., Laskin, O. L., Groopman, J. E., Volberding, P. A., Schooley, R. T., Jackson, G. G., Durack, D. T., Andrews, J. C., Nusinoff-Lehrman, S., & Barry, D. W. AZT Collaborative Working Group. (1989). Prolonged zidovudine therapy in patients with AIDS and advanced AIDS-related complex. AZT Collaborative Working Group. *JAMA, 262,* 2405–2410.

12. Seligmann, M. Concorde Coordinating Committee. (1994). Concorde: MRC/ANRS randomised double-blind controlled trial of immediate and deferred zidovudine in symptom-free HIV infection. *Lancet, 343,* 871–881.

13. Jariwalla, R. J., Niedzwiecki, A., & Rath, M. (2008). Micronutrients and nutrient synergy in immunodeficiency and infectious disease. In R. R. Watson & V. R. Preedy (Eds.), *Botanical medicine in clinical practice* (pp. 203–212). London: CAB International.

14. Clerici, M., Stocks, N. I., Zajac, R. A., Boswell, R. N., Lucey, D. R., Via, C. S., & Shearer, G. M. (1989). Detection of three distinct patterns of T helper cell dysfunction in asymptomatic, human immunodeficiency virus-seropositive patients. Independence of CD4 + cell numbers and clinical staging. *Journal of Clinical Investigation, 84*(6), 1892–1899.

15. Lane, H. C., & Fauci, A. S. (1985). Immunologic abnormalities in the acquired immunodeficiency syndrome. *Annual Review of Immunology, 3,* 477–500.

16. Morgan, D., Mahe, C., Mayanja, B., Okongo, J. M., Lubega, R., & Whitworth, J. A. (2002). HIV-1 infection in rural Africa: Is there a difference in median time to AIDS and survival compared with that in industrialized countries? *AIDS, 16*(4), 597–603.

17. Hogg, R. S., Yip, B., Chan, K. J., Wood, E., Craib, K. J., O'Shaughnessy, M. V., & Montaner, J. S. (2001). Rates of disease progression by baseline CD4 cell count and viral load after initiating triple-drug therapy. *JAMA, 286*(20), 2568–2577.

18. Paton, N. I., Sangeetha, S., Earnest, A., & Bellamy, R. (2006). The impact of malnutrition on survival and the CD4 count response in HIV-infected patients starting antiretroviral therapy. *HIV Medicine, 7*(5), 323–330.

19. Fischl, M. A., Richman, D. D., Grieco, M. H., Gottlieb, M. S., Volberding, P. A., Laskin, O. L., Leedom, J. M., Groopman, J. E., Mildvan, D., Schooley, R. T., Jackson, G. G., Durack, D. T., & King, D. The AZT Collaborative Working Group. (1987). The efficacy of azidothymidine (AZT) in the treatment of patients with AIDS and AIDS-related complex. A double-blind, placebo-controlled trial. *New England Journal of Medicine, 317,* 185–191.

20. Fischl, M. A., Parker, C. B., Pettinelli, C., Wulfsohn, M., Hirsch, M. S., Collier, A. C., Antoniskis, D., Ho, M., Richman, D. D., & Fuchs, E., et al. (1990). A randomized controlled trial of a reduced daily dose of zidovudine in patients with the acquired immunodeficiency syndrome. The AIDS Clinical Trials Group. *New England Journal of Medicine, 323*(15), 1009–1014.

21. Palella, F. J., Jr., Delaney, K. M., Moorman, A. C., Loveless, M. O., Fuhrer, J., Satten, G. A., Aschman, D. J., & Holmberg, S. D. (1998). Declining morbidity

and mortality among patients with advanced human immunodeficiency virus infection. HIV Outpatient Study Investigators. *New England Journal of Medicine, 338*(13), 853–860.

22. Jordan, J., Cahn, P., Goebel, F., Matheron, S., Bradley, C., & Woodcock., A. (2005). Abacavir compared to protease inhibitors as part of HAART regimens for treatment of HIV infection: Patient satisfaction and implications for adherence. *AIDS Patient Care and STDs, 19*(1), 9–18.

23. Frampton, J. E., Croom., & Katherine, F. (2006). Efavirenz/Emtricitabine/Tenofovir disoproxil fumarate: Triple combination tablet. *Drugs, 66*(11), 1501–1512.

24. Rodriguez, B., Sethi, A. K., Cheruvu, V. K., Mackay, W., Bosch, R. J., Kitahata, M., Boswell, S. L., Mathews, W. C., Bangsberg, D. R., Martin, J., Whalen, C. C., Sieg, S., Yadavalli, S., Deeks, S. G., & Lederman, M. M. (2006). Predictive value of plasma HIV RNA level on rate of CD4 T-cell decline in untreated HIV infection. *JAMA, 296*, 1523–1525.

25. Richman, D. D., Fischl, M. A., Grieco, M. H., Gottlieb, M. S., Volberding, P. A., Laskin, O. L., Leedom, J. M., Groopman, J. E., Mildvan, D., Hirsch, M. S., Jackson, G. G., Durack, D. T., & Nusinoff-Lehrman, S. The AZT Collaborative Working Group. (1987). The toxicity of azidothymidine (AZT) in the treatment of patients with AIDS and AIDS-related complex. A double-blind, placebo-controlled trial. *New England Journal of Medicine, 317*, 192–197.

26. Costello, C. (1988). Haematological abnormalities in human immunodeficiency virus (HIV) disease. *Journal of Clinical Pathology, 41*, 711–715.

27. Moore, R. D., Creagh-Kirk, T., Keruly, J., Link, G., Wang, M. C., Richman, D., & Chaisson, R. E. (1991). Long-term safety and efficacy of zidovudine in patients with advanced human immunodeficiency virus disease. Zidovudine Epidemiology Study Group. *Archives of Internal Medicine, 151*, 981–986.

28. Dainiak, N., Worthington, M., Riordan, M. A., Kreczko, S., & Goldman, L. (1988). 3'-Azido-3'-deoxythymidine (AZT) inhibits proliferation *in vitro* of human haematopoietic progenitor cells. *British Journal of Haematology, 69*, 299–304.

29. Carbonara, S., Fiorentino, G., Serio, G., Maggi, P., Ingravallo, G., Monno, L., Bruno, F., Coppola, S., Pastore, G., & Angarano, G. (2001). Response of severe HIV-associated thrombocytopenia to highly active antiretroviral therapy including protease inhibitors. *Journal of Infection, 42*, 251–256.

30. Mocroft, A., Kirk, O., Barton, S. E., Dietrich, M., Proenca, R., Colebunders, R., Pradier, C., dArminio Monforte, A., Ledergerber, B., & Lundgren, J. D. (1999). Anaemia is an independent predictive marker for clinical prognosis in HIV-infected patients from across Europe. EuroSIDA study group. *AIDS, 13*, 943–950.

31. Abrescia, N., D'Abbraccio, M., Figoni, M., Busto, A., Maddaloni, A., & De Marco, M. (2005). Hepatotoxicity of antiretroviral drugs. *Current Pharmaceutical Design, 11*, 3697–3710.

32. Fleischer, R., Boxwell, D., & Sherman, K. E. (2004). Nucleoside analogues and mitochondrial cytotoxicity. *Clinical Infectious Diseases, 38*, e79–e80.

33. Carr, A., Samaras, K., Burton, S., Law, M., Freund, J., Chisholm, D. J., & Cooper, D. A. (1998). A syndrome of peripheral lipodystrophy, hyperlipidaemia and insulin resistance in patients receiving HIV protease inhibitors. *AIDS, 12*(7), F51–F58.

34. Blanche, S., Tardieu, M., Rustin, P., Slama, A., Barret, B., Firtion, G., Ciraru-Vigneron, N., Lacroix, C., Rouzioux, C., Mandelbrot, L., Desguerre, I., Rötig, A., Mayaux, M. J., & Delfraissy, J. F. (1999). Persistent mitochondrial dysfunction and perinatal exposure to antiretroviral nucleoside analogues. *Lancet, 354*, 1084–1089.

35. Gerschenson, M., Erhart, S. W., Paik, C. Y., St Claire, M. C., Nagashima, K., Skopets, B., Harbaugh, S. W., Harbaugh, J. W., Quan, W., & Poirier, M. C. (2000). Fetal mitochondrial heart and skeletal muscle damage in Erythrocebus patas monkeys exposed *in utero* to 3'-azido-3'-deoxythymidine. *Aids Research and Human Retroviruses, 16*(7), 635–644.

36. De Souza, R. S. (2000). Effect of prenatal zidovudine on disease progression in perinatally HIV-1-infected infants. *Journal of Acquired Immune Deficiency Syndromes, 24*(2), 154–161.

37. Kuhn, L., Abrams, E. J., Weedon, J., Lambert, G., Schoenbaum, E. E., Nesheim, S. R., Palumbo, P., Vink, P. E., & Bulterys, M. (2000). Disease progression and early viral dynamics in human immunodeficiency virus-infected children exposed to zidovudine during prenatal and perinatal periods. *Journal of Infectious Diseases, 182*, 104–111.

38. World Health Organization. (2001). Global situation of the HIV/AIDS pandemic, end 2001, part I. *Weekly Epidemiologic Reviews, 76*(49), 381–384.

39. Kaiser, J. D., Campa, A. M., Ondercin, J. P., Leoung, G. S., Pless, R. F., & Baum, M. K. (2006). Micronutrient supplementation increases CD4 count in HIV-infected individuals on highly active antiretroviral therapy: A prospective, double-blinded, placebo-controlled trial. *Journal of Acquired Immune Deficiency Syndromes, 42*(5), 523–528.

40. Jariwalla, R. J. (2004). Nutrients as modulators of immune dysfunction and dyslipidemia in AIDS. In R. Watson (Ed.), *AIDS and heart disease* (pp. 163–175). New York, NY: Marcek Dekker Inc.

B. EFFECTS OF INDIVIDUAL VEGETABLES ON HEALTH

41. Muller, F., Svardal, A. M., Nordoy, I., Berge, R. K., Aukrust, P., & Froland, S. S. (2000). Virological and immunological effects of antioxidant treatment in patients with HIV infection. *European Journal of Clinical Investigation*, 30(10), 905–914.

42. Look, M. P., Rockstroh, J. K., Rao, G. S., Barton, S., Lemoch, H., Kaiser, R., Kupfer, B., Sudhop, T., Spengler, U., & Sauerbruch, T. (1998). Sodium selenite and N-acetylcysteine in antiretroviral-naive HIV-1-infected patients: A randomized, controlled pilot study. *European Journal of Clinical Investigation*, 28, 389–397.

43. Akerlund, B., Jarstrand, C., Lindeke, B., Sönnerborg, A., Akerblad, A. C., & Rasool, O. (1996). Effect of N-acetylcysteine (NAC) treatment on HIV-1 infection: a double-blind placebo-controlled trial. *European Journal of Clinical Investigation*, 50, 457–461.

44. Fuchs, J., Schofer, H., Milbradt, R., Freisleben, H. J., Buhl, R., Siems, W., & Grune, T. (1993). Studies on lipoate effects on blood redox state in human immunodeficiency virus infected patients. *Arzneimittelforschung*, 43, 1359–1362.

45. Abrams, B., Duncan, D., & Hertz-Picciotto, I. (1993). A prospective study of dietary intake and acquired immune deficiency syndrome in HIV-seropositive homosexual men. *Journal of Acquired Immune Deficiency Syndromes*, 8, 949–958.

46. Tang, A. M., Graham, N. M., Kirby, A. J., McCall, L. D., Willett, W. C., & Saah, A. J. (1993). Dietary micronutrient intake and risk of progression to acquired immunodeficiency syndrome (AIDS) in human immunodeficiency virus type 1 (HIV-1)-infected homosexual men. *American Journal of Epidemiology*, 138, 937–951.

47. Fawzi, W. W., Msamanga, G. I., Spiegelman, D., Wei, R., Kapiga, S., Villamor, E., Mwakagile, D., Mugusi, F., Hertzmark, E., Essex, M., & Hunter, D. J. (2004). A randomised trial of multivitamin supplements and HIV disease progression and mortality. *New England Journal of Medicine*, 351, 23–32.

48. FANTA. (2004). *HIV/AIDS: A guide for nutritional care and support.* (2nd ed.). Washington, DC: Academy for Educational Development Food and Nutrition Technical Assistance Project.

49. Jiamton, S., Pepin, J., Suttent, R., Filteau, S., Mahakkanukrauh, B., Hanshaoworakul, W., Chaisilwattana, P., Suthipinittharm, P., & Shetty, P. (2003). A randomized trial of the impact of multiple micronutrient supplementation on mortality among HIV-infected individuals living in Bangkok. *AIDS*, 17(17), 2461–2469.

50. Clerici, M., & Shearer, G. (1993). A TH1 to TH2 switch is a critical step in the etiology of HIV infection. *Immunology Today*, 14, 107–111.

51. Mannar, V., & Bellamy, C. (2004). *Vitamin and mineral deficiency: A global damage assessment report.* New York: UNICEF.

52. Harakeh, S., Jariwalla, R. J., & Pauling, L. (1990). Suppression of human immunodeficiency virus replication by ascorbate in chronically and acutely infected cells. *Proceedings of the National Academy of Sciences of the United States of America*, 87, 7245–7249.

53. Harakeh, S., & Jariwalla, R. J. (1991). Comparative study of the anti-HIV activities of ascorbate and thiol-containing reducing agents in chronically HIV-infected cells. *American Journal of Clinical Nutrition*, 54(6 Suppl.), 1231S–1235S.

54. Roederer, M., Staal, F. J., Raju, P. A., Ela, S. W., Herzenberg, L. A., & Herzenberg, L. A. (1990). Cytokine-stimulated human immunodeficiency virus replication is inhibited by N-acetyl-L-cysteine. *Proceedings of the National Academy of Sciences of the United States of America*, 87(12), 4884–4888.

55. Allard, J. P., Aghdassi, E., Chau, J., Tam, C., Kovacs, C. M., Salit, I. E., & Walmsley, S. L. (1998). Effects of vitamin E and C supplementation on oxidative stress and viral load in HIV-infected subjects. *AIDS*, 12, 1653–1659.

56. Herzenberg, L. A., De Rosa, S. C., Dubs, J. G., Roederer, M., Anderson, M. T., Ela, S. W., Deresinski, S. C., & Herzenberg, L. A. (1997). Glutathione deficiency is associated with impaired survival in HIV disease. *Proceedings of the National Academy of Sciences of the United States of America*, 94(5), 1967–1972.

57. Cathcart, R. F. (1984). Vitamin C in the treatment of the acquired immune deficiency syndrome (AIDS). *Medical Hypotheses*, 14, 423–433.

58. de la Asuncion, J. G., del Olmo, M. I., Saster, J., Millan, A., Pellin, A., Pallardo, F. V., & Vina, J. (1998). AZT treatment induces molecular and ultrastructural oxidative damage to muscle mitochondria. Prevention by antioxidant vitamins. *Journal of Clinical Investigation*, 102, 4–9.

59. Hurwitz, B. E., Klaus, J. R., Llabre, M. M., Gonzalez, A., Lawrence, P. J., Maher, K. J., Greeson, J. M., Baum, M. K, Shor-Posner, G., Skyler, J. S., & Schneiderman, N. (2007). Suppression of the human immunodeficiency virus type 1 viral load with selenium supplementation. *Archives of Internal Medicine*, 167, 148–154.

60. Jariwalla, R. J., Lalezari, J., Cenko, D., Mansour, S. E., Kumar, A., Gangapurkar, B., & Nakamura, D. (2008). Restoration of blood total glutathione status and lymphocyte function following α-lipoic acid supplementation in patients with HIV infection. *Journal of Alternative and Complementary Medicine.*, 14(2), 139–146.

Tomatoes, Tomato Products, and Lycopene in Prevention and Therapy of Prostate Diseases – Is There Evidence from Intervention Studies for Preventive and for Therapeutic Effects?

Sabine Ellinger

Department of Nutrition and Food Science – Nutritional Physiology, University of Bonn, Germany

1. INTRODUCTION

Prostate cancer (PCA) is the most common cancer in men in Europe [1] and in the USA [2]. In the USA, 186,320 new diagnosed cases (25% of all cancers) and 28,660 deaths from PCA are expected for 2008. Even though mortality from PCA has been decreasing [2], it takes second place in the USA [2] and third place in Europe [1] of all causes of cancer death in men. Benign prostate hyperplasia (BPH) is the most common benign neoplasm in men [3]. This raises the question if there are any dietary measures with proven evidence for the prevention and therapy of these prostate diseases.

Based on epidemiological data, regular consumption of tomatoes, tomato products (TTP), and isolated lycopene (the main carotenoid in TTP [4]) is postulated to contribute to the prevention of PCA [5] whereas epidemiological data for the prevention of BPH are scarce. Does the scientific literature provide evidence for a therapeutic benefit, investigated by intervention studies with patients suffering from BPH and PCA? There is a body of *in vitro* evidence that lycopene protects from DNA damage by increasing antioxidant protection. Moreover, lycopene induces apoptosis and phase II enzymes and inhibits the IGF-1 pathway which might prevent the development and progression of PCA [6]. Furthermore, lycopene reduces the synthesis of 5-α-dihydrotestosterone stimulating the growth of the prostate gland [7]. This would also be desirable for the prevention and therapy of BPH [8].

The protective effect of TTP and/or lycopene on the development of PCA can only be verified by intervention studies. Since an intervention study with the endpoint PCA would last several years, the benefit from diet or nutrients can only be assessed by suitable surrogate markers which are associated with PCA such as DNA damage and changes in the insulin-like growth factor I (IGF-1) signaling pathway [9]. The therapeutic value of TPP and/or lycopene for treatment of PCA was evaluated by consideration of intervention studies in which clinically or prognostically relevant parameters [prostate specific antigen (PSA), metastasis, symptoms, mortality] were measured. The same procedure was applied for the therapy of BPH. Since data on BPH prevention from intervention studies with healthy men on PSA as a surrogate marker for BPH are not available so far, epidemiological studies were considered instead.

Thus, the aim of this review was to judge the evidence of beneficial effects of TTP and lycopene in the prevention and therapy of BPH and PCA from human intervention studies considering the quantity and quality of the studies and the consistency of the results.

2. METHODS

The scientific literature was searched in Medline using the following key words: consumption/ingestion of tomatoes, tomato products, lycopene, prostate cancer, benign prostate hyperplasia, DNA damage, DNA strand breaks, 8-hydroxydesoxyguanosine (8-OHdG), oxidized (nucleo)bases, prostate specific antigen (PSA), IGF-1, prevention, therapy, intervention study.

2.1 Benign Prostate Hyperplasia

2.1.1 Prevention

The impact of lycopene-rich vegetables and lycopene on the prevention of BPH was investigated in two cohort studies [10,11] and in one case-control study [12]. Both cohort studies were performed in the USA with 24,829 [10] and 4770 [11] men and a follow-up of 9 [11] and 14 [10] years, respectively. The case-control study was done in Italy with 1369 patients and 1451 age-matched controls [12]. In all studies, fruit and vegetable consumption was assessed by a validated food frequency questionnaire [10–12]. The results of Rohrmann et al. show that neither total intake of lycopene-rich vegetables (sum of tomatoes, tomato juice, water melon, grapefruit, and pizza) nor the intake of lycopene affect the risk for BPH [10]. This could also be observed in the case-control study of Tavani et al. [12] for lycopene. In addition to surgical treatment [12], lower urinary tract symptoms and medical treatment served as criteria for BPH in these two studies [10]. In the cohort study of Kristal et al. [11], a tendency for lower risk could be observed if the incidence of BPH was defined by treatment with α-blockers, finasteride, surgical intervention, or clinical symptoms [international prostate symptom score (IPSS) ≥ 15]. If physicians' diagnosis of BPH was defined as the endpoint, the association became significant [11]. Since the average intake of lycopene was comparable in the studies of Tavani et al. (7.5 mg/d), Rohrmann et al. (8.6 mg/d) and Kristal et al. (6.9 mg/d), the different definitions of the endpoint may explain the divergent results. Thus, the question whether the ingestion of TTP and/or lycopene contributes to prevention of BPH cannot yet be evaluated.

2.1.2 Therapy

Table 23.1 provides an overview on intervention studies investigating the effect of TTP or lycopene intake on prognostically relevant parameters in patients with BPH.

The therapeutic benefit of tomato products was investigated only in a single study in which PSA level decreased by 11% after daily

TABLE 23.1 Effect of Tomato Paste or Lycopene Ingestion on Clinically and Prognostically Relevant Parameters in BPH Patients

	N	Supplementation	Additional Intake of Lycopene Per Day	Period of Intervention	Study Design	Parameter	Results
Edinger and Koff, 2006 [13]	43	Tomato paste, 50 g/d	13 mg	10 weeks	Not controlled	PSA	↓ Tomato paste
Mohanty et al., 2005 [14]	40	Lycopene	8 mg	12 months	Randomized controlled[a] parallel group design	PSA	↓ V, ↑ P[b]
Schwarz et al., 2008 [15]	40	Lycopene	15 mg	6 months	Double-blind placebo-controlled randomized parallel group design	PSA IGF-1 IGF-BP-3 Size of prostate gland Symptoms (IPSS points[c])	↓ V, ∅ P ∅ V, ∅ P ∅ V, ∅ P ↓ V, ↑ P ↓ V, ↑ P

[a]Patients in the control group were advised to renounce tomatoes and melons.
[b]Without statistical evaluation
[c]American Urological Association questionnaire to measure clinical symptoms and quality of life index.
IGF-1, insulin-like growth factor-1; IGFBP-3, insulin-like growth factor binding protein-3; IPSS, international prostate symptom score; P, placebo group; PSA, prostate specific antigen; V, treated group; ↓ significant decrease; ↑ significant increase; ∅ no significant changes.

ingestion of 50 g tomato paste for 10 weeks (equating to ingestion of 13 mg lycopene) [13]. However, the lack of a control group impairs the significance of this study. Since further studies are not available, the therapeutic effect of tomato product consumption in patients with BPH cannot be judged so far.

The impact of isolated lycopene was investigated in two intervention trials. In the study of Mohanty et al. [14], ingestion of 8 mg/d lycopene for 1 year by patients with manifest BPH and high grade prostatic intraepithelial neoplasia (HGPIN) decreased PSA by 42% in the treated group. In control patients, who were advised not to eat tomatoes and melons, PSA even increased. Even if the significance of these results cannot be evaluated objectively as statistical analysis is missing, the results of Mohanty et al. are confirmed by a randomized, double-blind, placebo-controlled study of Schwarz et al. (2008) in which daily ingestion of 15 mg lycopene for 6 months lowered PSA (mean change 11%) and prevented further enlargement of the prostate gland as assessed by digital rectal examination and trans-rectal ultrasonography [15]. Due to the consistent results and the excellent design of this powered study, suggestive evidence for a therapeutic benefit of lycopene in treatment of BPH seems to exist.

2.2 Prostate Cancer

2.2.1 Prevention

Effects on DNA Damage Prevention by Tomatoes and Tomato Products As shown in Table 23.2, reduced DNA damage in untreated leukocytes (DNA strand breaks *in vivo*) could only be observed in an

Table 23.2   Effect of Tomatoes and Tomato Product Ingestion on DNA Damage in Healthy Volunteers

	N	Supplementation	Additional daily intake of carotinoids/vitamins	Period	Study design	Biomarker	Results
Pool-Zobel et al., 1997 [16]	23	TJ 330 ml/d	40.0 mg Lycopene	2 wk	Not controlled	Ox. bases (leukocytes) SB *in vivo* SB vs. H_2O_2 *ex vivo*	Ø TJ; ↓ TJ; Ø TJ
Briviba et al., 2004 [17]	55	TE 3 capsules/d Placebo	4.88 mg Lycopene 0.48 mg Phytoene 0.44 mg Phytofluene 1.181 mg α-tocopherol	2 wk	Double-blind placebo-controlled randomized	SB *in vivo*	Ø TE; SB *in vivo* (mean damage) ↓ TE; % damaged cells, ↑ TE; % undamaged cells (only in non-smokers)
Riso et al., 2006 [18]	26	Soft drink with 6% TE 250 mL/d	5.7 mg Lycopene 3.7 mg Phytoene 2.7 mg Phytofluene 1.0 mg β-carotene 1.8 mg α-tocopherol	26 d	Double-blind placebo-controlled crossover	SB *in vivo*	Ø Soft drink with TE
Riso et al., 1999 [19]	10	TP 60 g/d Tomato free diet	1.5 mg Lycopene 0.6 mg β-carotene	3 wk	Controlled randomized crossover	SB *in vivo* SB vs. H_2O_2 ex vivo	Ø TP Ø Tomato free diet ↓ TP Ø Tomato free diet
Porrini and Riso, 2000 [20]	9	TP 25 g/d	7.0 mg Lycopene 0.3 mg β-carotene	2 wk	Not controlled	SB *in vivo* SB vs. H_2O_2 ex vivo	Ø TP ↓ TP
Porrini et al., 2005 [21]	26	Soft drink with 6% TE 250 mL/d	5.7 mg Lycopene 3.7 mg Phytoene 2.7 mg Phytofluene 1.0 mg β-carotene 1.8 mg α-tocopherol	26 d	Double-blind placebo-controlled crossover	SB vs. H_2O_2 *ex vivo*	↓ Soft drink with TE
Rao and Agarwal, 1998 [22]	19	TS 126 g/d (a) TS 126 g/d (b) TJ 540 mg/d TE 1 capsule/d (a) TE 1 capsule/d (b) Placebo	20.5 mg Lycopene 39.2 mg Lycopene 50.4 mg Lycopene 75.0 mg Lycopene 150.0 mg Lycopene	1 wk	Placebo-controlled randomized crossover	Ox. bases (leukocytes)	Ø TS (a) Ø TS (b) Ø TJ Ø TE (a) Ø TE (b)

Reference	N	Intervention	Carotenoids	Amount	Duration	Design	Outcome	Results
Riso et al., 2004 [24]	12	Mix of raw tomatoes, tomato puree, tomato sauce[b]	lycopene, β-carotene, Vitamin C	8.0 mg, 0.5 mg, 11.0 mg	3 wk	Not controlled	SB vs. Fe²⁺ ex vivo	↓ Mix of raw tomatoes, tomato puree, tomato sauce
Kiokias and Gordon, 2003 [25]	32	TE (24.6 mg) + further carotenoid-rich extracts[c] + fish oil fish oil[a]	Lycopene, β-carotene, Lutein, α-carotene	7.6 mg, 7.6 mg, 7.6 mg, 7.6 mg	3 wk	Double-blind placebo-controlled randomized crossover	Ox. bases (fasting urine, (8-OHdG:creatinine ratio)	∅ TE + fish oil, ↑ fish oil
Hadley et al., 2003 [26]	60	TS (concentrate) 300 mL/d	Lycopene, β-carotene, Lutein, Zeaxanthine	35 mg, 0.7 mg, 0.2 mg, 0.3 mg	15 d	Randomized	Ox. bases (fasting urine, 8-OHdG)	∅ TS (concentrate), ∅ TS (ready to serve), ∅ vegetable juice
		TS (ready to serve) 320 mL/d	Lycopene, β-carotene, Lutein, Zeaxanthine	23 mg, 0.5 mg, 0.2 mg, 0.3 mg				
		Vegetable juice 340 mL/d	Lycopene, β-carotene, Lutein, Zeaxanthine	25 mg, 1.4 mg, 0.4 mg, 0.3 mg				
Porrini et al., 2002 [28]	9	TP 25 g/d + spinach 150 g/d; spinach 150 g/d	Lycopene, β-carotene	7.0 mg, 0.3 mg	3 wk	Spinach as control	SB vs. H₂O₂ ex vivo	↓ TP + spinach, ↓ spinach

[a]Fish oil was given to induce oxidative challenge as supplementation with β-carotene in smokers was not considered ethical to use in these subjects.
[b]Subjects consumed on 3 days per week tomato sauce and tomatoes, respectively, and tomato puree twice per week.
[c]6.3 mg palm oil carotene extract, 2.0 mg marigold extract, 3.7 mg paprika extract, 3.7 mg bixin.
d, days; Ox. bases, oxidized DNA bases; SB, DNA-strand breaks in leukocytes; T, raw tomatoes; TE, tomato extract; TJ, tomato juice; TP, tomato puree; TS, tomato sauce; wk, week(s); ↓ significant reduction; ↑ significant increase; ∅ no significant change.

B. EFFECTS OF INDIVIDUAL VEGETABLES ON HEALTH

uncontrolled study after ingestion of tomato juice [16], whereas further studies with a controlled study design [17–19] did not show any changes in the extent of DNA damage after consumption of TTP [17–21]. The content of oxidized DNA bases in leukocytes did not change after daily consumption of tomato juice [16,22], spaghetti sauce, or tomato extract [22]. However, since DNA strand breaks *in vivo* as well as oxidized DNA bases in cells or tissues reflect only the steady state determined by the balance between damage and repair rates [23], conclusions from such studies on the effect of dietary measures on DNA damage cannot be drawn.

In contrast to these parameters, the resistance of DNA against strand breaks, induced by oxidants *ex vivo*, is a sensitive marker to evaluate changes in the antioxidant protection of DNA by an intervention [23]. Its resistance against strand breaks induced by oxidative challenge can be investigated by treatment of single cells in suspension (e.g. leukocytes) with oxidants *ex vivo* (e.g. hydrogen peroxide or ferrous ions). As shown by the majority of studies [19–21,24], the *ex vivo* resistance of leukocytes to DNA strand breaks increased after daily ingestion of tomatoes or tomato products for 2–4 weeks (Table 23.2). Only one study with an uncontrolled design did not show any changes [16]. Under conditions of oxidative stress induced by ingestion of fish oil, the 8-OHdG to creatinine ratio in urine, a reliable parameter for the rate of DNA damage (in contrast to oxidized bases in cells or tissue) [23], did not increase after ingestion of a carotenoid extract containing 60% tomato extract as observed in the control group [25]. Thus, supplementation seems to increase the resistance of cells against oxidative DNA damage *in vivo*. In contrast, Hadley et al. did not find any changes in urinary excretion of 8-OHdG [26]. However, they did not relate 8-OHdG to creatinine which has shown to be the main predictor of urinary 8-OHdG [27]. Therefore, the

validity of the results on urinary excretion of 8-OHdG in the study of Hadley et al. remains questionable.

Despite differences in sample size (9–60 participants), study design (placebo-/controlled [17–19,21,22,25,28] vs. uncontrolled [20, 24]), diet (no dietary guidelines [21,25] vs. restrictions on carotenoid intake [19,20,24,28]), and lycopene intake (1.5–8 mg/d), data are consistent that regular ingestion of tomato products increases the resistance of DNA to oxidative damage *ex vivo* [19–22,24] and *in vivo* [25]. Due to the correlation between DNA damage in leukocytes and in prostate tissue [29], similar protection of the DNA in prostate tissue can be expected. Since oxidative stress is involved in the development of PCA [30,31], regular consumption of TTP probably protects against development of PCA.

Effects on DNA Damage by Lycopene   In all studies, lycopene concentration in plasma [17,19–22,24–26] as well as in leukocytes [16,20,21,24] increased after ingestion of TTP with lycopene doses of 1.5–150 mg/d. Regarding non-lycopene carotenoids which were, however, ingested in comparably lower amounts (see Table 23.2), results are divergent [17,20, 21,24,25] except for phytofluene and phytoene which increased in plasma [17,21] and in leukocytes [21] as well as lycopene did.

To investigate whether lycopene is responsible for increased resistance of DNA to oxidative-induced damage after ingestion of TTP, isolated lycopene was supplemented in five studies (6.5–30 mg/d, 1–16 weeks) (Table 23.3). However, changes in the resistance of DNA to damage *ex vivo* [32–35] and in the content of oxidized nucleobases in leukocytes [36] and urine [32] did not occur if daily doses ≤15 mg/d were ingested. Only higher doses of lycopene (30 mg/d, 8 weeks) reduced H_2O_2-induced strand breaks in leukocytes which was accompanied by a decreased 8-OHdG to creatinine ratio

TABLE 23.3   Effect of Lycopene and Further Carotenoids on DNA Damage in Healthy Volunteers

	N	Supplementation	Additional Daily Intake of Carotenoids	Period	Study Design	Biomarker	Results
Devaraj et al., 2008 [32]	77	Lycopene Placebo	6.5 mg 15 mg 30 mg	8 wk	Double-blind placebo-controlled randomized parallel group comparison	SB vs H_2O_2 *ex vivo* Ox. bases (24 h urine, (8-OHdG: creatinine ratio)[a]	↓ 30 mg lycopene, ∅ 6.5 and 15 mg lycopene, ∅ placebo ↓ 30 mg lycopene ∅ placebo
Astley et al., 2004 [33]	28	Lycopene β-carotene Lutein Placebo	15 mg 15 mg 15 mg	4 wk	Double-blind placebo-controlled randomized crossover	SB *in vivo* SB vs H_2O_2 *ex vivo*	∅ lycopene, lutein, placebo, ↑ β-carotene ∅ All interventions
Torbergsen and Collins, 2000 [34]	8	Lycopene β-carotene Lutein	15 mg 15 mg 15 mg	1 wk	Crossover	SB *in vivo*	∅ All interventions
Zhao et al., 2006 [35]	37	Lycopene β-carotene Lutein Combination Placebo	12 mg 12 mg 12 mg 3×4 mg	8 wk	Double-blind placebo-controlled randomized parallel group comparison	SB vs H_2O_2 *ex vivo* SB *in vivo* SB vs H_2O_2 *ex vivo*	∅ lycopene, lutein, ↓ β-carotene ↓ All interventions ∅ All interventions
Collins et al., 1998 [36]	40	Lycopene α-/β-carotene Lutein Placebo	15 mg 15 mg 15 mg	16 wk	Placebo-controlled randomized parallel group comparison	Ox. Bases (leukocytes)	∅ All interventions

[a]Investigation of 8-OHdG to creatinine ratio only in the high dose and in the placebo group.
8-OHdG, 8-hydroxydesoxyguanosine; Ox. bases, oxidized DNA bases; SB, DNA strand breaks in leukocytes; wk, week(s); ↓ significant decrease; ↑ significant increase; ∅ no significant changes.

in 24 h urine [32]. Since ingestion of tomato products with lycopene doses above 30 mg/d for only one [22] or two weeks [16,26] did not show similar results, the longer period of supplementation in the study of Devaraj et al. [32] (8 weeks) may explain the beneficial effects.

Since DNA damage was not reduced in all these studies with well-controlled design after supplementation of isolated lycopene at moderate doses (12–15 mg/d) [32–36] which can be easily ingested with a standard serving of TTP (Table 23.3), further ingredients seem to be responsible for increased protection of cells against DNA damage after consumption of TTP. Therefore, similar trials were performed with β-carotene and lutein (Table 23.2). Except for β-carotene which reduced DNA damage in a single study [34], effects did not occur in further studies with a placebo-controlled design [33,35,36]. Even combined supplementation of lycopene, β-carotene, and lutein did not affect the resistance of leukocyte DNA to oxidative challenge [35]. Thus, in contrast to TTP, the ingestion of isolated carotenoids in comparable amounts does not protect against DNA damage induced by oxidative challenge *ex vivo*. Moreover, a reduction in DNA strand breaks *in vivo* [33–35] and in oxidized bases in leukocytes did not occur [36]. The weak contribution of lycopene, phytoene, and phytofluene concentration in leukocytes on DNA damage *ex vivo* (regression coefficient 0.3–0.34), determined after daily ingestion of tomato products [21], underlines that protection from DNA damage by TTP cannot be attributed to a single carotenoid, but rather to a broad spectrum of bioactive compounds (e.g. polyphenols [37,38] as well as vitamin C and E [39]) (see also Table 23.2) which may exert synergistic effects with carotenoids [40]. Thus, there is some evidence from experimental studies that protection from PCA can only be achieved by ingestion of TTP, and not by isolated ingestion of either lycopene, β-carotene, or lutein.

Effects on IGF Pathway by Tomato Products Within a double-blind, placebo-controlled trial, the concentration of lycopene and further carotenoids increased in plasma after ingestion of a tomato drink (250 mL/d, 26 days). Although the plasma levels of lycopene correlated inversely with those of IGF-1 (IGF-1 is accompanied by an increased risk for PCA [41]), IGF-1 concentration did not change after regular consumption of the tomato drink despite an increase in plasma lycopene. However, if only participants with a distinct increase in plasma lycopene level ($>0.25 \mu mol/$L and $>100\%$, respectively) were considered (high responders), a significant reduction of IGF-1 in this subgroup could be observed. However, the concentration of IGF binding protein 3 (IGFBP-3), which is known to reduce the activity of IGF-1 and the IGF-1 to IGFBP-3 ratio, did not change in high responders [18]. Since further studies are not available, the impact of regular consumption of TTP on

protection from PCA by modulation of the IGF pathway cannot yet be evaluated.

2.2.2 Therapy

Effects Of Tomato Products The effect of tomato consumption, including tomato products, on clinically and prognostically relevant parameters in patients with PCA has been investigated in four different studies so far (Table 23.4).

In the study of Chen et al. (2001) [42] (also published by Bowen et al. [29] and Kim et al. [43]), 32 patients consumed a pasta meal with 30 g tomato sauce (corresponding to 30 mg lycopene) daily for 3 weeks. Lycopene concentration increased in plasma ($+97\%$) and in prostate tissue ($+192\%$). PSA in serum as well as the content of oxidized nucleobases in leukocytes and in the prostate tissue decreased. After intervention, the concentration of oxidized nucleobases in the prostate tissue was lower compared to patients who did not receive any pasta meals. Furthermore, apoptotic cells in tumor areas increased which may suppress the progression of the disease [43]. Since a correlation between lycopene concentration and DNA damage in prostate tissue did not occur after intervention [29], the increased protection from DNA damage cannot be explained only by lycopene.

In the study of Kucuk et al. [44], 15 patients with PCA, also in stage T1/T2 disease, ingested tomato extract daily for 3 weeks. The lycopene concentration in prostate tissue was higher in the treated compared to the control group ($N = 11$) after intervention, but the PSA level as well as oxidized nucleobases in leukocytes did not change despite the intake of the same dose of lycopene (30 mg/d) as in the study of Chen et al. [42]. Since lycopene in plasma did not increase [44], the bioavailability of carotenoids from the tomato extract might have been lower than from the tomato sauce ingested with the pasta meal in the study of Chen et al. [42]. This may be the reason for the lack of changes in

TABLE 23.4 Effect of Tomato Products on Clinically and Prognostically Relevant Parameters in PCA Patients

	N	Supplementation	Additional Daily Lycopene Intake	Period	Stage of Disease	Study Design	Parameter	Results
Bowen et al., 2002 [29]; Chen et al., 2001 [42]; Kim et al., 2003 [43]	32	1 pasta meal/d with tomato sauce 200 g/d	30 mg	3 wk	T1/T2	Not controlled	PSA Ox. bases (leukocytes) Ox. bases (prostate gland) Apoptosis tumor tissue	↓ V ↓ V ↓ V, V <R[b] ↑ V
Kucuk et al., 2002 [44]	26	Tomato extract 2 capsules/d	30 mg	3 wk	T1/T2	Controlled[a] parallel group comparison	PSA Ox. bases (leukocytes) IGF-1 IGFBP-3	∅ V, ∅ C ∅ V, ∅ C ↓ V, ↓ C ↓ V, ↓ C
Grainger et al.,2008 [45]	41	Tomato products[c] or soy for 4 wk; thereafter tomato products + soy for 4 weeks for both groups	≥25 mg 43 ± 15 mg	8 wk	Relapse 50% hormone therapy, 50% hormone naive[d]	Not controlled randomized two arms parallel group design	PSA response (lower PSA PSA-doubling >9 months) IGF-1	25% Tomato products 43% Soy; T = S 30 vs. 48%, T = S ∅
Jatoi et al., 2007 [46]	46	Tomato puree or juice	30 mg	4 mo	Androgen-independent PCA	Not controlled	PSA response (↓ ≥50% vs. t_0 for ≥8 weeks) metastases	∅ ∅

[a]Controls should maintain dietary habits, but consume five portions of fruit and vegetables per day.
[b]Reference group: Seven patients who did not receive the pasta meal and who were only investigated once.
[c]E.g. juice, raw, catsup, salsa, pizza, sauce, soup, chilli.
[d]Both subgroups were evenly distributed between the two groups.
C, control group; IGF-1, insulin-like growth factor-1; IGFBP-3, insulin-like growth factor binding protein-3; mo, months; Ox. bases, oxidized DNA bases; PSA, prostate specific antigen; wk, weeks; R, reference group; V, treated group; ↓ significant decrease; ↑ significant increase; ∅ no significant changes.

oxidized nucleobases of leukocytes and in PSA after supplementation of tomato extract in the study of Kucuk et al. [44]. The decrease in IGF-1 and IGFBP-3 in both treated and control groups in the study of Kucuk et al. [44] is not surprising as patients from the control group were advised to consume five servings of fruit and vegetable per day which might also increase the intake of carotenoids. Thus, the effects achieved from tomato extract and high consumption of

fruit and vegetables seems to have comparable effects on the IGF pathway.

In contrast to Kucuk et al., Grainger et al. [45] did not observe any changes in IGF-1 after regular ingestion of diverse tomato products and a comparably higher mean ingestion of lycopene (43 mg vs. 30 mg), but a decrease of PSA. However, due to the lack of a control group, the effect of tomato products on IGF-1 and PSA remains unclear.

TABLE 23.5   Effect of Lycopene on Clinically and Prognostically Relevant Parameters in PCA Patients

	N	Supplementation	Additional daily Lycopene Intake	Period	Stage of Disease	Study Design	Parameter	Results
Clark et al., 2006 [47]	36	Lycopene	15 mg 30 mg 45 mg 60 mg 90 mg 120 mg	1 yr	Relapse[a] No metastases	Not controlled six arms randomized parallel group comparison	PSA response ($\downarrow \geq 50\%$ vs. t_0)	$\varnothing$ (all doses)
Ansari & Gupta, 2003 [48]	54	Orchidectomy + lycopene vs. orchidectomy without lycopene	4 mg	2 yr	Bone metastases (M1b, D2)	Two arms-randomized-controlled-parallel group comparison	PSA PSA response:	$\downarrow$ V, $\varnothing$ C
							• complete ($\downarrow$ to $<4\,\text{ng/mL}$)	V > C (78 vs 40%)
							• partial ($\downarrow \geq 50\%$ vs. t_0)	V vs. C $\varnothing$
							• progression ($\uparrow \geq 25\%$ vs. t_0)	V < C (7 vs 25%)
							Response bone metastases:	
							• complete (no metastases)	V > C (25 vs. 15%)
							• partial ($\downarrow \geq 50\%$ vs. t_0)	V vs. C $\varnothing$
							• progression (new metastases)	V < C (7 vs. 15%)
							Lower urinary tract symptoms (peak flow rate, voiding symptoms)	$\uparrow$ V, $\varnothing$ C
							Mortality rate (2 yr)	V < C (13 vs. 22%)

Ansari & Gupta, 2004 [49]	20	Lycopene	10 mg	3 mo	Hormone-resistant PCA	Not controlled	PSA response:	
							• complete (↓ to <4 ng/mL and disappearance of any signs of disease for ≥8 wk)	5%
							• partial (↓ ≥50% vs. t_0 for ≥8 wk without ↑ ECOG PS and ↓ BP)	30%
							• stable disease (↓ PSA <50% vs. t_0 or ↑ PSA <25% without ↑ ECOG PS and/ or BP for ≥8 wk)	50%
							• progression (↑ PSA ≥50% vs. t_0, metastasis and/or ↑ ECOG-PS and/ or ↑ BP)	15%
							LUTS	↓ 61%

[a]Triple PSA increase within 3 months.

N, number of participants; C, control group; BP, bone pain; ECOG PS, Eastern Cooperative Oncology Group performance status; LUTS, lower urinary tract symptoms; V, treated group; wk, week; ↓ significant decrease, ↑ significant increase, ⊘ no significant changes and differences, respectively.

B. EFFECTS OF INDIVIDUAL VEGETABLES ON HEALTH

PSA response and metastases did not change in the study of Jatoi et al. after daily supplementation of tomato puree or juice for 4 months, providing 30 mg/d lycopene [46]. As to Grainger et al., conclusions on the effect of tomato products on prognostically relevant parameters cannot be drawn due to the uncontrolled study design.

Because of the divergent results on prognostically relevant parameters and methodological weakness (lack of adequate control group) in all these studies [29,42–46], evidence for a therapeutic benefit of regular consumption of TTP in patients with PCA is scarce. Intervention studies with a randomized, (placebo-) controlled study are urgently necessary to judge the evidence.

Effects of Lycopene The effect of lycopene on prognostically relevant parameters in patients with PCA was investigated only in three intervention trials (Table 23.5). Clark et al. [47] measured PSA after ingestion of 15, 30, 45, 60, 90 or 120 mg/d lycopene for 1 year. However, a decline in PSA level of at least 50% versus baseline – according to Ansari and Gupta (2003) defined as partial PSA response [48] – did not occur. In a controlled, randomized trial from Ansari and Gupta [48] with orchidectomized patients suffering from bone metastases, mean PSA level decreased after regular ingestion of 4 mg/d lycopene for 2 years compared with controls who did not receive lycopene. Even if the percentage of patients with partial response on PSA and metastases (at least 50% reduction vs. baseline) was not different between the groups, a complete response (PSA reduction to values <4 ng/mL, no metastases detectable) occurred more often in the treated group than in the control group (78% vs. 40%), whereas progression of disease (PSA increase ≥25%: 7% vs. 25%; new metastases: 7% vs. 15%) was less frequent in the treated compared to the control group. This might also explain the lower mortality in treated patients than in controls (13% vs. 22%). The peak flow rate as well as further voiding symptoms improved in the treated group. The longer period of supplementation of lycopene in this study (2 years) compared to other studies (ingestion of tomato products or lycopene for weeks or months) with unclear or only partial effects [42,44] may be relevant for the results. Perhaps therapeutic effects of tomato products or lycopene can only be observed in the long term. Thus, daily intake of relatively low doses of lycopene (4 mg/d), which can also be easily ingested by consumption of TTP (Table 23.6), may be an efficient measure within adjuvant therapy of PCA. The question if patients with androgen-resistant PCA would also profit from such measures remains open, as the studies of Jatoi et al. [46] and Ansari and Gupta [49] were uncontrolled and did not allow final conclusions.

Thus, in contrast to the prevention of PCA, there is at least a little evidence that regular ingestion of lycopene in doses of 4 mg/d, which can easily be achieved by a standard

TABLE 23.6 Lycopene Intake by Consumption of Tomatoes and Tomato Products

	Standard Serving	Lycopene Intake (mg)[a]
Tomatoes	1 piece, medium, 65 g	2.0
Tomato puree	1 tablespoon, 20 g	3.3
Tomato catsup	1 tablespoon, 20 g	3.4
Tomato juice	1 glass, 200 g	18.7
Tomato soup	1 dish, 200 g	21.8
Tomato sauce	4 tablespoons, 60 g	1.0
Pizza	1 piece, medium, 500 g	16.3[b]
Pasta with tomato sauce	1 dish, 250 g	7.9

[a]Calculation based on carotenoid database of US. Department of Agriculture and National Cancer Institute (Holden et al., *J. Food Comp. Anal.* 1999) by consideration of commonly consumed servings (according to *Bundeslebensmittelschluessel* II.3.2).
[b]Mean value of diverse pizza.

serving of TTP (Table 23.6), may be useful in the therapy of PCA.

3. SUMMARY

The role of TTP in the prevention of BPH cannot yet be evaluated, but there is some evidence that regular consumption of TTP may protect against the development of PCA. Tomato products are probably useful in the therapy of BPH, whereas their benefit in therapy of PCA is not clear as well-controlled studies are lacking.

Supplementation of isolated lycopene does not protect against DNA damage and thus against development of PCA – in contrast to TTP. However, in the therapy of BPH and PCA, supplementation of lycopene in relatively low doses which are also ingested with common servings of TTP may be useful in the long term.

References

1. Ferlay, J., Autier, P., Boniol, M., Heanue, M., Colombet, M., & Boyle, P. (2007). Estimates of the cancer incidence and mortality in Europe in 2006. *Annals of Oncology, 18,* 581–592.
2. Jemal, A., Siegel, R., Ward, E., Hao, Y., Xu, J., Murray, T., & Thun, M. J. (2008). Cancer statistics. *CA – A Cancer Journal for Clinicians, 58,* 71–96.
3. Wei, J. T., Calhoun, E., & Jacobsen, S. J. (2008). Urologic diseases in america project: benign prostatic hyperplasia. *Journal of Urology, 179,* S75–S80.
4. Khachik, F., Carvalho, L., Bernstein, P. S., Muir, G. J., Zhao, D. Y., & Katz, N. B. (2002). Chemistry, distribution, and metabolism of tomato carotenoids and their impact on human health. *Experimental Biology and Medicine (Maywood), 227,* 845–851.
5. Etminan, M., Takkouche, B., & Caamano-Isorna, F. (2004). The role of tomato products and lycopene in the prevention of prostate cancer: a meta-analysis of observational studies. *Cancer Epidemiology Biomarkers & Prevention, 13,* 340–345.
6. Wertz, K., Siler, U., & Goralczyk, R. (2004). Lycopene: modes of action to promote prostate health. *Archives of Biochemistry and Biophysics, 430,* 127–134.
7. Obermuller-Jevic, U. C., Olano-Martin, E., Corbacho, A. M., Eiserich, J. P., van der Vliet, A., Valacchi, G., Cross, C. E., & Packer, L. (2003). Lycopene inhibits the growth of normal human prostate epithelial cells in vitro. *Journal of Nutrition, 133,* 3356–3360.
8. Kaplan, S. A. (2005). Lycopene: modes of action to promote prostate health. *Journal of Urology, 174,* 679.
9. Bowen, P. E. (2005). Selection of surrogate endpoint biomarkers to evaluate the efficacy of lycopene/tomatoes for the prevention/progression of prostate cancer. *Journal of Nutrition, 135,* 2068S–2070S.
10. Rohrmann, S., Giovannucci, E., Willett, W. C., & Platz, E. A. (2007). Fruit and vegetable consumption, intake of micronutrients, and benign prostatic hyperplasia in US men. *American Journal of Clinical Nutrition, 85,* 523–529.
11. Kristal, A. R., Arnold, K. B., Schenk, J. M., Neuhouser, M. L., Goodman, P., Penson, D. F., & Thompson, I. M. (2008). Dietary patterns, supplement use, and the risk of symptomatic benign prostatic hyperplasia: results from the prostate cancer prevention trial. *American Journal of Epidemiology, 167,* 925–934.
12. Tavani, A., Longoni, E., Bosetti, C., Maso, L. D., Polesel, J., Montella, M., Ramazzotti, V., Negri, E., Franceschi, S., & La Vecchia, C. (2006). Intake of selected micronutrients and the risk of surgically treated benign prostatic hyperplasia: a case-control study from Italy. *European Urology, 50,* 549–554.
13. Edinger, M. S., & Koff, W. J. (2006). Effect of the consumption of tomato paste on plasma prostate-specific antigen levels in patients with benign prostate hyperplasia. *Brazilian Journal of Medical and Biological Research, 39,* 1115–1119.
14. Mohanty, N. K., Saxena, S., Singh, U. P., Goyal, N. K., & Arora, R. P. (2005). Lycopene as a chemopreventive agent in the treatment of high-grade prostate intraepithelial neoplasia. *Urologic Oncology, 23,* 383–385.
15. Schwarz, S., Obermuller-Jevic, U. C., Hellmis, E., Koch, W., Jacobi, G., & Biesalski, H. K. (2008). Lycopene inhibits disease progression in patients with benign prostate hyperplasia. *Journal of Nutrition, 138,* 49–53.
16. Pool-Zobel, B. L., Bub, A., Muller, H., Wollowski, I., & Rechkemmer, G. (1997). Consumption of vegetables reduces genetic damage in humans: first results of a human intervention trial with carotenoid-rich foods. *Carcinogenesis, 18,* 1847–1850.
17. Briviba, K., Kulling, S. E., Moseneder, J., Watzl, B., Rechkemmer, G., & Bub, A. (2004). Effects of supplementing a low-carotenoid diet with a tomato extract for 2 weeks on endogenous levels of DNA single strand breaks and immune functions in healthy nonsmokers and smokers. *Carcinogenesis, 25,* 2373–2378.
18. Riso, P., Brusamolino, A., Martinetti, A., & Porrini, M. (2006). Effect of a tomato drink intervention on insulin-like growth factor (IGF)-1 serum levels in healthy

subjects. *Nutrition and Cancer – an International Journal*, *55*, 157–162.

19. Riso, P., Pinder, A., Santangelo, A., & Porrini, M. (1999). Does tomato consumption effectively increase the resistance of lymphocyte DNA to oxidative damage? *American Journal of Clinical Nutrition*, *69*, 712–718.

20. Porrini, M., & Riso, P. (2000). Lymphocyte lycopene concentration and DNA protection from oxidative damage is increased in women after a short period of tomato consumption. *Journal of Nutrition*, *130*, 189–192.

21. Porrini, M., Riso, P., Brusamolino, A., Berti, C., Guarnieri, S., & Visioli, F. (2005). Daily intake of a formulated tomato drink affects carotenoid plasma and lymphocyte concentrations and improves cellular antioxidant protection. *British Journal of Nutrition*, *93*, 93–99.

22. Rao, A. V., & Agarwal, S. (1998). Bioavailability and *in vivo* antioxidant properties of lycopene from tomato products and their possible role in the prevention of cancer. *Nutrition and Cancer-an International Journal*, *31*, 199–203.

23. Loft, S., & Poulsen, H. E. (2000). Antioxidant intervention studies related to DNA damage, DNA repair and gene expression. *Free Radical Research*, *33*(Suppl.), S67–S83.

24. Riso, P., Visioli, F., Erba, D., Testolin, G., & Porrini, M. (2004). Lycopene and vitamin C concentrations increase in plasma and lymphocytes after tomato intake. Effects on cellular antioxidant protection. *European Journal of Clinical Nutrition*, *58*, 1350–1358.

25. Kiokias, S., & Gordon, M. H. (2003). Dietary supplementation with a natural carotenoid mixture decreases oxidative stress. *European Journal of Clinical Nutrition*, *57*, 1135–1140.

26. Hadley, C. W., Clinton, S. K., & Schwartz, S. J. (2003). The consumption of processed tomato products enhances plasma lycopene concentrations in association with a reduced lipoprotein sensitivity to oxidative damage. *Journal of Nutrition*, *133*, 727–732.

27. Pilger, A., Germadnik, D., Riedel, K., Meger-Kossien, I., Scherer, G., & Rudiger, H. W. (2001). Longitudinal study of urinary 8-hydroxy-2'-deoxyguanosine excretion in healthy adults. *Free Radical Research*, *35*, 273–280.

28. Porrini, M., Riso, P., & Oriani, G. (2002). Spinach and tomato consumption increases lymphocyte DNA resistance to oxidative stress but this is not related to cell carotenoid concentrations. *European Journal of Nutrition*, *41*, 95–100.

29. Bowen, P., Chen, L., Stacewicz-Sapuntzakis, M., Duncan, C., Sharifi, R., Ghosh, L., Kim, H. S., Christov-Tzelkov, K., & van Breemen, R. (2002). Tomato sauce supplementation and prostate cancer: lycopene accumulation and modulation of biomarkers of carcinogenesis. *Experimental Biology and Medicine(Maywood)*, *227*, 886–893.

30. Lockett, K. L., Hall, M. C., Clark, P. E., Chuang, S. C., Robinson, B., Lin, H. Y., Su, L. J., & Hu, J. J. (2006). DNA damage levels in prostate cancer cases and controls. *Carcinogenesis*, *27*, 1187–1193.

31. Miyake, H., Hara, I., Kamidono, S., & Eto, H. (2004). Oxidative DNA damage in patients with prostate cancer and its response to treatment. *Journal of Urology*, *171*, 1533–1536.

32. Devaraj, S., Mathur, S., Basu, A., Aung, H. H., Vasu, V. T., Meyers, S., & Jialal, I. (2008). A dose-response study on the effects of purified lycopene supplementation on biomarkers of oxidative stress. *Journal of the American College of Nutrition*, *27*, 267–273.

33. Astley, S. B., Hughes, D. A., Wright, A. J., Elliott, R. M., & Southon, S. (2004). DNA damage and susceptibility to oxidative damage in lymphocytes: effects of carotenoids *in vitro* and *in vivo*. *British Journal of Nutrition*, *91*, 53–61.

34. Torbergsen, A. C., & Collins, A. R. (2000). Recovery of human lymphocytes from oxidative DNA damage; the apparent enhancement of DNA repair by carotenoids is probably simply an antioxidant effect. *European Journal of Nutrition*, *39*, 80–85.

35. Zhao, X., Aldini, G., Johnson, E. J., Rasmussen, H., Kraemer, K., Woolf, H., Musaeus, N., Krinsky, N. I., Russell, R. M., & Yeum, K. J. (2006). Modification of lymphocyte DNA damage by carotenoid supplementation in postmenopausal women. *American Journal of Clinical Nutrition*, *83*, 163–169.

36. Collins, A. R., Olmedilla, B., Southon, S., Granado, F., & Duthie, S. J. (1998). Serum carotenoids and oxidative DNA damage in human lymphocytes. *Carcinogenesis*, *19*, 2159–2162.

37. Shen, Y. C., Chen, S. L., & Wang, C. K. (2007). Contribution of tomato phenolics to antioxidation and down-regulation of blood lipids. *Journal of Agricultural and Food Chemistry*, *55*, 6475–6481.

38. Minoggio, M., Bramati, L., Simonetti, P., Gardana, C., Iemoli, L., Santangelo, E., Mauri, P. L., Spigno, P., Soressi, G. P., & Pietta, P. G. (2003). Polyphenol pattern and antioxidant activity of different tomato lines and cultivars. *Annals of Nutrition and Metabolism*, *47*, 64–69.

39. Campbell, J. K., Canene-Adams, K., Lindshield, B. L., Boileau, T. W., Clinton, S. K., & Erdman, J. W. Jr., (2004). Tomato phytochemicals and prostate cancer risk. *Journal of Nutrition*, *134*, 3486S–3492S.

40. Ellinger, S., Ellinger, J., & Stehle, P. (2006). Tomatoes, tomato products and lycopene in the prevention and

treatment of prostate cancer: do we have the evidence from intervention studies? *Current Opinion in Clinical Nutrition and Metabolic Care, 9,* 722–727.

41. Renehan, A. G., Zwahlen, M., Minder, C., O'Dwyer, S. T., Shalet, S. M., & Egger, M. (2004). Insulin-like growth factor (IGF)-I, IGF binding protein-3, and cancer risk: systematic review and meta-regression analysis. *Lancet, 363,* 1346–1353.

42. Chen, L., Stacewicz-Sapuntzakis, M., Duncan, C., Sharifi, R., Ghosh, L., van Breemen, R., Ashton, D., & Bowen, P. E. (2001). Oxidative DNA damage in prostate cancer patients consuming tomato sauce-based entrees as a whole-food intervention. *Journal of the National Cancer Institute, 93,* 1872–1879.

43. Kim, H. S., Bowen, P., Chen, L., Duncan, C., Ghosh, L., Sharifi, R., & Christov, K. (2003). Effects of tomato sauce consumption on apoptotic cell death in prostate benign hyperplasia and carcinoma. *Nutrition and Cancer-an International Journal, 47,* 40–47.

44. Kucuk, O., Sarkar, F. H., Djuric, Z., Sakr, W., Pollak, M. N., Khachik, F., Banerjee, M., Bertram, J. S., & Wood, D. P. Jr., (2002). Effects of lycopene supplementation in patients with localized prostate cancer. *Experimental Biology and Medicine (Maywood), 227,* 881–885.

45. Grainger, E. M., Schwartz, S. J., Wang, S., Unlu, N. Z., Boileau, T. W., Ferketich, A. K., Monk, J. P., Gong, M. C., Bahnson, R. R., DeGroff, V. L., & Clinton, S. K. (2008). A combination of tomato and soy products for men with recurring prostate cancer and rising prostate specific antigen. *Nutrition and Cancer-an International Journal, 60,* 145–154.

46. Jatoi, A., Burch, P., Hillman, D., Vanyo, J. M., Dakhil, S., Nikcevich, D., Rowland, K., Morton, R., Flynn, P. J., Young, C., & Tan, W. (2007). A tomato-based, lycopene-containing intervention for androgen-independent prostate cancer: results of a Phase II study from the North Central Cancer Treatment Group. *Urology, 69,* 289–294.

47. Clark, P. E., Hall, M. C., Borden, L. S., Jr., Miller, A. A., Hu, J. J., Lee, W. R., Stindt, D., D'Agostino, R., Jr., Lovato, J., Harmon, M., & Torti, F. M. (2006). Phase I-II prospective dose-escalating trial of lycopene in patients with biochemical relapse of prostate cancer after definitive local therapy. *Urology, 67,* 1257–1261.

48. Ansari, M. S., & Gupta, N. P. (2003). A comparison of lycopene and orchidectomy vs orchidectomy alone in the management of advanced prostate cancer. *BJU Int, 92,* 375–378discussion 378.

49. Ansari, M. S., & Gupta, N. P. (2004). Lycopene: a novel drug therapy in hormone refractory metastatic prostate cancer. *Urologic Oncology, 22,* 415–420.

24

Fruit, Vegetables, and Legumes Consumption: Role in Preventing and Treating Obesity

Ana B. Crujeiras, Estíbaliz Goyenechea, and J. Alfredo Martínez

Department of Nutrition and Food Sciences, Physiology and Toxicology, University of Navarra, Pamplona, Spain

1. INTRODUCTION

Eating patterns in Western industrialized countries are characterized by a high energy intake and chronic overconsumption of saturated fat, cholesterol, sugar, and salt [1]. Many chronic illnesses such as obesity, diabetes, cardiovascular or neurodegenerative diseases as well as cancer have been repeatedly associated with dietary unhealthy habits [1]. In contrast, low saturated fat intake and high fruit and vegetable intake have been found to be relevant in the prevention of health problems and in the reduction of chronic disease risks [1,2].

The prescription of nutritionally balanced low-energy diets is a common strategy for body weight reduction [3]. These hypocaloric diets are designed according to traditional nutrient recommendations to supply a balanced ratio of protein (10–20% energy), carbohydrate (50–65% energy), and fat (25–35% energy) in reduced quantities to provide an energy intake of 3350–6280 KJ (800–1500 kcal/d). However, the traditional nutritionally adequate low-energy diets frequently failed to promote stable weight loss, and the explanations for such limited success were mostly attributed to the 'poor adherence' to specific low-energy diets. In this context, dietary approaches based upon changes in the macronutrient distribution rather than food restriction to treat obesity are becoming increasingly popular because they might favorably affect weight loss and lipid profile [3]. Consequently, at the present time, clinical trials show that the enrichment of diets in foods with high fiber content and antioxidant properties such as fruits, vegetables, and legumes with the purpose of reducing the risk of illnesses associated with obesity could provide an additional healthy value by improving compliance and lowering risk.

Fruit, vegetables, and legumes are rich sources of a variety of nutrients, including minerals, vitamins, fiber, and some kinds of biologically active substance such as polyphenols, flavonoids, and carotenoids, among many others. These compounds may be important in dietary disease prevention, either alone or, in some cases, in combination with nuts and other plant-foods [4–6].

In this context, daily consumption of at least three to five servings of fruits and vegetables may inhibit or slow down chronic disease progression, due to their capacity to modulate biological processes by means of their nutrients and phytochemical compounds [7]. However, the putative mechanisms that may be responsible for the favorable effects of consumption of these plant derived-foods remain mainly undefined. In this chapter, some mechanisms concerning fruit, vegetables, and legumes consumption, currently proposed to prevent chronic disease risk, such as obesity and its comorbidities, and promote health status, are identified and reported.

2. OBESITY: STATE OF THE ART

Obesity and associated complications are creating a serious pandemic in the industrialized world and in transition countries. The prevalence of obesity has increased dramatically during the last decade not only in the USA but also in Europe and other regions [8]. The health burden of this metabolic disorder in developed countries represents an important economic load on the total health cost [9]. This dramatic economic outcome is apparently worsened by the fact that high body mass index (BMI) or obesity is an important independent risk factor for various chronic diseases, including the development of type II diabetes, hypertension, coronary

heart disease, and an increased incidence of several types of cancer [10,11].

Obesity in humans is also considered a state of chronic inflammation and oxidative stress [12,13]. An increasing amount of evidence suggests that oxidative stress is linked to pathophysiological mechanisms concerning multiple acute and chronic human diseases [14]. Therefore, oxidative stress has been proposed as a possible mechanism underlying the development of comorbidities in obesity [12]. Indeed, obesity is not a single disorder, but a complex multifactorial disease involving the interactions between environmental and genetic factors [15,16]. Among the environmental factors, together with a sedentary lifestyle and low physical activity patterns, diet appears be an important contributor in the development of obesity [3,17].

In this context, dietary patterns, including typical foods of the traditional Mediterranean diet such as vegetables, fruits, and legumes, may be more useful than nutrient-based methods for dietary therapy and in public health efforts [18]. These foods present special properties that make them useful to nutritional treatment by enhancing weight loss and decreasing the development of obesity comorbidities.

3. NUTRITIONAL VALUE AND PHYTOCHEMICALS CONTENT OF FRUITS, VEGETABLES, AND LEGUMES

Distinct types of fruit and vegetables differ widely in their nutrient content and, in recognition of this, national and international agencies recommend fruit and vegetable consumption from diverse groups, frequently based on nutrient content and form. In addition, preparation and storage conditions affect the nutrient content of fruit and vegetables. In this chapter, the nutritional values and phytochemical components of fruit, vegetables, and legumes are

TABLE 24.1 Fruit Content of Macronutrients, Fructose, Fiber, Vitamins, and Total Antioxidant Capacity (TAC) in European and American Diets

Fruit	CHO[b] (g/100 g)	Protein[b] (g/100 g)	Fat[b] (g/100 g)	Fructose[a] (g/100 g)	Fiber[b] (g/100 g)	Vit C[b] (mg/100 g)	Vit E[b] (mg/100 g)	Vit A[a] (μEq. retinol/100 g)	TAC[c] (mmol/kg)
Apple	10.5	0.3	Tr	5.0	2.3	12.4	0.4	4.0	3.23
Banana	20.8	1.2	0.27	3.8	2.5	11.5	0.23	18	2.28
Cherry	13.5	0.8	0.5	5.5	1.5	8.0	0.1	3.0	8.10
Grape (black)	15.5	0.6	0.7	8.0	0.4	4.0	0.7	3.0	11.09
Grape (white)	16.1	0.6	Tr	8.2	0.9	4.0	0.70	3.0	3.25
Kiwifruit	12.1	1.0	0.5	4.4	1.5	94	–	3.0	7.41
Melon	6	0.6	Tr	–	0.7	25	0.1	784	5.73
Orange	8.9	0.8	Tr	2.4	2.3	50.6	0.21	49.0	20.50
Peach	9.0	0.6	0.1	0.9	1.4	8.0	0.5	17.0	6.57
Pear	11.7	0.4	0.1	6.0	2.2	5.2	0.9	2.0	5.00
Pineapple	11.5	0.5	0.1	2.3	1.2	20.0	0.1	3.0	15.73
Plum	11.0	0.6	0.2	3.5	2.1	3.0	0.7	21.0	12.79
Strawberry	7.0	0.7	0.6	2.3	2.2	60	0.2	1	22.74
Watermelon	4.5	0.4	Tr	–	0.2	5	0.1	87	1.13

[a]Source: Belitz et al., 2004 [22]; Ansorena, 1999 [23].
[b]Source: Mataix, 2003 [24]
[c]Source: ferric reducing-antioxidant power (FRAP; mmol Fe^{+3}/kg); Pellegrini et al., 2003 [21].
Tr, traces.

summarized (Tables 24.1 and 24.2). However, foods are biochemically complex and contain compounds which may interact with each other. Therefore, the effect of a particular food in the human body cannot be completely described by the effects of single nutrients [19].

3.1 Fruits and Vegetables

High fruit and vegetable intake is commonly recommended because plant-foods contain a high proportion of water, are low in fats, provide a high content of fiber and fructose, and are good sources of vitamins and minerals [7]. Also, there are many minor components in such foods known as phytonutrients, which may elicit biological responses in mammalian systems that are consistent with reduced risk of one or more chronic diseases [20].

Fruit and vegetables can be consumed fresh, canned, or as juice. Both fruits and fruit juices are rich in vitamins and minerals like vitamin C, potassium, and folate. Total antioxidant capacity in fruit juices generally has a similar value to the whole fruit [21]. Therefore, individuals can consume fruit and vegetable juices to achieve the recommended amounts of vitamin A, vitamin C, folate, and potassium, while whole plant-foods offer some nutritional advantages. Fresh, canned, and frozen fruits and vegetables are generally a good source of fiber, but fruit juices are not often good suppliers.

Fruits and vegetables constitute an indispensable group of foods for balancing the human diet, especially by their contribution to fiber and vitamin supply (Table 24.1 and Table 24.2). Fruits and vegetables tend to be juicy because of their high content of water, usually ranging from 75 to 90%. Soluble

TABLE 24.2 Vegetables Content of Macronutrients, Fiber, Vitamins, and Total Antioxidant Capacity (TAC) in European and American Diets

Vegetables	CHO[a] (g/100 g)	Protein[a] (g/100 g)	Fat[a] (g/100 g)	Fiber[a] (g/100 g)	Vit C[a] (mg/100 g)	Vit E[a] (mg/100 g)	Vit A[a] (μEq. retinol/100 g)	TAC[b] (mmol/kg)
Artichoke	2.9	2.4	0.1	10.8	7.6	0.2	8.0	11.09
Asparagus	1.7	2.9	Tr	1.5	21.6	2.0	53.0	10.60
Broccoli	1.8	4.4	0.9	2.6	87.0	1.3	69.0	11.67
Carrot	7.3	0.9	0.2	2.9	6.0	0.5	1346.0	1.06
Cauliflower	3.1	2.2	0.2	2.1	67.0	0.2	0.0	4.27
Cucumber	1.9	0.7	0.2	0.5	6.0	0.1	2.0	0.71
Eggplant	2.7	1.2	0.2	1.4	5.0	0.1	5.2	3.77
Endive	3.6	1.5	0.2	1.3	10.0	1.0	25	3.24
Garlic	24.3	5.3	0.2	1.2	14.0	0.1	Tr	0.62[c]
Lettuce	1.4	1.5	0.6	1.5	12.2	0.5	29.0	4.94
Mushroom	0.5	1.8	Tr	1.9	4.0	0.1	0.0	16.39
Onion	5.3	1.4	Tr	1.8	6.9	0.4	1.4	5.28
Pepper (red bell)	3.7	0.9	0.2	1.4	131.0	0.8	67.5	20.98
Potato	16.1	2.5	0.2	1.8	18.0	0.1	0.0	3.67
Spinach	2.0	2.5	0.5	1.8	35.0	1.6	542.0	26.94
Swiss chard	4.5	2.0	0.4	0.8	20.0	0.1	183.0	11.60
Tomato	3.5	1.0	0.1	1.40	26.6	0.9	94.0	5.12
Zucchini	6.0	1.3	0.2	1.30	10.0	–	4.0	3.33

[a]Source: Mataix, 2003 [24].
[b]Source: ferric reducing-antioxidant power (FRAP; mmol Fe^{+3}/kg); Pellegrini et al., 2003 [21].
[c]Source: ferric reducing-antioxidant power (FRAP; mmol Fe^{+3}/kg); Nencini et al., 2007 [32].
Tr, traces.

substances that can be found in the moisture are sugars, salts, organic acids, minerals, water-soluble pigments, and vitamins [22]. Aside from water, carbohydrates are the main constituents in fruits, which include sugars, starches, and non-digestible carbohydrates [22,23]. In addition, fruits and vegetables provide bulk to the diet through their content of fiber, including celluloses, hemicelluloses, and pectin substances, which are not extensively digested because of the lack of enzymes capable of hydrolyzing such substances. Moreover, one serving of most fruits commonly contains 1 g or less of protein with a low amount of fat. Fruits are poor in calcium and phosphorus and, in general, are not particularly good sources of iron [22,23].

Vegetables contain an average of 1–3% nitrogen compounds, of which 35–80% is protein. The rest is constituted by amino acids, peptides, and other compounds. The lipid content of vegetables is generally low (0.1–0.9%). Generally, potassium is by far the most abundant mineral in vegetables, followed by calcium, sodium, and magnesium [22].

With regard to vitamins, most fruits are low in the B-vitamins, but citrus fruits, including oranges, lemons, and grapefruit, are excellent sources of ascorbic acid [22,23]. Yellow fruits such as peaches are a fairly good source of carotenoids, the precursor of vitamin A (Table 24.1). The vitamin content of vegetables may vary significantly depending on the cultivar and climate [22]. In spinach, for example, the ascorbic acid content varies from 40 to 155 mg/100 g fresh weight and potatoes contain 15–20 mg/100 g of vitamin C (Table 24.2), whereas carotenoids are occasionally found in large amounts in vegetables [22].

Furthermore, fruits and vegetables are able to supply different non-nutritional components such as polyphenols or other compounds, which may have healthy benefits and positive functions [21].

In this context, the antioxidant capacity of several substances occurring in plants has been documented in human intervention studies, although most of the work has been directed toward the effects of vitamins C and E and β-carotene [4,25,26]. Also, flavonoids, which are more potent antioxidants than vitamins C and E, have been receiving increasing attention [27]. These constituents operate additive and synergistically, contributing to the health benefits attributed to the diet [2,28,29].

Because different antioxidant compounds may act *in vivo* through different mechanisms, no single method can fully evaluate the total antioxidant capacity (TAC) of foods. Based on this assumption, several studies evaluated the TAC of individual foods by means of three assays [21]: Trolox equivalent antioxidant capacity (TEAC), total radical-trapping antioxidant parameter (TRAP), and ferric reducing-antioxidant power (FRAP). The TRAP assays evaluate the chain-breaking antioxidant potential and FRAP methods assess the reducing power of the sample [21], while TEAC assay measures the ability of antioxidants to quench a 2,2'-azinobis(3-ethylbenzothiazoline-6sulfonic acid) radical cation (ABTS$^+$) in both lipophilic and hydrophilic environments [23]. Based on TEAC data from Pellegrini et al. (2003), the TAC value of some commonly consumed fruits in the European and American diet [30,31] is reported (Table 24.1).

Analyzed berries, plums, and some varieties of apples have a relatively high TAC, which is likely to be associated with the high content of flavonoids such as anthocyanins. Oranges and grapes exhibited intermediate antioxidant capacity in agreement with the higher concentrations of phenolic compounds and vitamin C, while bananas, melon, and watermelon had low TAC values [21,33].

Taking into account vegetable TAC values (Table 24.2), it has been shown that spinach was the food with the greatest antioxidant capacity, followed by peppers, whereas cucumber and endive exhibited the lowest TAC values [21].

3.2 Legumes

Grain legumes have been eaten by humans in every country for thousands of years, but only lately have their consumption as a food and their effects on wellbeing begun to be investigated using suitably scientific approaches [6]. Thus, studies have now pointed out that legumes supply the diet with complex carbohydrates, soluble fibers, essential vitamins, and metals as well as with polyphenols such as flavonoids, isoflavones, and lignans, which might classify this food as a potential functional food [34]. However, with the exception of soyabeans, few nutritional studies have been carried out in the area of legumes.

In terms of economic importance, the Leguminosae constitute one of humanity's most important groups of plants. They are the major staple foods and the seed of legumes provides humankind with a highly nutritious food resource [6]. Moreover, they are cheaper than animal products. Therefore, they are consumed worldwide as major staple sources of protein especially in low income countries [34]. Indeed, grain legumes apparently provide specific therapeutic properties that have been described in traditional medicine books for phytotherapeutic use, being also an important element of the Mediterranean diet pattern [6]. Therefore, grain legumes may effectively contribute to a balanced diet and can prevent widespread illnesses including diabetes mellitus, cardiovascular diseases, and cancer [35].

Legumes include peas, beans, lentils, chickpeas, and other plants that are used as food in different culinary forms [36]. From the nutritional point of view, legumes are considered as a source of cholesterol-free proteins. The protein content of legumes (Table 24.3) ranges generally between 20% and 30% of energy, although the quality of legumes protein is often underestimated because of their relatively low content of sulfur amino acids [22]. However, this relatively low sulfur amino acids content (methionine, cysteine) may provide an advantage in terms of calcium retention [5]. The hydrogen ions produced from sulfur amino acid metabolism cause demineralization of bone and excretion of calcium in the urine, suggesting a legumes beneficial effect on bone health [5]. Moreover, the amounts of another essential amino acid, lysine, are much greater than in cereal grains, legume and cereal proteins being nutritionally complementary with respect to lysine and sulfur amino acid contents. In addition, legumes are an excellent source of dietary fiber (Table 24.3) which confers low glycemic values to these plant-foods.

Most legumes are very low in fat, generally containing about 5% of energy or less as fat (Table 24.3) with the exception of some chickpeas and soybeans, which may contain about 15% and 47% fat, respectively [5,22]. The legumes fat contains predominantly linoleic acid, n-3 fatty acid, and α-linolenic acid [5,22]. However, because of the low fat content of legumes, the contribution of this grain to linolenic acid intake is scarce.

Taking into account micronutrients, legumes are a good source of folate, iron, zinc, and calcium, although bioavailability of iron is poor [36]. In general, it is recognized that legumes are good source of complex B vitamins, such as thiamin, niacin, and folic acid, while vitamins A and E are found in low amounts [36].

Currently, legumes are being extensively investigated for their content of several components that traditionally have been considered to be antinutrients, such as trypsin inhibitors, hydrolase inhibitors, phytate, oligosaccharides, lectins, tannins, saponins, and isoflavones among others [5,6]. However, some of these antinutritive components have been related to beneficial effects on preventing chronic diseases [5].

The TAC of legumes has been less investigated than that of cereals, vegetables, and fruit. However, it has been demonstrated that lentils had the highest TAC (Table 24.3) among the legumes analyzed [37–39], while chickpeas

TABLE 24.3 Legumes Content of Macronutrients, Fiber, Vitamins, and Total Antioxidant Capacity (TAC) in European and American Diets

Legume	CHO[a] (g/100 g)	Protein[a] (g/100 g)	Fat[a] (g/100 g)	Fiber[a] (g/100 g)	Vit C[a] (mg/100 g)	Vit E[a] (mg/100 g)	Vit A[a] (μEq. retinol/100 g)	TAC[b] (mmol/kg)
Lentils	54.8	23.0	1.7	11.2	3.4	1.8	10.0	41.65
Garden peas	56.0	21.6	2.3	16.7	2.0	1.0	42.0	8.59
Beans	54.8	21.4	1.5	21.3	3.4	2.0	Tr	9.59
Chickpeas	55.8	20.5	5.5	13.6	4.1	3.1	33.0	5.50
Soyabean	15.8	35.0	18.0	15.7	Tr	2.9	2.0	8.20[c]

[a]Source: Mataix, 2003 [24].
[b]Source: ferric reducing-antioxidant power (FRAP; mmol Fe^{+3}/kg); Pellegrini et al., 2006 [37].
[c]Source: ferric reducing-antioxidant power (FRAP; mmol Fe^{+3}/kg); Halvorsen et al., 2002 [33].
Tr, traces.

showed the lowest TAC values (Table 24.3) [37–39].

Therefore, in addition to providing proteins and dietary fiber, legumes may serve as an excellent dietary source of natural antioxidants for disease prevention and health promotion, as well as their potential use as functional ingredients for processing into nutraceuticals and health food in the food industry [38,39].

4. BENEFITS OF FRUIT, VEGETABLES, AND LEGUMES CONSUMPTION: CLINICAL AND EPIDEMIOLOGICAL EVIDENCE

4.1 Promoting Weight Loss

Weight lowering in obesity conditions is a required target to improve health. In this context, an inverse association of fruit/vegetable consumption with weight gain has been revealed [40] (Table 24.4). Dietary patterns associated with a high intake of fruits and vegetables may reduce long-term risk of subsequent weight gain and obesity [41]. This increase in consumption of fruit and vegetables together with a reduction of fat intake can contribute to a diet lower in energy density, moderating weight gain and promoting weight maintenance [42,43]. Moreover, hypocaloric diets enriched with fruit seem to be involved in beneficial effects on cardiovascular disease associated risk by decreasing cholesterol plasma levels in addition to promoting weight loss [44,45] (Table 24.4). However, interventions prescribing a plant-based diet without a specific energy restriction do not appear to promote such changes in body weight [46]. Thus, several relatively small studies indicate that advice to increase fruit and vegetable intake, without other specific guidelines in the dietary pattern, was associated with no body weight change in spite of the spontaneous decrease in the proportion of fat in the diet [47]. Therefore,

in order to maintain body weight it is important to give advice on the form of food, as well as on the fruits and vegetables intake.

Regarding legume eating, epidemiological observations indicate that subjects with higher soy food consumption have lower body weight than those with lower soy intakes [66,67]. In fact, intervention studies suggest that obese individuals are more prone to lose weight with soy protein than with animal proteins as a major diet component. This effect is more relevant when a soy-based meal replacement is coupled with a high fruit and vegetables consumption [58]. Moreover, it has been demonstrated that a soy-based low-calorie diet decreases the body fat percentage as compared with a traditional low-calorie diet, although the effectiveness of weight loss was similar [65]. In the same way, a non soybean legumes-based hypocaloric diet was shown to induce higher weight loss than a conventional balanced hypocaloric diet in obese subjects [57]. Moreover, an energy-restricted diet containing high legume proteins improved weight loss by decreasing the fat mass content to a greater extent in comparison with a fatty fish and a control balanced diet without legume and fatty fish consumption [64] (Table 24.4).

4.2 Decreasing the Obesity Comorbidities Risk

Obesity is consistently considered as a risk factor associated with the origin or development of various chronic diseases, including coronary heart disease, hypertension, type II diabetes mellitus, and cancer, among other complications [10,11].

Epidemiological and experimental studies have shown a role for plant-food intake in the maintenance of health by decreasing the risk of chronic diseases [68]. So, current scientific evidence (Table 24.4) suggests that a vegetable-based and high-fiber diet is associated with

TABLE 24.4 Epidemiological and Experimental Evidence Concerning the Role of Plant Foods on Obesity and in Associated Complications

Study Design	Population	Major Finding	Study Reference
Intervention	Obese women ($N = 15$)	Fiber content from enriched **fruit** diets may be involved in the favorable effects on **cholesterol plasma levels**	[44]
Longitudinal	Diagnosed breast cancer women ($N = 285,526$)	Total or specific **vegetable** and **fruit** intake is not associated with risk for **breast cancer**	[48]
Case-control	Smoker subjects ($N = 20$)	Regular ingestion of modest amounts of **blueberries** may reduce the risk of **CVD** by decrease in lipid hydroperoxides	[49]
Cross-sectional	Postmenopausal women ($N = 586$)	Improvement overall survival after **breast cancer** diagnosis with **plant-based** and high-fiber diet	[50]
Longitudinal	Diagnosed oral premalignant lesions subjects ($N = 207$)	Reduced risk of oral **premalignant lesions** with higher consumption of citrus fruits and juices	[51]
Longitudinal	Women in the Nurses' Health Study (NHS) ($N = 34,467$)	Frequent consumption (5 servings/d) of **fruit** may reduce the risk of **colorectal adenomas**	[52]
Case-control	Healthy ($N = 385$) and diagnosed bladder cancer subjects ($N = 200$)	The effect of cigarette smoking on **bladder cancer** risk is reduced by **fruit** consumption	[53]
Intervention	Hypertensive patients ($N = 20$)	The consumption of a combination of **fruits** and **vegetables** may decrease the **cardiovascular risk** factors	[54]
Intervention	Hyperlipidemic patients ($N = 57$)	Fresh red **grapefruit** to generally accepted diets could be beneficial for hyperlipidemic patients suffering from **coronary atherosclerosis**	[55]
Intervention	Male Wistar rats ($N = 5$)	10 weeks' administration of the **fruit** extract revealed significant decrements of **blood glucose** levels after glucose loading	[56]
Cross-sectional	Men ($N = 5094$) and women ($N = 6613$)	Fiber or **fruit/vegetable** consumption is inversely associated with **weight gain**	[40]
Intervention	Obese women ($N = 15$)	Antioxidant substances of **fruit**, with the weight reduction, could increase the improvement of **cardiovascular risk factors related to obesity**	[45]
Intervention	Obese people ($N = 30$)	A hypocaloric diet including 4 d/wk **legume servings** empowers **weight loss** and mitigates **oxidative stress**	[57]
Longitudinal	Healthy participants ($N = 2006$)	Dietary patterns associated with a high intake of **fruits and vegetables** may reduce long-term risk of subsequent **weight gain and obesity**	[41]
Longitudinal and cross-sectional	Non-Hispanic white women ($N = 192$)	Increasing consumption of **fruit and vegetables** and reducing fat intake moderates **weight gain** and promotes **weight maintenance**	[42]
Intervention	Obese women ($N = 43$)	A **soy-based meal** replacement **coupled** with **fruit and vegetables** consumption is associated with a great **weight loss** and significant reductions in **body fat mass** and **circulating cholesterol**	[58]
Longitudinal	Type II diabetic patients	**Soy protein** consumption decrease **cardiovascular risk** factors	[59]

(Continued)

B. EFFECTS OF INDIVIDUAL VEGETABLES ON HEALTH

TABLE 24.4 (Continued)

Study Design	Population	Major Finding	Study Reference
Intervention	Families with obese children (N = 41)	**Fruit and vegetable** intake and low-fat dairy products, in an energy-restricted programme, is associated with a reduction in **BMI**, as well as with no relapse **in weight regain**	[43]
Longitudinal-controlled	PREDIMED cohort (N = 372)	**Mediterranean diet** could exert protective effects on **CVD** develop by decreasing the oxidative damage to LDL	[60]
Cross-sectional and longitudinal	SUVIMAX cohort (N = 13,017)	DASH dietary patterns (rich in **fruits and vegetables**) lower the **blood pressure** with aging	[61]
Intervention	Free-living adults (N = 45)	**Chickpeas** in the *ad libitum* diet improve serum **lipid profile** and **glycemic control**	[62]
Longitudinal	EPIC cohort (N>500,000)	Diet high in **vegetables, legumes,** and **fruit** was associated with a reduced risk of **all-cause mortality**, especially deaths due to **CVD**	[63]
Intervention	Obese men (N = 35)	High **legume-protein** hypocaloric diet is more effective than a normal hypocaloric diet for reducing **body weight** and improving some **CVD risk factors**	[64]
Intervention	Obese adults (N = 30)	**Soy-based low-calorie diets** decreased **blood lipids** and had a greater effect on reducing **body fat percentage**	[65]

CVD, cardiovascular disease; DASH, dietary approaches to stop hypertension; LDL, low-density lipoprotein.

improvements in overall survival after breast cancer diagnosis in postmenopausal women [50]. The risk of oral premalignant lesions is significantly reduced with higher consumption of fruits, particularly citrus fruit and juices [51], and frequent consumption of fruit may reduce the risk of colorectal adenoma [52]. Furthermore, the effect of cigarette smoking on bladder cancer risk was reduced by fruit consumption [53]. However, some cohort studies concluded that fruit and vegetable consumption has no effect in relation to overall cancer [48,68]. Actually, there is limited evidence for a cancer-preventive effect of the consumption of fruits and vegetables; nevertheless it is important to recognize that some cancers might be preventable by increasing fruit and vegetable intake [69].

Concerning the metabolic syndrome many studies have reported a benefit of fruit and vegetables intake on cardiovascular disease and diabetes mellitus [19]. In fact, in a case-control study, an inverse association has been found between the first acute myocardial infarction and the consumption of fruits among the Spanish Mediterranean diet [70]. Specific nutritional intervention studies have shown, for instance (Table 24.4), that fresh red grapefruit inclusion in balanced diets could be beneficial for hyperlipidemic patients suffering from coronary atherosclerosis [55]. Likewise, fruit consumption in smoking subjects could contribute to the prevention of cardiovascular disease [49] and a combination of fruit and vegetables prescribed intake may improve some cardiovascular risk factors in hypertensive patients [54]. Thus, the dietary patterns based on the DASH (Dietary Approaches to Stop Hypertension) diet pattern which emphasizes the consumption of fruits, vegetables, and low-fat dairy products and is reduced in saturated fat, total fat, and cholesterol substantially lowered blood pressure and low-density lipoprotein cholesterol [71]. Participants from the SU.VI.MAX

(Supplémentation en Vitamines et Minéraux Antioxydants) cohort who achieved the current daily fruit and vegetable intake recommendations within the DASH diet guidelines presented a lower increase in blood pressure with aging [61]. Moreover, in subjects following a Mediterranean diet, a decrease in the oxidative damage to LDL appears to be one of the protective effects against CVD developing according to a PREDIMED (Prevención Con Dieta Mediterránea) cohort trial [60] (Table 24.4). However, data from the SUN (Seguimiento University of Navarra) cohort evidenced that the protection afforded by fruit and vegetables against the risk of hypertension may be less apparent in relatively young people due to their exposure to healthy factors and the lower baseline risk [72].

High dietary consumption of fruit and vegetables also results in apparent lower risks of diabetes [19]. Thus, after 10 weeks' administration of a fruit extract, glucose tolerance tests revealed significant decreases of glycemic levels after glucose loading, supporting an advantageous association of fruit consumption with diabetes [56]. In addition, a prospective study in the EPIC (European Prospective Investigation into Cancer and Nutrition) cohort evidenced that a high vegetables, legumes, and fruit diet was associated with a reduced risk of all-cause mortality, especially deaths due to cardiovascular disease, but not for deaths due to cancer, underlying the recommendation for the diabetic population to eat large amounts of vegetables, legumes, and fruit [63].

Legumes have also been suggested to contribute in preventing cardiovascular disease and diabetes mellitus (Table 24.4). Indeed, epidemiological studies have shown that Asian people consuming soy in their staple diet present much lower mortality and morbidity from cardiovascular disease than their counterparts in Western countries [73]. For instance, soy protein consumption has been associated with a decrease in cardiovascular risk factors among type II diabetic patients [59]. Moreover, soy-based low-calorie diets significantly decrease blood lipid concentrations, suggesting a beneficial effect on preventing the risk of cardiovascular disease [59]. However, lentils, chickpeas, peas, and beans are the legumes more commonly consumed in Western countries [74]. In this context, the incorporation of chickpeas in the *ad libitum* diet of free-living adults for 12 weeks improves serum lipid profile and glycemic control [62]. Moreover, it has also been demonstrated that a non-soybean legumes-based hypocaloric diet induced a higher decrease in blood lipids concentrations [57,64] as well as lower lipid peroxidation markers related to obesity comorbidities [57], as compared to a conventional and balanced hypocaloric diet.

5. POTENTIAL MECHANISMS BY WHICH FRUIT, VEGETABLES, AND LEGUMES MAY PROTECT AGAINST OBESITY AND ITS COMORBIDITIES

A number of clinical trials involving plant-food intake have evidenced a beneficial effect in promoting health by decreasing weight gain and obesity-related comorbidities risk. The fiber content, minerals, and other compounds of these foods may be responsible for such a protective effect (Table 24.5). In addition, part of the beneficial effects of fruit, vegetables, and legumes has also been attributed to their high level of antioxidant compounds, which contribute by decreasing oxidative stress (Table 24.5).

5.1 Influence of Fiber Content on Glycemic Index

Fruit, together with vegetables, cereals, and legumes, are the major sources of dietary fiber, which after 30 years of research, is still a

TABLE 24.5 Potential Health Benefits of Selected Functional Nutrients or Ingredients and Their Main Food Sources

Functional Nutrients	Claimed Activities	Food Source		
		Fruits	**Vegetables**	**Legumes**
β-glucan	Inmunomodulation, cardioprotective		Mushrooms	
Ascorbic acid	Antioxidant, anticancer	Citrus fruit, kiwi	Tomatoes, potatoes, broccoli	
Carotenoids Carotenes	Antioxidant, anticancer	Tangerine, orange, yellow fruits, grapefruit, watermelon	Carrot, tomatoes, pumpkin, maize, potatoes, spinach, kale, chard, turnip, beet, broccoli, romaine lettuce	
Xanthophylls		Mango, tangerine, orange, peaches, grapefruit, kiwi, oranges, plum, honeydew melon, pear	Avocado, spinach, kale, turnip, pumpkin, romaine lettuce, squash, potatoes, rhubarb	Legumes (peas and others)
Catechins	Antioxidant, anticancer, cardioprotective	Fruits in general: berries, grapes	Fresh tea leaf and vegetables in general	
Fiber Soluble	Cardioprotective (positive influence on blood lipid profile; control of blood glucose)	Some fruits and fruit juices (particularly prune juice, plums, and berries)	Broccoli, carrots, and artichokes. Root vegetables such as potatoes, sweet potatoes, and onions	Peas, soybeans, other beans and lentils
Insoluble		Citrus fruit. The skins of some fruits and apples	Vegetables in general, such as green beans, tomatoes, cauliflower, zucchini (courgette), and celery	
Folic acid	Cardioprotective (decreasing homocystein levels)	Fruits in general	Leafy vegetables such as spinach, turnip greens, lettuces	Legumes in general and peas.
Fructooligo saccharides	Bone protection, iunmunomodulation, anticancer, improvement of gastrointestinal conditions	Fruits in general, especially bananas	Onion, garlic, asparagus, tomatoes, leeks, artichoke	
Isoflavones	Prevention of cardiovascular disease, antioxidant, anticancer, anti-inflammatory		Celery	Various legumes bean, soybeans, and chickpea
Lectins	Hypocholesterolemic			Various legumes: beans, soybeans, lentils, peas
Lignans	Prevention of cardiovascular disease, cancer, and osteoporosis	Strawberries, apricots	Cabbage, Brussels sprouts	Soybean, beans
Phytic acid	Antioxidant, anticancer			Fiber of legumes, soybeans

(Continued)

B. EFFECTS OF INDIVIDUAL VEGETABLES ON HEALTH

TABLE 24.5 (Continued)

Functional Nutrients	Claimed Activities	Food Source		
		Fruits	Vegetables	Legumes
Potassium	Cardioprotective (decreasing hypertension and stroke prevention)	Fruits in general, especially banana, plums, orange	Vegetables in general, especially potato, tomato, artichoke, acorn squash, spinach	Legumes in general
Polyphenols	Antioxidant, anti-inflammatory, anticancer, cardioprotective	Berries, grapes, pomegranates, and fruit skin	Vegetables	Most legumes and soybeans
Resveratrol	Antioxidant and anticancer	Grapes		
Saponins	Antioxidant, anticancer, cardioprotective		Most vegetables	Legumes in general, especially soybeans and peas
Tannins	Antioxidant, anticancer, anti-inflammatory	Pomegranates, berries, apple juice, grape juices		Most legumes, red-colored beans, chickpeas
Tocopherol	Antioxidant, anticancer, cardioprotective	Kiwi	Green leafy vegetables, spinach, carrot, avocado	Soybean

Source: Wildman, 2006 [78]; Linus Pauling Institute (http://lpi.oregonstate.edu) [79].

substantial key to a healthy diet according to current recommendation criteria [75].

Although there are as yet no conclusive data on recommendations of different types of fiber, it is still appropriate to prescribe a diet providing 20–35 g/d of fiber from different sources [75]. Both a high-fiber diet and the prescription of fiber are common in the primary and secondary care management of constipation, since they accelerate gastrointestinal flow [76]. Indeed, a reduced transit time has been associated with a protective role against colorectal cancer, by decreasing the likelihood of colonic carcinogen exposure [77].

Dietary fiber reaches the large bowel where it is attacked by colonic microflora, yielding short-chain fatty acids, hydrogen, carbon dioxide, and methane as fermentation products. Short-chain fatty acids are implicated in some beneficial functions for the human organism [80]. Insoluble fibers, cellulose and hemicelluloses, are known to slowly and selectively stimulate anaerobic bacterial fermentation into more distal areas of the colon [75]. The slow, sustained effect of metabolic activity and production of short-chain fatty acids, specifically butyrate, and consequent reduction in pH and conversion of bile acids into more distal regions, has been shown to have a strong physiological impact on biomarkers [81]. Mechanisms for beneficial effects of fiber might include changing the activity of exogenous carcinogens through metabolic modulation and/or detoxification. Moreover, modification of immune responses could be important in producing beneficial effects of dietary fiber [82].

Taking into account the features of metabolic syndrome, pectins have been proposed as one way in which plant-food consumption contributes to cardiovascular disease prevention [83]. Several mechanisms by which fiber lowers blood cholesterol have been reported. Thus,

B. EFFECTS OF INDIVIDUAL VEGETABLES ON HEALTH

there is evidence to suggest that some soluble fibers bind bile acids and cholesterol during intraluminal formation of micelles, resulting in liver cell cholesterol content reduction [84]. Another reported mechanism is the inhibition of hepatic fatty acid synthesis mediated by fermentation products such as acetate, butyrate, and propionate [85].

Body weight loss could be modulated by specific nutrients and macronutrient distribution included in energy-restricted diets [3]. Many conventional dietary approaches concerning weight management are based on reduction of the dietary fat intake in order to induce weight loss, which in some cases is achieved by increasing fruit and vegetable consumption [44]. In addition, the body weight reduction induced by a legumes-based hypocaloric diet was directly associated with the dietary fiber content [57]. In this regard, it has been suggested that fiber acts as a physiological obstacle to energy intake, displacing available calories and nutrients from the diet, increasing satiety and decreasing the absorption efficiency of the small intestine [86]. Indeed, populations reporting higher fiber diet intake demonstrate lower obesity rates [47,86]. Therefore, fruit, vegetables, and legumes can contribute against excessive body weight gain by means of their high fiber content besides their low fat content.

Moreover, dietary fiber has been shown to delay the absorption of carbohydrates after a meal and thereby decrease the insulinemic response to dietary carbohydrates [19,87]. In this context, fiber-rich foods such as fruits, vegetables, and legumes are often characterized by low glycemic index and glycemic load [88]. Indeed, the low glycemic index and glycemic load are another important functional aspect of whole plant-foods. The evidence suggests that the replacement of high glycemic index foods in the diet by plant-foods may have a wide range of beneficial public health consequences, including reduced risks of obesity, coronary heart disease, and development

of type II diabetes [19]. In the same manner, a nutritional pattern with a low glycemic index has been associated with lower risks of type II diabetes, and coronary heart disease and obesity in prospective studies [89]. Moreover, it has been demonstrated that there is increased satiety, delayed return of hunger, or decreased food intake after consumption of foods with a low, rather than high glycaemic index [90]. Thus, lower-GI energy-restricted diets achieved through a specific differential food selection can improve the homeostatic adaptations during obesity treatment, favoring weight loss and probably weight maintenance compared to higher-GI hypocaloric diets [17].

5.2 Influence of Fructose Content

Fruit provides high amounts of fructose and some investigations have shown that the fructose intake may be associated with the prevalence of obesity, type II diabetes mellitus, and non-alcoholic fatty liver disease development, since high fructose levels can serve as a relatively unregulated source of acetyl-CoA [91]. Indeed, studies in human subjects have demonstrated that fructose ingestion results in markedly increased rates of de novo lipogenesis [92]. However, those trials finding a positive relationship between fructose intake and non-healthy effects have been carried out by means of punctual interventions based upon high-fructose corn syrup intake [93] or incorporating high quantities of dairy fructose intake [94]. The oral administration of small amounts of fructose in animals and in humans, appears to have a specific action increasing the hepatic glucose absorption thereby stimulating glycogen synthesis and restoring the ability of hyperglycemia to regulate hepatic glucose production [95].

Some nutritionists and researchers consider fructose as a relatively safe form of sugar, being often recommended for people with diabetes and included in many weight-loss programs. This monosaccharide appears to facilitate weight loss by suppressing appetite during the postprandial period [96]. Moreover, some scientific evidences have found an increase in hepatic gluconeogenesis during caloric restriction mediated by fructose intake [97]. So, this monosaccharide would not be used for *de novo* lipogenesis when it is included in a hypocaloric diet [97], but this view is not always supported by scientific evidence.

Dietary fruits and vegetables produce beneficial effects as a source of fructose because of the relatively small amount of sugar, with the conjunction of fiber and antioxidants, with fructose in fruit and vegetables unlikely to cause any of the described disturbances [19]. Reinforcing this observation, nutritional studies showed that subjects treated with a fruit-rich hypocaloric diet reached the estimated weight loss and decreased circulating cholesterol levels [44,45,98].

5.3 Influence of Legumes Protein Content

It has now been recognized that food proteins are not only a source of constructive and fuel compounds as the amino acids, but they may also play bioactive roles by themselves and can be the precursors of biologically active peptides with various physiological functions [6]. As mentioned above, legume seeds are among the richest food sources of plant proteins and amino acids for human nutrition [36]. Although other legume components are considered the main responsible agents for their beneficial properties, some studies have attributed hypocholesterolemic effects and lipid homeostasis control to legume proteins, particularly soybean proteins [65] but also other legume seeds [62]. Moreover, some of the legume proteins can present anticarcinogenic [99,100] and antidiabetic properties [101]. In the same manner, peptides in peas, chickpeas,

and mung beans present numerous nutritional and bioactive properties to be used potentially in the formulation of therapeutic products for the treatment and prevention of human diseases [102].

In the context of obesity, increasing evidence from nutritional intervention studies suggests that protein is more satiating than carbohydrate or fat [103]. Thus, dietary soy protein and some of its constituents can favorably affect food intake and reduce excess body fat in obese animals and humans, and also reduce plasma lipids and fat accumulation in liver and adipose tissue, which may decrease the risks of atherosclerosis and lipotoxicity and possibly other obesity-related complications [104]. These beneficial effects that favorably affect energy balance and fat metabolism are mediated by regulating a wide spectrum of biochemical and molecular activities [104]. Thus, soy protein may operate by inhibiting hepatic fatty acid synthesis enzymes in the liver [105]; activating nuclear transcription factors that regulate the expression of genes involved in glucose homeostasis, lipid metabolism, and fatty acid oxidation [106]; or stimulating the production of adiponectin, a cytokine produced by fat cells that plays a key role in regulating in adipocyte differentiation and secretory function, and in enhancing insulin sensitivity [107].

Which component in legume protein is responsible for its hypolipidemic and anti-obesity effects is not entirely clear [104]. However, a specific direct action of certain polypeptides or subunits of legume protein such as grain legume α-amylase protein inhibitors, which may hinder the digestion of complex carbohydrates, thereby promoting or supporting weight loss, has been considered for its potential use in the prevention and therapy of obesity and diabetes [108]. In addition, it has been suggested that individual amino acids could have specific effects. Thus, dietary proteins with low ratios of methionine-glycine and lysine-arginine, such as soya protein without isoflavones, have a cholesterol-lowering effect by inducing the fecal excretion of bile acids and cholesterol among other hypocholesterolemic mechanisms [109]. Dietary supplementation with lysine has also been demonstrated to improve immune function and sensitive indicators of nutritional status [110], and lysine plus methionine supplements was associated with a marked inhibition of body weight gain in rabbits, in spite of adequate food intake [111]. Finally, studies have confirm that dietary methionine restriction, such as the consumption of legumes or a vegan diet, increases lifespan in rats and mice similar to caloric restriction, by suppressing mitochondrial superoxide generation and by decreasing systemic levels of insulin [112].

5.4 Influence of Antioxidant Properties

In aerobic organisms, reactive oxygen species (ROS) are generated constantly during mitochondrial oxidative metabolism [113]. These highly reactive compounds will potentially alter the structure and function of several cellular components, such as lipids, proteins, and nucleic acids [114]. In response to free radical overproduction, living organisms have developed an antioxidant defense network, which should prevent the harmful effects, removing these reactive species before damage, eliminating damaged molecules, and preventing mutations [114,115]. An excessive and/or sustained increase in free radical production associated with diminished efficacy of the cellular defense systems results in oxidative stress, which occurs in many pathological processes and may significantly contribute to disease onset [115]. In this context, it has been hypothesized that ROS play a key role in cardiovascular disease, cancer initiation, the aging process, inflammatory disease, and a variety of neurological disorders [116]. Oxidative stress has also been associated with diabetes mellitus and obesity [12,117].

To maintain the oxidative equilibrium, there is a requirement for the continuous supply of antioxidants, and the first source of antioxidant defense is mainly the diet [2,7]. In this context, fruits, vegetables, and legumes have been regarded as having considerable health benefits, due in particular to their antioxidant content, which can protect the human body against excessive cellular oxidation reactions [21,37].

Assuming that the antioxidant components of plant-food such as fruit, vegetables, and legumes may be responsible for the effects of such food, many studies have focused on vitamin C and carotenoids [118]. However, the results of supplementation studies with pure vitamins are not conclusive about their contribution to health promotion. Natural phytochemicals may not be effective or safe when consumed at high doses, even in a pure dietary supplement form [2]. Some findings suggest that long-term experimental antioxidant vitamin supplementation increases oxidative stress, which may be partly related to the direct prooxidant effect of vitamin radicals [119,120].

On the other hand, the antioxidant effect could be produced by the action of lesser known compounds or from a combination of different compounds occurring in the foods with direct or indirect antioxidant effects [118]. In this context, fructose has been proposed to produce specific effects on oxidative stress. Animal models fed with a high content of fructose have shown a significant increase in antioxidant capacity and prevention of lipid peroxidation [121]. This fruit monosaccharide induces uric acid synthesis due to its rapid metabolism by fructokinase [122]. Uric acid has been widely recognized in the scientific literature as a metabolic compound with high antioxidant power participating as an *in vivo* scavenger [123]. In addition to this, a study suggests that urate is responsible for the increase in antioxidant capacity after consuming apple as fruit [124]. In agreement with these observations, our research group found that the antioxidant capacity of plasma was associated with blood urate concentration in obese women, while keeping the levels in the acceptable metabolic range [45].

Furthermore, some studies have attributed antioxidative properties to fiber-enriched diets, since these compounds enhance the capacity to detoxify free radicals [125]. Numerous factors may explain the role and effects of dietary fiber on antioxidant capacity [125]. Thus, fiber alters fat absorption from the diet by impairing lipid hydrolysis, resulting in increased fat excretion. Moreover, fiber secondary metabolites that arise from bacterial fermentation in the colon may have antioxidant properties [125].

In 2001, the concept of antioxidant dietary fiber (AODF) was introduced, since this natural product contains significant amounts of natural antioxidants associated with the fiber matrix [126]. In this context, some fruits, such as grapes, are a suitable source to provide antioxidant dietary fiber [80]. Reinforcing this idea, a significant correlation between antioxidant power in plasma and dietary fiber plus fructose evidenced the beneficial effect of fruit intake on antioxidant capacity in obese women [45]. Hence, it is conceivable that some reported antioxidant health effects of phytochemicals from fruits can be also associated with the metabolic effect of fructose and fiber on antioxidant defenses.

Additionally, it has been described that the fruit-induced decrease in cholesterol levels and body weight was in parallel with oxidative stress improvement when evaluated by means of the pro-oxidant and antioxidant ratio in plasma, suggesting an indirect antioxidant effect of fruit intake mediated by weight loss and hypocholesterolemic induction [45]. These results are in agreement with those from other nutritional trials showing, after dietary restriction and weight loss in obese people, a decrease in oxidative damage to lipids and a direct influence of the hypocholesterolemic legumes diet-related effect on lipid peroxidation [57].

Legumes supply phytochemical products that also provide antioxidant properties (Table 24.5). In this context, it has been suggested that soy intake aids against oxidative stress by means of these bioactive products [127,128]. However, other kind of legumes more commonly consumed in European countries [74] can improve oxidative stress in obese people [57].

The relationship between obesity and its comorbidities is likely to be linked to a number of metabolic impairments accompanied by oxidative stress disturbances. Therefore, antioxidant-enriched diets including fruit, vegetables, and legumes could be applied in the nutritional therapy of obesity by improving oxidative stress damage and increasing the beneficial effects associated with weight loss.

6. SUMMARY

Obesity represents a metabolic disorder that is an important economic burden for the worldwide health economy due to its relationship with chronic diseases. A number of epidemiological and nutritional intervention studies have associated plant-food consumption with a decreased risk of suffering chronic diseases such as obesity and its comorbidities. Diets based on fruit, vegetables, and legumes aid reduction of body weight and decrease the damage associated with oxidative stress. Thus, the major beneficial effects on obesity are apparently mediated by biologically active substances found in fruits, vegetables, and legumes, besides their low fat and low calorie content. Other compounds such as fructose and fiber have been proposed to produce specific effects. Fructose could induce an antioxidant mechanism by means of uric acid synthesis, a potent scavenger, and fiber decreases lipid plasma levels and induces weight loss by inducing satiety and decreasing the intestinal fat absorption. Also, the low glycemic index and glycemic load is another important functional aspect of plant-foods, contributing to the regulation of glucose homeostasis and promoting weight maintenance. Therefore, the effects of fruit, vegetable, and legume consumption on weight loss, lipid profile, and glucose homeostasis could be mediated by the fiber, proteins, and low calorie content. Moreover, the antioxidant properties of bioactive plant-food compounds could prevent and improve chronic diseases and co-morbidities of obesity.

References

1. Engbers, L. H., van Poppel, M. N., Chin, A. P. M., & van Mechelen, W. (2006). The effects of a controlled worksite environmental intervention on determinants of dietary behavior and self-reported fruit, vegetable and fat intake. *BMC Public Health, 6,* 253.

2. Liu, R. H. (2003). Health benefits of fruit and vegetables are from additive and synergistic combinations of phytochemicals. *American Journal of Clinical Nutrition, 78,* 517S–520S.

3. Abete, I., Parra, M., Zulet, M., & Martnez, J. (2006). Different dietary strategies for weight loss in obesity: Role of energy and macronutrient content. *Nutrition Research Reviews, 19,* 1–19.

4. Lampe, J. W. (1999). Health effects of vegetables and fruit: Assessing mechanisms of action in human experimental studies. *American Journal of Clinical Nutrition, 70,* 475S–490S.

5. Messina, M. J. (1999). Legumes and soybeans: Overview of their nutritional profiles and health effects. *American Journal of Clinical Nutrition, 70,* 439S–450S.

6. Duranti, M. (2006). Grain legume proteins and nutraceutical properties. *Fitoterapia, 77,* 67–82.

7. Pajk, T., Rezar, V., Levart, A., & Salobir, J. (2006). Efficiency of apples, strawberries, and tomatoes for reduction of oxidative stress in pigs as a model for humans. *Nutrition, 22,* 376–384.

8. Bray, G. A., & Bellanger, T. (2006). Epidemiology, trends, and morbidities of obesity and the metabolic syndrome. *Endocrine, 29,* 109–117.

9. Gonzalez-Zapata, L. I., Ortiz-Moncada, R., & Alvarez-Dardet, C. (2007). Mapping public policy options responding to obesity: The case of Spain. *Obesity Reviews, 8,* 99–108.

10. Ceschi, M., Gutzwiller, F., Moch, H., Eichholzer, M., & Probst-Hensch, N. M. (2007). Epidemiology and pathophysiology of obesity as cause of cancer. *Swiss Medical Weekly, 137,* 50–56.

11. Smith, S. C., Jr., (2007). Multiple risk factors for cardiovascular disease and diabetes mellitus. *The American Journal of Medicine, 120,* 3–11.

12. Vincent, H. K., & Taylor, A. G. (2006). Biomarkers and potential mechanisms of obesity-induced oxidant stress in humans. *International Journal of Obesity (Lond.), 30,* 400–418.

13. Garcia-Diaz, D., Campion, J., Milagro, F. I., & Martinez, J. A. (2007). Adiposity dependent apelin gene expression: Relationships with oxidative and inflammation markers. *Molecular and Cellular Biochemistry, 305,* 87–94.

14. Dalle-Donne, I., Rossi, R., Colombo, R., Giustarini, D., & Milzani, A. (2006). Biomarkers of oxidative damage in human disease. *Clinical Chemistry, 52,* 601–623.

15. Marti, A., Martinez-Gonzalez, M. A., & Martínez, J. A. (2008). Interaction between genes and lifestyle factors on obesity. *The Proceedings of the Nutrition Society, 67,* 1–8.

16. Crujeiras, A. B., Parra, D., Goyenechea, E., & Martínez, J. A. (2008). Sirtuin gene expression in human mononuclear cells is modulated by caloric restriction. *European Journal of Clinical Investigation, 38,* 672–678.

17. Abete, I., Parra, D., & Martinez, J. A. (2008). Energy-restricted diets based on a distinct food selection affecting the glycemic index induce different weight loss and oxidative response. *Clinical Nutrition, 27,* 545–551.

18. Schroder, H., Marrugat, J., Vila, J., Covas, M. I., & Elosua, R. (2004). Adherence to the traditional mediterranean diet is inversely associated with body mass index and obesity in a Spanish population. *The Journal of Nutrition, 134,* 3355–3361.

19. Bazzano, L. (2005). 'Dietary Intake of Fruit and Vegetables and Risk of Diabetes Mellitus and Cardiovascular Disease.' World Health Organization (WHO). <http://www.who.int/dietphysicalactivity/publications/f&v_cvd_diabetes.pdf>.

20. Beecher, G. R. (1999). Phytonutrients' role in metabolism: Effects on resistance to degenerative processes. *Nutrition Reviews, 57,* S3–S6.

21. Pellegrini, N., Serafini, M., Colombi, B., Del Rio, D., Salvatore, S., Bianchi, M., & Brighenti, F. (2003). Total antioxidant capacity of plant foods, beverages and oils consumed in Italy assessed by three different *in vitro* assays. *The Journal of Nutrition, 133,* 2812–2819.

22. Belitz, H. D., Grosch, W., & Schieberle, P. (2004). *Food chemistry.* Berlin: Springer.

23. Ansorena, D. (1999). Frutas y frutos secos. In J. A. Martínez, & I. Astiasarán, (Eds.), *Alimentos:*

Composición y propiedades (pp. 219–241). Pamplona: Universidad de Navarra.

24. Mataix, J. (2003). 'Tabla de Composición de Alimentos Españoles.' Instituto de Nutrición y Tecnología de alimentos, Universidad de Granada, Granada.

25. Goralczyk, R., Bachmann, H., Wertz, K., Lenz, B., Riss, G., Buchwald Hunziker, P., Greatrix, B., & Aebischer, C. P. (2006). Beta-carotene-induced changes in RARbeta isoform mRNA expression patterns do not influence lung adenoma multiplicity in the NNK-initiated A/J mouse model. *Nutrition and Cancer, 54*, 252–262.

26. Takase, B., Etsuda, H., Matsushima, Y., Ayaori, M., Kusano, H., Hamabe, A., Uehata, A., Ohsuzu, F., Ishihara, M., & Kurita, A. (2004). Effect of chronic oral supplementation with vitamins on the endothelial function in chronic smokers. *Angiology, 55*, 653–660.

27. Scalzo, J., Mezzetti, B., & Battino, M. (2005). Total antioxidant capacity evaluation: Critical steps for assaying berry antioxidant features. *Biofactors, 23*, 221–227.

28. Saura-Calixto, F., & Goñi, I. (2006). Antioxidant capacity of the Spanish mediterranean diet. *Food Chemistry, 94*, 442–447.

29. Serafini, M., & Del Rio, D. (2004). Understanding the association between dietary antioxidants, redox status and disease: is the total antioxidant capacity the right tool? *Redox Report, 9*, 145–152.

30. Darmon, N., Darmon, M., Maillot, M., & Drewnowski, A. (2005). A nutrient density standard for vegetables and fruits: Nutrients per calorie and nutrients per unit cost. *Journal of the American Dietetic Association, 105*, 1881–1887.

31. Naska, A., Vasdekis, V. G., Trichopoulou, A., Friel, S., Leonhauser, I. U., Moreiras, O., Nelson, M., Remaut, A. M., Schmitt, A., Sekula, W., Trygg, K. U., & Zajkas, G. (2000). Fruit and vegetable availability among ten European countries: How does it compare with the 'five-a-day' recommendation? DAFNE I and II projects of the European Commission. *The British Journal of Nutrition, 84*, 549–556.

32. Nencini, C., Cavallo, F., Capasso, A., Franchi, G. G., Giorgio, G., & Micheli, L. (2007). Evaluation of antioxidative properties of Allium species growing wild in Italy. *Phytotherapy Research, 21*, 874–878.

33. Halvorsen, B. L., Holte, K., Myhrstad, M. C., Barikmo, I., Hvattum, E., Remberg, S. F., Wold, A. B., Haffner, K., Baugerod, H., Andersen, L. F., Moskaug, O., Jacobs, D. R., Jr., & Blomhoff, R. (2002). A systematic screening of total antioxidants in dietary plants. *The Journal of Nutrition, 132*, 461–471.

34. Oboh, G. (2006). Antioxidant properties of some commonly consumed and underutilized tropical legumes. *European Food Research Technology, 224*, 61–65.

35. Leterme, P. (2002). Recommendations by health organizations for pulse consumption. *The British Journal of Nutrition, 88*, 239–242.

36. Martínez, J. A., & Zulet, M. A. (1999). Leguminosas. In J. A. Martínez, & I. Astiasarán, (Eds.), *Alimentos: Composición y propiedades*. Pamplona: Universidad de Navarra.

37. Pellegrini, N., Serafini, M., Salvatore, S., Del Rio, D., Bianchi, M., & Brighenti, F. (2006). Total antioxidant capacity of spices, dried fruits, nuts, pulses, cereals and sweets consumed in Italy assessed by three different *in vitro* assays. *Molecular Nutrition & Food Research, 50*, 1030–1038.

38. Xu, B. J., & Chang, S. K. (2007). A comparative study on phenolic profiles and antioxidant activities of legumes as affected by extraction solvents. *Journal of Food Science, 72*, 159–166.

39. Xu, B. J., Yuan, S. H., & Chang, S. K. (2007). Comparative studies on the antioxidant activities of nine common food legumes against copper-induced human low-density lipoprotein oxidation *in vitro*. *Journal of Food Science, 72*, S522–S527.

40. Bes-Rastrollo, M., Martinez-Gonzalez, M. A., Sanchez-Villegas, A., de la Fuente Arrillaga, C., & Martinez, J. A. (2006). Association of fiber intake and fruit/vegetable consumption with weight gain in a Mediterranean population. *Nutrition, 22*, 504–511.

41. Vioque, J., Weinbrenner, T., Castello, A., Asensio, L., & Garcia de la Hera, M. (2008). Intake of fruits and vegetables in relation to 10-year weight gain among Spanish adults. *Obesity (Silver Spring), 16*, 664–670.

42. Savage, J. S., Marini, M., & Birch, L. L. (2008). Dietary energy density predicts women's weight change over 6 years. *The American Journal of Clinical Nutrition, 88*, 677–684.

43. Epstein, L. H., Paluch, R. A., Beecher, M. D., & Roemmich, J. N. (2008). Increasing healthy eating vs. reducing high energy-dense foods to treat pediatric obesity. *Obesity (Silver Spring), 16*, 318–326.

44. Rodriguez, M. C., Parra, M. D., Marques-Lopes, I., De Morentin, B. E., Gonzalez, A., & Martinez, J. A. (2005). Effects of two energy-restricted diets containing different fruit amounts on body weight loss and macronutrient oxidation. *Plant Foods for Human Nutrition, 60*, 219–224.

45. Crujeiras, A. B., Parra, M. D., Rodriguez, M. C., Martinez de Morentin, B. E., & Martinez, J. A. (2006). A role for fruit content in energy-restricted diets in improving antioxidant status in obese women during weight loss. *Nutrition, 22*, 593–599.

46. Thomson, C. A., Rock, C. L., Giuliano, A. R., Newton, T. R., Cui, H., Reid, P. M., Green, T. L., & Alberts, D. S. (2005). Longitudinal changes in body weight and

body composition among women previously treated for breast cancer consuming a high-vegetable, fruit and fiber, low-fat diet. *European Journal of Nutrition, 44,* 18–25.

47. Tohill, B. (2005). Dietary intake of fruit and vegetables and management of body weight. World Health Organization (WHO). http://www.who.int/dietphysicalactivity/publications/f&v_weight_management.pdf.

48. van Gils, C. H., Peeters, P. H., Bueno-de-Mesquita, H. B., Boshuizen, H. C., Lahmann, P. H., Clavel-Chapelon, F., Thiebaut, A., Kesse, E., Sieri, S., Palli, D., Tumino, R., Panico, S., Vineis, P., Gonzalez, C. A., Ardanaz, E., Sanchez, M. J., Amiano, P., Navarro, C., Quiros, J. R., Key, T. J., Allen, N., Khaw, K. T., Bingham, S. A., Psaltopoulou, T., Koliva, M., Trichopoulou, A., Nagel, G., Linseisen, J., Boeing, H., Berglund, G., Wirfalt, E., Hallmans, G., Lenner, P., Overvad, K., Tjonneland, A., Olsen, A., Lund, E., Engeset, D., Alsaker, E., Norat, T., Kaaks, R., Slimani, N., & Riboli, E. (2005). Consumption of vegetables and fruits and risk of breast cancer. *JAMA, 293,* 183–193.

49. McAnulty, S. R., McAnulty, L. S., Morrow, J. D., Khardouni, D., Shooter, L., Monk, J., Gross, S., & Brown, V. (2005). Effect of daily fruit ingestion on angiotensin converting enzyme activity, blood pressure, and oxidative stress in chronic smokers. *Free Radical Research, 39,* 1241–1248.

50. Jaiswal McEligot, A., Largent, J., Ziogas, A., Peel, D., & Anton-Culver, H. (2006). Dietary fat, fiber, vegetable, and micronutrients are associated with overall survival in postmenopausal women diagnosed with breast cancer. *Nutrition and Cancer, 55,* 132–140.

51. Maserejian, N. N., Giovannucci, E., Rosner, B., Zavras, A., & Joshipura, K. (2006). Prospective study of fruits and vegetables and risk of oral premalignant lesions in men. *American Journal of Epidemiology, 164,* 556–566.

52. Michels, K. B., Giovannucci, E., Chan, A. T., Singhania, R., Fuchs, C. S., & Willett, W. C. (2006). Fruit and vegetable consumption and colorectal adenomas in the Nurses' Health Study. *Cancer Research, 66,* 3942–3953.

53. Kellen, E., Zeegers, M., Paulussen, A., Van Dongen, M., & Buntinx, F. (2006). Fruit consumption reduces the effect of smoking on bladder cancer risk. The Belgian case control study on bladder cancer. *International Journal of Cancer, 118,* 2572–2578.

54. Adebawo, O., Salau, B., Ezima, E., Oyefuga, O., Ajani, E., Idowu, G., Famodu, A., & Osilesi, O. (2006). Fruits and vegetables moderate lipid cardiovascular risk factor in hypertensive patients. *Lipids in Health and Disease, 5,* 14.

55. Gorinstein, S., Caspi, A., Libman, I., Lerner, H. T., Huang, D., Leontowicz, H., Leontowicz, M., Tashma, Z., Katrich, E., Feng, S., & Trakhtenberg, S. (2006). Red grapefruit positively influences serum triglyceride level in patients suffering from coronary atherosclerosis: Studies *in vitro* and in humans. *Journal of Agricultural and Food Chemistry, 54,* 1887–1892.

56. Sugiura, M., Nakamura, M., Ikoma, Y., Yano, M., Ogawa, K., Matsumoto, H., Kato, M., Ohshima, M., & Nagao, A. (2006). The homeostasis model assessment-insulin resistance index is inversely associated with serum carotenoids in non-diabetic subjects. *Journal of Epidemiology, 16,* 71–78.

57. Crujeiras, A. B., Parra, D., Abete, I., & Martinez, J. A. (2007). A hypocaloric diet enriched in legumes specifically mitigates lipid peroxidation in obese subjects. *Free Radical Research, 41,* 498–506.

58. Anderson, J. W., Fuller, J., Patterson, K., Blair, R., & Tabor, A. (2007). Soy compared to casein meal replacement shakes with energy-restricted diets for obese women: Randomized controlled trial. *Metabolism, 56,* 280–288.

59. Azadbakht, L., Atabak, S., & Esmaillzadeh, A. (2008). Soy protein intake, cardiorenal indices, and C-reactive protein in type 2 diabetes with nephropathy: A longitudinal randomized clinical trial. *Diabetes Care, 31,* 648–654.

60. Fito, M., Guxens, M., Corella, D., Saez, G., Estruch, R., de la Torre, R., Frances, F., Cabezas, C., Lopez-Sabater Mdel, C., Marrugat, J., Garcia-Arellano, A., Aros, F., Ruiz-Gutierrez, V., Ros, E., Salas-Salvado, J., Fiol, M., Sola, R., & Covas, M. I. (2007). Effect of a traditional Mediterranean diet on lipoprotein oxidation: A randomized controlled trial. *Archives of Internal Medicine, 167,* 1195–1203.

61. Dauchet, L., Kesse-Guyot, E., Czernichow, S., Bertrais, S., Estaquio, C., Peneau, S., Vergnaud, A. C., Chat-Yung, S., Castetbon, K., Deschamps, V., Brindel, P., & Hercberg, S. (2007). Dietary patterns and blood pressure change over 5-y follow-up in the SU.VI.MAX cohort. *The American Journal of Clinical Nutrition, 85,* 1650–1656.

62. Pittaway, J. K., Robertson, I. K., & Ball, M. J. (2008). Chickpeas may influence fatty acid and fiber intake in an *ad libitum* diet, leading to small improvements in serum lipid profile and glycemic control. *Journal of the American Dietetic Association, 108,* 1009–1013.

63. Nothlings, U., Schulze, M. B., Weikert, C., Boeing, H., van der Schouw, Y. T., Bamia, C., Benetou, V., Lagiou, P., Krogh, V., Beulens, J. W., Peeters, P. H., Halkjaer, J., Tjonneland, A., Tumino, R., Panico, S., Masala, G., Clavel-Chapelon, F., de Lauzon, B., Boutron-Ruault, M. C., Vercambre, M. N., Kaaks, R., Linseisen, J., Overvad, K., Arriola, L., Ardanaz, E., Gonzalez, C. A., Tormo, M. J., Bingham, S., Khaw, K. T., Key, T. J.,

Vineis, P., Riboli, E., Ferrari, P., Boffetta, P., Bueno-de-Mesquita, H. B., van der, A. D., Berglund, G., Wirfalt, E., Hallmans, G., Johansson, I., Lund, E., & Trichopoulo, A. (2008). Intake of vegetables, legumes, and fruit, and risk for all-cause, cardiovascular, and cancer mortality in a European diabetic population. *The Journal of Nutrition, 138,* 775–781.

64. Abete, I., Parra, D., & Martínez, J. (2009). Legume-, fish-, or high-protein-based hypocaloric diets: Effects on weight loss and mitochondrial oxidation in obese men. *Journal of Medicinal Food, 12,* 100–108.

65. Liao, F. H., Shieh, M. J., Yang, S. C., Lin, S. H., & Chien, Y. W. (2007). Effectiveness of a soy-based compared with a traditional low-calorie diet on weight loss and lipid levels in overweight adults. *Nutrition, 23,* 551–556.

66. Goodman-Gruen, D., & Kritz-Silverstein, D. (2003). Usual dietary isoflavone intake and body composition in postmenopausal women. *Menopause, 10,* 427–432.

67. Yamori, Y. (2004). Worldwide epidemic of obesity: Hope for Japanese diets. *Clinical and Experimental Pharmacology & Physiology, 31,* 2–4.

68. Potter, J. D. (2005). Vegetables, fruit, and cancer. *Lancet, 366,* 527–530.

69. Vainio, H., & Weiderpass, E. (2006). Fruit and vegetables in cancer prevention. *Nutrition and Cancer, 54,* 111–142.

70. Martinez-Gonzalez, M. A., Fernandez-Jarne, E., Serrano-Martinez, M., Marti, A., Martinez, J. A., & Martin-Moreno, J. M. (2002). Mediterranean diet and reduction in the risk of a first acute myocardial infarction: An operational healthy dietary score. *European Journal of Nutrition, 41,* 153–160.

71. Miller, E. R., 3rd., Erlinger, T. P., & Appel, L. J. (2006). The effects of macronutrients on blood pressure and lipids: An overview of the DASH and OmniHeart trials. *Current Atherosclerosis Reports, 8,* 460–465.

72. Nunez-Cordoba, J. M., Alonso, A., Beunza, J. J., Palma, S., Gomez-Gracia, E., & Martinez-Gonzalez, M. A. (2008). Role of vegetables and fruits in Mediterranean diets to prevent hypertension. *European Journal of Clinical Nutrition,* DOI: 10.1038/ejcn.2008.22.

73. Heneman, K. M., Chang, H. C., Prior, R. L., & Steinberg, F. M. (2007). Soy protein with and without isoflavones fails to substantially increase postprandial antioxidant capacity. *The Journal of Nutritional Biochemistry, 18,* 46–53.

74. Schneider, A. V. (2002). Overview of the market and consumption of pulses in Europe. *The British Journal of Nutrition, 88,* 243–250.

75. Escudero Alvarez, E., & Gonzalez Sanchez, P. (2006). Dietary fibre. *Nutricion Hospitalaria, 21,* 60–72.

76. Wisten, A., & Messner, T. (2005). Fruit and fibre (Pajala porridge) in the prevention of constipation. *Scandinavian Journal of Caring Sciences, 19,* 71–76.

77. Lim, C. C., Ferguson, L. R., & Tannock, G. W. (2005). Dietary fibres as 'prebiotics': Implications for colorectal cancer. *Molecular Nutrition & Food Research, 49,* 609–619.

78. Wildman, R. (2006). *Handbook of nutraceuticals and functional foods.* New York: CRC Press.

79. Linus Pauling Institute (2008). 'Micronutrient research for optimum health.' <http://lpi.oregonstate.edu>.

80. Saura-Calixto, F. (1998). Antioxidant dietary fiber product: A new concept and a potential food ingredient. *Journal of Agricultural and Food Chemistry, 46,* 4303–4306.

81. Van Loo, J. A. (2004). Prebiotics promote good health: The basis, the potential, and the emerging evidence. *Journal of Clinical Gastroenterology, 38,* S70–S75.

82. Pool-Zobel, B. L., Selvaraju, V., Sauer, J., Kautenburger, T., Kiefer, J., Richter, K. K., Soom, M., & Wolfl, S. (2005). Butyrate may enhance toxicological defence in primary, adenoma and tumor human colon cells by favourably modulating expression of glutathione S-transferases genes, an approach in nutrigenomics. *Carcinogenesis, 26,* 1064–1076.

83. Marlett, J. A., McBurney, M. I., & Slavin, J. L. (2002). Position of the American Dietetic Association: Health implications of dietary fiber. *Journal of the American Dietetic Association, 102,* 993–1000.

84. Anderson, J. W., Allgood, L. D., Lawrence, A., Altringer, L. A., Jerdack, G. R., Hengehold, D. A., & Morel, J. G. (2000). Cholesterol-lowering effects of psyllium intake adjunctive to diet therapy in men and women with hypercholesterolemia: Meta-analysis of 8 controlled trials. *The American Journal of Clinical Nutrition, 71,* 472–479.

85. Aller, R., de Luis, D. A., Izaola, O., La Calle, F., del Olmo, L., Fernandez, L., Arranz, T., & Hernandez, J. M. (2004). Effect of soluble fiber intake in lipid and glucose levels in healthy subjects: A randomized clinical trial. *Diabetes Research and Clinical Practice, 65,* 7–11.

86. Slavin, J. L. (2005). Dietary fiber and body weight. *Nutrition, 21,* 411–418.

87. Anderson, J. W., O'Neal, D. S., Riddell-Mason, S., Floore, T. L., Dillon, D. W., & Oeltgen, P. R. (1995). Postprandial serum glucose, insulin, and lipoprotein responses to high- and low-fiber diets. *Metabolism, 44,* 848–854.

88. Jenkins, D. J., Wolever, T. M., Taylor, R. H., Barker, H., Fielden, H., Baldwin, J. M., Bowling, A. C., Newman, H. C., Jenkins, A. L., & Goff, D. V. (1981). Glycemic index of foods: A physiological basis for carbohydrate exchange. *The American Journal of Clinical Nutrition, 34,* 362–366.

89. Willett, W., Manson, J., & Liu, S. (2002). Glycemic index, glycemic load, and risk of type 2 diabetes. *The American Journal of Clinical Nutrition, 76*, 274S–280S.

90. Ludwig, D. S. (2000). Dietary glycemic index and obesity. *The Journal of Nutrition, 130*, 280–283.

91. Bray, G. A., Nielsen, S. J., & Popkin, B. M. (2004). Consumption of high-fructose corn syrup in beverages may play a role in the epidemic of obesity. *The American Journal of Clinical Nutrition, 79*, 537–543.

92. Elliott, S. S., Keim, N. L., Stern, J. S., Teff, K., & Havel, P. J. (2002). Fructose, weight gain, and the insulin resistance syndrome. *The American Journal of Clinical Nutrition, 76*, 911–922.

93. Tordoff, M. G., & Alleva, A. M. (1990). Effect of drinking soda sweetened with aspartame or high-fructose corn syrup on food intake and body weight. *The American Journal of Clinical Nutrition, 51*, 963–969.

94. Anderson, J. W., Story, L. J., Zettwoch, N. C., Gustafson, N. J., & Jefferson, B. S. (1989). Metabolic effects of fructose supplementation in diabetic individuals. *Diabetes Care, 12*, 337–344.

95. Petersen, K. F., Laurent, D., Yu, C., Cline, G. W., & Shulman, G. I. (2001). Stimulating effects of low-dose fructose on insulin-stimulated hepatic glycogen synthesis in humans. *Diabetes, 50*, 1263–1268.

96. Gaby, A. R. (2005). Adverse effects of dietary fructose. *Alternative Medicine Review, 10*, 294–306.

97. Hagopian, K., Ramsey, J. J., & Weindruch, R. (2005). Fructose metabolizing enzymes from mouse liver: Influence of age and caloric restriction. *Biochimica et Biophysica Acta, 1721*, 37–43.

98. Conceicao de Oliveira, M., Sichieri, R., & Sanchez Moura, A. (2003). Weight loss associated with a daily intake of three apples or three pears among overweight women. *Nutrition, 19*, 253–256.

99. de Lumen, B. O. (2005). Lunasin: A cancer-preventive soy peptide. *Nutrition Reviews, 63*, 16–21.

100. Xiao, R., Badger, T. M., & Simmen, F. A. (2005). Dietary exposure to soy or whey proteins alters colonic global gene expression profiles during rat colon tumorigenesis. *Molecular Cancer, 4*, 1.

101. Nordentoft, I., Jeppesen, P. B., Hong, J., Abudula, R., & Hermansen, K. (2008). Increased insulin sensitivity and changes in the expression profile of key insulin regulatory genes and beta cell transcription factors in diabetic KKAy-mice after feeding with a soy bean protein rich diet high in isoflavone content. *Journal of Agricultural and Food Chemistry, 56*, 4377–4385.

102. Aluko, R. E. (2008). Determination of nutritional and bioactive properties of peptides in enzymatic pea, chickpea, and mung bean protein hydrolysates. *Journal of AOAC International, 91*, 947–956.

103. Clifton, P. M., & Keogh, J. (2007). Metabolic effects of high-protein diets. *Current Atherosclerosis Reports, 9*, 472–478.

104. Velasquez, M. T., & Bhathena, S. J. (2007). Role of dietary soy protein in obesity. *International Journal of Medical Microbiology, 4*, 72–82.

105. Xiao, C. W., Wood, C., Huang, W., L'Abbe, M. R., Gilani, G. S., Cooke, G. M., & Curran, I. (2006). Tissue-specific regulation of acetyl-CoA carboxylase gene expression by dietary soya protein isolate in rats. *The British Journal of Nutrition, 95*, 1048–1054.

106. Mezei, O., Banz, W. J., Steger, R. W., Peluso, M. R., Winters, T. A., & Shay, N. (2003). Soy isoflavones exert antidiabetic and hypolipidemic effects through the PPAR pathways in obese Zucker rats and murine RAW 264.7 cells. *The Journal of Nutrition, 133*, 1238–1243.

107. Nagasawa, A., Fukui, K., Kojima, M., Kishida, K., Maeda, N., Nagaretani, H., Hibuse, T., Nishizawa, H., Kihara, S., Waki, M., Takamatsu, K., Funahashi, T., & Matsuzawa, Y. (2003). Divergent effects of soy protein diet on the expression of adipocytokines. *Biochemical and Biophysical Research Communications, 311*, 909–914.

108. Mosca, M., Boniglia, C., Carratu, B., Giammarioli, S., Nera, V., & Sanzini, E. (2008). Determination of alpha-amylase inhibitor activity of phaseolamin from kidney bean (*Phaseolus vulgaris*) in dietary supplements by HPAEC-PAD. *Analytica Chimica Acta, 617*, 192–195.

109. Gudbrandsen, O. A., Wergedahl, H., Liaset, B., Espe, M., & Berge, R. K. (2005). Dietary proteins with high isoflavone content or low methionine-glycine and lysine-arginine ratios are hypocholesterolaemic and lower the plasma homocysteine level in male Zucker fa/fa rats. *The British Journal of Nutrition, 94*, 321–330.

110. Hussain, T., Abbas, S., Khan, M. A., & Scrimshaw, N. S. (2004). Lysine fortification of wheat flour improves selected indices of the nutritional status of predominantly cereal-eating families in Pakistan. *Food and Nutrititon Bulletin, 25*, 114–122.

111. Giroux, I., Kurowska, E. M., & Carroll, K. K. (1999). Role of dietary lysine, methionine, and arginine in the regulation of hypercholesterolemia in rabbits. *The Journal of Nutritional Biochemistry, 10*, 166–171.

112. McCarty, M. F., Barroso-Aranda, J., & Contreras, F. (2008). The low-methionine content of vegan diets may make methionine restriction feasible as a life extension strategy. *Medical Hypotheses*, DOI:10.1016/j.mehy.2008.07.044.

113. Finkel, T., & Holbrook, N. J. (2000). Oxidants, oxidative stress and the biology of ageing. *Nature, 408*, 239–247.

B. EFFECTS OF INDIVIDUAL VEGETABLES ON HEALTH

114. Sies, H. (1997). Oxidative stress: Oxidants and antioxidants. *Experimental Physiology, 82*, 291–295.

115. Halliwell, B. (2006). Reactive species and antioxidants. Redox biology is a fundamental theme of aerobic life. *Plant Physioloogy, 141*, 312–322.

116. Mayne, S. T. (2003). Antioxidant nutrients and chronic disease: Use of biomarkers of exposure and oxidative stress status in epidemiologic research. *The Journal of Nutrition, 133*, 933–940.

117. Dandona, P., Aljada, A., Chaudhuri, A., Mohanty, P., & Garg, R. (2005). Metabolic syndrome: A comprehensive perspective based on interactions between obesity, diabetes, and inflammation. *Circulation, 111*, 1448–1454.

118. Crujeiras, A. B., Parra, M. D., & Martinez, J. A. (2007). Functional properties of fruit. *Food, 1*, 30–35.

119. Halliwell, B. (2000). The antioxidant paradox. *Lancet, 355*, 1179–1180.

120. Versari, D., Daghini, E., Rodriguez-Porcel, M., Sattler, K., Galili, O., Pilarczyk, K., Napoli, C., Lerman, L. O., & Lerman, A. (2006). Chronic antioxidant supplementation impairs coronary endothelial function and myocardial perfusion in normal pigs. *Hypertension, 47*, 475–481.

121. Girard, A., Madani, S., El Boustani, E. S., Belleville, J., & Prost, J. (2005). Changes in lipid metabolism and antioxidant defense status in spontaneously hypertensive rats and Wistar rats fed a diet enriched with fructose and saturated fatty acids. *Nutrition, 21*, 240–248.

122. Heuckenkamp, P. U., & Zollner, N. (1971). Fructose-induced hyperuricaemia. *Lancet, 1*, 808–809.

123. Glantzounis, G. K., Tsimoyiannis, E. C., Kappas, A. M., & Galaris, D. A. (2005). Uric acid and oxidative stress. *Current Pharmaceutical Design, 11*, 4145–4151.

124. Lotito, S. B., & Frei, B. (2004). The increase in human plasma antioxidant capacity after apple consumption is due to the metabolic effect of fructose on urate, not apple-derived antioxidant flavonoids. *Free Radical Biology & Medicine, 37*, 251–258.

125. Diniz, Y. S., Faine, L. A., Galhardi, C. M., Rodrigues, H. G., Ebaid, G. X., Burneiko, R. C., Cicogna, A. C., & Novelli, E. L. (2005). Monosodium glutamate in standard and high-fiber diets: Metabolic syndrome and oxidative stress in rats. *Nutrition, 21*, 749–755.

126. Jimenez-Escrig, A., Rincon, M., Pulido, R., & Saura-Calixto, F. (2001). Guava fruit (*Psidium guajava* L.) as a new source of antioxidant dietary fiber. *Journal of Agricultural and Food Chemistry, 49*, 5489–5493.

127. Jenkins, D. J., Kendall, C. W., Vidgen, E., Vuksan, V., Jackson, C. J., Augustin, L. S., Lee, B., Garsetti, M., Agarwal, S., Rao, A. V., Cagampang, G. B., & Fulgoni, V., 3rd. (2000). Effect of soy-based breakfast cereal on blood lipids and oxidized low-density lipoprotein. *Metabolism, 49*, 1496–1500.

128. Wiseman, H., O'Reilly, J. D., Adlercreutz, H., Mallet, A. I., Bowey, E. A., Rowland, I. R., & Sanders, T. A. (2000). Isoflavone phytoestrogens consumed in soy decrease F(2)-isoprostane concentrations and increase resistance of low-density lipoprotein to oxidation in humans. *The American Journal of Clinical Nutrition, 72*, 395–400.

25

Spinach and Carrots: Vitamin A and Health

Guangwen Tang

Jean Mayer US Department of Agriculture, Human Nutrition Research Center on Aging,
Tufts University, Boston, MA, USA

1. VITAMIN A FROM OUR DIETS

Vitami A is an essential nutrient for the promotion of general growth, maintenance of visual function, regulation of the differentiation of epithelial tissues and immune function, and embryonic development [1]. Vitamin A can only be supplied naturally from our diets, either as preformed vitamin A from foods of animal origin (e.g. liver, dairy products, eggs, etc.) or as provitamin A carotenoids from plants such as carrots, spinach, red yam, pumpkins, cantaloupe, etc.

The conversion of provitamin A β-carotene to vitamin A *in vivo* was discovered by Moore in 1929 [2]. Specifically, after intake of provitamin A carotenoids (mainly all-*trans* β-carotene), in addition to the absorption of intact β-carotene, (see Figure 25.1), retinol can be found to be formed from the provitamin A carotenoids in our circulation. In 2005, both central (β-carotene 15,15′-oxygenase, BCO1) and excentric (β-carotene 9,10-oxygenase, BCO2) cleavage enzymes have been reported in the small intestine, liver, skin, eye, and other tissues [3].

2. VITAMIN A EQUIVALENCE OF β-CAROTENE

As defined by the Dietary Reference Intakes, 2001 [4], the carotene over the retinol equivalency ratio (carotene:retinol, in μg:μg) of a low dose (less than 2 mg) of purified β-carotene in oil is approximately 2:1 (i.e. 2 μg of β-carotene in oil yields 1 μg of retinol). This ratio was derived from the relative amount of β-carotene required to correct abnormal dark adaptation in vitamin A-deficient individuals [5]. The efficiency of the absorption of β-carotene in food is lower than the absorption of β-carotene in oil. Previously, due to a relative absorption efficiency of about 33% of β-carotene from food sources, it was assumed that 3 μg of dietary β-carotene is equivalent to 1 μg of purified

381

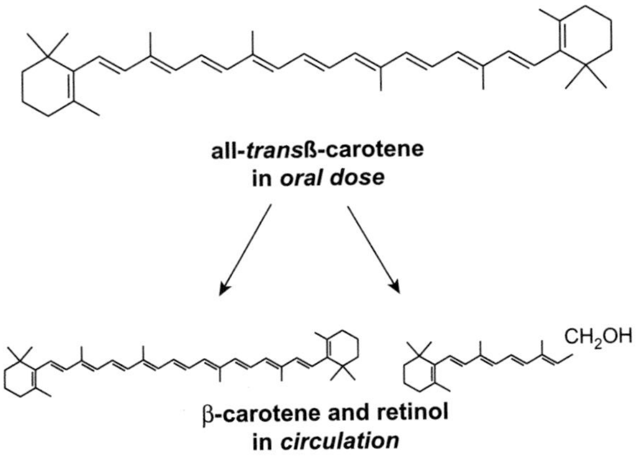

FIGURE 25.1 Retinol formation.

using RAE, the provitamin A carotenoids provide about 20% of our daily vitamin A intakes.

It has been common practice to assess the vitamin A value of a food from the amount of preformed vitamin A and provitamin A carotenoids contained in that food. In reality, many factors such as food matrices, food preparation, and the fat content of a meal affect the bioavailability of food carotenoids and the bioconversion of food carotenoids into vitamin A in humans. For example, did spinach β-carotene and carrot β-carotene show the same vitamin A equivalent when the same amounts of β-carotene were taken?

β-carotene in oil [6]. Therefore, the retinol equivalence (RE) for β-carotene from foods was estimated to be 6:1 (3 × 2:1). However, one report showed that 6 μg of β-carotene from a mixed diet is nutritionally equivalent to 1 μg of β-carotene in oil [4]. Therefore, the retinol activity equivalency (μg RAE) ratio for β-carotene from food is estimated to be 12:1 (6 × 2:1). Accordingly, based on the observation that the vitamin A activity of β-cryptoxanthin and α-carotene is approximately half of that for all-*trans* β-carotene [7,8] an RAE for dietary provitamin A carotenoids, other than β-carotene, is set at 24:1. Based on these considerations, the amount of vitamin A activity of provitamin A carotenoids, RAE yields half the amount when using RE.

3. DIETARY INTAKE OF VITAMIN A

Data published by CDC [9], from a survey conducted in the USA from 1999–2000 on a total of 8604 subjects, reported that the average dietary intake of preformed vitamin A was 938 RE, and the dietary intake of provitamin A carotenoids was 454 RE (assuming 6 μg of provitamin A is equivalent to 1 μg of retinol). Therefore, provitamin A carotenoids provide about 30% of our daily vitamin A intakes. If

4. VITAMIN A VALUE OF PLANT FOODS

As is well known, due to the inability to distinguish newly formed retinol from the body reserves, conversion of β-carotene to vitamin A cannot be investigated in well-nourished humans by supplementing *unlabeled* β-carotene (synthetic or in food). The problem can only be overcome by using an isotope technique which traces newly administered labeled nutrients, β-carotene and/or vitamin A. In order to achieve an accurate assessment of carotenoid bioabsorption, bioconversion, and a subsequent vitamin A value from a food source, food in which the carotenoids have been endogenously or intrinsically labeled with a stable isotope is required. This allows presentation of the carotenoids in their normal cellular compartments, while the isotopic label enables identification of those serum carotenoids (or derived retinol) which come from the specific food studied. Plant carotenoids have been intrinsically labeled with ^{2}H (deuterium) from ^{2}H$_2$O (heavy water). For ^{2}H$_2$O labeling, plants can easily be grown hydroponically [10] on a nutrient solution with a fixed ^{2}H atom percentage. Water with 25% atom excess of ^{2}H generates a range of isotopomers of carotenoids, with most abundant enrichment in the $^2H_{10}$ species (with 10 deuterium in a β-carotene molecule). After initiating the hydroponic growth in the ^2H_2O-enriched medium, spinach and carrots can be harvested in 32 and 60 days, respectively.

To prepare for an *in vivo* study on human volunteers, the spinach leaves (or sliced carrot root) were steamed in thin layers for 10 minutes. The cooked spinach leaves (or sliced carrot root) were immersed in cold water (1 liter water per 200 g vegetable) for 2 minutes, and then drained and pureed to be used for human ($N = 7$) consumption experiments, then placed in a sealed plastic container. The 300 g labeled spinach and 100 g labeled carrots each contained *ca*.11 mg (all-*trans*)-β-carotene, and it was assumed that α-carotene and (*cis*)-β-carotene, which were also present, have half the activity of (all-*trans*)-β-carotene.

Our investigation showed that the retinol equivalences were determined to be 21 µg spinach β-carotene to 1 µg retinol, and 15 µg carrot β-carotene to 1 µg retinol [11]. Therefore, similar amounts of β-carotene in various plant foods with different food matrices showed different levels of efficiency in converting to vitamin A *in vivo*.

Another human study on adults in Bangladesh [12] used a paired deuterated retinol dilution (DRD) test technique to measure the vitamin A pool size after 60-day supplementation with 750 RE/day as either retinyl palmitate, β-carotene, sweet potato, or Indian spinach, compared with a control containing no retinol or β-carotene. Vitamin A equivalency factors of 6:1 for β-carotene in oil, 10:1 for β-carotene in Indian spinach, and 13:1 for β-carotene in sweet potato were determined. It seems that the Indian spinach can be absorbed and converted to vitamin A more effectively by the population with relative low vitamin A status than the well-nourished USA population.

In 2007, Ribaya-Mercado et al. reported on a study that used mixed-vegetable intervention and the paired DRD test to measure the changes in vitamin A pool size of Filipino schoolchildren aged 9–12 years [13]. The results showed that the conversion factors of provitamin A carotenoids in the mixed-vegetable diets were better than 12:1 for β-carotene and 24:1 for other provitamin A carotenoids in the population with marginal vitamin A status. In addition, it reported that provitamin A carotenoid rich yellow and green leafy vegetables can be effectively ingested with minimal dietary fat (7 g fat per day or 2 g fat per meal). This observation provided further consideration on possible effects of vitamin A status and/or age on the bioconversion of dietary plant β-carotene to vitamin A.

Vitamin A equivalency of carrot provitamin A carotenoids (carrot is rich in both β-carotene and α-carotene) was studied in a human volunteer using deuterium-labeled vitamin A as an extrinsic standard [14]. In this study, a subject was given raw carrots containing 9.8 µmol (5 mg) β-carotene and 5.2 µmol α-carotene (2.8 mg), together with 7 µmol (2 mg) [$^2H_4$]-retinyl acetate, and the concentrations of β-carotene, α-carotene, and labeled and unlabeled retinyl esters in the triglyceride-rich lipoproteins (TRL) were measured at various time points for up to 7 hours. With the assumption that absorption of labeled retinyl acetate was about 80% of the dose, it was calculated that 0.8 µmol of the carrot β-carotene was absorbed intact and that 1.5 µmol of unlabeled retinyl esters were formed from the carrot dose. The mass equivalency of carrot β-carotene to vitamin A was, therefore, 13:1 (without considering the contribution from the 5.2 µmol of α-carotene). If the contribution of α-carotene were considered, assuming that α-carotene has half the activity of β-carotene, the ratio would be higher ($\sim$16:1).

Spinach and carrots are the major vegetables rich in provitamin A carotenoids and are commonly consumed. They represent dark green and yellow vegetables. The nutrient contents in raw spinach and carrots are presented in Table 25.1 and 25.2. The vitamin A content

TABLE 25.1   Nutrient Contents in Raw Spinach
NDB No. 11457
Spinach, raw
Spinacia oleracea
Refuse: 28% Large stems and roots

Nutrients and Units		Amount in 100 g of Edible Portion						Amount in Edible Portion of Common Measures of Food		
		Mean	Std. Error	Number of Data Points	Deriv Code	Source Code	Confidence Code	Measure 1	Measure 2	Measure 3
Proximates										
Water	g	91.40		1	JO	11		27.42	310.76	9.14
Energy	kcal	23		0	NC	4		7	79	2
Energy	kj	97		0	NC	4		29	330	10
Protein ($N \times 6.25$)	g	2.86	0.112	9		1		0.86	9.72	0.29
Total lipid (fat)	g	0.39	0.032	7		1		0.12	1.32	0.04
Ash	g	1.72	0.035	8		1		0.52	5.85	0.17
Carbohydrate, by difference	g	3.63		0	NC	4		1.09	12.34	0.36
Fiber, total dietary	g	2.2		1		1		0.7	7.5	0.2
Sugars, total	g	0.42		0	NC	4		0.13	1.43	0.04
Sucrose	g	0.07	0.036	8		1		0.02	0.23	0.01
Glucose (dextrose)	g	0.11	0.032	8		1		0.03	0.36	0.01
Fructose	g	0.15	0.070	8		1		0.04	0.50	0.01
Lactose	g	0.00		1		1		0.00	0.00	0.00
Maltose	g	0.00		1		1		0.00	0.00	0.00
Galactose	g	0.10		1		1		0.03	0.34	0.01
Starch	g									
Minerals										
Calcium, Ca	mg	99	4.996	9		1		30	337	10
Iron, Fe	mg	2.71	0.522	10		1		0.81	9.21	0.27
Magnesium, Mg	mg	79	4.794	7		1		24	269	8
Phosphorus, P	mg	49	3.479	7		1		15	167	5
Potassium, K	mg	558	28.703	10		1		167	1897	56
Sodium, Na	mg	79	10.835	10		1		24	269	8
Zinc, Zn	mg	0.53	0.039	7		1		0.16	1.80	0.05
Copper, Cu	mg	0.130	0.007	7		1		0.039	0.442	0.013
Manganese, Mn	mg	0.897	0.048	6		1		0.269	3.050	0.090
Selenium, Se	μg	1.0	0.335	5	A	1		0.3	3.4	0.1
Vitamins										
Vitamin C, total ascorbic acid	mg	28.1	4.129	7		1		8.4	95.5	2.8
Thiamin	mg	0.078	0.008	9		1		0.023	0.265	0.008
Riboflavin	mg	0.189	0.008	9		1		0.057	0.643	0.019

(Continued)

TABLE 25.1   (Continued)

Nutrients and Units		Amount in 100 g of Edible Portion						Amount in Edible Portion of Common Measures of Food		
		Mean	Std. Error	Number of Data Points	Deriv Code	Source Code	Confidence Code	Measure 1	Measure 2	Measure 3
Niacin	mg	0.724	0.032	9			1	0.217	2.462	0.072
Pantothenic acid	mg	0.065	0.008	6			1	0.020	0.221	0.007
Vitamin B6	mg	0.195	0.008	6			1	0.059	0.663	0.020
Folate, total	μg	194	35.597	6			1	58	661	19
Folic acid	μg	0		0	Z		7	0	0	0
Folate, food	μg	194	35.597	6			1	58	661	19
Folate, DFE	μg_DFE	194		0	NC		4	58	661	19
Choline, total	mg	18.0		0	BFSN		4	5.4	61.2	1.8
Betaine	mg	550.4		0	BFSN		4	165.1	1871.3	55.0
Vitamin B12	μg	0.00		0	Z		7	0.00	0.00	0.00
Vitamin B12, added	μg	0.00		0	Z		7	0.00	0.00	0.00
Vitamin A, RAE	μg_RAE	469		0	AS		1	141	1594	47
Retinol	μg	0		0	Z		7	0	0	0
Carotene, beta	μg	5626	766.716	5	A		1	1688	19130	563
Carotene, alpha	μg	0	0.000	4	A		1	0	0	0
Cryptoxanthin, beta	μg	0	0.000	4	A		1	0	0	0
Vitamin A, IU	IU	9377		0	AS		1	2813	31883	938
Lycopene	μg	0	0.000	7	A		1	0	0	0
Lutein − zeaxanthin	μg	12198	1930.873	7	A		1	3659	41474	1220
Vitamin E (alpha-tocopherol)	mg	2.03	0.152	7			1	0.61	6.91	0.20
Vitamin E, added	mg	0.00		0	Z		7	0.00	0.00	0.00
Tocopherol, beta	mg	0.00	0.000	7			1	0.00	0.00	0.00
Tocopherol, gamma	mg	0.18	0.036	7			1	0.05	0.61	0.02
Tocopherol, delta	mg	0.00	0.000	7			1	0.00	0.00	0.00
Vitamin D	IU									
Vitamin K (phylloquinone)	μg	482.9		1			1	144.9	1641.9	48.3

Blanks in the Mean column indicate lack of reliable data. The Number of Data Points column is the number of analyses upon which the mean is based. Number of Data Points of zero indicates the mean was either calculated (as for Energy) or estimated, usually from a recipe, another form of the same food, or similar food.

Common measures:

Measure 1 = 30 g: 1 cup

Measure 2 = 340 g: 1 bunch

Measure 3 = 10.0 g: 1 leaf

Calories factors: protein 2.44; **fat** 8.37; **carbohydrate** 3.57

Food group: 11 Vegetables and Vegetable Products USDA National Nutrient Database for Standard Reference, Release 21 (2008)

TABLE 25.2   Nutrient Contents in Raw Spinach and Carrots
NDB No. 11124
Carrots, raw
Daucus carota
Includes USDA commodity food A099
Refuse:11% Crown, tops and scrapings

Nutrients and Units		Amount in 100 g of Edible Portion						Amount in Edible Portion of Common Measures of Food		
		Mean	Std. Error	Number of Data Points	Deriv Code	Source Code	Confi- dence Code	Measure 1	Measure 2	Measure 3
Proximates										
Water	g	88.29	0.429	33	JO	11		113.01	97.12	107.71
Energy	kcal	41		0	NC	4		53	45	50
Energy	kj	173		0	NC	4		221	190	211
Protein (N × 6.25)	g	0.93	0.008	19	JO	11		1.18	1.02	1.13
Total lipid (fat)	g	0.24	0.018	26	JO	11		0.30	0.26	0.29
Ash	g	0.97	0.014	19	JA	6		1.24	1.06	1.18
Carbohydrate, by difference	g	9.58		0	NC	4		12.27	10.54	11.69
Fiber, total dietary	g	2.8		4	A	1		3.6	3.1	3.4
Sugars, total	g	4.74		0	NC	4		6.06	5.21	5.78
Sucrose	g	3.59	0.280	11	JA	6		4.60	3.95	4.38
Glucose (dextrose)	g	0.59	0.141	11	JA	6		0.76	0.65	0.72
Fructose	g	0.55	0.097	11	JA	6		0.71	0.61	0.67
Lactose	g	0.00	0.000	5	JA	6		0.00	0.00	0.00
Maltose	g	0.00	0.000	5	JA	6		0.00	0.00	0.00
Galactose	g	0.00		0	Z	7		0.00	0.00	0.00
Starch	g	1.43		2	A	1		1.83	1.58	1.75
Minerals										
Calcium, Ca	mg	33	1.120	75	JA	6		43	37	41
Iron, Fe	mg	0.30	0.014	75	JA	6		0.38	0.33	0.36
Magnesium, Mg	mg	12	0.367	75	JA	6		16	14	15
Phosphorus, P	mg	35	0.755	75	JA	6		45	38	42
Potassium, K	mg	320	8.418	76	JA	6		409	352	390
Sodium, Na	mg	69	3.358	81	JA	6		88	75	84
Zinc, Zn	mg	0.24	0.011	76	JA	6		0.31	0.27	0.30
Copper, Cu	mg	0.045	0.007	70	JA	6		0.058	0.050	0.055
Manganese, Mn	mg	0.143	0.006	66	JA	6		0.183	0.158	0.175
Selenium, Se	μg	0.1	0.038	39		1		0.1	0.1	0.1
Vitamins										
Vitamin C, total ascorbic acid	mg	5.9	1.130	21	JA	6		7.6	6.5	7.2

(Continued)

B. EFFECTS OF INDIVIDUAL VEGETABLES ON HEALTH

TABLE 25.2   (Continued)

Nutrients and Units		Amount in 100 g of Edible Portion						Amount in Edible Portion of Common Measures of Food		
		Mean	Std. Error	Number of Data Points	Deriv Code	Source Code	Confi-dence Code	Measure 1	Measure 2	Measure 3
Thiamin	mg	0.066	0.011	21	JA	6		0.085	0.073	0.081
Riboflavin	mg	0.058	0.013	19	JA	6		0.074	0.064	0.071
Niacin	mg	0.983	0.215	19	JA	6		1.259	1.082	1.200
Pantothenic acid	mg	0.273	0.145	9	JA	6		0.350	0.300	0.333
Vitamin B6	mg	0.138	0.030	19	JA	6		0.176	0.151	0.168
Folate, total	μg	19	5.175	19	JA	6		24	21	23
Folic acid	μg	0		0	Z	7		0	0	0
Folate, food	μg	19	5.175	19	JA	6		24	21	23
Folate, DFE	μg_DFE	19		0	NC	4		24	21	23
Choline, total	mg	8.8		0	AS	1		11.3	9.7	10.7
Betaine	mg	0.4		1	A	1		0.5	0.4	0.5
Vitamin B12	μg	0.00		0	Z	7		0.00	0.00	0.00
Vitamin B12, added	μ	0.00		0	Z	7		0.00	0.00	0.00
Vitamin A, RAE	μg_RAE	835		0	NC	4		1069	919	1019
Retinol	μg	0		0	Z	7		0	0	0
Carotene, beta	μg	8285	1082.194	197	JO	11		10605	9114	10108
Carotene, alpha	μg	3477	531.987	190	A	1		4451	3825	4242
Cryptoxanthin, beta	μg	0		45	A	1		0	0	0
Vitamin A, IU	IU	16706		0	NC	4		21384	18377	20382
Lycopene	μg	1	0.800	8	A	1		1	1	1
Lutein – zeaxanthin	μg	256	62.707	10	JO	11		328	282	313
Vitamin E (alpha-tocopherol)	mg	0.66	0.269	11	JO	11		0.85	0.73	0.81
Vitamin E, added	mg	0.00		0	Z	7		0.00	0.00	0.00
Tocopherol, beta	mg	0.01	0.005	11	JO	11		0.01	0.01	0.01
Tocopherol, gamma	mg	0.00	0.000	11	JO	11		0.00	0.00	0.00
Tocopherol, delta	mg	0.00	0.000	11	JO	11		0.00	0.00	0.00
Vitamin D	IU									
Vitamin K (phylloquinone)	μg	13.2		4	A	1		16.9	14.5	16.1
Lipids										
Fatty acids, total saturated	g	0.037		0	NC	4		0.047	0.040	0.045
4:0	g	0.000		0	Z	7		0.000	0.000	0.000

(*Continued*)

B. EFFECTS OF INDIVIDUAL VEGETABLES ON HEALTH

TABLE 25.2   (Continued)

Nutrients and Units		Amount in 100 g of Edible Portion						Amount in Edible Portion of Common Measures of Food		
		Mean	Std. Error	Number of Data Points	Deriv Code	Source Code	Confi-dence Code	Measure 1	Measure 2	Measure 3
6:0	g	0.000		0	Z	7		0.000	0.000	0.000
8:0	g	0.000		2	A	1		0.000	0.000	0.000
10:0	g	0.000		2	A	1		0.000	0.000	0.000
12:0	g	0.000		2	A	1		0.000	0.000	0.000
13:0	g									
14:0	g	0.000		2	A	1		0.000	0.000	0.000
15:0	g	0.000		2	A	1		0.000	0.000	0.000
16:0	g	0.035		2	A	1		0.044	0.038	0.042
17:0	g	0.000		2	A	1		0.000	0.000	0.000
18:0	g	0.002		2	A	1		0.002	0.002	0.002
20:0	g	0.000		2	A	1		0.000	0.000	0.000
22:0	g	0.000		2	A	1		0.000	0.000	0.000
24:0	g	0.000		2	A	1		0.000	0.000	0.000
Fatty acids, total monounsaturated	g	0.014		0	NC	4		0.018	0.016	0.018
14:1	g	0.000		2	A	1		0.000	0.000	0.000
15:1	g	0.000		2	A	1		0.000	0.000	0.000
16:1 undifferentiated	g	0.002		2	A	1		0.002	0.002	0.002
17:1	g	0.000		2	A	1		0.000	0.000	0.000
18:1 undifferentiated	g	0.012		2	A	1		0.016	0.014	0.015
20:1	g	0.000		2	A	1		0.000	0.000	0.000
22:1 undifferentiated	g	0.000		2	A	1		0.000	0.000	0.000
24:1 c	g									
Fatty acids, total polyunsaturated	g	0.117		0	NC	4		0.150	0.129	0.143
18:2 undifferentiated	g	0.115		2	A	1		0.148	0.127	0.141
18:3 undifferentiated	g	0.002		2	A	1		0.002	0.002	0.002
18:4	g	0.000		2	A	1		0.000	0.000	0.000
20:2 n-6 c,c	g	0.000		2	A	1		0.000	0.000	0.000
20:3 undifferentiated	g	0.000		2	A	1		0.000	0.000	0.000
20:4 undifferentiated	g	0.000		2	A	1		0.000	0.000	0.000
20:5 n-3	g	0.000		2	A	1		0.000	0.000	0.000
22:5 n-3	g	0.000		2	A	1		0.000	0.000	0.000

(Continued)

TABLE 25.2   (Continued)

Nutrients and Units		Amount in 100 g of Edible Portion						Amount in Edible Portion of Common Measures of Food		
		Mean	Std. Error	Number of Data Points	Deriv Code	Source Code	Confidence Code	Measure 1	Measure 2	Measure 3
22:6 n-3	g	0.000		2	A	1		0.000	0.000	0.000
Fatty acids, total trans	g	0.000		0	Z	7		0.000	0.000	0.000
Cholesterol	mg	0		0	Z	7		0	0	0
Phytosterols	mg									
Amino acids										
Tryptophan	g	0.012		0	JA	6		0.015	0.013	0.014
Threonine	g	0.191		0	JA	6		0.245	0.210	0.233
Isoleucine	g	0.077		0	JA	6		0.098	0.084	0.093
Leucine	g	0.102		0	JA	6		0.130	0.112	0.124
Lysine	g	0.101		0	JA	6		0.129	0.111	0.123
Methionine	g	0.020		0	JA	6		0.026	0.023	0.025
Cystine	g	0.083		0	JA	6		0.106	0.091	0.101
Phenylalanine	g	0.061		0	JA	6		0.078	0.067	0.074
Tyrosine	g	0.043		0	JA	6		0.054	0.047	0.052
Valine	g	0.069		0	JA	6		0.088	0.076	0.084
Arginine	g	0.091		0	JA	6		0.117	0.100	0.111
Histidine	g	0.040		0	JA	6		0.051	0.044	0.048
Alanine	g	0.113		0	JA	6		0.144	0.124	0.138
Aspartic acid	g	0.190		0	JA	6		0.243	0.209	0.231
Glutamic acid	g	0.366		0	JA	6		0.469	0.403	0.447
Glycine	g	0.047		0	JA	6		0.060	0.051	0.057
Proline	g	0.054		0	JA	6		0.069	0.060	0.066
Serine	g	0.054		0	JA	6		0.069	0.059	0.066
Hydroxyproline	g									
Others										
Alcohol, ethyl	g	0.0		0	Z	7		0.0	0.0	0.0
Caffeine	mg	0		0	Z	7		0	0	0
Theobromine	mg	0		0	Z	7		0	0	0

Blanks in the Mean column indicate lack of reliable data. The Number of Data Points column is the number of analyses upon which the mean is based. Number of Data Points of zero indicates the mean was either calculated (as for Energy) or estimated, usually from a recipe, another form of the same food, or similar food.

Common measures:
Measure 1 = 128 g: 1 cup chopped
Measure 2 = 110 g: 1 cup grated
Measure 3 = 122 g: 1 cup strips or slices
Calories factors: protein 2.78; **fat** 8.37; **carbohydrate** 3.84
Food group: 11 Vegetables and Vegetable Products USDA National Nutrient Database for Standard Reference, Release 21 (2008)

can be expressed as International Units (IU), where 1 IU is the biological equivalent of $0.3\,\mu g$ retinol, or of $0.6\,\mu g$ β-carotene. Both spinach and carrots are good sources of vitamin A and can supply significant amounts of vitamin A for our daily needs even though their vitamin A equivalencies are not as high as 6 to 1 (RE) or 12 to 1 (RAE).

5. VITAMIN A AND HEALTH

It is well known that vitamin A deficiency can result in visual malfunctions such as night blindness and xerophthalmia [15], and it can reduce immune responsiveness [16], causing an increased incidence and/or severity of respiratory infections, gastrointestinal infections [17], and measles [18]. On the other hand, it is less well known that excessive intakes of vitamin A can result in bone mineral loss [19,20], teratogenicity [21], or liver damage [22]. While vitamin A deficiency is rare in adults in industrialized countries, excessive intakes of vitamin A are of concern in well-nourished populations. Spinach and carrots are safe sources of vitamin A because the provitamin A carotenoids are not associated with specific adverse health effects. Their conversion to vitamin A decreases when body stores are full. A high intake of provitamin A carotenoids (which should not be encouraged) can turn the skin yellow, but this is not considered dangerous to our health.

6. SPINACH-CARROTS AND HEALTH

In addition to serving as dietary precursors to vitamin A, dietary provitamin A carotenoids may function as fat-soluble antioxidants. That is, increased consumption of foods rich in carotenoids, such as spinach, is associated with decreased risk of some degenerative diseases such as age-related macular degeneration [23].

From Table 25.1 we can see spinach and carrots are rich in other nutrients as well. Spinach is an excellent source of vitamin K, manganese, folate, magnesium, iron, vitamin C, vitamin B2, calcium, potassium, and vitamin B6. It is a very good source of dietary fiber, copper, phosphorus, and zinc as well.

Although best known for their high content of β-carotene and α-carotene, carrots also contain a phytonutrient called falcarinol that may be responsible for the recognized epidemiological association between frequently eating carrots and a reduced risk of cancers [24].

Spinach contains moderately high purine. Unless one has a risk for gout or health problems related to purine-related metabolism, there is no reason to avoid it in raw form as long as one keeps total portion sizes within practitioner's guidelines.

Spinach contains oxalate. Individuals who have previously formed calcium oxalate stones are recommended to limit or reduce oxalate intakes to prevent these stones.

In addition, spinach is among the foods on which pesticide residues have been most frequently found. Therefore, individuals wishing to avoid pesticide-associated health risks may want to avoid consumption of spinach unless it is grown organically. Also, spinach is one of the foods most commonly associated with allergic reactions.

In summary:

- Spinach and carrots provide about 20–30% of the vitamin A in the US diet.
- Vitamin A is produced in human metabolism by the breakdown of provitamin A carotenoids – the yellow pigments in spinach leaf (dark green leafy vegetables) and carrot roots (yellow colored vegetables).
- The conversion efficiency of vegetable β-carotene to vitamin A is poorer than previously thought. However, with

condensed contents of provitamin A carotenoids, dietary spinach and carrots can supply significant amounts of vitamin A for our daily needs.

- Over-consumption of vitamin A can be toxic to humans but over-consumption of provitamin A carotenoids is never toxic since carotenoids breakdown to vitamin A and is well controlled. Even though over-consumption of carotenoids rich foods, such as carrots, is not harmful, it is not encouraged.

References

1. Underwood, B. A., & Arthur, P. (1996). The contribution of vitamin A to public health. *FASEB Journal, 10,* 1040–1048.
2. Moore, T. (1929). Vitamin A and carotene. I. The association of vitamin A activity with carotene in the carrot root. *The Biochemical Journal, 23,* 803.
3. Lindqvist, A., He, Y. K., & Anderson, S. J. (2005). Cell type-specific expression of β-carotene 9′,10′-monooxygenase in human tissues. *Histochemistry Cytochemical, 53,* 1403–1412.
4. Food and Nutrition Board, Institute of Medicine. (2001). 'Dietary Reference Intakes for Vitamin A, Vitamin K, Arsenic, Boron, Chromium, Copper, Iodine, Iron, Manganese, Molybdenum, Nickel, Silicon, Vanadium, and Zinc.' A Report of the Panel on Micronutrients, Subcommittees on Upper Reference Levels of Nutrients and of Interpretation and Use of Dietary Reference Intakes, and the Standing Committee on the Scientific Evaluation of Dietary Reference Intakes. Washington, DC: National Academy Press.
5. Sauberlich, H. E., Hodges, H. E., Wallace, D. L., Kolder, H., Canham, J. E., Hood, J., Raica, N., & Lowry, L. K. (1974). Vitamin A metabolism and requirements in the human studied with the use of labeled retinol. *Vitamins and Hormones, 32,* 251–275.
6. NRC. (1989). *Recommended Dietary Allowances.* (10th ed.). Washington, DC: National Academy Press.
7. Bauernfeind, J. C. (1972). Carotenoid vitamin A precursors and analogs in foods and feeds. *Journal of Agricultural and Food Chemistry, 20,* 456–473.
8. Deuel, H. J., Greenberg, S. M., Straub, E., Fukui, T., Chatterjee, A., & Zechmeister, L. (1949). Stereochemical configuration and provitamin A activity. VII. Neocryptoxanthin U. *Archives of Biochemistry, 23,* 239–240.
9. Advanced Data from Vital and Health Statistics. (2004). Dietary Intake of Selected Vitamins for the United States Population: 1999–2000. Number 339, March 12, 2004.
10. Grusak, M. (1997). Intrinsic stable isotope labeling of plants for nutritional investigations in humans. *The Journal of Nutritional Biochemistry, 8,* 164.
11. Tang, G., Qin, J., Dolnikowski, G. G., Russell, R. M., & Grusak, M. G. (2005). Spinach or carrot can supply significant amounts of vitamin A as assessed by feeding with intrinsically deuterium-labeled vegetables. *The American Journal of Clinical Nutrition, 82,* 821–828.
12. Haskell, M. J., Jamil, K. M., Hassan, F., Peerson, J. M., Hossain, M. I., Fuchs, G. J., & Brown, K. H. (2004). Daily consumption of Indian spinach (*Basella alba*) or sweet potatoes has a positive effect on total-body vitamin A stores in Bangladeshi men. *The American Journal of Clinical Nutrition, 80,* 705–714.
13. Ribaya-Mercado, J. D., Maramag, C. C., Tengco, L. W., Dolnikowski, G. G., Blumberg, J. B., & Solon, F. S. (2007). Carotene-rich plant foods ingested with minimal dietary fat enhance the total-body vitamin A pool size in Filipino schoolchildren as assessed by stable-isotope-dilution methodology. *The American Journal of Clinical Nutrition, 85,* 1041–1049.
14. Parker, R. S., Swanson, J. E., You, C.. S., Edwards, A. J., & Huang, T. (1999). Bioavailability of carotenoids in human subjects. *Proceedings Nutrition Society, 58*(1), 155–162.
15. Sommer, A. (1982). *Nutritional Blindness: Xerophthalmia and keratomalacia.* New York: Oxford University Press.
16. Ross, A., & Stephensen, C. (1996). Vitamin A and retinoids in antiviral responses. *FASEB Journal, 10,* 979–985.
17. Usha, N., Sankaranarayanan, A., Walia, B., & Ganguly, N. (1991). Assessment of preclinical vitamin A deficiency in children with persistent diarrhoea. *Journal of Pediatric Gastroenterology and Nutrition, 13,* 168–175.
18. Sommer, A., & West, K. Jr., (1996). *Vitamin A Deficiency. Health, Survival, and Vision.* New York: Oxford University Press.
19. Melhus, H., Michaelsson, K., Kindmark, A., Bergstrom, R., Holmberg, L., Mallmin, H., Wolk, A., & Ljunghall, S. (1998). Excessive dietary intake of vitamin A is associated with reduced bone mineral density and increased risk for hip fracture. *Annals of Internal Medicine, 129,* 770–778.
20. Feskanich, D., Singh, V., Willett, W. C., & Colditz, G. A. (2002). Vitamin A intake and hip fractures among postmenopausal women. *JAMA, 287*(1), 47–54.

21. Rothman, K. J., Moore, L. L., Singer, M. R., Nguygen, U. D. T., Mannino, S., & Milunsky, B. (1995). Teratogenicity of high vitamin A intake. *The New England Journal of Medicine, 333*, 1369–1373.

22. Oren, R., & Ilan, Y. (1992). Reversible hepatic injury induced by long-term vitamin A ingestion. *The American Journal of Medicine, 93*, 703–704.

23. Seddon, J. M., Rosner, B., Sperduto, R. D., Yannuzzi, L., Haller, J. A., Blair, N. P., & Willett, W. (2001). Dietary fat and risk for advanced age-related macular degeneration. *Archives of Ophthalmology, 119*(8), 1191–1199.

24. Kobak-Larsen, M., Christensen, L. P., Vach, W., Ritskes-Hoitinga, J., & Brandt, K. (2005). Inhibitory effects of feeding with carrots or (−)-falcarinol on development of azoxymethane-induced preneoplastic lesions in the rat colon. *Journal of Agricultural and Food Chemistry, 53*(5), 1823–1827.

26

Spinach and Health: Anticancer Effect

Naoki Maeda[1], Hiromi Yoshida[1,2] and Yoshiyuki Mizushina[1,2]

[1]Laboratory of Food and Nutritional Sciences, Department of Nutritional Science, Kobe-Gakuin University, Nishi-ku, Kobe, Hyogo, Japan

[2]Cooperative Research Center of Life Sciences, Kobe-Gakuin University, Nishi-ku, Kobe, Hyogo, Japan

1. INTRODUCTION

In spite of the many advances in cancer treatment, chemotherapy for solid tumors is still greatly limited by a lack of selective anticancer drugs and by the recurrence of drug-resistant tumors. Finding a source of novel chemotherapeutics continues to be a focus of effort. Diets rich in vegetables are known to reduce cancer risk, suggesting edible plants as potential sources of anticancer agents.

Multiple organisms are known to contain at least 14 types of DNA polymerase (pol) [1], which catalyzes DNA replication, repair, and recombination [1,2]. Pol inhibitors could be employed as anticancer chemotherapy agents because they inhibit cell proliferation. Based on this idea, we have found many new pol inhibitors over the past 10 years from natural compounds, in particular food materials.

Of these, glycoglycerolipids in the class sulfoquinovosyl diacylglycerol (SQDG, 3-O-(6-deoxy-6-sulfono-α-D-glucopyranosyl)-bis-1,2-O-diacylglycerol) from a fern [3] and an alga [4,5]

are particularly potent inhibitors of eukaryotic pols. SQDG and its related glycoglycerolipids were chemically synthesized [6,7] and SQDG was the strongest pol inhibitor in the tested compounds [8–10]; therefore, SQDG has potential as an agent for cancer chemotherapy. Other glycoglycerolipids, such as monogalactosyl diacylglycerol (MGDG) and digalactosyl diacylglycerol (DGDG), are also from the chloroplast membrane in plants [11]; therefore, plant glycoglycerolipids could all be promising agents for cancer chemotherapy.

2. GLYCOGLYCEROLIPID CONTENTS OF VEGETABLES

The amounts of the major plant glycoglycerolipids in common vegetables were widely investigated, and three major compounds, MGDG (Figure 26.1A), DGDG (Figure 26.1B), and SQDG (Figure 26.1C), were analyzed by thin layer chromatography (TLC) [12]. Each of these compounds was completely purified by silica gel column chromatography, and their chemical structures were determined by ^{1}H-, ^{13}C- and DEPT (distortionless enhancement by polarization transfer) NMR (nuclear magnetic resonance) spectroscopic analyses. The glycoglycerolipids could theoretically have stereoisomers of two configurations, α- or β-type, between the sugar and the glyceride. Natural MGDG and SQDG had the β- and α-type, respectively, and DGDG had α- and β-types (Figure 26.1).

The amount of each glycoglycerolipid in 11 vegetables is shown in Table 26.1. Parsley had the largest amount of total glycoglycerolipids (793.3 mg/100 g of sample), whereas sweet pepper had the smallest (61.3 mg/100 g of sample). As for DGDG, parsley had the largest amount (305.3 mg/100 g of sample), whereas spinach had the most MGDG and SQDG (480.6 and

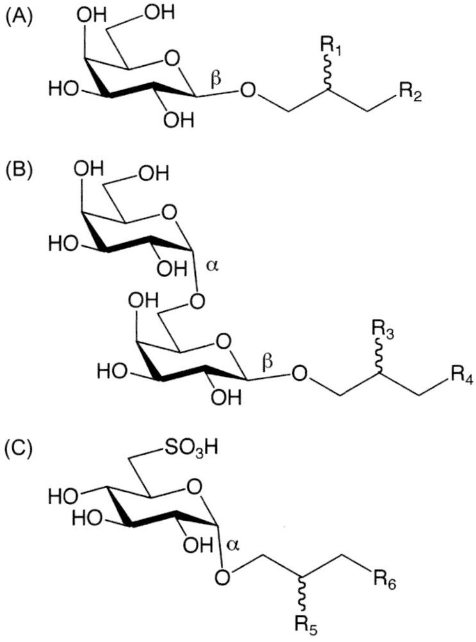

FIGURE 26.1 Chemical structures of major glycoglycerolipids from spinach.
(A) Monogalactosyl diacylglycerol (MGDG); (B) digalactosyl diacylglycerol (DGDG); (C) sulfoquinovosyl diacylglycerol (SQDG). R_1 to R_6 in these structures are fatty acids.

52.7 mg/100 g of sample, respectively) among the vegetables tested. Spinach also contained the second largest amount of total glycoglycerolipids (642.6 mg/100 g of sample). Spinach glycoglycerolipids had the strongest inhibitory effect on the activities of mammalian pols and human cancer cell proliferation in the tested vegetables [13]. These results suggest that spinach (*Spinacia oleracea* L.) could be a potent functional food for anticancer activity.

3. EFFECT OF EACH GLYCOGLYCEROLIPID FROM SPINACH ON THE ACTIVITIES OF DNA METABOLIC ENZYMES

The major glycoglycerolipids, MGDG, DGDG, and SQDG, were purified from spinach using silica gel column chromatography [12], and the inhibitory activities of pols and other DNA metabolic enzymes were investigated. Pols conduct cellular DNA synthesis [1], and eukaryotic cells reportedly contain three replicative types, pols α, δ, and ε, mitochondrial pol γ, and at least twelve repair types: pols β, δ, ε, ζ, η, θ, ι, κ, λ, μ, and σ and REV1 [2]. Since pols are essential for DNA replication, repair, and recombination and subsequently for cell division, inhibition of these enzymes will lead to cell death, especially under proliferation conditions. Inhibitors of eukaryotic pols should be considered potential agents for cancer chemotherapy. In the assay of pol activity, the substrates of pols were poly(dA)/oligo $(dT)_{12-18}$ and 2′-deoxythymidine 5′-triphosphate (dTTP) as the DNA template-primer and nucleotide substrate [i.e. 2′-deoxynucleoside 5′-triphosphate (dNTP)], respectively [14–16].

As shown in Figure 26.2, 100 µg/mL of MGDG and SQDG inhibited the activities of mammalian pols, including calf terminal deoxynucleotidyl transferase (i.e. DNA-independent pol), but DGDG had no such inhibitory effect. Interestingly, MGDG has no effect on pols β, λ and terminal deoxynucleotidyl transferase,

TABLE 26.1   The Major Glycoglycerolipid Composition of Vegetables (mg/100 g of Dried Vegetable)

Vegetable	MGDG	DGDG	SQDG	Total
Parsley	449.3 (0.566)[a]	305.3 (0.385)[b]	38.7 (0.049)[c]	793.3
Spinach	480.6 (0.748)	109.3 (0.170)	52.7 (0.082)	642.6
Green tea	255.3 (0.558)	176.7 (0.387)	25.0 (0.055)	457.0
Kale	196.0 (0.498)	157.3 (0.400)	40.0 (0.102)	393.3
Broccoli	113.3 (0.393)	149.3 (0.517)	26.0 (0.090)	288.6
Japanese honeywort	133.3 (0.549)	92.7 (0.382)	16.7 (0.069)	242.7
Chive	104.0 (0.441)	92.0 (0.390)	40.0 (0.169)	236.0
Cabbage	115.3 (0.581)	66.3 (0.335)	16.7 (0.084)	198.3
Common bean	117.3 (0.626)	41.3 (0.221)	28.7 (0.153)	187.3
Cucumber	36.7 (0.519)	34.0 (0.481)	ND	70.7
Sweet pepper	40.0 (0.653)	21.3 (0.347)	ND	61.3

Vegetables: parsley (*Petroselinum crispum* L.); spinach (*Spinacia oleracea* L.); green tea (*Thea sinensis* L.); kale (*Brassica oleracea* var. *acephala*); broccoli (*Brassica oleracea* var. *italica*); Japanese honeywort (*Cryptotaenia japonica*); chive (*Allium tuberosum* L.); cabbage (*Brassica oleracea* var. *capitata*); common bean (*Pisum sativum* L.); cucumber (*Cucumis sativus* L.); sweet pepper (*Capsicum annuum* var. *angulosum*).
[a]Ratio of MGDG in total glycoglycerolipids.
[b]Ratio of DGDG in total glycoglycerolipids.
[c]Ratio of SQDG in total glycoglycerolipids.
ND, not detected.

which are repair-related and/or recombination pols. The inhibitory effect of SQDG was more than 4-fold stronger than that of MGDG. These three glycoglycerolipids had no inhibitory effect on pols α and β from plant (cauliflower), prokaryotic pols such as *E. coli* pol I (Klenow fragment), T4 pol and *Taq* pol, and other DNA-metabolic enzymes such as human immunodeficiency virus type-1 (HIV-1) reverse transcriptase, T7 RNA polymerase, and bovine deoxyribonuclease I (DNase I) (Figure 26.2). These results suggest that MGDG and SQDG are selective mammalian pol inhibitors.

4. EXTRACTION AND ISOLATION OF THE GLYCOGLYCEROLIPID FRACTION FROM SPINACH

From the results of spinach glycoglycerolipids, an effective purification method of the glycoglycerolipid fraction from spinach was established, as shown in Figure 26.3. The major water-soluble substances were extracted from dried spinach (20 g) with 1000 mL of warm water (60°C). The tissue cake was added to 1000 mL of warm ethanol (60°C), and substances containing glycoglycerolipids were extracted (4.1 g) (i.e. ethanol extract). The ethanol extract was diluted with 1000 mL of 70% ethanol solution. The solution was subjected to Diaion HP-20 column chromatography (200 mL), a hydrophobic type of chromatography, washed with 1000 mL of 70% ethanol, eluted using 95% ethanol (1000 mL), and further eluted using chloroform (1000 mL). The 70% ethanol-washed solution, 95% ethanol-eluted solution, and chloroform-eluted solution were Fraction-I 'water-soluble fraction' (2.6 g), Fraction-II 'glycoglycerolipid fraction' (1.3 g), and Fraction-III 'fat-soluble fraction' (0.2 g), respectively.

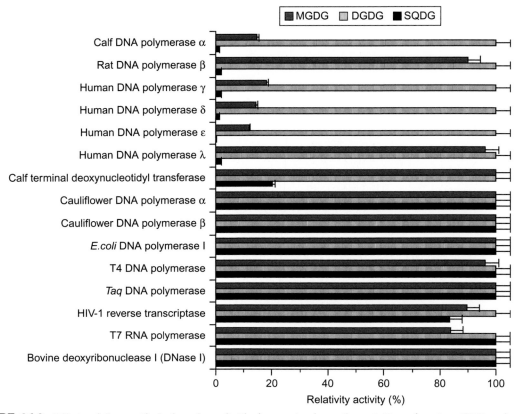

FIGURE 26.2 Effects of the purified glycoglycerolipids from spinach on the activities of various DNA polymerases and other DNA metabolic enzymes.

Purified MGDG, DGDG, and SQDG (100 μg/mL each) were incubated with each enzyme [14–16]. Enzyme activity in the absence of the compounds was taken as 100%. Data are shown as the means ± SEM of three independent experiments.

5. EFFECTS OF SPINACH FRACTIONS ON THE ACTIVITIES OF POL α AND HUMAN CANCER CELL GROWTH

The spinach ethanol extract and the three fractions by hydrophobic column chromatography were investigated for their inhibition of calf pol α, which is a replicative pol. As shown in Figure 26.4, the spinach glycoglycerolipid fraction (Fraction-II) dose-dependently inhibited the activity of pol α with an IC$_{50}$ value of 43.0 μg/ml, and the fat-soluble fraction (Fraction-III) slightly inhibited the activity

of pol α, although the water-soluble fraction (Fraction-I) did not show such an effect. Interestingly, the ethanol extract from spinach had no effect on pol α, although the extract contained pol-inhibitory glycoglycerolipids. There may be compounds that prevent pol α inhibitory activity by glycoglycerolipids in spinach extract.

To clarify the cytological effects of Fractions-I, II, and III from spinach, their influence on human cervix carcinoma (HeLa) survival was tested by MTT (3-(4,5-dimethylthiazol-2-yl)-2,5-diphenyl-2H-tetrazolium bromide) assay [17]. The cells were incubated with the spinach

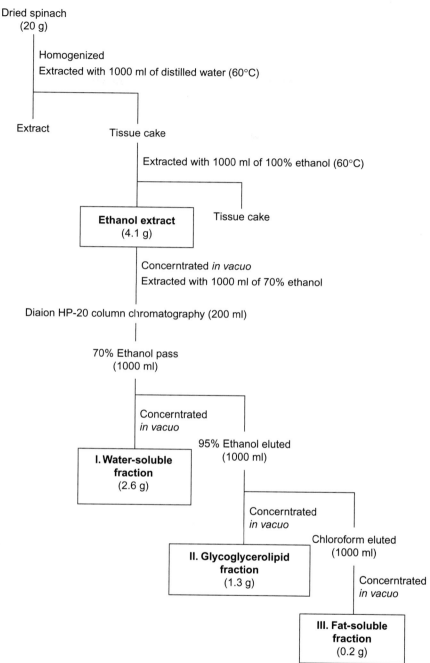

FIGURE 26.3 Purification method of the glycoglycerolipid fraction (Fraction-II) from dried spinach (*Spinacia oleracea* L.).

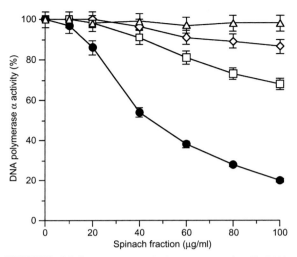

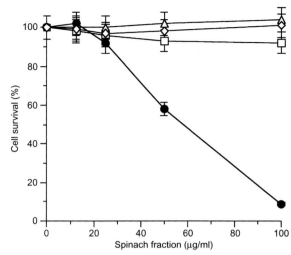

FIGURE 26.4 Inhibition of the activity of calf DNA polymerase α by spinach fractions.

Pol α activity (0.05 units, 5000 cpm) in the absence of the compounds was taken as 100% [15,16]. The spinach compounds tested and symbols used are as follows: ethanol extract (open diamond); water-soluble fraction (Fraction-I) (open triangle); glycoglycerolipid fraction (Fraction-II) (closed circle); fat-soluble fraction (Fraction-III) (open square). Data are shown as the means±SEM of three independent experiments.

FIGURE 26.5 Effect of spinach fractions on the growth of human cancer cells.

Dose-responsive curves of growth inhibition of human cervix carcinoma (HeLa) cells incubated with ethanol extract (open diamond), water-soluble fraction (Fraction-I) (open triangle), glycoglycerolipid fraction (Fraction-II) (closed circle), and fat-soluble fraction (Fraction-III) (open square) for 24 h. Cell proliferation was determined by MTT assay [17]. Data are shown as the means±SEM of four independent experiments.

fractions for 24 h. As shown in Figure 26.5, neither Fractions-I, III or the ethanol extract influenced cell growth. On the other hand, the spinach glycoglycerolipid fraction (Fraction-II) dose-dependently suppressed cell growth, and the LD_{50} value was 57.2 μg/mL. Since the value is approximately 1.3-fold the IC_{50} value on pol α activity, this inhibition must be mostly led by the function of pol α. A significant correlation was found between SQDG content and the inhibition of replicative pols such as pol α [13]. SQDG may be able to penetrate cancer cells and reach the nucleus, inhibiting the activities of pol α, and then the inhibition of pol α activity by SQDG may lead to cell growth suppression. Therefore, we concentrate on the properties of the glycoglycerolipid fraction from spinach in the latter part of this review.

6. HELA CELL GROWTH INHIBITORY PROPERTIES OF THE GLYCOGLYCEROLIPID FRACTION FROM SPINACH

Since the spinach glycoglycerolipid fraction inhibited the activities of mammalian pols and human cancer cell growth, the effect of the fraction on cell cycle regulation was investigated. When HeLa cells were treated with 57.2 μg/mL ($=LD_{50}$ value) of the glycoglycerolipid fraction for 48 h, the percentage of cells in the G1-phase increased (30.3 to 50.2%) and the percentage of cells in the G2/M-phase decreased (30.6 to 18.3%). S-phase cells were moderately decreased through incubation (39.1 to 31.5%). The cell cycle effect of the glycoglycerolipid fraction was investigated by the expression of cyclin proteins using Western blotting. Cyclin A and cyclin E proteins, which are regulated in the G1/S-phase [18–21], increased with glycoglycerolipid fraction treatment, but cyclin B, which is regulated in the G2/M-phase [22], decreased significantly. These results indicate that the glycoglycerolipid fraction induced G1- and S-phases arrest in human cancer cells, suggesting that HeLa cells treated with the glycoglycerolipid fraction, which inhibited the activity of replicative pols, overcome the S-phase block, divide, then enter a new G1-phase and stop cycling, being unable to replicate DNA and pass the G1/S checkpoint, and/or probably the fraction may act not so much as a DNA synthesis inhibitor (like, e.g., aphidicolin) but as mild anti-proliferative agents. Furthermore, DNA ladder formation was observed in HeLa cells treated with the LD_{50} value ($57.2 \mu g/mL$) of the fraction for 24h. These results suggest that both the inhibition of *in vivo* cell cycle arrest and the apoptotic effect occurred in HeLa cells by the glycoglycerolipid fraction from spinach, and the inhibition of mammalian pol activity by the fraction has a strong apoptotic effect on human cancer cells.

The spinach glycoglycerolipid fraction also dose-dependently suppressed the growth of other human cancer cell lines such as A549 (human lung cancer), BALL-1 (human leukemia), Molt-4 (human leukemia), and NUGC-3 (human gastric cancer), and the range of LD_{50} values was $58.5–81.7 \mu g/mL$ (Table 26.2). This fraction also inhibited the cell growth of all mice cancer cell lines (i.e. colon-26 and S-180), and the cancer cell lines from mice showed almost the same cell growth inhibitory results as cancer cell lines from humans; therefore, we used mouse colon-26 and S-180 cells in the latter part of this study.

7. ACUTE ORAL SAFETY TEST OF THE SPINACH GLYCOGLYCEROLIPID FRACTION

Analyses of the acute oral safety of the spinach glycoglycerolipid fraction were performed in ICR mice according to the standard protocols outlined by OECD 420. A sighting study at 300 mg/kg and another at 2000 mg/kg were performed. After checking for adverse effects, five female mice were used and the acute safety of orally administered glycoglycerolipid fraction at 2000 mg/kg for 14 days was assessed as the main study. At 2000 mg/kg of the spinach glycoglycerolipid fraction, no

TABLE 26.2 Inhibitory Effect of the Spinach Glycoglycerolipid Fraction from Spinach on Cultured Cancer Cell Growth

Cell Line	Type of Cancer	LD 50 Values ($\mu g/mL$)
A549	Human lung cancer	81.7 ± 9.9
BALL-1	Human B cell acute lymphoblastoid leukemia	59.0 ± 7.1
HeLa	Human cervix cancer	57.2 ± 6.9
Molt-4	Human T cell acute lymphoblastic leukemia	58.5 ± 7.1
NUGC-3	Human stomach cancer	79.8 ± 9.6
Colon-26	Mouse colon	74.5 ± 9.0
S-180	Mouse sarcoma	51.7 ± 6.5

Human cancer cells, such as A549, BALL-1, HeLa, Molt-4 and NUGC-3, and mouse cancer cells, such as colon-26 and S-180, were incubated with the glycoglycerolipid fraction for 24h. The rate of cell growth inhibition was determined by MTT assay [22]. Data are expressed as the mean $\pm$ SEM of five independent experiments.

TABLE 26.3   Acute Oral Safety of the Spinach Glycoglycerolipid Fraction in Mice

	Weight (g)						
	Body	**Brain**	**Heart**	**Lungs**	**Liver**	**Spleen**	**Kidneys**
Control	30.7 ± 1.6	0.475 ± 0.010	0.128 ± 0.001	0.193 ± 0.006	1.538 ± 0.136	0.132 ± 0.014	0.435 ± 0.025
Glycoglycerolipid fraction	31.5 ± 1.1	0.465 ± 0.016	0.139 ± 0.004	0.199 ± 0.006	1.643 ± 0.101	0.128 ± 0.012	0.460 ± 0.028

ICR mice treated with water or the spinach glycoglycerolipid fraction at 2000 mg/kg by oral administration in accordance with OECD 420. Fourteen days after administration, the body and major organ weights of mice were measured as the means ± SEM of five independent animals.

toxicity was noted at necropsy, indicated by a total lack of gross pathological alterations (Table 26.3).

8. PRELIMINARY MEDICATION OF TUMOR GROWTH INHIBITION BY THE SPINACH GLYCOGLYCEROLIPID FRACTION

The mice were randomized into two groups, and oral administration of the glycoglycerolipid fraction from spinach or phosphate buffer solution (PBS) (i.e. control) was started for 2 weeks. Then, 1×10^6 cells of the colon-26 cell line were implanted and the tumor volume of mouse solid tumor for 28 days after the implantation of cancer cells was measured ($N = 9$ of the each group). The spinach glycoglycerolipid fraction inhibited the tumor growth volume to 48.9% of the control tumor (Figure 26.6; Table 26.4). None of the mice showed any significant loss of body weight (control: 24.3 ± 0.4 g and the glycoglycerolipid fraction: 22.8 ± 0.4 g) and organopathy throughout the experimental period. Next, all mouse tumors were submitted for histopathological examination. PCNA (proliferating cell nuclear antigen), which is a marker protein for proliferating cells, is a reported biomarker for cancer [23]. The PCNA level expressed in tumor tissue was significantly decreased by orally

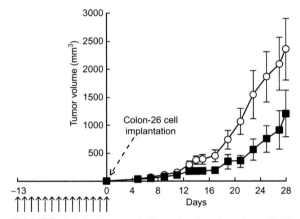

FIGURE 26.6   Effect of the spinach glycoglycerolipid fraction on the established colon cancer graft model for preliminary medication.

Vertical arrows show the injection timings of the glycoglycerolipid fraction (i.e. for 2 weeks, continuous intake before implantation). Control PBS (open circle), and the spinach glycoglycerolipid fraction (closed square) (20 mg/kg/day for 2 weeks). After the implantation of subcutaneous of colon-26 cells (1×10^6 cells), tumor volume was measured in all mice (BALB/c) for one month with no drug administration. All values are shown as the means ± SEM of five independent experiments.

administered spinach glycoglycerolipid fraction treatment (Table 26.4). Moreover, the tumor tissue of these mice was stained with hematoxylin and eosin (H&E) and mitoses counted, and a significant decrease of the mitotic index in the administered glycoglycerolipid fraction group compared with the control group was found (Table 26.4).

TABLE 26.4   Effect of the Spinach Glycoglycerolipid Fraction on Tumor Growth in Mice for Preliminary Medication

	Tumor Volume (mm³)	PCNA (%)	Mitosis (/HPF)
Control	2356.7 ± 550.1	64.0 ± 1.7	15.3 ± 1.7
Glycoglycerolipid fraction	1203.5 ± 416.0	55.0 ± 1.7	11.9 ± 0.4

The spinach glycoglycerolipid fraction at 20 mg/kg or PBS (i.e. control) was orally administered for 14 days, before colon-26 cells (1×10^6 cells) were implanted. The tumor volume of BALB/c mice on days 28 after cancer cell implantation was measured, and tumor tissue underwent histopathological analysis. Percentage of PCNA-positive cells and mitosis count in tumor tissue were used as an index of cell proliferation. Measurements are shown as the means $\pm$ SEM of nine independent experiments.

This animal model indicated not only tumor prevention but also some of the steps of metastasis. The major processes involved in metastasis are migration, intravasation, transport, extravasation, and metastatic colonization [24]. Metastatic colonization (i.e. adherence, growth and proliferation process of cancer cells) was shown in this study; therefore, the spinach glycoglycerolipid fraction must inhibit metastatic focus growth in tumor tissue.

9. ANTITUMOR ACTIVITY OF THE SPINACH GLYCOGLYCEROLIPID FRACTION IN MOUSE MODEL

Cells, 1×10^6 S-180, were implanted into the subcutaneous tissue of ICR mice. Four days after the implantation of sarcoma cells, the first group received daily PBS, and the second group received daily oral administration of the glycoglycerolipid fraction from spinach at 70 mg/kg. For 27 days after oral administration, the tumor volume and weight of all mice was measured once a day, and then the mice were killed humanely. As shown in Figure 26.7A, oral administration of the glycoglycerolipid fraction showed a superior antitumor effect. Oral administration of the fraction significantly decreased sarcoma tumor growth compared with the control mouse tumor (90%; $P < 0.05$), with tumor tissue almost lacking in mice. None of the mice showed any significant loss of body weight throughout the experimental period (Figure 26.7B). The *in vivo* antitumor effect of the glycoglycerolipid fraction induced no adverse drug reaction (i.e. no damage to major organs or drug-related animal death). These results suggest that the *in vitro* inhibition of cancer cell proliferation and *in vivo* induction of antitumor activity by the spinach fraction may be caused by the inhibition of replicative pols by the glycoglycerolipid.

10. DISCUSSION

Spinach plays an important role in nutrient supply. Spinach is a superior supplier of vitamin K, vitamin A, manganese, magnesium, folic acid, iron, vitamin C, vitamin B2, and potassium, and it includes a lot of dietary fiber, vitamin B6, vitamin E, and omega-3 fatty acids [25–28], which are essential for the maintenance, improvement, and regulation of human tissues. In addition, spinach includes major antioxidants of carotenoids and polyphenols (i.e. β-carotene, violaxanthin, neoxanthin, lutein, and phenolic acid) [29–34]. These antioxidant compounds or extracts including polyphenols, such as natural antioxidants (NAO) [34,35] of spinach, are related to disease prevention, including cancer [31,36]. Moreover, an epidemiological study suggested

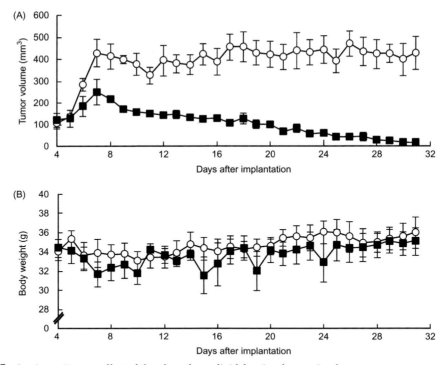

FIGURE 26.7   *In vivo* antitumor effect of the glycoglycerolipid fraction from spinach.
Four days after implantation of S-180 cells (1×10^6 cells) into ICR mice, we started orally administered PBS (control; open circle) and the glycoglycerolipid fraction (closed square) at 70 mg/kg for 27 days. All data are shown as the means $\pm$ SEM of four to five independent animals.

that spinach is associated with reduced cancer risk [37,38]. On the other hand, spinach includes oxalic acid and should be avoided by individuals with urinary calculus [39]. However, for many people, oxalic acid is not a problem.

The lipid composition of thylakoid membranes is highly conserved among higher plants such as spinach, algae, and cyanobacteria, comprised mainly of the following three glycoglycerolipids, MGDG, DGDG, and SQDG [40]. MGDG and DGDG are noncharged lipids, whereas SQDG possesses a negatively charged head group. Thylakoid membranes in plant chloroplasts and cyanobacterial cells are unique in possessing photosynthetic electron transport and photophosphorylation systems for the conversion of light to chemical energy. A mutant of *Chlamydomonas reinhardtii*, defective

in SQDG (*hf*-2), showed photosystem II (PSII) activity 40% lower than that of the wild type, an increase in sensitivity of PSII activity to 3-(3,4-dichlorophenyl)-1,1-dimethylurea (DCMU), and a lower growth rate [41–44]. In accordance with these observations, the incubation of isolated thylakoid membranes of *hf*-2 with SQDG *in vitro* reversed the lowered PSII activity. These results indicated that SQDG has the specific function of maintaining PSII properties.

The glycoglycerolipid fraction of spinach contains mainly MGDG, DGDG, and SQDG. SQDG and MGDG inhibit the activities of all mammalian pols and replicative pols, respectively [12], cancer cell proliferation [12,45], and angiogenesis [46]. Moreover, MGDG showed antipromotion [47,48] and anti-inflammatory

[49] actions. Furthermore, this fraction inhibited tumor growth *in vivo* with subcutaneous injection [50] or oral administration [51,52].

11. CONCLUSION

Spinach contains the most SQDG and MGDG in the glycoglycerolipid fraction of green vegetables tested, and a purification method of the spinach glycoglycerolipid fraction (Fraction-II) was established. This fraction inhibited the activities of mammalian replicative pols and suppressed the growth of human cancer cells. The water-soluble fraction (Fraction-I), fat-soluble fraction (Fraction-III), and ethanol extract from spinach had no influence, although the ethanol extract contained glycoglycerolipids such as SQDG and MGDG. Therefore, it is suggested that compounds preventing glycoglycerolipid bio-activity may be contained in the spinach ethanol extract. It is also important to purify the glycoglycerolipid fraction from spinach. In this review, we concluded that the spinach glycoglycerolipid fraction can inhibit mammalian pol activity, human cultured cancer cell growth, and *in vivo* solid tumor proliferation with oral administration. This fraction could help to prevent cancer, and be a functional food with anticancer activity.

ACKNOWLEDGMENTS

We are grateful for the donations of calf pol α, rat pol β, human pol γ, human pols δ, ε, and human pol λ by Dr M. Takemura of Tokyo University of Science (Tokyo, Japan), Dr A. Matsukage of Japan Women's University (Tokyo, Japan), Dr M. Suzuki of Nagoya University (Nagoya, Japan), Dr K. Sakaguchi of Tokyo University of Science (Chiba, Japan), and Dr O. Koiwai of Tokyo University of Science (Chiba, Japan), respectively. This work was supported in part by a Grant-in-aid for Kobe-Gakuin University Joint Research (A), and 'Academic Frontier' Project for Private Universities: matching fund subsidy from MEXT (Ministry of Education, Culture, Sports, Science and Technology), 2006–2010 (H. Y. and Y. M.). Y. M. acknowledges a Grant-in-Aid for Young Scientists (A) (No. 19680031) from MEXT, Grants-in-Aid from the Nakashima Foundation (Japan), Foundation of Oil & Fat Industry Kaikan (Japan), The Salt Science Research Foundation, No. 09S3 (Japan), and a Grant from the Industrial Technology Research Program from NEDO (Japan).

References

1. Hubscher, U., Maga, G., & Spadari, S. (2002). Eukaryotic DNA polymerases. *Annual Review of Biochemistry, 71*, 133–163.
2. Kornberg, A., & Baker, T. A. (1992). DNA replication, (2nd ed.). New York: W. H. Freeman Chaper 6, pp. 197–225.
3. Mizushina, Y., Watanabe, I., Ohta, K., Takemura, M., Sahara, H., Takahashi, N., Gasa, S., Sugawara, F., Matsukage, A., Yoshida, S., & Sakaguchi, K. (1998). Studies on inhibitors of mammalian DNA polymerase α and β: Sulfolipids from a pteridophyte, *Athyrium niponicum. Biochemical Pharmacology, 55*, 537–541.
4. Ohta, K., Mizushina, Y., Hirata, N., Takemura, M., Sugawara, F., Matsukage, A., Yoshida, S., & Sakaguchi, K. (1998). Sulfoquinovosyldiacylglycerol, KM043, a new potent inhibitor of eukaryotic DNA polymerases and HIV-reverse transcriptase type 1 from a marine red alga, *Gigartina tenella. Chemical and Pharmaceutical Bulletin, 46*, 684–686.
5. Ohta, K., Mizushina, Y., Hirata, N., Takemura, M., Sugawara, F., Matsukage, A., Yoshida, S., & Sakaguchi, K. (1999). Action of a new mammalian DNA polymerase inhibitor, sulfoquinovosyldiacylglycerol. *Biological & Pharmaceutical Bulletin, 22*, 111–116.
6. Hanashima, S., Mizushina, Y., Yamazaki, T., Ohta, K., Takahashi, S., Koshino, H., Sahara, H., Sakaguchi, K., & Sugawara, F. (2000). Structural determination of sulfoquinovosyldiacylglycerol by chiral syntheses. *Tetrahedron Letters, 41*, 4403–4407.
7. Hanashima, S., Mizushina, Y., Yamazaki, T., Ohta, K., Takahashi, S., Sahara, H., Sakaguchi, K., & Sugawara, F. (2001). Synthesis of sulfoquinovosylacylglycerols, inhibitors of eukaryotic DNA polymerase α and β. *Bioorganic & Medicinal Chemistry, 9*, 367–376.
8. Hanashima, S., Mizushina, Y., Ohta, K., Yamazaki, T., Sugawara, F., & Sakaguchi, K. (2000). Structure-activity

relationship of a novel group of mammalian DNA polymerase inhibitors, synthetic sulfoquinovosylacylglycerols. *Japanese Journal of Cancer Research: Gann, 91*, 1073–1083.

9. Murakami, C., Yamazaki, T., Hanashima, S., Takahashi, S., Ohta, K., Yoshida, H., Sugawara, F., Sakaguchi, K., & Mizushina, Y. (2002). Structure-function relationship of synthetic sulfoquinovosyl-acylglycerols as mammalian DNA polymerase inhibitors. *Archives of Biochemistry and Biophysics, 403*, 229–236.

10. Mizushina, Y., Xu, X., Asahara, H., Takeuchi, R., Oshige, M., Shimazaki, N., Takemura, M., Yamaguchi, T., Kuroda, K., Linn, S., Yoshida, H., Koiwai, O., Saneyoshi, M., Sugawara, F., & Sakaguchi, K. (2003). A sulphoquinovosyl diacylglycerol is a DNA polymerase ε inhibitor. *Biochemical Journal, 370*, 299–305.

11. Ishizuka, I., & Yamakiwa, T. (1985). In H. Wiegandt, (Ed.), *New comprehensive biochemistry* (Vol. 10). (p. 101). Amsterdam: Elsevier.

12. Murakami, C., Kumagai, T., Hada, T., Kanekazu, U., Nakazawa, S., Kamisuki, S., Maeda, N., Xu, X., Yoshida, H., Sugawara, F., Sakaguchi, K., & Mizushina, Y. (2003). Effects of glycolipids from spinach on mammalian DNA polymerases. *Biochemical Pharmacology, 65*, 259–267.

13. Kuriyama, I., Musumi, K., Yonezawa, Y., Takemura, M., Maeda, N., Iijima, H., Hada, T., Yoshida, H., & Mizushina, Y. (2005). Inhibitory effects of glycolipids fraction from spinach on mammalian DNA polymerase activity and human cancer cell proliferation. *Journal of Nutritional Biochemistry, 16*, 594–601.

14. Mizushina, Y., Yagi, H., Tanaka, N., Kurosawa, T., Seto, H., Katsumi, K., Onoue, M., Ishida, H., Iseki, A., Nara, T., Morohashi, K., Horie, T., Onomura, Y., Narusawa, M., Aoyagi, N., Takami, K., Yamaoka, M., Inoue, Y., Matsukage, A., Yoshida, S., & Sakaguchi, K. (1996). Screening of inhibitor of eukaryotic DNA polymerases produced by microorganisms. *Journal of Antibiotics (Tokyo), 49*, 491–492.

15. Mizushina, Y., Tanaka, N., Yagi, H., Kurosawa, T., Onoue, M., Seto, H., Horie, T., Aoyagi, N., Yamaoka, M., Matsukage, A., Yoshida, S., & Sakaguchi, K. (1996). Fatty acids selectively inhibit eukaryotic DNA polymerase activities in vitro. *Biochimica et Biophysica Acta, 1308*, 256–262.

16. Mizushina, Y., Yoshida, S., Matsukage, A., & Sakaguchi, K. (1997). The inhibitory action of fatty acids on DNA polymerase β. *Biochimica et Biophysica Acta, 1336*, 509–521.

17. Mosmann, T. (1983). Rapid colorimetric assay for cellular growth and survival: Application to proliferation and cytotoxicity assays. *Journal of Immunological Methods, 65*, 55–63.

18. Ohtsubo, M., Theodoras, A. M., Schumacher, J., Roberts, J. M., & Pagano, M. (1995). Human cyclin E, a nuclear protein essential for the G1-to-S phase transition. *Molecular and Cell Biology, 15*, 2612–2624.

19. Ohtsubo, M., & Roberts, J. M. (1993). Cyclin-dependent regulation of G1 in mammalian fibroblasts. *Science, 259*, 1908–1912.

20. Fang, F., & Newport, J. W. (1991). Evidence that the G1-S and G2-M transitions are controlled by different cdc2 proteins in higher eukaryotes. *Cell, 66*, 731–742.

21. Pagano, M., Pepperkok, R., Verde, F., Ansorge, W., & Draetta, G. (1992). Cyclin A is required at two points in the human cell cycle. *EMBO Journal, 11*, 961–971.

22. Nurse, P. (1990). Universal control mechanism regulating onset of M-phase. *Nature, 344*, 503–508.

23. Kubben, F. J., Peeters-Haesevoets, A., Engels, L. G., Baeten, C. G., Schutte, B., Arends, J. W., Stockbrugger, R. W., & Blijham, G. H. (1994). Proliferating cell nuclear antigen (PCNA): A new marker to study human colonic cell proliferation. *Gut, 35*, 530–535.

24. Fidler, I. J. (1990). Critical factors in the biology of human cancer metastasis: Twenty-eighth G.H.A. Clowes memorial award lecture. *Cancer Research, 50*, 6130–6138.

25. Schurgers, L. J., Shearer, M. J., Hamulyak, K., Stocklin, E., & Vermeer, C. (2004). Effect of vitamin K intake on the stability of oral anticoagulant treatment: Dose-response relationships in healthy subjects. *Blood, 104*, 2682–2689.

26. Tang, G., Qin, J., Dolnikowski, G. G., Russell, R. M., & Grusak, M. A. (2005). Spinach or carrots can supply significant amounts of vitamin A as assessed by feeding with intrinsically deuterated vegetables. *American Journal of Clinical Nutrition, 82*, 821–828.

27. Dainty, J. R., Bullock, N. R., Hart, D. J., Hewson, A. T., Turner, R., Finglas, P. M., & Powers, H. J. (2007). Quantification of the bioavailability of riboflavin from foods by use of stable-isotope labels and kinetic modeling. *American Journal of Clinical Nutrition, 85*, 1557–1564.

28. U.S. Department of Agriculture, Agricultural Research Service. (2005). USDA National Nutrient Database for Standard Reference, Release 21. Nutrient Data Laboratory Home Page, <http://www.nal.usda.gov/fnic/foodcomp/search/>.

29. Chitchumroonchokchai, C., Schwartz, S. J., & Failla, M. L. (2004). Assessment of lutein bioavailability from meals and a supplement using simulated digestion and caco-2 human intestinal cells. *Journal of Nutrition, 134*, 2280–2286.

30. Asai, A., Terasaki, M., & Nagao, A. (2004). An epoxide-furanoid rearrangement of spinach neoxanthin occurs in the gastrointestinal tract of mice and *in vitro*:

Formation and cytostatic activity of neochrome stereo-isomers. *Journal of Nutrition, 134*, 2237–2243.

31. Kotake-Nara, E., Kushiro, M., Zhang, H., Sugawara, T., Miyashita, K., & Nagao, A. (2001). Carotenoids affect proliferation of human prostate cancer cells. *Journal of Nutrition, 131*, 3303–3306.

32. Pool-Zobel, B. L., Bub, A., Muller, H., Wollowski, I., & Rechkemmer, G. (1997). Consumption of vegetables reduces genetic damage in humans: First results of a human intervention trial with carotenoid-rich foods. *Carcinogenesis, 18*, 1847–1850.

33. Chung, H. Y., Rasmussen, H. M., & Johnson, E. J. (2004). Lutein bioavailability is higher from lutein-enriched eggs than from supplements and spinach in men. *Journal of Nutrition, 134*, 1887–1893.

34. Bakshi, S., Bergman, M., Dovrat, S., & Grossman, S. (2004). Unique natural antioxidants (NAOs) and derived purified components inhibit cell cycle progression by downregulation of ppRb and E2F in human PC3 prostate cancer cells. *FEBS Letters, 573*, 31–37.

35. Lomnitski, L., Padilla-Banks, E., Jefferson, W. N., Nyska, A., Grossman, S., & Newbold, R. R. (2003). A natural antioxidant mixture from spinach does not have estrogenic or antiestrogenic activity in immature CD-1 mice. *Journal of Nutrition, 133*, 3584–3587.

36. Moeller, S. M., Jacques, P. F., & Blumberg, J. B. (2000). The potential role of dietary xanthophylls in cataract and age-related macular degeneration. *Journal of the American College Nutrition, 19*, 522S–527S.

37. Longnecker, M. P., Newcomb, P. A., Mittendorf, R., Greenberg, E. R., & Willett, W. C. (1997). Intake of carrots, spinach, and supplements containing vitamin A in relation to risk of breast cancer. *Cancer Epidemiology, Biomarkers & Prevention, 6*, 887–892.

38. Slattery, M. L., Benson, J., Curtin, K., Ma, K. N., Schaeffer, D., & Potter, J. D. (2000). Carotenoids and colon cancer. *American Journal of Clinical Nutrition, 71*, 575–582.

39. Taylor, E. N., & Curhan, G. C. (2007). Oxalate intake and the risk for nephrolithiasis. *Journal of the American Society of Nephrology, 18*, 2198–2204.

40. Benning, C. (1998). Biosynthesis and function of the sulfolipid sulfoquinovosyl diacylglycerol. *Annual Review of Plant Physiology and Plant Molecular Biology, 49*, 53–75.

41. Minoda, A., Sato, N., Nozaki, H., Okada, K., Takahashi, H., Sonoike, K., & Tsuzuki, M. (2002). Role of sulfoquinovosyl diacylglycerol for the maintenance of photosystem II in *Chlamydomonas reinhardtii. European Journal of Biochemistry, 269*, 2353–2358.

42. Sato, N., Aoki, M., Maru, Y., Sonoike, K., Minoda, A., & Tsuzuki, M. (2003). Involvement of sulfoquinovosyl diacylglycerol in the structural integrity and heat-tolerance of photosystem II. *Planta, 217*, 245–251.

43. Sato, N., Tsuzuki, M., Matsuda, Y., Ehara, T., Osafune, T., & Kawaguchi, A. (1995). Isolation and characterization of mutants affected in lipid metabolism of Chlamydomonas reinhardtii. *European Journal of Biochemistry, 230*, 987–993.

44. Dorne, A. J., Joyard, J., Block, M. A., & Douce, R. (1985). Localization of phosphatidylcholine in outer envelope membrane of spinach chloroplasts. *Journal of Cell Biology, 100*, 1690–1697.

45. Quasney, M. E., Carter, L. C., Oxford, C., Watkins, S. M., Gershwin, M. E., & German, J. B. (2001). Inhibition of proliferation and induction of apoptosis in SNU-1 human gastric cancer cells by the plant sulfolipid, sulfoquinovosyldiacylglycerol. *Journal of Nutritional Biochemistry, 12*, 310–315.

46. Matsubara, K., Matsumoto, H., Mizushina, Y., Mori, M., Nakajima, N., Fuchigami, M., Yoshida, H., & Hada, T. (2005). Inhibitory effect of glycolipids from spinach on *in vitro* and *ex vivo* angiogenesis. *Oncology Reports, 14*, 157–160.

47. Morimoto, T., Nagatsu, A., Murakami, N., Sakakibara, J., Tokuda, H., Nishino, H., & Iwashima, A. (1995). Anti-tumour-promoting glyceroglycolipids from the green alga, *Chlorella vulgaris. Phytochemistry, 40*, 1433–1437.

48. Wang, R., Furumoto, T., Motoyama, K., Okazaki, K., Kondo, A., & Fukui, H. (2002). Possible antitumor promoters in *Spinacia oleracea* (spinach) and comparison of their contents among cultivars. *Bioscience, Biotechnology and Biochemistry, 66*, 248–254.

49. Bruno, A., Rossi, C., Marcolongo, G., Di Lena, A., Venzo, A., Berrie, C. P., & Corda, D. (2005). Selective *in vivo* anti-inflammatory action of the galactolipid monogalactosyldiacylglycerol. *European Journal of Pharmacology, 524*, 159–168.

50. Maeda, N., Kokai, Y., Ohtani, S., Sahara, H., Hada, T., Ishimaru, C., Kuriyama, I., Yonezawa, Y., Iijima, H., Yoshida, H., Sato, N., & Mizushina, Y. (2007). Anti-tumor effects of the glycolipids fraction from spinach which inhibited DNA polymerase activity. *Nutrition Cancer, 57*, 216–223.

51. Maeda, N., Kokai, Y., Ohtani, S., Sahara, H., Kumamoto-Yonezawa, Y., Kuriyama, I., Hada, T., Sato, N., Yoshida, H., & Mizushina, Y. (2008). Anti-tumor effect of orally administered spinach glycolipid fraction on implanted cancer cells, colon-26, in mice. *Lipids, 43*, 741–748.

52. Maeda, N., Kokai, Y., Ohtani, S., Hada, T., Yoshida, H., & Mizushina, Y. (2009). Inhibitory effects of preventive and curative orally administered spinach glycoglycerolipid fraction on the tumor growth of sarcoma and colon in mouse graft models. *Food and Chemical, 112*, 205–210.

B. EFFECTS OF INDIVIDUAL VEGETABLES ON HEALTH

27

Potential Health Benefits of Rhubarb

Elisabetta M. Clementi[1] and Francesco Misiti[2]

[1]CNR, Istituto di Chimica del Riconoscimento Molecolare (ICRM), Rome, Italy
[2]Department of Health and Motor Sciences, University of Cassino, Cassino, (FR), Italy

1. GENERAL

1.1 History

Da huang, or Chinese rhubarb, is one of the most ancient and best known plants used in Chinese herbal medicine. Rhubarb and its wide range of uses were first documented in the *Divine Husbandman's Classic of Materia Medica*, which was written during the later Han Dynasty, around 200 AD. Rhubarb occurs in commerce under various names: Russian, Turkey, East Indian and Chinese; but the geographical source of all species is the same, the commercial names of the drug indicating only the route by which it formerly reached the European market. Previous to 1842, Canton was the only port of the Chinese Empire holding direct communication with Europe, so rhubarb mostly came by overland routes: the Russian rhubarb used to be brought by the Chinese to the Russian frontier town of Kiachta; the Turkey rhubarb received its name because it reached China by way of Asiatic Turkey, through the Levant; East Indian came by way of Singapore and other East Indian ports, and Chinese rhubarb was shipped from Canton [1]. Rhubarb first arrived in America in the 1820s, entering the country via Maine and Massachusetts and moving westwards with the settlers [2]. European herbalists recommended rhubarb as a laxative and diuretic and to treat kidney stones, gout, and liver diseases characterized by jaundice. Externally it was used to heal skin sores and scabs. Paradoxically, although larger doses were used as a laxative, small doses were used to treat dysenteric diarrhea [2]. The Chinese use rhubarb as an ulcer remedy and consider it a bitter, cold, dry herb used to clear 'heat' from the liver, stomach, and blood, to expel helminths, and to treat cancer, fever, upper intestinal bleeding (ulcers), and headache [3,4]. It is also used to treat toothaches [5].

1.2 Particular Common Names

These include: akar kalembak, chuòng diêp dai hoàng, dai hoàng, daioh, daiou, kot nam tao, rawind, Rhabarberwurzel, rhabarbarum, rhubarb, rhubard de Chine, rhubarb root, ta-huan, Canton rhubarb, Chinese rhubarb,

407

chong-gi huang, da-hunag, daio, Japanese rhubarb, medicinal rhubarb, racine de rhubarbee (French), rhaberber, rhei radix, rheun, rhizome rhei, shenshi rhubarb, tai hunag, Turkish rhubarb, turkey rhubarb [6].

1.3 Botany

Rhubarb is a perennial plant that grows from thick short rhizomes. The plants have large leaves that are somewhat triangular shaped with long fleshy petioles (Figure 27.1). The flowers are small, greenish-white to rose-red, and grouped in large compound leafy inflorescences. Three of the species (*R. officinale*, *R. palmatum*, and *R. tanguticum*) are highly regarded medicinal plants in China, and therefore widely cultivated.

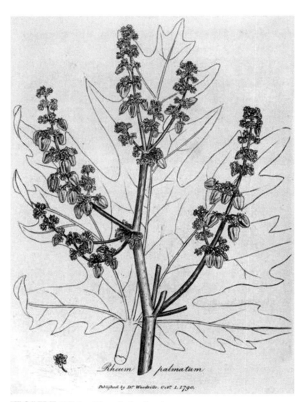

Rheum palmatum.

FIGURE 27.1 The rhubarb plant.
Royal Colleges of Physicians of London and Edinburgh Woodville, William (1802–1851).

Rheum is the technical name of the genus more familiarly known as rhubarb. It is said to be derived from Rha, the ancient name of the Volga, on whose banks the plants grow: but according to others it comes from the Greek *rheo* 'to flow,' in allusion to the purgative properties of the roots.

The genus is included in the Polygonaceae, a highly diversified genus with about 60 species, mainly distributed in the mountainous and desert regions of the Qinghai-Tibetan Plateau area and Asian interior [7]. These two adjacent areas are putatively centers of both origin and diversification of *Rheum*, due to its extremely diversified morphology and high endemism at both species and section level. Nine sections were recognized under *Rheum* by Losina-Losinskaya [7], who further suggested that Sect. *Palmata* is closely related to Sect. *Rheum*, and that both sections are primitive groups of the genus. Kao and Cheng [8] acknowledged only five of Losina-Losinskaya's sections and proposed two new sections: Sect. *Acuminata* based on the cordiform leaves of several species originally placed in Sect. *Rheum*; and the monotypic Sect. *Globulosa*, which has spherical inflorescences, but lacks distinct stems. To date, eight sections have been established and acknowledged under *Rheum*, according to Li [9], who further accepted Losina-Losinskaya's phylogenetic hypothesis of *Rheum* [7], although no new data were provided to support it. However, the phylogenetic relationships of some sections, e.g. Sect. *Nobilia* and Sect. *Globulosa*, are difficult to infer if exclusively based on their gross morphology, because they embrace unique morphologies which show no distinct connections with other sections [8,9]. Palynological research in 2001 has revealed diverse exine ornamentation in *Rheum* [10], but the variations in ornamentation are not consistent with the morphological classification. Some species with distinctly different morphology in different sections share similar types of pollen ornamentation while some species with very similar morphology have contrasting pollen ornamentation. For example, microechinate pollen ornamentation has been found in species of both Sect. *Rheum* and Sect. *Palmata*, while two species of Sect. *Nobilia* have very different ornamentation, pollen being densely microechinate and sparsely perforate in *R. globulosum*, but rugulate, verrucate, and high-relief in the other species, *R. nobile*. In addition, some papers have evidenced the existence of a relationship between chemical components pattern, genetic variations, and production sites [11,12]. This finding could represent an important factor in defining the appropriate use of rhubarb samples according to different therapeutic purposes.

1.4 Rhubarb as a Food Plant

Rhubarb is now grown in many areas and thanks to greenhouse production is available throughout much of the year. It is grown primarily for its fleshy petioles, commonly known as rhubarb sticks or stalks. In temperate climates rhubarb is one of the first food plants to be ready for harvest, usually in mid to late spring, and the season for field-grown plants lasts until September. Rhubarb is ready to be consumed as soon as it is harvested, and freshly cut stalks will be firm and glossy. The color of the rhubarb stalks can vary from the commonly associated deep red, through speckled pink, to simply green. The color results from the presence of anthocyanins, and varies according to both rhubarb variety and production technique. The color is not related to its suitability for cooking. The green-stalked rhubarb is more robust and has a higher yield, and the red-colored stalks are more popular with consumers. The stalks can be cooked in a variety of ways. Stewed, they yield a tart sauce that can be eaten with sugar and stewed fruit or used as filling for pies, tarts, and crumbles. This common use has led to the slang term for rhubarb, 'pie plant.' Cooked with strawberries or apples as a sweetener, or with stem or root

ginger, rhubarb makes excellent jam. It can also be used to make wine and as an ingredient in baked goods. Among species found in the wild, those most commonly used in cooking are the garden rhubarb (*R. rhabarbarum*) and *R. rhaponticum*, which, though a true rhubarb, bears the common name 'false rhubarb'. The many varieties of cultivated rhubarb more usually grown for eating are recognized as *Rheum × hybridum* in the Royal Horticultural Society's list of recognized plant names.

2. CHEMICAL COMPOSITION

The rhubarb plant is composed of three edible main parts: the leaves, the stalks or petioles, and rhizome or fleshy roots. The nutritional composition of rhubarb stalks for 100 g of edible portion is summarized in Table 27.1. The pleasantly acid taste (pH 3–3.6) is caused by the presence of malic, oxalic, and citric acids and saturated and unsaturated C_6 aldehydes and acids. In particular, because oxalic acid binds vital nutrients, long-term consumption of rhubarb can be problematic in individuals with kidney disorders, gout, and rheumatoid arthritis. Oxalic acid in fact combines with metals ions such as Ca^{2+}, Fe^{2+}, and Mg^{2+} to deposit crystals of the corresponding oxalates, which irritate the gut and kidneys. The median lethal dose (LD_{50}) for oxalic acid in rats is 375 mg/kg. So for a person of about 65 kg, about 25 g of pure oxalic acid is required to cause death. Rhubarb leaves are probably around 0.5% oxalic acid, so it would be necessary to eat quite a large serving of leaves, *ca.* 5 kg, to get that 24 g of oxalic acid. Some authors [13–15] state that the oxalic acid content of the petiole is approximately half that of the leaf blades, and that the oxalates of the leaves are predominantly potassium oxalate while calcium oxalate is predominant in the petiole.

The chemical composition of the roots is considerably different from that of the other parts

TABLE 27.1   Nutrition Factors in Rhubarb Stalks

Nutrient	Units	Value per 100 g of Edible Portion
pH		3 – 3.6
Water	g	93.61
Energy	kcal	21
Protein	g	0.90
Total lipid (fat)	g	0.20
Fiber	g	1.8
Sugars, total	g	1.10
Minerals		
Calcium, Ca	mg	86
Iron, Fe	mg	0.22
Magnesium, Mg	mg	12
Phosphorus, P	mg	14
Potassium, K	mg	288
Sodium, Na	mg	4
Zinc, Zn	mg	0.10
Copper, Cu	mg	0.021
Manganese, Mn	mg	0.196
Selenium, Se	μg	1.1
Vitamins		
Vitamin C, total ascorbic acid	mg	8.0
Thiamin	mg	0.020
Riboflavin	mg	0.030
Niacin	mg	0.300
Pantothenic acid	mg	0.085
Vitamin B6	mg	0.024
Folate, total	μg	7
Folate, food	μg	7
Folate, DFE	μg	7
Vitamin A, IU	IU	102
Vitamin A, RAE	μg	5
Vitamin E (α-tocopherol)	mg	0.38
Vitamin K (phylloquinone)	μg	41.0
Lipids		
Fatty acids, total saturated	g	0.053
14:0	g	0.001
16:0	g	0.046
18:0	g	0.004
Fatty acids, total monounsaturated	g	0.039
16:1 undifferentiated	g	0.001
18:1 undifferentiated	g	0.037
20:1	g	0.000
22:1 undifferentiated	g	0.000
Fatty acids, total polyunsaturated	g	0.099
18:2 undifferentiated	g	0.099
Other		
Carotene, beta	μg	61
Lutein + zeaxanthin	μg	170

Source: USDA National Nutrient Database for Standard Reference, Release 16 (July 2003).
DFE, dietary folate equivalents; IU, international units; RAE, retinol activity equivalents.

and not yet completely known, thus the rhizomes of this plant contain a series of biological active molecules which render the lyophilized and the dried rhizome particularly interesting for practical applications.

The most important constituents of rhubarb root are (Figure 27.2): i) hydroxyanthracene derivatives (anthraquinones) (3–12%), along with di-*O*,C-glycosides of the monomeric reduced forms (rheinosides A–D), and dimeric reduced forms (sennosides A–F); ii) tannins (5–10%); iii) stilbenes (1%); and iv) flavonoids (2–3%). In 2006, Komatsu et al. wrote a paper that evidenced a considerable variability in the chemical components in Rhei Rhizoma samples originating from different production areas [11]. Anthraquinones (Figure 27.3) are condensed *aromatic organic compound*s derivative of *anthracene*, insoluble in water but readily soluble in most organic solvents (*alcohol, nitrobenzene,* and *aniline*); these molecules are chemically fairly stable under normal conditions. The principal anthraquinones found in rhubarb include chrysophanol, emodin, aloeemodin, rhein, and physcion [16]. In Chinese traditional medicine, the anthraquinones have been used as a laxative, but pharmacological studies have ascribed to anthraquinones a new series of therapeutic properties besides confirming their laxative properties [17]. For example, chrysophanol has hemostatic and bactericidal

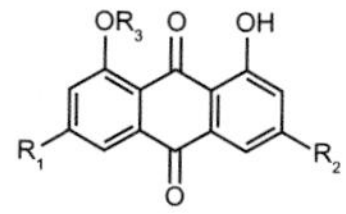

Anthraquinones

	R₁	R₂	R₃
1 Chrysophanol	H	CH₃	H
2 Emodin	OH	CH₃	H
3 Aloe-emodin	H	CH₂OH	H
4 Rhein	H	COOH	H
5 Physcion	OCH₃	CH₃	H

Anthraquinone glucosides

	R₁	R₂	R₃
6 Chrysophanol 8-O-β-D-glucopyranoside	H	CH₃	Glc.
7 Emodin 8-O-β-D-glucopyranoside	OH	CH₃	Glc.
8 Aloe-emodin 8-O-β-D-glucopyranoside	H	CH₃OH	Glc.
9 Rhein 5-O-β-D-glucopyranoside	H	COOH	Glc.
10 Physcion 8-O-β-D-glucopyranoside	OCH₃	CH₃	Glc.

Dianthrones

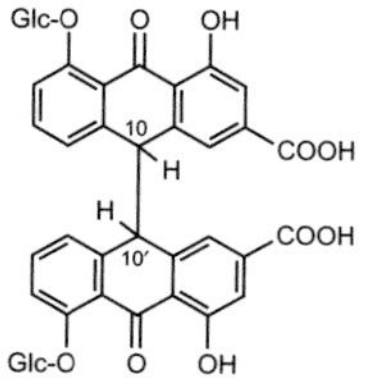

11 Sennoside A (10,10′-trans)
12 Sennoside B (10,10′-meso)

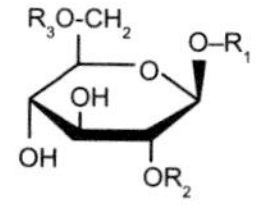

Phenylbutanones

	R₁	R₂	R₃
13 Lindleyin	L	H	G
14 Isolindleyin	L	G	H

Stilbenes

	R₁	R₂	R₃
15 Resveratrol 4′-O-β-D-glucopyranoside	S	H	H
16 Resveratrol 4′-O-β-D-(6″-O-galloyl)-glucopyranoside	S	H	G

Galloylglucoses

	R₁	R₂	R₃
21 1-O-Galloyl-β-D-glucose	H	G	H
22 6-O-Galloyl-β-D-glucose	H	H	G
23 1,6-Di-O-galloyl-β-D-glucose	G	H	G
24 1,2,6-Tri-O-galloyl-β-D-glucose	G	G	G

Acylglucoses

	R₁	R₂	R₃
25 1-O-Galloyl-2-O-cinnamoyl-β-D-glucose	G	C	H
26 1,6-Di-O-Galloyl-2-O-cinnamoyl-β-D-glucose	G	C	G
27 1,2-Di-O-Galloyl-6-O-cinnamoyl-β-D-glucose	G	G	C

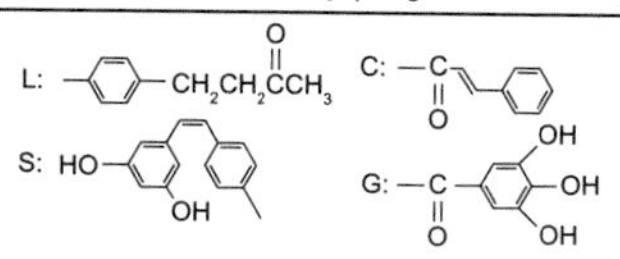

L: —CH₂CH₂CCH₃
C: —C
S: HO—
G: —C

Flavan-3-ols

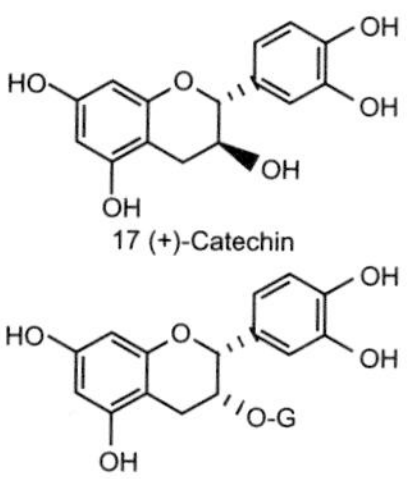

17 (+)-Catechin

18 (−)-Epicatechin 3-O-gallate

Procyanidins

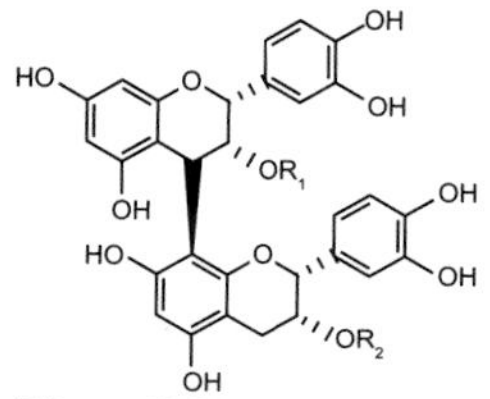

19 Procyanidin B-2 3′-O-gallate (R₁=H, R₂=G)
20 Procyanidin B-2 3,3′-di-O-gallate (R₁=R₂=G)

Phenol carboxylic acid

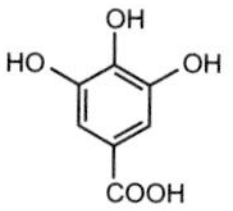

28 Gallic acid

Polymeric procyanidins

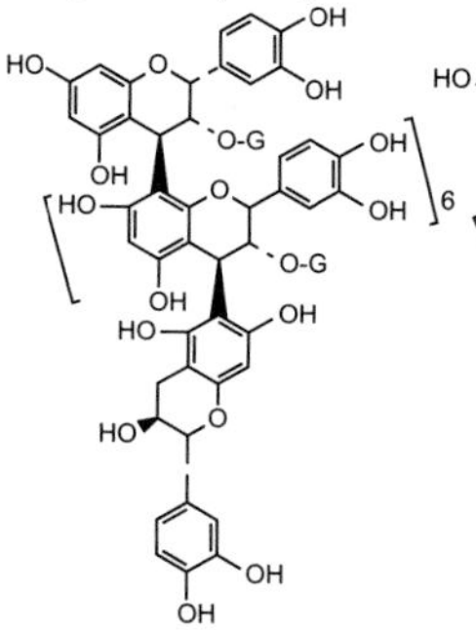

29 RG-tannin

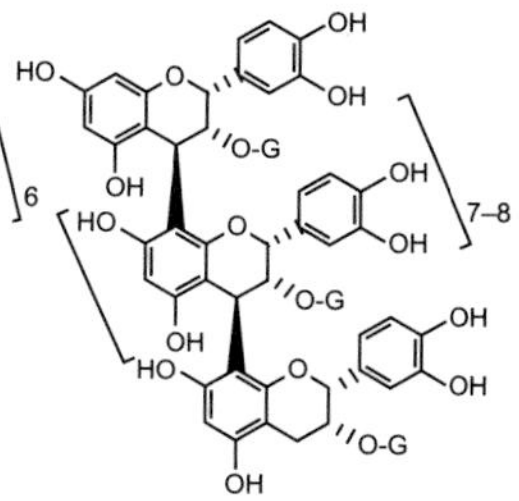

30 Rhatannin

FIGURE 27.2 Structure of the main chemical components purified from rhubarb roots. Data from Komatsu et al., 2006 [11].

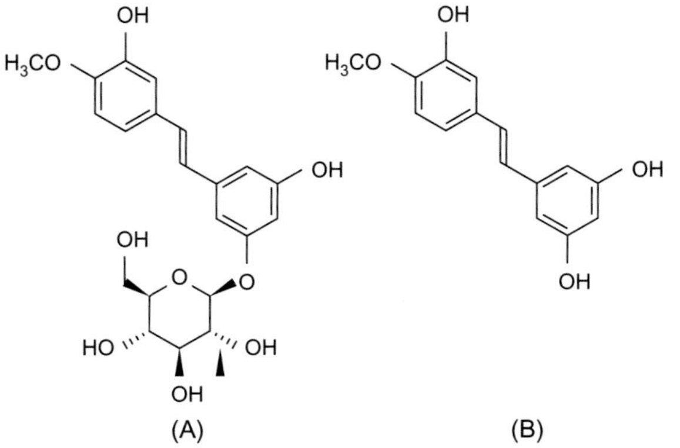

FIGURE 27.3 Chemical structures of rhaponticin (3,3′,5-trihydroxy-4′-methoxystilbene 3-O-d-glucoside) (A), and rhapontigenin (B), its aglycone metabolite.

properties, rhein is a potent antiviral, bactericidal and viricidal molecule, emodin is an anti-inflammatory, an antitumor agent, a vasorelaxant and viricide, and aloe-emodin has an antiherpetic and antileukemic action [18–23]. Other studies report that the major active principles producing the purgative action of rhubarb are the dimeric sennosides A–F [24]. The action of both the sennosides and rheinosides is limited to the large intestine, where they directly increase motor activity in the intestinal tract [24,25]. The *tannins* are *polyphenolic* compounds (Figure 27.2) containing *hydroxyls* and other suitable groups (such as *carboxyls*). Tannins comprise a large group of natural products widely distributed in plants; they have a great structural diversity and are usually divided into two basic groups: the hydrolyzable type, and the condensed type which are included in the polyphenolic compounds (see below). If ingested in excessive quantities, tannins inhibit partially the absorption of *minerals* such as *iron* and *calcium* [26] reducing the bioavailability of plant sources (i.e. iron non-heme), hence animal sources or heme iron absorption will not be affected by these molecules [27]. Plants that contain more than 10% tannins have potential adverse effects including upset stomach, renal damage, hepatic necrosis, and an increased risk

of esophageal and nasal cancer. Tannins are astringent and traditionally used topically for a variety of pathological skin conditions. Systematically, tannins have been used to treat diarrhea and inflamed mucous membranes. The principal tannin isolated from rhubarb root is gallotannin (Figure 27.2), a hydrolyzable molecule derived from a mixture of polygalloyl esthers of glucose, which have been shown to exert various biological effects ranging from anti-inflammatory to anticancer and antiviral implications [28,29]. The mechanisms underlying the anti-inflammatory effect of gallotannin include the scavenging of radicals [30] and inhibition of the expression of inflammatory mediators, such as some cytokines [29], inducible nitric-oxide synthase, and cyclooxygenase-2 [31]. Important tannin components isolated from rhubarb root are lindleyin and iso-lindleyin, molecules known to have several health benefits [32]. Lindleyin has estrogenic activity *in vitro* and *in vivo* and it is actually considered a novel phytoestrogen [33].

Stilbenes are *organic compounds* (Figure 27.2) that contain 1,2-diphenylethylene as a *functional group*. These molecules are synthesized via the phenylpropanoid pathway and share some structural similarities to estrogen. Stilbenes act as natural protective agents to defend the plant against viral and microbial attack, excessive ultraviolet exposure, and disease. Stilbenes from root rhubarb include: rhaponticin (3.5%), piceatannol-glucopyranoside (2.0%), desoxyrhaponticin (0.048%), isorhapontin (0.36%), rhapontigenin (0.58%), piceatannol (0.073%), desoxyrhapontigenin (0.015%), resveratrol (0.048%), rhaponticin 2-O-gallate (0.12%), and rhaponticin 6-O-gallate (0.087%). Rhaponticin has anti-allergic and antithrombotic properties, inhibits copper-induced lipid peroxidation [34,35], and has a neuroprotective effect against beta-amyloid induced neurotoxicity [36]. Moreover it has long been employed in Korea, Japan, and China as an oral hemostatic agent in treating oketsu, a disease characterized by poor circulation, pain, and chronic inflammation [37].

Rhaponticin and its aglycone rhapontigenin (Figure 27.3) have also been recommended by health professionals in Asian countries to treat and prevent allergies. Rhapontigenin and piceatannol were employed to assess their capacities to inhibit nitric oxide production in lipopolysaccharide-activated macrophages [38]. In mouse and rat experiments, anticancer, *anti-inflammatory*, blood-sugar-lowering and other beneficial cardiovascular effects of resveratrol have been reported [39–41]. The term *flavonoid* (or bioflavonoid) refers to a class of *plant metabolites* (Figure 27.2) which are derived from 2-(flavonoids) or 3-(isoflavonoids) or 4-phenyl-1, 4-*benzopyrone* (neoflavonoids). The principal flavo-noids purified from root rhubarb are catechin and epigallocatechin gallate (EGCg). They are reported to have various beneficial biological functions including antitumor [42], antioxidative, anti-inflammatory [43], and anti-obesity [44] activities. EGCg was reported to inhibit the differentiation of 3T3-L1 pre-adipocytes into adipocytes by suppressing the expression of peroxisome proliferator-activated receptor-γ2 and CCAAT/enhancer-binding protein-α, key transcription factors at an early stage of differentiation, and the expression of GLUT4 at a later stage [45]. Oral administration of EGCg to obese Zucker rats, a model of type II diabetes, significantly lowered blood glucose and insulin levels [46].

3. POTENTIAL HEALTH BENEFITS OF RHUBARB

3.1 Rhubarb and Alzheimer's Disease

Amyloid beta (1–42) peptide is considered responsible for the formation of senile plaques that accumulate in the brains of patients with Alzheimer's disease (AD). Researchers from Cassino University and CNR-ICRM (Chemistry of the Molecular Recognition Institute) in Italy studied *in vitro* the neuroprotective potentials of rhaponticin (3,3′,5-trihydroxy-4′-methoxy-stilbene 3-*O*-d-glucoside), a stilbene glucoside extracted from rhubarb roots (Rhei Rhizoma), and rhapontigenin, its aglycone metabolite (Figure 27.3A, B), against amyloid beta (1–42)-dependent toxicity. The obtained results show that rhapontigenin maintains significant cell viability in a dose-dependent manner and it exerts a protective effect on mitochondrial functionality. The protective mechanism mediated by the two stilbenes is related to their effect on bcl-2 gene family expression. Based on these studies, the authors suggest that rhaponticin and its main metabolite could be developed as agents for the management of AD [36].

3.2 Rhubarb and Cancer

Anthraquinones from rhubarb, *Rheum palmatum*, have been studied as anticancer molecules [47]. Emodin (3-methyl-1,6,8-trihydroxyanthraquinone) is one of the main active components contained in the root and rhizome of *Rheum palmatum* L. Emodin has been reported to exert antiproliferative effects in many cancer cell lines: HER-2/neu-overexpressing breast [47], lung cancer [48], leukemic HL-60 [49], human hepatocellular carcinoma [50], human cervical cancer, and prostate cancer cell lines through the activation of caspase-3 and up-regulation of TP-53 and p21 [49,50]. Moreover, emodin inhibits the kinase activity of p56lck, HER2/neu [47], and casein kinase II [51]. Therefore emodin significantly induces cytotoxicity in human myeloma cells through the elimination of myeloid cell leukemia 1 (Mcl-1) [52]. However, the actual molecular mechanisms of emodin-mediated tumor regression have not yet been fully defined. Rhein, another rhubarb anthraquinone, has the ability to induce cell death in tumor cells, inhibiting cell proliferation [53]. However, some test-tube studies showed that some rhubarb ingredients might have mutagenic activities, but the clinical link between the use of rhubarb and the development of gastric cancer was not clear [54]. A study by Huang et al. [54] suggests that the molecular effects of anthraquinones, though with different pathways, potentiate the antiproliferative effects of various chemotherapeutic agents. The most abundant anthraquinone of rhubarb, emodin, is capable of inhibiting cellular proliferation, inducing apoptosis, and preventing metastasis and these capabilities seem to act through tyrosine kinases, phosphoinositol 3-kinase (PI3K), protein kinase C (PKC), NF-kappa B (NF-κB), and mitogen-activated protein kinase (MAPK) signaling cascades [55]. Instead, the aloe-emodin antiproliferative property is related to p53 and its downstream p21 pathway. In one paper, a possible mechanistic explanation for the growth inhibitory effect of aloe-emodin includes cell cycle arrest and inducing differentiation [56]. Rhein, another major rhubarb anthraquinone, effectively inhibited the uptake of glucose in tumor cells, determining changes in membrane-associated functions, and led to cell death in the study.

3.3 Rhubarb and Pregnancy

Rhubarb showed a protective effect against high blood pressure during pregnancy (PIH) [57,58]. In a previous study, rhubarb was shown to reduce vascular endothelial cell damage significantly and alter the immune balance, which is effective in treating PIH.

3.4 Anti-atherogenic Effects

In traditional Chinese medicine some herbal prescriptions containing rhubarb have been used successfully to inhibit the development of atheromatous plaque formation: the molecular basis of this protective effect has been investigated and the results showed that the aqueous extract of rhubarb (AR) prevents the development of atherosclerosis through inhibiting vascular expression of pro-inflammatory and adhesion molecules via the regulation of the nitric oxide and endothelin system [59,60]. In particular, AR significantly reduces plasma low-density lipoprotein-cholesterol, increasing plasma high-density lipoprotein-cholesterol [61]. During purification of different components from rhubarb it was observed that piceatannol, a tetrahydroxystilbene extracted from *Rheum undulatum rizoma*, is the major mediator responsible for the vasorelaxing properties of rhubarb on endothelium-intact aorta and the vasorelaxant effects are mediated via the endothelium-dependent nitric oxide signalling pathway [35].

3.5 Rhubarb and Cholesterol

Several studies have shown that stalk fiber produced from rhubarb (*Rheum rhaponticum*) is potentially hypolipidemic [62]. In a study where diabetic rats were used, the rhubarb-fiber diet had no effect on the plasma cholesterol or triacylglycerol concentrations. The reported hypolipidemic effect of rhubarb [63] stalk fiber is probably due to the bile-acid-binding capacity of rhubarb fiber, which in turn upregulates cholesterol 7α-hydroxylase (cyp7a) activity, the first and the rate-limiting enzyme in the breakdown of cholesterol to bile acids.

3.6 Rhubarb and Antioxidant Effects

Antioxidative activities of components of rhubarb are shown in Table 27.2 [59] compared with probucol, a representative antioxidant that has been reported to prevent the progression of atherosclerotic lesions [64–66]. Among them, (−)-epicatechin 3-O-gallate showed strong antioxidative activity equal to that of probucol. The methanolic extracts from five kinds of rhubarb were found to show scavenging activity for 1,1-diphenyl-2-picryl hydrazyl (DPPH) radical and the superoxide radical [67].

TABLE 27.2 Antioxidative Activities of Components of Rhubarb, Compared With Probucol, a Representative Antioxidant

Components	Dose (μg/mL)
Anthraquinones free	
Aloe-emodin	ND
Rhein	3.65
Emodin	17.37
Chrysophanol	ND
Physcion	ND
Aloe-emodin 8-O-β-D-glucoside	ND
Rhein 8-O-β-D-glucoside	ND
Emodin 8-O-β-D-glucoside	ND
Chrysophanol 8-O-β-D-glucoside	ND
Glucoside anthrones	
• Sennoside A	>100
• Sennoside B	>100
Flavan-3-ols	
• Catechin	ND
• (–)-Epicatechin 3-O-gallate	0.35
Procyanidin B-2 3,3′-di-O-gallate	8.10
Naphtalene	ND
• 6-hydroxymusizin 8-O-β-glucoside	
Phenylbutanones	
• Lindleyin	1.25
• Isolindleyin	0.90
Stilbene	
• 3,5,4-Trihydroxystilbene 4′(6′-O-galloyl) glucose probucol	0.39

Antioxidative activities were found in seven of 18 rhubarb components. Among them, (−)-epicatechin 3-O-gallate showed strong antioxidative activity equal to that of probucol [59].
ND, not determinable.

3.7 Rhubarb and Renal Failure

In 1983, Yokozawa began a series of experiments to test the efficacy of rhubarb extract on rats with adenine-induced renal failure [68]. Through those experiments, the group found that an aqueous extract of rhubarb administered orally to rats after the induction of renal failure lowered BUN (blood urea nitrogen), SCr (plasma creatinine), methylguanidine, and guanidinosuccinic acid in a dose-dependent manner when compared to the controls. In 1991, the same group tested different tannins from rhubarb on rats with adenine-induced renal failure [69]. A decrease in SCr comparable to that observed after the administration of 270 mg rhubarb aqueous extract was found in rats given 5 and 10 mg of the rhubarb tannin (−)-epicatechin 3-O-gallate. On the contrary, another rhubarb tannin, procyanidin C-13,3′,3′-tri-O-gallate, produced significant increases in BUN and SCr causing aggravation of renal function [70]. Later, Zhang and el Nahas tested the efficacy of rhubarb in a subtotal nephrectomy model of chronic renal failure (CRF) [71]. Those rats consuming rhubarb extract in their drinking water (750 mg/kg per day) had significantly less proteinuria and less glomerulosclerosis than those rats not treated. Few clinical trials have been performed to determine the efficacy of rhubarb, and given the flaws in these studies, they can only be used to support further clinical trials. In a small study of patients with moderate to severe CRF ($N = 30$), Zhang et al. found the combination of captopril (25 mg, three times daily) and rhubarb extract (6–9 g/day) induced a non-significant improvement in renal function by normalizing BUN and SCr ($P > 0.05$) [72]. Kang et al. performed an uncontrolled observational study of 50 patients suffering from CRF [73]. While maintaining 'small doses of diuretics and hypertension pills' for 3 months, the main therapy was a decoction of 10 g rhubarb, 20 g dandelion (*Taraxacum officinale*), and 30 g oyster shell, administered orally or by retention enema. In the 1–3 years of follow-up, the BUN of 37 of the patients dropped from an average of 35 to 17.56 mmol/L, and pruritus and paresthesia were significantly decreased. The 13 remaining cases needed to be switched to full conventional therapy: seven had successful dialysis treatment and six died from complications of renal failure.

3.8 Rhubarb and Menstrual/Menopausal Problems

Rhubarb root has been used in traditional Chinese medicine to relieve menstrual problems. The herb stimulates the uterus and is thought to move stagnated blood, which also helps to relieve pains, cramps, and endometriosis correlated symptoms. Five known anthraquinones, chrysophanol, physcion, emodin, aloe-emodin, and rhein, showed estrogenic relative potency [74]. Moreover, a special extract of rhubarb (*Rheum rhaponticum*) has been [75] used to treat menopausal symptoms and the results have evidenced that rhubarb root extract administration significantly reduced the frequency and severity of hot flashes in perimenopausal women. Both these beneficial effects could be due to lindleyin [33], a tannin present in rhubarb root and to rhapontigenin and desoxyrhapontigenin, two stilbenes extracted from *Rheum rhaponticum* [76]. Lindleyin and rhapontigenin in fact bind '*in vitro*' specifically $ER\alpha$ and $ER\beta$, the principal estrogen receptors, demonstrating potential beneficial applications as selective ER modulators [77].

3.9 Antimicrobial Effect

It has been demonstrated that anthraquinone extracts were virucidal against HSV1 (herpes simplex virus type 1), measles, polio, and influenza virus *in vitro* [78–80]. Rhubarb extracts prevented cells from becoming infected with HSV1 [81]. In other studies, rhubarb was not active against HIV, vaccine viruses, polio virus, or hepatitis B virus (HBV) [20,82]. Other studies reported antibacterial activities [83,84].

3.10 Rhubarb and the Gastrointestinal System

In vitro data report that anthraquinone glycosides are hydrolyzed in the gut to aglycones which are reduced by bacteria to anthranols and anthrones. The laxative effect is due to: i) stimulation of colonic motility, which augments propulsion and accelerates colonic transit (which in turn reduces fluid absorption from the fecal mass; and ii) an increase in the paracellular permeability across the colonic mucosa probably owing to an inhibition of Na^+/K^+ exchanging ATPase or to an inhibition of chloride channels [85,86] which results in an increase in the water content in the large intestine [17,87,88]. Purgation is followed by an astringent effect owing to the tannins present [89].

In some animal studies rhubarb promoted electrical excitatory activity in the colon and duodenum which could be inhibited by atropine [90,91]. It has also been reported that rhubarb has excitatory actions on isolated gastric smooth muscle strips of guinea pig. The exciting action of rhubarb is partly mediated via the cholinergic M receptor, cholinergic N receptor, and L-type calcium channel [92]. Researchers also found that rhubarb extracts amplified the contraction amplitude of an isolated small intestinal smooth muscle of rabbit. The rate of change of contraction amplitude was elevated significantly after administration, while the frequency of contraction did not change [88].

3.11 Rhubarb and Ulcers

Rheum emodi extract was found to inhibit '*in vivo*' all strains of *Helicobacter pylori* (both sensitive and resistant) at a concentration of 10 mg/mL [93]. Furthermore, the microorganisms treated with rhubarb did not develop any resistance to root extract even after 10 subsequent passages, whereas it is well known that *H. pylori* strains usually acquire resistance to amoxicillin and clarithromycin after 10 sequential passages [94]. '*In vivo*' studies evidenced too that the alcoholic extract of *R. emodi*, containing especially anthraquinones derivatives, taken orally, may help prevent stomach ulcers in rats [93,95]. However, future studies concerning these properties must be carried out to justify the anti-ulcerogenic properties of rhubarb and the potential therapeutic applications.

3.12 Hepato-protective Effects

Rheum officinale Baill (Chinese rhubarb) is one of the most popular traditional herbal medicines that has pharmacological activities such as cathartic, anti-inflammatory, and antioxidative, and it is well known to promote excretion and decrease reabsorption of bilirubin, so it has been widely used alone or in combination with other crude drugs for the treatment of cholestatic hepatitis from ancient times in China [96]. The major active ingredient of the herb responsible for its hepato-protective effect is emodin, an anthraquinone. Ding et al. [21] reported that emodin has a protective effect on hepatocytes and a restoring activity on cholestatic hepatitis by anti-inflammation. The effects are mainly due to antagonizing pro-inflammatory cytokines and mediators, inhibiting oxidative damage, improving hepatic microcirculation, reducing impairment signals, and controlling neutrophil infiltration. Another important effect of rhubarb is to slow the development of liver fibrosis that, as the result of chronic liver injury, leads to the development of hepatocellular carcinoma and liver cirrhosis. Rhei rhizome extract significantly reduces liver fibrosis by the direct inhibition of stellate cells resulting in reduced type I procollagen mRNA, α-SMA (α-smooth muscle actin) and TIMP-1 and 2 (tissue inhibitors of metalloproteinases) expression [97].

4. ANALYTICAL PREPARATION

Although official and unofficial rhubarbs show significant difference in purgative effects, they are similar in physical appearance and are difficult to distinguish by conventional means. Therefore, a simple, rapid, and accurate method for the analysis of bioactive compounds in rhubarb is necessary for the quality control of crude rhubarb drugs and their pharmaceutical preparations. Qualitative and quantitative analyses of rhubarbs have been extensively pursued [98–102].

5. PRECAUTIONS AND SIDE EFFECTS

Chinese rhubarb should be prescribed only by a trained herbalist. It should not be taken by children under 12 years of age, or by pregnant women. Because it has been reported that anthranoid metabolites appear in breast milk, Rhizoma Rhei should not be used during lactation as there are insufficient data available to assess the potential for pharmacological effects in the breast-fed infant [2,103]. It should also not be used by persons with acute and chronic inflammatory diseases of the intestine, including Crohn's disease, appendicitis, and intestinal obstruction [104]. When rhubarb is used for its laxative-purgative qualities, the patient should be reminded that constipation is often caused by poor diet and lack of proper exercise. Correcting these patterns can improve bowel function without the use of any other therapy. People who use rhubarb root long-term for bowel problems may find that its effectiveness is decreased by extended use, and it can also cause electrolyte disturbances (hypokalemia, hypocalcemia), metabolic acidosis, malabsorption, weight loss, albuminuria, and hematuria [105,106]. Potassium deficiency can lead to disorders of heart function and muscular weakness, especially with concurrent use of cardiac glycosides, diuretics, or corticoadrenal steroids. Loss of potassium from the system can be decreased by combining the rhubarb root with licorice root. People with a history of renal stones or urinary problems should avoid rhubarb root, as should patients having blood thinning therapy with, for example, warfarin (Coumadin), clopidogrel (Plavix), aspirin, enoxaparin (Lovenox), dalteparin (Fragmin), or blood disease medication (sulfinpyrazone) [107,108]. It should be noted that rhubarb root can color the urine a deep yellow or even red. Melanotic pigmentation of the colonic mucosa (pseudomelanosis coli) has been observed in individuals taking anthraquinone laxatives for extended time periods [106,109]. The pigmentation is clinically harmless and usually reversible within 4–12 months after the drug has been discontinued [110]. It is possible to become intoxicated from an overdose of rhubarb, though the plant is generally safe to take in the recommended doses and manner. Signs of overdosage include vertigo, nausea, and vomiting, and severe abdominal cramps.

6. PREPARATIONS

Rhubarb root is usually taken from plants four or more years of age. It is dug up in the autumn, usually October, washed thoroughly, external fibers removed, and dried completely. The root is then pulverized and stored in a tightly closed container. Chinese rhubarb root usually comes from either China or Turkey. It can be purchased either in a powdered form or as a tincture. Medicinal uses supported by clinical data are reported for short-term treatment of occasional constipation [24,25,88]. The individually correct dosage is the smallest dosage necessary to maintain a soft stool. The average dose is 0.5–1.5 g of dried root. The powder is brought to a boil in a tea cup (240 mL) and then simmered at reduced heat for 10 minutes. It is advised to take 1 tablespoon (15 mL) at a time, up to 1 cup daily. [111]. The tincture can be taken in a dose of 1–2 mL three times a day. Note: 30–100 mg of hydroxyanthracene derivatives are equivalent to approximately 1.2–4.8 g of the dried root [112]. The total daily dose is 1–2 g of dried root. Provision of dosage information does not constitute a recommendation or endorsement, but rather indicates the range of doses commonly used in herbal practice. Doses may also vary according to the type and severity of the condition treated and individual patient conditions.

7. SUMMARY

Rhubarb root has been used for over two thousand years as a mild, yet powerful and effective laxative that empties the intestines and cleanses the bowels thoroughly. The anthraquinone glycosides are natural stimulants and produce a purging action, which make it useful for treating chronic constipation as reported by several clinical studies [87,113,114]. It is not recommended to use rhubarb preparations for medicinal uses in the presence of intestinal obstruction, appendicitis, abdominal pain, colitis, or Crohn's disease. It should be avoided or used with great caution by patients with a history of renal stones due its oxalate content [115]. It is not to be used by children under 12, by pregnant women, and nursing mothers. For healthy people, it is safe to use rhubarb in the recommended doses; however, using rhubarb can cause cramping and nausea. When using rhubarb, it is best to use it for no more than 2 weeks at a time. Long-term therapy can alter the body's normal balance of fluids and minerals, which can trigger irregular heart rhythms, kidney problems, fluid retention, promote bone deterioration, and produce a laxative dependence. Because of its potential to deplete potassium, it should be used cautiously by patients taking cardiac glycosides. Rhubarb leaves are *poisonous* because they contain high concentrations of oxalates, unlike the roots and petioles (stalks). It is recommended to use this herb only under the direction of a qualified professional.

References

1. Peigen, X., Livy, H., & Liwei, W. (1984). Ethnopharmacology study of Chinese rhubarb. *Journal of Ethnopharmacology, 10*, 275–293.

2. Castleman, M. (1991). *The healing herbs: The ultimate guide to the curative powers of nature's medicine.* pp. 305–307. Emmaus, PA: Rodale Press.

3. Borgia, M., Sepe, N., Borgia, R., & Ori-Bellometti, M. (1981). Pharmacological activity of an herbal extract: Controlled clinical study. *Current Therapeutic Research, 29*, 525–536.

4. Peirce, A. (1999). The American Pharmaceutical Association practical guide to natural medicine, New York: William Morrow and Company.

5. Duke, J. A. (1997). *Green Pharmacy.* p. 507. Emmaus, PA: Rodale Books.

6. Farnsworth, N. R. (Ed.), (1995). *NAPRALERT Database,* Chicago, IL: University of Illinois.

7. Losina-Losinskaya, A. S. (1936). The genus Rheum and its species. *Acta Instituti Botanici Academiae Scientiarum Unionis Rerum Publicarum Soveticarum Socialisticarum Ser, 1*, 5–141.

8. Kao, T. C., & Cheng, C. Y. (1975). Synopsis of the Chinese *Rheum. Acta Phytotaxonomica Sinica, 13*, 69–82.

9. Li, A. R. (1998). *Flora Republicae Popularis Sinicae.* Beijing: Science Press.

10. Yang, M. H., Zhang, D. M., Zheng, J. H., & Liu, J. Q. (2001). Pollen morphology and its systematic and ecological significance in *Rheum* (the Rhuburb genus, Polygonaceae) from China. *Nordic Journal of Botany, 21*, 411–418.

11. Komatsu, K., Nagayama, Y., Tanaka, K., Ling, Y., Cai, S. Q., Omote, T., & Meselhy, M. R. (2006). Comparative study of chemical constituents of rhubarb from different origins. *Chemical and Pharmaceutical Bulletin (Tokyo), 54*, 1491–1499.

12. Han, J., Ye, M., Xu, M., Qiao, X., Chen, H., Wang, B., Zheng, J., & Guo, D. A. (2008). Comparison of phenolic compounds of rhubarbs in the section *Deserticola* with *Rheum palmatum* by HPLC-DAD-ESI-MS^n. *Planta Medica, 74*, 873–879.

13. Pucher, G. W., Clark, H. E., & Vickery, H. B. (1937). The organic acids of rhubarb (rheum hybridum)*ii. the organic acid composition of the leaves. *The Journal of Biological Chemistry, 117*, 605–617.

14. Allsopp, A. (1937). Seasonal changes in the organic acids of rhubarb (Rheum hybridum). *The Biochemical Journal, 31*, 1820–1829.

15. Kasidas, G. P., & Rose, G. A. (1980). Oxalate content of some common foods: determination by an enzymatic method. *Journal of Human Nutrition, 34*, 255–266.

16. Koyama, J., Morita, I., & Kobayashi, N. (2007). Simultaneous determination of anthraquinones in rhubarb by high-performance liquid chromatography and capillary electrophoresis. *Journal of Chromatography A, 1145*, 183–189.

17. de Witte, P. (1993). Metabolism and pharmacokinetics of anthranoids. *Pharmacology, 47*(Suppl. 1), 86–97.

18. Huang, H. C., Lee, C. R., Chao, P. D., Chen, C. C., & Chu, S. H. (1991). Vasorelaxant effect of emodin, an anthraquinone from a Chinese herb. *European Journal of Pharmacology, 205*, 289–294.

19. Huang, Q., Shen, H. M., & Ong, C. N. (2004). Inhibitory effect of emodin on tumor invasion through suppression of activator protein-1 and nuclear factor-kappaB. *Biochemical Pharmacology, 68*, 361–371.

20. Li, Z., Li, L. J., Sun, Y., & Li, J. (2007). Identification of natural compounds with anti-hepatitis B virus activity from *Rheum palmatum* L. ethanol extract. *Chemotherapy, 53*, 320–326.

21. Ding, Y., Zhao, L., Mei, H., Zhang, S. L., Huang, Z. H., Duan, Y. Y., & Ye, P. (2008). Exploration of Emodin to treat alpha-naphthylisothiocyanate-induced cholestatic hepatitis via anti-inflammatory pathway. *European Journal of Pharmacology, 590*, 377–386.

22. He, T. P., Yan, W. H., Mo, L. E., & Liang, N. C. (2008). Inhibitory effect of aloe-emodin on metastasis potential in HO-8910PM cell line. *Journal of Asian Natural Products Research, 10*, 383–390.

23. Lin, C. W., Wu, C. F., Hsiao, N. W., Chang, C. Y., Li, S. W., Wan, L., Lin, Y. J., & Lin, W. Y. (2008). Aloe-emodin is an interferon-inducing agent with antiviral activity against Japanese encephalitis virus and enterovirus 71. *International Journal of Antimicrobial Agents, 32*, 355–359.

24. Nishioka, I. (1991). Biological activities and the active components of rhubarb.. *International Journal of Oriental Medicine, 16*, 193–212.

25. Reynolds, J. E. F. (Ed.), (1993). *Martindale, the Extra Pharmacopoeia,* 30th ed (p. 903). London: Pharmaceutical Press.

26. Brune, M., Rossander, L., & Hallberq, L. (1989). Iron absorption and phenolic compounds: importance of different phenolic structures. *European Journal of Clinical Nutrition, 43*, 547–557.

27. Hurrell, R. F., Reddy, M., & Cook, J. D. (1999). Inhibition of non-haem iron absorption in man by polyphenolic-containing beverages. *The British Journal of Nutrition, 81*, 289–295.

28. Van Molle, W., Vanden Berghe, J., Brouckaert, P., & Libert, C. (2000). Tumor necrosis factor-induced lethal hepatitis: Pharmacological intervention with

verapamil, tannic acid, picotamide and K76COOH. *FEBS Letters, 467,* 201–205.

29. Feldman, K. S., Sahasrabudhe, K., Lawlor, M. D., Wilson, S. L., Lang, C. H., & Scheuchenzuber, W. J. (2001). In vitro and in vivo inhibition of LPS-stimulated tumor necrosis factor-alpha secretion by the gallotannin beta-D-pentagalloylglucose. *Bioorganic and Medicinal Chemistry Letters, 11,* 1813–1815.

30. Hagerman, A. E., Dean, R. T., & Davies, M. J. (2003). Radical chemistry of epigallocatechin gallate and its relevance to protein damage. *Archives of Biochemistry and Biophysics, 414,* 115–120.

31. Lee, A. K., Sung, S. H., Kim, Y. C., & Kim, S. G. (2003). Inhibition of lipopolysaccharide-inducible nitric oxide synthase, TNF-alpha and COX-2 expression by sauchinone effects on I-kappa B alpha phosphorylation, C/EBP and AP-1 activation. *British Journal of Pharmacology, 139,* 11–20.

32. Tham, D. M., Gardner, C. D., & Haskell, W. L. (1998). Potential health benefits of dietary phytoestrogens: A review of the clinical, epidemiological and mechanistic evidence. *The Journal of Clinical Endocrinology and Metabolism, 83,* 2223–2235.

33. Usui, T., Ikeda, Y., Tagami, T., Matsuda, K., Moriyama, K., Yamada, K., Kuzuya, H., Kohno, S., & Shimatsu, A. (2002). Phytochemical Lindleyin, isolated from Rhei rhizome, mediates hormonal effects through estrogen receptors. *Journal of Endocrinology, 175,* 289–296.

34. Park, E. K., Choo, M. K., Yoon, H. K., & Kim, D. H. (2002). Antithrombotic and antiallergic activities of rhaponticin from Rhei Rhizoma are activated by human intestinal bacteria. *Archives of Pharmacal Research, 25,* 528–533.

35. Yoo, M. Y., Oh, K. S., Lee, J. W., Seo, H. W., Yon, G. H., Kwon, D. Y., Kim, Y. S., Ryu, S. Y., & Lee, B. H. (2007). Vasorelaxant effect of stilbenes from rhizome extract of rhubarb (*Rheum undulatum*) on the contractility of rat aorta. *Phytotherapy Research, 21,* 186–189.

36. Misiti, F., Sampaolese, B., Mezzogori, D., Orsini, F., Pezzotti, M., Giardina, B., & Clementi, M. E. (2006). Protective effect of rhubarb derivatives on amyloid beta (1-42) peptide-induced apoptosis in IMR-32 cells: A case of nutrigenomic. *Brain Research Bulletin, 71,* 29–36.

37. Matsuda, H., Tomohiro, N., Hiraba, K., Harima, S., Ko, S., Matsuo, K., Yoshikawa, M., & Kubo, M. (2001). Study on anti-Oketsu activity of rhubarb II. Antiallergic effects of stilbene components from Rhei undulati Rhizoma (dried rhizome of Rheum undulatum cultivated in Korea). *Biological and Pharmaceutical Bulletin, 24,* 264–267.

38. Kageura, T., Matsuda, H., Morikawa, T., Toguchida, I., Harima, S., Oda, M., & Yoshikawa, M. (2001). Inhibitors from rhubarb on lipopolysaccharide-induced nitric oxide production in macrophages: Structural requirements of stilbenes for the activity. *Bioorganic and Medicinal Chemistry, 9,* 1887–1893.

39. Kundu, J. K., & Surh, Y. J. (2008). Cancer chemopreventive and therapeutic potential of resveratrol: mechanistic perspectives. *Cancer Letters, 269,* 243–261.

40. Markus, M. A., & Morris, B. J. (2008). Resveratrol in prevention and treatment of common clinical conditions of aging. *Clinical Intervention Aging, 3,* 331–339.

41. Raval, A. P., Lin, H. W., Dave, K. R., Defazio, R. A., Della Morte, D., Kim, E. J., & Perez-Pinzon, M. A. (2008). Resveratrol and ischemic preconditioning in the brain. *Current Medicinal Chemistry, 15,* 1545–1551.

42. Yang, C. S., & Wang, Z. Y. (1993). Tea and cancer. *Journal of the National Cancer Institute, 85,* 1031–1049.

43. Tipoe, G. L., Leung, T. M., Hung, M. W., & Fung, M. L. (2007). Green tea polyphenols as an anti-oxidant and anti-inflammatory agent for cardiovascular protection. *Cardiovascular and Hematological Disorders Drug Targets, 7,* 135–144.

44. Klaus, S., Püeltz, S., Thöne-Reineke, C., & Wolfram, S. (2005). Epigallocatechin gallate attenuates diet-induced obesity in mice by decreasing energy absorption and increasing fat oxidation. *International Journal of Obesity, 29,* 615–623.

45. Furuyashiki, T., Nagayasu, H., Aoki, Y., Bessho, H., Hashimoto, T., Kanazawa, K., & Ashida, H. (2004). Tea catechin suppresses adipocyte differentiation accompanied by down-regulation of PPAR-γ2 and C/EBPα in 3T3-L1 cells. *Bioscience, Biotechnology, and Biochemistry, 68,* 2353–2359.

46. Dey, D., Mukherjee, M., Basu, D., Datta, M., Roy, S. S., Bandyopadhyay, A., & Bhattacharya, S. (2005). Inhibition of insulin receptor gene expression and insulin signaling by fatty acid: Interplay of PKC isoforms therein. *Cellular Physiology and Biochemistry, 16,* 217–228.

47. Wang, S. C., Zhang, L., Hortobagyi, G. N., & Hung, M. C. (2001). Targeting HER2: Recent developments and future directions for breast cancer patients. *Seminars in Oncology, 28,* 21–29.

48. Su, Y. T., Chang, H. L., Shyue, S. K., & Hsu, S. L. (2005). Emodin induces apoptosis in human lung adenocarcinoma cells through a reactive oxygen species-dependent mitochondrial signaling pathway. *Biochemical Pharmacology, 70,* 229–241.

49. Chen, Y. C., Shen, S. C., Lee, W. R., Hsu, F. L., Lin, H. Y., Ko, C. H., & Tseng, S. W. (2002). Emodin induces apoptosis in human promyeloleukemic HL-60 cells accompanied by activation of caspase 3 cascade but independent of reactive oxygen species production. *Biochemical Pharmacology, 64,* 1713–1724.

50. Shieh, D. E., Chen, Y. Y., Yen, M. H., Chiang, L. C., & Lin, C. C. (2004). Emodin-induced apoptosis through p53-dependent pathway in human hepatoma cells. *Life Sciences, 74*, 2279–2290.

51. Battistutta, R., Sarno, S., De Moliner, E., Papinutto, E., Zanotti, G., & Pinna, L. A. (2000). The replacement of ATP by the competitive inhibitor emodin induces conformational modifications in the catalytic site of protein kinase CK2. *The Journal of Biological Chemistry, 275*, 29618–29622.

52. Muto, A., Hori, M., Sasaki, Y., Saitoh, A., Yasuda, I., Maekawa, T., Uchida, T., Asakura, K., Nakazato, T., Kaneda, T., Kizaki, M., Ikeda, Y., & Yoshida, T. (2007). Emodin has a cytotoxic activity against human multiple myeloma as a Janus-activated kinase 2 inhibitor. *Molecular Cancer Therapeutics, 6*, 987–994.

53. Shi, P., Huang, Z., & Chen, G. (2008). Rhein induces apoptosis and cell cycle arrest in human hepatocellular carcinoma BEL-7402 cells. *The American Journal of Chinese Medicine, 36*, 805–813.

54. Huang, Q., Lu, G., Shen, H. M., Chung, M. C., & Ong, C. N. (2007). Anti-cancer properties of anthraquinones from rhubarb. *Medicinal Research Reviews, 27*, 609–630.

55. Huang, Q., Shen, H. M., Shui, G., Wenk, M. R., & Ong, C. N. (2006). Emodin inhibits tumor cell adhesion through disruption of the membrane lipid Raft-associated integrin signaling pathway. *Cancer Research, 66*, 5807–5815.

56. Guo, J. M., Xiao, B. X., Liu, Q., Zhang, S., Liu, D. H., & Gong, Z. H. (2007). Anticancer effect of aloe-emodin on cervical cancer cells involves G2/M arrest and induction of differentiation. *Guo Acta Pharmacol Sin, 28*, 1991–1995.

57. Zhang, Z. J., Cheng, W. W., & Yang, Y. M. (1994). Low-dose of processed rhubarb in preventing pregnancy induced hypertension. *Zhonghua Fu Chan Ke Za Zhi, 29*, 463–464, 509.

58. Wang, Z., & Song, H. (1999). Clinical observation on therapeutical effect of prepared rhubarb in treating pregnancy induced hypertension. *Zhongguo Zhong Xi Yi Jie He Za Zhi, 19*, 725–727.

59. Iizuka, A., Iijima, O. T., Kondo, K., Itakura, H., Yoshie, F., Miyamoto, H., Kubo, M., Higuchi, M., Takeda, H., & Matsumiya, T. (2004). Evaluation of Rhubarb using antioxidative activity as an index of pharmacological usefulness. *Journal of Ethnopharmacology, 91*, 89–94.

60. Liu, Y., Yan, F., Liu, Y., Zhang, C., Yu, H., Zhang, Y., & Zhao, Y. (2008). Aqueous extract of rhubarb stabilizes vulnerable atherosclerotic plaques due to depression of inflammation and lipid accumulation. *Phytotherapy Research, 22*, 935–942.

61. Ngoc, T. M., Hung, T. M., Thuong, P. T., Na, M., Kim, H., Ha do, T., Min, B. S., Minh, P. T., & Bae, K. (2008). Inhibition of human low density lipoprotein and high density lipoprotein oxidation by oligostilbenes from rhubarb. *Biological and Pharmaceutical Bulletin, 31*, 1809–1812.

62. Cheema, S. K., Goel, V., Basu, T. K., & Agellon, L. B. (2003). Dietary rhubarb (Rheum rhaponticum) stalk fibre does not lower plasma cholesterol levels in diabetic rats. *The British Journal of Nutrition, 89*, 201–206.

63. Goel, V., Cheema, S. K., Agellon, L. B., Ooraikul, B., & Basu, T. K. (1999). Dietary rhubarb (*Rheum rhaponticum*) stalk fibre stimulates cholesterol 7 alpha-hydroxylase gene expression and bile acid excretion in cholesterol-fed C57BL/6J mice. *The British Journal of Nutrition, 81*, 65–71.

64. Nagano, Y., Nakamura, T., Matsuzawa, Y., Cho, M., Ueda, Y., & Kita, T. (1992). Probucol and atherosclerosis in the watanabe heritable hyperlipidemic rabbit-long term antiatherogenic effects on established plaques. *Atherosclerosis, 92*, 131–140.

65. Braesen, J. H., Beisiegel, U., & Niendorf, A. (1995). Probucol inhibits not only the progression of atherosclerotic disease, but causes a different composition of atherosclerotic lesions in WHHL-rabbits. *Virchows Archiv, 426*, 179–188.

66. Oshima, R., Ikeda, T., Watanabe, K., Itakura, H., & Sugiyama, N. (1998). Probucol treatment attenuates the aortic atherosclerosis in Watanabe heritable hyperlipidemic rabbits. *Atherosclerosis, 137*, 13–22.

67. Matsuda, H., Morikawa, T., Toguchida, I., Park, J. Y., Harima, S., & Yoshikawa, M. (2001). Antioxidant constituents from rhubarb: Structural requirements of stilbenes for the activity and structures of two new anthraquinone glucosides. *Bioorganic and Medicinal Chemistry, 9*, 41–50.

68. Yokozawa, T., Zheng, P. D., Oura, H., Fukase, M., Koizumi, F., & Nishioka, I. (1983). Effect of extract from Rhei Rhizoma on adenine-induced renal failure in rats. *Chemical and Pharmaceutical Bulletin (Tokyo), 31*, 2762–2768.

69. Yokozawa, T., Fujioka, K., Oura, H., Nonaka, G., & Nishioka, I. (1991). Effects of rhubarb tannins on uremic toxins. *Nephron, 58*, 155–160.

70. Yokozawa, T., Fujioka, K., Oura, H., Nonaka, G., & Nishioka, I. (1993). Effects of rhubarb tannins on renal function in rats with renal failure. *Nippon Jinzo Gakkai Shi, 35*, 13–18.

71. Zhang, G., & el Nahas, A. M. (1996). The effect of rhubarb extract on experimental renal fibrosis. *Nephrology, Dialysis, Transplantation, 11*, 186–190.

72. Zhang, J. H., Li, L. S., & Zhang, M. (1990). Clinical effects of rheum and captopril on preventing

progression of chronic renal failure. *Chinese Medical Journal, 103*, 788–793.

73. Kang, Z., Bi, Z., Ji, W., Zhao, C., & Xie, Y. (1993). Observation of therapeutic effect in 50 cases of chronic renal failure treated with rhubarb and adjuvant drugs. *Journal of Traditional Chinese Medicine, 13*, 152–249.

74. Kang, S. C., Lee, C. M., Choung, E. S., Bak, J. P., Bae, J. J., Yoo, H. S., Kwak, J. H., & Zee, O. P. (2008). Anti-proliferative effects of estrogen receptor-modulating compounds isolated from *Rheum palmatum*. *Archives of Pharmacal Research, 31*, 722–726.

75. Heger, M., Ventskovskiy, B. M., Borzenko, I., Kneis, K. C., Rettenberger, R., Kaszkin-Bettag, M., & Heger, P. W. (2006). Efficacy and safety of a special extract of *Rheum rhaponticum* (ERr 731) in perimenopausal women with climacteric complaints: A 12-week randomized, double-blind, placebo-controlled trial. *Menopause, 13*, 724–726.

76. Wober, J., Möller, F., Richter, T., Unger, C., Weigt, C., Jandausch, A., Zierau, O., Rettenberger, R., Kaszkin-Bettag, M., & Vollmer, G. (2007). Activation of estrogen receptor-beta by a special extract of Rheum rhaponticum (ERr 731), its aglycones and structurally related compounds. *The Journal of Steroid Biochemistry and Molecular Biology, 107*, 191–201.

77. Cosman, F., & Lindsay, R. (1999). Selective estrogen receptor modulators: Clinical spectrum. *Endocrine Reviews, 20*, 418–434.

78. May, G., & Willuhn, G. (1978). Antiviral activity of aqueous extracts from medicinal plants in tissue cultures. *Arzneim-Forsch, 28*, 1–7.

79. Sydiskis, R. J., Owen, D. G., Lohr, J. L., Rosler, K. H., & Blomster, R. N. (1991). Inactivation of enveloped viruses by anthraquinones extracted from plants. *Antimicrobial Agents and Chemotherapy, 35*, 2463–2466.

80. Kurokawa, M., Ochiai, H., & Nagasaka, K. (1993). Antiviral traditional medicines against Herpes simplex virus (HSV 1), poliovirus *in vitro*. *Antiviral Research, 22*, 175–188.

81. Wang, Z., Wang, G., Xu, H., & Wang, P. (1996). Anti-herpes virus action of ethanol extract from the root and rhizome of rheum officinale Baill. *Chung-Kuo Chung Yao Tsa Chih – China Journal of Chinese Medica, 21*(3646), 384.

82. Chang, R., & Yeung, H. (1988). Inhibition of growth of human immunodeficiency virus *in vitro* by crud extracts Chinese medicinal herbs. *Antiviral Research, 9*, 163–175.

83. Lewis, W. S. (1977). *Medical botany: Plants affecting man's health*. New York: Wiley.

84. Cyong, J., Matsumoto, T., Arakawa, K., Kiyohara, H., Yamada, H., & Otsuka, Y. (1987). Anti-Bacterioides

fragilis substance from rhubarb. *Journal of Ethnopharmacology, 19*, 279–283.

85. Leng-Peschlow, E. (1986). Dual effect of orally administered sennosides on large intestine transit and fluid absorption in the rat. *The Journal of Pharmacy and Pharmacology, 38*, 606–610.

86. Yamauchi, K., Yagi, T., & Kuwano, S. (1993). Suppression of the purgative action of rhein anthrone, the active metabolite of sennosides A and B, by calcium channel blockers, calmodulin antagonists and indomethacin. *Pharmacology, 47*(Suppl. 1), 22–31.

87. Yamagishi, T., Nishizawa, M., Ikura, M., Hikichi, K., Nonaka, G., & Nishioka, I. (1987). New laxative constituents of rhubarb. Isolation and characterization of rheinosides A, B, C and D. *Chemical and Pharmaceutical Bulletin (Tokyo), 35*, 3132–3138.

88. Bisset, N. G. (1994). *Herbal Drugs and Phytopharmaceuticals*. p. 566. Stuttgart: Med Pharm CRC Press.

89. Grujic-Vasic, J., Bosnic, T., & Jovanovic, M. (1986). The examining of isolated tannins and their astringent effect. *Planta Medica, 52*, 548.

90. Jin, B. L., Ma, G. J., & Wang, X. L. (1989). Effect of rhubarb on electrical and contractive activities of the isolated intestine in rats (Chinese). *Chung-Kuo Chung Yao Tsa Chih – China Journal of Chinese Medica, 14*, 239–241, 256.

91. Yang, W. X., Jin, Z. G., & Tian, Z. S. (1993). Effect of dachengqi decoction and rhubarb on cellular electrical activities in smooth muscle of the guinea pig taenia coli (Chinese). *Chung-Kuo Chung His i Chieh Ho Tsa Chih, 13*, 33–35, 6.

92. Yu, M., Luo, Y. L., Zheng, J. W., Ding, Y. H., Li, W., Zheng, T. Z., & Qu, S. Y. (2005). Effects of rhubarb on isolated gastric muscle strips of guinea pigs. *World Journal of Gastroenterology, 11*, 2670–2673.

93. Ibrahim, M., Khan, A. A., Tiwari, S. K., Habeeb, M. A., Khaja, M. N., & Habibullah, C. M. (2006). Antimicrobial activity of *Sapindus mukorossi* and *Rheum emodi* extracts against H pylori: In vitro and *in vivo* studies. *World Journal of Gastroenterology, 12*, 7136–7142.

94. Sorberg, M., Hanberger, H., Nilsson, M., Bjorkman, A., & Nilsson, L. E. (1998). Risk of development of *in vitro* resistance to amoxicillin, clarithromycin, and metronidazole in *Helicobacter pylori*. *Antimicrobial Agents and Chemotherapy, 42*, 1222–1228.

95. Guo, K., Sun, L., & Lou, W. (1997). Four compounds of anthraquinones in *Rheum officinale* on *Helicobacter pylori* inhibition. *Chinese Pharmaceutical Journal, 32*, 278–279.

96. Hu, L. H. (1986). Experimental study of *Rheum officinale* Baill in treating severe hepatitis and hepatic coma. *Zhong Xi Yi Jie He Za Zhi, 6*, 41–42.

97. Jin, H., Sakaida, I., Tsuchiya, M., & Okita, K. (2005). Herbal medicine Rhei rhizome prevents liver fibrosis in rat liver cirrhosis induced by a choline-deficient L-amino acid-defined diet. *Life Science, 76,* 2805–2816.

98. Cai, Z., Lee, F. S., Wang, X. R., & Yu, W. J. (2002). A capsule review of recent studies on the application of mass spectrometry in the analysis of Chinese medicinal herbs. *Journal of Mass Spectrometry, 37,* 1013–1024.

99. Zhang, H. X., & Liu, M. C. (2004). Separation procedures for the pharmacologically active components in rhubarb. *Journal of Chromatography B, 812,* 175–181.

100. Koyama, J., Morita, I., Fujiyoshi, H., & Kobayashi, N. (2005). Simultaneous determination of anthraquinones, their 8-*β*-d-glucosides, and sennosides of *Rhei rhizoma* by capillary electrophoresis. *Chemical and Pharmaceutical Bulletin, 53,* 573–575.

101. Liu, L., Fan, L., Chen, H., Chen, X., & Hu, Z. (2005). Separation and determination of four active anthraquinones in Chinese herbal preparations by flow injection-capillary electrophoresis. *Electrophoresis, 26,* 2999–3006.

102. Ye, M., Han, J., Chen, H., Zheng, J., & Guo, D. (2007). Analysis of phenolic compounds in Rhubarbs using liquid chromatography coupled with electrospray ionization mass spectrometry. *Journal of the American Society Mass Spectrometry, 18,* 82–91.

103. Fok, T. F. (2001). Neonatal jaundice – traditional Chinese medicine approach. *Journal of Perinatology, 21* (Suppl. 1), S98–S100.

104. Siegers, C. P. (1992). Anthranoid laxatives and colorectal cancer. *Trends in Pharmacological Sciences, 13,* 229–231.

105. Godding, E. W. (1976). Therapeutics of laxative agents with special reference to the anthraquinones. *Pharmacology, 14*(Suppl. 1), 78–101.

106. Muller-Lissner, S. A. (1993). Adverse effects of laxatives: Facts and fiction. *Pharmacology, 47*(Suppl. 1), 138–145.

107. Massey, L. K., Roman-Smith, H., & Sutton, R. A. (1993). Effect of dietary oxalate and calcium on urinary oxalate and risk of formation of calcium oxalate kidney stones. *Journal of the American Dietetic Association, 93,* 901–906.

108. Yan, M., Zhang, L. Y., Sun, L. X., Jiang, Z. Z., & Xiao, X. H. (2006). Nephrotoxicity study of total rhubarb anthraquinones on Sprague Dawley rats using DNA microarrays. *Journal of Ethnopharmacology, 107,* 308–311.

109. Hardman, J. G., Goodman Gilman, A., Limbird, L., & Rall, T. W. (1990). *Goodman and Gilman's the pharmacological basis of therapeutics.* (8th ed). New York: McGraw Hill.

110. (1992). United States Pharmacopeia, drug information. Rockville, MD: Pharmacopeial Convention..

111. Hoffman, D., & Quayle, L. (1999). *The complete illustrated herbal: A safe and practical guide to making and using herbal remedies.* New York: Barnes and Noble Publishing.

112. Wichtl, M., (Ed.,) (1994). Rhei radix – Turkish rhubarb root (English translation by Norman Grainger Bisset). In: *Herbal drugs and phytopharmaceuticals.* (pp. 415–418). Stuttgart: CRC Press.

113. de Witte, P., & Lemli, L. (1990). The metabolism of anthranoid laxatives. *Hepatogastroenterology, 37,* 601–605.

114. Qu, Y., Wang, J. B., Li, H. F., Wang, Q., Xiao, X. H., & He, Y. Z. (2008). Study on relationship of laxative potency and anthraquinones content traditional Chinese drugs containing. *Zhongguo Zhong Yao Za Zhi, 33,* 806–808.

115. McGuffin, M., Hobbs, C., Upton, R., & Goldberg, A. (1997). *American Herbal Products Association's Botanical Safety Handbook.* p. 231. Boca Raton, FL: CRC Press.

28

Health Benefits of Fenugreek (*Trigonella foenum-graecum leguminosse*)

Tapan K. Basu and Anchalee Srichamroen

Department of Agriculture, Food and Nutritional Science, Faculty of Agricultural, Environmental and Life Sciences, University of Alberta, Edmonton, AB, Canada

1. INTRODUCTION

Fenugreek (*Trigonella foenum-graecum Leguminosse*) is a self-pollinating crop, which is a native plant of the Indian subcontinent and the Eastern Mediterranean region. The crop extends to central Asia and North Africa, and more recently it has been successfully grown in Central Europe, the UK, and North America. It belongs to the Papilionaceae section of the family Leguminosae. The species name, foenum-graecum, means 'Greek hay,' while the genus name, Trigonella, means 'little triangle' referring to the shape of its leaflet (Figure 28.1). Fenugreek plants have an erect growth habit with a height of 0.5–0.8 meters, mainly tap rooted with one main stem. Stems have alternating branches up to 0.4 meters with alternating compound pinnate trifoliate leaves 2–3 cm long. Flowers are axillary and cream in colour (1 cm long), developing into long slender (15 cm) green then yellow-brown pods. Each mature brown pod contains approximately 20 small yellow to brownish yellow seeds (Figure 28.2).

2. TRADITIONAL USES

Fenugreek is used for a variety of purposes. In the USA, fenugreek seed extract is the principal flavoring ingredient of simulated maple syrup [1]. It is also used as a tobacco flavoring ingredient, hydrolyzed vegetable protein flavor, perfume base, and a source of steroid sapogenin in drug manufacturing industries [2,3]. In India, the ground seeds are used in spice mixtures and as a condiment, and therefore constitute an important ingredient in chutneys and spice blends. The leaves are commonly consumed as a vegetable [4]. A poultice of leaves is often used on burns and swellings and applied to the scalp to treat baldness [5]. Indians also incorporate fenugreek into Methipak, a traditional food consumed during pregnancy and lactation.

In Egypt, the seeds are either eaten raw after sprouting or are used to make a confection following roasting, grinding, and cooking them with treacle and sesame seeds. Seeds are also sometimes added to maize and corn flour in bread making [6].

FIGURE 28.1   Fenugreek plant.

3. CHEMICAL COMPOSITION

Fenugreek leaves have been found to contain 25.0% protein, 25.9% starch, 12.9% neutral detergent fiber, 4.3% gum 10.8% ash, and 6.5% lipids. They have also been found to be a rich source of calcium, iron, β-carotene, and other vitamins [7,8]. The seeds contain 6–10% lipids (mainly as unsaturated fatty acids), 44–59% carbohydrates (mainly as galactomannans) and 20–30% protein (rich in arginine, alanine and glycine, but poor in lysine and methionine) [9]. The 4-hydroxy-isoleucine has been found to be a major free amino acid in the seeds. In comparison to other legumes, the seeds contain higher proportions of minerals (Ca, P, Fe, Zn, and Mn). Trigonelline is an important alkaloidal component of the seeds [10,11]. The seeds also contain some aromatic constituents such as n-alkanes, sesquiterpenes, and nonalactone. They have been found also to be rich in saponins. Three steroidal saponins, namely diosgenin, gitogenin, and tigogenin, have been

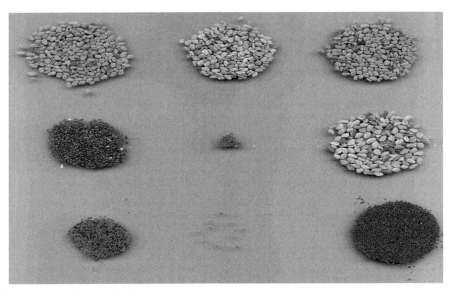

FIGURE 28.2   Seeds from fenugreek plants.

reported to be present in fenugreek seeds [12]. The seeds are also known to contain flavonoids, carotenoids, and coumarins.

4. POTENTIAL HEALTH BENEFITS

Fenugreek seeds have been known for a long time for their antidiabetic action. In 1948, Fourier [13] was first to observe that the administration of coarsely ground fenugreek seeds improved severe diabetes in human subjects. Since then, a number of studies have been conducted and the effects have been well documented in several investigations involving streptozotocin-induced diabetic rats, alloxan-diabetic dogs and type II diabetic subjects [14–17]. Initially, these antidiabetic properties were ascribed to the high fiber content of the seeds of fenugreek from Indian origin, affecting the intestinal absorptions of glucose and lipids [18,19]. However, research today has shown that not just the fiber, but many components of the seeds are involved in this effect. The components, such as 4-hydroxy-isoleucine and galactomannan, have shown insulinotropic properties [20–22] and inhibition of glucose absorption [16,18], respectively. Other components, such as saponins, were also found to increase cholesterol and bile salt excretion, and to have beneficial effects on many other health complications associated with diabetes such as polyuria, polydipsia, weakness, and weight loss [23]. The potential health benefits of fenugreek are generally thought to be mediated by its three active component groups, which include 4-hydroxy-isoleucine, galactomanans, and steroidal saponins.

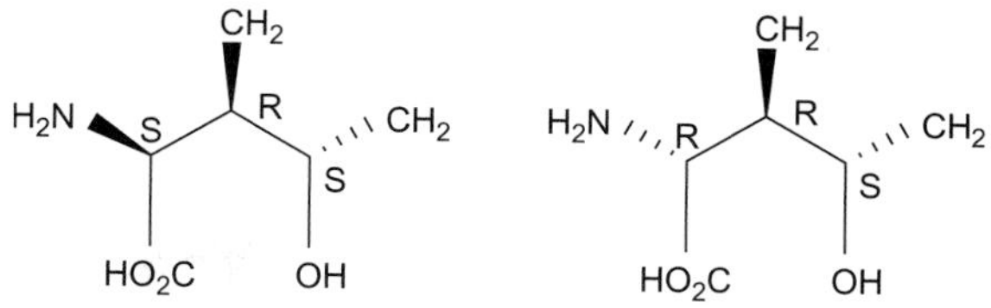

4.1 4-Hydroxy-Isoleucine (4-OH-Ile)

4-Hydroxy-isoleucine, a polar non-protein amino acid structurally related to branched-chain

FIGURE 28.3 Major and minor isomers of 4-hydroxy-isoleucine of fenugreek origin.

amino acids, is not present in mammalian tissues. It has been found only in some very specific plants, especially Trigonella species [20]. In fenugreek seeds it accounts for approximately 80% of the total content of free amino acids, 0.5% wt/wt [24]. The amino acid is present in the seeds in two diastereoisomers (Figure 28.3). The major one has been found to have a 2S, 3R, 4S configuration and represents up to 90% of total 4-OH-Ile content of the seeds. The minor possesses a 2R, 3R, 4S configuration. The major isomer, isolated and purified from fenugreek seeds, has been found to be a strong insulinotropic compound in both *in vitro* and *in vivo* studies [20,21]. Using islets of Langerhans from both rats and humans, 4-OH-Ile has thus been shown to be glucose dependent. In these *in vitro* studies, the amino acid appeared to be ineffective in stimulating insulin secretion at low (3 mmol/L) and basal (5 mmol/L) glucose concentrations, however, the insulin secretion was markedly induced as the glucose concentrations increased (6.6 –16.7 mmol/L). Such a response to glucose dependency is not shared by sulfonylureas, the widely used insulinotropic drugs to treat type II diabetes. Consequently, hypoglycemia remains a common untoward side effect of sulfonylurea treatment [25]. Thus, the novel insulinotropic property of 4-OH-Ile makes it worth further investigation to ascertain its potential clinical use, both as a purified insulinotropic compound and as a part of fenugreek seeds.

4.2 Galactomannans

Galactomannans belong to a family of seed gums and represent polymers of galactose and mannose [26]. Various soluble fibers such as guar gum, xanthine gum, gum acacia, locust bean gum (LBG), and fenugreek gum belong to this family. These galactomannans have different galactose:mannose (G:M) ratios and distributions of galactopyranosyl units along the mannan chains (Figure 28.4). The variability in galactose composition and distribution along the mannan main chain is responsible for variations in solubility and rheology of different sources [27,28]. The solubility and viscosity are increased when the G:M ratio increases from 1:4 in LBG to 1:3 in tara gum or to 1:2 in guar gum. In fenugreek galactomannan the G:M ratio is 1:1, making it superior in terms of its gel-forming characteristic over other galactomannans [23]. It has been found that by increasing the viscosity of digesta in the gut, these fibers, especially of fenugreek origin, delay the absorption of carbohydrates, thereby affecting postprandial plasma glucose response. Further, they have also been found to exert a hypoglycemic effect by increasing the short chain fatty acids mediated production of GLP-1 in the intestine [29]. Antidiabetic effects of GLP-1, including the ability to stimulate insulin secretion, inhibit glucagon production, and delay gastric emptying, have been documented in a number of studies [30]. When taken as supplements, these fibers have also been shown to be effective in lowering lipids and systolic blood pressure [31]. Fenugreek seeds contain up to 28% galactomannans. Considering the significant amounts of galactomannans in fenugreek seeds, it is of utmost importance that the effects of these in the control of postprandial glycemia and lipidemia in type II diabetic subjects be ascertained to possibly promote the clinical use of the seeds. Further, the systemic effects of their longitudinal intake on the molecular markers of carbohydrate and lipid metabolism are also of special interest.

Fenugreek has become a specialty crop in the Canadian Western Prairies. It has been found to be ideal for short-term crop rotation due to its high adaptability to dry climatic conditions, annual nature, and ability to fix atmospheric nitrogen in soil. It also provides excellent forage for cattle. The seeds of this Canadian-grown fenugreek have been found to contain 31.6% crude protein, 7.0% crude lipids, and 44.6% dietary fiber (25.8% insoluble and 18.8% soluble) (Table 28.1). These data were comparable to those of an Indian variety, except for the crude protein, which is significantly higher in the Canadian than in the Indian variety [32].

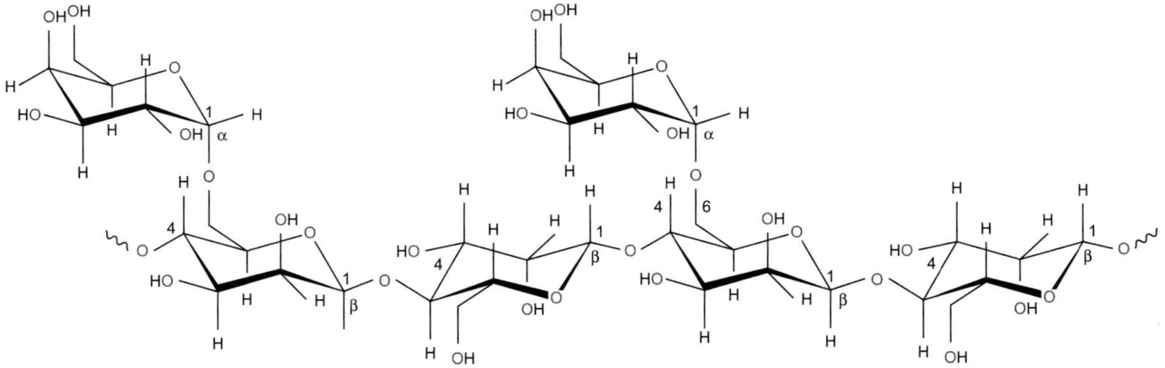

FIGURE 28.4 Galactomannan, consisting of polymers of galactose and mannose.

The galactomannan isolated and purified (91.4%) from Canadian-grown fenugreek seeds consists of a $\beta-(1,4)$ mannose backbone linking to a single α-(1,6) galactose side chain [32]. The high ratio of galactose to mannose (1:1.4) in fenugreek galactomannan makes it easy to dissolve in water and thus create a highly viscous solution at relatively low concentrations. Using this purified galactomannan, an *in vitro* study was carried out to examine its dose-related effects on the intestinal uptake of glucose in genetically determined lean JCR $(+/?)$ and obese JCR (cp/cp) rats [33]. The JCR (cp/cp) rats have been reported to be associated with the metabolic syndrome involving obesity, insulin resistance, hypertriglyceridemia, and secondary complications such as cardiovascular disease [34], and hence they are used as model animals for metabolic syndrome. There were significant differences ($P<0.001$) in body weight (g) between lean (326.9 ± 14.7) and obese (568.5 ± 14.7) rats, which were used in the *in vitro* study (Table 28.2). No significant difference was observed in the intestinal uptake of glucose between the two groups [33]. The uptake of glucose by the intestinal tissues, however, was significantly reduced in the presence of galactomannan; the effect was dose-related (0.1–0.5%, w/w), and this was true in both lean and obese rats. The inhibitory effect of galactomannan on glucose uptake was found to be in parallel with the degree of viscosity of the fiber solutions. The latter was determined after stirring for 60 min at a temperature-controlled (37 °C) fixed shear rate of 1.29 (L/s). In order to further examine these results, an *in vivo* study was carried out to examine the modifying effect of galactomannan extract from Canadian fenugreek source on glucose metabolism in high sucrose-fed rats [35]. The rational for feeding a high sucrose diet was that it increases triglycerides in both plasma and adipose tissue [36]. In rats fed this diet for 4 weeks, galactomannan (2.5% or 5.0%) reduced significantly the plasma triglycerides and cholesterol as well as hepatic cholesterol levels in association with a

TABLE 28.1 Proximate Composition of Fenugreek Seeds (% w/w, dry basis)

Component	Varieties	
	Canadian	Indian
Crude lipids	7.0 ± 0.3	9.6 ± 0.8
Crude proteins	31.6 ± 0.3^a	26.0 ± 0.3^b
Minerals	3.4 ± 0.1	3.2 ± 0.2
Soluble fiber	18.8 ± 0.2	17.5 ± 0.8
Insoluble fiber	25.8 ± 0.3^b	28.1 ± 0.1^a

Source Srichamroen, 2007 [32].

[a-b]Values in a row not sharing a common superscript differ significantly at $P<0.05$.

TABLE 28.2 Effect of Galactomannan on Food Intake, Weight Gain, and Adipose Tissue Weight

	Control	Low GAL (2.5%)	High GAL (5.0%)
Food intake (g/d)	26.8 ± 0.8^a	25.6 ± 0.8^a	22.6 ± 0.8^b
Body weight gain (g)	165.8 ± 9.7^a	157.4 ± 10.3^a	124.2 ± 10.3^b
Body weight gain/intake ratio	6.26 ± 0.1	6.10 ± 0.1	5.48 ± 0.1
Adipose tissue			
Abdominal adipose tissue (g)	5.90 ± 0.3^a	4.57 ± 0.3^b	2.58 ± 0.3^c
Perirenal adipose tissue (g)	1.38 ± 0.1	1.23 ± 0.1	0.91 ± 0.1

Source Srichamroen et al., 2009 [33].

Values represent means $\pm$ SEM of control ($N=9$), low GAL ($N=8$), high GAL ($N=8$).

[a-c]Values in a row not sharing a common superscript differ significantly at $P<0.05$.

reduction in plasma insulin concentrations (Table 28.3). It was of further interest that these animals had markedly lower sizes of abdominal adipose tissues compared with the control counterpart. These results can be explained by a decrease in insulin level leading to reduced lipogenesis. Both the *in vitro* and the *in vivo* results point to the significant potential of galactomannan of fenugreek origin in the management of metabolic syndrome, which is characterized by a cluster of hyperglycemia, hyperinsulinemia, hypertriglyceridemia, hyper-cholesterolemia, and increased abdominal fat.

4.3 Steroidal Saponins

Saponins are naturally occurring glycoside compounds found in a wide variety of food and forage plants and to a lesser extent in marine animals. The Asian fenugreek seeds have been found to contain steroidal saponins mainly in the form of diosgenin, which comprises approximately 5–6% of the seed [6]. The saponins mainly consist of an aglycone (e.g. sapogenin, sapogenol) linked to one or more sugars (Figure 28.5). The saponin (a bitter part) is hydrophobic, and the sugar part is hydrophilic. These structures give saponins their characteristic surface activity to form oil-in-water emulsions, and act as protective colloids. The structure also provides the ability to bind strongly to other components of the plant matrix. Hypocholesterolemic effects of the ethanol extracts of fenugreek seeds and also of purified saponins derived from these seeds have been reported in both *in vitro* and *in vivo* studies [19,37]. Such effects are thought to be

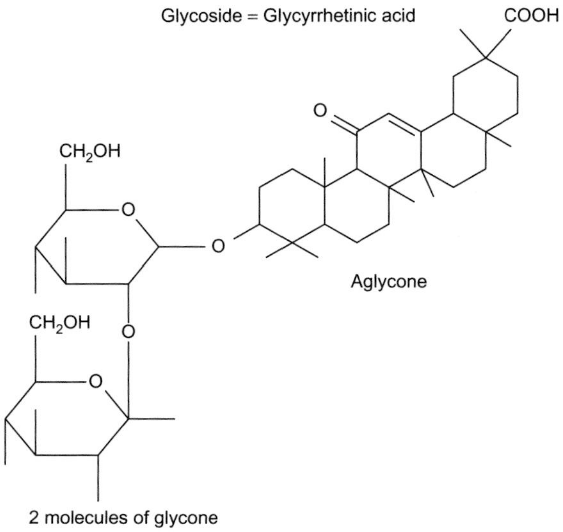

FIGURE 28.5 Saponin consisting of an aglycone linking to glucose molecules.

TABLE 28.3   Plasma Lipids and Insulin Status in Rats Fed a High Sucrose Diet Containing Low (2.5%) and High (5.0%) Galactomannan of Fenugreek Origin for 4 weeks

	Galactomannan		
	Control (N = 9)	Low (N = 8)	High (N = 8)
Triglyceride (mmol/L)	1.36 ± 0.2^a	0.79 ± 0.2^b	0.71 ± 0.2^b
Total cholesterol (mmol/L)	97.15 ± 0.6^a	75.27 ± 0.4^b	66.17 ± 0.5^b
LDL-cholesterol (mmol/L)	1.55 ± 0.1^a	$1.2 \pm 0.1^{a,b}$	0.75 ± 0.1^b
Free fatty acids (mmol/L)	16.16 ± 0.2^a	13.80 ± 0.2^b	15.95 ± 0.2^a
Insulin (pg/mL)	75.09 ± 0.3^a	60.12 ± 0.3^a	16.23 ± 0.3^b

Source Srichamroen et al., 2008 [35].
Values represent means $\pm$ SEM.
[a,b]Values in a row not sharing a common superscript differ significantly at $P<0.05$.

mediated by the saponins' ability to bind bile acids and thereby limit bile salt reabsorption in the gut, consequently accelerating cholesterol degradation and decreasing plasma cholesterol concentrations [38,39]. In addition to their hypocholesterolemic effects, the saponins have also been shown to have antioxidant properties. Such a response has been well documented for the saponins derived from soybeans, and they were found to have a protective effect against free radical-induced cellular damage [40]. However, the antioxidant effects of saponins from fenugreek seeds have not been determined. Therefore, to promote the potential use of fenugreek seeds as an antioxidant and hypocholesterolemic food product, it is of paramount importance that the effects of its intake on the antioxidant status and plasma lipid profiles of hypercholesterolemic subjects be further examined.

4.4 Additional Active Compounds in Fenugreek Seeds

Although considered to be the minor constituents, nicotinic acid, coumarin, and trigonelline are also believed to be involved in the hypoglycemic effects of the seeds [6]. A study [41] tested the effects of various pharmacological agents found in fenugreek seeds in normal and alloxan-diabetic rats. Nicotinic acid administration induced hypoglycemia lasting more than 5 hours after feeding in both normal and diabetic rats. The glucose-lowering effects of coumarin were also significant, persisting for more than 24 hours in both sets of animals. Trigonelline, the major alkaloid present in fenugreek, was also effective in causing a mild and transient hypoglycemia in diabetic rats. A high dose of scopoletine also caused a reduction in blood glucose in both sets of rats.

In addition to the above mentioned beneficial effects of fenugreek seeds or its isolated active compounds in maintenance of hyperglycemia

and hypercholesterolemia, the effects of the seeds on other factors regulating carbohydrate metabolism have also been reported. For instance, administration of fenugreek seed extract to experimental animals reduced the plasma concentration of triiodothyronine by inhibiting the conversion of thyroxine to triiodothyronine [42]. Reducing this conversion potentially prevents hyperthyroid-associated hypoglycemia. Further, supplementation of fenugreek seeds in the diets of diabetic rats reduced hepatic and renal output of glucose by decreasing the levels of glucose-6-phosphatase and fructose-1-6 bisphosphatase in these organs [43].

4.5 Prospects of Fenugreek in the Management of Diabetes

The antidiabetic properties of fenugreek were initially thought to be the result of the high fiber content of the seeds. However, it is now increasingly being realized that other seed components in addition to the fiber content may also account for its antidiabetic action. These components are believed to act synergistically in inhibiting glucose absorption and promoting pancreatic functions. Results of the extensive *in vitro* and *in vivo* studies have thus convincingly pointed to the fact that 4-hydroxyisoleucine, galactomannan, and saponin, the three potential active components isolated from the fenugreek seeds, lower blood glucose, cholesterol (including LDL-cholesterol), triglyceride, free fatty acids, and abdominal fat. Further, the galactomannan component of fenugreek has been shown to markedly reduce glycemic response in parallel with its insulin response to a glucose load. It has been known for many years that obesity is associated with insulin resistance, hyperinsulinemia, non-insulin-dependent diabetes mellitus (the type of diabetes that comprises over 90% of all diabetes), hyperlipidemia, and premature atherosclerosis, leading to increased morbidity and mortality from coronary arterial disease, stroke, and vascular disease. This cluster of associated diseases has been termed the 'insulin resistance syndrome,' 'syndrome X,' or 'metabolic syndrome.' It is increasingly being realized that the treatment of these multiple abnormalities is a central focus of management. Weight reduction associated with a decrease in total body and intra-abdominal fat results in a marked improvement in insulin resistance and in blood glucose and lipid profiles. The studies reported in this chapter show that fenugreek has the potential to lower blood glucose, lipid profiles, insulin resistance, and body weight (including the loss of abdominal fat), and thus fenugreek offers a viable choice in the treatment and prevention of diabetes and its complications. The conditions in which fenugreek with its low glycemic index could be potentially beneficial are shown in Figure 28.6.

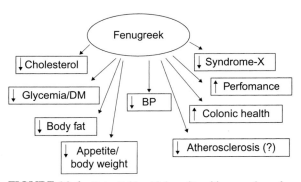

FIGURE 28.6   Potential health benefits of fenugreek seeds

TABLE 28.4   Elemental Similarities between Galactomannan in Fenugreek and the Antidiabetic Agent, Acarbose

	Galactomannan	Acarbose
Carbon (%)	48.1	46.5
Hydrogen (%)	6.3	6.7
Nitrogen (%)	1.7	2.2
Mol. wt	631	645
Mol. Formula	$C_{24}H_{43}O_{17}N_2$	$C_{25}H_{43}O_{18}N$

A synthetic oligosaccharide, acarbose, has been introduced for the treatment of diabetes [44]. It is an α-glucosidase inhibitor that improves glycemic control by reducing postprandial blood glucose excursions via competitive inhibition of the digestion and absorption of dietary carbohydrate. The saccharide has the molecular formula as $C_{25}H_{43}O_{18}N$ (mol. wt. 645). The soluble fiber from fenugreek seeds has shown a similar elemental analysis as $C_{24}H_{43}O_{17}N_2$ (mol. wt. 631), except that the amount of nitrogen appears to be two times that in acarbose (Table 28.4). Fenugreek is a legume plant, requiring nitrogen-fixing bacteria to grow; this may account for the increased amount of nitrogen in the seed compared with that in acarbose. Naturally occurring oligosaccharides containing a nitrogen atom are very rare. Hence, it would be an important finding if we can prove that the soluble fiber in fenugreek seeds is a potential source of oligosaccharide with hypoglycemic properties.

5. SUMMARY

Diabetes is a growing health concern. In the year 2025 the estimated diabetes incidence in North America was thought to exceed 24.5 million (U.S. Department of Health and Human Services, 2005). Generally, people diagnosed with diabetes undergo dietary and drug treatment regimens for the management of the disease. There are, however, borderline cases that remain untreated until there is an onset of the disease. These cases include gestational and non-gestational, associated with impaired blood glucose tolerance (e.g. fasting plasma glucose >6.2 mmol/L, and postprandial blood glucose response >11 mmol/L).

Central to the treatment of diabetes as well as the pre-diabetic state is the control of blood glucose levels to prevent or delay the associated complications of retinopathy, nephropathy, and cardiovascular disease. The results from major randomized trials have shown that glycemic control toward normal glucose levels can delay the onset and progression of these complications. To control these levels, a number of oral antihyperglycemic agents are currently available, including sulfonylureas, meglitinide analogues, biguanides, thiazolidinediones and α-glucosidase inhibitors. However, the detrimental side effects associated with their use such as weight gain, gastrointestinal disorders, an increased risk of cardiovascular morbidity and mortality, and a risk of hypoglycemia confound their use [45].

The evidence to date suggests that fenugreek could be a potential natural health product for the prevention and treatment of type II diabetics. This novel legume source provides soluble fiber along with other glucose-, cholesterol-, and triglyceride-lowering compounds. It would be a significant contribution to the daily management and stabilization of blood glucose and lipid levels for non-insulin-dependent diabetics. It is important to increase awareness of the public, dieticians, and other health professionals as to the unique properties of fenugreek and to recommend it for the prevention of hyperglycemia and hyperlipidemia. It is, however, important that the biological effects are firmly established through well-controlled clinical trials. Fenugreek seeds have been found to have a bitter taste. Thus, it is important that palatable food products in which the bitter flavor is effectively masked be developed in order to achieve sound consumer acceptability.

References

1. Abdel-Nabey, A. A., & Damir, A. A. (1990). Changes in some nutrients of fenugreek (*Trigonella foenum graecum*) seeds during boiling. *Plant Foods for Human Nutrition, 40*, 267–274.
2. Sorengarten, F., Jr., (1969). In *The Book of Spices* (pp. 250–253). PA: Livingston Publishing Company.
3. Leung, A. Y., & Foster, S. (1996). *Encyclopedia of Common Natural Ingredients Used in Food, Drugs and Cosmetics* (2nd ed.). New York: John Wiley & Sons pp. 243–245.
4. Patil, S. P., Niphadkar, P. V., & Bhat, M. M. (1997). Allergy to fenugreek (*Trigonella foenum graecum*). *Annals of Allergy Asthma Immunology, 78*, 297–300.
5. Fazli, F. R. Y., & Hardman, R. (1968). The spice, fenugreek (*Trigonella foenum graecum*): Its commercial varieties of seed as a source of diosgenin. *Tropical Science, 10*, 66–78.
6. Al-Habori, M., & Raman, A. (1998). Antidiabetic and hypocholesterolemic effects of fenugreek. *Phytotherapy Research, 12*, 233–242.
7. Gupta, K., Kumar, N., & Dahiya, D. S. (1998). Changes in structural carbohydrates and minerals in developing fenugreek (*Trigonella foenum graecum*) leaves. *International Journal of Tropical Agriculture, 16*, 221–227.
8. Gupta, U., Rati, E. R., & Joseph, R. (1998). Nutritional quality of lactic acid fermented bitter gourd and fenugreek leaves. *International Journal of Food Sciences and Nutrition, 49*, 101–108.
9. Sauvaire, Y., & Baccou, J. C. (1976). Nutritional value of the proteins of a leguminous seed: Fenugreek (*Trigonella foenum graecum*). *Nutrition Reports International, 14*, 527–537.
10. Mishkinsky, J., Joseph, B., & Sulman, F. (1967). Hypoglycaemic effect of trigonelline. *Lancet, 1*, 1311–1312.
11. Shani, J., Goldschmied, A., Ahronson, Z., & Sulman, F. G. (1974). Hypoglycaemic effect of *Trigonella foenum graecum* and *Lupinus termis* seeds and their major alkaoids in alloxan diabetic and normal rats. *Archives Internationales DE Pharmacodynamie Therapie, 210*, 27–36.
12. Dawidar, A. M., Saleh, A. A., & Elmotei, S. L. (1973). Steroid sapogenin constituents of fenugreek seeds. *Planta Medica, 24*, 367–370.
13. Fourier, F. (1948). Plantes medicinales et venereuses de France. *Paris, 111*, 495.
14. Ribes, G., Sauvaire, Y., Da Costa, C., Baccou, J. C., & Loubatieres-Mariani, M. M. (1986). Antidiabetic effects of subfractions from fenugreek seeds in diabetic dogs. *Proceedings of the Society for Experimental Biology and Medicine, 182*, 159–166.
15. Amin, R., Abdul-Ghani, A. S., & Suleiman, M. S. (1988). Effect of fenugreek and lupin seeds on the development of experimental diabetes in rats. *Planta Medica, 54*, 286–290.
16. Madar, Z., Abel, R., Samish, S., & Arad, J. (1988). Glucose-lowering effect of fenugreek in non-insulin dependent diabetes. *European Journal of Clinical Nutrition, 42*, 51–54.
17. Sharma, R., & Raghuram, T. C. (1990). Hypoglycemic effect of fenugreek seeds in non-insulin dependent diabetic subjects. *Nutrition Research, 10*, 731–739.

18. Madar, Z., & Shomer, I. (1990). Polysaccharide composition of a gel fraction derived from fenugreek and its effects on starch digestion and bile acid absorption in rats. *Journal of Agriculture and Food Chemistry, 38,* 1535–1539.

19. Stark, A., & Madar, Z. (1993). The effect of an ethanol extract derived from fenugreek (*Trigonella foenum graecum*) on bile acid absorption and cholesterol levels in rats. *British Journal of Nutrition, 69,* 277–287.

20. Sauvaire, Y., Petit, P., Broca, C., Manteghetti, M., Baissac, Y., Fernandea-Alvarez, J., Gross, R., Roy, M., Leconte, A., Gomis, R., & Ribes, G. (1998). 4-hydroxyisoleucine: A novel amino acid potentiator of insulin secretion. *Diabetes, 47,* 206–210.

21. Broca, C., Gross, R., Petit, P., Saauvaire, Y., Manteghetti, M., Tournier, M., Masiello, P., Gomis, R., & Ribes, G. (1999). 4-hydroxyisoleucine: Experimental evidence of its insulinotropic and antidiabetic properties. *American Journal Physiology, 277,* E617–E623.

22. Broca, C., Manteghetti, M., Gross, R., Baissac, Y., Jacob, M., Petit, P., Sauvaire, Y., & Ribes, G. (2000). 4-hydroxyisoleucine: Effects of synthetic and natural analogues on insulin secretion. *European Journal of Pharmacology, 390,* 339–345.

23. Evans, A. J., Hood, R. L., Oakenfull, D. G., & Sidhu, G. S. (1992). Relationship between structure and function of dietary fiber: A comparative study of three galactomannans on cholesterol metabolism in rats. *British Journal Nutrition, 68,* 217–229.

24. Gupta, D., Raja, J., & Baquer, N. Z. (1999). Modulation of some gluconeogenic enzyme activities in diabetic rat liver and kidney: Effect of antidiabetic compounds. *Indian Journal of Experimental Biology, 37,* 196–199.

25. Jennings, A. M., Wilson, R. M., & Ward, J. D. (1989). Symptomatic hypoglycemia in NIDDM patients treated with oral hypoglycaemic agents. *Diabetes Care, 12,* 203–208.

26. Buckeridge, M. S., Panegassi, V. R., Rocha, D. C., & Dietrich, S. M. C. (1995). Seed galactomannan in the classification and evolution of the leguminosae. *Phytochem, 38,* 871–875.

27. Maier, H., Anderson, M., Karl, C., Magnuson, K., & Whistler, R. L. (1993). Polysaccharides and their derivatives. In R. L. Whistler & J. N. BeMiller, (Eds.), *Industrial Gums* (pp. 181–227). San Diego: Academic Press.

28. Garti, N., Madar, Z., Aserin, A., & Sterneim, B. (1997). Fenugreek galactomannan as food emulsifiers. *Food Science and Technology, 30,* 305–311.

29. Groop, P. H., Aro, A., Stenman, S., & Groop, L. (1993). Long-term effects of guar gum in subjects with non-insulin-dependent diabetes mellitus. *American Journal of Clinical Nutrition, 58,* 513–518.

30. Massimino, S. P., McBurney, M. I., Field, C. J., Thomson, A. B. R., Keelan, M., Hayek, M. G., & Sunvold, G. D. (1998). Fermentable dietary fiber increases GLP-1 secretion and improves glucose homeostasis despite increased intestinal glucose transport capacity in healthy dogs. *Journal of Nutrition, 128,* 1786–1793.

31. Vuksan, V., Sievenpiper, J. L., Xu, Z., Wong, E. Y. Y., Jenkins, A. L., Beljan-Zdravkovic, U., Leiter, L. A., Josse, R. G., & Starvo, M. P. (2001). Konjac-mannan and American ginsing: Emerging alternate therapies for Type-2 diabetes mellitus. *Journal of American College of Nutrition, 20,* 370S–380S.

32. Srichamroen, A. (2007). Hypoglycemic and hypolipedimic effects of galactomannan from fenugreek (*Trigonella foenum graicum* L.) grown in Canada. PhD Thesis, Canada: University of Alberta..

33. Srichamroen, A., Thomson, A. B. R., Field, C. J., & Basu, T. K. (2009). *In vitro* intestinal glucose uptake is inhibited by galactomannan from Canadian fenugreek seed (*Trigonella foenum graecum* L) in genetically lean and obese rats. *Nutrition Research, 29,* 49–54.

34. O'Brien, S. F., & Russell, J. C. (1997). Insulin resistance and vascular wall function: Lessons from animal models. *Endocrinology and Metabolism (London), 4,* 155–162.

35. Srichamroen, A., Thomson, A. B. R., Field, C. J., & Basu, T. K. (2008). The modifying effects of galactomannan from Canadian-grown fenugreek (*Trigonella foenum-graecum* L.) on the glycemic and lipidemic status in rats. *Journal of Clinical Biochemistry and Nutrition, 43,* 167–174.

36. Aghelli, N., Kabir, M., Berni-Canani, S., Petijen, E., Boussairi, A., Luo, J., Bornet, F., Slama, G., & Rikzalla, S. (1998). Plasma lipids and fatty acid synthase activity are regulated by short-chain fructo-oligosaccharides in sucrose-fed insulin-resistant rats. *Journal of Nutrition, 128,* 1283–1288.

37. Petit, P. R., Sauvaire, Y. D., & Hillaire-Buys, D. M. (1995). Steroid saponins from fenugreek seeds: Extraction, purification, and pharmacological investigation on feeding behavior and plasma cholesterol. *Steroids, 60,* 674–680.

38. Sidhu, G. S., & Oakenfull, D. G. (1986). A mechanism for the hypocholesterolemic activity of saponins. *British Journal of Nutrition, 55,* 643–649.

39. Bhaumick, D., Acharya, S., & Basu, T. K. (2009). The potential benefits of saponin from Canadian-grown fenugreek (*Trigonella foenum graecum* L.) in modifying cholesterol status. *World Heart Journal, 1,* 263–270.

40. Francis, G., Kerem, Z., Makkar, H. P. S., & Becker, K. (2002). The biological action of saponins in animal systems: A review. *British Journal of Nutrition, 88,* 587–605.

41. Shani, J., Goldschmied, A., Ahronson, Z., & Sulman, F. G. (1974). Hypoglycaemic effect of *Trigonella foenum graecum* and *Lupinus termis* seeds and their major alkaloids in alloxan diabetic and normal rats. *Archives Internationales De Pharmacodynamie Et De Therapie, 210,* 27–36.

42. Panda, S., Tahiliani, P., & Kar, A. (1999). Inhibition of triiodothyronine production by fenugreek seed extract in mice and rats. *Pharmacological Research, 40,* 405–408.

43. Gupta, D., Raja, J., & Baquer, N. Z. (1999). Modulation of some gluconeogenic enzyme activities in diabetic rat liver and kidney: Effect of antidiabetic compounds. *Indian Journal of Experimental Biology, 37,* 196–199.

44. Johnston, P. S., Coniff, R. F., Hoogwerf, B. J., Santiago, J. V., Sunyer, F. X., & Krol, A. (1994). Effects of the carbohydrate inhibitor miglitol in sulphonulurea-treated NIDDM patients. *Diabetes Care, 17,* 20–29.

45. Camacho, P., Pitale, S., & Abraira, C. (2000). Beneficial and detrimental effects of intensive glycaemic control, with emphasis on type 2 diabetes mellitus. *Drugs Aging, 17,* 463–476.

29

Weight Loss Due to Fruit and Vegetable Use

Elena Rodríguez-Rodríguez, Ana M. López-Sobaler, and Rosa M. Ortega

Departamento de Nutrición, Facultad de Farmacia, UCM, Madrid, 28040, Spain

1. SUMMARY

Many overweight and obese people, and even people simply concerned about their weight, follow weight loss diets. However, there is a general lack of knowledge on the part of the public regarding the best steps to take in order to lose weight, and it is quite common for people to follow practices that, apart from not maintaining any weight lost, may lead to nutritional and health problems. This chapter discusses the most appropriate steps for achieving weight loss while maintaining a good nutritional status.

2. INTRODUCTION

The control of body weight is receiving growing interest for aesthetic and health reasons, but the number of overweight/obese persons continues to increase. Unfortunately, there

is a general lack of knowledge on the part of the public with respect to the most appropriate ways of achieving weight control [1], and people often follow inadequate diets that can lead to the appearance of mineral and vitamin deficiencies [2,3].

3. DEFINITION OF OVERWEIGHT AND OBESITY

A person who is overweight or obese suffers an abnormal or excessive accumulation of body fat [4]. The traditional method for determining whether a person is overweight or obese is to determine his/her body mass index (BMI): the person's body weight in kilograms divided by the square of his/her height in meters (kg/m^2) (Table 29.1). This highly reproducible, easy-to-use index reflects the amount of body fat in most people very well, and has been employed in a great many epidemiological studies. It is

437

TABLE 29.1 Body Mass Index (BMI) Intervals for Degree of Obesity in Adults [5]

Category	BMI Intervals (kg/m^2)
Underweight	<18.5
Normal weight	18.5–24.9
Grade I overweight	25–26.9
Grade II overweight (pre-obesity)	27–29.9
Type I obesity	30–34.9
Type II obesity	35–39.9
Type III obesity (morbid)	40–49.9
Type IV obesity (extreme)	≥50

also recommended for clinical use by different medical societies and international health organizations. However, the BMI is not a good indication of the amount of body fat carried by sportspersons or the elderly [5].

Other indicators of overweight/obesity have therefore been developed. For many years the ratio of the waist and hip circumferences (waist/hip ratio or WHR) has been used as an indicator of obesity, with central obesity said to exist when the WHR is >1 in men and >0.9 in women [6]. However, the latest trends recommend the use of the waist circumference alone to determine the excess of fat, rather than the WHR, given the positive correlation shown by the former with the abdominal fat load. A waist circumference of >95 cm in men, and >82 cm in women is associated with a greater risk of developing metabolic problems. When these circumferences reach >102 cm and >90 cm respectively, the risk of developing these problems becomes very high [5,7].

4. PREVALENCE OF OVERWEIGHT AND OBESITY

Obesity is one of the major public health problems of the twenty-first century. The latest WHO calculations [4] show that there has been a gradual increase in the incidence of overweight and obesity; in 2005 there were 1600 million overweight persons worldwide, plus another 400 million obese persons.

In Europe, the prevalence of obesity varies from one country to another. The highest values are seen in Central and Eastern Europe (5–23% for men and 7–36% for women). In 2007, obesity accounted for some 2–8% of health expenditure in Europe, and 10–13% of deaths [8,9].

Obesity in children is also increasing, as a consequence of changes in lifestyle. In Italy, Malta, and Greece, about 50% of children between 10 and 13 years of age are overweight/obese. However, this increase is not just seen in developed countries; developing countries are also experiencing the same problem [10]. Obesity during adolescence increases the chances of becoming an obese adult; the BMI of 14-year-olds is a good indicator of the likelihood of being obese in adulthood [11]. If action is not taken now, 2300 million people will be overweight by 2015, and 700 million more will be obese [4].

5. CAUSES OF THE INCREASE IN OBESITY

The increase in the incidence of obesity may have been favored by a growth in sedentary ways of life [12,13], but also by changes in food habits, especially in developed countries. There has been a gradual decline in the consumption of foods rich in carbohydrates and fiber, such as cereals and vegetables, and an increase in the consumption of foods rich in fat, refined sugar, and salt, and of those with high energy densities such as pastries, sugar-containing fizzy drinks, cured meats, and different types of fast food. This is especially true among members of the infant and juvenile populations [14].

Further, there is a trend toward consuming easy-to-prepare meals. The number of ready-made meals consumed has increased, as has the purchasing of foods from vending machines. Although this type of food may be a reasonable price it usually contains more fat and calories than food prepared at home [15]. The size of portions of this type of food has also been increasing, favoring an increase in the prevalence of obesity [16–18].

The number of meals taken outside the home has also increased [19], and this has only encouraged the appearance of obesity. Foods consumed outside the home may contain more fat (especially saturated fat) and contain less fiber and micronutrients than food prepared at home [20, 21]. The consumption of food at fast-food restaurants, such as hamburgers and chips, etc. (foods rich in fat and energy), has been associated with obesity [21–23]. Further, eating in restaurants can lead to the overconsumption of food, a consequence of the large portions served and the variety of appetizing – but high energy – foods offered [24,25].

Finally, it has been shown that under certain circumstances, people eat more than necessary, even though they have no appetite to satisfy. This may be seen at social events or when people are watching TV, etc. If repeated often, this can also lead to increases in body weight [26,27].

6. CONCERN ABOUT BODY WEIGHT IN THE POPULATION

Concern about one's body weight is not only healthy, it can also have an aesthetic impact. However, while many people who are not overweight want to lose some weight, there are many others who are overweight/obese who show no concern about their body weight or any desire to reduce their excess fat load [28].

Women, who constantly receive messages regarding patterns of beauty and the impact it has on their personal, social, and professional lives, are generally more concerned about their weight than men, and are more likely to follow diets or other strategies designed to help them lose weight [29,30]. This is true not only of women who are overweight/obese but also of those of normal and even low weight. Navia et al. [31] reported that, of 234 students aged 20–30, approximately half (total 47.9%; men 47.8%, women 47.9%) expressed a desire to lose weight. Yet among these, 30.4% of the men and 92.1% of the women students had BMIs between 18.5 and 25 kg/m^2; thus, even though they had normal BMIs, they still wanted to lose weight.

Despite this concern, there is a general lack of knowledge regarding the most adequate steps to take to achieve weight control, and the following of poorly designed, imbalanced diets is frequent. It is quite common, especially among adolescents, for people to embark on fasting, on eating very little food, on the use of food substitutes (powders or special drinks), to skip meals, and to smoke more cigarettes. Some even adopt methods such as self-induced vomiting and the taking of diet pills, laxatives, and diuretics [32]. This behavior and the following of imbalanced diets can lead to nutritional deficiencies [3], something that should be avoided.

7. INFLUENCE OF THE DIET ON WEIGHT CONTROL

The therapeutic objectives of weight loss in adults are the improvement or elimination of obesity-associated comorbidity and a reduction in the impact of future complications associated with weight excess. A number of weight loss strategies exist, including lifestyle changes (diet therapy, undertaking physical exercise, behavior modification), pharmacological treatment and bariatric (gastric bypass) surgery (which should be reserved for serious cases) [5]. This chapter looks at how diet can help achieve weight control, especially through the consumption of fruits and vegetables.

In order to design an appropriate weight loss diet for overweight/obese people it is necessary to bear in mind their gastronomic preferences, their economic possibilities, timetable, the climate, family and professional life, whether meals are taken alone or in company (watching TV or reading at the same time), the degree of physical activity undertaken, the person's clinical history, and other possible complications [33].

Many weight loss diets are based on the restriction of energy intake, yet weight loss appears to be regulated not just by the quantity of energy consumed but also by the composition of the diet [34]; it is therefore necessary to take this into account.

7.1 Energy Intake

The diets prescribed for the treatment of obesity in the general population should be slightly or moderately hypocaloric and balanced, and should, of course, be personalized. First, energy expenditure needs to be calculated with respect to the person's sex, age, and level of physical activity. This should allow a diet to be designed that provides about 500 kcal less than that required. Such a diet would allow a gradual weight loss that can be maintained over the long term [35].

In the loss and maintenance of weight it is also important to consider the distribution of energy intake over the day. A number of studies have shown that the greater the number of meals taken daily, the smaller the chance of becoming obese [36,37]. Navia and Perea [38] recommend that 4–5 meals be taken per day.

7.2 Diet Composition

Studies on overweight/obese people have found them to have food habits different to those of normal-weight people [39], including lower intakes of carbohydrates and fiber (mainly a lower consumption of fruits and vegetables) and higher intakes of fat. Obesity shows a positive correlation with fat intake and a negative correlation with carbohydrate intake [34–41]. This suggests that people who are interested in losing weight should, rather than simply reducing their energy intake, follow balanced diets with an energy profile close to that recommended but with 55–60% of all energy coming from carbohydrates, 10% from proteins, and <30% from fats [42]. It is recommended that the diet contain the minimum quantity recommended for each group; this will allow the energy profile to approach that recommended and improve nutritional status. Table 29.2 shows the number of servings and their sizes recommended for each food group. In order to help control body weight, the foods of least energy density within each group would be the best choices.

7.2.1 Milk and Milk Products

Weight control diets should include a relative increase in the consumption of skimmed milk/milk products since they contain less fat, and therefore less energy, than their whole milk counterparts [43,44]. Milk, fermented milk, and yoghurts would be more advisable than cured or semi-cured cheeses or milky desserts since the latter contain greater amounts of fat and should be taken in moderation [45].

7.2.2 Cereals and Pulses

There is evidence that, given their content in complex carbohydrates, fiber, and other minority components, the consumption of cereals [46,47] and pulses [48,49] favors weight loss. Cereals can be eaten either white or wholemeal, although the latter are recommended since they provide more fiber. This provides a sensation of being full and avoids constipation [50,51].

7.2.3 Fish, Meat, and Eggs

To favor weight loss, the meat consumed should be lean (chicken, turkey, rabbit, beef)

TABLE 29.2   Dietary Guidelines for Weight Control [38]

Food Groups	Servings/Day	Foods Included in the Group	Serving Size (g)
Dairy products*	2–3	Milk	200
		Yogurt	125
		Fermented milk	125
		Low-fat cheese	30–40
		High-fat cheese	15–30
Cereals and pulses	6	Bread	30–40
		Breakfast cereals	30–40
		Rice	100–150c
		Pasta	100–150c
		Pulses	100–150c
Meat, fish, and eggs	2–3	Meat and meat products	100–125r
		Fish and fish products	100–125r
		Egg	1 piece
Vegetables	3–4	Vegetables	100–200r
Fruits	2–3	Juices	150
		Fresh fruit	1 regular piece
Fats and oils	Moderate	Vegetable and seed oils, butter, margarine	Smallest possible
Sweet things	Moderate	Pastries, sugar, sweets	Smallest possible

*Skimmed or semi-skimmed.
r, Raw food weight, before the food is cooked.
c, Cooked food.
The number and weight of the servings indicated are an orientation and they have been established for one easy application to the dietetic daily practice

and all fat removed before cooking. Given their high fat content, meat derivatives should be avoided [52]. Some studies have shown that substituting meat for fish in hypocaloric diets can lead to greater weight losses; adding fish to the diet is therefore recommended [53,54]. Either blue or white fish can be eaten since, although blue fish contain more oil, they also have more omega-3 fatty acids. These can help protect against cardiovascular disease [55]. Fish canned in oil should be avoided, however, given its high energy content [56].

7.2.4 Fruits and Vegetables

Fruits and vegetables have different properties that are advantageous in weight loss diets:

a) Their energy density is low. These foods contain large quantities of water; their weight and volume is therefore high but their energy content low. Thus, more of these foods can be eaten for the same energy intake, reducing the hunger experienced when following diets that reduce the number of food servings [57]. Ello-Martin et al. [57] studied 97 obese women advised to reduce the energy density of their diets. These were randomly assigned to two groups: group RF, in which the fat intake was reduced, or group RF + FV in which the fat intake was reduced and the intake of water-rich foods, particularly fruit and vegetables, increased. At 1 year of follow-up the RF + FV group

had a lower dietary energy density
($P = 0.019$) and experienced less hunger
($P = 0.003$) than did the RF group. The
weight loss experienced by the RF + FV
group was consequently greater than that
achieved by the RF subjects [7.9 ± 0.9 kg
compared to 6.4 ± 0.9 kg ($P = 0.021$)].

b) They have high fiber contents. Many studies
have shown that increasing fiber
consumption increases satiety and reduces
hunger and energy intake [58]. Howarth
et al. [51] found that consuming an
additional 14 g/d of fiber for more than 2
days was associated with a 10% decrease in
energy intake and a loss of 1.9 kg over 3.8
months. Thus, the increase in fiber intake
associated with an increased consumption
of fruits and vegetables could help to
reduce energy intake and body weight.

c) They have low glycemic indices (GI) and are
often associated with low glycemic loads
(GL) (they cause only a slow increase in
blood glucose after their consumption);
several studies suggest that this facilitates
weight loss [59–61]. This may be due to a
lowering of postprandial insulin secretion or
because of their ability to produce a feeling
of fullness and thus reduce the amounts of
food eaten [62]. Given their low GI and GL
indices they are also associated with high
HDL-cholesterol levels and low triglyceride
and LDL-cholesterol levels. Their
consumption therefore reduces the risk of
cardiovascular disease [63,64].

d) They are rich in folic acid [56]. This vitamin
has been negatively associated with
overweight/obesity; the consumption of
these foods is therefore advisable in weight
loss diets [65,66]. In a study of 182 patients
with morbid obesity following lifestyle
changes combined with pharmacological
treatment, Martínez et al. [66] reported
serum folic acid to be the only independent
predictor of weight loss. An increase of
1 ng/mL improved the probability of a

successful treatment outcome by 28% and
serum folate was a better predictor of
success than fruit and vegetable intake.
However, this last variable was not found
to be a significant independent predictor
after its inclusion in a multiple linear
regression model.

These properties have inspired numerous
studies designed to confirm the effects of fruits
and vegetables on the regulation of food intake
and weight control. In those in which the only
step recommended to achieve weight loss was
to increase the consumption of fruits and vegetables, with no other advice at all, most of the
participants experienced no change in their
body weight [67–69]. However, when subjects
were also told to reduce their fat intake, the
majority lost weight by the end of the intervention [57,70,71]. Ortega et al. [71] studied 57
Spanish women with a BMI of 24–35 kg/m^2,
randomly assigned to one of two slightly hypocaloric diets for a 6-week period: diet V, in
which the consumption of greens and vegetables was increased, or diet C, in which the
consumption of cereals was increased. Weight
losses of 1.56 ± 0.93 kg and 1.02 ± 0.55 kg were
seen at 2 weeks with the C and V diets respectively, and of 2.8 ± 1.4 kg and 2.0 ± 1.3 kg at
6 weeks ($P < 0.05$). This shows that approximating the diet to the theoretical ideal by increasing the consumption of vegetables or cereals,
within the context of a hypocaloric diet, may
be of use in weight control.

The fact that fruit and vegetable consumption has a greater impact on weight control
when fat intake is also reduced might lead one
to think that any weight lost is simply a consequence of the reduction in fat intake. However,
part of the effect is due to the consumption
of fruits and vegetables. Singh et al. [72,73]
showed that adding fruits and vegetables to
a fat restriction diet resulted in significantly
more weight being lost than a fat restriction
diet alone.

It is important to bear in mind the form in which fruits and vegetables are consumed. Fruit juice, dried fruits, and vegetables cooked with oil or butter might have less influence on satiety than fruits and vegetables with lower energy densities, while adding appreciably to daily energy intake [74]. Weinrich et al. [75], who studied the relationships between selected dietary consumption and BMI in 204 African American men from southern states, found that cooking vegetables with butter ($P = 0.03$) was significantly associated with an increased BMI. In addition, Shi et al. [76] observed that a vegetable-rich food pattern was associated with a higher risk of obesity/central obesity in 2849 Chinese men and women aged 20 years; the problem was a high intake of energy due to the generous use of oil for stir-frying the vegetables.

Bearing these results in mind, it is essential to include quantities of fruits and vegetables in hypocaloric diets designed to achieve weight loss. In addition, some studies show that eating relatively more fruits and vegetables also improves dietary quality in addition to achieving weight loss [71]. The consumption of these foods can also help prevent diabetes [77], cancer [78], and cardiovascular disease [79]. Rodríguez-Rodríguez et al. [80] found that women who followed a diet rich in vegetables had lower plasma homocysteine levels, higher serum folate levels, and lower levels of LDL-cholesterol and non-HDL-cholesterol after 6 weeks.

Bearing in mind that reduced serum levels of antioxidants are present in obese people [81,82], the consumption of fruits and vegetables – which are rich in them [56] – may help control oxidative stress and consequently the inflammatory response associated with obesity [83,84]. This could help maintain and even improve the health of overweight persons.

7.3 Other Dietary Recommendations

a) The diet should be varied since this is the best way to ensure that all necessary nutrients are obtained in adequate amounts. Many epidemiological studies have shown that a varied diet is associated with a better nutrient intake profile and a reduction in morbidity and mortality [85].

b) Lipid consumption should be reduced. Foods should be cooked on a hotplate or steamed. Stews and fried foods, especially battered and breaded foods, should be avoided [86].

c) Food should be well chewed and eaten slowly to give time for satiety signals to develop. This will favor the digestion of food and prevent the appearance of meteorism [38].

d) The use of spices and herbs is recommended since, apart from making food more palatable [87], they may also increase the feeling of satiety and promote thermogenesis. They could therefore be of great use in the treatment of obesity [88]. It has been shown that the components of these ingredients reduce the inflammatory response of the adipose tissue in obese persons, perhaps improving the chronic inflammation associated with this problem [89].

e) Although salt has no calories and thus has little to do with weight loss, it should be used in moderation in order to reduce the high blood pressure so often seen in obese people [90]. In addition, salt makes food more palatable, which could lead to increased food consumption; this would favor an increase in weight [91].

8. PROMOTION OF FRUIT AND VEGETABLE CONSUMPTION

Nutrition education is vital in the prevention of obesity, yet there is a general lack of knowledge in nutrition matters among the population. This is particularly true of overweight/obese people [92]. This may be partly due to the many incorrect messages regarding nutrition transmitted by the media and sent out in publicity campaigns [38].

This general lack of information, coupled with incorrect information, leads to the appearance of inadequate food habits and therefore the swelling of the ranks of the overweight and obese. In fact, the consumption of fruits and vegetables by the population is much lower than that recommended, especially among adolescents, who also tend to restrict the intake of these foods when trying to lose weight [3,93]. The general public and overweight/obese persons in particular, therefore need to be given correct nutritional information [38]. Obesity prevention programs should encourage the acquisition of healthy food habits, including the consumption of fruit and vegetables, and promote physical exercise with the aim of reducing body weight and adiposity to desirable levels [94].

The 'Five a day' campaign, which began in the 1990s, aimed at increasing the consumption of fruits and vegetables. The consequence of this campaign has been a slight increase in their consumption, but a large percentage of the population are still not eating fruits or vegetables daily. Greater efforts will be required if people are to be convinced they should consume at least five portions of these foods daily [95].

Although obesity prevention programs should be directed toward the entire population, it is important to focus efforts on the infant population given the recent increase in obesity in this subgroup, and because the food and physical activity habits of infants can still be modified [96]. Different countries have tried different strategies for preventing infant obesity [96–99]. In Spain, the NAOS (Nutrición, Actividad Física y Prevención de la Obesidad) strategy aims to modify the composition of foods (reducing their sugar, fat, and salt contents), publish nutritional guides, promote physical activity within the educational setting, and examine the adequacy of the food available in school canteens and vending machines [96].

The main strategies that should be adopted to prevent infant obesity are [94]:

- The promotion of breast feeding;
- The promotion of a varied, balanced, and healthy diet;
- The promotion of increased fruit, vegetable, wholemeal cereal, and pulse consumption;
- The moderation of total fat intake;
- The moderation of the intake of prepared foods, sweets, and pastries;
- The promotion of gratifying, harmonious and continued physical exercise;
- The promotion of nutrition education at school, in the family, and in the community;
- The sensitization of social authorities and consumer education;
- The promotion of ethical–scientific coherence in TV publicity of foods and drinks.

By being constructive but not restrictive (therefore avoiding hunger and making a diet easier to follow), increasing the relative consumption of fruits and vegetables could be a good strategy for achieving weight control while improving food habits and nutritional status.

References

1. Ortega, R. M., Requejo, A. M., Quintas, M. E., Andrés, P., Redondo, M. R., & López-Sobaler, A. M. (1996). Desconocimiento sobre la relación dieta-control de peso corporal de un grupo de jóvenes universitarios. *Nutrición Clínica, 16,* 147–153.

2. Manore, M. M. (2000). Effect of physical activity on thiamine, riboflavin, and vitamin B-6 requirements. *American Journal of Clinical Nutrition, 72,* 598–606.

3. Neumark-Sztainer, D., Hannan, P. J., Story, M., & Perry, C. L. (2004). Weight-control behaviors among adolescent girls and boys: Implications for dietary intake. *Journal of the American Dietetic Association, 104,* 913–920.

4. WHO (World Health Organization). (2006). Obesity and overweight. WHO, Geneva. Fact sheet No.311. In: <http://www.who.int/mediacentre/factsheets/fs311/es/index.html/>.

5. Salas-Salvadó, J., Rubio, M. A., Barbany, M., & Moreno, B. (2007). Consenso SEEDO 2007 para la evaluación del sobrepeso y la obesidad y el establecimiento de criterios de intervención terapéutica. *Medicina Clínica, 128,* 184–196.

6. SEEDO (Sociedad Española para el Estudio de la Obesidad). (1996). Consenso español 1995 para la evaluación de la obesidad y para la realización de estudios epidemiológicos. *Medicina Clínica, 107,* 782–787.

7. Koster, A., Leitzmann, M. F., Schatzkin, A., Mouw, T., Adams, K. F., van Eijk, J. T., Hollenbeck, A. R., & Harris, T. B. (2008). Waist circumference and mortality. *American Journal of Epidemiology, 167,* 1465–1475.

8. WHO (World Health Organization). (2006). Obesity in Europe. In: <http://www.euro.who.int/obesity/>.

9. Branca, F., Nikogosian, H., & Lobstein, T. (2007). *The challenge of obesity in the WHO European region and the strategies for response.* Copenhagen: WHO.

10. Aranceta Bartrina, J., Serra Majem, Ll., Ribas Barba, L., & Pérez Rodrigo, C. (2001). Factores determinantes de la obesidad en la población infantil y juvenil española. In Ll. Serra Majem & J. Aranceta Bartrina (Eds.), *Obesidad infantil y juvenil. Estudio enKid* (pp. 109–128). Barcelona: Masson.

11. Laitinen, J., Power, C., & Järvelin, M. R. (2001). Family social class, maternal body mass index, childhood body mass index, and age at menarche as predictors of adult obesity. *American Journal of Clinical Nutrition, 74,* 287–294.

12. Tzotzas, T., & Krassas, G. E. (2004). Prevalence and trends of obesity in children and adults of South Europe. *Pediatric Endocrinology Reviews, 1*(Suppl. 3), 448–454.

13. Bastos, A., González, R., Molinero, O., & Salguero, A. (2005). Obesity, nutrition and physical activity. *Revista Internacional de Medicina y Ciencias de la Actividad Física y del Deporte, 18,* 140–153.

14. Tojo, R., & Leis, R. (2001). Obesidad infantil. Factores de riesgo y comorbilidades. In Ll. Serra Majem & J. Aranceta Bartrina (Eds.), *Obesidad infantil y juvenil. Estudio enKid* (pp. 39–54). Barcelona: Masson.

15. Summerfield, L. M. (2002). *Sobrepeso Delgadez y Obesidad.* Madrid: Thomson Editores.

16. Young, L. R., & Nestle, M. N. (2002). The contribution of expanding portion sizes to the US obesity epidemic. *American Journal of Public Health, 92,* 246–249.

17. Matthiessen, J., Fagt, S., Biltoft-Jensen, A., Beck, A. M., & Ovesen, L. (2003). Size makes a difference. *Public Health Nutrition, 6,* 65–72.

18. WHO (World Health Organization). (1998). *Obesity preventing and managing the global epidemic.* Geneva: WHO.

19. Nielsen, S. J., Siega-Riz, A. M., & Popkin, B. M. (2002). Trends in food locations and sources among adolescents and young adults. *Preventive Medicine, 35,* 107–113.

20. McCrory, M. A., Fuss, P. J., Saltzman, E., & Roberts, S. B. (2000). Dietary determinants of energy intake and weight regulation in healthy adults. *Journal of Nutrition, 130,* 276–279.

21. French, S. A., Story, M., Neumark-Sztainer, D., Fulkerson, J. A., & Hannan, P. (2001). Fast food restaurant use among adolescents: Associations with nutrient intake, food choices and behavioral and psychosocial variables. *International Journal of Obesity Relational Metabolic Disorders, 25,* 1823–1833.

22. French, S. A., Harnack, L., & Jeffery, R. W. (2000). Fast food restaurant use among women in the pound of prevention study: Dietary, behavioral and demographic correlates. *International Journal of Obesity Relational Metabolic Disorders, 24,* 1353–1359.

23. Satia, J. A., Galanko, J. A., & Siega-Riz, A. M. (2004). Eating at fast-food restaurants is associated with dietary intake, demographic, psychosocial and behavioural factors among African Americans in North Carolina. *Public Health Nutrition, 7,* 1089–1096.

24. Prentice, A. M., & Jebb, S. A. (2003). Fast foods, energy density and obesity: A possible mechanistic link. *Obesity Reviews, 4,* 187–194.

25. Close, R. N., & Schoeller, D. A. (2006). The financial reality of overeating. *Journal of the American College of Nutrition, 25,* 203–209.

26. Birch, C. D., Stewart, S. H., & Brown, C. G. (2007). Exploring differential patterns of situational risk for binge eating and heavy drinking. *Addictive Behaviors, 32,* 433–448.

27. Herman, C. P., & Polivy, J. (2007). Norm-violation, norm-adherence, and overeating. *Collegium Antropologicum, 31,* 55–62.

28. López-Sobaler, A. M., Ortega, R. M., Aparicio, A., Bermejo, L. M., & Rodríguez-Rodríguez, E. (2007). La preocupación por el peso corporal. Estudio a nivel nacional sobre errores y hábitos relacionados con el tema. In R. M. Ortega (Ed.), *Nutrición en población femenina. Desde la infancia a la edad avanzada* (pp. 39–49). Madrid: Ergon.

29. Skeie, G., & Klepp, K. I. (2002). Dieting among girls from Hordaland. *Tidsskrift for Den Norske Laegeforening, 122,* 1771–1773.

30. Anderson, D. A., Lundgren, J. D., Shapiro, J. R., & Paulosky, C. A. (2003). Weight goals in a college-age population. *Obesity Research, 11,* 274–278.

31. Navia, B., Ortega, R. M., Requejo, A. M., Mena, M. C., Perea, J. M., & López-Sobaler, A. M. (2003). Influence of the desire to lose weight on food habits, and knowledge of the characteristics of a balanced diet, in a group of Madrid university students. *European Journal of Clinical Nutrition, 57,* 90–93.

32. Neumark-Sztainer, D., Wall, M., Guo, J., Story, M., Haines, J., & Eisenberg, M. (2006). Obesity, disordered

B. EFFECTS OF INDIVIDUAL VEGETABLES ON HEALTH

eating, and eating disorders in a longitudinal study of adolescents: How do dieters fare 5 years later? *Journal of the American Dietetic Association, 106,* 559–568.

33. Díaz, J., Armero, M., Calvo, I., & Rico, M. A. (2000). Obesidad. In C. Martín, J. Díaz, T. Motilla, & P. Martínez, (Eds.), *Nutrición y dietética* (pp. 425–452). Madrid: Ediciones D.A.E.

34. Ortega, R. M., Requejo, A. M., & Andrés, P. (1999). Influencias dietéticas y control de peso corporal. *Nutrición y Obesidad, 2,* 4–13.

35. Russolillo, G., Astiasarán, I., & Martínez, J. A. (2003). *Intervención Dietética en la Obesidad.* Pamplona: EUNSA.

36. Toschke, A. M., Küchenhoff, H., Koletzko, B., & von Kries, R. (2005). Meal frequency and childhood obesity. *Obesity Research, 13,* 1932–1938.

37. Zizza, C., Siega-Riz, A. M., & Popkin, B. M. (2001). Significant increase in young adults' snacking between 1977–1978 and 1994–1996 represents a cause for concern! *Preventive Medicine, 32,* 303–310.

38. Navia, B., & Perea, J. M. (2006). Dieta y control de peso. In A. M. Requejo & R. M. Ortega (Eds.), *Nutriguía manual de nutrición clínica en atención primaria* (pp. 117–125). Madrid: Complutense.

39. Ortega, R. M., Andrés, P., Requejo, M., López-Sobaler, A. M., Redondo, R. M., & González-Fernández, M. (1996). Hábitos alimentarios e ingesta de energía y nutrientes en adolescentes con sobrepeso en comparación con los de peso normal. *Anales Españoles de Pediatría, 44,* 203–208.

40. Astrup, A. (2001). The role of dietary fat in the prevention and treatment of obesity. Efficacy and safety of low-fat diets. *International Journal of Obesity Relational Metabolic Disorders, 25,* 46–50.

41. Bray, G. A., Paeratakul, S., & Popkin, B. M. (2004). Dietary fat and obesity: A review of animal, clinical and epidemiological studies. *Physiology & Behavior, 83,* 549–555.

42. Lesi, C., Giaquinto, E., Valeriani, L., & Zoni, L. (2005). Diet prescription in obese patients. *Monaldi Archives for Chest Disease, 64,* 42–44.

43. Ortega, R. M., Requejo, A. M., Carcela, M., Pascual, M. J., & Montero, P. (1999). *Pautas Dietético-sanitarias Útiles en el Control de Peso.* Ayuntamiento de Madrid (Área de Salud y Consumo. Dirección de Servicios de Higiene y Salud Pública, Escuela de Sanidad y Consumo). Madrid: Universidad Complutense de Madrid.

44. WHO (World Health Organization). (2003). *Development of a WHO global strategy on diet, physical activity and health: European regional consultation.* Copenhagen: WHO.

45. Consejería de Sanidad de la Comunidad de Madrid. (2007). *Control de Peso de Forma Saludable.* Madrid: Dirección General de Salud Pública y Alimentación.

46. Mattes, R. D. (2002). Ready-to-eat cereal used as a meal replacement promotes weight loss in humans. *Journal of the American College of Nutrition, 21,* 570–577.

47. Melanson, K. J., Angelopoulos, T. J., Nguyen, V. T., Martini, M., Zukley, L., Lowndes, J., Dube, T. J., Fiutem, J. J., Yount, B. W., & Rippe, J. M. (2006). Consumption of whole-grain cereals during weight loss: Effects on dietary quality, dietary fiber, magnesium, vitamin B-6, and obesity. *Journal of the American Dietetic Association, 106,* 1380–1388.

48. Anderson, J. W., & Major, A. W. (2002). Pulses and lipaemia, short- and long-term effect: Potential in the prevention of cardiovascular disease. *British Journal of Nutrition, 88,* 263–271.

49. Abete, I., Parra, M. D., Martínez, B. E., Pérez, S., Rodríguez, M. C., & Martínez, J. A. (2005). Evaluación del efecto de una dieta hipocalórica, rica en legumbres, sobre la pérdida de peso y sobre marcadores de síndrome metabólico. *Nutrición Hospitalaria, 20,* 198.

50. Witkowska, A., & Borawska, M. H. (1999). The role of dietary fiber and its preparations in the protection and treatment of overweight. *Pol Merkur Lekarski, 6,* 224–226.

51. Howarth, N. C., Saltzman, E., & Roberts, S. B. (2001). Dietary fiber and weight regulation. *Nutrition Reviews, 59,* 129–139.

52. SENC (Sociedad Española de Nutrición Comunitaria). (2004). *Guía de Alimentación Saludable.* Madrid: Sociedad Española de Nutrición Comunitaria.

53. Mori, T. A., Bao, D. Q., Burke, V., Puddey, I. B., Watts, G. F., & Beilin, L. J. (1999). Dietary fish as a major component of a weight-loss diet: Effect on serum lipids, glucose, and insulin metabolism in overweight hypertensive subjects. *American Journal of Clinical Nutrition, 70,* 817–825.

54. Thorsdottir, I., Tomasson, H., Gunnarsdottir, I., Gisladottir, E., Kiely, M., Parra, M. D., Bandarra, N. M., Schaafsma, G., & Martinéz, J. A. (2007). Randomized trial of weight-loss-diets for young adults varying in fish and fish oil content. *International Journal of Obesity, 31,* 1560–1566.

55. Mozaffarian, D., Bryson, C. L., Lemaitre, R. N., Burke, G. L., & Siscovick, D. S. (2005). Fish intake and risk of incident heart failure. *Journal of the American College of Cardiology, 45,* 2015–2021.

56. Ortega, R. M., López-Sobaler, A. M., Requejo, A. M., & Andrés, P. (2008). *La Composición de los Alimentos. Herramienta básica para la valoración nutricional.* Madrid: Complutense.

57. Ello-Martin, J. A., Roe, L. S., Ledikwe, J. H., Beach, A. M., & Rolls, B. J. (2007). Dietary energy density in the treatment of obesity: A year-long trial comparing 2

weight-loss diets. *American Journal of Clinical Nutrition, 85,* 1465–1477.

58. Gustafsson, K., Asp, N. G., Hagander, B., & Nyman, M. (1995). Satiety effects of spinach in mixed meals: Comparison with other vegetables. *International Journal of Food Sciences and Nutrition, 46,* 327–334.

59. Colombani, P. C. (2004). Glycemic index and load-dynamic dietary guidelines in the context of diseases. *Physiology & Behavior, 83,* 603–610.

60. Brand-Miller, J. (2007). Effects of glycemic load on weight loss in overweight adults. *American Journal of Clinical Nutrition, 86,* 1249–1250.

61. Maki, K. C., Rains, T. M., Kaden, V. N., Raneri, K. R., & Davidson, M. H. (2007). Effects of a reduced-glycemic-load diet on body weight, body composition, and cardiovascular disease risk markers in overweight and obese adults. *American Journal of Clinical Nutrition, 85,* 724–734.

62. Brand-Miller, J. C., Holt, S. H., Pawlak, D. B., & McMillan, J. (2002). Glycemic index and obesity. *American Journal of Clinical Nutrition, 76,* 281–285.

63. Liu, S., Manson, J. E., Stampfer, M. J., Holmes, M. D., Hu, F. B., Hankinson, S. E., & Willett, W. C. (2001). Dietary glycemic load assessed by food-frequency questionnaire in relation to plasma high-density-lipoprotein cholesterol and fasting plasma triacylglycerols in postmenopausal women. *American Journal of Clinical Nutrition, 73,* 560–566.

64. Ma, Y., Li, Y., Chiriboga, D. E., Olendzki, B. C., Hebert, J. R., Li, W., Leung, K., Hafner, A. R., & Ockene, I. S. (2006). Association between carbohydrate intake and serum lipids. *Journal of the American College of Nutrition, 25,* 155–163.

65. Hirsch, S., Poniachick, J., Avendaño, M., Csendes, A., Burdiles, P., Smok, G., Diaz, J. C., & de la Maza, M. P. (2005). Serum folate and homocysteine levels in obese females with non-alcoholic fatty liver. *Nutrition, 21,* 137–141.

66. Martínez, J. J., Ruiz, F. A., & Candil, S. D. (2006). Baseline serum folate level may be a predictive factor of weight loss in a morbid-obesity-management programme. *British Journal of Nutrition, 96,* 956–964.

67. Maskarinec, G., Chan, C. L., Meng, L., Franke, A. A., & Cooney, R. V. (1999). Exploring the feasibility and effects of a high-fruit and -vegetable diet in healthy women. *Cancer Epidemiology Biomarkers & Prevention, 8,* 824–919.

68. Smith-Warner, S. A., Elmer, P. J., Tharp, T. M., Fosdick, L., Randall, B., Gross, M., Wood, J., & Potter, J. D. (2000). Increasing vegetable and fruit intake: Randomized intervention and monitoring in an at-risk population. *Cancer Epidemiology Biomarkers & Prevention, 9,* 307–317.

69. Whybrow, S., Harrison, C. L., Mayer, C., & James Stubbs, R. (2006). Effects of added fruits and vegetables on dietary intakes and body weight in Scottish adults. *British Journal of Nutrition, 95,* 496–503.

70. Ortega, R. M., López-Sobaler, A. M., Rodríguez Rodríguez, E., Bermejo, L. M., García González, L., & López Plaza, B. (2005). Response to a weight control program based on approximating the diet to its theoretical ideal. *Nutrición Hospitalaria, 20,* 393–402.

71. Ortega, R. M., Rodríguez-Rodríguez, E., Aparicio, A., Marín-Arias, L. I., & López-Sobaler, A. M. (2006). Responses to two weight-loss programs based on approximating the diet to the ideal: Differences associated with increased cereal or vegetable consumption. *International Journal for Vitamin and Nutrition Research, 76,* 367–376.

72. Singh, R. B., Rastogi, S. S., Niaz, M. A., Ghosh, S., Singh, R., & Gupta, S. (1992). Effect of fat-modified and fruit- and vegetable-enriched diets on blood lipids in the Indian Diet Heart Study. *American Journal of Cardiology, 70,* 869–874.

73. Singh, R. B., Dubnov, G., Niaz, M. A., Ghosh, S., Singh, R., Rastogi, S. S., Manor, O., Pella, D., & Berry, E. M. (2002). Effect of an Indo-Mediterranean diet on progression of coronary artery disease in high risk patients (Indo-Mediterranean Diet Heart Study): A randomised single-blind trial. *Lancet, 360,* 1455–1461.

74. Rolls, B. J., Ello-Martin, J. A., & Tohill, B. C. (2004). What can intervention studies tell us about the relationship between fruit and vegetable consumption and weight management? *Nutrition Reviews, 62,* 1–17.

75. Weinrich, S. P., Priest, J., Reynolds, W., Godley, P. A., Tuckson, W., & Weinrich, M. (2007). Body mass index and intake of selected foods in African American men. *Public Health Nursing, 24,* 217–229.

76. Shi, Z., Hu, X., Yuan, B., Hu, G., Pan, X., Dai, Y., Byles, J. E., & Holmboe-Ottesen, G. (2008). Vegetable-rich food pattern is related to obesity in China. *International Journal of Obesity, 32,* 975–984.

77. Harding, A. H., Wareham, N. J., Bingham, S. A., Khaw, K., Luben, R., Welch, A., & Forouhi, N. G. (2008). Plasma vitamin C level, fruit and vegetable consumption, and the risk of new-onset type 2 diabetes mellitus: The European prospective investigation of cancer–Norfolk prospective study. *Archives of Internal Medicine, 168,* 1493–1499.

78. Béliveau, R., & Gingras, D. (2007). Role of nutrition in preventing cancer. *Canadian Family Physician, 53,* 1905–1911.

79. He, F. J., Nowson, C. A., & MacGregor, G. A. (2006). Fruit and vegetable consumption and stroke: Meta-analysis of cohort studies. *Lancet, 367,* 320–326.

B. EFFECTS OF INDIVIDUAL VEGETABLES ON HEALTH

80. Rodríguez-Rodríguez, E., Ortega, R. M., López-Sobaler, A. M., Andrés, P., Aparicio, A., Bermejo, L. M., & García-González, L. (2007). Restricted-energy diets rich in vegetables or cereals improve cardiovascular risk factors in overweight/obese women. *Nutrition Research, 27*, 313–320.

81. Kuno, T., Hozumi, M., Morinobu, T., Murata, T., Mingci, Z., & Tamai, H. (1998). Antioxidant vitamin levels in plasma and low density lipoprotein of obese girls. *Free Radical Research, 28*, 81–86.

82. Galan, P., Viteri, F. E., Bertrais, S., Czernichow, S., Faure, H., Arnaud, J., Ruffieux, D., Chenal, S., Arnault, N., Favier, A., Roussel, A. M., & Hercberg, S. (2005). Serum concentrations of beta-carotene, vitamins C and E, zinc and selenium are influenced by sex, age, diet, smoking status, alcohol consumption and corpulence in a general French adult population. *European Journal of Clinical Nutrition, 59*, 1181–1190.

83. Staruchová, M., Volková, K., Lajdová, A., Misl'anová, C., Collins, A., Wsólová, L., Staruch, L., & Dusinská, M. (2006). Importance of diet in protection against oxidative damage. *Neuroendocrinology Letters, 27*, 112–115.

84. Thompson, H. J., Heimendinger, J., Sedlacek, S., Haegele, A., Diker, A., O'Neill, C., Meinecke, B., Wolfe, P., Zhu, Z., & Jiang, W. (2005). 8-Isoprostane F2alpha excretion is reduced in women by increased vegetable and fruit intake. *American Journal of Clinical Nutrition, 82*, 768–776.

85. Drewnowski, A., Henderson, S. A., Driscoll, A., Rolls, B. J., Drewnowski, A., Henderson, S. A., Driscoll, A., & Rolls, B. J. (1997). The dietary variety score: Assessing diet quality in healthy young and older adults. *Journal of the American Dietetic Association, 97*, 266–271.

86. Guallar-Castillón, P., Rodríguez-Artalejo, F., Fornés, N. S., Banegas, J. R., Etxezarreta, P. A., Ardanaz, E., Barricarte, A., Chirlaque, M. D., Iraeta, M. D., Larrañaga, N. L., Losada, A., Mendez, M., Martínez, C., Quirós, J. R., Navarro, C., Jakszyn, P., Sánchez, M. J., Tormo, M. J., & González, C. A. (2007). Intake of fried foods is associated with obesity in the cohort of Spanish adults from the European Prospective Investigation into Cancer and Nutrition. *American Journal of Clinical Nutrition, 86*, 198–205.

87. Billing, J., & Sherman, P. W. (1998). Antimicrobial functions of spices: Why some like it hot. *Quarterly Review of Biology, 73*, 3–49.

88. Westerterp-Plantenga, M., Diepvens, K., Joosen, A. M., Bérubé-Parent, S., & Tremblay, A. (2006). Metabolic effects of spices, teas, and caffeine. *Physiology and Behavior, 89*, 85–91.

89. Woo, H. M., Kang, J. H., Kawada, T., Yoo, H., Sung, M. K., & Yu, R. (2007). Active spice-derived components can inhibit inflammatory responses of adipose tissue in obesity by suppressing inflammatory actions of macrophages and release of monocyte chemoattractant protein-1 from adipocytes. *Life Sciences, 80*, 926–931.

90. Stanton, R. A. (2006). Nutrition problems in an obesogenic environment. *Medical Journal of Australia, 184*, 76–79.

91. Yeomans, M. R., Blundell, J. E., & Leshem, M. (2004). Palatability: Response to nutritional need or need-free stimulation of appetite? *British Journal of Nutrition, 92*, 3–14.

92. Ortega, R. M., & Lopez-Sobaler, A. M. (2005). How justifiable is it to distort the energy profile of a diet to obtain benefits in body weight control? *American Journal of Clinical Nutrition, 82*, 1140–1141.

93. Larson, N. I., Neumark-Sztainer, D., Hannan, P. J., & Story, M. (2007). Trends in adolescent fruit and vegetable consumption, 1999–2004: Project EAT. *American Journal of Preventive Medicine, 32*, 147–150.

94. Aranceta, J., & Serra, L. (2006). El sobrepeso y la obesidad como problema de salud pública. In Ll. Serra & J. Aranceta (Eds.), *Nutrición y Salud Pública. Métodos, bases científicas y aplicaciones*. Barcelona: Masson.

95. Kvaavik, E., Samdal, O., Trygg, K., Johansson, L., & Klepp, K. I. (2007). Five a day – Ten years later. *Tidsskrift for Den Norske Laegeforening, 127*, 2250–2253.

96. AESA (Agencia Española de Seguridad Alimentaria). (2005). *Estrategia NAOS. Invertir la tendencia de la obesidad*. Madrid: Ministerio de Sanidad y Consumo.

97. Lau, D. C., Douketis, J. D., Morrison, K. M., Hramiak, I. M., Sharma, A. M., & Ur, E. Obesity Canada Clinical Practice Guidelines Expert Panel. (2007). Canadian clinical practice guidelines on the management and prevention of obesity in adults and children [summary]. *Canadian Medical Association Journal, 176*, 1–13.

98. Nakamura, T. (2008). The integration of school nutrition program into health promotion and prevention of lifestyle-related diseases in Japan. *Asia Pacific Journal of Clinical Nutrition, 17*, 349–351.

99. Neumark-Sztainer, D., Haines, J., Robinson-O'Brien, R., Hannan, P. J., Robins, M., Morris, B., & Petrich, C. A. (2009). 'Ready. Set. ACTION!' A theater-based obesity prevention program for children: a feasibility study. *Health Education Research, 24*, 407–420.

B. EFFECTS OF INDIVIDUAL VEGETABLES ON HEALTH

30

Legumes and Cardiovascular Disease

Peter M. Clifton

CSIRO Preventative Health Flagship, CSIRO Human Nutrition, Adelaide SA, Australia

1. INTRODUCTION

Legumes (particularly beans) have been associated with protection from heart disease in the Mediterranean diet for many years, as have soybeans in Japan, but it has only been in recent years that epidemiology has shown more convincing links. Ken Carroll et al. first showed in 1978 in a small number of subjects that exchanging animal protein with soy protein lowers serum cholesterol [1]. This area has been an active focus of research ever since.

2. EPIDEMIOLOGY

The Puerto Rico heart study of 8218 urban and rural men showed that legume intake was related to a lower incidence of heart disease after adjustment for all other risk factors [2]. Guo et al. [3] showed that in China legume intake was related to both ischemic heart disease and stroke while cholesterol and smoking were not. Trichopoulou et al. [4] and Willett and colleagues [5] emphasized the high intake of legumes in the Greek and Mediterranean

diets and suggested these were protective. In the seven countries studied, legume intake on univariate analysis was inversely related to coronary heart disease (CHD) mortality ($r = -0.822$) [6]. In a study of dietary patterns in 45,000 US men, a protective pattern consisting of fruit, vegetables, legumes, whole grains, fish, and poultry was associated with a 30% reduction in CHD [7]. A similar observation was made in US women [8] and in young Finns [9]. In Japan, there was an inverse correlation between soy product intake (as total amount or soy protein) and heart disease mortality rate was statistically significant in women after controlling for covariates (and $r = -0.31$, $P = 0.045$) but not in men [10]. This was confirmed by Sasazuki et al. [11] in Fukuoko city. In the USA, 9632 men and women participated in the First National Health and Nutrition Examination Survey Epidemiologic Follow-up Study (NHEFS). Legume consumption was significantly and inversely associated with risk of CHD ($P = 0.002$ for trend) and CVD ($P = 0.02$ for trend) after adjustment for established CVD risk factors. Legume consumption four times or more per week compared with less than once a week was associated with a 22% lower risk of

CHD (relative risk, 0.78; 95% confidence interval, 0.68–0.90) and an 11% lower risk of CVD (relative risk, 0.89; 95% confidence interval, 0.80–0.98) [12]. Trichopoulou et al [13] have extensively studied the Greek population in relation to protective dietary components and found using a 10 point Mediterranean score that an increase in 2 points in adherence resulted in a 25% reduction in total mortality, and a 33% reduction in CHD mortality [14]. In those with prevalent CHD at baseline a similar reduction in mortality was found [15]. In the whole EPIC elderly cohort (aged>60 years at enrolment) those with CHD had an 18% reduction in mortality associated with a 2 point increase in score [16] while those without CHD had a decrease of 8% [17] or in another analysis a 14% decrease per SD of the diet score [18].

In the Japan Public Health Center-based (JPHC) study cohort, 40,462 Japanese (40–59 years old, without cardiovascular disease or cancer at baseline) were followed for 12 years. For women, the multivariable hazard ratios for soy intake ≥5 times per week versus 0–2 times per week were 0.64 for risk of cerebral infarction, 0.55 for risk of myocardial infarction and 0.31 for cardiovascular disease mortality. Similar but weaker inverse associations were observed between intake of miso soup and beans and risk of cardiovascular disease mortality. The multivariable hazard ratios for the highest versus the lowest quintiles of isoflavones in women were 0.35 for cerebral infarction, 0.37 for MI, and 0.87 for cardiovascular disease mortality. No significant association of dietary intake of soy, miso soup, and beans and isoflavones with cerebral infarction or myocardial infarction was present in men [19]. In the USA, nurses' adherence to a DASH diet [a score based on eight food and nutrient components (fruits, vegetables, whole grains, nuts and legumes, low-fat dairy, red and processed meats, sweetened beverages, and sodium)] was associated with a 24% reduction in CHD and an 18% reduction in

stroke [20]. A Mediterranean diet pattern was found to be beneficial in 214,284 men and 166,012 women in the National Institutes of Health (NIH)-AARP (also known as the American Association of Retired Persons) Diet and Health Study. A 9 point score was calculated which included vegetables, legumes, fruits, nuts, whole grains, fish, monounsaturated fat–saturated fat ratio, alcohol, and meat. People with the highest score had a 20–22% reduction in CHD and total mortality in both men and women [21]. A meta-analysis has been performed of eight studies relating adherence to a Mediterranean diet to mortality [22]. A 2 point increase in adherence has been associated with a 9% reduction in total mortality and 9% reduction in cardiovascular mortality.

In the EPIC study 10,500 people with diabetes at presentation were found to have a significant 27% reduction in all cause mortality (1346 deaths) in the highest quartile of legume consumption (32 g/day). This reduction was totally accounted for by a reduction in cardiovascular disease death with no effects on cancer or non-cancer, non-CVD deaths [23]. Twenty grams per day of legumes reduced CVD by 18%, which gave a stronger effect than 80 g/d of either fruit or vegetables.

3. INTERVENTIONS WITH LEGUMES

A very large number of studies have been performed, with seven meta-analyses available [24–31]. In 2008, Harland and Haffner [31] reported on research that had examined 30 studies containing 42 treatment arms ($N = 2913$), with an average soya protein intake of 26.9 g. LDL was significantly lowered by 0.22 mmol/L (about 6%, $P<0.001$) with no dose response. Taku and colleagues examined 11 studies specifically looking at isoflavones and found this component lowered LDL-cholesterol by 0.10 mmol/L while soy protein containing isoflavones lowered LDL-cholesterol by 0.18 mmol/L [29]. Soy isoflavones increase sterol regulatory binding protein 2, increasing the expression of the LDL receptor and enhancing LDL clearance [32].

The Portfolio Diet of David Jenkins and colleagues which contains 21.4 g/1000 kcal of soy protein along with plant sterols (1.0 g/1000 kcal), viscous fibers (9.8 g/1000 kcal), and almonds (14 g/1000 kcal) lowers LDL cholesterol by 28.6% and CRP by 28.2% over a 1 month period [33]. After 12 months this had been reduced to 12.8% with about 30% of participants having a reduction >20% [34]. Lukaczer et al. [35] found a lesser effect of a similar portfolio diet (30 g of soy protein and 4 g of phytosterols per day) in 59 postmenopausal women with LDL-cholesterol reductions of 14.8% at 3 months.

Chickpeas can lower LDL-cholesterol by 0.20 mmol/L over 5 weeks [36]. The dietary fiber component of chickpeas appeared to have the strongest effect [37]. Serra-Majem et al. in 2006 [38] reviewed 43 publications containing 35 studies of the Mediterranean diet and showed favorable effects in most. Lairon [39] has reviewed six published interventions with the Mediterranean diet, but it is difficult to be certain how much increase in legume intake occurred in these interventions as it is generally not recorded. In the Esposito study [40], 90 volunteers were asked to consume 400 g/d of whole grains which included legumes. In the nutritional report intervention, volunteers reported eating 274 g/d more of fruit, vegetables, nuts, and legumes and 100 g/d more whole grains. The number of people with the metabolic syndrome was reduced to 40 while in the control group 78 out of 90 still had the syndrome. Endothelial function improved in the intervention group. In the Medi-Rivage study [41], although it was recommended to increase legumes, there was only a minor increase in fiber and a decrease in carbohydrate and no record of any increase in legumes consumed.

The study itself showed no differences between control and Mediterranean diets. The Estruch paper [42] made no comment about any increase in legumes in the Predimed study, nor did the Fuentes Mediterranean intervention study [43].

Teede et al. [44] showed that soy protein elevated 24-hour ambulatory blood pressure compared with gluten. In a study [45] with 22 men in a randomized crossover design a mean intake of 91.9 g/d extruded dry beans had no significant effects on total serum cholesterol, LDL-cholesterol, triglycerides, apolipoprotein A or B, plasma fibrinogen, and plasma viscosity concentrations. HDL-cholesterol concentrations decreased in both the dry bean and control periods. Lipoprotein (a) concentrations increased with intake of extruded dry beans, but this increase was probably not due to an independent effect of extruded dry beans. Plasminogen activator inhibitor 1 levels were significantly lower after the intake of extruded dry beans compared to the control period. Similar lack of effect was found using six 440 g cans of baked beans/week in 20 mildly hypercholesterolemic men [46], and in 40 hypercholesterolemic subjects on a low fat diet [47], although positive effects have been found using beans by Shutler et al. [48], and Anderson et al. [49,50] in which total cholesterol was lowered from 10% to 19% and by Frühbeck et al. [51] and Duane [52] (9% lowering with 120 g mixed legumes/day). Pinto beans (120 g/d of dried beans) lower total cholesterol by 8% [53]. This area was reviewed by Geil and Anderson in 1994 [54]. A plant-based low fat diet with additional legumes, vegetables, and wholegrains lowered LDL-cholesterol by 0.18 mmol/L compared with the low fat diet in 120 healthy subjects [55].

4. CONCLUSIONS

Increasing fruit and vegetables will decrease heart disease risk, but a specific increase in legumes of the order of 120 g/d of dried beans will be even more effective and probably acts via lowering LDL-cholesterol.

References

1. Carroll, K. K., Giovannetti, P. M., Huff, M. W., Moase, O., Roberts, D. C., & Wolfe, B. M. (1978). Hypocholesterolemic effect of substituting soybean protein for animal protein in the diet of healthy young women. *American Journal of Clinical Nutrition, 31*, 1312–1321.

2. Garcia-Palmieri, M. R., Sorlie, P., Tillotson, J., Costas, R., Jr., Cordero, E., & Rodriguez, M. (1980). Relationship of dietary intake to subsequent coronary heart disease incidence: The Puerto Rico Heart Health Program. *American Journal of Clinical Nutrition, 33*, 1818–1827.

3. Guo, W., Li, J. Y., King, H., & Locke, F. B. (1992). Diet and blood nutrient correlations with ischemic heart, hypertensive heart, and stroke mortality in China. *Asia-Pacific Journal of Public Health, 6*, 200–209.

4. Trichopoulou, A., Bamia, C., Norat, T., Overvad, K., Schmidt, E. B., Tjønneland, A., Halkjaer, J., Clavel-Chapelon, F., Vercambre, M. N., Boutron-Ruault, M. C., Linseisen, J., Rohrmann, S., Boeing, H., Weikert, C., Benetou, V., Psaltopoulou, T., Orfanos, P., Boffetta, P., Masala, G., Pala, V., Panico, S., Tumino, R., Sacerdote, C., Bueno-de-Mesquita, H. B., Ocke, M. C., Peeters, P. H., Van der Schouw, Y. T., González, C., Sanchez, M. J., Chirlaque, M. D., Moreno, C., Larrañaga, N., Van Guelpen, B., Jansson, J. H., Bingham, S., Khaw, K. T., Spencer, E. A., Key, T., Riboli, E., & Trichopoulos, D. (2007). Modified Mediterranean diet and survival after myocardial infarction: The EPIC-Elderly study. *European Journal of Epidemiology, 22*, 871–881.

5. Willett, W. C., Sacks, F., Trichopoulou, A., Drescher, G., Ferro-Luzzi, A., Helsing, E., & Trichopoulos, D. (1995). Mediterranean diet pyramid: A cultural model for healthy eating. *American Journal of Clinical Nutrition, 61*(Suppl.), 1402S–1406S.

6. Menotti, A., Kromhout, D., Blackburn, H., Fidanza, F., Buzina, R., & Nissinen, A. (1999). Food intake patterns and 25-year mortality from coronary heart disease: Cross-cultural correlations in the Seven Countries Study. The Seven Countries Study Research Group. *European Journal of Epidemiology, 15*, 507–515.

7. Hu, F. B., Rimm, E. B., Stampfer, M. J., Ascherio, A., Spiegelman, D., & Willett, W. C. (2000). Prospective study of major dietary patterns and risk of coronary heart disease in men. *American Journal of Clinical Nutrition, 72*, 912–921.

8. Fung, T. T., Willett, W. C., Stampfer, M. J., Manson, J. E., & Hu, F. B. (2001). Dietary patterns and the risk of coronary heart disease in women. *Archives of Internal Medicine, 161*, 1857–1862.

9. Mikkilä, V., Räsänen, L., Raitakari, O. T., Marniemi, J., Pietinen, P., Rönnemaa, T., & Viikari, J. (2007). Major dietary patterns and cardiovascular risk factors from childhood to adulthood. The Cardiovascular Risk in Young Finns Study. *The British Journal of Nutrition, 98*, 218–225.

10. Nagata, C. (2000). Ecological study of the association between soy product intake and mortality from cancer and heart disease in Japan. *International Journal of Epidemiology, 29*, 832–836.

11. Sasazuki, S. Fukuoka Heart Study Group. (2001). Case-control study of nonfatal myocardial infarction in relation to selected foods in Japanese men and women. *Japanese Circulation Journal-English Edition, 65*, 200–206.

12. Bazzano, L. A., He, J., Ogden, L. G., Loria, C., Vupputuri, S., Myers, L., & Whelton, P. K. (2001). Legume consumption and risk of coronary heart disease in US men and women: NHANES I Epidemiologic Follow-up Study. *Archives of Internal Medicine, 161*, 2573–2578.

13. Trichopoulou, A., Lagiou, P., & Trichopoulos, D. (1994). Traditional Greek diet and coronary heart disease. *Journal of Cardiovascular Risk, 1*, 9–15.

14. Trichopoulou, A., Costacou, T., Bamia, C., & Trichopoulos, D. (2003). Adherence to a Mediterranean diet and survival in a Greek population. *The New England Journal of Medicine, 348*, 2599–2608.

15. Trichopoulou, A., Bamia, C., & Trichopoulos, D. (2005). Mediterranean diet and survival among patients with coronary heart disease in Greece. *Archives of Internal Medicine, 165*, 929–935.

16. Trichopoulou, A., Bamia, C., Norat, T., Overvad, K., Schmidt, E. B., Tjønneland, A., Halkjaer, J., Clavel-Chapelon, F., Vercambre, M. N., Boutron-Ruault, M. C., Linseisen, J., Rohrmann, S., Boeing, H., Weikert, C., Benetou, V., Psaltopoulou, T., Orfanos, P., Boffetta, P., Masala, G., Pala, V., Panico, S., Tumino, R., Sacerdote, C., Bueno-de-Mesquita, H. B., Ocke, M. C., Peeters, P. H., Van der Schouw, Y. T., González, C., Sanchez, M. J., Chirlaque, M. D., Moreno, C., Larrañaga, N., Van Guelpen, B., Jansson, J. H., Bingham, S., Khaw, K. T., Spencer, E. A., Key, T., Riboli, E., & Trichopoulos, D. (2007). Modified Mediterranean diet and survival after myocardial infarction: The EPIC-Elderly study. *European Journal of Epidemiology, 22*, 871–881.

17. Trichopoulou, A., Orfanos, P., Norat, T., Bueno-de-Mesquita, B., Ocké, M. C., Peeters, P. H., van der Schouw, Y. T., Boeing, H., Hoffmann, K., Boffetta, P.,

Nagel, G., Masala, G., Krogh, V., Panico, S., Tumino, R., Vineis, P., Bamia, C., Naska, A., Benetou, V., Ferrari, P., Slimani, N., Pera, G., Martinez-Garcia, C., Navarro, C., Rodriguez-Barranco, M., Dorronsoro, M., Spencer, E. A., Key, T. J., Bingham, S., Khaw, K. T., Kesse, E., Clavel-Chapelon, F., Boutron-Ruault, M. C., Berglund, G., Wirfalt, E., Hallmans, G., Johansson, I., Tjonneland, A., Olsen, A., Overvad, K., Hundborg, H. H., Riboli, E., & Trichopoulos, D. (2005). Modified Mediterranean diet and survival: EPIC-elderly prospective cohort study. *BMJ, 330*, 991.

18. Bamia, C., Trichopoulos, D., Ferrari, P., Overvad, K., Bjerregaard, L., Tjønneland, A., Halkjaer, J., Clavel-Chapelon, F., Kesse, E., Boutron-Ruault, M. C., Boffetta, P., Nagel, G., Linseisen, J., Boeing, H., Hoffmann, K., Kasapa, C., Orfanou, A., Travezea, C., Slimani, N., Norat, T., Palli, D., Pala, V., Panico, S., Tumino, R., Sacerdote, C., Bueno-de-Mesquita, H. B., Waijers, P. M., Peeters, P. H., van der Schouw, Y. T., Berenguer, A., Martinez-Garcia, C., Navarro, C., Barricarte, A., Dorronsoro, M., Berglund, G., Wirfält, E., Johansson, I., Johansson, G., Bingham, S., Khaw, K. T., Spencer, E. A., Key, T., Riboli, E., & Trichopoulou, A. (2007). Dietary patterns and survival of older Europeans: The EPIC-Elderly Study (European Prospective Investigation into Cancer and Nutrition). *Public Health Nutrition, 10*, 590–598.

19. Kokubo, Y., Iso, H., Ishihara, J., Okada, K., Inoue, M., & Tsugane, S. JPHC Study Group. (2007). Association of dietary intake of soy, beans, and isoflavones with risk of cerebral and myocardial infarctions in Japanese populations: The Japan Public Health Center-based (JPHC) study cohort I. *Circulation, 116*, 2553–2562.

20. Fung, T. T., Chiuve, S. E., McCullough, M. L., Rexrode, K. M., Logroscino, G., & Hu, F. B. (2008). Adherence to a DASH-style diet and risk of coronary heart disease and stroke in women. *Archives of Internal Medicine, 168*, 713–720.

21. Mitrou, P. N., Kipnis, V., Thiébaut, A. C., Reedy, J., Subar, A. F., Wirfält, E., Flood, A., Mouw, T., Hollenbeck, A. R., Leitzmann, M. F., & Schatzkin, A. (2007). Mediterranean dietary pattern and prediction of all-cause mortality in a US population: Results from the NIH-AARP Diet and Health Study. *Archives of Internal Medicine, 167*, 2461–2468.

22. Nöthlings, U., Schulze, M. B., Weikert, C., Boeing, H., van der Schouw, Y. T., Bamia, C., Benetou, V., Lagiou, P., Krogh, V., Beulens, J. W., Peeters, P. H., Halkjaer, J., Tjønneland, A., Tumino, R., Panico, S., Masala, G., Clavel-Chapelon, F., de Lauzon, B., Boutron-Ruault, M. C., Vercambre, M. N., Kaaks, R., Linseisen, J., Overvad, K., Arriola, L., Ardanaz, E., Gonzalez, C. A., Tormo, M. J., Bingham, S., Khaw, K. T., Key, T. J.,

Vineis, P., Riboli, E., Ferrari, P., Boffetta, P., Bueno-de-Mesquita, H. B., van der A., D. L., Berglund, G., Wirfält, E., Hallmans, G., Johansson, I., Lund, E., & Trichopoulou, A. (2008). Intake of vegetables, legumes, and fruit, and risk for all-cause, cardiovascular, and cancer mortality in a European diabetic population. *Journal of Nutrition, 138*, 775–781.

23. Sofi, F., Cesari, F., Abbate, R., Gensini, G. F., & Casini, A. (2008). Adherence to Mediterranean diet and health status: Meta-analysis. *BMJ, 337*, a1344doi: 10.1136/bmj.a1344. Review.

24. Anderson, J. W., Johnstone, B. M., & Cook-Newell, M. E. (1995). Meta-analysis of the effects of soy protein intake on serum lipids. *The New England Journal of Medicine, 333*, 276–282.

25. Weggemans, R. M., & Trautwein, E. A (2003). Relation between soy-associated isoflavones and LDL and HDL cholesterol concentrations in humans: A meta-analysis. *European Journal of Clinical Nutrition, 57*, 940–946.

26. Zhuo, X. G., Melby, M. K., & Watanabe, S. (2004). Soy isoflavone intake lowers serum LDL cholesterol: A meta-analysis of 8 randomized controlled trials in humans. *Journal of Nutrition, 134*, 2395–2400.

27. Zhan, S., & Ho, S. C. (2005). Meta-analysis of the effects of soy protein containing isoflavones on the lipid profile. *American Journal of Clinical Nutrition, 81*, 397–408.

28. Reynolds, K., Chin, A., Lees, K. A., Nguyen, A., Bujnowski, D., & He, J. A. (2006). Meta-analysis of the effect of soy protein supplementation on serum lipids. *American Journal of Cardiology, 98*, 633–640.

29. Taku, K., Umegaki, K., Sato, Y., Taki, Y., Endoh, K., & Watanabe, S. (2007). Soy isoflavones lower serum total and LDL cholesterol in humans: A meta-analysis of 11 randomized controlled trials. *American Journal of Clinical Nutrition, 85*, 1148–1156.

30. Sirtori, C. R., Eberini, I., & Arnoldi, A. (2007). Hypocholesterolaemic effects of soya proteins: Results of recent studies are predictable from the Anderson meta-analysis data. *The British Journal of Nutrition, 97*, 816–822.

31. Harland, J. I., & Haffner, T. A. (2008). Systematic review, meta-analysis and regression of randomised controlled trials reporting an association between an intake of circa 25 g soya protein per day and blood cholesterol. *Atherosclerosis, 20*, 13–27.

32. Torres, N., Torre-Villalvazo, I., & Tovar, A. R. (2006). Regulation of lipid metabolism by soy protein and its implication in diseases mediated by lipid disorders. *Journal of Nutrition Biochemical, 17*, 365–373.

33. Jenkins, D. J., Kendall, C. W., Marchie, A., Faulkner, D. A., Wong, J. M., de Souza, R., Emam, A., Parker, T. L., Vidgen, E., Lapsley, K. G., Trautwein, E. A.,

Josse, R. G., Leiter, L. A., & Connelly, P. W. (2003). Effects of a dietary portfolio of cholesterol-lowering foods vs lovastatin on serum lipids and C-reactive protein. *The Journal of the American Medical Association, 290*, 502–510.

34. Jenkins, D. J., Kendall, C. W., Faulkner, D. A., Nguyen, T., Kemp, T., Marchie, A., Wong, J. M., de Souza, R., Emam, A., Vidgen, E., Trautwein, E. A., Lapsley, K. G., Holmes, C., Josse, R. G., Leiter, L. A., Connelly, P. W., & Singer, W. (2006). Assessment of the longer-term effects of a dietary portfolio of cholesterol-lowering foods in hypercholesterolemia. *American Journal of Clinical Nutrition, 83*, 582–591.

35. Lukaczer, D., Liska, D. J., Lerman, R. H., Darland, G., Schiltz, B., Tripp, M., & Bland, J. S. (2006). Effect of a low glycemic index diet with soy protein and phytosterols on CVD risk factors in postmenopausal women. *Nutrition, 22*, 104–113.

36. Pittaway, J. K., Ahuja, K. D., Robertson, I. K., & Ball, M. J. (2007). Effects of a controlled diet supplemented with chickpeas on serum lipids, glucose tolerance, satiety and bowel function. *Journal of the American College of Nutrition, 26*, 334–340.

37. Pittaway, J. K, Robertson, I. K., & Ball, M. J. (2008). Chickpeas may influence fatty acid and fiber intake in an *ad libitum* diet, leading to small improvements in serum lipid profile and glycemic control. *Journal of the American Dietetic Association, 108*, 1009–1013.

38. Serra-Majem, L., Roman, B., & Estruch, R. (2006). Scientific evidence of interventions using the Mediterranean diet: A systematic review. *Nutrition Reviews, 64*, S27–S47.

39. Lairon, D. (2007). Intervention studies on Mediterranean diet and cardiovascular risk. *Molecular Nutrition & Food Research, 51*, 1209–1214.

40. Esposito, K., Marfella, R., Ciotola, M., Di Palo, C., Giugliano, F., Giugliano, G., D'Armiento, M., D'Andrea, F., & Giugliano, D. (2004). Effect of a mediterranean-style diet on endothelial dysfunction and markers of vascular inflammation in the metabolic syndrome: A randomized trial. *The Journal of the American Medical Association, 292*, 1440–1446.

41. Vincent-Baudry, S., Defoort, C., Gerber, M., Bernard, M. C., Verger, P., Helal, O., Portugal, H., Planells, R., Grolier, P., Amiot-Carlin, M. J., Vague, P., & Lairon, D. (2005). The Medi-RIVAGE study: Reduction of cardiovascular disease risk factors after a 3-month intervention with a Mediterranean-type diet or a low-fat diet. *American Journal of Clinical Nutrition, 82*, 964–971.

42. Estruch, R., Martinez-González, M. A., Corella, D., Salas-Salvadó, J., Ruiz-Gutiérrez, V., Covas, M. I., Fiol, M., Gómez-Gracia, E., López-Sabater, M. C., Vinyoles, E., Arós, F., Conde, M., Lahoz, C., Lapetra, J., Sáez, G., & Ros, E. PREDIMED Study Investigators. (2006). Effects of a Mediterranean-style diet on cardiovascular risk factors: A randomized trial. *Annals of Internal Medicine, 145*, 1–11.

43. Fuentes, F., López-Miranda, J., Sánchez, E., Sánchez, F., Paez, J., Paz-Rojas, E., Marin, C., Gómez, P., Jimenez-Pereperez, J., Ordovás, J. M, & Pérez-Jiménez, F. (2001). Mediterranean and low-fat diets improve endothelial function in hypercholesterolemic men. *Annals of Internal Medicine, 134*, 1115–1119.

44. Teede, H. J., Giannopoulos, D., Dalais, F. S., Hodgson, J., & McGrath, B. P. (2006). Randomised, controlled, cross-over trial of soy protein with isoflavones on blood pressure and arterial function in hypertensive subjects. *Journal of the American College of Nutrition, 25*, 533–540.

45. Oosthuizen, W., Scholtz, C. S., Vorster, H. H., Jerling, J. C., & Vermaak, W. J. (2000). Extruded dry beans and serum lipoprotein and plasma haemostatic factors in hyperlipidaemic men. *European Journal of Clinical Nutrition, 54*(5), 373–379.

46. Cobiac, L., McArthur, R., & Nestel, P. J. (1990). Can eating baked beans lower plasma cholesterol? *European Journal of Clinical Nutrition, 44*, 819–822.

47. Mackay, S., & Ball, M. J (1992). Do beans and oat bran add to the effectiveness of a low-fat diet? *European Journal of Clinical Nutrition, 46*, 641–648.

48. Shutler, S. M., Bircher, G. M., Tredger, J. A., Morgan, L. M., Walker, A. F., & Low, A. G. (1989). The effect of daily baked bean (*Phaseolus vulgaris*) consumption on the plasma lipid levels of young, normo-cholesterolaemic men. *The British Journal of Nutrition, 61*, 257–265.

49. Anderson, J. W., Story, L., Sieling, B., Chen, W. J., Petro, M. S., & Story, J. (1984). Hypocholesterolemic effects of oat-bran or bean intake for hypercholesterolemic men. *American Journal of Clinical Nutrition, 40*, 1146–1155.

50. Anderson, J. W., Gustafson, N. J., Spencer, D. B., Tietyen, J., & Bryant, C. A. (1990). Serum lipid response of hypercholesterolemic men to single and divided doses of canned beans. *American Journal of Clinical Nutrition, 51*, 1013–1019.

51. Frühbeck, G., Monreal, I., & Santidrián, S. (1997). Hormonal implications of the hypocholesterolemic effect of intake of field beans (*Vicia faba* L.) by young men with hypercholesterolemia. *American Journal of Clinical Nutrition, 66*, 1452–1460.

52. Duane, W. C. (1997). Effects of legume consumption on serum cholesterol, biliary lipids, and sterol metabolism in humans. *Journal of Lipid Research, 38*, 1120–1128.

53. Finley, J. W., Burrell, J. B., & Reeves, P. G. (2007). Pinto bean consumption changes SCFA profiles in fecal fermentations, bacterial populations of the lower

bowel, and lipid profiles in blood of humans. *Journal of Nutrition, 137,* 2391–2398.

54. Geil, P. B., & Anderson, J. W. (1994). Nutrition and health implications of dry beans: A review. *Journal of the American College of Nutrition, 13,* 549–558.

55. Gardner, C. D., Coulston, A., Chatterjee, L., Rigby, A., Spiller, G., & Farquhar, J. W. (2005). The effect of a plant-based diet on plasma lipids in hypercholesterolemic adults: A randomized trial. *Annals of Internal Medicine, 142,* 725–733.

B. EFFECTS OF INDIVIDUAL VEGETABLES ON HEALTH

ACTIONS OF INDIVIDUAL FRUITS IN DISEASE AND CANCER PREVENTION AND TREATMENT

31

Biological Effects of Pomegranate (*Punica granatum* L.), especially its Antibacterial Actions, Against Microorganisms Present in the Dental Plaque and Other Infectious Processes

Glauce S. B. Viana[1], Silvana Magalhães Siqueira Menezes[2], Luciana N. Cordeiro[2], and F. J. A. Matos[2]

[1]Rua Barbosa de Freitas, Fortaleza, Brazil
[2]Department of Physiology and Pharmacology, Federal University of Ceará (UFC), Fortaleza, Brazil

1. INTRODUCTION

Dentistry professionals have developed preventive and therapeutic regimens for two of the most common diseases of humans, caries and inflammatory periodontal disease, exacerbated by the accumulation of bacterial plaques on tooth surfaces. Dental plaque formation is a dynamic and complex process involving many stages, from the adsorption of salivary pellicles to bacterial accumulation and growth. Plaque removal is an important issue in health promotion.

Punica granatum L. (Punicaceae), known as pomegranate, is a small tree that is common in the Mediterranean region. Its fruit has a number of biological activities, such as antitumor, antibacterial, antidiarrheal, antifungal, and anti-ulcer, that have been reported with various extracts/constituents of different parts of this plant. From the fruit seeds, around 7% of a particular oil was isolated, presenting as its main constituents punicinic (33%), palmitic (10%), estearic (6%), nonedecanoic (6%), heneicosanoic (5%), tricosanoic (5%), 13-methylestearic (1.5%), and 4-methyl lauric (0.5%) acids. From the pericarp, elagic tannins were isolated, such as granatins A and B, punicalagin and punicalin (these two present as major constituents) (Figures 31.1, 31.2). This is the fruit part responsible for the

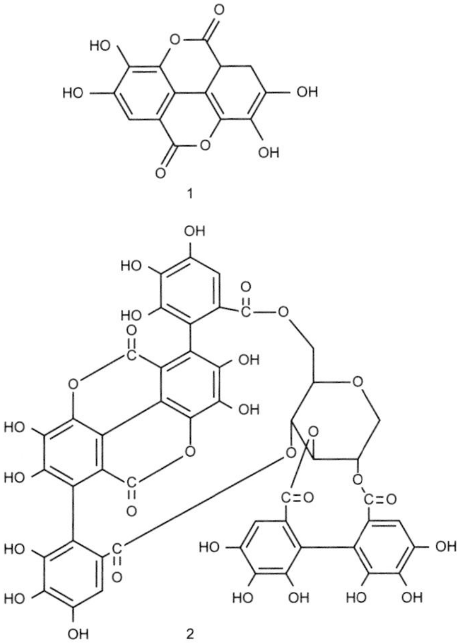

FIGURE 31.1 Structures of ellagic acid and punicalagin. Source Kwak H-M et al., 2005. *Archives of Pharmaceutical Research, 28,* 1328–1332; Saruwatari et al., 2008. *Journal of Medicinal Food, 11,* 623–628.

observed antibacterial action of the species, especially against *Staphylococcus aureus*, *Clostridium perfringens*, *Salmonella* sp., and *Serratia* sp. [1]. Another important action of the pericarp constituents is that against herpes simplex 2, responsible for genital herpes manifestations [2]. This action, added to the pericarp astringent effect as well as low toxicity, corroborates its large and diffuse use in Brazil, in mouth and throat inflammatory processes [3]. The root bark has also been used as an antihelminthic, due to the presence of alkaloids such as pelletierine, also called punicin, as well as N-methylpelletierine and pseudopelletierine. The plant alkaloids are toxic for humans, mainly due to their action on the CNS. The root bark also contains around 20% of tannins, in part combined with alkaloids [3]. In the plant leaves, ellagic acid and several other related tannins were identified.

The objective of this work was to review the biological effects of pomegranate (*Punica granatum*), especially those related to the fruit's

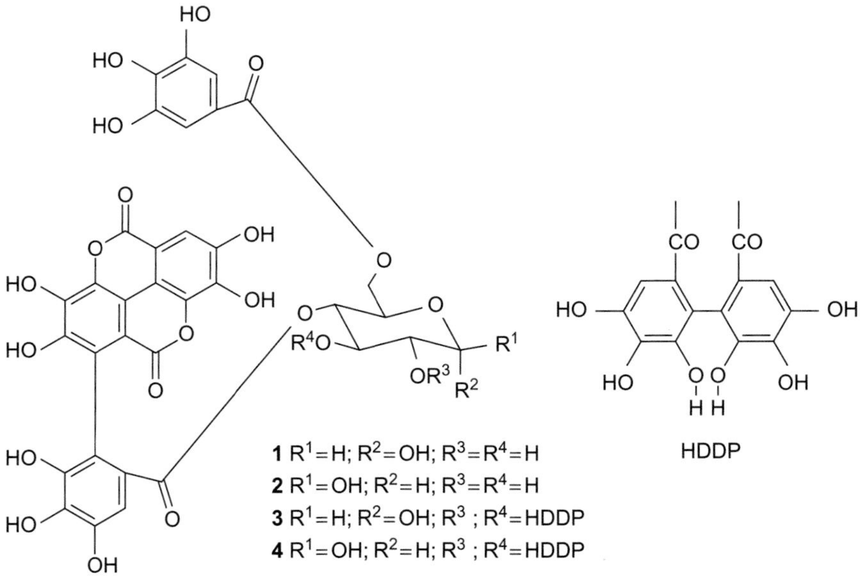

1 R^1=H; R^2=OH; R^3=R^4=H
2 R^1=OH; R^2=H; R^3=R^4=H
3 R^1=H; R^2=OH; R^3 ; R^4=HDDP
4 R^1=OH; R^2=H; R^3 ; R^4=HDDP

HDDP

FIGURE 31.2 Ellagitannins from the pericarp of *Punica granatum*. Source Machado et al., 2002. International *Journal of Antimicrobial Agents, 21,* 279–284.

antibacterial activity in general, as well as its potential against microorganisms present in dental plaque and other infectious and inflammatory processes. The popular use of pomegranate in Brazil and elsewhere makes this species potentially useful for government health programs, due to its adherence by regional communities, low toxicity, and already proven efficacy.

2. PERIODONTAL DISEASES, DENTAL PLAQUE AND OTHER ORAL INFECTIOUS DISEASES, AND THE USE OF NATURAL PRODUCTS

Periodontal diseases and dental caries are closely associated with the development of dental plaque, formed as a result of the complex interactions between teeth and adsorbed host and bacterial molecules, passive transport of oral bacteria, co-adhesion of successive bacterial strains, and the multiplication of associated microorganisms [4]. More than 300 bacterial species that inhabit the oral cavity participate in this process [5].

The pellicle formed as a result of selective adsorption of salivary proteins can influence early bacterial attachment and affect the acquisition of diseased or healthy microflora on the tooth. In order to colonize, bacteria must firstly adhere to the tooth surface and then resist the cleansing forces of flowing saliva, lips, and cheeks [6]. Following the attachment of pioneer colonizers, subsequent layers may form by cell-to-cell adherence and by the proliferation of adhering bacteria. Several bacterial and host factors are involved in the establishment of a balanced or climaxed microbial community [7].

Dental caries is a multifactorial pathology which involves a susceptible host, a cariogenic microbiota, and a cariogenic diet. These factors must occur simultaneously, during a certain period of time, for the occurrence of dental caries [8]. This disease is initiated by the bacterial adhesion and subsequent plaque formation on the tooth surface, followed by bacterial carbohydrate fermentation and organic acid formation. These acids diffuse into and ultimately demineralize the tooth [9]. *Streptococcus mutans* is considered the main etiological agent of caries of smooth surfaces, in human and in animal models. Previous studies showed that *mutans* streptococci are involved in the initial stages of dental demineralization [8,10].

Periodontal disease is one of the main causes of tooth loss. The presence of pathogenic plaques, together with host-related factors that modify the response to plaque bacteria, is the key determinant contributing to development of the disease [11]. Periodontitis is a common infectious disease, to which *Porphyromonas gingivalis* has been closely linked, in which the attachment tissues of the teeth and their alveolar bone housing are destroyed [12]. The role of dental plaque as the primary etiological agent in gingivitis has been demonstrated in classic studies of experimental gingivitis in adults and children [13]. The mechanical removal of supragingival plaques is an effective method for controlling plaque and gingival inflammation. However, most subjects lack the compliance and dexterity required for appropriate toothbrushing and flossing, and thus are not able to maintain an adequate standard of gingival health [14]. Hence, interest in alternative mouthrinses and toothpastes based on natural products and plant extracts has increased.

Other species of microorganisms that are present in the oral cavity, such as *Candida albicans* and *Staphylococcus aureus,* may also cause pathologies, particularly under specific conditions. Microorganisms of the Enterobacteriaceae family and of the genus *Pseudomonas* have been extensively studied, due to their pathogenic potential. Also, these bacteria are correlated to severe periodontal diseases [15,16].

The establishment and maintenance of oral microbiota are related not only to interbacterial co-aggregations but also to interactions of these bacteria with yeasts, such as *Candida albicans* [17].

C. ACTIONS OF INDIVIDUAL FRUITS IN DISEASE AND CANCER PREVENTION AND TREATMENT

Fungi are frequently isolated from several oral sites, including the tongue, buccal mucosa, palate, dental biofilm, subgingival microbiota, carious lesions, and prosthetic appliances [18]. Possible relations between *C. albicans* and periodontal disease, dentin and/or root caries have been suggested [19]. These studies showed that *C. albicans* has a similar capacity for colonizing hydroxyapatite as that of *S. mutans*, although using different mechanisms.

Exponential advancements in the field of cariology have re-emphasized the importance of prevention. There has been a change in thinking globally, with a growing tendency to go natural. The limitations of mechanical control of the dental biofilm originated several studies on the activity of chemical agents. These products can act by interfering in bacterial adhesion to the dental surface, reducing bacterial proliferation, or removing the pre-existing biofilm [10]. Review of the literature also showed that many studies are being conducted to identify therapeutic agents from natural sources, for the management of dental diseases. Most of the agents being evaluated are plant extracts, aiming at the management of periodontal diseases and dental caries. *Terminalia chebula* is one of the exceptions, in that its extract was already being used for prevention of dental caries. The aqueous extract of *T. chebula* is a potential anticariogenic mouthwash that increases the oral pH and buffering capacity and decreases the microbial count. For better efficiency, a suitable vehicle such as gel or varnish formulations has to be chosen [20].

The surface roughness of the dental enamel, caused by toothbrushing with abrasive dentifrices, is an important factor in oral health. It modifies the shape of the dental surface, facilitating the retention of bacterial plaque, which is a cause of dental caries and periodontal diseases [21]. It is considered that there is an increasing need to obtain substances that remove bacterial plaque without causing any other damage, such as an increase in surface roughness and, consequently, the process of dental abrasion [22].

The literature regarding the use of oil in dental brushing is not abundant [23]. Therefore, it is desirable to offer a clear understanding on the subject, which will demonstrate the removal of bacterial plaque (preventing caries and periodontal disease) without interfering with the salivary flow and buffer capacity (important in the process of dental demineralization), unbalancing oral microorganisms (enabling prolonged use), or causing abrasion of hard tissues of the oral cavity.

In Brazil, the population generally keeps its teeth up to an advanced age, when it is possible to detect problems of increased sensitivity and root caries resulting from the use of abrasive dentifrices (cervical abrasion). These lesions can be avoided by the replacement of abrasive dentifrices by abrasion-free dentifrices containing, for instance, a vegetable oil [22]. In one study, it was proven that, when pure almond oil was employed in toothbrushing, there was a reduction in dental plaque, as compared to the use of regular dentifrices containing abrasives and antiplaque agents [24].

There is an increasing number of investigations on new and natural substances, in order to evaluate their activity and possible application for controlling the dental biofilm. Scientific studies on the chemical and pharmacological properties of medicinal plants allow scientists to indicate their proper use. Low toxicity, when correctly employed, is considered one of the main advantages of treatment with medicinal plants [25]. Natural products have also proven to be an alternative to synthetic chemical substances. Nostro et al. [26] demonstrated that *Helichrysum italicum* extract interfered with the cariogenic properties of *S. mutans*, through reductions of superficial hydrophobicity, inhibiting adherence of cell growth to glass in 90–93%. While assessing the minimum inhibitory concentrations of adherence of different vegetable dyes and propolis, these authors observed that the capacity of the *H. italicum* extract to prevent bacterial adherence could be due to the effect of its flavonoid components, which have an antiglycosyltransferase activity.

Aqueous and alcoholic extracts from *Juglans regia* (Jangladaceae), used as chewing sticks to maintain oral hygiene, were tested for their ability to inhibit the growth and some physiological functions of *Streptococcus mutans*. Both extracts strongly inhibited the growth, *in vitro* adherence, acid production, and glucan-induced aggregation of *S. mutans* [27]. The antimicrobial activity of anacardic acids of the cashew nut shell oil from *Anacardium occidentale* L. (Anacardiaceae) was studied on the oral microorganisms *Streptococcus mutans* ATCC 25175, *Staphylococcus aureus* ATCC 12598, *Candida albicans* ATCC 10231, and *Candida utilis*. The anacardic acids presented antibacterial activity against the above microorganisms, but higher inhibitory activity occurred with gram-positive bacteria, such as *Streptococcus mutans,* known to be one of the main causes of tooth decay [28].

Alviano et al. [29] determined the antibacterial activities of crude extracts from *Cocos nucifera* L. (husk fiber), *Ziziphus joazeiro* Mart. (inner bark*)*, *Caesalpinia pyramidalis* Tul. (leaves), and *Aristolochia cymbifera* Mart. et Zucc. (rhizomes) against *Prevotella intermedia, Porphyromonas gingivalis, Fusobacterium nucleatum, Streptococcus mutans,* and *Lactobacillus casei.* The antioxidant activity and acute toxicity of these extracts were also evaluated. All oral bacteria tested (planktonic or in artificial biofilms) were more susceptible to, and rapidly killed in the presence of *A. cymbifera, C. pyramidalis,* and *C. nucifera* extracts than in the presence of the *Z. joazeiro* extract. In Thailand, the gum of *Cratoxylum formosum* Dyer, commonly known as mempat, is a natural agent that has been extensively used for caries prevention. The gum of *C. formosum* has high antimicrobial activity against *S. mutans,* and may become a promising herbal varnish against caries [30].

Prabu et al. [31] investigated the anti-*Streptococcus mutans* activity and the *in vitro*

effects of sub-minimal inhibitory concentrations of guaijaverin, isolated from *Psidium guajava* L., on the cariogenic properties of *Streptococcus mutans*. This study demonstrated that guaijaverin presented a growth-inhibitory action against *Streptococcus mutans*. Other natural products, such as those from *Arnica* and propolis, have been used for thousands of years in folk medicine for many purposes. They possess several biological activities, such as anti-inflammatory, antifungal, antiviral, and tissue regenerative, among others. Although the antibacterial activity of propolis has already been demonstrated, very few studies have been done on bacteria of clinical relevance in dentistry. Also, the antimicrobial activity of *Arnica montana* L. has not been extensively investigated. The antimicrobial activity, inhibition of adherence of *Streptococcus mutans,* and inhibition of formation of water-insoluble glucan by *A. montana* and propolis extracts were evaluated *in vitro*. *A. montana* (10%, w/v) and propolis (10%, w/v) extracts from the State of Minas Gerais, Brazil, were compared with controls. Twelve microorganisms were used as follows: *Candida albicans, Staphylococcus aureus, Enterococcus faecalis, Streptococcus sobrinus, S. sanguis, S. cricetus, S. mutans, Actinomyces naeslundii, A. viscosus, Porphyromonas gingivalis, P. endodontalis,* and *Prevotella denticola.* The propolis extract significantly inhibited all the microorganisms tested ($P<0.05$), showing the largest inhibitory zone for *Actinomyces* spp. While the propolis extract showed *in vitro* antibacterial activity, inhibition of cell adherence, and inhibition of water-insoluble glucan formation, the *Arnica* extract was only slightly active under those three conditions [32].

Feres et al. [33] determined, *in vitro*, the antimicrobial effect of plant extracts and propolis in saliva samples of 25 periodontally healthy subjects and 25 subjects with chronic periodontitis. Propolis showed significant antimicrobial properties in those samples from periodontally healthy or diseased subjects, suggesting that this substance may be used therapeutically in the future for inhibiting oral microbial growth. The methanol extracts of *Hamamelis virginiana* L. and *Arnica montana* L. and, to a lesser extent, *Althaea officinalis* L. were very active against anaerobic and facultative aerobic periodontal bacteria: *Porphyromonas gingivalis, Prevotella* spp., *Fusobacterium nucleatum, Capnocytophaga gingivalis, Veilonella parvula, Eikenella corrodens, Peptostreptococcus micros,* and *Actinomyces odontolyticus.* This study suggested the use of the alcohol extracts of *H. virginiana, A. montana,* and *A. officinalis* for topical medications in periodontal prophylactics [34].

The aqueous extracts of *Piper betle* L. and *Psidium guajava* L. were prepared and tested for their anti-adherence effect on the adhesion of early plaque settlers (*Streptococcus mitis, S. sanguinis,* and *Actinomyces* sp.). *P. guajava* was shown to have a slightly greater anti-adherence effect on *S. sanguinis* by 5.5% and *Actinomyces* sp. by 10% and a significantly higher effect on *S. mitis* (70%), as compared to *Piper betle* L. The three bacterial species are known to be highly hydrophobic, and that hydrophobic bonding seemed to be an important factor in their adherence activities. It is therefore suggested that the plant extracts, in expressing their anti-adherence activities, could have altered the hydrophobic nature of the bonding between the bacteria and the saliva-coated glass surfaces [35].

Cacao bean husk extract (CBH) has been shown to possess antibacterial and anti-glucosyltransferase activities, through its unsaturated fatty acids and epicatechin polymers, respectively. In the present study, the anti-plaque activities of CBH were examined *in vitro* and *in vivo*. The extract inhibited the adherence of *Streptococcus mutans* MT8148 to saliva-coated hydroxyapatite, and reduced the accumulation of artificial dental plaque by *S. mutans* MT8148 on orthodontic wire. The number of *mutans* streptococci in dental plaques was also significantly reduced when human dental plaque (from 21 children at 37°C for 1 h) was exposed to CBH. For the *in vivo* study, 28 volunteers

aged 19–29 years old rinsed their mouths with CBH, before and after each intake of food, and before sleeping at night, for 4 days, without using any other oral hygiene procedures. Plaque depositions and the numbers of *mutans* streptococci were reduced in the subjects, as compared to rinsing with 1% ethanol alone. These results indicated that CBH possesses significant antiplaque activity *in vitro* and *in vivo* [36].

Yamaguti-Sasaki et al. [37] showed *in vitro* assessments of the antibacterial potential of *Paullinia cupana* L. *var. sorbilis* extracts against *Streptococcus mutans*. Another study tested *in vitro* the bacteriostatic and bactericidal activities of a *Mahonia aquifolium* (Pursh) Nutt. extract and two of its major alkaloids, berberine chloride and oxyacanthine sulfate, against nine different oral bacteria. The most susceptible bacterium against all three test substances was *Porphyromonas gingivalis* [38].

Centella asiatica (L.) Urb. and *Punica granatum* L. (pomegranate) are medicinal plants that have been reported to promote tissue healing and modulate host responses. A preliminary study revealed positive clinical effects of an innovative preparation from the two herbal extracts, in the form of biodegradable chips, as a subgingival adjunct to scaling and root planing. The results of this research indicate that adjunctive local delivery of extracts from *C. asiatica,* in combination with *P. granatum,* significantly improved clinical signs of chronic periodontitis and IL-1β levels in maintenance patients [39]. A pomegranate fruit extract has been used worldwide for the treatment and prevention of arthritis and other inflammatory diseases. Thus, it has been shown that standardized extracts of the pomegranate fruit possess anti-inflammatory and cartilage-sparing effects *in vitro* [40]. Others have shown that the extract of pomegranate fruit constituents inhibits the proliferation of human cancer cells, and also modulates inflammatory subcellular signaling pathways and apoptosis when added directly to the culture medium [41,42].

C. ACTIONS OF INDIVIDUAL FRUITS IN DISEASE AND CANCER PREVENTION AND TREATMENT

3. THE ANTIBACTERIAL ACTIVITY OF *PUNICA GRANATUM* AND ITS BIOACTIVE CONSTITUENTS

The major class of compounds occurring in pomegranate is represented by polyphenols that include flavonoids, condensed tannins, and hydrolyzable tannins. Additionally, other components are alkaloids such as pelletitierine. Hydrolyzable tannins are the predominant polyphenols found in the pomegranate juice, accounting for around 90% of its antioxidant activity [43,44]. Evidences have shown that the antioxidant capacity of pomegranate juice is three times that of red wine and green tea, presumably due to the presence of hydrolyzable tannins in the rind, along with anthocyanins and ellagic acid derivatives. Flavonoid-rich fractions of pomegranate fruit extract have also been shown to exert an antiperoxidative effect, as its administration significantly decreased the concentrations of malondialdehydes and hydroperoxides, and enhanced the activities of catalase, superoxide dismutase, glutathione peroxidase, and glutathione reductase in the liver [45,46]. It has also been shown that pomegranate extract exerts a powerful influence in inhibiting the expression of inflammatory cytokines IL-1β and IL-6 in adjunctive periodontal therapy [39].

Furthermore, several groups reported that the consumption of pomegranate may have cholesterol-lowering and preventive effects against cardiovascular and chronic diseases *in vivo* [47,48]. In these studies, the major effect of the pomegranate extract was the reduction of oxidative stress, inhibition of the p38-mitogen-activated protein kinase (p38-MAPK) pathway, and inhibition of the activation of the transcription factor NF-κB. Activations of p38-MAPK and NF-κB are intimately associated with the increased gene expression of TNF-α, IL-1β, MCP1, inducible nitric oxide synthase (iNOS), and cyclooxygenase-2 (COX-2), that

are critical mediators of inflammation and pathogenesis of inflammatory and degenerative joint diseases [49,50]. Recently, a new piece of evidence also suggested that the extract from pomegranate may exert an anti-inflammatory effect by inhibiting the inflammatory cytokine-induced production of prostaglandins E2 (PGE2) and NO *in vivo* [43].

The edible part of pomegranate is rich in anthocyanins, a group of polyphenolic compounds that possess antioxidant and anti-inflammatory activities [44,51]. Other studies in animal models of cancer suggest that pomegranate extract may be anticarcinogenic [52,53], whereas studies in mice and humans indicate that it may also have a potential therapeutic and chemopreventive adjuvant effect in cardiovascular disorders [46]. Anthocyanins were shown to be effective inhibitors of lipid peroxidation, production of NO, and iNOS activity, in different model systems [46,54]. In a comparative study, anthocyanins from the pomegranate fruit were shown to possess higher antioxidant activity than vitamin E (α-tocopherol), ascorbic acid, and β-carotene [55]. Antioxidant activities of the three major anthocyanidins present in pomegranate (delphinidin, cyanidin, and pelargonidin) were also evaluated and shown to be potent antioxidants [56]. In related studies, prodelphinidins inhibited COX-2 and lipoxygenase activity and the production of prostaglandins E2. They also activated the synthesis of type II collagen in human chondrocytes [57,58].

The synergistic action of pomegranate constituents appears to be superior to that of single constituents [59]. Numerous studies on the antioxidant, anticarcinogenic, and anti-inflammatory properties of pomegranate constituents have been published, focusing on the treatment and prevention of cancer, cardiovascular disease, diabetes, dental conditions, erectile dysfunction, bacterial infections, antibiotic resistance, and ultraviolet radiation-induced skin damage [60]. In addition, the effects of pomegranate extracts and ellagic acid on the proliferation of prostate cancer cells were also studied [61].

The antiatherogenic activity of the pomegranate seed aril juice has been attributed to its antioxidant polyphenols, such as ellagitannin and punicalagin. Cerdá et al. [61] evaluated, in healthy humans, the bioavailability and metabolism of pomegranate juice ellagitannins, to assess their effects on several blood parameters and to compare the antioxidant activity of punicalagin with that of the *in vivo* generated metabolites, Their data showed that the potential systemic biological effects of pomegranate juice ingestion should be attributed to the colonic microflora metabolites rather than to the polyphenols present in the juice.

Punicalagin, ellagic acid, and the total pomegranate tannin extract were evaluated for antiproliferative activity on human oral, colon, and prostate tumor cells [41]. These authors showed that the pomegranate juice showed the greatest antiproliferative activity against all cell lines. The superior bioactivity of pomegranate, as compared to its purified polyphenols, illustrated the multifactorial effects and chemical synergy of the action of multiple compounds compared to single purified active ingredients. Furthermore, *in vitro* cytotoxic studies against three cell lines (a monkey kidney cell, a human larynx epithelial cancer cell, and a human lung carcinoma small cell line) showed that punicalagin is toxic only at higher concentrations [62].

Pomegranate may inhibit cell proliferation and apoptosis through the modulation of cellular transcription factors and signaling proteins. In previous studies, pomegranate juice and its ellagitannins inhibited proliferation and induced apoptosis in HT-29 colon cancer cells [42]. These authors examined the effects of pomegranate juice on inflammatory cell signaling proteins in these same cells, and found that not only the pomegranate juice, but also pomegranate tannin extract and punicalagin, significantly suppressed TNF-α-induced COX-2 protein expression. Additionally, they reduced

phosphorylation of the p65 subunit and binding of the NF-κB response element. The pomegranate juice also abolished TNF-α induced AKT activation, needed for NF-κB activity. These authors concluded that polyphenolic compounds present in pomegranate play an important role in the modulation of inflammatory cell signaling in colon cancer cells.

Besides being tested for antioxidant activity, several fractions and compounds from *Punica granatum*, such as ellagic acid, gallagic acid, punicallins, and punicalagins, were also recently evaluated for antiplasmodial and antimicrobial activities, in cell-based assays [63]. They showed that gallagic acid and punicalagins exhibited antiplasmodial activity, and pomegranate fractions presented antimicrobial activity when assayed against *Escherichia coli, Pseudomonas aeruginosa, Candida albicans, Cryptococcus neoformans,* and methicillin-resistant *Staphylococcus aureus* (MRSA). These compounds and the mixture of tannins presented also a high antioxidant activity.

Punicalagin isolated from the fruit of *Punica granatum* was identified as a potent immune suppressant, based on its inhibitory action on the activation of the nuclear factor of activated T cells (NFAT). Punicalagin down-regulated the mRNA and soluble protein expression of IL-2 from anti-CD3/anti-CD28-stimulated murine splenic CD4+ T cells, and suppressed a mixed leukocyte reaction, without exhibiting cytotoxicity to these cells. *In vivo* punicalagin treatment inhibited phorbol 12-myristate 13-acetate-induced chronic ear edema in mice, and decreased CD3+ T cell infiltration of the inflamed tissue [64].

Punicic acid [65] from pomegranate seed oil (PSO) and polyphenols [66–69] inhibits prostaglandin biosynthesis. The ethyl acetate extract of pomegranate fermented juice (W) inhibits soybean lipoxygenase (LOX) but not sheep cyclooxygenase (COX), while a phenolic-rich extract of pomegranate seed oil strongly inhibits lipoxygenase and cyclooxygenase [70]. Applied to mouse skin, whole pomegranate aqueous extract inhibits cyclooxygenase expression [47], while W, pomegranate peel extract (P), and pomegranate seed oil (PSO) each inhibit PC-3 human prostate cancer cell phospholipase A2 expression *in vitro*. These suppressive effects were supra-additively enhanced when two of W, P, and PSO, but especially all three pomegranate components were combined [71].

The acetone extracts of whole pomegranate fruits (WPFE) inhibited phosphorylation of several cytokines in UVB-irradiated keratinocytes, including mitogen activated protein kinases (MAPK). The extracts also diminished activation of NF-κB [52]. Inhibition of NF-κB, MAPK, and related cytokines by WPFE occurred *in vivo*, in mouse skin exposed to 12-O-tetradecanoylphorbol-13-acetate (TPA) [47], and in human chondrocytes induced by cytokine interleukin IL-1 [40] with up-regulation of MAPK-APK2, in PSO-treated human DU-145 prostate cancer cells [72]. The beneficial effect of pomegranate extract reduction of cytokine activity has been shown to occur in patients with periodontitis. Patients experiencing this form of oral inflammation received intragingival chips impregnated with pomegranate peel extract (and extract of *Centella asiatica*), which resulted in reduced inflammatory cytokines (IL-1β and IL-6) several months post-treatment [39].

Matrix metalloproteinases (MMP) are enzymes important in the maintenance of normal cellular architecture, assisting with the creation of interstitial spaces by destroying structural proteins, thereby facilitating multiple inflammatory processes [73–76]. Human chondrocyte MMPs were inhibited by WFPE [40], while P, and to a lesser extent W and the pomegranate seed extract (PSE), but not PSO, inhibited human dermal fibroblast MMP-1 [77].

Pomegranate juice has been demonstrated to present a high antioxidant activity, and it is effective in the prevention of atherosclerosis [44,70]. Pomegranate juice also displays a potent antiatherogenic action in atherosclerotic

mice and humans [78,79]. All these activities may be related to diverse phenolic compounds present in pomegranate juice, including punicalagin isomers, ellagic acid derivatives, and anthocyanins (delphinidin, cyanidin, and pelargonidin 3-glucosides and 3,5-diglucosides). These compounds are known for their properties in scavenging free radicals and inhibiting lipid oxidation *in vitro* [43,56].

Li et al. [80] found that pomegranate peel had the highest antioxidant activity among the peel, pulp, and seed fractions of 28 kinds of fruit commonly consumed in China, as determined by FRAP (ferric reducing antioxidant power) assays. In this study, antioxidants from pomegranate peel were extracted using a mixture of ethanol, methanol, and acetone, and the antioxidant properties of the extract were further investigated, as compared to the pulp extract. The results showed that the pomegranate peel extract had markedly higher antioxidant capacity than the pulp extract, in scavenging or preventive capacity against superoxide anions, hydroxyl and peroxyl radicals, as well as inhibiting $CuSO_4$-induced LDL oxidation. These authors concluded that the pomegranate peel extract appeared to have more potential as a health supplement rich in natural antioxidants than the pulp extract, and merits further intensive study.

Singh et al. [81] reported that the methanol extract of pomegranate peel has much higher antioxidant capacity than that of seeds, as demonstrated by using the β-carotene-linoleate and diphenylpicrylhydrazyl (DPPH) model systems. This pomegranate peel extract could effectively protect (after oral administration) against CCl_4 induced hepatotoxicity, in which reactive oxygen species (ROS) damage was intensively involved. Other examples of *in vivo* studies of beneficial effects of pomegranate antioxidant activity include: protection of rat gastric mucosa from ethanol or aspirin toxicity [82,83], protection of neonatal rat brain from hypoxia [84], prevention of male rabbit erectile tissue dysfunction [85], and abrogation of ferric nitrilotriacetate (Fe-NTA)-induced hepatotoxicity, evidenced by mitigated hepatic lipid peroxidation, actions of glutathione (GSH), catalase (CAT), glutathione peroxidase (GPX), glutathione reductase (GR), glutathione-*S*-transferase (GST), serum aspartate aminotransferase (AST), alanine aminotransferase (ALT), and alkaline phosphatase (ALP), bilirubin and albumin levels, hepatic ballooning degeneration, fatty changes, and necrosis [86]. Other authors related that treatment with antioxidant polyphenols contained in pomegranate juice may promote a sustained correction of the nitric oxide synthase NOSIII down-regulation induced by oxidized LDL in human coronary endothelial cells [87].

Ajaikumar et al. [83] evaluated *in vivo* antioxidant and anti-ulcer activity of the 70% methanolic extract of *Punica granatum* fruit rind. The administration of this extract (250 mg/kg and 500 mg/kg) showed inhibitions of 22 and 74, and 22 and 63%, in aspirin- and ethanol-induced gastric ulceration, respectively. In treated animal groups, the *in vivo* antioxidant levels, such as superoxide dismutase (SOD), catalase, glutathione (GSH), and glutathione peroxidase (GPx), were increased or found to be more or less equal to the normal values. The tissue lipid peroxidation level decreased in treated groups of animals, as compared to the control group. Histopathological examination of the stomach of ulcerated animals showed severe erosion of gastric mucosa, submucosal edema, and neutrophil infiltration. In general, the results of the present investigation revealed the gastroprotective activity of the extract through antioxidant mechanisms.

Pomegranate juice, which is rich in tannins, possesses antiatherosclerotic properties which could be related to its potent antioxidative characteristics. As some antioxidants were shown to reduce blood pressure, the effect of pomegranate juice consumption (50 mL, 1.5 mmol of total polyphenols per day, for 2 weeks) by hypertensive patients was studied on their blood pressure and on serum angiotensin converting enzyme (ACE) activity. A 36% decrement in serum ACE activity and a 5% reduction in systolic blood pressure were noted. A similar dose-dependent inhibitory effect (31%) of pomegranate juice on serum ACE activity was also observed *in vitro*. As the reduction in serum ACE activity, even with no decrement in blood pressure, was previously shown to attenuate atherosclerosis, pomegranate juice can offer wide protection against cardiovascular diseases that could be related to its inhibitory effect on oxidative stress and on serum ACE activity [88].

The ability of any chemotherapeutic agent to selectively inhibit proliferation of malignant, but not normal, cells is the hallmark of a promising anticancer therapeutic agent. Pomegranate peel extracts have been shown to retard proliferation of cells in several different human cancer cell lines [53,89,90]. In human breast cancer cells, for example, the effects of fermented pomegranate juice and pomegranate peel extract were most pronounced against estrogen-responsive MCF-7 cells, less pronounced against estrogen-negative MDA-MB-231 cells, and least pronounced against immortalized normal breast epithelial cells MCF-10A [91,92], strongly suggesting a spectrum of anticancer activity and not the presence of indiscriminate cytotoxic compounds. Pomegranate components have previously been shown to possess an ability for inhibiting the estrogenic action of 17-β-estradiol, an activity best explained through competitive binding to estrogen receptors by a number of non-steroidal estrogenic flavonoids, such as kaempferol, quercetin, naringenin, and luteolin [90].

In 2008, it was reported that pomegranate standardized extract decreased prostate cancer xenograft size, tumor vessel density, vascular endothelial growth factor (VEGF) peptide levels, and hypoxia-inducible factor 1 alpha (HIF-1α) expression, after 4 weeks of treatment in severe combined immunodeficient (SCID) mice. These results demonstrated that an ellagitannin-rich pomegranate extract can inhibit tumor-associated angiogenesis, as one of several potential mechanisms for slowing the growth of prostate cancer in chemopreventive applications [93].

Prostate cancer is the most common invasive malignancy and the second leading cause of cancer-related deaths among US males, with a similar trend in many Western countries. One approach to control this malignancy is its prevention through the use of agents present in the humans' consumed diet. Malik et al. [94] showed that pomegranate fruit extract (PFE) possesses remarkable antitumor-promoting effects in mouse skin. In this study, employing human prostate cancer cells, the anti-proliferative and pro-apoptotic properties of PFE were evaluated. PFE treatment of highly aggressive human prostate cancer PC3 cells resulted in a dose-dependent inhibition of cell growth and cell viability and induction of apoptosis. These authors suggest that pomegranate juice may have cancer chemopreventive as well as cancer chemotherapeutic effects against prostate cancer in humans.

In clinical studies, pomegranate seed aril juice administration led to a decrease in the rate of rise of prostate specific antigen (PSA) after primary treatment with surgery or radiation [95]. Others reported that the consumption of standardized pomegranate extract potently delayed the onset and reduced the incidence of collagen-induced arthritis in mice [43]. The severity of arthritis was also significantly lower in the group treated with the standardized extract. Histopathological data of this group demonstrated reduced joint infiltration by inflammatory cells, and the destruction of bone and cartilage were alleviated. Levels of IL-6 were significantly decreased. In mouse macrophages, the standardized extract abrogated multiple signal transduction pathways and downstream mediators implicated in the pathogenesis of rheumatoid arthritis.

Postprandial hyperglycemia plays an important role in the development of type II diabetes

and has been proposed as an independent risk factor for cardiovascular diseases. The flowering part of *Punica granatum* Linn. (Punicaceae) tree (PGF) has been recommended in the Unani literature as a remedy for diabetes. Li et al. [96] investigated the effects and action mechanisms of a methanolic extract from PGF on hyperglycemia, *in vivo* and *in vitro*. Oral administration of PGF extract markedly lowered plasma glucose levels in non-fasted Zucker diabetic fatty rats (a genetic model of obesity and type II diabetes), whereas it had little effect in the fasted animals, suggesting that it affected postprandial hyperglycemia in type II diabetes. In support of this conclusion, the extract was found to markedly inhibit the increase of plasma glucose levels after sucrose loading, but not after glucose loading in mice, and it had no effect on glucose levels in normal mice. *In vitro*, PGF extract demonstrated a potent inhibitory effect on α-glucosidase activity (IC$_{50}$: 1.8 μg/mL). The inhibition is dependent on the concentration of enzyme and substrate, as well as on the length of pretreatment with the enzyme. These findings strongly suggest that PGF extract improves postprandial hyperglycemia in type II diabetes and obesity, at least in part, by inhibiting intestinal α-glucosidase activity.

Modern uses of pomegranate-derived products now include treatment of acquired immune deficiency syndrome (AIDS) [97], in addition to cosmetic beautification [98,99] and enhancement [100], hormone replacement therapy [101], resolution of allergic symptoms [102], cardiovascular protection [103,104], oral hygiene[105], ophthalmic ointment [106], and as an adjunct therapy to increase bioavailability of radioactive dyes during diagnostic imaging [107,108].

In India, various medicinal species have been traditionally used since ancient times for the preservation of food products, as they have been regarded as having antiseptic and disinfectant properties. Of the 35 different Indian species tested, pomegranate seeds had potent antimicrobial activities against *Bacillus subtilis* (ATCC 6633), *Escherichia coli* (ATCC 10536), and *Saccharomyces cerevisiae* (ATCC10536) [109].

The interaction between *P. granatum* methanolic extract and five antibiotics (chloramphenicol, gentamicin, ampicillin, tetracycline, and oxacillin) against 30 clinical isolates of methicillin-resistant and methicillin-sensitive *Staphylococcus aureus* demonstrated that the pomegranate extract dramatically enhanced the activity of all antibiotics tested, with synergistic activity detected between the extract and the antibiotics tested [110].

The effect of the pomegranate extract on *Staphylococcus aureus* growth and subsequent enterotoxin production was studied. At a low concentration (0.01% v/v), the bacterial growth was delayed, and at a higher concentration (1% v/v) it was eliminated. At a 0.05% (v/v) pomegranate extract concentration, staphylococcal enterotoxin production was inhibited [111].

Fourteen extracts from Brazilian traditional medicinal plants used to treat infectious diseases were utilized to look for potential antimicrobial activity against multiresistant bacteria of medical importance. *Staphylococcus aureus* strains were susceptible to extracts of *P. granatum* and *Tabebuia avellanedae*. A mixture of ellagitannins isolated from *P. granatum* and two naphthoquinones isolated from *T. avellanedae* demonstrated antibacterial activities against all *S. aureus* strains tested. Semi-synthetic furanonaphthoquinones (FNQs) showed lower minimum inhibitory concentrations than those exhibited by naturally occurring naphthoquinones. The results indicate that these natural products can be effective potential candidates for the development of new strategies to treat MRSA infections [112]. Pomegranates also have a wide spectrum of antihelminthic and antidiarrheal properties. The use of pomegranate rind and root bark as a treatment for tapeworm infestation (*Latas tineas ventris*) was already recommended [113].

The establishment and maintenance of oral microbiota is related not only to interbacterial co-aggregations but also to interactions of these bacteria with yeasts, such as *Candida albicans* [114]. Vasconcelos et al. [115] investigated the antimicrobial effects of a pomegranate phytotherapeutic gel and miconazole (Daktarin® oral gel) against three standard streptococcal strains (*S. mutans* ATCC 25175, *S. sanguis* ATCC 10577, and *S. mitis* ATCC 9811), besides *S. mutans* clinically isolated and *Candida albicans*, either alone or in association. In experiments with three and four associated microorganisms, the *Punica granatum* L. gel had greater efficiency in inhibiting microbial adherence than miconazole. The results of this study suggested that this phytotherapeutic agent might be used in the control of adherence of different microorganisms in the oral cavity.

The antimicrobial activity of *P. granatum* has been widely investigated [116,117]. The findings of several studies, including some relating to inhibition of adherence, suggest that the phytotherapeutic use of this plant might be a viable option in controlling different microbial species. The major components of the *P. granatum* fruit extract are tannins and polyphenolics [118]. Studies have demonstrated the specific antimicrobial action of the plant on dental biofilm bacteria, i.e. disturbance of polyglycan synthesis, thus acting on the adherence mechanisms of these organisms to the dental surface. The results indicated that glucan synthesis and its antimicrobial action gave the pomegranate gel an effective control of the already formed biofilm, which is considered the primary etiological agent in caries disease and stomatitis [116,119].

There is a growing interest in using tannins as antimicrobial agents in caries prevention [120]. The action of tannins against bacteria and yeasts can be established by a relation between their molecular structure and their toxicity, astringent properties, or other mechanisms. The effect of tannins on microbial metabolism can be measured by their action on membranes.

They can cross the cell wall, composed of several polysaccharides and proteins, and bind to its surface. This adhesion can also help determine minimum inhibitory concentrations for yeasts and bacteria. Gebara et al. [121] observed that the inhibition of S. mutans and S. sobrinus adherence was a result of the inhibition of glucan synthesis by these substances. This study found a similar effect on the inhibition of adherence to glass of S. mutans, S. sanguis, S. mitis, and C. albicans by P. granatum in the presence of sucrose.

The hydroalcoholic extract (HAE) from Punica granatum fruits was tested against dental plaque microorganisms on 60 healthy patients using fixed orthodontic appliances. In this

TABLE 31.1 Tests of Sensitivity by Spread Plate Performed with *Punica granatum* Hydroalcoholic Extract (HAE) and Chlorhexidine on Selected Microorganisms

Microorganism	Control	HAE (mg/mL)			Chlorhexidine (%)		
		60	30	15	0.12	0.06	0.03
Shigella flexneri	−	−	−	−	−	−	−
Salmonella typhimurium	−	−	−	−	−	−	−
Staphylococcus aureus	−	+	+	+	+	+	+
Staphylococcus epidermidis	−	+	+	−	−	−	−
Staphylococcus B-hemolyticus	−	+	+	+	+	+	+
Streptococcus B-hemolyticus	−	+	+	−	−	+	+
Pseudomonas aeruginosa	−	+	−	−	+	−	−
Pseudomonas sp.	−	+	−	−	−	−	−
Klebsiella pneumoniae	−	+	+	+	+	+	+
Proteus vulgaris	−	+	+	+	+	+	+
Providencia alcalifaciens	−	−	−	−	−	−	−
Alcaligenes	−	−	−	−	−	−	−
Escherichia coli		+	+	+	+	+	+
Candida albicans	−	+	+	+			

+, sensitive; −, insensitive.
The initial concentration of *P. granatum* hydroalcoholic extract (PGHE) was 60 mg/mL and that of chlorhexidine was 0.12%. Control was distilled water.

TABLE 31.2 Effects of *Punica granatum* Hydroalcoholic Extract (HAE) on Colony Forming Units (CFU) from Dental Plaques

Group	CFU × 10⁵		
	Before Mouth-Rinse	After Mouth-Rinse	% Inhibition
HAE	154.4 ± 41.18	25.4 ± 7.76*	83.5
Chlor.	208.7 ± 58.81	44.0 ± 15.85*	79.0
Water	81.1 ± 10.12	71.9 ± 8.68	11.3

Values are means ± SEM of CFU from 20 patients per group, before and after mouth-rinses with HAE, chlorexidine (Chlor.) and distilled water (control). The experiment was performed as described in methods.
* $P < 0.05$ as compared to values before the mouth-rinse (Mann–Whitney U test).

C. ACTIONS OF INDIVIDUAL FRUITS IN DISEASE AND CANCER PREVENTION AND TREATMENT

study, HAE showed significant activity against microorganisms commonly present in the dental plaque (Table 31.1). The effect was observed immediately after the use of the mouthrinse

TABLE 31.3 Studies on Biological Effects of *Punica granatum* L. (Pomegranate)

Effect	Reference
Activity against genital herpes virus	[2]
Activity against herpes simplex virus type 1	[123]
Effect on rheumatoid arthritis	[43,124]
Effect on prostate cancer	[94,95,125]
Effect on candidosis	[117]
Antiplasmodial activity	[63]
Hypoglycemic activity	[127]
Effect on breast cancer	[91,92,128]
Inhibits methicillin-resistant *S. aureus*	[111,112]
Effect on colon cancer	[41,42,62]
Neuroprotective action	[129,130]
Beneficial effect on dental plaque	[122]
Inhibits microbial adherence	[115]
Antioxidant action	[44,70,88,131,134]
Inhibits lung tumorigenesis in mice	[132]
Antimicrobial activity	[63,133]
Reduces collagen-induced arthritis in mice	[43]
UVA and UVB-induced cell damage protection	[135]
Antihelminthic activity	[113]
Antidiarrheal activity	[113]
Antiatherogenic effect	[78,79]
Effective on a depressive state and bone loss	[137]
Protects against cardiovascular diseases	[138,139]
Prevention of erectile dysfunction	[59,85]
Gastroprotective activity	[83]
Reduces oxidative stress	[88]
Regeneration of epidermis in human skin cells	[136]
Prevention of male infertility	[59]
Prevention of Alzheimer's disease	[59]
Effect on periodontal disease	[39]
Molluscicidal activity	[126]

with HAE, when the number of colony forming units (CFU) was reduced by 84%. These results suggested that the mouthrinse aqueous extract from *P. granatum* could be beneficial in the prophylactic treatment of dental plaque bacteria [122]. Table 31.2 shows the effects of *P. granatum* HAE on the colony units (CFU) from dental plaques.

4. SUMMARY

The pharmacological actions of *P. granatum* chemical components suggest a wide range of potential clinical applications for the treatment of several diseases. It has been reported that pomegranate juice inhibited the progression of atherosclerotic lesions, improved stress-induced ischemia in patients with coronary heart diseases, and improved postprandial hyperglycemia and lipid profile in diabetic patients. Pomegranate has a wide spectrum of antibacterial, antiviral, and antihelminthic properties. A gel containing an extract of *P. granatum* was reported to be effective in dental diseases. Table 31.3 presents a list of recent studies on the biological effects of *P. granatum*. The intake of pomegranate has been demonstrated to have health-promoting and disease-preventing effects, combined with a low toxicity profile. These characteristics favor the inclusion of this plant in more detailed preclinical and mainly in clinical trials. Its potential may be utilized in prevention in cancer of prostate, breast, and colon, and skin tumors. Pomegranate seems to be especially useful in several types of diseases such as arthritis, among others, where chronic inflammation plays an essential role.

References

1. Aguiar, L. B. M. A., & Matos, F. J. A. (1983). Atividade antibiótica de plantas da flora nordestina – II. *Ciência e Cultura, 36*, 464.
2. Zhang, J., Zhan, B., Yao, X., Gao, Y., & Shong, J. (1995). Antiviral activity of tannin from the pericarp of

Punica granatum L. against genital herpes virus *in vitro*. *Zhongguo Zhong Yao Za Zhi, 20,* 556–558.

3. Sousa, M., Pinheiro, C., Matos, M. E. O., Matos, F. J., Lacerda, M. I., & Craveiro, A. A. (1991). Constituintes químicos de plantas medicinais brasileiras. Fortaleza: Universidade Federal do Ceará (pp. 385–388).

4. Marsh, P. D. (2006). Dental plaque as a biofilm and a microbial community: Implications for health and disease. *BMC Oral Health, 6,* 214.

5. Aas, J. A., Paster, B. J., Stokes, L. N., Olsen, I., & Dewhirst, F. E. (2005). Defining the normal bacterial oral cavity. *Journal of Clinical Microbiology, 43,* 5721–5732.

6. Olsson, J., Van Der Heijde, Y., & Holmberg, K. (1992). Plaque formation *in vivo* and bacterial attachment *in vitro* on permanently hydrophobic and hydrophilic surfaces. *Caries Research, 26,* 428–433.

7. Marsh, P. D. (1994). Microbial ecology of dental plaque and its significance in health and disease. *Advances in Dental Research, 8,* 263–271.

8. Hamada, S., & Slade, H. D. (1980). Biology, immunology and cariogenicity of *Streptococcus mutans*. *Microbiology Reviews, 44,* 331–384.

9. Scheie, A. A., Fejerskov, O., & Danielsen, B. (1994). The effects of xylitol-containing chewing gums on dental plaque and acidogenic potential. *International Dental Journal, 46,* 91–96.

10. Seif, T. (1997). *Cariología: Prevención, diagnóstico y tratamiento contemporáneo de la caries dental.* Caracas: AMOLSA.

11. Papanaou, P. N. (1999). Microbial markers of periodontal diseases. In H. J., Busscher, & L. V., Evans, (Eds.), *Oral Biofilms and Plaque Controls* (pp. 205–220). Amsterdam: Harwood Academic Publishers.

12. Page, R. C., Lantz, M. S., Darveau, R., Jeffcoat, M., Mancl, L., Houston, L, Braham, P., & Persson, G. R. (2007). Immunization of *Macaca fascicularis* against experimental periodontitis using a vaccine containing cysteine proteases purified from *Porphyromonas gingivalis*. *Oral Microbiology and Immunology, 22,* 162–168.

13. Ramires-Romito, A. C. D., Oliveira, L. B., Romito, G. A., Mayer, M. P. A., & Rodrigues, C. R. M. D. (2005). Correlation study of plaque and gingival indexes of mothers and their children. *Journal of Applied Oral Science, 13,* 227–231.

14. Axelsson, P., Lindhe, J., & Nystrom, B. (1991). Prevention of caries and periodontal disease. Results of a 15-year longitudinal study in adults. *Journal of Clinical Periodontology, 18,* 182–189.

15. Slots, J., Feik, D., & Rams, T. E. (1990). Prevalence and antimicrobial susceptibility of *Enterobacteriaceae, Pseudomonadaceae* and *Acinetobacter* in human periodontitis. *Oral Microbiology and Immunology, 5,* 149–154.

16. Slots, J., & Rams, T. E. (1991). New views on periodontal microbiota in special patient categories. *Journal of Clinical Periodontology, 18,* 411–420.

17. Jenkinson, H. F., Lala, H. C., & Shepherd, M. G. (1990). Coaggregation of *Streptococcus sanguis* and other streptococci with *Candida albicans*. *Infection and Immunity, 58,* 1429–1436.

18. Nikawa, H., Egusa, H., Makihira, S., Nishimura, M., Ishida, K., Furukawa, M., & Hamada, T. (2002). A novel technique to evaluate the adhesion of *Candida* species to gingival epithelial cells. *Mycoses, 46,* 384–389.

19. Makihira, S., Nikawa, H., Tamagami, M., Hamada, T., Nishimura, H., Ishida, K., & Yamashiro, H. (2002). Bacterial and *Candida* adhesion to intact and denatured collagen *in vitro*. *Mycoses, 45,* 389–392.

20. Carounanidy, U., Satyanarayanan, R., & Velmurugan, A. (2007). Use of an aqueous extract of *Terminalia chebula* as an anticaries agent: A clinical study. *Indian Journal of Dental Research, 18,* 152–156.

21. Heath, J., & Wilson, J. (1976). Abrasion of restorative materials by dentifrice. *Journal of Oral Rehabilitation, 3,* 121–138.

22. Aguiar, A. A. A., Saliba, N. A., Consani, S., & Sinhoreti, M. A. C. (2002). Óleo vegetal: Um substituto aos abrasivos dos dentifrícios! [abstract]. *Pesquisa Odontologica Brasileira, 16*(Suppl), 193.

23. Aguiar, A. A. A., & Moimaz, S. A. S. (2000). Redução da placa bacteriana dentária através da escovação com óleo de amêndoa [abstract]. *Pesquisa Odontologica Brasileira, 14*(Suppl), 128.

24. Aguiar, A. A. A., & Saliba, N. A. (2004). Toothbrushing with vegetable oil: A clinical and laboratorial analysis. *Brazilian Oral Research, 18,* 168–173.

25. Ccahuana-Vasquez, R. A., Santos, S. S. F., Koga-Ito, C. Y., & Jorge, A. O. C. (2007). Antimicrobial activity of *Uncaria tomentosa* against oral human pathogens. *Brazilian Oral Research, 21,* 46.

26. Nostro, A., Canntelli, M. A., Crisafi, G., Musolino, A. D., Procopio, F., & Alonzo, V. (2004). Modifications of hydrophobicity, *in vitro* adherence and cellular aggregation of *Streptococcus mutans* by *Helichrysum italicum* extract. *Letters in Applied Microbiology, 38,* 423–427.

27. Jagtap, A. G., & Karkera, S. G. (2000). Extract of *Juglandaceae regia* inhibits growth, *in-vitro* adherence, acid production and aggregation of *Streptococcus mutans*. *Journal of Pharmacy and Pharmacology, 52,* 235–242.

28. Lima, C. A. A., Pastore, G. M., & Lima, E. D. P. A. (2000). Study of the antibacterial activity of anacardic acids from the cashew *Anacardium occidentale* nut shell oil of the clone of cashew-midget-precocious ccp-76 and ccp-09 in five stages of maturation on oral microorganisms. *Ciência e Tecnologia de Alimentos, 20,* 358–362.

C. ACTIONS OF INDIVIDUAL FRUITS IN DISEASE AND CANCER PREVENTION AND TREATMENT

29. Alviano, W. S., Alviano, D. S., Diniz, C. G., Antoniolli, A. R., Alviano, C. S., Farias, L. M., Carvalho, M. A., & Souza, M. M. (2008). *In vitro* antioxidant potential of medicinal plant extracts and their activities against oral bacteria based on Brazilian folk medicine. *Archives of Oral Biology, 53,* 545–552.

30. Suddhasthira, T., Thaweboon, S., Dendoung, N., Thaweboon, B., & Dechkunakorn, S. (2006). Antimicrobial activity of *Cratoxylum formosum* on *Streptococcus mutans. Southeast Asian Journal of Tropical Medicine and Public Health, 37,* 1156–1159.

31. Prabu, G. R., Gnanamani, A., & Sadulla, S. G. (2006). A plant flavonoid as potential antiplaque agent against *Streptococcus mutans. Journal of Applied Microbiology, 101,* 487–495.

32. Koo, H., Gomes, B. P., Rosalen, P. L., Ambrosano, G. M., Park, Y. K., & Cury, J. A. (2000). *In vitro* antimicrobial activity of propolis and *Arnica montana* against oral pathogens. *Archives of Oral Biology, 45,* 141–148.

33. Feres, M., Figueiredo, L. C., Barreto, I. M., Coelho, M. H., Araujo, M. W., & Cortelli, S. C. (2005). *In vitro* antimicrobial activity of plant extracts and propolis in saliva samples of healthy and periodontally-involved subjects. *Journal of the International Academy of Periodontology, 7,* 90–96.

34. Iauk, L., Lo Bue, A. M., Milazzo, I., Rapisarda, A., & Blandino, G. (2003). Antibacterial activity of medicinal plant extracts against periodontopathic bacteria. *Phytotherapy Research, 17,* 599–604.

35. Razak, F. A., & Rahim, Z. H. (2003). The anti-adherence effect of *Piper betle* and *Psidium guajava* extracts on the adhesion of early settlers in dental plaque to saliva-coated glass surfaces. *Journal of Oral Science, 45,* 201–206.

36. Matsumoto, M., Tsuji, M., Okuda, J., Sasaki, H., Nakano, K., Osawa, K., Shimura, S., & Ooshima, T. (2004). Inhibitory effects of cacao bean husk extract on plaque formation *in vitro* and *in vivo. European Journal of Oral Sciences, 112,* 249–252.

37. Yamaguti-Sasaki, E., Ito, L. A., Canteli, V. C., Ushirobira, T. M, Ueda-Nakamura, T., Dias Filho, B. P., Nakamura, C. V., & de Mello, J. C. (2007). Antioxidant capacity and *in vitro* prevention of dental plaque formation by extracts and condensed tannins of *Paullinia cupana. Molecules, 12,* 1950–1963.

38. Rohrer, U., Kunz, E. M., Lenkeit, K., Schaffner, W., & Meyer, J. (2007). Antimicrobial activity of *Mahonia aquifolium* and two of its alkaloids against oral bacteria. *Schweizerischen Monatschrift Zahnmedizin, 117,* 1126–1131.

39. Sastravaha, G., Gassmann, G., Sangtherapitikul, P., & Grimm, W. D. (2005). Adjunctive periodontal treatment with *Centella asiatica* and *Punica granatum* extracts in supportive periodontal therapy. *Journal of the International Academy of Periodontology, 7,* 70–79.

40. Ahmed, S., Wang, N., Hafeez, B. B., Cheruvu, V. K., & Haqqi, T. M. (2005). *Punica granatum* L. extract inhibits IL-1beta-induced expression of matrix metalloproteinases by inhibiting the activation of MAP kinases and NF-kappa B in human chondrocytes *in vitro. Journal of Nutrition, 135,* 2096–2102.

41. Seeram, N. P., Adams, L. S., Henning, S. M., Niu, Y., Zhang, Y., Zhang, Y., Nair, M. G., & Heber, D. (2005). *In vitro* antiproliferative, apoptotic and antioxidant activities of punicalagin, ellagic acid and a total pomegranate tannin extract are enhanced in combination with other polyphenols as found in pomegranate juice. *Journal of Nutritional Biochemistry, 16,* 360–367.

42. Adams, L. S., Seeram, N. P., Aggarwal, B. B., Takada, Y., Sand, D., & Heber, D. (2006). Pomegranate juice, total pomegranate ellagitannins, and punicalagin suppress inflammatory cell signaling in colon cancer cells. *Journal of Agricultural and Food Chemistry, 54,* 980–985.

43. Shukla, M., Gupta, K., Rasheed, Z., Khan, K. A., & Haqqi, T. M. (2008). Bioavailable constituents/metabolites of pomegranate (*Punica granatum* L.) preferentially inhibit COX_2 activity *ex vivo* IL-I beta-induced PGE_2 production in human chondrocytes *in vitro. Journal of Inflammation, 5,* 9.

44. Gil, M. I., Tomas-Barberan, F. A., Hess-Pierce, B., Holcroft, D. M., & Kader, A. A. (2000). Antioxidant activity of pomegranate juice and its relationship with phenolic composition and processing. *Journal of Agricultural and Food Chemistry, 48,* 4581–4589.

45. Sudheesh, S., & Vijayalakshmi, N. R. (2005). Flavonoids from *Punica granatum* – potential antiperoxidative agents. *Fitoterapia, 76,* 181–186.

46. Aviram, M., Dornfield, L., & Coleman, R. (2002). Pomegranate juice flavonoids inhibit low-density lipoprotein oxidation in cardiovascular diseases: Studies in atherosclerotic mice and in humans. *Drugs under Experimental and Clinical Research, 28,* 49–62.

47. Afaq, F., Saleem, M., Krueger, C. G., Reed, J. D., & Mukhtar, H. (2005). Anthocyanin- and hydrolyzable tannin-rich pomegranate fruit extract modulates MAPK and NF-kappaB pathways and inhibits skin tumorigenesis in CD-1 mice. *International Journal of Cancer, 113,* 423–433.

48. Esmaillzadeh, A., Tahbaz, F., Gaieni, I., Alavi-Majd, H., & Azadbakht, L. (2006). Cholesterol-lowering effect of concentrated pomegranate juice consumption in type II diabetic patients with hyperlipidemia. *International Journal for Vitamin and Nutrition Research, 76,* 147–151.

49. Hayden, M. S., & Ghosh, S. (2004). Signaling to NF-κB. *Genes and Development, 18,* 2195–2224.

50. Scieven, G. S. (2005). The biology of p38 kinase: A central role in inflammation. *Current Topics in Medicinal Chemistry, 5,* 921–928.

51. Ahmed, S., Wang, N., Hafeez, B. B., Cheruvu, V. K., & Haqqi, T. M. (2005). *Punica granatum* L. extract inhibits IL-1β-induced expression of matrix metalloproteinases by inhibiting the activation of MAP kinases and NF-κB in human chondrocytes *in vitro. Journal of Nutrition, 135,* 2096–2102.

52. Afaq, F., Malik, A., Syed, D., Maes, D., Matsui, M. S., & Mukhtar, H. (2005). Pomegranate fruit extract modulates UV-B-mediated phosphorylation of mitogen-activated protein kinases and activation of nuclear factor kappa B in normal human epidermal keratinocytes. *Photochemistry and Photobiology, 81,* 38–45.

53. Kawaii, S., & Lansky, E. P. (2004). Differentiation-promoting activity of pomegranate (Punica granatum) fruit extracts in HL-60 human promyelocytic leukemia cells. *Journal Medicinal Food, 7,* 13–18.

54. Tsuda, T., Horio, F., & Osawa, T. (2002). Cyanidin 3-O-beta-D-glucoside suppresses nitric oxide production during zymogen treatment in rats. *Journal of Nutritional Science and Vitaminology, 48,* 305–310.

55. Youdim, K. A., McDonald, J., Kalt, W., & Joseph, J. A. (2002). Potential role of dietary flavonoids in reducing microvascular endothelium vulnerability to oxidative and inflammatory insults. *Journal of Nutritional Biochemistry, 13,* 282–288.

56. Noda, Y., Kaneyuki, T., Mori, A., & Packer, L. (2002). Antioxidant activities of pomegranate fruit extract and its anthocyanidins: Delphinidin, cyanidin, and pelargonidin. *Journal of Agricultural and Food Chemistry, 50,* 166–171.

57. Seeram, N. P., & Nair, M. G. (2002). Inhibition of lipid peroxidation and structure-activity related studies of the dietary constituents anthocyanins, anthocyanidins, and catechins. *Journal of Agricultural and Food Chemistry, 50,* 5308–5312.

58. Garbacki, N., Angenot, L., Bassleer, C., Damas, J., & Tits, M. (2002). Effects of prodelphinidins isolated from *Ribex nigrum* on chondrocytes metabolism and COX activity. *Archives Pharmacol, 365,* 434–441.

59. Jurenka, J. S. (2008). Therapeutic applications of pomegranate (*Punica granatum L*): a review. *Alternative Medicine Review, 13,* 128–144.

60. Bell, C., & Hawthorne, S. (2008). Ellagic acid, pomegranate and prostate cancer, a mine review. *Journal of Pharmacy and Pharmacology, 60,* 139–144.

61. Cerdá, B., Espín, J. C., Parra, S., Martinez, P., & Tomas-Barbéran, F. A. (2004). The potent *in vitro* antioxidante ellagitannins from pomegranate juice are metabolized into bioavailable but poor antioxidant hydroxyl-6H-dibenzopyran-6-one derivatives by the colonic microflora of healthy humans. *European Journal of Nutrition, 43,* 205–220.

62. Kulkarni, A. P., Mahal, H. S., Kapoor, S., & Aradhya, S. M. (2007). *In vitro* studies on the binding, antioxidant, and cytotoxic actions of punicalagin. *Journal of Agricultural Food Chemistry, 55,* 1491–1500.

63. Reddy, M. K., Gupta, S. K., Jacob, M. R., Khan, S. I., & Ferreira, D. (2007). Antioxidant, antimalarial and antimicrobial activities of tannin-rich fractions, ellagitannins and phenolic acids from *Punica granatum* L. *Planta Medica, 73,* 461–467.

64. Lee, S. I., Kim, B. S., Kim, K. S., Lee, S., Shin, K. S., & Lim, J. S. (2008). Immune-suppressive activity of punicalagin via inhibition of NFAT activation. *Biochemical and Biophysical Research Communications, 371,* 799–803.

65. Nugteren, D. H., & Christ-Hazelhof, E. (1987). Naturally occurring conjugated octadecatrienoic acids are strong inhibitors of prostaglandin biosynthesis. *Prostaglandins, 33,* 403–417.

66. Landolfi, R., Mower, R. L., & Steiner, M. (1984). Modification of platelet function and arachidonic acid metabolism by bioflavonoids. Structure-activity relations. *Biochemical Pharmacology, 33,* 1525–1530.

67. Welton, A. F., Tobias, L. D., Fiedler-Nagy, C., Anderson, W., Hope, W., Meyers, K., & Coffey, J. W. (1986). Effect of flavonoids on arachidonic acid metabolism. *Progress in Clinical and Biological Research, 213,* 231–242.

68. Wallace, J. M. (2002). Nutritional and botanical modulation of the inflammatory cascade – eicosanoids, cyclooxygenases, and lipoxygenases as an adjunct in cancer therapy. *Integrative Cancer Therapies, 1,* 7–37.

69. Morikawa, K., Nonaka, M., Narahara, M., Torii, I., Kawaguchi, K., Yoshikawa, T., Kumazawa, Y., & Morikawa, S. (2003). Inhibitory effect of quercetin on carrageenan-induced inflammation in rats. *Life Sciences, 74,* 709–721.

70. Schubert, S. Y., Lansky, E. P., & Neeman, I. (1999). Antioxidant and eicosanoid enzyme inhibition properties of pomegranate seed oil and fermented juice flavonoids. *Journal of Ethnopharmacology, 66,* 11–17.

71. Lansky, E. P., Jiang, W., Mo, H., Bravo, L., Froom, P., Yu, W., Harris, N. M., Neeman, I, & Campbell, M. J. (2005). Possible synergistic prostate cancer suppression by anatomically discrete pomegranate fractions. *Investigational New Drugs, 23,* 11–20.

72. Albrecht, M., Jiang, W., Kumi-Diaka, J., Lansky, E. P., Gommersall, L. M., Patel, A., Mansel, R. E., Neeman, I., Geldof, A. A., & Campbell, M. J. (2004). Pomegranate extracts potently suppress proliferation, xenograft growth, and invasion of human prostate cancer cells. *Journal of Medicinal Food, 7,* 274–283.

C. ACTIONS OF INDIVIDUAL FRUITS IN DISEASE AND CANCER PREVENTION AND TREATMENT

73. Shapiro, S. D. (1997). Mighty mice: Transgenic technology 'knocks out' questions of matrix metalloproteinase function. *Matrix Biology, 15*, 527–533.

74. Leppert, D., Lindberg, R. L., Kappos, L., & Leib, S. L. (2001). Matrix metalloproteinases: Multifunctional effectors of inflammation in multiple sclerosis and bacterial meningitis. *Brain Research Reviews, 36*, 249–257.

75. Okamoto, T., Akuta, T., Tamura, F., van Der Vliet, A., & Akaike, T. (2004). Molecular mechanism for activation and regulation of matrix metalloproteinases during bacterial infections and respiratory inflammation. *Biological Chemistry, 385*, 997–1006.

76. Salvi, G. E., & Lang, N. P. (2005). Host response modulation in the management of periodontal diseases. *Journal of Clinical Periodontology, 32*, 108–129. comment, 130–131.

77. Aslam, M. N., Fligiel, H., Lateef, H., Fisher, G. J., Ginsburg, I., & Varani, J. (2005). PADMA28: A multicomponent herbal preparation with retinoid-like dermal activity but without epidermal effects. *Journal of Investigative Dermatology, 124*, 524–529.

78. Aviram, M., Dornfeld, L., Rosenblat, M., Volkova, N., Kaplan, M., & Coleman, R. (2000). Pomegranate juice consumption reduces oxidative stress, atherogenic modifications to LDL, and platelet aggregation: Studies in humans and in atherosclerotic apolipoprotein E-deficient mice. *American Journal of Clinical Nutrition, 71*, 1062–1076.

79. Kaplan, M., Hayek, T., Raz, A., Coleman, R., Dornfeld, L., & Vaya, M. (2001). Pomegranate juice supplementation to atherosclerotic mice reduces macrophage lipid peroxidation, cellular cholesterol accumulation and development of atherosclerosis. *Journal of Nutrition, 131*, 2082–2089.

80. Li, Y., Guo, C., Yang, J., Wei, J., Xu, J., & Cheng, S. (2006). Evaluation of antioxidant properties of pomegranate peel extract in comparison with pomegranate pulp extract. *Food Chemistry, 96*, 254–260.

81. Singh, R. P., Murthy, K. N. C., & Jayaprakasha, G. K. (2002). Studies on the antioxidant activity of pomegranate peel and seed extracts using *in vitro* models. *Journal of Agricultural and Food Chemistry, 50*, 81–86.

82. Khennouf, S., Gharzouli, K., Amira, S., & Gharzouli, A. (1999). Effects of *Quercus ilex* and *Punica granatum* polyphenols against ethanol-induced gastric damage in rats. *Pharmazie, 54*, 75–76.

83. Ajaikumar, K. B., Asheef, M., Babu, B. H., & Padikkala, J. (2005). The inhibition of gastric mucosal injury by *Punica granatum* L. (pomegranate) methanolic extract. *Journal of Ethnopharmacology, 96*, 171–176.

84. Loren, D. J., Seeram, N. P., Schulman, R. N., & Holtzman, D. M. (2005). Maternal dietary supplementation with pomegranate juice is neuroprotective in an animal model of neonatal hypoxic-ischemic brain injury. *Pediatric Research, 57*, 858–864.

85. Azadzoi, K. M., Schulman, R. N., Aviram, M., & Siroky, M. B. (2005). Oxidative stress in arteriogenic erectile dysfunction: Prophylactic role of antioxidants. *Journal of Urology, 174*, 386–393.

86. Kaur, G., Jabbar, Z., Athar, M., & Alam, M. S. (2006). *Punica granatum* (pomegranate) flower extract possesses potent antioxidant activity and abrogates Fe-NTA induced hepatotoxicity in mice. *Food and Chemical Toxicology, 44*, 984–993.

87. Nigris, F., Williams-Ignarro, S., Botti, C., Sica, V., Ignarro, L. J., & Napoli, C. (2006). Pomegranate juice reduces oxidized low-density lipoprotein downregulation of endothelial nitric oxide synthase in human coronary endothelial cells. *Nitric Oxide, 15*, 259–263.

88. Aviram, M., & Dornfeld, L. (2001). Pomegranate juice consumption inhibits serum angiotensin converting enzyme activity and reduces systolic blood pressure. *Atherosclerosis, 158*, 195–198.

89. Settheetham, W., & Ishida, T. (1995). Study of genotoxic effects of antidiarrheal medicinal herbs on human cells *in vitro*. *Southeast Asian Journal of Tropical Medicine and Public Health, 26*, 306–310.

90. Mavlyanov, S. M., Islambekov, S. Y., Karimdzhanov, A. K., & Ismailov, A. I. (1997). Polyphenols of pomegranate peel show marked anti-viral and anti-tumor action. *Khimiia Prirodnykh Soedineni, 33*, 98–99.

91. Kim, N. D., Mehta, R., Yu, W., Neeman, I., Livney, T., Amichay, A., Poirier, D., Nicholls, P., Kirby, A., Jiang, W., Mansel, R., Ramachandran, C., Rabi, T., Kaplan, B., & Lansky, E. (2002). Chemopreventive and adjuvant therapeutic potential of pomegranate (*Punica granatum*) for human breast cancer. *Breast Cancer Research and Treatment, 71*, 203–217.

92. Toi, M., Bando, H., Ramachandran, C., Melnick, S. J., Imai, A., Fife, R. S., Carr, R. E., Oikawa, T., & Lansky, E. P. (2003). Preliminary studies on the anti-angiogenic potential of pomegranate fractions *in vitro* and *in vivo*. *Angiogenesis, 6*, 121–128.

93. Sartippour, M. R., Seeram, N. P., Rao, J. Y., Moro, A., Harris, D. M., Henning, S. M., Firouzi, A., Rettig, M. B., Aronson, W. J., Pantucky, A. J., & Heber, D. (2008). Ellagitannin-rich pomegranate extract inhibits angiogenesis in prostate cancer *in vitro* and *in vivo*. *International Journal of Oncology, 32*, 475–480.

94. Malik, A., Afaq, F., Sarfaraz, S., Adhami, V. M., Syed, D. N., & Mukhtar, H. (2005). Pomegranate fruit juice for chemoprevention and chemotherapy of prostate cancer. *PNAS, 102*, 14813–14818.

95. Heber, D. (2008). Multitargeted therapy of cancer by ellagitannins. *Cancer Letter*, May 12.

96. Li, Y., Wen, S., Kota, B. P., Peng, G., Li, G. Q., Yamahara, J., & Roufogalis, B. D. (2005). *Punica granatum* flower extract, a potent glucosidase inhibitor, improves postprandial hyperglycemia in Zucker diabetic fatty rats. *Journal of Ethnopharmacology, 99*, 239–244.

97. Lee, J., & Watson, R. R. (1998). Pomegranate: A role in health promotion and AIDS? In R.R., Watson, (Ed.), *Nutrients and Foods in AIDS* (pp. 179–192). Boca Raton, FL: CRC Press.

98. Kawamada, Y., & Shimada, T. (2002). Cosmetic or topical compositions containing *Punica granatum* extracts. Japan Kokai Tokkyo Koho, Japanese Patent: JP 2002234814 A2 20020823.

99. Moayadi, A. (2004). Mixtures of pomegranate seed oils for cosmetics. Japanese Patent: JP 2004083544 A2 20040318.

100. Curry, S. C. (2004). Breast enhancement system. US Patent 6,673,366.

101. Lansky, E. P. (2000). Pomegranate supplements prepared from pomegranate material including pomegranate seeds. US Patent 6,060,063.

102. Watanabe, K., and Hatakoshi, M. (2002). *Punica granatum* leaf extracts for inactivation of allergen. Japan Kokai TokkyoKoho (Japanese patent) JP 2002370996 A2 20021224, 5 pp.

103. Shiraishi, T., Abe, M., and Miyagawa, T. (2002). Cheese foods containing conjugated polyunsaturated fatty acid glycerides. Japanese Patent: JP 2002176913.

104. Aviram, M., and Dornfeld, L. (2003). Methods of using pomegranate extracts for causing regression in lesions due to arteriosclerosis in humans. US Patent 6,641,850.

105. Kim, M. M., and Kim, S. (2002). Composition for improving oral hygiene containing *Punica granatum* L. extract. Korean Patent: KR 2002066042.

106. Bruijn, C. D, Christ, F. R., and Dziabo, A. J. (2003). Ophthalmic, pharmaceutical and other healthcare preparations with naturally occurring plant compounds, extracts and derivatives. US Patent Application 20030086986.

107. Il'iasov, T. N. (1975). Comparative characteristics of barium suspension prepared on water and the extract from the pomegranate peel for examination of the large intestine. *Vestnik Rentgenologii i Radiologii, 5*, 83–84.

108. Amorim, L. F., Catanho, M. T. J. A., Terra, D. A., Brandão, K. C., Holanda, C. M. C. X., Jales-Junior, L. H., Brito, L. M., Gomes, M. L., De Melo, V. G., Bernardo-Filho, M., & Cavalcanti Jales, R. L. (2003). Assessment of the effect of *Punica granatum* (pomegranate) on the bioavailability of the radiopharmaceutical sodium pertechnetate (99mTc) in Wistar rats. *Cellular and Molecular Biology, 49*, 501–507.

109. Minakshi, De., Krishna, A., & Banerjee, A. B. (1999). Antimicrobial screening of some Indian species. *Phytotherapy Research, 13*, 616–618.

110. Braga, L. C., Leite, A. A., & Xavier, K. G. (2005). Synergistic interation between pomegranate extract and antibiotics against *Staphylococcus aureus*. *Canadian Journal of Microbiology, 51*, 541–547.

111. Braga, L. C., Shupp, J. W., Cummings, C., Jett, M., Takahashi, J. A., Carmo, L. S., Chartone-Souza, E., & Nascimento, A. M. (2005). Pomegranate extract inhibits *Staphylococcus aureus* growth and subsequente enterotoxin production. *Journal of Ethnopharmacology, 96*, 335–339.

112. Machado, T. B., Pinto, A. V., Pinto, M. C. F. R., Leal, I. C. R., Silva, M. G., Amaral, A. C. F., Kuster, R. M., & Netto-dos Santos, K. R. (2003). *In vitro* activity of Brazilian medicinal plants, naturally occurring naphtoquinones and their analogues, against methicillin resistant *Staphylococcus aureus*. *International Journal of Antimicrobial Agents, 21*, 279–284.

113. Chevalier, A. (1996). *Encyclopedia of Medicinal Plants*. London: Dorling Kinsdersley p. 257.

114. Jenkinson, H. F., Lala, H. C., & Shepherd, M. G. (1990). Coaggregation of *Streptococcus sanguis* and other streptococci with *Candida albicans*. *Infection and Immunity, 58*, 1429–1436.

115. Vasconcelos, L. C. S., Sampaio, F. C., Sampaio, M. C. C., Pereira, M. S. V., Higinao, J. S., & Peixoto, M. H. P. (2006). Minimum inhibitory concentration of adherence of *Punica granatum Linn* (pomegranate) gel against *S. mutans, S. mitis* and *C. albicans*. *Brazilian Dental Journal, 17*, 223–227.

116. Pereira, J. V., Pereira, M. S. V., Sampaio, F. C., Sampaio, M. C. C., Alves, P. M., Araújo, C. R. F., & Higino, J. S. (2006). *In vitro* antibacterial and anti-adherence effect of *Punica granatum Linn* extract upon dental biofilm microorganisms. *Brazilian Journal of Pharmacognosy, 16*, 88–93.

117. Vasconcelos, L. C. S., Sampaio, M. C. C., Sampaio, F. C., & Higino, J. S. (2003). Use of *Punica granatum Linn* as an antifungal agent against candidosis associated with denture stomatitis. *Mycoses, 46*, 192–196.

118. Haslam, E. (1996). Natural polyphenols (vegetables tannins) as drugs: Possible modes of action. *Journal of Natural Products, 59*, 205–215.

119. Kakiuchi, N., Hattori, M., & Nishizawa, M. (1986). Studies on dental caries prevention by traditional medicines. Inhibitory effect of various tannins on glucan synthesis by glycosyltransferase from *Streptococcus mutans*. *Chemical and Pharmaceutical Bulletin, 34*, 720–725.

C. ACTIONS OF INDIVIDUAL FRUITS IN DISEASE AND CANCER PREVENTION AND TREATMENT

120. Scalbert, A. (1991). Antimicrobial properties of tannins. *Chemistry, 30*, 3875–3883.

121. Gebara, E. C. E., Zardetto, C. G. D. C., & Mayer, M. P. A. (1996). *In vitro* study of the antimicrobial activity of natural substances against *S. mutans* and *S. sobrinus*. *Brazilian Oral Research, 10,* 251–256 (formerly Rev Odontol Univ São Paulo).

122. Menezes, S. M. S., Cordeiro, L. N., & Viana, G. S. B. (2006). *Punica granatum* (Pomegranate) Extract is active against dental plaque. *Journal of Herbal Pharmacotherapy, 6*, 79–92.

123. Li, Y., Ooi, L. S., Wang, H., But, P. P., & Ooi, V. E. (2004). Antiviral activity of medicinal herbs traditionally used in southern mainland China. *Phytotherapy Research, 9*, 718–722.

124. Shukla, M., Gupta, K., Rasheed, Z., Khan, K. A., & Haqqi, T. M. (2008). Consumption of hydrolysable tannins-rich pomegranate extract suppresses inflammation and joint damage in rheumatoid arthritis. *Nutrition, 7-8*, 733–743.

125. Rettig, M. B., Heber, D., An, J., Seeram, N. P., Rao, J. Y., Liu, H., Klatte, T., Belldegrun, A., Moro, A., Henning, S. M., Mo, D., Aronson, W. J., & Pantuck, A. (2008). Pomegranate extract inhibits androgen-independent prostate cancer growth through a nuclear factor-kappaB-dependent mechanism. *Molecular Cancer Therapeutics, 7*, 2662–2671.

126. Tripathi, S. M., & Singh, D. K. (2000). Molluscicidal activity of *Punica granatum* bark and Canna indica root. *Brazilian Journal of Medical and Biological Research, 33*, 1351–1355.

127. Das, A. K., Mandal, S. C., Banerjee, S. K., Sinhá, S., Saha, B. P., & Pal, M. (2001). Studies on the hypoglycaemic activity of *Punica granatum* seed in streptozotocin induced diabetic rats. *Phytotherapy Research, 15* (7), 628–629.

128. Mehta, R., & Lansky, E. P. (2004). Breast cancer chemopreventive properties of pomegranate (*Punica granatum*) fruit extracts in a mouse mammary organ culture. *European Journal of Cancer Prevention, 13*, 345–348.

129. Hartman, R. E., Shah, A., Fagan, A. M., Schwetye, K. E., Parsadanian, M., Schulman, R. N., Finn, M. B., & Holtzman, D. M. (2006). Pomegranate juice decreases amyloid load and improves behavior in a mouse model of Alzheimer's disease. *Neurobiology of Disease, 24*, 506–515.

130. West, T., Atzeva, M., & Holtzman, D. M. (2007). Pomegranate polyphenols and resveratrol protect the neonatal brain against hypoxic-ischemic injury. *Developmental Neuroscience, 29*, 363–372.

131. Sestili, P., Martinelli, C., Ricci, D., Fraternale, D., Bucchini, A., Giamperi, L., Curcio, R., Piccoli, G., & Stocchi, V. (2007). Cytoprotective effect of preparations from various parts of *Punica granatum* L. fruits in oxidatively injured mammalian cells in comparison with their antioxidant capacity in cell free systems. *Pharmacology Research, 56*, 18–26.

132. Khan, N., Afaq, F., Kweon, M. H., Kim, K., & Mukhtar, H. (2007). Oral consumption of pomegranate fruit extract inhibits growth and progression of primary lung tumors in mice. *Cancer Research, 67*, 3475–3482.

133. Naz, S., Siddigi, R., Ahmad, S., Rasool, S. A., & Sayeed, S. A. (2007). Antibacterial activity directed isolation of compounds from *Punica granatum*. *Journal of Food Science, 72*, 341–345.

134. Patel, C., Dadhaniya, P., Hingorani, L., & Soni, M. G. (2008). Safety assessment of pomegranate fruit extract: Acute and subchronic toxicity studies. *Food and Chemical Toxicology, 46*, 2735–2778.

135. Pacheco-Palencia, L. A., Noratto, G., Hingorani, L., Talcott, S. T., & Mertens-Talcott, S. U. (2008). Protective effects of standardized pomegranate (*Punica granatum* L.) polyphenolic extract in ultraviolet-irradiated human skin fibroblasts. *Pharmacology Research, 56*, 8434–8441.

136. Aslam, M. N., Lansky, E. P., & Varani, J. (2006). Pomegranates cosmetical source: Pomegranate fractions promote proliferation and procollagen synthesis and inhibit matrix metalloproteinase e-1 production in human. *Journal of Ethnopharmacology, 103*, 311–318.

137. Mori-Okamoto, J., Otawava-Hamamoto, Y., & Yamato, H. (2004). Pomegranate extract improves a depressive state and bone properties in menopausal syndrome model ovariectomized mice. *Journal of Ethnopharmacology, 92*, 93–101.

138. Sumner, M. D., Elliot-Eller, M., & Weidner, G. (2005). Effects of pomegranate juice consumption on myocardial perfusion in patients with coronary heart disease. *American Journal of Cardiology, 96*, 810–814.

139. Aviram, M., Rosenblat, M., & Gaitine, D. (2007). Pomegranate juice consumption for 3 years by patients with carotid artery stenosis reduces common carotid intima-media thickness blood pressure and LDL oxidation. *Clinical Nutrition, 23*, 423–433.

32

Açaí (*Euterpe oleracea* Mart.): A Macro and Nutrient Rich Palm Fruit from the Amazon Rain Forest with Demonstrated Bioactivities *In Vitro* and *In Vivo*

Alexander G. Schauss

Natural and Medicinal Foods Division, AIBMR Life Sciences, Puyallup, WA, USA

1. INTRODUCTION

Euterpe is a genus of native tropical palm trees found in the Amazon and a few Caribbean islands. There are three species producing edible fruit found widely dispersed through the Amazon: *Euterpe edulis, Euterpe precatoria*, and *Euterpe oleracea*.

E. edulis fruit harvest starts in December and runs through February in Maranhao state, the coastal region of Bahia state, and Espirito Santo. *E. precatoria* fruit is harvested in February and continues through June, with most of the fruit coming from Rondonia, Acre, and Mato Grosso. *E. oleracea* fruit starts its harvest as early as May, but usually in mid-June to late July, and continues into December in Para and Amapa states.

The Amazon River basin covers about 5,000,000 km^2, after having collected melting snowfall and rainfall throughout the year from the eastern slopes of the Andes mountain range, and rainfall from the Brazilian and Guiana highlands.

The Amazon floodplain, locally called the *varzea*, occupies about 150,000 km^2 within which *E. oleracea* palm trees are particularly abundant. In the varzea, an enormous amount of fresh water is pushed back and forth twice a day by tidal activity resulting in inundated lowlands. Above the varzea some 8 m higher in altitude is another type of forest called the terra-firme, which also is abundant in *E. oleracea* palm trees [1].

A large number of people living in the varzea and terra-firme make their living collecting forest products. In eastern Para state, Brazil, the forest is primarily varzea, consisting of hundreds of islands that receive over 2500 mm of rainfall each year. The rainy season begins in December and continues into June, after which evaporative transpiration exceeds rainfall. In 2007, the predominant agricultural product exported out of Para state was the fruit of *E. oleracea*, known locally as *Açaí*. This fruit represents 75% of all agricultural food products (including wood) harvested in 2007 [2]. In 2002, less than 20% of such products included açaí fruit.

From early to late May, depending on when the rainy season ends, this fruit provides a significant source of nutrition for millions of people living in the rain forest. In Belem, the largest city in Para state, with a population of nearly 1.3 million residents (Figure 32.1), the consumption of watered-down and sometimes sweetened açaí pulp can be as much as two liters a day during the months of July through December. However, due to numerous food processors of açaí operating during those dry months in and around Belem and neighboring states, consumption of açaí pulp is virtually year round, with hardly a day going by without a meal containing açaí.

Supplementing the diet with açaí are seasonal vegetables including manioc (*Manihot ultissima*) cultivated in the terra-firme for domestic use, which is as commonly consumed in the Amazon rain forest as potatoes, corn, and wheat are consumed in North America. When açaí is in season, many of the meals eaten by natives combine manioc with the pulp, particularly in fish dishes.

For natives that live among the thousands of islands found in eastern Amazonas state, Para and Amapa states, the açaí fruit is a major source of nutrition. Yet what makes this fruit remarkable is not just its nutritional profile and antioxidant properties, but how little pulp is found within the fruit due to the size of the seed in the fruit relative to the pulp. On average 87% of the fruit is a single seed. At maturity, the fruit is 1.5–2.0 cm wide. There is little variation in the size of the fruit. Hence, it takes a lot of fruit to provide enough pulp to feed the millions of people who rely on it as a major source of nutrition in the Amazon rain forest.

In a Brazilian government survey of açaí palm trees in the varzea, some islands were found to have over 7000 açaí palm trees per hectare (2.47 acres).

Natives that live on islands in Cameta County, Para state, Brazil, not only consume the açaí fruit daily with almost every meal during the dry season when it is readily available in almost any backyard, but they also harvest it daily to sell in marketplaces along tributaries of the Amazon River that wholesale it to processors who turn it into frozen pulp. This frozen pulp is then either sold as such or preserved in other ways to support the increasing export market for this fruit.

Those individuals who harvest açaí fruit are known locally as *ribeirinhos*, which means people who live on the river's riverines. Many of these ribeirinhos form cooperatives to bring fruit to market.

In the 1990s hardly anyone outside of the Amazon flood plains had heard of açaí fruit. What led to its popularity outside of Brazil was the discovery that it had the highest antioxidant capacity of any fruit, vegetable, or nut, other than a few spices, against the peroxyl and superoxide radical *in vitro*. This discovery led to açaí being called a 'superfood' by the media and discussed on television, radio, and print media. By 2008, it was the third most popular fruit juice product sold in the USA in grocery and natural products stores, behind orange juice and pomegranate juice.

2. BOTANY AND HORTICULTURE

E. oleracea is a perennial and riparian palm tree indigenous to the Amazon that primarily grows in the rain forests of Brazil and its northern bordering countries, particularly Suriname and the Guianas.

Also known as the cabbage palm or açaí palm, *E. oleracea* typically grows to 30–40 m, providing the canopy for the rain forest in the

FIGURE 32.1 Map of the Amazon River delta.

Amazonian flood plains. The palm tree grows in copses or groves called *rebolades*. These clumps of palm tree thickets average 6–12 or 10–18 trees in number, depending on their location, soil condition, and genetics, do not vary much in height or lifespan, but do vary in terms of dominance within the multi-stratus native forest. In some regions of the Amazon further inland and well away from the river's shorelines or its estuaries and above the terra-firme, *E. oleracea* can grow to a height of 80–100 m.

Within three years of emerging from the forest floor, *E. oleracea* begins to produce fruit suitable for consumption by humans. The fruit grows on inflorescences that hang just under the tree's fronds near its crown. At maturity and when ready for harvest, an inflorescence bears several hundred fruit, which depending on the weight of the stalk and the number of fruit can range from half a kilo to nearly two kilos in weight. The fruit weighs on average a few grams and is symmetrically round (1.5–2.0 cm), consisting of a hard endocarp (single seed) that constitutes ~87% of the fruit, and covered by a thin pulpy mesocarp, extracted to produce a juice or other food uses.

Fruit maturity is dictated by color. When fully ripe the skin of the fruit appears black. The most desirable fruit believed to produce the best pulp for juicing has a slightly powdery white film on its skin.

During its most productive years, from 3 to 10 years of age, a single palm tree can yield numerous inflorescences that produce ~1000 kilos of fruit. Keep in mind that the primary purpose for harvesting the fruit is to obtain its pulp for a juice. The seed is disposed of as it has limited use for nutritional purposes. Instead the seed is commonly utilized to produce a range of by-products such as handicrafts, animal feed, or composted organic fertilizer, as well as a fuel source to generate energy.

However, the seed contains phytochemicals that may have potentially useful biological activity. In a joint French–Brazilian study, a hydro-alcohol extract of the seed of açaí was produced and shown to have a potential in the treatment of cardiovascular diseases *in vivo* [3]. The extract induced a long-lasting endothelium-dependent vasodilation dependent on activation of the NO-cGMP pathway and possibly also involving endothelium-derived hyperpolarizing factor (EDHF) release. Further research on the seed is warranted, as the pulp has received almost all attention by the research community.

In the varzea, where the highest concentration of *E. oleracea* palms is found, fruit production declines markedly after 10 years, whereafter the palm tree may be cut down to obtain the core of the trunk near the crown, called the *palmito*, to sell as palm hearts. Each segment of palmito, also known as heart-of-palm, is from 25 to 100 cm in length, and can weigh from several kilos up to 25 kilos, depending on the age of the palm tree. These segments are processed by a cannery, into sliced segments, and canned. Sliced palmito is a popular vegetable or added as a garnish in salads.

As fruit mature on an inflorescence and begin to fall to the forest floor, fruit growing on adjacent inflorescences start changing color: the transition of the fruit's color is from white to a muted olive green, finally maturing to a shiny black skin. The cycle of maturing fruit on each inflorescence continues throughout the dry season until each inflorescence has borne and released mature fruit, a process that begins in May when the rainy season typically ends in the lower basin of the Amazon, continuing well into late November and early December just before the rainy season begins.

Once the rainy season begins, harvesting in the varzea shifts from gathering açaí to two other fruits: cupuacu (*Theobroma grandiflorum*), which grows on a tree and is related to cacao (*Theobroma cacao*), and the miriti palm fruit (*Mauritia flexuosa*), which can grow much taller than açaí palms in the varzea. Unlike cacao that contains xanthines (caffeine, theobromine, and theophylline), the fragrant cupuacu fruit contains theacrine rather than xanthines. Harvesting of both fruits ends about the time the rainy season ends and açaí harvesting starts again.

Harvesting açaí fruit is physically demanding and requires considerable skill, agility, and strength, for it grows near the crown of the palm tree. For harvesting to be commercially viable, cooperatives of sufficiently trained harvesters are needed, who have the means to move large quantities of fruit to market via watercraft. These cooperatives must include sufficient numbers of producers able to harvest large enough quantities of açaí fruit on scheduled days, and to do so in a manner that maintains quality standards and respect for hygiene. This is not easy to achieve given the challenges associated with the terrain and geography of the Amazon basin, including the twice daily tidal changes, and the need for infrastructure to support the movement of fruit from producers to consumers.

For example, in Para state, the leading producer of açaí fruit in the world, much of the açaí fruit that reaches markets by boat must consider the tides that change twice daily on the Amazon River and its tributaries and estuaries. Traveling by boat against the tide can add hours to a trip. Once fruit reaches the marketplace, governmental institutions must be in place to monitor sales and insure that only legally registered producers and their cooperatives are involved in the process. This is no small undertaking.

Research conducted on the fruit has determined that the açaí fruit can lose an appreciable amount of its nutritional and antioxidant qualities unless processed and refrigerated rapidly as it is very perishable after harvesting. ('Rapid' means reaching the processor within less than 24 hours.)

This explains why açaí fruit has not been seen in grocery stores or supermarkets outside of the Amazon. To move the fruit from harvesters to processors requires a reliable system of suppliers and industrial processors, and transportation resources capable of moving hundreds of tons of fruit daily throughout the dry season. (Technically there is no 'dry' season. All months have precipitation of at least 60 mm, but this is a small amount of rainfall compared to the extremes that can be experienced during the wet season.)

Small monoculture plantations can be found in Para state that are close to roads and markets. However, most açaí fruit still comes from harvesters living on islands in the rain forest.

Let me describe how elaborate and demanding a system is required to move açaí fruit to processors. During a trip to Brazil in 2008, the author traveled to Belem, on the estuary of the Tocantins and Para rivers, in Para state.

Traveling by car for two hours from the center of the city, the author reached one of the region's leading açaí fruit processors just as the last few tons of fruit were being unloaded off flatbed trucks. At 1 pm, we decided to trace the origin of the fruit that had just been unloaded to its source within a county that contains over 350 islands. These islands have numerous açaí harvesting cooperatives.

As we left the processor, to retrace the fruit's origin, we were unprepared for how seriously rough road conditions were. Once we left the outskirts of Belem, road conditions rapidly deteriorated. Within an hour began a never-ending series of speed bumps, often without any warning due to lack of signage as we passed through one village after another. With each mile the number and depth of potholes got worse until one dared not fall asleep for fear of hitting one's head on the ceiling of the car or biting one's tongue. After several hours the night set in, reducing visibility and making the ride that much more challenging.

After six hours the road suddenly ended at a river crossing requiring a ferry to take the car to the other side so as to continue the trip. The ferry waited until trucks heading to pick up tons of açaí fruit arrived. One hour later we were back on the road. It was not another hour before the road ended again at another river crossing, where we again waited for the ferry, this time carrying either trucks laden with açaí fruit, or either empty and heading to the same point on one of the estuary's ports that we were heading to. Two hours later, we arrived at the end of the road traversable by car, for here small boats carrying anywhere from 1000 to 2000 baskets, known as *hazas*, weighing 14 kilos each, are unloaded and the fruit transferred to the trucks.

At 11 pm we left this 'port' village by the river in a speedboat. For the next several hours we navigated between countless islands under a half moon to reach our destination at 3 am. There we met the head of the cooperative on the island whose açaí fruit we had just watched being processed 14 hours earlier in the outskirts of Belem.

On this particular island lived 112 families who earned their living for 8 months of the year harvesting the açaí fruit. The other four months they harvested miriti fruit.

Now repeat this journey several hundred times by numerous suppliers on any day during the dry season in this county alone, multiply that by several counties supplying açaí fruit in Para state, and one gets a sense of how expansive and complex this process has become as world demand for this fruit increases.

Demand for açaí fruit worldwide has been dramatic. More than 45 countries around the world offered açaí-based fruit juice and açaí fruit by-products, from cosmetics to shampoos, by the end of 2008.

What accounted for the dramatic increase in açaí consumption was the discovery of the pulp's remarkably high antioxidant capacity *in vitro* when compared to the antioxidant activity of fruits and berries such as cranberry, blackcurrant, black raspberry, and blueberries. The attention given to açaí fruit came at a time when the importance of increasing antioxidant-rich phytochemicals found in fruits, vegetables, nuts, and spices, was being promoted by the medical community and nutritionists, as the search for and popularization of 'superfoods' as part of the diet came into vogue.

3. NUTRITIONAL COMPOSITION AND PHYTOCHEMISTRY

A comprehensive report on the phytochemical, nutrient, and antioxidant capacity of açaí fruit was published by a consortium of government, university, and private industry scientists in two papers published in 2006 [4,5]. These findings are also summarized in a book dedicated to the açaí fruit [6].

The nutrient analysis of freeze-dried açaí pulp and skin is shown in Table 32.1. Table 32.2 provides the analysis of amino acids while Table 32.3 lists the fatty acids, as reported previously [4].

What challenged food scientists until the late 1990s, particularly in Brazil, was the lack of processing facilities able to produce commercial quantities of freeze-dried açaí, given its perishable nature.

The advantage of studying freeze-dried fruits and vegetables is well known to food scientists. Until the last decade, the means to study the açaí fruit's chemistry and antioxidant capacity in the Amazon, using sophisticated and validated analytical and assay methods with precision, had been lacking. Once freeze-dried samples became available to study the açaí fruit for its chemical and biological properties, inter-laboratory

TABLE 32.1 Nutrient Analysis of Freeze-dried Açaí Pulp and Skin

Analytes	Result	Unit per 100 g (DW [1])
Label analytes		
Calories	533.9	
Calories from fat	292.6	
Total fat	32.5	g
Saturated fat	8.1	g
Cholesterol	13.5	mg
Sodium	30.4	mg
Total carbohydrate	52.2	g
Dietary fiber	44.2	g
Sugars	1.3	g
Protein (F = 6.25)	8.1	g
Vitamin A	1002	IU
Vitamin C	<0.1	mg
Calcium	260.0	mg
Iron	4.4	mg
Contributing analytes		
Moisture	3.4	g
Ash	3.8	g

DW, dry weight; F, conversion factor.

TABLE 32.2 Analysis of Amino Acids from Freeze-dried Açaí

Amino Acids	Result
Aspartic acid	0.83%
Threonine	0.31%
Serine	0.32%
Glutamic acid	0.80%
Glycine	0.39%
Alanine	0.46%
Valine	0.51%
Methionine	0.12%
Isoleucine	0.38%
Leucine	0.65%
Tyrosine	0.29%
Phenylalanine	0.43%
Lysine	0.66%
Histidine	0.17%
Arginine	0.42%
Proline	0.53%
Hydroxyproline	<0.01%
Cystine	0.18%
Tryptophan	0.13%
Total	**7.59%**

TABLE 32.3 Fatty Acid Composition of Freeze-dried Açaí

Fatty Acids	Formula	Content
Saturated fatty acids		
Butynic	4:0	<0.1%
Caproic	6:0	<0.1%
Caprylic	8:0	<0.1%
Capric	10:0	<0.1%
Undecanoic	11:0	<0.1%
Lauric	12:0	0.1%
Tridecanoic	13:0	<0.1%
Myristic	14:0	0.2%
Pentadecanoic	15:0	<0.1%
Palmitic	16:0	24.1%
Margaric	17:0	0.1%
Stearic	18:0	1.6%
Nonadecanoic	19:0	<0.1%
Eicosanoic	20:0	<0.1%
Behenic	22:0	<0.1%
Tricosanoic	23:0	<0.1%
Lignoceric	24:0	<0.1%
Total		26.1%
Monounsaturated fatty acids		
Tridecenoic	13:1	<0.1%
Myristoleic	14:1	<0.1%
Pentadecenoic	15:1	<0.1%
Palmitoleic	16:1	4.3
Margaroleic	17:1	0.1
Oleic	18:1C	56.2
Elaidic	18:1T	<0.1%
Gadoleic	20:1	<0.1%
Erucic	22:1	<0.1%
Nervonic	24:1	<0.1%
Total		60.6%
Polyunsaturated fatty acids		
Linoleic	18:2	12.5%
Linolenic	18:3	0.8
Gamma linolenic	18:3G	<0.1%
Eicosadienoic	20:2	<0.1%
Eicosatrienoic	20:3	<0.1%
Homogamma linolenic	20:3G	<0.1%
Arachidonic	20:4	<0.1%
Eicosapentaenoic	20:5	<0.1%
Docosadienoic	22:2	<0.1%
Docosahexaenoic	22:6	<0.1%
Total		13.3%

reproducibility became possible, which had not been possible using sun-dried, spray-dried, or frozen samples, particularly given the existence of oxidase enzymes found in the fruit that contribute to its rapid degradation.

Although our research was performed on freeze-dried açaí, a significant amount of the açaí used in commercial products is produced using either spray-dried powders or frozen pulp. The former was found to show dramatically lower oxygen radical antioxidant capacity (ORAC) values compared to freeze-dried açaí. One assay found spray-dried açaí's ORAC was in the range of 50–70 Trolox equivalent per gram (μmol TE/g), whereas the freeze-dried açaí was above 1000 μmol TE/g.

At the time these results were reported by the leading ORAC assay laboratory and confirmed by the USDA's nutrient analysis and antioxidant laboratory using the same analytical and assay protocols, the ORAC score was reported to be the highest for any fruit vegetable or nut, other than for a few spices. This was true even when comparing various antioxidant-rich fruits based on moisture content and serving size (Table 32.4).

The phytochemical antioxidants in açaí are anthocyanins (ACNs), proanthocyanidins (PACs), and other major flavonoids, of which cyanidin 3-glucoside and cyanidin 3-runtinoside were found to be the predominant ACNs, along with numerous minor ACNs [4]. The total content of ACNs was 3.1919 mg/g dry weight. Polymers were found to be the major proanthocyanidins (PACs). The concentration of total PACs was calculated as 12.89 mg/g dry weight. Other flavonoids found in the açaí fruit include homoorientin, orientin, isovitexin, scoparin, and taxifolin deoxyhexose, along with a number of unknown flavonoids. The stilbene, resveratrol, was found at a very low concentration.

Freeze-dried açaí feels oily to touch. Analysis of the nutritional composition of açaí revealed that the predominant fatty acid was oleic acid, followed by palmitic acid and linoleic acid. Total unsaturated fatty acid was 74% of all fatty acids, in the range of olive and avocado oils.

Unfortunately, an examination of many açaí-based juices in the marketplace in the USA revealed that such juices are clarified and filtered, resulting in the loss of the mono- and poly-unsaturated fatty acids found naturally in the fruit. Combined with the presence of five sterols found in açaí fruit, including beta-sitosterol, campesterol, and signasterol, the loss of these fatty acids and sterols in terms of potential health benefits by excessive processing of açaí-based juices, may mislead consumers into believing that clarified and/or filtered juices contain the nutritional components found in unclarified açaí juices.

A pharmacokinetic study of clarified and unclarified açaí juices found that açaí juice had approximately 45% more total anthocyanins than clarified juice. Açaí juice with the pulp resulted in higher plasma antioxidant levels. A large portion of the anthocyanins was found bound to the water insoluble fraction of the pulp which would be digested in the gut and by its flora. Following

TABLE 32.4　Comparison of Antioxidant-rich Fruits and Berries by Serving Size, Adjusted for Moisture

Food	% Moisture	TAC Serving		TAC per Serving
		μmol TE/g	Size (g)	
Açaí pulp	60.0	410.80	145	59,566
Apples (Granny)	85.7	38.99	138	5381
Avocado (Hass)	72.0	19.33	173	3344
Blackberry	86.9	53.48	145	7701
Blueberry (Cult)	85.0	62.20	145	9019
Cranberry	87.1	94.56	95	8983
Raspberry	85.8	49.25	123	6058
Strawberry	91.1	35.77	166	5938

Hydrophilic-ORAC + lipophilic-ORAC = total antioxidant capacity (TAC); TE, Trolox Equivalent.
Source: Schauss, 2008, pp. 71–73 [6].

ingestion of açaí with the pulp, antioxidant compounds in serum were consistently higher during 4 hours of observation compared to clarified juice ($P<0.05$). Further, peak antioxidant levels in blood dropped sooner if the pulp had been removed. The maximum concentration of anthocyanins was 105% higher in açaí juice with pulp compared to clarified açaí juice. Finally, the half-life of açaí juice with pulp was 6.56 hours compared to 3.0 hours for clarified juice [7].

Using bioactive-guided fractionation of a methanol-soluble extract of açaí fruit led to the isolation of 14 isolates found to be active in a hydroxyl radical scavenging assay and seven isolates in a DPPH assay [8]. These assays identified dihydroconiferyl alcohol, (+)-lariciresinol, (+)-pinoresinol, (+)-syringaresinol, and protocatechuic acid, each of which exhibited cytoprotection of MCF-7 cells oxidatively stressed by hydrogen peroxide. A number of lignans were also reported representative of the aryltetrahydronaphthalene, dihydrobenzofuran, furofan, 8-O-4'-neolignan, and tetrahydrofuran structural types.

Nutritionally, the açaí fruit contains 19 amino acids, including the essential amino acids, along with virtually all vitamins, with the exception of phylloquinone and menaquinone, and a wide range of all essential minerals and numerous trace elements [4]. With regard to heavy metal content, many samples of açaí fruit tested have found extraordinarily low concentrations of cadmium, and mercury, in the range of 36 to 1 ppb, respectively.

Total carotenoids in açaí pulp have been studied and found to contribute to the antioxidant activity of the fruit along with the total anthocyanins [9].

4. ANTIOXIDANT CAPACITY AND ACTIVITY

The significance of açaí fruit's high antioxidant capacity *in vitro* in terms of potential health benefits is illustrated by comparing the antioxidant capacity of skin-bearing apples with açaí.

The antioxidant value of a small apple of 113 g (a quarter pound) is equivalent to 1400 mg of vitamin C. However, most of the antioxidant capacity of an apple does not come from its vitamin C content. Rather it comes from its combined phytochemical composition. Less than 0.4% of the apple's total antioxidant capacity (TAC) comes from vitamin C alone.

Apples have been shown to inhibit colon cell proliferation by 43% *in vitro*. Apples also have been shown to decrease mammary and other tumors *in vivo* in rats. The percent of inhibition increases the more apples are consumed. Just one apple inhibited 17% of tumors, but three apples inhibited 39%, while six apples inhibited tumor development by 44% [6].

An apple's hydrophilic and lipophilic antioxidant capacity (total ORAC) is between 22 and 43 μmol TE/g, depending on the color and variety of apple assayed based on USDA reported data. Adjusting for moisture content, the comparable dry weight of açaí has a total ORAC of 630 μmol TE/g, a value that is 45 times the antioxidant capacity of a quarter pound of apples.

The comparisons between açaí and various antioxidant rich fruits and vegetables have led to considerable speculation as to its potential benefits *in vivo*.

One study found that a fraction of compounds found in the açaí fruit had an antiproliferative effect on a line of human leukemia cells resulting in a significant increase in apoptosis of HL-60 leukemia cells [10].

Freeze-dried açaí fruit pulp has been reported to have exceptional activity against superoxide in the superoxide scavenging (SOD) assay, the highest of any food, at 1614 units/g [5].

An assay measuring the inhibition of reactive oxygen species (ROS) formation in freshly purified human neutrophils showed that the antioxidants in açaí are able to enter human cells in a fully functional form and perform their oxygen quenching function at a very low dose, in a H_2O_2-induced formation of ROS model, even at 0.1 pg/g [5].

The contents of anthocyanins, proanthocyanidins, and other polyphenol compounds in freeze-dried açaí were found to be much lower than those found in blueberry or other berries with elevated ORAC values [5]. The total phenolics in açaí were found to be only 13.9 mg/g gallic acid equivalents (GAE). The ratio between hydrophilic ORAC and total phenolics was reported in a study to vary dramatically from less than two to more than 100 for different groups of foods [11]. For most fruits and vegetables, this ratio is about 10. However, the ratio in açaí is 50, five times greater than that found for any other fruit. This 'unusual' ratio raises questions whether or not açaí contains much stronger antioxidants than found in other berries on an equal weight basis. Determining which antioxidants contributed to this unusual ratio warrants further work that is being carried out by the author and colleagues.

The ability of an orally ingested açaí-based fruit juice to inhibit ROS formation in humans has been investigated experimentally. A pilot study of 7 subjects and a randomized, double-blind, placebo-controlled, crossover trial involving 12 healthy adults, showed a significant increase in serum antioxidants at 1 hour and 2 hours, a correlating increase in antioxidant compounds, as well as a significant inhibition of lipid peroxidation ($P<0.01$) at post-consumption [12,13].

These results confirmed observed cell-based antioxidant protection of erythrocytes from oxidative damage ($P<0.001$) and reduced formation of ROS in polymorphonuclear cells ($P<0.003$), and reduced migration toward three different pro-inflammatory chemoattractants [fmlp ($P<0.001$), leukotriene B4 ($P<0.05$), and IL-8 ($P<0.03$)], using the same açaí-fruit based juice [12]. The reduction in pro-inflammatory chemoattractants supports earlier observations of açaí's potent inflammatory properties [5].

In a study that evaluated the effects of *E. oleracea* flowers, fruits, and fractions found therein on nitric oxide (NO) production and scavenging capacity in the expression of the inducible nitric oxide synthase (iNOS) enzyme, it was found that the fruit was most potent in inhibiting NO production and iNOS expression [14]. Fractionation of polyphenols found that the concentration of cyanidin 3-glucoside and cyanidin 3-rhamnoside contributed to the pronounced effect observed. The results suggest that fractions found in the fruit inhibit NO production and reduce levels of iNOS expression with implications in the treatment of cardiovascular disease.

The brain is especially susceptible to oxidative stress. It is rich in polyunsaturated fatty acids that are subject to lipid peroxidation given the disproportionate utilization of oxygen by the brain. To protect against oxidative damage enzymes are utilized by the brain, such as catalase and superoxide dismutase (SOD). Catalase converts hydrogen peroxide to water and molecular oxygen, while SOD converts superoxide to molecular oxygen and hydrogen peroxide which catalase turns into water and stable oxygen.

A study from Brazil reported evidence that açaí pulp might contain compounds that could protect the brain from free radical damage. The study pre-treated brain tissues taken from rodents and exposed the cerebral cortex, hippocampus, and cerebellum, to H_2O_2. Pre-treatment of these brain tissues with açaí decreased H_2O_2-induced damage of both proteins and lipids. The authors suggested, 'that açaí could prevent the development of age-related neurodegenerative diseases' [15].

To evaluate the impact of oral consumption of a fruit and berry blend on pain and range of motion (ROM), an open-label clinical pilot study was performed on subjects, many of which had osteoarthritis, 44–84 years of age, with limited ROM associated with pain that affected daily living. In this 12-week study, participants consumed a fruit juice whose predominant fruit was unclarified açaí pulp with other antioxidant fruits. Subjects were assessed at baseline and 2, 4, 8, and 12 weeks by structured nurse interviews, questionnaires on pain and Activities of Daily Living (ADL), blood samples, and ROM assessment. Pain was scored using a visual analogue scale (VAS). ROM assessment was performed using dual digital inclinometry. Consumption of the juice resulted in pain reduction ($P<0.01$), improved ROM ($P<0.05$), and ADL ($P<0.025$). Serum antioxidant status based on the CAP-e cell-based antioxidant protection assay [16] was significantly improved already after 2 weeks ($P<0.05$), and kept improving throughout the 12 weeks of study participation particularly in the lower back and knees of subjects ($P<0.0001$) [17]. A mild decrease in lipid peroxidation was also observed at 12 weeks ($P<0.16$).

A pilot and randomized, double-blind, placebo-controlled, crossover study was performed on the same juice used in the ROM study, to evaluate *in vitro* and *in vivo* antioxidant and anti-inflammatory capacity of the açaí-rich juice.

The cell-based antioxidant protection of erythrocytes (CAP-e) assay demonstrated that antioxidants in the açaí juice blend penetrated and protected cells from oxidative damage ($P<0.001$), while polymorphonuclear (PMN) cells showed reduced formation of ROS ($P<0.003$) and reduced migration toward three different pro-inflammatory chemoattractants: fmlp ($P<0.001$), leukotriene B4 ($P<0.05$), and IL-8 ($P<0.03$). Blood samples at baseline, 1 hour, and 2 hours following consumption of the juice or placebo were tested for antioxidant capacity using several antioxidant assays and the thiobarbituric acid reactive substances (TBARS) assay. A within-subject comparison post consumption showed an increase in serum antioxidants at 1 hour ($P<0.03$) and 2 hours ($P<0.015$), as well as inhibition of lipid peroxidation at 2 hours ($P<0.01$) [12].

The authors found of interest that the açaí-rich juice treatment of freshly purified human PMN cells *in vitro* affected random and fmlp-directed migratory behavior differently than PMN migration toward the inflammatory mediators LTB4 and IL-8.

This led to speculation that the anti-inflammatory properties of the juice *in vivo* may allow normal immune surveillance, while at the same time reducing inflammatory conditions.

Further, consumption of the juice under conditions of fasting-induced oxidative stress resulted in an increase in serum antioxidant compounds that were able to enter living cells and protect them from oxidative damage in 91% of study subjects, as evaluated by applying the serum samples to the CAP-e assay.

Based on these pilot and experimental studie, further research is warranted to evaluate whether the increased antioxidant protection and phytochemicals found in açaí pulp, added to unfiltered/unclarified açaí juice products, have a benefit for those afflicted with autoimmune and chronic inflammatory diseases, including diabetes mellitus.

References

1. Bohrerm, C. B. A., Goncalves, L. M. C. (1991). Vegetacao. In IBGE (Ed.), Geografia do Brasil, Vol. 3, Regiao Norte (pp. 137–168). Rio de Janeiro: Instituto Brasileiro de Geografia e Estatistica.
2. Mochiutti, S. (2007). *Potencial de Producao de Fructose de Açai em açaizais manejados no estuaries amazonico*. Brasilia: AMAPA.
3. Rocha, A. P. M., Carvalho, L. C. R. M., Sousa, M. A. V., Madeira, S. V. F., Sousa, P. J. C., Tano, T., Schini-Kerth, V. B., Resende, A. C., & Soares de Moura, R. (2007). Endothelium-dependent vasodilator effect of *Euterpe oleracea* Mart. (Açaí) extracts in mesenteric vascular bed of the rat. *Vascular Pharmacology, 46*, 97–104.
4. Schauss, A. G., Wu, X., Prior, R. L., Ou, B., Patel, D., Huang, D., & Kababick, J. P. (2006). Phytochemical and nutrient composition of the freeze-dried Amazonian palm berry, *Euterpe oleraceae* Mart. (Açaí). *Journal of Agricultural and Food Chemistry, 54*, 8598–8603.
5. Schauss, A. G., Wu, X., Prior, R. L., Ou, B., Huang, D., Owens, J., Agarwal, A., Jensen, G. S., Hart, A. N., & Shanbrom, E. (2006). Antioxidant capacity and other bioactivities of the freeze-dried Amazonian palm berry, *Euterpe oleraceae* Mart. (Açaí). *Journal of Agricultural and Food Chemistry, 54*, 8604–8610.

C. ACTIONS OF INDIVIDUAL FRUITS IN DISEASE AND CANCER PREVENTION AND TREATMENT

6. Schauss, A. G. (2009). *Açaí: An extraordinary antioxi-dant-rich palm fruit from the Amazon.* Tacoma: Biosocial Publications.

7. Mertens-Talcott, S., Rios, J., Jilma-Stohlawetz, P., Pacheco-Palencia, L. A., Meibohm, B., Talcott, S. T., & Derendorf, H. (2008). Pharmacokinetics of anthocya-nins and antioxidant effects after consumption of anthocyanin-rich açaí juice and pulp (*Euterpe oleracea* Mart.) in human healthy volunteers. *Journal of Agricultural and Food Chemistry, 56,* 7796–7802.

8. Chin, Y. W., Chai, H. B., Keller, W. J., & Kinghorn, A. D. (2008). Lignans and other constituents of the fruits of *Euterpe oleracea* (Açaí) with antioxidants and cytoprotective activities. *Journal of Agricultural and Food Chemistry, 56,* 7759–7764.

9. Dos Santos, G. M, Maia, G. A., de Sousa, P. H., da Costa, J. M., de Figueiredo, R. W., & do Prado, G. M. (2008). Correlation between antioxidant activity and bioactive compounds of açaí (*Euterpe oleracea* Mart) commercial pulps. *Archive Latinoamericanos De Nutricion, 58,* 187–192.

10. Del Pozo-Insfran, D., Percival, S. S., & Talcott, S. T. (2006). Açaí (*Euterpe oleracea*) polyphenolics in their glycoside and aglycone forms induce apoptosis of HL-60 leukemia cells. *Journal of Agricultural and Food Chemistry, 54,* 1222–1229.

11. Wu, X., Beecher, G., Holden, J., Haytowitz, D., Gebhardt, S. E., & Prior, R. L. (2004). Lipophilic and hydrophilic antioxidant capacities of common foods in the U.S. *Journal of Agricultural and Food Chemistry, 52,* 4026–4037.

12. Jensen, G. S., Patterson, K. M., Barnes, J., Certer, S. G., Wu, W., Scherwitz, L., Beaman, R., Endres, J. R., & Schauss, A. G. (2008). *In vitro* and *in vivo* antioxidant and anti-inflammatory capacity of an antioxidant-rich fruit and berry juice blend. Results of a pilot and ran-domized, double-blind, placebo-controlled, crossover study. *Journal of Agricultural and Food Chemistry, 56,* 8326–8333.

13. Schauss, A. G., Jensen, G., Wu, X., and Scherwitz, L. (2007). Increased antioxidant activity *in vivo* consuming Monavie, an açaí (*Euterpe oleracea*) berry fruit-based beverage. *'Proceedings of the Second International Symposium on Human Health Effects on Fruits and Vegetables,'* pp. 39–40, October 9, 2007, Houston, Texas.

14. Metheus, M. E., De Oliveira Fernandes, S. B., Silveira, C. S., Rodruues, V. P., de Sousa Menezes, F., & Fernandes, P. D. (2006). Inhibitory effects of *Euterpe oleracea* Mart. on nitric oxide production and iNOS expression. *Journal of Ethnopharmacology, 107,* 291–296.

15. Spada, P. D. S., Dani, C., Bortolini, G. V., Funchal, C., Henriques, J. A. P., & Salvador, M. (2009). Frozen fruit pulp of *Euterpe oleracea* Mart. (Açaí) prevents hydrogen peroxide induced damage in the cerebral cortex, cere-bellum and hippocampus of rats. *Journal of Medicinal Food,* in press.

16. Honzel, D., Carter, S. G., Redman, K. A., Schauss, A. G., Endres, J., & Jensen, G. S. (2008). Comparison of chemical and cell-based antioxidant methods for eval-uation of foods and natural products: Generating mul-tifaceted data by parallel testing using erythrocytes and polymorphonuclear cells. *Journal of Agricultural Food Chemistry, 56,* 8319–8325.

17. Jensen, G. S., Schauss, A. G., Beaman, R., & Ager, D. M. (2009). Pain reduction and improvement of range of motion after consumption of an antioxidant-rich fruit and berry juice blend. *Submitted for publication.*

33

Beneficial uses of Breadfruit (*Artocarpus altilis*): Nutritional, Medicinal and Other Uses

Neela Badrie and Jacklyn Broomes

Department of Food Production, Faculty of Science and Agriculture, University of the West Indies, St. Augustine, Republic of Trinidad and Tobago, West Indies

1. *ARTOCARPUS ALTILIS* (BREADFRUIT)

The breadfruit is a member of the genus Moraceae which contains about 50 species of trees [1]. It is grown throughout the tropical regions of the world yet only a few cultivars are found within most countries. Three species, *Artocarpus altilis* (Parkinson) Fosberg, *Artocarpus camansi* Blanco (breadnut), and *Artocarpus mariannensis* Trécul (dug-dug), plus natural hybrids (*A. altilis * A. mariannensis*) constitute the breadfruit complex [2]. However, *A. altilis* [synonyms *A. communis* J.R. Forst and G. Forst; *A. incisus* (Thunb.) L.f] is the most extensively dispersed species and exhibits great variability [3]. For this reason, this chapter will address mainly *A. altilis*, hereafter referred to as breadfruit (see Fig. 33.1). There are two distinct breadfruit varieties: seeded and seedless. The seeded breadfruit variety is considered to be the wild form [4].

The common name breadfruit in English is translated into Spanish as 'fruta de pan' (fruit), or 'arbol de pan,' 'arbol del pan' (tree), or 'pan de pobre'; into French as 'fruit à pain' (seedless), 'chataignier' (with seeds), 'arbre à pain' (tree); Portuguese, 'fruta pâo,' or 'pâo de massa'; Dutch, 'broodvrucht' (fruit), 'broodboom' (tree). In Venezuela, it may be called 'pan de año,' 'pan de todo el año,' 'pan de palo,' 'pan de name,' 'topán,' or 'túpan'; in Guatemala and Honduras, 'mazapán' (seedless), 'castaña' (with seeds); in Peru, 'marure'; in Yucatan, 'castaño de Malabar' (with seeds); in Puerto Rico, 'panapén' (seedless), 'panade pepitas' (with seeds). In Malaya and Java, it is 'suku' or 'sukun' (seedless); 'kulur,' 'kelur,' or 'kulor' (with seeds); in Thailand, 'sa-ke'; in the Philippines, 'rimas' (seedless); on Hawaii, 'ulu' [5,6].

1.1 Geographical Distribution

Breadfruit originated in the Western Pacific, with New Guinea and associated islands such as the Bismarck Archipelago being the centre of diversity for wild seeded forms of *Artocarpus altilis* (Parkinson) Fosberg. Zerega

FIGURE 33.1 A breadfruit tree located at The University Field Station, Valsayn, Trinidad. Photographer: Jacklyn Broomes.

1.2 Botany

The breadfruit tree is fast-growing with the capability of reaching 26 m in height and often buttressed at the base. There are many spreading branches (Fig. 33.1). The leaves, evergreen or deciduous depending on climatic conditions, are situated on thick, yellow petioles, entire at the base, then more or less deeply cut into 5–11 pointed lobes. They are bright green and glossy on the upper surface, with conspicuous yellow veins; dull, yellowish, and coated with minute, stiff hairs on the underside. The tree bears a multitude of tiny flowers with the male densely set on a drooping cylindrical or club-shaped spike, thick, yellowish at first and becoming brown. The female are massed in a somewhat rounded or elliptic, green, prickly head, which develops into the compound fruit. Generally the rind is green at first, turning yellowish-green, yellow or yellow-brown when ripe. The fruit is borne singly or in clusters of two or three at the branch tips. All parts of the tree, including the unripe fruit, are rich in milky, gummy latex [1,5,10].

et al. [3] stated that breadfruit was originally domesticated in Oceania. At the end of the sixteenth century, European explorers and naturalists traveling to Oceania quickly recognized the potential of breadfruit as a highly productive, cheap source of nutrition and introduced a limited number of cultivars to their tropical colonies [6]. The most famous of these attempts was led by William Bligh and culminated in the mutiny aboard the H.M.S. Bounty [7,8]. Today breadfruit is grown throughout the tropics but is especially important in Oceania and the Caribbean. In general, breadfruit is a crop for the hot, humid tropical lowlands and grows best at temperatures of 21–32°C [1]. Breadfruit has an annual rainfall requirement of 2000–3000 mm [9].

1.3 Propagation

In vitro technologies are increasingly important for preserving diversity, particularly for the vegetatively propagated species such as breadfruit and seeded breadfruit that produce recalcitrant seeds which do not store well [11]. Breadfruit is propagated vegetatively by root cuttings and root suckers whereas the seeded breadfruit is propagated by seed [12]. The roots grow on or slightly below the surface of the ground and will often produce a shoot, especially if cut or damaged. Air-layering or marcottage is another method which has shown good results [6]. Propagating by seed is not popular as seeds lose viability quickly and the germination percentage is low. Plant tissue culture techniques have allowed for mass clonal propagation, germplasm conservation and exchange, and improvement of crop species [13]. A protocol for the *in vitro* regeneration of breadfruit and seeded breadfruit has been established [12]. The addition of 2 g/L polyvinylpyrrolidone significantly reduced the oxidative browning of explants of both species. Aseptic cultures of mature meristem and shoot tip explants, juvenile shoot tip, and nodal explants were successfully established after exposure to antibiotic and fungicide pretreatments and antibiotic medium meristems. Approximately 40% of plantlets were successfully hardened under greenhouse conditions.

FIGURE 33.2   The maturing breadfruit. Photographer: Jacklyn Broomes.

1.4 Harvesting

Breadfruits are picked when maturity is indicated by the appearance of small drops of latex on the surface. In Jamaica, two harvest maturities, 'young and fit,' are recognized [8]. This corresponds with 'immature and mature' as described by Graham and Negron de Bravo [14] and Ragone and Wiseman [15]. Immature is characterized by light green skin, almost complete absence of latex, and closely packed fruit segments, indicating that they are not fully grown. Mature is characterized by a darker skin, with some browning and external dried latex, and the segments not closely packed (Fig. 33.2). Worrell and Carrington [16] described 'boilers' as breadfruit suitable for boiling and slightly immature with greener skin, rougher texture, and little or no latex flow, and 'roasters' as fully mature fruit for baking or roasting characterized by surface latex, paler skin, and fully flattened segments with no 'rough' feel.

1.5 Productivity and Yields

In the South Pacific, the trees yield 50–150 fruits per year. In southern India, normal production is 150–200 fruits annually [5,6].

Productivity varies between wet and dry areas. In the West Indies, a conservative estimate is 25 fruits per tree. Studies in Barbados indicate a reasonable potential of 16–32 tons per hectare while Roberts-Nkrumah [17] suggested 50 tons per hectare for selected varieties. Much higher yields have been forecasted but some experts view these as unrealistic.

1.6 Post-harvest Management

Breadfruit is harvested in the mature stage to facilitate preferred methods of preparation. A major constraint to breadfruit storage is its rapid respiration rate which causes the fruit to ripen in 2–3 days [18]. As the fruit ripens, it softens and starch is converted to sugar, thereby rendering the fruit unacceptable. At 26 or 28°C, breadfruit storage life is only 2 or 3 days [19,20]. Bernardin et al. [21] claimed that browning of the peel was the limiting factor in storage. In certain rural areas of Jamaica, it is the practice to store breadfruit underwater at ambient temperature. It is believed to extend their storage life, which was confirmed by Thompson et al. [19], who found that they did not begin to soften until after 14 days at 28°C, at which time they split, having absorbed too much water. Refrigerated storage recommendations are as follows:

- 12.5°C and 92–98% relative humidity (r.h.) for about 5 days [19];
- 13.3°C and 85–90% r.h. for 14–40 days [22];
- 13°C and 95% r.h. for 1–3 weeks [23].

Other methods of prolonging shelf-life under investigation are shrink-wrapping, coatings, packaging, and controlled atmosphere storage [20]. There is also interest in minimal processing of the fresh fruit to prolong its shelf-life.

2. USES OF SPECIFIC PARTS OF THE BREADFRUIT

The immature, mature, and ripe breadfruit may be incorporated into a variety of food uses. All parts (flesh, peel, core, and seeds) of edible mature and ripe fruits including the leaves of the breadfruit tree are fed to livestock [2,24]. The multipurpose breadfruit tree species provides medicine, construction materials, adhesives, insecticides, and animal feed, and is a primary component of traditional agro-forestry systems in Oceania [6]. All parts of the breadfruit trees are used medicinally, especially the latex, leaf tips, and inner bark. Breadfruit has been used as a trellis tree for yam (*Dioscorea* spp.). Honeybees visit male inflorescences and collect pollen, especially from fertile, seeded varieties. These bees also collect latex that oozes from the fruit surface [24]. Table 33.1 shows the non-food uses of the specific parts of breadfruit tree [6,7,25–28].

2.1 Food Uses

2.1.1 Different Food Preparation

The breadfruit's value as a food crop arises from its high potential productivity of 50 t/ha/yr [17]. Initially cultivated as a bread substitute due

TABLE 33.1 Uses of Parts of the Breadfruit Tree for Non-food Uses

Part of the Tree	Uses
Tree	Agro-forestry; canopy; erosion control; shelter for important pollinators or seed dispersers such as honeybees, birds, flying foxes; firewood; utensils
Timber	Construction materials, buildings, canoes, surfboards; drums, furniture and other objects, carvings, firewood, handicrafts, storage containers
Latex	Adhesive; caulking for canoes; birdlime; chewing gum; medicine; sheen on tapa cloth; mixed with colored earth for use as paints; mosquito repellent; catch fish with an arrow; glue to trap birds on sticks; repair canoes; stick feathers for traditional dance costumes; latex with oil to trap houseflies
Bark	Medicine
Bast (inner bark)	Cordage; tapa (bark cloth); cordage for fishing; animal harnesses, nets
Sap	Prepare bark for artistic painting; secure skin on 'kundu' drums
Male inflorescences	Candied and eaten; dried and used as mosquito repellent medicine, pollination, yellow/brown dye, repellent against flying insects
Fiber	Clothing; harnesses for buffaloes
Deeply pinnate lobed-leaves	Ornamental appeal; wrapping food for cooking or serving; livestock feed; medicine; dried leaves stipules used as a sanding cloth; fishing; kites; plates; organic composting; medicine
Fruits and seeds	Cooked fruits and seeds used for human consumption; uncooked for livestock feed

Source: references 6, 7, 24–29.

to similarities in taste and smell, breadfruit is now commonly used to replace starchy vegetables, pasta, or rice. The fruit is most often consumed when it is mature, but still firm. Wootton and Tumaalii [30] and Roberts-Nkrumah and Badrie [31] have noted that consumer preference for breadfruit consumption is at the mature but unripe stage. In a consumer study of the most common methods of cooking mature breadfruit in Trinidad, West Indies [31], given were 'oil-down' (83.6%), 'steaming' (62.3%), 'boiling' (54.1%), 'frying of thick slices' (53.5%), 'frying (chips) roasting' (44.8%), 'roasting then frying' (44.3%), 'soup-making' (23.5%), 'baking' (23.5%), 'pie-making' (21.3%), 'currying' (7.7%), and microwaving (4.4%). In the Caribbean, breadfruit 'oil down' is a one-pot meal of salted cured meat (chicken, beef, pig tail, pig snout), chicken, dumpling, breadfruit and callaloo (young dasheen leaves – *Colocasia esculenta*) and coconut water [32,33]. A breadfruit 'oil down' serving was found to be high in fat, protein, dietary fiber, phosphorus, sodium, and calcium, good in carbohydrates and potassium, but low in zinc, iron, and copper [34]. However, immature fruits can also be cooked by boiling, pickling, or marinating, imparting a flavor that is said to be similar to that of artichoke hearts [5]. Sliced fruit is sometimes fried to make chips [34] and male fruits can be candied [35]. A few cultivated varieties of breadfruit can be safely eaten without cooking. However, most varieties are purgative if eaten raw, and some are boiled twice and the water thrown away, to avoid unpleasant effects. The ripe fruit is somewhat creamy and sweet and a few cultivated varieties can be eaten raw or used in dessert recipes. In addition, fruit made into cereal, or pureed ripe fruits are regarded as good baby foods. Numerous processed breadfruit products are also available including flour, chips, and sliced fruits which are available frozen, dehydrated, or canned [6,36].

2.1.2 *Applications of Modified Starches in Foods*

The high carbohydrate content of breadfruit makes it a valuable source of starch (Table 33.2). Native starches have been used as raw material to prepare different products and are good texture stabilizers in food systems. However, native breadfruit starches exhibit limited industrial applications. These starches possess low shear stress resistance and thermal decomposition, in addition to high retrogradation and syneresis. Hence, breadfruit was isolated and modified by oxidation, acetylation, heat-moisture treatment and annealing [40]. All forms of starch modification reduced pasting temperature, peak viscosity, hot paste viscosity and cold paste viscosity of the native starch, except that heat-moisture-treatment increased the pasting temperature. Setback value reduced after modifications, indicating that modifications would minimize starch retrogradation [40].

2.2 Compositional and Nutritional Values

Nutritionists have promoted the breadfruit as one of the top 25 superfoods that are useful in the management of prevalent diet-related diseases in the Caribbean, such as diabetes and hypertension [41]. Breadfruit is consumed primarily for its nutritional benefits (see Table 33.2). The fruits and seeds are good sources of carbohydrate, protein, potassium, and calcium with fair amounts of ascorbic acid, niacin, and iron [6,42]. Compared to other staple starch foods, breadfruit is a better source of protein than cassava and is comparable to sweet potato and banana. In addition, the breadfruit seeds (Table 33.1) were found to contain more than 20% carbohydrates (between 26.6 and 38.2%), crude protein (7.9–8.1%) and fat (2.5–4.9%). The nutritional composition of breadfruit varies among cultivars and should aid in the selection of cultivars for different uses for fresh consumption and processed products [6].

TABLE 33.2   Proximate Composition of Breadfruit in Different Forms

Nutrient	Per 100 g Edible Portion						
	Fresh	Flour	Boiled	Roasted/ Baked	Fermented	Paste	Breadfruit Seeds
Water (%)	63.8–74.3	2.5–19.0	67.5–70.3	59.0–70.3	67.3–71.2	20.8	47.7–61.9
Protein (g)	0.7–3.8	2.9–5.0	0.95–1.2	0.8–2.2	0.7	6.3	7.9–8.1
Carbohydrate (g)	22.8–77.3	61.5–84.2	24.5–30.3	28.7–37.6	27.9	67.7	26.6–38.2
Fat (g)	0.26–2.36	1.93	0.24	0.11–0.39	1.13	2.2	2.5–4.9
Calcium (mg)	15.2–31.1	50	12.1–21.1	18.0–26.3	42.0	134	–
Potassium (mg)	352	1630	–	–	20–399	–	46.6–48.3
Phosphorus (mg)	34.4–79.0	90	27.3–37.9	42.7–91.7	–	164	–
Iron (mg)	0.29–1.4	1.9	0.27–0.49	0.68–1.56	0.73–1.18	0.83	186–189
Sodium (mg)	7.1	2.8	–	2.4–5.3	–	–	–
Magnesium (mg)	–	–	–	–	–	–	2.3
Thiamine (mg)	0.07–0.12	–	0.08	0.07–0.09	–	0.14	0.13–0.33
Riboflavin (mg)	0.03–0.1	0.2	0.05–0.07	0.06–0.1	–	0.12	0.08–0.10
Niacin (mg)	0.81–1.96	2.4	0.62–0.74	1.13–1.54	–	7.42	1.8–2.1
Ascorbic acid	19.0–34.4	22.7	2.9–3.2	1.0–2.6	4–20	–	1.9–22.6
β-carotene	0.01	–	–	–	0.04–0.29	–	–

2.2.1 Carbohydrates (Energy)

Since the eighteenth century, the seedless breadfruit has been considered an important nonconventional food product on account of its high caloric content [43]. It is an important energy food because of its starch and sugar content: 68% starch, dry-weight basis [14], and 15.5% starch fresh weight basis [44]. The levels of these vary according to the stage of ripeness at which the fruit is eaten. The carbohydrate content of the West Indian white heart cultivar showed the highest concentration to be arabinose (468 mg), sucrose (425 mg), and glucose (365 mg) [44].

2.2.2 Dietary Fiber

Breadfruit is rich in dietary fiber. The Secretariat of the Pacific Community (SPC) [45], stated that the boiled flesh of the unseeded mature breadfruit contains 2.5 g fiber/100 g (see Table 33.2) while Graham and Negron de Bravo [14] stated that 100 g of breadfruit flour contains 3.5 g of fiber. It is currently recommended that adults consume 20–35 g of dietary fiber per day. Two cups (500 g) of boiled breadfruit at lunch and dinner provide around 25 g of fiber, but a similar serving of white rice provides only 6.8 g (Fig. 33.3). Dietary fiber, principally the non-starch polysaccharides of the plant cell wall, is an important component of our diet. Fiber has uniquely significant physical effects in the gut and in addition through fermentation is a major determinant of large bowel function and bowel habit. A diet rich in fiber can help to control blood sugar in diabetics, reduce blood lipids (a risk for heart disease), and help to control weight. Its physical properties in the small bowel effect lipid absorption and the glycemic response. Fiber has some modest effects on appetite. These benefits feed through into a protective role in large bowel cancer, diabetes, and coronary heart disease. Table 33.3 illustrates some of the physiological and health benefits of dietary fiber within the diet [46,47].

2.2.3 Provitamin A and Carotenoids

The flesh of ripe, seeded breadfruit is particularly rich in provitamin A carotenoids. The amount of provitamin A carotenoids, the precursors to vitamin A, varies with ripeness. Consuming provitamin A carotenoids may help protect against infection, diabetes, heart disease, and cancer and help maintain good eye health and vision and strong blood. Two cups of ripe seeded breadfruit eaten at lunch and dinner provides 100% of the estimated daily vitamin A requirements for an adult. A wide range of provitamin A carotenoid levels was found in breadfruit cultivars, some

containing very high levels of 295–868 mg/100 g in breadfruit (edible portion). This compares to β-carotene content of 515–6360 mg/100 g in banana (*Musa* spp.) and 260–1651 mg/100 g in taro (*Colocasia esculenta*) [45]. Mature breadfruit has a creamy colored edible flesh and has low levels of carotenoid (23–40 mg/100 g) [48]. Persons may cook and eat breadfruit at the ripe soft sweet stage, which has a more yellow-colored flesh and a slightly fermented taste desired by many. Despite the yellow color of the ripe breadfruit samples, most were found to have low levels of provitamin A carotenoid. On the other hand, seeded breadfruit, which is particularly yellow, had a high provitamin A carotenoid content.

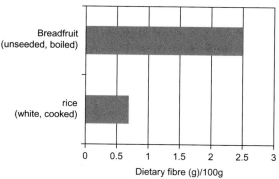

FIGURE 33.3 Dietary fiber (g)/100 g in breadfruit and white rice.
Source: Reference 45.

2.2.4 Vitamins and Minerals

Ragone [6] stated that breadfruit is a good source of iron, calcium, potassium, and riboflavin. The breadfruit also carries significant levels of B vitamins, niacin and thiamine which are essential to metabolism [45]. In addition, breadfruit is a good source of vitamin C which aids in the fighting of infection. In particular, as Figure 33.4 shows, a typical serving of one of the seeded types of breadfruit can meet daily needs for vitamin C. Rice is very low in both nutrients [45].

TABLE 33.3 Established Physiological Properties and Health Benefits of Dietary Fiber

Property	Mechanism	Related Conditions
Substrate for fermentation	Microbial growth stimulated	Bowel habit/constipation
	Short chain fatty acids	Diverticular disease
	Changes in nitrogen, bile acid, and xenobiotic metabolism	Colo-rectal cancer
Physical effects in small bowel	Gel properties	Glycemic response/diabetes
	Secondary effects on insulin secretion and gut hormones	Lipid absorption/coronary heart disease
Satiety and gastric emptying	Chewing of food	Short-term appetite reduced
	Delay in gastric emptying	

Source: Reference 47.

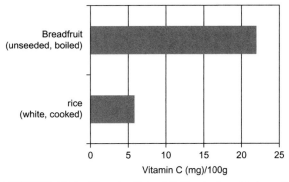

FIGURE 33.4 Vitamin C (mg)/100 g in breadfruit and white rice.
Source: Reference 45.

2.2.5 Protein

Table 33.1 shows that breadfruit supplies a limited amount of protein (0.7–3.8 g/100 g) while breadfruit seeds are a fair source of protein (7.9–8.1 g/100 g). All of the essential amino acids were detected in the breadfruit sample in varying amounts. The essential amino acids that were found in the greatest amounts were leucine (605 mg) and lysine (799 mg) during the ripe developmental stage [44]. These essential amino acids accounted for approximately 30% of the total amino acid content of the breadfruit sample, rendering it a relatively good source of essential amino acids in a diet.

2.2.6 Glycemic Index

The glycemic index (GI) is a classification of the glucose-raising potential of carbohydrate foods relative to glucose [49]. It relates to the blood glucose level after a meal. In response to low GI foods, there are several advantages including longer satiety, lower blood pressure, and lower plasma low-density lipoproteins (LDL)-cholesterol levels related to the less pronounced insulin response to low GI foods. Breadfruit has been designated an intermediate GI food (60 ± 9.0) along with other foods studied including yam, tannia, eddoes, cooking 'green' banana, and roti [50]. This is particularly important since the regular consumption of diets containing high-GI foods is associated with an increased risk for type II diabetes mellitus [51] and coronary heart disease (CHD) [52,53]. Small changes in diet GI are associated with a significant reduction in CHD risk [54] and diabetes risk [51], and improvements in insulin sensitivity and glycemic control [55–57]. The GI may therefore provide the rationale for choosing carbohydrate foods for meal plans created for individuals with diabetes [58].

2.2.7 Fatty Acids

Fatty acids are very important compounds in biological systems. First, they are important components of structural molecules such as phospholipids and are an important source of energy. In breadfruit, essential fatty acids detected at the highest concentration were linoleic acid, an omega-6 fatty acid (0.15 mg), and linolenic acid, an omega-3 fatty acid (2.13 mg), both at the ripe stage [44]. Higher proportions of linoleic and linolenic acids in the diet are inversely related to coronary artery disease.

The essential fatty acids combined here have proven to impart a regulatory function on the body's fatty acid metabolism. Fat metabolism is as important, if not more important, than our body's metabolism of proteins and carbohydrates, as evidenced by the drastic rise in fat-related degenerative diseases, such as vascular disease and stroke. Fatty acids are important structural molecules such as phospholipids and an important source of energy [44].

2.3 Medicinal Uses

2.3.1 Phenolic Compounds

Research on the chemical constituents of breadfruit (*A. altilis*) has isolated several classes of compounds such as various triterpenes and flavonoids [59–66]. It is documented that the genus *Artocarpus* is a rich source of prenylated phenolic compounds such as geranylated flavones [67]. A recent study reported the isolation of stilbenes (1–3), arylbenzofuran (4), flavanone (5), flavones (6–8), triterpenes (9 and 10), and sterols (11 and 12) from the fruits of breadfruit. Compounds 2, 6, 9, and 12 are reported from this plant for the first time [68]. One of the benzene rings in these phenolics often has a unique 2,4-di- or 2,4,5-trioxygenation pattern as exemplified by 20,40-dihydroxyflavones. Compounds 1–8 isolated from the fruits of *A. altilis* in the present study all possess the characteristic 2,4-dioxygenation pattern and can be used as taxonomic markers for the family *Moraceae*, especially the genus *Artocarpus* [68]. Using a two-dimensional counter-current chromatographic system, HPLC analysis for preparative isolation and purification of three prenylflavonoids from breadfruit (*A. altilis*) showed their purities as: (1) 98.7%, (2) 98.3%, and (3) 97.2% [69]. The phytochemical studies on the leaves [65], heartwood [70], and bud covers [71] of breadfruit (*A. altilis*) have revealed the presence of phenolic compounds such as stilbenes, chalcones, and flavones. It is to be noted that the leaves and bud covers mainly contain geranyl flavonoids [62,65], while the roots and stems provide various types of prenylated flavonoids [60,66]. The ethyl acetate soluble fraction of the methanol extract of the breadfruit leaves of *A. altilis* was subjected to repeated silica gel column chromatography which resulted in the isolation of nine compounds with the structural elucidation of five new geranyl dihydrochalcones (2, 4, 5, 8, 9), along with four known geranyl flavonoids (1, 3, 6, 7) [65].

The pharmacological studies have indicated that some flavonoids from breadfruit (*A. altilis*) have anti-inflammatory activities [71] and are able to inhibit 5-lipoxygenase of cultured mastocytoma cells [72], cathepsin K [62], and 5α-reductase [70]. However, characterization of the active principles responsible for these biological effects has not yet been determined.

Platelet aggregation is an important pathogenic factor in the development of atherosclerosis and associated thrombosis in humans [73]. Prenylflavonoids isolated from *A. altilis* have revealed significant antiplatelet effects [74]. Four flavonoids, dihydroartomunoxanthone (1), artomunoisoxanthone (2), cyclocomunomethonol (3), and artomunoflavanone (4) together with three known compounds, artochamins B (5), D, and artocommunol CC (6) were isolated from the cortex of the roots of *A. altilis*. The structures of 1–4 were determined by spectroscopic methods. The antiplatelet effects of the flavonoids, 1–3, 5 and 6 on human platelet-rich plasma (PRP) were evaluated. Of the compounds tested in human PRP, compounds 1, 5, and 6 showed significant inhibition of secondary aggregation induced by adrenaline. It is concluded that the antiplatelet effect of 1, 5, and 6 is mainly owed to an inhibitory effect on thromboxane formation [66]. It indicates that the antiplatelet effects of 1, 5, and 6 relate to an inhibitory effect on thromboxane formation [75–77]. Therefore, compounds 1, 5, and 6 showed promise as antithrombotic agents.

2.3.2 *Artocarpin*

Artocarpin (Ar) is an extract of heartwood of *A. altilis* which possesses potent 5α-reductase inhibitory effect. A study suggested that alginate/chitosan (ACS)–artocarpin (Ar) microparticles may be targeted for transfollicular delivery and may be a promising delivery system for the safe treatment of androgen-dependent disorders such as acne, seborrhea, hirsutism, and androgenic alopecia also in humans as the structure of the hair follicles of hamster and rat is not different regarding the uptake mechanisms, i.e. interruption of the stratum corneum around the hair shaft which makes a transfollicular particle uptake possible [68].

2.3.3 Lectin

Novel chitin-binding lectins have been extracted from the seeds of *A. altilis*. Most chitin-binding lectins have shown antifungal activity against phytopathogenic species, since chitin is the key component of the cell wall of these microorganisms. They have shown to affect fungal growth and development, disturbing the synthesis and/or deposition of chitin in the cell wall [78,79]. The isolated chitin-binding lectins from the seeds of *A. altilis*, frutackin, promoted hemagglutination and growth inhibition against fungi *F. moniliforme* and *S. cerevisiae* [80].

2.3.4 Breadfruit Starch in Tablets

Starch constitutes an important class of tablet disintegrant and is rated among the top ten pharmaceutical ingredients by the International Joint Conference on Excipients [81]. The high carbohydrate content of breadfruit makes it a valuable source of starch (see Table 33.2). Despite its presumed safety, acceptance as pharmaceutical grade starch would require detailed investigation and submission of supportive data ascertaining the safety, efficacy, stability, and compatibility with candidate drugs and other formulation additives [82].

A comparative evaluation of starch powder extracted from breadfruit as a tablet disintegrant was made with corn starch British Pharmacopoeia (BP). The tablets were tested for disintegration, dissolution, and physical qualities following the BP compendial procedures. As endo-disintegrant, however, only corn starch at 5% and 10% w/w and breadfruit at 10% passed the BP limit test. Generally, the rank order of effectiveness of disintegrant was corn $_{exo}$>breadfruit $_{exo}$>corn $_{endo}$>breadfruit $_{endo}$. Breadfruit starch powder appears to be a suitable substitute for official corn starch only as exo-disintegrant in paracetamol tablet formulation at 5% w/w concentration. In that form, it not only produced rapid tablet break-up and drug dissolution, it also improved tablet crushing strength. Breadfruit starch appeared to be suitable as a substitute for official corn starch as an exo-disintegrant in paracetamol tablet formulation (Table 33.4).

2.4 Folkloric Medicinal Uses

Breadfruit is one of the most widely used folk medicines. A number of authors [5,7,24,83–86] have reported on folkloric medicinal uses of the breadfruit (Table 33.5).

TABLE 33.4 Effect of Compression Force on Disintegration Times of Paracetamol Tablets Containing 5% w/w of Starch Disintegrant

		Disintegration Time (min)			
	No Disintegrant	Exo-disintegrant		Endo-disintegrant	
		Breadfruit	Corn	Breadfruit	Corn
0.5	50.30±11.25	0.83±0.22	0.57±0.09	15.75±1.45	4.47±0.09
1.0	61.86±10.18	3.09±0.50	1.01±0.12	21.33±8.16	8.62±0.34
1.5	64.82±8.06	16.35±3.85	3.18±0.61	32.28±0.93	13.09±0.95
2.0	68.41±2.46	41.75±9.07	14.02±3.66	36.99±2.19	18.26±1.36

Source: reference 84.

TABLE 33.5 Uses of Parts of the Breadfruit Tree for Folkloric Medicinal Uses

Part of the Tree Used	Uses
Crushed breadfruit	Poultice on tumors
Latex with/without mashed leaves	Antifungal; diluted latex for diarrhea; massage ointment to treat broken bones, sprains, and bruises; roasted leaves as a remedy for enlarged spleen; the latex is bandaged onto the spine to relieve sciatica; treatment of sores; relief of stomach aches
Root	Antimicrobial, antitumor, poultice for skin ailment, purgative
Bark	Antimicrobial, antitumor, headaches, cytotoxic activity in bioassays against leukemia
Sap	Ear infections
Male inflorescences	Toasted inflorescences for relief of toothache
Leaves	Leaf juice as ear drops, headaches, crushed leaves on tongue for treatment of thrush, roasted leaves for treatment of enlarged spleen, liver cirrhosis, hypertension, and diabetes, lower blood pressure, relief of asthma

Source: References 5, 7, 24, 84–86.

3. RESEARCH NEEDS

Although breadfruit is grown in around 90 countries, it has received limited commercial and research attention. It is an important crop for food, feed, and non-food uses. Despite the nutritional and medicinal benefits, the breadfruit remains underutilized in most tropical areas. This could be due to the lack of characterization, evaluation, and description of breadfruit germplasm, difficulty in obtaining planting materials, limited information and planting material for commercial orchard establishment and management, lack of a consistent supply of good-quality fruits, determination of the right stage of maturity of fruits, inaccessibility for harvesting due to the tallness of breadfruit trees, seasonality of the breadfruit, fruit rot, short-shelf life of fruits, lack of identification of suitable cultivars for processing, lack of appropriate technology for the development of new value-added food products, bulky nature of fruits making manual peeling difficult, ineffective approach for the isolation of bioactive constituents, lack of information on the medicinal benefits, and negative perception related to the colonial history [6,15,18,24].

4. SUMMARY

Creating an awareness of the nutritional and medicinal benefits of the breadfruit will help improve consumption levels [87]. Removing the social stigma and increasing awareness about its nutritional qualities are major challenges in the quest to transform breadfruit from its hidden identity into a crop that enhances livelihoods [28]. Breadfruit is an important source of significant nutrients and palatable foods. Breadfruit is consumed primarily for its nutritional benefits and as a major source of carbohydrates. The fruits and seeds are good sources of carbohydrates, protein, dietary fiber, fatty acids, pro-vitamin A, potassium, and calcium with significant amounts of ascorbic acid, niacin, and iron [6,42].

The FAO/UN expert consultation on carbohydrates in human nutrition recommends the consumption of foods that possess low glycemic index or slow digestion, i.e. promote slow release of glucose in the body. The GI may provide the rationale for choosing carbohydrate foods for meal plans created for individuals with diabetes [56,58].

Breadfruit is also useful medicinally as it is a provider of phenolic compounds, artocarpins, and lectins which carry different but important applications such as anti-inflammatory activities, antiplatelet effects, reductase inhibitory effects, and antifungal activity. The breadfruit starch may also be utilized as a tablet disintegrant [62,68,70–72,74,78,79,81].

4.1 Health and Medical Applications of Breadfruit

In terms of nutrition, breadfruit is an energy rich food. It provides up to 2.5 g dietary fiber/100 g, is particularly rich in provitamin A carotenoids, is a good source of iron, calcium, potassium, and riboflavin, and carries significant levels of B vitamins, niacin, and thiamine. In addition, breadfruit is a good source of vitamin C. Breadfruit is limited in protein, has an intermediate glycemic index, and carries linoleic and linolenic fatty acids.

Medicinally, breadfruit has phenolic compounds including triterpenes and flavonoids, artocarpin which has a reductase inhibitory effect, and lectin which has antifungal properties. Breadfruit starch is utilized as a tablet disintegrant.

References

1. Purseglove, J. W. (1968). *Tropical Crops: Dicotyledons.* 3, pp. 383–388. Bristol: J. W Arrowsmith Ltd, Winstoke Road.
2. Ragone, D. (2007). Breadfruit: diversity, conservation and potential. In M. B. Taylor, J. Woodend, & D. Ragone, (Eds.), *The First International Symposium on Breadfruit Research and Development* (pp. 19–22). EU-ACP Technical Centre for Agricultural and Rural Cooperation (CTA).
3. Zerega, N., Ragone, D., & Motley, T. J. (2005). Systematics and species limits of breadfruit (*Artocarpus*, Moraceae). *Systematic Botany*, 30, 603–615.
4. Bapat, V. A., & Mhatre, M. (2005). *Ficus carica* fig, *Artocarpus* spp. jackfruit and Breadfruit and Morus spp. Mulberry. In R. E. Litz, (Ed.), 'Biotechnology of Fruit and Nut Crops.' *Biotechnology in Agriculture Series, N. 29* (pp. 350–363). Wallingford, Oxfordshire, UK: CABI Publishing.
5. Morton, J. (1987). Breadfruit. *In* 'Fruits of Warm Climates' (J. F. Morton, Ed.), pp. 50–58. *Creative Resources Systems*, Miami, FL.
6. Ragone, D. (1997). *Breadfruit.* Artocarpus altilis *(Parkinson) Fosberg. Promoting the conservation and use of underutilized and neglected crops, 10.* Rome, Italy: Institute of Plant Genetics and Crop Plant Research, Gatersleben/International Plant Genetic Resources Institute (pp. 17–20).
7. Powell, J. M. (1976). Part 111. Ethnobotany. In K. Paijmans, (Ed.), *New Guinea Vegetation* (pp. 106–183). Canberra: Australian National University Press.
8. Thompson, A. K. (2003). Chapter 12, Postharvest technology of fruits and vegetables. In *Fruit and Vegetables: Harvesting, Handling and Storage* 2nd ed (pp. 174–175). Oxford: Blackwell Publishing.
9. Rajendran, R. (1992). *Artocarpus altilis* (Parkinson) Fosberg. In E. W. M. Verhij, & R. E. Coronel, (Eds.), *Plant Resources of South-East Asia, No. 2 Edible Fruits and Nuts* (pp. 83–86). Wageningen: Pudoc-DLO.
10. Ragone, D. 2003. Breadfruit. pp. 655–661. *In* B. Caballero, L. Trugo, and P. Finglas (eds). Encyclopedia of Food Sciences and Nutrition. Academic Press, San Diego.
11. Towill, L. E. (2005). Germplasm preservation. In N. R. Trigiano, & D. J. Gray, (Eds.), *Plant Development and Biotechnology* (pp. 277–284). London: CRC Press.
12. Rouse-Miller, J., and Duncan, E. J. (2007). *In vitro* regeneration of *Artocarpus camansi* and *Artocarpus altilis. In* 'Proceedings of the First International Symposium on Breadfruit Research and Development' (M. B. Taylor, J. Woodend and D. Ragone, Co-convenors), pp. 153–260, EU-ACP Technical Centre for Agricultural and Rural Cooperation (CTA). International Society for Horticultural Science, Leuven, Belgium.
13. Trigiano, N. R., & Gray, D. J. (2005). Plant Development and Biotechnology, London: CRC Press.
14. Graham, H. D., & Negron de Bravo, E. (1981). Composition of the breadfruit. *Journal of Food Science*, 46(2), 535–539.
15. Ragone, D., & Wiseman, J. (2008). Developing and applying descriptors for breadfruit germplasm *Acta Hort.* (ISHS) 757, 71–78.
16. Worrell, D. B., & Carrington, C. M. S. (1997). Breadfruit. In S. K. Mitra, (Ed.), *Postharvest Physiology and Storage of Tropical and Subtropical Fruit.* Oxford: CAB International.
17. Roberts-Nkrumah, N. (1998). A Preliminary Evaluation of the Imported Breadfruit Germplasm at the University Field Station, Trinidad. Caribbean Food Crops Society (34th) and the Jamaican Society for Agricultural Sciences (9th), July 12–18. Montego Bay, Jamaica.
18. Roberts-Nkrumah, N. (2007). An overview of breadfruit (*Artocarpus altilis*) in the Caribbean *Acta Hort.* (ISHS) 757, pp. 51–56.

19. Thompson, A. K., Been, B. O., & Perkins, C. (1974). Storage of fresh breadfruit. *Tropical Agriculture Trinidad, 51*, 407–415.

20. Maharaj, R., & Sankat, C. K. (1990). The shelf-life of breadfruit stored under ambient and refrigerated conditions. *Acta Horticulturae, 269*, 411–424.

21. Bernardin, J., Sankat, C. K., and Willemot, C. (1994). Breadfruit Shelf-life Extension by CaCl₂ Treatment, Hortscience 29, 6.

22. SeaLand. (1991). *Shipping Guide for Perishables.* Edison, NJ: SeaLand Corporate Marketing, 379 Thornall St.

23. Snowdon, A. L. (1992). 'A Colour Atlas of Postharvest Diseases and Disorders of Fruits and Vegetables. Volume 1. General Introduction and Fruits,' p. 302. Wolfe Scientific, London.

24. Ragone, D. (2006). *Artocarpus Altilis* (breadfruit). Moraceae (Mulberry Family) Species Profiles for Pacific Island Agroforestry. <www.traditionaltree.org> April, 2006, pp. 17.

25. Lorens, A. S., & Englberger, L. (2007). The importance and use of breadfruit cultivars in Pohnpei, Federated States of Micronesia *Acta Hort.* (ISHS) 757, 101–107.

26. Navarro, M., & Malres, S. (2007). Vanuatu breadfruit project: survey on botanical diversity and traditional uses of *Artocarpus altilis.* In M. B. Taylor, J. Woodend, & D. Ragone, (Eds.), *The First International Symposium on Breadfruit Research and Development.* EU-ACP Technical Centre for Agricultural and Rural Cooperation, CTA.

27. Redfern, T. (2007). Breadfruit improvement activities in Kirbati. *Acta Hort.* (ISHS) 757, 93–99.

28. Omobuwajo, T. O. (2007). Breadfruit as a key component of sustainable livelihoods in Nigeria: Prospects, opportunities and challenges. In M. B. Taylor, J. Woodend, & D. Ragone, (Eds.), *The First International Symposium on Breadfruit Research and Development* EU-ACP Technical Centre for Agricultural and Rural Cooperation, CTA.

29. Taylor, M. B., & Tuia, V. S. (2007). Breadfruit in the Pacific Regions. In M. B. Taylor, J. Woodend, & D. Ragone, (Eds.), *The First International Symposium on Breadfruit Research and Development.* EU-ACP Technical Centre for Agricultural and Rural Cooperation, CTA.

30. Wootten, M., & Tumaalii, F. (1984). Breadfruit production, utilization and composition. *A Review, Food Technology in Australia, 36*, 464–465.

31. Roberts-Nkrumah, L., & Badrie, N. (2004). Breadfruit consumption, cooking methods and cultivar preference among consumers in Trinidad, West Indies. *Food Quality and Preference, 16*, 267–274.

32. Mcintosh, C., & Manchew, P. (1993). The breadfruit in nutrition and health. *Tropical Fruit Newsletter, 6*, 5–6.

33. Badrie, N., Balfour, S., Ottley, K., & Chang-Yen, I. (2005). Nutrient composition of a commonly consumed West Indian meal of breadfruit (*Artocarpus altilis* Fosberg) oil down. *Journal of Nutrition in Recipe and Menu Development, 3*, 19–35.

34. Roberts, K., Badrie, N., & Roberts-Nkrumah, L. (2007). Colour and sensory characteristics of fried chips from three breadfruit (*Artocarpus altilis*) cultivars. In M. B. Taylor, J. Woodend, & D. Ragone, (Eds.), *The First International Symposium on Breadfruit Research and Development* (pp. 225–229). EU-ACP Technical Centre for Agricultural and Rural Cooperation (CTA).

35. Roberts-Nkrumah, L. (1993). Breadfruit in the Caribbean: a bicentennial review. Extension newsletter. Department of Agriculture, University of the West Indies (Trinidad and Tobago). *Tropical Fruits Newsletter, 24*, 1–3.

36. Bacchus-Taylor, G., & Akingbala, J. (2007). Breadfruit studies at the Food Science and Technology Unit, University of the West Indies, St. Augustine, Trinidad. In M. B. Taylor, J. Woodend, & D. Ragone, (Eds.), *The First International Symposium on Breadfruit Research and Development* (pp. 177–181). EU-ACP Technical Centre for Agricultural and Rural Cooperation (CTA).

37. Murai, M., Pen, P., & Miller, C. (1958). *Some Tropical Pacific Foods.* Honolulu: University of Hawaii.

38. Aalberg, W., Lovelace, C., Madhoji, K., & Parkinson, S. (1988). Davuke, the traditional Fijan method of pit preservation of staple carbohydrate foods. *Ecology of Food and Nutrition, 21*, 173–180.

39. Dalessandri, K., & Boor, K. (1994). World nutrition – the great breadfruit source. *Ecology of Food and Nutrition, 33*, 131–134.

40. Adebowale, K. O., Olu-Owolabi, B. I., Olawumia, E. K., & Lawal, O. S. (2005). Functional properties of native, physically and chemically modified breadfruit (*Artocarpus artilis*) starch. *Industrial Crops and Products, 21*, 343–351.

41. Magnus, H. (2005). The top 25 superfoods. *Cajanus, 30*, 3–13.

42. Ragone, D., & Cavaletto, C. G. (2006). Sensory evaluation of fruit quality and nutritional composition of 20 breadfruit (*Artocarpus, Moraceae*) cultivars. *Economic Botany, 60*, 335–346.

43. Omobuwajo, T. (2003). Compositional characteristics and sensory quality of biscuits, prawn crackers and fried chips produced from breadfruit. *Innovative Food Science and Emerging Technologies, 4*, 219–225.

44. Golden, K. D., and Williams, O. J. (2007). The amino acid, fatty acid and carbohydrate content of *Artocarpus altilis* (breadfruit): the white cultivar from the West Indies. *Acta Hort.* (ISHS) *757*, 201–208.

C. ACTIONS OF INDIVIDUAL FRUITS IN DISEASE AND CANCER PREVENTION AND TREATMENT

45. Secretariat of the Pacific Community. (2006). Breadfruit. A Publication of the Healthy Pacific Lifestyle Section of the Secretariat of the Pacific Community. Pacific Food leaflet No. 3. Website: <http://www.spc.int/lifestyle>.

46. Asp, N. G. (1992). Resistant starch. Physiological implications of the consumption of resistant starch in man. Proceedings from the Second plenary meeting of EURESTA: European FLAIR Concerted Action No. 11 (COST 911). *Eur. J. Clin. Nutr. 46*(Suppl. 2), 1–148.

47. Cummings, J., Edmond, L., & Magee, E. (2004). Dietary carbohydrates and health, do we still need the fibre concept? *Clinical Nutrition Supplementum, 1*, 5–17.

48. Dignan, C., Burlingame, B., Kumar, S., & Aalbersberg, W. (2004). The Pacific Islands Food Composition Tables, (2nd ed). Rome: FAO.

49. Wolever, T. M. S., Vorster, H. H., Björck, I., Brighenti, F., Mann, J. I., Ramdath, D. D., Granfeldt, Y., Holt, S., Perry, T. L., Venter, C., & Xiaomei, Wu. (2003). Determination of the glycaemic index of foods: Interlaboratory study. *European Journal of Clinical Nutrition, 57*, 475–482.

50. Ramdath, D., Isaacs, R., Teelucksingh, S., & Wolever, T. (2004). Glycaemic index of selected staples commonly eaten in the Caribbean and the effect of boiling vs crushing. *The British Journal of Nutrition, 91*, 971–977.

51. Salmeron, J., Manson, J. E., Stampfer, M. J., Colditz, G. A., Wing, A. L., & Willett, W. C. (1997). Dietary fiber, glycemic load, and risk of non-insulin-dependent diabetes mellitus in women. *The Journal of the American Medical Association, 227*, 472–477.

52. Ford, E. S., & Liu, S. (2001). Glycemic index and serum high-density lipoprotein cholesterol concentration among US adults. *Archives of Internal Medicine, 161*, 572–576.

53. Liu, S., & Manson, J. E. (2001). Dietary carbohydrates, physical activity, obesity, and the 'metabolic syndrome' as predictors of coronary heart disease. *Current Opinion in Lipidology, 12*, 395–404.

54. Liu, S., Willett, W., Stampfer, M., Hu, F., Franz, M., Sampson, L., Hennekens, C., & Manson, J. (2000). A prospective study of dietary glycemic load, carbohydrate intake and risk of coronary heart disease in US women. *The American Journal of Clinical nutrition, 71*, 1455–1461.

55. Frost, G., Leeds, A., Trew, G., Margara, R., & Dornhorst, A. (1998). Insulin sensitivity in women at risk of coronary heart disease and the effect of a low glycemic index diet. *Metabolism, 47*, 1245–1251.

56. Schakel, S., Schauer, R., Himes, J., Harnack, L. Van Heel, N. (2008). Development of a glycemic index for dietary assessment. *J. Food Composition and Analysis. 21*, S50–S55.

57. Brand-Miller, J., Wolever, T. M. S., Foster-Powell, K., & Colagiuri, S. (2003). *The New Glucose Revolution: The Authoritative Guide to the Glycemic Index.* New York: Marlowe & Company.

58. FAO/UN. (1998). Carbohydrates in human nutrition. *Report of a Joint FAO/WHO expert consultation Rome, 14–18 April 1997, FAO Food and Nutrition Paper no. 66.* Food and FAO, Rome: Agriculture Organization of the United Nations.

59. Altman, L. J., & Zito, S. W. (1976). Sterols and triterpenes from the fruit of *Artocarpus altilis. Phytochemistry, 15*, 829–830.

60. Lin, C. N., & Shieh, W. L. (1992). Pyranoflavonoids from *Artocarpus communis. Phytochemistry, 31*, 2922–2924.

61. Ashok, D.P., Alan, J.F., Lew, K., Priscilla,O., Paul, B. T., Bartholomew, J.V., and Randall, K.J. (2002). A new dimeric dihydrochalcone and a new prenylated flavone from the bud covers of *Artocarpus altilis*: Potent inhibitors of cathepsin K, *J. Nat. Prod. 65*, 624–627.

62. Patil, A. D., Freyer, A. J., Killmer, L., Offen, P., Taylor, P. B., Votta, B. J., & Johnson, R. K. (2002). A new dimeric dihydrochalcone and a new prenylated flavone from the bud covers of *Artocarpus altilis*: potent inhibitors of cathepsin K. *Journal of Natural Products, 65*, 624–627.

63. Chan, S. C., Ko, H. H., & Lin, C. N. (2003). New prenylflavonoids from *Artocarpus communis. Journal of Natural Products, 66*, 427–430.

64. Han, A. R., Kang, Y. J., Windono, T., Lee, S. K., & Seo, E. K. (2006). Prenylated flavonoids from the heartwood of *Artocarpus communis* with inhibitory activity on lipopolysaccharide-induced nitric oxide production. *Journal of Natural Products, 69*, 719–721.

65. Wang, Y., Deng, T. L., Lin, L., Pan, Y. J., & Zheng, X. X. (2006). Bioassayguided isolation of antiatherosclerotic phytochemicals from *Artocarpus altilis. Phytotherapy Research, 20*, 1052–1055.

66. Weng, J. R., Chan, S. C., Lu, Y. H., Hsien-Cheng Lin, H. C., Ko, H. H., & Chun-Nan Lin, C. N. (2006). Antiplatelet prenylflavonols from *Artocarpus communis. Phytochemistry, 67*, 824–829.

67. Hakim, E. H., Achmad, S. A., Juliawaty, L. D., Makmur, L., Syah, Y. M., Aimi, N., Kitajima, M., Takayama, H., and Ghisalberti, E. L. (2006). Prenylated flavonoids and related compounds on the Indonesian Artocarpus (Moraceae) *J. Nat. Med. 60*, 161–184.

68. Pitaksuteepong, T., Somsiri, A., & Waranuch, N. (2007). Targeted transfollicular delivery of artocarpin extract from *Artocarpus incisus* by means of microparticles. *European Journal of Pharmaceutics and Biopharmaceutics, 67*, 639–645.

69. Lu, Y., Cuirong Sun, C., Yu Wang, Y., & Pan, Y. (2007). Two-dimensional counter-current chromatography for

the preparative separation of prenylflavonoids from *Artocarpus altilis*. 4th International Conference on Countercurrent Chromatography, 4th International Conference on Countercurrent Chromatography. *Journal of Chromatography. A, 1151,* 31–36.

70. Shimizu, K., Kondo, R., Sakai, K., Buabarn, S., & Dilokkunanant, U. (2000). A geranylated chalcone with 5α-reductase inhibitory properties from *Artocarpus incisus. Phytochemistry, 54,* 737–739.

71. Wei, B. L., Weng, J. R., Chiu, P. H., Hung, C. F., Wang, J. P., & Lin, C. N. (2005). Anti-inflammatory flavonoids from *Artocarpus heterophyllus* and *Artocarpus communis. Journal of Agricultural and Food Chemistry, 53,* 3867–3871.

72. Koshihara, Y., Fujimoto, Y., & Inoue, H. (1988). A new 5-lipoxygenase selective inhibitor derived from *Artocarpus communis* strongly inhibits arachidonic acid-induced ear edema. *Biochemical Pharmacology, 37,* 2161–2165.

73. Ko, H. H., Hsieh, H. K., Liu, C. T., Lin, H. C., Teng, C. M., & Lin, C. N. (2004). Structure-activity relationship studies on chalcone derivatives: potent inhibitor of platelet aggregation. *The Journal of Pharmacy and Pharmacology, 54,* 1333–1337.

74. Lin, C. N., Shieh, W. L., Ko, F. N., & Teng, C. M. (1993). Antiplatelet activity of some prenylflavonoids. *Biochemical Pharmacology, 45,* 509–512.

75. Mitchell, J. R. A., & Sharp, A. A. (1964). Platelet clumping *in vitro. British Journal of Haematology, 10,* 78–93.

76. Mustand, J. F., Perry, D. W., Kinlough-Rathbone, R. L., & Packham, M. A. (1975). Factors responsible for ADP-induced release reaction of human platelets. *American Journal of Physiologic Imaging, 228,* 1757–1765.

77. Weiss, H. J. (1983). Antiplatelet drugs pharmacological aspects. In R. Alan, (Ed.), *Platelets: Pathophysiology and Antiplatelet Drug Therapy* (pp. 46–49). New York: Liss Inc.

78. Selitrennikoff, C. P. (2001). Antifungal proteins. *Applied and Environmental Microbiology, 67,* 2883–2894.

79. Ng, T. B. (2004). Antifungal proteins and peptides of leguminous and nonleguminous origins. *Peptides, 25,* 1215–1222.

80. Trindade, M. B., Lopes, J. L. S., Soares-Costa, A., Monteiro-Moreira, A. C., Moreira, R. A., Maria Luiza, V., Oliva, M. L. V., & Beltramini, L. M. (2006). Structural characterization of novel chitin-binding lectins from the genus *Artocarpus* and their antifungal activity. *Biochimica Et Biophysica Acta, 1764,* 146–152.

81. Shangraw, R. P. (1992). International harmonization of compendia standards for pharmaceutical excipients. In D. J. A. Crommelin, & K. Midha, (Eds.), *Topics in Pharmaceutical Sciences* (pp. 205–223). Stuttgart: MSP.

82. Adebayo, S. A., Eugenie Brown-Myrie, E., & Itiola, O. A. (2006). Comparative disintegrant activities of breadfruit starch and official corn starch. *Powder Technology, 181,* 98–103.

83. Kasahara, S., & Hemmi, S. (1986). *Medicinal Herb Index in Indonesia.* Jakarta: Eisai Indonesia, p. 184.

84. Young, R. E., Williams, L. A. D., Gardner, M. T., & Fletcher, C. K. (1993). An extract of the leaves of the breadfruit, *Artocarpus altilis* (Parkinson) Fosberg exerts a negative inotropic effect on rat myocardium. *Phytotherapy Research, 7,* 183–190.

85. Wang, Y., Kedi, X., Lin, L., Pan, Y., & Zheng, X. (2007). Geranyl flavonoids from the leaves of *Artocarpus altilis. Phytochemistry, 68,* 1300–1306.

86. Navarro, M, Malres, S., Labouisse, J. P. and Roupsard, O. (2007). Vanuatu breadfruit project: survey on botanical diversity and traditional uses of *Artocarpus altilis. Acta Hort.* (ISHS) *757,* 81–88.

87. Medagoda, I. (2007). Breadfruit in the Asian Region – Focus on Sri Lanka. *Acta Hort.* (ISHS) *757,* 65–69.

C. ACTIONS OF INDIVIDUAL FRUITS IN DISEASE AND CANCER PREVENTION AND TREATMENT

34

Bioactive Compounds in Mango (*Mangifera indica* L.)

Sônia Machado Rocha Ribeiro[1] *and Andreas Schieber*[2]

[1]Federal University of Viçosa, Department of Health and Nutrition, Viçosa, Minas Gerais State, Brazil
[2]University of Alberta, Department of Agricultural, Food and Nutritional Science, Edmonton, AB, Canada

1. INTRODUCTION

1.1 Importance of Mango in the Global Market

Mango (*Mangifera indica* L.) is a member of the *Anarcadiaceae* family which comprises more than 70 genera. Historical records suggest that its cultivation as a fruit tree originated in India around 4000 years ago. In the early period of domestication, mango trees probably yielded small fruits, but folk selection of superior seedlings over many hundreds of years would have resulted in the production of larger fruits [1].

Before 1970, mangoes were little known to consumers outside the tropics and the trade involving fresh fruit was non-existent. There was, nevertheless, in the subsequent years, a rapid expansion of mango production into non-traditional areas and the mango trade became well established as fresh fruit and processed products [2].

With a growing world production, the mango represents one of the most important tropical fruits and is produced worldwide. Mango production is, however, quite concentrated, since Asia accounts for approximately 77% of global mango production, and America and Africa account for the remaining 23% [3].

The mango is an important fruit for human nutrition in several parts of the world. It is a tropical fruit widely accepted by consumers throughout the world for its succulence, sweet taste and exotic flavor, being called the 'king of fruits' [4]. Mango flesh is consumed in varied forms in both ripe and unripe stages. It is mostly eaten fresh, but a vast range of processed foods and drinks can be prepared, such as pickles, beverages, vinegar, chutneys, and desserts, as well as dessert flavoring and meat tenderizer.

Along with the trade expansion of fresh mangoes, there has been an increase in world demand for processed mango products [5]. Fruit processing is one way to reduce losses at peak harvest periods and to maximize the fruit's great potential through varied products, including juices, nectar, flesh, and others. Besides, since the mango is a fruit with high nutritive value, it is quite advantageous from the human health standpoint to enable fruit supply throughout the year via processed products. Mango processing also allows the use of less appealing varieties which cannot be sold on the fresh fruit market. Not only can the mango flesh be used for human food, but the waste originating from mango processing is a source of both macro- and micronutrients.

This chapter reviews data from studies on nutrient and non-nutrient phytochemicals in mangoes, focusing on their contents, biological action, and antioxidant activity.

1.2 Mango as a Fruit with High Functional Potential

The concept of a healthy diet has changed over the years. It was believed that a healthy diet provided all nutrients at adequate levels. Currently, besides supplying nutrients at adequate quantities and quality, it is believed that a healthy diet should have additional attributes, contributing to protection against diseases. Such protection is achieved by the presence of bioactive compounds contained in 'functional foods,' which are defined as 'a food that may provide a health benefit beyond the traditional nutrients it contains' [6].

Despite these divergent concepts, there are opinions that a functional food can be a natural food [6]. Within this concept, the mango can be included in the category of functional foods, since it provides the human diet with macro- and micronutrients and contains a large pool of bioactive compounds that are relevant to improving health and reducing the risk of disease. Furthermore, other parts of mango are also rich in bioactive compounds and nutrients, and could be exploited as nutraceuticals or active ingredients in the provision industry.

1.2.1 Macronutrients

Macronutrient composition of mango flesh seems to differ very little among varieties. A study carried out in our laboratory evaluated the composition of four mango varieties showing that it contains low levels of lipids and proteins and approximately 15% of total carbohydrate. Like most fruits, mango flesh contributes little to the caloric supply of a diet (Table 34.1).

Agricultural residues of mango are also a source of nutrients. The macronutrient content of flour obtained from mango kernels (variety Ikanekpo, Nigeria) presented the following composition per kilogram: protein (66.1 g), fat (94.0 g), fiber (28.0 g), and starch (500.0 g) [7]. Although mango seed kernels have low protein contents, the composition of essential amino acids indicates a good quality protein. The pattern in limiting amino acids (methionine, cystine, isoleucine, and valine) seems to differ among cultivars [7,8]. Investigations on the compositional quality of mango seed kernel of Egyptian varieties (Zebda, Balady, and Succary) revealed all essential amino acids to be present at higher levels than those of FAO reference protein [9].

TABLE 34.1 Chemical Composition[a] of Mango Flesh from Four Varieties (g/100 g Fresh Weight)

Variety	Moisture	Protein	Lipid	Total Carbohydrate	Ash	Kcal(Kjoule)
Haden	83.61	0.64	0.15	15.31	0.29	65.15(272.77)
Tommy Atkins	84.38	0.55	0.07	14.67	0.29	61.51(257.53)
Palmer	81.96	0.59	0.09	17.02	0.34	71.25(298.31)
Ubá	83.17	0.50	0.14	15.87	0.32	66.74(279.43)

[a]Ribeiro, S. M. R. (2006). Mango (*Mangifera indica* L.) antioxidant potential: characterization and evaluation. Thesis (Doctor Science). Department of Molecular Biology and Biochemistry. Federal University of Viçosa, Brazil.

However, it needs to be noted that the presence of tannins [10] can reduce the biological value of the protein.

Mango kernel has fat contents ranging from 6 to 12% on a dry matter basis, and the profile of fatty acids shows high levels of stearic and oleic acids [9,11] with physical properties adequate for use by the food industry. Mango seed fat has been approved by European Union authorities as a cocoa butter substitute.

There are few studies on the content and quality of dietary fiber in mango flesh. It is likely that there are significant differences in the amounts and quality of fibers among mango varieties, since some varieties contain much higher amounts of fiber than others. This is one of the main characteristics influencing consumer preference for fresh consumption varieties, as low-fiber or fiberless flesh is mostly preferred. Keitt mango growing in Florida, in unripe and ripe states, had total dietary fiber contents of 1.6 and 1.4 g/100 g fruit, respectively, and a large proportion consisted of pectin [12]. Some studies have focused on the analysis of mango peel, because this can be considered a source of dietary fiber of excellent quality. Peels of Haden variety contain high amounts of soluble (281 g/kg of dry matter) and insoluble fiber (434 g/kg) [13], and a large fraction of the soluble fiber is pectin [14,15].

1.2.2 Micronutrients

Various studies have demonstrated that the mineral content in mango flesh is not high [16,17] and, therefore, mango flesh is not considered a good dietary source of these nutrients. In contrast, fiber from mango peel (variety Haden) had high contents of some minerals that are important for human nutrition, including calcium (4445 mg/kg), potassium (2910 mg/kg), magnesium (950 mg/kg), iron (175 mg/kg), and zinc (32.5 mg/kg) [13].

Mango flesh contains provitamin A carotenoids, with β-carotene being the most abundant carotenoid in many varieties [18,19]. This attributes an additional nutritive value to the fruit because β-carotene is the carotenoid that possesses the highest provitamin A activity. Therefore, mango consumption is very important for some populations in tropical regions, where the deficiency of vitamin A constitutes a public health problem [20]. A study using an *in vitro* model indicated that there are varietal differences in the content and bioavailability of β-carotene from mango, and the ingestion of flesh blended with milk is beneficial, increasing the bioavailability [21]. However, *in vivo* studies would be more appropriate to clarify this question.

Mango flesh contains ascorbic and dehydro-ascorbic acids [22,23], and the fruit can be considered an excellent source of vitamin C for the human diet, for two reasons: first, the flesh, the most commonly consumed form, provides favorable conditions for preservation of ascorbic acid when compared with other fruits that are predominantly consumed as juices or with cooked vegetables; second, organic acids, mainly citric and malic acids, can stabilize ascorbic acid through metal chelation [24]. In addition, phenolic compounds also present in mango flesh provide protection against ascorbate oxidation [25].

Therefore, the daily consumption of mango fruits by population groups of all life stages should be increased to meet the recommended dietary requirements of vitamins A and C.

2. BIOACTIVE COMPOUNDS IN MANGO

Apart from being important as a food, mango fruits as well as other parts of the plant are a source of bioactive compounds with potential health-promoting activity (Table 34.2).

All parts of mango trees have been used in traditional South Asian medicine: kernels, flowers, leaves, gum, bark, and peel. Diseases commonly treated with herbal remedies obtained

TABLE 34.2   Bioactive Compounds in Mango

Bioactive Compound	Part
Ascorbic and dehydroascorbic acids	Flesh
β-carotene	Flesh
Other carotenoids: ζ-carotene, mutato-chrome, α-cryptoxanthin, viola-xanthin, luteoxanthin, mutatoxanthin, auroxanthin	Flesh
Polyphenols: mangiferin, isomangi-ferin, homomangiferin, quercetin, kaempferol, anthocyanins	Flesh, bark, seed, peel, leaves, twigs
Phenolic acids: gallic, protocatechuic, ferulic, caffeic, coumaric, ellagic, 4-caffeoylquinic acids	Flesh, peel, seeds, kernel
Other phenols: Alk(en)ylresorcinols	Peel, sap
Fiber	Peel, seed, flesh
Terpenoids: α-pinene, β-pinene, β-myrcene, limonene, *cis*-ocimene, *trans*-ocimene, terpinene, α-guaiene, camphene, fenchene, α-humulene and others (lactones, aldehydes, acids, sesquiterpenes, esters and aliphatic alcohols)	Flesh, peel, sap
Antioxidant minerals: potassium, copper, zinc, manganese, iron, selenium	Flesh, peel, seed, stem bark

from parts of the mango tree include dysentery, diarrhea, urinary tract inflammation, rheumatism, and diphtheria. A number of these uses are supported by scientific evidence [26]. Vimang®, an extract obtained from the stem bark of mango trees, shows *in vitro* and *in vivo* anti-inflammatory and antioxidant activities and is currently produced on an industrial scale in Cuba [27]. Aqueous decoctions of mango flowers showed potential gastroprotective and ulcer-healing properties in the acute and subacute models of induced ulcer in mice and rats [28]. Extracts of mango leaves showed moderate larvicidal activity in experiments with *Culex quinquefasciatus* Say, the main mosquito vector of lymphatic filariasis, which is widely distributed in tropical regions [29]. The natural product used in Cuban traditional medicine, obtained from mango bark (MSBE), modulated the P450

enzymes in cultured cells, demonstrating inhibition of CYP1A2 and 2E1. The authors postulate that, by this mechanism, chemopreventive properties could be attributed to the natural product, since both P450 enzymes are involved in the bioactivation of mutagens and carcinogens [30]. One study has indicated that peel extracts from mango had thyroid stimulatory effects on animals with induced hypothyroidism, and reduced lipid peroxidation in heart, liver, and kidney tissues [31]. Extracts from mango kernel showed superoxide anion scavenging activity in a cell-free system [32], suggesting one possible bioactivity by antioxidant mechanism. Mangiferin, a xanthone present in mango, when administered at a dose of 100 mg/kg/d to rats subjected to experimental periodontitis, demonstrated an anti-inflammatory property, accelerating the processes of repairing and healing injured tissues [33].

Bioactive compounds present in fruits have attracted attention from both the consumer and the scientific community, considering strong epidemiological evidences that show the benefits of fruit intake in human disease prevention [34,35]. Mangoes contain several constituents which are included in the category of bioactive compounds with a great potential to modulate risk factors of diseases.

2.1 Ascorbic and Dehydroascorbic Acids

The term 'vitamin C' comprises the sum of ascorbic acid (AA) and dehydroascorbic acid (ADA) because ADA can be converted to AA in humans [36]. Similar to other fruits, mangoes differ in their ascorbic acid content because of genotype variations, climatic factors, agricultural practices, and ripening stage [37]. Literature reports indicate great variation in ascorbic acid contents, ranging from 9.79 to 186 mg per 100 g of mango flesh [22,23,38–42]. Besides other factors, such variation can be partially attributed to ripening stage, since ascorbic acid content declines during the maturation process. Therefore, products made from unripe or half-ripe mangoes usually have higher ascorbic acid content than those produced from ripe fruits. Apart from differences in the raw material, the large variations observed for vitamin C may be attributed also to differences in sample preparation and the analytical methods used for quantification. Another possible reason for the inconsistent reports is the fact that ADA contents were not always considered in previous investigations.

In addition to its function as scurvy preventing agent, ascorbic acid is considered a potent water-soluble antioxidant because the molecule can donate a hydrogen atom and form a relatively stable ascorbyl free radical, with a half-life of approximately 10^{-5} seconds [43]. The antioxidant effect of ascorbate is related to its capacity to remove reactive oxygen species (ROS) by reacting with superoxide radicals, hydrogen peroxide, hydroxyl radicals, and singlet oxygen [44]. It also removes reactive nitrogen species (RNS), preventing nitration reactions [45]. It is believed that ascorbate participates in the regeneration of tocopherol, and this explains the antioxidant synergism between the two nutrients. It is considered an antioxidant protector of cell soluble compartments, and helps to maintain tocopherol in its reduced form, being, therefore, considered a vitamin E 'regenerator.' By participating in reactions as reducing agent, ascorbate is oxidized, forming dehydroascorbate, which can be regenerated by specific reducing systems in the organism [46].

Intake levels of ascorbic acid can modify risk factors of cardiovascular diseases and cancer. A study by Block et al. has pointed out that the plasma level of vitamin C is inversely related to blood pressure in young black and white women [47], reinforcing previous evidence that a low vitamin C status may increase the risk of mortality from cancer and cardiovascular disease [48]. Evidence based on the prospective study indicated that higher plasma vitamin C levels are inversely associated with gastric cancer [49], and studies on the molecular mechanism of biological action suggest the role of vitamin C in the protection of the DNA against mutations [50].

2.2 Carotenoids

Mango flesh contains a wide pattern of carotenoids, including β-carotene, violaxanthin, cryptoxanthin, neoxanthin, luteoxanthin, and zeaxanthin [19,22,23,51–53]. There is also a wide variation in β-carotene (550 ~3210 μg/100 g) and total carotenoid (1159 ~3000 mg/100 g) contents in flesh of different mango varieties [19,21–23,51–57]. However, since many of these studies do not characterize the maturity of the fruit tested, it cannot be concluded whether the differences are attributable to varietal characteristics or related to other factors, including ripening stage. Incidence of sunlight may induce carotenogenesis as a fruit defense mechanism, protecting it against UV radiation injuries [58]. Mango production on a commercial scale uses sophisticated management to ensure fruit quality characteristics and some of these strategies may influence the content of bioactive compounds. For example, a CaO solution is used for fruit protection against injuries from excess sunlight. In theory, this practice can influence the fruit physiological response, decrease carotenogenesis, and thus reduce the carotenoid content in the fruit. There are indications of varietal differences in amounts of each minor carotenoid pattern [21].

Many hundreds of carotenoids are found in nature, but the five main ones found in human tissues are β-carotene, lutein, lycopene, β-cryptoxanthin, and α-carotene [59]. Carotenoids are important not only because of their provitamin A activity but also because of a number of other actions in biological systems. In animal tissues, because of their lipophilicity, carotenoids are distributed in apolar compartments, including membranes, lipoprotein particles (LDL and HDL), and serum, bound to a transport protein [60].

Carotenoids are efficient as ROS quenchers, specifically singlet oxygen and peroxyl radicals. The antioxidant properties of carotenoids explain their protective effects against diseases related to oxidative stress [61]. However, gene modulation seems to be the most relevant biological mechanism of carotenoid action in some pathological processes. Carotenoids affect gene expression regulation. Nuclear retinoic acid receptors bind to retinoic acid-responsive elements [62], and this event results in the expression of specific genes. Connexins are gene products that have been extensively investigated, since they increase cell-to-cell communication and thus affect cell proliferation. Cancer cells communicate poorly with normal cells and proliferate abnormally [63]. These carotenoid–gene interactions seem to explain part of the associations between high carotenoid intake and lower cancer incidence observed in epidemiological studies. These studies suggest a positive correlation between higher intake and tissue concentrations of carotenoids and lower risk of certain diseases, such as cardiovascular diseases, some types of cancer, osteoporosis, infectious diseases, cataract, and others [64]. One study with older adults has demonstrated that higher total plasma carotenoids were associated with a significantly lower risk of developing severe walking disability [65].

2.3 Phenolic Compounds

The presence of the phenolic compounds glucogallin and gallotanin in mango flesh and seeds, and mangiferin, isomangiferin, homomangiferin, fisetin, quercetin, isoquercitrin, astragalin, gallic acid, methyl gallate, digallic acid, β-glucogallin, and gallotanin in leaves, twigs, seeds, and fruits of 20 local varieties was described already in 1971 [66].

It has been reported that the total content of phenolic compounds in mango flesh ranges from 9.0 to 208.0 mg/100 g [22,42]. Peels and

kernels contain large amounts of extractable phenolics, and there are varietal differences in their contents [16]. A wide pattern of phenolic compounds has been described in the flesh, peels, and kernels of mangoes (Table 34.2) [67–72]. In particular, flavonols and xanthones have been identified and quantified (Table 34.2) [70–72]. The flavonols (quercetin, kaempferol, and rhamnetin) are present mostly as O-glycosides, whereas mangiferin is a C-glycoside and occurs both in its non-esterified form and conjugated with gallic acid.

There are varietal differences in the profile and content of flavonols and xanthones in mangoes originating from several countries [71], whereas the qualitative profile is relatively conservative. It seems that phenolic compounds are the main antioxidant constituents with greater variation in mango, because they are plant secondary metabolites and their contents differ not only by genetic characteristics and maturity stage but also agricultural practices. Studies with mango varieties demonstrating significant differences in the phenolic pattern raised the hypothesis that growing mangoes using simple management practices and pesticide-free technology enables the natural plant and fruit defense against environmental adversities. It results in the increased synthesis of secondary metabolites (phenolic compounds), and therefore improves functional properties of the fruits [70].

Quercetin and kaempferol belong to the flavonoid class and have been receiving great attention as bioactive compounds for many years. A wide range of biological activities, including antibacterial, antithrombotic, vasodilatory, anti-inflammatory, and anticarcinogenic effects mediated by different mechanisms, are associated with flavonoids [73]. Inverse relations were found between the dietary intake of some flavonoids and incidence of several chronic diseases [74,75].

Quercetin (3,3',4',5,7-pentahydroxyflavone) is a potent antioxidant [76] because its structure contains a double bond in the C-ring and the 4-oxo group, which are structural determinants that enhance its action as an antioxidant [77]. Studies have shown biological effects of quercetin on inhibition of protein kinases, DNA topoisomerases, and regulation of gene expression [78]. There is consistent evidence from these studies that quercetin may reduce the risk of cancer. Pretreatment of primary hippocampal cultures with quercetin attenuated β-amyloid induced cytotoxicity, protein oxidation, lipid peroxidation, and apoptosis, suggesting that quercetin may provide a promising approach for treatment of Alzheimer's disease and other oxidative stress-related neurodegenerative diseases [79].

Kaempferol (3,5,7,4'-tetrahydroxyflavone) has a chemical structure with various hydroxyl substitutions, transforming the molecule into a potent antioxidant [80]. Associations among kaempferol intake and reducing risk factors of chronic diseases have already been suggested [74]. A population study showed that higher intakes of kaempferol tended to lower ischemic heart disease mortality, and the incidence of brain vascular disease leading to hospitalization or death was also diminished [74].

Biological actions of mangiferin (1,3,6,7-tetrahydroxyxanthone-2-glucopyranoside) have been exhaustively studied, and several investigations have confirmed the bioactivity of xanthones. Studies conducted with VIMANG®, a formulation manufactured in Cuba that contains mangiferin as the main active ingredient, demonstrated protective effects on hepatic and brain tissues of mice against induced oxidative stress [81]. The inhibitory effect of mangiferin on carcinogenesis in rats [82], and against induced oxidative stress in cardiac and renal tissues of rats was also demonstrated [83]. Mangiferin showed antioxidant activity by eliminating the superoxide radical in *in vitro* tests, in which 100 μM of mangiferin was equivalent to the activity of 1 U/mL of superoxide dismutase, besides having other pharmacological

effects modulating gene expression related to the inflammatory response [84].

Studies on the immunomodulatory activity of mangiferin showed that xanthone modulates the expression of various genes critical for apoptosis regulation, viral replication, tumorigenesis, inflammation, and autoimmune diseases. These results suggest its possible value in the treatment of inflammatory diseases and/or cancer [85]. Mangiferin protected human lymphocytes from DNA lesions when exposed to gamma radiation, raising the possibility of its use in patients undergoing radiotherapy or people occupationally exposed to radiation [86].

Evidence now indicates that mangiferin is a promising chemopreventive [87], with bioactivity involving antioxidant action [88] and modulation of gene expression [89,90]. In another study, mangiferin was shown to provide protection against gastric injury induced by ethanol and indomethacin [91]. It must be emphasized that mangiferin is present in peels, bark, and leaves in higher concentrations than in the flesh and that xanthone derivatives have not been detected in some varieties [70]. Hence, the hypothesis has been raised that mangoes growing in more natural conditions may contain higher xanthone levels than others subjected to treatment against physical and microbial injuries. Singh [92] confirmed that the mangiferin content was higher in cultivars resistant to malformation syndrome associated with abnormal inflorescence and suggested that xanthone could be a potent inducer of plant natural defense.

Alk(en)ylresorcinols are phenolic lipids that are present in mango peels and sap. They have been demonstrated to possess antifungal and anti-inflammatory activities [93–95].

2.4 Terpenoids

Terpenoids are compounds belonging to the prenyl lipids class and represent probably the most widespread group of natural products. Monoterpenes and diterpenes, the main components of essential oils, can act as allelopathic agents, as attractants in plant–plant or plant–pathogen/herbivore interactions, or as repellents. In addition to carotenoids, which are tetraterpenoids, mangoes contain mono-, di- and triterpenoids, including ocimene, myrcene or limonene, terpinolene, and carene [96–99]. Several factors can affect the biosynthesis of aroma volatile compounds in mango [100–105].

There is currently a good prospect for exploiting the biological activity of terpenoids, as previous studies indicated that monoterpenes inhibited cell growth, cell cycle progression, and cyclin D1 gene expression in human cancer cell lines [106]. Monoterpenes would appear to act through multiple mechanisms in cancer chemoprevention and chemotherapy [107]. Mono- and diterpenes are effective antioxidants and studies have demonstrated their *in vitro* antioxidant activity [108,109]. Another study showed that combinations of rutin with terpinene can have synergistic effects by acting as hydrophilic and lipophilic antioxidants [110]. Studies have demonstrated terpenoids to act as chemopreventive agents [111]. Lupeol, a triterpene present in mango, has shown apoptogenic activity in mouse prostate by early increase of reactive oxygen species [112,113]. The anti-urolithic effect of lupeol and lupeol linoleate has been demonstrated in experimental hyperoxaluria [114]. These evidences indicated that terpenoids in mangoes have biological activity, contributing to raise the functional potential of the fruit, and the need for further studies to investigate their nutraceutical effects.

2.5 Fiber

Studies on mango flesh showed that a high proportion of the fiber fraction consists of pectin and its content in the peel is also quite high ([14,15]. Pectin is not hydrolyzed in humans by endogenous digestive enzymes but is fermented by the colon microflora [115], thus showing a prebiotic effect. Its biological activities have attracted interest in the last decades, because of their postulated positive effects on health such as cholesterol-lowering [116], cancer-preventing [117], and blood glucose-regulating properties [118,119]. Studies have since demonstrated that soluble fiber can stimulate protein turnover in intestines and liver [120].

The presence of pectin in mango flesh adds to it a functional attribute, and this points to the need for further investigation of fiber in mango flesh from different varieties. Fiber extracted from fruit agro-industrial residues can be used in the industry as a food ingredient [14,121,122]. Although the pectin extracted from mango peel had net yield similar to apple pectin, its low content of anhydro-galacturonic acid leads to a low jellifying capacity [15].

2.6 Antioxidant Minerals

In the area of human nutrition, selenium, copper, zinc, iron, and manganese are included in the group of antioxidant minerals and their deficiency in the body affects the activity of enzymes involved in protection against oxidative stress. Thus, copper, zinc, manganese, iron, and selenium have been considered essential minerals for the optimization of the antioxidant enzyme response.

Compared with other foods, mango flesh contains lower levels of antioxidant minerals such as copper, iron, manganese, and zinc [16,17]. Nevertheless, the mineral content should not be neglected because mango consumption is associated with the intake of numerous antioxidants acting synergistically. Stem bark of mango trees grown in Cuba presented high concentrations of copper, iron, selenium, and zinc. The authors suggested that these elements contribute to the antioxidant activity of this product [123].

3. TOTAL ANTIOXIDANT CAPACITY OF MANGOES

Bioactive compounds can protect against diseases via several mechanisms, but it is believed that the antioxidant activity is extremely important for protection against diseases related to oxidative stress [124]. Mango contains at least three classes of compounds, i.e. ascorbic acid, carotenoids, and phenolic components, that can support the antioxidant defense in humans. Despite the low content of the minerals copper, zinc, manganese, and iron in mango flesh, their importance should not be disregarded, as the fruit intake provides a set of antioxidants that may offer protection to the organism in a synergistic way.

The *in vivo* action of antioxidants demonstrates the synergism phenomenon, which is a cooperative action among several substances with antioxidant properties to protect oxidation targets [46,125]. Synergism occurs by co-antioxidant effect, involves more than one antioxidant with different reduction potentials and polarities participating in redox reactions in a system under pro-oxidant conditions, until a nonreactive product is formed, stabilizing the medium. Considering that mango contains this group of compounds, it can be assumed that the mango is a fruit with high antioxidant potential. Evidence suggests that a single antioxidant cannot replace a combination of antioxidants. Thus, a powerful antioxidant defense can be achieved in the biological media through mango consumption.

Total antioxidant capacity of foods has been suggested as a tool for investigating the health effects of antioxidants in mixed diets [126], and there is evidence that food selection based on total antioxidant capacity can modify antioxidant intake, system inflammation, and liver

function without altering markers of oxidative stress [127].

A comparative study carried out in our laboratory to investigate the antioxidant potential of four mango varieties indicated that the flesh extracts showed capacity to scavenge diphenylpicrylhydrazyl (DPPH) radicals in a dose-dependent manner with scores similar to or above the antioxidant standards gallic acid, butylated hydroxyanisole (BHA), and catechin at 100 ppm (Figure 34.1 A and B). There were significant differences in the antioxidant activity of the mango flesh extracts among varieties, for both reducing power and radical scavenging activity (RSA) tests (Figure 34.1 A and B). In all tested concentrations, the extract of mango variety Ubá showed scavenging activity significantly higher than the others, because the antioxidant contents in fruits of variety Ubá were higher [22], contributing to the positive effect in both antioxidant tests.

Other studies have demonstrated the total antioxidant capacity of mangoes [12,14,23,128]. During mango processing, the peels emerge as a byproduct and are usually discarded as waste. A number of valuable antioxidant compounds are contained in this residue [129]. For this reason, there are studies addressing the use of natural antioxidants obtained from mango agro-industry residues as food preservers in substitution to artificial antioxidants [130,131].

Furthering the research carried out in our laboratory, an *in vivo* study was performed to measure the antioxidant potential of Ubá mangoes because this variety presented high antioxidant activity in previous *in vitro* tests. Biological assays were carried out with Wistar rats in an induced oxidative stress model. The animals received acetaminophen, in a dose sufficiently high to induce oxidative stress in the liver as demonstrated in another study [132]. We performed an experiment supplementing the animals' diet with lyophilized mango flesh at 3%, which is a concentration equivalent to

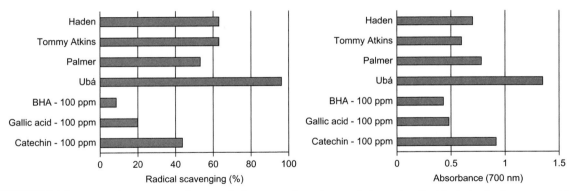

FIGURE 34.1 Antioxidant activity of mango flesh extracts from four varieties and antioxidant standards, measured by radical scavenging[1] (A) and reducing power[2]z (B) tests.

[1]Determined according to Blois, M. S. (1958). Antioxidant determinations by use of a stable free radical. *Nature* **181**, 3–14.

[2]According to Oyaizu, M. (1986). Studies on products of the browning reaction. Antioxidative activities of browning reaction products prepared from glucosamine. *Eiyogaku Zasshi* **44**, 307–315.

human consumption level. After having induced the oxidative stress, the animals were fed the diet containing mango at 3% in the subsequent 24 hours. The animals were then euthanized, and blood and liver were collected for analysis. A hepatoprotective effect was demonstrated with reduction of serum aminotransferases, mediated by antioxidant mechanism with decreased lipid peroxidation in liver homogenates. This finding confirmed that at concentrations similar to usual human consumption, mangoes provided protection to hepatic tissues against induced oxidative injury. These studies have demonstrated the potential bioactivity of compounds in mango flesh involving redox mechanisms.

4. SUMMARY

Mango is a fruit with high nutritional value, supplying the human diet with calories, fiber, vitamins, and minerals. Flesh and agro-industrial residues (peels and seeds) of mangoes contain several bioactive compounds, comprising nutrient and non-nutrient substances with biological properties that act mainly via redox mechanisms. Compounds contained in mango flesh can act as biological antioxidants maximizing the human antioxidant defense. Additive and synergistic effects of bioactive compounds from mangoes suggest that the fruit has great potential to improve health and reduce the risk of chronic diseases.

Despite the numerous bioactive compounds in mangoes, which may promote benefits to human health, the potential for allergenicity of the fruit has been shown. Conventional technological processing of mango into flesh-containing products does not allow complete elimination of the allergenic potency [133].

All mango varieties can supply the diet with nutrients, but considering that the contents of bioactive compounds are influenced by several factors, it was assumed that population groups with the same mango intake may be ingesting such compounds at different levels, not guaranteeing a comparable modulation potential of risk factors of diseases. Figure 34.2 summarizes some potential mango benefits for

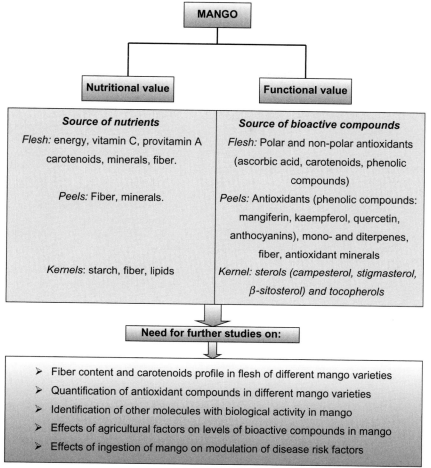

FIGURE 34.2 Nutrients and non-nutrient compounds in mangoes with health-promoting activity, and perspectives for future studies.

human health to be exploited, considering current knowledge.

References

1. Mukherjee, S. K. (1997). Introduction: Botany and importance. In R. E. Litz (Ed.), *The mango, production and uses*. Oxon, UK: CAB International.
2. Litz, R. E. (Ed.), (1997). *The mango, production and uses*, Oxon, UK: CAB International.
3. FAOSTAT. (2007). FAO Statistical Database – Agriculture. <http://faostat.fao.org/>. Accessed November 8, 2007.
4. Ramteke, R. S., Vilajayalaskshmi, M. R., & Eipeson, W. E. (1999). Processing and value addition to mangoes. *Indian Food Industry, 18*, 155–163.
5. Sauco, G. V. (2004). Mango production and world market: Current situation and future prospects. *Acta Horticulturae, 645*, 107–116.
6. Scrinis, G. (2008). Functional foods or functionally marketed foods? A critique of, and alternative to, the category of 'functional foods'. *Public Health Nutrition, 11*, 541–545.
7. Arogba, S. S. (1997). Physical, chemical and functional properties of Nigerian mango (*Mangifera indica*) kernel and its processed flour. *Journal of the Science of Food and Agriculture, 73*, 321–328.

8. Dhingra, S., & Kapoor, A. C. (1985). Nutritive value of mango seed kernel. *Journal of the Science of Food and Agriculture, 36*, 752–756.

9. Abdalla, A. E. M., Darwish, S. M., Ayad, E. H. E., & El-Hamahmy, R. M. (2007). Egyptian mango by-product 1. compositional quality of mango seed kernel. *Food Chemistry, 103*, 1134–1140.

10. Garg, N., & Tandon, D. K. (1997). Amylase activity of *A. oryzae* grown on mango kernel after certain treatments and aeration. *Indian Food Packer, 51*, 26–29.

11. Rukmini, C., & Vijayaraghavan, M. (1984). Nutritional and toxicological evaluation of mango kernel oil. *Journal of the American Oil Chemists'Society, 61*, 789–792.

12. Mahattanatawee, K., Manthey, J. A., Luzio, G., Talcott, S. T., Goodner, K., & Baldwin, E. A. (2006). Total antioxidant activity and fiber content of select Florida-grown tropical fruits. *Journal of Agricultural and Food Chemistry, 54*, 7355–7363.

13. Larrauri, J. A., Rupérez, P., Borroto, B., & Saura-Calixto, F. (1996). Mango peels as a new tropical fibre: Preparation and characterization. *Lebensmittel-Wissenschaft und –Technologie, 29*, 729–733.

14. Berardini, N., Knödler, M., Schieber, A., & Carle, R. (2005). Utilization of mango peels as a source of pectin and polyphenolics. *Innovative Food Science and Emerging Technologies, 6*, 443–453.

15. Sirisakulwat, S., Nagel, A., Sruamsiri, P., Carle, R., & Neidhart, S. (2008). Yield and quality of pectins extractable from the peels of Thai mango cultivars depending on fruit ripeness. *Journal of Agricultural and Food Chemistry, 56*, 10727–10738.

16. Ribeiro, S. M. R. (2006). Mango (*Mangifera indica*, L.) Antioxidant Potential of mangoes: Characterization and Evaluation. Thesis (Doctor Science) – Federal University of Viçosa, Viçosa, Brazil..

17. Leterme, P., Buldgen, A., Estrada, F., & Londoño, A. M. (2006). Mineral content of tropical fruits and unconventional foods of the Andes and the rain forest of Colombia. *Food Chemistry, 95*, 644–652.

18. Mercandante, A. Z., & Rodriguez-Amaya, D. B. (1998). Effects of ripening, cultivar differences, and processing on the carotenoid composition of mango. *Journal of Agricultural and Food Chemistry, 46*, 128–130.

19. Ornelas-Paz, J. J., Yahia, E. M., & Gardea-Bejar, A. (2007). Identification and quantification of xanthophylls esters, carotenes, and tocopherols in the fruit of seven Mexican mango cultivars by liquid chromatography-atmospheric pressure chemical ionization-time-of-flight mass spectrometry [LC-(APcl$^+$)-MS]. *Journal of Agricultural and Food Chemistry, 55*, 6628–6635.

20. WHO (World Health Organization) (1995). Global prevalence of vitamin A deficiency: micronutrient deficiency information system; Working Paper 2. WHO, Geneva (Document WHO/NUT/95.3).

21. Veda, S., Platel, K., & Srinivasan, K. (2007). Varietal differences in the bioaccessibility of β-carotene from mango (*Mangifera indica*) and papaya (*Carica papaya*) fruits. *Journal of Agricultural and Food Chemistry, 55*, 7931–7935.

22. Ribeiro, S. M. R., Queiroz, J. H., Lopes, M. E. L. R., Milagres, F. C., & Pinheiro-Sant'Ana, H. M. (2007). Antioxidant in mango (*Mangifera indica* L) pulp. *Plant Foods for Human Nutrition, 62*, 13–17.

23. Corral-Aguayo, R. D., Yahia, E. M., Carrilo-Lopez, A., & González-Aguilar, G. (2008). Correlation between some nutritional components and the total antioxidant capacity measured with six different assays in eight horticultural crops. *Journal of Agricultural and Food Chemistry, 56*, 10498–10504.

24. Nagy, S. (1980). Vitamin C contents of citrus and their products: A review. *Journal of Agricultural and Food Chemistry, 28*, 8–18.

25. Miller, N., & Rice-Evans, C. A. (1997). The relative contributions of ascorbic acid and phenolic antioxidants to the antioxidant activity of orange and apple fruit juices and blackcurrant drink. *Food Chemistry, 60*, 331–337.

26. Ross, I. A. (2003). *Medicinal plants of the world: Chemical constituents, traditional medicinal uses.* pp. 315–328. Totowa, NJ: Humana Press Inc.

27. Pardo-Andreu, G. L., Dorta, D. J., Delgado, R., Cavalheiro, R. A., Santos, A. S., Vercesi, A. E., & Curti, C. (2006). Vimang (*Mangifera indica* L. extract) induces permeability transition in isolated mitochondria, closely reproducing the effect of mangiferin, Vimang's main component. *Chemico-Biological Interactions, 159*, 141–148.

28. Lima, Z. P., Severi, J. A., Pellizon, C. H., Brito, A. R. M. S., Solis, P. N., Cáceres, A., Girón, L. M., & Hiruma-Lima, C. A. (2006). Can the aqueous decoction of mango flowers be used as an antiulcer agent? *Journal of Ethnopharmacology, 106*, 29–37.

29. Rahuman, A. A., Bagavan, A., Kamaraj, C., Vadivelu, M., Zahir, A. A., Elango, G., & Pandiyan, G. (2008). Evaluation of indigenous plant extracts against larvae of *Culex quinquefasciatus* Saty (Diptera: Culicidae). *Parasitology Research, 104*, 637–643.

30. Rodeiro, I., Donato, M. T., Lahoz, A., González-LavautLaguna, A., Castell, J. V., Delgado, R., & Gómez-Lechón, M. J. (2008). Modulation of P450 enzymes by Cuban natural products rich in polyphenolic compounds in rat hepatocytes. *Chemico-Biological Interactions, 172*, 1–10.

31. Parmar, H. S., & Kar, A. (2008). Protective role of *Mangifera indica, Cucumis melo* and *Citrullus vulgaris*

peel extracts in chemically induced hypothyroidism. *Chemico-Biological Interactions, 177*, 254–258.

32. Saito, N. K., Kohno, M., Yoshizazaki, F., & Niwano, K. (2008). Extensive screening for edible herbal extracts with potent scavenging activity against superoxide anions. *Plant Foods for Human Nutrition, 63*, 65–70.

33. Carvalho, R. R., Pellizon, C. H., Justulin, L., Jr., Felisbino, S. L., Vilegas, W., Bruni, F., Lopes-Ferreira, M., & Hiruma-Lima, C. A. (2009). Effect of mangiferin on the development of periodontal disease: Involvement of lipoxin A$_4$, anti-chemotaxic action in leucocyte rolling. *Chemico-Biological Interactions, 179*, 344–350.

34. Block, G., Patterson, B., & Subar, A. (1992). Fruit, vegetables, and cancer prevention: A review of the epidemiological evidence. *Nutrition and Cancer, 18*, 1–29.

35. Ames, B. M., Shigenaga, M. K., & Hagwn, T. M. (1993). Oxidants, antioxidants and the degenerative diseases of aging. *Proceedings of the National Academy of Sciences of the United States of America, 90*, 7915–7922.

36. Linster, C. L., & Van Shaftingen, E. (2007). Vitamin C. Biosynthesis, recycling and degradation in mammals. *FEBS Journal, 274*, 1–22.

37. Lee, S. K., & Kader, A. (2000). Preharvest and postharvest factors influencing vitamin C content of horticultural crops. *Postharvest Biology and Technology, 20*, 207–220.

38. Franke, A. A., Custer, L. J., Araraki, C., & Murphy, S. P. (2004). Vitamin C and flavonoid levels of fruits and vegetables consumed in Hawaii. *Journal of Food Composition and Analysis, 17*, 1–35.

39. Nisperos-Carriedo, M. O., Buslig, B. S., & Shaw, P. E. (1992). Simultaneous detection of dehydroascorbic, ascorbic, and some organic acids in fruits and vegetables by HPLC. *Journal of Agricultural and Food Chemistry, 40*, 1127–1130.

40. Vinci, G., Botré, F., Mele, G., & Ruggieri, G. (1995). Ascorbic acid in exotic fruits: A liquid chromatographic investigation. *Food Chemistry, 53*, 211–214.

41. Reys, L. F., & Cisneros-Zevallos, L. (2007). Electron-beam ionizing radiation stress effects on mango fruit (*Mangifera indica* L.) antioxidant constituents before and during postharvest storage. *Journal of Agricultural and Food Chemistry, 55*, 6132–6139.

42. Gil, M. I., Aguayo, E., & Kader, A. A. (2006). Quality changes and nutrient retention in fresh-cut versus whole fruits during storage. *Journal of Agricultural and Food Chemistry, 54*, 4284–4296.

43. Buettner, G. R. (1993). The pecking order of free radicals and antioxidants: Lipid peroxidation, alpha-tocopherol, and ascorbate. *Archives of Biochemistry and Biophysics, 300*, 534–543.

44. Weber, P., Bendich, A., & Schalch, W. (1996). Vitamin C and human health – a review of a recent data relevant to requirements. *International Journal for Vitamin and Nutrition Research, 66*, 19–30.

45. Tannenbauen, S. R., Wishnok, J. S., & Leaf, C. D. (1991). Inhibition of nitrosamine formation by ascorbic acid. *American Journal of Clinical Nutrition, 53*(Suppl. 1), 2475–2505.

46. Nordberg, J., & Arnér, E. S. J. (2001). Reactive oxygen species antioxidants, and the mammalian thioredoxin system. *Free Radical Biology and Medicine, 31*, 1287–1312.

47. Block, G., Jensen, C. D., Norkus, E. P., Hudes, M., & Crawford, P. B. (2008). Vitamin C in plasma is inversely related to blood pressure and change in blood pressure during the previous year in young black and white women. *Nutrition Journal, 7*, 37–39.

48. Loria, C. M., Klag, M. J., Cauldield, L. E., & Whelton, P. K. (2000). Low vitamin C status may increase the risk of mortality from cancer and cardiovascular disease. *American Journal of Clinical Nutrition, 72*, 139–145.

49. Jenabi, M., Riboli, E., Ferrari, P., Sabate, J., Slimani, N., & Nora, T., et al. (2006). Plasma and dietary vitamin C levels and risk of gastric cancer in the European Prospective Investigation into Cancer and Nutrition (EPIC-EURGAST). *Carcinogenesis, 27*, 2250–2257.

50. Halliwell, B. (2001). Vitamin C and genomic stability. *Mutation Research, 475*, 29–35.

51. Godoy, H. T., & Rodriguez-Amaya, D. B. (1989). Carotenoid composition of commercial mangoes from Brazil. *Lebensmittel-Wissenschaft und –Technologie, 22*, 100–103.

52. Chen, J. P., Tai, C. Y., & Chen, B. H. (2004). Improved liquid chromatographic method for determination of carotenoids in Taiwanese mango (*Mangifera indica*). *Journal of Chromatography A, 1054*, 261–268.

53. Pott, I., Breithaupt, D. E., & Carle, R. (2003). Detection of unusual carotenoid esters in fresh mango (*Mangifera indica* L. cv. 'Kent'). *Phytochemistry, 64*, 825–829.

54. Mercadante, A. Z., Rodriguez-Amaya, D. B., & Britton, G. (1997). HPLC and mass spectrometric analysis of carotenoids from mango. *Journal of Agricultural and Food Chemistry, 45*, 120–123.

55. Hulshof, P. J. M., Xu, C., Van De Bovenkamp, P., Muhital& West, C. E. (1997). Application of a validated method for determination of provitamin A carotenoids in Indonesian foods of different maturity and origin. *Journal of Agricultural and Food Chemistry, 45*, 1147–1179.

56. Ben-Amotz, A., & Fishler, R. (1998). Analysis of carotenoids with emphasis on 9-*cis*-β-carotene in vegetables and fruits commonly consumed in Israel. *Food Chemistry, 62*, 515–520.

C. ACTIONS OF INDIVIDUAL FRUITS IN DISEASE AND CANCER PREVENTION AND TREATMENT

57. Setiawan, B., Sulaeman, A., Giraud, D. W., & Driskell, J. A. (2001). Carotenoid content of selected Indonesian fruits. *Journal of Food Composition and Analysis, 14*, 169–176.

58. Lutz, C., Navakoudis, E., Seidlitz, H. K., & Kotzabasis, K. (2005). Stimulated solar irradiation with enhanced UV-B plastid-and thylakoid-associated polyamine changes for UV-B protection. *Biochimica Biophysica Acta, 1710*, 24–33.

59. Thurnham, D. I. (1994). Carotenoids: Functions and fallacies. *The Proceedings of the Nutrition Society, 53*, 77–87.

60. IOM (1998). Food and Nutrition Board. Dietary reference intakes: proposed definition and plan for review of dietary antioxidants and related compounds. Published: August 5, 1998. Available at: www.nap.edu. Accessed: May 15, 2002..

61. Krinsky, N. (2001). Carotenoids as antioxidants. *Nutrition, 17*, 815–817.

62. Chambon, P. A. (1996). A decade of molecular biology of retinoic acid receptors. *FASEB Journal, 10*, 940–943.

63. Burri, J. B. (2000). Carotenoids and gene expression. *Nutrition, 16*, 7–8.

64. Rao, A. V., & Rao, L. G. (2007). Carotenoids and human health. *Pharmacological Research, 55*, 207–216.

65. Lauretani, F., Semba, R. D., Bandinelli, S., Dayhoff-Brannigan, M., Lauretani, F., Corsi, A. M., Guralnik, J. M., & Ferrucci, L. (2008). Carotenoids as protection against disability in older persons. *Rejuvenation Research, 11*, 557–563.

66. El Ansari, M. A., Reddy, K. K., Sastry, K. N. S., & Nayudamma, Y. (1971). Polyphenols of *Mangifera indica*. *Phytochemistry, 10*, 2239–2241.

67. Schieber, A., Ulrich, W., & Carle, R. (2001). Characterization of polyphenols in mango puree concentrate by HPLC with diode array and mass spectrometric detection. *Innovative Food Science and Emerging Technologies, 1*, 161–166.

68. Schieber, A., Berardini, N., & Carle, R. (2003). Identification of flavonol and xanthone glycosides from mango (*Mangifera indica* L. Cv. 'Tommy Atkins') peels by high-performance liquid chromatography-electrospray ionization mass spectrometry. *Journal of Agricultural and Food Chemistry, 51*, 5006–5011.

69. Berardini, N., Schieber, A., Klaiber, I., Beifuss, U., Carle, R., & Conrad, J. (2005). 7-*O*-Methylcyanidin 3-*O*-β-D-galactopyranoside, a novel anthocyanin from mango (*Mangifera indica* L. cv. 'Tommy Atkins') peels. *Zeitschrift für Naturforschung, 60b*, 801–804.

70. Ribeiro, S. M. R., Barbosa, L. C. A., Queiroz, J. H., Knödler, M., & Schieber, A. (2008). Phenolic compounds and antioxidant capacity of Brazilian mango (*Mangifera indica* L.) varieties. *Food Chemistry, 110*, 620–628.

71. Berardini, N., Fezer, R., Conrad, J., Beifuss, U., Carle, R., & Schieber, A. (2005). Screening of mango (*Mangifera indica* L.) cultivars for their contents of flavonol *O*- and xanthone *C*-glycosides, anthocyanins, and pectin. *Journal of Agricultural and Food Chemistry, 53*, 1563–1570.

72. Barreto, J. C., Trevisan, M. T. S., Hull, W. E., Erben, G., de Brito, E. S., Pfundstein, B., Würtele, G., Spiegelhalder, B., & Owen, R. W. (2008). Characterization and quantification of polyphenolic compounds in bark, kernel, leaves, and peel of mango. *Journal of Agricultural and Food Chemistry, 56*, 5599–5610.

73. Middleton, E., Jr., Kandaswami, C., & Theoharides, T. C. (2000). The effects of plant flavonoids on mammalian cells: Implications for inflammation, heart disease, and cancer. *Pharmacological Reviews, 52*, 673–751.

74. Knekt, P., Kumpulainen, J., Järvinen, R., Rissanen, H., Heliövaara, M., Reunanen, A., Hakulinen, T., & Haromaa, A. (2002). Flavonoid intake and risk of chronic diseases. *American Journal of Clinical Nutrition, 76*, 560–568.

75. Knouhouser, M. L. (2004). Review: Dietary flavonoids and cancer risk: Evidence from human population studies. *Nutrition and Cancer, 50*, 1–7.

76. Paganga, G., Miller, N., & Rice-Evans, C. A. (1999). The polyphenolic content of fruit and vegetables and their antioxidant activities. What does a serving constitute? *Free Radical Research, 30*, 153–162.

77. Rice-Evans, C. A., Miller, N. J., & Paganga, G. (1996). Structure-antioxidant activity relationships of flavonoids and phenolic acids. *Free Radical Biology and Medicine, 20*, 933–956.

78. Moskaug, J.Ø., Carlsen, H., Myhrstad, M., & Blomhoff, R. (2004). Molecular imaging of the biological effects of quercetin and quercetin-rich foods. *Mechanisms of Ageing and Development, 125*, 315–324.

79. Ansari, A. A., Abdul, H. M., Joshi, G., Opii, W. O., & Butterfield, A. (2009). Protective effect of quercetin in primary neurons against A β(1-42): relevance to Alzheimer's disease. *Journal of Nutritional Biochemistry, 20*, 269–275.

80. Cao, G., Sofic, E., & Prior, R. (1997). Antioxidant and prooxidant behaviour of flavonoids: structure-activity relationships. *Free Radical Biology and Medicine, 22*, 749–760.

81. Sanchéz, G. M., Re, L., Giulian, A., Núñez-Sellés, A. J., Davison, G. P., & Léon-Fernández, O. S. (2000). Protective effects of *Mangifera indica* L. extract, mangiferin and selected antioxidants against TPA-induced biomolecules oxidation and peritoneal macrophage activation in mice. *Pharmacological Research, 42*, 565–573.

82. Yoshimi, N., Matsunaga, K., Katayama, M., Yamada, Y., Kuno, T., Qiao, Z., Hara, A., Yamahara, J., & Mori, H. (2001). The inhibitory effects of mangiferin, a naturally occurring glucosylxanthone, in bowel carcinogenesis of male F344 rats. *Cancer Letters, 163*, 163–170.

83. Muruganandan, S., Gupta, S., Kataria, M., Lal, J., & Gupta, P. K. (2002). Mangiferin protects the streptozotocin-induced oxidative damage to cardiac and renal tissues in rats. *Toxicology, 176*, 165–173.

84. Leiro, J. M., Álvarez, E., Arranz, J. A., Siso, I. G., & Orallo, F. (2003). *In vitro* effects of mangiferin on superoxide concentrations and expression of the inducible nitric oxide synthase, tumor necrosis factor-α and transforming growth factor-β genes. *Biochemical Pharmacology, 65*, 1361–1371.

85. Leiro, J., Arranz, J. A., Yáñez, M., Ubeira, F. M., Sanmartin, M. L., & Orallo, F. (2004). Expression profiles of genes involved in the mouse nuclear factor-kappa B signal transduction pathway are modulated by mangiferin. *International Immunopharmacology, 4*, 763–778.

86. Jagetia, G., & Venkatesha, V. (2006). Mangiferin protects human peripheral blood lymphocytes against γ-radiation-induced DNA strand breaks: A fluorescence analysis of DNA unwinding assay. *Nutrition Research, 26*, 303–311.

87. Rajendran, P., Ekambaram, G., & Sakthisekaran, D. (2008). Effect of mangiferin on benzo(a)pyrene induced lung carcinogenesis in experimental Swiss albino mice. *Natural Product Research, 22*, 672–680.

88. Rodriguez, J., Di Pierro, D., Gioia, M., Monaco, S., Delgado, R., Coletta, M., & Marini, S. (2006). Effects of a natural extract from *Mangifera indica* L., and its active compound mangiferin, on state and lipid peroxidation of red blood cells. *Biochimica et Biophysica Acta, 1760*, 1333–1342.

89. Wilkinson, A. S., Monbteich, G. R., Shaw, P. N., Lin, C. N., Gidley, M. J., & Roberts-Thomson, S. J. (2008). Effects of the mango components mangiferin and quercetin and the putative mangiferin metabolite norathyriol on the transactivation of peroxisome proliferator-activated receptor isoforms. *Journal of Agricultural and Food Chemistry, 56*, 3037–3042.

90. Bhatia, H. S., Candelario-Jalil, E., Oliveira, A. C., Olajide, O. A., Martinez-Sanchez, G., & Fiebich, B. L. (2008). Mangiferin inhibits cyclooxygenase-2 expression and prostaglandin E (2) production in activated rat microglia cells. *Archives of Biochemistry and Biophysics, 477*, 253–258.

91. Carvalho, A. C., Guedes, M. M., Souza, A. L., Trevisan, M. T., Lima, A. F., Santos, F. A., & Rao, V. S. (2007). Gastroprotective effect of mangiferin a xanthonoid from *Mangifera indica*, aginst gastric injury induced by ethanol and indomethacin in rodents. *Planta Medica, 73*, 1372–1376.

92. Singh, V. K. (2006). Physiological and biochemical changes with special reference to mangiferin and oxidative enzymes level in malformation resistant and susceptible cultivars of mango (*Mangifera indica* L.). *Scientia Horticulturae, 108*, 43–48.

93. Knödler, M., Berardini, N., Kammerer, D. R., Carle, R., & Schieber, A. (2007). Characterization of major and minor alk(en)ylresorcinols from mango (*Mangifera indica* L.) peels by high-performance liquid chromatography/atmospheric pressure chemical ionization mass spectrometry. *Rapid Communications in Mass Spectrometry, 21*, 945–951.

94. Cojocaru, M., Droby, S., Glotter, E., Goldman, A., Gottlier, H. E., Jacoby, B., & Prusky, D. (1986). (12-Heptadecenyl)-resorcinol, the major component of the antifungal activity in the peel of mango fruit. *Phytochemistry, 25*, 1093–1095.

95. Knödler, M., Conrad, J., Wenzig, E. M., Bauer, R., Lacorn, M., Beifuss, V., Carle, R., & Schieber, A. (2008). Anti-inflammatory 5-(11Z-heptadecenyl)- and 5-(8Z, 11Z-heptadecadienyl)-resorcinols from mango (*Mangifera indica* L.) peels. *Phytochemistry, 69*, 988–993.

96. Dang, K. T. H., Singh, Z., & Swinny, E. E. (2008). Edible coatings influence fruit ripening quality, and aroma biosynthesis in mango fruit. *Journal of Agricultural and Food Chemistry, 56*, 1361–1370.

97. Andrade, E. H. A., Maia, J. G. S., & Zoghbi, M. G. B. (2000). Aroma volatile constituents of Brazilian varieties of mango fruit. *Journal of Food Composition and Analysis, 13*, 27–33.

98. Sagar, S. P., Chidley, H. G., Kulkarni, R. S., Pujari, K. H., Giri, A. P., & Gupta, V. S. (2009). Cultivar relationships in mango based on fruit volatile profiles. *Food Chemistry, 114*, 363–372.

99. Jhon, K. S., Bhat, S. G., & Rao, U. J. S. P. (2003). Biochemical characterization of sap (latex) of a few Indian mango varieties. *Phytochemistry, 62*, 13–19.

100. Pino, J. A., Mesa, J., Muñoz, Y., Mati, M. P., & Mabbot, R. (2005). Volatile components from mango (*Mangifera indica* L.) cultivars. *Journal of Agricultural and Food Chemistry, 53*, 2213–2223.

101. Lalel, H. J. D., Singh, Z., & Tan, S. C. (2003). Aroma volatiles production during fruit ripening of 'Kensington Pride' mango. *Postharvest Biology and Technology, 27*, 323–336.

102. Lalel, H. J. D., Singh, Z., & Tan, S. C. (2003). Distribution of aroma volatile compounds in different parts of mango fruit. *Journal of Horticultural Science & Biotechnology, 78*, 131–138.

C. ACTIONS OF INDIVIDUAL FRUITS IN DISEASE AND CANCER PREVENTION AND TREATMENT

103. Lalel, H. J. D., Singh, Z., & Tan, S. C. (2003). Glycosidically-bound aroma volatile compounds in the skin and pulp of 'Kensington Pride' mango fruit at different stages of maturity. *Postharvest Biology and Technology, 29*, 205–218.

104. Lalel, H. J. D., Singh, Z., & Tan, S. C. (2003). Maturity stage at harvest affects fruit ripening, quality and biosynthesis of aroma volatile compounds in 'Kensington Pride' mango. *Journal of Horticultural Science & Biotechnology, 78*, 225–233.

105. Lalel, H. J. D., Singh, Z., & Tan, S. C. (2004). Ripening temperatures influence biosynthesis of aroma volatiles compounds in 'Kensington Pride' mango fruit. *Journal of Horticultural Science & Biotechnology, 79*, 146–157.

106. Bardon, S., Picard, K., & Martel, P. (1998). Monoterpenes inhibit cell growth, cell cycle progression, and cyclin D1, gene expression in human breast cancer cell lines. *Nutrition and Cancer, 32*, 1–7.

107. Crowell, P. L. (1999). Prevention and therapy of cancer by dietary monoterpenes. *The Journal of Nutrition, 129*, 775–778.

108. Foti, M. C., & Ingold, K. U. (2003). Mechanism of inhibition of lipid peroxidation by gamma-terpinene, an unusual and potentially useful hydrocarbon antioxidant. *Journal of Agricultural and Food Chemistry, 51*, 2758–2765.

109. Hwang, Y. P., & Jeong, H. G. (2008). The coffee diterpene kahweol induces heme oxygenase-1 via the PI3K and p38/Nrf2 pathway to protect human dopaminergic neurons from 6-hydroxydopamine-derived oxidative stress. *FEBS Letters, 582*, 2655–2662.

110. Grassmann, J. (2005). Terpenoids as plant antioxidants. *Vitamins and Hormones, 72*, 5005–5535.

111. Rabi, T., & Gupta, S. (2008). Dietary terpenoids and prostate cancer chemoprevention. *Frontiers in Bioscience, 13*, 3457–3469.

112. Prasad, S., Kabra, N., & Shukla, Y. (2008). Induction of apoptosis by lupeol and mango extract in mouse prostate and LNCaP cells. *Nutrition and Cancer, 60*, 120–130.

113. Chaturvedi, P. K., Bhui, K., & Shukla, Y. (2008). Lupeol: Connotations for chemoprevention. *Cancer Letters, 263*, 1–13.

114. Sudhahar, V., Veena, C. K., & Varalakshmi, P. (2008). Antiurolithic effect of lupeol and lupeol linoleate in experimental hyperoxaluria. *Journal of Natural Products, 71*, 1509–1512.

115. Gibson, G. R., & Roberfroid, M. (1995). Dietary modulation of human colonic microbiota: Introducing the concept of prebiotics. *The Journal of Nutrition, 125*, 1401–1412.

116. Brown, L., Rosner, B., Willet, W. W., & Sacks, F. M. (1999). Cholesterol-lowering effects of dietary fiber: A meta-analysis. *American Journal of Clinical Nutrition, 69*, 30–42.

117. Umar, S., Morris, A. P., Korouma, F., & Sellin, J. H. (2003). Dietary pectin and calcium inhibit colonic proliferation *in vivo* by differing mechanisms. *Cell Proliferation, 36*, 361–375.

118. Jenkins, D. J., Leeds, A. R., Gassul, M. A., Cochet, B., & Albert, G. M. (1977). Decrease in post prandial insulin and glucose concentrations by guar and pectin. *Annals of Internal Medicine, 86*, 20–23.

119. Kim, M. (2005). High-methoxyl pectin has greater enhancing effect on glucose uptake in intestinal perfused rats. *Nutrition, 21*, 372–377.

120. Pirman, T., Mosoni, L., Ramond, D., Ribeyre, M. C., Buffière, C., Salobir, J., & Mirand, P. P. (2008). Differential response of protein metabolism in splanchnic organs and muscle to pectin feeding. *The British Journal of Nutrition, 100*, 306–311.

121. Larrauri, J. A., Rupérez, P., & Saura-Calixto, F. (1995). Mango peels as new tropical fiber: Obtention and characterization. *Lebensmittel-Wissenschaft und –Technologie, 29*, 729–733.

122. Prasanna, V., Prabha, T. N., & Tharanathan, R. N. (2004). Pectic polysaccharides of mango (*Mangifera indica* L.): Structural studies. *Journal of the Science of Food and Agriculture, 84*, 1731–1735.

123. Núñez-Selléz, A. J., Rodriguez, M. D. D., Balseiro, E. G., Gonzalez, L. N., Nicolais, V., & Rastrelli, L. (2007). Comparison of major and trace elements concentrations in 16 varieties of Cuban mango stem bark (*Mangifera indica* L.). *Journal of Agricultural and Food Chemistry, 55*, 2176–2181.

124. Halliwell, B. (1999). Antioxidant defense mechanisms: From the beginning to the end (of the beginning). *Free Radical Research, 31*, 261–272.

125. Zhou, B., Wu, L., Yang, L., & Liu, Z. (2005). Evidence for α-tocopherol regeneration reaction of green tea polyphenols in SDS micelles. *Free Radical Biology and Medicine, 38*, 78–84.

126. Brighenti, F., Valtuena, S., & Pellegrini, N. (2005). Total antioxidant capacity of the diet is inversely and independently related to plasma concentration of high-sensitivity C-reactive protein in adult Italian subjects. *The British Journal of Nutrition, 93*, 619–625.

127. Valtueña, S., Pellegrini, N., Franzini, L., Bianchi, M. A., Ardigo, D., Del Rio, D., Piatti, P. M., Scazzina, F., Zavaroni, I., & Brighenti, F. (2008). Food selection based on total antioxidant capacity can modify antioxidant intake, systemic inflammation, and liver function without altering markers of oxidative stress. *American Journal of Clinical Nutrition, 87*, 1290–1297.

128. Ajila, C. M., Naidu, K. A., Bhat, S. G., & Prasada Rao, U. J. S. (2007). Bioactive compounds and

antioxidant potential of mango peel extract. *Food Chemistry, 105*, 982–988.

129. Ajila, M., Naidu, K. A., Bhat, S. G., & Prasada Rao, U. J. S. (2007). Valuable components of raw and ripe peels from two Indian mango varieties. *Food Chemistry, 102*, 1006–1011.

130. Abdalla, A. E. M., Darwish, S. M., Ayad, E. H. E., & El-Hamahmy, R. M. (2006). Egyptian mango by-product 2: Antioxidant and antimicrobial activities of extract and oil from mango seed kernel. *Food Chemistry, 103*, 1141–1152.

131. Amimoto, T., Matsura, T., Koyama, S., Nakanish, T., Yamada, K., & Kajiyama, G. (1995). Acetaminophen-induced hepatic injury in mice: The role of lipid peroxidation and effects of pretreatment with coenzyme Q10 and α-tocopherol. *Free Radical Biology and Medicine, 19*, 169–176.

132. Paschke, A., Kinder, H., Zunker, K., Wigotzki, M., Wessbecher, R., Viluf, I., & Steinhart, H. (2001). Charactrization of cross-reaction allergens in mango fruit. *Allergy, 56*, 237–242.

133. Dube, M., Zunker, K., Neidhart, S., Carle, R., Steinhart, H., & Paschke, A. (2004). Effect of technological processing on the allergenicity of mangoes (*Mangifera indica* L.). *Journal of Agricultural and Food Chemistry, 52*, 3938–3945.

C. ACTIONS OF INDIVIDUAL FRUITS IN DISEASE AND CANCER PREVENTION AND TREATMENT

35

Health Benefits of Bitter Melon (Momordica charantia)

Edralin A. Lucas[1], Gerard G. Dumancas[2], Brenda J. Smith[1], Stephen L. Clarke[1], and Bahram H. Arjmandi[3]

[1]Nutritional Sciences Department, Oklahoma State University, Stillwater, OK, USA
[2]Chemistry Department, Oklahoma State University, Stillwater, OK, USA
[3]Department of Nutrition, Food & Exercise Sciences, Florida State University, Tallahassee, FL, USA

1. DESCRIPTION AND COMPOSITION

Momordica charantia (MC), a member of the Cucurbitaceae family, is also known as bitter melon, bitter gourd, balsam pear, pare, or karela. It is a widely grown and consumed vegetable in Asia, East Africa, India, and South America. Momordica, which means 'to bite' refers to the jagged edges of the plant leaf that appear as if the leaf was bitten [1]. The plant grows as a climbing perennial and all parts of the plant, including its fruit, are bitter. The fruit resembles a warty gourd or cucumber with the young fruit being an emerald green color that turns orange-yellow when ripe (Fig. 35.1). The bitter taste of the fruit becomes more pronounced as the fruit ripens [1]. Cultivars of Chinese MC can be divided into three types: a) small fruit type, 10–20 cm long, 100–300 g in weight, usually dark green, and the fruit is extremely bitter; b) long fruit type, most commonly grown commercially in China, 30–60 cm long, 200–600 g in weight, light green in color with medium size protuberances, and only slightly bitter; and c) triangular fruit type, cone-shaped, 9–12 cm long, 300–600 g in weight, light to dark green with prominent tubercles, moderately to strongly bitter [2].

Nutritional analyses of MC indicate that this vegetable is rich in fiber, calcium, potassium, iron, and vitamins C and A (Table 35.1) [3]. Additionally, the pulp around the seeds of the mature ripe fruit is a good source of the carotenoid lycopene [4]. The bitter component of MC has been characterized as having four cucurbitane glycosides, momordicosides K and L, and momordicines I and II [5,6]. A few phytochemicals, including polypeptide-p (a 166 residue insulin mimetic peptide) [7], a steroid glycoside charantin [8], momordin Ic, oleanolic

525

TABLE 35.1 Nutrient Composition of Raw *Momordica charantia* [3]

Nutrient	Amount per 100 g
Proximates	
Water, g	94.03
Energy, kcal	17.0
Protein, g	1.0
Total lipid (fat), g	0.17
Ash, g	1.10
Carbohydrate (by difference), g	3.70
Total dietary fiber, g	2.80
Minerals	
Calcium (Ca), mg	19.0
Iron (Fe) , mg	0.43
Magnesium (Mg), mg	17.0
Phosphorus (P), mg	31.0
Potassium (K), mg	296.0
Sodium (Na), mg	5.0
Zinc (Zn), mg	0.80
Copper (Cu), mg	0.034
Manganese (Mn), mg	0.089
Selenium (Se), μg	0.20
Vitamins	
Vitamin C (total ascorbic acid), mg	84.0
Thiamin, mg	0.040
Riboflavin, mg	0.040
Niacin, mg	0.400
Pantothenic acid, mg	0.212
Vitamin B6, mg	0.043
Folate (total), μg	72.0
Vitamin A (RAE), μg RAE	24.0
Carotene (beta), μg	190
Carotene (alpha), μg	185
Vitamin A, IU	471
Lutein + zeaxanthin, μg	170

RAE, retinol activity equivalent.

FIGURE 35.1 Fruit of the *Momordica charantia* (Chinese phenotype) [95].

total phenolic content of MC obtained by subcritical water extraction is 48.177 mg gallic acid equivalents/g dry matter with gallic acid being the main phenolic acid in MC (0.6462 mg/g dry weight) [17].

The bioactive components responsible for many of the proposed health benefits of MC remain in question. Compounds isolated from the fruits and seeds of MC that are thought to contribute to the hypoglycemic property of MC include charantin and polypeptide p or plant insulin [7,8]. Other compounds found in MC that may contribute to its health benefits include other glycosides such as mormordin, and antioxidants such as vitamin C, carotenoids, flavanoids, and polyphenols [18,19]. Harinantenaina and colleagues [11] have demonstrated that cucurbutanoid compounds are the bioactive compounds in MC contributing to its hypoglycemic properties.

2. POTENTIAL CLINICAL APPLICATIONS

2.1 Diabetes

The potential for MC to modulate blood glucose has received the most attention from investigators searching for natural foods or compounds that may be useful in the treatment of diabetes. Different parts of the plant have been

acid-3-*O*-monodesmoside, and oleanolic acid-3-*O*-glucuronide have been identified in both the fruits and seeds of MC [9]. Many cucurbitane-type triterpenoids have been isolated from the fruits [5,10,11], seeds [12,13], stem [14,15], roots [16], and leaves and vines [6,14]. The shown to possess hypoglycemic properties in animal models, cell-based assays, and a limited number of clinical trials.

Most of the studies to date examining the hypoglycemic properties of MC have been conducted using animal models. MC can lower blood glucose in normal animals, in animals fed a high fat diet, and in streptozotocin (STZ)-, alloxan- and genetically-induced animal models of diabetes. Table 35.2 summarizes the different animal studies investigating the hypoglycemic properties of MC. The results of the animal work to date indicate that MC has profound effects in lowering fasting blood glucose [20–22], improving oral glucose tolerance test results [20–22], and increasing both liver and muscle glycogen stores [23,24].

Although numerous animal studies have been conducted investigating the hypoglycemic properties of MC, only limited data from controlled clinical trials are available to determine if the findings from the animal studies also hold true for humans (Table 35.3). Based on our review of the literature, only 11 human studies have been conducted assessing the efficacy of MC in lowering blood glucose. The results of these clinical trials also showed some glucose-lowering effects of MC but were not as convincing compared to those of the animal studies. Moreover, many of these human studies were plagued by inadequate sample size, poor study design, and differences in the MC preparation. As a result, the previously mentioned factors may have contributed to the inconsistent results obtained from human studies. Thus, it is imperative that well-designed, and adequately powered, placebo-controlled randomized clinical trials be conducted to determine the effective dose, methods of preparation, and safety of MC prior to making any recommendations for human use.

In addition to the animal and human studies, *in vitro* studies have proven quite useful in the determination of the components of MC responsible for its hypoglycemic effect. Further,

these cell-based studies have allowed for the elucidation of the mechanisms of action of MC on the liver, as well as in peripheral tissues (i.e. skeletal muscle, adipose tissues, intestine, and pancreas). For example, using an insulin-resistant liver cell line (FL83B) and myocyte (C2C12) cells, Cheng et al. [15] reported that extracts from the fruit, seed, or stem of MC all contain components contributing to the improved glucose uptake of these cells. Upon further screening of the components of the crude extract of MC stem, they identified that triterpenoids (CH10, CH63, and CH93) are capable of preventing cellular insulin resistance as evidenced by enhanced glucose uptake and phosphorylation of insulin resistant substrate-1 (IRS-1). Similarly, Yibchok-anun and colleagues [25] showed that protein extracts from the MC fruit pulp significantly enhanced glucose uptake in both 3T3-L1adipocytes and C2C12 myocytes. MC fruit extract has also been shown to stimulate both glucose and amino acid uptake in L6 myotubes [26] and 3T3-L1 adipocytes [27]. The increased uptake of glucose in skeletal muscle may be due in part to increased GLUT-4 transporter protein [26,28]. In addition to increased glucose uptake in the skeletal muscle and adipocytes, MC has been shown to increase hepatic glucose utilization and glycogen synthesis [29–31]. Increased activity of enzymes involved in glucose metabolism such as hepatic glucokinase, hexokinase, glucose-6-phosphatase, and glycogen synthase has also been observed in STZ-induced diabetic rats supplemented with MC extract [29,32,33]. An intriguing hypothesis is that MC contains bioactive components that act as ligands for peroxisome proliferator activated receptors (PPAR), ligand-activated transcription factors important in modulating both lipid and glucose metabolism [34,35].

Another proposed mechanism of action of MC is through its direct effect on the β cells of the pancreas and on the intestinal absorption of dietary glucose and amino acids.

TABLE 35.2   Animal Studies Investigating the Hypoglycemic/Antidiabetic Properties of *Momordica charantia* (MC)

Author(s)	Species and Strain Treatment Groups and MC Preparation and Study Duration	Results

Akhtar et al. [96]	Male albino rabbits
	Treatment groups: ($N = 6$/group)
	
	1. Normal
	  a. Control (1% carboxymethyl cellulose – CMC)
	  b. Dried fruit powder suspended in CMC (0.25, 0.5, 1.0, or 1.5 g/kg bw)
	2. Alloxan-induced diabetes – same treatment groups as above
	
	Treatment duration: one time before blood collection
		• In normal rabbits, dose starting at 0.5 g/kg bw reduced blood glucose
		• In diabetic rabbits, 1.0 and 1.5 g/kg bw reduced blood glucose but not at lower doses
		• For both normal and diabetic rabbits, the maximum effect was observed after 10 hours of administration

Karunanayake et al. [20]	Male Sprague-Dawley rats *Treatment groups*: ($N = 6$/group) 1. Control (saline –1 mL/100 g bw) 2. MC fruit juice (1 mL/100 g bw) 3. *Salacia reticulata* root aqueous extract 4. *Aegle marmelos* root aqueous extract *Treatment duration:* one time before blood collection	• MC showed the most profound effect both on fasting blood glucose and the oral glucose tolerance test (OGTT)
Welihinda and Karunanayake [23]	Male Sprague-Dawley rats *Treatment groups*: 1. Control (distilled water − 1 mL/100 g bw) 2. Fruit juice (1 mL/100 g bw) *Treatment duration:* one time	• MC increased glycogen content of the liver and muscle but did not alter triglyceride content of adipose tissue
Srivastava et al. [44]	Charles Foster rats *Treatment groups*: ($N = 10$/group) 1. Normal 2. Alloxan-diabetes 3. Alloxan-diabetes + fruit aqueous extract (2 mL) *Treatment duration:* 2 months	• MC significantly retarded the onset of retinopathy
Srivastava et al. [45]	Charles Foster rats (alloxan-induced diabetic) *Treatment groups*: ($N = 10$/group) 1. Control (saline) 2. Fruit aqueous extract (4 g/d) *Treatment duration:* 2 months	• Cataract formation was delayed in animals receiving MC extract compared to the control (3 months vs. 5 months)
Day et al. [31]	Male Thieller Original mice *Treatment groups*: 1. Normal ($N = 5$/group) a. Control	• In normal mice, extract A lowered the glycemic response to both oral and intraperitoneal glucose, without altering the insulin response

(Continued)

TABLE 35.2 (Continued)

Author(s)	Species and Strain Treatment Groups and MC Preparation and Study Duration	Results

	b. Fruit aqueous extract (extract A)
	2. Streptozotocin (STZ)-induced diabetes ($N = 3-5$/group)
	  a. Control
	  b. Fruit aqueous extract (extract A)
	  c. Residue after chloroform extraction of extract A (extract B)
	  d. Material after alkaline water wash of chloroform extract of A (extract C)
	  e. Material recovered by acid water wash of chloroform extract remaining after removal of extract C (extract D)
	Dose: (4 mL/kg bw)
	
	Treatment duration: one time
		• Extract A and B were both hypoglycemic, extract C was not, and extract D exhibited a slower hypoglycemic effect
		• Two types of hypoglycemic components are present in MC: a more rapidly effective water-soluble component and a more slowly acting component, possibly an alkaloid

Ali et al. [97]	Male Long-Evans rats
	Treatment groups:
	
	1. Normal
	  a. Water ($N = 8$)
	  b. Fruit pulp
	
	    • juice ($N = 11$)
	    • juice methanol extract ($N = 10$)
	    • juice saponin-free methanol extract ($N = 14$)
	  c. Seed methanol extract ($N = 13$)
	  d. Whole plant
	
	    • methanol extract ($N = 8$)
	    • saponin-free MeOH extract ($N = 7$)
	2. Insulin-dependent diabetes mellitus (IDDM)-STZ-induced diabetes
	  a. Water ($N = 6$)
	  b. Pulp juice ($N = 6$)
	  c. Seed MeOH extract ($N = 5$)
	3. Non insulin-dependent diabetes mellitus (NIDDM)-STZ-induced diabetes
	  a. Control ($N = 5$; H_2O)
	  b. Pulp saponin-free MeOH extract ($N = 5$)
	Dose: 250 mg/2 mL; saponin-free extract 150 mg/2 mL
	
	Treatment duration: one time
		• Fasting blood glucose in normal rats was significantly lowered by pulp juice, however, a more pronounced effect was observed with a saponin-free methanol extract of the pulp juice
		• In IDDM rats, pulp juice had no significant effect on either fasting or postprandial blood glucose level
		• In NIDDM rats, saponin-free methanol extract of the pulp juice has a significant hypoglycemic effect both in fasting or postprandial states
		• Methanol extract of seed and whole plant, saponin-free methanol extract of the whole plant has no hypoglycemic effect in normal or IDDM rats

Shibib *et al.* [29]	Male albino mice *Treatment groups:* ($N = 4-5$/group) 1. Normal	• Hypoglycemic effect of MC and *Coccinia indica* is mediated through (a) suppression of the key hepatic

(Continued)

TABLE 35.2   (Continued)

Author(s)	Species and Strain Treatment Groups and MC Preparation and Study Duration	Results

	2. STZ-induced diabetes
	  a. MC fruit alcohol extract
	  b. *Coccinia indica* leaves alcohol extract
	Dose: 200 mg/kg bw
		gluconeogenic enzymes glucose-6-phosphatase and fructose-1,6-bisphosphatase, and (b) an accelerated rate of glucose oxidation through the pentose phosphate pathway

	Treatment duration: one time	

Platel and Srinivasan [98]	Wistar rats
	Treatment groups: ($N = 10$/group)
	
	1. Normal
	  a. Control
	  b. Diet containing 0.5% freeze-dried fruit powder
	2. STZ-induced diabetes – same treatment groups as above
		• MC has no beneficial effect on blood glucose levels or diabetes-related metabolites

	Treatment duration: 6 weeks	
Sarkar et al. [30]	Female Wistar rats *Treatment groups:* **Experiment I** (glucose primed normal rat; $N = 6$/group) 1. Vehicle 2. Lyophilized alcohol fruit extract (500 mg/kg) 3. Tolbutamide (100 mg/kg) *Treatment duration:* one time **Experiment II** (normal rats; $N = 6$–8/group) 1. Vehicle 2. Lyophilized alcohol fruit extract (500 mg/kg) *Treatment duration:* 7 days **Experiment III** (STZ-induced diabetes; $N = 6$/group) 1. Vehicle 2. Lyophilized alcohol fruit extract (500 mg/kg) 3. Metformin (200 mg/kg)	• Efficacy of MC in reducing glucose concentration was 25–30% of tolbutamide • In STZ-induced diabetic rats, MC improved oral glucose tolerance by reducing plasma glucose 26% compared to 40–50% with metformin • MC caused a 4–5 fold increased in the rate of glycogen synthesis in the liver of normal fed rats
	Treatment duration: one time	
Ahmed et al. [36]	Male Wistar rats *Treatment groups:* ($N = 4$–5/group) 1. Normal 2. STZ-induced diabetes 3. STZ-induced diabetes + fruit juice (10 mL/kg bw) *Treatment duration:* 10 weeks	• MC increased the number of insulin-positive cells of the pancreas perhaps by preventing the death of β cells and/or by permitting the recovery of partially destroyed β cells

(Continued)

C. ACTIONS OF INDIVIDUAL FRUITS IN DISEASE AND CANCER PREVENTION AND TREATMENT

TABLE 35.2   (Continued)

Author(s)	Species and Strain Treatment Groups and MC Preparation and Study Duration	Results
Jayasooriya et al. [47]	Male Sprague-Dawley rats *Treatment groups:* **Experiment I** ($N = 5$/group) 1. Control 2. Lyophilized fruit powder (0.5, 1 or 3%) **Experiment II** ($N = 6$/group) 1. Control 2. Lyophilized fruit powder (1%) added to cholesterol-free or cholesterol-enriched diet *Treatment duration:* 14 days	• MC reduced blood glucose in rats fed cholesterol-free diets but not in those fed cholesterol-enriched diets • MC only marginally reduced serum triglycerides, total cholesterol, and phospholipids. However, MC elevated HDL-cholesterol both in the presence or absence of dietary cholesterol

Grover et al. [42]	Albino mice
	Treatment groups: ($N = 6$/group)
	
	1. Normal
	2. STZ-induced diabetes
	   a. No extract
	   b. Lyophilized MC fruit extract (200 mg/kg)
	   c. Lyophilized *Eugenia jambolana* (EJ) kernel (200 mg/kg)
	   d. *Mucuna pruriens* alcohol extract (MP; 200 mg/kg)
	   e. *Tinospora cordifolia* aqueous extract (TC; 400 mg/day)
	
	Treatment duration: 40 days
		• Plasma glucose concentrations in STZ-diabetic mice were reduced by MC, EJ, TC, and MP by 24.4, 20.8, 7.4, and 9.1%, respectively
		• All plants prevented polyuria and reduced urinary albumin levels with MC having the most significant effect
		• MC and EJ significantly prevented renal hypertrophy

Miura et al. [28]	KK-Ay mice *Treatment groups:* ($N = 5$–6/group) 1. Control 2. Fruit aqueous extract (100 mg/kg bw) *Treatment duration:* one time	• Three weeks after administration, MC reduced blood glucose and serum insulin • Protein content of glucose transporter isoform 4 (GLUT4) in the muscle plasma membrane was significantly higher in the MC-treated mice than in control animals
Vikrant et al. [99]	Albino rats *Treatment groups:* ($N = 8$/group) 1. Normal chow diet 2. Fructose-rich diet 3. Fructose-rich diet + MC fruit aqueous or alcoholic extract (100, 200, or 400 mg/day)	• Alcoholic extracts of MC and EJ did not reduce blood glucose whereas the highest dose of aqueous extract prevented the development of hyperglycemia and hyperinsulinemia

(Continued)

C. ACTIONS OF INDIVIDUAL FRUITS IN DISEASE AND CANCER PREVENTION AND TREATMENT

TABLE 35.2   (Continued)

Author(s)	Species and Strain Treatment Groups and MC Preparation and Study Duration	Results
	4. Fructose-rich diet + *Eugenia jambolana* (EJ) kernel aqueous or alcoholic extract (100, 200, or 400 mg/d) *Treatment duration:* 15 days	
Rathi et al. [33]	**Experiment 1** alloxan-injected (32 mg/kg) albino Wistar rats *Treatment groups:* (N = 8/group) 1. Normal 2. Diabetic 3. Diabetic + lyophilized aqueous fruit extract (50, 100, or 200 mg/kg/d) 4. Diabetic + alcohol fruit extract (50, 100, or 200 mg/kg/d) *Treatment duration:* 3 weeks **Experiment 2** alloxan-injected (120 mg/kg) albino Wistar rats *Treatment groups:* (N = 8/group) 1. Normal 2. Diabetic 3. Diabetic + lyophilized aqueous fruit extract (200 mg/kg/d) *Treatment duration:* 16 weeks **Experiment 3** STZ-injected (150 mg/kg) albino mice – same treatment group as Experiment 2 *Treatment duration:* 60 days	• Aqueous extract of MC was more effective than the alcoholic extract in reversing alloxan-induced hyperglycemia in rats • The percentage reduction in glucose levels in the three models of diabetes: mild (alloxan, 32 mg/kg), moderate (alloxan, 120 mg/kg), and severe (STZ, 150 mg/kg) was 69.25, 64.33, and 11.31%, respectively • The alterations in hepatic and skeletal muscle glycogen content and hepatic glucokinase, glucose-6-phosphatase, and phosphofructokinase levels in diabetic mice were partially restored by MC

Rathi et al. [43]	Albino rats
	Treatment groups: (N = 8/group)
	
	1. Normal
	2. STZ-induced diabetes
	  a. No extract
	  b. Lyophilized MC fruit aqueous extract (200 mg/kg)
	  c. Lyophilized *Eugenia jambolana* (EJ) kernel aqueous extract (200 mg/kg)
	  d. *Mucuna pruriens* alcoholic extract (MP; 200 mg/kg)
	  e. *Tinospora cordifolia* aqueous extract (TC; 400 mg/d)
	
	Treatment duration: 40 days
		• MC and EJ prevented the development of cataracts in addition to significantly reducing plasma glucose levels

(*Continued*)

C. ACTIONS OF INDIVIDUAL FRUITS IN DISEASE AND CANCER PREVENTION AND TREATMENT

TABLE 35.2 (Continued)

Author(s)	Species and Strain Treatment Groups and MC Preparation and Study Duration	Results
Chen et al. [21]	**Experiment 1** Female Sprague-Dawley (SD) rats fed low fat (LF) and high fat (HF) diet for 6 weeks *Treatment groups:* 1. LF ($N = 5$) 2. LF + 1.5% freeze-dried juice ($N = 4$) 3. HF ($N = 8$) 4. HF + 0.375, 0.75, or 1.5% freeze-dried juice ($N = 8-9$) *Treatment duration:* 9 weeks **Experiment 2** Male SD rats fed LF and HF diet for 4 weeks *Treatment groups:* 1. LF ($N = 7$) 2. HF/LF ($N = 8$) 3. HF/HF ($N = 8$) 4. HF/LF + MC ($N = 8$) 5. HF/HF + MC ($N = 8$) *Treatment duration:* 7 weeks	• OGTT was improved in rats fed HF diet supplemented with 0.75% MC • Highest dose MC had lower energy efficiency and animals tended to exhibit less visceral mass • MC supplementation to HF diet improved insulin resistance, lowered serum insulin, and reduced serum leptin levels

Virdi et al. [50]	Male Wistar rats
	Treatment groups:
	
	1. Normal
	   a. Dried whole fruit methanol extract
	   b. Dried whole fruit chloroform extract
	   c. Dried whole fruit aqueous extract
	   d. Glibenclamide (0.1 mg/kg bw/d)
	2. Alloxan-induced diabetes – same treatment as above
	
	Dose of extract: 20 mg/kg bw/d
	Treatment duration: 4 weeks
		• Aqueous extract of MC was more effective than either methanol or chloroform extracts and was as effective as glibenclamide in reversing alloxan-induced hyperglycemia
		• No toxicity to liver and kidney was observed with MC treatment

Chaturvedi et al. [22]	Male Horts Men albino rats
	Treatment groups: ($N = 5$/group)
	
	1. Normal
	2. Alloxan-induced diabetes
	   a. Control
	   b. Fruit methanol extract (140 mg/kg bw/d)
	
	Treatment duration: 30 days
		• MC methanol extract improved OGTT
		• The effect on OGTT was more pronounced when the test was performed on rats fed the extract on the day of the test compared to those that were not given MC extract

Singh et al. [39]	Male Wistar rats *Treatment groups:* ($N = 10$/group)	• MC juice reduced Na^+ and K^+-dependent glucose absorption by

(Continued)

C. ACTIONS OF INDIVIDUAL FRUITS IN DISEASE AND CANCER PREVENTION AND TREATMENT

TABLE 35.2 (Continued)

Author(s)	Species and Strain Treatment Groups and MC Preparation and Study Duration	Results

	1. Normal
	2. STZ-induced diabetes
	  a. Control
	  b. Fruit juice (10 mL/kg bw/d)
	
	Treatment duration: 10 weeks
		the brush border membrane vesicles of the jejunum; stimulated glucose uptake in L6 myotubes, and normalized structural abnormalities of peripheral nerves

Chaturvedi [100]	Male Horts Men albino rats *Treatment groups:* ($N = 5$/group) 1. Normal 2. Alloxan-induced diabetes 3. Alloxan-induced diabetes + fruit methanol extract (80, 100, or 120 mg/kg bw/d) *Treatment duration:* 45 days	• 100 and 120 mg/kg bw extract normalized blood glucose but not the 80 mg/kg bw dose

Sathishsekar and Subramanian [24]	Male albino Wistar rats
	Treatment groups: ($N = 6$/group)
	
	1. Normal
	2. STZ-induced diabetes
	  a. Control
	  b. Country variety seed aqueous extract (MCSEt1; 150 mg/kg bw/d)
	  c. Hybrid variety seed aqueous extract (MCSEt2; 150 mg/kg bw/d)
	  d. Glibenclamide (600 µg/kg bw/d)
	
	Treatment duration: 30 days
		• Both varieties significantly reduced blood glucose, HbA1c, lactate dehydrogenase, glucose-6-phosphatase, fructose-1-6 bisphosphatase, glycogen phosphorylase, and increased glycogen and activities of hexokinase and glycogen synthase
		• The hypoglycemic properties were more pronounced in MCSEt1 compared to MCSEt2 and glibenclamide

Shetty et al. [41]	Male Wistar rats
	Treatment groups:
	
	1. Normal ($N = 6$/group)
	  a. Starch containing diet
	  b. MC containing diet (10%)
	2. STZ-induced diabetes – same treatment group as above ($N = 12$/group)
	
	Treatment duration: 5 weeks
		• MC reduced fasting blood glucose by 30% in diabetic rats
		• Water consumption, urine volume, and urine glucose were significantly higher in diabetic controls compared to normal rats; MC feeding alleviated this elevation by about 30%
		• MC supplementation reduced renal hypertrophy

Shetty et al. [101]	Male Wistar rats
	Treatment groups:
	
	1. Normal ($N = 6$/group)
	  a. Starch containing diet
	  b. MC containing diet (10%)
		• MC prevented progression of diabetes by modulating the activities of intestinal and renal disaccharidases

(Continued)

C. ACTIONS OF INDIVIDUAL FRUITS IN DISEASE AND CANCER PREVENTION AND TREATMENT

TABLE 35.2 (Continued)

Author(s)	Species and Strain Treatment Groups and MC Preparation and Study Duration	Results
	2. STZ-induced diabetes – same treatment group as above ($N = 14$/group)	
	Treatment duration: 5 weeks	
Harinantenaina et al. [11]	Male ddY mice *Treatment groups:* alloxan-induced diabetes ($N = 3$–5/group) 1. Control 2. Glibenclamide (200 or 400 mg/kg) 3. Fruit ether fraction 4. Fruit ethyl acetate fraction 5. Isolated compound from MC extract (400 mg/kg) *compound 4*-5β,19-epoxy-3β,25-dihydroxycucurbita-6,23 (*E*)-diene *compound 5*-3β,7β,25-trihydroxycucurbita-5,23(*E*)-diene-19-al *compound 10 and 11*- sitosterol and stigmastadienol (aglycone of charantin) *Treatment duration:* one time	• Major pure cucurbutanoid constituent of MC exhibits *in vivo* hypoglycemic effect while the aglycones of charantin did not reveal any marked effect on blood glucose
Ojewole et al. [57]	Wistar rats *Treatment groups:* normal or STZ-induced diabetic rats ($N = 8$/group) 1. Water (control) 2. Protein extract of MC fruit (1, 5, or 10 mg/kg) *Treatment duration:* one time before blood collection	• Acute oral treatment of moderate to high doses of MC extract produced dose-dependent significant reductions in plasma glucose concentrations in both normal and diabetic rats
Yibchok-anun et al. [25]	Male Wistar rats *Treatment groups:* normal or STZ-induced diabetic rats ($N = 8$/group) 1. Control 2. Fruit acid ethanol extract (1, 5, 10 mg/kg) *Treatment duration:* one time before blood collection	• MC protein extract (5, 10 mg/kg) significantly and markedly reduced plasma glucose concentrations in both normal and diabetic rats • MC extract also increased plasma insulin concentrations by two-fold 4 hours following subcutaneous injection
Fernandes et al. [49]	Albino Wistar rats *Treatment groups:* alloxan-induced diabetic rats ($N = 8$/group) 1. Normal control 2. Alloxan-induced diabetes	• MC normalized blood glucose, glycosylated hemoglobin, lipids, insulin, total protein, and liver glycogen

(Continued)

C. ACTIONS OF INDIVIDUAL FRUITS IN DISEASE AND CANCER PREVENTION AND TREATMENT

TABLE 35.2 (Continued)

Author(s)	Species and Strain Treatment Groups and MC Preparation and Study Duration	Results
	a. Control b. Standardized MC extract (150, 300 mg/kg) c. Glibenclamide (4 mg/kg) *Treatment duration:* 30 days	
Oishi et al. [102]	Male Wistar rats *Treatment groups:* ($N = 6$/group) 1. Control 2. Water-soluble fraction of MC (100 or 1000 mg/kg bw) 3. Saponin fraction of MC (50 or 100 mg/kg bw) *Treatment duration:* 10 min before sucrose load	• The saponin fraction exhibited a stronger inhibitory effect on the elevation of the blood glucose level than the water-soluble fraction
Han et al. [103]	Female Kunming mice *Treatment groups:* ($N = 12$/group) 1. Control 2. Alloxan-induced hyperglycemic 3. Alloxan + Xiaoke pill (Chinese antidiabetes medicine; 200 mg/kg/d) 4. Alloxan + saponin fraction (100, 200, or 500 mg/kg/d) *Treatment duration:* 20 days	• Similar to Xiaoke pill, the highest dose of SF decreased blood glucose and increased insulin secretion and glycogen synthesis in alloxan-induced hyperglycemic mice
Nerurkar et al. [54]	Female C57BL/6 mice *Treatment groups:* ($N = 7$/group) 1. Control 2. High fat diet (HFD) 3. HFD + 1.5% freeze-dried MC fruit juice *Treatment duration:* 16 weeks	• MC juice not only improves glucose and insulin tolerance but also lowers plasma apo-B100 and apoB-48 in HFD-fed mice • Primary mechanism by which MC juice restores hepatic glucose and lipid metabolism is due to increased post-insulin receptor (IR) signal transduction linked to tyrosine phosphorylation of IRβ and IRS proteins leading to P13K activation

Shih et al. [56]	Male C57BL/6 mice
	Treatment groups:
	
	1. Low fat diet (N = 7)
	2. High fat diet (HFD) for 8 weeks
	  a. HFD
	  b. HFD + fruit water extract (0.5 or 1.0 g/kg bw/d)
		• MC ameliorated hyperglycemia and hyperleptinemia; reduced hemoglobin A1c (HbA1c) and plasma free fatty acid; increased adipose peroxisome proliferator-activated receptors (PPAR)γ and liver PPARα gene expression

(Continued)

TABLE 35.2 (Continued)

Author(s)	Species and Strain Treatment Groups and MC Preparation and Study Duration	Results
	c. HFD + fruit methanol extract (0.2 or 1.0 g/kg bw/d)	
	d. HFD + rosiglitazone (10 mg/kg bw/d)	
	Treatment duration: 4 weeks	
Singh et al. [104]	Albino Wistar rats (alloxan-induced diabetes) *Treatment groups:* (N = 18/group)	• All doses of extract significantly decreased blood glucose level at all time points – even after discontinuation of extract for 15 days
	1. Control	• MC extract improved granulation of pancreatic β cells and the size of the islets
	2. Whole fruit alcoholic extract (25, 50, or 100 mg/kg bw)	
	Treatment duration: 15 d, 30 d, 30 d + another 15 d normal diet (N = 6/time point)	
Sridhar et al. [105]	Male Wistar rats *Treatment groups:* (N = 8/group)	• MC extract improved glucose tolerance, insulin sensitivity, skeletal muscle insulin signaling, plasma lipids, and adiposity
	1. Control diet for 10 weeks	• MC improved insulin-stimulated insulin receptor substrate-1 (IRS-1) tyrosine phosphorylation compared to HFD alone
	2. High fat diet (HFD) for 10 weeks	
	3. HFD for 8 weeks and HFD + MC juice (10 mL/kg/d) for 2 weeks	
Tan et al. [106]	Male C57/BL6 mice fed normal or high fat diet for 7 weeks	• Momordicoside S significantly enhanced glucose removal from the circulation and promoted fatty acid oxidation
	Treatment groups: (N = 8/group)	
	1. Control	• Momordicoside S is more potent than the AMPK agonist, AICAR, in modulating glucose clearance during glucose tolerance test
	2. Momordicoside S (100 mg/kg)	
	3. Momordicoside T (10 mg/kg)	
	4. Metformin (200 mg/kg)	
	5. AICAR (500 mg/kg; AMPK agonist)	
	Treatment duration: one time before glucose tolerance test	

AICAR, 5-aminoimidazole-4-carboxamide-1-β-4-ribofuranoside; AMPK, AMP-activated protein kinase; bw, body weight; CMC, carboxymethyl cellulose; EJ, *Eugenia jambolana*; HFD, high fat diet; IDDM, insulin-dependent diabetes mellitus; MC, *Momordica charantia*; MP; *Mucuna pruriens*; NIDDM, non-insulin-dependent diabetes mellitus; OGTT, oral glucose tolerance test; STZ, streptozotocin ; TC, *Tinospora cordifolia*.

Ahmed and colleagues [36] demonstrated that oral feeding of MC juice prevented the death of pancreatic β cells and/or facilitated the recovery of partially destroyed β cells of STZ-induced diabetic rats. Using a pancreatic perfusion technique, Yibchok-anun and colleagues [25] showed that MC has stimulatory effects on insulin secretion, but not glucagon secretion, through its direct action on β cells of the pancreas. An aqueous extract of MC fruit was found to be a potent stimulator of insulin release from isolated β-cell-rich pancreatic islets obtained from obese hyperglycemic mice [37].

TABLE 35.3 Clinical Studies on Hypoglycemic Properties of *Momordica charantia* (MC)

Author(s)	Subject Characteristics	Treatment Groups and MC Preparation	Study Duration/ Mode of Administration	Results
Patel et al. [107]	Diabetic patients ($N = 10$)	Juice ($N = 4$) Powder ($N = 3$) Powder in capsule ($N = 2$) All three forms ($N = 1$)	4 oz 3 × day 16 g 3 × day 16 g 3 × day	• No significant effect on glucose tolerance test (GTT) by any of these MC preparations • Only one subject exhibited improvement of GTT after taking MC powder for 12 weeks
Baldwa et al. [108]	*Experimental group:* patients diagnosed with diabetes ($N = 9$) *Control group:* healthy volunteers ($N = 5$) and with overt diabetes ($N = 5$)	Purified protein extract from fruit and tissue cultures (v-insulin)	One time subcutaneous injection Dose dependent on fasting sugar level	• V-insulin has hypoglycemic effect in diabetic patients • Peak effect was seen after 4–12 hours of injection
Khanna et al. [7]	*Experimental group:* patients with juvenile diabetes ($N = 5$) and maturity onset diabetes ($N = 6$) *Control group:* patients with juvenile diabetes ($N = 6$) and maturity onset diabetes ($N = 2$)	Purified protein extract from fruit, seeds, and tissue cultures (polypeptide p)	One time subcutaneous injection Dose dependent on fasting sugar level	• Hypoglycemic effect on both groups • Peak effect is 4–8 hours in juvenile diabetes • Peak response in maturity onset diabetes not as readily determined
Leatherdale et al. [109]	Diabetic patients ($N = 9$)	Fruit juice Fried fruit	50 mL one time 0.23 kg for 8–11 wks daily	• Hypoglycemic effect during an oral glucose tolerance test (OGTT) • Smaller effect compared to juice on OGTT but has lower hemoglobin A1c (Hb1Ac).
Akhtar [110]	Diagnosed maturity onset diabetic patients ($N = 8$)	Dried and powdered whole fruit	Taken with milk 2 × per day (50 mg/ kg bw) for 7 days	• Mean blood sugar was significantly lower (after 50 g glucose load) after treatment compared to baseline
Welihinda, et al. [111]	Newly diagnosed maturity onset diabetic patients ($N = 18$)	Fruit juice	100 mL one time at 30 min before OGTT vs. water	• 73% of patient had improved glucose tolerance while 27% failed to respond
Grover and Gupta [112]	Non-insulin-dependent diabetes mellitus (NIDDM) subjects ($N = 14$)	Seeds	No information	• Postprandial glucose values: 350–380 and 150–180 mg/dL for with and without MC seeds, respectively

(Continued)

TABLE 35.3 (Continued)

Author(s)	Subject Characteristics	Treatment Groups and MC Preparation	Study Duration/ Mode of Administration	Results
	Insulin-dependent diabetes mellitus (IDDM) subjects ($N = 6$)			
Srivastava et al. [113]	Type II diabetics ($N = 12$)	Aqueous extract (100 g chopped MC in 100 mL) vs. 15 g of dried fruit powder	Single morning dose for the extract vs. 3 × daily for dried fruit powder for 3 weeks	• Significant reductions in blood glucose (54%) and HbA1c (17%) with aqueous extract • Non-significant reductions in blood glucose (25%) for dried fruit powder
Ahmad et al. [114]	Moderate NIDDM patients ($N = 100$)	Aqueous suspension of the vegetable pulp	Single dose [body wt (kg) × 2 dissolved in 150–200 mL H_2O] before OGTT	• Eighty six percent of the study participants showed a significant reduction in both fasting and postprandial glucose
Rosales and Fernando [115]	Type II diabetics on oral hypoglycemic agents and have not achieved adequate glycemic control ($N = 27$)	3 g of dried fruit and seeds and brewed as tea vs. Lipton tea	1 glass (200 mL) after each major meals, cross-over (12 weeks on each treatment)	• Significant reduction (0.63%) in HbA1c • No effect on weight and blood pressure and minor gastrointestinal side effects
Dans et al. [90]	Newly diagnosed or poorly controlled type II diabetics ($N = 40$)	Capsule (extract of fruit and seed) vs. placebo	Two capsules 3 × a day for 3 months	• Slight change in HbA1c (0.22% in favor) and no significant effects on fasting sugar, total cholesterol, weight and other clinical parameters

IDDM, insulin-dependent diabetes mellitus; MC, *Momordica charantia*; NIDDM, non-insulin-dependent diabetes mellitus; OGTT, oral glucose tolerance test.

Finally, another potential hypoglycemic mechanism of action of MC is by decreasing Na^+ and K^+-dependent glucose absorption at the brush border membrane [38,39] and inhibiting intestinal monosaccharide and amino acid uptake [40]. Moreover, MC has been shown to modulate activities of intestinal and renal disaccharidases in STZ-induced diabetic rats [41].

Complications associated with diabetes such as neuropathy and microvascular damage to the eyes and kidneys have also been reduced by supplementation with MC. For instance, MC supplementation in STZ-induced diabetic rats normalized structural abnormalities of peripheral nerves [39] and restored kidney function as measured by creatinine, albumin, urine volume, and gross tissue weight [41,42]. MC treatments of diabetic rats also significantly delayed the onset of retinopathy [43–45]. Taken together, these very promising findings demonstrate that MC may provide an alternative treatment for lowering blood glucose and reducing some of the clinical complications associated with diabetes mellitus; however, well-designed clinical trials establishing the safety and efficacy of MC are needed before recommendations for humans should be made.

2.2 Dyslipidemia, Obesity, and Hypertension

Obesity and dyslipidemia are major co-morbid factors contributing to the development of diabetes. Accordingly, MC has also been investigated primarily using animal models, for its role in preventing these two co-morbid conditions. In hamsters fed a high-cholesterol diet, supplementation with methanol fraction of MC was able to significantly reduce hypertriglyceridemia and hypercholesterolemia in a dose-dependent manner [46]. Marginal results, on the other hand, were obtained on the effect of this methanol fraction on liver triglyceride and total cholesterol levels [46]. However, the findings of Jayasooriya and colleagues [47] using freeze-dried MC fed to normolipidemic and hypercholesterolemic rats differed from those observed in hamsters using an MC methanol fraction [46]. Freeze-dried MC reduced hepatic cholesterol and triglyceride content but had little effect on serum lipid parameters, except for high density lipoprotein (HDL)-cholesterol, which was consistently increased by MC either in the presence or absence of dietary cholesterol [47]. Similarly, mice fed a high fat diet supplemented with MC exhibited reductions in both hepatic and skeletal muscle triglyceride concentrations [48]. MC fruit has also been shown to restore lipid profile (lowering of total cholesterol and triglycerides while elevating HDL-cholesterol levels) to nearly normal levels in animal models of diabetes [49]. Aqueous extract of both the MC fruit and seeds appears to be capable of reducing very low density lipoprotein (VLDL) in normal and diabetic rats [50]. In a study comparing the effects of three different varieties of MC (Koimidori, Powerful-Reishi, and Hyakunari), the Koimidori variety was found to be the most effective in lowering hepatic triglycerides in rats as compared to the other two varieties, suggesting a variety-dependent difference in their activity [51].

The mechanism(s) responsible for the hypolipidemic effects of MC remains to be determined. It has been reported that MC contains active components such as saponin and plant sterols that may possibly influence lipid metabolism [52]. Saponins have been shown to exhibit hypotriglyceridemic activity through the inhibition of triglyceride synthesis in the liver [53]. Others have also shown that MC juice normalizes apoB-100 and apoB-48 in mice fed a high fat diet, possibly through modulation of the insulin signaling pathway [54]. The primary mechanism of MC's action to restore hepatic glucose and lipid metabolism is due to increased post-insulin receptor (IR) signaling linked to phosphorylation of IRβ and IRS proteins ultimately leading to PI3K activation [54].

In addition to its hypoglycemic and hypolipidemic effects, long-term supplementation of MC has been shown to reduce visceral fat mass and energy efficiency in rats [21,48,55]. The anti-obesity effects of MC are attributed to enhanced sympathetic activity and lipolytic processes as suggested by elevated plasma catecholamines and free fatty acids [55]. MC was also shown to enhance mitochondrial uncoupling and lipid oxidation due to increased mitochondrial transport of long-chain fatty acids and activity of acyl-coA dehydrogenase, a key enzyme in fatty acid oxidation [48]. Similar to the previous reports by Chuang et al. [34] and Chao and Huang [35], Shih and colleagues [56] speculated that MC contains compounds that serve as ligands for PPARs and positively regulate hepatic fatty acid oxidation, thereby lowering blood lipid. Additionally, these compounds may also regulate adipocytokine production, resulting in improved insulin sensitivity. In fact, the most potent compound in MC capable of activating PPAR was isolated and identified to be 9cis,

11*trans*, 13*trans*-conjugated linolenic acid (9*c*, 11*t*, 13*t*-CLN) [34]. It remains to be determined whether this compound can alter visceral fat mass in other animal models of obesity and even humans, but given the success of other PPAR agonists as useful therapeutic agents, further investigation of this issue is warranted.

To our knowledge, only one group has reported the effects of MC on blood pressure. Ojewole and colleagues [57] found that MC extract significantly reduced systemic arterial blood pressure and heart rates of both normotensive and hypertensive rats in a dose-dependent manner. It is postulated that the hypotensive effect of MC may be due in part to altering adrenergic mechanisms by blocking peripheral α- and/or β-adrenoreceptor effects of endogenous noradrenaline and/or adrenaline on blood vessels and cardiac muscles. While these findings related to antihypertensive actions of MC are very intriguing, considerably more work is needed to corroborate these preliminary findings.

2.3 Cancer

A few studies have also been conducted to investigate the proposed anticancer properties of MC. Crude extract of MC reduced the occurrence of prostate carcinoma in rats [58] and inhibited tumor formation in mice given injections of tumor cells to induce lymphoma [59]. Development of mammary tumors was also significantly inhibited in mice given free access to drinking water containing 0.5% MC extract [60]. MC fruit extract also significantly reduced tumor incidence of benzo(a)pyrene induced fore-stomach tumorigenesis in mice [61]. A study by Guevara and colleagues [62] identified 3-*O*-[6′-*O*-palmitoyl-β-D-glucosyl-stigmasta-5,25(27)-dien as the major component and stearyl derivative as the minor component of MC extract responsible for the reduction in the number of micronucleated polychromatic erythrocytes induced by the mutagen mitomycin C in mice. Extracts of MC peel, pulp, seed, and whole fruit were also shown to inhibit mouse skin papillomagenesis [63]. Although the findings of the animal studies are promising, to date there are no human studies to demonstrate the inhibitory effect of MC on tumor formation. The only cancer-related human study with MC explored its role in cervical cancer patients undergoing radiotherapy [64]. Their findings showed that ingestion of MC is not beneficial for cervical cancer patients when treated with radiation but reduced the multidrug-resistant phenomenon, which may be useful for patients being treated with chemotherapy [64].

In vitro studies have also been utilized to investigate the anticancer properties of MC and identify the active component(s) and mechanism(s) of action. A study comparing extracts from MC, soybean, dokudami, and welsh onion demonstrated that MC is the most potent inhibitor of the P-glycoprotein (P-gp) activity in Caco-2 intestinal cells [65]. P-gp is expressed at the apical surface of intestinal epithelial cells and involved in the regulation of intestinal absorption of pharmacotherapeutics and other xenobiotic compounds [66]. A component of MC, 1-monopalmitin, is the most potent inhibitor of P-gp in MC [65]. P-gp also acts as an ATP-driven efflux pump that decreases intracellular drug accumulation, thereby potentially decreasing the effectiveness of chemotherapeutic agents [67,68]. Limtrakul and colleagues [69] demonstrated that MC leaf extract inhibited P-gdp-mediated drug efflux resulting in an increase in the intracellular accumulation and cytotoxicity of chemotherapeutic drugs in drug-resistant human cervical carcinoma.

Another anticancer mechanism of the action of MC is through inhibition of guanylate cyclase activity. The guanylate cyclase–cyclic GMP system is intimately involved in cell proliferation, DNA and RNA synthesis, and possibly malignant transformation [70,71]. A guanylate cyclase inhibitor was isolated from the ripe fruits and leaves of MC and shown to inhibit the growth of the rat prostatic adenocarcinoma *in vitro* and reduce [³H]thymidine incorporation into DNA [72,73]. Two ribosome inactivating proteins, momorcharin and momordin, were isolated from MC and shown to inhibit both RNA and protein synthesis. Ng and colleagues [74] showed that one of these proteins, momorcharin, suppressed human choriocarcinoma and melanoma. Free fatty acids from MC seed oil rich in *9cis,11trans-conjugated linoleic acid* were shown to induce apoptosis in Caco-2 cells through increased expression of growth arrest and DNA damage inducible gene 45 (GADD45), the tumor suppressor gene p53, and the gene encoding the nuclear hormone receptor PPARγ [75]. GADD45 and p53 are known to play important roles in growth inhibition and apoptosis in many types of cancer cells [76,77]. PPARγ is being investigated as a target molecule for cancer prevention [78,79]. Additional components of MC extract, α-eleostearic acid and its dihydroxy derivative, were shown to strongly inhibit the growth of some cancer and fibroblast cell lines such as HL60 leukemia and HT 29 colon carcinoma [80].

2.4 Immune Function and Other Effects

Inflammation has been implicated in the etiology of many chronic diseases, and MC exhibits anti-inflammatory properties. Extracts of MC reduced the expression of tumor necrosis factor (TNF)-α and other pro-inflammatory genes in RAW 264.7 macrophages [81]. The anti-inflammatory effect of MC was thought to be due to linoleoyl- and linoleoyl-lysophosphatidyl-choline [81]. MC was also shown to induce both intestinal and systematic anti-inflammatory responses both *in vitro* and *in vivo* by decreasing interleukin (IL)-7

secretion and the number of lymphocytes. Further, MC also increases the secretion of both transforming growth factor (TGF)-β and IL-10, increases populations of helper T (Th) cells and natural killer cells, and stimulates immunoglobulin production of lymphocytes [82]. Interestingly, at least in mice, MC pulp induced interferon-gamma (IFN-γ) production and may therefore provide a potential source of immunostimulatory therapy specific for Th1 cells and IFN-γ production [83]. MC protein (MRK29) isolated from MC ripe fruit and seed was shown to inhibit HIV-1 reverse transcriptase [84].

Momorcharin isolated from MC seeds is an extremely potent modulator of a variety of cell-mediated and humoral immune responses, both *in vivo* and *in vitro* [85]. The observed immunosuppressive effect of momorcharin proteins is not likely to be due to direct lymphocytotoxicity or to a shift in the kinetic parameter of the immune response, but rather may be through its selective cytostatic action on the immune cells [85]. Cytostatic activity prevents both concanavalin A-stimulated thymidine incorporation into human lymphocytic DNA and the subsequent induction of a cyclic AMP phosphodiesterase [86].

The antimicrobial activity of MC seed essential oil, with *trans*-nerolidol, apiole, *cis*-dihydrocarveol, and germacrene as its main constituents, was found to be more effective against *S. aureus* than against *E. coli* or *C. albicans* [87]. Lyophilized MC extracts prepared from accessions collected in Togo showed antiviral activity against Sindbis and herpes simplex type 1 viruses and anthelmintic activity against *Caenorhabditis elegans*. The triterpene glycosides momordicins I and II were found to be anthelmintic but not antiviral [88].

The mature fruits of MC were also tested for anti-ulcerogenic activity on various ulcer models in rats [89]. The olive oil extract of MC as well as dried powdered fruits in filtered honey showed significant and dose-dependent anti-ulcerogenic activity against ethanol-induced ulcerogenesis model in rats. A potent and dose-dependent inhibitory activity was also observed by the administration of ethanol and hexane extracts of the fruit against the same ulcer model. Furthermore, ethanol extract of the fruit showed significant activity against HCl-EtOH induced ulcerogenesis in indomethacin-pretreated rats and diethyldithiocarbamate-induced ulcer models [89].

3. SIDE EFFECTS AND TOXICITY

Despite several experiments demonstrating the potential health benefits of MC, adverse effects have also been reported. Table 35.4 summarizes the endocrine, gastrointestinal, genitourinary, hematologic, hepatic, and neurological adverse effects of MC. Some side effects reported in humans include diarrhea [90] and headaches [91]. Hypoglycemic coma and convulsions were also reported in children after administration of MC tea [52,91,92]. Vicine, a glycosidic compound isolated from MC, has been shown to induce favism [91]. Because of the limited human studies with MC, the severity of these side effects and the doses that produce these effects are poorly understood. As stated above, if we are to examine the efficacy and safety of MC supplementation, we must utilize well-designed and properly controlled clinical research trials.

In experimental animals, MC was found to be generally safe when ingested in low doses for up to a 2-month study period [1,50,93]. However, acute toxicity studies of El Batran and colleagues [94] reported that toxicity may occur at higher doses. MC juice was found to be more potent than the alcoholic extract (LD_{50} of MC juice and alcoholic extract is 91.9 and 362.34 mg/100 g body weight, respectively). A high dose MC alcoholic extract induced toxic symptoms in mice that included increased respiratory rate and strong heart beats [94].

TABLE 35.4 Adverse Effects of *Momordica charantia* in *Various Body Systems*

Body System	Effects
Endocrine	Hypoglycemic coma and convulsions in children after administration of MC tea [52,91,92]
Gastrointestinal	Report of bouts of diarrhea after taking MC for 2 months [90]
	Inhibition of protein synthesis in the intestinal wall due to the toxic lectin from MC seeds and outer rind. However, no correlation has been done with clinical signs or symptoms in humans [116]
Genitourinary	Isolated momorcharins from MC induced mid-term abortion and terminated early pregnancy in mice [85,117–119]
	Inhibition of spermatogenesis in dogs fed MC fruit extract for 60 days [120]
	Fertility dropped significantly in mice fed MC daily [121]
	MC juice caused uterine hemorrhage and death in two pregnant rabbits which was not observed in non-pregnant rabbits [122]
Hematologic	Individuals with glucose-6-phosphate dehydrogenase deficiency may develop favism, characterized by the onset of hemolytic anemia, after ingestion of MC seeds [52]
	Vicine, a glycosidic compound isolated from MC, is known to induce favism [91]
Hepatic	Increases in γ-glutamyltransferase and alkaline phosphatase levels in animals after oral intake of MC fruit juice and seed extract [123]
Neurological	Headaches after ingestion of MC seeds [52]

Post-mortem examination showed general congestion of internal organs, particularly the lung and heart [94]. High doses of MC juice extract caused abdominal convulsions, nervous disturbance, respiratory problems, and possibly death. Post-mortem examination showed congestion of lungs, heart, abdomen, and intestine; the heart was engorged with blood, and the mucous membrane of the eye was cyanosed [94]. These findings highlight the fact that certain extract preparations may not be safe and alternative preparations or means of incorporating reasonable amounts of MC into the diet should be considered.

4. SUMMARY

MC contains many diverse compounds that may contribute to its proposed health benefits. Numerous animal studies have been conducted to investigate its role in diabetes, dyslipidemia, adiposity, cancer, and inflammation. However, well-designed and controlled human studies are lacking to establish its efficacy. Moreover, standardization of MC preparation and dose needs to be undertaken, and information on side effects and toxicity in humans obtained.

References

1. Grover, J. K., & Yadav, S. P. (2004). Pharmacological actions and potential uses of *Momordica charantia*: A review. *Journal of Ethnopharmacology, 93*, 123–132.
2. Yang, S., & Walters, T. (1992). Ethnobotany and the economic role of cucurbitaceae of China. *Economic Botany, 46*, 349–367.
3. (2008). USDA National Nutrient Database for Standard Reference, Release 21. <http://www.nal.usda.gov/fnic/foodcomp/cgi-bin/list_nut_edit.pl/>.
4. Yen, G. C., & Hwang, L. S. (1985). Lycopene from the seeds of ripe bitter melon *(Momordica charantia)* as a potential red food colorant. II. Storage stability, preparation of powdered lycopene and food applications. *Journal of Chinese Agricultural Chemical Society, 23*, 151–161.
5. Okabe, H., Miyahara, Y., & Yamauchi, T. (1982). Studies on the constituents of *Momordica charantia L.* IV. Characterization of the new cucurbitacin glycosides of the immature fruits. (2). Structures of the bitter glycosides, momordicosides K and L. *Chemical and Pharmaceutical Bulletin (Tokyo), 30*, 4334–4340.
6. Yasuda, M., Iwamoto, M., Okabe, H., & Yamauchi, T. (1984). Structures of momordicines I, II, and III, the bitter principles in the leaves and vines of *Momordica charantia L. Chemical and Pharmaceutical Bulletin (Tokyo), 32*, 2044–2047.
7. Khanna, P., Jain, S. C., Panagariya, A., & Dixit, V. P. (1981). Hypoglycemic activity of polypeptide-p from a plant source. *Journal of Natural Products, 44*, 648–655.
8. Lotlikar, M. M., & Rajarama Rao, M. R. (1966). Pharmacology of a hypoglycaemic principle isolated from fruit of *Momordica charantia Linn. Industrial Journal of Pharmacology, 28*, 129–133.
9. Matsuda, H., Li, Y., Murakami, T., Matsumura, N., Yamahara, J., & Yoshikawa, M. (1998). Antidiabetic principles of natural medicines. III. Structure-related inhibitory activity and action mode of oleanolic acid glycosides on hypoglycemic activity. *Chemical and Pharmaceutical Bulletin (Tokyo), 46*, 1399–1403.
10. Murakami, T., Emoto, A., Matsuda, H., & Yoshikawa, M. (2001). Medicinal foodstuffs. XXI. Structures of new cucurbitane-type triterpene glycosides, goyaglycosides-a, -b, -c, -d, -e, -f, -g, and -h, and new oleanane-type triterpene saponins, goyasaponins I, II, and III, from the fresh fruit of Japanese *Momordica charantia L. Chemical and Pharmaceutical Bulletin (Tokyo), 49*, 54–63.
11. Harinantenaina, L., Tanaka, M., Takaoka, S., Oda, M., Mogami, O., Uchida, M., & Asakawa, Y. (2006). *Momordica charantia* constituents and antidiabetic screening of the isolated major compounds. *Chemical and Pharmaceutical Bulletin (Tokyo), 54*, 1017–1021.
12. Miyahara, Y., Okabe, H., & Yamauchi, T. (1981). Studies on the constituents of *Momordica charantia L.* II. Isolation and characterization of minor seed glycosides, Momordicosides C, D, and E. *Chemical and Pharmaceutical Bulletin (Tokyo), 29*, 1561–1566.
13. Okabe, H., Miyahara, Y., Yamauchi, T., Miyahara, K., & Kawasaki, T. (1980). Studies on the constituents of *Momordica charantia L.* I. Isolation and characterization of momordicosides A and B, glycosides of a penta-hydroxy-cucurbitane triterpene. *Chemical and Pharmaceutical Bulletin (Tokyo), 28*, 2753–2762.
14. Chang, C. I., Chen, C. R., Liao, Y. W., Cheng, H. L., Chen, Y. C., & Chou, C. H. (2006). Cucurbitane-type triterpenoids from *Momordica charantia. Journal of Natural Products, 69*, 1168–1171.
15. Cheng, H. L., Huang, H. K., Chang, C. I., Tsai, C. P., & Chou, C. H. (2008). A cell-based screening identifies compounds from the stem of *Momordica charantia* that overcome insulin resistance and activate AMP-activated protein kinase. *Journal of Agricultural and Food Chemistry, 56*, 6835–6843.
16. Chen, J., Tian, R., Qiu, M., Lu, L., Zheng, Y., & Zhang, Z. (2008). Trinorcucurbitane and cucurbitane triterpenoids from the roots of *Momordica charantia. Phytochemistry, 69*, 1043–1048.
17. Budrat, P., & Shotipruk, A. (2008). Extraction of phenolic compounds from fruits of bitter melon

(*Momordica charantia*) with subcritical water extraction and antioxidant activities of these extracts. *Chiang Mai Journal of Science, 35*, 123–130.

18. Jantan, I., Rafi, I. A., & Jalil, J. (2005). Platelet-activating factor (PAF) receptor-binding antagonist activity of Malaysian medicinal plants. *Phytomedicine, 12*, 88–92.

19. Anila, L., & Vijayalakshmi, N. R. (2000). Beneficial effects of flavonoids from *Sesamum indicum, Emblica officinalis* and *Momordica charantia. Phytotherapy Research, 14*, 592–595.

20. Karunanayake, E. H., Welihinda, J., Sirimanne, S. R., & Sinnadorai, G. (1984). Oral hypoglycaemic activity of some medicinal plants of Sri Lanka. *Journal of Ethnopharmacology, 11*, 223–231.

21. Chen, Q., Chan, L. L., & Li, E. T. (2003). Bitter melon (*Momordica charantia*) reduces adiposity, lowers serum insulin and normalizes glucose tolerance in rats fed a high fat diet. *The Journal of Nutrition, 133*, 1088–1093.

22. Chaturvedi, P., George, S., Milinganyo, M., & Tripathi, Y. B. (2004). Effect of *Momordica charantia* on lipid profile and oral glucose tolerance in diabetic rats. *Phytotherapy Research, 18*, 954–956.

23. Welihinda, J., & Karunanayake, E. H. (1986). Extra-pancreatic effects of *Momordica charantia* in rats. *Journal of Ethnopharmacology, 17*, 247–255.

24. Sathishsekar, D., & Subramanian, S. (2005). Beneficial effects of *Momordica charantia* seeds in the treatment of STZ-induced diabetes in experimental rats. *Biological and Pharmaceutical Bulletin, 28*, 978–983.

25. Yibchok-anun, S., Adisakwattana, S., Yao, C. Y., Sangvanich, P., Roengsumran, S., & Hsu, W. H. (2006). Slow acting protein extract from fruit pulp of *Momordica charantia* with insulin secretagogue and insulinomimetic activities. *Biological and Pharmaceutical Bulletin, 29*, 1126–1131.

26. Cummings, E., Hundal, H. S., Wackerhage, H., Hope, M., Belle, M., Adeghate, E., & Singh, J. (2004). *Momordica charantia* fruit juice stimulates glucose and amino acid uptakes in L6 myotubes. *Molecular and Cellular Biochemistry, 261*, 99–104.

27. Roffey, B. W., Atwal, A. S., Johns, T., & Kubow, S. (2007). Water extracts from *Momordica charantia* increase glucose uptake and adiponectin secretion in 3T3-L1 adipose cells. *Journal of Ethnopharmacology, 112*, 77–84.

28. Miura, T., Itoh, C., Iwamoto, N., Kato, M., Kawai, M., Park, S. R., & Suzuki, I. (2001). Hypoglycemic activity of the fruit of the Momordica charantia in type 2 diabetic mice. *Journal of Nutritional Science and Vitaminology (Tokyo), 47*, 340–344.

29. Shibib, B. A., Khan, L. A., & Rahman, R. (1993). Hypoglycaemic activity of *Coccinia indica* and *Momordica charantia* in diabetic rats: Depression of the hepatic gluconeogenic enzymes glucose-6-phosphatase and fructose-1,6-bisphosphatase and elevation of both liver and red-cell shunt enzyme glucose-6-phosphate dehydrogenase. *The Biochemical Journal, 292*(Pt 1), 267–270.

30. Sarkar, S., Pranava, M., & Marita, R. (1996). Demonstration of the hypoglycemic action of *Momordica charantia* in a validated animal model of diabetes. *Pharmacological Research, 33*, 1–4.

31. Day, C., Cartwright, T., Provost, J., & Bailey, C. J. (1990). Hypoglycaemic effect of *Momordica charantia* extracts. *Planta Medica, 56*, 426–429.

32. Sekar, D. S., Sivagnanam, K., & Subramanian, S. (2005). Antidiabetic activity of *Momordica charantia* seeds on streptozotocin induced diabetic rats. *Pharmazie, 60*, 383–387.

33. Rathi, S. S., Grover, J. K., & Vats, V. (2002). The effect of *Momordica charantia* and *Mucuna pruriens* in experimental diabetes and their effect on key metabolic enzymes involved in carbohydrate metabolism. *Phytotherapy Research, 16*, 236–243.

34. Chuang, C. Y., Hsu, C., Chao, C. Y., Wein, Y. S., Kuo, Y. H., & Huang, C. J. (2006). Fractionation and identification of 9c, 11t, 13t-conjugated linolenic acid as an activator of PPARalpha in bitter gourd (*Momordica charantia* L.). *Journal of Biomedical Science, 13*, 763–772.

35. Chao, C. Y., & Huang, C. J. (2003). Bitter gourd (*Momordica charantia*) extract activates peroxisome proliferator-activated receptors and upregulates the expression of the acyl CoA oxidase gene in H4IIEC3 hepatoma cells. *Journal of Biomedical Science, 10*, 782–791.

36. Ahmed, I., Adeghate, E., Sharma, A. K., Pallot, D. J., & Singh, J. (1998). Effects of *Momordica charantia* fruit juice on islet morphology in the pancreas of the streptozotocin-diabetic rat. *Diabetes Research and Clinical Practice, 40*, 145–151.

37. Welihinda, J., Arvidson, G., Gylfe, E., Hellman, B., & Karlsson, E. (1982). The insulin-releasing activity of the tropical plant *Momordica charantia. Acta Biologica Et Medica Germanica, 41*, 1229–1240.

38. Meir, P., & Yaniv, Z. (1985). An *in vitro* study on the effect of *Momordica charantia* on glucose uptake and glucose metabolism in rats. *Planta Medica*, 12–16.

39. Singh, J., Adeghate, E., Cummings, E., Giannikipolous, C., Sharma, A. K., & Ahmed, I. (2004). Beneficial effects and mechanism of action of *Momordica charantia* juice in the treatment of streptozotocin-induced diabetes mellitus in rat. *Molecular and Cellular Biochemistry, 261*, 63–70.

40. Mahomoodally, M. F., Fakim, A. G., & Subratty, A. H. (2004). *Momordica charantia* extracts inhibit uptake

C. ACTIONS OF INDIVIDUAL FRUITS IN DISEASE AND CANCER PREVENTION AND TREATMENT

of monosaccharide and amino acid across rat everted gut sacs *in vitro*. *Biological and Pharmaceutical Bulletin*, *27*, 216–218.

41. Shetty, A. K., Kumar, G. S., Sambaiah, K., & Salimath, P. V. (2005). Effect of bitter gourd (*Momordica charantia*) on glycaemic status in streptozotocin induced diabetic rats. *Plant Foods for Human Nutrition*, *60*, 109–112.

42. Grover, J. K., Vats, V., Rathi, S. S., & Dawar, R. (2001). Traditional Indian anti-diabetic plants attenuate progression of renal damage in streptozotocin induced diabetic mice. *Journal of Ethnopharmacology*, *76*, 233–238.

43. Rathi, S. S., Grover, J. K., Vikrant, V., & Biswas, N. R. (2002). Prevention of experimental diabetic cataract by Indian Ayurvedic plant extracts. *Phytotherapy Research*, *16*, 774–777.

44. Srivastava, Y., Venkatakrishna-Bhatt, H., Verma, Y., & Prem, A. S. (1987). Retardation of retinopathy by *Momordica charantia* L. (bitter gourd) fruit extract in alloxan diabetic rats. *Indian Journal of Experimental Biology*, *25*, 571–572.

45. Srivastava, Y., Venkatakrishna-Bhatt, H., & Verma, Y. (1988). Effect of *Momordica charantia Linn.* pomous aqueous extract on cataractogenesis in murrin alloxan diabetics. *Pharmacological Research Communications*, *20*, 201–209.

46. Senanayake, G. V., Maruyama, M., Sakono, M., Fukuda, N., Morishita, T., Yukizaki, C., Kawano, M., & Ohta, H. (2004). The effects of bitter melon (*Momordica charantia*) extracts on serum and liver lipid parameters in hamsters fed cholesterol-free and cholesterol-enriched diets. *Journal of Nutritional Science and Vitaminology (Tokyo)*, *50*, 253–257.

47. Jayasooriya, A. P., Sakono, M., Yukizaki, C., Kawano, M., Yamamoto, K., & Fukuda, N. (2000). Effects of *Momordica charantia* powder on serum glucose levels and various lipid parameters in rats fed with cholesterol-free and cholesterol-enriched diets. *Journal of Ethnopharmacology*, *72*, 331–336.

48. Chan, L. L., Chen, Q., Go, A. G., Lam, E. K., & Li, E. T. (2005). Reduced adiposity in bitter melon (*Momordica charantia*)-fed rats is associated with increased lipid oxidative enzyme activities and uncoupling protein expression. *The Journal of Nutrition*, *135*, 2517–2523.

49. Fernandes, N. P., Lagishetty, C. V., Panda, V. S., & Naik, S. R. (2007). An experimental evaluation of the antidiabetic and antilipidemic properties of a standardized *Momordica charantia* fruit extract. *BMC Complementary and Alternative Medicine*, *7*, 29.

50. Virdi, J., Sivakami, S., Shahani, S., Suthar, A. C., Banavalikar, M. M., & Biyani, M. K. (2003). Antihyperglycemic effects of three extracts from *Momordica charantia*. *Journal of Ethnopharmacology*, *88*, 107–111.

51. Senanayake, G. V., Maruyama, M., Shibuya, K., Sakono, M., Fukuda, N., Morishita, T., Yukizaki, C., Kawano, M., & Ohta, H. (2004). The effects of bitter melon (*Momordica charantia*) on serum and liver triglyceride levels in rats. *Journal of Ethnopharmacology*, *91*, 257–262.

52. Raman, A., & Lau, C. (1996). Anti-diabetic properties and phytochemistry of *Momordica charantia L* (Cucurbitaceae). *Phytomedicine*, *2*, 349–362.

53. Ng, T. B., Wong, C. M., Li, W. W., & Yeung, H. W. (1986). A steryl glycoside fraction from *Momordica charantia* seeds with an inhibitory action on lipid metabolism *in vitro*. *Biochemistry and Cell Biology*, *100*, 766–771.

54. Nerurkar, P. V., Lee, Y. K., Motosue, M., Adeli, K., & Nerurkar, V. R. (2008). *Momordica charantia* (bitter melon) reduces plasma apolipoprotein B-100 and increases hepatic insulin receptor substrate and phosphoinositide-3 kinase interactions. *The British Journal of Nutrition*, *100*, 751–759.

55. Chen, Q., & Li, E. T. (2005). Reduced adiposity in bitter melon (*Momordica charantia*) fed rats is associated with lower tissue triglyceride and higher plasma catecholamines. *The British Journal of Nutrition*, *93*, 747–754.

56. Shih, C. C., Lin, C. H., & Lin, W. L. (2008). Effects of *Momordica charantia* on insulin resistance and visceral obesity in mice on high-fat diet. *Diabetes Research and Clinical Practice*, *81*, 134–143.

57. Ojewole, J. A., Adewole, S. O., & Olayiwola, G. (2006). Hypoglycaemic and hypotensive effects of *Momordica charantia Linn* (Cucurbitaceae) whole-plant aqueous extract in rats. *Cardiovascular Journal of South Africa*, *17*, 227–232.

58. Fletcher, M. A., Caldwell, K., Claflin, A., & Malinin, T. (1980). Further characterization of a tumor cell growth inhibitor from the balsam pear. *Federation American Society Experimental Biology*, *39*, 414 (Abstract).

59. Jilka, C., Strifler, B., Fortner, G. W., Hays, E. F., & Takemoto, D. J. (1983). *In vivo* antitumor activity of the bitter melon (*Momordica charantia*). *Cancer Research*, *43*, 5151–5155.

60. Nagasawa, H., Watanabe, K., & Inatomi, H. (2002). Effects of bitter melon (*Momordica charantia L.*) or ginger rhizome (*Zingiber offifinale rosc*) on spontaneous mammary tumorigenesis in SHN mice. *The American Journal of Chinese Medicine*, *30*, 195–205.

61. Deep, G., Dasgupta, T., Rao, A. R., & Kale, R. K. (2004). Cancer preventive potential of *Momordica*

charantia L. against benzo(a)pyrene induced fore-stomach tumourigenesis in murine model system. *Indian Journal of Experimental Biology, 42,* 319–322.

62. Guevara, A. P., Lim-Sylianco, C., Dayrit, F., & Finch, P. (1990). Antimutagens from *Momordica charantia. Mutation Research, 230,* 121–126.

63. Singh, A., Singh, S. P., & Bamezai, R. (1998). *Momordica charantia* (Bitter Gourd) peel, pulp, seed and whole fruit extract inhibits mouse skin papillomagenesis. *Toxicology Letters, 94,* 37–46.

64. Pongnikorn, S., Fongmoon, D., Kasinrerk, W., & Limtrakul, P. N. (2003). Effect of bitter melon (*Momordica charantia Linn*) on level and function of natural killer cells in cervical cancer patients with radiotherapy. *Journal of the Medical Association of Thailand, 86,* 61–68.

65. Konishi, T., Satsu, H., Hatsugai, Y., Aizawa, K., Inakuma, T., Nagata, S., Sakuda, S. H., Nagasawa, H., & Shimizu, M. (2004). Inhibitory effect of a bitter melon extract on the P-glycoprotein activity in intestinal Caco-2 cells. *British Journal of Pharmacology, 143,* 379–387.

66. Thiebaut, F., Tsuruo, T., Hamada, H., Gottesman, M. M., Pastan, I., & Willingham, M. C. (1987). Cellular localization of the multidrug-resistance gene product P-glycoprotein in normal human tissues. *Proceedings of the National Academy of Sciences of the United States of America, 84,* 7735–7738.

67. Hunter, J., Jepson, M. A., Tsuruo, T., Simmons, N. L., & Hirst, B. H. (1993). Functional expression of P-glycoprotein in apical membranes of human intestinal Caco-2 cells. Kinetics of vinblastine secretion and interaction with modulators. *The Journal of Biological Chemistry, 268,* 14991–14997.

68. Larsen, A. K., Escargueil, A. E., & Skladanowski, A. (2000). Resistance mechanisms associated with altered intracellular distribution of anticancer agents. *Pharmacology and Therapeutics, 85,* 217–229.

69. Limtrakul, P., Khantamat, O., & Pintha, K. (2004). Inhibition of P-glycoprotein activity and reversal of cancer multidrug resistance by *Momordica charantia* extract. *Cancer Chemotherapy and Pharmacology, 54,* 525–530.

70. Zhu, B., Vemavarapu, L., Thompson, W. J., & Strada, S. J. (2005). Suppression of cyclic GMP-specific phosphodiesterase 5 promotes apoptosis and inhibits growth in HT29 cells. *Journal of Cellular Biochemistry, 94,* 336–350.

71. Yasuda, H., Hanai, N., Kurata, M., & Yamada, M. (1978). Cyclic GMP metabolism in relation to the regulation of cell growth in BALB/c3T3 cells. *Experimental Cell Research, 114,* 111–116.

72. Vesely, D. L., Graves, W. R., Lo, T. M., Fletcher, M. A., & Levey, G. S. (1977). Isolation of a guanylate cyclase inhibitor from the balsam pear (*Momordica charantia abreviata*). *Biochemical and Biophysical Research Communications, 77,* 1294–1299.

73. Claflin, A. J., Vesely, D. L., Hudson, J. L., Bagwell, C. B., Lehotay, D. C., Lo, T. M., Fletcher, M. A., Block, N. L., & Levey, G. S. (1978). Inhibition of growth and guanylate cyclase activity of an undifferentiated prostate adenocarcinoma by an extract of the balsam pear (*Momordica charantia abbreviata*). *Proceedings of the National Academy of Sciences of the United States of America, 75,* 989–993.

74. Ng, T. B., Liu, W. K., Sze, S. F., & Yeung, H. W. (1994). Action of alpha-momorcharin, a ribosome inactivating protein, on cultured tumor cell lines. *General Pharmacology, 25,* 75–77.

75. Yasui, Y., Hosokawa, M., Sahara, T., Suzuki, R., Ohgiya, S., Kohno, H., Tanaka, T., & Miyashita, K. (2005). Bitter gourd seed fatty acid rich in 9c,11t,13t-conjugated linolenic acid induces apoptosis and upregulates the GADD45, p53 and PPARgamma in human colon cancer Caco-2 cells. *Prostaglandins, Leukotrienes, and Essential Fatty Acids, 73,* 113–119.

76. Han, C., Demetris, A. J., Michalopoulos, G. K., Zhan, Q., Shelhamer, J. H., & Wu, T. (2003). PPARgamma ligands inhibit cholangiocarcinoma cell growth through p53-dependent GADD45 and p21 pathway. *Hepatology, 38,* 167–177.

77. Nagamine, M., Okumura, T., Tanno, S., Sawamukai, M., Motomura, W., Takahashi, N., & Kohgo, Y. (2003). PPAR gamma ligand-induced apoptosis through a p53-dependent mechanism in human gastric cancer cells. *Cancer Science, 94,* 338–343.

78. Sporn, M. B., Suh, N., & Mangelsdorf, D. J. (2001). Prospects for prevention and treatment of cancer with selective PPAR gamma modulators (SPARMs). *Trends in Molecular Medicine, 7,* 395–400.

79. Suh, N., Wang, Y., Williams, C. R., Risingsong, R., Gilmer, T., Willson, T. M., & Sporn, M. B. (1999). A new ligand for the peroxisome proliferator-activated receptor-gamma (PPAR-gamma), GW7845, inhibits rat mammary carcinogenesis. *Cancer Research, 59,* 5671–5673.

80. Kobori, M., Ohnishi-Kameyama, M., Akimoto, Y., Yukizaki, C., & Yoshida, M. (2008). Alpha-eleostearic acid and its dihydroxy derivative are major apoptosis-inducing components of bitter gourd. *Journal of Agricultural and Food Chemistry, 56,* 10515–10520.

81. Kobori, M., Nakayama, H., Fukushima, K., Ohnishi-Kameyama, M., Ono, H., Fukushima, T., Akimoto, Y., Masumoto, S., Yukizaki, C., Hoshi, Y., Deguchi, T., & Yoshida, M. (2008). Bitter gourd suppresses lipopolysaccharide-induced inflammatory responses. *Journal of Agricultural and Food Chemistry, 56,* 4004–4011.

C. ACTIONS OF INDIVIDUAL FRUITS IN DISEASE AND CANCER PREVENTION AND TREATMENT

82. Manabe, M., Takenaka, R., Nakasa, T., & Okinaka, O. (2003). Induction of anti-inflammatory responses by dietary *Momordica charantia* L. (bitter gourd). *Bioscience, Biotechnology, and Biochemistry*, *67*, 2512–2517.

83. Ike, K., Uchida, Y., Nakamura, T., & Imai, S. (2005). Induction of interferon-gamma (IFN-gamma) and T helper 1 (Th1) immune response by bitter gourd extract. *The Journal of Veterinary Medical Science*, *67*, 521–524.

84. Jiratchariyakul, W., Wiwat, C., Vongsakul, M., Somanabandhu, A., Leelamanit, W., Fujii, I., Suwannaroj, N., & Ebizuka, Y. (2001). HIV inhibitor from Thai bitter gourd. *Planta Medica*, *67*, 350–353.

85. Leung, S. O., Yeung, H. W., & Leung, K. N. (1987). The immunosuppressive activities of two abortifacient proteins isolated from the seeds of bitter melon (*Momordica charantia*). *Immunopharmacology*, *13*, 159–171.

86. Takemoto, D. J., Kaplan, S. A., & Appleman, M. M. (1979). Cyclic guanosine 3',5'-monophosphate and phosphodiesterase activity in mitogen-stimulated human lymphocytes. *Biochemical and Biophysical Research Communications*, *90*, 491–497.

87. Braca, A., Siciliano, T., D'Arrigo, M., & Germano, M. P. (2008). Chemical composition and antimicrobial activity of *Momordica charantia* seed essential oil. *Fitoterapia*, *79*, 123–125.

88. Beloin, N., Gbeassor, M., Akpagana, K., Hudson, J., de, S. K., Koumaglo, K., & Arnason, J. T. (2005). Ethnomedicinal uses of *Momordica charantia* (Cucurbitaceae) in Togo and relation to its phytochemistry and biological activity. *Journal of Ethnopharmacology*, *96*, 49–55.

89. Gurbuz, I., Akyuz, C., Yesilada, E., & Sener, B. (2000). Anti-ulcerogenic effect of *Momordica charantia* L. fruits on various ulcer models in rats. *Journal of Ethnopharmacology*, *71*, 77–82.

90. Dans, A. M., Villarruz, M. V., Jimeno, C. A., Javelosa, M. A., Chua, J., Bautista, R., & Velez, G. G. (2007). The effect of *Momordica charantia* capsule preparation on glycemic control in type 2 diabetes mellitus needs further studies. *Journal of Clinical Epidemiology*, *60*, 554–559.

91. Basch, E., Gabardi, S., & Ulbricht, C. (2003). Bitter melon (*Momordica charantia*): A review of efficacy and safety. *American Journal of Health-System Pharmacy*, *60*, 356–359.

92. Hulin, A., Wavelet, M., & Desbordes, J. M. (1988). Intoxication aigue par *Momordica charantia* (sorrossi). A propos de deux cas. *La Semaine Des Hopitaux*, *64*, 2847–2848.

93. Platel, K., Shurpalekar, K. S., & Srinivasan, K. (1993). Influence of bitter gourd (*Momordica charantia*) on growth and blood constituents in albino rats. *Nahrung*, *37*, 156–160.

94. Abd El Sattar El Batran, S., El-Gengaihi, S. E., & El Shabrawy, O. A. (2006). Some toxicological studies of *Momordica charantia* L. on albino rats in normal and alloxan diabetic rats. *Journal of Ethnopharmacology*, *108*, 236–242.

95. (2009). Fruit of *Momordica charantia*. <www.omafra.gov.on.ca/>.

96. Akhtar, M. S., Athar, M. A., & Yaqub, M. (1981). Effect of *Momordica charantia* on blood glucose level of normal and alloxan-diabetic rabbits. *Planta Medica*, *42*, 205–212.

97. Ali, L., Khan, A. K., Mamun, M. I., Mosihuzzaman, M., Nahar, N., Nur-e-Alam, M., & Rokeya, B. (1993). Studies on hypoglycemic effects of fruit pulp, seed, and whole plant of *Momordica charantia* on normal and diabetic model rats. *Planta Medica*, *59*, 408–412.

98. Platel, K., & Srinivasan, K. (1995). Effect of dietary intake of freeze dried bitter gourd (*Momordica charantia*) in streptozotocin induced diabetic rats. *Nahrung*, *39*, 262–268.

99. Vikrant, V., Grover, J. K., Tandon, N., Rathi, S. S., & Gupta, N. (2001). Treatment with extracts of *Momordica charantia* and *Eugenia jambolana* prevents hyperglycemia and hyperinsulinemia in fructose fed rats. *Journal of Ethnopharmacology*, *76*, 139–143.

100. Chaturvedi, P. (2005). Role of *Momordica charantia* in maintaining the normal levels of lipids and glucose in diabetic rats fed a high-fat and low-carbohydrate diet. *British Journal of Biomedical Science*, *62*, 124–126.

101. Shetty, A. K., Suresh Kumar, G., & Salimath, P. V. (2005). Bitter gourd (*Momordica charantia*) modulates activities of intestinal and renal disaccharidases in streptozotocin-induced diabetic rats. *Molecular Nutrition and Food Research*, *49*, 791–796.

102. Oishi, Y., Sakamoto, T., Udagawa, H., Taniguchi, H., Kobayashi-Hattori, K., Ozawa, Y., & Takita, T. (2007). Inhibition of increases in blood glucose and serum neutral fat by *Momordica charantia* saponin fraction. *Bioscience, Biotechnology, and Biochemistry*, *71*, 735–740.

103. Han, C., Hui, Q., & Wang, Y. (2008). Hypoglycaemic activity of saponin fraction extracted from *Momordica charantia* in PEG/salt aqueous two-phase systems. *Natural Product Research*, *22*, 1112–1119.

104. Singh, N., Gupta, M., Sirohi, P., & Varsha (2008). Effects of alcoholic extract of *Momordica charantia* (*Linn.*) whole fruit powder on the pancreatic islets of alloxan diabetic albino rats. *Journal of Environmental Biology*, *29*, 101–106.

105. Sridhar, M. G., Vinayagamoorthi, R., Suyambunathan, V. A., Bobby, Z., & Selvaraj, N.

(2008). Bitter gourd (*Momordica charantia*) improves insulin sensitivity by increasing skeletal muscle insulin-stimulated IRS-1 tyrosine phosphorylation in high-fat-fed rats. *The British Journal of Nutrition, 99,* 806–812.

106. Tan, M. J., Ye, J. M., Turner, N., Hohnen-Behrens, C., Ke, C. Q., Tang, C. P., Chen, T., Weiss, H. C., Gesing, E. R., Rowland, A., James, D. E., & Ye, Y. (2008). Antidiabetic activities of triterpenoids isolated from bitter melon associated with activation of the AMPK pathway. *Chemistry & Biology, 15,* 263–273.

107. Patel, J. C., Dhirawani, M. K., & Doshi, J. C. (1968). Karella in the treatment of diabetes mellitus. *Indian Journal of Medical Sciences, 22,* 30–32.

108. Baldwa, V. S., Bhandari, C. M., Pangaria, A., & Goyal, R. K. (1977). Clinical trials in patients with diabetes mellitus of an insulin-like compound obtained from plant sources. *Upsala Jurnal of Medical Sciences, 82,* 39–41.

109. Leatherdale, B. A., Panesar, R. K., Singh, G., Atkins, T. W., Bailey, C. J., & Bignell, A. H. (1981). Improvement in glucose tolerance due to *Momordica charantia* (karela). *British Medical Journal (Clinical Research Ed.), 282,* 1823–1824.

110. Akhtar, M. S. (1982). Trial of *Momordica charantia Linn* (Karela) powder in patients with maturity-onset diabetes. *The Journal of the Pakistan Medical Association, 32,* 106–107.

111. Welihinda, J., Karunanayake, E. H., Sheriff, M. H., & Jayasinghe, K. S. (1986). Effect of *Momordica charantia* on the glucose tolerance in maturity onset diabetes. *Journal of Ethnopharmacology, 17,* 277–282.

112. Grover, J. K., & Gupta, S. R. (1990). Hypoglycemic activity of seeds of *Momordica charantia. European Journal of Pharmacology, 183,* 1026–1027.

113. Srivastava, Y., Venkatakhrishna-Bhatt, H., Verma, Y., Venkaiah, K., & Raval, B. H. (1993). Anti-diabetic and adaptogenic properties of *Momordica charantia* extract: An experimental and clinical evaluation. *Phytotherapy Research, 7,* 285–289.

114. Ahmad, N., Hassan, M. R., Halder, H., & Bennoor, K. S. (1999). Effect of *Momordica charantia* (Karolla) extracts on fasting and postprandial serum glucose levels in NIDDM patients. *Bangladesh Medical Research Council Bulletin, 25,* 11–13.

115. Rosales, R., & Fernando, R. (2001). An inquiry to the hypoglycemic action of *Momordica charantia* among type 2 diabetic patients. *Philippine Journal of Internal Medicine, 39,* 213–216.

116. Marles, R., & Farnsworth, N. (1997). Antidiabetic plants and their active constituents: An update. *Phytomedicine, 2,* 137–189.

117. Ng, T. B., Chan, W. Y., & Yeung, H. W. (1992). Proteins with abortifacient, ribosome inactivating, immunomodulatory, antitumor and anti-AIDS activities from Cucurbitaceae plants. *General Pharmacology, 23,* 579–590.

118. Tam, P. P., Law, L. K., & Yeung, H. W. (1984). Effects of alpha-momorcharin on preimplantation development in the mouse. *Journal of Reproduction and Fertility, 71,* 33–38.

119. Chan, W. Y., Tam, P. P., & Yeung, H. W. (1984). The termination of early pregnancy in the mouse by beta-momorcharin. *Contraception, 29,* 91–100.

120. Dixit, V. P., Khanna, P., & Bhargava, S. K. (1978). Effects of *Momordica charantia L.* fruit extract on the testicular function of dog. *Planta Medica, 34,* 280–286.

121. Stepka, W., Wilson, K. E., & Madge, G. E. (1974). Anti-fertility investigation on Momordica. *Lloydia, 37,* 645.

122. Sharma, V. N., Sogani, R. K., & Arora, R. B. (1960). Some observations on hypoglycaemic activity of *Momordica charantia. The Indian Journal of Medical Research, 48,* 471–477.

123. Tennekoon, K. H., Jeevathayaparan, S., Angunawala, P., Karunanayake, E. H., & Jayasinghe, K. S. (1994). Effect of *Momordica charantia* on key hepatic enzymes. *Journal of Ethnopharmacology, 44,* 93–97.

C. ACTIONS OF INDIVIDUAL FRUITS IN DISEASE AND CANCER PREVENTION AND TREATMENT

36

Pomegranate in Human Health: An Overview

Ana Faria[1,2] *and Conceição Calhau*[1]

[1]Biochemistry Department (U38-FCT), Faculty of Medicine, University of Porto, Portugal
[2]Chemistry Investigation Centre (CIQ), Department of Chemistry, Faculty of Sciences, University of Porto, Portugal

The purpose of this chapter is to provide an update of current knowledge on basic principles and concepts, as well as clinical nutrition evidence of pomegranate effects on the prevention and treatment of disease.

Pomegranate (*Punica granatum L.*), an ancient and mystical fruit, affectionately known as the 'jewel of winter,' is the predominant member of two species comprising the Punicaceae family. It is native from the Himalayas in northern India to Iran but has been cultivated and naturalized, since ancient times, over the entire Mediterranean region. Now it is cultivated in Afghanistan, India, China, Japan, Russia, and the USA, particularly in Arizona and California.

While the pomegranate plant is considered either a small tree or a large shrub, its fruit is often deemed to be a large berry. This fruit is deep red, with leathery skin, grenade-shaped and crowned by the pointed calyx. It contains numerous arils, each surrounded by a translucent juice-containing sac. Thin acrid-tasting membranes extend into the interior of the fruit from the pericarp, providing a latticework for suspending the arils.

The fruit can be divided into three parts: the seeds (about 3% of the fruit weight), juice (about 30% of the fruit weight) and the peels, that also include the interior network of membranes [1].

Pomegranate has accompanied mankind over hundreds of years. This fruit has acquired symbolisms in Judaism, Christianity, Islam, Buddhism, and Zoroastrianism. It is a symbol of life, health, longevity, femininity, fecundity, knowledge, morality, immortality, and spirituality [2]. In alternative medicines, pomegranate is considered 'a pharmacy unto itself,' and is used as an antiparasitic agent [3], a 'blood tonic,' and to help in aphthae, diarrhea, and ulcers [4]. Modern uses of pomegranate include treatment of acquired immune deficiency syndrome, hormone replacement therapy, cardiovascular protection, oral hygiene, adjunct therapy to increase bioavailability of radioactive dyes during diagnostic imaging, as well as cosmetic beautification and enhancement [5–7]. Over the past few years, scientific investigations have given a credible basis for some of the traditional uses of pomegranate and the number of scientific papers concerning pomegranate and its health properties has increased greatly.

The relationship between the chemical constituents of pomegranate and their pharmacological effects is yet to be fully understood. However, significant progress has been made in the last few years in the establishment of pharmacological mechanisms responsible for this relationship and identification of chemicals responsible for the effects. Among the great variety of chemical components present in the pomegranate, ellagic acid, ellagitannins (including punicalagins), punic acid, anthocyanins, flavonols, flavan-3-ols, and flavones seem to be the ones responsible, at least in part, for most of the therapeutic benefits from the consumption of pomegranate [8,9].

1. CHEMICAL CONSTITUENTS OF POMEGRANATE FRACTIONS

1.1 Seed

The seeds comprise about 3% of the pomegranate weight. The seed polyphenol content is quite low and, consequently, the correlation between phenolic content and antioxidant activity is not clear [10,11]. In addition, the pomegranate seed oil, that comprises 12–20% of total seed weight, consists of approximately 80% conjugated octadecatrienoic fatty acids, with high content of 9*cis*, 11*trans*, 13*cis* acid (i.e. punicic acid) [12]. There are several studies focusing on biological effects of the seed oil. Minor components of the oil include sterols, steroids, and cerebrosides [13].

1.2 Peel

Pomegranate peel is characterized by the presence of flavonoids and tannins which have been associated with many of the biological properties demonstrated by the peel. Among seed, peel, and juice, the peel is the constituent which possesses higher antioxidant activity *in vitro*, in good correlation with the high content of polyphenols [11,14]. This part of the fruit contains the punicalagins, the ellagitannins typical of pomegranate. Punicalagin is unique to pomegranate and is part of a family of ellagitannins which include the minor tannins called punicalin and gallagic acid. All these ellagitannins have the ability to be hydrolyzed to ellagic acid resulting in a prolonged release of this acid into the blood. Punicalagin is a large polyphenol, with a molecular weight greater than 1000. During juice processing, the whole fruit is pressed and ellagitannins are extracted into pomegranate juice in significant quantities, reaching levels of over 2 g/L juice [8]. Among other fruit peels, such as bananas, mangos, and coconuts, pomegranate demonstrates the highest antioxidant capacity *in vitro* [15].

1.3 Juice

Pomegranate juice is the greatest contributor for pomegranate ingestion. The main antioxidant compounds in pomegranate juice are hydrolyzable tannins, with the contribution of anthocyanins (3-glucosides and 3,5-glucosides of delphinidin, cyanidin, and pelargonidin) and ellagic acid derivatives. There are some differences in phenolic composition between commercial juices and experimental ones [8]. Also, the use of the arils alone or the whole fruit to make juice has an enormous impact on polyphenol content and consequently on antioxidant capacity of the juice. Pomegranate access, as well as geographical region, harvesting, and season can also alter the fruit composition and consequently the juice composition [16].

2. POMEGRANATE IN HEALTH AND DISEASE

Pomegranate has been used for centuries as a therapeutic agent for the treatment of inflammatory diseases and disorders of the digestive tract by practitioners of the Ayurvedic and Unani systems of medicine.

Complementary/alternative medicine has become increasingly popular in the Western world over the last 10–15 years. The use of oral supplements falls within this framework. Most oral supplements are based on or include several botanical ingredients, many of which have long histories of traditional or folk medicine usage. Several of the available products derived from botanical sources are touted for their health benefits. On the other hand, it is true that in this age of targeted therapy, the failure of most current drug-discovery efforts to yield safe, effective, and inexpensive drugs has generated widespread concern. Successful drug development has been stymied by a general focus on target selection rather than clinical safety and efficacy. Thus, it has become necessary to rethink drug development strategies. The focus in this chapter will be on pomegranate derivates that have been used for preventive/therapeutic purposes.

Health beneficial effects of pomegranate fruit and derivatives such as its juice have been studied. The actual explosion of scientific interest in pomegranate as a medicinal and nutritional product is evident by a MEDLINE search.

It is important to understand why this is a hot topic. These days, an increasing proportion of people around the world are more health conscious than ever before. Perhaps this is due to an increased awareness of what seem to be mostly preventable diseases (i.e. obesity, heart disease, cancers, osteoporosis, arthritis, and type II diabetes mellitus) in part related to an expansion in educational vehicles. Perhaps another reason for a greater collective health consciousness is a continuous shift of the population toward a more-advanced age. Along with age comes an increased incidence of disease and thus individual focus upon prevention and treatment of such disorders. The notion that food may possess the ability to prevent disease and/or be used as treatment of ailments dates back a couple of millennia. Hippocrates proclaimed, some 2500 years ago, 'Let food be thy medicine and medicine be thy food.' Early evidence of the medicinal application of plants includes the archaeological discovery of a 60,000-year-old Neanderthal burial ground in present-day Iraq.

Many of these medicinal applications of plants are still popular in folk medicine today.

2.1 Biological Properties

Several studies have reported that diets high in fruits and vegetables can be associated a with markedly decreased risk of cardiovascular disease and cancer. This has been frequently related to high levels of antioxidants present in these foods.

In fact, although it may seem strange that oxygen can be damaging to living organisms, the knowledge that oxygen and oxygen-derived species can cause damage to cellular components is not new [17]. Defense against potentially harmful oxidant species, termed reactive oxygen species (ROS) and reactive nitrogen species (RNS), is needed. If the generation of reactive species is high and overcomes the effectiveness of the antioxidant defense system, a condition referred to as oxidative stress arises [17]. Damage to cellular components, mainly lipids, proteins, and nucleic acids occurs, impairing the functions of the molecules involved, which in turn results in metabolic insufficiencies or defects [18,19].

Oxidative stress is, thus, a serious problem and has been given growing attention as it has been found to constitute an etiological factor, and sometimes also a consequence, of diseases with increasing incidence in Western societies. Oxidative stress is a critical step in cancer, inflammatory, cardiovascular and neurodegenerative diseases, and aging [20,21]. Organisms have developed strategies to deal with oxidative stress which comprise different approaches. Endogenous protection is achieved by enzymes like superoxide dismutase, catalase, and glutathione peroxidase that catalytically remove free radicals and other reactive species. Another mechanism consists in the existence of low molecular weight agents like glutathione and α-tocopherol that scavenge ROS and RNS. There is an intimate relationship between nutrition and the antioxidant defense system, as some exogenous low molecular weight antioxidants may be supplied by the diet. These two main systems of antioxidant defense act in coordination, their levels being regulated by each other, protecting against oxidative stress events [22]. So, diet is indisputably the best possible manipulative source of antioxidant molecules that may help prevent oxidative damage to cellular components. The recognition of this fact has drawn attention to the identification of natural antioxidant sources for potential inclusion in human diet [23,24]. The oxidation protection actions that can be attributed to polyphenols are broad, from metal chelation, reactive species scavenging, and enzyme modulation to interference with cell signaling and molecular transcription [25–28]. On account of their antioxidant properties, these are regarded as anti-atherosclerotic, antitumoral, anti-inflammatory and anti-aging molecules [29–32].

Pomegranate has gained scientist attention in the past decade, and several reports have focused on antioxidant actions *in vitro*, *ex vivo*, and *in vivo* of pomegranate juice, seed oil, and peel. The antioxidant activity demonstrated by pomegranate has always been related to its chemical composition. However, it is true that in spite of the vast literature about the antioxidant 'current' focus on enhancing ROS elimination and inhibiting ROS generation, many other cellular processes are likely involved. Additionally, we have to consider that ROS are a potential double-edged sword in disease promotion and prevention.

In addition to its ancient historical uses, pomegranate use is justified by its anti-atherosclerotic capacity [33,34] as well as other benefits such as chemoprevention and chemotherapy of prostate cancer [30]. Seeram et al. [35] reported antiproliferative, apoptotic, and antioxidant activity of pomegranate tannin extract in an *in vitro* study. Other effects of pomegranate juice such as anti-inflammatory activity, reduction of platelet aggregation, reduction of atherogenic modifications of low density lipoprotein (LDL), reduction of macrophage oxidative state [8,29,36,37], anti-obesity effects [38], antidiabetic activity based on PPAR-α/-γ activator properties of pomegranate flower [39], and antilipidemic effects based on the inhibition of increased cardiac fatty acids uptake and oxidation in a diabetic condition [40] have also been reported. Most of these beneficial effects have been attributed to the high antioxidant capacity of this fruit juice [8,37,41], probably due to the presence of a complex mixture of phytochemicals.

The pomegranate seed presents a moderate biological activity. The antioxidant activity of the seed is limited in agreement with its polyphenol content. Guo et al. demonstrated a modest antioxidant activity of seed extracts of different varieties of pomegranate [11] using a chemiluminescence method. Although the antioxidant activity of pomegranate seed is limited, pomegranate seed oil has been object of many studies due to the presence of the conjugated punicic acid. There is a study in which the authors demonstrate that seed oil has strong antioxidant activity, close to that of butylated hydroxyanisole (BHA) and green tea

(*Thea sinensis*), and significantly greater than that of red wine (*Vitis vitifera*). Despite the polyphenol composition of this extract having not been clarified, the fatty acid composition is known, punicic acid percentage being over 65% of total oil [42]. In one study, seed oil supplementation of obese rats had a significant effect on thrombospondin-1 expression [43]. It was also found to promote regeneration of epidermis [44]. Kohno et al. suggest that this seed oil can suppress azoxymethane-induced colon carcinogenesis [45]. Yamasaki et al. tested the immune function and lipid metabolism of C57BL/6 N mice after supplementation with pomegranate seed oil and verified that B cell function *in vivo* could be enhanced [46]. Contradictory to these studies, Aviram et al. verified that pomegranate seed did not present any effect as to anti-atherogenic properties [47].

Pomegranate peel, the pomegranate component with the highest content in polyphenols, namely tannins and ellagitannins, is also the one that presents the highest antioxidant activity [11,48]. A study that compared peel and pulp hydro extract of Persian pomegranate cultivars demonstrated that peel has a higher capacity to chelate metals than the pulp extract [49]. In addition, other authors confirmed the stronger antioxidant properties of the peel when compared to the pulp extract, including scavenging or preventive capability against several reactive oxygen species and inhibition of LDL oxidation. Again, the authors attributed the high antioxidant activity of the peel extract to its high phenolic content [14]. In another work, a strong antioxidant activity was demonstrated by peel pomegranate extracts which also showed antimutagenic activity. A direct correlation between the antimutagenic and antioxidant activity has not been found, but the authors hypothesize that the quality and quantity of polyphenols and other bioactive compounds present in the different extracts were responsible for this lack of correlation [50].

Singh et al. studied the antioxidant activity of the seed and peel of pomegranate using *in vitro* models. The peel extract presented a higher phenolic weight percentage than the seed and a higher antioxidant activity in the models used [10]. The peel extract that presented higher antioxidant activity was tested using *in vivo* models. Treatment of rats with this extract and carbon tetrachloride (CCl_4) resulted in a decrease of lipid peroxidation and a protection of the hepatocyte histology, when compared to CCl_4 alone. Also, this treatment preserves catalase, peroxidase, and superoxide dismutase, enzymes involved in the endogenous antioxidant protection system [51]. Other authors studied the effect of chronic administration of pomegranate peel extract on liver fibrosis induced by bile duct ligation in rats. Plasma antioxidant capacity and hepatic reduced glutathione levels were similar to control levels in the group treated with pomegranate after liver fibrosis induced by bile duct ligation. Moreover, malonaldehyde levels and myeloperoxidase activity were also similar to control levels after pomegranate peel extract treatment [52].

A study by Seeram et al. in which the antioxidant potency of commonly consumed polyphenol-rich beverages in the USA was evaluated through several *in vitro* methodologies, showed that pomegranate juice was the tested beverage with the highest antioxidant capacity, followed by red wine [53]. The juice has been the most studied component because major pomegranate consumption is by juice ingestion. Several studies have confirmed the antioxidant capacity of the juice *in vitro*, *ex vivo*, as well as *in vivo*. As previously mentioned, there are *in vitro* comparative studies showing that the juice has less antioxidant capacity than the peel but more than the seeds [8,11]. However, the juice presents an antioxidant activity three times higher than red wine and green tea infusion, when tested with the same methodology [8]. This author also concluded that pomegranate juice processing influences polyphenol content and consequently antioxidant activity [54]. Commercial juices show higher antioxidant activity than handmade juices, probably due to the presence of tannins from the peel that are extracted to the juice during processing [8].

It is worth noting that the methodologies used *in vitro* are multiple and diffuse, and most of the time evaluate different capacities (radical scavenging, metal chelation, DNA damage). Even for the same ability, for example radical scavenging, there are different methods that can be used and that are suitable for lipophilic or hydrophilic antioxidants. To make this task even harder, some authors study only a specific extract of the peel or of the seed, which makes the comparison between components and between authors extremely difficult.

Nevertheless, some *ex vivo* and *in vivo* studies have confirmed pomegranate juice antioxidant activities. In an *ex vivo* study, hepatic oxidative stress was evaluated in mice after chronic pomegranate juice consumption. The authors found an improvement of protein and DNA oxidative state [55]. Another *ex vivo* study showed a high plasma antioxidant capacity in human volunteers after pomegranate juice consumption [56]. Pomegranate juice consumption by diabetic patients led to a significant reduction in serum lipid peroxides and thiobarbituric acid reactive substances (TBARS) levels, whereas serum sulfhydryl groups and serum paraoxonase 1 activity significantly increased compared to healthy subjects [57]. In another study, the effect of pomegranate juice consumption by atherosclerotic patients with carotid artery stenosis has been evaluated. After 12 months of pomegranate juice consumption, the patients' serum paraoxonase 1 activity was increased, whereas serum LDL basal oxidative state and LDL susceptibility to copper ion-induced oxidation were both significantly reduced compared to values obtained before consumption. Furthermore, serum total

antioxidant status was increased by 130% after 1 year of pomegranate juice consumption [58].

Because pomegranate is rich in polyphenols, and most of the biological properties of this fruit are attributed to polyphenols, several studies aimed to investigate the properties of these compounds in isolation. Punicalagin is unique to pomegranate and is part of the family of ellagitannins. It has been extensively studied for its antioxidant and biological properties, being reported as the substance responsible for over half of the juice's potent antioxidant activity [8]. Kulkarni et al. showed that the antioxidant action of punicalagin is expressed not only through its scavenging reactions but also by its ability to form metal chelates [9]. Other authors have confirmed not only the antioxidant capacity of punicalagins, but also other biological properties such as antiproliferative, apoptotic, anti-inflammatory, and hepatoprotective effects [35,59–61]. The metabolism of these compounds by gut flora forms urolithins such as urolithin A [62]. These metabolites have been identified in urine and plasma, in conjugated and free forms [63]. Since 2007 there has been an effort to clarify the role of these metabolites in prostate cancer [64,65].

2.2 Pomegranate on Cancer and Inflammation

A strong link has been shown between cancer and inflammation. In fact, it has been demonstrated that inflammation is a hallmark of several forms of cancer (prostate, colon, breast, etc). Sometimes inflammation precedes, and sometimes it is a consequence of tumor cell growth. As whole pomegranate, or a specific part of this fruit has strong antioxidant activity, this property may justify, at least in part, the benefits of its intake to prevent or treat several diseases, in particular cancer and inflammation [1]. As a matter of fact, diet factors with an impact on inflammation seem to be candidates for the intervention in cancers since chronic inflammation can serve as an important etiopathogenic factor for chronic diseases including cancer.

The fact that dietary compounds influence the susceptibility of human beings to cancer is widely accepted. One of the possible mechanisms responsible for these (anti)carcinogenic effects is modulation of biotransformation enzymes, thereby affecting the (anti)carcinogenic potential of other compounds. In line with this, our group investigated the effects of pomegranate consumption on metabolizing activity of phase-I enzymes, cytochromes P450 (CYPs), as a possible mechanism of that antitumoral action [66]. CYPs constitute a superfamily of hemethiolate isoenzymes of major importance in the oxidation of endogenous compounds, drugs, environmental pollutants, and dietary chemicals, and in the activation of procarcinogens [67]. Our results, obtained in mice, showed that pomegranate juice consumption decreased total hepatic CYP content as well as the expression of CYP1A2 and CYP3A. Thus, prevention of procarcinogen activation through CYP450 activity/expression inhibition may be involved in pomegranate juice protection against tumor initiation, promotion, and/or progression.

Another curious finding was that obtained by Mehta and Lansky [68]. These authors treated mouse mammary organ culture with pomegranate seed oil and observed a higher suppression of tumor occurrence with a lower tested concentration. This suggested that an optimal biological dose is more important than a maximally tolerated one, which is in agreement with a hormetic effect. There are different studies verifying an effect of pomegranate fractions (peel, juice, and seed oil) on the modulation of cell signaling molecules in the cell cycle machinery [69]. Also, effects on differentiation may possibly contribute to observed anticancer activity of pomegranate extracts [1]. There are in the literature reports on pomegranate effects

upon enzymes probably involved in anticancer effects. Carbonic anhydrase (CA) is one of them, CA inhibition strongly inhibiting cancer cell growth *in vitro* and *in vivo* [70,71]. Another enzyme activity that is inhibited by pomegranate is ornithine decarboxylase [72,73]; considering that polyamines regulate growth processes that may facilitate the growth of cancer, this may constitute a relevant anticancer effect of pomegranate. Aromatase, the estrogen synthase, and 17-β-hydroxysteroid dehydrogenase type 1, responsible for the reduction of estrone to the much more potent estrogenic 17-β-estradiol, are potently inhibited by fermented pomegranate juice and by pomegranate peel extract [74]. These effects justify other possible chemopreventive and adjuvant therapeutic applications of pomegranate in cancer, especially upon estrogen-dependent ones. In addition, some pomegranate components have been described to possess estrogenic activity, thus inhibiting the estrogenic action of 17-β-estradiol by competitive binding to estrogen receptors [74]. Estrogenic agonism could be important in inhibiting androgenic activity, especially in prostate cancer [75].

The capacity of any chemotherapeutic agent to selectively inhibit proliferation and/or induce apoptosis in malignant cells, preserving normal cells, is the hallmark of a successful anticancer therapeutic. A note of caution should be inserted here to point out that whole or complex pomegranate fractions possess antiproliferative, pro-apoptotic, and/or anti-angiogenic effects superior to those observed with their isolated key active compounds, suggesting therapeutic strategies that may depart from the orthodox preference for pure single agents. Dose-dependent antiproliferative and pro-apoptotic properties of pomegranate fruit extract have been demonstrated both *in vitro* and *in vivo* [30]. This inhibitory effect on cell growth and viability was shown to be mediated through the cki–cyclin–cdk network with up-regulation of p21 and p27 during G1-phase arrest, independent of p53. These effects were accompanied by down-regulation of the cyclins D1, D2, and E and cdks-2, -4, and -6, operative in the G1 phase of the cell cycle [30].

Skin cancer is an important and prevalent form of cancer, justifying the considerable attention now focused on the photopreventive impact of several natural compounds. Pomegranate has the potential to inhibit UVB-induced oxidative stress-mediated activation of signaling pathways in human epidermal cells, appearing as a promising agent against photo-carcinogenesis [76].

Inhibition of the initiation and development of new blood vessels (angiogenesis), essential to the supply of oxygen and nutrients to tumors, is a promising therapeutic approach for treating solid tumors. The anti-angiogenic potential of pomegranate fractions observed by Toi et al. in different breast cancer cell lines deserves further *in vivo* and clinical investigations [77]. The same group previously showed that pomegranate seed oil and fermented juice polyphenols retarded oxidation and prostaglandin synthesis, inhibited breast cancer cell proliferation and invasion, and promoted breast cancer cell apoptosis. The relevant effects on apoptosis seemed to be partially mediated by the caspase-3 pathway in the prostate cancer cell line [78] or through the mitochondrial pathway in colon cancer Caco-2 cells [79]. These results reinforce the pomegranate's potential as chemotherapeutic 'agent.'

Inflammation can result in persistent oxidative stress in the cancer cell environment, which may lend a survival advantage for the cancer. Thus, another area of potential therapeutic interest of pomegranate derives from its anti-inflammatory activity. Pomegranate, consumed as a whole fruit, or juice, or its different fractions, presents a large broad of effects: i) it inhibits cyclooxygenase (COX)-2 expression and consequently eicosanoid biosynthesis [37,80]; ii) it synergistically suppresses inflammatory cytokine expression [1,37]; iii) it inhibits

matrix metalloproteinases [44,81]. In addition, NFκB activation leads to inflammation and cell proliferation. Constitutive activation of NFκB, frequently identified in different cancer cell lines [82], up-regulates the transcription of collagenase gene, cell adhesion molecules, and cytokines (as TNF-α, IL-1, 2, 6, and 8) genes and regulates a number of downstream genes including COX-2 [83]. It is also known that NFκB regulates genes involved in the immune response, as well as cell cycle control and cell death in response to pro-inflammatory cytokines. NFκB expression is also associated with the transcription of genes involved in cell survival such as Bcl_x and inhibitors of apoptosis.

Any of these effects, and much more so their sum, attests to the anti-inflammatory activity and therefore the anticancer potential of this special fruit.

3. FOOD–DRUG INTERACTIONS AND SAFETY

The popularity of pomegranate juice has increased significantly after research studies showing putative beneficial health properties of its consumption. It has been of interest to determine whether pomegranate has any effect on cytochrome P450, the hepatic enzyme system involved in the metabolism of drugs and other xenobiotics, as well as some endogenous substrates. Drug interactions can frequently arise when drugs are co-administered and one drug inhibits the metabolic clearance of the second drug by inhibition of a specific CYP enzyme. Likewise, natural products can also affect activity of CYP enzymes; concomitant drug and food intake creates the opportunity for interactions that may change the oral bioavailability and resulting effectiveness or toxicity of a drug. Grapefruit juice and St John's Wort are notable examples of this type of interaction [84], resulting from a strong inhibition of CYP3A activity [85–87]. Some approaches using animal models led to the conclusion that pomegranate juice consumption inhibited CYP enzymes, thus altering drug pharmacokinetics [66,88,89]. Also, some authors showed that pomegranate juice has a CYP3A inhibitory activity using human microsomes [89,90]. In contrast, one human study has shown that pomegranate juice does not alter clearance of intravenous or oral midazolam, a CYP3A substrate [91]. Whether pomegranate juice consumption interacts with drugs needs further elucidation.

Some studies have been conducted in order to assess pomegranate safety. No toxic effects were reported after repeated punicalagin consumption by rats [92]. In agreement, another study also confirmed the nonexistence of toxic effects in rats consuming pomegranate juice in an acute and subchronic approach [93]. In addition, research with overweight human volunteers showed no adverse effects of pomegranate supplementation [94]. Nevertheless, although pomegranate consumption has not been reported to possess deleterious health effects, apparent innocuousness cannot be simply extended to enriched pomegranate extracts.

Therefore, an occasional, moderate consumption of pomegranate or pomegranate juice as a regular food probably carries no problems and is instead beneficial to health, but high dose ingestion, with 'pharmacological' intention, needs much more investigation to avoid placing people at serious health risk.

4. SUMMARY

Scientific evidence that NFκB activation is associated with heightened proliferation, increased angiogenesis, and resistance to apoptosis, suggests that the chemopreventive action of pomegranate compounds are significantly mediated through their NFκB inhibitory effects.

Punicalagin and ellagic acid seem to be two important components of pomegranate. Both compounds tested in isolation would provide results different from those obtained with the whole fruit, considering the interactions with all other pomegranate matrix compounds that could be strongly important for the final result.

Present evidence supports the interest of clinical trials to further assess chemopreventive and adjuvant therapeutic applications of pomegranate in human cancer.

References

1. Lansky, E. P., & Newman, R. A. (2007). Punica granatum (pomegranate) and its potential for prevention and treatment of inflammation and cancer. *Journal of Ethnopharmacology, 109*, 177–206.
2. Mahdihassan, S. (1984). Outline of the beginnings of alchemy and its antecedents. *American Journal Chinese Medicals, 12*, 32–42.
3. Naqvi, S. A. H., Khan, M. S. Y., & Vohora, S. B. (1991). Anti-bacterial anti-fungal and anthelmintic investigations on Indian medicinal plants.. *Fitoterapia, 62*, 221–228.
4. Caceres, A., Giron, L. M., Alvarado, S. R., & Torres, M. F. (1987). Screening of antimicrobial activity of plants popularly used in Guatemala for the treatment of dermatomucosal diseases. *Journal of Ethnopharmacology, 20*, 223–237.
5. Curry, S. C. (2004). Breast enhancement system. USA.
6. Mixture of vegetable oil for use in cleansing cosmetics e. g. Soap and shampoo, contains extract obtained from varieties of pomegranate seeds such as black large seeded shiraz from Iran and malas shirin. PERUSHAZAKURO YAKUHIN KK (PERU-Non-standard). Japanese patent no. JP2004083544-A.
7. Cosmetics for suppressing ageing effect contain ageing inhibitor which consists of essence of plant seed of pomegranate. Pola Chem Ind. Inc. (Pokk). Japanese patent no. JP2000143491-A.
8. Gil, M. I., Tomas-Barberan, F. A., Hess-Pierce, B., Holcroft, D. M., & Kader, A. A. (2000). Antioxidant activity of pomegranate juice and its relationship with phenolic composition and processing. *Journal of Agricultural and Food Chemistry, 48*, 4581–4589.
9. Kulkarni, A. P., Mahal, H. S., Kapoor, S., & Aradhya, S. M. (2007). *In vitro* studies on the binding, antioxidant, and cytotoxic actions of punicalagin. *Journal of Agricultural and Food Chemistry, 55*, 1491–1500.
10. Singh, R. P., Chidambara Murthy, K. N., & Jayaprakasha, G. K. (2002). Studies on the antioxidant activity of pomegranate (Punica granatum) peel and seed extracts using *in vitro* models. *Journal of Agricultural and Food Chemistry, 50*, 81–86.

11. Guo, S., Deng, Q., Xiao, J., Xie, B., & Sun, Z. (2007). Evaluation of antioxidant activity and preventing DNA damage effect of pomegranate extracts by chemiluminescence method. *Journal of Agricultural and Food Chemistry, 55*, 3134–3140.

12. Kaufman, M., & Wiesman, Z. (2007). Pomegranate oil analysis with emphasis on maldi-tof/ms triacylglycerol fingerprinting. *Journal of Agricultural and Food Chemistry, 55*, 10405–10413.

13. Tsuyuki, H., Itoh, S., & Nakatsukasa, Y. (1981). Studies on the lipids in pomegranate punica-granatum-var-nana seeds. *Bulletin of the College of Agriculture and Veterinary Medicine Nihon University, 6*, 141–148.

14. Li, Y. F., Guo, C. J., Yang, J. J., Wei, J. Y., Xu, J., & Cheng, S. (2006). Evaluation of antioxidant properties of pomegranate peel extract in comparison with pomegranate pulp extract. *Food Chemistry, 96*, 254–260.

15. Okonogi, S., Duangrat, C., Anuchpreeda, S., Tachakittirungrod, S., & Chowwanapoonpohn, S. (2007). Comparison of antioxidant capacities and cytotoxicities of certain fruit peels. *Food Chemistry, 103*, 839–846.

16. Mirdehghan, S. H., & Rahemi, M. (2007). Seasonal changes of mineral nutrients and phenolics in pomegranate (punica granatum l.) fruit. *Scientia Horticulturae, 111*, 120–127.

17. Halliwell, B. (1999). Antioxidant defence mechanisms: From the beginning to the end (of the beginning). *Free Radical Research, 31*, 261–272.

18. Halliwell, B., & Gutteridge, J. (2000). *Free radicals, other reactive species and disease. free radicals in biology and medicine*. Oxford: Oxford Scientific Publications pp. 617, 783.

19. Valko, M., Izakovic, M., Mazur, M., Rhodes, C. J., & Telser, J. (2004). Role of oxygen radicals in DNA damage and cancer incidence. *Molecular and Cellular Biochemistry, 266*, 37–56.

20. Kehrer, J. P. (1993). Free radicals as mediators of tissue injury and disease. *Critical Reviews in Toxicology, 23*, 21–48.

21. Storz, P. (2005). Reactive oxygen species in tumor progression. *Frontiers in Bioscience, 10*, 1881–1896.

22. Masella, R., Di Benedetto, R., Vari, R., Filesi, C., & Giovannini, C. (2005). Novel mechanisms of natural antioxidant compounds in biological systems: Involvement of glutathione and glutathione-related enzymes. *Journal of Nutritional Biochemistry, 16*, 577–586.

23. Benzie, I. F. (2003). Evolution of dietary antioxidants. *Comparative Biochemistry and Physiology Part A: Molecular and Integrative Physiology, 136*, 113–126.

24. Ross, J. A., & Kasum, C. M. (2002). Dietary flavonoids: Bioavailability, metabolic effects, and safety. *Annual Review of Nutrition, 22*, 19–34.

25. Heim, K. E., Tagliaferro, A. R., & Bobilya, D. J. (2002). Flavonoid antioxidants: Chemistry, metabolism and structure-activity relationships. *Journal of Nutritional Biochemistry, 13*, 572–584.

26. Rice-Evans, C. A., Miller, N. J., & Paganga, G. (1996). Structure-antioxidant activity relationships of flavonoids and phenolic acids. *Free Radical Biology and Medicine, 20*, 933–956.

27. Tapiero, H., Tew, K. D., Ba, G. N., & Mathe, G. (2002). Polyphenols: Do they play a role in the prevention of human pathologies? *Biomedicine and pharmacotherapy, 56*, 200–207.

28. Williams, R. J., Spencer, J. P., & Rice-Evans, C. (2004). Flavonoids: Antioxidants or signalling molecules? *Free Radical Biology and Medicine, 36*, 838–849.

29. Aviram, M., Dornfeld, L., Rosenblat, M., Volkova, N., Kaplan, M., Coleman, R., Hayek, T., Presser, D., & Fuhrman, B. (2000). Pomegranate juice consumption reduces oxidative stress, atherogenic modifications to LDL, and platelet aggregation: Studies in humans and in atherosclerotic apolipoprotein E-deficient mice. *American Journal of Clinical Nutrition, 71*, 1062–1076.

30. Malik, A., Afaq, F., Sarfaraz, S., Adhami, V. M., Syed, D. N., & Mukhtar, H. (2005). Pomegranate fruit juice for chemoprevention and chemotherapy of prostate cancer. *Proceedings of the National Academy of Sciences of the USA, 102*, 14813–14818.

31. Martin, I., & Grotewiel, M. S. (2006). Oxidative damage and age-related functional declines. *Mechanisms of Ageing and Development, 127*, 411–423.

32. Terman, A., & Brunk, U. T. (2006). Oxidative stress, accumulation of biological 'garbage', and aging. *Antioxidants and Redox Signaling, 8*, 197–204.

33. Aviram, M., & Dornfeld, L. (2001). Pomegranate juice consumption inhibits serum angiotensin converting enzyme activity and reduces systolic blood pressure. *Atherosclerosis, 158*, 195–198.

34. Kaplan, M., Hayek, T., Raz, A., Coleman, R., Dornfeld, L., Vaya, J., & Aviram, M. (2001). Pomegranate juice supplementation to atherosclerotic mice reduces macrophage lipid peroxidation, cellular cholesterol accumulation and development of atherosclerosis. *Journal of Nutrition, 131*, 2082–2089.

35. Seeram, N. P., Adams, L. S., Henning, S. M., Niu, Y., Zhang, Y., Nair, M. G., & Heber, D. (2005). *In vitro* antiproliferative, apoptotic and antioxidant activities of punicalagin, ellagic acid and a total pomegranate tannin extract are enhanced in combination with other polyphenols as found in pomegranate juice. *Journal of Nutritional Biochemistry, 16*, 360–367.

36. Rozenberg, O., Howell, A., & Aviram, M. (2006). Pomegranate juice sugar fraction reduces macrophage

oxidative state, whereas white grape juice sugar fraction increases it. *Atherosclerosis, 188,* 68–76.

37. Adams, L. S., Seeram, N. P., Aggarwal, B. B., Takada, Y., Sand, D., & Heber, D. (2006). Pomegranate juice, total pomegranate ellagitannins, and punicalagin suppress inflammatory cell signaling in colon cancer cells. *Journal of Agricultural and Food Chemistry, 54,* 980–985.

38. Lei, F., Zhang, X. N., Wang, W., Xing, D. M., Xie, W. D., Su, H., & Du, L. J. (2007). Evidence of anti-obesity effects of the pomegranate leaf extract in high-fat diet induced obese mice. *International Journal of Obesity (2005), 31,* 1023–1029.

39. Li, Y., Qi, Y., Huang, T. H., Yamahara, J., & Roufogalis, B. D. (2008). Pomegranate flower: A unique traditional antidiabetic medicine with dual PPAR-alpha/gamma activator properties. *Diabetes Obesity and Metabolis, 10,* 10–17.

40. Huang, T. H., Peng, G., Kota, B. P., Li, G. Q., Yamahara, J., Roufogalis, B. D., & Li, Y. (2005). Pomegranate flower improves cardiac lipid metabolism in a diabetic rat model: Role of lowering circulating lipids. *British Journal of Pharmacology, 145,* 767–774.

41. Kaur, G., Jabbar, Z., Athar, M., & Alam, M. S. (2006). *Punica granatum* (pomegranate) flower extract possesses potent antioxidant activity and abrogates Fe-NTA induced hepatotoxicity in mice. *Food and Chemical Toxicology, 44,* 984–993.

42. Schubert, S. Y., Lansky, E. P., & Neeman, I. (1999). Antioxidant and eicosanoid enzyme inhibition properties of pomegranate seed oil and fermented juice flavonoids. *Journal of Ethnopharmacology, 66,* 11–17.

43. de Nigris, F., Balestrieri, M. L., Williams-Ignarro, S., D'Armiento, F. P., Fiorito, C., Ignarro, L. J., & Napoli, C. (2007). The influence of pomegranate fruit extract in comparison to regular pomegranate juice and seed oil on nitric oxide and arterial function in obese Zucker rats.. *Nitric Oxide, 17,* 50–54.

44. Aslam, M. N., Lansky, E. P., & Varani, J. (2006). Pomegranate as a cosmeceutical source: Pomegranate fractions promote proliferation and procollagen synthesis and inhibit matrix metalloproteinase-1 production in human skin cells. *Journal of Ethnopharmacology, 103,* 311–318.

45. Kohno, H., Suzuki, R., Yasui, Y., Hosokawa, M., Miyashita, K., & Tanaka, T. (2004). Pomegranate seed oil rich in conjugated linolenic acid suppresses chemically induced colon carcinogenesis in rats. *Cancer Science, 95,* 481–486.

46. Yamasaki, M., Kitagawa, T., Koyanagi, N., Chujo, H., Maeda, H., Kohno-Murase, J., Imamura, J., Tachibana, H., & Yamada, K. (2006). Dietary effect of pomegranate seed oil on immune function and lipid metabolism in mice. *Nutrition, 22,* 54–59.

47. Aviram, M., Volkova, N., Coleman, R., Dreher, M., Reddy, M. K., Ferreira, D., & Rosenblat, M. (2008). Pomegranate phenolics from the peels, arils, and flowers are antiatherogenic: Studies *in vivo* in atherosclerotic apolipoprotein E-deficient (E 0) mice and *in vitro* in cultured macrophages and lipoproteins. *Journal of Agricultural and Food Chemistry, 56,* 1148–1157.

48. Guo, C. J., Yang, J. J., Wei, J. Y., Li, Y. F., Xu, J., & Jiang, Y. G. (2003). Antioxidant activities of peel, pulp and seed fractions of common fruits as determined by FRAP assay. *Nutrition Research, 23,* 1719–1726.

49. Hajimahmoodi, M., Oveisi, M. R., Sadeghi, N., Jannat, B., Hadjibabaie, M., Farahani, E., Akrami, M. R., & Namdar, R. (2008). Antioxidant properties of peel and pulp hydro extract in ten Persian pomegranate cultivars. *Pakistan Journal of Biological Sciences, 11,* 1600–1604.

50. Negi, P. S., Jayaprakasha, G. K., & Jena, B. S. (2003). Antioxidant and antimutagenic activities of pomegranate peel extracts. *Food Chemistry, 80,* 393–397.

51. Chidambara Murthy, K. N., Jayaprakasha, G. K., & Singh, R. P. (2002). Studies on antioxidant activity of pomegranate (*Punica granatum*) peel extract using in vivo models. *Journal of Agricultural and Food Chemistry, 50,* 4791–4795.

52. Toklu, H. Z., Dumlu, M. U., Sehirli, O., Ercan, F., Gedik, N., Gokmen, V., & Sener, G. (2007). Pomegranate peel extract prevents liver fibrosis in biliary-obstructed rats. *Journal of Pharmacy and Pharmacology, 59,* 1287–1295.

53. Seeram, N. P., Aviram, M., Zhang, Y., Henning, S. M., Feng, L., Dreher, M., & Heber, D. (2008). Comparison of antioxidant potency of commonly consumed polyphenol-rich beverages in the United States. *Journal of Agricultural and Food Chemistry, 56,* 1415–1422.

54. Alper, N., Bahceci, K. S., & Acar, J. (2005). Influence of processing and pasteurization on color values and total phenolic compounds of pomegranate juice. *Journal of Food Processing and Preservation, 29,* 357–368.

55. Faria, A., Monteiro, R., Mateus, N., Azevedo, I., & Calhau, C. (2007). Effect of pomegranate (*Punica granatum*) juice intake on hepatic oxidative stress. *European Journal of Nutrition, 46,* 271–278.

56. Mertens-Talcott, S. U., Jilma-Stohlawetz, P., Rios, J., Hingorani, L., & Derendorf, H. (2006). Absorption, metabolism, and antioxidant effects of pomegranate (*Punica granatum* L.) polyphenols after ingestion of a standardized extract in healthy human volunteers. *Journal of Agricultural and Food Chemistry, 54,* 8956–8961.

57. Rosenblat, M., Hayek, T., & Aviram, M. (2006). Anti-oxidative effects of pomegranate juice (PJ) consumption by diabetic patients on serum and on macrophages. *Atherosclerosis, 187,* 363–371.

C. ACTIONS OF INDIVIDUAL FRUITS IN DISEASE AND CANCER PREVENTION AND TREATMENT

58. Aviram, M., Rosenblat, M., Gaitini, D., Nitecki, S., Hoffman, A., Dornfeld, L., Volkova, N., Presser, D., Attias, J., Liker, H., & Hayek, T. (2004). Pomegranate juice consumption for 3 years by patients with carotid artery stenosis reduces common carotid intima-media thickness, blood pressure and LDL oxidation. *Clinical Nutrition, 23*, 423–433.

59. Lin, C. C., Hsu, Y. F., Lin, T. C., & Hsu, H. Y. (2001). Antioxidant and hepatoprotective effects of punicalagin and punicalin on acetaminophen-induced liver damage in rats. *Phytotherapy Research, 15*, 206–212.

60. Lin, C. C., Hsu, Y. F., & Lin, T. C. (1999). Effects of punicalagin and punicalin on carrageenan-induced inflammation in rats. *American Journal of Chinese Medicine, 27*, 371–376.

61. Lin, C. C., Hsu, Y. F., Lin, T. C., Hsu, F. L., & Hsu, H. Y. (1998). Antioxidant and hepatoprotective activity of punicalagin and punicalin on carbon tetrachloride-induced liver damage in rats. *Journal of Pharmacy and Pharmacology, 50*, 789–794.

62. Cerda, B., Periago, P., Espin, J. C., & Tomas-Barberan, F. A. (2005). Identification of urolithin A as a metabolite produced by human colon microflora from ellagic acid and related compounds. *Journal of Agricultural and Food Chemistry, 53*, 5571–5576.

63. Seeram, N. P., Henning, S. M., Zhang, Y., Suchard, M., Li, Z., & Heber, D. (2006). Pomegranate juice ellagitannin metabolites are present in human plasma and some persist in urine for up to 48 hours. *Journal of Nutrition, 136*, 2481–2485.

64. Seeram, N. P., Zhang, Y. J., Sartipipour, M., Henning, S. M., Lee, R. P., Harris, D. M., Moro, A., and Heber, D. (2007). Pharmacokinetics and tissue disposition of urolithin A, an ellagitannin-derived metabolite, in mice. *Faseb Journal, 21*, 842.10.

65. Seeram, N. P., Aronson, W. J., Zhang, Y., Henning, S. M., Moro, A., Lee, R. P., Sartippour, M., Harris, D. M., Rettig, M., Suchard, M. A., Pantuck, A. J., Belldegrun, A., & Heber, D. (2007). Pomegranate ellagitannin-derived metabolites inhibit prostate cancer growth and localize to the mouse prostate gland. *Journal of Agricultural and Food Chemistry, 55*, 7732–7737.

66. Faria, A., Monteiro, R., Azevedo, I., & Calhau, C. (2007). Pomegranate juice effects on cytochrome p450s expression: *In vivo* studies. *Journal of Medicinal Food, 10*, 643–649.

67. Patterson, L. H., & Murray, G. I. (2002). Tumour cytochrome p450 and drug activation. *Current Pharmaceutical Design, 8*, 1335–1347.

68. Mehta, R., & Lansky, E. P. (2004). Breast cancer chemopreventive properties of pomegranate (*Punica granatum*) fruit extracts in a mouse mammary organ culture. *European Journal of Cancer Prevention, 13*, 345–348.

69. Shukla, S., & Gupta, S. (2004). Molecular mechanisms for apigenin-induced cell-cycle arrest and apoptosis of hormone refractory human prostate carcinoma DU145 cells. *Molecular Carcinogenesis, 39*, 114–126.

70. Pastorekova, S., Parkkila, S., Pastorek, J., & Supuran, C. T. (2004). Carbonic anhydrases: Current state of the art, therapeutic applications and future prospects. *Journal of Enzyme Inhibition and Medicinal Chemistry, 19*, 199–229.

71. Satomi, H., Umemura, K., Ueno, A., Hatano, T., Okuda, T., & Noro, T. (1993). Carbonic anhydrase inhibitors from the pericarps of *Punica granatum* L. *Biological & Pharmaceutical Bulletin, 16*, 787–790.

72. Afaq, F., Malik, A., Syed, D., Maes, D., Matsui, M. S., & Mukhtar, H. (2005). Pomegranate fruit extract modulates UV-B-mediated phosphorylation of mitogen-activated protein kinases and activation of nuclear factor kappa B in normal human epidermal keratinocytes paragraph sign. *Photochemistry and Photobiology, 81*, 38–45.

73. Hora, J. J., Maydew, E. R., Lansky, E. P., & Dwivedi, C. (2003). Chemopreventive effects of pomegranate seed oil on skin tumor development in cd1 mice. *Journal of Medicinal Food, 6*, 157–161.

74. Kim, N. D., Mehta, R., Yu, W., Neeman, I., Livney, T., Amichay, A., Poirier, D., Nicholls, P., Kirby, A., Jiang, W., Mansel, R., Ramachandran, C., Rabi, T., Kaplan, B., & Lansky, E. (2002). Chemopreventive and adjuvant therapeutic potential of pomegranate (*Punica granatum*) for human breast cancer. *Breast Cancer Research and Treatment, 71*, 203–217.

75. Zhu, Y. S., Cai, L. Q., Huang, Y., Fish, J., Wang, L., Zhang, Z. K., & Imperato-McGinley, J. L. (2005). Receptor isoform and ligand-specific modulation of dihydrotestosterone-induced prostate specific antigen gene expression and prostate tumor cell growth by estrogens. *Journal of Andrology, 26*, 500–508 discussion 509–510.

76. Syed, D. N., Afaq, F., & Mukhtar, H. (2007). Pomegranate derived products for cancer chemoprevention. *Seminars in Cancer Biology, 17*, 377–385.

77. Toi, M., Bando, H., Ramachandran, C., Melnick, S. J., Imai, A., Fife, R. S., Carr, R. E., Oikawa, T., & Lansky, E. P. (2003). Preliminary studies on the anti-angiogenic potential of pomegranate fractions *in vitro* and *in vivo*. *Angiogenesis, 6*, 121–128.

78. Albrecht, M., Jiang, W., Kumi-Diaka, J., Lansky, E. P., Gommersall, L. M., Patel, A., Mansel, R. E., Neeman, I., Geldof, A. A., & Campbell, M. J. (2004). Pomegranate extracts potently suppress proliferation,

xenograft growth, and invasion of human prostate cancer cells. *Journal of Medicinal Food, 7,* 274–283.

79. Larrosa, M., Tomas-Barberan, F. A., & Espin, J. C. (2006). The dietary hydrolysable tannin punicalagin releases ellagic acid that induces apoptosis in human colon adenocarcinoma caco-2 cells by using the mitochondrial pathway. *Journal of Nutritional Biochemistry, 17,* 611–625.

80. Shukla, M., Gupta, K., Rasheed, Z., Khan, K. A., & Haqqi, T. M. (2008). Bioavailable constituents/metabolites of pomegranate (*Punica granatum* L) preferentially inhibit COX2 activity *ex vivo* and il-1beta-induced pge2 production in human chondrocytes *in vitro*. *Journal of Inflammation (London, England), 5,* 9.

81. Okamoto, T., Akuta, T., Tamura, F., van Der Vliet, A., & Akaike, T. (2004). Molecular mechanism for activation and regulation of matrix metalloproteinases during bacterial infections and respiratory inflammation. *Biological Chemistry, 385,* 997–1006.

82. Heber, D. (2008). Multitargeted therapy of cancer by ellagitannins. *Cancer Letters, 269,* 262–268.

83. Conner, E. M., & Grisham, M. B. (1996). Inflammation, free radicals, and antioxidants. *Nutrition, 12,* 274–277.

84. Sparreboom, A., Cox, M. C., Acharya, M. R., & Figg, W. D. (2004). Herbal remedies in the United States: Potential adverse interactions with anticancer agents. *Journal of Clinical Oncology, 22,* 2489–2503.

85. Hidaka, M., Fujita, K., Ogikubo, T., Yamasaki, K., Iwakiri, T., Okumura, M., Kodama, H., & Arimori, K. (2004). Potent inhibition by star fruit of human cytochrome P450 3A (CYP3A) activity. *Drug Metabolism and Disposition, 32,* 581–583.

86. Obach, R. S. (2000). Inhibition of human cytochrome P450 enzymes by constituents of St. John's Wort, a herbal preparation used in the treatment of depression. *Journal of Pharmacology and Experimental Therapeutics, 294,* 88–95.

87. Wang, Z., Gorski, J. C., Hamman, M. A., Huang, S. M., Lesko, L. J., & Hall, S. D. (2001). The effects of St John's Wort (Hypericum perforatum) on human cytochrome P450 activity. *Clinical Pharmacology and Therapeutics, 70,* 317–326.

88. Hidaka, M., Okumura, M., Fujita, K., Ogikubo, T., Yamasaki, K., Iwakiri, T., Setoguchi, N., & Arimori, K. (2005). Effects of pomegranate juice on human cytochrome P450 3A (CYP3A) and carbamazepine pharmacokinetics in rats. *Drug Metabolism and Disposition, 33,* 644–648.

89. Nagata, M., Hidaka, M., Sekiya, H., Kawano, Y., Yamasaki, K., Okumura, M., & Arimori, K. (2007). Effects of pomegranate juice on human cytochrome P450 2C9 and tolbutamide pharmacokinetics in rats. *Drug Metabolism and Disposition, 35,* 302–305.

90. Kim, H., Yoon, Y. J., Shon, J. H., Cha, I. J., Shin, J. G., & Liu, K. H. (2006). Inhibitory effects of fruit juices on CYP3A activity. *Drug Metabolism and Disposition, 34,* 521–523.

91. Farkas, D., Oleson, L. E., Zhao, Y., Harmatz, J. S., Zinny, M. A., Court, M. H., & Greenblatt, D. J. (2007). Pomegranate juice does not impair clearance of oral or intravenous midazolam, a probe for cytochrome P450-3A activity: Comparison with grapefruit juice. *Journal of Clinical Pharmacology, 47,* 286–294.

92. Cerda, B., Ceron, J. J., Tomas-Barberan, F. A., & Espin, J. C. (2003). Repeated oral administration of high doses of the pomegranate ellagitannin punicalagin to rats for 37 days is not toxic. *Journal of Agricultural and Food Chemistry, 51,* 3493–3501.

93. Patel, C., Dadhaniya, P., Hingorani, L., & Soni, M. G. (2008). Safety assessment of pomegranate fruit extract: Acute and subchronic toxicity studies. *Food and Chemical Toxicology, 46,* 2728–2735.

94. Heber, D., Seeram, N. P., Wyatt, H., Henning, S. M., Zhang, Y., Ogden, L. G., Dreher, M., & Hill, J. O. (2007). Safety and antioxidant activity of a pomegranate ellagitannin-enriched polyphenol dietary supplement in overweight individuals with increased waist size. *Journal of Agricultural and Food Chemistry, 55,* 10050–10054.

C. ACTIONS OF INDIVIDUAL FRUITS IN DISEASE AND CANCER PREVENTION AND TREATMENT

37

Kiwifruit and Health

Denise C. Hunter, Margot A. Skinner, A. Ross Ferguson, and Lesley M. Stevenson

The New Zealand Institute for Plant & Food Research Ltd, Auckland, New Zealand

1. INTRODUCTION

The word 'kiwifruit' is now widely used for plants in the genus *Actinidia* Lindl., and the fruit they produce. *Actinidia* species predominantly come from south-western China, but a small number are found in countries adjoining China [1]. More than 90% of the kiwifruit in international trade are of a single cultivar, *A. deliciosa* (A. Chev.) C.F. Liang and A.R. Ferguson 'Hayward' ('green' kiwifruit), and the most commonly traded yellow-fleshed kiwifruit is *A. chinensis* Planch. 'Hort16A' (sold as ZESPRI® GOLD Kiwifruit) and often called 'gold' kiwifruit. Very small amounts of a third species, *A. arguta* (Sieb. et Zucc.) Planch. ex Miq. ('hardy' kiwifruit, or 'baby kiwi'), are also produced.

In China, the *Actinidia* species are generally referred to as *mihoutao*, a name used since the beginning of the Tang Dynasty (618–906 AD). According to Chinese pharmacopoeia, *mihoutao* were used to aid digestion, reduce irritability, relieve rheumatism, prevent kidney or urinary tract stones, and cure hemorrhoids, dyspepsia, and vomiting [2]. They were also used for the treatment of many types of cancer, especially breast cancer and cancers of the digestive system [3,4]. The bioactive constituents in kiwifruit that are recognized in traditional Chinese medicine include polysaccharides [5,6], alkaloids [7,8], saponins [9,10], and organic acids [11,12]. As populations age, there is increasing emphasis on maintenance of health and well-being, and more interest in developing functional foods based on traditional therapies. Kiwifruit have long been used in China for medicinal purposes and their potential for use in modern therapy deserves detailed study.

Kiwifruit are often promoted for their high vitamin C content, which probably contributes to the health benefits observed. *A. deliciosa* 'Hayward' and *A. chinensis* 'Hort16A' typically contain 85 and 100 mg/100 g (fresh weight) respectively, and a single fruit can provide 100% of the recommended daily intake [13]. Vitamin C levels vary among and within *Actinidia* species [14,15]. The health benefits attributed to vitamin C include antioxidant, anti-atherogenic, and anti-carcinogenic activity, as well as immunomodulation [16]. However, kiwifruit also contain other vitamins and minerals that may contribute to possible health benefits, including folate, potassium, and magnesium, as well as dietary fiber and phytochemicals [13]. Typically, green-fleshed kiwifruit (*A. deliciosa*) contain chlorophylls *a* and *b*, and carotenoids such as β-carotene, lutein, violaxanthin, and 9′-*cis*-neoxanthin; yellow-fleshed kiwifruit (e.g. *A. chinensis* or *A. macrosperma* C.F. Liang) also contain carotenoids, but little or no chlorophyll [17]; and red-fleshed genotypes (e.g. *A. chinensis* 'Hongyang') contain anthocyanins [18]. In addition, 'Hayward' leaf tissue and juice prepared from the fruit have been shown to contain polyphenolics, specifically flavonols [19,20]. Phytochemicals are well recognized for their antioxidant and cell protection activities, and they are becoming increasingly recognized for immunomodulatory properties. However, the phytochemical composition of kiwifruit can vary greatly according to the particular species or genotype. Many experimental studies give inadequate details of the species or cultivar of kiwifruit used. Most studies, especially those from outside China, have probably used fruit of only *A. deliciosa* 'Hayward', but even then the composition of 'Hayward' fruit could be affected by varying growing conditions, the maturity or ripeness of the fruit, or different periods in coolstore. Some authors do not specify the plant part analyzed.

2. HEALTH BENEFITS FROM THE *ACTINIDIA* SPECIES

Consumers are becoming more interested in the health benefits of food, particularly fruits [21]. Studies using *in vitro* and animal models, and evidence from human epidemiological studies provide supporting evidence for health claims. However, the most convincing evidence is ultimately that provided by clinical intervention trials, the type of evidence required to validate any health claims made about functional foods.

2.1 Cell Protection

Cells are continuously exposed to reactive oxygen species (ROS), produced during normal metabolism, and at higher concentrations as a result of disease, smoking, alcohol consumption, poor diet, and environmental pollution. ROS can potentially cause oxidative damage to DNA, proteins, and lipids. Dietary antioxidants are important in helping to protect against the oxidative damage common in degenerative diseases. Numerous studies have concluded that diets rich in fruits and vegetables reduce the risk from degenerative diseases [22], including cardiovascular disease [23] and cancer [24]. Antioxidant capacity of fruits and vegetables could come from vitamin C, vitamin E, carotenoids, and phenolic compounds.

The antioxidant capacity of dietary components is usually measured using chemical assays, including the Trolox™ equivalent antioxidant capacity (TEAC) assay, the ferric reducing antioxidant power (FRAP) assay, the DPPH (2,2′-diphenyl-1-picrylhydrazyl) free radical scavenging potential assay, the oxygen radical absorbance capacity (ORAC) assay, and the total radical absorption potential (TRAP) assay. Fruits and fruit extracts need to be screened using at least two of these assays, since various fruit components perform differently in individual assays [25]. The antioxidant capacity of kiwifruit is only average when compared with that of other fruits. By the ORAC assay, kiwifruit ranked fifth highest out of 13 fruits on a fresh weight basis [26]. Similar results were obtained using a modified TEAC assay [27]. The antioxidant capacity of 'Hayward' kiwifruit consistently ranges from 6.0 to 9.2 μmol Trolox™ equivalent/g fresh weight, when assayed by the ORAC method [26,28–30]; the antioxidant activity of 'Hort16A' kiwifruit, is reportedly 12.1 μmol TE/g fresh weight [28]. However, other non-commercial genotypes of kiwifruit may have a higher antioxidant capacity, with *A. eriantha* and *A. latifolia* rating highest in a study of eight genotypes, when analyzed using several assays [31]. The antioxidant capacity of different kiwifruit cultivars correlates well with their total polyphenol content [31,32] and vitamin C content [31].

The potential antioxidant capacity of fruit of *Actinidia* species may depend on the part of the fruit that is analyzed. Guo and colleagues found that the skin of a kiwifruit (species and cultivar undefined) contained three times more antioxidant activity than the flesh [33]. However, the skin makes up only a small part of the total fresh weight of the fruit and although the flesh of 'Hayward' kiwifruit is eaten, the coarse and fuzzy skin typically is not. The skin of 'Hort16A' is edible, even if not particularly palatable, but is not eaten by most consumers. Of the different types of kiwifruit available commercially, only the skin of *A. arguta* is normally eaten. Therefore, the antioxidant content of the skin is probably of little practical significance, unless future consumers are encouraged to consume it because it is 'healthy.'

The relevance of chemical-based antioxidant assays in predicting antioxidant activity in a biological setting has been questioned. The ability of fruit antioxidants to protect against oxidative stress needs to be demonstrated in cell-based assays or *in vivo*. This has been done for kiwifruit in only a few studies. For example, juice concentrate from 'Hayward' extracts protected Jurkat T cells from cell death induced by hydrogen peroxide-mediated oxidative stress [34]. Similarly, chemical-induced oxidative damage was found to be reduced in rats fed a diet supplemented with 15 and 30% kiwifruit (species and cultivar undefined) [35]. In a rat model of chronic liver injury, an ethanolic extract from *A. rubricaulis* roots, stems, and leaves fed to rats (0.5 or 1.0 g/kg/d) during induction of liver fibrosis significantly increased the levels of hepatic glutathione reductase, superoxide dismutase, and glutathione peroxidase [36]. Intense physical exercise is often associated with increased production of free radicals, and these may contribute to

muscle damage and fatigue during exercise. An extract of 'Hort16A' fruit mediated a temporary reduction in muscle fatigue in isolated mouse soleus muscle, and increased performance; a comparable effect was observed with the addition of superoxide dismutase in the short term (15 min), although the effect of superoxide dismutase was more prolonged (up to 55 min) [37,38].

Ingestion of kiwifruit juice improves scavenging of ROS generated in human plasma [39]. Ten trial participants were supplied with 150 mL kiwifruit juice (species and cultivar not defined) and blood was collected at intervals after ingestion. The antioxidant capacity of the plasma was evaluated by inhibition against ROS generation, as measured by the prevention of oxidation of 2',7'-dichlorodihydrofluorescein (DCHF) to its fluorescent product, dichlorofluorescein (DCF). Ingestion of kiwifruit juice enhanced the antioxidant capacity of the plasma within 30 min, and this was sustained for up to 90 min [39]. 'Hayward' kiwifruit (300 g) also significantly increased whole plasma antioxidant capacity (as measured by ORAC) for 5 h post-ingestion, during which time a steady increase in plasma vitamin C levels was also observed [40].

Taken together, these findings indicate that regular consumption of fruits and fruit juices, including kiwifruit, could be important in protecting against oxidative stress produced under normal and stressful conditions. However, the effect of long-term supplementation on plasma antioxidant activity and the implications this has on specific health targets remain to be investigated.

2.2 Cancer

Many cancers common in the Western world, such as colon, prostate, and breast cancers, have been related to dietary mutagens, including compounds in cooked meat, *N*-nitroso compounds, and fungal toxins, and to dietary habits, including high meat and saturated fat consumption [41]. Alcohol and tobacco may also contribute to the risk. Dietary antimutagens may slow the progression toward cancer, and there are numerous mechanisms by which they might act. They could inhibit mutagen uptake by adsorbing mutagens and promoting fecal bulking, thereby reducing transit time and contact time of the mutagen in the gut [41]. 'Hayward' kiwifruit contain relatively high amounts of lignin, which contributes to their insoluble dietary fiber content [42]. Lignin is hydrophobic and higher lignin content increases adsorption of heterocyclic aromatic amines, potential mutagenic compounds, and potential carcinogens following absorption and metabolism in the liver [43]. The high lignin content of kiwifruit may therefore act as a dietary antimutagen.

Dietary antimutagens can also inhibit endogenous formation of mutagens. Vitamins C and E, and plant polyphenols inhibit endogenous production of N-nitroso compounds from amino precursors [44]. Ethanolic extracts of kiwifruit also prevent mutagenic activity *in vitro* of N-nitrosoamines [45]. The prebiotic effect of dietary fiber may modulate colonic bacteria populations and this could reduce production of mutagens following fermentation of dietary chemicals by gut bacteria [41]. Dietary antioxidants, including vitamins C and E, and polyphenols, might also protect DNA, either directly by scavenging radical species, or indirectly through promoting endogenous antioxidant activity [41]. Juice from *A. sinensis* (*A. chinensis* or *A. deliciosa*) inhibited nitrosation reactions by scavenging nitrite, reportedly due to the vitamin C content and potential breakdown product of vitamin C, 3-hydroxy-3 pyranone [46]. However, although kiwifruit contain fiber, high levels of vitamin C, and phytochemicals, there is little evidence that kiwifruit specifically have antimutagenic activity.

Increased susceptibility to mutagenicity is positively correlated with incidence of cancer.

DNA repair has been used as a marker for susceptibility to mutagenicity [47]. Lymphocyte DNA repair was increased in healthy volunteers supplemented with β-carotene or lycopene, but this might have been due to a direct antioxidant effect, rather than to any antimutagenic effect [48]. Kiwifruit contain β-carotene, and similar results have been achieved by feeding subjects with kiwifruit. In a pilot trial two groups of six healthy subjects were given dietary and physical activity advice, and one group was given daily one kiwifruit ('Hayward', L.R. Ferguson, pers. comm.) for every 30 kg of body weight [49]. Leukocytes from subjects eating kiwifruit daily were better able to repair DNA after a peroxide challenge *ex vivo*. However, this was not specifically shown to be due to carotenoid content, and the vitamin C, folate, and fiber content of the fruit may also have contributed to the improvement. In a randomized crossover trial, Collins and colleagues fed 14 subjects one, two, or three kiwifruit (species and cultivar undefined) for three weeks [50]. Kiwifruit consumption was associated with a decrease in endogenous oxidation of DNA pyrimidines and purines, suggesting that a small intake of kiwifruit, as is manageable in a normal diet, may provide protection against DNA damage that might otherwise lead to mutations and cancer [50].

Toxicity to cancer cells provides a further mechanism by which kiwifruit might assist prevention or treatment of cancer. Such cytotoxicity is usually determined *in vitro*. For example, selected fractions from the peel of *A. deliciosa* fruit were shown to be cytotoxic to two human oral tumor cell lines (HSC-2 and HSG), but not to a 'normal' oral cell line (human gingival fibroblast) [51]. Roots of *Actinidia* species also contain effective compounds. For example, two novel polyoxygenated triterpenoids from the roots of *A. valvata*, a species recognized in traditional Chinese medicine as having antitumor activity, were cytotoxic to human hepatocellular carcinoma cell lines BEL-7402 and

SMMC-7721 [52]. Similarly, four novel phenolic compounds isolated from the roots of *A. chinensis* were cytotoxic to the cell lines P-388 (murine leukemia cell line) and A-549 (human lung carcinoma cell line) [53]. A chlorophyll-like compound isolated from a solvent extract from *A. arguta*, presumably from the vegetative parts, was also cytotoxic to a range of tumor cells [54]. This chlorophyll derivative was more toxic to leukemia cells than to solid tumor or 'normal' fibroblast cells; it rapidly blocked cellular DNA synthesis and induced apoptosis in human leukemia Jurkat T cells [54]. These findings indicate that extracts from *Actinidia* species may be toxic to cancer cells, but it should be noted that these effects have only been demonstrated *in vitro*, and not in a clinical setting.

Many chemotherapeutic agents cause serious myelotoxic side effects, such as anemia, leukopenia, and thrombocytopenia [55]. Various growth factors and some traditional medicines have been used to stimulate bone marrow proliferation in cancer patients [56,57]. Methanolic extracts from *A. arguta* stems stimulated both proliferation and colony formation of bone marrow (a measure of the potential of progenitor cells to develop into myeloid cells) in isolated bone marrow cells from mouse femurs [58]. The stimulation was largely due to (+)-catechin and (−)-epicatechin. These findings were confirmed *in vivo* using a mouse model of reduced bone marrow cellularity [58]; catechin, found in *A. arguta* stems [58], and also in small quantities in juice of 'Hayward' fruit [19], stimulated bone marrow proliferation and could therefore be effective in reducing the toxicity of chemotherapeutic agents.

2.3 Gut Health, Digestion, and Immune Function

Traditionally, kiwifruit were used to aid digestion; and kiwifruit have now been widely recognized for their strong laxative properties, and this has been confirmed by clinical studies. Healthy elderly subjects fed 'Hayward' kiwifruit daily for three weeks, one fruit per 30 kg of body weight, reported an improvement in laxation parameters, such as frequency and ease of defecation, stool bulk, and stool softness [59]. It was concluded that kiwifruit, eaten in these realistic quantities, could be useful in maintaining regularity in elderly people who otherwise have no major bowel problems. A subsequent trial indicated that 'Hayward' kiwifruit can relieve chronic constipation [60]. Importantly, there appeared to be no adverse effects, with no participants reporting diarrhea. The laxative effects of kiwifruit have been ascribed to their content of dietary fiber [59,60], approximately 2–3% fresh weight [61,62] and about 10% of recommended daily intake [13]. The cell walls of kiwifruit are unusual in that during ripening they swell much more than those of other fruits [63], and this may reflect an exceptionally high water-holding capacity, important for fecal bulking and enhancement of laxation [59]. Other kiwifruit components suggested as having laxative properties include the proteolytic enzyme actinidin, and non-digestible oligosaccharides. Further studies are required to elucidate their role in the promotion of laxation by kiwifruit [59].

Gut health and immune function are strongly influenced by microbial colonization of the gastrointestinal tract. Gut microflora act as an effective barrier against opportunistic and pathogenic microorganisms [64], and stimulating proliferation of beneficial bacteria with prebiotic components in the diet (dietary fiber) could reduce colonization by harmful bacteria. Dietary fiber is the 'non-digestible food ingredient that beneficially affects the host by selectively stimulating the growth and/or activity of one or a limited number of bacteria in the colon' [65]. Kiwifruit are a reasonable source of dietary fiber [66], mostly cellulose, pectin polysaccharides, and hemicelluloses such as glucuronoxylans and xyloglucans [67]. Kiwifruit polysaccharides have the potential to prevent adhesion of enteropathogens, and enhance adhesion of probiotic bacteria to Caco-2 (colon epithelial-derived) cells [68]. Aqueous extracts and extracts from the edible flesh of 'Hayward' and 'Hort16A' also promoted the growth of lactic acid bacteria and reduced the growth of *Escherichia coli* using a fecal batch fermentation model [69]. Changes to gut flora populations may also modulate immune function, influencing immunoglobulin (Ig) levels and cytokine production [70]. The contribution of such modulation to clinical and health outcomes remains uncertain, but the dietary fiber present in kiwifruit may contribute to the immunomodulatory activity demonstrated by various kiwifruit cultivars.

Antimicrobial activity of kiwifruit may protect the plants against infection. Animals that eat the fruit or other parts of kiwifruit plants might also be protected. For example, antifungal proteins isolated from green and gold kiwifruit are active against the plant fungal pathogens *Botrytis cinerea* and *Fusarium oxysporum*, respectively [71,72]. Antimicrobial activity against animal and human microbial pathogens has also been demonstrated. An essential oil prepared from fresh leaves of *A. macrosperma* had antibacterial activity *in vitro*, the activity being mild against *Staphylococcus aureus* and *Bacillus subtilus*, and strong against *Escherichia coli* [73]. Furthermore, the oil had higher fungicidal activity against *Candida albicans*, *Aspergillus fumigatus*, and *Microsporum canis* than the positive control clotrimazole [73]. Similarly, extracts from the skin, pulp, seeds, leaves, and stems of *A. chinensis* were bacteriostatic [74]. These reports suggest that ingestion of kiwifruit may provide an additional line of defense against infection. Furthermore, the use of kiwifruit in the development of antimicrobial products may slow the progression of bacterial

and fungal resistance to traditional drugs across populations.

Nutrient status is an important contributing factor to immune competence, and essential nutrients for efficient immune function include essential amino acids, linolenic acid, folic acid, vitamins A, B6, B12, C, and E, and the minerals zinc, copper, iron, and selenium [75]. Kiwifruit (*A. deliciosa* and *A. chinensis*) provide a reasonable source of copper, vitamin E, and folate (*A. deliciosa* 'Hayward': 18, 11, and 10% US recommended daily intake (RDI) respectively; *A. chinensis* 'Hort16A': 19, 11, and 8% US RDI, respectively), and are an excellent source of vitamin C [13]. In addition, kiwifruit also contain phytochemicals, in particular carotenoids, and some evidence suggests that higher carotenoid intake is associated with reduced incidence of infection [76,77].

Immune function can be broadly divided into innate immunity, defined as a nonspecific first line of defense, and adaptive immunity, a specific response that requires an element of memory. Markers of innate immunity that can be used for determining the effect of nutrition and functional foods on immune function include phagocytosis, oxidative burst, and natural killer (NK) cell activity; markers of adaptive immunity include lymphocyte proliferation and activation, cytokine production, and circulating immunoglobulins (Ig) [78].

Most studies examining the influence of kiwifruit on immune function have used murine models. For example, supplementation of the diet with up to 30% kiwifruit extract (plant part, cultivar, and species not defined) stimulated phagocytosis in Kunming mice, and enhanced levels of the serum immunoglobulins IgA, IgG, and IgM [79]. *A. macrosperma* has been used in Chinese medicine for treatment of cancer and stimulation of the immune system [12]. The mode of action was studied *in vivo* by feeding aqueous extracts of *A. macrosperma* (plant part not specified) to mice with the S180 ascites tumor. Tumor weight was not affected, but lymphocyte proliferation, NK cytotoxic activity, and phagocytosis were significantly stimulated [12]. Mice (BALB/c) were also used in studies which showed that aqueous and supercritical fluid extracts of 'Hort16A' and 'Hayward' fruit enhanced both innate and acquired immunity [80]. The extracts enhanced nonspecific NK cell activity and cytokine production (interferon γ; IFN-γ), and enhanced antigen-specific antibody production following vaccination. The effect of kiwifruit on an antigen-specific systemic antibody and gut-associated responses in mice has also been studied. Feeding ZESPRI® GOLD Kiwifruit puree, a processed product prepared from 'Hort16A,' enhanced a gut-associated adaptive immune response to a model protein, ovalbumin, in C57Bl/6 mice following oral immunization with OVA [32]. The puree stimulated antigen-specific antibody production (total Ig, and IgG) and antigen-specific proliferation of mesenteric lymph node (MLN) cells. Carotenoids and possibly water-soluble polysaccharides were suggested to be implicated in modulation of the adaptive immune response.

Adaptive immune responses can be broadly classified as resulting from one of two pathways: T helper Th1 and Th2 responses. These pathways are characterized by the production of specific cytokines and antibodies, with IFN-γ, IL-12, and IgG2a associated with a TH1 response, and IL-4, IL-5, IL-10, IL-13, IgG1, and IgE associated with a Th2 response. Both pathways were affected by consumption of the ZESPRI® GOLD Kiwifruit puree, when investigated using the murine model described above. Mice fed the puree produced higher levels of IgG1, which is effective at fixing complement and dealing with bacterial and viral infections, and interleukin (IL)-5, indicating a predominantly Th2-type response [37,38]. However, the mice also produced higher levels of IgG2c and IgG2b antibodies, typical of a

Th1-type response, suggesting a general stimulation of the humoral response.

Imbalance of these pathways toward a predominant Th2 type response has been implicated in the onset of some allergic responses and hyperresponsiveness. Extracts of dried fruit of *A. arguta* have shown promise as antiallergenic agents. In an OVA-sensitized mouse model that promotes a Th2 type response, feeding with the extracts was shown to decrease significantly the level of Th2 cytokines, and increase the level of Th1 cytokines, and this was also accompanied by decreased levels of plasma IgE and IgG1 and increased IgG2a [81]. Consumption of the extracts down-regulated the GATA-binding protein 3 (GATA-3) and induced T-box transcription factor (T-bet). These transcription factors control the differentiation of Th1 and Th2 cells [81]. The potential of *A. arguta* fruit extracts as orally active immune modulators for the therapy of allergic diseases was further tested using a mouse model for atopic dermatitis, a chronic inflammatory skin disease. The extracts reduced the severity of the dermatitis induced in NC/Nga mice. There was an associated decrease in plasma IgE, IgG1, and IL-4 levels, and an increase in IgG2a and IL-12 [82]. Splenic levels of IL-4, IL-5, and IL-10 also fell, whereas those of IFN-γ and IL-12 increased.

Bronchial asthma is a chronic inflammatory disease in which there is relative overproduction of Th2 cytokines relative to Th1 cytokines, and one of the results is accumulation of eosinophils in airways [83]. Induced bronchial inflammation in mice was reduced by feeding with extracts of *A. polygama* fruit, which appeared to have both anti-inflammatory and anti-asthmatic activity. Accumulation of eosinophils into airways and the levels of IL-4, IL-5, IL-13, and IgE in bronchoalveolar lavage fluid were reduced [83]. Although the active compound in *A. polygama* was not identified, a human study indicates that intake of fruits rich in vitamin C (citrus fruit and kiwifruit) reduced wheezing symptoms, and even eating the fruits as infrequently as once a week was found to alleviate symptoms [84].

Traditionally, *A. polygama* fruit have been used to treat rheumatic diseases and inflammation [85]. Extracts of *A. polygama* fruit have been tested and found to be effective in inhibiting acute inflammation and reducing edema formation in the carrageenan-induced hind paw edema model in rats [85], a model considered to predict reliably the anti-inflammatory efficacy of orally active anti-inflammatory agents. *In vitro* assays indicated that the anti-inflammatory activity was due to reduced expression of inducible nitric oxide synthase (iNOS), and cyclooxygenase 2 (COX-2) [85]. The active agent was identified as α-linolenic acid [86], which acts by down-regulating enzymes such as iNOS and COX-2 and expression of the inflammatory cytokine tumor necrosis factor-α (TNF-α) through blocking of the nuclear factor NFκB [87]. NFκB is involved in the regulation of transcription of many genes involved in inflammation and its inhibition by linolenic acid offers promising therapeutic possibilities for treatment of rheumatoid arthritis and inflammatory bowel diseases, including Crohn's disease. Other *Actinidia* species also contain compounds that can inhibit TNF-α production. Aqueous and ethyl acetate kiwifruit extracts were tested *in vitro* for their ability to inhibit TNF-α production initiated by ligation of microbial products to pattern recognition receptors (PRR) found on immune cells [88]. Inhibition varied according to the type of extract, the fruit used ('Hayward' or 'Hort16A') and the particular PRR selectively stimulated through the use of different ligands [88]. Inhibition of TNF-α production has also been achieved through the addition of aqueous extracts ('Hayward' or 'Hort16A') to human monocyte cultured cells (THP-1 cells) or peripheral whole blood cells, stimulated with lipopolysaccharide or IL-1β [89]. *In vitro*, the 'Hort16A' extract inhibited TNF-α production

more effectively, but when tested in the *ex vivo* whole blood model, the efficacy of the extracts was less clear because variations in the response from different blood donors became evident. Overall, the varying results between studies probably reflect the different phytochemical constituents of the different kiwifruit species [88], possibly the varying contents of α-linolenic acid [90], but demonstrate the potential anti-inflammatory properties of kiwifruit in general.

2.4 Cardiovascular Disease

Modification of the diet by increased consumption of fruit and vegetables has been associated with reduced risk of cardiovascular disease (CVD) [91,92]. The result is decreased mortality from cardiovascular disease and also reduced risk of ischemic stroke [91]. This has been ascribed to phytochemicals in fruit and vegetables such as carotenoids, flavonoids, polyphenols, and other antioxidants [92,93].

Crude ethanolic and aqueous extracts of kiwifruit (probably 'Hayward') have been shown to have antioxidant, antihypertensive, hypocholesterolemic, and fibrinolytic activity using *in vitro* assays [94]. Confirmation of the antioxidant activity of kiwifruit suggests that consumption of kiwifruit could reduce oxidation of cholesterol, thereby lowering the formation of atherosclerotic lesions. Antihypertensive activity was assessed by inhibition of angiotensin I-converting enzyme (ACE) which is important in the regulation of blood pressure [94]. There is increasing interest in the possibility of reducing hypertension through ACE inhibitors in food. Hypocholesterolemic activity was assayed by inhibition of 3-hydroxy-3-methyl-glutaryl coenzyme A (HMG-CoA), a key enzyme in the biosynthesis of cholesterol. Kiwifruit extracts were only weak inhibitors [94]. The extracts also mildly stimulated fibrinolytic activity. Fibrinolysis is the process by which fibrin in blood clots is dissolved, and increased fibrino-lytic activity is associated with a reduced risk of thromboembolic and cardiovascular disease. These results indicate that kiwifruit have poten-tial as cardiovascular protectants through vari-ous mechanisms and although some effects observed were relatively minor, cumulatively they may represent a significant reduction in overall risk. However, it is important to acknowledge the limitation of *in vitro* studies, and the *in vivo* effects of kiwifruit consumption depend on the absorption and metabolism of bioactives in the fruit.

Adhesion and aggregation of platelets at the site of injury in atherosclerotic vessel walls are very important in the pathogenesis of CVD [95]. Drugs such as aspirin are commonly used to inhibit platelet aggregation, and treatment with aspirin reduces myocardial infarction, stroke, and death. However, aspirin has only relatively weak antiplatelet activity and, more-over, can cause serious side effects in some patients. *In vitro* data suggest that natural pro-ducts, including fruits, may also inhibit platelet activity by a different mechanism to aspirin, with tomatoes and kiwifruit having particu-larly high inhibitory activity [95]. This has been confirmed *in vivo* and consumption of two or three kiwifruit per day by healthy volunteers was shown to inhibit platelet aggre-gation and also to raise plasma antioxidant levels and lower triglyceride levels [96]. Regular consumption of kiwifruit could there-fore be of great benefit in preventing and halt-ing the processes that lead to CVD [96].

Obesity is a strong risk factor for diseases such as diabetes and CVD. Inhibitors of pan-creatic lipase, a key enzyme for lipid absorp-tion, could be useful for treatment of obesity. Ursolic acid and a coumaryl triterpene isolated from *A. arguta* roots are useful inhibitors of pancreatic lipase activity *in vitro* [97]. Although much less potent than Orlistat, the widely used anti-obesity drug, they warrant further study.

2.5 Other Health Benefits from Kiwifruit

In addition to the 'mainstream' health tar-gets commonly used to define the benefits from functional foods and fruits, a small number of studies demonstrate more novel bioactive effects of kiwifruit. These often arise from use of less common *Actinidia* species or of fractions from commercial kiwifruit cultivars. For exam-ple, polysaccharide extracts from *A. chinensis* stimulated proliferation of keratinocyte and fibroblast cells *in vitro*, as well as stimulating collagen production from stable dermal equiva-lents [98]; this finding led to the conclusion that kiwifruit could have potential as pharmacologi-cally active agents in dermatology. Another example is the antihyperglycemic activity of leaf extracts from *A. deliciosa* which inhibited the carbohydrate-hydrolyzing enzymes α-amy-lase and α-glucosidase [99]; kiwifruit might have a use in treatment of hyperglycemia or dia-betes. Less convincing is the suggestion that kiwifruit juice (species and cultivar not speci-fied, but probably 'Hayward') might be useful in dislodging meat bolus obstructions [100], which had previously been suggested to be a result of the proteolytic enzyme (actinidin) con-tent [101]. Whether kiwifruit juice could actu-ally be used clinically for removal of bolus obstructions does not seem to have been tested.

Some of the proteins and polypeptides in kiwifruit may have specific biological effects and could have pharmacological benefits. A good example is kissper, a small, cysteine-rich peptide, present in large amounts in ripe kiwi-fruit (probably 'Hayward') [102]. When inges-ted, kissper might affect gastrointestinal physiology and, it is conjectured, might have potential for the treatment of disorders such as cystic fibrosis, in which ion transport mechan-isms are disturbed. Further research and clini-cal trials are required to determine whether the many bioactive compounds in kiwifruit have any real effects when the fruit are consumed.

3. KIWIFRUIT ALLERGIES, AND OTHER DETRIMENTAL HEALTH EFFECTS

As kiwifruit consumption has become more common around the world, so too has the incidence of kiwifruit allergy. Allergic responses to kiwifruit range from localized oral allergy syndrome to life-threatening anaphylaxis [103], and several clinical subgroups have been established [104]. The allergenicity of kiwifruit is not in doubt but is currently ill-defined and poorly understood [105,106]. Three of the allergens identified, actinidin, thaumatin, and kiwellin, are proteins often present in kiwifruit in large amounts and their identification as allergens may simply reflect this abundance, rather than their allergenicity [105]. Actinidin content and protease activity vary in different kiwifruit. Fruit of some *A. arguta* cultivars contain more actinidin than 'Hayward', whereas actinidin was not detectable or was barely detectable in two *A. rufa* selections [107] and some cultivars of *A. chinensis*, including 'Hort16A' [108]. Further work on kiwifruit allergens is required, as future commercial kiwifruit cultivars should preferably be less allergenic than the cultivars now available.

Kiwifruit contains small amounts of oxalate, and this is commonly associated with the 'catch' in the throat sometimes experienced when fresh kiwifruit and especially processed kiwifruit products are eaten [109,110]. Oxalate is present in cultivars of *A. deliciosa* (e.g. 'Hayward') as insoluble, fine needle-like, calcium oxalate raphide crystals [109], but in other *Actinidia* species the crystal shape varies [111]. Fruit of *A. chinensis* contain 18–45 mg/100 g fresh weight oxalate [14], similar to levels in mature fruit of *A. deliciosa* and *A. eriantha* [112]. High dietary intake of oxalate may be a major risk factor for calcium oxalate kidney stone formation [113], and oxalate consumption also has an antinutritional effect, reducing the bioavailability of Ca^{2+}, Mg^{2+}, and Fe^{2+} [114]. Normal dietary intake of oxalate is estimated to be 50–200 mg per day [113], and other foods, such as spinach, have considerably higher levels of total oxalate (1959 mg/100 g fresh weight) [113] than kiwifruit. Although kiwifruit contain oxalate, the amount is probably not of concern to healthy individuals who maintain a well-balanced diet; the effect of raphides on kiwifruit palatability is more significant.

4. SUMMARY

Dieticians along with other health professionals have the responsibility of providing consumers with scientifically supported knowledge to help them make informed dietary decisions. The scientific information supporting the unique health benefits of kiwifruit is growing rapidly. There is evidence from *in vitro* cell studies and animal models. Such evidence must be validated by human intervention trials.

Two intervention trials have been described that provide evidence for improved oxidation status of the blood after consumption of kiwifruit, but it is currently not clear how this influences the health of the individual. Oxidative stress may play a role in the pathogenesis of cardiovascular disease, the leading cause of death in developed countries. However, there is no evidence to support the use of antioxidant supplements to prevent mortality in healthy subjects or patients with various diseases [115]. It remains to be proven whether the compounds with antioxidant activity that are present in foods such as kiwifruit provide a health benefit.

A pilot intervention trial demonstrated that kiwifruit protect DNA from damage [49]. This damage may be associated with mutagenesis and cancer but further evidence is required before the consumption of kiwifruit for any tangible health benefit in this area can be recommended to consumers.

One intervention trial provided evidence for decreased platelet aggregation after consumption of kiwifruit [96]. This could result in an alteration in the natural course of atherosclerosis and reduce the risk of coronary arteriole disease, myocardial infarction, and stroke.

The strongest evidence for a health benefit of kiwifruit is in the area of intestinal well-being. Intestinal well-being is an ill-defined state often equated with an absence of symptoms. Bowel habit is a useful overall biomarker of gut function, particularly colonic function. Kiwifruit have been shown in two intervention trials to contribute to gastrointestinal well-being by their positive effects on laxation in targeted groups of subjects, the elderly [59] and those suffering from constipation [60].

A diet rich in fruits and vegetables offers health and wellness benefits that go beyond basic nutrition [116]. Consumption of green kiwifruit ('Hayward') contributes by having positive effects on cardiovascular and gut health. As supporting evidence accumulates for other cultivars, such as 'Hort16A,' it can be used to justify progression to human intervention trials. Validation of health claims that might be made about kiwifruit in the future may be based on consumption of the whole fruit, when the fruit is palatable (e.g. 'Hort16A'), or processed functional products, when the fruit is not palatable (e.g. *A. polygama*). With the possibility that these cultivars may modulate our immune system in a positive way, in the future we can look forward to further ways that kiwifruit will contribute to our health.

References

1. Ferguson, A. R. (1990). The genus Actinidia. In I. J. Warrington & G. C. Weston (Eds.), *Kiwifruit: science and management* (pp. 15–35). Auckland: Ray Richards Publisher and New Zealand Society for Horticultural Science Inc.

2. Ferguson, A. R. (1990). The kiwifruit in China. In I. J. Warrington & G. C. Weston (Eds.), *Kiwifruit: science and management* (pp. 155–164). Auckland: Ray Richards Publisher and New Zealand Society for Horticultural Science.

3. Motohashi, N., Shirataki, Y., Kawase, M., Tani, S., Sakagami, H., Satoh, K., Kurihara, T., Nakashima, H., Mucsi, I., Varga, A., & Molnár, J. (2002). Cancer prevention and therapy with kiwifruit in Chinese folklore medicine: a study of kiwifruit extracts. *Journal of Ethnopharmacology, 81*, 357–364.

4. Ai, X.-P., Li, Y.-G., & Zhang, M.-X. (1982). *Chinese medicines against cancers.* Heilong-jiang Province: Heilongjiang Science Technology Publishers.

5. Yan, J. Q., Wang, J. Y., & Zhao, M. (1995). *Actinidia chinensis* Planch. polysaccharide superoxide free radical hydroxyl free radical electron spin resonance. *Chinese Journal of Biochemical Pharmaceutics, 16*, 12–14.

6. Zhu, C. C., & Xu, G. J. (1996). *Actinidia rubricaulis* var. *coriacea* polysaccharides Arps – 2 Arps – 3. *Journal of Chinese Medicinal Materials, 19*, 623–625.

7. Sakan, T., Fujino, A., Murai, F., Butsugan, Y., & Suzui, A. (1959). On the structure of actinidine and matatabilactone, the effective components of *Actinidia polygama. Bulletin of the Chemical Society of Japan, 32*, 315.

8. Tang, S. H., & Zhang, K. M. (1997). *Actinidia chinensis* Planch actinidine composition resistance cancer. *Journal of Jishou University, 18*, 69–71.

9. Qing, Y. M., Chen, X. H., Cai, M. T., Hang, C. S., & Liu, H. X. (1999). Studies on the triterpenoids constituents of indochina *Actinidia* root (*Actinidia indochinensis*). *Journal of Chinese Medicinal Materials, 30*, 323–326.

10. Jin, Y. R., Gui, M. Y., & Ma, B. R. (1998). Chemical constituents of *Actindia arguta* roots. *Chinese Pharmacology Journal, 33*, 402–404.

11. Heatherbell, D. A. (1975). Identification and quantitative analysis of sugars and non-volatile organic acids in Chinese gooseberry fruit (*Actindia chinensis* Planch). *Journal of the Science of Food and Agriculture, 26*, 815–820.

12. Lu, Y., Fan, J., Zhao, Y.-P., Chen, S.-Y., Zheng, X.-D., Yin, Y.-M., & Fu, C.-X. (2007). Immunomodulatory activity of aqueous extract of *Actinidia macrosperma. Asia-Pacific Journal of Clinical Nutrition, 16*(Suppl 1), 261–265.

13. Ferguson, A. R., & Ferguson, L. R. (2002). Are kiwifruit really good for you? *Acta Horticulturae, 610*, 131–138.

14. Rassam, M., & Laing, W. A. (2005). Variation in ascorbic acid and oxalate levels in the fruit of *Actinidia chinensis* tissues and genotypes. *Journal of Agricultural and Food Chemistry, 53*, 2322–2326.

15. Nishiyama, I., Yamashita, Y., Yamanaka, M., Shimohashi, A., Fukuda, T., & Oota, T. (2004). Varietal difference in vitamin C content in the fruit of kiwifruit and other *Actinidia* species. *Journal of Agricultural and Food Chemistry, 52,* 5472–5475.

16. Naidu, K. A. (2003). Vitamin C in human health and disease is still a mystery? An overview. *Nutrition Journal, 2,* 7.

17. McGhie, T. K., & Ainge, G. D. (2002). Color in fruit of the genus *Actinidia*: carotenoid and chlorophyll compositions. *Journal of Agricultural and Food Chemistry, 50,* 117–121.

18. Montefiori, M., McGhie, T. K., Costa, G., & Ferguson, A. R. (2005). Pigments in the fruit of red-fleshed kiwifruit (*Actinidia chinensis* and *Actinidia deliciosa*). *Journal of Agricultural and Food Chemistry, 53,* 9526–9530.

19. Dawes, H. M., & Keene, J. B. (1999). Phenolic composition of kiwifruit juice. *Journal of Agricultural and Food Chemistry, 47,* 2398–2403.

20. Webby, R. F. (1990). Flavonoid complement of cultivars of *Actinidia deliciosa* var. *deliciosa,* kiwifruit. *New Zealand Journal of Crop and Horticultural Science, 18,* 1–4.

21. Crawford, K., & Mellentin, J. (2008). *Successful superfruit strategy: how to build a superfruit business.* Cambridge: Woodhead Publishing.

22. Ames, B. N., Shigenaga, M. K., & Hagen, T. M. (1993). Oxidants, antioxidants, and the degenerative diseases of aging. *Proceedings of the National Academy of Sciences of the United States of America, 90,* 7915–7922.

23. Joshipura, K. J., Hu, F. B., Manson, J. E., Stampfer, M. J., Rimm, E. B., Speizer, F. E., Colditz, G., Ascherio, A., Rosner, B., Spiegelman, D., & Willett, W. C. (2001). The effect of fruit and vegetable intake on risk for coronary heart disease. *Annals of Internal Medicine, 134,* 1106–1114.

24. Block, G., Patterson, B., & Subar, A. (1992). Fruit, vegetables, and cancer prevention: a review of the epidemiological evidence. *Nutrition and Cancer, 18,* 1–29.

25. Ozgen, M., Reese, R. N., Tulio, A. Z. J., Scheerens, J. C., & Miller, A. R. (2006). Modified 2,2-azino-bis-3-ethylbenzothiazoline-6-sulfonic acid (ABTS) method to measure antioxidant capacity of selected small fruits and comparison to ferric reducing antioxidant power (FRAP) and 2,2'-diphenyl-1-picrylhydrazyl (DPPH) methods. *Journal of Agricultural and Food Chemistry, 54,* 1151–1157.

26. Wang, H., Cao, G.-H., & Prior, R. L. (1996). Total antioxidant capacity of fruits. *Journal of Agricultural and Food Chemistry, 44,* 701–705.

27. Chun, O. K., Kim, D.-O., Smith, N., Schroeder, D., Han, J. T., & Lee, C. Y. (2005). Daily consumption of phenolics and total antioxidant capacity from fruit and vegetables in the American diet. *Journal of the Science of Food and Agriculture, 85,* 1715–1724.

28. United States Department of Agriculture. (2007). Oxygen Radical Absorbance Capacity (ORAC) of Selected Foods – 2007 (United States Department of Agriculture, Ed.) pp. 1–34, Agricultural Research Service, Beltsville.

29. Wu, X.-L., Beecher, G. R., Holden, J. M., Haytowitz, D. B., Gebhardt, S. E., & Prior, R. L. (2004). Lipophilic and hydrophilic antioxidant capacities of common foods in the United States. *Journal of Agricultural and Food Chemistry, 52,* 4026–4037.

30. Wu, X.-L., Gu, L.-W., Holden, J., Haytowitz, D. B., Gebhardt, S. E., Beecher, G. R., & Prior, R. L. (2004). Development of a database for total antioxidant capacity in foods: a preliminary study. *Journal of Food Composition Analysis, 17,* 407–422.

31. Du, G., Li, M., Ma, F., & Liang, D. (2009). Antioxidant capacity and the relationship with polyphenol and vitamin C in *Actinidia* fruits. *Food Chemistry, 113,* 557–562.

32. Hunter, D. C., Denis, M., Parlane, N. A., Buddle, B. M., Stevenson, L. M., & Skinner, M. A. (2008). Feeding ZESPRI™ GOLD Kiwifruit puree to mice enhances serum immunoglobulins specific for ovalbumin and stimulates ovalbumin-specific mesenteric lymph node cell proliferation in response to orally administered ovalbumin. *Nutrition Research (NY), 28,* 251–257.

33. Guo, C.-J., Yang, J.-J., Wei, J.-Y., Li, Y.-F., Xu, J., & Jiang, Y.-G. (2003). Antioxidant activities of peel, pulp and seed fractions of common fruits as determined by FRAP assay. *Nutrition Research (NY), 23,* 1719–1726.

34. Hunter, D. C., Zhang, J., Stevenson, L. M., & Skinner, M. A. (2008). Fruit-based functional foods II: the process for identifying potential ingredients. *International of Journal of Food Science and Technology, 43,* 2123–2129.

35. Ma, A.-G., Zhang, Y., Han, X.-X., Lan, J., & Gao, Y.-H. (2006). Effect of kiwifruit extract on antioxidant activity and erythrocyte membrane fluidity in rats. *The FASEB Journal, 20,* A429.

36. Liao, J.-C., Lin, K.-H., Cheng, H.-Y., Wu, J.-B., Hsieh, M.-T., & Peng, W.-H. (2007). *Actinidia rubricaulis* attenuates hepatic fibrosis induced by carbon tetrachloride in rats. *American Journal of Chinese Medicine, 35,* 81–88.

37. Skinner, M. A., Hunter, D. C., Denis, M., Parlane, N., Zhang, J., Stevenson, L. M., & Hurst, R. (2007). Health benefits of ZESPRI™ GOLD Kiwifruit: effects on muscle performance, muscle fatigue and immune responses (abstract). *Asia Pacific Journal of Clinical Nutrition, 16,* S31.

38. Skinner, M. A., Hunter, D. C., Denis, M., Parlane, N., Zhang, J., Stevenson, L. M., & Hurst, R. (2007). Health

benefits of ZESPRI™ GOLD Kiwifruit: effects on muscle performance, muscle fatigue and immune responses. *Proceedings of the Nutrition Society of New Zealand, 32,* 49–59.

39. Ko, S.-H., Choi, S.-W., Ye, S.-K., Cho, B.-L., Kim, H.-S., & Chung, M.-H. (2005). Comparison of the antioxidant activities of nine different fruits in human plasma. *Journal of Medicinal Food, 8,* 41–46.

40. Prior, R. L., Gu, L., Wu, X., Jacob, R. A., Sotoudeh, G., Kader, A. A., & Cook, R. A. (2007). Plasma antioxidant capacity changes following a meal as a measure of the ability of a food to alter *in vivo* antioxidant status. *Journal of the American College of Nutrition, 26,* 170–181.

41. Ferguson, L. R., Philpott, M., & Karunasinghe, N. (2004). Dietary cancer and prevention using antimutagens. *Toxicology, 198,* 147–159.

42. Bunzel, M., & Ralph, J. (2006). NMR characterization of lignins isolated from fruit and vegetable insoluble dietary fiber. *Journal of Agricultural and Food Chemistry, 54,* 8352–8361.

43. Funk, C., Braune, A., Grabber, J. H., Steinhart, H., & Bunzel, M. (2007). Model studies of lignified fiber fermentation by human fecal microbiota and its impact on heterocyclic aromatic amine adsorption. *Mutation Research/Fundamental and Molecular Mechanisms of Mutagenesis, 624,* 41–48.

44. Bartsch, H., Ohshima, H., & Pignatelli, B. (1988). Inhibitors of endogenous nitrosation mechanisms and implications in human cancer prevention. *Mutation Research/Fundamental and Molecular Mechanisms of Mutagenesis, 202,* 307–324.

45. Ikken, Y., Morales, P., Martinez, A., Marin, M. L., Haza, A. I., & Cambero, M. I. (1999). Antimutagenic effect of fruit and vegetable ethanolic extracts against N-nitrosamines evaluated by the Ames test. *Journal of Agricultural and Food Chemistry, 47,* 3257–3264.

46. Normington, K. W., Baker, I., Molina, M., Wishnok, J. S., Tannenbaum, S. R., & Puju, S. (1986). Characterization of a nitrite scavenger, 3-hydroxy-2-pyranone, from Chinese wild plum juice. *Journal of Agricultural and Food Chemistry, 34,* 215–217.

47. Collins, A. R., & Gaivao, I. (2007). DNA base excision repair as a biomarker in molecular epidemiology studies. *Molecular Aspects of Medicine, 28,* 307–322.

48. Torbergsen, A. C., & Collins, A. R. (2000). Recovery of human lymphocytes from oxidative DNA damage; the apparent enhancement of DNA repair by carotenoids is probably simply an antioxidant effect. *European Journal of Nutrition, 39,* 80–85.

49. Rush, E., Ferguson, L. R., Cumin, M., Thakur, V., Karunasinghe, N., & Plank, L. (2006). Kiwifruit consumption reduces DNA fragility: a randomized controlled pilot study in volunteers. *Nutrition Research (NY), 26,* 197–201.

50. Collins, A. R., Harrington, V., Drew, J., & Melvin, R. (2003). Nutritional modulation of DNA repair in a human intervention study. *Carcinogenesis, 24,* 511–515.

51. Motohashi, N., Shirataki, Y., Kawase, M., Tani, S., Sakagami, H., Satoh, K., Kurihara, T., Nakashima, H., Wolfard, K., Miskolci, C., & Molnár, J. (2001). Biological activity of kiwifruit peel extracts. *Phytotherapy Research, 15,* 337–343.

52. Xin, H.-L., Yue, X.-Q., Xu, Y.-F., Wu, Y.-C., Zhang, Y.-N., Wang, Y.-Z., & Ling, C.-Q. (2008). Two new polyoxygenated triterpenoids from *Actinidia valvata.* *Helvetica Chimica Acta, 91,* 575–580.

53. Chang, J., & Case, R. (2005). Cytotoxic phenolic constituents from the root of *Actinidia chinensis.* *Planta Medica, 71,* 955–959.

54. Park, Y. H., Chun, E. M., Bae, B. A., Seu, Y. B., Song, K. S., & Kim, Y. H. (2000). Induction of apoptotic cell death in human Jurkat T cells by a chlorophyll derivative (Cp-D) isolated from *Actinidia arguta* Planchon. *Journal of Microbiology and Biotechnology, 10,* 27–34.

55. Gentile, P., & Epremian, B. E. (1987). Approaches to ablating the myelotoxicity of chemotherapy. *Critical Reviews in Oncology/Hematology, 7,* 71–87.

56. Danova, M., & Aglietta, M. (1997). Cytokine receptors, growth factors and cell cycle in human bone marrow and peripheral blood hematopoietic progenitors. *Haematologica, 82,* 622–629.

57. Hisha, H., Yamada, H., Sakurai, M. H., Kiyohara, H., Li, Y., Yu, C.-Z., Takemoto, N., Kawamura, H., Yamaura, K., Shinohara, S., Komatsu, Y., Aburada, M., & Ikehara, S. (1997). Isolation and identification of hematopoietic stem cell-stimulating substances from Kampo (Japanese herbal) medicine, Juzen-Taiho-To. *Blood, 90,* 1022–1030.

58. Takano, F., Tanaka, T., Tsukamoto, E., Yahagi, N., & Fushiya, S. (2003). Isolation of (+)-catechin and (−)-epicatechin from *Actinidia arguta* as bone marrow cell proliferation promoting compounds. *Planta Medica, 69,* 321–326.

59. Rush, E. C., Patel, M., Plank, L. D., & Ferguson, L. R. (2002). Kiwifruit promotes laxation in the elderly. *Asia Pacific Journal of Clinical Nutrition, 11,* 164–168.

60. Chan, A. O., Leung, G., Tong, T., & Wong, N. Y. (2007). Increasing dietary fiber intake in terms of kiwifruit improves constipation in Chinese patients. *World Journal of Gastroenterology, 13,* 4771–4775.

61. Fourie, P. C., & Hansmann, C. F. (1992). Fruit composition of four South African-grown kiwifruit cultivars. *New Zealand Journal of Crop and Horticultural Science, 20,* 449–452.

C. ACTIONS OF INDIVIDUAL FRUITS IN DISEASE AND CANCER PREVENTION AND TREATMENT

62. Rupérez, P., Bartolomé, A. P., & Fernández-Serrano, M. I. (1995). Dietary fibre in Spanish kiwifruit. *European Journal of Clinical Nutrition, 49*(Suppl 3), S274–S276.

63. Hallett, I. C., MacRae, E. A., & Wegrzyn, T. F. (1992). Changes in kiwifruit cell wall ultrastructure and cell packing during postharvest ripening. *International Journal of Plant Science, 153*, 49–60.

64. Cummings, J. H., Antoine, J.-M., Azpiroz, F., Bourdet-Sicard, R., Brandtzaeg, P., Calder, P. C., Gibson, G. R., Guarner, F., Isolauri, E., Pannemans, D., Shortt, C., Tuijtelaars, S., & Watzl, B. (2004). PASSCLAIM – Gut health and immunity. *European Journal of Nutrition, 43* (Suppl 2), II/118–II/173.

65. Gibson, G. R., & Roberfroid, M. B. (1995). Dietary modulation of the human colonic microbiota: introducing the concept of prebiotics. *Journal of Nutrition, 125*, 1401–1412.

66. Lund, E. D., Smoot, J. M., & Hall, N. T. (1983). Dietary fiber content of eleven tropical fruits and vegetables. *Journal of Agricultural and Food Chemistry, 31*, 1013–1016.

67. Martin-Cabrejas, M. A., Esteban, R. M., López-Andreu, F. J., Waldron, K., & Selvendran, R. R. (1995). Dietary fiber content of pear and kiwi pomaces. *Journal of Agricultural and Food Chemistry, 43*, 662–666.

68. Ying, D. Y., Parkar, S., Luo, X. X., Seelye, R., Sharpe, J. C., Barker, D., Saunders, J., Pereira, R., & Schröder, R. (2007). Microencapsulation of probiotics using kiwifruit polysaccharide and alginate chitosan. *Acta Horticulturae, 753*, 801–808.

69. Molan, A. L., Kruger, M. C., & Drummond, L. N. (2007). The ability of kiwifruit to positively modulate key markers of gastrointestinal function. *Proceedings of the Nutrition Society of New Zealand, 32*, 66–71.

70. Lim, C. C., Ferguson, L. R., & Tannock, G. W. (2005). Dietary fibres as 'prebiotics': Implications for colorectal cancer. *Molecular Nutrition and Food Research, 49*, 609–619.

71. Wang, H., & Ng, T. B. (2002). Isolation of an antifungal thaumatin-like protein from kiwi fruits. *Phytochemistry, 61*, 1–6.

72. Xia, L., & Ng, T. B. (2004). Actinchinin, a novel antifungal protein from the gold kiwi fruit. *Peptides, 25*, 1093–1098.

73. Lu, Y., Zhao, Y. P., Wang, Z. C., Chen, S. Y., & Fu, C. X. (2007). Composition and antimicrobial activity of the essential oil of *Actinidia macrosperma* from China. *Natural Product Research, 21*, 227–233.

74. Basile, A., Vuotto, M. L., Violante, U., Sorbo, S., Martone, G., & Castaldo-Cobianchi, R. (1997). Antibacterial activity in *Actinidia chinensis, Feijoa sellowiana* and *Aberia caffra. International Journal of Antimicrobial Agents, 8*, 199–203.

75. Calder, P. C., & Kew, S. (2002). The immune system: a target for functional foods? *The British Journal of Nutrition, 88*(Suppl 2), S165–S176.

76. Cser, M. A., Majchrzak, D., Rust, P., Sziklai-László, I., Kovács, I., Bocskai, E., & Elmadfa, I. (2004). Serum carotenoid and retinol levels during childhood infections. *Annals of Nutrition and Metabolism, 48*, 156–162.

77. van der Horst-Graat, J. M., Kok, F. J., & Schouten, E. G. (2004). Plasma carotenoid concentrations in relation to acute respiratory infections in elderly people. *British Journal of Nutrition, 92*, 113–118.

78. Albers, R., Antoine, J.-M., Bourdet-Sicard, R., Calder, P. C., Gleeson, M., Lesourd, B., Semartin, S., Sanderson, I. R., Van Loo, J., Vas Dias, F. W., & Watzl, B. (2005). Markers to measure immunomodulation in human nutrition intervention studies. *British Journal of Nutrition, 94*, 452–481.

79. Ma, A.-G., Han, X.-X., Zhang, Y., Gao, Y.-H., & Lan, J. (2006). Effect of kiwifruit extract supplementation on levels of serum immunoglobulins and phagocytosis activity in mice. *The FASEB Journal, 20*, A1057.

80. Shu, Q., De Silva, U. M., Chen, S., Peng, W.-D., Ahmed, M., Lu, G.-J., Yin, Y.-J., Liu, A.-H., & Drummond, L. (2008). Kiwifruit extract enhances markers of innate and acquired immunity in a murine model. *Food and Agricultural Immunology, 19*, 149–161.

81. Park, E.-J., Kim, B. C., Eo, H. K., Park, K. C., Kim, Y. R., Lee, H. J., Son, M. W., Chang, Y.-S., Cho, S.-H., Kim, S. Y., & Jin, M. R. (2005). Control of IgE and selective T_H1 and T_H2 cytokines by PG102 isolated from *Actinidia arguta. Journal of Allergy and Clinical Immunology, 116*, 1151–1157.

82. Park, E. J., Park, K. C., Eo, H., Seo, J., Son, M., Kim, K. H., Chang, Y. S., Cho, S. H., Min, K. U., Jin, M., & Kim, S. (2007). Suppression of spontaneous dermatitis in NC/Nga murine model by PG102 isolated from *Actinidia arguta. Journal of Investigative Dermatology, 127*, 1154–1160.

83. Lee, Y.-C., Kim, S.-H., Seo, Y.-B., Roh, S.-S., & Lee, J.-C. (2006). Inhibitory effects of *Actinidia polygama* extract and cyclosporine A on OVA-induced eosinophilia and bronchial hyperresponsiveness in a murine model of asthma. *International Immunopharmacology, 6*, 703–713.

84. Forastiere, F., Pistelli, R., Sestini, P., Fortes, C., Renzoni, E., Rusconi, F., Dell'Orco, V., Ciccone, G., & Bisanti, L. and SIDRRIA Collaborative Group Italy. (2000). Consumption of fresh fruit rich in vitamin C and wheezing symptoms in children. *Thorax, 55*, 283–288.

85. Kim, Y. K., Kang, H. J., Lee, K. T., Choi, J. G., & Chung, S. H. (2003). Anti-inflammation activity of *Actinidia polygama*. *Archives of Pharmacal Research, 26,* 1061–1066.

86. Ren, J., Han, E. J., & Chung, S. H. (2007). *In vivo* and *in vitro* anti-inflammatory activities of α-linolenic acid isolated from *Actinidia polygama* fruits. *Archives of Pharmacal Research, 30,* 708–714.

87. Ren, J., & Chung, S. H. (2007). Anti-inflammatory effect of α-linolenic acid and its mode of action through the inhibition of nitric oxide production and inducible nitric oxide synthase gene expression via NFκB and mitogen activated protein kinase pathways. *Journal of Agricultural and Food Chemistry, 55,* 5073–5080.

88. Philpott, M., Mackay, L., Ferguson, L. R., Forbes, D., & Skinner, M. A. (2007). Cell culture models in developing nutrigenomics foods for inflammatory bowel disease. *Mutation Research/Fundamental and Molecular Mechanism of Mutagenesis, 622,* 94–102.

89. Farr, J. M., Hurst, S. M., & Skinner, M. A. (2007). Anti-inflammatory effects of kiwifruit. *Proceedings of the Nutrition Society of New Zealand, 32,* 20–25.

90. Wang, M. (2007). Distribution of important flavour precursors in *Actinidia*: long chain fatty acids. *Acta Horticulturae, 753,* 427–432.

91. Miura, K., Greenland, P., Stamler, J., Liu, K., Daviglus, M. L., & Nakagawa, H. (2004). Relation of vegetable, fruit, and meat intake to 7-year blood pressure change in middle-aged men: The Chicago Western Electric Study. *American Journal of Epidemiology, 159,* 572–580.

92. Lauretani, F., Semba, R. D., Dayhoff-Brannigan, M., Corsi, A. M., Di Iorio, A., Buiatti, E., Bandinelli, S., Guralnik, J. M., & Ferrucci, L. (2008). Low total plasma carotenoids are independent predictors of mortality among older persons. The InCHIANTI study. *European Journal of Nutrition, 47,* 335–340.

93. Emsaillzadeh, A., & Azadbakht, L. (2008). Dietary flavonoid intake and cardiovascular mortality. *The British Journal of Nutrition, 100,* 695–697.

94. Jung, K.-A., Song, T.-C., Han, D.-S., Kim, I.-H., Kim, Y.-E., & Lee, C.-H. (2005). Cardiovascular protective properties of kiwifruit extracts *in vitro*. *Biological and Pharmaceutical Bulletin, 28,* 1782–1785.

95. Duttaroy, A. K. (2007). Kiwifruits and the cardiovascular health. *Acta Horticulturae, 753,* 819–824.

96. Duttaroy, A. K., & Jørgensen, A. (2004). Effects of kiwi fruit consumption on platelet aggregation and plasma lipids in healthy human volunteers. *Platelets, 15,* 287–292.

97. Jang, D. S., Lee, G. Y., Kim, J., Lee, Y. M., Kim, J. M., Kim, Y. S., & Kim, J. S. (2008). A new pancreatic lipase inhibitor isolated from the roots of *Actinidia arguta*. *Archives of Pharmacal Research, 31,* 666–670.

98. Deters, A. M., Schröder, K. R., & Hensel, A. (2005). Kiwi fruit (*Actinidia chinensis* L.) polysaccharides exert stimulating effects on cell proliferation via enhanced growth factor receptors, energy production, and collagen synthesis of human keratinocytes, fibroblasts, and skin equivalents. *Journal of Cellular Physiology, 202,* 717–722.

99. Shirosaki, M., Koyama, T., & Yazawa, K. (2008). Anti-hyperglycemic activity of kiwifruit leaf (*Actinidia deliciosa*) in mice. *Bioscience, Biotechnology, and Biochemistry, 72,* 1099–1102.

100. Thomas, L., Low, C., Webb, C., Ramos, E., Panarese, A., & Clarke, R. (2004). Naturally occurring fruit juices dislodge meat bolus obstruction *in vitro*. *Clinical Otolaryngology, 29,* 694–697.

101. Karanjia, N. D., & Rees, M. (1993). The use of Coca-Cola in the management of bolus obstruction in benign oesophageal stricture. *Annals of the Royal College of Surgeons of England, 75,* 94–95.

102. Ciardiello, M. A., Meleleo, D., Saviano, G., Crescenzo, R., Carratore, V., Camardella, L., Gallucci, E., Micelli, S., Tancredi, T., Picone, D., & Tamburrini, M. (2008). Kissper, a kiwi fruit peptide with channel-like activity: Structural and functional features. *Journal of Peptide Science, 14,* 742–754.

103. Lucas, J. S. A., Lewis, S. A., & Hourihane, J. O. B. (2003). Kiwi fruit allergy: A review. *Pediatric Allergy and Immunology, 14,* 420–428.

104. Alemán, A., Sastre, J., Quirce, S., de las Heras, M., Carnés, J., Fernández-Caldas, E., Pastor, C., Blázquez, A. B., Vivanco, F., & Cuesta-Herranz, J. (2004). Allergy to kiwi: A double-blind, placebo-controlled food challenge study in patients from a birch-free area. *Journal of Allergy and Clinical Immunology, 113,* 543–550.

105. Lucas, J. S. A., & Atkinson, R. G. (2008). What is a food allergen? *Clinical and Experimental Allergy, 38,* 1095–1099.

106. Lucas, J. S. A., Nieuwenhuizen, N. J., Atkinson, R. G., MacRae, E. A., Cochrane, S. A., Warner, J. O., & Hourihane, J. O. B. (2007). Kiwifruit allergy: actinidin is not a major allergen in the United Kingdom. *Clinical and Experimental Allergy, 37,* 1340–1348.

107. Nishiyama, I., Fukuda, T., & Oota, T. (2004). Varietal differences in actinidin concentration and protease activity in the fruit juice of *Actinidia arguta* and *Actinidia rufa*. *Journal of the Japanese Society for Horticultural Science, 73,* 157–162.

108. Bublin, M., Mari, A., Ebner, C., Knulst, A., Scheiner, O., Hoffmann-Sommergruber, K., Breiteneder, H., &

C. ACTIONS OF INDIVIDUAL FRUITS IN DISEASE AND CANCER PREVENTION AND TREATMENT

Radauer, C. (2004). IgE sensitization profiles toward green and gold kiwifruits differ among patients allergic to kiwifruit from 3 European countries. *Journal of Allergy and Clinical Immunology, 114*, 1169–1175.

109. Perera, C. O., Hallett, I. C., Nguyen, T. T., & Charles, J. C. (1990). Calcium oxalate crystals: the irritant factor in kiwifruit. *Journal of Food Science, 55*, 1066–1069.

110. Walker, S., & Prescott, J. (2003). Psychophysical properties of mechanical oral irritation. *Journal of Sensory Studies, 18*, 325–346.

111. Watanabe, K., & Takahashi, B. (1998). Determination of soluble and insoluble oxalate contents in kiwifruit (*Actinidia deliciosa*) and related species. *Journal of the Japanese Society for Horticultural Science, 67*, 299–305.

112. Rassam, M., Bulley, S. M., & Laing, W. A. (2007). Oxalate and ascorbate in *Actinidia* fruit and leaves. *Acta Horticulturae 753*, 479–484.

113. Siener, R., Hönow, R., Seidler, A., Voss, S., & Hesse, A. (2006). Oxalate contents of species of the Polygonaceae, Amaranthaceae and Chenopodiaceae families. *Food Chemistry, 98*, 220–224.

114. Massey, L. K. (2003). Dietary influences on urinary oxalate and risk of kidney stones. *Frontiers in Bioscience, 8*, s584–s594.

115. Bjelakovic, G., Nikolova, D., Gluud, L. L., Simonetti, R. G., & Gluud, C. (2008). Antioxidant supplements for prevention of mortality in healthy participants and patients with various diseases (Review). *Cochrane Database of Systematic Reviews, 2*, 1–188.

116. Van Duyn, M. A. S., & Pivonka, E. (2000). Overview of the health benefits of fruit and vegetable consumption for the dietetics professional: Selected literature. *Journal of the American Dietetic Association, 100*, 1511–1521.

38

Bioactive Chemicals and Health Benefits of Grapevine Products

Marcello Iriti and Franco Faoro

Dipartimento di Produzione Vegetale, Università di Milano and Istituto di Virologia Vegetale, Dipartimento Agroalimentare, CNR, Milano, Italy

> We have wine even in this country, for our soil grows grapes and the sun ripens them, but this drinks like nectar and ambrosia all in one
>
> The Odyssey, Book IX, Homer

1. INTRODUCTION

Fruits and vegetables represent important components of the Mediterranean diet, a complex of dietary habit characterized by frequent consumption of whole grains, legumes, fish, olive oil for seasoning, low intake of red meat, yoghurt, and cheese, and associated with regular, moderate wine drinking (two glasses/day). According to the modern guidelines, dietary patterns including 400–600 g (up to 8 servings of 80 g) of fruits and vegetables per day, with a reduced calorie content and an increased nutrient density, as well as diets deficient in saturated fats and refined carbohydrates and abounding in fiber, significantly lower the incidence of insulin resistance (metabolic) syndrome, type II diabetes, and obesity, risk factors for cardiovascular diseases, cerebrovascular accidents, and cancers. Currently, the Mediterranean dietary style is certainly the diet most similar to the nutritional habits that determined the evolution of the genus *Homo* during the Palaeolithic era. Nonetheless, some of the previously mentioned risk factors, for instance excessive red meat consumption (≥ 7 times per week), plays an important role in the nutritional etiology of many malignancies [1].

As attested by Theophrastus and Hesiod, viticulture and winemaking were widely practiced in ancient Greece, though it is widely believed that winemaking began earlier, in the Neolithic Period (8500–4000 BC). Besides religious, social, and academic (at the symposium) contexts wherein wine was introduced, its medical uses were studied by Greek physicians, as reported by Hippocrates (460–370 BC). He recommended wine to cure fever and as analgesic, antiseptic, diuretic, tonic, and digestive [2]. The Romans also attributed therapeutic properties to wine, and Galen (129–200 AD), in particular, provided a detailed description on the medical uses of wine in his practice. In the care of gladiators, he used wine as an antiseptic for wound healing and as an analgesic for surgery. Other illnesses that Roman physicians treated by wine included depression, memory loss, constipation, diarrhea, gout, halitosis, snakebites, tapeworms, urinary tract ailments, and vertigo [3].

In this chapter, we first provide a brief description of the complex grape chemistry, i.e. the great variety of metabolites synthesized by the plant and stored in different berry tissues; second, we describe the most relevant biological activities of grapevine products and grape chemicals, reported in both human and animal *in vivo/in vitro* experimental models.

2. GRAPE CHEMISTRY

Secondary metabolites (phytochemicals) in grapevine (*Vitis vinifera* L.) occur in wood, leaves, stems (rachis and pedicels), and berries [4]. Although the berries are consumed as fruit and, along with stems, for winemaking, the leaves are normally not used as an edible vegetable by humans, with the exception of some typical Greek (dolmadakia) and Middle Eastern dishes of stuffed grapevine leaves rolled with rice. Therefore, in this section, we will emphasize the phytochemicals present in berry tissues.

Grape chemistry is rather complex and some thousands of compounds have been identified in the genus *Vitis*, included in the three main classes of natural products, phenylpropanoids, isoprenoids, and alakaloids, widely distributed both in plant foods and medicinal herbs [5,6]. In general, secondary metabolites exert a functional role in plant/grapevine ecology, mainly as phytoalexins, compounds involved in defense against pathogens and phytophages, as well as intolerance to detrimental abiotic conditions; for instance, adverse climate, high UV irradiance, exposure to excess light (photo-oxidation), water deficit, and anthropogenic pollutants [7,8].

2.1 Phenylpropanoids

The phenylpropanoid pathway starts from the aromatic amino acid phenylalanine (Phe, with the phenylpropanoid moiety C_6–C_3) and leads to derivatives with one, two, or more aromatic rings (C_6), each ring with a characteristic substitution pattern, and with different modifications of the propane residue of Phe (C_3) (Figure 38.1). Hydroxycinnamic acids (C_6–C_3) include *p*-coumaric, caffeic, ferulic, and sinapic acids, with different degrees of hydroxylation and methylation of C_6 (Figure 38.1). The cleavage of a C_2 fragment from the aliphatic side chain of *p*-coumaric acid leads to hydroxybenzoic acids (C_6–C_1), such as salicylic, vanillic, gallic, and syringic acids (Figure 38.1) [9].

The condensation of three C_2 residues with an activated hydroxycinnamic acid produces two classes of metabolites with a second aromatic ring linked to the phenylpropanoid moiety, the flavonoids (C_6–C_3–C_6) (Figure 38.2) and the stilbenes (C_6–C_2–C_6) (Figure 38.3) [10]. The basic flavonoid chemical structure is the flavan nucleus, consisting of 15 carbon atoms arranged in three rings: two benzene rings (A and B) combined by an oxygen-containing pyran ring (C) (Figure 38.2). The main classes of flavonoids in grape (flavanones, flavones, flavonols, flavanols, and anthocyanidins) differ in the level of oxidation and saturation of the C ring, while individual compounds within a class vary in the substitution pattern of the A and B rings [11].

Among flavonoids, anthocyanidins are the most abundant pigments in grape berry skin. Their conjugated derivatives (glycones), anthocyanins, mainly bound to sugars, hydroxycinnamates, or organic acids, are water-soluble

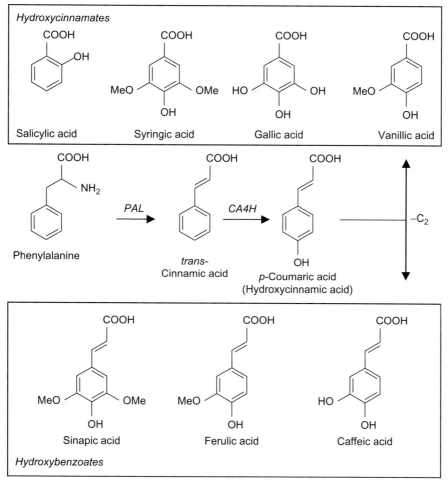

FIGURE 38.1 The simple phenols, hydroxycinnamates and hydroxybenzoates, are originated from the phenylpropanoid pathway that starts with deamination of phenylalanine by phenylalanine ammonia-lyase (PAL), leading to *trans*-cinnamic acid, in turn hydroxylated to *p*-coumaric acid via cinnamic acid 4-hydroxylase (CA4H).

pigments conferring blue, dark blue, red, and purple hues to flowers, fruits, and other plant organs. Anthocyanins of *Vitis* are structurally based on five aglycones/anthocyanidins – malvidin, cyanidin, delphinidin, peonidin, and petunidin – which differentiate on the basis of number and position of their hydroxyl groups and their degree of methylation (Figure 38.2) [12,13]. Flavonols include mainly kaempferol, quercetin, and myricetin aglycones. Apigenin is the main flavone in grape, whereas flavanols provide catechins, the monomeric units for proanthocyanidin biosynthesis (Figure 38.2) (see later in this section). Molecules belonging to the stilbene family are also abundant in berry skin cells, and possess the basic chemical structure based on the *trans*-resveratrol skeleton (Figure 38.3). Stilbenes comprise piceids, pterostilbenes, and viniferins that are glucosides, dimethylated derivatives, and oligomers of resveratrol, respectively (Figure 38.3) [14,15].

C. ACTIONS OF INDIVIDUAL FRUITS IN DISEASE AND CANCER PREVENTION AND TREATMENT

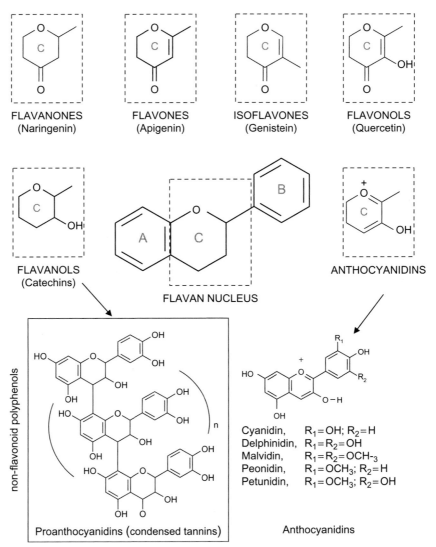

FIGURE 38.2 General flavonoid chemical structure: the different classes of flavonoids vary in the level of oxidation and substitution of the C ring, while individual compounds within a class differ in the substitution pattern of A and B rings, as in the case of anthocyanidins. Polymerization of flavonoid units, particularly flavanols, leads to proanthocyanidins.

Proanthocyanidins, also known as condensed tannins or simply tannins, are both oligomeric and polymeric compounds arising from flavanol condensation. In grapes, the most common monomers include catechin epimers [(+)-catechin and (−)-epicatechin], whose polymerization degree ranges mainly between 3 and 11 (Figure 38.2) [16]. In *Vitis*, proanthocyanidins are mainly present in seed, skin, and stem tissues of the bunch. In seeds, proanthocyanidins represent the major fraction of the total polyphenol extract and are characterized by a lower degree of polymerization than those present in berry skin. However, skin proanthocyanidins are more easily extracted during winemaking, because of their localization in the vacuole and cell wall, thus conferring important organoleptic properties to wine, such as astringency, bitterness, browning, turbidity, and color stability [17–19].

To summarize, flavonoids, stilbenes, and proanthocyanidins are collectively grouped in polyphenols (Figure 38.4), the name indicating both the compounds with a second aromatic ring and those arising from the polymerization of flavonoidic/catechin units.

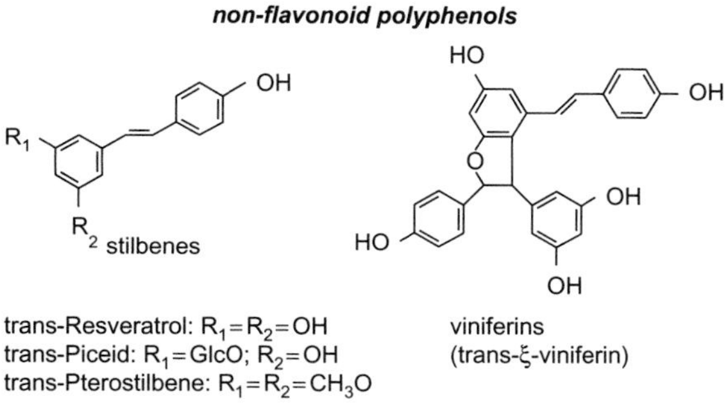

non-flavonoid polyphenols

trans-Resveratrol: $R_1 = R_2 = OH$
trans-Piceid: $R_1 = GlcO$; $R_2 = OH$
trans-Pterostilbene: $R_1 = R_2 = CH_3O$

viniferins
(trans-ξ-viniferin)

FIGURE 38.3 The stilbene family includes non-flavonoid compounds that possess a skeleton based on the *trans*-resveratrol structure.

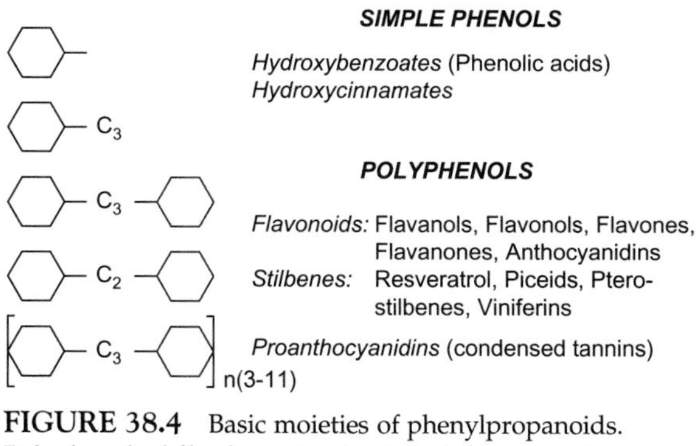

SIMPLE PHENOLS

Hydroxybenzoates (Phenolic acids)
Hydroxycinnamates

POLYPHENOLS

Flavonoids: Flavanols, Flavonols, Flavones, Flavanones, Anthocyanidins
Stilbenes: Resveratrol, Piceids, Pterostilbenes, Viniferins
Proanthocyanidins (condensed tannins)

FIGURE 38.4 Basic moieties of phenylpropanoids. Polyphenols differ from simple phenols because of a second aromatic ring, whereas proanthocyanidins are oligomeric or polymeric derivatives of flavonoids with polymerization degree, ranging from 3 to 11 in grape.

2.2 Isoprenoids

Isoprenoids, or terpenoids, are a huge and diversified group of lipidic compounds, deriving from acetyl-coenzyme A (CoA) via the intermediate molecule mevalonate and by the activity of key enzymes hydroxymethylglutaryl-CoA (HMG-CoA) synthase and reductase (Figure 38.5) [20]. Several hundreds of volatile compounds have been identified in grapes, whose aroma is largely due to C_{10} monoterpenes, representatives of isoprenoids and major components of essential oils. They include the acyclic alcohols geraniol, nerol, linalool, citronellol, and homotrienol and the monocyclic α-terpineol (Figure 38.5) [21–23]. In human breast adenocarcinoma (MCF-7) cells, geraniol was shown to inhibit cell cycle progression and proliferation, via a mevalonate-independent pathway (similarly to statins, some isoprenoids, including monoterpenes, can inhibit HMG-CoA reductase activity and, consequently, mevalonate accumulation, thus inhibiting cell growth) [24].

Carotenoids are isoprenoid tetraterpenes (C_{40}) accumulating in ripening grape berries, whose oxidation produces volatile fragments, the C_{13}-norisoprenoids (Figure 38.5) [23]. These are strongly odoriferous compounds, such as β-ionone (aroma of viola), β-damascenone (aroma of honey and exotic fruits), β-damascone (aroma of rose and fruits) and β-ionol (aroma of flowers and fruits) (Figure 38.5) [21,25–28]. In particular, β-ionone, a compound with an end ring analogue of β-carotene, exhibits a potent anticancer activity, as reported in both animal models and cell lines. In 2008 a dose-dependent inhibition of carcinogenesis by dietary β-ionone was observed in a rat mammary gland cancer model induced by 7,12-dimethylbenz[a]anthracene (DMBA) [29]. Moreover, in previous studies, β-ionone inhibited the growth of MCF-7, human gastric adenocarcinoma (SGC-7901), human colon adenocarcinoma (Caco-2), and human promyelocytic leukemia (HL-60) cell lines through different

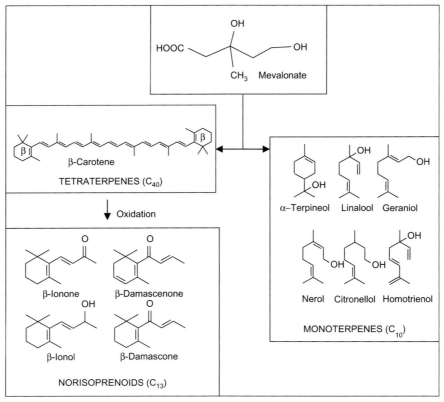

FIGURE 38.5   Major classes of grape isoprenoids, mevalonate being the precursor.

molecular mechanisms: arrest of cell cycle, induction of apoptosis, and inhibition of cell proliferation, invasiveness, and metastasis [24,30–32]. Because of the similarity between the chemical structure of β-ionone and retinoids (β-carotene, vitamin A, retinoic acids), it was suggested that β-ionone can act as a ligand agonist of the retinoid receptors, a class of nuclear receptors regulating cell growth, differentiation, and apoptosis, as demonstrated by the dose-dependent up-regulation of the retinoid receptor mRNA in the human colon cancer HCT-166 cell line [33].

2.3 Alkaloids

Indole alkaloids, deriving from the aromatic amino acid tryptophan (Trp), have been detected both in grape and grape products. Tetrahydro-β-carbolines are tricyclic indole derivatives occurring in low amounts in grapes, grape juice, and wine. In particular, carboxylic acids of tetrahydro-β-carbolines, widespread in fruits, arise from Trp via a nonenzymatic Pictet–Spengler condensation, i.e. the indole nucleus cyclization with carbonyl substrates, aldehydes typically (Figure 38.6). They occur at the ng/g and μg/L levels in grapes and juice/wine, respectively, contributing to the antioxidant capacity of these products [34–37]. Additionally, tetrahydro-β-carbolines are synthesized in mammalian brain tissues via the endogenous condensation of Trp, or 5-hydroxytryptophan, with formaldehyde or acetaldehyde [38]. In the central nervous system, they display a wide spectrum of psychoactive properties, as neurotransmitters and

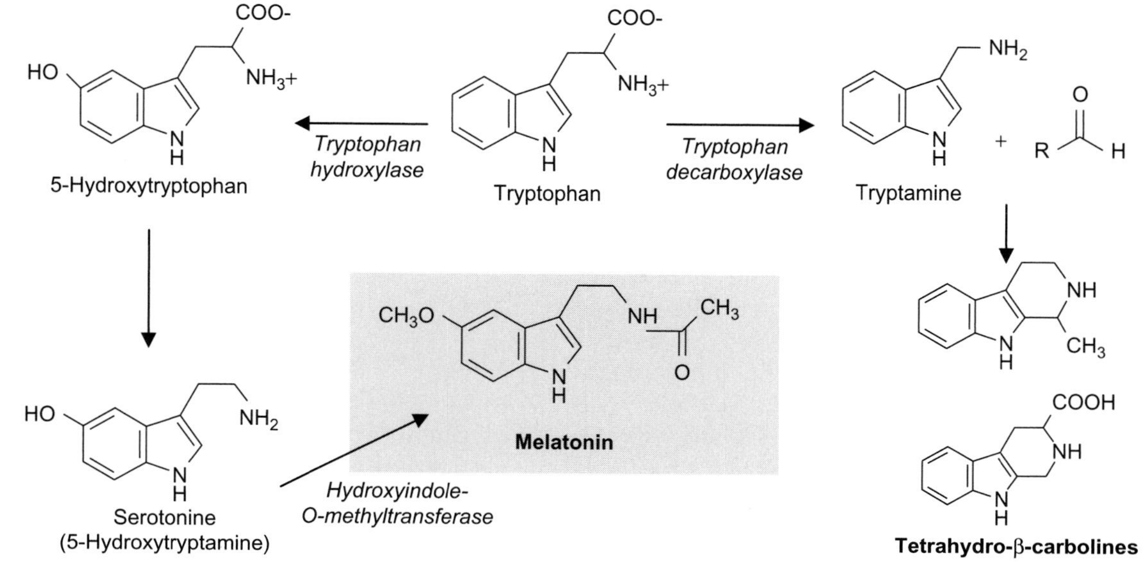

FIGURE 38.6 Main indole alkaloids of grape from tryptophan.
Tetrahydro-β-carbolines arise via a non-enzymatic condensation with carbonyl substrates (aldehydes) and indole nucleus cyclization, while melatonin is synthesized through the intermediates 5-hydroxytryptophan and serotonin.

neuromodulators: tetrahydro-β-carbolines function as potent reversible inhibitors of the enzyme monoamine oxidase (MAO), besides binding to the benzodiazepine, imidazoline, serotonin (5-hydroxytryptamine, 5-HT), and dopamine receptors and inhibiting 5-HT (re)uptake [39,40]. MAO catalyzes the oxidative deamination of biogenic amines, including neurotransmitters (dopamine, 5-HT, tryptamine, norepinephrine), vasoactive dietary (tyramine) and xenobiotic amines, thus being implicated in neurological disorders, psychiatric conditions, and depression [41,42]. In addition to the reported psychopharmacological effects, tetrahydro-β-carbolines are potent antioxidant and anticancer agents, active by different biochemical and molecular mechanisms, such as induction of apoptosis, inhibition of DNA topoisomerase I and II, and of cyclin-dependent kinases [35,43,44]. However, the precise effects of these compounds on human health are still debated, and the amount of dietary tetrahydro-β-carbolines effectively accumulated in biological tissues and fluids is still unknown,

despite their occurrence in many commercial fermented and smoked foodstuffs, such as vinegar, beer, cheese, yogurt, bread, and fish [45,46].

Melatonin (N-acetyl-5-methoxytryptamine) has been discovered in the berry skin of grapes [47,48]. The essential amino acid Trp is the precursor of all 5-methoxyindoles, or indoleamines/tryptamines, including melatonin, through the intermediate 5-HT and the activity of hydroxyindole-O-methyltransferase (Figure 38.6) [49]. Melatonin was long thought to be a neurohormone found exclusively in vertebrates, until its detection in bacteria, protozoans, algae, plants, fungi, and invertebrates [50]. Ever since, melatonin has been found in edible plants, medicinal herbs, and seeds, though its physiological and pathophysiological function *in planta* is still unclear. Nevertheless, a hormone-like role has been putatively attributed to melatonin in some plant species [51–53]. Among the examined grape cultivars, Nebbiolo and Croatina contain the highest melatonin levels, 0.9 and 0.8 ng/g, respectively, whereas the lowest concentration has been detected in Cabernet Franc (0.005 ng/g) [47]. As expected, melatonin has been found in wine too, with concentrations ranging from 0.05 to 0.5 ng/g, and, interestingly, in humans, serum melatonin level increases significantly after red wine intake (1 h after a single 100 mL supplementation) [54,55]. Melatonin is also a powerful antioxidant. It possesses an electron-rich aromatic indole ring and easily acts as an electron donor for molecules lacking it, thereby reducing and repairing electrophilic radicals [56]. After oxidation by a free radical, for instance hydroxyl anion radical ($^{\bullet}OH^{-}$), melatonin generates a resonance-stabilized nitrogen-centred radical, the indolyl (or melatonyl) cation radical. The latter, after a further quenching of superoxide anion radical ($^{\bullet}O_2^{-}$), forms the stable, nontoxic N^1-acetyl-N^2-formyl-5-methoxykynuramine, itself a powerful antioxidant and an anti-inflammatory able to improve mitochondrial metabolism and to inhibit cyclooxygenase (COX) 2 [57,58]. Intriguingly, it was postulated that a reaction between melatonin and peroxidases is present in plant tissues, able to improve the production of kynuramines [59]. Finally, melatonin can also counteract the cell oxidative burden indirectly, by stimulating the production of ROS detoxifying enzymes, specifically glutathione peroxidase, glutathione reductase, and superoxide dismutase [60].

3. GRAPEVINE CHEMICAL/PRODUCT BIOACTIVITIES: FOCUS ON POLYPHENOLS

Health-promoting effects of grapevine products strictly depend on the potpourri of the chemicals present in grape tissues, phenylpropanoids, isoprenoids, and alkaloids, thus supporting the assumption that no particular compound is by itself responsible for the health benefits widely attributed to these plant foods and beverages. However, in the last decades, a lot of studies focused mainly on polyphenols, considered as the archetypes of grapevine product bioactivities. In this section, we also emphasize polyphenolic properties, treating separately the findings of grape research.

3.1 Oxidative Stress and Structure–Radical Scavenging Activity Relationship of Polyphenols

At biochemical level, oxidative stress can be defined as a disturbance in the cell oxidation/reduction (redox) status. Metabolism of aerobic organisms unavoidably and continuously produces partially reduced oxygen intermediates, more reactive than molecular oxygen in its ground state, including both radical and nonradical forms collectively termed as reactive oxygen species (ROS) (Figure 38.7). During respiration, leakage of electrons from the mitochondrial transport chain leads to the single-electron reduction of molecular oxygen and to

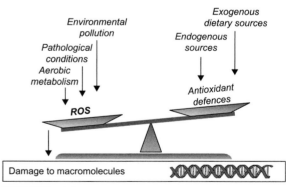

FIGURE 38.7 Homeostasis of the cell oxidation/reduction state.
Disturbance in the pro-oxidant–antioxidant balance may cause an overaccumulation of reactive oxygen species (ROS), thus leading to an oxidative stress, harmful for biomacromolecules.

the consequent formation of the superoxide anion radical ($\bullet O_2^-$) (Figure 38.8).

In addition to the sources of ROS under stable physiological conditions, other sites of ROS production exist. During an inflammation and immune response, the activated phagocytic white cells (neutrophils, macrophages, and monocytes) generate $\bullet O_2^-$ from molecular oxygen by an NADPH oxidase. This radical is then transformed into other ROS, mainly hydrogen peroxide (H_2O_2) and hydroxyl anion radicals ($\bullet OH^-$), involved in direct toxicity toward microbes, a process known as the respiratory burst. External factors can also detrimentally affect aerobic organisms, such as diseases, cigarette smoking, exposure to radiations, pollutants, and xenobiotic metabolism by the cytochrome P450 oxidase detoxifying system, all conditions that contribute to exacerbate the oxidative burden by generating significant amounts of ROS (Figure 38.7) [61–64].

Pathological conditions mechanistically linked to oxidative stress include inflammation, atherogenesis, and carcinogenesis. Thus it is not surprising that foods rich in antioxidant compounds play an essential role in the prevention of cancer, cardiovascular diseases, neurodegenerative disorders such as Parkinson's and Alzheimer's diseases, and premature aging (Figure 38.9) [61,62,64]. Due to the high reactivity of ROS, their uncontrolled production can cause injury to the nearest biomacromolecules (lipids, proteins, nucleic acids, and carbohydrates), if the pro-oxidant/antioxidant balance is not preserved (Figure 38.7). In particular, lipid peroxidation is a free radical-mediated reaction which damages both polyunsaturated fatty acids, in cell membranes, and plasma lipoprotein particles, such as low density lipoproteins (LDL) (Figure 38.8) [65,66]. To overcome these and other side effects of aerobic life and to protect vulnerable

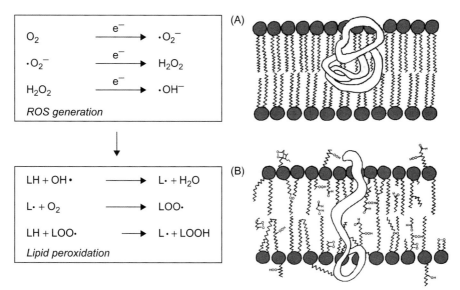

FIGURE 38.8 The oxidative burst generates reactive oxygen species (ROS) (A) that cause lipid peroxidation with consequent alterations of the bilayer of the cell membrane (B).
Lipid peroxidation is a chain reaction starting by the extraction of an H atom from polyunsatured fatty acids (PUFAs) and forming a fatty acid radical ($L^\bullet$); the latter reacts with O_2 to give a lipid peroxyl radical ($LOO^\bullet$) and, finally, a lipid hydroperoxide ($LOOH$).

C. ACTIONS OF INDIVIDUAL FRUITS IN DISEASE AND CANCER PREVENTION AND TREATMENT

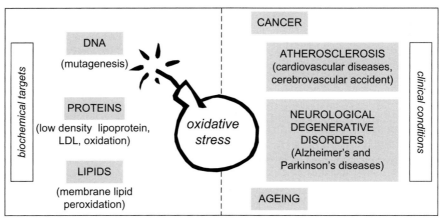

FIGURE 38.9   Biochemical targets and clinical conditions mechanistically linked to oxidative stress.

targets, organisms evolved sophisticated strategies, collectively termed antioxidant defenses, that counteract the imbalance of the cell redox homeostasis and keep ROS levels under the cytotoxic threshold (Figure 38.7) [67]. Antioxidant defenses also comprise vitamins and nutraceuticals, i.e. nonenzymatic scavengers abundant in plant foodstuffs and introduced by diets, including polyphenol-rich grape products [68–71]. Any compound capable of quenching ROS, without itself undergoing conversion to a destructive radical species, can be considered as an antioxidant, as in the case of dietary flavonoids [72,73].

As already introduced, the basic flavonoid chemical structure is the flavan nucleus, consisting of 15 carbon atoms arranged in three rings (C_6–C_3–C_6): two aromatic rings (A and B) connected by a three-carbon-atom heterocyclic ring, an oxygen-containing pyran ring (C) (Figure 38.2) [11]. Flavonoids act as antioxidants by donating electrons and stopping radical chains (Figure 38.10). This activity is attributed to the phenolic hydroxyls, increasing with the number of OH groups in A and B rings (Figure 38.10) [74]. The structural requirements considered to be essential for effective radical scavenging by flavonoids are the presence of a 3′,4′-dihydroxy group (o-diphenolic

group, a catechol structure) on the B ring, and a double bond between C_2 and C_3 ($\Delta^{2,3}$) conjugated with a keto function at C_4 of the C ring (Figure 38.2–38.10). Hydroxyl groups on the B ring donate hydrogen and an electron to radical species, stabilizing them and giving rise to a relatively stable flavonoid radical (Figure 38.10). The C_2–C_3 double bond and the 4-keto group are responsible for electron delocalization from the B ring (Figure 38.10). Hydroxyl groups in positions 3 and 5, in combination with 4-oxo function and C_2–C_3 double bond, contribute to further enhance the radical scavenging activity (Figure 38.10) [75–78].

3.2 Antioxidant and Antimutagenic Activity of Polyphenols

Probably the most investigated biological activity of polyphenols is their antioxidant power, though they also possess a plethora of more or less correlated properties, such as antimutagenic, anti-inflammatory, antitumoral, antihypertensive, cardio- and neuroprotective activities [6,11].

As reported above, polyphenols supplied by diet exert health benefits by ROS scavenging, and this has also been shown in studies based

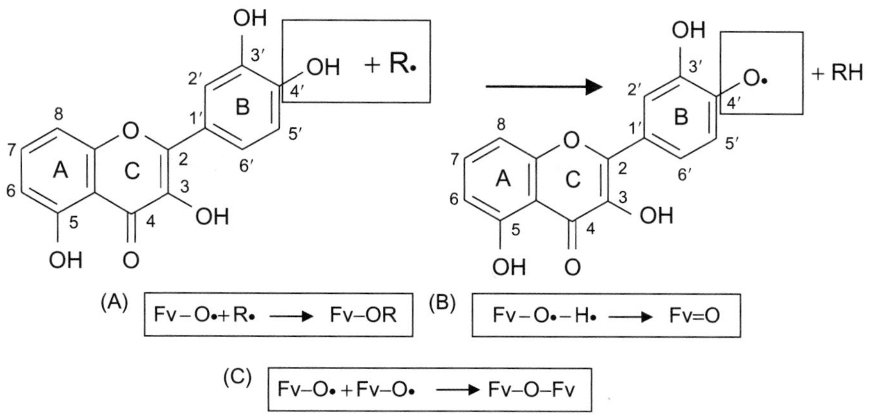

FIGURE 38.10 The activity of flavonoids as radical scavengers is attributed to the phenolic hydroxyls and increases with the number of − OH groups in A and B rings.

An − OH can react with a free radical ($R^\bullet$) to give a flavonoid radical (Fv-$O^\bullet$); afterwards three different termination reactions can occur: a coupling reaction between a flavonoid radical with another radical species (A); a dimerization with another flavonoid radical (B); a further loss of a hydrogen atom from the flavonoid radical (C).

on the supplementation of grapevine products [70,79]. In one study, the antioxidant capacity of red grape was evaluated in HepG2 (human hepatocellular liver carcinoma) cell line and positively correlated to the total phenolic content and to the oxygen radical absorbance capacity (ORAC) values of grape extracts. The authors concluded that increasing fruit consumption represents a suitable strategy to counteract oxidative stress and to reduce the risk of cancer [80]. Dietary polyphenols also contribute to elevate the antioxidant capacity of human blood. Daily consumption of grape juice (10 mL/kg body weight) for 2 weeks resulted in an increased resistance of LDL to *ex vivo* oxidation, comparable to the value obtained after α-tocopherol (400 IU) supplementation [81]. These results further confirmed the data previously reported by Day and colleagues, showing that the daily intake of grape juice (125 mL), for 1 week, significantly reduced the LDL oxidizability. They also showed an 8% increase in plasma antioxidant capacity measured 1 h after the grape juice supplementation as ferric-reduced antioxidant potential (FRAP) [82]. In a short-term study,

the acute intake of a phenolic-rich juice (400 mL), with grape as major ingredient, improved the antioxidant status in healthy subjects, determined both in serum and urine by FRAP. In the same work, the authors showed that the phenolic compounds of the juice were bioavailable, as revealed by the increase of phenolics that could bind the lipid fraction of serum and by their rise in the urinary excretion, with a maximum reached 2 h after consumption [83]. In pre- and postmenopausal women, the whole-body oxidative stress was significantly reduced after the daily supplementation of a lyophilized grape powder (36 g) for 4 weeks, by reducing the levels of urinary F_2-isoprostanes, biomarkers of oxidative stress [84]. Consuming black grape (1 g/kg body weight) exerted similar effects, compared with juice and powder, significantly raising the plasma antioxidant potential of healthy volunteers 4 h after ingestion [85].

Oxidative DNA damage, leading to modifications of DNA bases, is related to mutagenesis, carcinogenesis, and aging (Figure 38.9). Daily grape juice supplementation (480 mL), for 8 weeks, reduced DNA strand breaks in peripheral lymphocytes, as detected by the single cell gel electrophoresis (comet assay, a powerful tool in mutagenesis studies), beside decreasing the amount of released ROS [86]. Similarly, treatment of human lymphocytes with a grape seed extract reduced the frequency of micronuclei (an assay for the detection of DNA damage) by 40% and the production of malonyldialdehyde (a biomarker of lipid peroxidation) by 30%, while increasing the activities of the antioxidant enzymes catalase and glutathione S-transferase by 10 and 15%, respectively [87]. In another study, the antimutagenic activity of both aqueous and methanolic extract from two Greek grape varieties (red and white) was assessed against the ROS-induced DNA damage, using the *Salmonella*/reversion assay and the oxidant mutagens bleomycin and H_2O_2. Unexpectedly, both polyphenol-rich fractions and single polyphenols (resveratrol, catechin, epicatechin, quercetin, gallic and protocatechuic acids) from red grape extracts did not affect the bleomycin- and H_2O_2-induced mutagenicity, either positively or negatively, though some slight pro-oxidant and mutagenic effects were reported for resveratrol and quercetin at the assayed concentrations [88]. In agreement with these results, the reported data on the antimutagenic activity of individual grape phytochemicals are still elusive. In purified calf thymus DNA treated with oxidants, resveratrol exhibited a bimodal response on the formation of 8-hydroxy-2-deoxyguanosine (8-OH-dG, a biomarker of oxidative DNA damage), with a slight pro-oxidant effect, at lower concentrations, and an antioxidant activity at higher concentrations, reducing the 8-OH-dG accumulation in a dose-dependent manner. This biomarker causes G→T and A→C transversions during DNA replication, resulting in carcinogenesis [89]. Intriguingly, melatonin, besides being more effective than resveratrol, reversed the pro-oxidant DNA damage induced by low concentrations of resveratrol, when added in combination,

showing a synergistic action [90]. Furthermore, pretreatment with resveratrol prevented the accumulation of DNA strand breaks induced by tobacco smoke condensate in cell lines of different histogenetic origin, as assessed by the comet test [91]. Accordingly, in animal cell cultures, resveratrol failed to induce DNA damage, though it slightly increased chromosomal aberrations at the highest assayed doses [92].

In conclusion, it seems that the protective effect against ROS-induced oxidative DNA damage cannot be attributed to polyphenols singly present in grape, but rather to the synergism among polyphenols themselves and/or between them and other types of bioactive chemicals.

3.3 Cancer Chemoprevention

Prevention has become as important as therapy to control cancer, in order to reduce both cancer morbidity and mortality. Cancer chemoprevention is the strategy of preventing, arresting, or reversing carcinogenesis by means of chemopreventive agents, dietary therapeutics effective at each step of neoplastic progression. Therefore, chemopreventive agents can be divided into blocking agents, that arrest the initiation stage of malignancy, and suppressing agents, that act on tumor promotion and progression by inhibiting the malignant transformation of initiated cells. Again, chemopreventive agents can avoid the time-dependent tumor resistance (chemoresistance) to chemotherapeutic agents, and their nonspecific toxicity toward non-target cells [93–99].

Cancer is a multistage and multifactorial disease, the second leading cause of death worldwide after heart disease, whose risk and incidence augments with age. In addition to genetic factors, environmental and nutritional factors play a major role in cancer etiology. In industrialized developed countries, breast, prostate, and colorectal cancers predominate, because of a diet rich in animal foods and refined carbohydrates and deficient in plant foods. Conversely, in developing countries, where diet is largely based on cereal/starchy foods, esophageal, stomach, and liver cancers have a higher incidence [100].

An intricate network of signaling pathways is involved in cancer pathogenesis, regulating the (im)balance between cell growth-promoting and growth-inhibiting mechanisms (Figure 38.11). At a molecular level, the interactions with both transcription factors and receptors have been proposed as putative mechanisms for the reported anticarcinogenic activitiy of (grape) polyphenols (Figure 38.11). Furthermore, chemopreventive dietary agents can promote apoptosis in premalignant and malignant cells by modifying different stages of the apoptotic process, including the cell redox status (Figure 38.11) [101].

Chemopreventive properties of grape products are more likely attributable to the combined effect of their bioactive components, rather than to one or a few specific molecules, although resveratrol represents the most studied example of grape biologically active compound [5,6]. In his seminal and pioneering study, Pezzuto and his group [b] reported for the first time the chemopreventive potential of resveratrol in different assays representing the three carcinogenesis stages. They also showed that resveratrol inhibited the development of preneoplastic lesions in a mouse mammary gland culture treated with the carcinogen DMBA, as well as tumorigenesis in a mouse skin cancer model [102]. Nevertheless, it must

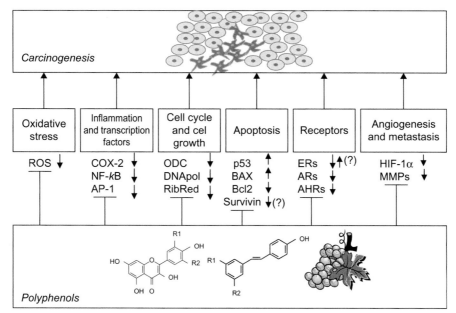

FIGURE 38.11 Mechanisms of cancer prevention of polyphenols by inhibiting ($\downarrow$) or enhancing ($\uparrow$): reactive oxygen species (ROS) production; important transcription factors (NFκB, AP-1) and enzymes involved in inflammation (cyclooxygenase-2, COX-2); androgen and estrogens receptors (ARs, ERs); aryl hydrocarbon receptors (AHRs); pro-apoptotic (p53, BAX) and antiapoptotic proteins (Bcl-2, Survivin); ornithine decarboxylase (ODG) and other enzymes regulating cell growth (DNA polymerase, ribonucleotide reductase); hypoxia inducing factor 1α (HIT-1α) and matrix metalloproteinases (MMPs) involved in metastasis and angiogenesis (see text for details).

be underlined that, when consuming a food or beverage, we ingest its potpourri of chemicals.

Cancer and chronic inflammation are causally linked, as demonstrated by the increased gene expression and/or activity of enzymes synthesizing arachidonate-derived proinflammatory mediators, and by the augmented production of these mediators (prostanoids) in various cancers [103,104]. Hence, inhibition of cyclooxygenases (COXs), particularly of inducible COX-2 isozyme, and blockage of prostaglandin cascade are relevant and effective mechanisms to counteract multistage carcinogenesis (Figure 38.11) [105]. Among polyphenols, resveratrol's capability of blocking various components of the proinflammatory cascade has been known for a long time, as reported in phorbol ester-treated human mammary epithelial cells, where resveratrol inhibited both COX-2 gene transcription and enzymatic activity [106]. Anti-inflammatory properties of resveratrol were then reported in a variety of models and at different biochemical levels. This compound significantly inhibited the expression of COX-2 in: i) mouse peritoneal macrophages treated with LPS, 12-O-tetradecanoylphorbol-13-acetate (TPA) or H_2O_2 [107]; ii) RAW 264.7 macrophages stimulated with LPS plus interferon γ (INF-γ) [108]; and iii) mouse skin treated with TPA [109]. $In\ vivo$ (F344 rat), resveratrol at concentrations between 1 and 2 mg/kg body weight suppressed both the number and size of N-nitrosomethylbenzylamine (NMBA)-induced esophageal tumors per rat, by targeting COXs and prostaglandin E_2 (PGE$_2$) [110]. Resveratrol and other stilbenes also decreased COX-2 activity and reduced the production of PGE$_2$ in peripheral blood leukocytes treated with LPS plus INF-γ [111].

Resveratrol was shown to inhibit two important transcription factors, NFκB, involved in signaling pathways mediating inflammation, oncogenesis (including angiogenesis and metastasis), apoptosis and (together with anthocyanins) the activator protein 1 (AP-1), regulating the expression of genes involved in cell adaptation, differentiation, and proliferation (Figure 38.11) [112–116].

The interaction with both androgen (ARs) and estrogen receptors (ERs), belonging to the nuclear steroid hormone receptor family, represents another molecular mechanism involved in resveratrol-mediated chemoprevention (Figure 38.11). A decrease in cell proliferation was reported in the androgen-responsive prostate cancer cell lines LNCaP treated with resveratrol and quercetin, due to the inhibition of both expression and function of ARs [117,118]. ARs represent essential mediators of androgen activity, controlling the transcription of androgen-inducible genes, such as prostate-specific antigen (PSA). Therefore they are implicated in the development of hormone-responsive prostate cancer. Moreover, the growth of androgen-unresponsive prostate cancer cells was also inhibited by resveratrol, though to a lesser extent than that of the androgen-responsive cell lines [119]. Estrogens regulate the transcription of target genes by binding to different intracellular estrogen receptors (ERα and ERβ) with tissue and ligand specificity, influencing the growth, differentiation, and function of target tissues and playing a pivotal role in breast cancer. As a phytoestrogen, resveratrol was shown to possess either estrogen agonist and antagonist activity, thus raising some controversy regarding its therapeutic application against estrogen-responsive breast cancers [120–122]. Phytoestrogens are diphenolic plant metabolites that exert estrogen agonist/antagonist activity because of their structural similarities to natural and synthetic estrogen steroids. They are either hormone-like compounds, with inherent estrogenic activity, or can be converted by intestinal flora to weakly estrogenic compounds. Other phytoestrogens include lignans and isoflavones, present in whole cereals and legumes respectively, and classified as selective estrogen receptor modulators (SERMs) [123].

Inhibition of both aromatase (estrogen synthetase) activity and expression by grape seed extract represents another mechanism of breast cancer suppression (Figure 38.11), as demonstrated in an aromatase-transfected MCF-7 breast cancer xenograft model [124]. Aromatase is a cytochrome P450 enzyme, which converts C_{19} androgens to aromatic C_{18} estrogens, expressed at higher levels in breast cancer than in normal tissues. Therefore, its overexpression in breast cancerous cells can influence the tumoral progression itself, because of the major role of estrogens in breast cancer development [125].

Aryl hydrocarbon receptor (AHR) is a cytosolic protein that translocates to the nucleus upon ligand binding. Metabolic activation of aryl hydrocarbons (AH) results from their binding to AHR that, after migration, activates the transcription of the CYP1A1 gene, encoding for the cytochrome P450 (CYP450) isozyme CYP1A1. CYP450 enzymes are involved in the metabolism of a variety of xenobiotics, including carcinogens such as AH, and are overexpressed in a variety of tumors (Figure 38.11). The metabolized active forms of carcinogens can subsequently interact with DNA, thus causing mutations. Resveratrol was shown to exert a strong inhibitory effect on AH-induced CYP1A1 expression, both at the mRNA and protein level, as well as on other CYP450 isozymes, such as CYP1A2 and CYP3A4 [126–128].

Besides the above-mentioned signaling networks regulated by polyphenols, the induction of apoptosis is another molecular mechanism involved in their antiproliferative effects (Figure 38.11) [101]. In a variety of tumor cell lines, among them leukemia cells, it was reported that resveratrol activates the mitochondrial-dependent apoptotic pathway by the up-regulation of pro-apoptotic p53 and Bax proteins and the down-regulation of the death inhibitory protein Bcl-2 [129–134]. The gene p53 is an important oncosuppressor whose mutations, as well as the loss of p53 protein function, are related to more than half of human cancers [101]. However, in different cell lines, polyphenols induce apoptosis by mechanisms other than p53 gene modulation [135,136]. Suppression of anti-apoptotic survivin may be another pro-apoptotic mechanism promoted by grapevine polyphenols, as reported for green tea polyphenols, which decreased both mRNA and protein expression of survivin (Figure 38.11) [137]. Survivin is a member of the inhibitor of apoptosis protein (IAP) family, overexpressed in several human neoplasms [138].

The inhibition of ribonucleotide reductase (the enzyme that catalyzes the reduction of ribonucleotides into deoxyribonucleotides), of DNA polymerase, of ornithine decarboxylase (ODC, a key enzyme of polyamine synthesis greatly involved in cancer growth), and the promotion of cell cycle arrest are key processes further contributing to the chemopreventive potential of polyphenols (Figure 38.11) [139–146].

Grape seed extract showed promising efficacy also against two important processes involved in cancer progression, angiogenesis and metastasis, inhibited in prostate and breast carcinoma, respectively [147,148]. Because of increased metabolic activity and oxygen consumption of rapidly proliferating cells, solid tumors are likely to maintain an intratumoral hypoxic environment, which, in turn, induces a set of hypoxia-responsive genes in order to allow tumor cell adaptation. The expression of these genes is regulated by the hypoxia inducible factor (HIF), a major regulator of cellular oxygen homeostasis [149]. Resveratrol was reported to inhibit the expression of HIF-1α in human ovarian cancer cells OVCAR-3, as well as of the vascular endothelial growth factor (VEGF), an HIF-regulated angiogenic factor [150]. Angiogenic factors promote neovascularization of interstitial stroma, the tissue surrounding the primary tumor site, a process which supplies nutritional requirements to proliferating neoplastic cells and facilitates their access to the vascular system (intravasation).

Before penetrating blood vessel endothelium, from the primary tumor site, and gaining access to the blood stream, cancer cells must invade local tissues by degrading extracellular matrix (ECM) components and, ultimately, traverse the basement membrane. Once in circulation, these cells can form metastatic colonies at secondary locations. Resveratrol was reported to inhibit the invasiveness of diverse cancer cells by reducing the expression and activity of matrix metalloproteinase (MMP)-2 and MMP-9, involved in ECM degradation [151–153].

Apart from the beneficial effects exerted by polyphenols mainly on prostate, breast, and blood cancers, as discussed above, other types of tumors can benefit from regular, moderate consumption of grape products. Carcinomas of the digestive tract are common and their risk increases with age. In gastric cancer cells, resveratrol was shown to inhibit cell proliferation and to induce apoptosis [154–156]. Resveratrol and red wine extracts also inhibited the growth of 15 clinical isolates of *Helicobacter pylori*, the primary etiological determinant of gastric cancer [157]. Grape polyphenols, mainly quercetin, were shown to suppress the formation of aberrant crypt foci, in animal models of carcinogenesis, by modulating both cell proliferation and apoptosis [158]. The crypt is the fundamental unit of epithelial proliferation in the colonic mucosa, where genetically damaged stem cells are removed from the epithelium by apoptosis, before they undergo clonal expansion. Hence, increased apoptosis in the proliferating zone of the colonic crypt provides a protective mechanism against crypt cell hyperproliferation and neoplasia [158]. In both cancerous and noncancerous human colon tissues, black grape extracts modified the activity of enzymes involved in DNA turnover, adenosine deaminase, 5' nucleotidase, and xanthine oxidase, thus depriving cancer cells of nucleotides for proliferation [159]. Similar results were reported on cancerous and noncancerous human urinary bladder tissues [160].

Moreover, the growth of human colorectal carcinoma cells was inhibited by a grape seed extract rich in proanthocyanidins [161], and a red grape dietary fiber (obtained from seeds) induced epithelial hypoplasia with a decrease in the depth of crypts in both rat cecal and distal colonic mucosa as well as a decrease in crypt density and mucosal thickness [162].

Photochemoprevention by botanical agents may prevent skin cancer at various stages of carcinogenesis, as shown in different models. Topical application of resveratrol and apigenin to SKH-1 hairless mice, prior to UV exposure, effectively prevented radiation-induced carcinogenesis [163]. Mechanisms involved in photochemoprevention by resveratrol include the inhibition of NFκB signaling, as demonstrated in UV-exposed normal human epidermal keratinocytes [164]. In SKH-1 mice, grape seed proanthocyanidins also prevented UV-induced oxidative stress, decreased lipid peroxidation, and inhibited the activation of NFκB signaling and, finally, photocarcinogenesis [165,166]. As for other cell lines, resveratrol induced apoptosis in two human melanoma cell lines by activating a MAP kinase pathway [167].

The paradigm that health benefits arising from fruit consumption are due to the efficacy of all the biologically active phytocomponents has been emphasized on pancreatic cancer cells. Pancreatic cancer is a highly aggressive malignancy, currently treated with limited success by conventional therapeutics and with an extremely poor prognosis [168]. A mixture of isoflavone 10 nM + cucurmin 500 nM + epigallocatechin-3-gallate 125 nM + resveratrol 125 nM inhibited by 40%, up to 72 h, the cell growth of BxPC-3 cells, a human pancreatic cancer cell line, via a mechanism partly due to the inactivation of NFκB. The authors concluded that a combined treatment with phytochemicals induces a greater inhibition of cell growth than that obtained after treatment with single compounds [169]. Finally, in human pancreatic cancer cell line Panc02 inoculated into

C57BL/6 mice, both resveratrol and quercetin suppressed pancreatic cancer via different mechanisms, namely apoptosis induction (by caspase 3 and 8), cell cycle arrest at G1 phase, and inhibition of tumor cell migration (invasiveness) through the ECM barrier [170].

3.4 Atherogenesis, Hypertension, and Cardioprotection

Atherosclerosis is a chronic, inflammatory, fibroproliferative process of large and medium-sized arteries. It results in the progressive formation of fibrous plaques that impair the blood flow inside the vessels. In the affected artery, atherosclerotic lesions resulting from an eccentric thickening of the intima can either promote an occlusive thrombosis or produce a gradual stenosis of the arterial lumen. In the first case, thrombus formation due to the disruption of the lesion surface can lead to infarction of the organ supplied by the afflicted vessel, such as in a heart attack, when a coronary artery is suddenly blocked, or in a thrombotic stroke, when a cerebral artery is damaged. In the second case, the stenosis of the vessel limits the blood supply to local tissues, leading to a progressive and gradual injury of the affected organ [171].

Endothelial dysfunction, oxidative modification of LDL, platelet aggregation, and inflammation are key factors in atherogenesis and hypertension. Endothelial cells exert multiple physiological functions, maintaining the integrity of the vascular wall and representing a permeable barrier through which diffusion and active transport of several substances occur. Furthermore, endothelial cells constitute a non-thrombotic and non-adherent surface for platelets and leukocytes; they regulate the vascular tone by producing nitric oxide (NO), eicosanoids, endothelins, and cytokines [172].

In cardiovascular diseases, an important inflammatory process takes place after leukocyte mobilization. Phospholipase A2 leads to the hydrolysis of plasma membrane phospholipids, mainly phosphatidylcholine and phosphatidylethanolamine, and to the subsequent release of arachidonic acid. This acid is the substrate of COX and lipoxygenase (LOX), the enzymes involved in the synthesis of prostanoids (prostaglandins and thromboxanes) and leukotrienes, respectively, the inflammation mediators collectively grouped in eicosanoids. Among COX-derived platelet modulators, thromboxanes, mainly thromboxane A2 (TXA2), potentiate platelet reactivity, whereas prostacyclins help to maintain platelets in a quiescent state [173]. Polyphenols, mainly quercetin and resveratrol, were shown to inhibit COX and LOX activities and eicosanoid synthesis (Table 38.1, Figure 38.12) [107,174–178].

A great deal of evidence suggests an inverse relationship between grape product consumption and cardiovascular disease. Clinical trials demonstrated that grape and grape juice

TABLE 38.1 Mechanisms of Grape Polyphenol-Promoted Cardiovascular Protective Effects

Mechanisms	References
Inhibition of inflammation and eicosanoid synthesis	[84,107,174–178,207]
Improvement of endothelial function	[186–190]
Decrease of LDL	[84,196]
Increase of HDL	[84,196]
Inhibition of LDL oxidation	[81,82,187]
Inhibition of platelet aggregation and thrombosis	[177,179,180,191–195]
Improvement of vasorelaxation	[193,199–202]
Inhibition of ET-1	[204]
Improvement of fibrinolysis	[209]
Inhibition of VSMC proliferation and vascular hyperplasia	[210–212]

LDL, low density lipoproteins; HDL, high density lipoproteins; ET-1, endothelin-1; VSCM, vascular smooth muscle cells.

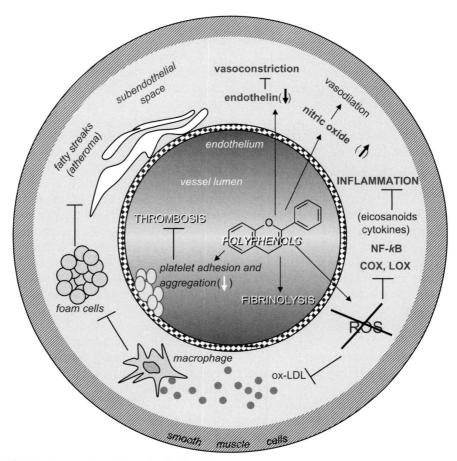

FIGURE 38.12　Cardioprotective effects of polyphenols.
Molecular mechanisms by which polyphenolic compounds may counteract endothelial dysfunction and atherosclerosis include: prevention of low density lipoprotein (LDL) oxidation and atheromatous plaque deposition; suppression of cyclooxygenases (COX), lipoxygenases (LOX), nuclear factor κB (NFκB) and pro-inflammatory eicosanoids and cytokines; inhibition of platelet aggregation and thrombus formation; enhancement of fibrinolysis; decrease of endothelin-mediated vasoconstriction; release of nitric oxide and vasodilation.

consumption improved endothelial function, reduced platelet aggregation and platelet–endothelial cell interactions, decreased blood pressure, and positively influenced biomarkers such as LDL and HDL (high density lipoproteins) (Table 38.1, Figure 38.12). Additionally, the prevention of LDL oxidation, as previously described, further contributed to the cardioprotective effect of grapevine products (Table 38.1, Figure 38.12) [179–186].

In patients with coronary heart disease, ingestion of red grape juice (8 mL/kg body weight) daily for 2 weeks improved endothelial function, increasing the flow-mediated dilation of the brachial artery to 6.5%, compared to baseline values measured by high-resolution brachial artery ultrasonography (a measure of endothelium-dependent vasodilation). The authors also reported a reduction of LDL susceptibility to oxidation in the same subjects

[187,188]. Similar results were subsequently reported, in hypercholesterolemic patients supplied with red grape juice (500 mL/d) for 2 weeks, with a 6.7% increase of the brachial artery flow-mediated dilation, related to the baseline [189]. The improvement of arterial endothelial dilation was also shown in subjects with coronary heart disease, peaking 60 min after the acute intake of a red grape phenolic extract (600 mg in 20 mL of water) [190].

Inhibition of platelet activity after grape juice supplementation was extensively reported. Red grape juice reduced platelet-mediated thrombus formation in stenosed canine coronary arteries, abolishing the cyclic (thrombogenic) flow reduction in coronary blood flow [191]. *In vitro* incubation of platelets with diluted red grape juice reduced their aggregation, decreased $\cdot O_2^-$ production, and enhanced the release of platelet-derived nitric oxide (NO). In turn, the release of NO by platelets contributes to further inhibit their own aggregation and their recruitment to a growing thrombus [192]. The same results were reported in plasma samples from healthy subjects daily supplemented with grape juice (7 mL/kg), for 2 weeks [193]. Intriguingly, orange and grapefruit juices failed to inhibit platelet function in healthy patients, probably because of the lower amount of polyphenols in these products than in red grape juice, as asserted by the authors [194]. In fact, in that study only the daily intake of grape juice (5–7.5 mL/kg) for 7–10 days reduced by 77% the baseline values of the platelet aggregation response to 1.0 mg/L of collagen [194]. Moreover, the collagen-mediated platelet aggregation was greater when grape seed and grape skin extracts were supplied individually than when used in combination, both in human and dog blood samples [195].

In hemodialysis patients, daily supplementation with red grape juice (100 mL) for 2 weeks ameliorated the lipoprotein profile by decreasing plasma concentration of LDL and increasing that of HDL, besides reducing the plasma level of oxidized LDL and monocyte chemoattractant protein 1 (MCP-1), an inflammatory biomarker associated with cardiovascular disease risk. In the same study, a rapid absorption of quercetin was reported, with the maximum plasma concentration reaching 3 h after the grape juice ingestion [196]. Also the levels of tumor necrosis factor α (TNF-α), a plasma proinflammatory cytokine released by both endothelial cells and leukocytes, decreased in pre- and postmenopausal women after the supplementation of a lyophilized grape powder (36 g daily) for 4 weeks. Vascular inflammation and endothelial activation induced by TNF-α play a critical role in atherosclerosis. Plasma triglyceride concentrations were also reduced in the same patients [84].

Blood pressure parameters and atherosclerosis were also ameliorated by grape juice intake. Supplementing hypertensive men with red grape juice (5.5 mL/kg body weight), daily for 8 weeks, reduced on average by 7.2 and 6.2 mm Hg the systolic and diastolic blood pressure, respectively, compared to the baseline values, and by 3.5 and 3.2 mm Hg, respectively, compared to a calorie-matched placebo [197]. By using a hamster model of atherosclerosis, it was shown that grape juice intake reduced by 10% the aorta area covered by foam cells, i.e. macrophages internalizing oxidized LDL, in animals supplied with a cholesterol/saturated fat diet for 10 weeks [71].

Interestingly, another antithrombotic effect of red wine polyphenols is their capacity to prevent the rebound phenomenon of platelet hyperaggregability observed after acute alcohol consumption and responsible for ischemic strokes or sudden deaths occurring after episodes of drunkenness. This rebound effect was not observed after acute wine intake, because of the bioactive components present in this product with respect to other alcoholic beverages [198].

Other molecular mechanisms by which polyphenols may counteract endothelial dysfunction, thus playing a role in the etiopathogenesis and pathophysiology of atherosclerosis, include: i) the release of NO by the endothelial nitric oxide synthase (eNOS); ii) the decrease of endothelin 1 (ET-1) production; and iii) the suppression of NFκB expression (Table 38.1, Figure 38.12).

Polyphenols, mainly anthocyanins, trigger an endothelium-dependent, NO-mediated vasorelaxation, whereas delphinidin inhibits the apoptosis of endothelial cells [199,200]. NO exerts vasodilating, antithrombotic, and antiproliferative effects, besides inhibiting leukocyte adhesion to the vascular wall. Alcohol-free red wine polyphenol extract increases eNOS expression and subsequent NO release [201]. Additionally, incubation of endothelial cells with red wines up-regulated eNOS mRNA and protein expression with a production of bioactive NO up to three times higher than that reported in control cells [202]. Among ROS, $\bullet O^{2-}$ can react rapidly with the endothelium-derived NO, leading to the strong oxidant peroxynitrite ($ONOO^-$) and reducing the amount of bioactive NO available for vasodilation. Therefore, antioxidant power represents another cardioprotective mechanism of polyphenols by scavenging ROS and reducing NO breakdown [203].

In other works, procyanidins, a class of proanthocyanidins, blocked the production of ET-1 by suppressing *ET-1* gene transcription in cultured bovine aortic endothelial cells. ET-1 is a highly potent vasoconstrictor, which also promotes leukocyte adhesion, monocyte chemotaxis, and smooth muscle cell proliferation, in addition to facilitating LDL uptake by the endothelial cells [204]. ET-1 acts as the natural counterpart to endothelium-derived NO which, besides its arterial blood pressure-raising effect, induces vascular and myocardial hypertrophies, an independent risk factor for cardiovascular morbidity and mortality [205].

As explained in the previous section, NFκB is a pleiotropic transcription factor subjected to a redox regulation and involved in different signaling pathways and processes, including vascular inflammation. Therefore, it can be evoked by an oxidative stress and inhibited by antioxidants [206]. In human coronary endothelial cells, resveratrol (10^{-6} mol/L) reduced by 50% the activation of NFκB mediated by TNF-α, in addition to attenuating the monocyte adhesiveness to endothelium [207]. Disruption of the cytokine-activated NFκB signaling pathway exerts a vasculoprotective action by attenuating vascular inflammation and preventing atherogenesis. Additionally, a spectrum of different genes expressed in atherosclerosis was shown to be up-regulated by NFκB, including those encoding for TNF-α, MCP-1, interleukin 1 (IL-1), vascular cell adhesion molecules (VCAM), and intracellular adhesion molecules (ICAM) [206].

Enhancement of fibrinolysis and inhibition of vascular smooth muscle cell (VSMC) proliferation and migration represent two additional processes involved in cardiovascular protective effects of polyphenols (Table 38.1, Figure 38.12). Clinical and epidemiological studies suggest that impairment of the fibrinolytic system contributes significantly to atherothrombosis development [208]. It has been demonstrated that grape polyphenols (catechin, epicatechin, quercetin, and resveratrol) increase mRNA levels of plasminogen activators (PAs) in human umbilical vein endothelial cells, independently of ethanol [209]. PAs convert plasminogen to plasmin, the latter able to degrade fibrin within a thrombus and eventually leading to clot dissolution. The abnormal proliferation of VSMC in the arterial intima plays an important role in the pathogenesis of atherosclerosis. In an endothelial denudation model, rabbits fed with a high-dose resveratrol diet developed less intimal hyperplasia than control rabbits, with a considerably reduced number of SMCs in resveratrol-treated animals [210]. In agreement with these results, bovine aortic SMC proliferation was inhibited after treatment with grape polyphenols (dealcoholized red wine, red wine polyphenol extract, or resveratrol), according to a dose-response relation [211]. In stroke-prone, hypertensive rats, resveratrol inhibited proliferation and collagen synthesis in VSMCs at concentration attainable in plasma at therapeutic doses (from 0.01 to $1.0\,\mu M$) [212].

3.5 Neuroprotection and Aging

Neurodegeneration is a process involved in both neuropathological conditions and brain aging. Although the brain accounts for less than 2% of the body weight, it consumes about 20% of the oxygen available through respiration. Therefore, because of its high oxygen demand, the brain is the most susceptible organ to oxidative damage [213–215]. Additionally, the high amount of polyunsaturated fatty acids (PUFAs) present in neuronal membranes makes the brain tissues particularly susceptible to lipid peroxidation reactions (Figure 38.8), resulting in the formation of cytotoxic aldehydes, such as malondialdhyde (MDA) and 4-hydroxynonenal (HNE) [216]. This oxidative burden can be effectively counteracted by the cell antioxidant defenses (see Section 3.1), including ascorbic acid (vitamin C), whose concentration in brain is the highest among all the body tissues. Flavonoids may also play an important role as neuroprotectants, by virtue of their free radical scavenging power (Figure 38.10) [217,218]. Another important property for a neuroprotective agent regards its ability to cross the blood–brain barrier (BBB) in order to reach the target sites of the central nervous system. A limited number of studies, both *in vitro* and on animal models focused on the ability of flavonoids to cross the endothelial cell layer of the BBB depending on the compound lipophilicity and on the activity of specific transporters [219–223].

According to the World Health Organization (WHO), neurodegenerative diseases (NDs) in 2020 will represent the eighth greatest cause of disease burden for developed countries, and, by the middle of the century they will become the world's second leading cause of death, overtaking cancer [224]. Many important NDs include amyloidogenic diseases, characterized by conformational changes (misfolding) and aggregation of proteins and peptides inside or outside cells. Major amyloid-related diseases include Alzheimer's disease (AD), Parkinson's disease (PD), Huntington's disease, prion disease, and type II diabetes [225]. Familial forms of these diseases due to a mutation of the gene coding for the abnormally aggregating protein represent a minority of cases, most NDs occurring sporadically and arising through interactions among genetic and environmental factors [226].

AD is a progressive, degenerative disorder which accounts for 65% of all age-related dementias, with an estimated prevalence between 1 and 5% among people aged 65, doubling every four years to reach about 30% at 80 years [227]. Histopathology reveals that one of the major hallmarks of AD is the abundant protein deposits in neurons that trigger neuronal degeneration. These deposits result from the extracellular and intracellular accumulation of amyloid β (Aβ) peptides and phosphorylated tau (Pτ) protein, respectively. Aβ and Pτ aggregation leads to the formation and deposition of senile plaques and neurofibrillary tangles, respectively, which promote inflammation and neuronal cell death. Among nongenetic factors influencing AD, studies strongly supported the hypothesis that certain dietary habits, such as those of Mediterranean diets, may play a beneficial role in the relative risk for AD clinical dementia [228]. Conversely, some other dietary factors such as high caloric intake in the form of saturated fat may be involved in the nutritional etiology of NDs and promote AD neuropathology [229,230].

In the Canadian Study of Health and Aging, a prospective analysis of risk factors for AD was conducted by a large-scale cohort study on a representative sample of the Canadian population aged 65 years or older. Regular alcohol consumption (beer, wine, and spirits, at least weekly) was associated with a reduced risk of AD, with wine intake reducing the risk by 50%. Interestingly, the daily coffee consumption was observed to be significantly associated with a lower risk of AD, differently from tea intake, though both beverages are important sources of polyphenolic compounds [231]. Similar results were reported in the *Kame* Project, another large population-based prospective study based on a cohort of Japanese Americans. In this study, a significant, inverse relation between fruit/vegetable juices and AD was observed. Furthermore, the risk for AD was significantly reduced among people who consumed fruit and vegetable juices three or more times per week, compared with those who drank these juices less than once per week. Again, tea was not associated with the risk of AD, whereas the association between wine (or sake) intake and AD was inverse but not statistically significant, probably because of the few subjects drinking wine included in the *Kame* Project cohort [232]. In another prospective community study, in the Bordeaux area, wine consumption (3–4 standard glasses) was associated with more than an 80% reduced risk of dementia and a 75% reduced risk of AD [233].

In different cell lines stably transfected with human amyloid-β protein precursor (APP), resveratrol was shown to promote the intracellular degradation of Aβ peptides via a mechanism involving the proteasome, without direct inhibition of the enzymes β- and γ-secretases implicated in Aβ protein synthesis [234]. Neuroprotective effects of three major grape polyphenolic constituents (resveratrol, quercetin, and catechin) were assessed in cultured mixed (glial/neuronal) cells of rat hippocampus, a brain area severely affected in both AD

and ischemia. Hippocampal cell treatment with polyphenols reduced both the cytotoxicity induced by the NO free radical donor sodium nitroprusside (SNP) and by intracellular ROS accumulation [235]. In a mouse model of Alzheimer's disease, the moderate consumption of Cabernet Sauvignon promoted the non-amyloidogenic processing of APP mediated by α-secretase, thereby preventing or delaying the generation of Aβ peptides [236]. In 2008, a study showed that a grape seed polyphenolic extract significantly prevented Aβ protein oligomerization, by inhibiting the Aβ protein aggregation into high molecular weight oligomeric Aβ species, both *in vitro* and in Tg2576 mice. Besides, when orally administered to these animals, the extract attenuated the cognitive deterioration typical of Alzheimer's disease [237].

Polyphenols were described to inhibit the formation of amyloid fibril assembly *in vitro* and to reduce its cytotoxicity. The mechanism involved in this process is based on structural constraints and aromatic interactions which direct polyphenols to the amyloidogenic core. All the efficient polyphenol inhibitors are composed of at least two phenolic rings with two to six atom linkers, and a minimum of three OH groups in the aromatic rings. It seems that these structural features are essential for the non-covalent interaction with β-sheet structures, common to all amyloidogenic structures [238]. It was also reported that red wine polyphenols exert anti-amyloidogenic and fibril-destabilizing effects in a dose-dependent manner, besides scavenging the free radical species involved in β-amyloid neurotoxicity [239–241]. Another study attributed the anti-amyloidogenic activity of polyphenols to their particular C_6–linkers–C_6 structure, able to inhibit Aβ fibril aggregation. Malvidin, malvidin glucoside, and resveratrol were the most efficient fibril inhibitors whereas phenolic acids (coumaric, caffeic, hydroxybenzoic acids) showed only a weak inhibitory activity, probably because of their C_6–C_3 structure, thus pointing out the structure–activity relationship of polyphenols against Aβ fibril formation [242].

Vascular dementia (VaD) is another neurological disorder due to brain vascular atrophy, characterized by decreased brain perfusion and causing a mild cognitive impairment. Foods and beverages rich in flavonoids have been advocated as preventive agents to counteract dementia [217,228]. In particular, flavanols (i.e. catechin and epicatechin) were reported to increase brain blood flow and perfusion, delaying the onset of brain vascular atrophy and the development of mild cognitive impairment [243]. In 2008, a systematic review and a meta-analysis were carried out to evaluate the correlation between the incidence of dementia or cognitive decline in the elderly and alcohol consumption. Results suggested that low to moderate amounts of wine consumption are associated with a 38 and 32% reduced risk for dementia and AD, respectively. Although for VaD and cognitive decline the results showed a similar trend, they were not statistically significant [244].

Parkinson's disease (PD) is a movement disorder considered the most frequent neurodegenerative disease after AD, caused by the degeneration of dopaminergic neurons in the substantia nigra. One of the main pathological hallmarks of PD is the aggregation of the intracellular protein α-synuclein to form intracytoplasmic inclusions (Lewy bodies) in these neurons [226]. Currently, oxidative and nitrosative stress is believed to be one of the leading causes of neuronal degeneration in PD [245,246]. The daily administration of resveratrol (50 or 100 mg/kg) for 1 or 2 weeks to adult male mice significantly prevented the nigrostriatal dopaminergic neuron depletion after acute treatment with the neurotoxin 1-methyl-4-phenyl-1,2,3,6-tetrahydropyridine (MPTP) injected intraperitoneally [247]. In general, protective effects of polyphenols against PD can be ascribed to their anti-amyloidogenic and antioxidant activities [238].

Cerebrovascular accidents include mainly ischemic (occlusive) and hemorrhagic strokes. Ischemic stroke represents one of the leading causes of mortality and permanent disability in adults worldwide. It results from occlusion of a major cerebral artery by a thrombus or embolism, with the subsequent loss of blood flow and decrease in the supply of oxygen and nutrients to the affected brain region. Hypertension is the main cause of hemorrhagic stroke, exacerbating the risk of cerebral hemorrhage [228,248]. In the Copenhagen City Heart Study, wine intake on a monthly, weekly, or daily basis was associated with a lower risk of stroke, compared with no wine intake. No association between beer or spirit consumption and risk of stroke was reported, thus suggesting that some wine components, in addition to ethanol, may be responsible for the beneficial effect of wine intake on the risk of stroke [249]. The Framingham Study evaluated the association between the type of alcoholic beverages and incidences of ischemic stroke, showing a protective effect of wine consumption among subjects aged 60–69 years [250]. Data from a case-control study on young women were consistent with the above reported studies and, in general, beer and spirit intake failed to exert neuroprotective effects [249–251].

Other mechanisms by which polyphenols retard the aging process and delay the onset of aging-related diseases resemble those induced by caloric restriction (CR), suggesting that these compounds and CR share quite similar molecular pathways (Figure 38.13) [252,253]. A moderate reduction in calorie intake by 20–40% significantly extends the lifespan in a wide spectrum of organisms, ranging from bacteria to primates, a process mediated by a class of silent information regulator (SIR) proteins, the sirtuins [254–256]. In mammals, the seven members of the sirtuin family (SIRT1–7) represent novel therapeutic targets to treat age-associated and neurodegenerative diseases, being implicated in a variety of cellular functions,

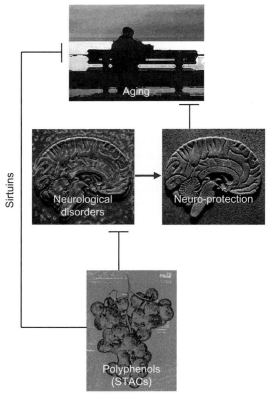

FIGURE 38.13 Resveratrol, quercetin, and other polyphenols are sirtuin activating compounds (STACs), able to
activate SIRT1, which in turn regulates some physiological
processes affected during aging.
Polyphenols also exert a neuroprotective effect by counteracting the onset of neurological disorders.

ranging from gene silencing, over the control
of cell cycle and apoptosis, to energy metabolism [257]. Two different enzymatic activities
have been reported for mammalian sirtuins, a
nicotinamide adenine dinucleotide (NAD$^+$)-
dependent deacetylase activity or an adenosine
diphosphate (ADP)-ribosyl transferase activity,
and they have diverse subcellular localizations
(nucleus, nucleolus, cytosol, and mitochondria)
[258,259]. Therefore, sirtuin activating compounds (STACs) represent a promising class of
therapeutics, including resveratrol, quercetin,
and other polyphenols able to activate SIRT1,
which in turn regulates some physiological

processes affected during aging and modified
by CR (Figure 38.13) [260–262]. SIRT1 deacety-
lates many substrates, including histones and
non-histone substrates, such as the tumor suppressor protein p53 and the transcription factor
NFκB, besides regulating the activity of the
nuclear receptor PPAR-γ (peroxisome proliferator-activated receptor γ) and PGC-1α
(PPAR-γ co-activator 1α), master regulators of
adipogenesis, fat storage in white adipose tissue, liver metabolism, and muscle cell differentiation [254].

4. GRAPEVINE PRODUCTS AND ORAL HEALTH

The burden of oral diseases is particularly
high for the disadvantaged and poor population
groups in both developing and developed countries. Oral diseases such as dental caries, periodontal disease, tooth loss, oral mucosa lesions
and oropharyngeal cancers, human immunodeficiency virus/acquired immunodeficiency syndrome (HIV/AIDS)-associated oral disease, and
orodental trauma represent some of the major
public health problems worldwide [263].

Among the tumors of head and neck, carcinomas of the oral cavity, mainly of squamous-cell
type, comprise an important group of malignancies whose incidence is increasing all over the
world. Cancer of the upper aerodigestive tract,
including cancers of lip, tongue, oral cavity,
and pharynx, ranks as the seventh most common cause of cancer worldwide and represents
2–4% of all diagnosed cancers, associated with
a poor prognosis and survival rate if not
diagnosed early [264–266]. Administration of
proanthocyanidins showed suppressed proliferation of human oral squamous-cell carcinoma
in a dose-dependent manner [267]. More interestingly, proanthocyanidins inhibited the proliferation of cell carcinoma also after transfection
with human papillomavirus (HPV), another
putative risk factor for oral cancer [267]. It has

also been reported that, in the two well-characterized oral squamous-cell carcinoma cell lines CAL-27 and SCC-25, the antiproliferative effect of a grape seed extract was correlated to the dramatic up-regulation of mRNA expression of caspases 2 and 8, proteases considered the executioners of apoptosis [268]. In human tongue squamous-cell carcinoma cells SCC-9, resveratrol inhibited angiogenesis by reducing HIF-1α protein accumulation and VEGF expression, promoting the proteasomal degradation of HIF-1α without affecting HIF-1α mRNA expression [269].

It is noteworthy that alcohol consumption, as well as tobacco smoking, accounts for most oral cancers, and thus it would be possible to prevent about 75% of such cancer cases by avoiding alcohol abuse and smoking. Alcohol may act as a solvent for carcinogens and enhance the permeability of oral mucosa to carcinogens themselves, such as those from tobacco. Besides, the ethanol metabolite acetaldehyde has been identified as a tumor-causing substance [270,271]. However, the carcinogenic effect of alcohol may depend on drinking habits, though the effect of beverage type on the risk of developing oral cancer remains controversial. On the whole, it seems that heavy alcohol consumption is associated with a major incidence of oral cancer, the latter being higher among spirits consumers than wine drinkers [272–274]. Therefore, it is possible that red wine, by virtue of its polyphenolic content, may have a beneficial effect on the risk of cancers of the upper aerodigestive tract, especially in the context of a Mediterranean diet [275,276]. In opposition, Maserejian and colleagues observed similar increases in the development of oral premalignant lesions after consumption of spirits, wine, and beer, although the association with wine was not statistically significant [277].

Chronic periodontitis is a local inflammatory disease mediating the destruction of periodontal (tooth-supporting) tissues, triggered by bacterial infection and leading to bone resorption and tooth loss. Gram-negative periodontopathogenic bacteria that are diffused in subgingival sites, including *Porphyromonas gingivalis*, *Tannerella forsythia*, *Treponema denticola*, *Fusobacterium nucleatum*, *Actinobacillus actinomycetemcomitans*, induce a local immune and inflammatory response by releasing bacterial products such as lipopolysaccharides (LPS). In order to directly poison the pathogens, the activated immune cells produce a variety of molecules via the respiratory burst, mainly ROS and reactive nitrogen species (RNS). However, increased tissue levels of these reactive molecular species create an oxidative stress which, finally, results in cell death and tissue destruction [278]. The antioxidant properties of grape seed proanthocyanidins were investigated in an *in vitro* model of murine macrophages (RAW 264.7) stimulated with LPS isolated from periodontopathogens. Treatment of immune cells with grape seed proanthocyanidin extract, at noncytotoxic concentrations, decreased both ROS and NO production as well as iNOS (inducible nitric oxide synthase) protein expression [279].

In 2008, antimicrobial constituents of raisins were assayed against two oral pathogens, *Streptococcus mutans* and *Porphyromonas gingivalis*, associated with caries and periodontitis, respectively. Oleanolic acid and its derivatives, oleanolic aldehyde, 5-(hydroxymethyl)-2-furfural, and rutin were the most effective compounds, inhibiting the bacterial growth at concentrations ranging from 4 to 500 μg/mL [280].

5. NEW PERSPECTIVES IN GRAPE RESEARCH: MELATONIN

Undoubtedly, polyphenols represent the prototype of the health-promoting effects associated with grapevine product intake, though a discovery of a new bioactive metabolite in grape, melatonin, adds a new element to further comprehend the pharmaconutritional properties of grape products, besides opening new perspectives in the field of grape research [48].

In mammals, melatonin is synthesized mostly in the pineal gland, predominantly during the night-time, though it can also be produced in other organs, such as retina, gastrointestinal tract, lymphocytes, and bone marrow cells. Conversely, light has an inhibitory effect on melatonin biosynthesis, initiated by the uptake of L-tryptophan from the circulation into the pineal gland and occurring within the pinealocytes. Once synthesized, melatonin is not stored in the pineal cells but is released into the bloodstream with a circadian rhythm from which it reaches other body fluids, including saliva, cerebrospinal fluid, bile, semen, and amniotic fluid. The circadian rhythm of melatonin secretion is generated by the biological clock, situated in the suprachiasmatic nucleus (SCN) of the hypothalamus, via a neuronal pathway that begins in the retina and involves the retinohypothalamic tract [281]. In mammals, melatonin acts principally by activating two high-affinity, G-protein-coupled membrane receptors designated MT_1 and MT_2. These receptors have the highest density in the central nervous system (CNS), particularly in adenohypophysis and SCN, and in the cardiovascular system, localized mainly on endothelial and vascular smooth muscle cells of heart, coronary arteries, and aorta. The third receptor type, MT_3, with a lower affinity, is not coupled with G-protein. Pathophysiological conditions regulated by melatonin via a receptor-mediated mechanism include the control of the sleep/wake cycle, regulation of reproductive development, and bone metabolism [282]. Apart from these receptor-mediated processes, melatonin exerts a series of receptor-independent functions mainly due to its powerful antioxidant activity. Melatonin can directly scavenge free radical species, both ROS and RNS, stimulate the activity of antioxidant enzymes, and inhibit pro-oxidant pathways [283].

Some plants, as well as non-mammalian vertebrates and some invertebrates, also synthesize melatonin, and it has been suggested that phytomelatonin, i.e. melatonin from edible plants, can promote health benefits by virtue of its antioxidant potential and/or by modulating the receptor-dependent pathways [284,285]. In this view, phytomelatonin could contribute to ameliorate the physiological functions regulated in humans by endogenous melatonin, and contribute to counteract mutagenesis, carcinogenesis, cardiovascular diseases, neurological disorders, and aging [286–288].

In particular, it was proven that melatonin exerts an anti-amyloidogenic activity due to its structural interaction with $A\beta$ peptides [289–291]. It was also reported that melatonin had a protective effect in animal experimental models of PD, both *in vitro* and *in vivo*, because of its strong antioxidant capacity able to prevent nigral dopaminergic cell damage [292–294]. Melatonin improves cardiovascular health: cardioprotective mechanisms include vasodilation, central and peripheral anti-adrenergic action. In the CNS, melatonin enhances GABA-ergic signaling and reduces sympathetic output, finally resulting in a blood pressure decrease [295]. Leukocytes synthesize melatonin and possess both membrane and nuclear melatonin receptors, thus indicating a relation between melatonin and immune cell production and function [296]. The anti-inflammatory activity of melatonin is mainly due to its ability to reduce NFκB binding to DNA, probably preventing its translocation to the nucleus [297]. Anticancer activity of melatonin involves different mechanisms, among which are the decrease of uptake and metabolism of fatty acids in cancer cells, the reduction of telomerase activity and restoration of chromosome instability, the anti-estrogenic activity relevant in some hormone-responsive tumors, and the inhibition of ET-1 and angiogenesis [281,282,286].

In a pioneering study, the bioavailability of melatonin from edible plants was demonstrated as well as the competitive binding of phytomelatonin to melatonin receptors in mammal brain. In particular, feeding chicks with plant products rich in melatonin increased their plasma melatonin levels, whereas (phyto)melatonin extracted from plants inhibited the binding of labeled melatonin to cell membrane receptors in rabbit brain [298]. In 2008, it was shown that, in humans, serum melatonin concentration increased significantly 1 h after a single 100 mL red wine administration. In that work, the melatonin content in wines ranged from 50 to 80 pg/mL, and the serum melatonin concentration in volunteers augmented significantly from 10 to 12 pg/mL. White wine, in general containing lesser amounts of melatonin, failed to raise the serum melatonin levels [299].

6. SUMMARY

At the end of this survey, one could suppose that grapevine products, particularly red wine, represent a sort of panacea against a plethora of pathological conditions, including the most important causes of morbidity and mortality in developed countries. As a matter of fact, the health benefits arising from a regular and appropriate consumption of these foodstuffs can be mostly ascribed to the several hundreds of bioactive components present therein, and so the attempt of determining which of these compounds are specifically responsible for the reported beneficial effects represents a very hard, and possibly inappropriate task. Furthermore, in the last decades research has focused mainly on a relatively small number of molecules, above all resveratrol.

It must also be taken into account that each fruit or vegetable possesses health-promoting properties not only due to its polyphenol content, but also in virtue of its components other than polyphenols, namely vitamins, mineral salts, dietary fibers, macronutrients (carbohydrates, lipids, proteins), and other phytochemicals. Thus, a single plant food by itself can questionably improve health status even if regularly consumed; and an appropriate dietetic rule is that of diversifying as much as possible the foods consumed, as in the context of a Mediterranean diet.

Finally, a better comprehension of the bioavailability of dietary phytochemicals is critical in order to correctly evaluate their bioactivity, to interpret the experimental results, and to design new approaches. However, biokinetic data supporting their absorption, distribution, metabolism, and excretion in the human body are still fragmentary despite the enormous amount of indications on their bioactivities. Dietary phytochemicals have to be absorbed to exert their health benefits, and several human studies indeed reported direct evidence of absorption and urinary excretion of these compounds after intake. Nevertheless, the high variability of the data concerning their bioavailability, at least for polyphenols, depends on the complexity of the food (fruits and vegetables) matrix and on the chemical structure of the specific compounds [300–302].

ACKNOWLEDGMENTS

The authors warmly thank Dr Giovanni Lodi for critical reading of the section on oral health.

References

1. Simopoulos, A. P. (2001). The Mediterranean diets: What is so special about the diet of Greece? The scientific evidence. *Journal of Nutrition, 131*, S3065–S3073.
2. Johnson, H. (1989). *Vintage: The Story of Wine*. Simon and Schuster. pp. 35–46, New York.
3. Phillips, R. (2000). *A Short History of Wine*. Harper and Collins, pp. 57–63, New York.
4. Mikeš, O., Vrchotová, N., Tríska, J., Kyseláková, M., & Šmidrkal, J. (2008). Distribution of major polyphenolic compounds in vine grapes of different cultivars growing in south Moravian vineyards. *Czechoslovak Journal of Food Science, 26*, 182–189.

5. Iriti, M., & Faoro, F. (2006). Grape phytochemicals: A *bouquet* of old and new nutraceuticals for human health. *Medical Hypotheses, 67*, 833–838.

6. Pezzuto, J. M. (2008). Grapes and human health: A perspective. *Journal of Agricultural and Food Chemistry, 56*, 6777–6784.

7. Jeandet, P., Douillet-Breuil, A.-C., Bessis, R., Debord, S., Sbaghi, M., & Adrian, M. (2002). Phytoalexins from the Vitaceae: Biosynthesis, phytoalexin gene expression in transgenic plants, antifungal activity and metabolism. *Journal of Agricultural and Food Chemistry, 50*, 2731–2741.

8. Iriti, M., & Faoro, F. (2004). Plant defense and human nutrition: The phenylpropanoids on the menù. *Current Topics Nutraceutical. Research, 2*, 47–65.

9. Herrmann, K. (1989). Occurrence and content of hydroxycinnamic acid and hydroxybenzoic acid compounds in foods. *Critical Reviews in Food Science and Nutrition, 28*, 315–347.

10. Iwashina, T. (2000). The structure and distribution of the flavonoids in plants. *Journal of Plant Research, 113*, 287–299.

11. Pietta, P.-G. (2000). Flavonoids as antioxidants. *Journal of Natural Products, 63*, 1035–1042.

12. Mazza, G. (1995). Anthocyanins in grapes and grape products. *Critical Reviews in Food Science and Nutrition, 35*, 341–371.

13. Clifford, M. N. (2000). Anthocyanins – Nature, occurrence and dietary burden. *Journal of the Science of Food and Agriculture, 80*, 1063–1072.

14. Langcake, P. (1981). Disease resistance of *Vitis* spp. and the production of stress metabolites resveratrol, ε-viniferin, α-viniferin and pterostilbene. *Physiologia Plantarum Pathologica, 18*, 213–226.

15. Waterhouse, A. L., & Lamuela-Raventos, R. M. (1994). The occurrence of piceid, a stilbene glucoside, in grape berries. *Phytochemistry, 37*, 571–573.

16. Santos-Buelga, C., & Scalbert, A. (2000). Proanthocyanidins and tannin-like compounds – nature, occurrence, dietary intake and effects on nutrition and health. *Journal of the Science of Food and Agriculture, 80*, 226–230.

17. Pieur, C., Rigaud, J., Cheynier, V., & Moutounet, M. (1994). Oligomeric and polymeric procyanidins from grape seed. *Phytochemistry, 36*, 781–784.

18. Souquet, J. M., Cheynier, V., Brossaud, F., & Moutounet, M. (1996). Polymeric proanthocyanidins from grape skins. *Phytochemistry, 43*, 509–512.

19. Gavetta, B., Fuzzati, N., Griffino, A., Lolla, E., Pace, R., Ruffilli, T., & Peterlongo, F. (2000). Characterization of proanthocyanidins from grape seeds. *Fitoterapia, 71*, 162–175.

20. Holstein, S. A., & Hohl, R. J. (2004). Isoprenoids: Remarkable diversity of form and function. *Lipids, 34*, 293–309.

21. Pisarnitskii, A. F. (2001). Formation of wine aroma: Tones and imperfections caused by minor components. *Applied Biochemistry Microbiologica, 37*, 552–560.

22. Sejer Pedersen, D., Capone, D. L., Skouroumounis, G. K., Pollnitz, A. P., & Sefton, M. A. (2003). Quantitative analysis of geraniol, nerol, linalool and α-terpineol in wine. *Analytical and Bioanalytical Chemistry, 375*, 517–522.

23. Iriti, M., & Faoro, F. (2006). Lipids biosynthesis in spermatophyta. In J. A. Teixeira da Silva, (Ed.), *Floriculture, Ornamental and Plant Biotechnology: Advances and Topical Issues* (Vol. 1), (pp. 359–372). London: Global Science Books.

24. Duncan, R. E., Lau, D., El-Sohemy, A., & Archer, M. C. (2004). Geraniol and β-ionone inhibit proliferation, cell cycle progression and cyclin-dependent kinase 2 activity in MCF-7 breast cancer cells independent of effects on HMG-CoA reductase activity. *Biochemical Pharmacology, 68*, 1739–1747.

25. Kotseridis, Y., Baumes, R. L., Bertrand, A., & Skouroumounis, G. K. (1999). Quantitative determination of β-ionone in red wines and grapes of Bordeaux using a stable isotope dilution assay. *Journal of Chromatography A, 848*, 317–325.

26. Kotseridis, Y., Baumes, R. L., & Skouroumounis, G. K. (1999). Quantitative determination of free and hydrolytically liberated β-damascenone in red grapes and wines using a stable isotope dilution assay. *Journal of Chromatography A, 849*, 245–254.

27. Baumes, R., Wirth, J., Bureau, S., Gunata, Y., & Razungles, A. (2002). Biogeneration of C_{13}-norisoprenoid compounds: Experiment supportive for an apocarotenoid pathway in grapevines. *Analytica Chimica Acta, 458*, 3–14.

28. Pineau, B., Barbe, J.-C., Van Leeuwen, C., & Dubourdieu, D. (2007). Which impact for β-damascenone on red wine aroma? *Journal of Agricultural and Food Chemistry, 55*, 4103–4108.

29. Liu, J.-R., Sun, X.-R., Dong, H.-W., Sun, C.-H., Sun, W.-G., Chen, B.-Q., Song, Y.-Q., & Yang, B.-F. (2008). β-ionone suppresses mammary carcinogenesis, proliferative activity and induces apoptosis in the mammary gland of the Sprague-Dawley rat. *International Journal of Cancer, 122*, 2689–2698.

30. Mo, H., & Elson, C. E. (1999). Apoptosis and cell-cycle arrest in human and murine tumor cells are initiated by isoprenoids. *Journal of Nutrition, 129*, 804–813.

31. Liu, J.-R., Yang, B.-F., Chen, B.-Q., Yang, Y.-M., Dong, H.-W., & Song, Y.-Q. (2004). Inhibition of β-ionone on SGC-7901 cell proliferation and upregulation of

metalloproteinase-1 and -2 expression. *World Journal of Gastroenterology, 10,* 167–171.

32. Liu, J.-R., Chen, B.-Q., Yang, B.-F., Dong, H.-W., Sun, C.-H., Wang, Q., Song, G., & Song, Y.-Q. (2004). Apoptosis of human gastric adenocarcinoma cells induced by β-ionone. *World Journal of Gastroenterology, 10,* 348–351.

33. Janakiram, N. B., Cooma, I., Mohammed, A., Steele, V. E., & Rao, C. V. (2008). β-Ionone inhibits colonic aberrant crypt foci formation in rats, suppresses cell growth and induces retinoid X receptor-α in human colon cancer cells. *Molecular Cancer Therapeutics, 7,* 181–190.

34. Herraiz, T. (1999). 1-Methyl-1,2,3,4-tetrahydro-β-carbo-line-3-carboxilic acid and 1,2,3,4-tetrahydro-β-carbo-line-3-carboxilic acid in fruits. *Journal of Agricultural and Food Chemistry, 47,* 4883–4887.

35. Herraiz, T., & Galisteo, J. (2002). Tetrahydro-β-carbo-line alkaloids that occur in foods and biological systems act as radical scavengers and antioxidants in the ABTS assay. *Free Radical Research, 36,* 923–928.

36. Herraiz, T., & Galisteo, J. (2003). Tetrahydro-β-carbo-line alkaloids occur in fruits and fruit juice. Activity as antioxidant and radical scavengers. *Journal of Agricultural and Food Chemistry, 51,* 7156–7161.

37. Herraiz, T. (2007). Identification and occurrence of β-carboline alkaloids in raisins and inhibition of mono-amine oxidase (MAO). *Journal of Agricultural and Food Chemistry, 55,* 8534–8540.

38. Brossi, A., Focella, A., & Teitel, S. (1973). Alkaloids in mammalian tissues. 3. Condensation of L-tryptophan and L-5-hydroxytryptophan with formaldehyde and acetaldehyde. *Journal of Medicinal Chemistry, 16,* 418–420.

39. Buckholtz, N. S. (1980). Neurobiology of tetrahydro-β-carbolines. *Life Sciences, 27,* 893–903.

40. Airaksinen, M. M, & Kari, I. (1981). β-Carbolines, psy-choactive compounds in the mammalian body. 1. Occurrence, origin and metabolism. *Medical Biology, 59,* 21–34.

41. Yamada, M., & Yasuhara, H. (2004). Clinical pharma-cology of MAO inhibitors: Safety and future. *Neurotoxicology, 25,* 215–221.

42. Youdim, M. B. H., Edmondson, D., & Tipton, K. F. (2006). The therapeutic potential of monoamine oxidase inhibitors. *Nature Rreviews Neuroscience, 22,* 197–217.

43. Pari, K., Sundari, C. S., Chandani, S., & Balasubramanian, D. (2000). β-Carbolines that accumu-late in human tissues may serve a protective role against oxidative stress. *The Journal of Biological Chemistry, 275,* 2455–2462.

44. Chen, Q., Chao, R., Chen, H., Hou, X., Yan, H., Zhou, S., Peng, W., & Xu, A. (2004). Antitumor and neurotoxic effects of novel harmine derivatives and structure-activity relationship analysis. *International Journal of Cancer, 114,* 675–682.

45. Herraiz, T. (1996). Occurrence of tetrahydro-β-carbo-line-3-carboxylic acids in commercial foodstuffs. *Journal of Agricultural and Food Chemistry, 44,* 3057–3065.

46. Gonzales, G. F., & Gonzales-Castañeda, C. (2008). The methyltetrahydro-β-carbolines in Maca (*Lepidium meye-nii*). *Evidence-based Compliance Alternative Medicine,* doi: 10.1093/ecam/nen041.

47. Iriti, M., Rossoni, M., & Faoro, F. (2006). Melatonin content in grape: Myth or panacea? *Journal of the Science of Food and Agriculture, 86,* 1432–1438.

48. Iriti, M. (2008). Melatonin in grape, not just a myth, maybe a panacea. *Journal of Pineal Research,* doi: 10.1111/j.1600-079X.2008.00616.x.

49. Axelrod, J., & Weissbach, H. (1961). Purification and properties of hydroxyindole-O-methyltransferase. *The Journal of Biological Chemistry, 236,* 211–213.

50. Hardeland, R., & Poeggeler, B. (2003). Non-vertebrate melatonin. *Journal of Pineal Research, 34,* 233–241.

51. Kolár, J., & Macháčková, I. (2005). Melatonin in higher plants: Occurrence and possible functions. *Journal of Pineal Research, 39,* 333–341.

52. Hernández-Ruiz, J., Cano, A., & Arnao, M. B. (2005). Melatonin acts as a growth-stimulating compound in some monocot species. *Journal of Pineal Research, 39,* 137–142.

53. Arnao, M. B., & Hernández-Ruiz, J. (2007). Melatonin promotes adventitious- and lateral root regeneration in etiolated hypocotyls of *Lupinus albus* L. *Journal of Pineal Research, 42,* 147–152.

54. Marcolini, L., Saracino, M. A., & Bugamelli, F. (2008). HPLC-F analysis of melatonin and resveratrol isomers in wine using a SPE procedure. *Journal of Separation Science, 31,* 1007–1014.

55. Guerrero, J. M., Martínez-Cruz, F., & Elorza, F. L. (2008). Significant amount of melatonin in red wine: Its con-sumption increases blood melatonin levels in humans. *Food Chemistry,* doi: 10.1016/j.foodchem.2008.02.007.

56. Allegra, M., Reiter, R. J., Tan, D.-X., Gentile, C., Tesoriere, L., & Livrea, M. A. (2003). The chemistry of melatonin's interaction with reactive species. *Journal of Pineal Research, 34,* 1–10.

57. Burkhardt, S., Reiter, R. J., Tan, D.-X., Hardeland, R., Cabrera, J., & Karbownik, M. (2001). DNA oxidatively damaged by chromium(III) and H_2O_2 is protected by melatonin, N^1-acetyl-N^2-formyl-5-methoxykynuramine, resveratrol and uric acid. *The International Journal of Biochemistry and Cell Biology, 33,* 775–783.

58. Mayo, J. C., Sainz, R. M., Tan, D.-X., Hardeland, R., Leon, J., Rodriguez, C., & Reiter, R. J. (2005).

C. ACTIONS OF INDIVIDUAL FRUITS IN DISEASE AND CANCER PREVENTION AND TREATMENT

Anti-inflammatory actions of melatonin and its metabolites, N1-acetyl-N2-formyl-5-methoxykynuramine (AFMK) and N1-acetyl-5-methoxykynuramine (AMK), in macrophages. *Journal of Neuroimmunology, 165,* 139–149.

59. Ximenes, V. F, Fernandes, J. R, Bueno, V. B, Catalani, L. H, Oliveira, G. H., & Machado, R. G. P. (2007). The effect of pH on horseradish peroxidase-catalyzed oxidation of melatonin: Production of N1-acetyl-N2-formyl-5-methoxykynuramine versus radical-mediated degradation. *Journal of Pineal Research, 42,* 291–296.

60. Antolín, I., Rodríguez, C., Sáinz, R. M., Mayo, J. C., Uría, H., Kotler, M. L., Rodríguez-Colunga, M. J., Tolivia, D., & Menéndez-Peláez, A. (1996). Neurohormone melatonin prevents cell damage: Effect on gene expression for antioxidant enzymes. *The FASEB Journal: Official Publication of the Federation of American Societies for Experimental Biology, 10,* 882–890.

61. Ames, B. N., Shigenaga, M. K., & Hagen, T. M. (1993). Oxidants, antioxidants, and the degenerative diseases of aging. *The Proceedings of the National Academy of Sciences of the United States of America, 90,* 7915–7922.

62. Halliwell, B. (1997). Antioxidants and human disease: A general introduction. *Nutrition Reviews, 55,* S44–S49.

63. Dahlgren, C., & Karlsson, A. (1999). Respiratory burst in human neutrophils. *Journal of Immunological Methods, 232,* 3–14.

64. Lee, J., Koo, N., & Min, D. B. (2004). Reactive oxygen species, aging, and antioxidant nutraceuticals. *Comprehensive Review Food Science Food Safety, 3,* 21–33.

65. Wiseman, H. (1996). Dietary influences on membrane function: Importance in protection against oxidative damage and disease. *Journal of Nutrition Biochem, 7,* 2–15.

66. Tavolini, C., Juliano, L., Piu, F., Franconi, F., & Cabrini, L. (2000). Resveratrol inhibition of lipid peroxidation. *Free Radical Research, 33,* 105–114.

67. Yu, B. P. (1994). Cellular defenses against damage from reactive oxygen species. *Physiological Reviews, 74,* 139–162.

68. Cao, G., Sofic, E., & Prior, R. L. (1997). Antioxidant and prooxidant behavior of flavonoids: Structure-activity relationships. *Free Radical Biology and Medicine, 22,* 749–760.

69. Prior, R. L, & Cao, G. (2000). Antioxidant phytochemicals in fruits and vegetables: Diet and health implications. *Horticultural Science, 35,* 588–592.

70. Vinson, J. A., Yang, J. H., Proch, J., & Liang, X. (2000). Grape juice, but not orange juice, has *in vitro, ex vivo* and *in vivo* antioxidant properties. *Journal of Medicinal Food, 2,* 167–171.

71. Vinson, J. A., Teufel, K., & Wu, N. (2001). Red wine, dealcoholized red wine and specially grape juice, inhibit atherosclerosis in a hamster model. *Atherosclerosis, 156,* 42–67.

72. Bors, W., Heller, W., Michel, C., & Saran, M. (1990). Flavonoids as antioxidants: Determination of radical-scavenging efficiencies. *Methods in Enzymology, 186,* 343–355.

73. Bors, W., Michel, C., & Stettmaier, K. (1997). Antioxidant effects of flavonoids. *Biofactors, 6,* 399–402.

74. Frankel, E. N. (1999). Food antioxidants and phytochemicals: Present and future perspectives. *Fett/Lipid, 101,* 450–455.

75. Heim, K. E., Tagliaferro, A. R., & Bobilya, D. J. (2002). Flavonoid antioxidants: Chemistry, activity and structure-activity relationships. *Journal of Nutrition Biochemistry, 13,* 572–584.

76. Amic, D., Davidovic-Amic, D., Bešlo, D., & Trinajstic, N. (2003). Structure-radical scavenging activity relationships of flavonoids. *Croatica Chimica Acta, 76,* 55–61.

77. Soobrattee, M. A., Neergheen, C. S., Luximon-Ramma, A., Aruoma, O. I., & Bahorun, T. (2005). Phenolics as potential antioxidant therapeutic agents: Mechanism and actions. *Mutation Research, 579,* 200–213.

78. Seyoum, A., Asres, K., & El-Fiky, F. K. (2006). Structure-radical scavenging activity relationships of flavonoids. *Phytochemistry, 67,* 2058–2070.

79. Wang, H., Cao, G., & Prior, R. L. (1996). Total antioxidant capacity of fruits. *Journal of Agricultural and Food Chemistry, 44,* 701–705.

80. Wolfe, K. L., Kang, X., He, X., Dong, M., Zhang, Q., & Liu, R. H. (2008). Cellular antioxidant activity of common fruits. *Journal of Agricultural and Food Chemistry, 56,* 8418–8426.

81. O'Byrne, D. J., Devaraj, S., Grundy, S. M., & Jialal, I. (2002). Comparison of the antioxidant effects of Concord grape juice flavonoids and α-tocopherol on markers of oxidative stress in healthy adults. *The American Journal of Clinical Nutrition, 76,* 1367–1374.

82. Day, A. P., Kemp, H. J., Bolton, C., Hartog, M., & Stansbie, D. (1997). Effect of concentrated red grape juice consumption on serum antioxidant capacity and low-density lipoprotein oxidation. *Annals of Nutrition and Metabolism, 41,* 353–357.

83. García-Alonso, J., Ros, G., Vidal-Guevara, M. L., & Periamo, M. J. (2006). Acute intake of phenolic-rich juice improves antioxidant status in healthy subjects. *Nutrition Research, 26,* 330–339.

84. Zern, T. L., Wood, R. J., Greene, C., West, K. L., Liu, Y., Aggarwal, D., Shachter, N. S., & Fernández, M. L. (2005). Grape polyphenols exert a cardioprotective effect in pre- and postmenopausal women by lowering plasma lipids and reducing oxidative stress. *Journal of Nutrition, 135,* 1911–1917.

85. Durak, I., Köseoglu, M. H., Kaçmaz, M., Büyükkoçak, S., Çimen, M. Y. B., & Öztürk, H. S. (1999). Black grape enhances plasma antioxidant potential. *Nutrition Research, 19*, 973–977.

86. Park, Y. K., Park, E., Kim, J.-S., & Kang, M.-H. (2003). Daily grape juice consumption reduces oxidative DNA damage and plasma free radical levels in healthy Koreans. *Mutation Research, 529*, 77–86.

87. Stankovic, M., Tesević, V., Vajs, V., Todorovic, N., Milosavljevic, S., & Godevac, D. (2008). Antioxidant properties of grape seed extract on human lymphocyte oxidative defence. *Planta Medica, 74*, 730–735.

88. Stagos, D., Kazantzoglou, G., Theofanidou, D., Kakalopoulou, G., Magiatis, P., Mitaku, S., & Kouretas, D. (2006). Activity of grape extracts from Greek varieties of Vitis vinifera against mutagenicity induced by bleomycin and hydrogen peroxide in *Salmonella typhimurium* strain TA102. *Mutation Research, 609*, 165–175.

89. Cheng, K. C., Cahill, D. S., Kasai, H., Nishimura, S., & Loeb, N. A. (1992). 8-Hydroxyguanine, an abundant form of oxidative DNA damage, causes G-T and A-C substitutions. *The Journal of Biological Chemistry, 267*, 166–172.

90. López-Burillo, S., Tan, D.-X., Mayo, J. C., Sainz, R. M., Manchester, L. C., & Reiter, R. J. (2003). Melatonin, xanthurenic acid, resveratrol, EGCG Vitamin C and alpha-lipoic acid differentially reduce oxidative DNA damage induced by fenton reagents: A study of their individual and synergistic actions. *Journal of Pineal Research, 34*, 269–277.

91. Sgambato, A., Ardito, R., Faraglia, B., Boninsegna, A., Wolf, F. I., & Cittadini, A. (2001). Resveratrol, a natural phenolic compound, inhibits cell proliferation and prevents oxidative DNA damage. *Mutation Research, 496*, 171–180.

92. De Salvia, R., Festa, F., Ricordy, R., Particone, P., & Cozzi, R. (2002). Resveratrol affects in a different way primary versus fixed DNA damage induced by H_2O_2 in mammalian cells *in vitro*. *Toxicology Letters, 135*, 1–9.

93. Wattenberg, L. W. (1985). Chemoprevention of cancer. *Cancer Research, 45*, 1–8.

94. Greewald, P. (2002). Cancer chemoprevention. *British Medical Journal, 324*, 714–718.

95. Young, G., & Le Leu, R. (2002). Preventing cancer: Dietary lifestyle or clinical intervention? *Asia Pacific Journal of Clinical Nutrition, 11*, S618–S631.

96. Suhr, Y.-J. (2003). Cancer chemoprevention with dietary phytochemicals.. *Nature Reviews Cancer, 3*, 768–780.

97. Garg, A. K., Buchholz, T. A., & Aggarwal, B. B. (2005). Chemosensitization and radiosensitization of tumors by plant polyphenols. *Antioxidants and Redox Signaling, 7*, 1630–1647.

98. Fresco, P., Borges, F., Diniz, C., & Marques, M. P. M. (2006). New insights on the anticancer properties of dietary polyphenols. *Medicinal Research Reviews, 26*, 747–766.

99. Hail, N., Cortes, M., Drake, E. N., & Spallholz, J. E. (2008). Cancer chemoprevention: A *radical* perspective. *Free Radical Biology and Medicine, 45*, 97–110.

100. La Vecchia, C., & Borsetti, C. (2007). Diet and cancer risk in Mediterranean countries. *Hungarica Medical Journal, 1*, 13–23.

101. Khan, N., Adhami, V. M., & Mukhtar, H. (2008). Apoptosis by dietary agents for prevention and treatment of cancer. *Biochemical Pharmacology*, doi: 10.1016/j.bcp.2008.07.015.

102. Jang, M., Cai, L., Udeani, G. O., Slowing, K. V., Thomas, C. F., Beecher, C. W., Fong, H. H., Farnsworth, N. R., Kinghorn, A. D., Mehta, R. G., Moon, R. C., & Pezzuto, J. M. (1997). Cancer chemopreventive activity of resveratrol, a natural product derived from grapes. *Science, 275*, 218–220.

103. Kundu, J. K., & Surh, Y.-J. (2005). Breaking the relay in deregulated cellular signal transduction as a rationale for chemoprevention with anti-inflammatory phytochemicals. *Mutation Research, 591*, 123–146.

104. Aggarwal, B. B., Shishodia, S., Sandur, S. K., Pandey, M. K., & Sethi, G. (2006). Inflammation and cancer: How hot is the link? *Biochemical Pharmacology, 72*, 1605–1621.

105. Subbaramaiah, K., & Donnenberg, A. J. (2003). Cyclooxygenase 2: A molecular target for chemoprevention and treatment. *Trends in Pharmacological Sciences, 24*, 96–102.

106. Subbaramaiah, K., Chung, W. J., Michaluart, P., Telang, N., Tanabe, T., Inoue, H., Jang, M., Pezzuto, J. M., & Donnenberg, A. J. (1998). Resveratrol inhibits cyclooxygenase 2 transcription and activity in phorbol ester-treated human mammary epithelial cells. *The Journal of Biological Chemistry, 273*, 21875–21882.

107. Martinez, J., & Moreno, J. J. (2000). Effect of resveratrol, a natural polyphenolic compound, on reactive oxygen species and prostaglandin production. *Biochemical Pharmacology, 59*, 865–870.

108. Murakami, A., Matsumoto, K., Koshimizu, K., & Ohigashi, H. (2003). Effects of selected food factors with chemopreventive properties on combined lipopolysaccharide- and interferon-gamma-induced IκB degradation in RAW264.7 macrophages. *Cancer Letters, 195*, 17–25.

109. Kundu, J. K., Shin, Y. K., Kim, S. H., & Surh, Y.-J. (2006). Resveratrol inhibits phorbol ester-induced

C. ACTIONS OF INDIVIDUAL FRUITS IN DISEASE AND CANCER PREVENTION AND TREATMENT

expression of COX-2 and activation of NF-κB in mouse skin by blocking IκB kinase activity. *Carcinogenesis, 27*, 1465–1474.

110. Li, Z. G., Hong, T., Shimada, Y., Komoto, I., Kawabe, A., Ding, Y., Kaganoi, J., Hashimoto, Y., & Imamura, M. (2002). Suppression of N-nitrosomethyl-benzylamine (NMBA)-induced esophageal tumorigenesis in F344 rats by resveratrol. *Carcinogenesis, 23*, 1531–1536.

111. Richard, N., Porath, D., Radspieler, A., & Schwager, A. (2005). Effects of resveratrol, piceatannol, tri-acetoxystilbene and genistein on the inflammatory response on human peripheral blood leukocytes. *Molecular Nutrition and Food Research, 49*, 431–442.

112. Holmes-McNary, M., & Baldwin, A. S. (2000). Chemopreventive properties of trans-resveratrol are associated with inhibition of activation of the IkappaB kinase. *Cancer Research, 60*, 3477–3483.

113. Manna, S. K, Mukhopadhyay, A., & Aggarwal, B. B. (2000). Resveratrol suppresses TNF-induced activation of nuclear transcription factors NF-B, activator protein-1, and apoptosis: Potential role of reactive oxygen intermediates and lipid peroxidation. *Journal of Immunology, 164*, 6509–6519.

114. Dong, Z. (2003). Molecular mechanism of the chemopreventive effect of resveratrol. *Mutation Research, 523–524*, 145–150.

115. Pervaiz, S. (2003). Resveratrol: From grapevines to mammalian biology. *The FASEB Journal: Official Publication of the Federation of American Societies for Experimental Biology, 17*, 1975–1985.

116. Hou, D.-X., Kai, K., Li, J. J., Lin, S., Terahara, N., Wakamatsu, M., Fujii, M., Young, M. R., & Colburn, N. (2004). Anthocyanidins inhibit activator protein 1 activity and cell transformation: Structure-activity relationship and molecular mechanisms. *Carcinogenesis, 25*, 29–36.

117. Mitchell, S. H., Zhu, W., & Young, C. Y. (1999). Resveratrol inhibits the expression and function of the androgen receptor in LNCaP prostate cancer cells. *Cancer Research, 59*, 5892–5895.

118. Xing, N., Chen, Y., Mitchell, S. H., & Young, C. Y. (2001). Quercetin inhibits the expression and function of the androgen receptor in LNCaP prostate cancer cells. *Carcinogenesis, 22*, 409–414.

119. Hsieh, T. C., & Wu, J. M. (1999). Differential effects on growth, cell cycle arrest, and induction of apoptosis by resveratrol in human prostate cancer cell lines. *Experimental Cell Research, 249*, 109–115.

120. Lu, R., & Serrero, G. (1999). Resveratrol, a natural product derived from grape, exhibits antiestrogenic activity and inhibits the growth of human breast cancer cells. *Journal of Cellular Physiology, 179*, 297–304.

121. Basly, P., Marre-Fournier, F., Le Bail, J. C., Habrioux, G., & Chulia, A. J. (2000). Estrogenic/antiestrogenic and scavenging properties of (E)- and (Z)-resveratrol. *Life Sciences, 66*, 769–777.

122. Bhat, K. P., Lantvit, D., Christov, K., Mehta, R. G., Moon, R. C., & Pezzuto, J. M. (2001). Estrogenic and antiestrogenic properties of resveratrol in mammary tumor models. *Cancer Research, 61*, 7456–7463.

123. Ososki, A. L., & Kennelly, E. J. (2003). Phytoestrogens: A review of the present state of reaserch. *Phytother. Research, 17*, 845–869.

124. Kijima, I., Phung, S., Hur, G., Kwok, S. L., & Chen, S. (2006). Grape seed extract is an aromatase inhibitor and a suppressor of aromatase expression. *Cancer Research, 66*, 5960–5967.

125. Sun, X. Z., Zhou, D., & Chen, S. (1997). Autocrine and paracrine actions of breast tumor aromatase. A three-dimensional cell culture study involving aromatase transfected MCF-7 and T-47D cells. *The Journal of Steroid Biochemistry and Molecular Biology, 63*, 29–36.

126. Ciolino, H. P., & Yeh, G. C. (1999). Inhibition of aryl hydrocarbon-induced cytochrome P-450 1A1 enzyme activity and CYP1A1 expression by resveratrol. *Molecular Pharmacology, 56*, 760–767.

127. Chan, W. K., & Delucchi, A. B. (2000). Resveratrol, a red wine constituent, is a mechanism-based inactivator of cytochrome P450 3A4. *Life Sciences, 67*, 3103–3112.

128. Chang, T. K, Chen, J., & Lee, W. B. (2001). Differential inhibition and inactivation of human CYP1 enzymes by trans-resveratrol: Evidence for mechanism-based inactivation of CYP1A2. *The Journal of Pharmacology and Experimental Therapeutics, 299*, 874–882.

129. Clement, M. V., Ponton, A., & Pervaiz, S. (1998). Apoptosis induced by hydrogen peroxide is mediated by decreased superoxide anion concentration and reduction of intracellular milieu. *FEBS Letters, 440*, 13–18.

130. Huang, C., Ma, W., Goranson, A., & Dong, Z. (1999). Resveratrol suppresses cell transformation and induces apoptosis through a p53-dependent pathway. *Carcinogenesis, 20*, 237–242.

131. Suhr, Y. J., Hurh, Y. J., Kang, Y. J., Lee, E., Kong, G., & Lee, S. J. (1999). Resveratrol, an antioxidant present in red wine, induces apoptosis in human promyelocytic leukemia (HL-60) cell. *Cancer Letters, 140*, 1–10.

132. Dorrie, J., Gerauer, H., Wachter, Y., & Zunino, S. J. (2001). Resveratrol induces extensive apoptosis by depolarizing mitochondrial membranes and

C. ACTIONS OF INDIVIDUAL FRUITS IN DISEASE AND CANCER PREVENTION AND TREATMENT

activating caspase-9 in acute lymphoblastic leukemia cells. *Cancer Research, 61*, 4731–4739.

133. Tinhofer, I., Bernhard, D., Senfter, M., Anether, G., Loeffler, M., Kroemer, G., Kofler, R., Csordas, A., & Greil, R. (2001). Resveratrol, a tumor-suppressive compound from grapes, induces apoptosis via a novel mitochondrial pathway controlled by Bcl-2. *The FASEB Journal : Official Publication of the Federation of American Societies for Experimental Biology, 15*, 1613–1615.

134. Roman, V., Billard, C., Kern, C., Ferry-Dumazet, H., Izard, J. C., Mohammad, R., Mossalayi, D. M, & Kolb, J. P. (2002). Analysis of resveratrol-induced apoptosis in human B-cell chronic leukaemia. *British Journal of Haematology, 117*, 842–851.

135. Soleas, G. J., Goldberg, D. M., Grass, L., Levesque, M., & Diamandis, E. P. (2001). Do wine polyphenols modulate p53 gene expression in human cancer cell lines? *Clinical Biochemistry, 34*, 415–420.

136. Schneider, Y., Fischer, B., Coelho, D., Roussi, S., Gosse, F., Bischoff, P., & Raul, F. (2004). Z)-3,5,4'-Tri-O-methyl-resveratrol induces apoptosis in human lymphoblastoid cells independently of their p53 status. *Cancer Letter, 211*, 155–161.

137. Tang, Y., Zhao, D. Y., Elliot, S., Zhao, W., Curiel, T. J., & Beckman, B. S. (2007). Epigallocathechin-3 gallate induces growth inhibition and apoptosis in human breast cancer cells through surviving suppression. *International Journal of Oncology, 31*, 705–711.

138. Johnson, M. E., & Howerth, E. W. (2004). Survivin: A bifunctional inhibitor of apoptosis protein. *Veterinary Pathologica, 41*, 599–607.

139. Fontecave, M., Lepoivre, M., Elleingand, E., Gerez, C., & Guittet, O. (1998). Resveratrol, a remarkable inhibitor of ribonucleotide reductase. *FEBS Letters, 421*, 277–279.

140. Ragione, F. D., Cucciolla, V., Borriello, A., Pietra, V. D., Racioppi, L., Soldati, G., Manna, C., Galletti, P., & Zappia, V. (1998). Resveratrol arrests the cell division cycle at S/G2 phase transition. *Biochemistry and Biophysics Research Communications, 250*, 53–58.

141. Hsieh, T. C., Burfeind, P., Laud, K., Backer, J. M., Traganos, F., Darzynkiewicz, Z., & Wu, J. M. (1999). Cell cycle effects and control of gene expression by resveratrol in human breast carcinoma cell lines with different metastatic potentials. *International Journal of Oncology, 15*, 245–252.

142. Bernhard, D., Tinhofer, I., Tonko, M., Hubl, H., Ausserlechner, M. J., Greil, R., Kofler, R., & Csordas, A. (2000). Resveratrol causes arrest in the S-phase prior to Fas-independent apoptosis in CEM-C7H2 acute leukemia cells. *Cell Death Differentiation, 7*, 834–842.

143. Joe, A. K., Liu, H., Suzui, M., Vural, M. E., Xiao, D., & Weinstein, I. B. (2002). Resveratrol induces growth inhibition, S-phase arrest, apoptosis, and changes in biomarker expression in several human cancer cell lines. *Clinical Cancer Research, 8*, 893–903.

144. Schneider, Y., Vincent, F., Duranton, B., Badolo, L., Gosse, F., Bergmann, C., Seiler, N., & Raul, F. (2000). Anti-proliferative effect of resveratrol, a natural component of grapes and wine, on human colonic cancer cells. *Cancer Letter, 158*, 85–91.

145. Tsan, F., White, J. E., Maheshwari, J. G., & Chikkappa, G. (2002). Anti-leukemia effect of resveratrol. *Leukemia and Lymphoma, 43*, 983–987.

146. Agarwal, C., Sharma, Y., Zhao, J., & Agarwal, R. (2000). A polyphenolic fraction from grape seeds causes irreversible growth inhibition of breast carcinoma MDA-MB468 cells by inhibiting mitogen-activated protein kinases activation and inducing G1 arrest and differentiation. *Clinical Cancer Research, 6*, 2921–2930.

147. Singh, R. P., Tyagi, A. K., Dhanalakshmi, S., Agarwal, R., & Agarwal, C. (2004). Grape seed extract inhibits advanced human prostate tumor growth and angiogenesis and upregulates insulin-like growth factor binding protein-3. *International Journal of Cancer, 108*, 733–740.

148. Mantena, S. K., Baliga, M. S., & Katiyar, S. K. (2006). Grape seed proanthocyanidins induce apoptosis and inhibit metastasis of highly metastatic breast carcinoma cells. *Carcinogenesis, 27*, 1682–1691.

149. Liao, D., & Johnson, R. S. (2007). Hypoxia: A key regulator of angiogenesis in cancer. *Cancer Metastasis Reviews, 26*, 281–290.

150. Cao, Z., Fang, J., Xia, C., Shi, X., & Jiang, B. H. (2004). *Trans*-3,4,5'-trihydroxystilbene inhibits hypoxia-inducible factor 1 alpha and vascular endothelial growth factor expression in human ovarian cancer cells. *Clinical Cancer Research, 10*, 5253–5263.

151. Sun, C. Y., Hu, Y., Guo, T., Wang, H. F., Zhang, X. P., He, W. J., & Tan, H. (2006). Resveratrol as a novel agent for treatment of multiple myeloma with matrix metalloproteinase inhibitory activity. *Acta Pharmacologica Sinica, 27*, 1447–1452.

152. Yu, H., Pan, C., Zhao, S., Wang, Z., Zhang, H., & Wu, W. (2007). Resveratrol inhibits tumor necrosis factor-α-mediated matrix metalloproteinase-9 expression and invasion of human hepatocellular carcinoma cells. *Biomedicine & Pharmacotherapy*, doi:10.1016/j.biopha.2007.09.006.

153. Tang, F. Y., Chiang, E. P., & Sun, Y. C. (2008). Resveratrol inhibits heregulin-β1-mediated matrix metalloproteinase-9 expression and cell invasion in

C. ACTIONS OF INDIVIDUAL FRUITS IN DISEASE AND CANCER PREVENTION AND TREATMENT

human breast cancer cells. *Journal of Nutrition Biochemistry*, 19, 287–294.

154. Holian, O., Wahid, S., Atten, M. J., & Attar, B. M. (2002). Inhibition of gastric cancer cell proliferation by resveratrol: Role of nitric oxide. *The American Journal of Physiology*, 282, G809–G816.

155. Atten, M. J., Godoy-Romero, E., Attar, B. M., Milson, T., Zopel, M., & Holian, O. (2005). Resveratrol regulates cellular PKC alpha and delta to inhibit growth and induce apoptosis in gastric cancer cells. *Investigational. New Drugs*, 23, 111–119.

156. Riles, W. L., Erickson, J., Nayyar, S., Atten, M. J., Attar, B. M., & Holian, O. (2006). Resveratrol engages selective apoptotic signals in gastric adenocarcinooma cells. *World Journal of Gastroenterology*, 12, 5628–5634.

157. Mahady, G. B., & Pendland, S. L. (2000). Resveratrol inhibits the growth of *Helicobacter pylori* in vitro. *The American Journal of Gastroenterology*, 95, 1849.

158. Johnson, I. T. (2002). Anticarcinogenic effects of diet-related apoptosis in the colorectal mucosa. *Food and Chemical Toxicology*, 40, 1171–1178.

159. Durak, I., Çetin, R., Devrim, E., & Ergüder, I. B. (2005). Effects of black grape extract on activities of DNA turn-over enzymes in cancerous and noncancerous human colon tissues. *Life Sciences*, 76, 2995–3000.

160. Durak, I., Biri, H., Ergüder, I. B., Devrim, E., Senocak, Ç., & Avci, A. (2007). Effects of garlic and black grape extracts on the activity of adenosine deaminase from cancerous and noncancerous human urinary bladder tissues. *Med. Chem. Research*, 16, 259–265.

161. Kaur, M., Singh, R. P., Gu, M., Agarwal, R., & Agarwal, C. (2006). Grape seed extract inhibits *in vitro* and *in vivo* growth of human colorectal carcinoma cells. *Clinical Cancer Research*, 12, 6194–6202.

162. López-Oliva, M. E., Agis-Torres, A., García-Palencia, P., Goñi, I., & Muñoz-Martínez, E. (2006). Induction of epithelial hypoplasia in rat cecal and distal colonic mucosa by grape antioxidant dietary fiber. *Nutrition Research*, 26, 651–658.

163. Birt, D. F, Mitchell, D., Gold, B., Pour, P., & Pinch, H. C. (1997). Inhibition of ultraviolet light induced skin carcinogenesis in SKH-1 mice by apigenin, a plant flavonoid. *Anticancer Research*, 17, 85–91.

164. Adhami, V. M., Afaq, F., & Ahmad, N. (2003). Suppression of ultraviolet B exposure-mediated activation of NF-κB in normal human keratinocytes by resveratrol. *Neoplasia*, 5, 74–82.

165. Mittal, A., Elmets, C. A., & Katiyar, S. K. (2003). Dietary feeding of proanthocyanidins from grape seeds prevents photocarcinogenesis in SKH-1 hairless

mice: Relationship to decreased fat and lipid peroxidation. *Carcinogenesis*, 24, 1379–1388.

166. Sharma, S. D., Meeran, S. M., & Katiyar, S. K. (2007). Dietary grape seed proanthocyanidins inhibit UVB-induced oxidative stress and activation of mitogen-activated protein kinases and nuclear factor-κB signaling in in vivo SKH-1 hairless mice. *Molecular Cancer Therapeutics*, 6, 995–1005.

167. Niles, R. M., McFarland, M., Weimer, M. B., Redkar, A., Fu, Y.-M., & Meadows, G. G. (2003). Resveratrol is a potent inducer of apoptosis in human melanoma cells. *Cancer Letter*, 190, 157–163.

168. Wang, Z., Desmoulin, S., Banerjee, D., Li, Y., Deraniyagala, R. L., Abbruzzese, J., & Sarkar, F. H. (2008). Synergistic effects of multiple natural products in pancreatic cancer cells. *Life Sciences*, 83, 293–300.

169. Jemal, A., Siegel, R., Ward, E., Hao, Y., Xu, J., Murray, T., & Thun, M. J. (2008). Cancer statistics, 2008. *CA: A Cancer Journal for Clinicians*, 58, 71–96.

170. Cui, G., Zhang, Y., Liu, C., Skinner, S., & Qin, Y. (2008). Pancreatic cancer suppression by natural polyphenols. *Scholarship Research Exchange*, doi:10.3814/2008/540872.

171. Berliner, J. A., Navab, M., Fogelman, A. M., Frank, J. S., Demer, L. L., Edward, P. A., Watson, A. D., & Lusis, A. J. (1995). Atherosclerosis: Basic mechanisms. Oxidation, inflammation, and genetics. *Circulation*, 91, 2488–2496.

172. da Luz, P. L., & Coimbra, S. R. (2004). Wine, alcohol and atherosclerosis: Clinical evidences and mechanisms. *Brazilian Journal of Medical and Biological Research*, 37, 1275–1295.

173. Smith, W. L. (1989). The eicosanoids and their biochemical mechanisms of action. *The Biochemical Journal*, 259, 315–324.

174. Kimura, Y., Okuda, H., & Arichi, S. (1985). Effects of stilbenes on arachidonate metabolism in leukocytes. *Biochimica et Biophysica Acta*, 834, 268–275.

175. Ferrandiz, M. L., & Alcaraz, M. J. (1991). Anti-inflammatory activity and inhibition of arachidonic acid metabolism by flavonoids. *Agents Actions*, 32, 283–288.

176. Laughton, M. J., Evans, P. J., Moroney, M. A., Hoult, J. R., & Halliwell, B. (1991). Inhibition of mammalian 5-lipoxygenase and cyclo-oxygenase by flavonoids and phenolic dietary additives. Relationship to antioxidant activity and to iron ion-reducing ability. *Biochemical Pharmacology*, 42, 1673–1681.

177. Pace-Asciak, C. R., Hahn, S., Diamandis, E. P., Soleas, G., & Goldberg, D. M. (1995). The red wine phenolics, trans-resveratrol and quercetin, block human platelet aggregation and eicosanoid synthesis:

Implications for protection against coronary heart diseases. *Clinical Chimica Acta, 236,* 207–219.

178. Kim, H. P., Mani, I., Iversen, L., & Ziboh, V. A. (1998). Effects of naturally-occurring flavonoids and biflavonoids on epidermal cyclooxygenase and lipoxygenase from guinea-pigs. *Prostaglandins, Leukotrienes, and Essential Fatty Acids, 58,* 17–24.

179. Bertelli, A. A., Giovannini, L., Giannessi, D., Migliori, M., Bernini, W., Fregoni, M., & Bertelli, A. (1995). Antiplatelet activity of synthetic and natural resveratrol in red wine. *International Journal of Tissue Reactions, 17,* 1–3.

180. Bertelli, A. A., Giovannini, L., Bernini, W., Migliori, M., Fregoni, M., Bavaresco, L., & Bertelli, A. (1996). Antiplatelet activity of cis-resveratrol. *Drugs Experimental Clinical Research, 22,* 61–63.

181. Di Castelnuovo, A., Rotondo, S., Iacoviello, L., Donati, M. B., & de Gaetano, G. (2002). Meta-analysis of wine and beer consumption in relation to vascular risk. *Circulation, 105,* 2836–2844.

182. Bradamante, S., Barenghi, L., & Villa, A. (2004). Cardiovascular protective effects of resveratrol. *Cardiovascular Drug Reviews, 22,* 169–188.

183. Dell'Agli, M., Buscialà, A., & Bosisio, E. (2004). Vascular effects of wine polyphenols. *Cardiovascular Research, 63,* 593–602.

184. Stoclet, J.-C., Chataigneau, T., Ndiaye, M., Oak, M.-H., El Bedoui, J., Chataigneau, M., & Schini-Kerth, V. B. (2004). Vascular protection by dietary polyphenols. *European Journal of Pharmacology, 500,* 299–313.

185. Cordova, A. C., Jackson, L. S. M., Berke-Schlessel, D. W., & Sumpio, B. E. (2005). The cardiovascular protective effect of red wine. *Journal of the American College of Surgeons, 200,* 428–439.

186. Perez-Vizcaino, F., Duarte, J., & Andriantsitohaina, R. (2006). Endothelial function and cardiovascular disease: Effects of quercetin and wine polyphenols. *Free Radical Research, 40,* 1054–1065.

187. Stein, J. H., Keevil, J. G., Wiebe, D. A., Aeschlimann, S., & Folts, J. D. (1999). Purple grape juice improves endothelial function and reduces the susceptibility of LDL cholesterol to oxidation in patients with coronary artery disease.. *Circulation, 100,* 1050–1055.

188. Chou, E. J., Keevil, J. G., Aeschlimann, S., Wiebe, D. A., Folts, J. D., & Stein, J. H. (2001). Effect of ingestion of purple grape juice on endothelial function in patients with coronary heart disease. *The American Journal of Cardiology, 88,* 553–555.

189. Coimbra, S. R., Lage, S. H., Brandizzi, L., Yoshida, V., & da Luz, P. L. (2005). The action of red wine and purple grape juice on vascular reactivity is independent of plasma lipids in hypercholesterolemic patients. *Brazilian Journal of Medical and Biological Research, 38,* 1339–1347.

190. Lekakis, J., Rallidis, L. S., Andreadou, I., Vamvakou, G., Kazantzoglou, G., Magiatis, P., Skaltsounis, A.-L., & Kremastinos, D. (2005). Polyphenolic compounds from red grapes acutely improve endothelial function in patients with coronary heart disease. *European Journal of Cardiovascular Prevention and Rehabilitation, 12,* 596–600.

191. Demrow, H. S., Slane, P. R., & Folts, J. D. (1995). Administration of wine and grape juice inhibits *in vivo* platelet activity and thrombosis in stenosed canine coronary arteries. *Circulation, 91,* 1182–1188.

192. Freedman, J. E., Loscalzo, J., Barnard, M. R., Alpert, C., Keaney, J. F., & Michelson, A. D. (1997). Nitric oxide released from activated platelets inhibits platelet recruitment. *Journal of Clinical Investigation, 100,* 350–356.

193. Freedman, J. E., Parker, C., Li, L., Perlman, J. A., Frei, B., Ivanov, V., Deak, L. R., Iafrati, M. D., & Folts, J. D. (2001). Select flavonoids and whole juice from purple grapes inhibit platelet function and enhance nitric oxide release. *Circulation, 103,* 2792–2798.

194. Keevil, J. G., Osman, H. E., Reed, J. D., & Folts, J. D. (2000). Grape juice, but not orange juice or grapefruit juice, inhibits human platelet aggregation. *Journal of Nutrition, 130,* 53–56.

195. Shanmuganayagam, D., Beahm, M. R., Osman, H. E., Krueger, G. G., Reed, J. D., & Folts, J. D. (2002). Grape seed and grape skin extracts elicit a greater anti-platelet effect when used in combination than when used individually in dogs and humans. *Journal of Nutrition, 132,* 3592–3598.

196. Castilla, P., Echarri, R., Dávalos, A., Cerrato, F., Ortega, H., Teruel, J. L., Lucas, M. F., Gómez-Coronado, D., Ortuño, J., & Lasunción, M. A. (2006). Concentrated red grape juice exerts antioxidant, hypolipidemic, and antiinflammatory effects in both hemodialysis patients and healthy subjects. *The American Journal of Clinical Nutrition, 84,* 252–62.

197. Park, Y. K., Kim, J.-S., & Kang, M.-H. (2004). Concorde grape juice supplementation reduces blood pressure in Korean hypertensive men: Double-blind, placebo controlled intervention trial. *BioFactors, 22,* 145–147.

198. Ruf, J. C., Berger, J. L., & Renaud, S. (1995). Platelet rebound effect of alcohol withdrawal and wine drinking in rat. Relation to tannins and lipid peroxidation. *Arteriosclerosis, Thrombosis, and Vascular Biology, 1,* 140–144.

199. Martin, S., Giannone, G., Andriantsitohaina, R., & Martinez, M. C. (2003). Delphinidin, an active

C. ACTIONS OF INDIVIDUAL FRUITS IN DISEASE AND CANCER PREVENTION AND TREATMENT

compound of red wine, inhibits endothelial cell apoptosis via nitric oxide pathway and regulation of calcium homeostasis. *British Journal of Pharmacology, 139,* 1095–1102.

200. Fumagalli, F., Rossoni, M., Iriti, M., Di Gennaro, A., Faoro, F., Borroni, E., Borgo, M., Scienza, A., Sala, A., & Folco, G. (2006). From field to health: A simple way to increase the nutraceutical content of grape as shown by NO-dependent vascular relaxation. *Journal of Agricultural and Food Chemistry, 54,* 5344–5349.

201. Leikert, J. F., Rathel, T. L., Wohlfart, T., Cheynier, V., Vollmar, A. M., & Dirsch, V. M. (2002). Red wine polyphenols enhance endothelial nitric oxide synthase expression and subsequent nitric oxide release from endothelial cells. *Circulation, 106,* 1614–1617.

202. Wallerath, T., Deckert, G., Ternes, T., Anderson, H., Li, H., Witte, K., & Forstermann, U. (2002). Resveratrol, a polyphenolic phytoalexin present in red wine, enhances expression and activity of endothelial nitric oxide synthase. *Circulation, 106,* 1652–1658.

203. Huie, R. E, & Padmaja, S. (1993). The reaction of NO with superoxide. *Free Radical Research, 18,* 195–199.

204. Corder, R., Douthwaite, J. A., Lees, D. M., Khan, N. Q., Viseu Dos Santos, A. C., Wood, E. G., & Carrier, M. J. (2001). Endothelin-1 synthesis reduced by red wine. *Nature, 414,* 863–864.

205. Kiely, D. G., Cargill, R. I., & Struthers, A. D. (1997). Cardiopulmonary effects of endothelin-1 in man. *Cardiovasc. Research, 33,* 378–386.

206. Brand, K., Page, S., Walli, A. K., Neumeier, D., & Bauerle, P. A. (1997). Role of nuclear factor NF-κB in atherogenesis. *Experimental Physiology, 82,* 297–304.

207. Csiszar, A., Smith, K., Labinskyy, N., Orosz, Z., Rivera, A., & Ungvari, Z. (2006). Resveratrol attenuates TNF-alpha-induced activation of coronary arterial endothelial cells: Role of NF-kappaB inhibition. *American Journal of Physiology. Heart and Circulatory Physiology, 291,* H1694–H1699.

208. Hamsten, A., & Eriksson, P. (1995). Fibrinolysis and atherosclerosis. *Baillieres Clinical Haematologica, 8,* 345–363.

209. Abou-Agag, L. H., Aikens, M. L., & Tabengwa, E. M. (2001). Polyphenolics increase *t*-PA and *u*-PA gene transcription in cultured human endothelial cells. *Alcoholism, Clinical and Experimental Research, 25,* 155–162.

210. Zou, J., Huang, Y., & Cao, K. (2000). Effect of resveratrol on intimal hyperplasia after endothelial denudation in an experimental rabbit model. *Life Sciences, 68,* 153–163.

211. Araim, O., Ballantyne, J., Waterhouse, A., & Sumpio, B. E. (2002). Inhibition of vascular smooth muscle cell proliferation with red wine and red wine polyphenols. *Journal of Vascular Surgery, 35,* 1226–1232.

212. Mizutani, K., Ikeda, K., & Yamori, Y. (2000). Resveratrol inhibits AGEs-induced proliferation and collagen activity in vascular smooth muscle cells from stroke-prone spontaneously hypertensive rats. *Biochemical and Biophysical Research Communications, 274,* 61–67.

213. Halliwell, B. (1992). Reactive oxygen species and central nervous systems. *Journal of Neurochemistry, 59,* 1609–1623.

214. Floyd, R. A. (1999). Antioxidants, oxidative stress and degenerative neurological disorders. *Proceedings of the Society for Experimental Biology and Medicine, 222,* 236–245.

215. Barja, G. (2004). Free radicals and aging. *Trends Neuroscience, 23,* 209–216.

216. Smith, K. J., Kapoor, R., & Felts, P. A. (1999). Demyelination: The role of reactive oxygen and nitrogen species. *Brain Pathologica, 9,* 69–92.

217. Commenges, D., Scotet, V., Renaud, S., Jacqmin-Badda, H., Barberger-Gateau, P., & Dartigues, J. F. (2000). Intake of flavonoids and risk of dementia. *European Journal of Epidemiology, 16,* 357–363.

218. Schmitt-Schillig, S., Schaffer, S., Weber, C. C., Eckert, G. P., & Müller, W. E. (2005). Flavonoids and the aging brain. *Journal of Physiology and Pharmacology, 56,* 23–36.

219. Peng, H. W., Cheng, F. C., Huang, Y. T., Chen, C. F., & Tsai, T. H. (1998). Determination of naringinin and its glucoronide conjugate in rat plasma and brain tissue by high-performance liquid chromatography. *Journal of Chromatography, 714,* 369–374.

220. Tsai, T. H, & Chen, Y. F. (2000). Determination of unbound hesperetin in rat blood and brain by microdialysis coupled to microbore liquid chromatography. *Journal of Food Drug Analysis, 8,* 331–336.

221. Abd El Mohsen, M. M., Kuhnle, G., & Rechner, A. R. (2002). Uptake and metabolism of epicatechin and its access to the brain after oral ingestion. *Free Radical Biology and Medicine, 33,* 1693–1702.

222. Youdim, K. A., Dobbie, M. S., & Kuhnle, G. (2003). Interaction between flavonoids and the blood-brain barrier: In vitro studies. *Journal of Neurochemistry, 85,* 180–192.

223. Youdim, K. A., Shukitt-Hale, B., & Joseph, J. A. (2004). Flavonoids and the brain: Interactions at the blood-brain barrier and their physiological effects on the central nervous systems. *Free Radical Biology & Medicine, 37,* 1683–1693.

C. ACTIONS OF INDIVIDUAL FRUITS IN DISEASE AND CANCER PREVENTION AND TREATMENT

224. Menken, M., Munsat, T. L., & Toole, J. F. (2000). The global burden of disease study: Implications for neurology. *Archives of Neurology, 57,* 418–420.

225. Agorogiannis, E. I., Agorogiannis, G. I., Papadimitriou, A., & Hadjigeorgiou, G. M. (2004). Protein misfolding in neurodegenerative diseases. *Neuropathology and Applied Neurobiology, 30,* 215–224.

226. Elbaz, A., Dufouil, C., & Alpérovitch, A. (2007). Interaction between genes and environment in neurodegenerative diseases. *Comptes Rendus Biologies, 330,* 318–328.

227. Ritchie, K., & Lovestone, S. (2002). The dementias. *Lancet, 360,* 1759–1766.

228. Pinder, R. M., & Sandler, M. (2004). Alcohol, wine and mental health: Focus on dementia and stoke. *Journal of Psychopharmacology, 18,* 449–456.

229. Scarmeas, N., Luchsinger, J. A., Mayeux, R., & Stern, Y. (2007). Mediterranean diet and Alzheimer disease mortality. *Neurology, 69,* 1084–1093.

230. Pasinetti, G. M. (2008). Can diet modifications play a preventive role in the onset of Alzheimer's diseases? *Aging Health, 4,* 1.

231. Lindsay, J., Laurin, D., Verreault, R., Hébert, R., Helliwell, B., Hill, G. B., & McDowell, I. (2002). Risk factors for Alzheimer's disease: A prospective analysis from the Canadian Study of Health and Aging. *American Journal of Epidemiology, 156,* 445–453.

232. Dai, Q., Borenstein, A. R., Wu, Y., Jackson, J. C., & Larson, E. B. (2006). Fruit and vegetable juices and Alzheimer's disease: The *Kame* Project. *The American Journal of Medicine, 119,* 751–759.

233. Orgogozo, J. F., Dartigues, J. F., Lafont, S., Letenneur, L., Commenges, D., Salamon, R., Renaud, S., & Breteler, M. B. (1997). Wine consumption and dementia in the elderly: A prospective community study in the Bordeaux area. *Revue Neurologique, 153,* 185–192.

234. Marambaud, P., Zhao, H., & Davies, P. (2005). Resveratrol promotes clearance of Alzheimer's disease amyloid-β peptides. *The Journal of Biological Chemistry, 280,* 37377–37382.

235. Bastianetto, S., Zheng, W.-H., & Quirion, R. (2000). Neuroprotective abilities of resveratrol and other red wine constituents against nitric oxide toxicity in cultured hippocampal neurons. *British Journal of Pharmacology, 131,* 711–720.

236. Wang, J., Ho, L., Zhao, W., Seror, I., Humala, N., Dickstein, D. L., Thiyagarajan, M., Percival, S. S., Talcott, S. T, & Pasinetti, G. M. (2006). Moderate consumption of Cabernet Sauvignon attenuates Aβ neuropathology in a mouse model of Alzheimer's diseases. *The FASEB Journal: Official Publication of the Federation of American Societies for Experimental Biology, 20,* 2313–2320.

237. Wang, J., Ho, L., Zhao, W., Ono, K., Rosensweig, C., Chen, L., Humala, N., Teplow, D. B., & Pasinetti, G. M. (2008). Grape-derived polyphenolics prevent Aβ oligomerization and attenuate cognitive deterioration in a mouse model of Alzheimer's disease. *The European Journal of Neuroscience, 28,* 6388–6392.

238. Porat, Y., Abramowitz, A., & Gazit, E. (2006). Inhibition of amyloid fibril formation by polyphenols: Structural similarity and aromatic interactions as a common inhibition mechanism. *Chemical Biology and Drug Design, 67,* 27–37.

239. Conte, A., Pellegrini, S., & Tagliazucchi, D. (2003). Synergistic protection of PC12 cells from beta-amyloid toxicity by reveratrol and catechin. *Brain Research Bulletin, 62,* 29–38.

240. Ono, K., Yoshiike, Y., Takashima, A., Hasegawa, K., Naiki, H., & Yamada, M (2003). Potent anti-amyloidogenic and fibril-destabilizing effects of polyphenols *in vitro*: Implications for the prevention and therapeutics of Alzheimer's disease. *Journal of Neurochemistry, 87,* 172–181.

241. Savaskan, E., Olivieri, G., Meier, E., Seifritz, E., Wirz-Justice, A., & Muller-Spahn, F. (2003). Red wine ingredient resveratrol protects from beta-amyloid neurotoxicity. *Gerontology, 49,* 380–383.

242. Rivière, C., Richard, T., Vitrac, X., Mérillon, J.-M., Valls, J., & Monti, J.-P. (2008). New polyphenols active on β-amyloid aggregation. *Bioorganic and Medicinal Chemistry Letters, 18,* 828–831.

243. Patel, A. K., Rogers, J. T., & Huang, X. (2008). Flavanols, mild cognitive impairment and Alzheimer's dementia. *International Journal of Clinical and Experimental Medicine, 1,* 181–191.

244. Peters, R., Peters, J., Warner, J., Beckett, N., & Bulpitt, C. (2008). Alcohol, dementia and cognitive decline in the elderly: A systematic review. *Age Ageing, 37,* 505–512.

245. Hunot, S., Boissiere, F., Faucheux, B., Brugg, B., Mouatt-Prigent, A., Agid, Y., & Hirsch, E. C. (1996). Nitric oxide synthase and neuronal vulnerability in Parkinson's disease. *Neuroscience, 72,* 355–363.

246. Jenner, P. (2003). Oxidative stress in Parkinson's disease. *Annals of Neurology, 53,* S26–S36.

247. Blanchet, J., Longpré, F., Bureau, G., Morisette, M., DiPaolo, T., Bronchti, G., & Martinoli, M.-G. (2008). Resveratrol, a red wine polyphenol, protects dopaminergic neurons in MPTP-treated mice. *Progress in Neuro-Psychopharmacology and Biological Psychiatry, 32,* 1243–1250.

248. Czlonkowski, A., Mirowska-Guzel, D., & Czlonkowska, A. (2007). Therapy of stroke – from

C. ACTIONS OF INDIVIDUAL FRUITS IN DISEASE AND CANCER PREVENTION AND TREATMENT

experimental studies to clinical trials. *Pharmacological Reports, 59,* 123–128.

249. Truelsen, T., Grønbæk, M., Schnohr, P., & Boysen, G. (1998). Intake of beer, wine and spirits and risk of stroke. The Copenhagen City Heart Study. *Stroke, 29,* 2467–2472.

250. Djoussé, L., Curtis Ellison, R., Beiser, S., Scaramucci, A., D'Agostino, R. B., & Wolf, P. A. (2002). Alcohol consumption and risk of ischemic stroke. The Framingham Study. *Stroke, 33,* 907–912.

251. Malarcher, A. M., Giles, W. H., Croft, J. B., Wozniak, M. A., Wityk, R. J., Stolley, P. D., Stern, B. J., Sloan, M. A., Sherwin, R., Price, T. R., Macko, R. F., Johnson, C. J., Earley, C. J., Buchholz, D. W., & Kittner, S. J. (2001). Alcohol intake, type of beverage and risk of cerebral infarction in young women. *Stroke, 32,* 77–83.

252. Howitz, K. T., Bitterman, K. J., Cohen, H. Y., Lamming, D. W., Lavu, S., Wood, J. G., Zipkin, R. E., Chung, P., Kisielewski, A., Zhang, L.-L., Scherer, B., & Sinclair, D. A. (2003). Small molecule activators of sirtuins extend *Saccharomyces cerevisiae* lifespan. *Nature, 425,* 191–196.

253. Wood, J. G., Rogina, B., Lavu, S., Howitz, K., Helfand, S. L., Tartar, M., & Sinclair, D. A. (2004). Sirtuin activators mimic caloric restriction and delay ageing in metazoans. *Nature, 430,* 686–689.

254. Bordone, L., & Guarente, L. (2005). Calorie restriction, SIRT1 and metabolism: Understanding longevity. *Nature Reviews. Molecular Cell Biology, 6,* 298–305.

255. Chen, D., & Guarente, L. (2006). SIR2: A potential target for calorie restriction mimetics. *Trends in Molecular Medicine, 13,* 64–71.

256. Haigis, M. C., & Guarente, L. P. (2006). Mammalian sirtuins – emerging roles in physiology, aging and calorie restriction. *Genes Development, 20,* 2913–2921.

257. Outeiro, T. F., Marques, O., & Kazantsev, A. (2008). Therapeutic role of sirtuins in neurodegenerative disease. *Biochimica et Biophysica Acta, 1782,* 363–369.

258. Michan, S., & Sinclair, D. (2007). Sirtuins in mammals: Insights into their biological function. *The Biochemical Journal, 404,* 1–13.

259. Yamamoto, H., Schoonjans, K., & Auwerx, J. (2007). Sirtuin functions in health and disease. *Molecular Endocrinology, 21,* 1745–1755.

260. Porcu, M., & Chiarugi, A. (2005). The emerging therapeutic potential of sirtuin-interacting drugs: From cell death to lifespan extension. *Trends in Pharmacological Sciences, 26,* 94–103.

261. Anekonda, T. S. (2006). Resveratrol – A boon for Alzheimer's disease? *Brain Research Review, 52,* 316–326.

262. Lavu, S., Boss, O., Elliott, P. J., & Lambert, P. D. (2008). Sirtuins – Novel therapeutic targets to treat age-associated diseases. *Nature Reviews. Drug Discovery, 7,* 841–853.

263. Petersen, P. E., Buorgeois, D., Ogawa, H., Estupinan-Day, S., & Ndiaye, C (2005). The global burden of oral diseases and risks to oral health. *Bulletin WHO, 83,* 661–669.

264. Jane-Salas, E., Chimenos-Kustner, E., Lopez-Lopez, J., & Rosello-Llabres, X. (2003). Importance of diet in the prevention of oral cancer. *Medicina Oral, 8,* 260–268.

265. Davies, L., & Welch, H. G. (2006). Epidemiology of head and neck cancer in the United States. *Otolaryngology and Head and Neck Surgery, 135,* 451–457.

266. Pavia, M., Pileggi, C., Nobile, C. G., & Angelillo, I. F. (2006). Association between fruit and vegetable consumption and oral cancer: A meta-analysis of observational studies. *The American Journal of Clinical Nutrition, 83,* 1126–1134.

267. King, M., Chatelain, K., Farris, D., Jensen, D., Pickup, J., Swapp, A., O'Malley, S., & Kingsley, K. (2007). Oral squamous cell carcinoma proliferative phenotype is modulated by proanthocyanidins: A potential prevention and treatment alternative for oral cancer. *BMC complementary and Alternative Medicine, 7,* 22.

268. Chatelain, K., Phippen, S., McCabe, J., Teeters, C. A., O'Malley, S., & Kingsley, K. (2008). Cranberry and grape sees extracts inhibit the proliferative phenotype of oral squamous cell carcinomas. *Evidence-based Complementary and Alternative Medicine,* doi: 10.1093/ecam/nen047.

269. Zhang, Q., Tang, X., Lu, Q. Y., Zhang, Z. F., Brown, J., & Le, A. D. (2005). Resveratrol inhibits hypoxia-inducible accumulation of hypoxia-inducible factor-1α and VGEF expression in human tongue squamous cell carcinoma and hepatoma cells. *Molecular Cancer Therapeutics, 4,* 1465–1474.

270. Pöschl, G., & Seitz, H. K. (2004). Alcohol and cancer. *Alcohol and Alcoholism, 39,* 155–165.

271. Mehrotra, R., & Yadav, S. (2006). Oral squamous cell carcinoma: Etiology, pathogenesis and prognostic value of genomic alterations. *Indian Journal of Cancer, 43,* 60–66.

272. Castellsagué, X., Quintana, M. J., Martínez, M. C., Nieto, A., Sánchez, M. J., Juan, A., Monner, A., Carrera, M., Agudo, A., Quer, M., Muñoz, N., Herrero, R., Franceschi, S., & Bosch, F. X. (2004). The role of type of tobacco and type of alcoholic beverage in oral carcinogenesis. *International Journal of Cancer, 108,* 741–749.

C. ACTIONS OF INDIVIDUAL FRUITS IN DISEASE AND CANCER PREVENTION AND TREATMENT

273. Petti, S., & Scully, C. (2005). Oral cancer: The association between nation-based alcohol-drinking profiles and oral cancer mortality. *Oral Oncology, 41*, 828–834.

274. Güneri, P., Çankaya, H., Yavuzer, A., Güneri, E. A., Erisen, L., Özkul, D., Nehir, El.S., Karakaya, S., Arican, A., & Boyacioglu, H. (2005). Primary oral cancer in a Turkish population sample: Association with sociodemographic features, smoking, alcohol, diet and dentition. *Oral Oncology, 41*, 1005–1012.

275. Bosetti, C., Gallus, S., Trichopoulou, A., Talamini, R., Franceschi, S., Negri, E., & La Vecchia, C. (2003). Influence of the Mediterranean diet on the risk of cancers of the upper aerodigestive tract. *Cancer Epidemiology, Biomarkers and Prevention, 12*, 1091–1094.

276. Rossi, M., Garavello, W., Salamini, R., Negri, E., Borsetti, C., Dal Maso, L., Lagiou, P., Tafani, A., Polesel, J., Barman, L., Ramazzotti, V., Franceschi, S., & La Vecchia, C. (2007). Flavonoids and the risk of oral and pharyngeal cancer: A case-control study from Italy. *Biomarkers Prev Cancer Epidemiology, Biomarkers and Prevention, 16*, 1621–1625.

277. Maserejian, N. N., Joshipura, K. J., Rosner, B. A., Giovannucci, E., & Zavras, A. I. (2006). Prospective study of alcohol consumption and risk of oral premalignant lesions in men. *Cancer Epidemiology, Biomarkers and Prevention, 15*, 774–781.

278. Kinane, D. F. (2000). Causation and pathogenesis of periodontal disease. *Periodontology, 25*, 8–20.

279. Houde, V., Grenier, D., & Chandad, F. (2006). Protective effects of grape seed proanthocyanidins against oxidative stress induced by lipopolysaccharides of periodontopathogens. *Journal of Periodontology, 77*, 1371–1379.

280. Rivero-Cruz, J. F., Zhu, M., Kinghorn, A. D., & Wu, C. D. (2008). Antimicrobial constituents of Thompson seedless raisins (*Vitis vinifera*) against selected oral pathogens. *Phytochemistry Letter, 1*, 151–154.

281. Anisimov, V. N., Popovich, I. G., Zabezhinski, M. A., Anisimov, S. V., Vesnushkin, G. M, & Vinogradova, I. A. (2006). Melatonin as antioxidant, geroprotector and anticarcinogen. *Biochimica et Biophysica Acta, 1757*, 573–589.

282. Reiter, R. J., Tan, D.-X., Manchester, L. C., Pilar Terron, M., Flores, L. J., & Koppisepi, S. (2007). Medical implications of melatonin: Receptor-mediated and receptor-independent actions. *Advances in Medical Sciences, 52*, 11–28.

283. Reiter, R. J., Tan, D.-X., Pilar Terron, M., Flores, L. J., & Czarnocki, Z. (2007). Melatonin and its metabolites: New findings regarding their production and their radical scavenging action. *Acta Biochimica Polonica, 54*, 1–9.

284. Hardeland, R., & Poeggeler, B. (2003). Non-vertebrate melatonin. *Journal of Pineal Research, 34*, 233–241.

285. Kolár, J., & Macháčková, I. (2005). Melatonin in higher plants: Occurrence and possible functions. *Journal of Pineal Research, 39*, 333–341.

286. Reiter, R. J. (2004). Mechanisms of cancer inhibition by melatonin. *Journal of Pineal Research, 37*, 213–214.

287. Karasek, M. (2007). Does melatonin play a role in aging processes? *Journal of Physiology and Pharmacology, 58*, 105–113.

288. Szczepanik, M. (2007). Melatonin and its influence on immune system. *Journal of Physiology and Pharmacology, 58*, 115–124.

289. Pappolla, M., Bozner, P., Soto, C., Shao, H., Robakis, N. K., Zagorski, M., Frangione, B., & Ghiso, J. (1998). Inhibition of Alzheimer beta-fibrillogenesis by melatonin. *The Journal of Biological Chemistry, 273*, 7185–7188.

290. Pappolla, M. A., Chyan, Y., Poeggeler, B., Frangione, B., Wilson, G., Ghiso, J., & Reiter, R. J. (2000). An assessment of the antioxidant and the antiamyloidogenic properties of melatonin: Implications for Alzheimer's disease. *Journal of Neural Transmission, 107*, 203–231.

291. Marambaud, P., Zhao, H., & Davies, P. (2005). Resveratrol promotes clearance of Alzheimer's disease amyloid-β peptides. *The Journal of Biological Chemistry, 280*, 37377–37382.

292. Acuña-Castroviejo, D., Coto-Montes, A., Gaia Monti, M., Ortiz, G. G., & Reiter, R. J. (1997). Melatonin is protective against MPTP-induced striatal and hippocampal lesions. *Life Sciences, 60*, PL23–PL29.

293. Ortiz, G. G., Crespo-López, M. E., Moran-Moguel, C., Garcia, J. J., Reiter, R. J., & Acuña-Castroviejo, D. (2001). Protective role of melatonin against MPTP-induced mouse brain cell DNA fragmentation and apoptosis *in vivo*. *Neuroendocrinology Letter, 22*, 101–108.

294. Antolin, I., Mayo, J. C., Sainz, R. M., del Brio, M. L., Herrera, F., Martin, V., & Rodriquez, C. (2002). Protective effect of melatonin in a chronic experimental model of Parkinson's disease. *Brain Research, 943*, 163–173.

295. Paulis, L., & Šimko, F. (2007). Blood pressure modulation and cardiovascular protection by melatonin: Potential mechanisms behind. *Physiology Research, 56*, 671–684.

296. Carrillo-Vico, A., Calvo, J. R., Abreu, P., Lardone, P. J., Garcia-Maurino, S., Reiter, R. J., & Guerrero, J. M. (2004). Evidence of melatonin synthesis by human lymphocytes and its physiological significance: Possible role as intracrine, autocrine, and/or paracrine substance. *The FASEB Journal: Official*

C. ACTIONS OF INDIVIDUAL FRUITS IN DISEASE AND CANCER PREVENTION AND TREATMENT

Publication of the Federation of American Societies for Experimental Biology, 18, 537–539.

297. Mayo, J. C., Sainz, R. M., Tan, D.-X., Hardeland, R., Leon, J., Rodriguez, C., & Reiter, R. J. (2005). Anti-inflammatory actions of melatonin and its metabolites, N1-acetyl-N2-formyl-5-methoxykynuramine (AFMK) and N1-acetyl-5-methoxykynuramine (AMK), in macrophages. *Journal of Neuroimmunology, 165*, 139–149.

298. Hattori, A., Migitaka, H., Iigo, M., Itoh, M., Yamamoto, K., Ohtani-Kaneko, R., Hara, M., Suzuki, T., & Reiter, R. J. (1995). Identification of melatonin in plants and its effects on melatonin levels and binding to melatonin receptors in vertebrates. *Biochemistry and Molecular Biology International, 35*, 627–634.

299. Guerrero, J. M., Martínez-Cruz, F., & Elorza, F. L. (2008). Significant amount of melatonin in red wine: Its consumption increases blood melatonin levels in humans. *Food and Chemical*, doi: 10.1016/j.foodchem.2008.02.007.

300. Manach, C., Scalbert, A., Morand, C., Rémésy, C., & Jimenez, L. (2004). Polyphenols – Food sources and bioavailability. *The American Journal of Clinical Nutrition, 79*, 727–747.

301. Manach, C., Williamson, G., Morand, C., Scalbert, A., & Remesy, C. (2005). Bioavailability and bioefficacy of polyphenols in humans. I. Review of 97 bioavailability studies. *The American Journal of Clinical Nutrition, 81*, 230S–242S.

302. Williamson, G., & Manach, C. (2005). Bioavailability and bioefficacy of polyphenols in humans. II. Review of 93 intervention studies. *The American Journal of Clinical Nutrition, 81*, 243S–255S.

C. ACTIONS OF INDIVIDUAL FRUITS IN DISEASE AND CANCER PREVENTION AND TREATMENT

39

Soursop (*Annona muricata* L.): Composition, Nutritional Value, Medicinal Uses, and Toxicology

Neela Badrie[1] and Alexander G. Schauss[2]

[1]Department of Food Production, Faculty of Science and Agriculture, University of the West Indies, St. Augustine, Republic of Trinidad and Tobago, West Indies
[2]Natural and Medicinal Products Research, AIBMR Life Sciences, Puyallup, Washington, USA

1. INTRODUCTION

Annona is a genus of tropical fruit trees belonging to the family *Annonaceae*, of which there are approximately 119 species. Seven species and one hybrid are grown for domestic/commercial use [1]. *Annona muricata* L. is known as soursop in English-speaking countries and is referred to by numerous common names (Table 39.1) [2–6]. After the arrival of the Spanish in the Americas, the Annona species were distributed throughout the tropics [7]. Soursop trees are widespread in the tropics and frost-free subtropics of the world [8,9] and are found in the West Indies, North and South America, lowlands of Africa, Pacific islands, and Southeast Asia. The soursop fruit and other parts of the tree are considered to be underutilized. Information on the composition, nutritional value, medicinal uses, and toxicology of the soursop fruit and plant is limited and scattered.

2. BOTANY AND HORTICULTURE

The soursop is an upright, low-branching tree reaching 8 to 10 meters [7,10]. The tree has green, glossy evergreen leaves [4]. The flowers appear anywhere on the trunk or any branch [11]. It is usually grown from seeds [12] which can be stored for several months before planting. Germination of seeds usually takes 3 weeks, but under suboptimal conditions can be delayed for up to 2–3 months. Alternatively, propagation of the *Annona* species is achieved by cuttings for rapid multiplication of new genotypes and for the elimination of viral and disease infection [13]. With the exception of a few cultivars, clonal propagation of the *Annona* species by cutting or

TABLE 39.1   Some Common Names for Soursop by Country

Country/Language	Common Names
Argentina	Anona de puntitas; anona de broquel
Bolivia	Sinini
Brazil	Araticum, araticum-do-grande; coraç ão-da-rainha; condessa; graviola; jaca-do-pará; jaca-de-pobre; fruta-do-conde
Cameroon	Soursop
Caribbean (English-speaking)	graviola; Jamaica soursop, prickly custard apple, soursop
Cambodia	Guayabano
Dominican Republic	Guanábana
El Salvador	Guanaba dulce, guanabana azucaron
French-speaking countries	Corossolier; cacheimantier; épineux; corossol; corossol épineux; cachiman épineux; grand corossol; sapotille
French Polynesia	Tapotapo papaa
Germany	Sauersak
Guam	Laguana; laguaná; laguanaba; labuanaha
Guatemala	Huanábano; huanabana, huanaba
Indian	Mamphal
Indonesia	Sirsak; nangka belanda; nangka seberang; zuurzak
Jamaica	Jamaica soursop
Laos	Khan thalot
Malaysia	Durian belanda; durian blanda, durian benggala, durian maki; durian makkah; seri kaya belanda
Marshall islands	Jojaab
Mexico	Catuche; catucho; cabeza de negro; catuch; catucho; guanábana; guanábano; polvox; tak-ob; caduts-at; xunápill; llama de tehuantepec; zopote de viejas (Rep. Mexico)
Netherlands Antilles	Zunrzak, sorsaka
North Vietnam	Corossol; grand corossol; corossol epineux; cachiman epineux
Papua New Guinea	Saua sap
Peru	Guanábano; guanábana
Philippines	Babaná; guyabano; gwabana
Portuguese (Brazil)	Graviola; araticum do grande; jaca do para
Samoa	Sasalapa
Sierra Leone	Soursap; soursapi, soursop
Surinam	Zunrzak
Thailand	Thu-rian-khack, thurian-thet, thurian-khaek
Tonga	Apele initia
United States of America	Soursop (Hawaii and Puerto Rico); languana (Guam)
Venezuela	Catoche; catuche
Vietnam	Mang câù-xiûm
Yoruba	Ebo

Source: references 2–6.

air layering has not been very successful [14]. Vegetative propagation of rootstocks or cultivars of known agronomic potential could eliminate tree to tree variability in growth and productivity [15]. However, the seedling rootstocks are highly variable in vigor and disease resistance and consequently scion growth and productivity are also variable. They are considered as minor tropical fruits due to strict environmental requirements for tree planting and the short post-harvest life of their fruits [16].

2.1 Fruit Description

The soursop tree (Fig. 39.1, 39.2, 39.3) produces dark green, spiny aggregate fruits made up of berries fused together with associated flower parts [5]. The oval or heart-shaped and frequently irregular lopsided composite soursop fruit is derived from the fusion of many fruitlets and can weigh more than 4 kg [3,11,17,18]. The fruit pulp consists of white fibrous juicy segments surrounding an elongated receptacle [19]. In each fertile segment there is a single oval, smooth hard, black seed {1/2}–{3/4} in (1.25–2 cm) long [3]. A fruit may contain as few as 5 or up to 200 or more seeds

FIGURE 39.1 Soursop tree. Photograph by N. Badrie.

FIGURE 39.2 Soursop tree with a young soursop fruit. Photograph taken by N. Badrie.

FIGURE 39.3 Soursop tree with maturing fruit. Photograph taken by A. Boodoo.

[3,19]. The reticulated leathery looking skin has short spines [19]. Its inner surface is cream-colored and granular and separates easily from the mass of white, fibrous juicy segments which surround the central soft pithy core [3]. In Puerto Rico, the seedling soursops are roughly divided into three general classifications: sweet, subacid, and acid. These are subdivided as round, heart-shaped, oblong, or angular and finally classed according to flesh consistency which varies from soft and juicy to firm and comparatively dry [3].

2.2 Physiological Changes

By measurement of the respiration rate, soursop was classified as climacteric [20]. Climacteric fruits are often harvested in an immature state and ripened post-harvest. It is classified as a multiple climacteric fruit owing to the berries that make up a single fruit being of different maturities and thus ripening at different times [20]. Mature fruit produced a biphasic respiratory climacteric, with CO_2 production reaching 100 mL/kg/h and then 350 mL/kg/h at 25–30°C. Peak ethylene production (250–350 mL/kg/h) occurred between the two respiratory maxima. The respiratory climacteric of the harvested immature fruit tends to be higher and later than that of mature fruit [21].

As the fruit reaches maturity, there is a slight paling as the green chlorophyll peel color changes to a slight yellowing green [21–24] which may reflect a loss of chlorophyll with carotenoids contributing more to overall peel color. During the late ripening stage the peel becomes dark brown [25], due possibly to chloroplast breakdown releasing polyphenol oxidases causing oxidation and polymerization of phenols. Maturity can be detected as the skin becomes smooth [17] as the density of the spurs on the fruit surface reaches a minimum value (6–7 per $12\,cm^2$) [21]. Also, the mature fruit is soft to the touch [3]. Within 2 days of harvesting of soursop fruit, changes in the molecular weight of the starch have been recorded [26] while at 6 days from harvest only a small fraction of the starch remains [4]. Concomitant with the decline of ethanol fraction and starch is an increase in ascorbic acid, total soluble solids, total soluble sugars, and total titratable acidity [25]. Total soluble solids increase to about 16°Brix. The total sugars begin increasing 1–2 days from harvest at the same time as the increase in respiration. Ethanol-soluble sugars increase rapidly as ripening progresses [25,26]. Glucose and fructose reach a maximum concentration 5 days after harvest.

There is a marked decrease in pH from 5.5 to 3.7 over the 3 days of ripening [25]. The decrease in pH is correlated by an increase in titratable acidity [25]. The malic acid content begins to increase within 2 days of harvest and there is a 3-fold increase during ripening [4]. This increase in malic acid contributes to the acidic flavor of the fruit and the decrease in pulp pH [24].

Soursop volatile production varies during ripening. Prior to the rise in respiration rate, there are few volatiles [25]. Volatile production begins to increase drastically with ethylene production within 2 days after the increase in the rate of respiration [25]. The volatiles and ethylene peak about 4 days after harvest and mark the optimum eating stage [25]. Total ethanol-soluble phenol first increases (10%), then declines to 50% of preclimacteric levels with the greatest decline occurring after the climactic peak [24]. The decline in phenols probably leads to a loss of astringency during ripening [24] and leads to a bland flavor of the slightly overripe fruit [4]. Very overripe fruit has an off-flavor due to low phenols [24], lower organic acids [25], and some fermentation [4].

The physical, chemical, and biochemical changes related to softening during maturation of Crioula soursop fruits (L.) were studied in Brazil [27]. The soursop fruits were harvested when physiologically mature and were stored at $26.3\pm0.6°C$ and $88\pm12\%$ relative humidity. During days 1, 2, 3, 4, and 5, no significant variation in the soluble pectin content was found. Weight loss reached 5% in the fifth day, without causing fruit shriveling. Reductions in starch and in total pectin contents occurred during the period of greatest enzymatic activity, respectively, of amylase and of polygalacturonase, and cell wall β-galactosidase. The most significant changes in the contents of starch and total pectin, in pectin solubilization, and in the activity of the enzymes amylase, pectin methylesterase, polygalacturonase, and cell wall β-galactosidase occurred from the second to the fourth day after harvest.

2.3 Post-harvest Issues

A preliminary evaluation of the post-harvest problems associated with tropical fruits in Costa Rica indicated that the post-harvest loss was 75.8% [28]. The main problems were due to deficient field practices and lack of knowledge on the fruit quality parameters by fruit growers. Also, inadequate handling during the marketing process increases the loss.

Annona fruits are climacteric with a short shelf-life [16]. The soursop fruits are harvested when full grown and still firm but slightly yellow-green. If they are allowed to soften on the tree, they will fall and crush. Ripe fruits are very soft and easily bruised [2] which shortens the post-harvest life. Figure 38.4 shows a ripe bruised soursop fruit. The immature fruits when ripened off the tree do not develop full flavor and aroma, thus ripening at room temperature is recommended [4,29]. Soursop fruits are very susceptible to chilling temperatures. The symptoms of chilling injury in soursop are skin darkening, failure to ripen, pulp discoloration, poor flavor and aroma, and increase in rot [4]. Fully ripe soursop fruits can be held 2–3 days longer by refrigerator [17]. Storage of soursop fruits at 10°C for 1 day leads to noticeable loss of flavor and aroma [30]. The trends in respiration and ethylene production relate to changes in composition during ripening. A low level of ethylene production occurs during the preclimacteric post-harvest stage [4]. Ethylene production increases about the fourth day after harvest from 0.2 to 0.9 μL/kg/h, 48 hours after respiratory climacteric is initiated [4]. Ethylene production peaks (290 μL/kg/h) at about the same time as respiration rate reaches a plateau of 108 μL/kg/h at day 6 [24]. Studies of the

FIGURE 39.4 A ripe bruised soursop fruit. Photograph taken by N. Badrie.

ripening process in Hawaii have determined that the optimum stage for eating is 5–6 days after harvest, at the peak of ethylene production. Thereafter, the flavor is less pronounced and a faint off-odor develops. All damaged soursop fruits must be removed as they might become sources of ethylene gas and will increase the rate of ripening. Also, all immature soursop fruits should be removed as their high respiration rate will affect the rate of ripening of mature soursop fruits. The short harvest period and rapid fruit ripening have stimulated the introduction of cultivars having different maturity periods in order to even out market supply [16].

2.4 Pests and Diseases

Production problems in Brazil have included low fruit set due to poor pollination and adverse climatic conditions and the attack of several devastating pests and diseases [31]. Soursop plantations in north east Brazil have been attacked by several pests, especially the fruit borer, *Cerconota anonella* Sepp, the seed borer, *Bephratelloides maculicollis* Bondar, the stem borer, *Cratosomus* spp., the leafminer, *Prinomerus anonicola* Bondar, and some species of Membracidae, Coccidae, Diaspididae, and Aphididae [32–34].

The most important pests were the fruit borer *Cerconota anonella* and the 'irapua' bee *Trigona spinips*, the maximum fruit damage being 62 and 53%, respectively [35]. The seed borer *Bephratelloides maculicollis* caused the least fruit damage. The damage from different species of Membracidae, Coccidae, Diaspididae and Aphididae varied from 16 to 20%. The leafminer *Prinomerus anonicola* was continuously present during the 3-year period with plant damage varying from 17 to 22%. Other less important pests were stem borer *Cratosomus* sp., ants, stink bugs, and defoliators that together caused on average 19–32% plant damage.

The principal pest of the soursop in the West Indies is the mealy bug (*Maconellicoccus hirsutus*) which may occur in masses on the soursop fruits, where the tree is often infested with scale insects. Sometimes it may be infected by a lacewing bug. The soursop fruit is subject to attack by soursop fruit flies, and red spiders are a problem in dry climates. In Trinidad, the damage done to soursop flowers by *Thecla ortygnus* seriously limits the cultivation of the fruit [3]. Most of the flowers and young fruits of the *Annona* species are susceptible to anthracnose (*Colletotrichum gloeosporioides* Penz.) and its incidence increases in the warm humid environments [3,36–40]. Black canker caused by *Phomopsis anonacearum* occurs in the wet season, purple spots occurring at or near the distal end [38]. As the lesions enlarge the surface becomes hard and cracked. Minute spore-containing black bodies (pycnidia) develop. Botryodiplodia rot is caused by *Botryodiplodia theobromae*, first purple then later pimpled with black pycnidia. The flesh is rapidly invaded, becoming brown and corky [41].

3. TYPES OF ENZYMES

The enzymes pectinase, catalase, and peroxidase have been detected in soursop pulp [42,43]. Pectinesterase (PE) is one of the most heat-resistant enzymes present in soursop fruit [44] which could lead to gelation and precipitation of pectin in puree and juice with subsequent loss of cloud. Two forms of pectinesterase were purified using the techniques of ammonium sulfate fractionation, ion-exchange chromatography, and gel filtration [43]. PE I had a specific activity of approximately 4 units/mg (43-fold), that of PE II was 6.4 units/mg (229-fold). These pectinesterases (PE I and PE II) had approximate molecular weights of 29,100 and 24,100, respectively, as estimated by gel filtration, and 31,000 and 28,000, respectively, as estimated by sodium dodecyl sulfate polyacrylamide electrophoresis. The optimum temperature for enzyme activity was shown to be 60°C for both PE I and PE II. The thermostability of the two forms of PE showed that PE 1 was more stable than PE11 at pH 7.5 [45]. The changes in temperature required to increase the inactivation rate 10-fold (Z value) were calculated at 8.5 and 8.6% for PE I and PE II, respectively. Both enzymes also tested positive for their ability to destabilize soursop juice cloud at 5 and 30°C. Cloud destabilization by PE I occurred the fastest (large decrease in absorbance 660 nm) in the natural juice at 30°C.

Optimum pasteurization for inactivation of pectinesterase activity and vitamin C retention was at 78.8°C for 69 s at pH 3.7 [46]. Also, a sensory panel found that the taste, smell, and overall acceptability of pasteurized soursop puree were not significantly different from those of the control. However, the color and appearance of the pasteurized puree were improved. In another study, the minimum temperature for the inactivation of pectinase was 60°C for 1.5 min; inactivation of catalase was accomplished at 85°C for 2 min or 90°C for 1 min [42]. Sánchez-Nieva and colleagues [47] found that during processing of soursop puree the minimum temperature required to inactivate the peroxidase was 65.5°C.

Pectinesterase (pectinase) enzyme was found to increase the yield of soursop juice extracted by 41% [48]. Optimum yield of the soursop juice was with 0.075% (Pectinex Ultra SPK) at 2 h incubation time. Sensory results indicated that the enzyme-extracted juice had better flavor characteristics than a commercial juice. Also, a significant increase of acidity and °Brix level resulted with 0.05% pectinase at 1 h incubation time.

Amylase was detected during ripening and its activity increases 18-fold as the fruit ethylene increases [25]. A similar increase was found for two wall-degrading enzymes: polygalacturonase (43-fold) and cellulose (7-fold)

[4]. This suggests that degradative events provide the substrates for the synthesis of ethanol-soluble sugars, organic acids, and volatiles [49,50].

Polyphenol oxidase (PPO) is found in fruits and is the enzyme responsible for them turning brown. Soursop constitutes a rich source of PPO [51]. This enzyme (PPO) catalyzes the ortho-hydroxylation of phenols and the oxidation of catechols to orthoquinone in the presence of O_2 [51]. High correlation was found between PPO activity and phenolic compounds. It is the rapid polymerization of O-quinones that is the cause of enzymic browning of soursop during maturation. The values for polyphenol oxidase (PPO), total phenols and ascorbic acid, browning intensity, and pH in soursop pulp decreased during maturation. The least browning intensity has been found in fully ripe Crioula soursop fruits [52]. Optimum pH was 7.5 (fully developed immature, mature unripe, and ripe fruits) or 7.0 (fully ripe fruits).

The optimum pH and temperature for PPO activity from ripe soursop were 7.5 and 32°C, respectively [53]. Among the chemical inhibitors, ascorbic acid and SO_2 were almost equally effective but ascorbic acid EDTA showed a very small inhibitory effect. PPO inhibition of 61.9–71.5% was achieved using 0.28 mM and 0.84 mM concentrations of ascorbic acid, while for 55.4–70.2% inhibition 0.78 mM and 2.34 mM concentrations of SO_2 were needed. Heating the enzyme extract at 70C, 80C, 90C, and 95°C for 6 s reduced 15.2, 49.5, 84.8 and 95.1% PPO activity, respectively. At the natural pH of pulp (4.3), SO_2 at concentration of 3.9 mM reduced specific activity of PPO from about 60 to about 8.4 U/mg protein in partially purified enzyme extract (PPEE), while the specific activity of 22.5 U/mg protein was retained in the pulp. Heating PPEE at 80°C at pH 4.3 (natural pH of pulp) for about 14 s decreased specific enzyme activity from 60.0 to about 11.0 U/mg and heating during 8 min decreased to 9.0 U/mg protein.

4. COMPOSITIONAL CHARACTERISTICS OF SOURSOP FRUIT

The soursop fruit consists of 67.5% edible pulp, 20% peel, 8.5% seeds, and 4% core by weight [29]. The white edible pulp contains 80–81% water, 1% protein, 18% carbohydrate, 3.43% titratable acidity, 24.5% non-reducing sugar, and vitamins B1, B2, and C [54,55]. Table 39.2 shows the high percentage of moisture, low fat, and low protein per 100 g edible value [3]. Some physicochemical characteristics were refractive indices of 1.335 for the seeds and 1.356 for the pulp, pH values of 8.34 for the soursop seed and 4.56 for the pulp, and soluble solids contents of 1.5°Brix for the soursop seed and 15°Brix for the pulp [55]. Table 39.3

TABLE 39.2 Food Value per 100 g of Edible Portion* of Soursop

Calories	53.1–61.3
Moisture	82.8 g
Protein	1.00 g
Fat	0.97 g
Carbohydrates	14.63 g
Fiber	0.79 g
Ash	60 g
Calcium	10.3 mg
Phosphorus	27.7 mg
Iron	0.64 mg
Vitamin A (β-carotene)	0
Thiamine	0.11 mg
Riboflavin	0.05 mg
Niacin	1.28 mg
Ascorbic acid	29.6 mg
Tryptophan	11 mg
Methionine	7 mg
Lysine	60 g

Source: reference 3.
*Analyses made at the Laboratorio FIM de Nutricion, Havana, Cuba.

TABLE 39.3 Physical and Chemical Characteristics of Soursop Pulp and Puree

Physicochemical Analyses	Pulp	Puree
Titratable acidity (g/100 g) as citric acid	1.02 ± 0.43	0.61 ± 0.02
Ascorbic acid (mg/100 g)	20.9 ± 1.84	9.83 ± 0.26
Pectinesterase (PE) activity (units/g)	32.1 ± 2.40	15.2 ± 0.15
Cloud stability (at 660 nm)	0.94 ± 0.06	0.65 ± 0.00
Total soluble solids (°Brix)	11.00 ± 0.4	8.00 ± 0.00
pH	3.70 ± 0.06	3.70 ± 0.04
Viscosity (cp)	25.60 ± 0.14	19.20 ± 0.16
Color 'L'	65.89 ± 1.01	61.01 ± 0.35
Color 'a'	− 2.24 ± 0.10	− 3.16 ± 0.00
Color 'b'	6.04 ± 0.55	3.83 ± 0.10
Fructose, g/100 g	3.60 ± 0.27	3.09 ± 0.22
Glucose, g/100 g	2.97 ± 0.24	2.90 ± 0.29
Sucrose, g/100 g	1.02 ± 0.28	0.99 ± 0.05

Source: reference 46.

TABLE 39.4 Nutrient Composition (%) and Citrate (g/L) of Soursop Fruit

Type of Analysis	Unripe Soursop Mesocarp Powder	Unripe Soursop Epicarp Powder	Ripe Soursop Juice
Bound moisture	2.14	4.72	75.0
Total carbohydrate	84.82	84.94	12.52
Glucose	NT	NT	6.14
Protein	7.34	6.26	2.91
Lipid	1.68	0.62	3.25
Ash	4.02	3.43	NT
Fiber	4.33	41.15	0
Citrate	NT	NT	8.82

Source: reference 59.
NT, not tested.

shows some physicochemical properties of soursop pulp and soursop puree [46].

The second most abundant component of soursop pulp next to water is the sugars, which constitute about 67.2–69.9% of the total solids [56]. The reducing sugars, glucose and fructose, were 81.9–93.6% of the total sugar content. Using gas–liquid chromatography (GLC), fructose, D-glucose, and sucrose contents of soursop pulp were 1.80, 2.27, and 6.57%, respectively, to make a total sugar content of 10.48% [57].The soursop fruit contains 12% sugar (mostly glucose and some fructose), pectin, potassium, sodium, calcium chloride, and citrate [58]. Table 39.4 shows that the citrate concentration in soursop juice was higher (8.82 g/L) than in WHO/UNICEF Oral Rehydration Salt (ORS) preparation standard (2.9 g/L) in the form of trisodium citrate dehydrate [59].

The fiber content of soursop pulp was reported as 0.78 [60] and 0.95% [61]. The alcohol-insoluble solids were mainly pectin, which in ripe fruit is 0.91% on a fresh weight basis [62]. The fraction declines from 12.0 to 4.0% on a dry weight basis from preclimacteric to climacteric stages [24].

The wet weight of soursop pulp has been reported to be 0.055 gN/100 g [63]. Ninety-one percent of this amount was contributed by the acid and neutral free amino acids [2]. Eleven free amino acids were identified by paper chromatography and four other unidentified ninhydrin-positive components were detected. Proline and γ-aminobutyric acid were the most abundant free amino acids [63]. Other amino acids detected, in order of relative amount, were glutamic acid, aspartic acid, serine glycine, alanine, citrulline, cysteine (or cystine), arginine, and lysine [4].

A study of the pre-harvest deterioration of soursop and its effect on nutrient composition was conducted in Ibadan, southwestern Nigeria [64]. Four fungal pathogens including *Botryodiplodia theobromae*, *Fusarium* sp., *Rhizopus stolonifer*, and *Aspergillus niger* were found to be associated with the pre-harvest deteriorating soursop. Nutrient analysis revealed that the freshly harvested non-infected soursop fruits had 78.49–78.68% moisture content, 14.88–14.91% carbohydrates, 1.20–1.24% crude protein, 0.89–0.90% ash, 19.15–19.35% dry matter, 1.39–1.41% potassium, and 0.63–0.65% sodium at five tested locations. Comparable values have also been documented by Rice and colleagues [54]. An approximate 39% reduction in the carbohydrate contents was observed in the infected freshly harvested fruits [64]. This was probably due to the degradative activities of the pathogens leading to reduction of the quality of the fruit. The infected fruits had about 20 and 11% loss in crude protein and dry matter, respectively. However, the ash and moisture contents of the infected fruits were higher than those of the non-infected ones.

TABLE 39.5 Some Toxicants in the Seed and Seed Coat of Soursop

Parameter	Seed	Seed Coat
Tannin (mg/100 g)	2.6	4.9
Phytate (mg/100 g)	620.5	188.0
Cyanide (mg/kg)	3.7	10.8

Source: reference 65.

5. COMPOSITIONAL CHARACTERISTICS OF SOURSOP SEEDS

The seeds of soursop fruit are rich in oil and protein and low in toxicants (tannins, phytate, and cyanide) and therefore could be harnessed in human and animal nutrition [65]. Soursop seeds contained 22.10% pale-yellow oil and 21.43% protein [66]. The oil was bland in taste and possessed an acid value of 0.93, a saponification value of 227.48, an iodine value of 111.07, and an acetyl value of 66.77. The oil consisted of 28.07% saturated and 71.93% unsaturated fatty acids [66]. The seeds yield a yellowish-brown 70% unsaturated oil which consisted of 12–33% linoleic, 41–58% oleic, 16% palmitic, and 5% stearic with a trace of myristic [67,68]. This oil may have economic value [69] as an edible oil, if some possibly toxic components can be removed [4]. Table 39.5 shows some toxicants in the seed and seed coat of soursop [65]. The seed has a higher content of magnesium and zinc than the pulp, and the pulp has a higher content of potassium and calcium than the peel or seeds [4]. The seeds also contained 0.2% water-soluble ash, 0.79% titratable acidity, and 17.0 mg calcium/100 g.

6. PROCESSING OPTIONS AND FOOD USES

The soursop fruit's major use is in the juice industry in the tropics, and the seeds are an abundant by-product of this industry [66,70]. In El Salvador, two types of soursops were distinguished: guanaba azucarón (sweet) eaten raw and used for drinks, and guanaba acida (very sour) used only for drinks [3]. In the Dominican Republic, the guanabana dulce (sweet soursop) is popular due to its low acidity.

Soursop is suitable for processing because of its high sugar content and delicate flavor [56,71,72]. However, the industrial processing of soursop is limited. The soursop is sold as fresh or frozen pulp, strained soursop juice, and frozen concentrates, which have been preserved as various juice blends, ice creams, sherberts, nectars, syrups, shakes, jams, jellies, preserves, yoghurts, and ice creams [2,3,73–76]. It is also a raw material for powders, fruits, bars, and flakes [77]. It can be made into a fruit jelly with the addition of some gelatin or used in the preparation of beverages, sherberts, ice creams, and syrups [2].In Cuba and Brazil, a refreshing drink is prepared by mixing the fruit with milk and sugar (champola), while in Puerto Rico it is generally mixed with water (carato) [2,3]. Morton [3] reported that a canned soursop soft drink which contained 12–15% pulp stored well for a year or more. Miller and colleagues [78] provided several recipes for soursop sherbert and mousse. Soursop and blended soursop-tamarind drinks were prepared by varying soursop pulp addition (10–15%) to give 15°Brix and by increasing the amount of tamarind pulp in the ratios of soursop:tamarind pulp of 6:4 to 6:8 [79].

Franco-Betancourt and Alvarez Reguera [42] developed a method for canning soursop drinks by blending with sugar cane juice (guarapo) or papaya juice. These drinks were prepared by mixing 20% soursop juice and 80% sugarcane juice, heating to 100°C, canning, and heat processing at 100°C for 15 min. The formulation for a soursop–papaya drink was 20% soursop pulp, 30% papaya pulp, 50% sugar syrup of 25°Brix. The pH of the papaya pulp was adjusted to below 4.5 before preparing the drinks. In the subtropics, the soursop fruit has been pulverized and strained, then mixed with rum, brandy, or milk to make a beverage.

The immature fruits with soft seeds have been cooked as vegetables [80–82], cooked in coconut milk [80], or used in soups [83]. Seeds are roasted or fried in Brazil [82].

6.1 Extraction of Soursop Pulp

The percentage of soursop pulp recovery has varied from 62 to 82.5% which could be due to the type of equipment, extraction method, cultivar, cultural practices, and number of seeds per fruit [84]. Prior to peeling, the soursop fruit is washed with chlorinated water to assist in the removal of dirt and debris [4].The removal of the outer skin is difficult since the peel is fragile, breaking into small pieces. Lye peeling causes the pulp to develop a reddish-brown discoloration [85] as well as lye contamination of the pulp [29]. The viscosity of the pulp varies with the stage of fruit, ripening season, and growing area [29]. The amount of water to add is determined by the final pulp viscosity desired, with sugar normally added to bring the total soluble solids to 11 to 15°Brix and citric acid to restore the pH to about 3.7 [29,86].

The extraction of soursop pulp is achieved by sieving to separate the pulp from the seeds [84] or by a processor [72]. The seeded soursop may be pressed in a colander or sieve or squeezed in cheesecloth to extract the rich, creamy juice. If an electric blender is used, the soursop seeds must first be removed since they are somewhat toxic and none should be accidentally ground up in the juice. In Puerto Rican processing factories, the hand-peeled and cored fruits are passed through a mechanical pulper having nylon brushes that press the pulp through a screen, separating it from the seeds and fiber [87].

Enzyme liquefaction of the soursop pulp could possibly facilitate its mechanical extraction and the subsequent stabilization process by tangential microfiltration. Among the common preparations tested, Rapidase Pomaliq gave a liquefaction rate of the insoluble purified cell walls of 30% after 2 h of incubation for 500 ppm of fresh soursop pulp [88]. The soluble pectin of the soursop is characterized by a very low galacturonic acid content (24%) and a high degree of esterification (70%).

6.2 Drying

Pressure drop and minimum fluidization velocity were experimentally studied in a vibro-fluidized bed of inert particles subjected to different vibration intensities during drying of soursop pulp [89]. Pulps were initially concentrated, resulting in pastes with different soluble solids content, and a constant fraction of maltodextrin was fixed in the final pulp samples. Maltodextrin was added to the pulp in order to prevent stickiness between particles and the consequent bed collapse. Increasing pulp apparent viscosity caused a considerable increase in the vibro-fluidized bed pressure drop during pulp drying and, as a consequence, resulted in a larger value of minimum vibro-fluidization velocity. On the other hand, the negative effect of increasing apparent viscosity could be attenuated by increasing the fluidized bed vibration intensity, which could prevent stickiness between particles.

6.3 Pasteurization and Nectar Processing

In the preparation of soursop nectar, the process involved dispersing the soursop fruit pulp in water, removing the seeds by screening, treating the pulp in a screw press or paddle finisher with 0.02 inch perforated screens to remove fiber, the addition of water to lower viscosity, the addition of citric acid to pH 3.7 and sugar to 15°Brix, and flash-pasteurizing. Cans of the juice kept well at room temperature for

at least a year. The ascorbic acid content of the pulp is approximately 9.0 mg/100 g [29].

Sánchez-Nieva and colleagues [29] determined the effect of dilution of the extracted soursop pulp on the viscosity and soluble solids (°Brix) content of the pulp dispersion. The total soluble solids should be between 6 and 8°Brix. To obtain the correct acid–sugar, the pH of the nectar should be 3.7, corresponding to a total acidity of 0.4%; the total soluble solids should be between 11-15°Brix. The nectars were pasteurized at 90.6°C and canned in plain tin cans.

Ascorbic acid is to be added to the pasteurized puree at a rate of 0.5–1.5 g/0.45 kg as this improves the retention of the nectar's flavor and serves as an antioxidant to control polyphenol oxidase-mediated pulp darkening of the fruit [90].

Pasteurization at 79°C for 69 s improved the sensory color, flavor, appearance, and overall acceptability of soursop puree [77]. Also the pasteurized puree packed in laminated aluminum foil at 4°C had the highest score for all sensory attributes evaluated over the lacquered can and high-polyethylene plastic bottle. Flash-pasteurized and canned nectar can be kept for up to a year at 30°C without noticeable loss in quality [29,79,91].

Soursop nectar was processed from pasteurized unstored or pasteurized frozen pulp. Nectars of pH 3.6–3.7 with 0.1% xanthan gum were produced from either 6° or 8°Brix pulp and increased to 13° or 15°Brix by addition of sucrose [92].

6.4 Concentration

In the production of soursop concentrate, Torres and Sánchez [93] indicated the following steps: 1) fruit selection; 2) fresh water rinse; 3) hand peeling; 4) seed removal; 5) pulp scalding (1 min); 6) cooling; 7) soluble solids determination: 8) addition of 0.1% sodium benzoate; 9) blending (10 min); 10) sieving; 11) sugar addition; 12) air elimination and pulp concentration; 13) packing of pulp in container; 14) covering; 15) cooling; 16) labeling; and 17) storage.

The pH of marmalade ranged from 3.1 to 3.3 and contained 60% concentrated soursop pulp and 31% added sugar [90]. The following steps have been suggested to prepare marmalade: 1) scalding or boiling of the pulp; 2) homogenization; 3) water addition; 4) sugar addition; 5) cooking for 30 minutes: 6) sugar addition; 7) cooking for 45 minutes; 8) addition of fruit pieces; 9) final cooking for 10 minutes; 10) pouring into container while warm and label [93].

6.5 Fermentation – Soursop Nectar Yogurt

Soursop nectar was added at 0, 5, 10 and 15% v/v in stirred yogurts [94]. Yogurts with 10 and 15% soursop nectar had the highest ($P<0.05$) overall quality scores (12.60/20 and 12.75/20, respectively) but differed ($P<0.05$) in flavor and aroma from plain yogurt and 5% soursop yogurt. Most panelists would consider purchase of 10 and 15% soursop yogurts over 0 and 5% soursop yogurts. These yogurts provided high percentage daily values of zinc, phosphorus, and calcium and a good level of protein.

6.6 Canning

The seeded soursop has been canned [3]. A study of the thermal diffusivity (α) of soursop pulp by the use of heat penetration curves in eight Z short cans indicated no difference in the α values for the pulps from ripe and unripe soursop pulps at the 5% level of significance [95]. The values for α for the unfrozen soursop pulp are 16 times those of frozen pulp, so this is important when designing

processes for freezing, thawing, or pasteurization of soursop pulp.

6.7 Oil Extraction

The soursop seed is about 4% of the whole Fruit [94]. The fat content (ether extract) of the soursop seeds was 22.57% which could supplement animal feed concentrate. The seed coat had negligible fat content. Essential oils extracted from soursop (pulp) have industrial applications and the oils are also thought to improve the flavor of processed fruit products [1].

7. FLAVOR COMPONENTS AND QUALITY CHANGES

The ripe fruit has a very pleasant, distinctive, sub-acid, aromatic, juicy flesh [17]. In an early study, some volatiles reported were methyl alcohol and ethyl alcohol, acetaldehyde, amyl caproate, geranyl caproate, and citral in ripe fruit [62]. The principal aroma components were esters with approximately 80% of compounds found in the 2-methylbutane solvent extract by low-temperature–high-vacuum distillation [96]. Using solid-phase microextraction (SPME), 21 compounds were identified of which 12 were esters [97]. Table 39.6 shows the prevailing compounds which were identified in soursop aroma after SPME [97]. They were α-unsaturated methyl esters of the type R-CH$_5$CH-COOCH$_3$ (Rethyl, butyl, hexyl) as methyl crotonate (Rethyl), methyl 2-hexenoate (Rbutyl), and methyl 2-octenoate (Rhexyl) as well as aliphatic esters of butyric and caproic acids.

Methyl hexanoate (~31%) and methyl hex-2-enoate (~27%) were the two most abundant components and amounted to ~0.7 mg/kg of fruit [96]. Trans-β-farnesene (6.5%) and some other terpenes were present at less than 1%.

TABLE 39.6 Some Compounds Identified in Soursop Aroma after Solid Phase Microextraction

Group	Compounds
Alcohols	1-Butanol
	3-Hexen-1-ol
Aldehydes	Nonyl aldehyde
	Decyl aldehyde
Esters	Ethyl acetate
	Methyl butyrate
	Methyl crotonate
	Ethyl butyrate
	Butyl acetate
	Ethyl crotonate
	Methyl caproate
	Methyl 2-hexenoate
	Ethyl caproate
	Ethyl 2-hexenoate
	Methyl caprylate
	Methyl 2-octenoate
Ketones	1-Phenyl-2-pentanone
Terpenic compounds	Camphene
	β-Mircene
	Limonene
	Ocimene
	β-Linalool
	α-Terpineol
	Geraniol
Other	Diacetyl
	Acetic acid
	2,5-Dihydro-2,5-dimethoxyfuran
	2,4,5-Trimethyl-1,3-dioxolane
	γ-Octalactone
	Palmitic acid

Source: reference 97.

The β-farnesene imparts a floral odor [4]. Forty-four main components were separated of which 24, comprising nearly 96% of the sample, are positively identified [96]. No peak seemed to represent any specific element characteristic of ripe soursop flavor [4]. Using gas chromatographic/spectroscopic (GC/FID and GC/MS), esters of aliphatic acids were dominant odor compounds (~51%), with 2-hexenoic acid ethyl ester (8.6%), 2-octenoic acid methyl ester (5.4%) and 2-butenoic acid methyl ester (2.4%) in essential oil extracted from Cameroon soursop pulp [98]. In addition, mono- and sesquiterpenes such as β-caryophyllene (12.7%), 1, 8-cineole (9.9%), linalool (7.8%), α-terpineol (2.8%), linalyl propionate (2.8%), linalyl propionate (2.2%) and calarene (2.2) are highly concentrated in the essential oil. The volatile components of soursop which were isolated by simultaneous steam distillation/solvent extraction and GC/MS analysis identified 41 compounds [99]. The major volatiles were methyl 3-phenyl-2-propenoate, hexadecanoic acid, methyl (E)-2-hexenoate, and methyl 2-hydroxy-4-methyl valerate.

The principal component present in unripe soursop fruit is hexen-1-ol [100]. The volatile compounds in soursop pulp were obtained by a liquid–liquid continuous extraction procedure from the aqueous solution of blended soursop pulp and analyzed by gas chromatography (GC) and GC-mass spectrometry [101]. Twelve volatiles were identified by comparing their mass spectra and Kovats retention indexes with those of standard compounds; five were identified tentatively from MS data only. Eight are being reported for the first time. (Z)-3-Hexen-1-ol was the main volatile present in mature-green fruit, while Me (E)-2-hexenoate, Me (E)-2-butenoate, Me butanoate, and Me hexanoate were the four main volatiles present in ripe fruit. Concentrates of these five volatiles decreased and several other unidentified volatiles appeared when the fruit became overripe.

The components which were responsible for the intense odor of fresh soursop pulp from the Cameroon were investigated [102]. Esters of aliphatic acids dominated (~51%) with 2-hexenoic acid methyl ester (23.9%), 2-hexenoic acid ethyl ester (8.6%), 2-octenoic acid methyl ester (5.4%), and 2-butenoic acid methyl ester (2.4%) as main compounds in the essential oil of the fresh soursop pulp. Additional mono- and sesquiterpenes such as β-car-yophyllene (12.7%), 1, 8-cineole (9.9%), linalool (7.8%), α-terpineol (2.8%), linalyl propionate (2.2%), and calarene (2.2%) were highly concentrated in the essential oil of the fresh fruit. Another study by Pelissier and colleagues [103] also showed that the essential oils which were extracted from the peel and pulp of soursop were rich in aliphatic esters (69%), with the most abundant being methyl 2-hexenoate (41%) as determined by hydrodistillation and GC-MS. The peel oil contained fewer aliphatic esters (21%) than pulp oil. Peel oil was rich in monoterpenes (16%), such as α-phellandrene (8.5%), sesquiterpenes (6.5%), and alkanes (13%).

The non-volatile organic acid fraction of soursop consists of a mixture of about 2 parts malic acid, 1 part citric acid, and traces of iso-citric acid [62]. The ratio between total malic and citric acids changes during ripening, with oxalic acid also being present [25]. Hydrocyanic acid is found within the bark of soursop trees; smaller amounts are present in the roots and leaves and a trace found in the fruit [104].

8. QUALITY CHANGES IN PROCESSED PRODUCTS

The pure white color of soursop pulp is very stable [2]. A reddish-brown discoloration in soursop during lye peeling was noted [29] and rapid browning in an overripe soursop [42]. The flavor is somewhat volatile so pasteurized products are less attractive than fresh ones, and the off-white color can become an unpleasant gray unless oxidation is prevented [105].

The soluble solids of a pasteurized soursop juice were more stable at refrigeration (4°C) than at ambient temperature (28°C) [106]. In another storage study, soursop nectar was stored at 4°C for 12 weeks and remained unchanged in browning, unlike the soursop nectar which was stored at 30°C [92]. Also, nectars

processed from stored pulp became browner in appearance than those from unstored pulp.

The effects of processing temperature, the addition of sugar to 45° to 59°Brix, and the addition of four levels of ascorbic acid ranging from 0.5 to 1.5 g/0.45 kg on the quality, shelf-life, and ascorbic acid retention of frozen soursop was studied by Sánchez-Nieva and colleagues [47]. Pulps enriched by the addition of 0.5 to 1.5 g ascorbic acid/0.45 kg retained their flavor better than the unenriched sweetened and unsweetened pulps.

9. FOLKLORIC USES

The traditional use of *A. muricata* is recorded worldwide in folk medicine systems, including Barbados [107], Borneo (Kalimantan, Indonesia) [108], Brazil [109–111], Cook Islands [112], Curacao [113], Dominica [114], Guatemala [115], Guam [116], Guyana [117], Haiti [118], Jamaica [119], Madagascar [120], Malaysia [121,122], Peru [123], Suriname [124], Togo [125], the West Indies [126], and the Amazon [127].

The genus Annonaceae comprises about 70 species, of which nine are indigenous to India. *A. muricata* is widely used in traditional Indian medicine for the 'treatment of kidney troubles, fever, nervousness, ulcers and wounds' and possesses antispasmodic, antidysenteric, and parasiticidal activity [128,129]. In terms of the plants parts, the 'leaves are used as suppurative, febrifuge; its bark as tonic; roots as antispasmodic, parasiticidal; flowers as bechic; unripe fruit as antiscrobutic; and seeds as insecticidal, astringent, fish-poison.' Within Ayurvedic medicine in India, the plant is used as a bitter, tonic, abortifacient, febrifuge, for scorpion stings, high blood pressure, and as a respiratory stimulant.

In Trinidad and Tobago, the leaves of *A. muricata* are used to treat hypertension [130].

Taylor [131] reports that in 'Jamaica, Haiti, and the West Indies, the fruit and/or fruit juice is used for fevers, parasites and diarrhea, and as a lactagogue; the bark or leaf is used as an antispasmodic, sedative, and nervine for heart conditions, coughs, gripe, difficult childbirth, asthma, asthenia, hypertension, and parasites.'

In Brazil, the fruit, its juice, and crushed seeds are used as a vermifuge and anthelmintic, while the fruit and its juice alone are used to increase mother's milk (lactagogue) and as an astringent for diarrhea and dysentery. A tea from the leaf is used to treat arthritis pain, rheumatism, and neuralgia. In the Peruvian Amazon the leaves and root bark are used as an antispasmodic, a sedative, and for diabetes. In the Guianas, a tea is prepared from either the leaf or bark as a cardiotonic or sedative [131].

Many of these folkloric uses have been scientifically validated since the 1940s. The bark and extracts of *A. muricata* have been shown by scientists to be hypotensive, antispasmodic, anticonvulsant, vasodilatory, a smooth muscle relaxant, and to possess cardiodepressant activity in animals [132,133].

10. MEDICINAL USES

Soursop is used as an antispasmodic, emetic, and sudorific in herbal medicine. A decoction of the leaves is used to kill head lice and bed bugs while a tea from the leaves is well known in the West Indies to have sedative properties. The juice of the fruit is taken orally for hematuria, liver complaints, and urethritis.

An extract of the stem bark of *A. muricata* has been shown to exhibit antistress activity. Studies have found that an ethanol extract of the stem bark of soursop induces a significant reduction in stress levels in stress-induced rats, suggesting a role for the extract as an adaptogenic agent. The plant extract was found to

reduce the stress-induced rise in brain 5-hydroxytryptamine (5-HT) and 5-hydroxyindole acetic acid (5-HIAA) levels, by increasing the levels of the MAO enzyme, which was observed to have lowered 5-HT and 5-HIAA [134]. Administration of the ethanol extract has also been demonstrated to significantly inhibit the cold immobilization stress-induced increase in lipid peroxidation in the liver and brain of rats [129].

11. BIOACTIVITY AND TOXICOLOGY

Analysis of defatted seeds from ripe soursop led to the isolation of annonaceous acetogenins including annonacin, isoannonacin, isoannonain-10-one, goniothalamicin, and gigantetrocin [135–137] which showed cytotoxicities for the A-549 lung carcinoma, MCF-7 breast carcinoma, and HT-29 colon adenocarcinoma human tumor cell lines [138–140]. These single ring acetogenins are not considered as biologically potent as nonadjacent bistetrahydrofuran ring compounds [141]. Bioactivity-directed fractionation of the leaves of soursop led to the isolation of two additional acetogenins, muricoreacin and murihexocin, that also showed significant cytotoxicities in six human tumor cells, including PACA-2 pancreatic carcinoma and PC-3 prostate adenocarcinoma cell lines [142].

Unlike animals or aquatic species, when plants are threatened the only defense they have is chemistry (although thorns and other defensive strategies can be effective against some predators) against microbes, insects, and herbivores. Two annonaceous species, *A. muricata* and *Asimina triloba* Dunal ('paw paw'), are abundant fruit trees that both yield acetogenins. When discovered, these bioactive compounds were found to be a new class of natural pesticides, explaining why these plants put energy into their creation. One of these annonaceous acetogenins, bullatacin, showed 300 times the potency of taxol against L1210 murine leukemia [143,144]. Bullatacin and several other acetogenins, a number of which are only found in soursop, do so by inhibiting adenosine triphosphate (ATP) production. This has profound implications for the development of complex acetogenin drugs that could be very effective against multidrug-resistant tumors [145].

Certain isoquinoline derivatives are known to be toxic to dopaminergic neurons and inhibitors of the mitochondrial respiratory chain. The aqueous extract of leaves and the extract of the root bark of *A. muricata* and infusions and decoctions of the fruit have been shown in both *in vitro* and *in vivo* experiments to be a potentially toxic inhibitor of the mitochondrial respiratory chain [146,147]. Experimental studies have confirmed annonacin, an isoquinoline derivative, the major acetogenin found in soursop, as the toxic agent responsible for this effect [148–150]. *A. muricata* and other plants of Annonaceae contain potential neurotoxins, particularly the isoquinolinic alkaloids and acetogenins, structurally homogenous fatty acid derivatives known as polyketides specific to Annonaceae. This class of polyketides are among the most potent inhibitors of complex I of the mitochondrial respiratory chain known in nature, some 50-fold more potent than the class complex I inhibitor MPP+ and two times more potent than rotenone in inducing neuronal death. Inhibition of the mitochondrial respiratory chain impairs energy production, as demonstrated by using purified annonacin extracted from soursop [analyzed by matrix associated laser desorption ionization-time of flight (MALDITOF) mass spectrometry] and then given to animals [150].

The significance of these findings relates to the abnormally high rate of atypical parkinsonism found on islands such as Guam in the Northern Mariana islands, New Caledonia, western New Guinea, the Kii peninsula of Japan, and the French West Indian island of Guadeloupe in the Caribbean, where epidemiological evidence suggests a close association of the disease with regular consumption of soursop fruit, infusions, and decoctions [151–155].

In Guadeloupe atypical parkinsonian patients represent two-thirds of all cases of parkinsonism. In European countries and the USA, atypical parkinsonism accounts for around 5% of such cases. The atypical form's clinical entity is described as 'a unique combination of levodopa-resistant parkinsonism, tremor, myoclonus, hallucinations, REM sleep behavior disorder and fronto-subcortical dementia' [156]. Chronic administration of annonacin to rats via Alzet osmotic minipumps has shown that annonacin is able to reproduce the characteristic lesions found in the brains of humans with atypical parkinsonism [156].

In a revealing report published in 2004, researchers studying the atypical parkinsonism in Guadeloupe quantified the amount of annonacin in the fruit and in commercial products. The average fruit contained 15 mg of annonacin, a can of commercial nectar had 36 mg, and a cup of infusion or decoction 14 μg [157]. It was estimated that an adult who consumed one fruit or can of nectar a day would ingest in 1 year the amount of annonacin found to induce brain lesions in rats that received purified annonacin via intravenous infusion. While soursop fruit also contains alkaloids that are toxic *in vitro* to dopaminergic and other neurons, annonacin was found to be toxic in nanomolar concentrations, whereas micromolar concentrations of the alkaloids were needed [150].

What is interesting is the virtual disappearance over the years of the disproportionate incidence of atypical parkinsonism previously reported in Guam and New Guinea. It has been suggested that the significant reduction in incidences may be due to changes in diet in New Guinea and in Guam, in particular owing to the adoption of Western diets and the abandonment of native foods, such as soursop, despite their continued availability.

Studies in animals have shown that rotenone neurotoxicity is similar in toxicity to that demonstrated for annonacea acetogenins. Caparros-Lefebvre and colleagues have suggested that simultaneous exposure to acetogenins and rotenone can produce a synergistic toxicity on neurons that may explain its persistence in some islands and virtual disappearance in others [158]. Rotenone is used in solution as a pesticide and insecticide and works by interfering with the electron transport chain in mitochondria. The compound is found in a number of plants growing in the tropics and subtropics, including the seed of the Jicama plant (*Tephrosia virginana*), the Tuba plant (*Derris elliptica*) and several Derris species (*Derris involuta, Derris thyrsiflora,* and *Derris walchiil*), which are favored by natives as a tender root vegetable.

Given the amount of evidence showing a relationship between the acetogenins and atypical parkinsonism, chronic oral intake of the fruit, and the derivatives ingested that are produced from the plant, caution is warranted and further research encouraged. Should there be an increased risk of neurotoxicity associated with acetogenins with other plant toxins, it is important to determine what it is so that dietary advice can be given to minimize risk of neurodegenerative outcomes.

12. SUMMARY

Soursop fruit is useful as a processed product due to its high pulp recovery and many flavor compounds, particularly rich volatiles. Some constraints to processing are: 1) short storage life of the soursop fruit; 2) fragile peel; 3) uneven ripening of soursop fruit which makes the selection for processing tedious; 4) loss of flavor by thermosensitive processing methods; and 5) the need to inactivate the enzymes in soursop pulp. One hundred grams of raw soursop fruit yields 66 calories, 3.3 g dietary fiber, 14 mg calcium, 278 mg potassium, 20.6 mg vitamin C, 27 mg phosphorus, and 16.8 g carbohydrate [82]. A serving size (225 g) or 1 cup soursop is very low in saturated fat (0.1 g; 1% DV), low in sodium (32 mg; 1%DV), has no cholesterol (0 mg; 0% DV), high dietary fiber (7.4 g : 30% DV), and very high vitamin C (77% DV), but is very high in sugars (30.5 g) based on a 2000 calorie diet (Calorie Count, 2009 [159]). It is a good source of vitamin B (0.07 mg/100 g pulp) and vitamin C (20 mg/100 g pulp) and a poor to fair source of calcium and phosphorus [81].The citrate concentration in soursop juice is higher (8.82 g/L) than in WHO/UNICEF Oral Rehydration Salt (ORS) preparation standard (2.9 g/L) in the form of presidium citrate dehydrate.

Soursop is a good source of nutrition, yet *A. muricata,* including its fruit, contains annonacin, the most abundant acetogenin, which has been experimentally demonstrated to be toxic *in vitro* and *in vivo* to dopaminergic and other neurons. Epidemiological evidence in several regions of the world has linked consumption of the fruit to an increased risk of developing atypical parkinsonism. The absence of family histories of parkinsonism and the cross-ethnic origins found among islands around the world led to the suggestion that consumption of soursop fruit and other consumables derived from this plant places those who consume the fruit at possible risk. That risk is associated with cross-interactions with compounds found in other foods is suggested by the continued consumption of soursop in places like the North Marianna Islands (e.g. Guam) and the virtual disappearance of atypical parkinsonism in recent decades. Could it be that dietary changes among such islanders have made accounts for the disappearance of this neurodegenerative disease? And if so, which food combinations, if any, with soursop, should be avoided? A clearer understanding of the risks associated with chronic intake of soursop is warranted given the presence of acetogenins and other alkaloids in the fruit so that competent and reliable dietary advice can be given. Pharmacokinetic and metabonomic studies in animals and humans would be helpful, particularly the latter, to determine which enzymes are affected that may play a part in the heightened risk of atypical parkinsonism. Animal studies do not, however, suggest that occasional consumption of the fruit places consumers at risk. How often one can safely eat the fruit or commercial products derived thereof over a given period remains debatable until more answers come from ongoing research on this tropical fruit.

References

1. International Centre for Underutilized Crops. (2002). *Fruits for the future. Newsletter 5. March. Institute of Irrigation and Development Studies.* Southampton, UK: University of Southampton http://www.icuc-iwmi.org/files/News/Resources/Factsheets/annona.pdf. Accessed 15th August, 2008.

2. Bueso, C. E. (1980). Soursop, tamarind and chironja. In S. Nagy, & P. S. Shaw, (Eds.), *Tropical and Subtropical Fruits: Composition, Properties and Uses* (pp. 375–406). Westport, CT: AVI Publishing.

3. Morton, J. F. (1987). Soursop. In *Fruits of Warm Climates* (pp. 75–80). Greensboro, NC: Media Incorporated.

4. Paull, R. E. (1998). Soursop. In P. E. Shaw, H. T. Chan, Jr. & S. Nagy, (Eds.), *Tropical and Subtropical Fruits* (pp. 386–400). Auburndale, FL: Agscience.

5. Thompson, A. K. (2003). Postharvest technology of fruits and vegetables. In A. K. Thompson (Ed.), *Fruit and Vegetables: Harvesting, Handling and Storage* (2nd ed.). (pp. 115–369). Chapter 12, Oxford: Blackwell Publishing.

6. Blench, R., & Dendo, M. (2007). A History of fruits on the SE Asian MainLand. Paper presented at the EUREAA, Bourgon, 26 September 2006, p. 1–26. Cambridge, UK. http://www.rogerblench.info/RBOP.htm Accessed 08/30/09..

7. Popenoe, W. (1920). *Manual of Tropical and Subtropical Fruits.* New York: Hafner Press.

8. Morton, J. (1973). *Fruits of Warm Climates. Media.* Winterville, NC: Creative Resource Systems, Inc (pp. 75–80).

9. Samson, J. A. (1980). *Tropical Fruits.* (2nd Ed). NY: Longman Scientific & Technical, Longman Inc p. 336. (Tropical Agriculture Series).

10. Mowry, H., Toy, L. R., & Wolfe, H. S. (1953). Miscellaneous tropical and subtropical Florida fruits. *Florida Agricultural Experiment Station Bulletin,* 156.

11. Salazar, C. G. (1965). Some tropical fruits of minor economic importance in Puerto Rico. *Review Agriculture Puerto Rico,* 52, 135–156 (In Spanish).

12. Lopes, J. G. V., De Almeida, J. I., & Assuncao, M. V. (1981). Preservacao do poder germinatiro de sementes de graviola (*Annona muricata* L.) sob diferentes temperaturas e tipos de embalagens. *Proc. Trop. Reg. Am. Soc. Horticultural Society,* 25, 275–280.

13. Frey, K. J. (1981). *Plant Breeding II.* Amsterdam: The Iowa State University Press p. 497.

14. Rasai, S., George, A. P., & Kantharajah, A. S. (1995). Tissue culture of Annona spp. (cherimoya, atemoya, sugar apple and soursop) a review. *Scientia Horticulturae,* 62, 1–14.

15. George, A. P., & Nissen, R. J. (1987). Propagation of Annona species: a review. *Scientia Horticulturae,* 33, 75–85.

16. Encina, C. L. (2005). Annona spp. Atemoya, cherimoya, soursop and sugar apple. In: R. E. Litz (Ed.), *Biotechnology of Fruits and Nut Crops.* In Biotechnology in Agriculture Series. No, 29, chapter 3.1, (pp. 74–87). Wallingford, Oxfordshire, UK: CABI Publishing.

17. Morton, J. F. (1966). The soursop or guanabana (*Annona muricata* Linn). *Proceedings Fla State Horticulture Society,* 79, 355–366.

18. Coronel, R. E. (1983). *Promising Fruits in the Philippines.* Los Banos, Philippines: College of Agriculture, University of the Philippines.

19. Paull, R. E. (1998). Soursop. In P. E. Shaw, & H. T. Chan, (Eds.), *Tropical and Subtropical Fruits* (pp. 386–400). Auburndale, FL: Agscience.

20. Biale, J. B., & Barcus, D. E. (1970). Respiratory patterns in tropical fruits of the Amazon Basin. *Tropical Science,* 12, 93–104.

21. Worrell, D. B., Carrington, C. M. S., & Huber, D. J. (1994). Growth, maturation and ripening of soursop (*Annona muricata* L.) fruit. *Scientia Horticulturae,* 57, 7–15.

22. Zayas, J. C. (1944). Fruits in Cuba. *Review Agriculture,* 27, 23–140.

23. Araque, R. (1967). The soursop. Rural Welfare Council. Consejo de Bienestar Rural Crop Service. No. 13. Caracas, Venezuela (Spanish).

24. Paull, R. E. (1982). Postharvest variation in composition of soursop (*Annona muricata* L.) fruit in relation to respiration and ethylene production. *Journal of American Society Horticulture Science,* 107, 582–585.

25. Paull, R. E., Deputy, J. C., & Chen, N. J. (1983). Changes in organic acids, sugars and headspace volatiles during fruit ripening of soursop (*Annona muricata*

L.). *Journal of American Society Horticulture Science, 108*, 931–934.

26. Paull, R. E. (1990). Soursop ripening – starch breakdown. *Acta Horticulturae, 269*, 277–281.

27. De Lima, M. A. C., Alves, R. E., & Filgueiras, H. A. C. (2006). Changes related to softening of soursop during postharvest maturation. *Pesquisa Agropecuaria Brasileira, 41*(12), 1707–1713.

28. Arauz, L. F., & Mora, D. (1983). Evaluacion prelimnar de los problemas postcosecha en seis frutas tropicales de Costa Rica. *Agronomic Costau, 7*(1/2), 43–53.

29. Sánchez-Nieva, F., Igaravidez, L., & Lopez-Ramos, B. (1953). The preparation of soursop nectar. *University of Puerto Rico Agriculture Experimental Stational Technical Paper, 11*, 5–19.

30. Paull, R. E. (1990). Chilling injury of crops of tropical and subtropical origin. In C. Y. Wang, (Ed.), *Chilling Injury of Horticultural Crops* (pp. 17–36). Boca Raton, FL: CRC Press.

31. FAO. (1986). *Food and Fruit Bearing Forest Species.* No. 3: Examples from Latin America. Forestry Paper 44/3. Rome: FAO..

32. Peña, J. E., & Bennet, F. D. (1995). Arthropods associated with *Annona* spp in the Neotropics. Gainesville, FL. 32611. *University of Florida Entomology of Florida, 2*, 329–349.

33. Junqueira, N. T. D., Cunha, M. M. da., Oliveira, M. A. S., & Pinto, A. C. de Q. (1996). FRUPEX. Graviola para exportação : aspectos fitos sanitários (Soursop for Export: Phytosanitary Aspects). Brasilia. EMBRAPA- SPI, 67P. FRUPEX. Publicações Técnicas, **3**, 67 pp..

34. Braga Sobrinho, R., Mesquita, A. L. M., & Bandeira, C. T. (1998). Avanços tecnológicos no manejo integrado de pragas da gravioleira. (Technologic Advances in the Soursop Integrated Pest Management). Simpósio Avanços tecnológicos na Agroindústria Topical, pp. 143–145. EMBRAPA-CNPAT. Fortaleza-CE. Anais.

35. Braga Sobrinho, R., Bandeira, C. T., & Mesquita, A. L. M. (1999). Occurrence and damage of soursop pests in northeast Brazil. *Crop Protection, 18*, 539–541.

36. Alvarez-Garcia, L. A. (1949). Anthracnose of the *Annonaceae* in Puerto Rico. *University of Puerto Rico Journal of Agriculture, 33*, 27–43.

37. Persley, D. M., Pegg, K. G., & Syme, J. R. (1989). Fruit and Nut Crop: a Disease Management Guide. Information Series QI 88018. Dept of Primary Industries, Brisbane, Queensland, Australia.

38. Snowdon, A. L. (1991). *A Color Atlas of Post-harvest Diseases and Disorders of Fruits and Vegetables. Vol. 1.* General Introduction and Fruits. Boca Raton, FL: CRC Press.

39. Pennisi, A. M., & Agosteo, G. E. (1994). Foliar alterations by *Colleotrichum gloeosporioides* on Annona. *Informatore Fitopatologico, 44*, 63–64.

40. Zarate-Reyes, R. D. (1995). Diseases of the soursop, *Annona muricata* L., in Colombia: characteristics, management and control. *Fitopatologia Colombiana, 19*, 68–74.

41. Lutchmeah, R. S. (1988). Botryodiplodia theobromae causing fruit rot of *Annona muricata* in Mauritius. *Plant Pathology, 37*, 152.

42. Franco-Betancourt, J. J., & Alvarez Reguera, J. (1980). Juices of Cuban Fruits. Cuban Institute Technology Research. No. 11. La Habana, Cuba (Spanish).

43. Arbaisah, S. M., Asbi, B. A., Junainah, A. H., & Jamilah, B. (1997). Purification and properties of pectinesterase from soursop (*Annona muricata*) pulp. *Food Chem., 59*(1), 33–40.

44. Garces, M. (1969). Tropical fruit technology. *Adv. Food Res., 17*, 153–214.

45. Arbaisah, S. M., Asbi, B. A., Junainah, A. H., Jamilah, B., & Kennedy, J. F. (1997). Soursop pectinesterases: thermostability and effect on cloud stability of soursop juice. *Carbohydrate Polymers, 34*, 177–182.

46. Umme, A., Asbi, B. A., Sahnah, Y., Junainah, A. H., & Jamilah, B. (1997). Characteristics of soursop natural puree and determination of optimum conditions for pasteurization. *Food Chem., 58*, 119–124.

47. Sánchez-Nieva, F., Hernandez, F. I., & M.De George, L. (1970). Frozen soursop puree. *Journal of Agriculture University, Puerto Rico, 54*, 220–226.

48. Yusof, S., & Ibrahim, N. (1994). Quality of soursop juice after pectinase enzyme treatment. *Food Chem., 51*, 83–88.

49. Bruinsma, J., & Paull, R. E. (1984). Respiration during post-harvest development of soursop fruit. *Annona muricata* L. *Plant Physiology, 76*, 131–138.

50. Bruinsma, J., & Paull, R. E. (1984). Respiration and ethylene production in post-harvest soursop fruit (*Annona muricata* L.). In *Ethylene and Plant Development Proceedings*London: 39th Easter School, Nottingham University. Butterworths Scientific.

51. Sobral Bezerra, V., de Lima Filho, J. L., Montenegro, M. C. B. S. M., & Lins da Silva, A. N. A. V. (2003). Flow-injection amperometric determination of dopamine in pharmaceuticals using a polyphenol oxidase biosensor obtained from soursop pulp. *J. Pharm. Biomed. Anal., 33*, 1025–1031.

52. Lima de Oliviera, S., Barbosa Guerra, N., Sucupira Maciel, M.. I., & Souza Livera, A. V. (1994). Polyphenoloxidase activity, polyphenols concentration and browning intensity during soursop (*Annona muricata,* L.) maturation. *J. Food Sci., 59*, 1050–1052.

53. Bora, P. S., Holschuh, H. J., & da Silva Vasconcelos, M. A. (2004). Characterization of polyphenol oxidase of soursop (*Annona muricata* L.) fruit and a comparative study of its inhibition in enzyme extract and in pulp. *Ciencia y Tecnologia Alimentaria (Ourense, Spain), 4*, 267–273.

C. ACTIONS OF INDIVIDUAL FRUITS IN DISEASE AND CANCER PREVENTION AND TREATMENT

54. Rice, R. P., Rice, L. W., & Tindal, H. D. (1991). *Fruit and Vegetable Production in Warm Climate*. Oxford: Macmillan Education Ltd.

55. Onimawo, I. A. (2002). Proximate composition and selected chemical properties of the seed, pulp and oil of soursop (*Anona muricata*). *Plant Foods Hum. Nutr.*, 57 (2), 165–171.

56. Sánchez-Nieva, F., Hernandez, F. I., & M.De George, L. (1970). Frozen soursop puree. *Journal of Agriculture University, Puerto Rico*, 54, 220–226.

57. Chan, H. T., & Lee, C. W. Q. (1975). Identification and determination of sugars in soursop, rose apple, mountain apple and Surinam cherry. *J. Food Sci.*, 40, 892–893.

58. WHO. (1991). Diarrhoea. World Health Organization. *Medicare J.*, 4(10), 22–25.

59. Enweani, I. B., Obroku, J., Enahoro, T., & Omoifo, C. (2004). The biochemical analysis of soursop (*Annona muricata* L.) and sweetsop (*A.squamosa* L.) and their potential use as oral rehydration therapy. *Food, Agriculture and Environment*, 2, 39–43.

60. Axtmayer, J. H., & Cook, D. H. (1942). Manual of Bromatology. Pan-Am. Sanit. Bur Pub, 186, (Spanish), Washington, DC.

61. Wenkam, N. S., & Miller, C. D. (1965). *Composition of Hawaii Fruit*. Hawaii, USA: Hawaii Agriculture Experimental Station, University of Hawaii Bulletin 135.

62. Nelson, E. K., & Curl, A. L. (1940). *The non-volatile acids and flavor of the soursop*. Puerto Rico: Rep. Puerto Rico Exp. Station. US Dept of Agriculture pp. 88–91.

63. Ventura, M. M., & Hollanda-Lima, I. (1961). Ornithine cycle amino acids and other free amino acids in fruits *Annona squamosa* L. and *Annona muricata* L. *Phyton*, 17, 39–47.

64. Amusa, N. A., Ashaye, O. A., Oladapo, M. O., & Kafaru, O. O. (2003). Pre-harvest deterioration of soursop (*Annona muricata*) at Ibadan Southwestern Nigeria and its effect on nutrient composition. *African Journal of Biotechnology*, 2, 23–25.

65. Fasakin, A. O., Fehintola, E. O., Obijole, O. A., & Oseni, O. A. (2008). Compositional analyses of seed of soursop, Annona muricata L. as a potential animal feed supplement. *Scientific Research and Essay*, 3, 521–523.

66. Awan, J. A., Kar, A., & Udoudoh, P. J. (1980). Preliminary studies on the seeds of *Annona muricata* Linn. *Plant Foods Hum. Nutr.*, 30, 163–168.

67. Asenjo, C. F., & Goyco, J. A. (1943). Puerto Rican fatty oils. II. The characteristics and composition of Guanabana seed oil. *J. Am. Chem. Soc.*, 65, 208–209.

68. Aecco de Castro, F., Arraes Maia, G., Holanda, L. F. F., Guedes, Z. B. L., & J. De Anchieta Moura, F. E. (1984). Physical and chemical characteristics of soursop fruit. Pesq. Agropec. bras. *Brasilia*, 19, 361–365 (Portuguese).

69. Ngiefu, C. K., Paquot, C., & Vieux, A. (1976). Oil-bearing plants of Zaire. II. Botanical families providing oils of medium unsaturation. *Oleagineux*, 31, 545–547.

70. Koesriharti (1991). Edible fruits. And nuts. In: E. W. Verheij & R. E. Coronel, (Eds.), *Plant Resources of South-East Asia*, No 2, pp. 75–78. Pudoc Wageningen, Prosea Foundation: Bogor, Indonesia, and Pudoc-DLO: S Wageningen, Netherlands.

71. George, G. P. (1984). Annonaceae. In P. E. Page, (Ed.), *Tropical Fruit for Australia* (pp. 35–41). Brisbane, Australia: Queensland Department of Primary Industries.

72. Mororó, R. C., Freire, E. S., & Sacramento, C. K. (1997). Processamento de Graviola para Obtenção de Polpa (Portuguese). In A. R. São José, I. V. Boas, O. M. Morais, & T. N. H. Rebouças, (Eds.), *Anonáceas Poduoção e Mercado (Pinha, Graviola, Atemóia e Cherimólia)* (pp. 263–274). Vitoria da Conquista, Brasil: Universidade Estadula do Sudoeste da Bahia.

73. Umme, A., Salmah, Y., Jamilah, B., & Asbi, B. A. (1999). Microbial and enzymatic changes in natural soursop puree during storage. *Food Chem.*, 65, 315–322.

74. Lutchmedial, M., Ramlal, R., Badrie, N., & Chang-Yen, I. (2005). Nutritional and sensory quality of stirred soursop (*Annona muricata* L.) yoghurt. *International Journal of Food Sciences*, 55, 407–414.

75. Pinto, A. C.de. Q., Cordeiro, M. C. R., de Andrade, S. R. M., Ferreira, F. R., Filgueiras, H. A.de. C., Alves, R. E., & Kinpara, D. I. (2005). *Annona species. International Centre for Underutilized Crops*. Southampton, UK: University of Southampton.

76. Gratão, A. C. A., Silveira, V., Jr., & J. Telis-Romero, J. (2007). Laminar flow of soursop pulp juice through concentric annuli ducts: friction factors and rheology. *Journal of Food Engineering*, 78, 1343–1354.

77. Umme., A., Bambang, S., Salmah, Y., & B. Jamilah, B. (2001). Effect of pasteurization on sensory quality of natural soursop puree under different storage conditions. *Food Chem.*, 75, 293–301.

78. Miller, C. D., Bazore, K., & Barton, M. (1965). *Fruits of Hawaii*. Honolulu: University of Hawaii Press.

79. Benero, J. R., Collazo-Riviera, A. L., & De George, L. M. (1974). Studies on the preparation and shelf-life of soursop, tamarind and blended soursop-tamarind soft drinks. *Journal of Agriculture University of Puerto Rico*, 58, 99–104.

80. Cantillang, P. H. (1976). Fruits: Guayabano is in great demand. *Greenfields*, 6(7), 25–26.

81. FAO (2001). Part 1 Species with potential for commercial development. In: 'Underutilized Tropical Fruits of Thailand' (S. Subhadrabandhu, Ed.). FAO Corporate

C. ACTIONS OF INDIVIDUAL FRUITS IN DISEASE AND CANCER PREVENTION AND TREATMENT

Repository. Food and Agriculture Organization of the United Nations Regional Office for Asia and Pacific, Bangkok, Thailand. RAP Publication. 2001/26, December 2001, <http://www.fao.org/docrep/004/ab777e/ab777e07.htm>#. Accessed 30th November, 2008.

82. GetJamaica.Com (2008). Jamaican food – soursop. <http://www.getjamaica.com/Jamaican%20Food%20%20Jamaican%20Soursop.asp>. Accessed March 17, 2008.

83. Ochse, J. J., & Bakhuizen van den Brink, R. C. Jr., (1931). *Vegetables of the Dutch East Indies*. Buitenzorg, Java: Department of Agriculture, Industry and Commerce of the Netherlands East Indies.

84. Nakasone, H. Y., & Paull, R. E. (1998). Annonas. In H. Y. Nakasone, & R. E. Paull, (Eds.), *Tropical Fruits* (pp. 45–75). London, UK: CAB International.

85. Benero, J. R., Rodriguez, A. J., & Roman de Sandoval, A. (1971). A soursop pulp extraction procedure. *Journal of Agriculture University of Puerto Rico*, 55, 518–519.

86. Benero, J. R., Collazo e Riviera, A. L., & De George, L. M. (1974). Studies on the preparation and shelf-life of soursop, tamarind and blended soursop-tamarind soft drinks. *Journal of Agriculture University of Puerto Rico*, 58, 99–104.

87. Cloudforest Café, (2008). *A Detailed Annona Report*. <www.cloudforest.com>. Accessed February 1st, 2009.

88. Vaillant, F., Millan, P., Tchiliguirian, C., & Reynes, M. (1998). Preliminary characterization of the cell wall polysaccharides of the soursop pulp and study of their enzyme degradation. *Fruits*, 53, 257–270.

89. Telis-Romero, J., Beristain, C. I., Gabas, A. L., & Telis, V. R. N. (2007). Effect of apparent viscosity on the pressure drop during fluidized bed drying of soursop pulp. *Chemical Engineering and Processing*, 46, 684–694.

90. de Oliveira, S. L., Guerra, N.b., Maciel, M. I. S., & Livera, A. V. S. (1994). Polyphenoloxidase activity, polyphenols concentration and browning intensity during soursop (*Annona muricata*) maturation. *Journal of Food Science*, 59, 1050–1052.

91. Payumo, E. M., Pilac, L. M., & Mnaiguis, P. L. (1965). The preparation and storage properties of canned guwayabanom (*Annona muricata* L.) concentrate. *Phillipine Journal of Science*, 94, 161–169.

92. Peters, M., Badrie, N., & Comissiong, E. (2001). Processing and quality evaluation of soursop (*Annona muricata* L.). *Journal of Quality*, 24, 361–374.

93. Torres, W. E., & Sánchez, L. A. (1992). *Fruticultura Colombiana Guanábano. (Spanish)*. Bógota, Colombia: Instituto Colombiano Agropecuario. ICA Manual de Asistencia Técnica 57.

94. Fasakin, A. O., Fehintola, E. O., Obijole, O. A., & Oseni, O. A. (2008). Compositional analyses of the seed of soursop, *Annona muricata* L., as a potential aminal feed. *Science Research Essays* 3(10), 521–523.

95. Jaramillo-Flores, M. E., & Hernández-Sánchez, H. (2000). Thermal diffusivity of soursop (*Annona muricata* L.) pulp. *Journal of Food Engineering*, 46, 139–143.

96. MacLeod, A. J., & Pieris, N. M. (1981). Volatile flavor components of soursop (*Annona muricata*). *J. Agric. Food Chem.*, 29, 488–490.

97. Augusto, F., Valente, A. L. P., dos Santos Tada, E., & Rivellino, S. R. (2000). Screening of Brazilian fruit aromas using solid-phase microextraction–gas chromatography–mass spectrometery. *J. Chromatogr. A*, 873, 117–127.

98. Jirovetz, L., Buchbauer, G., & Nagassoum, M. B. (1998). Essential oil compounds of the *Annona muricata* fresh fruit pulp from Cameroon. *J. Agri Food Chem.*, 46, 3719–3720.

99. Pino, J. A., Agüero, J., & Marbot, R. (2001). Volatile components of soursop (*Annona muricata* L.). *Journal of Essential Oil Research*, 13, 140–141.

100. Zhang, X. R. (1991). *Identification and changes of volatile constituents of soursop (*Annona muricata* L.) during ripening*. Master's thesis. Hawaii, USA: University of Hawaii.

101. Iwaoka, W. T., Zhang, X., Hamilton, R. A., Chia, C. L., & Tang, C. S. (1993). Identifying volatiles in soursop and comparing their changing profiles during ripening. *Horticultural Science*, 28, 817–819.

102. Jirovetz, L., Butchbauer, G., & Ngassoum, M. B. (1998). Essential oil compounds of *Annona muricata* fresh pulp from the Cameroon. *J. Agric. Food Chem.*, 46, 3719–3720.

103. Pelissier, Y., Marion, C., Kone, D., Lamaty, G., Menut, C., & Bessiere, J. M. (1994). Volatile components of *Annona muricata* L. *Journal of Essential Oil Research.*, 6, 411–414.

104. Quisumbing, E. (1951). *Medicinal plants of the Philippines*. Philippines: Philippine Department of Agriculture Natural Resource, Manila Technical Bulletin.

105. Arkcoll, D. (1990). New crops from Brazil. In J. Janick, & J. E. Simon, (Eds.), *Advances in New Crops*, (pp. 367–371). Portland, OR: Timber Press.

106. Abbo, E. S., Olurin, T. O., & Odeyemi, G. (2006). Studies on the storage stability of soursop (*Annona muricata* L.). *African Journal of Biotechnology*, 5, 1808–1812.

107. Morton, J. F. (1980). Caribbean and Latin American folk medicine and its influence in the United States. *Quarterly Journal of Crude Drug Research*, 18, 57–75.

C. ACTIONS OF INDIVIDUAL FRUITS IN DISEASE AND CANCER PREVENTION AND TREATMENT

108. Leaman, D. J. (1995). Malaria remedies of the Kenyah of the Apo Kayan, East Kalimantan, Indonesia Borneo: a quantitative assessment of local consensus as an indicator of biological efficacy. *J. Ethnopharmacol.*, *49*, 1–16.

109. Branch, L. C., & Da Silva, I. M. F. (1983). Folk medicine of Alter Do Chao, Para, Brazil. *Acta Amazonica*, *13*, 737–797.

110. Mors, W. B., Rizzini, C. T., & Pereria, N. V. (2000). In R. A. DeFilipps, (Ed.), *Medicinal Plants of Brazil* Algonac, MI: Reference Publications.

111. Santos, A. F., & Sant'Ana, A. E. G. (2001). Molluscicidal properties of some species of Annona. *Phytomedicine*, *8*, 115–120.

112. Holdsworth, D. K. (1990). Traditional medicinal plants of Rarotonga, Cook Islands. Part 1. *International Journal of Crude Drug Research*, *28*, 209–218.

113. Morton, J. F. (1968). A survey of medicinal plants of Curacao. *Economic Botany*, *22*, 87–102.

114. Hodge, W. H., & Taylor, D. (1956). The ethnobotany of the Island Caribs of Dominica. *Webbia*, *12*, 513–644.

115. Caceres, A., Lopez, B. R., Giron, M. A., & Logemann, H. (1991). Plants used in Guatemala for the treatment of dermatophytic infections. 1. Screening for antimycotic activity of 44 plant extracts. *J. Ethnopharmacol.*, *31*, 263–276.

116. Haddock, R. L. (1974). Some medicinal plants of Guam including English and Guamanian common names. Report of the Regional and Technical Meeting on Medicinal Plants, Papeete, Tahiti, November, 1973, South Pacific Commission, Noumea, New Caledonia, p. 79.

117. Grenand, P., Moretti, C., & Jacquemin, H. (1987). Pharmacopees traditionnels en Guyane: Creoles, Palikur, Wayapi. Collection Mem., No 108, Paris, France, pp. 569–575..

118. Weiniger, B., Rousier, M., Daguilh, R., Henrys, J. H., & Anton, R. (1986). Popular medicine of the central plateau of Haiti. 2. Ethnopharmacological inventory. *J. Ethnopharmacol.*, *17*, 13–30.

119. Asprey, G. F., & Thornton, P. (1955). Medicinal plants of Jamaica. III. *West Ind. Med. J.*, *4*, 69–92.

120. Novy, J. W. (1997). Medicinal plants of the eastern region of Madagascar. *J. Ethnopharmacol.*, *55*, 119–126.

121. Ahmad, F. B., & Holdsworth, D. K. (1994). Medicinal plants of Sabah, Malaysia. Part II. The Muruts. *International Journal of Pharmacognosy*, *32*, 378–383.

122. Ilham, M., Yaday, M., & Norhanom, A. W. (1995). Tumour promoting activity of plants used in Malaysian traditional medicine. *Nat. Prod. Sci.*, *1*, 31–42.

123. De Feo, V. (1992). Medicinal and magical plants in the northern Peruvian Andes. *Fitoterapia*, *63*, 417–440.

124. Hasrat, J. A., De Bruyne, T., De Backer, J. P., Vauquelin, G., & Vlietinck, A. J. (1997). Isoquinoline derivates isolated from the fruit of Annona muricata as 5-HTergic and 5-HT1α receptor agonists in rats: unexploited antidepressive (lead) products. *J. Pharmeut. Pharmacol.*, *49*, 1145–1149.

125. Gbeassor, M., Kedjagni, A. Y., Koumaglo, K., De Souza, C., Agbo, K., Aklikokou, K., & Amegbo, K. A. (1990). *In vitro* antimalarial activity of six medicinal plants. *Phytother. Res.*, *4*, 115–117.

126. Ayensu, E. S. (1978). Medicinal Plants of the West Indies. Unpublished manuscript.

127. Schultes, R. E., & Raffauf, R. F. (1990). *The Healing Forest: Medicinal and Toxic Plants of the Northwest Amazonia*. Portland, OR: Discorides Press.

128. Anonymous. (1985). In *Wealth of India, Revised. Raw materials* (Volume 1A). (pp. 278–295). New Delhi: CSIR.

129. Padma, P., Chansouria, J. P. N., & Khosa, R. L. (1997). Effect of alcohol extract of *Annona muricata* on cold immobilization stress induced tissue lipid peroxidation. *Phytother. Res.*, *11*, 326–327.

130. Lans, C. A. (2006). Ethnomedicines used in Trinidad and Tobago for urinary problems and diabetes mellitus. *J. Ethnobiol. Ethnomed.*, *2*, 45–55.

131. Taylor, L. (2002). Graviola (*Annona muricata*). In *Herbal Secrets of the Rainforest* 2nd ed. Roseville, CA: Prima Publishing.

132. Meyer, T. M. (1941). The alkaloids of Annona muricata. *De Ing Nederland-Indie*, *8*, 64–70.

133. Feng, P. C., Haynes, L. J., Magnus, K. E., Plimmer, J. R., & Sherratt, H. S. (1962). Pharmacological screening of some West Indian medicinal plants. *J. Pharmeut. Pharmacol.*, *14*, 556–561.

134. Padma, P., Chansauria, J. P. N., Khosa, R. L., & Ray, A. K. (2001). Effect of Annona muricata and Polyalthia ceradoides on brain neurotransmitters and enzyme monoamine oxidase following cold immobilization stress. *Journal of Natural Remedies*, *1*, 144–146.

135. Meyer, B. N., Ferrigni, N. R., Putnam, J. E., Jacobson, L. B., Nichols, D. E., & McLaughlin, J. L. (1982). Brine shrimp: a convenient general bioassay for active plant constituents. *Planta Med.*, *45*, 31–34.

136. Rieser, M. J., Kozlowski, J. F., Wood, K. V., & McLaughlin, J. L. (1991). On the mode of baker's yeast reduction of enantiomeric 4-acyl butanolides. *Tetrahedron Letter*, *32*, 1137.

137. Reiser, M. J., Fang, X. P., Rupprecht, J. K., Hui, Y. H., Smith, D. L., & McLaughlin, J. L. (1993). Bioactive single-ring acetogenins from seed extracts of Annona muricata. *Planta Med.*, *59*, 91–92.

138. Giard, D. J., Aaronson, S. A., Todaro, G. J., Arnstein, P., Kersey, J., Dosik, H., & Parks, W. P. (1973). *In vitro*

cultivation of human tumors: establishment of cell lines derived from a serious of solid tumors. *J. Natl Cancer Inst.*, *55*, 1417–1423.

139. Soule, H. D., Vazquez, L. J., Albert, A. S., & Brennan, M. (1973). A human cell line from a pleural effusion derived from a breast carcinoma. *J. Natl Cancer Inst.*, *51*, 1409–1416.

140. Fogh, J., & Trempe, G. (1975). New human tumor cell lines. In J. Fogh, (Ed.), *Human Tumor Cells In Vitro* (pp. 115–159). New York: Plenum Press.

141. Rupprecht, J. K., Hui, Y. H., & McLaughlin, J. L. (1990). Annonaceous acetogenins: a review. *J. Nat. Prod.*, *53*, 237–278.

142. Kim, G. S., Zeng, L., Alali, F., Rogers, L. L., Wu, F. E., Sastrodihardio, S., & McLaughlin, J. L. (1998). Muricoreacin and murihexocin C, mono-tetrahydro-furan acetogenins, from the leaves of *Annona muri-cata*. *Phytochemistry*, *49*, 565–571.

143. Ahammadsahib, K. I., Hollingworth, R. M., McGovren, J. P., Hui, Y. H., & McLaughlin, J. L. (1993). Mode of action of bullatacin: A potent antitumor and pesticidal annonaceous acetogenin. *Life Sci.*, *53*, 1113–1120.

144. McLaughlin, J. L., & Rogers, L. L. (1998). The use of biological assays to evaluate botanicals. *Drug Inf. J.*, *32*, 513–524.

145. Oberlies, N. H., Chang, C., & McLaughlin, J. L. (1997). Structure-activity relationships of diverse Annonaceous acetogenins against multidrug resistant human mammary adenocarcinoma (MCF-7/Adr) cells. *Journal of Medicinal Chemistry*, *40*, 2102–2106.

146. Cave, A., Figadere, B., Laurens, A., & Cortes, D. (1997). Acetogenins from Annonaceae. In W. Herz, D. W. Kirby, R. E. Moore, W. Steglich, & C. Tamm, (Eds.), *Progress in the Chemistry of Organic Natural Products* (pp. 81–88). New York: Springer.

147. Lannuzel, A., Michel, P. P., Caparros-Lefebvre, D., Abaul, J., Hocquemiller, R., & Ruberg, M. (2002). Toxicity of annonaceae for dopaminergic neurons: potential role in atypical parkinsonism in Guadeloupe. *Mov. Disord.*, *17*, 84–90.

148. Lannuzel, A., Michel, P. P., Hoglinger, G. U., Champy, P., Jousset, A., & Medja, F. (2003). The mitochondrial complex I inhibitor annonacin is toxic to mesencephalic dopaminergic neurons by impairment of energy metabolism. *Neuroscience*, *121*, 287–296.

149. Champy, P., Hoglinger, G. U., Feger, J., Gleye, C., Hocquemiller, R., & Laurens, A. (2004). Annoncin, a lipophilic inhibitor of mitochondrial complex I, induces nigral and striatal neurodegeneration in rats: possible relevance for atypical parkinsonism in Guadeloupe. *J. Neurochem.*, *88*, 63–69.

150. Lannuzel, A., Hoglinger, G. U., Champy, P., Michel, P. P., Hirsch, E. C., & Ruberg, M. (2006). Is atypical parkinsonism in the Caribbean caused by the consumption of Annonacea?. *J. Neural Transmission (Suppl).*, *70*, 153–157.

151. Caparros-Lefebvre, D., & Elbaz, A. (1999). A possible correlation of atypical parkinsonism in the French West Indies with consumption of topical plants: a case-control study. Caribbean Parkinsonism Study Group. *Lancet*, *354*, 281–286.

152. Steele, J. C., Morris, H. R., Lees, A. J., Perez-Tur, J., & McGeer, P. L. (1999). Atypical parkinsonism in the French West Indies. *Lancet*, *354*, 1474.

153. Caparros-Lefebvre, D. (2004). Atypical parkinsonism in New Caledonia: comparison with Guadeloupe and association with Annonaceae consumption. *Mov. Disord.*, *19*, 604.

154. Caparros-Lefebvre, D., & Steele, J. (2005). Atypical parkinsonism on Guadeloupe, comparison with the parkinsonism-dementia complex of Guam, and environmental toxic hypotheses. *Environmental Toxicology and Pharmacology*, *19*, 407–413.

155. Lannuzel, A., Hoglinger, G. U., Verhaeghe, S., Gire, L., Belson, S., Escobar-Khondiker, M., Poullain, P., Oertel, W. H., Hirsch, E. C., Dubois, B., & Ruberg, M. (2007). Atypical parkinsonism in Guadeloupe: a common risk factor for two closely related phenotypes? *Brain*, *130*, 816–827.

156. Lannuzel, A., Ruberg, M., & Michel, P. P. (2008). Atypical parkinsonism in the Caribbean island of Guadeloupe: etiological role of the mitochondrial complex I. *Mov. Disord.*, *23*, 2122–2128.

157. Champy, P., Melot, A., Eng, V. G., Gleye, C., Fall, D., Hoglinger, G. U., Ruberg, M., Lannuzel, A., Laprevote, O., Laurens, A., & Hocquemiller, R. (2005). Quantification of acetogenins in *Annona muricata* linked to atypical parkinsonism in Guadeloupe. *Mov. Disord.*, *20*, 1629–1633.

158. Caparros-Lefebvre& Elbaz, D. (2006). A Possible correlation of atypical parkinsonism in the French West Indies with consumption of topical plants: a case-control study. Caribbean Parkinsonism Study Group. *Lancet*, *354*, 281–286.

159. Calorie Count (2009). *Calories in soursop.* <http://caloriecount.about.com/calories-soursop-i9315>. Accessed 13th July, 2009.

C. ACTIONS OF INDIVIDUAL FRUITS IN DISEASE AND CANCER PREVENTION AND TREATMENT

40

Carotenoids in Vegetables: Biosynthesis, Occurrence, Impacts on Human Health, and Potential for Manipulation

Dean A. Kopsell[1] and David E. Kopsell[2]

[1]Plant Sciences Department, The University of Tennessee, Knoxville, TN, USA
[2]Department of Agriculture, Illinois State University, Normal, IL, USA

1. INTRODUCTION

Mounting scientific evidence provides the association between dietary choices and chronic disease expression. Dietary guidelines, now in place, are designed to prevent the onset of such chronic diseases as tissue-specific cancers, cardiovascular diseases, and osteoporosis. The cornerstone of recommended dietary guidelines is increased consumption of fruits and vegetables. Current United States Department of Agriculture (USDA) dietary guidelines recommend eating 7–9 servings of fruits and vegetables per day. However, average adult consumption in the USA is only 4.4 servings per day, with an estimated 42% of Americans eating <2 daily servings for fruits and vegetables (see: http://www.healthierus.gov/dietaryguidelines). Consumption of vegetables provides the human

diet with many essential vitamins and minerals important for health maintenance. Vegetables also contain secondary metabolite phytochemicals, which provide benefits beyond normal health maintenance and nutrition and play active roles in chronic disease reductions. One important class of phytochemicals is the carotenoids. Carotenoids are lipid-soluble pigments found in all photosynthetic organisms. Among the naturally occurring plant pigments, carotenoids are widely distributed, demonstrate a high degree of structural diversity, and possess large variations in biological functions [1,2]. There are over 600 carotenoids found in nature, with 40 dietary carotenoids regularly consumed as part of a typical human diet [3]. The many health benefits attributed to carotenoid intake include prevention of certain cancers [4–6], cardiovascular diseases [7], and aging eye diseases

[8,9], as well as enhanced immune system functions [10,11]. Pro-vitamin A activity is the classical biological function of carotenoids in mammalian systems.

Research into carotenoid enhancement of vegetable crops to benefit human health has paralleled efforts to increase consumption of fruits and vegetables in the diet. Release of carotenoid compounds from the membranes of plant tissues facilitates intestinal absorption; however, changes in carotenoid chemistry by biotic and abiotic factors can influence bioavailability. Current methods to assess bioavailability include serum measurements and *in vitro* digestion models. Programs designed to improve carotenoid levels in vegetable crop tissues must successfully link plant physiology with accurate bioavailability assessments in human subjects. The focus of this chapter will be discussions of the current knowledge of carotenoid chemistry, bioavailability assessments, the impacts of certain carotenoids on human health and disease suppression, how pre- and post-harvest cultural practices can alter carotenoid levels and influence potential bioavailability, and current vegetable carotenoid enhancement efforts.

2. STRUCTURAL CHEMISTRY AND THE PLANT CAROTENOID BIOSYNTHETIC PATHWAY

Carotenoids are C_{40} isoprenoid polyene compounds that form lipid-soluble yellow, orange, and red pigments [12,13]. They are considered secondary plant metabolites and are divided into two main structural groups: the oxygenated xanthophylls such as lutein (3R,3′R,6′R β,ε-carotene-3, 3′diol), zeaxanthin (3,3′R-β,β-carotene-3,3′diol), and violaxanthin (3S,5R,6S,3′S,5′R,6′S-5,5,5′, 6′-depoxy-5,6,5′,6′-tetrahydro-β, β-carotene-3,3′ diol), and the hydrocarbon carotenes such as β-carotene (β,β-carotene), α-carotene (6′R, β,ε-carotene), and lycopene (ψ,ψ-carotene) [12]. In higher plants, carotenoid compounds are synthesized and localized in cellular plastids and are associated with light-harvesting complexes in the thylakoid membranes, or present as semicrystalline structures derived from the plastids [14,15].

The carotenoid biosynthetic pathway in plants was elucidated in the mid-1960s [15]. Carotenoids are produced in the plastids and are derived via the isopentenyl diphosphate biochemical pathway. In the first step in biosynthesis, isopentenyl diphosphate is isomerized to dimethylallyl diphosphate, which becomes the substrate for the C_{20} geranylgeranyl diphosphate. The enzyme geranylgeranyl diphosphate synthase catalyzes the formation of geranylgeranyl diphosphate from isopentenyl diphosphate and dimethylallyl diphosphate [16,17]. The first step unique to carotenoid biosynthesis is the condensation of two molecules of geranylgeranyl diphosphate to form the first C_{40} carotenoid, the colorless phytoene pigment, via phytoene synthase [1]. Two structurally similar enzymes, phytoene desaturase and ξ-carotene desaturase, make the conversions of phytoene to lycopene via several important intermediates [2,16,17]. These desaturase enzymes create the chromophore present in the carotenoid pigments, and change the colorless phytoene into the pink-colored lycopene. The carotenoid pathway then branches at the cyclization reactions of lycopene to produce carotenoids with either two β-rings (e.g. β-carotene, zeaxanthin, anteraxanthin, violaxanthin, and neoxanthin) or carotenoids with one β-ring and one ε-ring (e.g. α-carotene and lutein) [16,18]. The pathway advances with the additions of oxygen moieties, which convert the hydrocarbons, α-carotene and β-carotene, into the oxygenated subgroup referred to as the xanthophylls. Further steps in xanthophyll synthesis include epoxidation reactions. The reversible epoxidation/de-epoxidation reaction converting violaxanthin back to zeaxanthin via the intermediate antherxanthin is collectively referred to as the violaxanthin cycle and is vital for energy dissipation from incoming solar radiation [15,17]

(Fig. 40.1). Currently, genes and cDNAs for the major enzymes functioning in carotenoid biosynthesis have been cloned from plant, algal, and microbial sources [16].

Within the plant thylakoid membranes of chlorophyll organelles, carotenoids are found in close association with the specific protein complexes of photosystem I and photosystem II. Carotenoids function to help harvest light energy, mostly in the blue-green wavelength range, which is transferred to the photosynthetic reaction centers. In the photosystem II complex, β-carotene is highly concentrated close to the reaction center, while lutein is present in several light-harvesting antennae components [19,20]. When the absorption of light radiation exceeds the capacity of photosynthesis, excess excitation energy can result in the formation of triplet excited chlorophyll (^{3}Chl) and reactive singlet oxygen (^{1}O$_2$). Carotenoid pigments protect photosynthetic structures by quenching excited ^{3}Chl to dissipate excess energy, and by binding ^{1}O$_2$ to inhibit oxidative damage [20–23]. The antioxidant capacity attributed to the carotenoid chemical structure conveys the ability of these compounds to protect plant photosynthetic machinery in time of excess excitation, and is also responsible for the antioxidant ability of carotenoids in the prevention of chronic diseases in humans. Xanthophylls may also be involved in structural stabilization of light-harvesting complexes and reduction of lipid peroxidation [21].

The conjugated double-bond systems of the carotenoids create the light-absorbing chromophores which result in the distinctive colors associated with carotenoid plant pigments [16]. In the pigment–protein complexes of photosynthetic organisms, the all-*trans* configurations of carotenoids are the major component of the light-harvesting complexes, while the photosynthetic reaction centers contain 15-*cis* carotenoid configurations [24–26]. The all-*trans* carotenoids in the light-harvesting complexes provide efficient singlet-energy transfer to

chlorophyll molecules, and thus participate mainly in light-harvesting. The 15-*cis* carotenoids show a preference for isomerization toward the all-*trans* configurations upon excitation, and thus are better suited for photoprotective functions in the reaction centers [24]. Therefore, there is a physiological basis for carotenoid isomerization in plant tissues.

3. EPOXIDATION AND ISOMERIZATION OF CAROTENOID STRUCTURES

In the human body, oxidants produced during normal metabolism and immune defense against infectious and/or chemical agents are responsible for damage to cellular tissues, DNA structures, and proteins [27,28]. This harmful oxidative damage is considered the major cause of aging and degenerative diseases such as cancer, cardiovascular disease, immune-system decline, and cataract. Compounds such as ascorbate, α-tocopherol, and carotenoids are examples of antioxidants possessing the ability to quench reactive oxygen species [28]. The physical properties of carotenoid molecules, especially the conjugated carbon–carbon double-bond system, permit the quenching of ^{1}O$_2$. The localization of carotenoid molecules in biological tissues will also influence their ability to encounter and scavenge free radicals. *In vivo* antioxidant activity is determined by carotenoid structure and concentration, as well as the nature and concentration of the reactive oxygen species present. In fact, carotenoids are one of the most potent biological quenchers of reactive oxygen species [29].

The antioxidant activity of carotenoids comes from the susceptibility of 5,6 and 5′,6′ double bonds in their cyclic end groups to undergo epoxidation with ^{1}O$_2$ [30,31]. Based on their chemistry, epoxide isomers would lack antioxidant activity because they are unable to bind more ^{1}O$_2$, and some may even have pro-oxidant

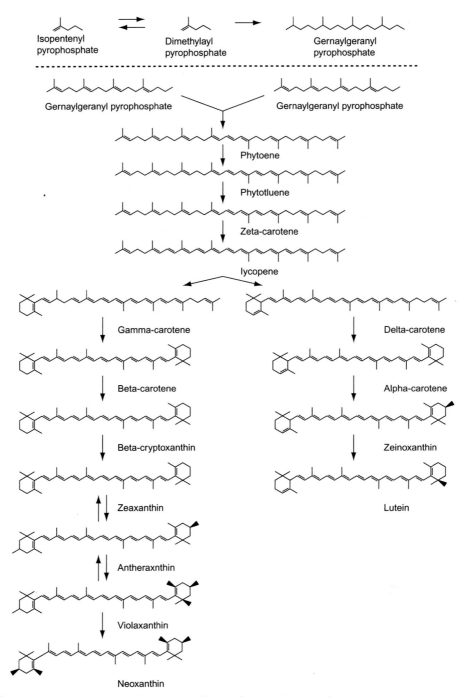

FIGURE 40.1   A simplified version of the carotenoid biosynthetic pathway in plants.

TABLE 40.1   Carotenoid Pigments Identified and Quantified in the Edible Tissues of Major Vegetable Crops

Commodity	Carotenoid Pigments Identified in Edible Tissues	Reference
Beans, green	all-*trans* β-carotene, all-*trans* lutein, 9-*cis* lutein, 9′-*cis* lutein, 13-*cis* lutein, all-*trans* lutein epoxide, 9′-*cis* neoxanthin, neolutein, all-*trans* violaxanthin, all-*trans* zeaxanthin, 9-*cis* zeaxanthin, 13-*cis* zeaxanthin	[36,109]
Broccoli	all-*trans* β-carotene, all-*trans* lutein, 9-*cis* lutein, 9′-*cis* lutein, 13-*cis* lutein, all-*trans* and *cis* lutein epoxide, neolutein, all-*trans* neoxanthin, 9′-*cis* neoxanthin, violaxanthin, all-*trans* zeaxanthin, 9-*cis* zeaxanthin, 13-*cis* zeaxanthin	[36,109]
Cabbage	β-carotene, lutein, lutein epoxide, neoxanthin, violaxanthin, zeaxanthin	[36,48,80]
Carrot	all-*trans* α-carotene, all-*trans* β-carotene, lutein, lycopene	[47,93,96]
(Sweet) Corn	α-carotene, β-carotene, β-cryptoxanthin, all-*trans* lutein, 9-*cis* lutein, 9′-*cis* lutein, 13-*cis* lutein, all-*trans* zeaxanthin, 9-*cis* zeaxanthin	[36,56]
Kale/collards	all-*trans* β-carotene, all-*trans* lutein, 9-*cis* lutein, 9′-*cis* lutein, 13-*cis* lutein, all-*trans* and *cis* lutein epoxide, neolutein, all-*trans* neoxanthin, 9′-*cis* neoxanthin, violaxanthin, all-*trans* zeaxanthin, 9-*cis* zeaxanthin, 13-*cis* zeaxanthin	[36,48,49,109]
Lettuce	all-*trans* β-carotene, all-*trans* lutein, 9-*cis* lutein, 9′-*cis* lutein, 13-*cis* lutein, all-*trans* and *cis* lutein epoxide, neolutein, all-*trans* neoxanthin, 9′-*cis* neoxanthin, violaxanthin, all-*trans* zeaxanthin, 9-*cis* zeaxanthin, 13-*cis* zeaxanthin	[36,55,109]
Pepper	α-carotene, β-carotene, β-cryptoxanthin, capsanthin, lutein, zeaxanthin	[86]
Spinach	all-*trans* β-carotene, all-*trans* lutein, 9-*cis* lutein, 9′-*cis* lutein, 13-*cis* lutein, all-*trans* and *cis* lutein epoxide, neolutein, all-*trans* neoxanthin, 9′-*cis* neoxanthin, violaxanthin, all-*trans* zeaxanthin, 9-*cis* zeaxanthin, 13-*cis* zeaxanthin	[36,96,109]
Squash	α-carotene, β-carotene, β-cryptoxanthin, lutein, neurosporene, neoxanthin, phytofluene, violaxanthin	[108]
Tomato (raw)	all-*trans* β-carotene, all-*trans* γ-carotene, all-*trans* δ-carotene, ξ-carotene, all-*trans* lutein, all-*trans* lycopene, neurosporene, phytoene, phytofluene, lycopene-5,6 diol	[37,67,110]
Watermelon	β-carotene, phytofluene, all-*trans*-lycopene, *cis*-lycopene	[68]

Source: Table reprinted in part from *Trends in Plant Science*, Vol. 11/No. 10, D.A. Kopsell and D.E. Kopsell, Assessing bioavailability of carotenoids in vegetable crops, pp. 499–507, Copyright 2006, with permission from Elsevier.

activity. Epoxide forms of carotenoids, with a bound oxygen to the 5,6 or 5′,6′ position are present in the edible tissues of many different vegetable crops (Table 40.1). The ratio of lutein epoxide:all-*trans*-lutein was 1:1.5 in cabbage (*Brassica oleracea* L. var. *capitata*), 1:2 in broccoli (*B. oleracea* L. var. *botrytis*), 1:6 in spinach (*Spinacia oleracea* L.), and 1:23 in kale (*B. oleracea* L. var. Acephala) [32]. Low doses of lycopene from tomato (*Lycopersicon esculentum* Mill.) products significantly increased serum lycopene levels and reduced lipid peroxidation *in vivo* [33]. However, the complexity of measuring *in vivo* antioxidant behavior, the variability associated with carotenoid content of vegetables, and the nutritional status of subjects used in human dietary intervention studies will affect interpretation of results [14,31].

All carotenoids exhibit *cis-trans* isomerization, and both isomeric groups can be found in vegetable crops (Table 40.1) [32,34–36]. It is the presence of the conjugated double-bonds in the carotenoid structure that causes considerable isomerization to occur. The *trans* isomers are more stable and most commonly found in intact foods. Phytoene exists predominately as the 15-*cis* isomer; however, the predominate isomer of lycopene is the all-*trans* geometric form,

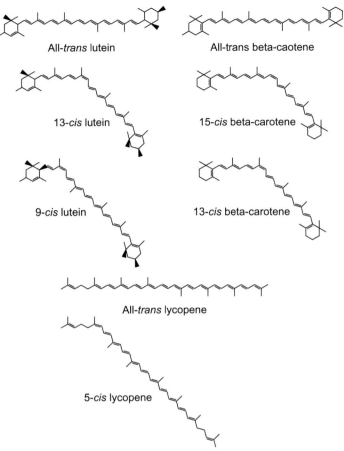

All-*trans* lutein

All-trans beta-caotene

13-*cis* lutein

15-*cis* beta-carotene

9-*cis* lutein

13-*cis* beta-carotene

All-*trans* lycopene

5-*cis* lycopene

FIGURE 40.2 Structures of some common all-*trans* and *cis* carotenoid isomers found in vegetable crops.
The occurrence and properties of carotenoid isomers in food crops can influence intestinal absorption. Source: reprinted from *Trends in Plant Science*, Vol. 11/No. 10, D.A. Kopsell and D.E. Kopsell, Assessing bioavailability of carotenoids in vegetable crops, pp. 499–507, Copyright 2006, with permission from Elsevier.

indicating isomerase activity is present in higher plants to mediate *cis* to *trans* conversions [15]. All-*trans* carotenoids in plants are susceptible to photo, thermal, and chemical isomerization [24]. This can result in post-harvest cultural activities such as harvest, transport, and environmental storage conditions having a tremendous effect on the stability and preservation of carotenoid compounds in edible vegetable crop tissues.

Carotenoid *cis-trans* isomers differ in their intestinal absorption in humans (Fig. 40.2).

Human blood plasma contains mostly all-*trans* carotenoids, but some plasma carotenoids can be found as high as 50% in the *cis* form [37]. As mentioned previously, the majority of lycopene found in fresh and processed tomatoes exists in the all-*trans* form [38,39]. However, lycopene in human and animal tissues exits predominantly as *cis*-isomers, indicating possible preference for *cis*-lycopene in intestinal absorption [40,41]. In contrast, greater excretions of *cis*-β-carotene and lower excretions of *trans*-β-carotene were measured in human subjects after ingestion of both raw and processed carrots (*Daucus carota* L. var. *sativa*). Such data may indicate an absorption preference for all-*trans* β-carotene [42]. Differences in intestinal absorption among the carotenoid isomers have been established; however, very little is currently known about the biological significance of these different isomers in human health [43].

4. VEGETABLE CAROTENOIDS AND THEIR IMPACT ON HUMAN HEALTH

Epidemiological data clearly support the positive association between higher dietary intake of foods high in carotenoids and greater carotenoid tissue concentrations with lower risks of certain chronic diseases. Many of these disease suppressing abilities can be attributed to the antioxidant properties of carotenoids. One of the most important physiological functions of carotenoids in human nutrition is to act as vitamin A precursors. Pro-vitamin A carotenoid compounds support the maintenance of healthy epithelial cell differentiation, normal reproductive performance, and visual functions [44]. Both pro-vitamin A carotenoids (β-carotene, α-carotene, and cryptoxanthins) and non-provitamin A carotenoids (lutein, zeaxanthin, lycopene) function as free radical scavengers, enhance the immune response, suppress cancer development, and protect eye tissues [45]. Humans cannot synthesize carotenoids, and therefore must rely on dietary sources to provide sufficient levels. Fruits and vegetables are primary sources of carotenoids in the human diet and their consumption has been associated with numerous health benefits [27,30]. Studies indicate that a high intake of a variety of vegetables, providing a mixture of carotenoids, was more strongly associated with reduced cancer and eye disease risk than intake of individual carotenoid supplements [9].

The colors visible in vegetable crop tissues come from the presence of anthocyanin, betalain, chlorophyll, and carotenoid pigments. Vegetable crop species differ in the composition and concentrations of these pigments present. Carotenoid pigments are responsible for the brilliant reds, yellows, and oranges that become visible as fruit tissues of vegetable crops ripen and mature. Carotenoids are still present in leafy tissues, although they are masked by the high concentrations of the green chlorophyll pigments. There is a wealth of scientific information which demonstrates that vegetable carotenoid concentrations are influenced by both genetic and environmental factors, both pre- and postharvest. In this respect, it would be misleading to list the reported carotenoid concentrations identified thus far in vegetable crops. For example, simply selecting one spinach cultivar vs. another may result in greater than a 2-fold difference in leaf tissue lutein concentrations, based solely on genetic propensity [46]. It is therefore advisable to consult current web data published by the USDA-Agricultural Research Service Nutrient Database Laboratory located in Beltsville, MD. The database is routinely updated and maintains a searchable menu of the nutritional content, including the major carotenoids, of the top foods and food products sold in the USA (see: http://www.nal.usda.gov/fnic/foodcomp/search/).

4.1 Vegetable Sources of β-*carotene and Its Impact on Human Health*

β-carotene participates as an accessory pigment in light absorption and energy dissipation in photosynthesis, as well as general antioxidant functions. Therefore, β-carotene can be found in leaf, fruit, and even root tissues of many vegetable crops (Table 40.1). The root crop carrot (*Daucus carrota* L.) has some of the highest concentrations of β-carotene. β-carotene levels in carrot can range from 3.2 to 6.1 mg/100 g fresh weight [47]. However, cruciferous leafy vegetable crops can have concentrations equal to, or higher than carrot. Values in kale and collards are reported to range from 3.8 to 10.0 mg β-carotene/100 g fresh weight [48,49].

Vitamin A deficiency is the single most important cause of childhood blindness in developing countries around the world, and subclinical levels can contribute significantly to increased child mortality [44]. Vitamin A can be consumed in the diet as both preformed retinoids from animal tissues and as pro-vitamin A carotenoids found in plant tissues. The major sources for vitamin A for most populations worldwide are the plant-based pro-vitamin A carotenoids. The same factors that may limit carotenoid bioavailability (see Section 5) also affect vitamin A status. The bioconversion of pro-vitamin A carotenoids is mediated by a predominantly cytosolic enzyme, β-carotene 15,15'-deoxygenase, present in the intestinal mucosa, the liver, and the corpus luteum [44]. The activity of this enzyme is most efficient for β-carotene; however, other carotenoids can be metabolized to yield retinal (α-carotene and cryptoxanthins). After conversion of the pro-vitamin A carotenoids, vitamin A is transported and stored in the liver mainly as retinyl esters. Metabolism of vitamin A occurs following esterification, conjugation, oxidation, and/or isomerization reactions, after which retinal forms participate in maintenance of epithelial cell differentiation, reproductive performance, and visual functions [44].

The chemical structure of β-carotene makes it an efficient *in vitro* neutralizer of singlet oxygen ($^1O_2$), and to a lesser extent, an effective agent at reducing lipid peroxidation. Early research that demonstrated both its antioxidant and antigenotoxic properties resulted in β-carotene being one of the most extensively studied cancer chemopreventative agents in research supported by the National Cancer Institute [50]. Unexpectedly, results from three separate clinical trials revealed that β-carotene supplements, either alone or in combination with vitamin E, administered for cancer prevention actually increased incidences of lung cancers in heavy smokers and asbestos workers [51,52]. Based on the results showing pro-oxidant behavior in the presence of tobacco smoke, it appears plausible that β-carotene could be causing stimulation of pre-existing latent tumors under these conditions, rather than initiating tumorigenesis [50]. However, research showed β-carotene supplementation in animal models significantly increased phase I carcinogen enzymes in the lung, including the cytochrome P450s of CYP1A1, CYP1A2, CYP3A, CYP2B1, and CYP2A [53]. It is hypothesized that high β-carotene supplementation may increase tissue oxidative stress, or could act synergistically with known carcinogenic chemicals (present in tobacco smoke) in CYP induction. In a current review on the subject, Paolini et al. [50] clearly demonstrate that detrimental effects of individual β-carotene supplements are possible when subject individuals are exposed to environmental mutagens and carcinogens. The authors still encourage a diet high in fruit and vegetables, but warn of the possible dangers in consuming high concentrations of one or more isolated supplements.

4.2 Vegetable Sources of Lutein and Zeaxanthin and Their Impacts on Human Health

The roles that lutein and zeaxanthin carotenoids play in photosynthesis, excess light energy dissipation, and general antioxidant functions cause them to be ubiquitous in leaf and fruit tissues of vegetable crops (Table 40.1). The highest concentrations of lutein can be found in dark green leafy vegetables such as kale, collards, and spinach. Values in kale and spinach can range from 5.8 to 12.9 mg lutein/100 g fresh weight [46,48]. Significant positive correlations exist between chlorophyll and carotenoids pigments in leafy vegetable crops [48,54,55]. Thus, the darker green the leafy tissues, the greater the concentrations of lutein that will be expected to be present. Zeaxanthin accompanies lutein in both leaf and fruit tissues; however, it is usually only found in minor concentrations in most vegetables. The exceptions are sweet corn kernel (*Zea mays* var. *rugosa* L.) and tabasco pepper (*Capsicum frutescens*), where zeaxanthin can accumulate to levels of 0.5 and 2.0 mg/100 g fresh weight, respectively [56,57].

Lutein and zeaxanthin are two of seven major carotenoids found in human blood serum; however, they are the only carotenoids present in the retina and lens of the eye [58]. In the retina, lutein and zeaxanthin are selectively deposited and are chiefly responsible for the yellow pigmentation collectively referred to as macular pigment [59]. Lutein and zeaxanthin have been found in human retinal and lens tissues, and in other areas of the eye where concentrations of long-chain polyunsaturated fatty acids and the potential for tissue oxidation are the highest. The yellow pigments are postulated to participate in photoprotection, and diminished macular pigment may be related to retinal damage [60,61]. Possible modes of action for photoprotection of the retinal carotenoids include their ability to filter harmful short-wave UV (blue) light and their function as antioxidants. Factors that increase macular pigment include increased antioxidant ingestion, high fruit and vegetable consumption, high dietary carotenoid intake and the resultant elevated serum carotenoid concentration, normal body mass index, and history of no tobacco use. Many of these same factors are also associated with a decreased risk of developing age-related macular degeneration and suggest there may be a causal relationship [62,63]. However, a direct correlation between macular pigment levels and development of macular eye diseases has not been established [60,64], although strong associative relationships are reported. Strong epidemiological associations also demonstrate that increased intakes of lutein and zeaxanthin, but not β-carotene, α-carotene, lycopene, or β-crytoxanthin, are associated with decreased risks of developing cataracts [65].

Research by our group revealed differences in serum carotenoids and responses to macular pigment optical density evaluations in human subjects administered different doses of lutein from mono-molecular supplements and whole food sources (spinach). Serum carotenoid levels increased from baseline after ingestion of 10 mg or 30 mg lutein supplements, or spinach fortified with 8 or 12 mg lutein per 100 g fresh weight. Significant increases in macular pigment optical density (indicative of photoprotection) from baseline to the end of a 12-week intervention occurred in subjects administered 30 mg lutein supplements and spinach with 12 mg lutein per 100 g fresh weight [46]. Conclusions from this study support earlier research showing the ability to modify both serum carotenoid concentrations and macular pigment through increased consumption of carotenoid-rich plant foods. Unique to the study is the demonstration that serum carotenoid concentrations and macular pigment optical density were affected by the concentrations of spinach tissue lutein. Results may imply a dose-dependent nutritional impact of carotenoid enhancement in plant tissues, and emphasize the importance of phytochemical enhancement efforts in fruit and vegetable crops.

Xanthophyll carotenoids can also possess antimutagenic and anticarcinogenic properties. Mechanisms may include selective modulation of cellular apoptosis, inhibition of angiogenesis, increased gap junction intercellular communications, induction of cellular differentiation, and the prevention of oxidative damage [65]. Lutein serum status is inversely associated with cytochrome CYP1A2 activity, a hepatic enzyme responsible for activation of several human carcinogens [66]. Although human studies regarding the impacts of dietary lutein and zeaxanthin on the risk of breast, lung, colorectal, prostate, and other cancers have been mostly inconclusive, there have been positive results from cell bioassays and animal models that may support a protective role of the xanthophylls [65]. In many of the studies reported in the review by Ribaya-Mercado and Blumberg [65], it is clear that greater cancer protection may be afforded by consuming a variety of vegetables (supplying both carotenes and xanthophylls) as compared to only consuming foods rich in one particular carotenoid.

4.3 Vegetable Sources of Lycopene and Its Impact on Human Health

Tomatoes and tomato-based products such as juices, pastes, and sauces are frequently consumed in the diets of populations around the world. The red color of the tomato is due to the concentration of its major carotene, lycopene, which accumulates as the fruit ripens. Minor concentrations of β-carotene, and the colorless precursors phytoene and phytofluene can be found in ripe tomato fruits; however, lycopene accounts for >90% of the total carotenoids [1]. Lycopene concentrations for whole tomatoes can reach levels of 5.0 mg/100 g fresh weight; however, fruit lycopene concentrations are highly dependent on ripeness stages [67]. Processed tomato products can range from 2.2 to 10.7 mg lycopene/100 g fresh weight [67]. Another major dietary contributor of lycopene is watermelon. An evaluation of lycopene concentrations among commercial seeded and seedless watermelons reported average values to be 6.1 mg lycopene/100 g fresh weight [68]. The predominate pigment among the watermelon cultivars was all-*trans*-lycopene, with minor amounts of *cis*-lycopene, β-carotene, and phytofluene present in the fruit.

The activity of carotenoids in health maintenance and disease suppression comes from their unique chemical structures. Lycopene is a highly unsaturated straight-chain hydrocarbon which lacks any terminal β- or ε-ionic ring structures, and thus lycopene lacks pro-vitamin A activity. Similar to the other carotenoids, lycopene is subject to oxidative, thermal, and photo-degradation, causing structural isomerizations. These structural changes have significant impact on the degree of intestinal absorption, described in further detail later in this chapter. In general, the absorption of dietary lycopene in humans is between 10 and 30% [39]. It has been demonstrated that lycopene absorption from tomato-based products is greater than from fresh tomatoes, which may be attributed to the greater release of plant membrane-bound lycopene and/or higher incidences of *cis*-lycopene isomers during food processing activities. There are many different forms of lycopene present in vegetable crop tissues (Table 40.1). The presence of *cis*- and *trans*-isomers, different oxidized forms, and polar metabolites of lycopene have been isolated and identified from blood serum and tissues in animal models [69]. To date, there is very little information as to the metabolism and biological functions of lycopene *in vivo* in human systems [43]. Lycopene is believed to convey *in vivo* antioxidant properties, function in gene regulation, modulate hormonal and immune metabolism, participate in gap junction communication, and influence carcinogen metabolism, as well as affecting enzymes in the pathways associated with phase II drug metabolism [43].

The impact of lycopene on cancer prevention has clearly been established. The main body of research has focused on the influence of increased lycopene intake on the suppression of prostate cancer in adult men. Giovannucci et al. [70] first established the link between consumption of tomatoes and tomato-based products and reduced prostate cancer risks. The authors followed up that work with a comparative study of data gathered from more than 70 different epidemiological studies which established the role of dietary lycopene in the reduction of not only prostate cancer, but mammary, cervical, ovarian, and liver cancer as well [71]. Antioxidants, such as β-carotene and lycopene, can reduce oxidative stress associated with the onset of osteoporosis. It has been demonstrated that dietary lycopene intake and serum lycopene levels may be associated with a reduced incidence of osteoporosis in postmenopausal women [72]. The study showed that lycopene intake reduced protein oxidation and specific bone resorption markers in subject participants. The antioxidant functions of lycopene can also reduce incidences of cardiovascular disease. Epidemiological studies have demonstrated that serum lycopene can reduce the oxidation of low-density lipoproteins (LDL) in subjects who consume tomato-based products or lycopene supplements [73]. The largest case study to date, conducted in Europe, showed increased adipose tissue lycopene status to decrease the risk of cardiovascular disease in a dose-dependent relationship [74]. More information on the association between dietary lycopene and other carotenoids in the reduction of cardiovascular disease prevalence can be found in a review of the subject [75]/ce:cross-ref>.

5. FACTORS THAT IMPACT CAROTENOID BIOAVAILABILITY

The bioavailability of carotenoids from plant foods is highly variable and is influenced by the crop species, the composition of carotenoid structures present in the food matrix, release of carotenoids from the food matrix, amount consumed and absorption in the intestinal tract, transportation within the lipoprotein serum fractions, the potential for *in vivo* biochemical conversions, tissue-specific depositions, as well as nutritional status of the ingesting host [12,14,76]. Carotenes are entirely lipophilic molecules located in the hydrophobic cores of plant membranes. Similarly, xanthophylls are largely hydrophobic molecules with their polar groups at opposite ends of a non-polar carbon skeleton [77]. Because of their lipophilic nature, biotic or abiotic activities that expose carotenoid molecules to potential oxidation, degradation, or isomerization will ultimately have an influence on carotenoid biochemistry and bioavailability.

Carotenoid localization in plant membranes may also influence their release from the food matrix. Results from a 12-week dietary intervention of two different spinach types, differing only in measured carotenoid concentrations, showed that lutein was more bioavailable from the spinach matrix than β-carotene, based on serum profiles in the subject participants [46]. One possible explanation for differences in bioavailability between lutein and β-carotene from the spinach matrix may come from differences in molecular orientation in plant membranes between carotenes and xanthophylls. Hydrophobic interactions and lack of polar end groups localize carotenes, such as β-carotene, within the hydrophobic core of biomembranes at several different orientations. The positions of xanthophyll carotenoids (lutein and zeaxanthin) are oriented to span the entire membrane, allowing for the positioning of polar groups outside of the hydrophobic core, or in the polar head-group region of the membrane. Differences in molecule positioning may be expected to affect release of carotene and xanthophyll carotenoids from the matrix of plant tissues, and thus affect bioavailability.

Carotenoid accumulation in vegetable crops appears to be shaped by a plant species' physiological, genetic, and biochemical attributes, as well as environmental growth factors such as light, temperature, and fertility [48,78–81]. Significant differences among vegetable crop species for carotenoid accumulations have been reported [8,82,83]. Significant genetic variation within species has been found for carrot [47], corn (*Zea mays* L.) [56], kale [48,54,78], lettuce (*Lactuca* species) [55], potato (*Solanum tuberosum* subsp. *tuberosum* L.) [84], pepper (*Capsicum* species) [85,86], and edible green soybean (*Glycine max* L.) [87]. Genetic variation both within and among vegetable crop species is of importance since concentrations of carotenoid present in the food matrix can influence serum carotenoids levels following intestinal absorption [46].

The increased coloration in vegetable and fruit tissues associated with maturity is often indicative of increases in carotenoid concentrations [57,85,86]. Carotenoid concentrations increase in leaf tissues with maturity [49,88] and decrease during senescence [1,89]. Manipulation of cultural growing conditions and time of harvest would therefore be influential on carotenoid concentrations of fruit and vegetable crops. Environmental growing conditions will also have a large influence on the accumulation of carotenoid in plant foods. Carotenoid accumulations have been shown to increase and decrease in response to environmental manipulations, with results differing among plant species. Changes in the growing air temperature [90], irradiance level [91], irradiance photoperiod [92], and nutritional fertility [93–95] all affect plant carotenoid accumulations.

The first step in carotenoid bioavailability is release from the food matrix. Absorption of carotenoids from raw, uncooked vegetables can be extremely low, in some cases less than 1–2%. Food processing activities, such as heating or pureeing, will act to increase intestinal absorption. Thermal processing, mincing, or liquefying results in changes to carotenoid chemistry, most likely through isomerization or oxidation reactions [35,42,76,96,97]. However, frozen or low-temperature storage generally preserves carotenoid concentrations by reducing potential enzymatic oxidation [96,97]. Processing activities usually increase bioavailability through increased release of bound carotenoids from the food matrix; however, thermal breakdowns of the exposed carotenoid compounds may adversely affect bioavailability in some food crops.

Absorption of carotenoids in humans is passive and follows digestive pathways similar to lipids. Once released from plant tissues, protein- or membrane-bound carotenoids must be dissolved in a hydrophobic domain (oils, fats, or bulk lipid emulsions). Carotenoid absorption requires the presence of dietary fat in the small intestine, which stimulates the release of emulsifying bile acids by the gallbladder [12]. Further studies provide evidence of increased absorption when carotenoids are ingested with dietary lipids [98,99]. Due to their hydrophobic nature, carotenoids in the mostly aqueous environment in plant foods must be transferred to bulk lipids or intestinal micelles in the digesta [14]. After release from the food matrix, carotenoids are assimilated and oriented into lipid micelles before uptake by intestinal mucosal cells. Once in the enterocyte, carotenoids are incorporated into chylomicrons, which are eventually delivered to the bloodstream, and ultimately to the liver. The distribution of carotenoids among the various transporting lipoprotein classes is influenced by their physical structures. The hydrocarbon carotenes are transported mainly in low-density lipoproteins, while the more polar xanthophylls are evenly distributed between the low-density and high-density lipoprotein serum factions [44]. Carotenoid compounds can remain in the liver, or get transferred to low-density or high-density lipoproteins before eventual tissue-specific deposition [12].

Carotenoid bioavailability is most often assessed in blood serum after ingestion in dietary trials or cohort studies. The relatively simple analysis quantifies carotenoid changes in serum at various time intervals following ingestion of whole foods or supplements. Some caveats to interpreting serum carotenoid bioavailability include: 1) serum responses to single oral doses of carotenoids are highly variable; 2) carotenoids measured in serum signify an equilibrium between intestinal absorption, breakdowns, tissue uptake, and tissue release; and 3) there are high concentrations of endogenous carotenoids (α-carotene, β-carotene, lycopene, lutein) already present in serum [45]. Other studies have also utilized *in vitro* Caco-2 human intestinal cell lines to assess carotenoid bioavailability [100–102]. In these studies, pure carotenoid compounds and whole food samples are brought through an *in vitro* digestion model and reacted with Caco-2 human intestinal cells. Absorption potential is measured using standard high performance liquid chromatography (HPLC) carotenoid analysis. Current results demonstrate that *in vitro* Caco-2 cells can accurately predict carotenoid bioavailability from supplements and whole foods [100–102]. Increases in serum carotenoid concentrations typically result after ingestion of carotenoids from whole food or mono-molecular supplements [33,42,45,76,98, 99,102]. *In vitro* and *in vivo* studies show that carotenoid bioavailability is influenced by source (whole food vs. supplement), degree of processing, interactions with other carotenoid compounds, the degree of isomerization before, during and after absorption, transit time in the intestine, and the nutritional status of the human subjects [14].

6. ENHANCEMENT EFFORTS TO INCREASE VEGETABLE CROP CAROTENOIDS

Carotenoid enhancement efforts in plant foods have advanced through both traditional breeding and molecular approaches. Some reviews have chronicled molecular advances in carotenoid pathway manipulation to improve biosynthesis and partitioning [13,15,103]. Successful approaches have centered on modification of the biosynthetic pathway to change the flux and end products, increasing preexisting carotenoids, and engineering carotenogenic behavior in tissues completely devoid of carotenoid activity [13]. Identification and

cloning of genes and cDNAs in the carotenoid pathway have facilitated genetic manipulations. Most studies have demonstrated that phytoene synthase holds greatest control over fluxes in the carotenoid pathway; however, success has also been achieved overexpressing phytoene desaturase enzymes [15]. A 2–3-fold increase in tomato fruit-specific carotenoid accumulation was achieved by utilizing a bacterial phytoene synthase (*crt*B) and a tissue-specific promoter [104]. Genetic strategies have also increased β-carotene production in canola (*B. napus* L.) using seed-specific promotion [105], xanthophyll carotenoids in potato using tuber-specific promotion and suppression of epoxidation reactions [106], and most notably, xanthophyll and carotene production in rice (*Oryza sativa* L.) endosperm ('Golden Rice') using exogenous cyclase and desaturase enzymes [107]. There are obvious advantages to genetic engineering techniques, but they must be accompanied by a clear understanding of how these manipulations will ultimately affect plant physiology. Advances in carotenoid biosynthesis have clearly been facilitated through molecular approaches, but the question still remains if the general public will accept genetic modification of whole, unprocessed foods.

Programs designed to increase carotenoid bioavailability in food crops must begin with a firm understanding of carotenoid chemistry and biosynthesis. Before selecting candidate crops for enhancement efforts, care must be taken to outline specific research impacts, whether combating widespread vitamin A deficiency or fighting aging eye diseases. Understanding how carotenoids in the food matrix respond to food processing activities and any subsequent changes to carotenoid chemistry will be of primary importance. Lastly, there is a necessity for collaborative activities to assess bioavailability in model systems or human subjects to gauge progress in enhancement efforts. Epidemiological data firmly establish the link between increased carotenoid intakes and chronic disease suppression;

however, properly designed dietary intervention studies with varying levels of vegetable carotenoid treatments are needed to clearly demonstrate the *in vivo* health attributes of higher consumption of vegetable crops high in carotenoids.

7. SUMMARY

Dietary guidelines recommend consuming 7–9 servings of fruits and vegetables daily, which has been positively associated with reduced chronic disease risk. Specifically, carotenoid compounds in fruits and vegetables provide improved health maintenance. Research demonstrates the antioxidant activity of β-carotene, lutein, zeaxanthin, and lycopene in promoting disease suppression, and their activity is affected by the amount consumed, conditions of the food matrix, intestinal absorption, and biometabolism. Genetic and environmental effects strongly influence carotenoid content in fruits and vegetables. Therefore, current carotenoid enhancement efforts and assessment of carotenoid consumption on human health need to consider many complicated and interrelated factors.

References

1. Gross, J. (1991). *Pigments in Vegetables: Chlorophylls and Carotenoids.* New York: AVI/Van Nostrand Reinhold.
2. DellaPenna, D. (1999). Carotenoid synthesis and function in plants: insights from mutant studies in *Arabidopsis thaliana.* In H. A. Frank, A. J. Young, G. Britton, & R. J. Cogdell, (Eds.), *The Photochemistry of Carotenoids* (Vol. 8). (pp. 21–37). The Netherlands: Kluwer, Dordrecht.
3. Bendich, A. (1993). Biological functions of carotenoids. In L. M. Canfield, N. I. Krinsky, & J. A. Olsen, (Eds.), *Carotenoids in Human Health* (Vol. 691). (pp. 61–67). New York: New York Academy of Sciences.
4. Tang, L. L., Jin, T. Y., Zeng, X. B., & Wang, J. S. (2005). Lycopene inhibits the growth of human androgen-independent prostate cancer cells *in vitro* and in BALB/c nude mice. *The Journal of Nutrition, 135,* 287–290.
5. Seifried, H. E., McDonald, S. S., Anderson, D. E., Greenwald, P., & Milner, J. A. (2003). The antioxidant conundrum in cancer. *Cancer Review, 63,* 4295–4298.
6. Finley, J. W. (2005). Proposed criteria for assessing the efficacy of cancer reduction by plant foods enriched in carotenoids, glucosinolates, polyphenols and selenocompounds. *Annals of Botany, 95,* 1075–1096.
7. Granado, F., Olmedilla, B., & Blanco, I. (2003). Nutritional and clinical relevance of lutein in human health. *The British Journal of Nutrition, 90,* 487–502.
8. Sommerburg, O., Keunen, J. E. E., Bird, A. C., & van Kuijk, F. J. G. M. (1998). Fruits and vegetables that are sources for lutein and zeaxanthin: the macular pigment in human eyes. *British Journal of Ophthalmologica, 82,* 907–910.
9. Johnson, E. J., Hammond, B. R., Yeum, K. J., Qin, J., Wang, X. D., Castaneda, C., Snodderly, D. M., & Russell, R. M. (2000). Relation among serum and tissue concentrations of lutein and zeaxanthin and macular pigment density. *The American Journal of Clinical Nutrition, 71,* 1555–1562.
10. Garcia, A. L., Ruhl, R., Herz, U., Koebnick, C., & Schweigert, F. J. (2003). Retinoid- and carotenoid-enriched diets influence the ontogenesis of the immune system in mice. *Immunology, 110,* 180–187.
11. Hughes, D. A. (1999). Effects of carotenoids on human immune function. *The Proceedings of the Nutrition Society, 58,* 713–718.
12. Zaripheh, S., & Erdman, J. W. (2002). Factors that influence the bioavailability of xanthophylls. *The Journal of Nutrition, 132,* 531S–534S.
13. Sandmann, G. (2001). Carotenoid biosynthesis and biotechnological applications. *Archives of Biochemistry and Biophysics, 385,* 4–12.
14. Faulks, R. M., & Southon, S. (2005). Challenges to understanding and measuring carotenoid bioavailability. *Biochemica Biophysica Acta, 1740,* 95–100.
15. Fraser, P. D., & Bramley, P. M. (2004). The biosynthesis and nutritional uses of carotenoids. *Progress in Lipid Research, 43,* 228–265.
16. Cunningham, F. X., & Gantt, E. (1998). Genes and enzymes of carotenoid biosynthesis in plants. *Annual Review of Plant Physiology and Plant Molecular Biology, 49,* 577–583.
17. Bramley, P. M. (2002). Regulation of carotenoid formation during tomato fruit ripening and development. *Journal of Experimental Botany, 53,* 2107–2113.
18. Cunningham, F. X. (2002). Regulation of carotenoid synthesis and accumulation in plants. *Pure and Applied Chemistry. Chimie Pure et Appliquee, 74,* 1409–1417.
19. Niyogi, K. K., Bjorkman, O., & Grossman, A. R. (1997). The roles of specific xanthophylls in

photoprotection. *Proceedings of the National Academy of Sciences, 94*, 14162–14167.

20. Demmig-Adams, B., & Adams, W. W. (1996). The role of xanthophylls cycle carotenoids in the protection of photosynthesis. *Trends in Plant Science, 1*, 21–26.

21. Frank, H. A., & Cogdell, R. J. (1996). Carotenoids in photosynthesis. *Photochemistry and Photobiology, 63*, 257–264.

22. Paulsen, H. (1999). Carotenoids and the assembly of light-harvesting complexes. In H. A. Frank, A. J. Young, G. Britton, & R. J. Cogdell, (Eds.), *The Photochemistry of Carotenoids* (Vol. 8). (pp. 123–135). The Netherlands: Kluwer, Dordrecht.

23. Tracewell, C. A., Vrettos, J. S., Bautista, J. A., Frank, H. A., & Brudvig, G. W. (2001). Carotenoid photooxidation in photosystem II. *Archives of Biochemistry and Biophysics, 385*, 61–69.

24. Koyama, Y., & Fujii, R. (1999). *Cis-trans* carotenoids in photosynthesis: configurations, excited-state properties and physiological functions. In H. A. Frank, A. J. Young, G. Britton, & R. J. Cogdell, (Eds.), *The Photochemistry of Carotenoids* (Vol. 8). (pp. 161–188). The Netherlands: Kluwer, Dordrecht.

25. Koyama, Y., Kanaji, M., & Shimamura, T. (1988). Configurations of neurosporene isomers isolated from the reaction center and light-harvesting complex or Rhodobacter sphaeroides G1C. A resonance Raman, electronic absorption, and 1H-NMR study. *Photochemistry and Photobiology, 48*, 107–114.

26. Bialek-Bylka, G. E., Tomo, T., Satoh, K., & Koyama, Y. (1995). 15-*Cis-β*-carotene found in the reaction center of spinach photosystem II. *FEBS Letters, 363*, 137–140.

27. Mortensen, A., Skibsted, L. H., & Truscott, T. G. (2001). The interaction of dietary carotenoids with radical species. *Archives of Biochemistry and Biophysics, 385*, 13–19.

28. Ames, B. N., Shigenaga, M. K., & Hagen, T. M. (1993). Oxidants, antioxidants, and the degenerative diseases of aging. *Proceedings of the National Academy of Science, 90*, 7915–7922.

29. DiMascio, P., Kaiser, S., & Sies, H. (1989). Lycopene as the most efficient biological carotenoid singlet oxygen quencher. *Archives of Biochemistry and Biophysics, 274*, 532–538.

30. Grusak, M. A., & DellaPenna, D. (1999). Improving the nutrient composition of plants to enhance human nutrition and health. *Annual Review of Plant Physiology and Plant Molecular Biology, 50*, 133–161.

31. Young, A. J., & Lowe, G. M. (2001). Antioxidant and prooxidant properties of carotenoids. *Archives of Biochemistry and Biophysics, 385*, 20–27.

32. Khachik, F., Beecher, G. R., & Whittaker, N. F. (1986). Separation, identification, and quantification of the major carotenoids and chlorophyll constituents in extracts of several green vegetables by liquid chromatography. *Journal of Agricultural and Food Chemistry, 34*, 603–616.

33. Rao, A. V., & Shen, H. (2002). Effect of low dose lycopene intake on lycopene bioavailability and oxidative stress. *Nutrition Research, 22*, 1125–1131.

34. Emenhiser, C., Simunovic, N., Sander, L. C., & Schwartz, S. J. (1996). Separation of geometrical carotenoid isomers in biological extracts using a polymeric C_{30} column in reverse-phase liquid chromatography. *Journal of Agricultural and Food Chemistry, 44*, 3887–3893.

35. Updike, A. A., & Schwartz, S. J. (2003). Thermal processing of vegetables increases *cis* isomers of lutein and zeaxanthin. *Journal of Agricultural and Food Chemistry, 51*, 6184–6190.

36. Humphries, J. M., & Khachik, F. (2003). Distribution of lutein, zeaxanthin, and related geometrical isomers in fruits, vegetables, wheat, and pasta products. *Journal of Agricultural and Food Chemistry, 51*, 1322–1327.

37. Khachik, F., Carvalho, L., Bernstein, P. S., Muir, G. J., Zhao, D. Y., & Katz, N. B. (2002). Chemistry, distribution, and metabolism of tomato carotenoids and their impact on human health. *Experimental Biology and Medicine, 227*, 845–851.

38. Clinton, S. K., Emenhiser, C., Schwartz, S. J., Bostwick, D. G., Williams, A. W., Moore, B. J., & Erdman, J. W. (1996). *Cis-trans* lycopene isomers, carotenoids, and retinol in the human prostate. *Cancer Epidemiology Biomarkers & Prevention, 5*, 823–833.

39. Gärtner, C., Stahl, W., & Sies, H. (1997). Lycopene is more bioavailable from tomato paste than from fresh tomatoes. *The American Journal of Clinical Nutrition, 66*, 116–122.

40. Wu, K., Schwartz, S. J., Platz, E. A., Clinton, S. K., Erdman, J. W., Ferruzzi, M. G., Willett, W. C., & Giovannucci, E. L. (2003). Variations in plasma lycopene and specific isomers over time in a cohort of U.S. men. *The Journal of Nutrition, 133*, 1930–1936.

41. Boileau, T. W. M., Boileau, A. C., & Erdman, J. W. (2002). Bioavailability of all-*trans* and *cis*-isomers of lycopene. *Experimental Biology and Medicine, 227*, 914–919.

42. Livny, O., Reifen, R., Levy, I., Madar, Z., Faulks, R., Southon, S., & Schwartz, B. (2003). *β*-carotene bioavailability from differently processed carrot meals in human ileostomy volunteers. *European Journal of Nutrition, 42*, 338–345.

43. Rao, A. V., & Rao, L. G. (2007). Carotenoids and human health. *Pharmacological Research, 55*, 207–216.

44. Combs, G. F. (1998). Vitamin A. In R. Suja, (Ed.), *The Vitamins: Fundamental Aspects in Nutrition and Health* (2nd ed.) (pp. 107–153). San Diego: Academic Press.

C. ACTIONS OF INDIVIDUAL FRUITS IN DISEASE AND CANCER PREVENTION AND TREATMENT

45. Yeum, K. J., & Russell, R. M. (2002). Carotenoid bioavailability and bioconversion. *Annual Review of Nutrition, 22*, 483–504.

46. Kopsell, D. A., Kopsell, D. E., Lefsrud, M. G., Wenzel, A. J., Gerweck, C., & Curran-Celentano, J. (2006). Spinach cultigen variation for tissue carotenoid concentrations influence human serum carotenoid levels and macular pigment optical density following a twelve-week dietary intervention. *Journal of Agricultural and Food Chemistry, 54*, 7998–8005.

47. Nicolle, C., Simon, G., Rock, E., Amouroux, P., & Remesy, C. (2004). Genetic variability influences carotenoid, vitamin, phenolic, and mineral content in white, yellow, purple, orange, and dark-orange carrot cultivars. *Journal of the American Society for Horticultural Science, 129*, 523–529.

48. Kopsell, D. A., Kopsell, D. E., Lefsrud, M. G., Curran-Celentano, J., & Dukach, L. E. (2004). Variation in lutein, β-carotene, and chlorophyll concentrations among *Brassica oleracea* cultigens and seasons. *HortScience, 39*, 361–364.

49. de Azevedo, C. H., & Rodriguez-Amaya, D. B. (2005). Carotenoid composition of kale as influenced by maturity, season and minimal processing. *Journal of the Science of Food and Agriculture, 85*, 591–597.

50. Paolini, M. P., Abdel-Rahman, S. Z., Sapone, A., Pedulli, G. F., Perocco, P., Cantelli-Forti, G., & Legator, M. S. (2003). β-carotene: a cancer chemopreventive agent or a co-carcinogen? *Mutation Research, 543*, 195–200.

51. Omenn, G. S., Goodman, G. E., Thornquist, M. D., Balmes, J., Cullen, M. R., Glass, A., Keogh, J. P., Meyskens, F. L., Valanis, B., Williams, J. H., Barnhart, S., Cherniack, M. G., Brodkin, C. A., & Hammar, S. (1996). Risk factors from lung cancer and from intervention effects in CARET. *Journal of the National Cancer Institute, 88*, 1550–1558.

52. Omenn, G. S., Goodman, G. E., Thornquist, M. D., Balmes, J., Cullen, M. R., Glass, A., Keogh, J. P., Meyskens, F. L., Valanis, B., Williams, J. H., Barnhart, S., & Hammar, S. (1996). Effects of a combination of β-carotene and Vitamin A on lung cancer and cardiovascular disease. *The New England Journal of Medicine, 334*, 1150–1155.

53. Paolini, M. P., Cantelli-Forti, G., Perocco, P., Pedulli, G. F., Abdel-Rahman, S., & Legator, M. S. (1999). Cocarcinogenic effect of β-carotene. *Nature, 398*, 760–761.

54. Mercadante, A. Z., & Rodriguez-Amaya, D. R. (1991). Carotenoid composition of a leafy vegetable in relation to some agricultural variables. *Journal of Agricultural and Food Chemistry, 39*(6), 1094–1097.

55. Mou, B. (2005). Genetic variation of β-carotene and lutein contents in lettuce. *Journal of the American Society for Horticultural Science, 130*, 870–876.

56. Kurlich, A. C., & Juvik, J. A. (1999). Quantification of carotenoid and tocopherol antioxidants in *Zea mays*. *Journal of Agricultural and Food Chemistry, 47*, 1948–1955.

57. Russo, V. M., & Howard, L. R. (2002). Carotenoids in pungent and non-pungent peppers at various developmental stages grown in the field and glasshouse. *Journal of the Science of Food and Agriculture, 82*, 615–624.

58. Bone, R. A., Landrum, J. T., Friedes, L. M., Gomez, C. M., Kilburn, M. D., Menendez, E., Vidal, I., & Wang, W. L. (1997). Distribution of lutein and zeaxanthin stereoisomers in the human retina. *Experimental Eye Research, 64*, 211–218.

59. Khachik, F., Bernstein, P. S., & Garland, D. L. (1997). Identification of lutein and zeaxanthin oxidation products in human and monkey retinas. *Investigative Ophthalmology & Visual Science, 38*, 1082–1811.

60. Mares-Perlman, J. A., & Klein, R. (1999). Diet and age-related macular degeneration. In A. Taylor, (Ed.), *Nutritional and Environmental Influences on the Eye* (pp. 181–214). Boca Raton, FL: CRC Press.

61. Wooten, B. R., Hammond, B. R., Land, R. I., & Snodderly, D. M. (1999). A practical method for measuring macular pigment optical density. *Investigative Ophthalmology & Visual Science, 40*, 2481–2489.

62. Schalch, W. (2003). Lutein and zeaxanthin, the xanthophylis of the macula lutea. *Agro Food Industry Hi-Tech, 14*, 23–27.

63. Hammond, B. R., Wooten, B. R., & Snodderly, D. M. (1998). Preservation of visual sensitivity of older subjects: association with macular pigment density. *Investigative Ophthalmology & Visual Science, 39*, 397–406.

64. Landrum, J. T., & Bone, R. A. (2001). Lutein, zeaxanthin, and the macular pigment. *Archives of Biochemistry and Biophysics, 385*, 28–40.

65. Ribaya-Mercado, J. D., & Blumberg, J. B. (2004). Lutein and zeaxanthin and their potential roles in disease prevention. *Journal of the American College of Nutrition, 23*, 567S–587S.

66. Le Marchand, L., Franke, A. A., Custer, L., Wilkens, L. R., & Cooney, R. V. (1997). Lifestyle and nutritional correlates of cytochrome CYP1A2 activity: inverse associations with plasma lutein and alpha-tocopherol. *Pharmacogenetics, 7*, 11–19.

67. Nguyen, M., Francis, D., & Schwartz, S. J. (2001). Thermal isomerization susceptibility of carotenoids in different tomato varieties. *Journal of the Science of Food and Agriculture, 81*, 910–917.

68. Perkins-Veazie, P., Collins, J. K., Davis, A. R., & Robbins, W. (2006). Carotenoid content of 50 watermelon cultivars. *Journal of Agricultural and Food Chemistry, 54*, 2593–2597.

C. ACTIONS OF INDIVIDUAL FRUITS IN DISEASE AND CANCER PREVENTION AND TREATMENT

69. Jain, C. K., Agarwal, S., & Rao, A. V. (1999). The effects of dietary lycopene on bioavailability, tissue distribution, *in-vivo* antioxidant properties and colonic preneoplasia in rats. *Nutrition Reviews, 19,* 1383–1391.

70. Giovannucci, E., Ascherio, A., Rimm, E. B., Stampfer, M. J., Colditz, G. A., & Willett, W. C. (1995). Intake of carotenoids and retinol in relation to risk of prostate cancer. *Journal of the Natitonal Cancer Institute, 87,* 1767–1776.

71. Giovannucci, E. (1999). Tomatoes, tomato-based products, lycopene, and cancer: review of the epidemiologic literature. *Journal of the Natitonal Cancer Institute, 91,* 317–331.

72. Rao, L. G., Mackinnon, E. S., Josse, R. G., Murray, T. M., Strauss, A., & Rao, A. V. (2007). Lycopene consumption decreases oxidative stress and bone resorption markers in postmenopausal women. *Osteoporosis International, 18,* 109–115.

73. Agarwal, S., & Rao, A. V. (1998). Tomato lycopene and low density lipoprotein oxidation: a human dietary intervention study. *Lipids, 33,* 981–984.

74. Kohlmeier, L., Kark, J. D., GomezGracia, E., Martin, B. C., Steck, S. E., Kardinaal, A. F. M., Ringstad, J., Thamm, M., Masaev, V., Riemersma, R., MartinMoreno, J. M., Huttunen, J. K., & Kok, F. J. (1997). Lycopene and myocardial infarction risk in the EURAMIC study. *American Journal of Epidemiology, 146,* 618–626.

75. Voutilainen, S., Nurmi, T., Mursu, J., & Rissanen, T. H. (2006). Carotenoids and cardiovascular health. *The American Journal of Clinical Nutrition, 83,* 1265–1271.

76. Castenmiller, J. J. M., West, C. E., Linssen, J. P. H., van het Hof, K. H., & Voragen, A. G. J. (1999). The food matrix of spinach is a limiting factor in determining the bioavailability of β-carotene and to a lesser extent of lutein in humans. *The Journal of Nutrition, 129,* 349–355.

77. Gruszecki, W. I. (1999). Carotenoids in membranes. In H. A. Frank, A. J. Young, G. Britton, & R. J. Cogdell, (Eds.), *The Photochemistry of Carotenoids* (Vol. 8). (pp. 363–379). The Netherlands: Kluwer, Dordrecht.

78. Kurilich, A. C., Tsau, G. J., Brown, A., Howard, L., Klein, B. P., Jeffery, E. H., Kushad, M., Wallig, M. A., & Juvik, J. A. (1999). Carotene, tocopherol, and ascorbate in subspecies of *Brassica oleracea*. *Journal of Agricultural and Food Chemistry, 47,* 1576–1581.

79. Goldman, I. L., Kader, A. A., & Heintz, C. (1999). Influence of production, handling, and storage on phytonutrient content of foods. *Nutrition Reviews, 57,* S46–S52.

80. Kopsell, D. A., McElroy, J. S., Sams, C. E., & Kopsell, D. E. (2007). Genetic variation in carotenoids among diploid and amphidiploids rapid-cycling *Brassica* species. *HortScience, 42,* 461–465.

81. Kopsell, D. E., Kopsell, D. A., Randle, W. M., Coolong, T. W., Sams, C. E., & Curran-Celentano, J. (2003). Kale carotenoids remain stable while flavor compounds respond to changes in sulfur fertility. *Journal of Agricultural and Food Chemistry, 51,* 5319–5325.

82. Klein, B. P., & Perry, A. K. (1982). Ascorbic acid and vitamin A activity in selected vegetables from different geographic areas of the United States. *Journal of Food Science, 47,* 941–948.

83. Kimura, M., & Rodriguez-Amaya, D. B. (2003). Carotenoid composition of hydroponic leafy vegetables. *Journal of Agricultural and Food Chemistry, 51,* 2603–2607.

84. Nesterenko, S., & Sink, K. C. (2003). Carotenoid profiles of potato breeding lines and selected cultivars. *HortScience, 38,* 1173–1177.

85. Simonne, A. H., Simonne, E. H., Eitenmiller, R. R., Mills, H. A., & Green, N. R. (1997). Ascorbic acid and provitamin A contents in unusually colored bell peppers (*Capsicum annuum* L. *Journal of Food Composition and Analysis, 10,* 299–311.

86. Howard, L. R., Talcott, S. T., Brenes, C. H., & Villalon, B. (2000). Changes in phytochemical and antioxidant activity of selected pepper cultivars (*Capsicum* species) as influenced by maturity. *Journal of Agricultural and Food Chemistry, 48,* 1713–1720.

87. Simonne, A. H., Smith, M., Weaver, D. B., Vail, T., Barnes, S., & Wei, C. I. (2000). Retention and changes of soy isoflavones and carotenoids in immature soybean seeds (Edamame) during processing. *Journal of Agricultural and Food Chemistry, 48,* 6061–6069.

88. de Azevedo, C. H., & Rodriguez-Amaya, D. B. (2005). Carotenoids of endive and New Zealand spinach as affected by maturity, season and minimal processing. *Journal of Food Composition and Analysis, 18,* 845–855.

89. Lefsrud, M. G., Kopsell, D. A., Wenzel, A. J., & Sheehan, J. (2007). Changes in kale (*Brassica oleracea* L. var. Acephala) carotenoid and chlorophyll pigment concentrations during leaf ontogeny. *Scientia Horticulturae, 112,* 136–141.

90. Lefsrud, M. G., Kopsell, D. A., Kopsell, D. E., & Curran-Celentano, J. (2005). Air temperature affects biomass and carotenoid pigment accumulation in kale and spinach grown in a controlled environment. *HortScience, 40,* 2026–2030.

91. Lefsrud, M. G., Kopsell, D. A., Kopsell, D. E., & Curran-Celentano, J. (2006). Irradiance levels affect growth parameters and carotenoid pigments in kale and spinach grown in a controlled environment. *Physiologia Plantarium, 127*(4), 624–631.

92. Lefsrud, M. G., Kopsell, D. A., Augé, R. M., & Both, A. J. (2006). Biomass production and pigment

C. ACTIONS OF INDIVIDUAL FRUITS IN DISEASE AND CANCER PREVENTION AND TREATMENT

accumulation in kale grown under increasing photo-periods. *HortScience, 41,* 603–606.

93. Hochmuth, G. J., Brecht, J. K., & Bassett, M. J. (1999). Nitrogen fertilization to maximize carrot yield and quality on a sandy soil. *HortScience, 34,* 641–645.

94. Chenard, C. H., Kopsell, D. A., & Kopsell, D. E. (2005). Nitrogen concentration affects nutrient and carotenoid accumulation in parsley. *Journal Plant Nutrition, 28,* 285–297.

95. Kopsell, D. A., Kopsell, D. E., & Curran-Celentano, J. (2007). Carotenoid pigments in kale are influenced by nitrogen concentration and form. *Journal of the Science of Food and Agriculture, 87,* 900–907.

96. Kopas-Lane, L. M., & Warthesen, J. J. (1995). Carotenoid photostability in raw spinach and carrots during cold storage. *Journal of Food Science, 60,* 773–776.

97. Rodriguez-Amaya, D. B. (1999). Changes in carote-noids during processing and storage of foods. *Archos Latinoamericanos de Nutrición, 49,* 38S–47S.

98. Unlu, N. Z., Bohn, T., Clinton, S. K., & Schwartz, S. J. (2005). Carotenoid absorption from salad and salsa by humans is enhanced by the addition of avocado or avocado oil. *The Journal of Nutrition, 135,* 431–436.

99. Brown, M. J., Ferruzzi, M. G., Nguyen, M. L., Cooper, D. A., Eldridge, A. L., Schwartz, S. J., & White, W. S. (2004). Carotenoid bioavailability is higher from salads ingested with full-fat than with fat-reduced salad dressings as measured with electro-chemical detection. *The American Journal of Clinical Nutrition, 80,* 396–403.

100. Chitchumroonchokchai, C., Schwartz, S. J., & Failla, M. L. (2004). Assessment of lutein bioavailability from meals as a supplement using simulated diges-tion and Caco-2 human intestinal cells. *The Journal of Nutrition, 134,* 2280–2286.

101. Liu, C. S., Glahn, R. P., & Liu, R. H. (2004). Assessment of carotenoid bioavailability of whole foods using a Caco-2 cell culture model coupled with an *in vitro* digestion. *Journal of Agricultural and Food Chemistry, 52,* 4330–4337.

102. Reboul, E., Borel, P., Mikail, C., Abou, L., Charbonnier, M., Caris-Veyrat, C., Goupy, P., Portugal, H., Lairon, D., & Amiot, M. J. (2005). Enrichment of tomato paste with 6% tomato peel increases lycopene and β-carotene bioavailability in men. *The Journal of Nutrition, 135,* 790–794.

103. Naik, P. S., Chanemougasoundharam, A., Khurana, S. M. P., & Kalloo, G. (2003). Genetic manipulation of carotenoid pathway in higher plants. *Current Science, 85,* 1423–1430.

104. Fraser, P. D., Romer, S., Shipton, C. A., Mills, P. B., Kiano, J. W., Misawa, N., Drake, R. G., Schuch, W., & Bramley, P. M. (2002). Evaluation of transgenic tomato plants expressing an additional phytoene synthase in a fruit specific manner. *Proceeding of the Natitonal Academy Science, 99,* 1092–1097.

105. Shewmaker, C. K., Sheehy, J. A., Daley, M., Colburn, S., & Ke, D. Y. (1999). Seed-specific over-expression of phytoene synthase: increase in carote-noids and other metabolic effects. *The Plant Journal, 20,* 401–412.

106. Römer, S., Lubeck, J., Kauder, F., Steiger, S., Adomat, C., & Sandmann, G. (2002). Genetic engineering of a zeaxanthin-rich potato by antisense inactivation and co-suppression of a carotenoid epoxidation. *Metabolic Engineering, 4,* 263–272.

107. Burkhardt, P. K., Beyer, P., Wunn, J., Kloti, A., Armstrong, G. A., Schledz, M., von Lintig, J., & Potrykus, I. (1997). Transgenic rice (*Oryza sativa*) endosperm expressing daffodil (*Narcissus pseudonar-cissus*) phytoene synthase accumulates phytoene, a key intermediate of provitamin A biosynthesis. *The Plant Journal, 11,* 1071–1078.

108. Evangelina, G., Montengrao, M., Nazareno, M., & de Mishima, B. L. (2001). Carotenoid composition and vitamin A value of an Argentinian squash (*Cucurbita moschata*). *Archivos Latinoamericanos de Nutrición, 51,* 395–399.

109. Larsen, E., & Christensen, L. P. (2005). Simple saponi-fication method for the quantitative determination of carotenoid in green vegetables. *Journal of Agricultural and Food Chemistry, 53,* 6598–6602.

110. Tonucci, L. H., Holden, J. M., Beecher, G. R., Khachik, F., Davis, C. S., & Mulokozi, G. (1995). Carotenoid content of thermally processed tomato-based food products. *Journal of Agricultural and Food Chemistry, 43,* 579–586.

C. ACTIONS OF INDIVIDUAL FRUITS IN DISEASE AND CANCER PREVENTION AND TREATMENT

41

Apigenin and Cancer Chemoprevention

Sanjeev Shukla[1,2] *and Sanjay Gupta*[1,2,3]

[1]Department of Urology, Case Western Reserve University
[2]University Hospitals Case Medical Center
[3]Case Comprehensive Cancer Center, Cleveland, Ohio, USA

1. INTRODUCTION

Epidemiological studies have demonstrated an association between higher fruit and vegetable intake and lower risk of human diseases, but it is unclear which bioactive agents are responsible, or if the associations are a function of synergistic effects of the whole food. Fruits and vegetables are a rich source of flavonoids. In general, flavonoids are part of a large group of polyphenolic compounds found in foods of plant origin including vegetables, fruits, legumes, and beverages like tea and wine. There are approximately 5000 flavonoids which are generally categorized into the following subclasses: flavonols, flavanones, flavones, flavan-3-ols, isoflavones, and anthocyanidins (Figure 41.1). The various mechanisms of biological effects of flavonoids have been demonstrated by laboratory experiments, including antioxidation, induction of detoxification enzymes and inhibition of bioactivation enzymes, estrogenic and anti-estrogenic activity, antiproliferation, cell cycle arrest and apoptosis, promotion of differentiation, regulation of host immune function, and inhibition of angiogenesis and metastasis. Data from *in vitro* studies and animal models suggest that flavonoids have the ability to influence important cellular and molecular mechanisms related to carcinogenesis. Based on cumulative reports the association between flavonoid intake and lower risk of several cancers has been documented.

The Zutphen study was conducted in the Netherlands where a cohort of 878 men were followed beginning in 1960 for 25 years. In this study the incidence and mortality from all causes of cancer and the mortality from alimentary or respiratory tract cancers were considered – associated with the intake of five flavonoids: myricetin, quercetin, kaempferol, luteolin, and apigenin. The results supported the findings that high intake of flavonoids from vegetables and fruits was inversely associated with risk of cancer [1]. In a large cohort study the association between flavonoid intake and human cancers was investigated, following 9959 Finnish men from 1967 to 1991 [2]. This study provides stronger evidence for a

663

Structure	Flavonoids	Source
Flavones 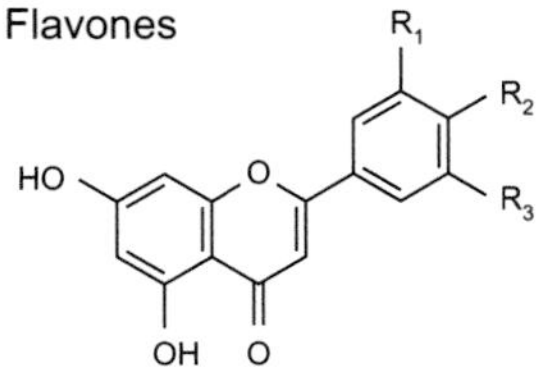	R1=H, R2=OH: Apigenin R1=R2=OH: Luteolin	Celery, parsley, thyme, onions etc. Red pepper, onions, lettuce, berries etc.
Flavonols 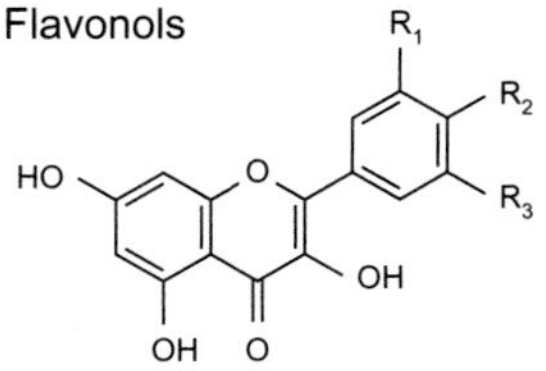	R2=OH, R1=R3=H: Kaempferol R1=R2=OH, R3=H: Quercetin R1=R2=R3=OH: Myricetin	Black tea etc. Olive oil, apple peels, kale etc.
Isoflavones 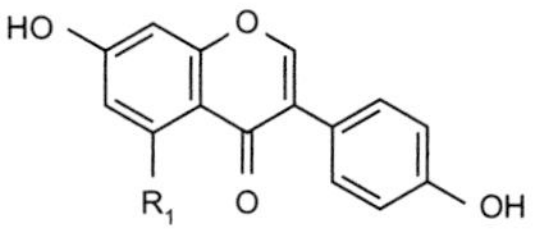	R1=H: Daidzein R1=OH: Genistein	Soybeans, legumes etc. Soybeans, legumes etc.
Flavanols 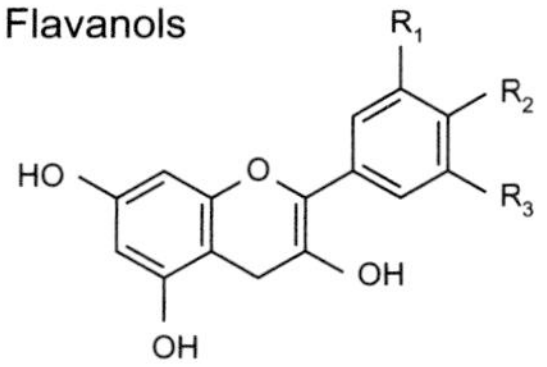	R1=R2=OH, R3=H:Catechins R1=R2=R3=OH: Gallocatechin	Green tea etc. Green tea etc.
Flavanones 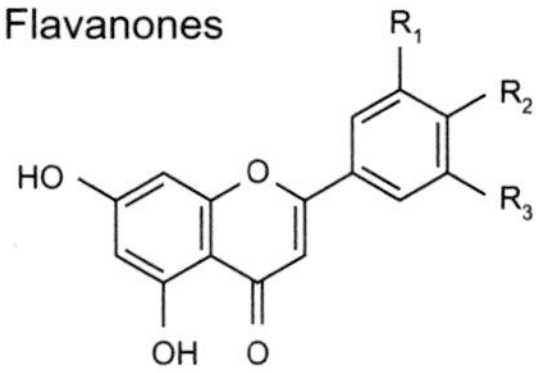	R1=H, R2=OH: Naringenin R1=R2=OH: Eriodictyol R1=OH, R2=OCH3: Hesperetin	Citrus fruits, grape fruits etc. Tomatoes, mint, citrus fruits etc. Citrus fruits, grape fruits etc.
Anthocyanins 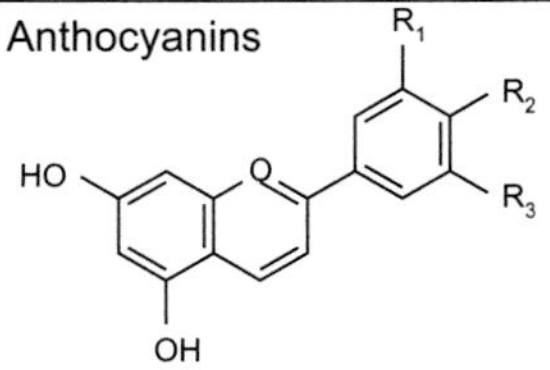	R1=H, R2=H: Pelargonidin R1=OH, R2=H: Cyanidin R1=R2=OH: Delphinidin R1=OCH3, R2=OH: Petunidin R1=R2=OCH3: Malvidin	Aubergines, radishes etc. Red wine, beans, berries etc. Pomegranate, cherries, berries etc. Cherries, strawberries etc. Cherries, berries, blackcurrants etc.

FIGURE 41.1   Chemical structure and source of some commonly occurring plant flavonoids.

protective role of flavonoids against lung cancer and other malignant neoplasms. In a study by Gates et al., associations between dietary flavonoid intakes of five common dietary flavonoids of the Zutphen study were investigated [3]. The results confirmed that dietary intake of these flavonoids reduced the risk of ovarian cancer. Similar findings were observed in a case-control

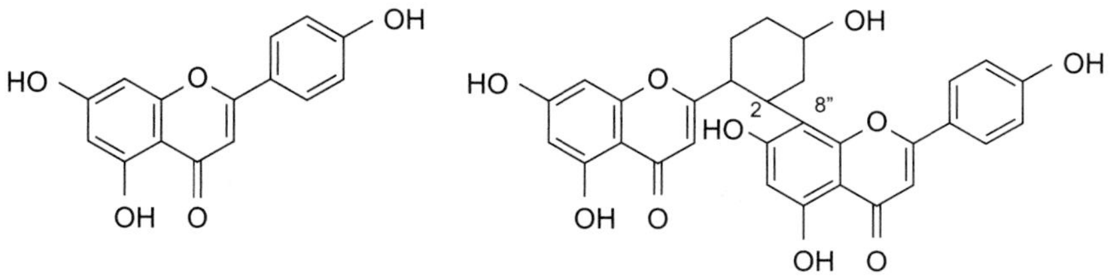

FIGURE 41.2 Chemical structure of apigenin and biapigenin.

study in Italy. The study investigated the relation of six classes of flavonoids with ovarian cancer risk, using the data from a multicentric case-control study between 1992 and 1999 that included 1031 cases with histologically confirmed epithelial ovarian cancer and 2411 hospital controls, essentially confirming the previous findings. Interestingly, another case-control study on intake of flavonoids and breast cancer risk investigating six principal classes of flavonoids was conducted in Italy between 1991 and 1994 on 2569 women with incident histologically confirmed breast cancer, and 2588 hospital controls [4]. This study documented an inverse association between flavones and breast cancer risk.

One study investigated the relation of flavonoids with the recurrence risk of neoplasia with resected colorectal cancer patients. Eighty seven patients, 36 with resected colon cancer and 51 after polypectomy, were divided into two groups: 31 patients received a flavonoid mixture (daily standard dose 20 mg apigenin and 20 mg epigallocathechin-gallate) and were compared with a 56 matched control group. Fourteen patients receiving flavonoid had no cancer recurrence in this group, and only one adenoma developed. Conversely, the cancer recurrence rate of the 15 matched untreated controls was 20% (3 of 15) and adenomas evolved in 4 of those patients (27%). The combined recurrence rate for neoplasia was 7% (1 of 14) in the treated patients and 47% (7 of 15) in the controls [5]. This study demonstrated that flavonoid intervention can reduce the recurrence

rate of neoplasia in patients with sporadic colorectal neoplasia. These observational and case-control reports of the beneficial effects of plant flavone intake are encouraging and demonstrate flavones as a class of beneficial compounds which possess health-promoting and disease-preventing dietary effects including efficacy in cancer prevention. An added benefit is that these compounds are associated with minimal toxicity and could be developed for chemoprevention protocols. Furthermore, most of the chemopreventive effects have been attributed to apigenin, a naturally occurring plant flavone that has demonstrated promising cancer chemopreventive/therapeutic properties.

2. APIGENIN – STRUCTURE, PROPERTIES AND SOURCES

Apigenin is a flavonoid belonging to the flavone structural class and chemically known as 4',5,7,-trihydroxyflavone, with molecular formula $C_{15}H_{10}O_5$. Apigenin has a low molecular weight (MW 270.24), structurally forming yellow needles in pure form. The melting point of apigenin is 347.5; it is practically insoluble in water, moderately soluble in hot alcohol, and is soluble in dilute potassium hydroxide and dimethyl sulfoxide [6]. It is incompatible with strong oxidizing agents. Apigenin also exists as a dimer, biapigenin, mainly isolated from the buds and flowers of *Hypericum perforatum* (Figure 41.2).

Apigenin is abundantly present in common fruits, plant-derived beverages, and vegetables such as parsley, onions, oranges, tea, chamomile, wheat sprouts, and some seasonings. One of the most common sources of apigenin consumed as single ingredient herbal tea is chamomile, prepared from the dried flowers from *Matricaria chamomilla* [7]. This is an annual herbaceous plant indigenous to Europe and western Asia that has been naturalized in Australia, Britain, and the USA. Also known as German chamomile, Hungarian chamomile, mayweed, sweet false chamomile, or wild chamomile, the plant is cultivated in Germany, Hungary, Russia, and other southern and eastern European countries for the flower heads. Infusions of chamomile contain maximum concentrations of apigenin ranging from 0.8 to 1.2% and essential oils which have aromatic, flavoring, and coloring properties. Chamomile is consumed in the form of tea at the rate of over 1 million cups per day. Other sources for apigenin include beverages such as wine and beer brewed from natural ingredients. Apigenin is commonly present as a constituent in red wine [8]. Like red wine, beer also provides a good source of apigenin [9]. In natural sources, apigenin is present as apigenin-7-*O*-glucoside and various acylated derivatives [10].

3. APIGENIN – INTAKE THROUGH DIET

There are no reports in the literature that have estimated the intake of apigenin alone in healthy adult individuals. According to an estimate, average intakes of flavonoids as flavonols and flavones have ranged from 6 mg per day in Finland to 64 mg/d in Japan, with intermediate intakes in the USA (13 mg/d), Italy (27 mg/d), and the Netherlands (33 mg/d). These estimates were based on analysis of five plant flavonoids: quercetin, kaempferol, myricetin, luteolin, and apigenin in composite food samples for the population analyzed in the Seven Countries Study [11]. In another study on the Hungarian population the intake of flavonoid was lower than in Dutch (23 mg/d), Danish (28 mg/d), and Finnish citizens (55 mg/d). The intake of five flavonoids in 17 different diets was estimated [12]. Lowest intake (1–9 mg/d) was estimated in South African diets, whereas the highest flavonoid intake (75–81 mg/d) was from a Scandinavian diet. In addition to flavonoid intake, dietary sources of the flavonols and flavones varied between different countries, with major contributions from tea in Japan (95%) and the Netherlands (64%), red wine and beer in Italy (46%), and vegetables and fruits in Finland (100%) and the USA (80%). For those regularly consuming wine and beer, apigenin will likely be the most important source of flavonoids. In Australia, tea remains the major dietary flavonoid with apparent dietary consumption up to 351 mg/person/d, of which 75% were flavan-3-ols [13].

4. APIGENIN – ABSORPTION AND METABOLISM

Although flavonoids have poor bioavailability, there are reports that fully methylated flavones have high metabolic stability as well as good intestinal absorption [14,15]. Studies with methylated apigenin (5,7,4′-trimethoxyflavone) in rats exhibited elevated intestinal absorption and high oral bioavailability and tissue accumulation greater than natural apigenin [16].

Apigenin in its natural form is present in foods mostly as glucoside conjugates and various acylated derivatives, which are more water soluble than the parent compound and remain mostly unmodified by various cooking methods. The moiety with which apigenin is conjugated is an important determinant of its absorption and bioavailability, since these attributes may require enzymatic cleavage by mammalian or microbial glucosidases [17]. Studies have shown that human absorption of quercetin glycoside from onions is far better than that of pure quercetin [18,19]. Consequently it seems likely that apigenin in its natural form bound to β-glucosides may be in its optimal bioavailable form.

Upon reaching the gut, apigenin is extensively metabolized via the dual recycling scheme involving both enteric and enterohepatic recycling [20,21]. Apigenin has been shown to be rapidly metabolized via UDP glucuronosyltransferase UGT1A1 into glucuroside and sulfate conjugates which are more readily transported in the blood and excreted in bile or urine [22]. In a study, absorption and excretion of apigenin were determined in 11 healthy subjects after ingestion of a single oral bolus of 2 g parsley. The maximum apigenin plasma concentration of 127 ± 81 nmol/L was observed after 7.2 ± 1.3 h; levels became undetectable after 28 h. The average apigenin content measured in 24-hour urine collections was 144 ± 110 nmol/24 h, corresponding to $0.22 \pm 0.16\%$ of the ingested dose [23]. These results suggest that only a small portion of dietary apigenin reaches the human circulation. A single oral dose ingestion of radiolabeled apigenin by rats demonstrated that 51% of the radioactivity appeared in urine, 12% in feces, 1.2% in blood, 0.4% in the kidneys, 9.4% in the intestine, 1.2% in liver, and 24.8% in the rest of the body within 10 days. The radioactivity appeared in blood 24 hours after oral apigenin intake. The kinetics of apigenin in blood was characterized by a prolonged elimination half-life of 91.8 h, which is relatively high compared to other dietary flavonoids [24]. In another study, a single oral dose of 200 mg/kg *Chrysanthemum morifolium* extract to rats showed that apigenin absorbed more efficiently than luteolin and both had a slow elimination phase [25]. These results suggest that although the bioavailability of apigenin is limited, the slow pharmacokinetics may allow this flavonoid to accumulate in various tissues and exert its salutary effects.

5. APIGENIN – MODE OF ACTION

Apigenin has gained particular interest over the years as a beneficial and health-promoting agent because of its low intrinsic toxicity and because of its striking effects on normal versus cancer cells, compared with other structurally related flavonoids [26]. There is very little evidence to date to suggest that apigenin promotes adverse metabolic reactions *in vivo* when consumed as part of a normal diet. Other studies have also confirmed that apigenin possesses 1) antioxidant, 2) antimutagenic, 3) anticarcinogenic, 4) anti-inflammatory, 5) antineoplastic proliferation and 6) antineoplasic progression properties [27].

Apigenin has been shown to possess antimutagenic properties in a setting of nitropyrene-induced genotoxicity in Chinese hamster ovary cells [28]. Apigenin has also been shown to inhibit benzo[a]pyrene and 2-aminoanthracene-induced bacterial mutagenesis [28]. Laboratory studies have demonstrated that apigenin promotes metal chelation, scavenges free radicals, and stimulates phase II detoxification enzymes in cell culture and in *in vivo* tumor models [29]. Exposure to apigenin prior to a carcinogenic insult has been shown to afford a protective effect in murine skin and colon cancer models [30,31]. Apigenin is a strong inhibitor of ornithine decarboxylase, an enzyme that plays a major role in tumor promotion [32]. In addition, apigenin has been shown to increase the intracellular concentration of glutathione, enhancing the endogenous defense against oxidative stress [33].

The anticarcinogenic effects of apigenin have been demonstrated in a skin carcinogenesis model. Topical application of apigenin inhibited dimethyl benzanthracene-induced skin tumors [32]. In addition, apigenin administration diminished the incidence of UV light-induced cancers and increased tumor-free survival in similar experiments [30].

The anti-inflammatory properties of apigenin are evident in studies that have shown suppression of lipopolysaccharide (LPS)-induced cyclooxygenase-2 and nitric oxide synthase-2 activity and expression in mouse macrophages [34]. Analyses of structure–activity relationships of 45 flavones, flavonols, and their related compounds showed that luteolin, ayanin, apigenin, and fisetin are the strongest inhibitors of IL-4 production [35]. This work was further confirmed in ovalbumin immunized BALB/C mice where production of IL-4 was down-regulated by apigenin [36]. In another study, apigenin treatment resulted in suppression of tumor necrosis factor (TNF) α-induced nuclear factor (NF)-κB activation in human umbilical vein endothelial cells [37].

Several studies have demonstrated that apigenin exerts a broad range of molecular signaling effects [38]. Apigenin has been reported to inhibit protein kinase C activity, mitogen activated protein kinase (MAPK), transformation of C3HI mouse embryonic fibroblasts, and the downstream oncogenes in v-Ha-ras-transformed NIH3T3 cells [39,40]. Apigenin is a well-known inhibitor of protein tyrosine kinases and has been shown to block peroxisome proliferation regulated kinase (ERK), a MAPK in isolated hepatocytes [41]. In 2007, we reported that apigenin-mediated inhibition of cell proliferation is due to modulations in MAPK, PI3K-Akt in human prostate cancer cells [42]. Apigenin has further been shown to down-regulate the expression of the Na^+/Ca^{2+} exchanger, a protein important for calcium extrusion in neonatal rat cardiac myocytes [43]. Apigenin treatment has been shown to decrease the levels of phosphorylated EGFR tyrosine kinase and of other MAPK and their nuclear substrate c-myc, which causes apoptosis in anaplastic thyroid cancer cells [44]. Furthermore, apigenin has been shown to

inhibit the expression of casein kinase (CK)-2 in both human prostate and breast cancer cells [45,46].

It has been demonstrated that apigenin exerts its effects on the cell cycle. Exposure of a wide array of malignant cells, including epidermal cells and fibroblasts, to apigenin induces a reversible G2/M and G0/G1 arrest by inhibiting p34 (cdc2) kinase activity, accompanied by increased p53 protein stability [47,48]. Apigenin has also been shown to induce WAF1/p21 levels resulting in cell cycle arrest and apoptosis in androgen-responsive human prostate cancer, LNCaP cells and androgen-refractory DU145 cells, regardless of the Rb status and p53 dependence or p53 independence [49,50]. In addition, apigenin has been shown to induce apoptosis in a wide range of malignant cells [51–53]. Apigenin treatment has been shown to alter the Bax/Bcl-2 ratio in favor of apoptosis, associated with release of cytochrome c and induction of apoptotic protease activating factor (Apaf)-1, which leads to caspase activation and poly (ADP-ribose) polymerase (PARP) cleavage [50].

Apigenin has shown promise in inhibiting tumor cell invasion and metastases by regulating protease production [54]. Apigenin under *in vivo* conditions is also effective in inhibiting TNF-α induced intracellular adhesion molecule-1 up-regulation in cultured human endothelial cells [55]. *In vivo* studies have also shown that apigenin inhibits melanoma lung metastases by impairing interaction of tumor cells with endothelium [56]. Furthermore, exposure of endothelial cells to apigenin results in suppression of the expression of vascular endothelial growth factor (VEGF), an important factor in angiogenesis via degradation of hypoxia-inducible factor (HIF)-1α protein [57]. Apigenin has also been shown to inhibit the expression of HIF-1α and VEGF via the PI3K/Akt/p70S6K1 and HDM2/p53 pathways in human ovarian cancer cells [58].

Studies demonstrated that apigenin is an effective inhibitor of aromatase and 17β-hydroxysteroid dehydrogenase activities in human placental microsomes, with resulting effects on steroid metabolism [59]. Oral administration of apigenin was shown to cause a significant increase in uterine weight and overall uterine concentration of estrogen receptor (ER)-α in female mice [60], and also suppresses prostate and breast cancer cell growth through estrogen receptor β1 [61]. Apigenin has been shown to decrease intracellular and secreted levels of prostate-specific antigen (PSA) in androgen-responsive human prostate cancer LNCaP cells [49]. It has also been shown that oral administration of apigenin suppresses the levels of insulin-like growth factor (IGF)-1 in prostate tumor xenografts and increases levels of insulin-like growth factor binding protein (IGFBP)-3, a binding protein that sequesters IGF-1 in vascular circulation [62]. Further studies have demonstrated that apigenin exposure to human prostate carcinoma DU145 cells caused increase in protein levels of E-cadherin and inhibited nuclear translocation of β-catenin and its retention to the cytoplasm [63]. These studies imply that apigenin may have the potential to inhibit hormone-related cancers as well.

A 2007 study demonstrated effects of apigenin on the immune system in C57BL/6 mice. Apigenin feeding for 2 weeks resulted in significant suppression of total immunoglobulin (Ig) E levels whereas levels of IgG, IgM, and IgA were not affected [36]. Apigenin feeding further resulted in decreased production of regulated-on-activation normal T cell expressed and secreted (RANTES) and soluble tumor necrosis factor receptor I in mouse serum.

Other important targets of apigenin include heat shock proteins [57], telomerase [64], fatty acid synthase [65], matrix metalloproteinases [66], aryl hydrocarbon receptor activity [67], HER2/neu [68], and casein kinase 2 alpha [69], all of which have relevance to the development of various human diseases.

6. APIGENIN – ROLE IN HUMAN DISEASES

For centuries, apigenin has been utilized as a traditional or alternative medicine against various human diseases. For example, passion flower, which contains high levels of apigenin, has been used effectively to treat asthma, intransigent insomnia, Parkinson's disease, neuralgia, and shingles [70]. Apigenin is a major constituent of chamomile, which is recognized for its antiphlogistic, antispasmodic, and antibacterial effects. Chamomile tea, 3–4 cups a day, has been used for centuries as a folk medicine remedy for relieving indigestion or calming gastritis [70]. In addition, chamomile preparations are widely used in skin care products to reduce cutaneous inflammation and other dermatological diseases [71]. Alcoholic tincture of the whole flowering plant of chamomile has been used topically as a rinse, gargle, cream, ointment, or bath additive. It has also been used as a vapor inhalant [70]. Studies providing objective evidence of the beneficial effects of apigenin on human health are quite limited, although a few epidemiological and clinical studies support the concept that apigenin is beneficial in counteracting coronary artery disease, gastrointestinal irritation, and dermatological disorders, in alleviating labor pain, and in providing antidepressant, calming, and relaxing effects. The putative effects of apigenin on various human diseases/disorders are shown in Table 41.1.

7. APIGENIN – ROLE IN CANCER PREVENTION

Over the years, apigenin has been increasingly recognized as a cancer chemopreventive agent. Interest in the possible cancer preventive activity of apigenin has increased owing to reports of potent antioxidant and

TABLE 41.1 Human Studies Examining the Effects of Apigenin in Various Disorders/Diseases

Number of Individuals	Source of Delivery	Doses	Period	Outcome	References
87 patients; 36 with resected colon cancer and 51 after polypectomy	Orally	20 mg apigenin and 20 mg epigallocate-chin-gallat in 31 patients	Daily dose 3–4 years surveillance and colonoscopy	Flavonoid-treated patients with resected colon cancer ($N = 14$), there was no cancer recurrence and one adenoma developed	[5]
Incidence of epithelial ovarian cancer among 66,940 women in the Nurses' Health Study. Analysis included 347 cases diagnosed between 1984 and 2002	Dietary intake including fruits, vegetables and tea	Five common dietary flavonoids: apigenin myricetin, kaempferol, quercetin, and luteolin	Daily	Dietary intake of certain flavonoids may reduce ovarian cancer risk	[3]
255 patients presenting with acute diarrhea, ages 6 months to 6 years	Orally	Apple pectin chamomile extract		Shortening the course of the disease and relieving associated symptoms	[146]
88 healthy breast-feeding colicky infants, ages 21–60 weeks	Beverage	Herbal mixture (71.1 mg/kg/d chamomile) or placebo	One week	Significantly reduced crying time compared with placebo. 85% reduction in treatment group vs. 49% in placebo group	[147]
55 postmenopausal women complaining of hot flushes and refusing hormonal therapy	Orally	Chewable tablets of Chamomilla (Climax)	5 tablets daily	Decrease in number and intensity of hot flushes from baseline to completion of treatment	[148]
Japanese young males	Beverage	One serving chamomile tea or placebo (hot water)	Single dose	Decreased heart rate and ratings of sadness and depression after drinking tea	[149]
161 patients administered before intubation on postoperative sore throat and hoarseness	Spray	111 mg chamomile extract/normal saline in placebo	Single dose	No significant difference was observed. 42 out of 80 patients in chamomile scored no postoperative sore throat in post-anesthesia care. There was significant correlation between use of oral airway and sore throat in the post-anesthesia care unit after 24 h of operation	[150]
8058 mothers in childbirth	Skin/inhalation aromatherapy	Chamomile and clary sage essential oils	Single dose	Effective in alleviating pain	[151]
79 children aged 6–66 months with acute, noncomplicated diarrhea	Liquid	Chamomile/ apple pectin preparation or placebo	Three day	Diarrhea ended sooner in treatment group (85%) than	[152]

(Continued)

C: ACTIONS OF INDIVIDUAL FRUITS IN DISEASE AND CANCER PREVENTION AND TREATMENT

TABLE 41.1 (Continued)

Number of Individuals	Source of Delivery	Doses	Period	Outcome	References
				(58%). Duration was also attenuated	
164 double-blind placebo control clinical trial 5FU-induced oral mucositis	Orally	Chamomile or placebo mouth wash	Thrice daily for 14 days	No stomatitis difference between patients randomized to either protocol arm	[153]
24 healthy subjects; erythema induced by the radiation and cell phone tape stripping	Topically applied	Chamomile cream	Single application	Erythema was suppressed in 24 h	[154]
68 healthy infants (2–8 weeks)	Beverage	150 mL herbal mixture including chamomile or placebo	One week	Significantly higher colic improvement score compared with placebo. Eliminated colic in 57% of treatment group vs. 26% in placebo group	[155]
22 eye patients	Oil inhalation	Chamomile/crushed green pepper or placebo	Two doses	Significantly increased the latency for all images and shifted mood rating and frequency judgment in positive direction	[156]
14 patients wound healing after dermabrasion of tattoos	Topically applied chamomile	Chamomile extract	Single	Significantly decreased the wound area and increased drying of the wound	[157]
12 patients heart disease hospitalized for cardiac catheterization	Beverage	2 cups chamomile tea	Single dose	Significant decrease in mean brachial artery pressure from baseline	[158]

anti-inflammatory activities. Support for this notion comes from a study where consumption of flavonoid-free diets by healthy human volunteers led to a decrease in markers of oxidative stress in blood, *viz.* plasma antioxidant vitamins, erythrocyte superoxide dismutase (SOD) activity, and lymphocyte DNA damage commonly associated with enhanced disease risk [72].

Many of the biological effects of apigenin in numerous mammalian systems *in vitro* as well as *in vivo* are related to its antioxidant effects and its role in scavenging free radicals. Furthermore, it exhibits antimutagenic, anti-inflammatory, antiviral, and purgative effects [73]. The actions of apigenin in inhibiting the cell cycle, diminishing oxidative stress, improving the efficacy of detoxification enzymes,

inducing apoptosis, and stimulating the immune system are quite limited [73–75]. One human study demonstrated that apigenin was absorbed systemically by a subject fed a diet high in parsley; this subject was found to have elevated levels of the antioxidant enzymes erythrocyte glutathione reductase and superoxide dismutase [76]. Activities of erythrocyte catalase and glutathione peroxidase, however, were found to be unchanged. Other biological effects induced by flavonoids include reduction of plasma levels of low density lipoproteins, inhibition of platelet aggregation, and reduction of cell proliferation [73–75,77]. This is apparent from another cross-sectional study conducted in Japan in which total intake of flavonoids among women was found to be inversely correlated with plasma total cholesterol and low density lipoprotein concentration, after adjustment for age, body mass index, and total energy intake [78]. The effects of flavonoids on the hematological system were assessed in a 7-day study of 18 healthy men and women, examining the effects of a daily dietary supplement providing quercetin (377 ± 10 mmol from onions) and apigenin (84 ± 6 mg from parsley) on platelet aggregation and other hemostatic variables. The authors observed no significant changes in collagen- or ADP-induced platelet number, factor VII, plasminogen, PAI-1 activity or fibrinogen concentrations [79]. These inherent properties of flavonoids categorize them as a class of beneficial compounds which possess health-promoting and disease-preventing dietary effects.

8. APIGENIN – ROLE IN HUMAN CANCERS

8.1 Breast Cancer

Studies have demonstrated antiproliferative effects of apigenin on human breast cancer cell lines with different levels of HER2/neu expression. Apigenin exhibited potent growth inhibitory activity in HER2/neu over-expressing breast cancer cells but was much less effective in inhibiting growth of cells expressing basal levels of HER2/neu [68]. Induction of apoptosis was also observed in HER2/neu over-expressing breast cancer cells in a dose- and time-dependent manner after apigenin treatment [80]. The cell survival pathway involving phosphatidylinositol 3-kinase (PI3K) and Akt/PKB is known to play an important role in inhibiting apoptosis in HER2/neu expressing breast cancer cells. Apigenin has been shown to inhibit Akt function in tumor cells by directly inhibiting PI3K activity and consequently inhibiting Akt kinase activity [81].

Additionally, inhibition of HER2/neu auto-phosphorylation and trans-phosphorylation resulting from depleting HER/neu protein *in vivo* was observed after apigenin treatment. Further studies from the same group showed that exposure of HER2/neu expressing breast cancer cells to apigenin resulted in induction of apoptosis by depleting HER2/neu protein and, in turn, suppressing the signaling of the HER2/HER3-PI3K/Akt pathway. Apoptosis in breast cancer cells exposed to apigenin was induced through cytochrome c release and rapid induction of DNA fragmentation factor 45. Apigenin has also been shown to down-regulate the levels of cyclin D1, D3, and cdk4 and increase p27 protein levels in breast cancer cells [80].

It has been reported that peptide hormones and protein kinase C (PKC)-activating phorbol ester (PMA) protect cells from apoptosis through activation of cellular signaling pathways such as the MAPK and PI3K pathways [81]. Additional studies have demonstrated suppression of TNF-α-induced apoptosis by treatment with PMA in MCF-7 breast carcinoma cells [81]. The ability of apigenin to block PMA-mediated cell survival was correlated with suppression of PMA-stimulated AP-1 activity, providing evidence of the ability of apigenin to affect cell survival pathways and offering an explanation for its antitumor activity.

The effect of apigenin on protease-mediated invasiveness was evaluated in estrogen-insensitive breast tumor cell line MDA-MB231, showing that apigenin strongly inhibited tumor cell invasion in a dose-dependent manner [54]. Apigenin inhibits growth and induces G2/M arrest by modulating cyclin-cdk regulators and ERK MAP kinase activation in breast carcinoma cells [82]. The growth inhibitory effects of apigenin were observed on MCF-7 cells that express two key cell cycle regulators, wild-type p53 and the retinoblastoma tumor suppressor protein (Rb), and in MDA-MB-468 cells which are mutant for p53 and Rb negative. Apigenin-mediated cell growth inhibition along with

G2/M arrest was accompanied by significant decrease in cyclin B1 and cdk1 protein levels, resulting in a marked inhibition of cdk1 kinase activity. Furthermore, apigenin treatment reduced the protein levels of cdk4, cyclin D1, and A, and inhibited Rb phosphorylation, but did not affect the protein levels of cyclin E, cdk2, or cdk6. Studies have shown that apigenin induces G(2)/M phase cell cycle arrest in SK-BR-3 cells which is via regulation of cdk1 and p21 (Cip1) pathway. In addition, apigenin treatment resulted in ERK MAP kinase phosphorylation and activation in MDA-MB-468 cells [83].

Further effects of apigenin and other phytoestrogens on DNA synthesis (estimated by thymidine incorporation analysis) were evaluated in estrogen-dependent MCF-7 cells in the presence of estradiol (E2), tamoxifen, insulin, or epidermal growth factor [84,85]. The results show that apigenin was capable of inhibiting E2-induced DNA synthesis in these cells. Overall, the effects of apigenin and other phytoestrogens in the presence of E2 or growth factors were variable and concentration dependent.

A study to characterize the estrogenic and anti-estrogenic activities of flavonoids was performed in the ER-positive MCF-7 human breast cancer cell line using an ER-dependent reporter gene assay and an ER competition binding assay [86]. In these studies apigenin was shown to possess anti-estrogenic activities which may be mediated through ER binding-dependent and independent mechanisms. A later study suggested that apigenin targets both ERalpha-dependent and ERalpha-independent pathways on estrogen-responsive, anti-estrogen-sensitive MCF7 breast cancer cells and has a growth inhibitory effect on two MCF7 sublines with acquired resistance to either tamoxifen or fulvestrant [87]. These anti-estrogenic activities were deemed to be biologically significant in the regulation of breast cancer cell proliferation.

The combined effects of multiple flavonoids on breast cancer resistance protein (BCRP) were demonstrated: several plant flavonoids including apigenin were used alone or in combinations to evaluate the potential interactions for BCRP inhibition [88]. Apigenin and other flavonoids were shown to inhibit the BCRP protein, which was highly efficacious in combination at equimolar concentrations. Another study compared the endocrine disruption activities of compounds in materials used to package foods including bisphenol derivatives and plant flavonoids, including apigenin, on human breast cancer cell lines MCF-7 which is ER(+) and MDA-MB453 which is AR(+) and GR(+). These studies suggested that natural compounds had a biphasic effect: at high concentrations they act as GR agonists and in low concentrations they may act as partial androgen receptor (AR) agonists [89]. In another study apigenin and genistein have been shown to stimulate the proliferation of MCF-7 and T47D cells [estrogen receptor alpha (ERalpha-positive)], but do not stimulate the proliferation of an ERalpha-negative cell line (MDA-MB-435 cells) [90]. These studies indicate that estrogenicity of the phytochemicals is quantitatively important in inducing cell proliferation or inhibiting aromatase, suggesting that perhaps a more cautionary approach should be taken before they are taken as food supplements [91].

Another study has shown that plant flavonoids can induce apoptosis in human breast and prostate cancer cells, an effect that is associated with their ability to inhibit the activity of fatty acid synthase, a key metabolic enzyme that catalyzes the synthesis of long-chain fatty acids over-expressed in neoplastic and malignant cells [65]. In this study at least six plant-derived flavonoids, including apigenin, had marked inhibitory effects on cancer cell growth and survival which appear to be related to their ability to inhibit fatty acid synthesis. A later observation confirmed that extra virgin olive oil derived apigenin content was able to suppress the expression of lipogenic enzyme fatty acid synthase in SKBR and MCF-7/HER2 cells [92]. The proteasomal chymotrypsin-like activity was inhibited by the apigenin and induces apoptosis by the activation of caspase 3, 7 and poly (ADP-ribose) polymerase cleavage, in cultured MDA-MB-231 cells and also in MDA-MB-231 xenografts [25]. Further studies have shown that apigenin inhibits hepatocyte growth factor-induced MDA-MB-231 cell invasiveness and metastasis by blocking Akt, ERK, and JNK phosphorylation and also inhibits clustering of β-4-integrin function at the actin-rich adhesive site [93].

8.2 Cervical Cancer

The first report about apigenin in human cervical carcinoma HeLa cells demonstrated that apigenin inhibited growth through an apoptotic pathway. Apigenin inhibited cell growth, caused G1 phase growth arrest, and induced apoptosis which was p53 dependent and associated with a marked increase in the expression of p21/WAF1 protein and with the induction of Fas/APO-1 and caspase-3 expression. Apigenin also decreased the expression of Bcl-2 protein, an anti-apoptotic factor [94].

Further studies demonstrated that apigenin can interfere with cell proliferation, cell survival, and gap junctional coupling. Exposure of noninvasive wild-type HeLa cells and their connexin43 (Cx43)-transfected counterparts to apigenin resulted in a significant and reversible inhibition of translocation of both cell types. The effect of apigenin on cell proliferation was less pronounced especially at low apigenin concentration, whereas its influence on cell motility correlated with the reduction of the invasive potential of HeLa Cx43 cells [95]. Another study with medicinal herb feverfew (*Tanacetum parthenium*) extract containing parthenolide, camphor, luteolin, and apigenin showed antiproliferative activity against SiHa human cervical cancer cells [96].

8.3 Colon Cancer

In various human colon carcinoma cell lines apigenin treatment resulted in cell growth inhibition and G2/M cell cycle arrest, which was associated with inhibition of p34 (cdc2) kinase, and with reduced accumulation of p34 (cdc2) and cyclin B1 proteins [97]. Additional studies were performed in individual and interactive influences of seven apigenin analogues on cell cycle, cell number, and cell viability in human colonic carcinoma cell lines [98]. These findings indicate that the induction of cell cycle arrest by five of seven tested apigenin analogues and the additive induction by the combination of flavonoids at low doses cooperatively protect against colorectal cancer through conjoint blocking of cell cycle progression.

An important effect of apigenin is to increase the stability of the tumor suppressor p53 gene in normal cells [47]. It is speculated that apigenin may play a significant role in cancer prevention by modifying the effects of p53 protein. Exposure of p53-mutant cancer cells to apigenin results in inhibition of cell growth and alteration of the cell cycle as demonstrated in a study in which apigenin treatment resulted in growth inhibition and G2/M phase arrest in two p53-mutant cancer cell lines, HT-29 and MG63 [99]. These effects were associated with a marked increase in the protein expression of p21/WAF1 in a dose- and time-dependent manner. These results suggest that there is a p53-independent pathway for apigenin in p53-mutant cell lines, which induces p21/WAF1 expression and growth inhibition. Further assessment was that adenomatous polyposis coli (APC) dysfunction may be critical for apigenin to induce cell cycle arrest in human colon cancer HT29-APC cells (mutated APC), but apigenin enhances APC expression and apoptosis in cells [100]. Studies with 5,6-dichloro-ribifuranosylbenzimidazole (DRB) and apigenin represented induced sensitization of colon cancer cells to TNF-α-mediated apoptosis [101]. Inhibition of CK2 in HCT-116 and HT-29 cells with the use of two specific CK2 inhibitors, DRB and apigenin, resulted in a synergistic reduction in cell survival when used in conjunction with TNF-α [101]. Chemopreventive activity of apigenin may be mediated by its ability to modulate the MAPK cascade. Apigenin induced a dose-dependent phosphorylation of both ERK and p38 kinase but had little effect on the phosphorylation of c-jun amino terminal kinase (JNK) [31]. Further studies on apigenin suggest that it inhibits ornithine decarboxylase (ODC) activity and the formation of aberrant crypt foci in two different mouse models systems; azoxymethane (AOM)-induced CF-1 mice and Min mice with mutant adenomatous polyposis coli (APC) gene [102].

The interactions between sulforaphane and apigenin resulted in the induction of UGT1A1 and GSTA1, the phase II detoxifying enzymes, in CaCo-2 cells [103]. Apigenin was shown to induce UGT1A1 transcription but not GSTA1; sulforaphane induced both UGT1A1 and GSTA1 transcription in both dose- and time-dependent manner. The combination of sulforaphane and apigenin resulted in a synergistic induction of UGT1A1 mRNA expression, although this interaction was not seen for GSTA1, suggesting that different signal transduction pathways regulate the expression of detoxification enzymes. Additional studies suggest that apigenin is more potent than tricin or quercetin in down-regulating inducible COX-2 expression in HCEC cells [104].

8.4 Hematological Cancer

Apigenin was also tested to ascertain its effects on human leukemia cells. Apigenin was shown to be markedly more effective than other tested flavonoids in inducing apoptosis in these cells [51]. Further studies have shown that apigenin and quercetin both inhibit topoisomerase-catalyzed DNA irregularities that are involved

in many aspects of leukemia cell DNA metabolism including replication and transcription reactions. Another study suggests that treatment with apigenin in different leukemia cell lines resulted in selective antiproliferative and apoptotic effects in monocytic and lymphocytic leukemias; this selective apoptosis is mediated by induction of protein kinase C delta [105]. Apigenin inhibits platelet function through several mechanisms including blockade of TxA$_2$ receptors (TPs). The inhibitory effect of apigenin in the presence of plasma might in part rely on TxA$_2$ receptor antagonism. This was demonstrated through a clear increase in the *ex vivo* antiplatelet effect of aspirin in the presence of apigenin, which encourages the idea of the combined use of aspirin and apigenin in patients in whom aspirin fails to properly suppress the TxA$_2$ pathway [106].

A flavonoids-based study was carried out mainly on apigenin, quercetin, kaempferol, and myricetin for their proteasome-inhibitory and apoptosis-inducing abilities in human leukemia cells. The authors reported that apigenin and quercetin were much more potent than kaempferol and myricetin in: 1) inhibiting chymotrypsin-like activity of purified 20S proteasome and of 26S proteosome; 2) accumulating putative ubiquitinated forms of two proteasome target proteins, Bax and IκBα; and 3) inducing activation of caspase-3 and cleavage of poly (ADP-ribose) polymerase in Jurkat T cells [107]. Furthermore, the proteasome-inhibitory abilities of these compounds correlated with their apoptosis-inducing potencies.

Further structurally related flavonoids, such as apigenin, quercetin, myricetin, and kaempferol, were able to induce apoptosis in human promyelocytic leukemia HL-60 cells [51]. Treatment of cells with flavonoids caused rapid induction of caspase-3 activity and stimulated proteolytic cleavage of poly (ADP-ribose) polymerase. These flavonoids induced loss of mitochondrial transmembrane potential, elevation of reactive oxygen species production, release of mitochondrial cytochrome c into the cytosol, and subsequent induction of procaspase-9, with apigenin having the highest potency in inducing apoptotic effects. Olive leaf (*Olea europaea* L.) extract from seven principal Tunisian olive varieties, Chemchali, Chemlali, Chétoui, Gerboui, Sayali, Zalmati, and Zarrazi, having one common compound apigenin-7-O-glucoside in all the extracts, has been shown to reduce nitroblue tetrazolium (a differentiation marker) in HL-60 cells [108]. Another study evaluated the potential of 22 flavonoids and related compounds by testing their apoptotic activity in leukemic U937 cells [109]. In these studies, apigenin and several other flavones but not the isoflavones or flavanones tested were shown to induce apoptosis in U937 cells.

The protective effects of four flavonoids, quercetin, rutin, luteolin, and apigenin, were evaluated by measuring the extent of H_2O_2-induced DNA damage in murine leukemia L1210 cells [110]. The results show that apigenin, at low concentrations, was marginally effective in reducing the extent of DNA damage. However, at high concentrations apigenin induced DNA single strand breaks, indicating its ability to serve as a pro-oxidant. Another study evaluated the role of dietary bioflavonoids in inducing cleavage in the MLL gene, which may contribute to infant leukemia [111]. Apigenin was shown to induce DNA cleavage in primary progenitor hematopoietic cells from healthy newborns and adults and in cell lines by targeting topoisomerase II, an enzyme that alters the DNA topology. It is not known whether this *in vitro* study can be extrapolated to human situations, because of the dose and bioavailability issue.

8.5 Lung Cancer

The effects of apigenin on lung cancer cells were evaluated. Apigenin inhibited A549 lung cancer cell proliferation and VEGF transcriptional activation in a dose-dependent manner [112]. Apigenin inhibited VEGF transcriptional activation through the HIF-1 binding site, and specifically decreased HIF-1α, but not HIF-1β subunit expression in these cells. In a signaling pathway that mediates VEGF transcriptional activation, apigenin inhibited AKT and p70S6K1 activation. Lung cancer cells SQ-5 incubated with apigenin exhibited significantly greater radiosensitivity and apoptosis than cells without apigenin [113]. Another report demonstrates that apigenin exhibited an inhibitory effect on hepatocyte growth factor-induced Akt phosphorylation in lung carcinoma A549 cells [114]. In addition, the exposure of nude mice with lung cancer to apigenin inhibited HIF-1α and VEGF expression in the tumor tissues, suggesting an inhibitory effect of apigenin on angiogenesis. Another study demonstrated the efficacy of apigenin administration against experimental Lewis lung carcinomas (LLC), C-6 gliomas and DHDK 12 colonic cancers *in vivo* [115]. Tumor-bearing mice received 50 mg/kg/d of apigenin in three different galenical formulations during 12 days in 8-hourly intervals. No *in vivo* response was observed, in contrast to the high *in vitro* sensitivity of LLC, C-6, DHDK 12, and endothelial cells to apigenin; complete growth suppression occurs *in vitro* at concentrations beyond 30 μg/mL. Studies in B57BL/6N mice injected with B16-BL6 tumor cells demonstrated that the number of tumor cells adhering to lung vessels was significantly diminished in animals treated with a single dose of apigenin and quercetin [56].

8.6 Ovarian Cancer

In human ovarian cancer cells, apigenin-inhibited VEGF expression was observed at the transcriptional level through expression of HIF-1α via the PI3K/AKT/p70S6K1 and HDM2/p53 pathways. Apigenin has also been shown to inhibit tube formation by endothelial cells *in vitro* [58,116]. Additionally, apigenin inhibited the activity of MAPK and PI3K in human ovarian carcinoma HO-8910PM cells [117]. Apigenin inhibits expression of focal adhesion kinase (FAK), migration, and invasion of human ovarian cancer A2780 cells. Further *in vivo* experiments also showed that apigenin inhibited spontaneous metastasis of A2780 cells implanted onto the ovary of nude mice [118]. A prospective study investigated the association between intake of five common dietary flavonoids, *viz.* myricetin, kaempferol, quercetin, luteolin, and apigenin, and incidence of epithelial ovarian cancer among 66,940 women in the Nurses' Health Study. The researchers calculated each participant's intake of these flavonoids from dietary data collected at multiple time points, and used Cox proportional hazards regression to model the incidence rate ratio (RR) of ovarian cancer for each quintile of intake. There was significant reduction in ovarian cancer risk after dietary intake of certain flavonoids [3]. Furthermore, study of the potency and safety of selectively oncolytic adenoviruses for treatment of advanced ovarian cancers is necessary. Apigenin has been shown to reduce the replication of adenovirus, which could provide a safety switch in case of replication-associated side effects encountered in these patients. Apigenin was also found to be useful for the treatment of systemic adenoviral infections in immunosuppressed patients [119].

8.7 Prostate Cancer

The effects of selected bioflavonoids including apigenin were compared on the proliferation of androgen-independent human prostate cancer PC-3 cells, which show complete growth retardation after apigenin exposure [120]. The effects of bioflavonoids on the activity and phosphotyrosine content of oncogenic proline-directed protein kinase FA (PDPK FA) in human prostate carcinoma cells have also been studied. Long-term treatment of human prostate carcinoma cells with low concentrations of quercetin, apigenin, and kaempferol potently induced tyrosine dephosphorylation and concurrently inactivated oncogenic PDPK FA in a concentration-dependent manner [121].

Apigenin has the capability to significantly reduce cell number and induce apoptosis in PWR-1E, LNCaP, PC-3, and DU145 cells [122]. The PC-3 and DU145 cells were less susceptible to apigenin-induced apoptosis than LNCaP and PWR-1E cells. The induction of apoptosis by apigenin is caspase-dependent. Apigenin generates reactive oxygen species, causes loss of mitochondrial Bcl-2 expression, increases mitochondrial permeability, causes cytochrome c release, and induces cleavage of caspase 3, 7, 8, and 9 and the concomitant cleavage of the inhibitor of apoptosis protein, cIAP-2. The over-expression of Bcl-2 in LNCaP B10 cells reduces the apoptotic effects of apigenin. A study demonstrated correlation between the activity of casein kinase (CK) 2 and certain growth properties of prostate cancer cells [45]. Apigenin exposure led to inhibition of CK2 activity in both hormone-sensitive LNCaP cells and hormone-refractory PC-3 cells but only the hormone-sensitive LNCaP cells responded with apoptosis. These studies suggest that a high CK2 activity is not essential for growth or protection against apoptosis in hormone-refractory prostate cancer cells.

We evaluated the growth-inhibitory effects of apigenin on normal human prostate epithelial cells (NHPE), virally transformed normal human prostate epithelial PZ-HPV-7 cells, and human prostate adenocarcinoma CA-HPV-10 cells [26]. Apigenin treatment to NHPE and PZ-HPV-7 resulted in almost identical growth-inhibitory responses of low magnitude whereas significant decrease in cell viability was observed in CA-HPV-10 cells. Further we reported that apigenin inhibits the growth of androgen-responsive human prostate carcinoma LNCaP cells and described the molecular basis for this observation. The cell growth inhibition achieved by apigenin treatment resulted in a significant decrease in AR protein expression along with a decrease in intracellular and secreted forms of PSA [49]. Apigenin treatment of LNCaP cells resulted in G1 arrest in cell cycle progression which was associated with a marked decrease in the protein expression of cyclin D1, D2, and E and their activating partner cdk2, 4 and 6 with concomitant induction of WAF1/p21 and KIP1/p27. The induction of WAF1/p21 appears to be transcriptionally upregulated and is p53 dependent. In addition, apigenin inhibited hyperphosphorylation of the pRb protein in these cells. Next we studied apigenin-mediated inhibitory effects in androgen-refractory human prostate carcinoma DU145 cells which have mutations in the tumor suppressor gene p53 and pRb. Exposure of DU145 cells to apigenin resulted in a dose- and time-dependent inhibition of growth, colony formation, and G1 phase arrest of the cell cycle [50]. Apigenin exposure also resulted in alteration in Bax/Bcl2 ratio in favor of apoptosis, which was associated with the release of cytochrome c and induction of apoptotic protease activating factor-1 (Apaf-1). This effect was found to result in a significant increase in cleaved fragments of caspase-9, -3, and poly (ADP-ribose) polymerase (PARP). Apigenin exposure also resulted in down-modulation of the constitutive expression of NFκB/p65 and NFκB/p50 in the nuclear fraction which correlated with an increase in the expression of IκBα in the cytosol. In another study we examined whether apigenin was effective in inhibiting the expression of NFκB, a gene that regulates several cell survival and anti-apoptotic genes. Exposure of PC-3 cells to apigenin inhibited DNA binding and reduced nuclear levels of the p65 and p50 subunits of NFκB with concomitant decrease in

IκBα degradation, IκBα phosphorylation, and IκKα kinase activity [123]. In addition, apigenin exposure inhibited TNF-α-induced activation of NFκB via the IκBα pathway, thereby sensitizing the cells to TNF-α-induced apoptosis. The inhibition of NFκB activation correlated with a decreased expression of NFκB-dependent reporter gene and suppressed expression of NFκB-regulated genes, specifically, Bcl2, cyclin D1, cyclooxygenase-2, matrix metalloproteinase 9, nitric oxide synthase-2, and VEGF. Furthermore, we investigated the *in vivo* growth-inhibitory effects of apigenin on androgen-sensitive human prostate carcinoma 22Rv1 tumor xenografts subcutaneously implanted in athymic male nude mice [62]. Apigenin feeding resulted in dose-dependent inhibition of tumor growth which was associated with increased accumulation of human IGFBP-3 in mouse serum. Apigenin consumption by these mice also resulted in simultaneous decrease in serum IGF-1 levels and induction of apoptosis in tumor xenografts, evidence favoring the concept that the growth-inhibitory effects of apigenin involve modulation of IGF-axis signaling in prostate cancer. Further studies with pharmacological intervention of apigenin showed a direct growth-inhibitory effect on human prostate tumors implanted in athymic nude mice. Oral feeding of apigenin resulted in dose-dependent: 1) increase in the protein expression of WAF1/p21, KIP1/p27, INK4a/p16, and INK4c/p18; 2) down-modulation of the protein expression of cyclins D1, D2, and E, and cyclin-dependent kinases (cdk), cdk2, cdk4, and cdk6; 3) decrease in retinoblastoma phosphorylation at serine 780; 4) increase in the binding of cyclin D1 toward WAF1/p21 and KIP1/p27; and 5) decrease in the binding of cyclin E toward cdk2 in both types of tumors [124]. Studies with apigenin in LNCaP and PC-3 cells demonstrated G0–G1 phase arrest, decrease in total retinoblastoma (Rb) protein, and its phosphorylation at Ser780 and Ser807/811 in dose- and time-dependent fashion. Apigenin treatment caused increased phosphorylation of ERK1/2 and JNK1/2 and this sustained activation resulted in decreased ELK-1 phosphorylation and c-FOS expression thereby inhibiting cell survival. Interestingly, apigenin caused a marked reduction in cyclin D1, D2, and E and their regulatory partners cdk 2, 4, and 6, operative in the G0–G1 phase of the cell cycle. This was accompanied by a loss of RNA polymerase II phosphorylation, suggesting the effectiveness of apigenin in inhibiting transcription of these proteins [42]. In another study using TRansgenic Adenocarcinoma of Mouse Prostate (TRAMP) model, we demonstrated that oral administration of apigenin at doses of 20 and 50 μg/mouse/d, 6 days per week for 20 weeks, significantly decreased tumor volumes of the prostate as well as completely abolishing distant-site metastases to lymph nodes, lungs, and liver [63]. Administration of apigenin resulted in increased levels of E-cadherin and decreased levels of nuclear β-catenin, c-Myc, and cyclin D1 in the dorso-lateral prostates of TRAMP mice. These studies indicate that apigenin is effective in suppressing prostate carcinogenesis in an *in vivo* model, at least in part, by blocking β-catenin signaling. Furthermore, we demonstrated that apigenin at different doses resulted in ROS generation, which was accompanied by rapid glutathione depletion, disruption of mitochondrial membrane potential, cytosolic release of cytochrome c, and apoptosis in human prostate cancer 22Rv1 cells [125]. There was accumulation of a p53 fraction to the mitochondria, which was rapid and occurred between 1 and 3 h after apigenin treatment. *In vivo*, 22Rv1 xenograft studies confirmed that apigenin administration resulted in p53-mediated induction of apoptosis in 22Rv1 tumors. These results indicated that apigenin-induced apoptosis in 22Rv1 cells is initiated by a ROS-dependent disruption of the mitochondrial membrane potential through transcriptional-dependent and -independent p53 pathways.

The mechanism(s) of apigenin action on the IGF/IGF-IR (insulin-like growth factor receptor 1 protein) signaling pathway in human prostate cancer DU145 cells markedly reduced IGF-I-stimulated cell proliferation and induced apoptosis [126]. This effect of apigenin might be partially due to reduced autophosphorylation of IGF-IR. Inhibition of p-Akt by apigenin resulted in decreased phosphorylation of GSK-3β. In another study using human prostate cancer PC-3 cells we further demonstrated that apigenin-mediated dephosphorylation of Akt resulted in inhibition of its kinase activity, which was confirmed by reduced phosphorylation of pro-apoptotic proteins BAD and glycogen synthase kinase-3, essential downstream targets of Akt [127]. These results suggest that Akt inactivation and dephosphorylation of BAD is a critical event, at least in part, in apigenin-induced decreased cell survival and apoptosis. One report in 2008 suggested that hypoxia induced a time-dependent increase in the level of HIF-1α subunit protein in PC3-M cells, which were markedly decreasing HIF-1α expression after apigenin treatment under both normoxic and hypoxic conditions [128]. Apigenin prevented the activation of the HIF-1 and its downstream target gene VEGF.

8.8 Skin Cancer

Studies have shown that apigenin is effective in the prevention of UVA/B-induced skin carcinogenesis in SKH-1 mice [30]. Topical application of apigenin has been shown to inhibit UV-mediated induction of ornithine decarboxylase activity, reduce tumor incidence, and increase tumor-free survival in mice. Several other studies have provided evidence that apigenin prevents UV-induced skin tumorigenesis by inhibiting the cell cycle and cyclin-dependent kinases [48]. Exposure of mouse keratinocytes to apigenin induced G2/M cell cycle arrest and accumulation of the p53 tumor suppressor protein with increased expression of p21/WAF1. This arrest was accompanied by inhibition of p34 (cdk2) kinase protein level and activity, which was found to be independent of p21/WAF1 [129]. In human diploid fibroblasts, apigenin produced G1 cell cycle arrest by inhibiting cdk2 kinase activity and inducing p21/WAF1. A short-term *in vivo* system was established to evaluate topical formulations of apigenin and to determine whether apigenin is effective when delivered as a topical preparation to the local skin lesions. It was observed that topical application of apigenin was capable of targeting local tissue [130]. Further study demonstrated the *in vivo* and *in vitro* percutaneous absorption of apigenin using different vehicles [131]. Some observations suggest that apigenin suppresses the UVB-induced increase in COX-2 expression (a key enzyme which converts the arachidonic acid to prostaglandins, and its over-expression results in carcinogenesis) and mRNA in mouse and even in human keratinocyte cell lines [132,133]. Through these studies it was apparent that delivery of apigenin into viable epidermis appears to be a necessary property for an apigenin formulation to be effective in skin cancer prevention.

The combined effects of quercetin and apigenin were evaluated on inhibition of melanoma growth, invasiveness, and metastatic potential, and demonstrated that *in vivo* administration of apigenin and quercetin was effective in inhibiting melanoma lung tumor metastasis in a B16-BL6 murine melanoma metastasis model, an effect that was postulated to be due to the impairment of endothelial interactions in malignant cells [134].

8.9 Thyroid Cancer

The effects of some selected flavonoids including apigenin were investigated on human thyroid carcinoma cell lines, UCLA NPA-87-1 (NPA) (papillary carcinoma), UCLA RO-82W-1 (WRO) (follicular carcinoma), and UCLA RO-81A-1 (ARO) (anaplastic carcinoma) [44]. Among the flavonoids tested, apigenin was the most potent inhibitor of the proliferation of these cell lines. Further study demonstrated that the inhibitory effect of apigenin on ARO cell proliferation was associated with inhibition of both EGFR tyrosine autophosphorylation and phosphorylation of its downstream effector MAPK [44]. Subsequent evaluation was done on the effects of flavonoids on iodide transport and growth of the human follicular thyroid cancer cell line (FTC133) which was stably transfected with the human Na (+)/I (−) symporter (hNIS) [135]. It was observed that apigenin inhibited NIS mRNA expression, a finding that may have therapeutic implications in the radioiodide treatment of thyroid carcinoma.

8.10 Endometrial cancer

The genomic aberrations in endometrial cancer cells were identified after treatment with phytoestrogenic compounds, including apigenin, using array-based comparative genomic hybridization [136]. Over 20% of the array genes involving insulin metabolism were modulated in the cancer cells treated with β-estradiol, compared to those treated with the same concentration of apigenin, suggesting that it may play a role in the treatment of endometrial cancer and in the treatment of postmenopausal women.

8.11 Gastric Cancer

Growth-inhibitory potential and apoptosis were evaluated after apigenin treatment on human gastric carcinoma SGC-7901 cells. Exposure of these cells to apigenin resulted in dose-dependent inhibition of the growth and clone formation of SGC-7901 cells by inducing apoptosis [137].

8.12 Liver Cancer

Initial studies on plant flavonoids have shown that structural analogues designated the flavonoid 7-hydroxyl group are potent inhibitors of the human P-form phenolsulfotransferase, which is of major importance in the metabolism of many drugs, resulting in either inactivation and rapid renal elimination of the highly ionized sulfuric acid ester conjugates formed or, in some instances, formation of conjugates with increased pharmacological activity [138]. Introduction of a prenyl group into the molecule increases the hydrophobicity which would be expected to improve their biochemical and pharmacological properties through enhanced affinity for the lipophilic membrane. C8-prenylation of apigenin enhances the cytotoxicity and induces apoptotic cell death in H4IIE hepatoma cells without affecting antioxidative properties [139]. Further inhibitory effects of luteolin and apigenin were investigated on human hepatocellular carcinoma HepG2 cells. The results indicate that both flavonoids exhibited cell growth-inhibitory effects which were due to cell cycle arrest and down-regulation of the expression of cdk4 with induction of p53 and p21, respectively [140]. Apigenin reduced cell viability, and induced apoptotic cell death in HepG2 cells. Additionally, it evoked a dose-related elevation of intracellular ROS level. Treatment with various inhibitors of NADPH oxidase significantly blunted both the generation of ROS and induction of apoptosis by apigenin. These results suggest that ROS generated through the activation of NADPH oxidase may play an essential role in the apoptosis induced by apigenin in HepG2 cells [83]. In *in vivo* studies protective effects of apigenin were observed against *N*-nitroso-diethylamine-induced and

phenobarbitol promoted hepatocarcinogenesis in Wistar albino rats [141]. Apigenin treatment of these rats at 25 mg/kg body weight for 2 weeks provided protection against the oxidative stress and DNA damage caused by the carcinogen. Combination therapy of gemcitabine and apigenin enhanced antitumor efficacy in pancreatic cancer cells (MiaPaca-2, AsPC-1). *In vitro*, the combination treatment resulted in growth inhibition and apoptosis through the down-regulation of NFκB activity with suppression of Akt activation [141]. Further, *in vivo*, combination therapy augmented tumor growth inhibition through the down-regulation of NFκB activity with the suppression of Akt in tumor tissue. The combination of gemcitabine and apigenin enhanced antitumor efficacy and apoptosis induction [114].

8.13 Adrenal Cortical Cancer

Laboratory studies of adrenocortical cancers have revealed aberrations in a wide variety of signaling pathways and enzymes including aromatase, a key enzyme in the synthesis of estrogen from androgens. The effects of various flavonoids on the catalytic and promoter specific expression of aromatase were investigated in H295R human adenocortical cancer cells. Plant flavonoids were shown to be potent aromatase inhibitors, a finding associated with increased intracellular cAMP concentrations [142]. The effects of plant flavonoids were assessed on cortisol production in H295R cells. Their results indicate that cells exposed to apigenin demonstrate decreased cortisol production and 3β-HSD II and P450c21 activity [143].

8.14 Neuroblastoma

Effects of apigenin were investigated on various human neuroblastoma cell lines [144]. Apigenin treatment has been shown to result in inhibition of colony-forming ability and survival, and induction of apoptosis in human neuroblastoma cells. The mechanism of action of apigenin seems to involve p53, as it increased the levels of p53 and the p53-induced gene products p21WAF1/CIP1 and Bax. Furthermore, apigenin induced cell death and apoptosis of neuroblastoma cells expressing wild-type but not mutant p53. Apigenin was shown to increase caspase-3 activity and PARP cleavage in these cells. Further studies with apigenin in neuroblastoma SH-SY5Y cells resulted in increased apoptosis, which was associated with increases in intracellular free (Ca^{2+}) and Bax/Bcl2 ratio, mitochondrial release of cytochrome c, and activation of caspase-9, calpain, caspase-3, and caspase-12 as well [145].

9. SUMMARY

Epidemiological studies considerably support the notion that diets rich in plant flavones are associated with a number of health benefits, including a reduction of the risk of developing certain cancers. Apigenin is one of the most bioactive plant flavones and is widely distributed in common fruits, beverages, and vegetables. Its intake in the most popular form is as herbal tea. Laboratory studies have demonstrated that apigenin possesses anti-inflammatory, antioxidant, and anticancer properties. Apigenin affects several critical pathways and/or targets which are associated with several health disorders, including cancer. Apigenin can also help in improving cardiovascular conditions, stimulate the immune system, and provide some protection against cancer. Establishing whether or not therapeutic effects of apigenin are beneficial to patients will require research and generation of scientific evidence. However, based on the above highlighted findings, apigenin has potential for further investigation and development as a cancer chemopreventive and/or therapeutic agent.

ACKNOWLEDGMENTS

The original work from the authors' laboratory outlined in this chapter was supported by United States Public Health Service Grants RO1 CA108512 and RO1 AT002709 and funds from Cancer Research and Prevention Foundation to SG.

Dr Sanjay Gupta is Carter Kissell Associate Professor in the Department of Urology and holds secondary appointment in the Department of Nutrition at Case Western Reserve University and Division of General Medical Sciences at Case Comprehensive Cancer Center, Cleveland, Ohio, USA.

Dr Sanjeev Shukla is Instructor in the Department of Urology at Case Western Reserve University, Cleveland, Ohio, USA.

References

1. Hertog, M. G., Feskens, E. J., Hollman, P. C., Katan, M. B., & Kromhout, D. (1994). Dietary flavonoids and cancer risk in the Zutphen Elderly Study. *Nutrition and Cancer, 22*, 175–184.
2. Knekt, P., Järvinen, R., Seppänen, R., Hellövaara, M., Teppo, L., Pukkala, E., & Aromaa, A. (1997). Dietary flavonoids and the risk of lung cancer and other malignant neoplasms. *American Journal of Epidemiology, 146*, 223–230.
3. Gates, M. A., Tworoger, S. S., Hecht, J. L., De Vivo, I., Rosner, B., & Hankinson, S. E. (2007). A prospective study of dietary flavonoid intake and incidence of epithelial ovarian cancer. *International Journal of Cancer, 121*, 2225–2232.
4. Rossi, M., Negri, E., Lagiou, P., Talamini, R., Dal Maso, L., Montella, M., Franceschi, S., & La Vecchia, C. (2008). Flavonoids and ovarian cancer risk: A case-control study in Italy. *International Journal of Cancer, 123*, 895–898.
5. Hoensch, H., Groh, B., Edler, L., & Kirch, W. (2008). Prospective cohort comparison of flavonoid treatment in patients with resected colorectal cancer to prevent recurrence. *World Journal of Gastroenterology, 14*, 2187–2193.
6. Chemical Sources International (2000). All chemical suppliers for apigenin. Chem. Sources Chemical Search [http://kw1.innova.net].
7. McKay, D. L., & Blumberg, J. B. (2006). A review of the bioactivity and potential health benefits of chamomile tea (*Matricaria recutita* L.). *Phytotherapy Research, 20*, 519–530.

8. Bevilacqua, L., Buiarelli, F., Coccioli, F., & Jasionowska, R. (2004). Identification of compounds in wine by HPLC-tandem mass spectrometry. *Annali di Chimica, 94*, 679–689.

9. Gerhauser, C. (2005). Beer constituents as potential cancer chemopreventive agents. *European Journal of Cancer, 41*, 1941–1954.

10. Svehlikova, V., Bennett, R. N., Mellon, F. A., Needs, P. W., Piacente, S., Kroon, P. A., & Bao, Y. (2004). Isolation, identification and stability of acylated derivatives of apigenin 7-O-glucoside from chamomile (*Chamomilla recutita* [L.] Rauschert). *Phytochemistry, 65*, 2323–2332.

11. Hertog, M. G., Kromhout, D., Aravanis, C., Blackburn, H., Buzina, R., Fidanza, F., Giampaoli, S., Jansen, A., Menotti, A., & Nedeljkovic, S. (1995). Flavonoid intake and long-term risk of coronary heart disease and cancer in the seven countries study. *Archives of Internal Medicine, 155*, 381–386.

12. de Vries, J. H., Janssen, P. L., Hollman, P. C., van Staveren, W. A., & Katan, M. B. (1997). Consumption of quercetin and kaempferol in free-living subjects eating a variety of diets. *Cancer Letters, 114*, 141–144.

13. Johannot, L., & Somerset, S. M. (2006). Age-related variations in flavonoid intake and sources in the Australian population. *Public Health Nutrition, 8*, 1045–1054.

14. Wen, X., & Walle, T. (2006). Methylated flavonoids have greatly improved intestinal absorption and metabolic stability. *Drug Metabolism and Disposition, 34*, 1786–1792.

15. Wen, X., & Walle, T. (2006). Methylation protects dietary flavonoids from rapid hepatic metabolism. *Xenobiotica, 36*, 387–397.

16. Walle, T., Ta, N., Kawamori, T., Wen, X., Tsuji, P. A., & Walle, U. K. (2007). Cancer chemopreventive properties of orally bioavailable flavonoids – methylated versus unmethylated flavones. *Biochemical Pharmacology, 73*, 1288–1296.

17. Ross, J. A., & Kasum, C. M. (2002). Dietary flavonoids: bioavailability, metabolic effects, and safety. *Annual Review of Nutrition, 22*, 19–34.

18. Hollman, P. C., van Trijp, J. M., Buysman, M. N., van der Gaag, M. S., Mengelers, M. J., de Vries, J. H., & Katan, M. B. (1997). Relative bioavailability of the antioxidant flavonoid quercetin from various foods in man. *FEBS Letters, 418*, 152–156.

19. Aziz, A. A., Edwards, C. A., Lean, M. E., & Crozier, A. (1998). Absorption and excretion of conjugated flavonols, including quercetin-4′-O-beta-glucoside and isorhamnetin-4′-O-beta-glucoside by human volunteers after the consumption of onions. *Free Radical Research, 29*, 257–269.

20. Chen, J., Lin, H., & Hu, M. (2003). Metabolism of flavonoids via enteric recycling: Role of intestinal disposition. *Journal of Pharmacology and Experimental Therapeutics, 304*, 1228–1235.

21. Chen, J., Wang, S., Jia, X., Bajimaya, S., Lin, H., Tam, V. H., & Hu, M. (2005). Disposition of flavonoids via recycling: Comparison of intestinal versus hepatic disposition. *Drug Metabolism and Disposition, 33*, 1777–1784.

22. Walle, U. K., & Walle, T. (2002). Induction of human UDP-glucuronosyltransferase UGT1A1 by flavonoids – structural requirements. *Drug Metabolism and Disposition, 30*, 564–569.

23. Meyer, H., Bolarinwa, A., Wolfram, G., & Linseisen, J. (2006). Bioavailability of apigenin from apiin-rich parsley in humans. *Annals of Nutrition and Metabolism, 50*, 167–172.

24. Gradolatto, A., Basly, J. P., Berges, R., Teyssier, C., Chagnon, M. C., Siess, M. H., & Canivenc-Lavier, M. C. (2005). Pharmacokinetics and metabolism of apigenin in female and male rats after a single oral administration. *Drug Metabolism and Disposition, 33*, 49–54.

25. Chen, T., Li, L. P., Lu, X. Y., Jiang, H. D., & Zeng, S. (2007). Absorption and excretion of luteolin and apigenin in rats after oral administration of *Chrysanthemum morifolium* extract. *Journal of Agricultural and Food Chemistry, 55*, 273–277.

26. Gupta, S., Afaq, F., & Mukhtar, H. (2001). Selective growth-inhibitory, cell-cycle deregulatory and apoptotic response of apigenin in normal versus human prostate carcinoma cells. *Biochemical and Biophysical Research Communications, 287*, 914–920.

27. Birt, D. F., Walker, B., Tibbel, M. G., & Bresnick, E. (1986). Anti-mutagenesis and anti-promotion by apigenin, robinetin and indole-3-carbinol. *Carcinogenesis, 7*, 959–963.

28. Kuo, M. L., Lee, K. C., & Lin, J. K. (1992). Genotoxicities of nitropyrenes and their modulation by apigenin, tannic acid, ellagic acid and indole-3-carbinol in the Salmonella and CHO systems. *Mutation Research, 270*, 87–95.

29. Middleton, E. J. R., Kandaswami, C., & Theoharides, T. C. (2000). The effects of plant flavonoids on mammalian cells: Implications for inflammation, heart disease, and cancer. *Pharmacological Reviews, 52*, 673–751.

30. Birt, D. F., Mitchell, D., Gold, B., Pour, P., & Pinch, H. C. (1997). Inhibition of ultraviolet light induced skin carcinogenesis in SKH-1 mice by apigenin, a plant flavonoid. *Anticancer Research, 17*, 85–91.

31. Van Dross, R., Xue, Y., Knudson, A., & Pelling, J. C. (2003). The chemopreventive bioflavonoid apigenin modulates signal transduction pathways in keratinocyte and colon carcinoma cell lines. *Journal of Nutrition, 133*, 3800S–3804S.

C: ACTIONS OF INDIVIDUAL FRUITS IN DISEASE AND CANCER PREVENTION AND TREATMENT

32. Wei, H., Tye, L., Bresnick, E., & Birt, D. F. (1990). Inhibitory effect of apigenin, a plant flavonoid, on epidermal ornithine decarboxylase and skin tumor promotion in mice. *Cancer Research, 50*, 499–502.

33. Myhrstad, M. C., Carlsen, H., Nordstrom, O., Blomhoff, R., & Moskaug, J. O. (2002). Flavonoids increase the intracellular glutathione level by transactivation of the gamma-glutamylcysteine synthetase catalytic subunit promoter. *Free Radical Biology and Medicine, 32*, 386–393.

34. Liang, Y. C., Huang, Y. T., Tsai, S. H., Lin-Shiau, S. Y., Chen, C. F., & Lin, J. K. (1999). Suppression of inducible cyclooxygenase and inducible nitric oxide synthase by apigenin and related flavonoids in mouse macrophages. *Carcinogenesis, 20*, 1945–1952.

35. Kawai, M., Hirano, T., Higa, S., Arimitsu, J., Maruta, M., Kuwahara, Y., Ohkawara, T., Hagihara, K., Yamadori, T., Shima, Y., Ogata, A., Kawase, I., & Tanaka, T. (2007). Flavonoids and related compounds as anti-allergic substances. *Allergology International, 56*, 113–123.

36. Yano, S., Umeda, D., Yamashita, T., Ninomiya, Y., Sumida, M., Fujimura, Y., Yamada, K., & Tachibana, H. (2007). Dietary flavones suppresses IgE and Th2 cytokines in OVA-immunized BALB/c mice. *European Journal of Nutrition, 46*, 257–263.

37. Choi, J. S., Choi, Y. J., Park, S. H., Kang, J. S., & Kang, Y. H. (2004). Flavones mitigate tumor necrosis factor-alpha-induced adhesion molecule upregulation in cultured human endothelial cells: Role of nuclear factor-kappa B. *Journal of Nutrition, 4*, 1013–1019.

38. Williams, R. J., Spencer, J. P., & Rice-Evans, C. (2004). Flavonoids: antioxidants or signalling molecules? *Free Radical Biology and Medicine, 36*, 838–849.

39. Lee, S. F., & Lin, J. K. (1997). Inhibitory effects of phytopolyphenols on TPA-induced transformation, PKC activation, and c-jun expression in mouse fibroblast cells. *Nutrition and Cancer, 28*, 177–183.

40. Lin, J. K., Chen, Y. C., Huang, Y. T., & Lin-Shiau, S. Y. (1997). Suppression of protein kinase C and nuclear oncogene expression as possible molecular mechanisms of cancer chemoprevention by apigenin and curcumin. *Journal of Cellular Biochemistry Supplement, 28–29*, 39–48.

41. Mounho, B. J., & Thrall, B. D. (1999). The extracellular signal-regulated kinase pathway contributes to mitogenic and antiapoptotic effects of peroxisome proliferators *in vitro*. *Toxicology and Applied Pharmacology, 159*, 125–133.

42. Shukla, S., & Gupta, S. (2007). Apigenin-induced cell cycle arrest is mediated by modulation of MAPK, PI3K-Akt, and loss of cyclin D1 associated retinoblastoma dephosphorylation in human prostate cancer cells. *Cell Cycle, 6*, 1102–1114.

43. Carrillo, C., Cafferatam, E. G., Genovese, J., O'Reilly, M., Roberts, A. B., & Santa-Coloma, T. A. (1998). TGF-beta1 up-regulates the mRNA for the Na + /Ca2 + exchanger in neonatal rat cardiac myocytes. *Cellular and Molecular Biology, 44*, 543–551.

44. Yin, F., Giuliano, A. E., & Van Herle, A. J. (1999). Signal pathways involved in apigenin inhibition of growth and induction of apoptosis of human anaplastic thyroid cancer cells (ARO). *Anticancer Research, 19*, 4297–4303.

45. Hessenauer, A., Montenarh, M., & Gotz, C. (2003). Inhibition of CK2 activity provokes different responses in hormone-sensitive and hormone-refractory prostate cancer cells. *International Journal of Oncology, 22*, 1263–1270.

46. Landesman-Bollag, E., Song, D. H., Romieu-Mourez, R., Sussman, D. J., Cardiff, R. D., Sonenshein, G. E., & Seldin, D. C. (2001). Protein kinase CK2: Signaling and tumorigenesis in the mammary gland. *Molecular and Cellular Biochemistry, 227*, 153–165.

47. Plaumann, B., Fritsche, M., Rimpler, H., Brandner, G., & Hess, R. D. (1996). Flavonoids activate wild-type p53. *Oncogene, 13*, 1605–1614.

48. Lepley, D. M., & Pelling, J. C. (1997). Induction of p21/WAF1 and G1 cell-cycle arrest by the chemopreventive agent apigenin. *Molecular Carcinogenesis, 19*, 74–82.

49. Gupta, S., Afaq, F., & Mukhtar, H. (2002). Involvement of nuclear factor-kappa B, Bax and Bcl-2 in induction of cell cycle arrest and apoptosis by apigenin in human prostate carcinoma cells. *Oncogene, 21*, 3727–3738.

50. Shukla, S., & Gupta, S. (2004). Molecular mechanisms for apigenin-induced cell-cycle arrest and apoptosis of hormone refractory human prostate carcinoma DU145 cells. *Molecular Carcinogenesis, 39*, 114–126.

51. Wang, I. K., Lin-Shiau, S. Y., & Lin, J. K. (1999). Induction of apoptosis by apigenin and related flavonoids through cytochrome c release and activation of caspase-9 and caspase-3 in leukaemia HL-60 cells. *European Journal of Cancer, 35*, 1517–1525.

52. Iwashita, K., Kobori, M., Yamaki, K., & Tsushida, T. (2000). Flavonoids inhibit cell growth and induce apoptosis in B16 melanoma 4A5 cells. *Bioscience Biotechnology and Biochemistry, 64*, 1813–1820.

53. Hirano, T., Oka, K., & Akiba, M. (1989). Antiproliferative effects of synthetic and naturally occurring flavonoids on tumor cells of the human breast carcinoma cell line, ZR-75-1. *Research Communications in Chemical Pathology and Pharmacology, 64*, 69–78.

54. Lindenmeyer, F., Li, H., Menashi, S., Soria, C., & Lu, H. (2001). Apigenin acts on the tumor cell invasion

process and regulates protease production. *Nutrition and Cancer, 39,* 139–147.

55. Panes, J., Gerritsen, M. E., Anderson, D. C., Miyasaka, M., & Granger, D. N. (1996). Apigenin inhibits tumor necrosis factor-induced intercellular adhesion molecule-1 upregulation *in vivo. Microcirculation, 3,* 279–286.

56. Piantelli, M., Rossi, C., Iezzi, M., La Sorda, R., Iacobelli, S., Alberti, S., & Natali, P. G. (2006). Flavonoids inhibit melanoma lung metastasis by impairing tumor cells endothelium interactions. *Journal of Cellular Physiology, 207,* 23–29.

57. Osada, M., Imaoka, S., & Funae, Y. (2004). Apigenin suppresses the expression of VEGF, an important factor for angiogenesis, in endothelial cells via degradation of HIF-1alpha protein. *FEBS Letters, 575,* 59–63.

58. Fang, J., Xia, C., Cao, Z., Zheng, J. Z., Reed, E., & Jiang, B. H. (2005). Apigenin inhibits VEGF and HIF-1 expression via PI3K/AKT/p70S6K1 and HDM2/p53 pathways. *FASEB Journal, 19,* 342–353.

59. Le Bail, J. C., Laroche, T., Marre-Fournier, F., & Habrioux, G. (1998). Aromatase and 17 beta-hydroxy-steroid dehydrogenase inhibition by flavonoids. *Cancer Letters, 133,* 101–106.

60. Hiremath, S. P., Badami, S., Hunasagatta, S. K., & Patil, S. B. (2000). Antifertility and hormonal properties of flavones of *Striga orobanchioides. European Journal of Pharmacology, 391,* 193–197.

61. Mak, P., Leung, Y. K., Tang, W. Y., Harwood, C., & Ho, S. M. (2006). Apigenin suppresses cancer cell growth through ERbeta. *Neoplasia, 8,* 896–904.

62. Shukla, S., Mishra, A., Fu, P., MacLennan, G. T., Resnick, M. I., & Gupta, S. (2005). Up-regulation of insulin-like growth factor binding protein-3 by apigenin leads to growth inhibition and apoptosis of 22Rv1 xenograft in athymic nude mice. *FASEB Journal, 19,* 2042–2044.

63. Shukla, S., MacLennan, G. T., Flask, C. A., Fu, P., Mishra, A., Resnick, M. I., & Gupta, S. (2007). Blockade of beta-catenin signaling by plant flavonoid apigenin suppresses prostate carcinogenesis in TRAMP mice. *Cancer Research, 67,* 6925–6935.

64. Menichincheri, M., Ballinari, D., Bargiotti, A., Bonomini, L., Ceccarelli, W., D'Alessio, R., Fretta, A., Moll, J., Polucci, P., Soncini, C., Tibolla, M., Trosset, J. Y., & Vanotti, E. (2004). Catecholic flavonoids acting as telomerase inhibitors. *Journal of Medicinal Chemistry, 47,* 6466–6475.

65. Brusselmans, K., Vrolix, R., Verhoeven, G., & Swinnen, J. V. (2005). Induction of cancer cell apoptosis by flavonoids is associated with their ability to inhibit fatty acid synthase activity. *Journal of Biological Chemistry, 280,* 5636–5645.

66. Kim, M. H. (2003). Flavonoids inhibit VEGF/bFGF-induced angiogenesis *in vitro* by inhibiting the matrix-degrading proteases. *Journal of Cellular Biochemistry, 89,* 529–538.

67. Reiners, J. J., Jr., Clift, R., & Mathieu, P. (1999). Suppression of cell cycle progression by flavonoids: dependence on the aryl hydrocarbon receptor. *Carcinogenesis, 20,* 1561–1566.

68. Way, T. D., Kao, M. C., & Lin, J. K. (2004). Apigenin induces apoptosis through proteasomal degradation of HER2/neu in HER2/neu-overexpressing breast cancer cells via the phosphatidylinositol 3-kinase/Akt-dependent pathway. *Journal of Biological Chemistry, 279,* 4479–4489.

69. Kim, J. S., Eom, J. I., Cheong, J. W., Choi, A. J., Lee, J. K., Yang, W. I., & Min, Y. H. (2007). Protein kinase CK2alpha as an unfavorable prognostic marker and novel therapeutic target in acute myeloid leukemia. *Clinical Cancer Research, 13,* 1019–1028.

70. <http://www.altnature.com/gallery/>. Alternative Herbal Index.

71. Graf, J. (2000). Herbal anti-inflammatory agents for skin disease. *Skin Therapy Letter, 5,* 3–5.

72. Kim, H. Y., Kim, O. H., & Sung, M. K. (2003). Effects of phenol-depleted and phenol-rich diets on blood markers of oxidative stress, and urinary excretion of quercetin and kaempferol in healthy volunteers. *Journal of the American College of Nutrition, 22,* 217–223.

73. Yang, C. S., Landau, J. M., Huang, M. T., & Newmark, H. L. (2001). Inhibition of carcinogenesis by dietary polyphenolic compounds. *Annual Review of Nutrition, 21,* 381–406.

74. O'Prey, J., Brown, J., Fleming, J., & Harrison, P. R. (2003). Effects of dietary flavonoids on major signal transduction pathways in human epithelial cells. *Biochemical Pharmacology, 66,* 2075–2088.

75. Thiery-Vuillemin, A., Nguyen, T., Pivot, X., Spano, J. P., Dufresnne, A., & Soria, J. C. (2005). Molecularly targeted agents: Their promise as cancer chemopreventive interventions. *European Journal of Cancer, 41,* 2003–2015.

76. Nielsen, S. E., Young, J. F., Daneshvar, B., Lauridsen, S. T., Knuthsen, P., Sandstrom, B., & Dragsted, L. O. (1999). Effect of parsley (*Petroselinum crispum*) intake on urinary apigenin excretion, blood antioxidant enzymes and biomarkers for oxidative stress in human subjects. *The British Journal of Nutrition, 81,* 447–455.

77. Surh, Y. J. (2003). Cancer chemoprevention with dietary phytochemicals. *Nature Reviews Cancer, 3,* 768–780.

78. Arai, Y., Watanabe, S., Kimira, M., Shimoi, K., Mochizuki, R., & Kinae, N. (2000). Dietary intakes of

flavonols, flavones and isoflavones by Japanese women and the inverse correlation between quercetin intake and plasma LDL cholesterol concentration. *Journal of Nutrition, 130*, 2243–2250.

79. Janssen, K., Mensink, R. P., Cox, F. J., Harryvan, J. L., Hovenier, R., Hollman, P. C., & Katan, M. B. (1998). Effects of the flavonoids quercetin and apigenin on hemostasis in healthy volunteers: Results from an *in vitro* and a dietary supplement study. *American Journal of Clinical Nutrition, 67*, 255–262.

80. Way, T. D., Kao, M. C., & Lin, J. K. (2005). Degradation of HER2/neu by apigenin induces apoptosis through cytochrome c release and caspase-3 activation in HER2/neu-overexpressing breast cancer cells. *FEBS Letters, 579*, 145–152.

81. Weldon, C. B., McKee, A., Collins-Burow, B. M., Melnik, L. I., Scandurro, A. B., McLachlan, J. A., Burow, M. E., & Beckman, B. S. (2005). PKC-mediated survival signaling in breast carcinoma cells: A role for MEK1-AP1 signaling. *International Journal of Oncology, 26*, 763–768.

82. Yin, F., Giuliano, A. E., Law, R. E., & Van Herle, A. J. (2001). Apigenin inhibits growth and induces G2/M arrest by modulating cyclin-CDK regulators and ERK MAP kinase activation in breast carcinoma cells. *Anticancer Research, 21*, 413–420.

83. Choi, E. J., & Kim, G. H. (2008). Apigenin causes G (2)/M arrest associated with the modulation of p21 (Cip1) and Cdc2 and activates p53-dependent apoptosis pathway in human breast cancer SK-BR-3 cells. *Journal of Nutritional Biochemistry, 20*, 285–290.

84. Wang, C., & Kurzer, M. S. (1997). Phytoestrogen concentration determines effects on DNA synthesis in human breast cancer cells. *Nutrition and Cancer, 28*, 236–247.

85. Wang, C., & Kurzer, M. S. (1998). Effects of phytoestrogens on DNA synthesis in MCF-7 cells in the presence of estradiol or growth factors. *Nutrition and Cancer, 31*, 90–100.

86. Collins-Burow, B. M., Burow, M. E., Duong, B. N., & McLachlan, J. A. (2000). Estrogenic and antiestrogenic activities of flavonoid phytochemicals through estrogen receptor binding-dependent and -independent mechanisms. *Nutrition and Cancer, 38*, 229–244.

87. Long, X., Fan, M., Bigsby, R. M., & Nephew, K. P. (2008). Apigenin inhibits antiestrogen-resistant breast cancer cell growth through estrogen receptor-alpha-dependent and estrogen receptor-alpha-independent mechanisms. *Molecular Cancer Therapeutics, 7*, 2096–2108.

88. Zhang, S., Yang, X., & Morris, M. E. (2004). Combined effects of multiple flavonoids on breast cancer resistance protein (ABCG2)-mediated transport. *Pharmaceutical Research, 21*, 1263–1273.

89. Stroheker, T., Picard, K., Lhuguenot, J. C., Canivenc-Lavier, M. C., & Chagnon, M. C. (2004). Steroid

activities comparison of natural and food wrap compounds in human breast cancer cell lines. *Food and Chemical Toxicology, 42*, 887–897.

90. Seo, H. S., DeNardo, D. G., Jacquot, Y., Laïos, I., Vidal, D. S., Zambrana, C. R., Leclercq, G., & Brown, P. H. (2006). Stimulatory effect of genistein and apigenin on the growth of breast cancer cells correlates with their ability to activate ER alpha. *Breast Cancer Research and Treatment, 99*, 121–134.

91. Van Meeuwen, J. A., Korthagen, N., de Jong, P. C., Piersma, A. H., & Van den Berg, M. (2007). (Anti) estrogenic effects of phytochemicals on human primary mammary fibroblasts, MCF-7 cells and their co-culture. *Toxicology and Applied Pharmacology, 221*, 372–383.

92. Menendez, J. A., Vazquez-Martin, A., Oliveras-Ferraros, C., Garcia-Villalba, R., Carrasco-Pancorbo, A., Fernandez-Gutierrez, A., & Segura-Carretero, A. (2008). Analyzing effects of extra-virgin olive oil polyphenols on breast cancer-associated fatty acid synthase protein expression using reverse-phase protein microarrays. *International Journal of Molecular Medicine, 22*, 433–439.

93. Lee, S. H., Ryu, J. K., Lee, K. Y., Woo, S. M., Park, J. K., Yoo, J. W., Kim, Y. T., & Yoon, Y. B. (2008). Enhanced anti-tumor effect of combination therapy with gemcitabine and apigenin in pancreatic cancer. *Cancer Letters, 259*, 39–49.

94. Zheng, P. W., Chiang, L. C., & Lin, C. C. (2005). Apigenin induced apoptosis through p53-dependent pathway in human cervical carcinoma cells. *Life Sciences, 76*, 1367–1379.

95. Czyz, J., Madeja, Z., Irmer, U., Korohoda, W., & Hulser, D. F. (2005). Flavonoid apigenin inhibits motility and invasiveness of carcinoma cells *in vitro*. *International Journal of Cancer, 114*, 12–18.

96. Wu, C., Chen, F., Rushing, J. W., Wang, X., Kim, H. J., Huang, G., Haley-Zitlin, V., & He, G. (2006). Antiproliferative activities of parthenolide and golden feverfew extract against three human cancer cell lines. *Journal of Medical Food, 9*, 55–61.

97. Wang, W., Heideman, L., Chung, C. S., Pelling, J. C., Koehler, K. J., & Birt, D. F. (2000). Cell-cycle arrest at G2/M and growth inhibition by apigenin in human colon carcinoma cell lines. *Molecular Carcinogenesis, 28*, 102–110.

98. Wang, W., VanAlstyne, P. C., Irons, K. A., Chen, S., Stewart, J. W., & Birt, D. F. (2004). Individual and interactive effects of apigenin analogs on G2/M cell-cycle arrest in human colon carcinoma cell lines. *Nutrition and Cancer, 48*, 106–114.

99. Takagaki, N., Sowa, Y., Oki, T., Nakanishi, R., Yogosawa, S., & Sakai, T. (2005). Apigenin induces cell cycle arrest and p21/WAF1 expression in a p53-independent pathway. *International Journal of Oncology, 26*, 185–189.

100. Chung, C. S., Jiang, Y., Cheng, D., & Birt, D. F. (2007). Impact of adenomatous polyposis coli (APC) tumor supressor gene in human colon cancer cell lines on cell cycle arrest by apigenin. *Molecular Carcinogenesis, 46,* 773–782.

101. Farah, M., Parhar, K., Moussavi, M., Eivemark, S., & Salh, B. (2003). 5,6-Dichloro-ribifuranosylbenzimidazole- and apigenin-induced sensitization of colon cancer cells to TNF-alpha-mediated apoptosis. *American Journal of Physiology Gastrointestinal and Liver Physiology, 285,* 919–928.

102. Au, A., Li, B., Wang, W., Roy, H., Koehler, K., & Birt, D. (2006). Effect of dietary apigenin on colonic ornithine decarboxylase activity, aberrant crypt foci formation, and tumorigenesis in different experimental models. *Nutrition and Cancer, 54,* 243–251.

103. Svehlikova, V., Bennett, R. N., Mellon, F. A., Needs, P. W., Piacente, S., Kroon, P. A., & Bao, Y. (2004). Isolation, identification and stability of acylated derivatives of apigenin 7-O-glucoside from chamomile (*Chamomilla recutita* [L.] Rauschert). *Phytochemistry, 65,* 2323–2332.

104. Al-Fayez, M., Cai, H., Tunstall, R., Steward, W. P., & Gescher, A. J. (2006). Differential modulation of cyclooxygenase-mediated prostaglandin production by the putative cancer chemopreventive flavonoids tricin, apigenin and quercetin. *Cancer Chemotherapy and Pharmacology, 58,* 816–825.

105. Vargo, M. A., Voss, O. H., Poustka, F., Cardounel, A. J., Grotewold, E., & Doseff, A. I. (2006). Apigenin-induced-apoptosis is mediated by the activation of PKCdelta and caspases in leukemia cells. *Biochemical Pharmacology, 72,* 681–692.

106. Navarro-Núñez, L., Lozano, M. L., Palomo, M., Martinez, C., Vicente, V., Castillo, J., Benavente-Garcia, O., Diaz-Ricart, M., Escolar, G., & Rivera, J. (2008). Apigenin inhibits platelet adhesion and thrombus formation and synergizes with aspirin in the suppression of the arachidonic acid pathway. *Journal of Agricultural and Food Chemistry, 56,* 2970–2976.

107. Chen, D., Daniel, K. G., Chen, M. S., Kuhn, D. J., Landis-Piwowar, K. R., & Dou, Q. P. (2005). Dietary flavonoids as proteasome inhibitors and apoptosis inducers in human leukemia cells. *Biochemical Pharmacology, 69,* 1421–1432.

108. Abaza, L., Talorete, T. P., Yamada, P., Kurita, Y., Zarrouk, M., & Isoda, H. (2007). Induction of growth inhibition and differentiation of human leukemia HL-60 cells by a Tunisian gerboui olive leaf extract. *Bioscience Biotechnology and Biochemistry, 71,* 1306–1312.

109. Monasterio, A., Urdaci, M. C., Pinchuk, I. V., Lopez-Moratalla, N., & Martinez-Irujo, J. (2004). Flavonoids induce apoptosis in human leukemia U937 cells through caspase- and caspase-calpain-dependent pathways. *Journal of Nutrition Cancer, 50,* 90–100.

110. Horvathova, K., Novotny, L., & Vachalkova, A. (2003). The free radical scavenging activity of four flavonoids determined by the comet assay. *Neoplasma, 50,* 291–295.

111. Strick, R., Strissel, P. L., Borgers, S., Smith, S. L., & Rowley, J. D. (2000). Dietary bioflavonoids induce cleavage in the MLL gene and may contribute to infant leukemia. *Proceedings of the National Academy of Sciences of the United States of America, 97,* 4790–4795.

112. Li, Z. D., Liu, L. Z., Shi, X., Fang, J., & Jiang, B. H. (2007). Benzo[a]pyrene-3,6-dione inhibited VEGF expression through inducing HIF-1alpha degradation. *Biochemical and Biophysical Research Communications, 357,* 517–523.

113. Watanabe, N., Hirayama, R., & Kubota, N. (2007). The chemopreventive flavonoid apigenin confers radiosensitizing effect in human tumor cells grown as monolayers and spheroids. *Journal of Radiation Research (Tokyo), 48,* 45–50.

114. Lee, W. J., Chen, W. K., Wang, C. J., Lin, W. L., & Tseng, T. H. (2008). Apigenin inhibits HGF-promoted invasive growth and metastasis involving blocking PI3K/Akt pathway and beta 4 integrin function in MDA-MB-231 breast cancer cells. *Toxicology and Applied Pharmacology, 226,* 178–191.

115. Engelmann, C., Blot, E., Panis, Y., Bauer, S., Trochon, V., Nagy, H. J., Lu, H., & Soria, C. (2002). Apigenin – strong cytostatic and anti-angiogenic action *in vitro* contrasted by lack of efficacy *in vivo*. *Phytomedicine, 9,* 489–495.

116. Fang, J., Zhou, Q., Liu, L. Z., Xia, C., Hu, X., Shi, X., & Jiang, B. H. (2007). Apigenin inhibits tumor angiogenesis through decreasing HIF-1alpha and VEGF expression. *Carcinogenesis, 28,* 858–864.

117. Zhu, F., Liu, X. G., & Liang, N. C. (2003). Effect of emodin and apigenin on invasion of human ovarian carcinoma HO-8910PM cells *in vitro*. *Ai Zheng, 22,* 358–362.

118. Hu, X. W., Meng, D., & Fang, J. (2008). Apigenin inhibited migration and invasion of human ovarian cancer A2780 cells through focal adhesion kinase. *Carcinogenesis, 29,* 2369–2376.

119. Kanerva, A., Raki, M., Ranki, T., Särkioja, M., Koponen, J., Desmond, R. A., Helin, A., Stenman, U. H., Isoniemi, H., Höckerstedt, K., Ristimäki, A., & Hemminki, A. (2007). Chlorpromazine and apigenin reduce adenovirus replication and decrease replication associated toxicity. *Journal of Gene Medicine, 9,* 3–9.

120. Knowles, L. M., Zigrossi, D. A., Tauber, R. A., Hightower, C., & Milner, J. A. (2000). Flavonoids suppress androgen-independent human prostate tumor proliferation. *Nutrition and Cancer, 38,* 116–122.

121. Lee, S. C., Kuan, C. Y., Yang, C. C., & Yang, S. D. (1998). Bioflavonoids commonly and potently induce tyrosine dephosphorylation/inactivation of oncogenic proline-directed protein kinase FA in human prostate carcinoma cells. *Anticancer Research, 18,* 1117–1121.

122. Morrissey, C., O'Neill, A., Spengler, B., Christoffel, V., Fitzpatrick, J. M., & Watson, R. W. (2005). Apigenin drives the production of reactive oxygen species and initiates a mitochondrial mediated cell death pathway in prostate epithelial cells. *Prostate, 63,* 131–142.

123. Shukla, S., & Gupta, S. (2004). Suppression of constitutive and tumor necrosis factor alpha-induced nuclear factor (NF)-kappaB activation and induction of apoptosis by apigenin in human prostate carcinoma PC-3 cells: correlation with down-regulation of NF-kappaB-responsive genes. *Clinical Cancer Research, 10,* 3169–3178.

124. Shukla, S., & Gupta, S. (2006). Molecular targets for apigenin-induced cell cycle arrest and apoptosis in prostate cancer cell xenograft. *Molecular Cancer Therapeutics, 5,* 843–852.

125. Shukla, S., & Gupta, S. (2008). Apigenin-induced prostate cancer cell death is initiated by reactive oxygen species and p53 activation. *Free Radical Biology and Medicine, 44,* 1833–1845.

126. Shukla, S., & Gupta, S. (2009). Apigenin suppresses insulin-like growth factor I receptor signaling in human prostate cancer: An *in vitro* and *in vivo* study. *Molecular Carcinogenesis, 48,* 243–252.

127. Kaur, P., Shukla, S., & Gupta, S. (2008). Plant flavonoid apigenin inactivates Akt to trigger apoptosis in human prostate cancer: An *in vitro* and *in vivo* study. *Carcinogenesis, 29,* 2210–2217.

128. Mirzoeva, S., Kim, N. D., Chiu, K., Franzen, C. A., Bergan, R. C., & Pelling, J. C. (2008). Inhibition of HIF-1 alpha and VEGF expression by the chemopreventive bioflavonoid apigenin is accompanied by Akt inhibition in human prostate carcinoma PC3-M cells. *Molecular Carcinogenesis, 47,* 686–700.

129. McVean, M., Xiao, H., Isobe, K., & Pelling, J. C. (2000). Increase in wild-type p53 stability and transactivational activity by the chemopreventive agent apigenin in keratinocytes. *Carcinogenesis, 21,* 633–639.

130. Li, B., & Birt, D. F. (1996). *In vivo* and *in vitro* percutaneous absorption of cancer preventive flavonoid apigenin in different vehicles in mouse skin. *Pharmaceutical Research, 13,* 1710–1715.

131. Li, B., Pinch, H., & Birt, D. F. (1996). Influence of vehicle, distant topical delivery, and biotransformation on the chemopreventive activity of apigenin, a plant flavonoid, in mouse skin. *Pharmaceutical Research, 13,* 1530–1534.

132. Tong, X., Van Dross, R. T., Abu-Yousif, A., Morrison, A. R., & Pelling, J. C. (2007). Apigenin prevents UVB-induced cyclooxygenase 2 expression: coupled mRNA stabilization and translational inhibition. *Molecular and Cellular Biology, 27,* 283–296.

133. Van Dross, R. T., Hong, X., Essengue, S., Fischer, S. M., & Pelling, J. C. (2007). Modulation of UVB-induced and basal cyclooxygenase-2 (COX-2) expression by apigenin in mouse keratinocytes: role of USF transcription factors. *Molecular Carcinogenesis, 46,* 303–314.

134. Caltagirone, S., Rossi, C., Poggi, A., Ranelletti, F. O., Natali, P. G., Brunetti, M., Aiello, F. B., & Piantelli, M. (2000). Flavonoids apigenin and quercetin inhibit melanoma growth and metastatic potential. *International Journal of Cancer, 87,* 595–600.

135. Schroder-van der Elst, J. P., van der Heide, D., Romijn, J. A., & Smit, J. W. (2004). Differential effects of natural flavonoids on growth and iodide content in a human Na + /I- symporter-transfected follicular thyroid carcinoma cell line. *European Journal of Endocrinology, 150,* 557–564.

136. O'Toole, S. A., Sheppard, B. L., Sheils, O., O'Leary, J. J., Spengler, B., & Christoffel, V. (2005). Analysis of DNA in endometrial cancer cells treated with phytoestrogenic compounds using comparative genomic hybridisation microarrays. *Planta Medica, 71,* 435–439.

137. Wu, K., Yuan, L. H., & Xia, W. (2005). Inhibitory effects of apigenin on the growth of gastric carcinoma SGC-7901 cells. *World Journal of Gastroenterology, 11,* 4461–4464.

138. Eaton, E. A., Walle, U. K., Lewis, A. J., Hudson, T., Wilson, A. A., & Walle, T. (1996). Flavonoids, potent inhibitors of the human P-form phenolsulfotransferase: Potential role in drug metabolism and chemoprevention. *Drug Metabolism and Disposition, 24,* 232–237.

139. Watjen, W., Weber, N., Lou, Y. J., Wang, Z. Q., Chovolou, Y., Kampkotter, A., Kahl, R., & Proksch, P. (2007). Prenylation enhances cytotoxicity of apigenin and liquiritigenin in rat H4IIE hepatoma and C6 glioma cells. *Food and Chemical Toxicology, 45,* 119–124.

140. Yee, S. B., Lee, J. H., Chung, H. Y., Im, K. S., Bae, S. J., Choi, J. S., & Kim, N. D. (2003). Inhibitory effects of luteolin isolated from *Ixeris sonchifolia* Hance on the

proliferation of HepG2 human hepatocellular carcinoma cells. *Archives of Pharmacal Research, 26,* 151–156.

141. Jeyabal, P. V., Syed, M. B., Venkataraman, M., Sambandham, J. K., & Sakthisekaran, D. (2005). Apigenin inhibits oxidative stress-induced macromolecular damage in N-nitrosodiethylamine (NDEA)-induced hepatocellular carcinogenesis in Wistar albino rats. *Molecular Carcinogenesis, 44,* 11–20.

142. Sanderson, J. T., Hordijk, J., Denison, M. S., Springsteel, M. F., Nantz, M. H., & van den Berg, M. (2004). Induction and inhibition of aromatase (CYP19) activity by natural and synthetic flavonoid compounds in H295R human adrenocortical carcinoma cells. *Toxicological Sciences, 82,* 70–79.

143. Ohno, S., Shinoda, S., Toyoshima, S., Nakazawa, H., Makino, T., & Nakajin, S. (1999). Effects of flavonoid phytochemicals on cortisol production and on activities of steroidogenic enzymes in human adrenocortical H295R cells. *Journal of Steroid Biochemistry and Molecular Biology, 80,* 355–363.

144. Torkin, R., Lavoie, J. F., Kaplan, D. R., & Yeger, H. (2005). Induction of caspase-dependent, p53-mediated apoptosis by apigenin in human neuroblastoma. *Molecular Cancer Therapeutics, 4,* 1–11.

145. Das, A., Banik, N. L., & Ray, S. K. (2006). Mechanism of apoptosis with the involvement of calpain and caspase cascades in human malignant neuroblastoma SH-SY5Y cells exposed to flavonoids. *International Journal of Cancer, 119,* 2575–2585.

146. Becker, B., Kuhn, U., & Hardewig-Budny, B. (2006). Double-blind, randomized evaluation of clinical efficacy and tolerability of an apple pectin-chamomile extract in children with unspecific diarrhea. *Arzneimittelforschung, 56,* 387–393.

147. Savino, F., Cresi, F., Castagno, E., Silvestro, L., & Oggero, R. A. (2005). Randomized double-blind placebo-controlled trial of a standardized extract of *Matricariae recutita, Foeniculum vulgare* and *Melissa officinalis* (ColiMil) in the treatment of breastfed colicky infants. *Phytotherapy Research, 19,* 335–340.

148. Kupfersztain, C., Rotem, C., Fagot, R., & Kaplan, B. (2003). The immediate effect of natural plant extract, *Angelica sinensis* and *Matricaria chamomilla* (Climex) for the treatment of hot flushes during menopause. A preliminary report. *Clinical and Experimental Obstetrics and Gynecology, 30,* 203–206.

149. Nakamura, H., Moriya, K., Oda, S., Yano, E., & Kakuta, H. (2002). Changes in the parameters of autonomic nervous system and emotion spectrum calculated from encephalogram after drinking chamomile tea. *Aroma Research, 3,* 251–255.

150. Kyokong, O., Charuluxananan, S., Muangmingsuk, V., Rodanant, O., Suornsug., K., & Punyasang, W. (2002). Efficacy of chamomile-extract spray for prevention of post-operative sore throat. *Journal of the Medical Association of Thailand, 85,* S180–S185.

151. Burns, E., Blamey, C., Ersser, S. J., Lloyd, A. J., & Barnetson, L. (2000). The use of aromatherapy in intrapartum midwifery practice – an observational study. *Complementary Therapies in Nursing and Midwifery, 6,* 33–34.

152. de la Motte, S., Bose-O'Reilly, S., Heinisch, M., & Harrison, F. (1997). Double-blind comparison of an apple pectin-chamomile extract preparation with placebo in children with diarrhea. *Arzneimittelforschung, 47,* 1247–1249.

153. Fidler, P., Lorinzi, C. L., O'Fallon, J. R., Leitch, J. M., Lee, J. K., Hayes, D. L., Novotny, P., Clemens-Schutjer, D., Bartel, J., & Michalak, J. C. (1996). Prospective evaluation of a chamomile mouthwash for prevention of 5-FU-induced oral mucositis. *Cancer, 77,* 522–525.

154. Korting, H. C., Schafer-Korting, M., Hart, H., Laux, P., & Schmid, M. (1993). Anti-inflammatory activity of *hamamelis* distillate applied topically to the skin. Influence of vehicle and dose. *European Journal of Clinical Pharmacology, 44,* 315–318.

155. Weizman, Z., Alkrinawi, S., Goldfarb, D., & Bitran, C. (1993). Efficacy of herbal tea preparation in infantile colic. *Journal of Pediatrics, 122,* 650–652.

156. Roberts, A., & Williams, J. M. G. (1992). The effect of olfactory stimulation on fluency, vividness of imagery and associated mood – a preliminary study. *The British Journal of Medical Psychology, 6,* 197–199.

157. Glowania, H. J., Raulin, C., & Swoboda, M. (1987). Effect of chamomile on wound healing – a clinical double-blind study. *Zeitschrift fur Hautkrankheiten, 62,* 1267–1271.

158. Gould, L., Goswami, M. K., Reddy, C. V., & Gomprecht, R. F. (1973). The cardiac effects of tea. *Journal of Clinical Pharmacology, 13,* 469–474.

C: ACTIONS OF INDIVIDUAL FRUITS IN DISEASE AND CANCER PREVENTION AND TREATMENT

42

Goitrogen in Food: Cyanogenic and Flavonoids Containing Plant Foods in the Development of Goiter

Amar K. Chandra

University College of Science and Technology, University of Calcutta, Kolkata, West Bengal, India

1. INTRODUCTION

Goitrogens are naturally occurring substances that can interfere with the function of the thyroid gland. Goitrogens get their name from the term 'goiter' which means the enlargement of the thyroid gland. If the thyroid gland has difficulty synthesizing thyroid hormone, it may enlarge to compensate for this inadequate hormone production. Goitrogens cause difficulty for the thyroid in making its hormone.

Goiter is usually the most obvious sign of iodine deficiency; however, brain damage, mental retardation, reproductive failure, and childhood mortality are more serious consequences. Iodine deficiency also affects the socioeconomic development of a community. The role of iodine deficiency as an environmental factor in the development of goiter is established. However, there are observations that indicate the existence of factors other than iodine deficiency. Iodine deficiency does not always cause endemic goiter. Iodine supplementation does not always result in complete eradication of goiter. Moreover, there are epidemiological and experimental evidences that concomitant exposure to other naturally occurring antithyroid agents magnifies the severity of endemic goiter. The foods that contain goitrogens/antithyroid substances and are responsible for the exaggeration, persistence, and development of goiter and associated disorders are reviewed in this chapter.

2. HISTORICAL REVIEW

Large goiters developed in rabbits being used to study experimental syphilis as reported by Chesney and co-workers [1,2]. These experiments and subsequent tests by others proved that the cabbage, which was the main item in the diets in these experiments, was goiter-producing. Some later experiments confirmed this, while others did not; it also became clear that cabbage grown in different parts of the world and during different seasons varied widely in goitrogenic properties. By the early 1930s it had been established that some foods were goitrogenic. After failing to produce goiter in experimental animals with a diet of cabbage, Hercus and Purves [3] in New Zealand demonstrated that the seeds of cabbage, rape, and mustard contained a goitrogenic principle. It was noted further that additional iodine prevented 'cabbage goitre' because it only partly inhibited the effects of active substance in the seeds.

A series of investigations in New Zealand confirmed the goitrogenic properties of the seeds of a number of *Brassica* that paved the way for the isolation of an active goitrogenic compound, L-5-vinyl-2-thio-oxazolidone, from the seeds of a number of *Brassica* and from the roots of swedes and turnips [4]. This was slightly more goitrogenic for humans than propylthiouracil and had about one-fifth the potency of the same drug for the rat. Tests with ^{131}I indicated that its action on the thyroid gland was similar to that of antithyroid drugs of the aminobenzine and thioamide series that blocked the organic binding of iodine.

Greer [5] proved that vinyl-oxazolidone (goitrin) was present in the seeds and roots as an inactive precursor which requires enzyme action for conversion to the active form. Cooking destroyed the enzyme and so the goitrogenic properties. Further, Greer [6] found that the progoitrin content of the seeds of *Brassica* varied from batch to batch and from season to season even for the same variety. However, progoitrin had not been demonstrated in cabbage or other species of *Brassica* used in the early experiments to produce cabbage goiter. Therefore some other substances must have been responsible for the results obtained by Chesney and co-workers.

Goiter occurs in many types of livestock and in some instances this has been shown to be associated with a particular diet. Linseed meal when fed to pregnant ewes produced large goiters in newborn lambs in New Zealand, which was proved to be due to thiocyanate derived from the detoxification of a cyanide-producing glycoside, linamarin, present in the seeds [7]. These effects could be eliminated by adding iodine in the diet. A severe outbreak of goiter, with heavy neonatal mortality in lambs, was reported from New Zealand in kale-fed ewes. The effects could be controlled with additional iodine. That the goitrogenic action of kale may be due to a thyroid-blocking agent, like thiouracil, has been suggested by the New Zealand Department of Agriculture [8]. Whittem, while investigating an outbreak of goiter in newborn lambs in Tasmania in 1956, concluded that this was due to the consumption by the pregnant ewes in the last month of pregnancy of particularly lush growths of the common crow's foot (*Erodium cicutarium*) and the long stork's bill (*Erodium botrys*). The overgrowth of these plants was associated with a warm humid autumn and winter; however, the causal agent was not isolated [9].

Based on the results of a number of studies on glycosides obtained from the red outer skin-covering of some edible nuts including peanuts, Moudgal and co-workers [10,11] in the University of Madras, concluded that the goitrogenic action of these nuts is probably due to a double action: the interference with the uptake of inorganic iodine, and the blockage of the organic binding of iodine by tyrosine. These actions were only partially inhibited by supplementation of iodine.

The results of these studies showed that there are certain plants and their seeds that are actually or potentially goitrogenic for humans and domestic animals and some experimental animals. It appears that there are at least two different kinds of compounds. One produces goiter by preventing the uptake of iodine by the thyroid gland, e.g. thiocyanate, the other group produces goiter by blocking the organic binding of iodine, e.g. goitrin and flavonoids.

3. GOITROGENS IN FOOD

The foods that have been associated with disrupted thyroid hormone production in humans are broadly classified into two categories: cyanogenic plant foods, and flavonoids containing plant foods. To gain insight into the antithyroid agents present in those plants, information concerning their chemical origin and biological effect when eaten is given as follows.

4. CYANOGENIC PLANT FOODS

Cyanide in trace amounts is almost ubiquitous in the plant kingdom and occurs mainly in the form of cyanogenic glycosides and thioglycosides or glucosinolates. Both are nitrogen-containing secondary metabolites sharing a number of common features. They derive biogenetically from amino acids and occur as glycosides which are stored in the vacuole. They function as prefabricated defense compounds that are activated by the action of a β-glucosidase in case of emergency, releasing the deterrent: toxic cyanide from cyanogens or isothiocyanates from glucosinolates. The goitrogenic substances produced from cyanogenic plants are thiocyanate, isothiocyanates, and thiooxazolidone.

4.1 Cyanogenic Glycosides

4.1.1 Structure and Occurrence of Cyanogenic Glycosides

Cyanogens are derivatives of 2-hydroxynitriles which form glycosides with β-D-glucose. More than 60 cyanogenic glycosides are known and are widely distributed among plants. More than 2600 cyanogenic taxa have been reported, especially among members of Rosaceae, Leguminosae, Gramineae, and Araceae [12,13]. A few insects, such as the zygaenidae (Lepidoptera), feed on plants with cyanogens and

sequester these compounds but are also capable of synthesizing the same cyanogens independently themselves [14]. Cyanogenic glycosides are easily hydrolyzed by dilute acid to nitriles which are further decomposed to aldehydes or ketones and hydrogen cyanide (HCN).

4.1.2 Metabolism of Inorganic Cyanide

Ingested cyanide is rapidly absorbed from the upper gastrointestinal tract. It also passes readily through skin, and HCN gas is rapidly absorbed from the lungs. In the body the principal metabolic pathway is by reaction with thiosulfate to form thiocyanate and sulfite, the reaction being catalyzed by rhodonase, an enzyme which may be described as a sulfur transferase [15]. The enzyme is widespread in living tissues, having its highest concentrations in the liver, kidney, thyroid, adrenal, and pancreas [16]. The thiocyanate thus produced is excreted in the urine. Thiocyanate is slowly oxidized to sulfate, but this does not appear to be an important factor in body tissues [17]. In another pathway cyanogen reacts with 3-mercaptopyruvate in the presence of another sulfur transferase to produce thiocyanate [18]. However, these reactions require the presence of adequate amounts of cystine as a sulfur donor. The metabolic disposal of inorganic cyanide is summarized in Figure 42.1.

Finally, the part played by hydroxocobalamin (vitamin B12) remains to be clearly defined. It is well known that hydroxocobalamin takes up cyanide as cyanocobalamin and readily liberates it on exposure to light [19]. This could merely be a secondary effect of the presence of cyanide in the body, but there is some evidence that hydroxocobalamin may play a more active role in cyanide detoxification. The administration of hydroxocobalamin has been shown to protect mice against cyanide poisoning to a remarkable extent [20]. Urinary thiocyanate excretion was found to be increased not only by the ingestion of cyanide [21] but also by vitamin B12 deficiency [22]. The latter finding could be

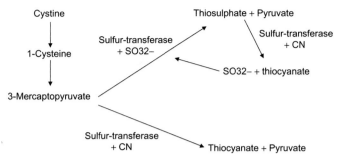

FIGURE 42.1 Metabolic disposal of inorganic cyanide.

interpreted as evidence of interference with cyanide detoxification.

In addition to diet, people are continually exposed to small doses of cyanide due to a polluted atmosphere and particularly in cigarette smoke [23,24]. The presence of trace amounts in the body may perhaps be regarded as physiological and may act as a brake in cellular oxidative processes.

4.1.3 Dietary Sources of Cyanogenic Glycosides

Relatively higher concentrations of cyanogenic glycosides are found in certain grasses, pulses, root crops, and fruit kernels. Most of these are consumed by animals while a few have a practical importance in human nutrition. The important glycosides that have been identified in edible species of plants are amygdalin, dhurrin, and linamarin.

Amygdalin was identified first in bitter almonds and in the kernel of other fruits. Dhurrin occurs in sorghum and other grasses. Linamarin is the glycoside of pulses, linseed, and cassava. The other dietary sources of glycosides that are often consumed by people are sweet potato and yam; maize and millets (especially *Sorghum vulgare*); bamboo and sugar cane; peas and beans; kernel of almond, lemon, lime, apple, pear, cherry, apricot, prune, and plum.

In addition to these plants, various leaves, particularly of the *Sorbus* and *Prunus* species, may cause animal poisoning because these are important cattle food. Linseed may yield up to 53 mg of HCN per 100 g and arrow grass up to 77 mg/100 g. A significant yield of cyanogen may also be obtained from certain varieties of *Vicia sativa* (common vetch) [25].

In grasses and cane, the highest yield is obtained from the tip of the immature plant, particularly if growing in rich soil [26,27]. In *P. lunatus* and in linseed the cyanide occurs in all parts of the plant throughout life, as well as in the seed. Systematic cultivation greatly reduces the cyanide content, but probably no variety can be produced that is entirely free of cyanogens.

Cyanogen may be found in the range of 2 mg/100 g in *P. vulgaris* (haricot or navy bean, red pea), *Vigna sinensis* (black-eye pea), and *Pisum sativum* (garden pea).

In cassava, significant amounts of cyanogens occur in the black variety, but there is no clear differentiation between bitter and sweet strains. The cyanogens occur in all parts of the plant; however, its importance lies in its concentration in the cortex of the root, where it is more plentiful than in the pulp.

4.1.4 Mechanism of Action of Thiocyanate on Goiter Formation

The thiocyanate may not be present in the plant as the free ion, but as a glycoside which is converted in the animal body to a thiocyanate. Thiocyanate and thiocyanate-like compounds arising from cyanogenic glycosides, glucosinolates, or thiocyanate interfere with iodine metabolism in the thyroid gland by reducing iodide uptake, stimulating iodide efflux, or replacing iodide by thiocyanate [28,29]. Moreover, thiocyanate and thiocyanate-like compounds mainly inhibit the iodide-concentrating mechanism of the thyroid gland by inhibiting the unidirectional clearance of iodide from the gland [30].

Thyroid peroxidase (TPO), the key enzyme in thyroid hormone biosynthesis, regulates organification, iodination, and coupling reactions to form thyroid hormones (T_3 and T_4) in the thyroid gland [31]. Thiocyanate, a monovalent anion having a molecular size corresponding to that of iodide, inhibits the incorporation of iodide into thyroglobulin by competing with iodide at the thyroid peroxidase level [29]. It inhibits iodide oxidation, i.e. conversion of I^- leads to I_2 by inhibiting TPO activity [32]. Moreover, thiocyanate also increases the formation of an essentially insoluble iodinated thyroglobulin within the thyroid in an iodine-depleted condition [33]. At low doses, thiocyanate has been shown to inhibit the uptake of iodide [34]; however, at higher doses thiocyanate in addition inhibits the incorporation of iodine into thyroglobulin [29].

A diet rich in thiocyanate (or glucosinolates) but deprived of iodide causes decreased levels of T_3 and T_4 [35,36]. Inhibited TPO activity associated with decreased concentration of iodide in the thyroid gland may be the probable reason for the reduced synthesis of thyroid hormones as reflected by thyroid hormone levels in thiocyanate (or glucosinolate) fed rats. Low circulating level of T_4 and T_3 stimulate the secretion of TSH from the pituitary by feedback mechanism and if this condition is continued for a longer duration, enlargement of the thyroid gland occurs or goiter develops [37,38].

4.2 Glucosinolates (Thioglucosides)

Glucosinolates resemble cyanogens in many respects, but contain sulfur as an additional atom. When hydrolyzed, glucosinolates liberate D-glucose, sulfate, and an unstable aglycone, which may form an isothiocyanate (common name 'mustard oil') as the main product under certain conditions, or a thiocyanate, a nitrile, or cyano-epithioalkane. Isothiocyanates are responsible for the distinctive, pungent flavour and odor.

4.2.1 Structure and Occurrence of Glucosinolates

More than 80 different glucosinolates have been found in higher dicotyledonous plants in the order Capparales, that includes the families Capparidaceae, Cruciferae (syn. Brassicaceae), Resedaceae, Moringaceae, Tropacolaceae, and others [39].

4.2.2 Dietary Sources of Glucosinolate/ Thioglucosides

Most of the thioglucosides have been characterized in part through the isothiocyanate formed from hydrolysis because its formation appears favored over the nitrile and the thiocyanate. The isothiocyanate lends itself to identification by formation of crystalline thiourea derivatives by reaction with ammonia. Also, isothiocyanates from many thioglucosides are easily recognized because of their pungency.

Sinigrin, sinalbin, progoitrin, and epi-progoitrin are trivial names for specific thioglucosides found in common plants. The first two were so named before much was known about their chemistry. Progoitrin was so named because it is the precursor of goitrin. The thioglucosides present in common plants used as food condiments and feed are summarized in Table 42.1.

4.2.3 Chemically Identified Goitrogens in Thioglucoside-Containing Plants

Goitrin and Related Oxazolidinethiones Goitrin is a potent antithyroidal agent. Progoitrin occurs in many crucifers in small amounts in their leaves and roots. The

TABLE 42.1   Domesticated Cruciferous Plants Containing Thioglucoside

Name[a]	Part of Plant[b]	Common Name of Thioglucoside	Amount[c]
Used as food			
Brassica oleraceae	L	Sinigrin	L
Cabbage, kale, brussels sprouts,		Glucobrassicin	L
cauliflower, kohlrabi		Progoitrin	S
		Gluconapin	S
		Neoglucobrassicin	S
Brassica campestris	R		
Turnips		Progoitrin	S
		Gluconasturtiin	S
	L,R,S	(R)–2–hydroxy–4–pentenyl-glucosinolate	S
Brassica napus	R,L	Progoitrin	L
Rutabaga		Glycobrassicin	S
		Neoglucobrassicin	S
Lepidium sativum	L	Glucotropaeolin	L
Garden cress			
Raphanus sativus	R	4-Methylthio-3-butenyl glucosinolate	L
Radish		Glucobrassicin	S
Used as condiments			
Armoracia lapathifolia	R	Sinigrin	L
A. rusticana			
Horseradish		Gluconasturiin	S
Brassica cartinata	S	Sinigrin	L
Ethiopian rapeseed			
B. juncea	S	Sinigrin	L
Indian or brown mustard			
B. nigra	S	Sinigrin	L
Black mustard			
Sinapis alba	S	Sinalbin	L
White mustard			
S. arvensis	S	Sinigrin	L
Charlock			
Used as feed			
B. campestris	S	Gluconapin	S
Rape, turnip rape,		Progoitrin	S
Polish rape, rubsen,		Glucobrassicanapin	S
naverte		Glucoalyssin	S
		Glucoraphanin	S
B. napus	S	Progoitrin	L
Rape, Argentine rape, winter rape		Gluconapin	L
		Glucobrassicanapin	S

(*Continued*)

C. ACTIONS OF INDIVIDUAL FRUITS IN DISEASE AND CANCER PREVENTION AND TREATMENT

TABLE 42.1 (Continued)

Name[a]	Part of Plant[b]	Common Name of Thioglucoside	Amount[c]
		Gluconasturtiin	S
		Glucoiberin	S
		Sinalbin	S
Crambe abyssinica	S	epi-Progoitrin	L
Crambe, Abyssinian		Sinigrin	S
kale		Gluconapin	S
		Gluconasturtiin	S

[a]Taxonomy of the Brassica and related genera.
[b]L, leaves or stalk; R, roots; S, seed.
[c]Amounts of each thioglucosides in relation to total thioglucoside content: L, large; S, small. [Ref: Van Etten C.H. (1969) Goitrogens. In Toxic Constituents of Plant Food Stuffs (I. E. Lienear, Ed.) Academic Press, New York.]

inhibitory effect of goitrin was first detected by measuring the reduced uptake by the thyroid of radioactive iodine (I^{131}) when the animals were fed the compound [40]. Hyperplasia and hypertrophy of the thyroid gland was noticed in goitrin-fed rats and this action is slightly less than that of thiourea, an antithyroid drug [41]. Ettlinger [42] synthesized the optically active (R) and (S) forms and the racemate of goitrin. Greer [43] showed that both the forms are equally potent as an antithyroid agent. The compound possessed 2 and 133% of the activity of the antithyroid drug propylthiouracil when tested with rats and with man respectively.

There are oxazolidinethiones in plants other than those formed from progoitrin and epi-progoitrin. The 5,5-dimethyloxazolidine-2-thione isolated from hare's ear mustard (Conringia cochlearia) by Hopkins [44] has about the same antithyroid activity as goitrin as tested by Astwood et al. [45]. Kjaer and Gmelin [46] isolated barbarin [(−)-5-phenyloxazolidine-2-thione] from various species of Barbarea and Reseda luteola. These species contain the precursor thioglucoside, glucobarbarin, in edible green parts of the plants. Barbarin has about half the antithyroid activity of goitrin. Kjaer and Thomsen [47] have isolated and identified other oxazolidinethiones from the Cleame genus in the

family of Capparidaceae and two from Sisymbrium austriacum [48,49].

Thiocyanate Ions and Organic Isothiocyanates Langer and his co-workers extensively studied the relationship between goiter caused by ingestion of the foliage of Brassica, viz., cabbage, and the inorganic thiocyanate found in these vegetables. It was found that the edible part of cabbage and related plants contained thiocyanate ion ranging from 0.7 to 10.2 mg/100 g fresh material. It was demonstrated that ingestion of these Brassica vegetables by animals and humans causes a rise of thiocyanate ion in the blood followed by its appearance in the urine. The thiocyanate level was found to drop as soon as the eating of Brassica plants was discontinued [50]. Michajlovskij and Langer [51] found that the thiocyanate content of the Brassica plants was highest in spring and varied little in relation to the regions where grown, but did show a variation from plant to plant within a single field. Quantitative estimation of the thiocyanate content of the plant in relation to the extent of thyroid inhibition indicated other goitrogens might be present in Brassica [52].

Langer [53] fed allyl isothiocyanate through a stomach tube to rats, on a low iodine containing ration, at levels approximately those found in an amount of cabbage that the animals might consume at one feeding. The thiocyanate level in the blood serum increased following this experiment. Further studies demonstrated that single doses of 2–4 mg of allyl isothiocyanate given to rats on a low iodine diet depressed the uptake of radioiodine by the thyroid significantly [54]. In regions where iodine content of the diet is low, benign goiter may be accentuated by eating excessive amounts of *Brassica* vegetables.

Conversion of mustard oil to thiocyanate ion under the action of myrosinase present in plants is a known mechanism of detoxification of isothiocyanate [55]. Their goitrogenic effect may be due to this conversion of thiocyanate ion, a known goitrogen. Bachelard and Trikojus [56] reported that cheiroline (an isothiocyanate) inhibits radioactive iodine uptake by the thyroid when injected subcutaneously into rats.

Other Possible Thioglucoside Products Nitriles formed from thioglucosides may act as goitrogens. Nitriles cause thyroid enlargement in rabbits given daily doses of the compound. Detoxification of nitriles to form thiocyanate ions is the probable reason [55]. However, the latter workers were not always able to confirm the reported goitrogenic properties of organic nitriles.

An organic polysulfide fraction from cabbage and other *Brassicas* when fed to rats caused histological changes in the thyroid like those caused by feeding cabbage [57]. Jirousek and Starka [58] were able to isolate a 'trithionen' (1,2-dithiocyclopent-4-en-3-thione) from this fraction. It was further noted that the sulfhydryl content of the polysulfides in the fraction was higher in *Brassica* plant juices from endemic goiter areas with no iodine deficiency [59].

4.2.4 Mechanism of Action of Isothiocyanates, Goitrin, and Related Compounds

Thiocyanates or thiocyanate-like compounds inhibit the iodine-concentrating mechanism of the thyroid and their goitrogenic activity can be overcome by iodine administration.

Isothiocyanates act on the thyroid mainly by their rapid conversion to thiocyanate. However, isothiocyanates not only use the thiocyanate metabolic pathway but also react spontaneously with amino groups forming thiourea derivatives, which produce a thiourea-like antithyroid effect. The thionamide or thiourea-like goitrogens interfere in the thyroid gland with the organification of iodine and formation of active thyroid hormones and their action cannot usually be antagonized by iodine. The naturally occurring goitrin is a representative of this category, besides, goitrin is not degraded like other thioglucosides.

The actual concentration of thiocyanate or isothiocyanates in a given foodstuff may not represent its true goitrogenic potential, nor does the absence of these compounds negate a possible antithyroid effect, because inactive precursors can be converted into goitrogenic agents both in the plant itself and in the animal body after its ingestion [60].

Mustard oil glycoside yields thiocyanate under the action of myrosinase, an enzyme present in plants. But ingestion of pure progoitrin, a naturally occurring thioglucoside, elicits antithyroid activity in rats and humans in the absence of this enzyme because it is partially converted in the animal body into the more potent goitrin by microflora of the intestine [61].

Small aliphatic disulfides, the major components of onion and garlic, have been shown to exert a marked thiourea-like antithyroid activity [62].

5. FLAVONOIDS CONTAINING FOODS

Flavonoids are important stable organic constituents of a wide variety of plants. They are universally present in vascular plants and in a large number of food plants. Because of their widespread occurrence in edible plants such as fruits, vegetables, and grains, flavonoids are an integral part of the human diet [63]. They are present in high concentrations in polymeric (tannins) and oligomeric (pigments) forms in various staple foods of developing countries, such as millet, sorghum, beans, and ground nuts.

5.1 Structure and Occurrence of Flavonoids

Flavonoids occur as aglycones, glycosides, and methylated derivatives. The flavonoid aglycone consists of a benzene ring (A) condensed with a six-membered ring (C), which in the 2-position carries a phenyl ring (B) as a substituent. The six-membered ring condensed with the benzene ring is either a α-pyrone (flavonols and flavonanes) or its dihydroderivatives (flavanols and flavanones). The position of the benzenoid substitute divides the flavonoid class into flavonoids (2-position) and isoflavonoids (3-position). Flavonols differ from flavonones by a hydroxy group in 3-position and C_2–C_3 double bonds [64]. Flavonoids are often hydroxylated in positions with an S (3,5,7,2′,3′,4′,5′). Methylethers and acetylesters of the alcohol group are known to occur in nature. When glycosides are formed, a glycosidic linkage is normally located in positions 3 or 7 and the carbohydroxy can be L-rhamnose, D-glucose, glucorhamnose, glactose, and arabinose [65].

5.2 Dietary Sources of Flavonoids

Flavonoids occur in nature in almost all plant families, and are found in considerable quantities in fruits, vegetables, grains, cola, tea, coffee, cocoa, beer, and red wine (Table 42.2) [67–69]. The daily dietary intake of mixed flavonoids is estimated to be in the range of 500–100 mg [70], but can be as high as several grams in those supplementing their diets with

TABLE 42.2 Flavonoid Source and Level in Commonly Consumed Foods

Flavonoid	Common Sources	Levels (mg/kg wet weight)
Quercetin	Onion	284–486
	Apple	21–72
	Kale	110
	Red wine	4–16 mg/L
	Green/black teas	10–25 mg/L
Genistein/diadzein	Tofu	159–306 (genistein)
		57–117 (diadzein)
		21 (genistein)
		6.8–7.2 (diadzein)
	Soy milk formula	19–23 (genistein)
		15–19 (diadzein)
Coumestrol	Alfalfa sprouts	47
	Clover sprouts	281
Apigenin/luteolin	Millet (fonio)	150 (apigenin)
		350 (lutcolin)
Maringin	Grapefruit and juice	100–800 mg/L

Source: reference 66.

flavonoids or flavonoid-containing herbal preparations in the USA.

Major groups of flavonoids in the human diet are the flavonoids, proanthocyanidins (that include catechins), isoflavonoids, flavones, and flavanones (Figure 42.2) [71,72] (Figure 42.1). Lignans, the closely related compounds (resorcyclic acid lactones), have similar properties to those of flavonoids. Quercetin, a flavonol, is the most predominant flavonoid in the human diet and human consumption is in the range of 4–68 mg per day based on epidemiological studies in the USA, Europe, and Asia [69,73,74]; the Japanese population had the highest levels of flavonol intake, mainly due to their green tea consumption.

Quercetin is found in high concentrations in commonly consumed foods such as onions, apples, kale [67], red wine, and green and

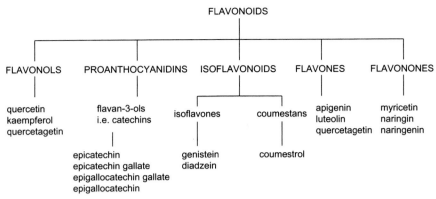

FIGURE 42.2   Family of major flavonoid groups.

black teas [69]. The bioavailability of quercetin varies greatly in foods due to absorption differences in conjugated forms. The absorption of quercetin is much higher than the absorption of quercetin without its sugar moiety [75]. Besides, the quercetin glucosides in onions are more readily absorbed than those found in tea [76] and the rutinosides in apples [77].

Tea and red wine have high levels of catechins, which are condensed tannins that contribute to the astringency of these beverages. Catechins comprise approximately 30% of unprocessed tea leaves, most of which is epigallocatechin-3-gallate (EGCG).

Isoflavones, coumestans, and lignans, generally known as 'phytoestrogens' are non-steroidal compounds with weak estrogenic activity. These compounds may compete with endogenous hormones, as well as inhibit a number of enzymes involved in estrogen metabolism.

Genistein and diadzein are the primary isoflavones with estrogenic properties. They are found in legumes such as lentils, chickpeas, soybeans, and other soy-based products such as tofu, miso, soy milk, and temph [68,78,79]. Daily dietary intake of soy protein in heavy soy consumers such as in Asian populations is estimated to be 20–80 g, whereas the average Western dietary intake is approximately 1–3 g [80]. Assuming a total isoflavone concentration of soy-based products to be in the range of

1–3 mg/g, thus expected daily isoflavone intake is in the ranges of 20–210 mg for Asians and 1–9 mg in other populations.

Coumestans, such as coumestrol, are found in high concentrations in alfalfa and clovers [79] and have been recognized, along with formononetin, as the cause of infertility in grazing herbivores.

Lignans are found in many fruits, vegetables, tea beverages, and cereal grains [81]. They contain very high concentrations of the lignan 2,3-bis (3-methoxy-4-hydroxybenzyl) butane-1,4-diol) (secoisolar iciresinol) [82], which is a precursor of the primary mammalian lignans, enterolactone and enterodiol [83]. Lignans are known for their antioxidant activity.

Flavones do not possess estrogenic activity; however, the flavones, apigenin and luteolin, act as potent inhibitors of aromatase and 17β-hydroxysteroid oxidoreductase enzymes involved in estrogen metabolism [84,85]. Studies have also demonstrated that the flavones and several glycocylflavones are potent goitrogens [86,87].

5.3 Mechanism of Action

Flavonoids, which exert a thiourea-like antithyroid effect, are polyhydroxyphenols with a C_6–C_3–C_6 structure. Following ingestion by mammals, flavonoids glycosides are hydrolyzed by intestinal microbial glycosidases to flavonoid aglycans which may be absorbed and metabolized by mammalian tissues or may be further processed in the intestinal microorganisms by B ring hydroxylation and middle ring fusion to produce various compounds such as phenolic acids, phloroglucinol, and gallic acid. All these have marked antithyroidal effects. For instance, flavonoids glycosides can inhibit the activity of the enzyme thyroid peroxidase (TPO) from 0.2 to 2.0 times more effectively than the thiourea-like antithyroid drug, propylthiouracil. Their corresponding aglycones are 0.5–14 times more potent and the middle ring fission metabolite, phloroglucinol, 38 times more potent.

Flavonoids not only inhibit the activity of TPO but also inhibit the activity of deiodinase enzymes resulting in the inhibition of peripheral metabolism of thyroid hormones. In addition, polymers of the flavonoid phloretin interact with thyroid stimulating hormone (TSH), preventing its action at the thyroid cell. Therefore, this class of compounds alters thyroid hormone economy in a complex way [60].

6. CYANOGENIC PLANT FOOD IN THE DEVELOPMENT OF GOITER

6.1 Cassava

The goitrogenic role of cassava has been evidenced through studies carried out in severe goiter endemic areas of Zaire, Idjwi Island, Kivu, and Ubangi. These studies were initially devoted to the adaptation mechanisms of the thyroid function to iodine deficiency. The existence in these areas of marked iodine deficiency has been extensively documented and most of the observed abnormalities of thyroid function have been satisfactorily explained on this basis. However, some findings have not been consistent with generally accepted views concerning the pathophysiology and management of endemic goiter and associated disorders. For instance, in Idjwi Island, very high (60–70%) or

very low (<5%) prevalences of goiter were observed in communities with similar low iodine intake levels. In both goitrous and non-goitrous regions of Idjwi Island, urinary iodine level was extremely low and nearly identical high values were observed for ^{131}I thyroid uptake. The discrepancy between goiter prevalence and iodine deficiency was first reported by Delange et al. [88] and subsequently confirmed by Thilly et al. [89].

At this stage of studies, despite the lack of definite evidence supporting the causative role of cassava in endemic goiter, based on experimental findings in the animal model, clinical studies were initiated in Ubangi in northeastern Zaire. Endemic goiter in this population of more than two million people is even more widespread than on Idjwi Island. Cassava intake is very high. In this goitrous population mean serum thiocyanate concentration of 1.0 mg/dL was observed, and some subjects had a serum thiocyanate level of 2–3 mg/dL, i.e. 10 times normal value. Plasma and urinary thiocyanate concentrations were compared in a large number of people. The concentrations increased in parallel up to a serum level of 1.0–1.2 mg/dL, then a renal adaptive mechanism prevented further increase in serum thiocyanate concentration.

Various cassava preparations were listed, calculating their cyanide content and their effect on thiocyanate load. Ingestion of large amounts of cassava for 3 days produced a significant increase in serum and urinary thiocyanate concentrations. Complete cessation of cassava intake induced a marked decrease in these values. These observations confirmed that cassava is responsible for abnormal thiocyanate loads in the population.

The goitrogenic action of cassava has been demonstrated in a comparative study of cassava intake and preparation method in different ethnic populations of Ubangi. In one population (Group I), cassava is consumed principally as cassava flour mixed with corn and ground, boiled cassava leaves. These preparations contain an average of 13.5 mg and 8.5 mg HCN/kg respectively. The other populations (Group II and Group III) consume cassava that is soaked and then mashed into a puree. This preparation contains about 3.5 mg HCN/kg. Group III populations eat large quantities of fish and thus have a higher iodine intake than the rest of the population and they are not affected by severe endemic goiter.

Serum and urinary thiocyanate levels in the Group I population are 1.5–2 times higher than in Group II and Group III populations (Table 42.3) [90]. In Groups I and II iodine intake is identical, but the prevalence of goiter is 60% in Group I and only 34% in Group II. This difference is associated with thiocyanate excretions of 2.57 mg/dL in Group I and 1.32 mg/dL in Group II. Urinary thiocyanate concentrations are similar in Group II and Group III, although Group III has a lower prevalence of goiter (17%) due to the higher iodine intake.

The overall results clearly indicate that the prevalence of goiter is linked to the balance between thiocyanate and iodine intake. In each community, prevalence is directly proportional to thiocyanate load and inversely proportional to iodine supply. On the basis of this relationship, an index corresponding to the ratio between urinary thiocyanate and iodine

concentration has been calculated. The indices are 2.1 (Group I), 0.75 (Group II), and 0.39 (Group III) corresponding to 60, 34, and 17% goiter prevalence, respectively.

6.2 Cyanogenic Plants of India

Consumption of cyanogenic foods has been considered as one of the etiological factors in certain instances for the persistence of endemic goiter in India. Thus cyanogenic plants were collected from the different states of the country where endemic goiter was prevalent during the post salt iodization period. Their cyanogenic glucosides, glucosinolates and thiocyanate, were estimated (Table 42.4).

Thyroid peroxidase activity (TPO) of human thyroid was assayed from the microsomal fraction following I_3^- from iodide. The anti-TPO activities of the plants were assayed after adding raw, boiled, and cooked extracts in the assay medium with and without iodide (Table 42.5).

Relative antithyroidal potency of the plant extracts was also evaluated in terms of the concentration (IC_{50}) necessary to produce 50% inhibition of TPO activity. PTU equivalence of the plant foods was also determined (Table 42.6).

The results showed that Indian cyanogenic plants had *in vitro* antithyroidal activity. Boiled

TABLE 42.3 Epidemiological and Metabolic Indicators for Three Groups Investigated

Group	Prevalence of Goiter (%) (Mean ± SE)	Prevalence of Cretinism (%) (Mean ± SE)	6 hour[131] Thyroid Uptake (%) (Mean ± SE)	Serum levels			Urinary levels		Urinary SCN/l Ratio (mg/μg) (Mean ± SE)
				T_4 (μg/dL) (Mean ± SE)	TSH (μU/mL) (Mean ± SE)	SCN (mg/dL) (Mean ± SE)	SCN (mg/dL) (Mean ± SE)	^{127}(μg/dL) (Mean ± SE)	
I	60 ± 4 (8)	6.1 ± 1.0 (8)	55 ± 2 (142)	4.4 ± 0.1 (408)	27.0 ± 4.0 (409)	1.24 ± 0.03 (406)	2.57 ± 0.10 (379)	1.9 ± 0.1 (254)	2.10 ± 0.21 (250)
II	34 (2)	1.1 (2)	42 ± 3 (46)	6.4 ± 0.4 (43)	3.8 ± 0.6 (44)	0.42 ± 0.04 (44)	1.32 ± 0.11 (49)	2.5 ± 0.3 (49)	0.75 ± 0.08 (49)
III	17 ± 5 (3)	0 (3)	23 ± 1 (75)	6.8 ± 0.2 (77)	3.3 ± 0.5 (76)	0.78 ± 0.04 (76)	1.60 ± 0.07 (77)	4.9 ± 0.3 (77)	0.39 ± 0.02 (77)

[a]Figures in parentheses are the number of villages or subjects.

TABLE 42.4 Distributions of Cyanogenic Glucosides, Glucosinolates and Thiocyanate Content in Selected Fresh Plant Foods

Plant foods	Cyanogenic Glucosides	Glucosinolates	Thiocyanate
Brassica family			
Cauliflower (Brassica oleracea var. botrytis)	1.82 ± 0.4	17.28 ± 1.6	5.04 ± 0.5
Cabbage (Brassica oleracea var. capitara)	1.6 ± 0.3	15.7 ± 1.3	11.6 ± 1.7
Mustard (Brassica juncea)	0.24 ± 0.01	4.0 ± 0.3	50.5 ± 2.9
Turnip (Brassica rapa)	1.3 ± 0.5	4.6 ± 0.8	20.1 ± 0.9
Radish (Raphanus sativus)	1.28 ± 0.4	2.64 ± 0.2	13.28 ± 0.9
Non-Brassica family			
Bamboo shoot (Bambusa arundinacea)	551.05 ± 72	9.57 ± 0.5	24.31 ± 5.2
Cassava (Manihot)	304.31 ± 41	4.32 ± 0.8	12.95 ± 2

Values are mean ± SD of six observations, expressed in terms of mg/kg wet weight.

TABLE 42.5 In vitro TPO Activity ($\Delta Od/Min/Mg$ Protein) of Raw, Boiled and Cooked Plant Extracts Without and With Extra Iodide

Plants	Raw Extract		Boiled Extract		Cooked Extract	
	Without Extra KI	With Extra KI	Without Extra KI	With Extra KI	Without Extra KI	With Extra KI
Control	1.62 ± 0.054	–	–	–	–	–
Cauliflower	0.409 ± 0.014 (74.75)	1.295 ± 0.024 (20.04)	0.160 ± 0.007 (90.12)	0.569 ± 0.02 (64.88)	0.405 ± 0.001 (75)	0.405 ± 0.006 (75)
Cabbage	0.562 ± 0.016 (65)	1.368 ± 0.012 (15)	0.333 ± 0.024 (79.44)	0.655 ± 0.011 (59.57)	0.550 ± 0.018 (66.05)	0.545 ± 0.017 (66.36)
Bamboo shoot	0.248 ± 0.02 (84.69)	0.562 ± 0.164 (65.31)	0.243 ± 0.024 (85)	0.655 ± 0.013 (59.17)	0.253 ± 0.018 (84.38)	0.896 ± 0.037 (44.69)
Cassava	0.488 ± 0.009 (69.88)	1.063 ± 0.01 (34.38)	0.472 ± 0.04 (70.86)	0.792 ± 0.023 (51.11)	0.482 ± 0.024 (70.25)	0.884 ± 0.017 (45.43)
Mustard	0.329 ± 0.019 (79.69)	1.062 ± 0.005 (34.44)	0.281 ± 0.015 (88.65)	0.483 ± 0.037 (70.19)	0.295 ± 0.013 (81.79)	0.881 ± 0.017 (45.62)
Turnip	0.470 ± 0.033 (70.99)	1.362 ± 0.005 (15.93)	0.294 ± 0.014 (81.85)	0.487 ± 0.008 (69.94)	0.412 ± 0.018 (74.57)	0.479 ± 0.019 (70.43)
Radish	0.655 ± 0.02 (59.57)	1.458 ± 0.008 (10)	0.328 ± 0.018 (79.75)	0.473 ± 0.02 (70.80)	0.401 ± 0.042 (75.25)	0.407 ± 0.04 (74.88)

extracts showed highest anti-TPO potency followed by cooked and raw extracts. Excess iodide reversed the anti-TPO activity of the plant foods to a certain extent but could not neutralize it [91].

6.2.1 Bamboo Shoot

The young shoots or sprouts of common bamboo Bambusa arundinacea (Retz.) Willd., family Graminaceae are used as a staple food as well as pickle and chutney in developing countries including India. Bamboo shoot has been considered as one of the etiological factors for the persistence of endemic goiter in Manipur, a northeastern state of India [92]. The influence of bamboo shoot feeding with and without iodine supplementation in rat food for different durations on thyroid status have been evaluated determining urinary iodine and thiocyanate concentrations, thyroid

weight, *in vivo* TPO activity, and thyroid hormone profiles [93].

Thyroid weight, TPO activity, and total serum thyroid hormone levels of bamboo shoot fed animals for 45 and 90 days, respectively, were determined and compared with control. Significant increase in thyroid weight as well as higher excretion of thiocyanate and iodine along with marked decrease in thyroid peroxidase activity, T_4 and T_3 levels were observed in the bamboo shoot fed group (Table 42.7).

Chronic bamboo shoot consumption gradually developed a state of hypothyroidism. Extra iodide reduced the antithyroidal effect of bamboo shoot to an extent but could not cancel it because of excessive cyanogenic glucoside, glucosinolate and thiocyanate present in it.

6.2.2 Radish

The effect of chronic or prolonged consumption of radish (*Raphanus sativus* Linn.) under varying state of iodine intake on morphological and functional status of thyroid in albino rats was evaluated by thyroid gland

TABLE 42.6 Concentration of fresh Cyanogenic Plant Foods Producing 50% Inhibition (IC_{50}) of Thyroid Peroxidase Activity

Plant foods	$IC_{50}(\mu g)$	PTU equivalence
PTU	0.9 ± 0.2	100
Cauliflower	51.25 ± 0.52	1.76
Cabbage	66.25 ± 1.42	1.36
Bamboo shoot	27.5 ± 0.77	3.27
Cassava	42.5 ± 1.35	2.12
Mustard	45.00 ± 0.67	2
Turnip	60.00 ± 1.10	1.5
Radish	51.88 ± 1.2	1.73

IC_{50}, inhibition constant$_{50}$.
Values are mean $\pm$ SD of six observations.
PTU-6-n-propyl-2-thiouracil.

TABLE 42.7 Bamboo Shoot Induced Alteration on Urinary Thiocyanate and Iodine and Thyroid Status Under Varying Iodide Intake in Albino Rats for (a) 45 days and (b) 90 days. Values are mean $\pm$ SD of ten observations

Groups		Urinary thiocyanate m moles/l	Urinary iodine μ moles/l	Thyroid weight mg/100g body weight	TPO activity ΔOD/min/mg protein	Serum T4 μg/dl	Serum T3 ng/dl
Control	A	0.16 ± 0.01	24.94 ± 1.06	7.23 ± 0.15	3.45 ± 0.38	3.90 ± 0.06	123.63 ± 3.94
	B	0.19 ± 0.05	32.06 ± 0.73	8.94 ± 0.94	5.34 ± 0.27	4.21 ± 0.28	143.12 ± 1.49
Bamboo shoot fed	A*	1.24 ± 0.12	35.4 ± 2.18	10.83 ± 0.2	2.62 ± 0.21	3.5 ± 0.11	107.59 ± 1.50
	B**	2.02 ± 0.13	41.12 ± 1.11	13.07 ± 0.96	0.39 ± 0.08	3.39 ± 0.09	94.42 ± 0.84
KI supplemented	A*	0.17 ± 0.01	178.48 ± 1.7	5.85 ± 0.21	7.51 ± 0.31	4.14 ± 0.09	134.52 ± 3.04
	B**	0.21 ± 0.09	187.66 ± 1.37	7.69 ± 0.32	9.96 ± 1.74	4.75 ± 0.29	152.19 ± 1.84
Bamboo shoot fed	A*	1.58 ± 0.17	190.61 ± 0.38	9.5 ± 0.41	3.1 ± 0.18	3.73 ± 0.08	123.53 ± 1.39
KI supplemented	B**	2.65 ± 0.16	197.03 ± 0.93	11.56 ± 0.93	2.04 ± 0.29	3.69 ± 0.11	126.67 ± 1.09

Values bearing superscripts are significantly different by ANOVA at $P < 0.05$.
*When compared with (A) 45 days control.
**When compared with (B) 90 days control.

morphology and histology, thyroid peroxidase activity, serum triiodothyronine, thyroxine, and thyrotropin levels. The consumption pattern of iodine and goitrogens of cyanogenic origin was evaluated by measuring urinary iodine and thiocyanate levels respectively. After chronic feeding, increased thyroid weight, decreased thyroid peroxidase activity, reduced thyroid hormone profiles, and elevated level of thyrotropin were observed resembling a relative state of hypoactive thyroid gland in comparison to control even after supplementation of adequate iodine (Table 42.8) [94,95].

Similar studies conducted with cabbage (*Brassica oleracea* var. capitata Linn.), cauliflower (*Brassica oleracea* var. botrytis Linn.), mustard (*Brassica juncea*), maize (*Zea mays*), and cassava (*Manihot esculenta* Linn.) found that all of these plant foods have potent antithyroidal activity even in the presence of iodine (Figure 42.3).

7. FLAVONOIDS CONTAINING FOOD IN GOITER DEVELOPMENT

7.1 Soyabean

Numerous surveys have shown that iodine intake is low in goiter areas. There are two types of endemic goiter. In mountainous areas it occurs as a diffuse parenchymatous goiter in adults with degenerative forms. The characteristic type in plain areas is a diffuse colloid goiter. The complete explanation of these differences probably depends on more than a simple iodine deficiency.

Soyabeans have a high nutritive value and probably provide as complete a diet as can be obtained from any single vegetable source. In spite of this McCarrison [96] had reported that soyabean would produce a goiter in rats that was not prevented by iodine. Subsequent animal studies reported similar findings [97,98]. Shepard et al. [99] reported the development

TABLE 42.8 Uncooked/fresh and Cooked Radish Induced Alterations on Urinary Thiocyanate and Iodine Thyroid Weight and Functional Status Under Varying Iodine Intake in Albino Rats for 90 days. Values are mean $\pm$ SD of ten observations; figures in parenthesis are % change as compared to control group.

Groups	Thyroid weight (mg/ 100g Body weight)	Urinary Iodine (μg/dL)	Urinary Thiocyanate (mg/dL)	TPO Activity (ΔOD/min/ mg Protein)	Serum T4 (μg/dL)	Serum T3 (ng/dL)	Scrum TSH (ng/dL)
Control	8.89 $\pm$ 0.34	416.29 $\pm$ 4.74	1.2 $\pm$ 0.13	5.45 $\pm$ 0.28	4.32 $\pm$ 0.16	141.5 $\pm$ 1.2	0.59 $\pm$ 0.13
Fresh radish	11.17 $\pm$ 0.51* (25.6%)	442.75 $\pm$ 4.08* (6.25%)	6.25 $\pm$ 0.19* (420.8%)	2.25 $\pm$ 0.19* (58.7%)	3.61 $\pm$ 0.11* (16.4%)	117.05 $\pm$ 2.27* (17.3%)	2.28 $\pm$ 0.14* (286.4%)
Cooked radish	10.29 $\pm$ 0.52* (15.7%)	423.22 $\pm$ 5.45 (1.68%)	5.67 $\pm$ 0.18* (372.5%)	3.82 $\pm$ 0.14* (29.9%)	3.67 $\pm$ 0.18* (15.0%)	123.67 $\pm$ 2.06* (12.6%)	1.83 $\pm$ 0.1* (210.2%)
KI supplemented	7.84 $\pm$ 0.24	2396.24 $\pm$ 3.89	1.46 $\pm$ 0.19	6.24 $\pm$ 0.52	4.82 $\pm$ 0.13	152.04 $\pm$ 3.02	0.31 $\pm$ 0.11
Fresh radish fed KI supplemented	10.08 $\pm$ 0.36** (28.6%)	2539.02 $\pm$ 5.38** (5.96%)	8.82 $\pm$ 0.14** (504.1%)	2.82 $\pm$ 0.14** (54.8%)	3.72 $\pm$ 0.22** (22.8%)	120.55 $\pm$ 2.31** (20.7%)	1.26 $\pm$ 0.16** (306.5%)
Cooked radish fed KI supplemented	9.69 $\pm$ 0.62** (23.6%)	2441.68 $\pm$ 6.03** (1.87%)	7.48 $\pm$ 0.27** (412.3%)	4.48 $\pm$ 0.27** (28.2%)	3.9 $\pm$ 0.11** (19.1%)	129.48 $\pm$ 1.58** (14.8%)	0.92 $\pm$ 0.12** (196.8%)

P values:
Values bearing superscripts are significantly different by ANOVA at <0.05.
*When compared to control group.
**When compared to KI supplemented group.

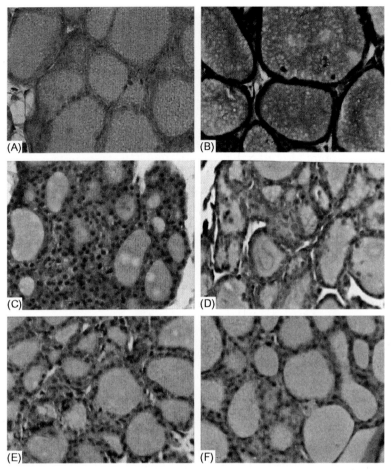

FIGURE 42.3 Photomicrographs of thyroid gland in (A) control, (B) KI supplemented, (C) fresh radish, (D) fresh radish fed KI supplemented, (E) cooked radish, and (F) cooked radish fed KI supplemented after 90 days treatment (H&E, × 20) Source: Chandra et al., 2004 [93].

of goiter in infants after feeding soya milk since birth, while Bruce et al. [100] found no difference in thyroid function when soyabeans were supplemented in women. Klein et al. [101] demonstrated that the energy metabolism status and thyroid hormone levels of growing rats are augmented in response to different proteins including soya proteins. The goitrogenic/antithyroidal potential of soyabean is still conflicting. The observations based on feeding of soyabeans of Indian origin in adult female albino rats by replacing one-third portion of

the normal diet for a period of 60 days are summarized [102].

A marked increase in thyroid weight was observed in the soyabean fed group of rats as compared to the control group. Further, there was a significant increase in serum protein content in the experimental group. However, there was no significant difference in urinary iodine concentration (Table 42.9).

The histological changes that occurred in the soya fed and control group of rats are shown in Figure 42.4. In the soyabean fed group the follicles are lined with high cuboidal/columnar cells showing hypertrophy and hyperplasia. Some follicles are invaded by epithelial cells.

TABLE 42.9 Changes in Thyroid Weight, Serum Protein Content and Urinary Iodine after Soyabean Feeding

Group	Thyroid Weight (mg/100 g Body Weight)	Serum Protein Content (g/dL)	Urinary Iodine Level (μmol/L)
Control	8.9 ± 0.5	3.7 ± 0.4	32.1 ± 0.9
Soyabean fed	11.4 ± 1.3	4.6 ± 0.5	30.7 ± 1.1

Values are mean $\pm$ SD of eight observations.

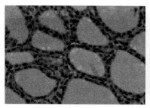 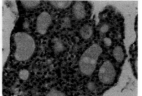

CONTROL   SOYABEAN FED

FIGURE 42.4 Photomicrographs of thyroid gland of control and soyabean fed rats showing follicles lined with cuboidal/columnar cells showing hypertrophy and hyperplasia, while in control groups the follicles are lined with low cuboidal cells and filled with colloid (H&E, $\times$ 400) Source: Mukhopadhyay et al., 2004 [102].

There was a great increase in the number of small follicles but no such changes were found in the control group.

In *in vitro* studies, inhibition of thyroid peroxidase activity (TPO) was found after addition of graded doses of soyabean extract. In an *in vivo* study, a marked inhibition of TPO activity was observed in a soyabean fed group of rats as compared to control group (Figures 42.5, 42.6).

Serum T_3 and T_4 values were also compared between the experimental and control group (Table 42.10). There was a significant increase ($P < 0.001$) in T_4 level but the increase in T_3 level was not significant in the soyabean fed group as compared to the control group of rats.

The conflicting results of morphological and functional aspects of thyroid under the influence of dietary soyabean are available from earlier studies for the lack of a logistic approach on the problem. In those studies iodine nutrition status and protein nutrition were not evaluated concomitantly with soyabean feeding because the antithyroidal potency of dietary goitrogens of soyabean is dependent on those body's nutritional indices. The results of this investigation show that soyabeans caused an enlargement of thyroid glands resembling a goitrous state, and inhibited TPO activity both *in vivo* and *in vitro*, but most strikingly elevated

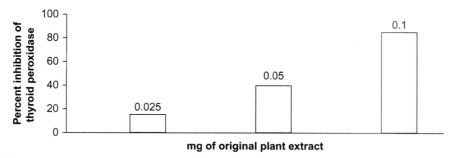

Inhibition of thyroid peroxidase activity by aqueous extract of soyabean at different concentrations

FIGURE 42.5 Inhibition of TPO activity by aqueous extract of soyabean. Source: Mukhopadhyay et al., 2004 [102].

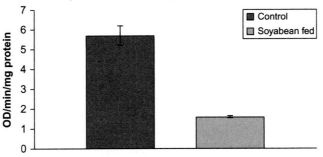

FIGURE 42.6 Comparison of *in vivo* thyroid peroxidase activity in control and soyabean fed rats. Source: Mukhopadhyay et al., 2004 [102].

TABLE 42.10 Changes in Serum Total T_3 and T_4 Following Soyabean Feeding

Group	Serum total T_3 ng/dL	Serum total T_4 mg/dL
Control	153.4 ± 1.9	4.8 ± 0.5
Soyabean fed	155.7 ± 1.2	5.8 ± 0.8

Values are mean ± SD of eight observations.

serum T_4 levels and T_3 levels to an extent. Genistein, the primary flavonoid present in soyabeans, is responsible for this inhibition [103] and the possible mechanism of inhibition is due to direct contact of thyroid peroxidase with isoflavone and H_2O_2 but not with iodine [104]. This study suggests that a goitrogenic/antithyroid effect of soyabean is expressed in a protein deficient state or in a state of iodine deficiency. Adequate supplies of dietary protein and iodine may prevent goitrogenic/antithyroidal potency of soyabeans [102].

7.2 Green Tea

Green tea leaves contain catechin (flavon-3-ol derivatives), including (−)-epicatechin (EC), (−)-epigallocatechin (EGC), (−)-epigallocatechin gallate (EGCg), (−)-epicatechin gallate (ECg), (−)-gallocatechin gallate (GCg), (−)-catechin gallate (Cg), and (+)-catechins (CT) [105]. Among others, Divi and Doerge [106]

reported that catechins inhibit thyroid peroxidase activity (TPO) ($IC_{50} = 36.4 ± 3.86\ \mu M$), though *in vivo* long-term effects on thyroid functions by oral administration have not been reported. However, polyphenone-60 (P-60), water extracts of green tea leaves, when orally administered for 2, 4, and 8 weeks in male rats, caused changes in the thyroid, testis and prostate gland, and related hormone concentrations – TSH, LH, FSH, testosterone, T_3, and T_4 [107].

It is evident from those observations that a 5% oral dose of polyphenone-60 (P-60), green tea extract catechins in the diet to male rats for 2–8 weeks induced goiters and decreased weights of the body, testis, and prostate gland. Endocrinologically, elevating plasma thyroid stimulating hormone (TSH), LH, and testosterone levels and decreasing T_3 and T_4 levels were induced by this treatment. The changes observed *in vivo* were in conformity with antithyroid effects and aromatase inhibition effects of P-60 and its constituents (Figure 42.7; Tables 42.11, 42.12).

7.3 Millets

Pearl millet [*Pennisetum millet* (L.) Leeke] is one of the most important food crops for the rural poor of the semiarid tropics, an ecological zone that includes the portion of Africa and Asia and almost encircles the earth.

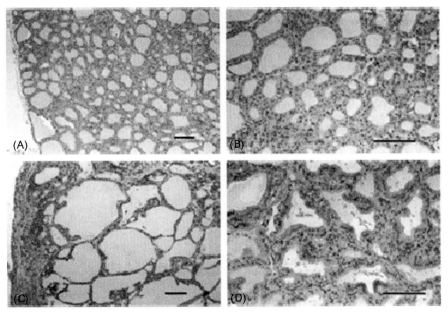

FIGURE 42.7 Thyroid gland of a male rat of the control group (A, B) and of the group administered a 5% polyphenone-60 (P-60) containing diet (C, D), for 8 weeks

Epidemiological evidence suggests that millet may play a role in the genesis of endemic goiter in these areas. *In vivo* and *in vitro* studies revealed that a millet diet rich in C-glycosylflavones produces goitrogenic and antithyroidal effects similar to those of certain other antithyroid agents and small doses of methyl mercapto imidazole (MMI).

In areas of iodine deficiency where millet is a major component of the diet, its ingestion may contribute to the genesis of endemic goiter [108]. Of importance in this regard is the increase in antithyroid activity of extracts of millet that occurs when they are boiled or stored, since boiling, and perhaps storage, of millet precedes its ingestion in the natural setting.

Chemical changes analogous to those that occur with the glycosylflavonoids after their ingestion in mammals, such as hydrolysis of glycosides to aglycones, B-ring methylation and C-ring fission, which result in a marked increase in antithyroid activity, as well as additive antithyroid effects, may also occur under the above circumstances [109].

7.4 Babussu and Mandioca

Among the southern edges of the Amazon Basin, particularly in the state of Maranhao in northern Brazil, abandoned fields and croplands have been taken over by palm trees. The palm tree, known as babassu (*Orbignya phalerata*), is a stubborn competitor that keeps other species from gaining a foothold. Thus, the immense forests of babassu in Maranhao are a basic source of food, fuel, shelter, and income [110]. In Maranhao, goiter prevalence among school children was still 38% in 1986 despite an adequate iodine intake in most of the population. Therefore, the question arose as to whether or not the ingestion of babassu alone or mixed with mandioca contributed to the persistence of goiter in this area of Brazil. The potential antithyroid effect of babassu and

TABLE 42.11   Effects of Polyphenone-60[b] Treatment in Male Rats for 2, 4 or 8 Weeks[a,b]

Feeding Period (Weeks)	Dose (%)	Body Weight[c] (g)	Thyroid Weight[c] (g)	Testis Weight[c] (g)	Prostate Gland Weight (mg)
2	0	139±5	9.2±2.6	1.72±0.1.21	48.1±6.0
	1.25	144±5	(6.6±1.9)	(1.23±0.08)	(34.6±5.6)
	5.0	124±5**	9.9±1.3	1.82±0.11	57.8±10.7
			(6.9±1.1)	(1.26±0.05)	(40.0±6.7)
			14.8±4.1*	1.48±0.01*	38.4±4.3
			(11.8±3.0) **	(1.43±0.26)	(30.9±2.7)
4	0	195±9	10.8±2.1	2.38±0.33	101.4±29.4
	1.25	203±11	(5.5±1.1)	(1.21±0.13)	(51.61±3.1)
	5.0	166±8**	11.9±1.7	2.61±0.10	125.4±21.3
			(5.9±1.1)	(1.29±0.08)	(61.5±8.7)
			28.6±12.2**	2.15±0.26	71.1±31.0
			(17.0±6.7)**	(1.29±0.12)	(42.3±17.4)
8	0	247±35	12.4±1.5	2.89±0.08	206.5±47.7
	1.25	249±14	(5.1±1.0)	(1.19±0.13)	(83.2±12.2)
	5.0	209±14*	12.9±2.0	2.85±0.05	194.0±41.6
			(5.2±0.5)	(1.17±0.07)	(77.6±15.4)
			44.3±11.7**	2.52±0.17*	125.4±34.3*
			(21.4±6.5)**	(1.21±0.05)	(59.3±12.74)*

[a]Rats were fed polyphenone-60 (P-60) in the diet at 0, 1.25, or 5%. Five animals/group. The values in the parentheses show the relative weight per 100 g body weight. The statistical significance of data was calculated using Dennett's test. Values represent mean ± SD.
[b]% of P-60 concentration in diet.
[c]Significantly different ($P < 0.05$) from control (0%).
*Significantly different ($P < 0.05$) from control (0%).
**Significantly different ($P < 0.01$) from control (0%).

mandioca has been assessed by means of *in vitro* and *in vivo* studies [111]. The edible parts of babassu produced significant *in vivo* antithyroid effects in rats on a high iodine intake as well as distinct and reproducible antithyroid and anti-TPO activities in an *in vitro* system, their action being similar to that of the thionamide-like antithyroid drugs propylthiouracil and methimazole. The antithyroid action of aqueous extract of mandioca flour *in vivo* and *in vitro* was also evident by significant and pronounced inhibition of the iodide organification process; however, in contrast to babassu, methanol and aqueous extracts of mandioca flour caused little inhibition *in vitro* in the TPO system. Little or no effect was produced by babassu or mandioca on thyroid iodide transport by thyroid slices or *in vivo* in the rat, indicating that neither thiocyanate nor perchlorate-like compounds are responsible for their antithyroid effects.

Results of this study provide direct experimental evidence, both *in vivo* and *in vitro*, of antithyroid effects of babassu and mandioca, supporting the hypothesis that the staple food is responsible, at least in part, for the persistence of goiter in the iodine-supplemented endemic regions of Maranhao in Brazil. Finally, research is needed to isolate and identify those compounds or agents responsible for the antithyroid effects of babassu and mandioca (Figure 42.8).

TABLE 42.12 Effects of Polyphenone-60 (P-60)[a] Treatment in Male Rats on Plasma TSH, T_4, T_3, Testosterone (TTN), LH, and FSH Level for 2, 4 or 8 Weeks[b,c]

Feeding period (weeks)	Dose (%)	TSH[f] (ng/mL)	T_4^f (ng/mL)	T_3^f (ng/mL)	Ttn[f] (ng/mL)	LH[f] (ng/mL)	FSH (ng/ml)
2	0	16.2±2.1	3.3±0.2	1.3±0.1	–*	–	–
(N = 5)	1.25	15.3±1.3	3.4±0.2	1.6±0.1	–	–	–
	5.0	14.8±2.4	2.5±0.3**	1.4±0.2	–	–	–
4	0	16.7±1.1	2.9±0.3	1.5±0.1	4.3±1.7	–	–
(N = 5)	1.25	16.6±0.9	2.9±0.4	1.3±0.1	4.9±2.3	–	–
	5.0	17.9±0.6	2.1±0.5	1.6±0.2	7.5±2.2*	–	–
8	0	18.2±0.7	2.9±0.2	1.4±0.1	1.7±0.6	9.3±2.3*	84.8±27.5*
(N = 7)	1.25	18.8±0.4	2.8±0.3	1.4±0.1	3.6±1.6	11.7±2.3*	88.8±12.5*
	5.0	19.2±0.7*	2.0±0.4*	1.1±0.2*	7.2±2.5**	13.6±2.9*	101.7±17.3*

[a]Percent of P-60 concentration in diet.
[b]Rats were fed P-60 in diet at 0, 1.25, or 5%.
[c]The statistical significance of data was calculated using Student's t-test.
All measurements were performed in triplicate, and values represent mean ± SD.
*Significantly different ($P<0.05$) from control (0%).
**Significantly different ($P<0.01$) from control (0%).

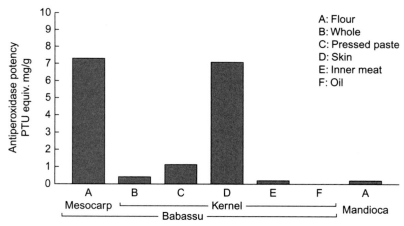

FIGURE 42.8 *In vitro* antithyroid peroxidase (TPO) of methanol extracts of edible parts of babassu and mandioca flour in the purified porcine TPO system.
Source: Gaitan et al., 1994 [111].

8. SUMMARY

All of these observations reveal that a number of plant foods usually consumed by the people of several regions contain goitrogenic/antithyroidal substances that interfere with thyroid hormone synthesis, acting at different levels in the thyroid gland and causing goiter and associated iodine deficiency disorders. Even supplementation of iodine fails to counteract the adverse effects of those dietary goitrogens on thyroid hormone synthesis, as has been observed in several epidemiological and experimental studies. Therefore, goitrogens that are present in food have an important role in the regulation of thyroid gland morphology and functions; their action is not only mediated through interference of iodine nutrition but also by other indirect mechanisms.

In spite of adequate iodine intake, endemic goiter, the most common manifestation of iodine deficiency disorders, is prevalent in many regions because of the consumption of goitrogenic foods. Therefore, to identify the region-specific environmental goitrogens and to find out measures to counteract their action is important, to prevent and control this public health problem.

Cyanogenic glycosides, glucosinolates, and flavonoids found in plants are the precursor molecules of goitrogenic substances. Cassava, cabbage, cauliflowers, radish, mustard, turnip, sweet potato, are bamboo shoots are the common dietary sources of cyanogenic glucosides and glucosinolates while soyabeans, millet, tea, onion, apple, kale, red wine, tofu, and grapefruit juice are the sources of flavonoids-containing plant foods.

References

1. Chesney, A. M., Clawson, T. A., & Webster, B. (1928). Endemic goiter in rabbits. I. Incidence and characteristics. *Johns Hopkins Hospital Bulletin, 43,* 261–277.

2. Webster, B., & Chesney, A. M. (1930). Studies in the etiology of simple goiter. *The American Journal of Pathology, 6,* 275–284.

3. Hercus, C. E., & Purves, H. D. (1936). Studies on endemic and experimental goitre. *Journal of Hygiene Cambridge, 36,* 182–203.

4. Astwood, E. B., Greer, M. A., & Ettlinger, M. G. (1949). 1-5-Vinyl-2-thiooxazolidone, an antithyroid compound from yellow turnip and from Brassica seeds. *The Journal of Biological Chemistry, 181,* 121–130.

5. Greer, M. A. (1956). Isolation from rutabaga seed of progoitrin, the precursor of the naturally occurring antithyroid compound, goitrin (L-5-vinyl-2-thiooxazolidone). *Journal of the American Chemical Society, 78,* 1260–1261.

6. Greer, M. A. (1957). Goitrogenic substances in food. *The American Journal of Clinical Nutrition, 5,* 440–444.

7. Care, A. D. (1954). Goitrogenic properties of linseed. *Nature, 173,* 172–173.

8. New Zealand Department of Agriculture (1957). Annual Report of the Director General of Agriculture for the year ended 31 March, 1957.

9. Clements, F. W. (1960). Naturally occurring goitrogens. *British Medical Bulletin, 16,* 133–137.

10. Moudgal, N. R., Srinivasan, V., & Sarma, P. S. (1957). Studies on goitrogenic agents in food. II. Goitrogenic action of arachidoside. *The Journal of Nutrition, 61,* 97–101.

11. Moudgal, N. R., Raghupathy, E., & Sarma, P. S. (1958). Studies on goitrogenic agents in food. III. Goitrogenic action of some glycosides isolated from edible nuts. *The Journal of Nutrition, 66,* 291–303.

12. Conn, E. E. (1980). Secondary plant products. In E. A. Bell, & B. V. Charlwood, (Eds.), *Encyclopedia of Plant Physiology* (Vol. 8), (pp. 461–492). New York: Springer.

13. Seigler, D. S. (1991). The chemical participants. In G. A. Rosenthal, & M. R. Berenbaum, (Eds.), *Herbivores – Their interaction with secondary metabolites* (Vol. I), (pp. 35–77). New York: Academic Press.

14. Nahrstedt, A. (1989). The significance of secondary plant metabolites for interactions between plants and insects. *Planta Medica, 55,* 333–338.

15. Lang, K. (1933). Die Rhodanbildung in Tierkörpor. *The Biochemical Journal, 259,* 243–256.

16. Rosenthal, O. (1948). The distribution of rhodanese. *Federation Proceedings, 7,* 181–182.

17. Clemedson, C. J., Sorbo, B., & Ullberg, S. (1960). Autoradiographic observations on injected S^{35} Thiocyanate and C^{14} cyanide in mice. *Acta Physiologica Scandinavica, 48,* 358–382.

18. Fiedler, H., & Wood, J. L. (1956). Specificity studies on the β-mercaptopyruvate-cyanide trans sulfuration system. *The Journal of Biological Chemistry, 222,* 387–397.

19. Wokes, F., Baxter, N., Horstord, J., & Preston, B. (1953). Effect of light on vitamin B_{12}. *The Biochemical Journal, 53,* xix–xx.

20. Mushett, C. W., Kelley, K. L., Boxer, G. E., & Rickards, J. C. (1952). Antidotal efficacy of vitamin B_{12a}, (hydro-cobalamin) in experimental cyanide poisoning. *Proceedings of the Society for Experimental Biology and Medicine, 81,* 234–237.

21. Hartmann, F., & Wagner, K. H. (1949). Studies uber die wirkung von dl-Methionine in stoffwechsel der erkranten leber II. Die BeeinFlussung des Rhodanstoff wechsels bei leiberkranken durch Methionine. *Deutsches Archiv fur Klinische Medizin, 196,* 432–434.

22. Wokes, F., Badenoch, J., & Sinclair, H. M. (1955). Human dietary deficiency of vitamin B_{12}. *The American Journal of Clinical Nutrition, 3,* 375–382.

23. Osborne, J. S., Adamek, S., & Hobbs, M. E. (1956). Some components of the gas phase of cigarette smoke. *Analytical Chemistry, 28,* 211–215.

24. Anonymous (1964). Smoking and health: Report of the advisory committee to the surgeon general of health services. US Dept Health, Education and Welfare, Public Health Service, Publ. No. 1103, p. 60.

25. Montgomery, R. D. (1964). Observations on the cyanide content and toxicity of tropical pulses. *The West Indian Medical Journal, 13,* 1–11.

26. Couch, J. F. (1934). Poisoning of Livestock by Plants that Produce Hydrocyanic Acid. *US Department of Agriculture,* 2–4, Leaflet No. 88.

27. Bagchi, K. N., & Ganguli, H. D. (1943). Toxicology of young shoots of common bamboos (*Bambusa arundinacea Willd*). *Indian Medical Gazette, 78,* 40–42.

28. Greer, W.L. (1978). Mechanism of action of antithyroid sulfurated compounds. In: S. C. Werner and S. H. Ingbar, (Eds.), (4th ed.) (p. 41). New York: Harper and Row Publishers.

29. Ermans, A. M., & Bourdoux, P. (1989). Antithyroid sulfurated compounds. In E. Gaitan, (Ed.), *Environmental Goitrogens* (pp. 15–31). Boca Raton, FL: CRC Press.

30. Mitchell, M. L., & O'Rourke, M. E. (1960). Response of thyroid gland to thiocyanate and thyrotropin. *The Journal of Clinical Endocrinology and Metabolism, 20,* 47.

31. Taurog, A. (1970). Thyroid peroxidase and thyroxine biosynthesis. *Recent Progress in Hormone Research, 26,* 189–247.

32. Virion, A., Deme, D., Pommier, J., & Nunez, J. (1980). Opposite effect of thiocyanate on tyrosine iodination and thyroid hormone synthesis. *European Journal of Biochemistry, 112,* 1–7.

33. Van Middlesworth, L. (1985). Thiocyanate feeding with low iodine diet causes chronic iodine retention in the thyroid of mice. *Endocrinology, 116,* 665–670.

34. Wolf, J. (1964). Transport of iodide and other anions in the thyroid gland. *Physiological Reviews, 44,* 45–90.

35. Lakshmy, R., Srinivas Rao, P., Sesikeran, B., & Suryaprakash, P. (1995). Iodine metabolism in response to goitrogen induced altered thyroid status under conditions of moderate and high intake of iodines. *Hormone and Metabolic Research, 27,* 450–454.

36. Schone, F., Leiterer, M., Hartung, H., Jahreis, G., & Tischendorf, F. (2001). Rapeseed glucosinolates and iodine in cows affect the milk iodine concentration and the iodine status of biglets. *The British Journal of Nutrition, 85,* 659–670.

37. Chandra, A. K., Ghosh, D., Mukhopadhyay, S., & Tripathy, S. (2004). Effect of bamboo shoot, *Bambusa arundinacea* (Retz.) wild on thyroid status under conditions of varying iodine intake in rats. *Indian Journal of Experimental Biology, 42,* 781–786.

38. Chandra, A. K., Ghosh, D., Mukhopadhyay, S., & Tripathy, S. (2006). Effect of radish (*Raphanus sativas* Linn.) on thyroid status under conditions of varying iodine intake in rats. *Indian Journal of Experimental Biology, 44,* 653–661.

39. Larsen, P. O. (1981). Secondary plant products. In P. K. Stumpf, & E. E. Conn, (Eds.), *The Biochemistry of Plants* (Vol. VII) (pp. 502–526). New York: Academic Press.

39a. Van Etten, C. H. (1969). Goitrogens. In I. E. Lienear, (Ed.), *Toxic Constituents of Plant Food Stuffs.* New York: Academic Press.

40. Astwood, E. B., Greer, M. A., & Ettlinger, M. A. (1949). *l*-5-Vinyl-2-thiooxazolidone, an antithyroid compound from yellow turnip and from Brassica seeds. *The Journal of Biological Chemistry, 181,* 121–130.

41. Carroll, K. K. (1949). Isolation of an antithyroid compound from rapeseed (*Brassica napus*). *Proceedings of the Society for Experimental Biology and Medicine, 71,* 622–624.

42. Ettlinger, M. G. (1950). Synthesis of the natural antithyroid factor. *l*-5-Vinyl-2-thiooxazolidone. *Journal of the American Chemical Society, 72,* 4792–4796.

43. Greer, M. A. (1962). The natural occurrence of goitrogenic agents. *Recent Progress in Hormone Research, 18,* 187–219.

44. Hopkins, C. Y. (1938). A sulfur-containing substance from the seed of *Conringia orientalis*. *Canadian Journal of Research, 168,* 341–344.

45. Astwood, E. B., Bissell, A., & Hughes, A. M. (1945). Further studies on the chemical nature of compounds which inhibit the function of the thyroid gland. *Endocrinology, 37,* 456–481.

46. Kjaer, A., & Gmelin, R. (1957). *iso* Thiocyanates. XXVIII. A new *iso*thiocyanate glucoside (glucobarbin) furnishing (−)-5-phenyl-2-oxazolidinethione upon enzymic hydrolysis. *Acta Chemica Scandinavica, 11,* 906–907.

47. Kjaer, A., & Thomsen, H. (1962). *iso* Thiocyanates. XLII. Glucocleomin, a new natural glucoside, furnishing (−)-5-ethyl-5-methyl-2-oxazolidinethione on enzymic hydrolysis. *Acta Chemica Scandinavica, 16,* 591–598.

48. Kjaer, A., & Christensen, B. W. (1959). *iso* Thiocyanates. XXXVI. (−)-4-methyl-2-oxazolidinethione, the enzymic hydrolysis product of a glucoside (glucosisymbrin) in seeds of *Sisymbrium austriacum* Jacq. *Acta Chemica Scandinavica, 13,* 1575–1584.

49. Kjaer, A., & Christensen, B. W. (1962). Isothiocyanates XL. Glucosisaustricin, a novel glucoside present in seeds of *Sisymbrium austriacum*. *Acta Chemica Scandinavica, 16,* 71–77.

C. ACTIONS OF INDIVIDUAL FRUITS IN DISEASE AND CANCER PREVENTION AND TREATMENT

50. Langer, P., & Michajlovskiji, N. (1958). The relation between thiocyanate formation and the goitrogenic effects of foods. II. The thiocyanate content of foods, the chief cause of thiocyanate excretion in urine of man and animals. *Zeitschrift fur Physiologische Chemie, 312,* 31–36.

51. Michajlovskij, N., & Langer, P. (1959). Studien über Beziehungen zwischen Rhodanidbildung und kropfbildender Eigenschaft von Nahrungsmitteln, III. *Zeitschrift Journal of Physiologische Chemie, 317,* 30–33.

52. Langer, P., & Stole, V. (1964). The relation between thiocyanate formation and goitrogenic effects of foodstuffs. V. Comparison of the effect of white cabbage and thiocyanate on the rat thyroid gland. *Zeitschrift fur Physiologische Chemie, 335,* 216–222.

53. Langer, P. (1964). The relation between thiocyanate formation and goitrogenic effect of foods. VI. Thiocyanate activity of allylisothiocyanate, one of the most frequently occurring mustard oils in plants. *Journal of Physiological Chemistry, 339,* 33–35.

54. Langer, P., & Stole, V. (1965). Goitrogenic activity of allylisothiocyanate – a widespread natural mustard oil. *Endocrinology, 76,* 151–155.

55. Greer, M. A. (1950). Nutrition and goiter. *Physiological Reviews, 30,* 513–548.

56. Bachelard, H. S., & Trikojus, V. M. (1960). Plant thioglycosides and the problem of endemic goitre in Australia. *Nature, 185,* 80–82.

57. Jirousek, L. (1956). On the antithyroid substances in cabbage and other *Brassica* plants. *Naturwissenschaften, 43,* 328–329.

58. Jirousek, L., & Starka, L. (1958). Concerning the occurrence of trithiones (1,2-dithiocyclopenta-4-ene-3-thione) in *Brassica* plants. *Naturwissenschaften, 45,* 386–387.

59. Jirousek, L., Reisenauer, R., & Hovarka, J. (1959). Polarographic determination of sulphydryl substances in biological materials. IV. Concerning the possibility of a relationship between sulphydryl substances and endemic goitre. *Endokrinologie, 37,* 269–276.

60. Gaitan, E. (1986). Environmental goitrogens other than iodine deficiency. *IDD Newsletter, 2,* 11–13.

61. Oginsky, E. L., Stein, A. E., & Greer, M. A. (1965). Myrosinase activity in bacteria as demonstrated by the conversion of progoitrin to goitrin. *Proceedings of the Society for Experimental Biology and Medicine, 119,* 360–364.

62. Gaitan, E. (1990). Goitrogens in food and water. *Annual Review of Nutrition, 10,* 21–39.

63. Gaitan, E. (1996). Flavonoids and the thyroid. *Nutrition, 12,* 127–129.

64. Harsteen, B. (1983). Flavonoids. A class of natural products of high pharmacological potency. *Biochemical Pharmacology, 32,* 1141–1148.

65. Middleton, E. (1984). The flavonoids. *Trends in Pharmacological Sciences, 5,* 335–338.

66. Skibola, C. F., & Smith, C. F. (2000). Potential health impacts of excessive flavonoid intake. *Free Radical Biology and Medicine, 29,* 375–383.

67. Hertog, G., Hollman, P., & Katan, M. (1992). Content of potentially anticarcinogenic flavonoids of 28 vegetables and 9 fruits commonly consumed in The Netherlands. *Journal of Agricultural and Food Chemistry, 40,* 2379–2383.

68. Dwyer, J. T., Goldin, B. R., Saul, N., Gualtieri, L., Barakat, S., & Adlercreutz, H. (1994). Tofu and soy drinks contain phytoestrogens. *Journal of the American Dietetic Association, 94,* 739–743.

69. Hertog, M. G., Hollman, P. C., Katan, M. B., & Kromhout, D. (1993). Intake of potentially anticarcinogenic flavonoids and their determinants in adults in The Netherlands. *Nutrition and Cancer, 20,* 21–29.

70. Kuhnau, J. (1976). The flavonoids, a class of semiessential food components: their role in human nutrition. *World Review of Nutrition and Dietetics, 24,* 117–191.

71. Harborne, J. B., Mabry, T. J., & Mabry, H. (1975). *The Flavonoids* (Vol. Part 1). New York: Academic Press.

72. Harborne, J. B., Mabry, T. J., & Mabry, H. (1975). *The Flavonoids* (Vol. Part 2). New York: Academic Press.

73. Knekt, P., Jarvinen, R., Seppanen, R., Hellovara, M., Teppo, L., Pukkala, E., & Aromaa, A. (1997). Dietary flavonoids and the risk of lung cancer and other malignant neoplasms. *American Journal of Epidemiology, 146,* 223–230.

74. Rimm, E. B., Katan, M. B., Ascherio, A., Stampfer, M. J., & Willett, W. C. (1996). Relation between intake of flavonoids and risk for coronary heart disease in male health professionals. *Annals of Internal Medicine, 125,* 384–389.

75. Hollman, P. C., Bijsman, M. N., Van Gameren, Y., Cnossen, E. P. J., de Vries, J. H., & Katan, M. B. (1999). The sugar moiety is a major determinant of the absorption of dietary flavonoid glycosides in man. *Free Radical Research, 31,* 569–573.

76. de Vries, J. H., Janssen, P. L., Hollman, P. C., Van Staveren, W. A., & Katan, M. B. (1997). Consumption of quercetin and kaempferol in free-living subjects eating a variety of diets. *Cancer Letters, 114(1,2),* 141–144.

77. Hollman, P. C., van Trijp, J. M., Buysman, M. N., van der Gaag, M. S., Mengelers, M. J., de Vries, J. H., & Katan, M. B. (1997). Relative bioavailability of the antioxidant quercetin from various foods in man. *FEBS Letters, 418(1,2),* 152–156.

78. Coward, L., Barnes, N., Setchell, K., & Barnes, S. (1993). The isoflavones genistein and daidzein in

C. ACTIONS OF INDIVIDUAL FRUITS IN DISEASE AND CANCER PREVENTION AND TREATMENT

soybean foods from American and Asian diets. *Journal of Agricultural and Food Chemistry, 41,* 1961–1967.

79. Franke, A. A., Custer, L. J., Cerna, C. M., & Narala, K. (1995). Rapid HPLC analysis of dietary phytoestrogens from legumes and from human urine. *Proceedings of the Society for Experimental Biology and Medicine, 208,* 18–26.

80. Barnes, S., Peterson, T. G., & Coward, L. (1995). Rationale for the use of genistein-containinng soy matricies in chemoprevention trials for breast and prostate cancer. *Journal of Cellular Biochemistry, 22,* 181–187.

81. Thompson, L. U., Robb, P., Serraino, M., & Cheung, F. (1991). Mammalian lignan production from various foods. *Nutrition and Cancer, 16,* 43–52.

82. Obermeyr, W. R., Musoer, S. M., Betz, J. M., Casey, R. E., Pohland, A. E., & Page, S. W. (1995). Chemical studies of phytoestrogens and related compounds in dietary supplements: flax and chaparral. *Proceedings of the Society for Experimental Biology and Medicine, 208,* 6–12.

83. Nesbitt, P. D., Lam, Y., & Thompson, L. W. (1999). Human metabolism of mammalian lignan precursors in raw and processed flaxseed. *The American Journal of Clinical Nutrition, 69,* 549–555.

84. Wang, C., Makela, T., Hase, T., Adlercreutz, H., & Kurzer, M. S. (1994). Lignans and flavonoids inhibit aromatase enzyme in human preadipocytes. *The Journal of Steroid Biochemistry and Molecular Biology, 50,* 205–212.

85. Le Bail, J. C., Laroche, T., Marre-Fournier, F., & Habrioux, G. (1998). Aromatase and 17beta-hydroxysteroid dehydrogenase inhibition by flavonoids. *Cancer Letters, 133,* 101–106.

86. Gaitan, E., Cooksey, R. C., Legan, J., & Linsay, R. H. (1995). Antithyroid effects *in vivo* and *in vitro* of vitexin : a C-glycosylflavone in millet. *Journal of Clinical Endocrinologia e Metabologia, 80,* 1144–1147.

87. Sartelet, H., Serghat, S., Lobstein, A., Ingenbleek, Y., Anton, R., Petitfrere, E., Aguie-Aguie, G., Martiny, L., & Haye, B. (1996). Flavonoids extracted from fonio millet *(Digitaria exilis)* reveal potent antithyroid properties. *Nutrition, 12,* 100–106.

88. Delange, F., Thilly, C., & Ermans, A. M. (1968). Iodine deficiency, a permissive condition in the development of endemic goiter. *The Journal of Clinical Endocrinology and Metabolism, 28,* 114–116.

89. Thilly, C. H., Delange, F., & Ermans, A. M. (1972). Further investigations of iodine deficiency in the etiology of endemic goiter. *The American Journal of Clinical Nutrition, 25,* 30–40.

90. Ermans, A. M., Bourdoux, P., Kinthaert, J., Langasse, R., Luvivila, K., Mafuta, M., Thilly, C. H., & Delange, F. (1982). Role of cassava in the etiology of endemic goiter and cretinism. In F. Delange, & R. Ahluwalia, (Eds.), *Cassava Toxicity and Thyroid: Research and public health issues* (pp. 9–16). International Development Research Centre (IDRC 207e).

91. Chandra, A. K., Mukhopadhyay, S., Lahiri, D., & Tripathy, S. (2004). Goitrogenic content of cyanogenic plant foods of Indian origin and their anti-thyroidal activity *in vitro. The Indian Journal of Medical Research, 119,* 180–185.

92. Chandra, A. K., Mukhopadhyay, S., Ghosh, D., & Tripathy, S. (2006). Effect of radish (*Raphanus sativus* Linn.) on thyroid status under conditions of varying iodine intake in rats. *Indian Journal of Experimental Biology, 44,* 653–661.

93. Chandra, A. K., Ghosh, D., Mokhopadhyaya, S., & Tripathy, S. (2004). Effect of bamboo shoot, *Bambusa arundinacea* (Retz.) Willd. on thyroid status under conditions of varying iodine intake in rats. *Indian Journal of Experimental Biology, 42,* 780–781.

94. Chandra, A. K., Ghosh, D., Mukhopadhyay, S., & Tripathy, S. (2006). Effect of cassava (*Manihot esculenta Crantz*) on thyroid status under conditions of varying iodine intake in rats. *African Journal of Traditional Complementary and Alternative Medicines, 3,* 87–99.

95. Chandra, A. K., Singh, L. H., Tripathy, S., Debnath, A., & Khanam, J. (2006). Iodine nutritional status of school children in selected areas of Imphal West district, Manipur, North East India. *Indian Journal of Pediatrics, 73,* 795–798.

96. McCarrison, R. (1933). A paper on food and goitre. *British Medical Journal, 2,* 671–675.

97. Sharpless, G. R. (1938). A new goiter producing diet for the rat. *Proceedings of the Society for Experimental Biology and Medicine, 38,* 166–168.

98. Wilgus, H. S., Jr., Gassner, F. X., Patton, A. H., & Gustavson, R. G. (1941). The goitrogenicity of soyabeans. *The Journal of Nutrition, 22,* 43–52.

99. Shepard, T. H., Gordon, E. P., Krischvink, J. F., & McLean, C. M. (1960). Soybean goiter. *The New England Journal of Medicine, 262,* 1099–1103.

100. Bruce, B., Spiller, G. A., & Holloway, L. (2000). Soy isoflavones do not have an antithyroid effect in postmenopausal women over 64 years of age. *FASEB Journal, 9,* 214–218.

101. Klein, M., Schadereit, R., & Kuchemneister, U. (2000). Energy metabolism and thyroid hormone levels of growing rats in response to different dietary proteins – soy protein or casein. *Archiv fur Tierernahrung, 53,* 99–125.

102. Mukhopadhyay, S., Ghosh, D., Tripathy, S., Jana, S., & Chandra, A. K. (2004). Evaluation of possible

C. ACTIONS OF INDIVIDUAL FRUITS IN DISEASE AND CANCER PREVENTION AND TREATMENT

goitrogenic and anti thyroid activity of soyabean (Glycine max) of Indian origin. *Indian Journal of Physiology and Allied Sciences, 58,* 1–12.

103. Divi, R. L., Chang, H. C., & Doerge, D. R. (1997). Anti-thyroid isoflavones from soybean. *Biochemical Pharmacology, 54,* 1087–1096.

104. Chang, H. C., & Doerge, D. R. (2000). Dietary genistein inactivates rat thyroid peroxidase *in vivo* without an apparent hypothyroid effect. *Toxicology and Applied Pharmacology, 168,* 244–252.

105. Miura, S., Watanabe, J., Tomita, T., Sano, M., & Tomita, I. (1994). The inhibitory effects of tea polyphenols (flavan-3-ol derivatives) on Cu^{2+} mediated oxidative modification of low density lipoprotein. *Biological and Pharmaceutical Bulletin, 17,* 1567–1572.

106. Divi, R. L., & Doerge, D. R. (1996). Inhibition of thyroid peroxidase by dietary flavonoids. *Chemical Research in Toxicology, 9,* 16–23.

107. Satoh, K., Sakamoto, Y., Ogata, A., Nagai, F., Mikuriya, H., Numazawa, M., Yamada, K., & Aoki, N. (2002). Inhibition of aromatase activity by green tea extract catechins and their endocrinological effects of oral administration in rats. *Food and Chemical Toxicology, 40,* 925–933.

108. Gaitan, E., Raymond, H. L., Riechert, R. D., Ingbar, S. H., Cooksey, R. C., Legan, J., Meydrech, E. F., Hill, J., & Kubota, K. (1989). Antithyroid and goitrogenic effects of millet: Role of C-glycosyl-flavones. *The Journal of Clinical Endocrinology and Metabolism, 68,* 707–714.

109. Lindsay, R. H., Gaitan, E., & Cooksey, R. C. (1989). Pharmacokinetics and intrathyroidal effects of flavonoids. In E. Gaitan, (Ed.), *Environmantal Goitrogenesis.* Boca Raton, FL: CRC Press.

110. Anderson, A., & Anderson, S. (1985). A tree of life grows in Brazil. *Natural History, 12,* 41–46.

111. Gaitan, E., Cooksey, R. C., Legan, J., Lindsay, R. H., Ingbar, S. H., & Medeiros-Neto, G. (1994). Antithyroid effects in vivo and in vitro of babassu and mandioca: a staple food in goiter areas of Brazil. *European Journal of Endocrinology, 131,* 138–144.

C. ACTIONS OF INDIVIDUAL FRUITS IN DISEASE AND CANCER PREVENTION AND TREATMENT

43

Nutritional and Health Benefits of Root Crops

Trinidad P. Trinidad, Aida C. Mallillin, Anacleta C. Loyola, Rosario R. Encabo, and Rosario S. Sagum

Food and Nutrition Research Institute, Department of Science and Technology, Bicutan, Taguig, Metro Manila 1631 Philippines

1. INTRODUCTION

The Philippines is one of the major producers of root crops in Asia along with Thailand, Indonesia, India, China, and Vietnam. Root crops such as sweet potato (kamote) and cassava are used as staples by many people in the Philippines, e.g. Batanes, Cebu, Lanao, Zamboanga, and Sulu. Root crops are rich in carbohydrates specifically dietary fiber, protein, and fats as well as vitamins and minerals.

Dietary fiber comes from the family of carbohydrates, a non-starch polysaccharide that is not digested in the small intestine but may be fermented in the colon into short chain fatty acids (SCFA) such as acetate, propionate, and butyrate. SCFA contributes 1.5–2.0 kcal/g dietary fiber [1]. It enhances water absorption in the colon, thus preventing constipation.

Dietary fiber at levels of 10–20% in food/meal may interfere with the absorptive processes in the small intestine and colon. Some fibers (e.g. cellulose) inhibit the absorption of minerals (e.g. iron, zinc, calcium) by forming insoluble complexes or by entrapping the minerals due to their viscosity and fibrous structure in the small intestine [2–4]. However, studies have shown that some fibers that reach the colon may be fermented and release the minerals for potential absorption [2–4]. Also, the product of fiber fermentation, the short chain fatty acids, e.g. acetate, propionate, and butyrate, can in turn enhance the absorption of minerals in the colon [5–7]. Thus, dietary fiber may not reduce the availability of minerals but may shift their absorption from the small intestine to the colon. Root crops and legumes are good sources of dietary fiber (5–30%). The dietary fiber present in root crops is mostly galactomannans and a fermentable fiber [8–10]. On the other hand, root crops also contain phytic and tannic acid which can inhibit mineral absorption through formation of insoluble iron complexes [11,12].

Dietary fiber has been shown to have important health implications in the prevention of risks of chronic diseases such as cancer, cardiovascular diseases, and diabetes mellitus. Propionate has been shown to inhibit the activity of the enzyme HMG CoA reductase, the limiting enzyme for cholesterol synthesis [13]. Dietary fiber has the ability to bind with bile acids and prevents their reabsorption in the liver, thus inhibiting cholesterol synthesis [13]. Butyrate enhances cell differentiation, thus preventing tumor formation in the colon [14].

Over time, the dietary fiber's viscous and fibrous structure can control the release of glucose in the blood, thus helping with the proper control and management of diabetes mellitus and obesity [15,16]. The glycemic index (GI), a classification of foods based on their blood glucose response relative to a starchy food, e.g. white bread, or standard glucose solution, has been proposed as a therapeutic principle for diabetes mellitus by slowing carbohydrate absorption [17,18]. Low GI foods, e.g. high dietary fiber food, has been shown to reduce postprandial blood glucose and insulin responses and improve the overall blood glucose and lipid concentrations in normal individuals [19], and patients with diabetes mellitus [20–22]. The utilization of root crops as a functional food will not only solve the problem of chronic diseases now prevailing in almost all countries but also encourage the industry and farmers to produce value-added or healthy products from root crops. This will increase the production and promotion of root crops.

The general objective of this study is to determine the nutritional and health benefits of the dietary fiber component of root crops, and the specific objectives are as follows: a) to determine the dietary fiber composition and fermentability characteristics of root crops; b) to determine the mineral availability from root crops; c) to determine glycemic index (GI) of root crops from non-diabetic and diabetic subjects; and d) to determine the cholesterol-lowering effect of root crops in moderately raised serum cholesterol levels of humans.

2. DIETARY FIBER AND FERMENTABILITY CHARACTERISTICS OF ROOT CROPS

Six root crops were studied as follows: ube (*Dioscorea alata*), gabi (*Colocasia esculenta*), tugi (*Dioscorea esculenta*), potato (*Solanum tuberosum*), camote (*Ipomoea batatas*), and cassava (*Manihot esculenta*). Root crops were cooked and freeze dried before analysis. The proximate analysis of root crops was determined using AOAC methods, and analyzed for total, soluble, and insoluble fiber using AOAC Methods against a standard wheat flour (NBS Standard Reference Material 1567 A, Gaithersburg, MD, USA) [23–26].

2.1 *In Vitro* Fermentation

Dietary fiber isolate from each root crop was fermented *in vitro* using human fecal inoculum [27], and 0.5 g of the fiber isolate was weighed in a serum bottle. Forty (40) mL of fermentation media (mix 2 L deionized water, 1 L 0.5 M sodium bicarbonate buffer solution, 1 L macro-mineral solution, 5 mL of 0.1% resazurin), and 2 mL reducing solution (a mixture of 1.25 g cysteine-HCl + 50 pellets of potassium hydroxide in 100 mL deionized water and 1.25 g sodium sulfide in 100 mL deionized water) were added to the serum bottle and flushed with carbon dioxide until colorless. The bottles were sealed with rubber stoppers and an aluminum seal and stored at 4°C. The next day the bottles were placed in a water bath for 1–2 hours at 37°C. A 10 mL fecal inoculum (1:15 dilution of fresh feces from a human volunteer) was added into each bottle and the mixture was incubated for 24 hours at 37°C. The fermented digest was filtered and read using high pressure liquid

TABLE 43.1 Proximate Analysis of Root Crops and Legumes, g/100 g; Sample (Mean ± SEM)

Test Food	Moisture	Ash	Fat	Protein	Carbohydrates*
Kamote	3.3 ± 0.[c]	3.0 ± 0.0[e]	0.6 ± 0.2[b]	3.4 ± 0.0[d]	89.6 ± 0.3[c]
Gabi	2.3 ± 0.[d]	5.6 ± 0.1[a]	0.4 ± 0.0[b]	1.1 ± 0.0[f]	90.7 ± 0.2[b]
Potato	7.1 ± 0.1[b]	4.7 ± 0.1[b]	1.0 ± 0.1[a]	10.0 ± 0.1[a]	77.1 ± 0.4[e]
Tugi	15.9 ± 0.[a]	3.4 ± 0.1[d]	0.2 ± 0.1[c]	4.8 ± 0.1[c]	75.6 ± 0.8[f]
Ube	3.4 ± 0.1[c]	4.0 ± 0.1[c]	0.4 ± 0.0[b]	5.2 ± 0.1[b]	87.0 ± 0.3[d]
Cassava	3.3 ± 0.[c]	2.9 ± 0.0[f]	0.3 ± 0.0[c]	2.4 ± 0.0[e]	91.1 ± 0.1[a]

[a–f]Denotes significant differences between test foods at $P<0.05$.
*Calculated by difference.

TABLE 43.2 Total, Soluble and Insoluble Fiber Content of Root Crops, g/100 g; Sample (Mean ± SEM)

Test Food	Total Dietary Fiber	Soluble Fiber	Insoluble Fiber
Kamote	8.1 ± 0.1[d]	4.8 ± 0.4[a]	3.2 ± 0.2[d]
Gabi	13.5 ± 0.1[a]	3.6 ± 0.1[b]	9.5 ± 0.2[a]
Potato	7.6 ± 0.2[e]	3.6 ± 0.1[b]	4.0 ± 0.2[c]
Tugi	10.3 ± 0.1[c]	3.6 ± 0.1[b]	7.0 ± 0.2[b]
Ube	11.8 ± 0.2[b]	4.4 ± 0.1[a]	7.4 ± 0.2[b]
Cassava	4.6 ± 0.2[f]	1.4 ± 0.4[c]	3.2 ± 0.1[d]

[a–f]Denotes significant differences between test foods at $P<0.05$.

chromatography (HPLC) to measure short chain fatty acids against a volatile acid standard mix (SUPELCO, Philadelphia, USA).

Table 43.1 shows the proximate analysis of root crop. Tugi has significantly greater moisture (15.9 ± 0.5) content among the root crops ($P<0.05$). Gabi (5.6 ± 0.1) has significantly greater ash content among the root crops ($P<0.05$). Ube has the highest protein content, kamote the highest fat content, and gabi and cassava the highest carbohydrates content among the root crops. The dietary fiber content of root crops ranged from 4.6 to 13.5 g/100 g while legumes ranged from 20.9 to 46.9 g/ 100 g, suggesting that both root crops and legumes are good sources of dietary fiber (Table 43.2). The ratio of insoluble to soluble fiber in root crops ranged from 1:1 to 2:1.

Significant amounts of short chain fatty acids were produced after fermentation of the fiber isolate of root crops with tugi and cassava having the greatest production of SCFA (Table 43.3). The best sources of acetate among the root crops were tugi (2.5 ± 0.4) and cassava (2.4 ± 0.1), for propionate, tugi (1.8 ± 0.2), and for butyrate, tugi and cassava (0.8 ± 0.0; Table 43.4).

3. MINERAL AVAILABILITY, *IN VITRO*, FROM ROOT CROPS

3.1 Test Foods

The following root crops were used in the study: ube (*Dioscorea alata*), gabi (*Colocasia esculenta*), tugi (*Dioscorea esculenta*), potato (*Solanum tuberosum*), kamote (*Ipomoea batatas*), and cassava (*Manihot esculanta*). Root crops were bought in local markets in Metro Manila. The root crops were boiled in water to cook, and freeze dried.

3.2 Analytical Methods

Total iron, zinc, and calcium were analyzed by wet digestion method. A 0.1 g sample of freeze-dried test food was weighed and dissolved in 1 mL 36N sulfuric acid (Analytical

TABLE 43.3   Short Chain Fatty Acids from Fiber Isolates of Root Crops, mmole/g (Mean $\pm$ SEM)

Test Food	Acetate	Propionate	Butyrate	Total
Kamote	1.1 ± 0.2^{bx}	0.4 ± 0.0^{cz}	0.5 ± 0.0^{by}	1.9 ± 0.2^{c}
Gabi	1.1 ± 0.1^{bx}	0.2 ± 0.0^{dy}	0.2 ± 0.0^{cy}	1.5 ± 0.1^{d}
Potato	0.2 ± 0.0^{cy}	0.1 ± 0.1^{dey}	0.6 ± 0.1^{bx}	1.0 ± 0.2^{e}
Tugi	2.5 ± 0.4^{ax}	1.8 ± 0.2^{ay}	0.8 ± 0.0^{az}	5.1 ± 0.6^{a}
Ube	0.2 ± 0.0^{cx}	0.1 ± 0.0^{ey}	0.1 ± 0.1^{cx}	0.4 ± 0.1^{f}
Cassava	2.4 ± 0.1^{ax}	0.7 ± 0.1^{by}	0.8 ± 0.0^{ay}	3.9 ± 0.2^{b}

[a-f]Denotes significant differences between test foods at $P<0.05$.
x–z Denotes significant differences between short chain fatty acids at $P<0.05$.

TABLE 43.4   Mineral Content Of Root Crops, mg/100 g Sample

Root crop	Total Iron	Total Zinc	Total Calcium
Ube	$0.94 \pm 0.06^{ey*}$	2.27 ± 0.24^{cx}	0.84 ± 0.05^{ey}
Gabi	6.95 ± 0.96^{ax}	6.03 ± 0.81^{ax}	4.10 ± 0.25^{by}
Tugi	5.89 ± 0.19^{ax}	1.30 ± 0.08^{ez}	2.84 ± 0.17^{cy}
Potato	1.26 ± 0.20^{cx}	1.32 ± 0.06^{ex}	1.24 ± 0.04^{dx}
Kamote	1.08 ± 0.07^{dz}	3.50 ± 0.64^{by}	9.86 ± 1.28^{ax}
Cassava	5.05 ± 0.14^{by}	1.92 ± 0.03^{dz}	8.21 ± 0.48^{ax}

[a-e]Denotes significant differences between test foods at $P<0.05$.
x–zDenotes significant differences between minerals for each test food at $P<0.05$.
*Mean $\pm$ SEM.

Reagent, Ajax Chemical, Auburn, NSW, Australia), and 3 mL 30% hydrogen peroxide (Analytical Reagent, Ajax Chemical, Auburn, NSW, Australia), and made to 50 mL volume with double deionized (DDI) water. For calcium, the digested sample was volumed to 50 mL with 10 mM lanthanum chloride (Analytical Reagent, Asia Pacific Specialty Chemical Limited, NSW, Australia). The resulting solution was read in the atomic absorption spectrophotometer (Buck Scientific, East Norwalk, Connecticut, USA). Total dietary fiber, phytic acid, and tannic acid were also analyzed from the test foods using AOAC methods [28–30].

3.3 In Vitro Determination of Mineral Availability [2–4]

A 20 g freeze-dried sample of the test food was weighed. Eighty grams of DDI water was added, and the mixture was homogenized. The pH of the solution was adjusted to pH 2.0 and incubated with pepsin–HCL [8 g pepsin (Sigma Chemical Company, St. Louis, MO, USA), 600 units/mg solid hog stomach in 50 mL of 0.1 M HCl] for 3 hours. A 20 g aliquot of the pepsin digest was weighed in a dialysis tubing (Spectrapor 1, width 23 mm, 6000–8000 MW cut off) containing 5 mL of pancreatin–bile mixture [1 g pancreatin (porcine pancreas Grade VI, Sigma Chemical Company, St. Louis, MO, USA) plus 6.25 g bile extract (porcine, Sigma Chemical Company) in 250 mL of 0.1 M sodium bicarbonate (Analytical Reagent, Mallinckrodt, NY, USA)], and dialyzed against sodium bicarbonate (pH 7.5) for 12 hours. The dialysate was collected every 3 hours and replaced with 100 mL DDI water. The dialysate was read in the atomic absorption spectrophotometer for mineral content. This simulated the amount of mineral released that can be potentially absorbed in the small intestine. The retentate was freeze dried. A 0.5 g of the retentate was weighed and fermented using human fecal inoculum. The fermented digest was placed in a dialysis bag and dialyzed against DDI water for 6 hours. Similarly, the dialysates were collected every 3 hours and replaced with 100 mL DDI water. The dialysates were read in an atomic absorption spectrophotometer for mineral content. This simulated the amount of mineral released that can be potentially absorbed in the colon.

Table 43.4 shows the total mineral content of the test foods. Among the root crops, gabi, tugi, and cassava were the best sources of iron while gabi was the best source of zinc. Kamote and cassava were the best sources of calcium.

Table 43.5 shows the dietary fiber, phytic acid, and tannic acid content of the test foods. The best source of dietary fiber among root crops was gabi but it has the highest phytic acid content. Ube has the highest tannic acid content followed by tugi.

Dialyzable mineral as percent of total mineral content of root crops and legumes is used as a measure of mineral availability. Table 43.6 shows the mineral availability from the test foods. Iron availability from root crops showed that gabi>cassava>tugi and potato>ube and kamote ($P<0.05$); while zinc availability from root crops showed that kamote>potato> ube>tugi>cassava> gabi ($P<0.05$). Calcium availability from root crops showed that kamote and cassava>ube>gabi and tugi> potato ($P<0.05$).

4. GLYCEMIC INDEX (GI) OF ROOT CROPS AND LEGUMES IN NON-DIABETIC AND DIABETIC PARTICIPANTS

4.1 Food Samples

Five root crops – kamote [*Ipomoea batatas* (L.) Lam], gabi [*Colacasia esculenta* (L.) Scholl], tugi (*Diocorea esculenta*), cassava (*Manihot esculenta* Crantz), and ube (*Dioscorea* spp.) – were used as test foods in the study. The test foods were prepared at the Nutrient Availability Section, Food and Nutrition Research Institute,

TABLE 43.5 Dietary Fiber, Phytic Acid, and Tannic Acid Content of Root Crops

Root Crop	Dietary Fiber g/100 g S	Phytic Acid mg/100 g S	Tannic Acid mg/100 g S
Ube	$11.8 \pm 0.2^{b*}$	265.0 ± 6.2^{d}	932.6 ± 1.8^{a}
Gabi	13.5 ± 0.1^{a}	657.2 ± 14.8^{a}	355.3 ± 0.4^{d}
Tugi	10.3 ± 0.1^{c}	334.1 ± 5.5^{c}	505.2 ± 0.6^{b}
Potato	7.6 ± 0.2^{e}	459.6 ± 0.7^{b}	397.4 ± 3.5^{c}
Kamote	8.1 ± 0.1^{d}	264.5 ± 10.2^{d}	200.4 ± 5.8^{e}
Cassava	4.6 ± 0.2^{f}	197.8 ± 1.0^{e}	124.5 ± 1.0^{f}

$^{a-f}$Denotes significant differences between test foods at $P<0.05$.
*Mean $\pm$ SEM.

TABLE 43.6 Mineral Availability from Test Foods (% of Total Mineral Content)

Root Crop	Iron	Zinc	Calcium
Ube	$3.1 \pm 0.1^{dz*}$	45.5 ± 4.6^{cx}	20.2 ± 5.9^{by}
Gabi	24.4 ± 2.7^{ax}	10.7 ± 0.5^{fz}	15.4 ± 1.9^{by}
Tugi	16.7 ± 2.0^{cy}	32.2 ± 5.8^{dx}	15.4 ± 3.2^{by}
Potato	11.6 ± 4.1^{cy}	54.6 ± 1.7^{bx}	7.7 ± 1.7^{cz}
Kamote	3.0 ± 0.3^{dz}	71.3 ± 13.6^{ax}	30.8 ± 2.4^{ay}
Cassava	20.7 ± 0.1^{by}	13.61 ± 0.02^{ez}	30.1 ± 1.0^{ax}

$^{a-e}$Denotes significant differences between test foods at $P<0.05$.
$^{x-z}$denote significant differences between minerals at $P<0.05$.
*Mean $\pm$ SEM.

Department of Science and Technology. Root crops were boiled in water, cooled, and fed to subjects. The control/standard food was white bread prepared in a bread maker (Regal Kitchen Pro Collection, KCTNO4584, made in China, serviced in USA) following the formulation of Wolever et al. [22] as follows: 334 g flour (Wooden Spoon All Purpose Flour, Pillmico Mauri Food Corporation, Kwalan Cove, Iligan City, Philippines), 4 g salt, 5 g yeast, 7 g sucrose, and 330 mL water per 250 g carbohydrate loaf. Crust ends were not used for the test meals.

4.2 Study Participants

Eleven non-diabetic and nine diabetic subjects (type II) were physically examined by a medical doctor and evaluated by an endocrinologist on the basis of the following criteria:

- non-diabetic subjects: body mass index (BMI) 20–25 kg/m^2 (WHO criteria for normal to overweight), fasting blood sugar (FBS) 4–7 mol/L, age 35–60 years, no physical defect, and non-smokers;
- diabetic subjects: BMI 20–25 kg/m^2 (WHO criteria for normal to overweight), FBS 7.5–11.0 mol/L, age 35–60, no intake of drugs, no complications, and non-smokers.

Each subject was interviewed for physical activity and was asked to fill up a 3-day food intake recall form. Participants with common food intake (pattern) and physical activity were included in the study. The diabetic subjects were managed through dietary consultations and advice. The subjects signed voluntary consent forms approved by the National Human Ethics Committee, Philippine Council for Health Research and Development, Department of Science and Technology, Metro Manila, Philippines.

4.3 Protocol of the Study

Using the randomized cross-over design, the control and test foods were fed in random order on separate occasions after an overnight fast. The control and test foods contained 50 g available carbohydrates. The participants were told to fast overnight (10–12 hours) prior to the start of the study. Feedings of white bread and test foods were repeated twice. A standard glucose (Medic Orange 50 Glucose Tolerance Test Beverage Product No. 089, 50 g glucose/240 mL; Medic Diagnostic Laboratory, Pasig City, Philippines) was given once to non-diabetic participants only (for safety purposes) to determine the glycemic index of white bread.

Blood samples, approximately 0.3–0.4 mL, were collected through finger prick before and after feeding in a 4 mm diameter and 10 cm long capillary tubing (PYREX, Corning, New York, USA) and sealed (Jockel Seal Sticks Cement, AH Thomas Catalogue No. 2454 W15, USA). For non-diabetic participants, samples were collected at 0 hour and every 15 minutes after feeding for 1 hour and every 30 minutes for the next hour, while for diabetic participants, samples were collected at 0 hour and at 30 minute intervals after feeding for a period of 3 hours. The serum was separated from the blood using a refrigerated centrifuge after all the blood was collected (Eppendorf Centrifuge, Eppendorf, Germany), and analyzed for glucose levels on the same day using a clinical chemistry analyzer (ARTAX Menarini Diagnostics, Firenze, Italy) after calibration with the glucose standard (Glucofix Reagent 1-Menarini Diagnostics, Firenze, Italy). The area under the glucose response curve for each food, ignoring areas below the fasting level, was calculated geometrically [21,22]. The glycemic index of each food was expressed as the percent of the mean glucose response of the test food divided by the standard food taken by the same subject and was determined by the following formula:

$$\text{Glycemic index} = \frac{\text{IAUC* of the test food} \times 100}{\text{IAUC of the standard food}}$$

Table 43.7 shows the characteristics of the subjects. The BMI of the non-diabetic participants was found to be 25.0 ± 0.6 kg/m^2 and the diabetics, 26.9 ± 1.2 kg/m^2. There were no significant differences between the non-diabetic and diabetic subjects for age and BMI. There were significant differences observed between study participants for the fasting blood sugar: as expected, those of diabetics

*Incremental area under the glucose response curve.

TABLE 43.7 Characteristics of the Subjects, Mean $\pm$ SEM

Subjects	No. of Subjects	Age (Years)	BMI (kg/m²)	FBS (mmol/L)
Non-diabetic	11	46.5 ± 1.5^a	25.0 ± 0.6^a	5.2 ± 0.1^b
Diabetic	9	48.3 ± 1.8^a	26.9 ± 1.2^a	9.1 ± 0.7^a

BMI, body mass index = weight/height²; FBS, fasting blood sugar.
[a,b]Denotes significant differences between subjects at $P<0.05$.

TABLE 43.8 Glycemic Index* of Local Root Crops in Non-Diabetic and Diabetic Subjects, Mean $\pm$ SEM

Test Food	Serving Size (g)	Non-Diabetic	Diabetic
Kamote	62	34 ± 1^{bx}	34 ± 1^{bx}
Gabi	65	32 ± 1^{bx}	32 ± 1^{bx}
Tugi	76	37 ± 1^{ax}	36 ± 1^{ax}
Cassava	58	30 ± 1^{cx}	30 ± 1^{cx}
Ube	66	27 ± 1^{dx}	23 ± 1^{dy}

[a–d]Denotes significant differences between test foods at $P<0.05$.
[xy]Denotes significant differences between normal and diabetic subjects at $P<0.05$.
*50 g available carbohydrate.

were significantly greater than the non-diabetics (Table 43.7; $P<0.05$). Normal readings of the standard glucose used in the clinical chemistry analyzer ranged from 4.0 to 6.4 mmol/L glucose. The initial blood glucose obtained from the non-diabetics for all test foods did not exceed 6.0 mmol/L. All of the participants were able to consume all test foods (white bread 2 $\times$, test foods 2 $\times$, and standard glucose 1 $\times$ in non-diabetics) on 27 occasions.

Table 43.8 shows the glycemic index (GI) of the test foods. The GI of the standard white bread determined from the incremental area under the glucose response curve (IAUC) from white bread divided by the IAUC from the standard glucose, was found to be 93 ± 9 for non-diabetic participants. Local root crops are considered to be low GI foods (GI <55;

Table 43.8; $P<0.05$). Ube has the lowest GI in non-diabetic (GI = 27) and diabetic participants (GI = 23) and was found to be significantly different (Table 43.8; $P<0.05$). Tugi has the highest GI in both non-diabetic (GI = 37) and diabetic participants (GI = 36) and was not found to be significantly different (Table 43.8).

5. CHOLESTEROL-LOWERING EFFECT IN HUMANS WITH MODERATELY RAISED SERUM CHOLESTEROL LEVELS

5.1 Study Participants

The participants were selected based on the following criteria: moderately raised cholesterol level (250–300 mg cholesterol/100 g serum), 30–60 years old, no drug intake for cholesterol-lowering, and no complications. They were interviewed to obtain data on their usual 3-day food intake, physical activity, and smoking habits. The total number of subjects in the study was 20, 18 females and 2 males.

5.2 Protocol of the Study

The study was conducted in a double-blind randomized cross-over design in a 14-week period (3.5 months), consisting of four 2-week experimental periods, each experimental period separated by a 2-week washout period for a total of three washout periods. The subjects were made to fast overnight (10–12 hours of fasting) prior to the study. They were weighed, their blood pressure was measured, and a sample of blood from their forearm vein was taken. The subjects were given the test foods to consume every day to ensure compliance of the subjects, except on Fridays when three test foods were given to include Saturday and Sunday intakes. The subjects recorded their respective food intakes for the duration of the experimental study. On the 15th day, blood was drawn from the subjects after an overnight fast. Blood samples were taken into plain glass tubes from the forearm vein, left to clot at room temperature, centrifuged, and the serum separated. Total cholesterol, HDL cholesterol, and triglycerides were measured in a clinical chemistry analyzer (ARTAX, Menarini Diagnostic, Florence, Italy) against standards (Cholesterol, Cholesterol Standard, 200 mg/dL, Sentinel CH, Milan, Italy; HDL Cholesterol, Cholesterol Standard, 50 mg/dL, Sentinel CH, Milan, Italy; Triglycerides, Glycerol Standard, 200 mg/dL, Sentinel CH, Milan, Italy). The amount of LDL was estimated from the formula used by Wolever et al. [31] as follows:

$$LDL = (total\ cholestrol - HDL - triglyceride/2.2)$$

Subjects served as their own control. The subjects signed voluntary consent forms approved by the National Human Ethics Committee, Philippine Council for Health Research and Development, Department of Science and Technology, Metro Manila, Philippines.

Non-significant results in total, LDL, and HDL cholesterol as well as triglycerides in all root crops studied were observed after 14 days except for tugi and cassava. Tugi had significantly increased HDL cholesterol by 16% while cassava significantly decreased LDL cholesterol levels by 7% ($P<0.05$; Table 43.9).

6. DISCUSSION

The study showed that root crops are good sources of dietary fiber and fermentable in the colon. The soluble fiber in root crops may have a physiological effect related to glucose and cholesterol metabolism [21,22]. Generally, soluble fibers have a tendency to have long-term effects associated with better glycemic control and lower insulin requirements due to increased insulin sensitivity or changes in insulin receptors as well as cholesterol-lowering effects. In terms of short chain fatty acid production, a good source of butyrate may have some important role in the prevention of risk of colon cancer, e.g. tugi and cassava.

Mineral availability from root crops differed and may be attributed to the mineral content, mineral–mineral interaction, and the presence of phytic and tannic acid in the test foods [32–40]. Among root crops, iron availability increased with increasing amount of iron (Tables 43.4 and 43.6; $r = 0.92$, $N = 12$; $P<0.01$). Similar results were also observed with calcium (Tables 43.4 and 6; $r = 0.85$, $N = 12$; $P<0.05$) but not for zinc. Zinc availability decreased with increasing iron content of root crops (Tables 43.4 and 43.6; $r = -0.87$, $N = 12$; $P<0.01$). An increase in zinc availability affected iron and calcium availability, e.g. ube, tugi, potato, and kamote (Table 43.6, $P<0.05$);

TABLE 43.9 Serum Total, HDL, and LDL Cholesterol, and Triglycerides Levels of Humans Fed With Root Crops

Root crop	Total Cholesterol		HDL		LDL		Triglycerides	
	Before	After	Before	After	Before	After	Before	After
Kamote	283±7	271±7	41±2	37±2	205±8	202±8	177±12	163±14
Cassava	255±6	247±5	31±2	29±2	199±6*	185±6	153±13	168±16
Tugi	256±6	261±7	27±2	32±2*	200±6	202±7	141±13	132±13
Ube	270±6	270±6	ND	ND	ND	ND	136±11	142±13
Gabi	222±5	251±7	ND	ND	ND	ND	137±11	158±13

ND, no data.
*Denotes significant differences between before and after feeding of root crops at $P<0.05$.

an increase in iron availability affected zinc and calcium availability, e.g. gabi (Table 43.6, $P<0.05$). An increase in calcium availability affected iron availability, e.g. ube, kamote, cassava. The phytic acid present in gabi (Table 43.5) may have affected zinc and calcium availability but not iron. The phytic acid content of potato (Table 43.5) decreased calcium availability but not zinc and iron availability (Table 43.6). The iron, zinc, and calcium content of potato did not differ significantly. The study showed that phytic acid did not affect iron availability from root crops. Ube has the highest tannic acid content (Table 43.5) and resulted in low iron availability but did not affect zinc and calcium availability. However, we do not believe that the high tannic acid content of ube has decreased iron availability (3.1%; Table 43.6) because kamote with similar iron availability (3.0%; Table 43.6) has a tannic acid content of 200.4 mg/100 g (Table 43.5).

Equal carbohydrates portions (50 g available carbohydrates) of the different test foods do not have the same glycemic effect in non-diabetic and diabetic subjects (Table 43.8). The amount of dietary fiber present in the test foods may have caused significant differences in the GI of root crops (Table 43.8). The increasing levels of dietary fiber, viscosity, cooking, particle size, form of food, and starch structure may all be attributed to slower nutrient absorption and delayed transit time [41–43]. Considering the dietary fiber present in whole foods, insoluble fiber was related more strongly to the GI than soluble fiber content [44]. The insoluble fiber content of root crops were significantly greater than that of soluble fiber except for kamote (Table 43.2; $P<0.05$). The fat content of root crops ranged from 0.2 to 1.0 g/100 g and may not affect the glycemic responses of the study participants from the foods ingested. A sufficient amount of fat in food (23 g fat/kg) can cause an early (0–90 min) decrease in glucose response but does not affect the overall glucose response to food [45]. The differences in glucose responses

between the two groups (non-diabetics and diabetics) may be due to rates of digestion and absorption in relation to the food ingested [44].

Root crops have been shown to be potentially hypocholesterolemic foods, e.g. tugi, cassava. The hypocholesterolemic property of dietary fiber is associated with the water-soluble fractions of fiber fermentable in the colon, e.g. galactomannans, uronic acid, glucomannans, galacturonic acids. However, various water-soluble fibers may differ in their ability to reduce serum cholesterol [46–48]. Taro or gabi has been shown to contain 85% galacturonic acid dry weight [49]. Sweet potato (kamote) is a significant source of dietary fiber as its pectin content can be as high as 5% of the fresh weight or 20% of the dry matter at harvest [50]. The Lipid Research Clinics Coronary Primary Prevention Trial predicted that for every 1% decrease in serum cholesterol concentration, there is a decreased risk of coronary heart disease (CHD) of 2% [51]. There was no significant increase or decrease in HDL cholesterol levels of all subjects except for tugi. The concentration of serum HDL cholesterol is affected by alcohol intake and body mass index (BMI) [52]. However, not all subjects drink alcohol and the BMIs were not significantly different between subjects in the duration of the experimental period. The study on cholesterol-lowering effect was a short-term (acute) study. A longer-term nutrition intervention study may give a more conclusive result.

7. SUMMARY

In conclusion: a) root crops are rich sources of dietary fiber that is fermentable and produces short chain fatty acids such as acetate, propionate, and butyrate; b) mineral availability from the root crops studied differed and may be attributed to the mineral content, mineral–mineral interaction, and the presence of phytic and tannic acid in the test foods; c) root crops are considered to be low GI foods (GI <55); and d) root crops have shown a potential hypocholesterolemic effect. The study may have a significant role in the proper control and management of chronic diseases, e.g. obesity, diabetes mellitus, cancer, and cardiovascular diseases and can be a scientific basis for considering root crops as functional foods. The utilization of root crops as a functional food will not only solve the problem of chronic diseases now prevailing in almost all countries but also encourage the industry and farmers to produce value-added or healthy products from root crops. This will increase production and promotion of root crops.

ACKNOWLEDGMENT

The authors wish to thank Mr Zoilo B. Villanueva, Josefina A. Desnacido, and Paz S. Lara, Nutritional Biochemistry Section, FNRI, DOST, for their technical assistance.

References

1. Roberfroid, M. (1997). Health benefits of non-digestible oligosaccharides. In D. Kritchevsky, & C. Bonfield, (Eds.), *Dietary fiber in health and disease. Advances in experimental biology* (Vol. 427). New York: Plenum Press.
2. Trinidad, T. P., Wolever, T. M. S., & Thompson, L. U. (1996). Availability of calcium for absorption in the small intestine and colon from dies containing available and unavailable carbohydrates: An *in vitro* assessment. *International Journal of Food Sciences and Nutrition, 47*, 83–88.
3. Trinidad, T. P., Wolever, T. M. S., & Thompson, L. U. (1996). Calcium availability from dairy foods as affected by dietary fibers. *Trace Elements and Electrolytes Journal, 13*, 123–125.
4. Thompson, L. U., Trinidad, T. P., & Jenkins, D. J. A. (1991). Methods for determining the minerals available for absorption from the small intestine and colon. *Trace Elements in Man and Animal, 7*, 12–25.
5. Trinidad, T. P., Wolever, T. M. S., & Thompson, L. U. (1996). Effect of acetate and propionate on calcium absorption from the rectum and distal colon of humans. *The American Journal of Clinical Nutrition, 63*, 574–578.
6. Trinidad, T. P., Wolever, T. M. S., & Thompson, L. U. (1993). Interactive effects of calcium and short chain fatty acids on absorption in the distal colon of man. *Nutrition Research, 13*, 417–425.
7. Wolever, T. M. S., Trinidad, T. P., & Thompson, L. U. (1995). Short chain fatty acids from the human distal colon: Interaction between acetate, propionate and calcium. *Journal of the American College of Nutrition, 14*, 293–298.
8. Champ, M., Brillouet, J. M., & Rouau, X. (1983). Nonstarchy polyssacharide of *Phaseolus vulgaris, Lens esculenta* and *Cicer arietinum* seeds. *Journal of Agricultural and Food Chemistry, 34*, 326–329.
9. Holloway, W. D., Monro, J. A., Gurnsey, J. C., Pomare, E. W., & Stace, N. H. (1985). Dietary fiber and other constituents of some Tongan foods. *Journal of Food Science, 50*, 1756–1757.
10. Howarth, N. C., Saltzman, E., McCrory, M. A., Greenberg, A. S., Dwyer, J., Ausman, L., Kramer, D. G., & Roberts, S. B. (2003). Fermentable and nonfermentable fiber supplements did not alter hunger, satiety or body weight in a pilot study of men and women consuming self-selected diets. *Journal of Nutrition, 133*, 3141–3144.
11. Brune, M., Rossander, L., & Hallberg, L. (1989). Iron absorption and phenolic compounds: Importance of different phenolic structures. *European Journal of Clinical Nutrition, 43*, 547–548.
12. Kelsay, L. (1987). Effect of fiber, phytic acid and oxalic acid in the diet on mineral bioavailability. *American Journal of Gastroenterology, 82*, 983–986.
13. Chen, W. J. L., Anderson, J. W., & Jenkinds, D. J. A. (1984). Propionate may mediate the phypocholesterolemic effects of certain soluble plant fibers in cholesterol-fed rats. *Proceedings of the Society for Experimental Biology and Medicine, 175*, 215–218.
14. (1994). *Dietary fiber.* ILSI Europe Concise Monograph Series, pp. 15–19. Brussels, Belgium: ILSI Press.
15. Jenkins, D. J. A., Ghafari, A., Wolever, T. M. S., Taylor, R. H., Barker, H. M., Filden, H., Jenkins, A. L., & Bowling, A. C. (1982). Relationship between the rate of digestion of foods and post-prandial glycemia. *Diabetologia, 22*, 450–455.
16. Creutzfeldt, W. (1983). Introduction. In W. Creutzfeldt, & R. U. Folsch, (Eds.), *Delaying absorption as a therapeutic principle in metabolic diseases* (pp. 1–). New York: Thieme-Stratton.
17. Jenkins, D. J. A., Cuff, D., & Wolever, T. M. S. (1987). Digestibility of carbohydrate foods in an ileostomate: Relationship to dietary fiber, *in vitro* digestibility and

glycemic response. *American Journal of Gastroenterology, 82,* 709–717.

18. Collier, G. R., Giudici, S., Kalmusky, J., Wolever, T. M. S., Helman, G., Wesson, V., Ehrlich, R. M., & Jenkins, D. J. A. (1988). Low glycemic index starchy foods improve glucose control and lower serum cholesterol in diabetic children. *Diabetes, Nutrition and Metabolism, 1,* 1–9.

19. Fontvieille, A. M., Acosta, M., Rizkalla, S. W., Bornet, F., David, P., Letanoux, M., Tehobroutsky, G., & Slama, G. (1988). A moderate switch from high to low glycemic index foods for three weeks improves the metabolic control of type I (IDDM) diabetic subjects. *Diabetes, Nutrition and Metabolism, 1,* 139–143.

20. Brand, J. C., Calagiuri, S., Crossman, S., Allan, A., Roberts, D. C. K., & Truswel, A. S. (1991). Low glycemic index foods improve long-term glycemic control in NIDDM. *Diabetes Care, 14,* 95–101.

21. Wolever, T. M. S., Jenkins, D. J. A., Vuksan, V., Jenkins, A. L., Wong, G. S., & Josse, R. G. (1992). Beneficial effect of low glycemic index diet in overweight NIDDM subjects. *Diabetes Care, 15,* 562–566.

22. Wolever, T. M. S., Jenkins, D. J. A., Vuksan, V., Jenkins, A. L., Buckly, G. C., Wong, G. S., & Josse, R. G. (1992). Beneficial effect of a low glycemic index diet in type 2 diabetes. *Diabetic Medicine, 9,* 451–458.

23. (1984). Official Methods of Analysis, Association of Official Analytical Chemists, 14th ed., Aoac International, p. 249.

24. Aoac (1984). Official Methods of Analysis, Association of Official Analytical Chemists, 14th ed., p. 251.

25. Sullivan, D.M., and Carpenter, D.E., Eds (1993). *Methods of analysis for nutrition labeling,* Aoac International, p. 402.

26. Aoac (1995). Official Methods of Analysis 991.43 Supplement, Association of Official Analytical Chemists.

27. McBurney, M. I., & Thompson, L. U. (1987). *In vitro* fermentation of purified fiber supplements. *Journal of Food Science, 54,* 347–350.

28. Aoac (1995). Official Methods of Analysis 991.43 Supplement, Association of Official Analytical Chemists pp. 249, 251.

29. Association of Official Analytical Chemists. (1986). Phytate in foods: Anion exchange method. *Journal of the Association of Official Analytical Chemists, 69,* 2.

30. Earp, C. F., Ring, S. H., & Rooney, L. W. (1981). Evaluation of several methods to determine tannins in sorghum with varying kernel characteristics. *Cereal Chemistry, 58,* 134–138.

31. Wolever, T. M. S., Jenkins, D. J. A., Mueller, S., Boctor, D. L., Ransom, T. P. P., Patten, R., Chao, E. S.

M., McMillan, K., & Fulgoni, V. III, (1994). Method of administration influences the serum cholesterol-lowering efect of psyllium. *American Journal of Clinical Nutrition, 59,* 1055–1059.

32. Cook, J. D., Dassenko, S. A., & Whittaker, P. (1991). Calcium supplementation: Effect on iron absorption. *American Journal of Clinical Nutrition, 53,* 106–111.

33. Davidsson, L., Almgren, A., Sandstrom, B., & Hurrell, R. F. (1995). Zinc absorption in adult humans: The effect of iron fortification. *The British Journal of Nutrition, 74,* 417–425.

34. Davidsson, L., Kastenmayer, P., & Hurrell, R. F. (1994). Sodium iron EDTA as a food fortificant: The effect on the absorption and retention of zinc and calcium in women. *American Journal of Clinical Nutrition, 60,* 231–237.

35. Dawson-Hughes, B., Seligson, F. H., & Hughes, V. A. (1986). Effects of calcium carbonate and hydroxyapatite on zinc and iron retention in postmenopausal women. *American Journal of Clinical Nutrition, 44,* 83–88.

36. Deehr, M. S., Dallal, G. E., Smith, K. H., Taulbee, J. D., & Dawson-Hughes, B. (1990). Effect of different calcium sources on iron absorption in post-menopausal women. *American Journal of Clinical Nutrition, 51,* 95–99.

37. Gleerup, A., Rossander-Hulthen, L., Gramatkovski, E., & Hallberg, L. (1995). Iron absorption from the whole diet: comparison of the effect of two different distributions of daily calcium intake. *American Journal of Clinical Nutrition, 61,* 97–104.

38. Forbes, R. M., Erdman, J. W., Jr., Parker, H. M., Kondo, H., & Keteisen, H. M. (1983). Bioavailability of zinc in coagulated soy protein (tofu) to rats and effect of dietary calcium at a constant phytate:zinc ratio. *Journal of Nutrition, 113,* 205–210.

39. Brune, M., Rossander, L., & Hallberg, L. (1989). Iron absorption and phenolic compounds: Importance of different phenolic structures. *American Journal of Clinical Nutrition, 43,* 547–548.

40. Gillooly, M., Bothwell, T. H., & Trrance, J. D. (1983). The effect of organic acid, phytates and polyphenols on absorption of iron from vegetables. *The British Journal of Nutrition, 49,* 331–342.

41. Anderson, J. W. (1990). Hypercholesterolemic effect of oat bran. In I. Furda, & C. J. Brine, (Eds.), *New development in dietary fiber: Physiological, physicochemical and analytical apect* (pp. 17–36). New York: Plenum Press.

42. Wolever, T. M. S., Jenkins, D. J. A., Kalmusky, J., Giordano, C., Giudici, S., Josse, R. G., & Wong, G. S. (1986). Glycemic response to pasta: Effect of food form, cooking and protein enrichment. *Diabetes Care, 9,* 401–404.

C. ACTIONS OF INDIVIDUAL FRUITS IN DISEASE AND CANCER PREVENTION AND TREATMENT

43. Jenkins, D. J. A., Wolever, T. M. S., Leeds, A. R., Gassull, M. A., Dilawari, J. B., Goff, D. V., Metz, G. I., & Alberti, K. G. M. M. (1978). Dietary fibers, fiber analogues and glucose tolerance: importance of viscosity. *British Medical Journal, 2,* 1744–1746.

44. Jenkins, D. J. A., Ghafari, A., Wolever, T. M. S., Taylor, R. H., Barker, H. M., Filden, H., Jenkins, A. L., & Bowling, A. C. (1982). Relationship between the rate of digestion of foods and post-prandial glycemia. *Diabetologia, 22,* 450–455.

45. Peters, A. L., & Davidson, M. B. (1983). Protein and fat effects on glucose response and insulin requirements in subjects with IDDM. *American Journal of Clinical Nutrition, 58,* 555–560.

46. Kirby, R. W., Anderson, J. W., Sieling, B., Rees, E. D., Chen, D. J. L., Miller, R. E., & Kay, R. M. (1981). Oat bran intake selectively lowers serum LDL cholesterol concentration of hypercholesterolimic men. *American Journal of Clinical Nutrition, 34,* 824–829.

47. Jenkins, D. J. A., Leeds, A. R., Newton, C., & Cummings, J. H. (1975). Effect of pectin, guargum and wheat fiber on serum cholesterol. *Lancet, 1,* 1116–1117.

48. Bell, L. P., Hectorn, K. J., Reynolds, H., & Hunninghake, D. B. (1990). Cholesterol-lowering effects of soluble fiber cereals as part of a prudent diet for patients with mild to moderate hypercholesterolemia. *American Journal of Clinical Nutrition, 52,* 1020–1026.

49. Kawakita, K., & Kojima, M. (1983). The isolation and properties of a factor in taro tuber that agglutinates spores of Ceratocytis fimbriata, black rot fungus. *Plant and Cell Physiology, 24,* 141–149.

50. Collins, W.W., and Walter, W.M. Jr. (1982). Potential for increasing the nutritional value of sweet potatoes. In R.L. Villareal & T.F. Griggs (Eds.) *Sweet Potatoes. Proceedings of the First International Symposium* (pp. 355–363). AVRDC Pub No. 82–192. Tainan, Taiwan, China: Hon Wen Printing, Woks..

51. Lipid Research Clinic Program. (1984). The lipid research clinics coronary primary prevention trial results: II. The relationship of reduction in incidence of coronary heart disease to cholesterol lowering. *JAMA, 251,* 365–373.

52. Bolton-Smith, C., Woodward, M., Smith, W. C. S., & Tunstall-Pedoe, H. (1991). Dietary and non-dietary predictors of serum total and HDL-cholesterol in men and women: results from The Scottish Heart Health Study. *International Journal of Epidemiology, 20,* 95–104.

C. ACTIONS OF INDIVIDUAL FRUITS IN DISEASE AND CANCER PREVENTION AND TREATMENT

Index